MW01409831

A COMPENDIUM OF ARMAMENTS AND MILITARY HARDWARE

A COMPENDIUM OF ARMAMENTS AND MILITARY HARDWARE

CHRISTOPHER CHANT

Routledge & Kegan Paul
London and New York

First published in 1987 by
Routledge & Kegan Paul Ltd
11 New Fetter Lane, London EC4P 4EE

Published in the USA by
Routledge & Kegan Paul Inc.
in association with Methuen Inc.
29 West 35th Street, New York, NY 10001

Phototypeset in Linotron Times, 9 on 10pt
by Input Typesetting Ltd, London
and printed in Great Britain
by R J Acford, Chichester

© *Imp Publishing Services Ltd 1987*

No part of this book may be reproduced in any form without permission from the publisher
except for the quotation of brief passages
in criticism

Library of Congress Cataloging in Publication Data

Chant, Christopher.
  A compendium of armaments and military hardware.

  Includes index.
  1. Weapons systems—Catalogs. I. Title.
UF500.C42   1987   355.8'2'0216   86–31326

British Library CIP Data also available
ISBN 0–7102–0720–4

CONTENTS

Introduction	vii
SECTION 1 LAND WEAPONS	**1**
Main battle tanks and medium tanks	3
Light tanks	21
Armoured cars, reconnaissance vehicles and scout cars	29
Armoured personnel carriers and infantry fighting vehicles	39
Artillery (self-propelled and towed)	73
AA systems (self-propelled and towed)	100
Multiple rocket systems	116
Unguided anti-tank weapons	126
Table 1.1 Machine-guns	130
Table 1.2 Mortars	132
SECTION 2 WARSHIPS	**135**
Submarines	
Nuclear-powered ballistic missile	137
Nuclear-powered cruise missile	143
Ballistic missile	146
Cruise missile	146
Nuclear-powered attack	147
Attack	159
Aircraft-carriers	181
Battleships and battle-cruisers	190
Guided-missile cruisers	191
Guided-missile destroyers	202
Destroyers	228

Guided-missile frigates	241
Frigates	268
Corvettes	286
Fast attack craft	294
Amphibious warfare vessels	317
Table 2.1 Mine warfare vessels	330
Table 2.2 Naval guns	332
Table 2.3 Surface skimmers	335
Table 2.4 Torpedoes	336

SECTION 3   AIRCRAFT — 339

Table 3.1 Flying training aircraft	483

SECTION 4   MISSILES — 485

Strategic	487
Cruise	496
Surface-to-surface	498
Air-to-surface	502
Anti-ship	509
Anti-submarine	518
Surface-to-air	521
Air-to-air	536
Anti-tank	544
Index	553

INTRODUCTION

The weapons deployed by modern armies, navies and air forces are complex and highly capable, with military applications that may range from a single highly specialized role to a diversity across a spectrum of related tasks. Such weapons are produced in large quantities by an increasing number of countries, and are rarely even in the first stages of production before variants begin to appear, either modifying the task of the baseline model or upgrading its capabilities with improved weapons, sensors and powerplant. The nature of these weapons conditions to a very great degree the capabilities of the user force, so it is impossible to arrive at any genuine assessment of any service's worth without at least a basic understanding of the weapons it operates and has ordered. And given the fact that modern military operations are increasingly of the combined-arms type (with land, sea and air forces all playing a part in the overall plan), it is ever more important to understand the basic capabilities of the weapons used by each of the services.

In recent years there has been a great upsurge in the number and quality of the military reference books available to the specialist and general reader. The practical limitations of these books have been triple: they are in general inordinately expensive if they are of any significance, almost invariably concentrate on only a single aspect of the military arena, and in the process present so large a quantity of information about every weapon and its development that it is difficult for the reader to pick out the salient technical and operational facts, especially in comparison with those of other weapons. The *Compendium of Armaments and Military Hardware* is designed to overcome these limitations by presenting in a single volume the primary military features of the main modern weapons operated by the world's more important armed forces. In this volume the reader, both specialist and interested, can find the key features (accurate up to late 1986) of the weapons that unfortunately play so crucial a part in current world events, with particular emphasis on sensor/data-processing/countermeasures electronics and weapons/ammunition. Only present types are covered, though an exception to this general principle has been made for a small number of major weapons currently under final development or earmarked for a major role in the future. Weapons of which virtually nothing is known have been omitted.

The volume is divided into four basic sections, dealing respectively with land weapons, warships, aircraft and missiles. It is clear that these systems cannot operate in any effective manner without a vast array of support equipment of various types, but space constraints have held the contents of the volume to the 'cutting edge' equipment in current service, this concept being stretched to include amphibious warfare vessels, tactical and strategic transport aircraft, etc. The information is designed to provide basic physical data (overall dimensions, weights, etc.), armament (type, calibre, disposition and ammunition), combat-related electronics (sensors, data-processing equipment, countermeasures, etc.), propulsion (powerplant and fuel, where appropriate) and performance (speed, range, etc.). Brief notes then describe the current variants and the ways in which these differ from the baseline model. Where two countries are shown, as in USA/South Korea, the former denotes the country of origin, and the latter the operating country where substantial modification has been made to the basic design.

The author has used a question mark to indicate an unknown or unrevealed quantity: thus in the 'Benjamin Franklin' class SSBN the reference '? UUM-44A SUBROC tube-launched missiles' indicates that this missile is used in the ships of the class, but that it is uncertain how many missiles are carried.

Section 1 Land Weapons
The land weapons section is organized along role lines, the disposition of entries within any one role being alphabetically by country of origin and within the entries for any one country by tactical importance. The author has had to use his discretion about the precise location of some weapon systems, and hopes that the reader will agree with his choice. The land weapons are arranged as follows:

1 main battle tanks and medium tanks (over 25 tonnes in weight);
2 light tanks (under 25 tonnes in weight);
3 tank destroyers designed to tackle other tanks without concern for other battlefield targets except in emergencies;
4 reconnaissance vehicles (both tracked and wheeled) designed to reconnoitre but also to fight in emergencies;
5 mechanized infantry combat vehicles and armoured personnel carriers (both tracked and wheeled), the former designed to transport troops and support them on the battlefield, the latter intended only to transport the troops;
6 tube artillery (of both the towed and the self-propelled varieties);
7 anti-aircraft weapons (both missile and gun, including towed and self-propelled varieties);
8 multiple-launch rocket systems; and

9 individual anti-tank weapons.

Tables give brief details of the world's most important mortars and machine-guns.

Section 2 Warships

The ship section is also organized in terms of role, and then alphabetically by class within any particular role specialization. It is difficult to classify some ships into a specific role type (especially where modern cruisers, destroyers, frigates and corvettes are concerned, the author having worked to the general premise that corvettes displace between 600 and 1,250 tons, frigates between 1,250 and 3,500 tons, destroyers between 3,500 and 7,000 tons, and cruisers upwards of 7,000 tons unless a class's particular capabilities make such a tonnage assessment invalid), and it is hoped that the reader will approve of the author's disposition of such vessels. The ships are arranged as follows:

1. nuclear-powered submarines carrying a primary armament of ballistic missiles for strategic attack of land targets (SSBN);
2. nuclear-powered submarines carrying a primary armament of cruise missiles for strategic/operational attack of high-value naval targets (SSGN);
3. conventionally-powered submarines carrying a primary armament of ballistic missiles for strategic attack on land targets (SSB);
4. conventionally-powered submarines carrying a primary armament of cruise missiles for strategic/operational attack on high-value naval targets (SSG);
5. nuclear-powered attack (fleet) submarines (SSN);
6. conventionally-powered attack (patrol) submarines (SS);
7. aircraft-carriers and helicopter-carriers (CV);
8. battleship (BB);
9. nuclear-powered battle-cruiser (CBN), a difficult classification bearing little resemblance to the conventional definition of such vessels, but intermediate in size and capability between the battleship and the cruiser;
10. guided-missile cruisers, both nuclear-powered (CGN) and conventionally-powered (CG) carrying a primary missile armament; with the exception of the Soviet 'Sverdlov' class, the world's elderly gun cruisers are not covered as being largely irrelevant to modern warfare;
11. guided-missile destroyers with a primary armament of medium/long-range missiles for offensive rather than self-defence purposes (DDG);
12. destroyers with a primary armament of guns, but often carrying self-defence missiles (DD);
13. guided-missile frigates with a primary armament of medium-range missiles for offensive rather than self-defence purposes (FFG);
14. frigates with a primary armament of guns, but often carrying self-defence missiles (FF);
15. corvettes with an armament of missiles and guns (FFL);
16. fast attack craft with a speed of 25 kts or more and armed with missiles, guns or torpedoes (FAC); and
17. amphibious assault vessels designed for the landing of ground forces over an enemy beach.

Tables give brief details of the world's most important air-cushion vehicles (hovercraft and surface-effect craft), minesweepers and mine-countermeasures vessels, naval guns and torpedoes.

In the *Complement* subsection for each type, a single number indicates the whole crew, two numbers the officers and other ranks, and the numbers the officers, petty officers, and other ranks. Ranges are quoted in statute (not nautical) miles.

Section 3 Aircraft

The aircraft section is organized along lines different from the other sections, being arranged alphabetically by manufacturer and within each manufacturer's section alphabetically by designation or name. A table gives brief details of the world's most important pure trainer aircraft.

Section 4 Missiles

The missile section is organized by role, within each role alphabetically by country of origin, within each country by manufacturer, and within each manufacturer by designation or name. The missiles are arranged as follows:

1. strategic missiles;
2. cruise missiles;
3. surface-to-surface missiles;
4. air-to-surface missiles;
5. anti-ship missiles;
6. anti-submarine missiles;
7. surface-to-air missiles; and
9. anti-tank missiles.

The pace of developments in the military field is ferociously fast, and the author is keenly aware that any book about modern weapons is inevitably out of date in some areas even before it is published. Information about Soviet weapons is also difficult to assess, the primary sources for such information in the West being defence agencies which tend to over- or under-state Soviet capabilities for a variety of reasons. The author thus hopes that the reader will bear these factors in mind, but remains confident that the essence of the weapons described has been conveyed accurately and with reasonable assessment.

SECTION 1

LAND WEAPONS

MAIN BATTLE TANKS AND MEDIUM TANKS

TAMSE/Thyssen Henschel TAM
(West Germany/Argentina)
Type: medium tank
Crew: 4
Combat weight: 30500 kg (67,240 lb)
Dimensions: length, gun forward 8.23 m (27.00 ft) and hull 6.775 m (22.23 ft); width 3.12 m (10.24 ft); height to turret top 2.42 m (7.94 ft)
Armament system: one 105-mm (4.13-in) Rheinmetall Rh-105-30 rifled gun with 50 rounds, two 7.62-mm (0.3-in) FN-MAG machine-guns (one co-axial and one AA) with 6,000 rounds, and four smoke-dischargers on each side of the turret; the turret is electro-hydraulically powered, the main gun is stabilized in elevation and azimuth, and an optical fire-control system is used
Armour: welded steel
Powerplant: one 537–kW (720-hp) MTU diesel engine with 650 litres (143 Imp gal) of internal fuel; two 200-litre (44-Imp gal) auxiliary tanks can be carried at the rear
Performance: speed, road 75 km/h (46.6 mph); range, road 550 km (342 miles) on internal fuel and 900 km (559 miles) with auxiliary fuel; fording 1.4 m (4.6 ft) without preparation, 2.25 m (7.4 ft) with preparation and 4.0 m (13.1 ft) with snorkel; gradient 65%; vertical obstacle 1.0 m (39.4 in); trench 2.5 m (8.2 ft); ground clearance 0.44 m (17.3 in)

Variant
TAM: this is the baseline medium tank model developed in West Germany to Argentine requirements, using the hull of the Marder MICV as the structural and automotive basis, and now in production in Argentina; the same basis has been used for the VCTP infantry fighting vehicle, designed to operate with the TAM; it is not known with certainty whether the main armament is the Rh-105 as stated above or the British ROF L7A3 of the same calibre

ENGESA EE-T1 Osorio
(Brazil)
Type: main battle tank
Crew: 4
Combat weight: 35000 kg (77,160 lb)
Dimensions: length, gun forward 10.03 m (32.91 ft) and hull 7.08 m (23.23 ft); width 3.256 m (10.68 ft); height to turret top 2.36 m (7.74 ft)
Armament system: one 105-mm (4.13-in) L7 rifled gun with 50 rounds, and two 7.62-mm (0.3-in) FN-MAG machine-guns (one co-axial and one AA) with 5,000 rounds or one 7.62-mm FN-MAG co-axial machine-gun with 3,000 rounds and one 0.5-in (12.7-mm) AA Browning machine-gun with 600 rounds, and four smoke-dischargers on each side of the turret; the turret is electrically powered, the main gun is stabilized in elevation and azimuth, and an OIP optical fire-control system with laser rangefinder and stabilized optics
Armour: welded steel
Powerplant: one 746-kW (1,000-hp) NWM TBD 234-V12 diesel engine with ? litres (? Imp gal) of internal fuel
Performance: speed, road 70 km/h (43.5 mph); range, road 550 km (342 miles); fording 1.2 m (3.9 ft) without preparation and 2.0 m (6.6 ft) with preparation; gradient 60%; side slope 40%; vertical obstacle 1.15 m (45.3 in); trench 3.0 m (9.8 ft); ground clearance 0.46 m (18.1 in)

Variants
EE-T1: this is the basic MBT model, and optional features are NBC protection, a more advanced fire-control system, an automatic fire-extinguishing system, and a land navigation system
EE-T2: this is the upgunned model, the main armament being a GIAT 120-mm smooth-bore gun with 40 rounds; this model has a combat weight of 39000 kg (85,979 lb) and a length, gun forward, of 9.995 m (32.79 ft)

Type 80
(China)
Type: main battle tank
Crew: 4
Combat weight: 38000 kg (83,774 lb)
Dimensions: length, gun forward 9.328 m (30.60 ft) and hull 6.325 m (20.75 ft); width 3.372 m (11.06 ft); height to top of AA machine-gun 2.874 m (9.43 ft)
Armament system: one 105-mm (4.13-in) rifled gun with 44 rounds, one 7.62-mm (0.3-in) Type 59T co-axial machine-gun with 2,250 rounds, one 12.7-mm (0.5-in) Type 54 AA machine-gun with 500 rounds, and four smoke-dischargers on each side of the turrent; the turrent is electro-hydraulically powered, the main gun in stabilized in elevation and azimuth, and an ISFCS-212 image-stabilized fire-control system is fitted, this last incorporating stabilized optics, a laser rangefinder, gun and vehicle sensors and a ballistic computer; the type can also generate smoke by injecting fuel into the exhaust system
Armour: cast and welded steel
Powerplant: (probably) one 545–kW (731-hp) Type 12150L-7BW turbocharged diesel with 1400 litres (308 Imp gal) of internal fuel, plus provision for ? litres (? Imp gal) of external fuel
Performance: speed, road 60 km/h (37.3 mph); range,

road 430 km (267 miles); fording 0.8 m (2.6 ft) without preparation and 5.0 m (16.4 ft) with preparation; gradient 60%; vertical obstacle 0.8 m (31.5 in); trench 2.7 m (8.9 ft); ground clearance 0.48 m (18.9 in)

Variants
Type 80: this new Chinese MBT is due to enter production in the later 1980s as the mainstay of the growing armour force fielded by the Chinese into the next century; the type has a wholly new hull, a turret modelled on that of the Type 69, and a rifled 105-mm (4.13-in) gun with an advanced fire-control system; standard equipment includes an NBC system and night vision equipment, and the comparatively high power-to-weight ratio offers better cross-country performance than that of earlier Chinese MBTs

Type 69-II
(China)
Type: main battle tank
Crew: 4
Combat weight: 36500/37000 kg (80,467/81,570 lb)
Dimensions: length, gun forward 8.657 m (28.40 ft) and hull 6.243 m (20.48 ft); width 3.298 m (10.82 ft); height with AA gun at full elevation 3.909 m (12.825 ft)
Armament system: one 100-mm (3.94-in) rifled gun with ? rounds, two 7.62-mm (0.3-in) Type 59T machine-guns (one bow and one co-axial) with ? rounds, and one 12.7-mm (0.5-in) Type 54 AA machine-gun with ? rounds; the turret is electro-hydraulically powered, the main gun is stabilized in elevation and azimuth, and a Tank Simplified Fire-Control System is fitted, this last incorporating a laser rangefinder, optical sights and a ballistic computer; an optional fit for export models is the TSFCS-C which adds sensors for variables such as cross wind, ambient temperature and propellant-charge temperature; the type can also generate smoke by injecting fuel into the exhaust system
Armour: cast and welded steel varying in thickness from 20 to 203 mm (0.79 to 8 in)
Powerplant: one 435-kW (583-hp) Model 12150L-7BW diesel engine with ? litres (? Imp gal) of internal fuel
Performance: speed, road 50 km/h (31.1 mph); range, road 440 km (273 miles); fording 1.4 m (4.6 ft); gradient 32 degrees, side slope 30 degrees; vertical obstacle 0.8 m (31.5 in); trench 2.7 m (8.9 ft); ground clearance 0.425 m (16.75 in)

Variants
Type 69-I: introduced in the early 1980s, the Type 69-I is in essence an improved Type 59 MBT with superior smooth-bore armament allied to a new fire-control system and night vision equipment; the type also features a slightly more powerful diesel powerplant; standard features are NBC protection and an engine-compartment fire-suppression system
Type 69-II: further improved model optimized for the export market and fitted with a more accurate rifled gun
Type 69-II Mk B: this is the command version of the Type 69-II with an additional radio set; a further development with two radios of the same type is the **Type 69-II Mk C**
Type 653: armoured recovery model of the Type 69 with a welded superstructure, winch, dozer blade and hydraulic crane; the type has a five-man crew and a combat weight of 38000 kg (83,774 lb)
Type 80: self-propelled twin 57-mm AA mounting based conceptually on the Soviet ZSU-57-2 but using the hull of the Type 69-II MBT; the type has a combat weight of 31000 kg (68,342 lb) and has a 433-kW (580-hp) diesel, and is limited to clear-weather operations by its lack of anything but optical sights; the Chinese have also developed a twin 37-mm AA mounting on the same chassis, though the designation for this useful system remains unknown; the two-man power-operated turrent has an optical fire-control system for the twin P793 Type B cannon, each with a cyclic rate of 380 rounds per minute and muzzle velocity of 1000 m (3,281 ft) per second; the effective AA range is 4000 m (4,375 yards), and the practical surface-to-surface range 3500 m (3,830 yards); combat weight is 34200 kg (75,397 lb)
Type 84: armoured vehicle-launched bridge variant of the Type 69, with the turret replaced by a hydraulically-operated folding bridge able to span a gap of 16.0 m (52.5 ft); combat weight is 38500 kg (84,876 lb)

Type 59
(China)
Type: main battle tank
Crew: 4
Combat weight: 36000 kg (79,365 lb)
Dimensions: length, gun forward 9.00 m (29.53 ft) and hull 6.17 m (20.24 ft); width 3.27 m (10.73 ft); height to turret top 2.40 m (7.87 ft)
Armament system: one 100-mm (3.94-in) Type 59 rifled gun with 34 rounds, two 7.62-mm (0.3-in) Type 59T machine-guns (one co-axial and one in the glacis) with 3,500 rounds, and one 12.7-mm (0.5-in) Type 54 AA machine-gun with 200 rounds; the turret is electro-hydraulically powered, the main gun is stabilized in elevation but not in azimuth, and an optical fire-control system is fitted though the type can be retrofitted with the TSFCS-C fire-control system (see Type 69-II); the type can also generate smoke by injecting fuel into the exhaust system
Armour: cast and welded steel varying in thickness from 20 mm (0.79 in) to 203 mm (8 in)
Powerplant: one 388-kW (520-hp) V-12 diesel engine with 960 litres (211 Imp gal) of internal fuel, plus

provision for 400 litres (88 Imp gal) of auxiliary fuel
Performance: speed, road 50 km/h (31.1 mph); range, road 400 km (249 miles) on internal fuel and 600 km (373 miles) with auxiliary fuel; fording 1.4 m (4.6 ft) without preparation and 5.5 m (18.0 ft) with preparation; gradient 60%; vertical obstacle 0.79 m (31 in); trench 2.68 m (8.8 ft); ground clearance 0.425 m (16.7 in)

Variant
Type 59: this is the Chinese derivative of the Soviet T-54, and is currently the main gun tank fielded by the armies of China, and the model has also been exported to Albania, Congo, Kampuchea, North Korea, Pakistan, Sudan, Tanzania and Vietnam; recently many tanks have been upgraded with Western guns (105-mm/4.13-in L7A3), laser rangefinders and night-vision devices

GIAT AMX-48 Leclerc
(France)
Type: main battle tank
Crew: 3
Combat weight: 50000 kg (110,229 lb)
Dimensions: length, gun forward ? m (? ft) and hull 6.60 m (21.65 ft); width ? m (? ft); height 2.30 m (7.55 ft)
Armament system: one 120-mm (4.72-in) GIAT smooth-bore gun with 40 rounds, one 20-mm F2 co-axial cannon with ? rounds, one 7.62-mm (0.3-in) NF1 AA machine-gun with ? rounds, and three smoke-dischargers on each side of the turret; the turret is electro-hydraulically powered, the main gun is stabilized in elevation and azimuth, and a Sagem digital fire-control system is fitted; this last combines optical sights, a low-light-level TV sight, various sensors, a laser rangefinder and a ballistic computer
Armour: welded steel with superimposed reactive armour
Powerplant: one 1120-kW (1,502-hp) Poyaud V8X-1500 Hyperbar diesal engine with ? litres (? Imp gal) of internal fuel
Peformance: not revealed

Variants
Leclerc: this is planned as the main French MBT for the 1990s and beyond; the prototype appeared early in 1987, and is characterized by its low overall height, very high power-to-weight ratio (offering the probability of high cross-country performance), and the latest reactive armour, which is a layer of small explosive plates over the basic steel armour, designed to detonate and so snuff out the hypervelocity gas jet of incoming HEAT rounds; the type has full protection, and is provided with a snorkel kit for deep wading

GIAT AMX-40
(France)
Type: main battle tank
Crew: 4
Combat weight: 43000 kg (94,797 lb)
Dimensions: length, gun forward 10.04 m (32.94 ft) and hull 6.80 m (22.31 ft); width 3.36 m (11.02 ft); height to turret top 2.38 m (7.81 ft)
Armament system: one 120-mm (4.72-in) GIAT smooth-bore gun with 37 rounds, one 20-mm F2 co-axial cannon with 580 rounds, one 7.62-mm (0.3-in) NF1 machine-gun with 2,170 rounds, and three smoke-dischargers on each side of the turret; the turret is electro-hydraulically powered, the main gun is stabilized in elevation and azimuth, and a COTAC fire-control system is fitted; this last combines optical sights, a low-light-level TV sight, various sensors, a laser rangefinder and a ballistic computer
Armour: welded and laminate
Powerplant: one 820-kW (1,100-hp) Poyaud V12X diesel engine with 1100 litres (242 Imp gal) of internal fuel
Performance: speed, road 65 km/h (40.4 mph); range, road 550 km (342 miles); fording 1.3 m (4.3 ft) without preparation and 2.3 m (7.5 ft) with preparation; gradient 60%; side slope 30%; vertical obstacle 1.0 m (39.4 in); trench 3.2 m (10.5 ft); ground clearance 0.45 m (17.7 in)

Variant
AMX-40: this is a French development vehicle serving primarily as an export type but also as prototype for a new-generation of French MBTs, current MBTs being obsolescent in terms of protection and fire-control compared with American, British and West German MBTs; the type has full NBC protection

GIAT AMX-32
(France)
Type: main battle tank
Crew: 4
Combat weight: 43000 kg (94,797 lb)
Dimensions: length, gun forward 9.45 m (31.0 ft) and hull 6.59 m (21.62 ft); width 3.24 m (10.63 ft); height to turret top 2.29 m (7.51 ft)
Armament system: one 105-mm (4.13-in) GIAT F1 rifled gun with 47 rounds, one 20-mm GIAT F2 co-axial cannon with 480 rounds, one 7.62-mm (0.3-in) GIAT NF1 cupola-mounted AA machine-gun with 2,170 rounds, and three smoke-dischargers on each side of the turret; the turret is electro-hydraulically powered, the main gun is stabilized in elevation and azimuth, and a COTAC fire-control system is fitted; this last combines optical sights, a low-light-level TV, various sensors, a laser rangefinder and a ballistic computer
Armour: conventional welded and cast steel, spaced and composite types

Powerplant: one 596-kW (800-hp) Hispano-Suiza HS 110-2-SR multi-fuel engine with 920 litres (202 Imp gal) of internal fuel
Performance: speed, road 65 km/h (40.4 mph); range, road 530 km (329 miles); fording 1.3 m (4.3 ft) without preparation and 2.2 m (7.2 ft) with preparation; gradient 60%; side slope 30%; vertical obstacle 0.9 m (35.4 in); trench 2.9 m (9.5 ft); ground clearance 0.44 m (17.3 in)

Variant
AMX-32: this is a development and export vehicle intended to combine the best features of the AMX-30 series with lighter weight and greater power (producing a better power-to-weight ratio for improved performance and mobility) and composite armour for better protection

GIAT AMX-30B2
(France)
Type: main battle tank
Crew: 4
Combat weight: 37000 kg (81,570 lb)
Dimensions: length, gun forward 9.48 m (31.1 ft) and hull 6.59 m (21.62 ft); width 3.10 m (10.17 ft); height to turret top 2.29 m (7.51 ft)
Armament system: one 105-mm (4.13-in) GIAT F1 rifled gun with 47 rounds, one 20-mm GIAT F1 co-axial cannon with 480 rounds, one 7.62-mm (0.3-in) GIAT NF1 cupola-mounted AA machine-gun with 2,050 rounds, and two smoke-dischargers on each side of the turret; the turret is electro-hydraulically powered, the main gun is stabilized in elevation and azimuth, and a COTAC fire-control system is fitted; this last combines optical sights, a low-light-level TV, various sensors, a laser rangefinder and a ballistic computer
Armour: welded and cast steel
Powerplant: one 537-kW (720-hp) Hispano-Suiza HS 110 multi-fuel engine with 900 litres (198 Imp gal) of internal fuel
Performance: speed, road 65 km/h (40.4 mph); range, road 450 km (280 miles); fording 1.2 m (3.9 ft) without preparation and 2.2 m (7.2 ft) with preparation; gradient 60%; side slope 30%; vertical obstacle 0.9 m (35.4 in); trench 2.9 m (9.5 ft); ground clearance 0.44 m (17.3 in)

Variants
AMX-30: baseline French MBT introduced in the mid-1960s with a 522-kW (700-hp) Hispano-Suiza HS 110 engine plus 970-litre (213-Imp gal) fuel capacity, a combat weight of 36000 kg (79,365 lb), a 12.7-mm (0.5-in) co-axial machine-gun and a less sophisticated fire-control system
AMX-30B2: designation of improved model introduced in 1979 with the COTAC integrated fire-control system, new gearbox and many detail improvements
AMX-30S: lower-powered version designed for desert operations and fitted with appropriate special equipment (sand shields etc)
AMX-30D: armoured recovery vehicle based on the chassis of the AMX-30 with a dozer blade, crane, 35000-kg (77,160-lb) main winch and 3500-kg (7,716-lb) auxiliary winch; there are three models of the ARV, the **AMX-30D** able to lift only 12000 kg (26,455 lb) except directly to the front of the vehicle, the **AMX-30D1** able to lift 15000 kg (33,069 lb) through an arc of 240 degrees, and the **AMX-30D(S)** desert version based on the AMX-30S
AMX-30R: this is the SAM version designed for the battlefield engagement of low-level targets ranging from stationary helicopters to supersonic attack aircraft; the twin-tube launcher (for Roland 1 and 2 missiles) and associated radar arrangement are similar to those of the West German Roland/Marder system, but French identification friend or foe equipment is fitted
AMX-30SR: AA variant delivered to Saudi Arabia from the late 1970s with a TG 230A turret carrying two HSS-831A cannon, 1,500 rounds and improved Oeil Vert search and fire-control radar
AMX-30 SABRE: AA model with the SABRE turret carrying two M693 (F2), HSS-820 or MK 20 Rh 202 20-mm cannon and an optional fire-control system
AMX-30 VLB: this vehicle-launched bridge version can span a gap of some 20 m (65.6 ft) with a 22-m (72.2 ft) scissors-type bridge launched from the rear of the turretless vehicle
AMX-30 EBG: combat engineer tractor version of the AMX-30 with a dozer blade, 20000-kg (44,092-lb) capacity winch, auger, 142-mm (5.59-in) demolition gun, and four mine-launcher tubes
Shahine, SICA: AA missile version of the AMX-30 series, developed in parallel with the AMX-30 SA AA gun system during the later 1970s; there are two components to the system, an acquisition unit with Thomson-CSF pulse-Doppler radar able to detect targets at a range of 18.5 km (11.5 miles) and handle simultaneously 18 of the 40 computer-registered targets, and a fire unit with three ready-to-fire Matra R.460 SAMs on each side of the 17-km (10.6-mile) range fire-control/missile-control radar; each type of unit has back-up TV sensors for fair-weather operation in the event of radar failure; reload missiles are carried on a separate vehicle and loaded with the aid of a crane

Krauss-Maffei/Krupp MaK Leopard 2
(West Germany)
Type: main battle tank
Crew: 4
Combat weight: 55150 kg (121,583 lb)
Dimensions: length, gun forward 9.668 m (31.72 ft)

and hull 7.722 m (25.33 ft); width over skirts 3.70 m (12.14 ft); height to turret top 2.46 m (8.07 ft)
Armament system: one 120-mm (4.72-in) Rheinmetall Rh-120 smooth-bore gun with 42 rounds, two 7.62-mm (0.3-in) MG3A1 machine-guns (one co-axial and one cupola-mounted) with 4,750 rounds, and eight smoke-dischargers on each side of the turret; the turret is electro-hydraulically powered, the main gun is stabilized in elevation and azimuth, and an EMES 15 fire-control system is fitted; this last combines optical sights, thermal sights, a laser rangefinder and a ballistic computer
Armour: spaced multi-layer type
Powerplant: one 820-kW (1,100-hp) MTU MB 873 Ka501 diesel engine with 1200 litres (264 Imp gal) of internal fuel
Performance: speed, road 72 km/h (44.7 mph); range, road 550 km (342 miles); fording 1.0 m (3.3 ft) without preparation, 2.25 m (7.4 ft) with preparation and 4.0 m (13.1 ft) with snorkel; gradient 60%; side slope 30%; vertical obstacle 1.1 m (43.3 in); trench 3.0 m (9.8 ft); ground clearance 0.63 m (24.8 in)

Variants
Leopard 2: baseline MBT in service with the West German army since 1979 and ordered by several other nations; the type is an extremely capable MBT, and features such modern systems as NBC protection, passive night-vision equipment and automatic fire-extinguishing
Leopard 2 AEV: armoured engineer vehicle derivative of the basic type currently under development
Bergepanzer 3: armoured recovery vehicle derivative of the basic type currently under development
Leopard 2 (Netherlands): version for Dutch service with FN-MAG machine-guns and various items of Dutch equipment
Leopard 2 (Switzerland): version for Swiss service with detail modifications and items of Swiss equipment

Krauss-Maffei/Krupp MaK Leopard 1A4
(West Germany)
Type: main battle tank
Crew: 4
Combat weight: 42400 kg (93,474 lb)
Dimensions: length, gun forward 9,543 m (31.31 ft) and hull 7.09 m (23.26 ft); width with skirts 3.37 m (11.06 ft); height to top of commander's periscope 2.764 m (9.07 ft)
Dimensions: one 105-mm (4.13-in) ROF/Rheinmetall L7A3 rifled gun with 60 rounds, two 7.62-mm (0.3-in) MG3A1 machine-guns (one co-axial and one cupola-mounted) with 5,500 rounds, and four smoke-dischargers on each side of the turret; the turret is electro-hydraulically powered, the main gun is stabilized in elevation and azimuth, and an EMES 12 fire-control system is fitted; this combines optical sights, a computer-controlled rangefinder and a ballistic computer
Armour: steel, varying in thickness from 10 mm (0.39 in) to 70 mm (2.76 in)
Powerplant: one 619-kW (830-hp) MTU MB 838 CaM500 multi-fuel engine with 985 litres (217 Imp gal) of internal fuel
Performance: speed, road 65 km/h (40.4 mph); range, road 600 km (373 miles); fording 1.2 m (3.9 ft) without preparation, 2.25 m (7.4 ft) with preparation and 4.0 m (13.1 ft) with snorkel; gradient 60%; side slope 30%; vertical obstacle 1.15 m (45.3 in); trench 3.0 m (9.8 ft); ground clearance 0.44 m (17.3 in)

Variants
Leopard 1: baseline West German MBT, which began to enter service in 1965; since that time these early vehicles have been upgraded to **Leopard 1A1** (thermal sleeve on the gun, gun-stabilization system, rubber skirts etc) and **Leopard 1A1A1** (additional turret and mantlet spaced armour plus detail improvements) standards
Leopard 1A2: similar to the Leopard 1A1A1 without spaced mantlet and turret armour but with an improved NBC system, a turret of improved steel, passive night sights and other detail modifications
Leopard 1A3: based on the Leopard 1A1 and Leopard L1A2 but with a new turret of welded spaced armour and wedge-shaped mantlet
Leopard 1A4: based on the Leopard 1A3 but with an integrated fire-control system; the type has full NBC protection and night-vision devices
Leopard AS 1: variant of the Leopard 1A3 for Australia with tropical kit and a SABCA integrated fire-control system with laser rangefinder and seven sensors
Leopard 1 (Belgium): Belgian version with FN-MAG machine-guns, detail modifications and SABCA fire-control system
Leopard 1 (Canada): Canadian version of the Leopard 1A3 with SABCA fire-control system
Leopard 1 (Denmark): Danish version of the Leopard 1 with only very small differences from the West German model
Leopard 1 (Greece): Greek version of the Leopard 1A3 with only very small differences from the West German model
Leopard 1 (Italy): Italian version produced by OTO Melara with only small alterations compared with the West German model
Leopard 1 (Netherlands): Dutch model with different radios, machine-guns, smoke-dischargers etc
Leopard 1 (Norway): Norwegian model with only very small differences from the West German model
Leopard 1 (Turkey): Turkish model with only very small differences from the West German Leopard 1A3 model
Leopard 1 ARV: armoured recovery vehicle variant

with a front-mounted dozer blade, a crane, a 35000-kg (77,160-lb) capacity winch and other equipment
Leopard 1 AEV: armoured engineer vehicle variant derived from the Leopard 1 ARV, but has an auger instead of the spare powerpack, a dozer blade fitted with scarifiers for the ripping-up of roads, a stock of explosives, and a heat exchanger system
Leopard 1 AVLB: armoured vehicle-launched bridge variant to span a gap of some 20 m (65.6 ft) with a 22-m (72.2-ft) bridge

Israeli Ordnance Corps Merkava Mk 1
(Israel)
Type: main battle tank
Crew: 4
Combat weight: 60000 kg (132,275 lb)
Dimensions: length, gun forward 8.63 m (28.31 ft) and hull 7.45 m (24.44 ft); width 3.70 m (12.14 ft); height to turret top 2.64 m (8.66 ft)
Armament system: one 105-mm (4.13-in) IMI M68 rifled gun with 92 rounds, three 7.62-mm (0.3-in) FN-MAG machine-guns (one co-axial and two roof-mounted) with 10,000 rounds, and one 60-mm (2.36-in) mortar with 30 rounds; the turret is electro-hydraulically powered, the main gun is stabilized in elevation and azimuth, and an Elbit fire-control system is fitted; this last combines optical sights, various sensors, a laser rangefinder and a ballistic computer
Armour: cast and welded steel
Powerplant: one 900-hp (671-kW) Teledyne Continental AVDS-1790-6A diesel engine with 900 litres (198 Imp gal) of internal fuel
Performance: speed, road 46 km/h (28.6 mph); range, road 400 km (249 miles); fording 1.38 m (4.5 ft) without preparation and 2.0 m (6.6 ft) with preparation; gradient 60%; side slope 38%; vertical obstacle 0.95 m (37.4 in); trench 3.0 m (9.8 ft); ground clearance 0.47 m (18.5 in)

Variants
Merkava Mk 1: this extremely capable MBT began to enter service with the Israeli army in 1979, and features full NBC protection, an explosion-suppressing system and night-vision equipment
Merkava Mk 2: entering service in 1983, this version of the Merkava has stabilized optics with slaved main armament, modifications to ease maintenance, Blazer reactive armour and improved fire-suppression equipment
Merkava Mk 3: due for service before 1990, the Merkava Mk 3 has armour 100% improved over that of the Mk 1, a 120-mm (4.72-in) smooth-bore gun, a 1200-hp (895-kW) powerpack for much enhanced performance and mobility, and hydro-pneumatic rather than helical spring suspension

OTO Melara OF-40
(Italy)
Type: main battle tank
Crew: 4
Combat weight: 45500 kg (100,309 lb)
Dimensions: length, gun forward 9.22 m (30.25 ft) and hull 6.893 m (22.61 ft); width with skirts 3.51 m (11.52 ft); height to turret top 2.42 m (7.94 ft)
Armament system: one 105-mm (4.13-in) OTO Melara OTO 105/52 rifled gun with 57 rounds, two 7.62-mm (0.3-in) FN-MAG machine-guns (one co-axial and one hatch-mounted AA) with 5,700 rounds, and four smoke-dischargers on each side of the turret; the turret is electro-hydraulically powered, the main gun has optional stabilization in elevation and azimuth, and an Officine Galileo fire-control system is fitted; this last combines stabilized optics, various sensors, a laser rangefinder and a ballistic computer
Armour: welded steel
Powerplant: one 619-kW (830-hp) MTU MB838 CaM500 multi-fuel engine with 1000 litres (220 Imp gal) of internal fuel
Performance: speed, road 65 km/h (40.4 mph); range, road 600 km (373 miles); fording 1.2 m (3.9 ft) without preparation, 2.25 m (7.4 ft) with preparation and 4.0 m (13.1 ft) with snorkel; gradient 60%; side slope 30%; vertical obstacle 1.1 m (43.3 in); trench 3.0 m (9.8 ft); ground clearance 0.44 m (17.3 in)

Variants
OF-40 Mk 1: this is the baseline MBT with such standard features as NBC protection and passive night-vision devices
OF-40 Mk 2: improved model with improved reliability, main armament stabilization and the upgraded Officine Galileo OG14L2 fire-control system
OF-40 ARV: armoured recovery vehicle using the standard OF-40 chassis, and fitted with a spade, 20000-kg (44,092-lb) capacity crane and 30000-kg (66,138-lb) capacity winch

Mitsubishi Type 88
(Japan)
Type: main battle tank
Crew: 4
Combat weight: 43000 kg (94,797 lb)
Dimensions: length, hull 7.50 m (24.61 ft); width 3.50 m (11.48 ft); height 2.40 m (7.87 ft)
Armament system: one 120-mm (4.72-in) Japan Iron Works smooth-bore gun with ? rounds, one 7.62-mm (0.3-in) co-axial machine-gun with ? rounds, and one 0.5-in (12.7-mm) Browning M2HB AA machine-gun with ? rounds; the turret is electro-hydraulically powered, the main gun is stabilized in elevation and azimuth, and an advanced fire-control system is fitted;

this last combines stabilized optical sights, a low-light-level TV, various sensors, a laser rangefinder and a ballistic computer
Armour: multiple-layer steel with ceramic pockets
Powerplant: one 1125-kW (1,509-hp) Mitsubishi 8ZG turbocharged diesel engine with ? litres (? Imp gal) of internal fuel
Performance: speed, road 70 km/h (43.5 mph); range, road 550 km (311 miles); fording not revealed; gradient not revealed; side slope not revealed; vertical obstacle not revealed; trench not revealed; ground clearance not revealed

Variant
Type 88: due to enter service in 1989 or 1990, this MBT is designed as an advanced successor to the Type 61

Mitsubishi Type 74
(Japan)
Type: main battle tank
Crew: 4
Combat weight: 38000 kg (83,774 lb)
Dimensions: length, gun forward 9.41 m (30.87 ft) and hull 6.70 m (21.98 ft); width 3.18 m (10.43 ft); height to turret top 2.48 m (8.14 ft)
Armament system: one 105-mm (4.13-in) ROF L7 rifled gun with 55 rounds, one 7.62-mm (0.3-in) Type 74 co-axial machine-gun with 4,500 rounds, one 0.5-in (12.7-mm) Browning M2HB roof-mounted machine-gun with 660 rounds, and three smoke-dischargers on each side of the turret; the turret is electrically powered, the main gun is stabilized in elevation and azimuth, and a Mitsubishi fire-control system is fitted; this last combines optical sights, a laser rangefinder and a ballistic computer
Armour: cast and welded steel
Powerplant: one 559-kW (750-hp) Mitsubishi 10ZF Type 22 WT diesel engine with 950 litres (209 Imp gal) of internal fuel
Performance: speed, road 53 km/h (33.0 mph); range, road 300 km (186 miles); fording 1.0 m (3.3 ft) without preparation and 2.0 m (6.6 ft) with preparation; gradient 60%; side slope 40%; vertical obstacle 1.0 m (39.4 in); trench 2.7 m (8.9 ft); ground clearance adjustable from 0.2 to 0.65 m (7.9 to 25.6 in)

Variants
Type 74: baseline Japanese MBT which entered service in 1974; the type has interesting variable-height hydro-pneumatic suspension that permits the crew not only to vary the type's ground clearance to suit the terrain, but also to incline the vehicle up/down and sideways; an NBC system is standard, and it is believed that the maximum speed is higher than officially admitted

Type 78 ARV: based on the chassis of the Type 74 MBT, this armoured recovery vehicle has a dozer blade and a side-mounted crane

Mitsubishi Type 61
(Japan)
Type: main battle tank
Crew: 4
Combat weight: 35000 kg (77,160 lb)
Dimensions: length, gun forward 8.19 m (26.87 ft) and hull 6.30 m (20.57 ft); width 2.95 m (9.68 ft); height to turret roof 2.49 m (8.17 ft)
Armament system: one 90-mm (3.54-in) Type 61 rifled gun with ? rounds, one 0.3-in (7.62-mm) Browning M1919A4 co-axial machine-gun with ? rounds, and one 0.5-in (12.7-mm) Browning M2HB cupola-mounted machine-gun with ? rounds, and three smoke-dischargers on each side of the turret; the turret is hydraulically powered, the main gun is unstabilized in either elevation and azimuth, and an optical fire-control system is fitted
Armour: cast and welded steel varying in thickness from 15 mm to 64 mm (0.59 in to 2.52 in)
Powerplant: one 447-kW (60-hp) Mitsubishi Type 12 HM 21 WT diesel engine with ? litres (? Imp gal) of internal fuel
Performance: speed, road 45 km/h (28.0 mph); range, road 200 km (124 miles); fording 0.99 m (3.25 ft) without preparation; gradient 60%; side slope ?; vertical obstacle 0.685 m (27 in); trench 2.489 m (8.17 ft); ground clearance 0.4 m (15.75 in)

Variants
Type 61: this type first saw service in 1962; it is now obsolete as a result of its poor performance and armament, and its lack of any adequate fire-control capability and even NBC protection
Type 67 AVLB: this armoured vehicle-launched bridge uses the chassis of the Type 61 MBT and can span a 10-m (32.8-ft) gap with a 12-m (39.4-ft) bridge
Type 67 AEV: this armoured engineer vehicle again uses the chassis of the Type 61 MBT, and is equipped with a dozer blade and a light-capacity crane
Type 70 ARV: this armoured recovery vehicle is another type developed on the chassis of the Type 61 MBT, and it is equipped with an A-frame, a dozer blade and a winch

Bofors Stridsvagn 103B
(Sweden)
Type: main battle tank
Crew: 3
Combat weight: 39700 kg (87,522 lb)
Dimensions: length overall 8.99 m (29.49 ft) and hull 7.04 m (23.10 mft); width 3.63 m (11.91 ft); height to top of cupola 2.14 m (7.02 ft)

Armament system: one 105-mm (4.13-in) Bofors L74 rifled gun with 50 rounds, three 7.62-mm (0.3-in) FFV ksp58 machine-guns (two co-axial and one AA) with 2,750 rounds, and four smoke-dischargers on each side of the hull; the gun has no stabilization, and a simple fire-control system with stabilized optical sights is fitted
Armour: welded steel
Powerplant: one 240-bhp (179-kW) Rolls-Royce K60 multi-fuel engine and one 490-shp (365-kW) Boeing 553 gas turbine with 960 litres (211 Imp gal) of internal fuel
Performance: speed, road 50 km/h (31 mph); range, road 390 km (242 miles); fording 1.5 m (4.9 ft) without preparation and with preparation amphibious; gradient 58%; side slope 70%; vertical obstacle 0.9 m (35.4 in); trench 2.3 m (7.5 ft); ground clearance 0.4 m (15.75 in)

Variants
Strv 103B: designation of the full-production model with permanent dozer blade and flotation collar; some vehicles has twin Lyran flare launchers; the type is not fitted with an NBC system; this is a highly unusual MBT, the main armament being fixed in the turretless hull, elevation being achieved by varying the comparative height of the road wheels, and traverse by slewing the entire vehicle
Strv 103B (Modernised): designation of the improved standard to which all Strv 103Bs are being brought; this has a 275-hp (205-kW) Detroit Diesel 6V-53T engine in place of the Rolls-Royce unit, and a Bofors Aerotronics fire-control system with a laser range-finder and ballistic computer; in 1986 the type was designated **Strv 103C**

Federal Construction Pz 68
(Switzerland)
Type: main battle tank
Crew: 4
Combat weight: 39700 kg (87,522 lb)
Dimensions: length, gun forward 9.49 m (31.14 ft) and hull 6.98 m (22.90 ft); width 3.14 m (10.30 ft); height to top of cupola 2.75 m (9.02 ft)
Armament system: one 105-mm (4.13-in) PzKan 61 (ROF L7A1) rifled gun with 56 rounds, two 7.5-in (0.295-in) MG-51 machine-guns (one co-axial and one AA) with 5,400 rounds, and three smoke-dischargers on each side of the turret; the turret is electro-hydraulically operated, the main gun is stabilized in elevation and azimuth, and a simple optical fire-control system is fitted
Armour: cast steel varying in thickness from 20 to 60 mm (0.79 to 2.36 in)
Powerplant: one 492-kW (660-hp) MTU MB 837 diesel engine with 710 litres (156 Imp gal) of internal fuel
Performance: speed, road 55 km/h (34 mph); range, road 350 km (217 miles); fording 1.1 m (3.6 ft) without preparation and 2.3 m (7.5 ft) with preparation; gradient 70%; side slope 30%; vertical obstacle 1.0 m (39.4 in); trench 2.6 m (8.5 ft); ground clearance 0.41 m (16 in)

Variants
Pz 68: designation of improved version of the Pz 61 with gun-stabilization system, ammunition resupply hatch, upgraded powerplant and better tracks; later redesignated **Pz 68 Mk 1**
Pz 68 Mk 2: improved Pz 68 Mk 1 with thermal sleeve for main armament
Pz 68 Mk 3: improved Pz 68 Mk 2 with larger turret
Pz 68 Mk 4: product-improved Pz 68 Mk 3
Brückenlegepanzer 68: armoured vehicle-launched bridge based on the Pz 68 chassis with a bridge measuring 18.23 m (59.8 ft) in length

Federal Construction Pz 61
(Switzerland)
Type: main battle tank
Crew: 4
Combat weight: 38000 kg (83,774 kg)
Dimensions: length, gun forward 9.43 m (30.94 ft) and hull 6.78 m (22.24 ft); width 3.06 m (10.04 ft); height to top of cupola 2.72 m (8.92 ft)
Armament system: one 105-mm (4.13-in) PzKan 61 (ROF L7A1) rifled gun with 56 rounds, two 7.5-mm (0.295-in) MG-51 machine-guns (one co-axial and one AA) with 5,400 rounds, and three smoke-dischargers on each side of the turret; the turret is electro-hydraulically powered, the main gun lacks stabilization in either elevation or azimuth, and a simple optical fire-control system is fitted
Armour: cast steel varying in thickness from 20 to 60 mm (0.79 to 2.36 in)
Powerplant: one 470-kW (630-hp) MTU MB 837 diesel engine with 760 litres (167 Imp gal) of internal fuel
Performance: speed, road 50 km/h (31 mph); range, road 300 km (186 miles); fording 1.1 m (3.6 ft) without preparation; gradient 70%; side slope 30%; vertical obstacle 0.75 m (29.5); trench 2.6 m (8.5 ft); ground clearance 0.42 m (16.5 in)

Variants
Pz 61: this was the first indigenously designed Swiss tank, and began to enter service in 1965; the initial models had a 20-mm co-axial cannon, but survivors have been modified to the **Pz 61 AA9** standard described above; the tank has an NBC system but only an optical fire-control system
Entpannungspanzer 65: armoured recovery vehicle based on the Pz 61, and fitted with a 15000-kg (33,069–kg) A-frame and a 25000-kg (55,115-lb) winch; the crew is five, and combat weight 38000 kg (83,774 lb)

T-80 MBT (USSR): see T-64 MBT

T-72
(USSR)
Type: main battle tank
Crew: 3
Combat weight: 41000 kg (90,388 lb)
Dimensions: length, gun forward 9.24 m (30.31 ft) and hull 6.95 m (22.80 ft); width with skirts 4.75 m (15.58 ft); height to top of cupola 2.37 m (7.78 ft)
Armament system: one 125-mm (4.92-in) 2A46 Rapira 3 smooth-bore gun with 40 rounds (including AT-8 Kobra missiles), one 7.62-mm (0.3-in) PKT co-axial machine-gun with 3,000 rounds, one 12.7-mm (0.5-in) NSVT AA machine-gun with 500 rounds, and 12 smoke-dischargers (five on the right of the turret and seven on the left); the type can also generate smoke by injecting fuel into the exhaust; the turret is electrically operated, the main gun is stabilized in elevation and azimuth, and an advanced fire-control system is fitted; this last combines optical sights and a ballistic computer
Armour: cast, welded and composite
Powerplant: one 580-kW (778-hp) V-46 diesel engine with 1000 litres (220 Imp gal) of internal fuel, plus provisional for 400 litres (88 Imp gal) of external fuel
Performance: speed, road 60 km/h (37 mph); range, road 480 km (298 miles) on internal fuel and 700 km (435 miles) with external fuel; fording 1.4 m (4.6 ft) without preparation and 5.5 m (18.0 ft) with snorkel; gradient 60%; side slope 40%; vertical obstacle 0.85 m (33.5 in); trench 2.8 m (9.2 ft); ground clearance 0.47 m (18.5 ft)

Variants
T-72: baseline model of one of the Soviets' most important MBTs, believed to have entered service in 1971 and to be a development of the T-64 with simplified systems, torsion-bar rather than hydro-mechanical suspension, and a different powerplant; all versions have an NBC system and night-vision devices; another modification from the T-64 series is the use of a cassette rather than basket autoloading system with the charges and projectiles in a two-row arrangement on the turret floor
T-72 M1980/1: improved version of the T-72 with new fire-control system using a laser rangefinder, improved armour and additional stowage on the outside of the vehicle; this version is designated M1980 in the USA but as the **T-74** by Western European nations
T-72 M1980/2: improved version of the T-72
T-72 M1981/1: improved version of the T-72
T-72 M1981/2: improved version of the T-72 with new fabric side skirts
T-72 M1981/3: improved version of the T-72
T-72K: command version of the T-72
T-72M1: late-production version of the T-72, believed to be designated **T-74** by the USSR; the **T-74M** version has additional overhead armour plus anti-radiation lining material to reduce the effects of enhanced-radiation (neutron) weapons and of electro-magnetic pulse
T-72MK: command version of the T-72M, believed to be designated **T-74K** by the USSR
BREM-1: armoured recovery vehicle variant

T-64A
(USSR)
Type: main battle tank
Crew: 3
Combat weight: 38000 kg (83,774 lb)
Dimensions: length, gun forward 9.10 m (29.86 ft) and hull 6.40 m (21.0 ft); width with skirts 4.64 m (15.22 ft); height to top of cupola 2.30 m (7.55 ft)
Armament system: one 125-mm (4.92-in) 2A46 Rapira 3 smooth-bore gun with 40 rounds, one 7.62-mm (0.3-in) PKT co-axial machine-gun with 3,000 rounds, and one 12.7-mm (0.5-in) NSVT AA machine-gun with 500 rounds; the turret is electrically powered, the main gun is stabilized in elevation and azimuth, and a fire-control system is fitted; the last combines optical sights and a ballistic computer; the type can generate smoke by injecting fuel into the exhaust system
Armour: cast and composite
Powerplant: one 559-kW (750-hp) five-cylinder diesel engine with 1000 litres (220 Imp gal) of internal and 400 litres (88 Imp gal) of external fuel
Performance: speed, road 70 km/h (43.5 mph); range, road 450 km (280 miles) on internal fuel and 700 km (435 miles) with external fuel; fording 1.4 m (4.6 ft) without preparation and 5.5 m (18.0 ft) with snorkel; gradient 60%; side slope 40%; vertical obstacle 0.915 m (36 in); trench 2.72 m (8.9 ft); ground clearance 0.377 m (14.8 in)

Variants
T-64: baseline variant of the Soviet army's current mainstay MBT, which began to enter service in 1968 and is armed, according to some analysts, with a 2A20 U-5TS Rapira 1 115-mm (4.53-in) main gun or, according to others, with a short-barrel version of the 2A46 Rapira 3 125-mm (4.92-in) gun; standard equipment includes an NBC protection pack and night-vision devices; the type has suffered from numerous mechanical problems, but is nonetheless significant for its advanced capabilities and high-quality fire-control system
T-64A: current model with a 125-mm (4.72-in) Rapira 3 main gun and automatic loader; it is believed that the type has suffered severe powerplant, transmission and suspension problems, which may explain why, in conjunction with its advanced capabilities, it has never been exported and then rapidly complemented by the T-72 series; there are several variants within the T-64A series, these being notable for revised

stowage, the addition of smoke-dischargers and the provision of less complex skirts; internally, one of the most important features of the T-64A is the provision of a basket-type autoloader for the main armament, with the semi-combustible propellant casings arranged in a ring round the inside of the turret ring and the projectiles in a ring arrangement in trays on the turret floor

T-64B: variant of the T-64A with a combined gun/missile-launcher main armament for standard smooth-bore ammunition and the tube-launched AT-8 Kobra guided missile; this latter is probably in the same 125-mm calibre and the standard projectiles, and is fired with a muzzle velocity of 150 m (492 ft) per second before the solid-propellant rocket motor ignites to boost cruise speed to 1800 km/h (1118 mph) for a 3000-m (3,280-yard) flight of 7 seconds; the missile is a laser beam-rider used in conjunction with a fairly large laser on the turret roof

T-64K: command version of the T-64

T-80: much improved version of the T-64B with the mechanical problems eliminated and (probably) provision for the Kobra tube-launched missile in only a small proportion of the overall total strength; perhaps the most important modifications are the replacement of the T-64's troublesome five-cylinder diesel with a 671-kW (900-shp) gas turbine engine and the readoption of conventional rubber-rimmed road wheels in place of the T-64's resilient steel wheels; the type is fitted with conventional armour (the Soviets having apparently opted for the manufacturing ease and cheapness of this type over more capable yet considerably more expensive and weighty composite armour), though it is likely that the Soviets will soon introduce reactive add-on armour to counter the worst effects of HEAT warheads; other modifications include an improved laser rangefinder

T-62A
(USSR)
Type: main battle tank
Crew: 4
Combat weight: 40000 kg (88,183 lb)
Dimensions: length, gun forward 9.335 m (30.63 ft) and hull 6.63 m (21.75 ft); width 3.30 m (10.83 ft); height 2.395 m (7.86 ft)
Armament system: one 115-mm (4.53-in) U-5T smooth-bore gun with 40 rounds, one 7.62-mm (0.3-in) PKT co-axial machine-gun with 3,000 rounds, and one 12.7-mm (0.5-in) DShKM AA machine-gun with 250 rounds; the turret is electro-hydraulically powered, the main gun is stabilized in elevation and azimuth, and an optical fire-control system is fitted; the type can generate smoke by injecting fuel into the exhaust system
Armour: cast and welded steel varying in thickness from 14 to 242 mm (0.59 to 9.53 in)

Powerplant: one 432-kW (580-hp) V-55 diesel engine with 675 litres (148 Imp gal) of internal fuel, plus provision for 285 litres (63 Imp gal) of external fuel; there is also provision for 400 litres (88 Imp gal) of supplementary external fuel
Performance: speed, road 50 km/h (31 mph); range, road 450 km (280 miles) on internal fuel and 650 km (404 miles) with external fuel; fording 1.4 m (4.6 ft) without preparation and 5.5 m (18.0 ft) with snorkel; gradient 60%; vertical obstacle 0.8 m (31.5 in); trench 2.85 m (9.35 ft); ground clearance 0.43 m (17 in)

Variants
T-62: baseline MBT developed from the T-54/T-55 series and introduced to service from 1962; this model lacks the 12.7-mm (0.5-in) AA machine-gun but, like all other T-62 variants, has NBC and fire-suppression systems
T-62A: improved T-62 with revised turret shape and 12.7-mm (0.5-in) AA machine-gun
T-62K: command variant of the T-62 series
T-62M: T-62A fitted with the track and drive sprocket of the T-72
T-62 Flamethrower: this has a flame gun fitted co-axially with the main armament; the flame gun has a range of 100 m (110 yards)
T-62 (Egypt): these vehicles have been regunned with ROF 105-mm (4.13-in) rifled guns
M1977 ARV: this armoured recovery vehicle is based on the chassis of the T-62, and is a limited-capability vehicle restricted to towing operations

T-55
(USSR)
Type: main battle tank
Crew: 4
Combat weight: 36000 kg (79,365 lb)
Dimensions: length, gun forward 9.00 m (29.53 ft) and hull 6.45 m (21.16 ft); width 3.27 m (10.73 ft); height to turret top 2.40 m (7.87 ft)
Armament system: one 100-mm (3.94-in) D-10T2S rifled gun with 43 rounds, one 7.62-mm (0.3-in) SGMT co-axial machine-gun with 3,500 rounds, and one 12.7-mm (0.5-in) DShKM AA machine-gun with 250 rounds; the turret is electro-hydraulically powered, the main gun is stabilized in elevation and azimuth, and an optical fire-control system is fitted; the type can generate smoke by injecting fuel into the exhaust system
Armour: steel varying in thickness from 20 to 150 mm (0.79 to 5.91 in)
Powerplant: one 432-kW (580-hp) V-55 diesel engine with 960 litres (211 Imp gal) of internal fuel and provision for 200 litres (44 Imp gal) of external fuel
Performance: speed, road 50 km/h (31 mph); range, road 500 km (311 miles) on internal fuel and 600 km (373 miles) with external fuel; fording 1.4 m (4.6 ft) without preparation and 4.546 m (14.9 ft) with

snorkel; gradient 58%; vertical obstacle 0.8 m (31.5 in); trench 2.7 m (8.85 ft); ground clearance 0.425 m (16.7 in)

Variants

T-55: the T-55 is modelled closely on the T-54, and entered service in the late 1950s; compared with the T-54 the T-55 lacks the loader's hatch and associated 12.7-mm (0.5-in) DShKM AA machine-gun, has a more powerful engine and features a main gun stabilized in elevation and azimuth; the type was later retrofitted with the AA machine-gun, and then designated **T-55(M)**; an NBC system is standard, having been retrofitted to this model but incorporated from scratch in later models

T-55A: an upgraded version of the T-55, the T-55A eliminates the 7.62-mm (0.3-in) machine-gun in the glacis plate, and replaces the SGMT co-axial machine-gun with a more modern PKT weapon of the same 7.62-mm calibre; when retrofitted with the 12.7-mm (0.5-in) AA machine-gun this model is known as the **T-55A(M)**

T-55 Flamethrower: this vehicle has a short-barrelled flame gun located co-axially with the main armament, and can project its flame to a range of some 150/200 m (165/220 yards)

T-55-T: this is an armoured recovery vehicle based on the T-55, with a rear platform to accommodate a spare engine; other items fitted on this model are a rear spade and a jib able to lift 1000 kg (2,205 lb); the Czechs produced a basically similar model with improved protection for the commander plus an armament of one 7.62-mm (0.3-in) machine-gun; the Polish version of the same basic model is **WZT-2**

MTU-20: Soviet armoured vehicle-launched bridge version of the T-55 with a gap-spanning capability of 18 m (59.1 ft) with a 20-m (65.6-ft) bridge

MT-55: Czech armoured vehicle-launched bridge model able to span a 16-m (52.5-ft) gap with an 18-m (59.1-ft) scissors bridge

BLG-60: East German armoured vehicle-launched bridge model able to span a 20-m (65.6-ft) gap with a 21.6-m (70.9-ft) scissors bridge

IMR: Soviet combat engineer vehicle derivative of the T-55 with the turret replaced by a fully traversible hydraulic crane fitted with pincers for the removal of trees and other items; there is also a straight dozer blade at the front

T-55 (Egypt): these are being regunned with the ROF L7A3 105-mm (4.13-in) rifled gun

T-55 (India): these have a 0.5-in (12.7-mm) Browning M2HB machine-gun in the AA role

T-55 (Israel): of the numerous T-55s captured by the Israelis, many have been revised with 105-mm (4.13-in) M68 rifled main guns, 0.3-in (7.62-mm) Browning co-axial machine-guns and 0.5-in (12.7-mm) Browning AA machine-guns; the vehicles have been considerably modified with Israeli and US equipment, especially in the communications and fire-control fields

T-55 (Romania): these modified vehicles have six rather than five roadwheels, new sideskirts, and a revised 12.7-mm (0.5-in) AA machine-gun armament; it is believed that these vehicles have the designation M-77 or TR-77

T-54

(USSR)

Type: main battle tank
Crew: 4
Combat weight: 36000 kg (79,365 lb)
Dimensions: length, gun forward 9.00 m (29.53 ft) and hull 6.45 m (21.16 ft); width 3.27 m (10.73 ft); height to turret top 2.40 m (7.87 ft)
Armament system: one 100-mm (3.94-in) D-10T rifled gun with 34 rounds, two 7.62-mm (0.3-in) SGMT machine-guns (one co-axial and one in glacis plate) with 3,000 rounds, and one 12.7-mm (0.5-in) DShKM AA machine-gun with 500 rounds; the turret is electro-hydraulically powered, the main gun lacks stabilization in either elevation or azimuth, and an optical fire-control system is fitted; the type can generate smoke by injecting fuel into the exhaust system
Armour: cast and welded steel varying in thickness from 20 to 203 mm (0.79 to 8 in)
Powerplant: one 388-kW (520-hp) V-54 diesel engine with 812 litres (179 Imp gal) of internal fuel, plus provision for external fuel
Performance: speed, road 48 km/h (30 mph); range, road 400 km (249 miles) on internal fuel and 600 km (373 miles) with external fuel; fording 1.4 m (4.6 ft) without preparation and 4.55 m (14.9 ft) with snorkel; gradient 58%; vertical obstacle 0.8 m (31.5 in); trench 2.7 m (8.85 ft); ground clearance 0.425 m (16.7 in)

Variants

T-54: the initial T-54 series entered service in 1948, and appeared in two basic forms; the earlier of these had two turret cupolas, could not be fitted with a snorkel for deep wading, and possessed an oddly shaped rear to the turret; the later model still has two cupolas (that on the right being fitted with a 12.7-mm/0.5-in DShKM AA machine-gun), but the turret was more normally shaped and could be fitted with a snorkel; when retrofitted with infra-red driving lights the type is designated **T-54(M)**, and the model is also built in China as the Type 59; standard features of current examples of the T-54 series are night-vision devices and NBC protection, while a major tactical disadvantage is the lack of gun depression so requiring the type to use hull-up positions where the benefits of its low silhouette and good ballistic shape are fully needed; the original D-10T gun was unstabilized

T-54A: introduced in the mid-1950s, this variant witnessed the adoption of a bore evacuator for the

D-10TG main armament, which was also stabilized in elevation; the retrofitting of infra-red driving lights resulted in the designation **T-54A(M)**
T-54B: this variant was adopted for service in 1957/8, and was built as standard with infra-red driving lights; the main armament is the D-10T2S gun stabilized in both elevation and azimuth
T-54C: this is basically the T-54B with the gunner's cupola replaced by a hatch without the machine-gun
T-54-T: this is the armoured recovery vehicle derivative of the basic T-54, and has a rear platform for the carriage of a spare engine, a rear spade and a small jib capable of lifting 1000 kg (2,205 lb)
T-54(A): this is an armoured recovery vehicle developed in East Germany on the basis of the T-54, and apart from a full range of specialist tools the type has a push/pull bar and a dismountable crane with a lifting capacity of 1000 kg (2,205 lb); chemical and nuclear agent detectors are fitted, but the type's capabilities are limited by lack of a winch or spade
T-54(B): this is essentially the T-54(A) with attachment points for tow ropes
T-54(C): this is another East German armoured recovery vehicle, and is fitted with a rear platform for a spare engine, a rear spade, a front-mounted dozer blade and a jib crane able to lift 20000 kg (44,092 lb)
WZT-1: armoured recovery vehicle developed in and used by Poland on the basis of the T-54
MTU: Soviet-developed armoured vehicle-launched bridge based on the T054 chassis, and able to span an 11-m (36.1-ft) gap with a 12.3-m (40.35-ft) bridge
T-54 (India): Indian T-54s have received the same basic modifications as the T-55s operated by the Indian army
T-54 (Israel): T-54s captured by the Israelis have undergone the same type of modification programme as Israel's T-55 tanks

T-34/85
(USSR)
Type: medium tank
Crew: 5
Combat weight: 32000 kg (70,547 lb)
Dimensions: length, gun forward 8.076 m (26.5 ft) and hull 6.19 m (20.31 ft); width 2.997 m (9.83 ft); height 2.743 m (9.0 ft)
Armament system: one 85-mm (3.35-in) ZIS-S53 rifled gun with 56 rounds, and two 7.62-mm (0.3-in) DTM machine-guns (one co-axial and one in glacis plate) with 2,394 rounds; the turret is electrically or manually powered, the main gun lacks stabilization in either elevation or azimuth, and an optical fire-control system is fitted
Armour: cast and welded steel varying in thickness from 18 to 90 mm (0.71 to 3.54 in)
Powerplant: one 373-kW (500-hp) V-2-34 or V-2-34M diesel engine with 590 litres (130 Imp gal) of internal fuel, plus provision for external fuel
Performance: speed, road 55 km/h (34 mph); range, road 300 km (186 miles); fording 1.32 m (4.33 ft); gradient 60%; vertical obstacle 0.73 m (28.75 in); trench 2.5 m (8.2 ft); ground clearance 0.38 m (15 in)

Variants
T-34/85: introduced in 1943, the T-34/85 was developed as an upgunned T-34/76, and the type was produced in vast numbers during and after world War II; many thousands were remanufactured in the 1960s, and these still constitute an important part of the arsenals of many smaller clients of the USSR; there were numerous variants on the basic chassis, but the only ones still in substantial service are various armoured recovery vehicles such as the **T-35-T(A)** turretless towing vehicle, the **T-34-T(B)** turretless ARV with a spare engine on an aft platform plus a small crane and winch, the **SKP-5** crane vehicle able to lift 5000 kg (11,023 lb) and the **WPT-34** with a large forward superstructure, a platform-mounted spare engine and rear-mounted spade and winch

ROF Khalid MBT (UK): see ROF FV4201 Chieftain MBT

ROF Shir MBT (UK): see ROF FV4201 Chieftain MBT

ROF Leeds FV4030/4 Challenger
(UK)
Type: main battle tank
Crew: 4
Combat weight: 136,640 lb (61980 kg)
Dimensions: length, gun forward 37.89 ft (11.55 m) and hull 27.53 ft (8.39 m); width 11.54 ft (3.518 m); height to top of commander's sight 9.48 ft (2.89 m)
Armament system: one 120-mm (4.72-in) ROF L11A5 rifled gun with between 44 and 52 rounds, two 7.62-mm (0.3-in) machine-guns (one co-axial L8A2 and one AA L37A2) with 4,000 rounds, and five smoke-dischargers on each side of the turret; the turret is electrically powered, the main gun is stabilized in elevation and azimuth, and a Marconi Improved Fire-Control System is fitted; this last combines stabilized day/night optical sights, a laser ranger, various sensors and a ballistic computer
Armour: laminate type
Powerplant: one 1,200-bhp (895-kW) Rolls-Royce Condor 12V 1200 diesel engine with about 210 Imp gal (955 litres) of internal fuel
Performance: speed, road 37 mph (59.5 km/h); range, road not known; fording 3.5 ft (1.07 m) without preparation; gradient 58%; side slope 27%; vertical obstacle 36 in (0.91m); trench 10.33 ft (3.15 m); ground clearance 19.7 in (0.5 m)

Variants
FV4030/4 Challenger: this is an evolutionary development of the Chieftain MBT via the FV4030/3 Shir 2 ordered by Iran but then cancelled after the fall of the Shah in 1979; the type began to enter British service in 1983 and is fitted with full NBC protection

ROF Leeds FV4201 Chieftain Mk 5
(UK)
Type: main battle tank
Crew: 4
Combat weight: 121,250 lb (55000 kg)
Dimensions: length, gun forward 35.42 ft (10.795 m) and hull 24.67 ft (7.518 m); width over skirts 11.50 ft (3.504 m); height overall 9.50 ft (2.895 m)
Armament system: one 120-mm (4.72-in) ROF L11A5 rifled gun with 64 rounds, two 7.62-mm (0.3-in) machine-guns (one co-axial L8A1 and one AA L37A1) with 6,000 rounds, and six smoke-dischargers on each side of the turret; the turret is electrically powered, the main gun is stabilized in elevation and azimuth, and a Marconi Improved Fire-Control System is fitted; this last combines stabilized day/night optical sights, a laser rangefinder, various sensors and a ballistic computer
Armour: cast and welded steel plus (as a retrofit) 'Stillbrew' appliqué armour comprising layers of composite armour under a steel shell on the frontal arc
Powerplant: one 750-bhp (559-kW) Leyland L60 No.4 Mk 8A multi-fuel engine with 210 Imp gal (955 litres) of internal fuel
Performance: speed, road 30 mph (48 km/h); range, road 310 miles (499 km); fording 3.5 ft (1.07 m) without preparation; gradient 70%; side slope 30%; vertical obstacle 36 in (0.91 m); trench 10.33 ft (3.15 m); ground clearance 20 in (0.51 m)

Variants
Chieftain Mk 1: training model of the Chieftain, of which 40 were built in 1965/1966 with a 585-bhp (436-kW) engine; of these vehicles most were later upgraded, the **Chieftain Mk 1/2** being the Mk 1 brought up to Mk 2 standard, the **Chieftain Mk 1/3** having a new powerpack, and the **Chieftain Mk 1/4** being the Mk 1 with a new powerpack and modified 0.5-in (12.7-mm) ranging machine-gun
Chieftain Mk 2: initial service model, which began to enter British army regimental use in 1967 with a 650-bhp (485-kW) engine
Chieftain Mk 3: second service model, essentially an improved Mk 2 with improved auxiliary power unit and main engine plus a number of detail modifications to the secondary armament and trackwork
Chieftain Mk 3/S: development of the Mk 3 with turret air-breathing and other modifications pioneered in the **Chieftain Mk 3/G** and **Chieftain Mk 3/2** prototypes
Chieftain Mk 3/3: basically the Mk 3 fitted with a longer-range ranging machine-gun, provision for a Barr & Stroud laser rangefinder, a low-loss air-cleaning system, turret air-breathing, a new NBC pack and 720-bhp (537-kW) engine; the export version for Iran was the **Chieftain Mk 3/3P**
Chieftain Mk 4: experimental model with greater fuel capacity
Chieftain Mk 5: improved version of Mk 3/3 with better engine and transmission system, and detail modifications to the fire-control system etc; the export models were the **Chieftain Mk 5/2K** for Kuwait and the **Chieftain Mk 5/5P** for Iran
Chieftain Mk 6: Mk 2 with a modified ranging machine-gun and new powerpack
Chieftain Mk 7: Mk 3 and Mk 3/S with modified ranging machine-gun and powerpack
Chieftain Mk 8: Mk 3/3 with Mk 7 modifications
Chieftain Mk 9: Mk 6 with the Improved Fire-Control System
Chieftain Mk 10: Mk 7 with IFCS
Chieftain Mk 11: Mk 8 with IFCS
Chieftain Mk 12: Mk 5 with IFCS
Chieftain 800: designation of the private-venture export variant produced by Royal Ordnance with an 800-bhp (596-kW) Rolls-Royce CV12 TCE diesel and improved transmission
Chieftain 900: comparable to the Chieftain 800 but with a 900-bhp (671-kW) version of the same CV12 engine (Rolls-Royce Condor 900E) and fitted with Chobham laminate armour for a considerably higher standard of protection; the combat weight is 123,450 lb (55997 kg)
Chieftain ARRV: based on the Chieftain MBT chassis and hull, the Chieftain Armoured Recovery and Repair Vehicle is designed specifically for operations in support of the Challenger MBT, whose powerpack is too heavy for the standard FV434 armoured repair vehicle used with the Chieftain
FV4204 Chieftain ARV: based on the chassis and hull of the Chieftain Mk 5, the Chieftain Armoured Recovery Vehicle has two winches and a crane, the main winch being rated at 90,000 lb (40824 kg) when used with the front-mounted dozer blade, and the crane at 12,800 lb (5806 kg)
FV4205 Chieftain AVLB: the Chieftain Armoured Vehicle-Launched Bridge can span gaps of up to 75 ft (22.86 m) with the 80-ft (24.38-m) No.8 Tank Bridge, or 40 ft (12.19 m) when fitted with the 44-ft (13.41-m) No.9 Tank Bridge
Chieftain Mk 6 AVLB: this improved Armoured Vehicle-Launched Bridge is based on the chassis and hull of the Chieftain Mks 1/4 gun tanks and fitted with the Mk 8 Bridge and Pierson mine plough system
Khalid: built for Jordan, this vehicle was originally developed for Iran as the **FV4030/2 Shir 1**, in essence a Chieftain Mk 5 with improved fire-control and automotive features; the Shir 1, of which 125 had been

ordered, was cancelled in 1979 together with the much improved FV4030/3 Shir 2 (1,225 ordered); the Khalid is the Shir 1 with detail modifications to suit the type to Jordanian requirements, and the powerplant is a 1,200-bhp (895-kW) Rolls-Royce Condor 12V 1200 diesel; the Khalid weighs 127,865 lb (58000 kg), and its dimensions include a hull length of 27.53 ft (8.39 m), a width of 11.54 ft (3.52 m) and a height of 9.88 ft (3.01 m)

ROF Leeds/ROF Woolwich/Leyland/Vickers FV4017 Centurion Mk 13
(UK)
Type: main battle tank
Crew: 4
Combat weight: 114,240 lb (51819 kg)
Dimensions: length, gun forward 32.33 ft (9.85 m) and hull 25.67 ft (7.82 m); width 11.12 ft (3.39 m); height to top of cupola 9.87 ft (3.01 m)
Armament system: one 105-mm (4.13-in) ROF L7A2 rifled gun with 64 rounds, one 0.5-in (12.7-mm) L21A1 co-axial ranging machine-gun with 600 rounds, two 7.62-mm (0.3-in) machine-guns (one co-axial L8A1 and one AA L37A1), with 4,750 rounds, and six smoke-dischargers on each side of the turret; the turret is electrically powered, the main gun is stabilized in elevation and azimuth, and an optical fire-control system is fitted
Armour: cast and welded steel varying in thickness from 0.66 to 6 in (17 to 152 mm)
Powerplant: one 650-bhp (485-kW) Rolls-Royce Meteor Mk IVB petrol engine with 228 Imp gal (1037 litres) of internal fuel
Performance: speed, road 21.5 mph (34.6 km/h); range, road 118 miles (190 km); fording 4.75 ft (1.45 m) without preparation and 9.0 ft (2.74 m) with preparation; gradient 60%; side slope 30%; vertical obstacle 36 in (0.91 m); trench 11.0 ft (3.35 m); ground clearance 20 in (0.51 m)

Variants
Centurion Mk 5: the Chieftain series began to enter service in the period immediately after World War II, but the Mk 5 is the earliest model to remain in service, and was designed by Vickers; it differs from earlier models primarily in having a 20-pdr (3.28-in/83.4-mm) gun with 65 rounds in place of the original 17-pdr (3-in/76.2-mm) weapon; developments of the Mk 5 are the uparmoured **Centurion Mk 5/1** and the **Centurion Mk 5/2** with the 105-mm (4.13-in) L7 gun
Centurion Mk 6: uparmoured Mk 5 with L7 gun and additional fuel capacity; developments are the **Centurion Mk 6/1** with infra-red night-vision equipment and stowage basket on the turret rear, and the **Centurion Mk 6/2** with a ranging machine-gun
Centurion Mk 7: Leyland-developed **FV4007** model with 20-pdr gun and 61 rounds plus a fume extractor; developments are the **FV4012 Centurion Mk 7/1** with thicker armour and the **Centurion Mk 7/2** with 105-mm gun
Centurion Mk 8: essentially the Mk 7 with improved main gun mounting and contra-rotating commander's cupola; developments are the uparmoured **Centurion Mk 8/1** and the upgunned **Centurion Mk 8/2** with 105-mm gun
Centurion Mk 9: upgunned and uparmoured Mk 7 fitted with the 105-mm gun and designated **FV4015**; developments are the **Centurion Mk 9/1** with infra-red driving lights and stowage basket on the turret rear, and the **Centurion Mk 9/2** with a ranging machine-gun for the main armament
Centurion Mk 10: Mk 8 with improved armour and 105-mm gun with 70 rounds, and designated **FV4017**; developments are the **Centurion Mk 10/1** with infra-red driving lights and a stowage basket on the turret rear, and the **Centurion Mk 10/2** with a ranging machine-gun for the main armament
Centurion Mk 11: Mk 6 with ranging machine-gun, infra-red driving lights and stowage basket on turret rear
Centurion Mk 12: Mk 9 with ranging machine-gun, infra-red driving lights and stowage basket on turret rear
Centurion Mk 13: Mk 10 with ranging machine-gun and infra-red driving lights
FV4002 Centurion AVLB: based on the chassis and hull of the Centurion Mk 5, the Centurion Armoured Vehicle-Launched Bridge can span a gap of 45 ft (13.72 m)
FV4003 Centurion AVRE: based on the chassis of the Centurion Mk 5, the Centurion Armoured Vehicle Royal Engineers has a turret-mounted 165-mm (6.5-in) demolition gun, and also a front-mounted dozer blade
FV4006 Centurion ARV Mk 2: the Centurion Armoured Recovery Vehicle Mk 2 is a standard gun tank with its turret replaced by a welded superstructure for the commander; the use of rear-mounted spades gives a winch capacity of 68,340 lb (31000 kg)
FV4018 Centurion BARV: this Centurion Beach Armoured Recovery Vehicle has a revised superstructure allowing the vehicle to operate in water 9.5 ft (2.90 m) deep
Centurion (Denmark): these are currently to Mk 5 and Mk 5/1 standard, with a 0.5-in (12.7-mm) Browning AA machine-gun; they are to be fitted with an Ericsson gunner's sight with laser rangefinder
Centurion (Israel): Israeli vehicles have been very substantially improved to **Upgraded Centurion** standard with a 750-hp (559-kW) Continental AVDS-1790-2A diesel and automatic gearbox, additional fuel capacity, redesigned ammunition stowage for the 105-mm M68 rifled gun (allowing the carriage of 72 rounds), and a 0.5-in (12.7-mm) Browning M2HB machine-gun mounted over the barrel of the main armament and fired by remote control; the result is

basically a Centurion Mk 5 with extra ammunition, a road speed of 48 km/h (26.7 mph) and a road range of 600+ km (373+ miles); other modifications developed and adopted by the Israelis are applique active armour of the Blazer type, and an advanced fire-control system
Centurion (Jordan): like the Israeli Centurion Mk 5s, these vehicles are being re-engined with the Continental AVDS-1790 diesel, and are being fitted with an advanced SABCA fire-control system imported from Belgium; the revised MBTs are designated **Tariq** in local service
Centurion (South Africa): South African Centurion tanks are being remanufactured with a new engine/transmission assembly, a locally produced 105-mm rifled gun and six smoke-dischargers on each side of the turret; the name **Olifant** has been given to these much improved MBTs
Centurion (Sweden): in Swedish service the Centurion has the designation **Strv 101** and **Strv 102**, and these vehicles are being modified with a new Bofors fire-control system with laser rangefinder, and with the Continental AVDS-1790-2DC diesel with automatic transmission; thus modified the vehicles are designated **Strv 104**
Centurion (Switzerland): Swiss Centurions are designated **Pz 55** (Mk 5) and **Pz 57** (Mk 7), and are essentially similar to the British versions

Vickers Main Battle Tank Mk 7
(UK)
Type: main battle tank
Crew: 4
Combat weight: 120,460 lb (54641 kg)
Dimensions: length, gun forward 35.92 ft (10.95 m), and hull 25.33 ft (7.72 m); width 11.25 ft (3.42 m); height overall 9.83 ft (2.99 m)
Armament system: one 120-mm (4.72-in) ROF L11A5 rifled gun or one 120-mm (4.72-in) Rheinmetall Rh-120 smooth-bore gun with 38 rounds, one co-axial 7.62-mm (0.3-in) RSAF Chain Gun with 3,000 rounds, one optional 7.62-mm (0.3-in) AA machine-gun with ? rounds, and six smoke-dischargers on each side of the turret; the turret is electrically powered, the main gun is stabilized in elevation and azimuth, and a Marconi Centaur 1 fire-control system is fitted; this last combines optical/thermal sights, a laser rangefinder, various sensors and a ballistic computer
Armour: welded steel and applique Chobham laminate
Powerplant: one 1119-kW (1,500-hp) MTU MB 873 Ka501 turbocharged diesel engine with 253 Imp gal (1150 litres) of internal fuel
Performance: speed, road 45 mph (72 km/h); range, road 310 miles (500 km); fording 2.6 ft (0.8 m) without preparation and 5.6 ft (1.7 m) with preparation; gradient 60%; side slope 30%; vertical obstacle 43 in (1.1 m); trench 9.85 ft (3.0 m); ground clearance 18 in (0.457 m)

Variants
Main Battle Tank Mk 7: originally known as the **Vickers Valiant** when first offered unsuccessfully with a Vickers-designed hull, the Main Battle Tank Mk 7 was developed in response to Middle Eastern requirements by Vickers in association with Krauss-Maffei, the former supplying the universal turret and overall design/integration control and the latter the hull and automotive components based very closely on those of the Leopard 2; the type is now ready for production should orders be forthcoming

Vickers Main Battle Tank Mk 3
(UK)
Type: main battle tank
Crew: 4
Combat weight: 85,120 lb (36610 kg)
Dimensions: length, gun forward 32.11 ft (9.79 m) and hull 24.81 ft (7.56 m); width 10.4 ft (3.17 m); height to turret top 8.12 ft (2.48 m)
Armament system: one 105-mm (4.13-in) ROF L7A1 rifled gun with 50 rounds, one 0.5-in (12.7-mm) L6A1 co-axial ranging machine-gun with 600 rounds, two 7.62-mm (0.3-in) machine-guns (one co-axial L8A2 and one optional AA) with 3,000 rounds, and six smoke dischargers on each side of the turret; the turret is electrically powered, the main gun is stabilized in elevation and azimuth, and a GEC-Marconi EC517 fire-control system is fitted; the last combines day/night optical sights, a laser rangefinder and a ballistic computer
Armour: welded steel varying in thickness from 0.67 to 3.15 in (17 to 80 mm)
Powerplant: one 720-bhp (537-kW) General Motors 12-71T diesel engine with 220 Imp gal (1000 litres) of internal fuel
Performance: speed, road 31 mph (50 km/h); range, road 375 miles (603 km); fording 3.75 ft (1.14 m); gradient 60%; side slope 30%; vertical obstacle 36 in (0.91 m); trench 8.0 ft (2.44 m); ground clearance 17 in (0.43 m)

Variants
Vickers Main Battle Tank Mk 1: this private-venture tank was developed in the early 1960s in response to a request from the Indian army; the resulting MBT Mk 1 was accepted for Indian service and placed in large-scale production in that country as the **Vijayanta** with a 650-bhp (485-kW) Leyland L60 Mk 4B multi-fuel engine and a main armament ammunition capacity of 44 rounds; some Indian-produced models are being fitted with the SFCS 600 fire-control system; some Viyayanta MBTs have been converted into self-propelled guns with a Soviet 130-mm (5.12-in) M1946 gun plus 30 rounds

Vickers Main Battle Tank Mk 3: this is the upgraded model with General Motors diesel engine for greater range, additional armour and improved fire-control capability; the type is in service with Nigeria as the **Eagle** and with Kenya, and Kuwaiti Mk 1s have been brought up to this standard by the retrofitting of General Motors diesel engines

Vickers Armoured Bridgelayer: based on the chassis of the Mk 3, this AVLB carries a 44-ft (13.41-m) bridge

Vickers Armoured Recovery Vehicle: based on the chassis of the Mk 3, this ARV has a 55,115-lb (25000-kg) capacity winch and (on some vehicles only) an 8,820-lb (4000-kg) crane

Vickers Vijayanta MBT (UK/India): see Vickers MBT Mk 3

General Dynamics M1 Abrams
(USA)
Type: main battle tank
Crew: 4
Combat weight: 120,250 lb (54,545 kg)
Dimensions: length, gun forward 32.04 ft (9.77 m) and hull 25.98 ft (7.92 m); width 11.98 ft (3.65 m); height to turret top 7.8 ft (2.375 m)
Armament system: one 105-mm (4.13-in) M68E1 rifled gun with 55 rounds, two M240 7.62-mm (0.3-in) machine-guns (one co-axial and one loader's AA) with 11,400 rounds, one 0.5-in (12.7-mm) Browning M2HB machine-gun (commander's AA) with 1,000 rounds, and six smoke-dischargers on each side of the turret; the turret is electro-hydraulically powered, the main gun is stabilized in elevation and azimuth, and an advanced fire-control system is fitted; this last combines stabilized day/night optical/thermal sights, a laser rangefinder, various sensors and a ballistic computer; the type can also generate smoke by injecting fuel into the exhaust system
Armour: Chobham laminate
Powerplant: one 1,500-hp (1118-kW) Avco Lycoming AGT-1500 gas turbine with 504 US gal (1908 litres) of internal fuel
Performance: speed, road 45 mph (72.4 km/h); range, road 310 miles (498 km); fording 4.0 ft (1.22 m) without preparation and 7.8 ft (2.375 m) with preparation; gradient 60%; side slope 40%; vertical obstacle 48 in (1.22 m); trench 9.0 ft (2.74 m); ground clearance 17 in (0.43 m)

Variants
M1 Abrams: this is the successor to the M60 series as the US Army's main battle tank, and has been designed with the European theatre in mind; the type began to enter service in 1980 with standard features such as an NBC pack and advanced fire-control for the fully-stabilized main armament, and offers significant improvements over the M60 series in armament, protection, mobility, reliability, availability, maintainability and durability

M1A1 Abrams: this much improved model of the Abrams began to enter service in 1986, and differs from the original model principally in its armament, which is the 120-mm (4.72-in) Rheinmetall Rh-120 smooth-bore gun produced in the USA as the M256; 40 rounds of ammunition are provided for this weapon, and other improvements to the M1A1 lie in the fields of vision, fire-control, ballistic and NBC protection, and refuelling rate; the combat weight is 126,000 lb (57154 kg) and the length, gun forward, is 32.25 ft (9.828 m)

General Dynamics M60A3
(USA)
Type: main battle tank
Crew: 4
Combat weight: 113,535 lb (51500 kg)
Dimensions: length, gun forward 30.96 ft (9.44 m) and hull 22.79 ft (6.95 m); width 11.91 ft (3.63 m); height 10.73 ft (3.27 m)
Armament system: one 105-mm (4.13-in) M68E1 rifled gun with 63 rounds, one 7.62-mm (0.3-in) M240 co-axial machine-gun with 5,950 rounds, one 0.5-in (12.7-mm) M85 AA machine-gun with 940 rounds, and six smoke-dischargers on each side of the turret; the turret is electro-hydraulically powered, the main gun is stabilized in elevation and azimuth, and an advanced fire control system is fitted; this last combines day/night optical sights, a laser rangefinder, various sensors and a ballistic computer; the type can also generate smoke by injecting fuel into the exhaust system
Armour: welded steel
Powerplant: one 750-bhp (559-kW) Continental AVDS-1790-2D diesel engine with 375 US gal (1420 litres) of internal fuel
Performance: speed, road 30 mph (48.3 km/h); range, road 300 miles (483 km); fording 4.0 ft (4.22 m) without preparation and 7.87 ft (2.4 m) with preparation; gradient 60%; side slope 30%; vertical obstacle 36 in (0.91 m); trench 8.5 ft (2.59 m); ground clearance 17.7 in (0.45 m)

Variants
M60: this initial variant began to supplant the M48 series as the US Army's main battle tank from 1960 and, while generally comparable in dimensions and performance to later models, it lacks the gun-stabilization system of the M60A3, has a mechanical device for computing ballistic solutions and possesses a turret of inferior ballistic qualities; the type has a combat weight of 102,000 lb (46267 kg) and full NBC protection
M60A1: this model began to enter service in 1962, and is an improvement over the M60 in having a sharper-fronted turret of superior ballistic qualities,

and stowage for 63 rather than 60 rounds of 105-mm (4.13-in) ammunition; the type has a combat weight of 108,000 lb (48989 kg)
M60A3: this is essentially a product-improved M60A1 with improved fire-control (with a laser rangefinder and digital computer), smoke-dischargers, a more reliable engine and an add-on gun-stabilization system; the type began to enter service in 1978 and is being upgraded with appliqué reactive armour to defeat HEAT warheads
M60 AVLB: this Armoured Vehicle-Launched Bridge uses the chassis of the M60 and is fitted with a 63-ft (19.2-m) scissors-type bridge able to span a gap of 60 ft (18.29 m)
M728 CET: this Combat Engineer Tractor is armed with a 165-mm (6.5-in) M125 demolition gun, and its specialized equipment includes a front-mounted dozer blade, a folding A-frame and a 25,000-lb (11340-kg) capacity winch
M60 (Israel): Israel has about 1,000 M60A1 and M60A3 tanks, and these have generally been upgraded with Blazer reactive armour, a new commander's cupola with externally-mounted 7.62-mm (0.3-in) machine-gun, a remotely-fired 0.5-in (12.7-mm) Browning machine-gun attached over the barrel of the main armament, and other modifications

Chrysler (General Dynamics) M48A5 Patton
(USA)
Type: main battle tank
Crew: 4
Combat weight: 108,000 lb (48989 kg)
Dimensions: length, gun forward 30.53 ft (9.31 m) and hull 20.06 ft (6.42 m); width 11.91 ft (3.63 m); height 10.12 ft (3.09 m)
Armament system: one 105-mm (4.13-in) M68E1 rifled gun with 54 rounds, three 7.62-mm (0.3-in) M60D machine-guns (one co-axial and two AA) with 10,000 rounds, and smoke-dischargers; the turret is electro-hydraulically powered, the main gun lacks stabilization in either elevation or azimuth, and an optical fire-control system is fitted; the type can also generate smoke by injecting fuel into the exhaust system
Armour: welded and cast steel varying in thickness from 0.5 to 4.72 in (12.7 to 120 mm)
Powerplant: one 750-bhp (559-kW) Continental AVDS-1790-2D diesel engine with 375 US gal (1420 litres) of internal fuel
Performance: speed, road 30 mph (48.3 km/h); range, road 310 miles (499 km); fording 4.0 ft (1.22 m) without preparation and 8.0 ft (2.44 m) with preparation; gradient 60%; side slope 30%; vertical obstacle 36 in (0.91 m); trench 8.5 ft (2.59 m); ground clearance 16.5 in (0.39 m)

Variants
M48: this was the original production model of the Patton tank, and entered service in 1952 as the replacement for the M47 series; the type has a combat weight of 99,000 lb (44906 kg), possesses a main armament of one 90-mm (3.54-in) gun with 60 rounds, a co-axial armament of one 0.3-in (7.62-mm) machine-gun with 5,900 rounds and an AA armament of one 0.5-in (12.7-mm) machine-gun with 180 rounds, and is powered by a 810-bhp (604-kW) Continental AVDS-1790 petrol engine with an internal fuel supply of 200 US gal (757 litres) for a range of only 70 miles (113 km)
M48A1: improved model remedying some of the M48's lesser failings without real concern for the major problems of a petrol engine and catastrophically short range, though provision is made for jettisonable external tanks
M48A2: this improved M48A1 adds a T-shape muzzle brake to the 90-mm (3.54-in) main armament, increases internal tankage to 335 US gal (1268 litres) and makes provision for jettisonable external tanks; main armament ammunition capacity is also increased to 64 rounds
M48A2C: M48A2 with improved fire-control equipment
M48A3: major model produced by rebuilding M48A1 and M48A2 tanks with the 750-bhp (559-kW) AVDS-1790-2A diesel and internal tankage for 375 US gal (1420 litres), track-return rollers reduced in number from five to three, and improved fire-control; the powerplant modification increases range to 290 miles (467 km)
M48A5: produced from 1975 by conversion of M48A1 and M48A3 tanks, the M48A5 introduces a host of internal and external modifications to improve the basic type's serviceability and habitability, and also saw the introduction of turret-mounted smoke-dischargers and of the 105-mm (4.13-in) main armament plus revised secondary and AA fits; almost identical modifications programmes are being undertaken by Greece, South Korea and Turkey
M48 Armored Vehicle-Launched Bridge: this AVLB has a 63-ft (19.2-m) scissors-type bridge capable of spanning a 60-ft (18.3-m) gap
M48 (Israel): Israel operates numbers of M48 tanks of M60-type capability with a 12-cylinder Continental diesel, 105-mm (4.13-in) main armament and an Israeli-designed commander's cupola
M48 (Spain): Spanish M48s have been upgraded considerably in a local modification programme resulting in the addition of the suffix 'E' to the basic designation; these modifications are directed generally to the powerplant (diesel) and to the main armament and its fire-control system
M48 (West Germany): the West German army uses a number of M48 series tanks brought up to **M48A2GA2** standard with M48A5-type modifications based on the 105-mm (4.13-in) L7A1 gun complete with thermal sleeve and 46 rounds of ammunition,

two 7.62-mm (0.3-in) MG3 machine-guns with 4,750 rounds, and four smoke-dischargers on each side of the turret

Detroit Tank Plant/American Locomotive M47M
(USA)
Type: medium tank
Crew: 4
Combat weight: 103,205 lb (46814 kg)
Dimensions: length, gun forward 28.06 ft (8.55 m) and hull 20.56 ft (6.27 m); width 11.12 ft (3.39 m); height to turret top 9.69 ft (2.95 m)
Armament system: one 90-mm (3.54-in) M36 rifled gun with 79 rounds, one 0.3-in (7.62-mm) M1919A4 co-axial machine-gun with 4,125 rounds, and one 0.5-in (12.7-mm) M2HB AA machine-gun with 440 rounds; the turret is hydraulically powered, the main gun lacks stabilization in either elevation or azimuth, and an optical fire-control system is fitted
Armour: welded steel varying in thickness from 0.5 to 4 in (12.7 to 102 mm)
Powerplant: one 750-bhp (559-kW) Continental AVDS-1790-2A diesel engine with 390 US gal (1476 litres) of internal fuel
Performance: speed, road 30 mph (48 km/h); range, road 375 miles (603 km); fording 4.0 ft (1.22 m); gradient 60%; side slope 30%; vertical obstacle 36 in (0.91 m); trench 8.5 ft (2.59 m); ground clearance 18.5 in (0.47 m)

Variants
M47: the M47 entered service with the US Army in 1953, but was not judged a great success, largely because it was powered by an 810-bhp (604-kW) Continental AVDS-1790 5B petrol engine with a mere 231 US gal (875 litres) of internal fuel, which provided a range of only 80 miles (129 km) for this five-crew tank; main armament ammunition amounted to 71 rounds, and the M47 also has a bow-mounted 0.3-in (7.62-mm) M1919A4 machine-gun
M47E1: this Spanish-developed variant has an AVDS-1790-2A diesel, an improved Cadillac Gage turret and a 7.62-mm (0.3-in) MG42 co-axial machine-gun
M47E2: this Spanish-developed variant is similar to the M47E1 but is fitted with 105-mm (4.13-in) M68 or Rh-105 main armament
M47M: this model was developed as a retrofit package for Iran, and provided the basic M47 with the powerplant, fire-control and fuel tankage of the M60A1 MBT
M47 (Israel): known as the **Rhino**, this Israeli-developed model has the powerpack and 105-mm (4.13-in) M68 rifled gun (plus 63 rounds) of the M60A1
M47 (South Korea): South Korean development has produced an **M47 Armoured Recovery Vehicle**, which has a turret-mounted winch in place of the 90-mm (3.54-in) gun, and an A-frame at the front of the hull

M4A3E8 Sherman
(USA)
Type: medium tank
Crew: 5
Combat weight: 71,175 lb (32285 kg)
Dimensions: length, gun forward 24.67 ft (7.5128 m) and hull 20.58 ft (6.273 m); width 8.75 ft (2.667 m); height to turret top 11.25 ft (3.43 m)
Armament system: one 3-in (76-mm) M1A1/2 rifled gun with 71 rounds, two 0.3-in (7.62-mm) Browning M1919A4 machine-guns (one co-axial and one bow) with 6,250 rounds, and one 0.5-in (12.7-mm) Browning M2HB AA machine-gun with 600 rounds; the turret is hydraulically powered, the main gun lacks stabilization in either elevation or azimuth, and an optical fire-control system is fitted
Armour: cast and welded steel varying in thickness from 1 to 3 in (25.4 to 76.2 mm)
Powerplant: one 450-hp (336-kW) Ford GAA petrol engine with 168 US gal (636 litres) of internal fuel
Performance: speed, road 30 mph (48 km/h); range, road 100 miles (161 km); fording 3 ft (0.91 m); gradient 60%; vertical obstacle 24 in (0.61 m); trench 7.4 ft (2.26 m); ground clearance 17 in (0.43 m)

Variants
M4: this initial model of the Sherman was introduced in October 1941 and is armed with a 75-mm (2.95-in) gun in a cast turret on a welded hull with a 353-hp (263-kW) Continental R-975-C1 radial engine; many versions of the Sherman remain in service despite their age, most having been upgraded to greater or lesser extent (generally with a diesel powerplant such as the 335-kW/449-kW Poyaud 520 V8S2 or Detroit Diesel 8V-71T) to provide continued utility (as well as greater range and enhanced reliability) in less advanced corners of the world
M4A1: introduced in December 1941 as a derivative of the M4 with a cast hull
M4A2: introduced in December 1941 as a version of the M4 powered by the 375-hp (277-kW) General Motors 6046 coupled powerplant
M4A3: introduced in January 1943 as a derivative of the M4 with the 450-hp (336-kW) Ford GAA-III powerplant
M4A4: introduced in February 1942 with welded hull and turret, and powered by a 375-hp (280-kW) Chrysler five-bank engine
M4A6: version of the M4A4 with the 450-hp (336-kW) Caterpillar RD-1820 diesel
M4 (76 mm): unofficial designation of earlier M4s retrofitted with the 76-mm (3-in) M1 gun and some 70 rounds (fired at a muzzle velocity of 2,600 ft/792 m per second) instead of the 75-mm M3 with some

90 rounds (fired at a muzzle velocity of 2,030 ft/619 m per second); the slightly larger gun offers significantly enhanced armour-penetration capabilities (4 in/102 mm rather than 3.15 in/80 mm at 1,000 yards/914 m)
Sherman Firefly: designation of M4A3 chassis fitted with the British 17-pdr anti-tank gun and 42 rounds; the gun has a muzzle velocity of 2,900 ft (884 m) per second, allowing its projectile to penetrate 4.72 in (120 mm) of armour at 500 yards (457 m)
M32: armoured recovery vehicle developed in World War II and based on the Sherman chassis with an A-frame and 60,000-lb (27216-kg) capacity winch
M74: armoured recovery vehicle developed after World War II and based on the M4A3 with a dozer blade, A-frame and 90,000-lb (40824-kg) capacity winch

LIGHT TANKS

Steyr SK105 Kürassier
(Austria)
Type: light tank
Crew: 3
Combat weight: 17500 kg (38,580 lb)
Dimensions: length, gun forward 7.76 m (25.47 ft) and hull 5.58 m (18.31 ft); width 2.50 m (8.20 ft); height to top of the commander's cupola 2.53 m (8.30 ft)
Armament system: one 105-mm (4.13-in) GIAT 105/57 rifled gun with 44 rounds, one 7.62-mm (0.3-in) MG74 co-axial machine-gun with 2,000 rounds, and three smoke-dischargers on each side of the turret; the turret is hydraulically powered, the main gun lacks stabilization in either elevation or azimuth, and the day/night optical/thermal fire-control system is supported by a laser rangefinder
Armour: welded steel varying in thickness from 8 to 40 mm (0.31 to 1.57 in)
Powerplant: one 239-kW (320-hp) Steyr 7FA diesel engine with 400 litres (88 Imp gal) of internal fuel
Performance: speed, road 65.3 km/h (40.6 mph); range, road 520 km (323 miles); fording 1.0 m (3.3 ft); gradient 75%; side slope 40%; vertical obstacle 0.8 m (31.5 in); trench 2.41 m (7.9 ft); ground clearance 0.4 m (15.75 in)

Variants
SK105 Kürassier: based on the chassis of the Saurer 4K 7FA APC, the Kürassier was developed in the second half of the 1960s and entered production in 1971; the type has individual NBC protection
SK105 Kürassier II: developed model of the Kürassier with a fully-stabilized upper turret and ballistic computer
4K 7FA SB 20 Bergepanzer Greif: armoured recovery vehicle development of the Kürassier, and using the same hull fitted with a 6000-kg (13,228-lb) capacity crane, a 20000-kg (44,092-lb) capacity winch and a front-mounted dozer blade
4KH 7FA-Pi Pionierpanzer: armoured engineer vehicle development of the Kürassier, basically similar to the Greif apart from having a larger dozer blade, an 8000-kg (17,637-lb) capacity winch and an excavator instead of the crane

Bernardini X1A2
(Brazil)
Type: light tank
Crew: 3
Combat weight: 19000 kg (41,887 lb)
Dimensions: length, gun forward 7.10 m (23.29 ft) and hull 6.50 m (21.33 ft); width 2.60 m (8.53 ft); height to turret top 2.45 m (8.04 ft)
Armament system: one 90-mm (3.54-mm) Cockerill/ENGESA rifled gun with 66 rounds, one 7.62-mm (0.3-in) co-axial machine-gun with 2,500 rounds, one 0.5-in (12.7-mm) M2HB AA machine-gun with 750 rounds, and three smoke-dischargers on each side of the turret; the turret is hydraulically powered, the main gun lacks stabilization in either elevation or azimuth, and an optical fire-control system is fitted (a laser rangefinder being optional)
Armour: welded steel
Powerplant: one 224-kW (300-bhp) Saab-Scania DS-11 diesel engine with 600 litres (132 Imp gal) of internal fuel
Performance: speed, road 55 km/h (34.2 mph); range, road 600 km (373 miles); fording 1.3 m (4.3 ft); gradient 70%; side slope 30%; vertical obstacle 0.7 m (27.6 in); trench 2.1 m (6.9 ft); ground clearance 0.5 m (19.7 in)

Variants
X1A: developed in the early 1970s, this Brazilian light tank was based on the chassis of the M3A1 Stuart light tank (built during World War II by the USA) but fitted with a 209-kW (280-bhp) Saab-Scania diesel and a new turret accommodating the French 90-mm (3.54-in) GIAT F1 rifled gun; the vehicle weighs 15000 kg (33.069 lb)

X1A1 Carcara: developed X1A with a stretched hull and an extra suspension group
X1A2: full-production version of the X1A1 with new rather than rebuilt chassis
XLP-10: this is the armoured vehicle-launched bridge derivative of the X1A series, and can span a gap of 10 m (32.8 ft)

Type 63
(China)
Type: light tank
Crew: 4
Combat weight: 18000 kg (39,683 lb)
Dimensions: length, gun forward 8.27 m (27.13 ft) and hull 7.16 m (23.36 ft); width 3.25 m (10.66 ft); height to top of cupola 2.19 m (7.19 ft)
Armament system: one 85-mm (3.35-in) rifled gun with 56 rounds, one 7.62-mm (0.3-in) co-axial machine-gun with 3,000 rounds, and one 12.7-mm (0.5-in) Type 54 AA machine-gun with 1,200 rounds; the turret is powered, the main gun lacks stabilization in either elevation or azimuth, and an optical fire-control system is fitted
Armour: cast and welded steel varying in thickness from 10 to 14 mm (0.39 to 0.55 in)
Powerplant: one 388-kW (520-hp) V-12 diesel engine with 545 litres (120 Imp gal) of internal fuel
Performance: speed, road 50 km/h (31.1 mph) and water 9 km/h (5.6 mph) driven by two waterjets; range, road 240 km (149 miles); fording amphibious; gradient 60%; side slope 30%; vertical obstacle 1.0 m (39.4 in); trench 3.0 m (9.8 ft); ground clearance 0.37 m (14.6 in)

Variant
Type 63: this is a derivative of the Soviet PT-76 amphibious light tank (built in China in small numbers as the Type 60) using the same armament and similar automotive components as the T-62 light tank, and propelled in the water by twin waterjets; the tank lacks NBC protection and night vision capability

Type 62
(China)
Type: light tank
Crew: 4
Combat weight: 21000 kg (46,296 lb)
Dimensions: length, gun forward 7.90 m (25.92 ft) and hull 5.55 m (18.21 m); width 2.86 m (9.38 ft); height 2.55 m (8.37 ft)
Armament system: one 85-mm (3.35-in) rifled gun with 47 rounds, one 7.62-mm (0.3-in) co-axial machine-gun with 1,750 rounds, and one 12.7-mm (0.5-in) Type 54 AA machine-gun with 1,250 rounds; the turret is hydraulically powered, the main gun lacks stabilization in either elevation or azimuth, and an optical fire-control system is fitted
Armour: cast and welded steel
Powerplant: one 283-kW (380-hp) diesel engine with 730 litres (160.6 Imp gal) of internal fuel
Performance: speed, road 60 km/h (37.3 mph); range, road 500 km (311 miles); fording 1.3 m (4.25 ft); gradient 60%; side slope 30%; vertical obstacle 0.7 m (27.6 in); trench 2.55 m (8.4 ft); ground clearance 0.42 m (16.5 in)

Variant
Type 62: this is little more than a scaled-down Type 59 MBT intended for operations in adverse terrain; the type lacks an NBC system, amphibious capability and night-vision devices

Creusot-Loire AMX-13/90
(France)
Type: light tank
Crew: 3
Combat weight: 15000 kg (33,069 lb)
Dimensions: length, gun forward 6.36 m (20.87 ft) and hull 4.88 m (16.01 ft); width 2.50 m (8.20 ft); height to commander's hatch 2.30 m (7.55 ft)
Armament system: one 90-mm (3.54-in) GIAT CN90 (F3) rifled gun with 34 rounds, two 7.5- or 7.62-mm (0.295- or 0.3-in) machine-guns (one co-axial and one AA) with 3,600 rounds, and two smoke-dischargers on each side of the turret; the turret is hydraulically powered, the main gun lacks stabilization in either elevation or azimuth, and an optical fire-control system is fitted
Armour: welded steel varying in thickness from 10 to 25 mm (0.39 to 1 in)
Powerplant: one 186-kW (250-hp) SOFAM 8Gxb petrol engine with 480 litres (105.6 Imp gal) of internal fuel
Performance: speed, road 60 km/h (37.3 mph); range, road 400 km (249 miles); fording 0.6 m (23.6 in); gradient 60%; side slope 30%; vertical obstacle 0.65 m (2.1 ft); trench 1.6 m (5.25 ft); ground clearance 0.37 m (14.6 in)

Variants
AMX-13 Modèle 51/75: this first production model of one of the world's most successful light tanks appeared in 1952, and is armed with a 75-mm (2.95-in) main gun fed from two six-round revolving magazines in the bustle of the oscillating turret; the version currently in service with the French army is also fitted with a pair of SS.11 wire-guided anti-tank missiles, one on each side of the turret; a version of the same basic concept is available for export with six Milan anti-tank missiles, three on each side of the turret
AMX-13/75: designed during the 1950s for North African operations, this variant has a manually-loaded 75-mm (2.95-in) gun in an FL-11 turret
AMX-13/90: this upgunned variant appeared in the early 1960s, and is similar to the Modèle 51/75 apart

from its main armament, a 90-mm (3.54-in) weapon in the same FL-10 turret; the same loading system is used, but whereas the Modèle 51/75 has only 12 rounds (requiring external reloading after this supply has been used) the AMX-13/90 carries 34 rounds

AMX-13/105: this model was developed as an upgunned export derivative with a 105-mm (4.13-in) gun in an FL-12 turret, and alternative SOFAM petrol-engine or 280-hp (209-kW) Detroit Diesel 6V-53T diesel-engine powerplant, the latter boosting speed to 64 km/h (40 mph) and range to 550 km (342 miles)

Char de Depannage Modèle 55: this is the armoured recovery variant of the AMX-13 series, and features two winches (the main one having a pull of 15000 kg/ 33,069 lb), an A-frame and four rear-mounted spades

Char Poseur de Pont AMX-13: this is the armoured vehicle-launched bridge derivative, and carries a 14.01-m (45.96-ft) folding bridge

PT-76
(USSR)
Type: light tank
Crew: 3
Combat weight: 14000 kg (30,864 lb)
Dimensions: length, gun forward 7.625 m (25.02 ft) and hull 6.91 m (22.67 ft); width 3.14 m (10.30 ft); height, early model 2.255 m (7.40 ft) and late model 2.195 m (7.20 ft)
Armament system: one 76.2-mm (3-in) D-56T rifled gun with 40 rounds, one 7.62-mm (0.3-in) SGMT co-axial machine-gun with 1,000 rounds, one optional 12.7-mm (0.5-in) DShKM AA machine-gun with ? rounds; the turret is electrically powered, the main gun lacks stabilization in either elevation or azimuth, and an optical fire-control system is fitted; the type can generate smoke by injecting fuel into the exhaust system
Armour: welded steel varying in thickness from 5 to 17 mm (0.2 to 0.67 in)
Powerplant: one 179-kW (240-hp) Model V-6 diesel engine with 250 litres (55 Imp gal) of internal fuel and provision for 180 litres (40 Imp gal) of external fuel
Performance: speed, road 44 km/h (27.3 mph) and water 10 km/h (6.2 mph) driven by two waterjets; range, road 450 km (280 miles) with maximum fuel and water 65 km (40 miles); fording amphibious; gradient 70%; vertical obstacle 1.1 m (43.3 in); trench 2.8 m (9.2 ft); ground clearance 0.37 m (14.6 in)

Variants
PT-76 Model 1: introduced in 1952, this is the main light tank used by the Soviet forces (army and naval infantry) for reconnaissance despite its small size, very thin armour, and lack of NBC protection and night-fighting capability; this initial model is apparently no longer in service, and was distinguishable by its multi-baffle muzzle brake and lack of bore evacuator
PT-76 Model 2: this introduced a double-baffle muzzle brake and forward-located bore evacuator
PT-76 Model 3: similar to the Model 2 but without a bore evacuator
PT-76B: this last variant has a fully-stabilized D-56TM main armament; the NATO reporting designation is **PT-76 Model 4**
Type 60: PT-76 built in China

Vickers/FMC Battle Tank Mk 5
(UK/USA)
Type: light tank
Crew: 4
Combat weight: 43,540 lb (19750 kg)
Dimensions: length, gun forward 8.657 m (28.25 ft) and hull 20.33 ft); 6.20 m); width 8.83 ft (2.69 m); height to turret roof 8.60 ft (2.62 m)
Armament system: one 105-mm (4.13-in) M68 rifled gun with 41 rounds, one co-axial 7.62-mm (0.3-in) L8A2 machine-gun with 5,000 rounds, one 0.5-in (12.7-mm) Browning M2HB AA heavy machine-gun with 1,000 rounds, and six smoke-dischargers on each side of the turret; the turret is electrically powered, the main armament in stabilized in both elevation and azimuth, and a Marconi fire-control system is fitted
Armour: cast and welded aluminium and steel
Powerplant: one 552-hp (412-KW) Detroit Diesel Model 6V-92 TA6 diesel engine with ? US gal (? litres) of internal fuel
Performance: speed, road 44 mph (70 km/h); range, road 300 miles (483 km); fording 4.33 ft (1.32 m) without preparation; gradient 60%; side slope 40%; vertical obstacle 2.5 ft (0.76 m); trench 7.0 ft (2.13 m); ground clearance 16 in (40.6 cm)

Variants
Battle Tank Mk 5: introduced in 1986, this export derivative of the CCVL produced jointly by Vickers (turrent, armament, fire-control system) and FMC (hull and automotive system); the result is a tank with the weight of a light tank but the armament and protection of an MBT
FMC CCVL: this is a competitor in the US Army's competition for a light combat vehicle needed by its light divisions; the FMC turret of the purely US model has a 105-mm (4.13-in) M68A1 rifled gun with a low-recoil system produced by Rheinmetall and an automatic loader plus 19-round magazine; another 24 rounds are stowed in the hull to replenish the loader; the gunner has a stabilized day/night sight plus laser rangefinder; the CCVL has a combat weight of 42,800 lb (19414 kg), and its principal dimensions are lengths of 30.75 ft 9.37 with the gun forward and 20.33 ft (6.20 m) for the hull, a width of 8.83 ft (2.69 m) and a height of 7.75 ft (2.36 m)

Cadillac Gage Stingray
(USA)
Type: light tank
Crew: 4
Combat weight: 42,000 lb (19051 kg)
Dimensions: length, gun forward 30.67 ft (9.35 m) and hull 20.67 ft (6.30 m); width 8.92 ft (2.71 m); height overall 8.33 ft (2.55 m)
Armament system: one 105-mm (4.13-in) Royal Ordnance L7A1 rifled gun with ? rounds, one co-axial 7.62-mm (0.3-in) machine-gun with ? rounds, one 0.5-in (12.7-mm) Browning M2HB AA machine-gun with ? rounds, and ? smoke-dischargers mounted on each side of the turret; the turret is hydraulically operated, the main gun is stabilized in elevation and azimuth, and a Marconi Digital Fire-Control System is fitted; this last combines day/night optical/thermal sights, a laser rangefinder, various sensors and a ballistic computer
Armour: welded steel
Powerplant: one 535-hp (399-kW) Detroit Diesel Model 8V-92TA turbocharged diesel engine with ? US gal (? litres) of fuel
Performance: speed, road 43 mph (69 km/h); range, road 300 miles (483 km); fording 4.0 ft (1.22 m); gradient 60%; side slope 30%; vertical obstacle 30 in (0.76 m); trench 5.6 ft (1.69 m); ground clearance ? in (? m)

Variant
Stingray: produced as a private venture, the Stingray is Cadillac Gage's company complement to the Commando series of wheeled AFVs, offering good firepower and mobility though only slight protection

Cadillac (General Motors) M41A3 Walker Bulldog
(USA)
Type: light tank
Crew: 4
Combat weight: 51,800 lb (23496 kg)
Dimensions: length, gun forward 29.94 ft (8.21 m) and hull 19.09 ft (5.82 m); width 10.50 ft (3.20 m); height to top of cupola 8.94 ft (2.73m)
Armament system: one 76-mm (3-in) M32A1 rifled gun with 65 rounds, one 0.3-in (7.62-mm) M1919A4 co-axial machine-gun with 5,000 rounds, and one 0.5-in (12.7-mm) M2HB AA machine-gun with 2,175 rounds; the turret is electro-hydraulically powered, the main gun lacks stabilization in either elevation or azimith, and an optical fire-control system is fitted
Armour: cast and welded steel varying in thickness from 0.35 to 1.25 in (9.25 to 31.75 mm)
Powerplant: one 500-bhp (373-kW) Avco Lycoming AOSI-895-5 petrol engine with 140 US gal (530 litres) of internal fuel
Performance: speed, road 45 mph (72.4 km/h); range, road 100 miles (161 km); fording 40 in (1.02 m); gradient 60%; side slope 30%; vertical obstacle 2.33 ft (0.71 m); trench 6.0 ft (1.83 m); ground clearance 17.75 in (0.45 m)

Variants
M41: this successor to the M24 Chaffee began to enter service in 1951, and is powered by the 500-bhp (373-kW) Continental AOS-895-3 petrol engine; 76-mm (3-in) ammunition stowage is 57 rounds
M41A1: improved M41
M41A2: M41A1 with simplified gun and turret control systems, stowage for 65 rounds of 76-mm (3-in) ammunition, and fuel-injected AOSI-895-5 engine
M41A3: improved M41A2
M41 (Brazil): the Brazilian army is converting its M41 tanks to **M41B** standard in an extensive modernization programme, which involves the replacement of the original engine with a 302-kW (405-hp) Saab-Scania DS-14A 04 diesel (producing a speed of 70 km/h/43.5 mph), the revision of the fuel tankage and the boring out of the original 76-mm (3-in) gun to 90-mm (3.54-in) calibre (under the revised designation Ca 76/90 M32) so that Cockerill 90-mm ammunition can be fired; other modifications include the provision of four smoke-dischargers on each side of the turret, and the incorporation of a night-vision system; the **M41C** for the Brazilian marine corps is similar to the M41B but is powered by a DS-14 OA diesel engine and has a new fire-control system for the Ca 76/90 M32 BR1 gun able to fire an APFSDS round
M41 (Spain): Spanish vehicles are being modified to **M41E** standard with the 400-hp (298-kW) General Motors 8V-71T turbocharged diesel engine and (possibly later) a 90-mm (3.54-in) gun; the use of a diesel powerplant increases range to 300 miles (483 km)
M41 (Thailand): the M41 light tanks operated by the Thai army are being upgraded to **M41GTI** (German Tank Improvement) standard by two West German concerns; this programme involves the replacement of the original petrol engine with a 330-kW (442-hp) MTU MB 833 Aa501 diesel engine, the enlarging of total fuel capacity to 800 litres (176 Imp gal), the fitting of a fire-suppression system and the incorporation of many suspension improvements; the revised vehicle has a maximum road speed of 60 km/h (37.3 mph) and a range of 600 km (373 miles); the original main armament is retained, but the secondary armament now comprises two 7.62-mm (0.3-in) Heckler und Koch HK21E machine-guns, and other modifications are a stabilized gun-control system used in conjunction with the MOLF 41 fire-control system (including thermal imaging capability and a laser rangefinder), the addition of an NBC system, and the incorporation of dual-role dischargers for smoke and anti-personnel grenades; the combat weight is 24500 kg (54,012 lb)
M41/Cockerill: the Belgian company Cockerill is

supplying kits by which M41s can be retrofitted with a 90-mm (3.54-in) Cockerill Mk IV gun
Type 64: Taiwanese reworking of the M41A2/A3 with stronger armour (including a bolt-on outer layer of laminate steel), a Taiwanese development of the original main armament with higher performance, and upgraded secondary armament; further improvements may include a modern fire-control system (with a laser rangefinder and ballistic computer) and night-vision equipment

Cadillac (General Motors) M24 Chaffee
(USA)
Type: light tank
Crew: 5
Combat weight: 40,500 lb (18371 kg)
Dimensions: length, gun forward 18.00 ft (5.49 m) and hull 16.50 ft (5.03 m); width 9.67 ft (2.95 m); height to top of commander's cupola 8.08 ft (2.46 m)
Armament system: one 75-mm (2.95-in) M6 rifled gun with 48 rounds, two 0.3-in (7.62-mm) M1919A4 machine-guns (one bow and one co-axial) with 3,750 rounds, and one 0.5-in (12.7-mm) M2HB AA machine-gun with 440 rounds; the turret is hydraulically powered, the main gun lacks stabilization in either elevation or traverse, and an optical fire-control system is fitted
Armour: cast and welded steel varying in thickness from 0.5 to 1.5 in (12.7 to 38.1 mm)
Powerplant: two 110-hp (82-kW) Cadillac 44T24 petrol engines with 110 US gal (416 litres) of internal fuel
Performance: speed, road 34 mph (55 km/h); range, road 175 miles (281 km); fording 3.33 ft (1.02 m); gradient 60%; side slope 30%; vertical obstacle 36 in (0.91 m); trench 8.0 ft (2.44 m); ground clearance 18.0 in (0.46 m)

Variants
M24: the initial production model of this replacement for the M3 and M5 series was delivered in 1944, but the type is no longer in service in its original form
M24 (Norway): the Norwegians have converted a number of M24s to a more advanced standard, under the designation **NM-116**, with a 250-hp (186-kW) Detroit Diesel 6V-53T turbocharged diesel engine and a new armament layout, comprising a French 90-mm (3.54-in) GIAT D-925 main gun with 41 rounds, two 0.5-in (12.7-mm) M2HB machine-guns (one co-axial and one AA) with 500 rounds and four smoke-dischargers on each side of the turret; other modifications have been made to the safety and fire-control systems; the Norwegians have also developed an armoured recovery vehicle variant, the **BK 710** with an 11000-kg (24,250-lb) capacity crane replacing the turret
M24 (Taiwan): this is essentially similar to the Norwegian model, but the Taiwanese have also developed a flamethrower version in which the flame gun replaces the bow machine-gun

TANK DESTROYERS AND ANTI-TANK VEHICLES

GIAT AMX-10 HOT tracked anti-tank vehicle (France): see GIAT AMX-10 tracked infantry fighting vehicle

Renault VCAC HOT wheeled anti-tank vehicle (France): see Renault VAB VTT wheeled armoured personnel carrier

Panhard VCR/TH wheeled anti-tank vehicle (France): see Panhard VCR/TT wheeled armoured personnel carrier

Henschel/Hanomag (Thyssen Henschel) Jagdpanzer Jaguar 1
(West Germany)
Type: tracked tank destroyer (missile)
Crew: 4
Combat weight: 23000 kg (50,705 lb)
Dimensions: length, hull 6.61 m (21.69 ft); width 3.12 m (10.24 ft); height including missiles 2.54 m (8.33 ft)
Armament system: one Euromissile K3S single-tube launcher system with 20 HOT anti-tank missiles, two 7.62-mm (0.3-in) MG3 machine-guns (one bow and one AA) with 3,200 rounds, and eight smoke-dischargers on the hull roof
Armour: welded and spaced steel varying in thickness from 12 to 50 mm (0.47 to 1.97 in)
Powerplant: one 373-kW (500-hp Daimler-Benz MB 837 diesel engine with 470 litres (103 Imp gal) of internal fuel
Performance: speed, road 70 km/h (43.5 mph); range, road 400 km (249 miles); fording 1.4 m (4.6 ft) without preparation and 2.1 m (6.9 ft) with preparation; gradient 60%; side slope 30%; vertical obstacle 0.75 m (29.5 in); trench 2.0 m (6.6 ft); ground clearance 0.44 m (17.3 in)

Variant
Jagdpanzer Jaguar 1: this vehicle is based on the same chassis as the Jagdpanzer Kanone tank destroyer and Marder MICV, and the type was built

from 1967 with the designation **Jagdpanzer Rakete** and a primary armament of two SS.11 wire-guided missile launchers; between 1978 and 1983 316 of the 370 vehicles were rebuilt to Jaguar 1 standard with the HOT system; the type has also been evaluated with a four-tube Euromissile HCT launcher for HOT missiles, but no orders have been placed for this variant; standard equipment includes an NBC system and night-vision devices

Henschel/Hanomag (Thyssen Henschel) Jagdpanzer Kanone (JPz 4-5)
(West Germany)
Type: tracked tank destroyer (gun)
Crew: 4
Combat weight: 27500 kg (60,626 lb)
Dimensions: length, overall 8.75 m (28.71 ft) and hull 6.238 m (20.47 ft); width 2.98 m (9.78 ft); height without machine-gun 2.085 m (6.84 ft)
Armament system: one 90-mm (3.54-in) rifled gun with 51 rounds, two 7.62-mm (0.3-in) MG3 machine-guns (one co-axial and one AA) with 4,000 rounds, and eight smoke-dischargers on the hull roof; the type is fitted with an optical fire-control system
Armour: welded and spaced steel varying in thickness from 12 to 50 mm (0.47 to 1.97 in)
Powerplant: one 373-kW (500-hp) Daimler-Benz MB 837 diesel engine with 470 litres (103 Imp gal) of internal fuel
Performance: speed, road 70 km/h (43.5 mph); range, road 400 km (249 miles); fording 1.4 m (4.6 ft) without preparation and 2.1 m (6.9 ft) with preparation; gradient 60%; side slope 30%; vertical obstacle 0.75 m (29.5 in); trench 2.0 m (6.6 ft); ground clearance 0.44 m (17.3 in)

Variants
Jagdpanzer Kanone: this is basically the same vehicle as the Jaguar 1 but armed with a 90-mm (3.54-mm) rifled gun rather than with a missile system, German experience in World War II having convinced them that low-silhouette gun-armed vehicles were (and still are) highly effective tank-killers; 750 production vehicles were produced between 1965 and 1967, and some thought has been given to converting at least some of the survivors to mount a 105-mm (4.13-in) rifled or 120-mm (4.72-mm) smooth-bore gun; standard equipment includes an NBC system and night-vision devices
Jagdpanzer Jaguar 2: produced by conversion between 1983 and 1985, these 162 vehicles are similar to the Jaguar 1 but have a TOW anti-tank missile launcher and TAS-4 night sight rather than the main gun
JPK 90: this is the Belgian version of the Jagdpanzer Kanone, from which it differs in having the power train, suspension and tracks of the Marder MICV, the SABCA fire-control system (in simplified form) of the Leopard 1 MBT, Lyran flare launchers, passive night-vision devices, and FN MAG rather than MG3 machine-guns

Porsche Wiesel/TOW tracked anti-tank vehicle
(West Germany): see Porsche Wiesel airportable tracked armoured vehicle

OTO Melara VCC-1/TOW tracked anti-tank vehicle
(Italy): see OTO Melara IAFV Mk 1 tracked infantry armoured fighting vehicle

Komatsu Type 60 Model C
(Japan)
Type: tracked tank destroyer (recoilless rifle)
Crew: 3
Combat weight: 8000 kg (26,247 lb)
Dimensions: length 4.30 m (14.11 ft); width 2.23 m (7.32 ft); height 1.59 m (5.22 ft)
Armament system: two 106-mm (4.17-in) Type 60 recoilless rifles on an elevating mounting with eight rounds, and one 0.5-in (12.7-mm) ranging machine-gun
Armour: welded and riveted steel to a maximum thickness of 12 mm (0.47 in)
Powerplant: one 112-kW (150-hp) Komatsu SA4D105 diesel engine with 77 litres (17 Imp gal) of internal fuel
Performance: speed, road 55 km/h (34 mph); range, road 130 km (81 miles); fording 1.0 m (3.3 ft); gradient 60%; side slope 30%; vertical obstacle 0.55 m (21.7 in); trench 1.8 m (5.9 ft); ground clearance 0.35 m (13.8 in)

Variants
Type 60 Model A: initial production model from 1960 with an 89-kW (120-hp) Komatsu 6T 120-2 diesel and a maximum speed of 45 km/h (28 mph)
Type 60 Model B: 1967 development with stronger construction
Type 60 Model C: definitive production model from 1975 to 1979 with more powerful engine; none of the variants has NBC protection

DAF PWAT wheeled anti-tank vehicle
(Netherlands): see DAF YP-408 wheeled armoured personnel carrier

BRAVIA Chaimite V-700 wheeled anti-tank vehicle
(Portugal): see BRAVIA Chaimite V-200 wheeled armoured personnel carrier

ENASA VEC wheeled anti-tank vehicle
(Spain): see ENASA VEC wheeled reconnaissance vehicle

Hägglund & Söner Infanterikanonvagn 91
(Sweden)
Type: tracked tank destroyer (gun)
Crew: 4

Combat weight: 16300 kg (35,935 lb)
Dimensions: length, gun forward 8.84 m (29.00 ft) and hull 6.41 m (21.03 ft); width 3.00 m (9.84 ft); height to top of commander's cupola 2.32 m (7.61 ft)
Armament system: one 90-mm (3.54-in) Bofors KV 90 S 73 rifled gun with 59 rounds, two 7.62-mm (0.3-in) machine-guns (one co-axial and one AA) with 4,250 rounds, two Lyran flare launchers on the turret roof, and six smoke-dischargers on each side of the turret; the turret is electro-hydraulically powered, the main gun lacks stabilization in either elevation or azimuth, and an optical fire-control system is fitted
Armour: welded steel
Powerplant: one 246-kW (330-hp) Volvo-Penta TD 120 A diesel engine with 400 litres (88 Imp gal) of internal fuel
Performance: speed, road 65 mph (40.4 mph) and water 6.5 km/h (4 mph) driven by its tracks; range, road 500 km (311 miles); fording amphibious; gradient 60%; side slope 30%; vertical obstacle 0.8 m (31.5 in); trench 2.8 m (9.2 ft); ground clearance 0.37 m (14.6 in)

Variants
Ikv 91: developed in the later 1960s by Hägglund & Söner, this is a powerful tank destroyer with a capable fire-control system, and produced for the Swedish army between 1975 and 1978; the type is designed for arctic conditions and has NBC protection, but lacks night-vision capability
Ikv 91–105: much improved version developed as a private venture on the basis of the Ikv 91 but fitted with the 105-mm (4.13-in) Rheinmetall Rh-105-11 low-impulse rifled gun and stowage for 40 rounds (under special circumstances the ammunition capacity can be increased to 50 rounds); length with gun forward is 9.72 m (31.89 ft); the type is also provided with a choice of two passive fire-control systems, one based on a low-light-level TV camera plus an infra-red scanner, and the other on a thermal imaging sight; the combat weight of this model has increased to 18000 kg (39,683 lb), but the type is still amphibious, and can be fitted with twin propellers for better waterborne performance

Hägglund & Söner Pvrbv 551
(Sweden)
Type: tracked tank-destroyer (missile)
Crew: 4
Combat weight: 9700 kg (21,384 lb)
Dimensions: length 4.81 m (15.78 ft); width 2.54 m (8.33 ft); height over raised missile launcher 2.98 m (9.78 ft)
Armament system: one single-tube TOW elevating launcher with ? missiles
Armour: welded steel
Powerplant: one 136-hp (101-kW) Ford Model 2658E petrol engine with 260 litres (57 Imp gal) of internal fuel
Performance: speed, road 41 km/h (25.5 mph); range, road 350 km (217 miles); fording 0.9 m (2.95 ft); gradient 60%; vertical obstacle not revealed; trench 1.5 m (4.9 ft); ground clearance 0.33 m (13 in)

Variant
Pvrbv 551: this vehicle is being produced by conversion of obsolete Ikv 102 and Ikv 103 infantry cannons, the whole powerplant and transmission being changed, the hull being extended to the rear, the protection and communications being upgraded and, perhaps most importantly of all, the glacis-mounted gun being removed (and the gap plated over) in favour of an elevating launcher for TOW anti-tank missiles

ASU-85
(USSR)
Type: airborne tracked tank destroyer (gun)
Crew: 4
Combat weight: 15500 kg (34,171 lb)
Dimensions: length, overall 8.49 m (27.85 ft) and hull 6.00 m (19.69 ft); width 2.80 m (9.19 ft); height without AA machine-gun 2.10 m (6.89 ft)
Armament system: one 85-mm (3.35-in) D-70 rifled gun with 40 rounds, one co-axial 7.62-mm (0.3-in) SGMT machine-gun with 2,000 rounds, one 12.7-mm (0.5-in) DShKM AA heavy machine-gun, and (some vehicles) smoke-dischargers on the hull rear; an optical fire-control system is fitted
Armour: welded steel varying in thickness from 8 to 40 mm (0.315 to 1.57 in)
Powerplant: one 179- or 209-kW (240- or 280-hp) V-6 diesel engine with 250 litres (55 Imp gal) of internal fuel
Performance: speed, road 45 km/h (28 mph); range, road 260 km (162 miles); fording 1.1 m (3.6 ft); gradient 70% vertical obstacle 1.1 m (43.3 in); trench 2.8 m (9.2 ft); ground clearance 0.4 m (15.75 in)

Variant
ASU-85: developed in the late 1950s as the primary self-propelled tank destroyer of the Soviet airborne forces, the ASU-85 is based on the chassis and automotive system of the PT-76 light tank; standard equipment includes an NBC system and night-vision devices

ASU-57
(USSR)
Type: airborne tracked tank destroyer (gun)
Crew: 3
Combat weight: 3350 kg (7,385 lb)
Dimensions: length overall 4.995 m (16.39 ft) and hull 3.48 m (11.42 ft); width 2.086 m (6.84 ft); height 1.18 m (3.87 ft)

Armament system: one 57-mm Ch-51 or Ch-51M rifled gun with 40 rounds, and (some vehicles) one 7.62-mm (0.3-in) machine-gun; an optical fire-control system is fitted
Armour: welded steel to a thickness of 6 mm (0.24 in)
Powerplant: one 41-kW (55-hp) M-20E petrol engine with 140 litres (31 Imp gal) of internal fuel
Performance: speed, road 45 km/h (28 mph); range, road 250 km (155 miles); fording 0.7 m (2.3 ft); gradient 60%; side slope 30%; vertical obstacle 0.5 m (19.7 in); trench 1.4 m (4.6 ft); ground clearance 0.2 m (8 in)

Variant
ASU-57: developed in the early 1950s, the ASU-57 is a small and very low light assault gun/tank transporter designed for use by the Soviet airborne forces; the vehicle is very simple in concept and operation, and lacks modern features such as night-vision devices and NBC protection

BMP-2 tracked anti-tank vehicle (USSR): see BMP-2 tracked mechanized infantry fighting vehicle

BMP-1 tracked anti-tank vehicle (USSR): see BMP-1 tracked mechanized infantry fighting vehicle

BMD tracked anti-tank vehicle (USSR): see BMD tracked airborne combat vehicle

BRDM-2/'Swatter' wheeled anti-tank vehicle (USSR): see BRDM-2 wheeled scout car

BRDM-2/'Sagger' wheeled anti-tank vehicle (USSR): see BRDM-2 wheeled scout car

BRDM-2/'Spandrel' wheeled anti-tank vehicle (USSR): see BRDM-2 wheeled scout car

BRDM-1/'Snapper' wheeled anti-tank vehicle (USSR): see BRDM-1 wheeled scout car

BRDM-1/'Swatter' wheeled anti-tank vehicle (USSR): see BRDM-1 wheeled scout car

BRDM-1/'Sagger' wheeled anti-tank vehicle (USSR): see BRDM-1 wheeled scout car

BTR-40/'Sagger' wheeled anti-tank vehicle (USSR/East Germany): see BTR-40 wheeled armoured personnel carrier

GKN Sankey FV438 tracked anti-tank vehicle (UK): see GKN Sankey FV432 tracked armoured personnel carrier

GKN Sankey FV432 Carl Gustav tracked anti-tank vehicle (UK): see GKN Sankey FV 432 tracked armoured personnel carrier

Alvis FV102 Striker tracked anti-tank vehicle (UK): see Alvis FV103 Spartan tracked armoured personnel carrier

Daimler FV712 Ferret Mk 5 wheeled anti-tank vehicle (UK): see Daimler FV711 Ferret Mk 4 wheeled scout car

FMC AIFV tracked anti-tank vehicle (USA): see FMC AIFV tracked armoured infantry fighting vehicle

FMC M901 Improved TOW Vehicle tracked anti-tank vehicle (USA): see FMC M113A1 tracked armoured personnel carrier

General Motors Canada LAV-25 wheeled anti-tank vehicle (USA/Canada): see General Motors Grizzly wheeled armoured personnel carrier

Cadillac Gage Commando V-150/TOW wheeled anti-tank vehicle (USA): see Cadillac Gage Commando V-150/20-mm Turret wheeled armoured personnel carrier

M36 Jackson
(USA)
Type: tracked tank destroyer (gun)
Crew: 5
Combat weight: 61,000 lb (27670 kg)
Dimensions: length, gun forward ? ft (? m) and hull 19.50 ft (5.94 m); width 10.00 ft (3.05 m); height 10.33 ft (3.15 m)
Armament system: one 90-mm (3.54-in) M3 rifled gun with 45 rounds and one 0.5-in (12.7-mm) Browning M2HB AA heavy machine-gun with 1,050 rounds in a hydraulically-powered turret
Armour: cast and welded steel up to a maximum thickness of 1.5 in (38.1 mm)
Powerplant: one 450-hp (336-kW) Ford GAA petrol engine with 192 US gal (727 litres) of internal fuel
Performance: speed, road 26 mph (42 km/h); range, road 155 miles (249 km); fording 3.0 ft (0.91 m); gradient 60%; vertical obstacle 18 in (0.46 m); trench 7.5 ft (2.23 m); ground clearance 17.4 in (0.44 m)

Variant
M36: a development of the legendary M4 Sherman medium tank, the M36 was introduced into service during 1943, and some examples remain in limited service to this day

M18 Hellcat
(USA)
Type: tracked tank destroyer (gun)
Crew: 5
Combat weight: 37,577 lb (17036 kg)
Dimensions: length, gun forward 21.83 ft (6.65 m) and hull 17.50 ft (5.33 m); width 9.42 ft (2.87 m); height 8.42 ft (2.57 m)
Armament system: one 76-mm (3-in) M1A rifled gun with 45 rounds and one 0.5-in (12.7-mm) Browning M2HB AA heavy machine-gun with 800 rounds in a hydraulically-powered turret
Armour: cast and welded steel varying in thickness from 0.3 to 1 in (7.9 to 25.4 mm)
Powerplant: one 400-hp (298-kW) Continental R-975-C4 petrol engine with 169 US gal (640 litres) of internal fuel
Performance: speed, road 55 mph (88 km/h); range, road 105 miles (169 km); fording 4.0 ft (1.22 m); gradient 60%; vertical obstacle 36 in (0.91 m); ground clearance 14.5 in (0.37 m)

Variant
M18: delivered in 1944, the Hellcat was designed round a capable anti-tank weapon and a high-speed chassis with good cross-country performance, and the type remains in limited service to this day

ARMOURED CARS, RECONNAISSANCE VEHICLES AND SCOUT CARS

ENGESA EE-9 Cascavel Mk IV
(Brazil)
Type: wheeled armoured car
Crew: 3
Combat weight: 12000 kg (26,455 lb)
Dimensions: length, gun forward 6.29 m (20.64 ft) and hull 5.25 m (17.22 ft); width 2.59 m (8.50 ft); height to top of commander's cupola 2.60 m (8.53 ft)
Armament system: one 90-mm (3.54-in) EC-90-III rifled gun with 44 rounds, one 7.62-mm (0.3-in) co-axial machine-gun with 2,400 rounds, one optional 0.5-in (12.7-mm) M2HB or 7.62-mm (0.3-in) AA machine-gun with ? rounds, and two or three smoke-dischargers on each side of the turret; the turret is electrically powered, the main gun lacks stabilization in either elevation or azimuth, and an optical fire-control system is fitted; this last can be upgraded optionally with a laser rangefinder
Armour: welded steel
Powerplant: (typically) one 212-hp (158-kW) Detroit Diesel 6V-53N diesel engine with 360 litres (79 Imp gal) of internal fuel, and driving a 6 × 6 layout
Performance: speed, road 100 km/h (62 mph); range, road 750 km (466 miles); fording 1.0 m (3.3 ft); gradient 60%; side slope 30%; vertical obstacle 0.6 m (23.6 in); ground clearance 0.375 m (14.75 in)

Variants
EE-9 Cascavel Mk I: this was the initial production model of the Cascavel, with deliveries beginning in 1974; armament was a 37-mm gun and the powerplant a Mercedes-Benz diesel; most have been upgraded with the ET-90 turret armed with a 90-mm (3.54-in) gun
EE-9 Cascavel Mk II: export version of the Mk I but with automatic rather than manual transmission and a French-supplied Hispano-Suiza H 90 turret with 90-mm (3.54-in) gun
EE-9 Cascavel Mk III: developed model with Mercedes-Benz diesel plus automatic transmission, and the new ET-90 turret with Brazilian-built 90-mm (3.54-in) gun
EE-9 Cascavel Mk IV: improved model with different powerplant; fitted with the ET-90 turret, this was the first variant to have a central tyre pressure-regulation system; an integrated fire-control system is under development to provide the type with the capability of making full use of its powerful main gun
EE-9 Cascavel Mk V: identical with the Mk IV apart from the powerplant, in this instance a 142-kW (190-hp) Mercedes-Benz OM 352A diesel

ENGESA EE-3 Jararaca
(Brazil)
Type: wheeled scout car
Crew: 3
Combat weight: 5500 kg (12.125 lb)
Dimensions: length 4.125 m (13.53 ft); width 2.13 m (6.99 ft); height to top of AA gun mount 1.97 m (6.46 ft)
Armament system: one pintle-mounted 0.5-in (12.7-mm) Browning M2HB machine-gun with ? rounds
Armour: welded steel
Powerplant: one 90-kW (121-hp) Mercedes-Benz OM 314A turbocharged diesel engine with 135 litres (30 Imp gal) of internal fuel, and driving a 4 × 4 layout
Performance: speed, road 100 km/h (62.1 mph); range, road 750 km (466 miles); fording 0.8 m (2.6 ft); gradient 60%; side slope 30%; vertical obstacle 0.4 m (15.75 in); ground clearance 0.315 m (12.4 in)

Variant
EE-3 Jararaca: introduced to the export market in the early 1980s, the Jararaca has proved moderately successful, and the company is now developing a number of alternative armament fits, such as a 20-mm cannon, two turret-mounted 7.62-mm (0.3-in) machine-guns, a 60-mm (2.36-in) mortar, and a launcher for the Milan anti-tank missile

GIAT AMX-10RC
(France)
Type: wheeled armoured reconnaissance vehicle
Crew: 4
Combat weight: 15880 kg (35,009 lb)
Dimensions: length, gun forward 9.15 m (30.02 ft) and hull 6.357 m (20.86 ft); width 2.95 m (9.68 ft); height to turret top 2.215 m (7.27 ft)
Armament system: one 105-mm (4.13-in) GIAT F2 rifled gun with 38 rounds, one 7.62-mm (0.3-in) co-axial machine-gun with 4,000 rounds, and two smoke-dischargers on each side of the turret; the turret is electro-hydraulically powered, the main gun lacks stabilization in either elevation or azimuth, and a COTAC fire-control system is fitted; this last combines optical sights, a low-light-level TV, a laser rangefinder, various sensors and a ballistic computer
Armour: welded aluminium
Powerplant: one 194-kW (260-hp) Renault (Hispano-Suiza) HS 115 diesel engine with ? litres (? Imp gal) of internal fuel, and driving a 6 × 6 layout
Performance: speed, road 85 km/h (53 mph) and water 7.2 km/h (4.5 mph) driven by two waterjets; range, road 800 km (497 miles); fording amphibious; gradient 60%; side slope 40%; vertical obstacle 0.7 m (27.6 in); trench 1.15 m (3.8 ft); ground clearance adjustable between 0.2 and 0.6 m (7.8 and 23.6 in)

Variants
AMX-10RC: this is basically the reconnaissance version of the AMX-10P infantry fighting vehicle, and is fitted with the TK 105 turret; the type began to enter French service in 1979; the vehicle has an NBC system, and the advanced fire-control/observation system permits effective engagements by day or night; the latest production batches have the 224-kW (300-hp) Baudouin 6F11 SRX diesel, which is more fuel-efficient than the Hispano-Suiza unit and thus provides greater range on the same fuel capacity
AMX-10RAC: derivative of the AMX-10RC with a TS 90 turret armed with a 90-mm (3.54-in) GIAT CS Super gun and 37 rounds
AMX-10 RAA: proposed AA version with the Thomson-CSF/SAMM SABRE turret with two 30-mm cannon and associated fire-control system
AMX-10 RTT: proposed APC version for a crew of 3+10 and armed with a 0.5-in (12.7-mm) Browning M2HB machine-gun

Panhard EBR
(France)
Type: wheeled armoured car
Crew: 4
Combat weight: 13500 kg (29,762 lb)
Dimensions: length, gun forward 6.15 m (20.18 m) and hull 5.56 m (18.24 m); width 2.42 m (7.94 m); height to turret top (on eight wheels) 2.32 m (7.61 ft)
Armament system: one 90-mm (3.54-in) GIAT rifled gun with 43 rounds, one 7.5-mm (0.295-in) co-axial machine-gun with 2,000 rounds, and two smoke-dischargers on each side of the turret; the turret is hydraulically powered, the main gun lacks stabilization, and an optical fire-control system is fitted
Armour: cast and welded steel varying in thickness from 10 to 40 mm (0.39 to 1.57 in)
Powerplant: one 149-kW (200-hp) Panhard 12 H 6000 petrol engine with 380 litres (84 Imp gal) of internal fuel, and driving an 8 × 8 layout
Performance: speed, road 105 km/h (65.2 mpg); range, road 650 km (404 miles); fording 1.2 m (3.9 ft); gradient 60%; vertical obstacle 0.4 m (15.75 in); trench 2.0 m (6.6 ft); ground clearance on four wheels 0.33 m (13.0 in) and on eight wheels 0.41 m (16.1 in)

Variants
EBR: this heavy armoured car is being phased out of service with the French army, and is unusual in having front and rear drivers and alternative 4 × 4 or 8 × 8 drives, the four central tyreless wheels being lifted off the ground unless needed for extra traction in adverse conditions; the main armament system is the same as that of later AMX-13 light tanks, featuring a 90-mm (3.54-in) gun in an FL-11 oscillating turret
EBR ETT: this is the armoured personnel carrier version of the EBR, built in very small numbers; 12 infantry can be carried in addition to the crew of three, and armament comprises a turret-mounted 7.5-mm (0.295-in) machine-gun

Renault VBC 90
(France)
Type: wheeled armoured car
Crew: 3
Combat weight: 12800 kg (28,219 lb)
Dimensions: length, gun forward 8.152 m (26.75 ft) and hull 5.495 m (18.03 ft); width 2.49 m (8.17 m); height to turret top 2.55 m (8.37 ft)
Armament system: one 90-mm (3.54-in) GIAT CS Super (F4) rifled gun with 45 rounds, two 7.62-mm (0.3-in) machine-guns (one co-axial and one AA) with 4,000 rounds, and two smoke-dischargers on each side of the turret; the turret is manually powered with electrical assistance, the main gun lacks stabilization in either elevation or azimuth, and a SOPTAC or simplified COTAC fire-control system is fitted; the

SOPTAC type is an optical system with the option of laser rangefinding, and the COTAC type includes optical sights, a laser rangefinder, various sensors and a ballistic computer
Armour: cast and welded steel
Powerplant: one 171-kW (230-bhp) Renault VI MIDS 06.20.45 diesel engine with ? litres (? Imp gal) of internal fuel, and driving a 6 × 6 layout
Performance: speed, road 92 km/h (57.2 mph); range, road 1000 km (621 miles); fording 1.2 m (3.9 ft); gradient 60%; side slope 30%; vertical obstacle 0.6 m (23.6 in); trench 1.0 m (3.3 ft); ground clearance ? m (? in)

Variant
VBC 90: this is a powerful armoured car normally fitted with the GIAT TS 90 turret as described above; optional turrets are the Hispano-Suiza Lynx 90 with 90-mm (3.54-in) GIAT F1 gun, the Hispano-Suiza MARS turret with 90-mm gun, the SAMM TTB 190 turret with 90-mm CS Super gun, the GIAT turret fitted with 81-mm (3.2-in) CL81 breech-loaded smooth-bore mortar, and the ESD TA-20 turret with two 20-mm cannon; optional extras are an NBC system and a front-mounted winch

Panhard Sagaie 2
(France)
Type: wheeled armoured car
Crew: 3
Combat weight: 10000 kg (22,046 lb)
Dimensions: length, gun forward 7.97 m (26.15 ft) and hull 5.57 m (18.27 ft); width 2.70 m (8.86 ft); height 2.30 m (7.55 ft)
Armament system: one 90-mm (3.54-in) GIAT CS Super (F4) rifled gun with 32 or 35 rounds, two 7.62-mm (0.3-in) machine-guns (one co-axial and one AA), and two or four smoke-dischargers on each side of the turret; the turret is manually powered, the main gun lacks stabilization in either elevation or azimuth, and an optical fire-control system is fitted
Armour: welded steel
Powerplant: two 73-kW (98-hp) Peugeot XD 34T turbocharged diesel engines, or two 108-kW (145-hp) Peugeot petrol engines with ? litres (? Imp gal) of fuel, and driving a 6 × 6 layout
Performance: speed, road 100 km/h (62 mph); range, road 600 km (373 miles); fording 1.2 m (3.9 ft); gradient 50%; side slope 30%; vertical obstacle 0.8 m (31.5 in); trench 0.8 m (31.5 in)

Variant
Sagaie 2: this is basically an improved version of the Sagaie 1 with an uparmoured SAMM TTB 90 turret in place of the Sagaie 1's TS 90 type; other improvements are an uprated powerplant and a hull enlarged in length and width; a wide diversity of optional equipment can be installed; as on the Sagaie 1, the central pair of wheels can be lifted for road travel

Panhard ERC 90 F4 Sagaie
(France)
Type: wheeled armoured car
Crew: 3
Combat weight: 8100 kg (17,857 lb)
Dimensions: length, gun forward 7.693 m (25.24 ft) and hull 5.098 m (16.73 ft); width 2.495 m (8.19 ft); height 2.254 m (7.40 ft)
Armament system: one 90-mm (3.54-mm) GIAT CS Super (F4) rifled gun with 20 rounds, two 7.62-mm (0.3-in) machine-guns (one co-axial and one AA) with 2,000 rounds, and two smoke-dischargers on each side of the turret; the turret is manually powered, the main gun lacks stabilization in either elevation or azimuth, and an optical fire-control system is fitted
Armour: welded steel
Powerplant: one 116-kW (155-hp) Peugeot petrol engine with 242 litres (53 Imp gal) of internal fuel, and driving a 6 × 6 layout
Performance: speed, road 100 km/h (62 mph) and water 9.5 km/h (5.9 mph) driven by two waterjets or 4.5 km/h (2.8 mph) driven by its wheels; range, road 800 km (497 miles); fording amphibious; gradient 60%; side slope 30%; vertical obstacle 0.8 m (31.5 in); trench 1.1 m (3.6 ft); ground clearance 0.344 m (13.5 in)

Variants
ERC 90 F4 Sagaie: this is the baseline model that began to enter service from 1979; optional extras are an NBC system, waterjet propulsion, air conditioning, ground navigation system, and extra ammunition capacity; like the other ERC models, this variant is now available with an 86-kW (115-hp) Mercedes-Benz OM 617A diesel, which boosts range to 1000 km (621 miles)
ERC 90 F4 Sagaie TTB 90: similar to the standard model, this variant has a TTB 190 rather than TS 90 turret, though the CS Super (F4) gun is retained together with 23 rounds of ammunition; gun stabilization and a better fire-control system are also available as extras
ERC 90 F1 Lynx: similar to the standard model but fitted with a Hispano-Suiza Lynx 90 turret with 90-mm (3.54-in) GIAT F1 gun and 21 rounds; optional extras are the same as those of the Sagaie
ERC 60-20 Serval: fitted with a revised turret, this model has a primary armament of one 60-mm (2.36-in) LR breech-loaded mortar with 50 rounds and one 20-mm cannon with 300 rounds
ERC 60/12 Mangouste: similar to the Serval, this model has a 60-mm (2.36-in) LR mortar with 60

rounds and one 0.5-in (12.7-mm) machine-gun with 1,200 rounds
EMC 81: mortar-carrier variant fitted with an open-topped turret and 81-mm (3.2-in) breech-loaded mortar, for which 78 rounds are provided
ERC Lanza: variant fitted with a SAMM TTB 125 turret armed with a 25-mm Hughes Chain Gun and 7.62-mm (0.3-in) co-axial machine-gun
ERC S 530 A: AA variant with a SAMM S 530 A turret with twin 20-mm cannon
ERC TA-20: AA variant with ESD TA-20 turret with twin 20-mm cannon
ERC 20 Kriss: AA variant fitted with SAMM TAB 220 turret with twin 20-mm cannon

Panhard AML H 90
(France)
Type: light wheeled armoured car
Crew: 3
Combat weight: 5500 kg (12,125 lb)
Dimensions: length, gun forward 5.11 m (16.77 ft) and hull 3.79 m (12.43 ft); width 1.97 m (6.46 m); height overall 2.07 m (6.79 m)
Armament system: one 90-mm (3.54-in) GIAT F1 rifled gun with 21 rounds, two 7.62-mm (0.3-in) machine-guns (one co-axial and the other AA) with 2,000 rounds, and two smoke-dischargers on each side of the turret; the turret is manually powered, the main gun lacks stabilization in elevation and azimuth, and an optical fire-control system is fitted
Armour: welded steel varying in thickness from 8 to 12 mm (0.315 to 0.47 in)
Powerplant: one 67-kW (90-hp) Panhard Model 4HD air-cooled petrol engine, with 156 litres (34 Imp gal) of fuel, and driving a 4 × 4 layout
Performance: speed, road 90 km/h (56 mph); range, road 600 km (373 miles); fording 1.1 m (3.6 ft) and with preparation amphibious; gradient 60%; side slope 30%; vertical obstacle 0.3 m (11.8 in); trench 0.8 m (2.6 ft) with one channel or 3.1 m (10.2 ft) with four channels; ground clearance 0.33 m (13 in)

Variants
AML H 90: the AML series began to enter service in 1961, and has since then been produced in large numbers for the domestic and export markets; the H 90 version is fitted with a powerfully armed Hispano-Suiza turret, which originally had provision for two SS.11 or ENTAC wire-guided anti-tank missiles, one located on each side of the turret
AML Lynx 90: successor to the H 90 version, the Lynx 90 variant is the current production model, and features a more advanced turret with provision for passive night-vision equipment, laser rangefinder and powered traverse; also available is a different powerplant, the 71-kW (95-hp) Peugeot XD 3T diesel, for a range of 1000km (621 miles)

AML HE 60-7: the HE 60 turret has twin 7.62-mm machine-guns on the left and a breech-loaded 60-mm (2.36-in) HB 60 mortar on the right, with 3,800 rounds and 53 rounds respectively, though these numbers are altered to 3,200 and 32 when additional radio equipment is fitted for the command role
AML HE 60-20: variant of the HE 60-7 type with a 20-mm cannon and 500 rounds in place of the twin 7.62-mm machine-gun installation
AML HE 60-20 Serval: improved version of the HE 60-20 type with powerful M693 or KAD-B16 cannon with 300 rounds
AML S 530: the S 530 turret has two M621 20-mm cannon with 600 rounds
AML TG 120: the TG 120 turret has a 20-mm cannon with 240 rounds and a co-axial 7.62-mm machine-gun
EPR: this is the scout car model, basically the AML with its turret replaced by a ring-mounted 12.7-mm (0.5-in) heavy machine-gun; variants currently proposed are the **ERA** raider vehicle (one 20-mm cannon and two 7.62-mm machine-guns, or one 20-mm cannon and two lateral 7.62-mm machine-guns, one Milan anti-tank launcher with six missiles plus two lateral 7.62-mm machine-guns), the **EPF** border vehicle (one 12.7-mm or two 7.62-mm machine-guns plus a searchlight) and the **EPA** airfield protection vehicle (three 7.62-mm machine-guns and a searchlight)
AML 20: model with open-top TL-20 SO turret carrying one 20-mm M693 cannon and 1,000 rounds
AML Eclairage: scout car model with open-top turret fitted with one M693 20-mm cannon and one 7.62-mm machine-gun (1,050 and 2,000 rounds respectively)
Eland Mk 1: South African licence-built version of the AML HE 60-7 with a 60-mm mortar and two 7.62-mm machine-gun
Eland Mk 2: improved version of the Eland Mk 1 with better engine and detail modifications
Eland Mk 5: designation of Eland Mk 1s upgraded to the latest standards
Eland Mk 6: designation of Eland Mk 2s upgraded to the latest standards
Eland Mk 7: current production model, produced with either the HE 60-7 turret or a more powerful turret featuring one 90-mm (3.54-in) GIAT F1 rifled gun, one 7.62-mm co-axial machine-gun and one 7.62-mm AA machine-gun; other improvements are a quick-change engine, better suspension and transmission, and upgraded protection; this version is also known as the **Eland 90**

Panhard VBL
(France)
Type: wheeled reconnaissance/anti-tank vehicle
Crew: 2 (reconnaissance version) or 3 (anti-tank version)
Combat weight: 3590 kg (7,914 lb) with NBC system

Dimensions: length 3.70 m (12.14 ft); width 2.02 m (6.63 ft); height overall 2.14 m (7.02 ft)
Armament system: one pintle-mounted 7.62-mm (0.3-in) machine-gun with 3,000 rounds, or one pintle-mounted 0.5-in (12.7-mm) machine-gun with 1,200 rounds, or one 7.62-mm machine-gun with 3,000 rounds plus one Milan anti-tank launcher with six missiles
Armour: welded steel varying in thickness from 5 to 11 mm (0.2 to 0.43 in)
Powerplant: one 78-kW (105-hp) Peugeot XD 3T turbocharged diesel engine with ? litres (? Imp gal) of fuel, and driving a 4 × 4 layout
Performance: speed, road 95 km/h (59 mph); range, road 750 km (466 miles) with standard fuel or 1000 km (621 miles) with onboard fuel cans; fording 0.9 m (2.95 ft) and with preparation amphibious driven by a single propeller; gradient 50%; side slope 30%; vertical obstacle ? m (? in); ground clearance 0.37 m (14.6 in)

Variants
VBL: the French army is to use the VBL as a two-seat front-line intelligence-gathering/reconnaissance vehicle, and as a three-man anti-tank machine; Panhard has proposed at least 22 derived models for a host of roles.
VBL ULTRAV M11: VBL optimized for jungle operations and carrying a one-man SAMM BTM 208 turret armed with one 0.5-in (12.7-mm) and one 7.62-mm (0.3-in) machine-guns

Thyssen Henschel Spähpanzer Luchs
(West Germany)
Type: wheeled reconnaissance vehicle
Crew: 4
Combat weight: 19500 kg (42,989 lb)
Dimensions: length 7.743 m (25.40 ft); width 2.98 m (9.78 ft); height to machine-gun rail 2.905 m (9.53 ft)
Armament system: one 20-mm Rheinmetall MK 20 Rh 202 cannon with 375 rounds, one 7.62-mm (0.3-in) MG3 machine-gun with 100 ready-use rounds, and four smoke-dischargers on each side of the turret; the turret is electro-hydraulically powered, the main gun lacks stabilization in either elevation or azimuth, and an optical fire-control system is fitted
Armour: welded steel
Powerplant: one 291-kW (390-hp) Daimler-Benz OM 403A turbocharged multi-fuel engine with 500 litres (110 Imp gal) of fuel, and driving an 8 x 8 layout
Performance: speed, road 90 km/h (56 mph) and water 9 km/h (5.6 mph) driven by two propellers; range, road 800 km (497 miles); fording amphibious; gradient 60%; side slope 30%; vertical obstacle 0.6 m (23.6 in); trench 1.9 m (6.2 ft); ground clearance 0.44 m (17.3 in)

Variant
Spähpanzer Luchs: the Luchs is the West German army's standard amphibious reconnaissance vehicle, and started to enter service in 1975; the type has a full NBC system and passive night-vision devices, and can be driven equally rapidly forwards and backwards

Porsche Wiesel
(West Germany)
Type: airportable tracked armoured vehicle
Crew: 2 (cannon-armed model) or 3 (TOW-armed model)
Combat weight: 2600 kg (8,530 lb)
Dimensions: length, gun forward (20-mm cannon version) 3.51 m (11.52 ft) and hull 3.263 m (10.71 ft); width 1.82 m (5.97 ft); height to hull top 1.252 m (4.11 ft)
Armament system: one 20-mm Rheinmetall MK 20 Rh 202 cannon with 400 rounds, or one TOW anti-tank launcher with eight missiles; the turret is manually operated, the main gun lacks stabilization in either elevation or azimuth, and an optical fire-control system is fitted
Armour: welded steel
Powerplant: one 73-kW (98-hp) VW-Audi petrol engine with 80 litres (17.6 Imp gal) of fuel
Performance: speed, road 85 km/h (53 mph); range, road 300 km (186 miles); fording not revealed; gradient 60%; side slope 30%; vertical obstacle 0.4 m (15.75 in); trench 1.5 m (4.9 ft); ground clearance 0.3 m (11.9 in)

Variants
Wiesel/E6-II: designed for use by West Germany airborne formations, and thus portable in transport aircraft and medium-lift helicopters, the Wiesel was developed by Porsche in the late 1970s but then cancelled, only to be reinstated for the West German army in the mid-1980s; this cannon-armed model has a one-man Keller und Knappich E6-II turret with a dual-feed cannon
Wiesel/TOW: derivative of the basic model with a TOW overhead launcher and eight rounds; a number of other variants have been proposed, including a command vehicle, a radar-equipped battlefield surveillance vehicle, an ammunition resupply vehicle, a recovery vehicle, a SAM vehicle, a reconnaissance vehicle, an ambulance and an armoured personnel carrier

FUG
(Hungary)
Type: wheeled scout car
Crew: 2+4
Combat weight: 7000 kg (15,432 lb)
Dimensions: length 5.79 m (12.76 ft); width 2.50 m

(8.20 m); height with turret (OT-65A) 2.25 m (7.38 ft)
Armament system: one pintle-mounted 7.62-mm (0.3-in) SGMB machine-gun with 1,250 rounds
Armour: steel up to a maximum of 13 mm (0.51 in)
Powerplant: one 75-kW (100-hp) Csepel D.414.44 diesel engine with 200 litres (44 Imp gal) of fuel, and driving a 4 x 4 layout
Performance: speed, road 87 km/h (54 mph) and water 9 km/h (5.6 mph); range, road 600 km (373 miles); fording amphibious; gradient 60%; side slope 30%; vertical obstacle 0.4 m (15.75 in); trench 1.2 m (3.9 ft); ground clearance 0.34 m (13.4 in)

Variants
FUG: this light scout car entered service with the Hungarian army in 1964, and features a two-man crew with accommodation in the rear for four infantrymen; the type lacks NBC protection, and variants include an ambulance and an NBC reconnaissance vehicle; in Czech service the type is known as the **OT-65**
OT-65A: this Czech-produced variant has the turret of the OT-62B APC installed over the hull; the turret is armed with a 7.62-mm vz 59T machine-gun and (on the outside of the right-hand side of the turret) an 82-mm (3.23-in) recoilless rifle

RAMTA RAM V-2L
(Israel)
Type: wheeled multi-role AFV
Crew: 2+8
Combat weight: 6000 kg (13,228 lb)
Dimensions: length 5.42 m (17.78 ft); width 2.03 m (6.66 ft); height overall 2.20 m (7.22 ft)
Armament system: one pintle-mounted 0.5-in (12.7-mm) Browning M2HB heavy machine-gun with 1,500 rounds, or one pintle-mounted 40-mm grenade-launcher with six ammunition boxes, or three pintle-mounted 7.62-mm (0.3-in) Browning M1919A4 machine-guns with 5,000 rounds, or one rocket-launcher with six rounds
Armour: steel to a maximum thickness of 8 mm (0.315 in)
Powerplant: one 98-kW (132-bhp) Deutz diesel engine with 160 litres (35 Imp gal) of fuel, and driving a 4 x 4 layout
Performance: speed, road 96 km/h (60 mph); range, road 950 km (590 miles); fording 1.0 m (3.3 ft); gradient 60%; side slope 35%; vertical obstacle 0.8 m (31.5 in); ground clearance 0.31 m (12.2 in)

Variants
RAM V-1: this first model began to enter service in 1980, and is the short-wheelbase model of the open-top V-1 series with a crew of two plus six infantrymen, a combat weight of 5400 kg (11,905 lb) and a length of 5.02 m (16.47 ft); the type can be used as an IFV (three 7.62-mm machine-guns with 5,000 rounds and one anti-tank launcher with six rockets), or as an ICV (three 7.62-mm machine-guns with 5,000 rounds, one 52-mm/2.05-in mortar with 36 rounds, and two anti-tank rockets), or as a close-range anti-tank vehicle (one M40 106-mm/4.17-in recoilless rifle with 18 rounds) or as a long-range anti-tank vehicle (one TOW launcher with 16 missiles and two 7.62-mm machine-guns with 2,500 rounds)
RAM V-1L: long-wheelbase model of the V-1 with crew of two plus eight infantrymen, combat weight of 5600 kg (12,346 lb) and length of 5.42 m (17.78 ft); the same armament options as the V-1 are available, together with an AA option with two 20-mm cannon and 720 rounds in a TCM-20 turret
RAM V-2: short-wheelbase model of the enclosed variant, with a crew of two plus six infantrymen, a combat weight of 5700 kg (12,566 lb) and a length of 5.02 m (16.47 ft); the same type of armament options as the V-1 are available, together with one 0.5-in (12.7-mm) Browning machine-gun with 1,500 rounds
RAM V-2L: long-wheelbase model of the V-2, with the whole range of armament options available for the V-2

Fiat/OTO Melara Tipo 6616
(Italy)
Type: wheeled armoured car
Crew: 3
Combat weight: 8000 kg (17,637 lb)
Dimensions: length 5.37 m (17.62 ft); width 2.50 m (8.20 ft); height to turret roof 2.035 m (6.68 ft)
Armament system: one 20-mm Rheinmetall MK 20 Rh 202 cannon with 400 rounds, one co-axial 7.62-mm (0.3-in) machine-gun with 1,000 rounds, and three smoke-dischargers on each side of the turret; the turret is electrically powered, the main gun lacks stabilization in either elevation or azimuth, and an optical fire-control system is fitted
Armour: steel varying in thickness from 6 to 8 mm (0.24 to 0.315 in)
Powerplant: one 119-kW (160-hp) Fiat modello 8062.24 diesel engine with 150 litres (33 Imp gal) of fuel, and driving a 4 x 4 layout
Performance: speed, road 100 km/h (62 mph) and water 5 km/h (3.1 mph) driven by its wheels; range, road 700 km (435 miles) and water 20 km (12.5 miles); fording amphibious; gradient 60%; side slope 30%; vertical obstacle 0.45 m (17.7 in); ground clearance 0.37 m (14.6 in)

Variant
Tipo 6616: derived from the Tipo 6614 APC, the Tipo 6616 began to enter service in the mid-1970s, and has secured small export sales in addition to limited domestic use for paramilitary roles; the type has also

been offered with a two-man turret fitted with a 90-mm (3.54-in) Cockerill Mk III rifled gun plus 36 rounds

Komatsu Type 87
(Japan)
Type: wheeled reconnaissance and patrol vehicle
Crew: 5
Combat weight: 13500 kg (29,762 lb)
Dimensions: length 5.37 m (17.62 ft); width 2.48 m (8.14 ft); height 2.38 m (7.80 ft)
Armament system: one 25-mm Oerlikon KBA cannon, one co-axial 7.62-mm (0.3-in) Type 74 machine-gun, and three smoke-dischargers on each side of the turret; the turret is electrically powered, the main gun lacks stabilization in elevation and azimuth, and an optical fire-control system is fitted
Armour: steel
Powerplant: one 227-kW (305-hp) Isuzu 10PBI diesel engine with ? litres (? Imp gal) of fuel, and driving a 6 x 6 layout
Performance: speed, road 100 km/h (62 mph); range, road not known; fording 1.0 m (3.3 ft); gradient 60%; vertical obstacle 0.6 m (23.6 in); trench 1.5 m (4.9 ft); ground clearance 0.45 m (17.7 in)

Variants
Type 82: this is a specialist eight-crew command and communications version of the Type 87, weighing 23600 kg (29,982 lb) and having an overall length of 5.72 m (18.77 ft); the armament comprises one 7.62-mm and one 0.5-in (12.7-mm) machine-guns, each ring-mounted; the type is entering service in the mid-1980s
Type 87: also entering service in the mid-1980s, this is the reconnaissance and patrol version of the Type 82 with a revised superstructure to accommodate the cannon-armed turret

Eland wheeled armoured car (South Africa): see Panhard AML H 90 wheeled armoured car

ENASA VEC
(Spain)
Type: wheeled reconnaissance vehicle
Crew: 5
Combat weight: 13750 kg (30,313 lb)
Dimensions: length 6.25 m (20.50 ft); width 2.50 m (8.20 ft); height to hull top 2.00 m (6.56 ft)
Armament system: one 20-mm Rheinmetall MK 20 Rh 202 cannon (in a Fiat/OTO Melara Tipo 6616 turret), or one 20-mm Oerlikon KAA cannon (in a GAD-BOA turret), or one 25-mm Oerlikon KBA-B cannon (in a GBD-COA turret), plus 7.62-mm (0.3-in) and/or 0.5-in (12.7-mm) Browning machine-guns and three smoke-dischargers on each side of the turret; the turret is electrically powered, the main gun lacks stabilization in either elevation or azimuth, and an optical fire-control system is fitted
Armour: welded steel
Performance: one 228-kW (306-hp) Pegaso modelo 9157/8 turbocharged diesel engine with 300 litres (66 Imp gal) of fuel, and driving a 6 x 6 layout
Performance: speed, road 100 km/h (62 mph); range, road 800 km (497 miles); fording amphibious; gradient 60%; side slope 30%; vertical obstacle not revealed; trench not revealed; ground clearance 0.4 m (15.75 in)

Variant
VEC: this derivative of the BMR-600 IFV began to enter Spanish service in the early 1980s, and is designed as a cavalry scout vehicle with good protection and a wide diversity of armament fits; options include an AA version with turreted cannon, and an anti-tank version with wire-guided missiles

MOWAG SPY
(Switzerland)
Type: wheeled reconnaissance vehicle
Crew: 3
Combat weight: 7500 kg (16,534 lb)
Dimensions: length 4.52 m (14.83 ft); width 2.50 m (8.20 ft); height to top of hull 1.66 m (5.45 ft)
Armament system: one 0.5-in (12.7-mm) Browning M2HB machine-gun with 3,000 rounds, one co-axial 0.3-in (7.62-mm) Browning M1919 machine-gun with 2,000 rounds, and three smoke-dischargers on each side of the turret; the turret is manually operated, the main gun lacks stabilization in either elevation or azimuth, and an optical fire-control system is fitted
Armour: welded steel
Powerplant: one 205-hp (153-kW) Detroit Diesel V-8 diesel engine with 200 litres (44 Imp gal) of fuel, and driving a 4 x 4 layout
Performance: speed, road 110 km/h (68 mph); range, road 700 km (435 miles); fording 1.3 m (4.3 ft); gradient 60%; side slope 35%; vertical obstacle 0.5 m (19.7 in); ground clearance 0.39 m (15.4 in)

Variant
SPY: using many components of the Piranha AFV series, the SPY is a light reconnaissance vehicle that began to enter service with an undisclosed Far Eastern country in the mid-1980s

BRDM-2
(USSR)
Type: wheeled scout car
Crew: 4
Combat weight: 7000 kg (15,432 lb)
Dimensions: length 5.75 m (18.86 ft); width 2.35 m (7.71 ft); height 2.31 m (7.58 ft)
Armament system: one 14.5-mm (0.57-in) KPVT

heavy machine-gun with 500 rounds, and one co-axial 7.62-mm (0.3-in) PKT machine-gun with 2,000 rounds; the turret is manually powered, the main gun lacks stabilization in either elevation or azimuth, and an optical fire-control system is fitted
Armour: welded steel varying in thickness from 2 to 14 mm (0.08 to 0.55 in)
Powerplant: one 104-kW (140-hp) GAZ-41 petrol engine with 290 litres (64 Imp gal) of fuel, and driving a 4 × 4 layout
Performance: speed, road 100 km/h (62 mph) and water 10 km/h (6.2 mph) driven by one waterjet; range, road 750 km (466 miles); fording amphibious; gradient 60%; vertical obstacle 0.4 m (15.75 in); trench 1.25 m (4.1 ft); ground clearance 0.43 m (16.9 in)

Variants
BRDM-2: developed in the late 1950s as successor to the BRDM-1, the BRDM-2 differs from its predecessor in having enclosed armament and a more powerful engine located at the rear for better road, cross-country and water performance; there are four semi-retractable chain-driven belly wheels (two on each side) for additional traction in difficult going; standard equipment includes an NBC system and waterjet propulsion
BRDM-2-rkh: this is the nuclear and chemical reconnaissance version of the baseline model, and is distinguishable by the racks for automatically positioned marker poles
BRDM-2U: command version with a hatch in place of the turret, and additional generating capacity for the extra radio equipment carried
BRDM-2/'Swatter': anti-tank version with the turret replaced by an overhead-covered retractable arm launcher for four AT-2 'Swatter-B' IR-homing missiles; four reloads are carried inside the hull; the variant appeared in the early 1970s and, like other missile-equipped variants, has a crew reduced to two or three so that reload missiles can be accommodated inside the hull
BRDM-2/'Sagger': anti-tank version with the turret replaced by an overhead-covered retractable arm launcher for six AT-3 'Sagger' wire-guided missiles; eight reloads are carried inside the hull; the variant appeared in the early 1970s
BRDM-2/'Spandrel': anti-tank version with its turret surmounted by five launcher tubes for AT-5 'Spandrel' semi-automatic command to line of sight missiles; another 10 missiles are carried inside the hull; the variant appeared in the mid-1970s
BRDM-2/SA-9 'Gaskin': introduced in 1968, this is a self-propelled SAM system mounted on the chassis of the BRDM-2 reconnaissance vehicle and designed to provide regimental-level short-range air defence against low-level air threats; threat detection is entrusted to a battery-level truck-mounted 'Gun Dish' radar, which passes target information to the 'Gaskin' launch vehicles, which each have an erector/launcher in place of the normal turret; this erector/launcher carries two missile container/launchers on each side of the central pedestal, and four reload rounds are carried on the outside of the launch vehicle's hull; the engagement envelope for the IR-homing missiles are minimum/maximum ranges of 500/8000 m (550/8,750 yards) and minimum/maximum altitudes of 20/5000 m (65/16,405 ft)

BRDM-1
(USSR)
Type: wheeled scout car
Crew: 5
Combat weight: 5600 kg (12,346 lb)
Dimensions: length 5.70 m (18.70 ft); width 2.25 m (7.38 ft); height without armament 1.90 m (6.23 ft)
Armament system: one pintle-mounted 7.62-mm (0.3-in) SGMB machine-gun with 1,250 rounds
Armour: welded steel to a maximum thickness of 10 mm (0.39 in)
Powerplant: one 67-kW (90-hp) GAZ-40P petrol engine with 150 litres (33 Imp gal) of fuel, and driving a 4 x 4 layout
Performance: speed, road 80 km/h (50 mph) and water 9 km/h (4.5 mph) driven by one waterjet; range, road 500 km (311 miles); fording amphibious; gradient 60%; vertical obstacle 0.4 m (15.75 in); trench 1.22 m (4.0 ft); ground clearance 0.315 m (12.4 in)

Variants
BRDM-1: this amphibious scout car became standard in Warsaw Pact armies from 1959, and was notable for its four chain-driven semi-retractable belly wheels for improved performance in adverse conditions, its exposed main armament and its lack of NBC protection
BRDM-U: command version with extra antennae
BRDM-rkh: nuclear and chemical reconnaissance version with racks for the automatic marking of safe lanes with special poles
BRDM-1/'Snapper': anti-tank version with the superstructure extended aft for the location of a fixed triple launcher for AT-1 'Snapper' missiles; reserve missiles are stored in the hull; crew is reduced to two or three
BRDM-1/'Swatter': anti-tank version with a fixed quadruple launcher for AT-2 'Swatter-A' missiles; crew is reduced to two or three
BRDM-1/'Sagger': anti-tank version with elevating and overhead-covered six-rail launcher for AT-3 'Sagger' missiles; this variant has a crew of two or three, and first appeared in 1965

Alvis FV601(C) Saladin Mk 2
(UK)
Type: wheeled armoured car
Crew: 3
Combat weight: 25,550 lb (11589 kg)
Dimensions: length, gun forward 17.33 ft (5.284 m) and hull 16.16 ft (4.93 m); width 8.33 ft (2.54 m); height to turret roof 7.19 ft (2.19 m)
Armament system: one 76-mm (3-in) ROF L5A1 rifled gun with 42 rounds, two 0.3-in (7.62-mm) Browning M1919 machine-guns (one co-axial L3A3 and one AA L3A4) with 2,750 rounds, and six smoke-dischargers on each side of the turret; the turret is electrically powered, the main armament lacks stabilization in either elevation or azimuth, and an optical fire-control system is fitted
Armour: welded steel varying in thickness from 8 to 32 mm (0.315 to 1.25 in)
Powerplant: one 170-bhp (127-kW) Rolls-Royce B80 Mk 6A petrol engine with 53 Imp gal (241 litres) of fuel, and driving a 6 x 6 layout
Performance: speed, road 45 mph (72 km/h); range, road 250 miles (400 km); fording 3.5 ft (1.07 m) without preparation and 7.0 ft (2.13 m) with preparation; gradient 46%; vertical obstacle 18 in (0.46 m); trench 5.0 ft (1.52 m); ground clearance 16.75 in (0.43 m)

Variants
FV601C Saladin Mk 2: this British armoured car entered service in 1959, and is still a useful vehicle despite its lack of NBC protection or night-vision equipment
FV601(D): used by the West German border police as the **Geschützer Sonderwagen III**, this version of the Saladin lacks the co-axial machine-gun

Alvis FV101 Scorpion
(UK)
Type: tracked combat reconnaissance vehicle
Crew: 3
Combat weight: 17,800 lb (8074 kg)
Dimensions: length 15.73 ft (4.79 m); width 7.33 ft (2.235 m); height overall 6.90 ft (2.102 m)
Armament system: one 76-mm (3-in) ROF L23A1 rifled gun with 40 rounds, one co-axial 7.62-mm (0.3-in) L43A1 machine-gun with 3,000 rounds, and four smoke-dischargers on each side of the turret; the turret is manually powered, the main gun lacks stabilization in either elevation or azimuth, and a day/night optical fire-control system is fitted
Armour: welded aluminium
Powerplant: one 190-hp (142-kW) Jaguar J60 No.1 Mk 100B petrol engine with 93 Imp gal (423 litres) of fuel
Performance: speed, road 50 mph (80 km/h) and water 4 mph (6.4 km/h) driven by its tracks; range, road 400 miles (644 km); fording 3.5 ft (1.07 m) without preparation and amphibious with preparation; gradient 60%; vertical obstacle 19.75 in (0.5 m); trench 6.75 ft (2.06 m); ground clearance 14 in (0.356 m)

Variants
FV101 Scorpion: baseline model of an extensive family whose models are derived generally from the Spartan APC version of the CVR(T), the Scorpion was designed by Alvis Ltd in the 1950s and 1960s, and began to enter British service in 1972 as the air-transportable **Combat Vehicle Reconnaissance (Tracked)**, a reconnaissance and anti-tank machine to complement the Fox wheeled reconnaissance vehicle; the type is fitted as standard with NBC protection
FV107 Scimitar: this is the version of the Scorpion with a 30-mm Raiden cannon instead of the low-velocity 76-mm gun
Scorpion 90: private-venture development of the basic Scorpion with a 90-mm (3.54-in) Cockerill Mk III rifled gun and 35 rounds, used in conjunction with a more capable fire-control system incorporating a laser rangefinder; this vehicle has a combat weight of 18,900 lb (8573 kg) and a length, gun forwards, of 17.33 ft (5.283 m)
Improved Scorpion: private-venture development with a 200-bhp (149-kW) Perkins T6.3544 diesel; this increases combat weight to 18,210 lb (8260 kg), with consequent degradation of performance with the notable exception of range, which increases to 520+ miles (837+ km); the improvement package has been incorporated in the Scorpion 90s built for Malaysia

ROF Leeds FV721 Fox
(UK)
Type: wheeled combat reconnaissance vehicle
Crew: 3
Combat weight: 13,495 lb (6121 kg)
Dimensions: length, gun forwards 16.67 ft (5.08 m); width 7.00 ft (2.134 m); height to turret top 6.5 ft (1.98 m)
Armament system: one 30-mm ROF L21 Raiden cannon with 99 rounds, one 7.62-mm (0.3-in) L8A1 co-axial machine-gun with 2,600 rounds, and four smoke-dischargers on each side of the turret; the turret is manually powered, the main gun lacks stabilization in either elevation or azimuth, and a day/night optical fire-control system is fitted
Armour: welded aluminium
Powerplant: one 190-bhp (142-kW) Jaguar J60 No.1 Mk 100B petrol engine with 32 Imp gal (145.5 litres) of fuel, and driving a 4 x 4 layout
Performance: speed, road 65 mph (104 km/h) and water 3.25 mph (5.2 km/h) driven by its wheels; range, road 270 miles (434 km); fording 3.3 ft (1.0 m) without preparation and amphibious with prep-

aration; gradient 58%; vertical obstacle 19.75 in (0.5 m); trench 4.0 ft (1.22 m) with channels; ground clearance 12 in (0.3 m)

Variants
FV107 Fox: a development of the Ferret scout car into an all-aluminium light armoured car, the Fox was designed by Daimler in the late 1960s and entered British service in 1973 as the **Combat Vehicle Reconnaissance (Wheeled)**
Panga: developed by Royal Ordnance, Leeds, the Panga is an export derivative of the Fox with a one-man turret armed with a 0.5-in (12.7-mm) Browning M2HB heavy machine-gun with or without a co-axial 7.62-mm L37 machine-gun; other armament options are a Hughes 25-mm Chain Gun with 250 rounds and a 7.62-mm machine-gun with 1,500 rounds, or a twin launcher for Milan anti-tank missiles coupled with a 7.62-mm machine-gun

Daimler FV711 Ferret Mk 4
(UK)
Type: wheeled scout car
Crew: 2
Combat weight: 11,900 lb (3598 kg)
Dimensions: length 13.00 ft (3.96 m); width 7.00 ft (2.134 m); height 6.67 ft (2.03 m)
Armament system: one 0.3-in (7.62-mm) Browning M1919 (L3A4) machine-gun with 2,500 rounds, and three smoke-dischargers on each side of the turret; the turret is manually powered, the gun lacks stabilization in either elevation or azimuth, and an optical fire-control system is fitted
Armour: welded steel varying in thickness from 8 to 16 mm (0.315 to 0.63 in)
Powerplant: one 129-bhp (96-kW) Rolls-Royce B60 Mk 6A petrol engine with 21 Imp gal (96 litres) of fuel, and driving a 4 x 4 layout
Performance: speed, road 50 mph (80 km/h); range, road 190 miles (306 km); fording 3.0 ft (0.91 m) without preparation and amphibious with preparation; gradient 46%; vertical obstacle 16 in (0.41 m); trench 4.0 ft (1.22 m) with channels; ground clearance 17 in (0.43 m)

Variants
FV701(C) Ferret Mk 1: this was the initial two-seat liaison scout car, and entered service in 1953 as the second production model; it lacked the turret and was armed with a 0.303-in Bren Gun and 450 rounds; the **FV704 Ferret Mk 1/2** was a three-seat version with an armoured roof
FV702(E) Ferret Mk 2: turret-armed reconnaissance version of the Mk 1, developed into the **Ferret Mk 2/2** with a collar to raise the turret slightly, the **FV701(H) Ferret Mk 2/3** with detail improvements, the Ferret Mk 2/3 with extra armour, the **Ferret Mk 2/5** Mk 2 brought up to Mk 2/4 standard, the **FV703 Ferret Mk 2/6** with a single launcher for three Swingfire anti-tank missiles, and the **Ferret Mk 2/7** version of the Mk 2/6 without the Swingfire system
Ferret Mk 3: designation of Ferret Mk 1/1s brought up to Mk 4 standard but fitted with a machine-gun turret
FV711 Ferret Mk 4: designation of earlier Ferrets upgraded with stronger suspension and a folded flotation screen
FV712 Ferret Mk 5: designation of Mk 4s with Swingfire system

Short Brothers Shorland Mk 4
(UK)
Type: wheeled armoured patrol car
Crew: 3
Combat weight: 7,400 lb (3357 kg)
Dimensions: length 15.08 ft (4.597 m); width 5.83 ft (1.778 m); height 7.50 ft (2.286 m)
Armament system: one 7.62-mm (0.3-in) L7A3 machine-gun with 1,500 rounds, and four smoke-dischargers on each side of the turret; the turret is manually powered, the gun lacks stabilization in either elevation or azimuth, and an optical fire-control system is fitted
Armour: welded steel
Powerplant: one 91-bhp (68-kW) Rover petrol engine with 28 Imp gal (128 litres) of internal fuel, and driving a 4 x 4 layout
Performance: speed, road 65 mph (104.6 km/h); range, road 320 miles (515 km); vertical obstacle 9 in (0.23 m); ground clearance 12.75 in (0.32 m)

Variants
Shorland Mk 1: derived from the long-wheelbase Land Rover and delivered from 1965 mainly for internal security duties, the Shorland Mk 1 has a 67-bhp (50-kW) four-cylinder engine and 7.25-mm (0.29-in) armour
Shorland Mk 2: improved model with four-cylinder 77-bhp (57-kW) engine and 8.25-mm (0.32-in) armour
Shorland Mk 3: improved version with V-6 91-bhp (68-kW) engine
Shorland Mk 4: improved version with V-8 engine and upgraded armour protection
Shorland SB 403: anti-hijack derivative for use at airports and fitted with armoured glass rather than the standard glass windows with armoured shutters

Cadillac Gage Commando Scout
(USA)
Type: wheeled scout car
Crew: 1+1 or 1+2
Combat weight: 14,500 lb (6577 kg)
Dimensions: length 15.42 ft (4.67 m); width 6.75 ft (2.06 m); height 7.33 ft (2.235 m)
Armament system: two 7.62-mm (0.3-in) machine-

guns with 4,800 rounds, or two 0.5-in (12.7-mm) Browning M2HB heavy machine-guns with 2,200 rounds, or one TOW launcher with six missiles, or one 106-mm (4.17-in) M40 recoilless rifle with 15 rounds; the turret is manually powered, the gun lacks stabilization in either elevation or azimuth, and an optical fire-control system is fitted
Armour: welded steel
Powerplant: one 149-hp (111-kW) Cummins V-6 diesel engine with 55 US gal (208 litres) of fuel, and driving a 4 x 4 layout
Performance: speed, road 55 mph (88 km/h); range, road 500 miles (805 km); fording 3.8 ft (1.17 m); gradient 60%; side slope 30%; vertical obstacle 24 in (0.61 m)

Variant
Commando Scout: introduced to service in 1983 by Indonesia, the Commando Scout is a very useful light AFV with extensive armament options suiting it for reconnaissance and light anti-tank operations; the type can also be used in the command role with an extended roof an armament comprising one 7.62-mm machine-gun

ARMOURED PERSONNEL CARRIERS AND INFANTRY FIGHTING VEHICLES

TAMSE/Thyssen Henschel VCTP
(West Germany/Argentina)
Type: tracked armoured personnel carrier
Crew: 2+10
Combat weight: 27500 kg (60,626 lb)
Dimensions: length 6.79 m (22.28 ft); width 3.28 m (10.76 m); height 2.45 m (8.04 ft)
Armament system: one 20-mm Rheinmetall MK 20 Rh 202 cannon and one 7.62-mm (0.3-in) AA machine-gun in the electro-hydraulically powered turret, one remotely-controlled 7.62-mm (0.3-in) machine-gun, and four smoke-dischargers on each side of the hull
Armour: welded steel
Powerplant: one 535-kW (717-hp) MTU diesel engine with 652 litres (143 Imp gal) of internal fuel plus provision for 400 litres (88 Imp gal) of auxiliary fuel
Performance: speed, road 75 km/h (46.6 mph); range, road 570 km (354 miles) on internal fuel or 915 km (569 miles) with auxiliary fuel; fording 1.0 m (3.3 ft); gradient 60%; vertical obstacle 1.0 m (39.4 in); trench 2.5 m (8.2 ft); ground clearance 0.44 m (17.3 in)

Variants
VCTP: this is in essence a less refined version of the West German Marder MICV with a more powerful engine; there is provision for the embarked troops to use their weapons from inside the vehicle
VCTM: mortar-carrier version of the VCTP with a roof hatch to allow the firing of the 120-mm (4.72-in) 120LR mortar located in the rear compartment; 49 mortar bombs are carried, and local defence is provided by a single 7.62-mm (0.3-in) pintle-mounted machine-gun

Steyr 4K 7FA G 127
(Austria)
Type: tracked armoured personnel carrier
Crew: 2+8
Combat weight: 14800 kg (32,628 lb)
Dimensions: length, hull 5.87 m (19.26 ft); width 2.50 m (8.20 ft); height without armament 1.61 m (5.29 ft)
Armament system: one pintle-mounted 0.5-in (12.7-mm) Browning M2HB heavy machine-gun with ? rounds, and four smoke-dischargers
Armour: welded steel
Powerplant: one 239-kW (320-hp) Steyr 7FA diesel engine with 360 litres (79 Imp gal) of internal fuel
Performance: speed, road 65 km/h (40 mph); range, road 520 km (323 miles); fording 1.0 m (3.3 ft); gradient 75% side slope 40%; vertical obstacle 0.8 m (31.5 in); trench 2.1 m (6.9 ft); ground clearance 0.42 m (16.5 in)

Variants
4K 7FA G 127: this is essentially an uparmoured development of the 4K 4FA with automative components of the SK 105 series; the first production models were completed in 1977, and options include night-vision equipment, individual NBC protection and provision for the embarked troops to use their personal weapons from inside the vehicle
4K 7FA G 20: updated version of the 4K 7FA G 127 with a 20-mm cannon in a one-man Oerlikon turret
4K 7FA-KUPz 1/90: this is the fire-support variant of the series, a five-man vehicle equipped with a TS 90 turret carrying the 90-mm (3.54-in) GIAT CS Super (F4) rifled gun and 64 rounds of ammunition, plus

2,000 rounds of ammunition for the co-axial and AA 7.62-mm machine-guns
4K 7FA-FU: this is the command model with a crew of seven and additional radio equipment
4K 7FA-San: ambulance model with a crew of two (driver and medical attendant) plus provision for two litters and four sitting wounded
4K 7FA GrW 81: mortar-carrier version with an 81-mm (3.2-in) mortar and 78 bombs, the mortar firing through a hatch in the roof; the crew is five
4K 7FA FLA 1/2.20: AA version fitted with the French TA 20/RA 20 turret carrying two 20-mm cannon
4K 7FA FLA 3/2.30: AA version fitted with the French SABRE turret carrying two 30-mm cannon
Leonidas: designation of the 4K 7FA built under licence in Greece, and identical in all respects to the Austrian-built model, though the type may be retrofitted with a Greek-designed turret accommodating a 20-mm cannon or 0.5-in (12.7-mm) Browning M2HB heavy machine-gun

Steyr 4K 4FA G 127
(Austria)
Type: tracked armoured personnel carrier
Crew: 2+8
Combat weight: 12500 kg (27l557 lb)
Dimensions: length, hull 5.40 m (17.72 ft); width 2.50 m (8.20 ft); height to top of machine-gun 2.10 m (6.89 ft)
Armament system: one pintle-mounted 0.5-in (12.7-mm) Browning M2HB heavy machine-gun, and four smoke-dischargers
Armour: welded steel varying in thickness from 8 to 35 mm (0.315 to 1.38 in)
Powerplant: one 186-kW (250-hp) Steyr 4FA diesel engine with 184 litres (40.5 Imp gal) of internal fuel
Performance: speed, road 65 km/h (40.4 mph); range, road 370 km (230 miles); fording 1.0 m (3.3 ft); gradient 75%; side slope 50%; vertical obstacle 0.8 m (31.5 in); trench 2.2 m (7.2 ft); ground clearance 0.42 m (16.5 in)

Variants
4K 4F: original production model with deliveries beginning in 1961; this variant is powered by the 186-kW (250-hp) Saurer 2P diesel; as in the standard version of the later 4K 7FA series, the embarked troops cannot fire their personal weapons from inside the vehicle, but must emerge from roof hatches to fire on the move
4K 3FA: modified version of the 4K 4F with a 171-kW (230-hp) diesel engine
4K 3FA-Ful: brigade commander's command and control vehicle
4K 3FA-FuA: artillery command vehicle
4K 3FA-Fu/FLA: AA command vehicle
4K 3FA-FS: wireless and teleprinter version
4K 4FA-G 127: main production variant
4K 4FA-G 20: upgunned version with a 20-mm Oerlikon 204 GK cannon and 100 rounds of ready-use ammunition in a GAD-AOA turret; this version has a combat weight of 15000 kg (33,069 lb)
4K 4FA-San: ambulance model with the same crew and casualty parameters as the 4K 7FA San
4K 4FA GrW 81: mortar-carrier version with an 81-mm (3.2-in) mortar

BN Constructions SIBMAS
(Belgium)
Type: wheeled armoured personnel carrier
Crew: 3+11
Combat weight: 14500 to 16500 kg (31,966 to 36,376 lb) depending on specific role
Dimensions: length 7.32 m (24.02 ft); width 2.50 m (8.20 ft); height to turret top 2.77 m (9.09 ft)
Armament system: see Variants (below)
Armour: welded steel
Powerplant: one 239-kW (320-hp) MAN D2566 MK diesel engine with 400 litres (88 Imp gal) of internal fuel, and driving a 6 x 6 layout
Performance: speed, road 100 km/h (62 mph) and water 11 km/h (6.8 mph) driven by two propellers or 4 km/h (2.5 mph) driven by its wheels; range, road 1000 km (621 miles); fording amphibious; gradient 70%; side slope 40%; vertical obstacle 0.6 m (23.6 in); trench 1.5 m (4.9 ft); ground clearance 0.4 m (15.75 in)

Variants
SIBMAS: available from the early 1980s, this useful wheeled APC has good range and payload, and has provision for the embarked troops to use their personal weapons from inside the hull; the type is offered with a wide variety of turreted armament options including the **AFSV-90** armoured fire-support vehicle with a Cockerill CM-90 turret with 90-mm (3.54-mm) Cockerill Mk III rifled gun and two 7.62-mm (0.3-in) machine-guns (one co-axial and the other AA), an ESD turret with two 20-mm AA cannon, a Lynx 90 turret with 90-mm GIAT F1 rifled gun and two 7.62-mm machine-guns (one co-axial and the other AA), a Serval 60/20 turret with a 60-mm (2.36-in) breech-loaded mortar plus a 20-mm cannon and a 7.62-mm machine-gun, a Vickers turret with two 7.62-mm machine-guns, a TG 120 (TTB 120) turret with a 20-mm cannon and a co-axial 7.62-mm machine-gun, and an S 365 (BTM 208) turret with 0.5-in (12.7-mm) and 7.62-mm machine-guns; other variants on offer are a command post vehicle, an ambulance vehicle for four litters and three sitting wounded, a cargo vehicle with a payload of 4000+ kg (8,818+ lb), and a recovery vehicle with front and rear spades plus a winch and a crane

Beherman Demoen BDX
(Belgium)
Type: wheeled armoured personnel carrier
Crew: 2+10
Combat weight: 10700 kg (23,589 lb)
Dimensions: length 5.05 m (16.57 ft); width 2.50 m (8.20 ft); height to top of machine-gun turret 2.84 m (9.32 ft)
Armament system: two 7.62-mm (0.3-in) L7 or MAG machine-guns in a manually-powered turret
Armour: welded steel varying in thickness from 9.5 to 12.7 mm (0.37 to 0.5 in)
Powerplant: one 180-hp (134-kW) Chrysler petrol engine with 248 litres (55 Imp gal) of internal fuel, and driving a 4 x 4 layout
Performance: speed, road 100 km/h (62 mph); range, road 900 km (559 miles); fording amphibious; gradient 60%; side slope 40%; vertical obstacle 0.4 m (15.75 in); ground clearance 0.4 m (15.75 in)

Variants
BDX: this is the improved and considerably developed Belgian production version of the Irish **Timoney Mk 5** APC, and deliveries began in 1978 mainly for paramilitary duties; there is provision for the embarked troops to use their personal weapons from inside the vehicle, and the type can be fitted with a turret for two Milan anti-tank missiles, or a Vickers turret for two 7.62-mm machine-guns, or a Vickers turret for one 7.62-mm machine-gun
Timoney Mk 6: this is a developed version of the Timoney Mk 5 (produced in Belgium as the BDX), with improved armour, lengthened wheelbase and a Detroit Diesel 4-53T engine boosting range to 1000 km (621 miles) as well as reducing the fire risk; there are many armament options, the most common being the Creusot-Loire TLi turret with one 0.5-in (12.7-mm) and one 7.62-mm machine-guns
Vickers Valkyr: joint prototype by Vickers of the UK and Beherman-Demoen of Belgium to enhance the BDX by improving the driver's position, bettering the ballistic protection, strengthening the suspension so that a greater range of armament installations can be offered, and incorporating a number of detail modifications

ENGESA EE-11 Urutu Model III
(Brazil)
Type: wheeled armoured personnel carrier
Crew: 1+13
Combat weight: 13000 kg (28,660 lb)
Dimensions: length 6.15 m (20.18 ft); width 2.59 m (8.50 ft); height to top of machine-gun mount 2.72 m (8.92 ft)
Armament system: one pintle-mounted 0.5-in (12.7-mm) Browning M2HB heavy machine-gun (but see Variants), and two or three smoke-dischargers
Armour: welded layered steel
Powerplant: one 212-hp (158-kW) Detroit Diesel 6V-53N diesel engine with 380 litres (83.6 Imp gal) of internal fuel, and driving a 6 x 6 layout
Performance: speed, road 90 km/h (56 mph) and water 8 km/h (5 mph) driven by two propellers; range, road 850 km (528 miles) and water 60 km (37 miles); fording amphibious; gradient 60%; side slope 30%; vertical obstacle 0.6 m (23.6 in); ground clearance 0.375 m (14.75 in)

Variants
EE-11: featuring great commonality of components with the EE-9 armoured car, the basic Urutu was designed in the early 1970s and entered service with the Brazilian army in 1975 as the **CTRA**; there is provision for the embarked troops to use their personal weapons from inside the vehicle, and the armament described above is standard for the APC, but alternatives include a ring-mounted 7.62-mm (0.3-in) machine-gun, a Hagglunds 20-mm cannon turret, a Brandt 60-mm (2.36-in) mortar turret, the turret of the British Scorpion reconnaissance vehicle with 76-mm (3-in) rifled gun, the ENGESA EC-90 turret with 90-mm (3.54-in) rifled gun (producing a variant known as the **Urutu AFSV**), the ENGESA ET-MD turret with 0.5-in and 7.62-mm machine-guns, and the ENGESA ET-20 turret with 20-mm Oerlikon cannon and 7.62-mm machine-gun; the EE-11 has so far appeared in four basic models with different engines, the **Urutu Model I** having a 130-kW (174-hp) Mercedes-Benz with manual transmission (and uniquely in the series no central tyre pressure regulation system), the **Urutu Model II** having a 142-kW (190-hp) Mercedes-Benz with automatic transmission, the **Urutu Model III** having the General Motors diesel with automatic transmission, and the **Urutu Model IV** having a 142-kW (190-hp) Mercedes-Benz with automatic transmission; unnamed variants include an 81-mm (3.2-in) mortar-carrier, an ambulance model for six to eight sitting wounded or fewer litters, a 2000-kg (4,409-lb) cargo model, a command model, a recovery model with a crane and specialist tools, and a riot-control model

General Motors Canada Grizzly
(Canada)
Type: wheeled armoured personnel carrier
Crew: 3+6
Combat weight: 26,000 lb (11794 kg)
Dimensions: length 18.69 ft (5.97 m); width 8.17 ft (2.49 m); height overall 8.63 ft (2.63 m)
Armament system: one 0.5-in (12.7-mm) Browning M2HB heavy machine-gun with 1,000 rounds, one 7.62-mm (0.3-in) C1 co-axial machine-gun with 4,400 rounds, and four smoke-dischargers on each side of the manually-powered turret
Armour: welded steel varying in thickness from 8 to 10 mm (0.315 to 0.39 in)

Powerplant: one 215-hp (160-kW) Detroit Diesel 6V-53 diesel engine with 54 US gal (204 litres) of internal fuel, and driving a 6 x 6 layout
Performance: speed, road 62 mph (10 km/h) and water 6.2 mph (10 km/h) driven by two propellers; range, road 603 miles (970 km); fording amphibious; gradient 60%; side slope 30%; vertical obstacle 15 in (0.38 m); ground clearance 15.5 in (0.39 m)

Variants
Grizzly: wheeled armoured personnel carrier of the Canadian **Armored Vehicle General Purpose** series (derived from the MOWAG Piranha family evolved in Switzerland) and entered service in 1979; the armament is fitted in a Cadillac Gage 1-m turret of the type used on the Commando Scout, and there is provision for the embarked troops to fire their personal weapons from inside the vehicle
Cougar: wheeled fire-support vehicle member of the family, with a crew of three and the turret of the British Scorpion reconnaissance vehicle (one 76-mm/3-in L23A1 rifled gun with 40 rounds and one 7.62-mm co-axial machine-gun) and a fire-control system incorporating a laser rangefinder
Husky: wheeled maintenance and recovery vehicle member of the family, with a crew of three and a 3,850-lb (1746-kg) capacity crane in place of the turret
LAV-25: entering service with the US Marine Corps is the Light Armored Vehicle-25, an 8 x 8 derivative of the Armored Vehicle General Purpose series fitted with an Arrowpointe turret containing a 25-mm M242 Hughes Chain Gun (210 rounds), an M240 7.62-mm co-axial machine-gun, and an M60 7.62-mm AA medium machine-gun or 0.5-in (12.7-mm) Browning M2HB AA heavy machine-gun, and four smoke-dischargers on each side of the turret, which is hydraulically powered and provides main gun stabilization in both elevation and azimuth; the LAV-25 has a crew of 3+6, a combat weight of 28,400 lb (12882 kg), a length of 20.97 ft (6.39 m) and a powerplant comprising one 275-hp (205-kW) Detroit Diesel 6V-53T turbocharged diesel engine; the LAV-25 is being produced also as a logistics vehicle, a mortar-carrier (81-mm/3.2-in weapon with 90 bombs or 107-mm/4.2-in weapon with 80 bombs), a recovery vehicle with crane and winch, an anti-tank vehicle with a retractable twin TOW launcher and 12 missiles, a command vehicle, and (possibly) an assault gun version with a turret-mounted 90-mm (3.54-in) Cockerill Mk III, 75-mm (2.95-mm) ARES or 60-mm (2.36-in) HMVS weapon

Cardoen VTP-2
(Chile)
Type: wheeled armoured personnel carrier
Crew: 2+10
Combat weight: 7800 kg (17,196 lb)
Dimensions: length 5.37 m (17.62 ft); width 2.32 m (7.61 ft); height 2.22 m (7.28 ft)
Armament system: one pintle-mounted 7.62-mm (0.3-in) or 0.5-in (12.7-mm) machine-gun
Armour: welded steel, either 6 or 8 mm (0.24 or 0.315 in) thick
Powerplant: one 89-kW (120-hp) Mercedes-Benz OM 352 diesel engine with 150 litres (33 Imp gal) of internal fuel, and driving a 4 x 4 layout
Performance: speed, road 100 km/h (62 mph); range, road 600 km (373 miles); fording not revealed; gradient 70%; side slope 30%; vertical obstacle 0.5 m (19.7 in); ground clearance 0.44 m (17.3 in)

Variant
VTP-2: based on the chassis and automotive components of the Mercedes-Benz Unimog high-mobility truck, the VTP-2 was developed in the early 1980s, and is available with open or enclosed bodies, and with armament options that include a 20-mm cannon in an Oerlikon GAD-AOA turret, a mortar, anti-tank missiles and turreted AA cannon; the basic vehicle has provision for the embarked troops to fire their personal weapons from inside the hull

NORINCO Type 63
(China)
Type: tracked armoured personnel carrier
Crew: 4+10
Combat weight: 12500 kg (27,557 lb)
Dimensions: length 5.385 m (17.67 ft); width 2.997 m (9.83 ft); height overall 2.61 m (8.56 ft)
Armament system: one pintle-mounted 12.7-mm (0.5-in) Type 54 heavy machine-gun with 1,250 rounds
Armour: welded steel
Powerplant: one 194-kW (260-hp) diesel engine with 450 to 480 litres (99 to 106 Imp gal) of internal fuel
Performance: speed, road 50 km/h (31 mph) and water 7 km/h (4.3 mph) propelled by its tracks; range, road 425 km (264 miles); fording amphibious; gradient 60%; side slope 40%; vertical obstacle 0.8 m (31.5 in); trench 2.5 m (8.2 ft); ground clearance 0.43 m (17 in)

Variants
Type 63: known to the Chinese army as the **Model 531** and otherwise described as the **K-63**, **M1967** and **M1970**, the Type 63 is believed to be based loosely on the chassis and automotive components of the Type 63 light tank; there is only one firing port in each side of the hull, and the type lacks night-vision equipment and an NBC system; the vehicle is also available in psychological warfare (with four loudspeakers), command post, ambulance and mortar-carrier versions; the chassis has also been used for two multiple rocket-launcher models and one self-propelled howitzer

H-1: considerably improved version of the Type 63 with a longer and slightly wider hull, a 239-kW (320-hp) Deutz diesel engine imported from West Germany, and a new suspension system; the type has a crew of 3+8, and there is provision for the embarked troops to fire their personal weapons from inside the vehicle

NORINCO-Vickers NVH-1: designation of an advanced infantry fighting vehicle being developed on the basis of the H-1 with a Vickers two-man turret; this latter is to be offered in manual and powered forms, and with a variety of fire-control systems for a main armament fit that can be varied between the ROF L21 Rarden 30-mm cannon with 120 rounds, the Mauser Model F (MK 30) 30-mm cannon and the Oerlikon KBB 25-mm cannon with 520 rounds; the co-axial armament will be a 7.62-mm (0.3-in) McDonnell Douglas Chain Gun; the combat weight will be some 16000 kg (35,273 lb) and the road speed 65 km/h (40.4 mph), and the dimensions will include a length of 6.125 m (20.10 ft), a width of 3.06 m (10.04 ft) and height 2.77 m (9.09 ft) with the Rarden turret; the type will be amphibious with the aid of a flotation screen, being driven in the water by its tracks

NORINCO-Vickers NVH-4: upgraded version of the NVH-1 with a lenghtened hull (requiring an extra road wheel on each side) for buoyancy without the flotation screen

122-mm SP howitzer: this is a Chinese development of the Type 63 APC chassis with a 122-mm (4.8-in) Type 54 howitzer mounted at the rear in an exposed position; the elevation limits are less than −3 degrees to +63.5 degrees, and traverse 10 degrees left and right

OT-64A
(Czechoslovakia)
Type: wheeled armoured personnel carrier
Crew: 2+18
Combat weight: 14300 kg (31,526 lb)
Dimensions: length 7.44 m (24.40 ft); width 2.55 m (8.37 ft); height to hull top 2.06 m (6.76 ft)
Armament system: one pintle-mounted 7.62-mm (0.3-in) PKT machine-gun with 1,250 rounds
Armour: welded steel to a maximum thickness of 10 mm (0.39 in)
Powerplant: one 134-kW (180-hp) Tatra 928-14 diesel engine with 320 litres (70 Imp gal) of internal fuel, and driving an 8 x 8 layout
Performance: speed, road 94.4 km/h (58.7 mph) and water 9 km/h (5.6 mph) driven by two propellers; range, road 710 km (441 miles); fording amphibious; gradient 60%; side slope 30%; vertical obstacle 0.5 m (19.7 in); trench 2.0 m (6.6 ft); ground clearance 0.46 m (18 in)

Variants
OT-64A: developed from the late 1950s by Czechoslovakia and Poland in the basis of the Tatra 813 truck, the OT-64A entered service in 1964, the Czech vehicles being unarmed and the Polish **SKOT** vehicles having a pintle-mounted PKT machine-gun; notable features are the size and capacity of the type (though there is only limited capability for the embarked troops to fire their personal weapons from inside the hull), the standard NBC system, full amphibious capability and a central tyre pressure regulation system; some OT-64As have been seen two launchers for four 'Sagger' anti-tank missiles

OT-64B: used only by Poland as the **SKOT-2**, this model has a pintle-mounted and shielded 7.62-mm PKT or 12.7-mm (0.5-in) DShKM machine-gun

OT-64C(1): development of the basic model with a 14.5-mm (0.57-in) KVPT machine-gun (500 rounds) and co-axial 7.62-mm PKT machine-gun (2,000 rounds) in a centrally-located and manually powered turret similar to that of the Soviet BRDM-2; the type is known to the Poles as the **SKOT-2A**

OT-64C(2): version used by Poland as the **SKOT-2AP** with a revised turret featuring a curved top

OT-64R-2 and **OT-64R-3:** command versions of the basic model

SKOT-WPT: designation of the Polish recovery and repair version, fitted with a light crane and armed with a single 7.62-mm machine-gun

Kadar Fahd
(Egypt)
Type: wheeled armoured personnel carrier
Crew: 2+10
Dimensions: length 6.00 m (19.69 ft); width 2.45 m (8.03 ft); height 2.10 m (6.89 ft)
Armament system: one turret-mounted 20-mm cannon, or one pintle-mounted 0.5-in (12.7-mm) or 7.62-mm (0.3-in) machine-gun
Armour: welded steel
Powerplant: one 125-kW (168-hp) Mercedes-Benz OM 352A diesel engine with ? litres (? Imp gal) of internal fuel, and driving a 4 x 4 layout
Performance: speed, road 90 km/h (56 mph); range, road 800 km (497 miles); fording 0.7 m (2.3 ft); gradient 70%; side slope 30%; vertical obstacle 0.5 m (19.7 in); trench 0.8 m (31.5 in)

Variant
Fahd: this Egyptian APC is based on the chassis and automotive system of the Daimler-Benz LAP 1117/32 truck; the type can carry the Milan anti-tank missile system, and variants optimized for the command post, recovery, logistics, ambulance and rocket-launcher roles are under development

SISU XA-180
(Finland)
Type: wheeled armoured personnel carrier
Crew: 2+10

Dimensions: length 7.35 m (24.11 ft); width 2.89 m (9.48 ft); height 2.47 m (8.10 ft)
Armament system: none
Armour: welded steel
Powerplant: one 183-kW (245-hp) Valmet turbocharged diesel engine with ? litres (? Imp gal) of internal fuel, and driving a 6 x 6 layout
Performance: speed, road 105 km/h (65 mph) and water 10 km/h (6.2 mph) driven by two propellers; range, road 800 km (497 miles); fording amphibious; ground clearance 0.45 m (17.7 in)

Variant
XA-180: this Finnish APC, entering service in the mid-1980s, is based on the chassis and automotive components of the Valmet SA-150 VK 4 x 4 truck, and can carry a payload of 6500 kg (14,330 lb) on roads, or of 3000 kg (6,514 lb) across country or swimming; the type has only one port for the use of a personal weapon by embarked troops; it is likely that command, mortar and AA variants will be developed

GIAT AMX-10P
(France)
Type: tracked mechanized combat infantry vehicle
Crew: 3+8
Combat weight: 14200 kg (31,305 lb)
Dimensions: length 5.778 m (18.06 ft); width 2.78 m (9.12 ft); height 2.57 m (8.43 ft)
Armament system: one 20-mm GIAT M693 (F2) cannon with 800 rounds, one 7.62-mm (0.3-in) co-axial machine-gun with 2,000 rounds, and two smoke-dischargers on each side of the electrically-powered turret
Armour: welded aluminium
Powerplant: one 209-kW (280-hp) Hispano-Suiza HS 115 supercharged diesel engine with 528 litres (116 Imp gal) of internal fuel
Performance: speed, road 65 km/h (40.4 mph) and water 7 km/h (4.3 mph) driven by two waterjets; range, road 600 km (373 miles); fording amphibious; gradient 60%; side slope 30%; vertical obstacle 0.7 m (27.6 in); trench 1.6 m (5.2 ft); ground clearance 0.45 m (17.7 in)

Variants
AMX-10P: designed from the mid-1960s as replacement for the AMX VCI, the AMX-10P is a useful MICV rather than a plain APC, and began to enter French service in 1973, since when substantial export orders have been received to bolster production swelled by a proliferation of variants; the type has an NBC system as standard and is fully amphibious, but the embarked troops can fire their personal weapons only through a pair of firing ports in the electrically-operated rear ramp; the standard Toucan II two-man turret can be replaced by a one-man Toucan I or Capre 20 turret, and the M693 cannon can be altered for other 20-mm weapons; there is also a driving training version of this baseline version with an observation position in place of the armament turret
AMX-10P 25: upgraded model first shown in 1983, and featuring a one-man Dragar turret with dual-feed 25-mm GIAT 811 cannon plus 175 rounds of HE and 45 rounds of AP ammunition, as well as a co-axial 7.62-mm machine-gun; the vehicle weighs 14300 kg (31,526 lb) and has a crew of 3+8
AMX-10P Marine: version developed for the Indonesian marines with improved waterjet propulsion for a speed of 10 km/h (6.2 mph), and a rear-mounted CIBI 50 turret with one 0.5-in (12.7-mm) Browning M2HB heavy machine-gun and 120 rounds; the crew is 2+13
AMX-10 Sanitaire: unarmed ambulance model with accommodation for a crew of three (driver and two attendants) plus three litters or one litter and four sitting wounded
AMX-10 ECH: repair version with Toucan I turret and 6000-kg (13,228-lb) capacity crane
AMX-10 HOT: anti-tank version with a two-man Lancelot turret for four loaded missile launchers, between 16 and 20 reload missiles being accommodated in the hull; the crew comprises the driver and two loading numbers in the hull, and the commander and gunner in the turret
AMX-10PC: command post version with additional radios and generating capacity for a crew of six (driver, two officers, one NCO and two signallers)
AMX-10P RATAC: radar version with RATAC artillery radar in place of the turret; the crew comprises the driver, commander, two radar operators and radio operator
AMX-10 RAV: ammunition resupply version for use with the self-propelled 155-mm (6.1-in) GCT vehicle
AMX-10RC: reconnaissance version (see relevant section)
AMX-10 SAO: artillery observation version with the Toucan II turret replaced by a two-man turret with long-range laser rangefinder and external 7.62-mm machine-gun; the five-man crew comprises the driver and two radio operators in the hull, and two artillery specialists in the turret
AMX-10 SAT: artillery survey vehicle with specialist survey equipment and navigation system
AMX-10 TM: tractor version for the 120-mm (4.72-in) Brandt MO-120-RT-61 rifled mortar, for which the vehicle carries 60 bombs as well as a crew of six and the Toucan I turret (20-mm cannon and 7.62-mm machine-gun)
AMX-10 VOA: artillery observer version of the AMX-10PC with two-man turret for observation (day and night) and local defence; the four-man crew comprises the driver and radio operator in the hull, and the commander and observer in the turret

AMX-10P VFA: regimental-level version with the ATILA artillery fire-control system with computer
AMX-10P VLA: artillery liaison version lacking the computerized elements of the ATILA system
AMX-10 SAF: high-echelon version of the ATILA artillery fire-control system
AMX-10 TMC-81: advanced fire-support version with an 81-mm (3.2-in) Brandt CL 81 smooth-bore mortar cannon with 108 HE and 10 APFSDS rounds; the type has a crew of four
AMX-10 PAC 90: export fire-support version with the TS 90 turret carrying the 90-mm (3.54-in) GIAT CS Super (F4) rifled gun (30 rounds), a co-axial 7.62-mm machine-gun (2,000 rounds) and a 7.62-mm or 0.5-in (12.7-mm) AA machine-gun (1,200 or 1,000 rounds respectively); this version has a combat weight of 14800 kg (32,628 lb), lengths of 7.22 m (23.69 ft) with gun forward and 5.87 m (19.26 ft) for the hull, plus increased width and height; advanced sights and/or fire-control systems can be fitted as required, and the crew is 3+4

Creusot-Loire AMX VCI
(France)
Type: tracked infantry combat vehicle
Crew: 3+10
Dimensions: length 5.70 m (18.70 ft); width 2.67 m (8.76 ft); height to turret top 2.41 m (7.91 ft)
Armament system: one 20-mm cannon, or one 7.5-mm (0.295-in) machine-gun, or one 7.62-mm (0.3-in) machine-gun in a manually-powered turret, or one pintle-mounted 0.5-in (12.7-mm) Browning M2HB heavy machine-gun
Armour: welded steel varying in thickness from 10 to 30 mm (0.30 to 1.18 in)
Powerplant: one 186-kW (250-hp) SOFAM 8Gxb petrol engine with 410 litres (90 Imp gal) of internal fuel
Performance: speed, road 60 km/h (37.3 mph); range, road 350 km (217 miles); fording 1.0 m (3.3 ft); gradient 60%; vertical obstacle 0.65 m (25.6 in); trench 1.6 m (5.2 ft); ground clearance 0.48 m (18.9 in)

Variants
AMX VCI: this APC was developed in the early part of the 1950s on the basis of the AMX-13 light tank, and the first production examples appeared in 1957 under the designation **TT 12 Ch Modèle 56**, later changed to **AMX VTP** and finally to AMX VCI; as evident from the Armament system section of the specification, the AMX VCI can be fitted with a wide diversity of barreled armament, either pintle- or turret-mounted; there is also provision for the embarked troops to fire their personal weapons from inside the vehicle; later-production examples were fitted as standard with an NBC system, and vehicles of this series can be engined (in construction or as a retrofit) with a diesel powerplant, either the 280-hp (209-kW) Detroit Diesel 6V-53T or the 201-kW (270-hp) Baudouin 6 F 11 SRY, the latter of Belgian origin; the use of a diesel engine reduces the fire risk, increases speed marginally, and boosts range by some 100 km (62 miles)
VCG: combat engineer vehicle with a front-mounted dozer blade and A-frame, a winch, specialist equipment and a crew of 10
VTT/Cargo: freight model with a payload of 3000 kg (6,614 lb)
VTT/LT: artillery fire-control model of the VTT/PC with a crew of seven and specialist equipment
VTT/PC: command post model with a crew of six to eight and additional features such as map tables
VTT/PM: mortar-carrier with an 81-mm (3.2-in) mortar and 128 bombs, or a 120-mm (4.72-in) mortar and 60 bombs, both types firing through a roof hatch; the crew is six
VTT/RATAC: artillery radar vehicle with radar mounted above the roof
VTT/TB: ambulance model with accommodation for three litters or four sitting wounded in addition to the driver, doctor and two attendants
VTT/VCA: support vehicle for the 155-mm (6.1-in) Mk F3 self-propelled gun, carrying part of the gun crew and 25 rounds, and towing a trailer with another 30 rounds

Renault VAB VTT
(France)
Type: wheeled armoured personnel carrier
Crew: 2+10
Combat weight: 13000 kg (28,660 lb)
Dimensions: length 5.98 m (19.62 ft); width 2.49 m (8.17 ft); height to hull top 2.06 m (6.76 ft)
Armament system: one pintle-mounted 7.62-mm (0.3-in) machine-gun
Armour: welded steel
Powerplant: one 175-kW (235-hp) MAN D2356 HM72 diesel engine or one 171-kW (230-hp) Renault VI MIDS 06.20.45 turbocharged diesel engine with 300 litres (66 Imp gal) of internal fuel, and driving a 4 x 4 layout
Performance: speed, road 92 km/h (57 mph) and water 7 km/h (4.3 mph) driven by two waterjets; range, road 1000 km (621 miles); fording amphibious; gradient 60%; side slope 35%; vertical obstacle 0.6 m (23.6 in); ground clearance 0.4 m (15.75 in)

Variants
VAB VTT: this is the French army's standard wheeled APC, and was designed in the early 1970s in 4 x 4 and 6 x 6 layouts in response to a requirement for such a vehicle as the primary transport of motorized formations, mechanized formations being allocated

the tracked AMX-10P; production began in 1976 for this purpose, the French army opting for the 4 x 4 layout; standard equipment includes NBC protection and active/passive night-vision equipment; there is provision for the embarked troops to fire their personal weapons from inside the vehicle; it is likely that the 6 x 6 version, which is proving successful in the export market, will also be adopted for the French army, this type having a combat weight of 14200 kg (31,305 lb) and a 1-m (3.3-ft) trench-crossing capability; various alternative armament installations are possible

VAB VCI: infantry combat vehicle with a crew of 3+8 and a one-man Toucan I turret with a 20-mm M621 (F1) cannon (700 rounds) and a co-axial 7.62-mm machine-gun (2,000 rounds); the French air force has adopted a 4 x 4 version as the **VIB**

VAB Dragar: derivative of the basic version with the Dragar one-man turret armed with a 25-mm cannon and co-axial 7.62-mm (0.3-in) machine-gun

VAB T 25: derivative of the basic version with the T 25 two-man turret armed with a 25-mm cannon and co-axial 7.62-mm (0.3-in) machine-gun

VAB Echelon: repair vehicle with specialized repair kit and a TLi 52A machine-gun turret

VAB PC: command post model for the French army with extra radios and additional mapboards; optional fits produce the **VAB RASIT** with ground-surveillance radar, the **VAB RATAC** with artillery radar, the **VAB FDC** fire-direction centre and the **VAB FOO** forward observation vehicle, the latter two as part of the ATILA artillery fire-control system

VAB Sanitaire: ambulance model with accommodation for 10 sitting wounded or four litters

VAB SATCP: under development in the mid-1980s for the French army is this specialized all-weather low-level air-defence model of the VAB, the HML turret over the hull rear accommodating six ready-to-fire Mistral short-range SAMs (three on each side of the turret) and the surveillance system, which can be an ESD Rodeo pulse-Doppler radar, a Thomson-CSF TRS 2600 pulse-Doppler radar, or a SAT Vipère infra-red system; six reload missiles are carried in the hull; the turret can also take four smoke-dischargers and a pintle-mounted 7.62-mm (0.3-in) machine-gun for local defence

VCAC HOT Mephisto: 4 x 4 French army anti-tank version with the Euromissile Mephisto retractable launcher with four ready-to-fire HOT missiles plus eight reloads in the hull

VCAC HOT UTM 800: 6 x 6 anti-tank export version with the Euromissile UTM 800 turret for four ready-to-fire HOT missiles, another 16 missiles being held in the hull

VAB VDAA TA 20: AA gun version of the VAB fitted with the TA 20 turret carrying two 20-mm cannon, one RA-20S radar and the associated fire-control system and ammunition

VAB VDAA TA 25: AA gun version of the VAB fitted with the TA 25 turret carrying two 25-mm Oerlikon KBB dual-feed cannon, Contraves Gun King fire-control system with laser rangefinder, and ESD Rodeo or RA-20S radar

VMO: internal-security model with light armament and an obstacle-clearing blade

VTM 120: tractor for the 120-mm (4.72-in) Brandt rifled mortar, with accommodation for 70 bombs and six crew

Berliet VXB-170

(France)
Type: wheeled armoured personnel carrier
Crew: 1+11
Combat weight: 12700 kg (27,998 lb)
Dimensions: length 5.99 m (19.65 ft); width 2.50 m (8.20 ft); height without armament 2.05 m (6.73 ft)
Armament system: one pintle-mounted 7.62-mm (0.3-in) machine-gun, or one 7.62-mm (0.3-in) machine-gun and one 40-mm grenade-launcher in a manually-powered turret, and two smoke-dischargers on each side of the hull
Armour: welded steel to a 7-mm (0.28-in) basis
Powerplant: one 127-kW (170-hp) Berliet V 800 M diesel engine with 220 litres (48 Imp gal) of internal fuel, and driving a 4 x 4 layout
Performance: speed, road 85 km/h (53 mph) and water 4 km/h (2.5 mph) driven by its wheels; range, road 750 km (466 miles); fording amphibious; gradient 60%; side slope 30%; vertical obstacle 0.3 in (11.8 in); ground clearance 0.45 m (17.7 in)

Variant

VXB-170: introduced in the early 1970s, the VXB has secured only limited sales, generally to forces with a paramilitary role; the type has provision for the embarked troops to fire their personal weapons from inside the vehicle, and options include an NBC system, night-vision devices, a front-mounted obstacle-clearing blade, and a winch

Panhard VCR/TT

(France)
Type: wheeled armoured personnel carrier
Crew: 3+9
Combat weight: 7900 kg (17,416 lb)
Dimensions: length 4.875 m (15.99 ft); width 2.478 m (8.13 ft); height including armament 2.56 m (8.40 ft)
Armament system: one pintle-mounted 20-mm cannon or 7.62-mm (0.3-in) machine-gun, and two smoke-dischargers on each side of the hull
Armour: welded steel varying in thickness between 8 and 12 mm (0.315 to 0.49 in)
Powerplant: one 116-kW (155-hp) Peugeot PRV petrol engine with 242 litres (53 Imp gal) of internal fuel, and driving a 6 x 6 layout
Performance: speed, road 100 km/h (62 mph) and

water 4 km/h (2.5 mph) driven by the wheels; range, road 800 km (497 miles); fording amphibious; gradient 60%; side slope 30%; vertical obstacle 0.8 m (31.5 in); trench 1.1 m (3.6 ft); ground clearance 0.315 m (12.4 in)

Variants
VCR/TT: this series of wheeled fighting vehicles has been developed from the basis of the ERC 6 x 6 armoured car, and features a retractable central pair of wheels for improved road performance without compromise of cross-country mobility; the VCR/TT is the standard APC and can be fitted with a wide range of armament (including some turreted weapons, generally causing a reduction in embarked troop accommodation to six) and optional equipment such as a winch and an NBC system; there is limited provision for the embarked troops to fire their personal weapons from inside the vehicle
VCR/AT: repair model with specialist equipment
VCR/IS: ambulance model with accommodation for three crew and four litters or two litters and six sitting wounded
VCR/PC: command post model with a crew of six and special fittings in the hull for radios, mapboards etc
VCR/TH: anti-tank version with the Euromissile UTM 800 launcher for four ready-to-launch HOT missiles, another 10 missiles being accommodated in the hull; this variant has a remotely-controlled 7.62-mm machine-gun on the roof for local protection, and a crew of four
VCR/TT 2: upgraded version of the basic model, available with two 73-kW (98-hp) Peugeot XD 3T diesel or 108-kW (145-hp) Peugeot V6PRV petrol engines, the former providing a range of 1000 km (621 miles) and the latter of 600 km (373 miles); the embarked troop strength varies according to the armament fitted, a typical complement when fitted with a single 0.5-in or 7.62-mm machine-gun being 3+12; fitted with the SAMM BTE 105 cupola with 0.5-in machine-gun, the VCR/TT 2 has a combat weight of 9600 kg (21,164 lb) and water propulsion by a pair of propellers

Panhard VCR (4 × 4)
(France)
Type: wheeled armoured personnel carrier
Crew: 3+9
Combat weight: 7800 kg (17,196 lb)
Dimensions: length 4.875 m (15.99 ft); width 2.478 m (8.13 ft); height to hull top 2.13 m (6.99 ft)
Armament system: one 20-mm cannon, or one or two 7.62-mm (0.3-in) machine-guns, or one 60-mm (2.36-in) breech-loaded mortar, or one Milan anti-tank launcher
Armour: welded steel varying in thickness from 8 to 12 mm (0.315 to 0.47 in)
Powerplant: one 116-kW (155-hp) Peugeot petrol engine with 242 litres (53 Imp gal) of internal fuel, and driving a 4 x 4 layout
Performance: speed, road 100 km/h (62 mph) and water 7.2 km/h (4.5 mph) driven by two waterjets; range, road 800 km (497 miles); fording amphibious; gradient 60%; side slope 30%; vertical obstacle 0.4 m (15.75 in); ground clearance 0.315 m (12.4 in)

Variant
VCR (4 x 4): this is a simple development of the VCR 6 x 6 series, and can be used in a similar diversity of roles with modified armament fits and interior configurations; the type has provision for the embarked troops to fire their personal weapons from inside the vehicle

Panhard M3
(France)
Type: wheeled armoured personnel carrier
Crew: 2+10
Combat weight: 6100 kg (13,448 lb)
Dimensions: length 4.45 m (14.60 ft); width 2.40 m (7.87 ft); height to top of turret 2.48 m (5.47 ft)
Armament system: one 20-mm cannon, or one 0.5-in (12.7-mm) heavy machine-gun, or one or two 7.62-mm (0.3-in) machine-guns, one 60-mm (2.36-in) breech-loaded mortar, one Milan anti-tank launcher
Armour: welded steel varying in thickness from 8 to 12 mm (0.315 to 0.47 in)
Powerplant: one 67-kW (90-hp) Panhard 4 HD petrol engine with 165 litres (36 Imp gal) of internal fuel, and driving a 4 x 4 layout
Performance: speed, road 90 km/h (56 mph) and water 4 km/h (2.5 mph) driven by the wheels; range, road 600 km (373 miles); fording amphibious; gradient 60%; side slope 30%; vertical obstacle 0.3 m (11.8 in); ground clearance 0.35 m (13.8 in)

Variants
M3: designed in the late 1960s, the M3 has sold successfully in Africa, some Moslem countries and elsewhere since the type entered production in 1971; the automotive components are almost identical with those of the AML armoured car series; the type has provision for the embarked troops to fire their personal weapons from inside the vehicle, and a very great range of armament fits, of both turreted and ring-mounted types, is possible
M3/VAT: repair version with specialist equipment and a crew of four
M3/VDA: air-defence version with a turret mounting two 20-mm Hispano-Suiza 820 SL cannon (plus 600 rounds), RA-20 radar and a fire-control system; the guns can elevate from −5 to +85 degrees, and the turet can traverse through 360 degrees; the cyclic rate is either 200 or 1,000 rounds per minute per barrel, and effective AA altitude 1500 m (4,920 ft) with a

variety of ammunition types; this model weighs 7200 kg (15,873 lb) and has a crew of three

M3/VLA: engineer version with a front-mounted dozer blade and a crew of six

M3/VPC: command post version with mapboards and extra radios

M3/VTS: ambulance version with a crew of three (driver and two attendants) plus six sitting wounded, or four litters, or two litters and three sitting wounded

Buffalo: M3 reworked with better stowage and a choice between the 108-kW (145-hp) Peugeot V6 PRV petrol engine or 71-kW (95-hp) Peugeot XD 3T turbocharged diesel engine; a variety of armament installations is possible, and options for the fully amphibious vehicle include a front-mounted winch and air conditioning; the basic model can also be outfitted as an internal security vehicle, ambulance, mortar carrier, command vehicle etc

ACMAT TPK 4.20 VSC
(France)
Type: wheeled armoured personnel carrier
Crew: 2+8
Combat weight: 7300 kg (16,094 lb)
Dimensions: length 5.98 m (19.62 ft); width 2.10 m (6.89 ft); height to hull top 2.21 m (7.23 ft)
Armament system: one pintle-mounted 0.5-in (12.7-mm) or 7.62-mm (0.3-in) machine-gun, or one 81-mm (3.2-in) mortar
Armour: welded steel 5.8 mm (0.23 in) thick
Powerplant: one 125-hp (93-kW) Perkins 6.3544 diesel engine with 360 litres (79 Imp gal) of internal fuel, and driving a 4 x 4 layout
Performance: speed, road 95 km/h (59 mph); range, road 1600 km (994 miles); fording 0.8 m (2.6 ft); gradient 60%; ground clearance 0.27 m (10.75 in)

Variants
TPK 4.20 VSC: based on the VLRA reconnaissance vehicle but fitted with an armoured body, the VSC is notable for its very considerable range; there is no provision for the embarked troops to fire their personal weapons from inside the vehicle, which has an open-topped configuration

VBL: armoured car version with a fully-enclosed troop compartment surmounted by a one-man turret for a variety of light armament fits

Thyssen Henschel/Krupp MaK Schützenpanzer Neu M-1966 Marder
(West Germany)
Type: tracked mechanized infantry combat vehicle
Crew: 4+6
Combat weight: 28200 kg (62,169 lb)
Dimensions: length 6.79 m (22.28 ft); width 3.24 m (10.63 ft); height to turret top 2.86 m (9.38 ft)
Armament system: one 20-mm Rheinmetall MK 20 Rh 202 cannon with 1,250 rounds, two 7.62-mm (0.3-in) MG3 machine-guns (one co-axial and the other in a remotely-controlled podded installation at the rear of the hull) with 5,000 rounds, one Milan anti-tank launcher, and three smoke-dischargers on each side of the turret; the turret is electro-hydraulically powered, the main gun lacks stabilization in either elevation or azimuth, and a day/night optical fire-control system is fitted
Armour: welded steel
Powerplant: one 447-kW (600-hp) MTU MB 833 Ea500 diesel engine with 652 litres (143 Imp gal) of internal fuel
Performance: speed, road 75 km/h (46.6 mph); range, road 520 km (323 miles); fording 1.5 m (4.9 ft) without preparation and 2.5 m (8.2 ft) with preparation; gradient 60%; vertical obstacle 1.0 m (3.3 ft); trench 2.5 m (8.2 ft); ground clearance 0.45 m (17.7 in)

Variants
Marder: designed by Henschel and then built from 1969 by Rheinstahl (which had taken over Henschel) and MaK, the Marder is the West German army's standard MICV and a pioneering vehicle of its type; the Marder was at first built without the Milan anti-tank system, which was later added on the right-hand side of the cannon turret; standard systems include an NBC system and night-vision equipment, and there is provision for the embarked troops to fire their personal weapons from inside the vehicle

Marder A1: upgraded version produced by retrofit of 670 vehicles from 1982; modifications include a double-belt feed for the cannon, improved night-vision capabilities and better habitability; this model weighs 30000 kg (66,138 lb) and has accommodation for only five infantrymen

Marder A1A: upgraded version produced by retrofit of 1,466 vehicles from 1982 to a standard similar to the Marder A1 but without improved passive night-vision capability

Marder 25: version with the 20-mm cannon replaced by a 25-mm Mauser Model E cannon, which will be retrofitted from the late 1980s

Radarpanzer TUR: battlefield surveillance version with a turret replaced by a hydraulically extending arm supporting the antenna of the 30-km (18.6-mile) range Siemens surveillance radar; the vehicle weighs 32000 kg (70,547 lb) and has a crew of four; the armament comprises two 7.62-mm machine-guns, and four smoke-dischargers are located on each side of the hull rear

Thyssen Henschel Transportpanzer 1 Fuchs
(West Germany)
Type: wheeled armoured personnel carrier
Crew: 2+10

Combat weight: 17000 kg (37,478 lb)
Dimensions: length 6.76 m (22.19 ft); width 2.98 m (9.78 ft); height to hull top 2.30 m (7.55 ft)
Armament system: one ring-mounted 20-mm Rheinmetall MK 20 Rh 202 cannon or 7.62-mm (0.3-in) MG3 machine-gun, and six smoke dischargers on the left of the hull
Armour: welded steel
Powerplant: one 239-kW (320-hp) Mercedes-Benz OM 402A turbocharged diesel engine with 390 litres (86 Imp gal) of internal fuel, and driving a 6 x 6 layout
Performance: speed, road 105 km/h (65 mph) and water 10.5 km/h (6.5 mph) driven by two propellers; range, road 800 km (497 miles); fording amphibious; gradient 70%; vertical obstacle not revealed; trench not revealed; ground clearance 0.41 m (16 in)

Variants
Transportpanzer 1: deliveries of this standard wheeled APC began in 1979, and the type is designed for maximum operational flexibility; when cargo is being carried, the normal load is 4000 kg (8,818 lb) reducing to 2000 kg (4,409 lb) when swimming; standard equipment includes an NBC system and night-vision equipment, but there is no provision for the embarked troops to fire their personal weapons from inside the vehicle; the basic version can also be configured as an engineer vehicle (for the carriage of mines, demolition charges and other specialist kit), as a supply carrier (for the battlefield movement of front-line supplies) and as an ambulance for four litters or two litters and four sitting wounded
TPz 1 Eloka: electronic warfare version fitted with the EK 33 jamming equipment; unlike other TPz 1 variants this model is not amphibious
TPz 1 FüFu: command and communications version with mapboards and extra radio equipment
ABC Erkundsgruppe: NBC reconnaissance version with specialist sampling and analysis equipment
Panzeraufklärungsradargerät: battlefield surveillance version with RASIT radar on an elevating arm above the front of the hull

Thyssen Henschel Condor
(West Germany)
Type: wheeled armoured personnel carrier
Crew: 3+9
Combat weight: 12000 kg (26,455 lb)
Dimensions: length 6.05 m (19.85 ft); width 2.47 m (8.10 ft); height to turret top 2.79 m (9.15 ft)
Armament system: one 20-mm Rheinmetall MK 20 Rh 202 cannon with 220 rounds, one co-axial 7.62-mm (0.3-in) machine-gun with 500 rounds, and three smoke-dischargers on each side of the manually-powered turret
Armour: welded steel
Powerplant: one 125-kW (168-hp Daimler-Benz OM 352A turbocharged diesel engine with 260 litres (57 Imp gal) of internal fuel, and driving a 4 x 4 layout
Performance: speed, road 100 km/h (62 mph) and water 8 km/h (5 mph) driven by a single propeller; range, road 900 km (559 miles); fording amphibious; gradient 60%; side slope 30%; vertical obstacle 0.55 m (21.7 in); ground clearance 0.475 m (18.7 in)

Variant
Condor: designed as a successor to the highly successful UR-416 series, the Condor is a versatile type capable of serving as an APC, ambulance and command post; the armament above can be varied to twin turret-mounted 7.62-mm (0.3-in) machine-guns or twin turret-mounted launchers for anti-tank missiles such as HOT; optional equipment includes an NBC system and habitability items; the type has provision for the embarked troops to fire their personal weapons from inside the vehicle

Thyssen Henschel TM 170
(West Germany)
Type: wheeled armoured personnel carrier
Crew: 2+10
Combat weight: 11200 kg (24,691 lb)
Dimensions: length 6.12 m (20.08 ft); width 2.45 m (8.04 ft); height without armament 2.32 m (7.61 ft)
Armament system: a wide diversity of armament fits is available, including a pintle-mounted 7.62-mm (0.3-in) machine-gun, twin turret-mounted 7.62-mm machine-guns, a turret-mounted 20-mm cannon, or twin turret-mounted launchers for anti-tank missiles
Armour: welded steel
Powerplant: one 125-kW (168-hp) Daimler-Benz OM 352A turbocharged diesel engine with 175 litres (38.5 Imp gal) of internal fuel, and driving a 4 x 4 layout
Performance: speed, road 100 km/h (62 mph) and water 9 km/h (5.6 mph) driven by two waterjets; range, road 700 km (435 miles); fording amphibious; gradient 80%; vertical obstacle not revealed; ground clearance 0.48 m (18.9 in)

Variant
TM 170: this vehicle is the largest Thyssen Henschel APC, and is designed for internal security rather than front-line duties; as such it can be fitted with a wide assortment of armament fits and optional extras, and there is provision for the embarked troops to fire their personal weapons from inside the vehicle

Thyssen Henschel TM 125
(West Germany)
Type: wheeled armoured personnel carrier
Crew: 2+10
Combat weight: 7600 kg (16,755 lb)
Dimensions: length 5.54 m (18.18 ft); width 2.46 m

(8.07 ft); height without armament 2.02 m (6.61 ft)
Armament system: a wide diversity of armament fits is available, including a pintle-mounted 7.62-mm (0.3-in) machine-gun, twin turret-mounted 7.62-mm machine-guns, a turret-mounted 20-mm cannon, or twin turret-mounted launchers for anti-tank missiles
Armour: welded steel
Powerplant: one 93-kW (125-hp) Daimler-Benz OM 352 supercharged diesel engine with 175 litres (38.5 Imp gal) of internal fuel, and driving a 4 x 4 layout
Performance: speed, road 85 km/h (53 mph) and water 8 km/h (5 mph) driven by two propellers; range, road 700 km (435 miles); fording amphibious; gradient 80%; side slope 40%; vertical obstacle 0.55 m (21.7 in); ground clearance 0.46 m (18.1 in)

Variant
TM 125: like the larger TM 170, the TM 125 is based on a commercial chassis and powerplant for maximum cost effectiveness and reliability, and is designed for internal security rather than front-line duties; there is provision for the embarked troops to fire their personal weapons from inside the vehicle

Thyssen Henschel UR-416
(West Germany)
Type: wheeled armoured personnel carrier
Crew: 2+8
Combat weight: 7600 kg (16,755 lb)
Dimensions: length 5.21 m (17.09 ft); width 2.30 m (7.55 ft); height with turret 2.52 m (8.27 ft)
Armament system: one 7.62-mm (0.3-in) pintle-mounted machine-gun
Armour: welded steel
Powerplant: one 89-kW (120-hp) Daimler-Benz OM 352 diesel engine with 150 litres (33 Imp gal) of internal fuel, and driving a 4 x 4 layout
Performance: speed, road 85 km/h (53 mph); range, road 700 km (435 miles); fording 1.4 m (4.6 ft); gradient 70%; vertical obstacle 0.55 m (21.7 in); ground clearance 0.44 m (17.3 in)

Variant
UR-416: the basic APC model can be fitted with a variety of armaments ranging from a single pintle-mounted machine-gun to a single turret-mounted 20-mm cannon, and there is provision for the embarked troops to fire their personal weapons from inside the vehicle; but the UR-416 is optimized for the internal security role in guises such as an ambulance vehicle (eight sitting wounded or two litters and four sitting wounded in addition to the two crew), anti-tank vehicle with a 90-mm (3.54-in) recoilless rifle, command and communications vehicle, reconnaissance vehicle, and workshop vehicle

PSZH-IV
(Hungary)
Type: wheeled armoured personnel carrier
Crew: 3+6
Combat weight: 7500 kg (16,534 lb)
Dimensions: length 5.70 m (17.70 ft); width 2.50 m (8.20 ft); height 2.30 m (7.55 ft)
Armament system: one 14.5-mm (0.57-in) KPVT heavy machine-gun with 500 rounds and one co-axial 7.62-mm (0.3-in) PKT machine-gun with 2,000 rounds in a manually-powered turret
Armour: welded steel up to a maximum thickness of 14 mm (0.55 in)
Powerplant: one 775-kW (100-hp) Csepel D.414.44 diesel engine with 200 litres (44 Imp gal) of internal fuel, and driving a 4 x 4 layout
Performance: speed, road 80 km/h (50 mph) and water 9 km/h (5.6 mph) driven by two waterjets; range, road 500 km (311 miles); fording amphibious; gradient 60%; vertical obstacle 0.4 m (15.75 in); trench 0.6 m (2 ft) with channels; ground clearance 0.42 m (16.5 in)

Variant
PSZH-IV: this small Hungarian wheeled APC appeared in the mid-1960s, and is used by other Warsaw Pact armed forces (Czechoslovakia and East Germany) mainly for internal security and border patrol; the Czech designation is **OT-66**; standard equipment includes infra-red driving lights, an NBC system and a central tyre pressure regulation system

NIMDA Shoet II
(Israel)
Type: wheeled armoured personnel carrier
Crew: 2+10
Combat weight: 9700 kg (21,384 lb)
Dimensions: length 6.64 m (21.78 ft); width 2.20 m (7.22 ft); height 2.08 m (6.82 ft)
Armament system: up to four 0.5-in (12.7-mm) or 7.62-mm (0.3-in) machine-guns pintle-mounted round the top of the vehicle
Armour: welded steel varying in thickness from 8 to 14 mm (0.315 to 0.44 in)
Powerplant: one 172-hp (128-kW) Detroit Diesel 6V-53 diesel engine with 170 litres (37.4 Imp gal) of internal fuel, and driving a 6 x 6 layout
Performance: speed, road 90 km/h (56 mph); range, road 400 km (249 miles); fording not revealed; gradient 60%; side slope 30%; vertical obstacle not revealed; ground clearance not revealed

Variants
Shoet II: developed as a private venture, the Shoet II has an open-topped troop compartment which allows the vehicle to be transformed rapidly into an

anti-tank vehicle (106-mm/4.17-in recoilless rifle or anti-tank missile launcher), an anti-aircraft vehicle (guns or missiles), a mortar carrier (81-mm/3.2-in mortar and machine-guns), a reconnaissance vehicle (61-mm/2.4-in light mortar and machine-guns), an ambulance and a recovery vehicle; standard equipment includes night-vision devices, a central tyre pressure regulation system, and fuel tank protection
Shoet III: improved version with more advanced operational features and other significant modifications

OTO Melara IAFV Mk 1
(Italy)
Type: tracked infantry armoured fighting vehicle
Crew: 2+7
Combat weight: 11600 kg (25,573 lb)
Dimensions: length 5.041 m (16.54 ft); width 2.686 m (8.81 ft); height including machine-gun 2.552 m (8.37 ft)
Armament system: one pintle-mounted 0.5-in (12.7-mm) Browning M2HB machine-gun with 1,050 rounds, and one pintle-mounted 7.62-mm (0.3-in) MG3 machine-gun with 1,000 rounds
Armour: welded aluminium armour with additional steel armour over the fronts and sides
Powerplant: one 215-bhp (160-kW) Detroit Diesel 6V-53 diesel engine with 360 litres (79 Imp gal) of internal fuel
Performance: speed, road 64.4 mph (40 mph) and water 5 km/h (3.1 mph) driven by its tracks; range, road 550 km (342 miles); fording amphibious; gradient 60%; side slope 30%; vertical obstacle 0.61 m (24 in); trench 1.68 m (5.5 ft); ground clearance 0.41 m (16 in)

Variants
AIFV Mk 1: this is an Italian upgrading of the US M113A1 APC with additional armour, improved firepower, the capability for the embarked troops to fire their personal weapons from inside the vehicle, and better seating; the type is known in the Italian army as the **VCC-1 Cavallino**, and improved versions are the **AIFV Mk 2** with a remotely-controlled 0.5-in machine-gun, and the **AIFV Mk 3** with a turreted 20-mm cannon; standard equipment includes infra-red driving lights, but no NBC system is fitted
VCC-1/TOW: version developed for Saudi Arabia with an Emerson TOW turret; this is an elevating and armoured launcher/sight unit with two ready-to-fire missiles; another 10 missiles are held in the hull
VCC-2: improved VCC-1 with appliqué armour on the front and sides

Fiat/OTO Melara Tipo 6614
(Italy)
Type: wheeled armoured personnel carrier
Crew: 1+10
Combat weight: 8500 kg (18,739 lb)
Dimensions: length 5.86 m (19.23 ft); width 2.50 m (8.20 ft); height including machine-gun mounting 2.18 m (7.15 ft)
Armament system: one pintle-mounted 0.5-in (12.7-mm) Browning M2HB heavy machine-gun
Armour: welded steel varying in thickness from 6 to 8 mm (0.24 to 0.315 in)
Powerplant: one 119-kW (160-hp) Fiat 8062.24 supercharged diesel engine with 142 litres (31.25 Imp gal) of internal fuel, and driving a 4 x 4 layout
Performance: speed, road 100 km/h (62 mph) and water 4.5 km/h (2.8 mph) driven by its wheels; range, road 700 km (435 miles); fording amphibious; gradient 60%; side slope 30%; vertical obstacle 0.4 m (15.75 in); ground clearance 0.37 m (14.6 in)

Variants
Tipo 6614: designed as a private venture by Fiat and OTO Melara, this is a very useful second-line APC with provision for the embarked troops to use their personal weapons from inside the hull; in place of the 0.5-in machine-gun, single or twin 7.62-mm (0.3-in) turreted gun(s) can be carried, and optional equipment includes night-vision equipment, air conditioning, smoke-dischargers and a front-mounted winch
KM900: designation of the Tipo 6614 built under licence in South Korea
KM901: designation of role-specific variants (ambulance, command post, mortar-carrier etc) of the KM900

Mitsubishi Type 73
(Japan)
Type: tracked armoured personnel carrier
Crew: 3+9
Combat weight: 13300 kg (29,321 lb)
Dimensions: length 5.80 m (19.03 ft); width 2.80 m (9.19 ft); height including machine-gun 2.20 m (7.22 ft)
Armament system: one pintle-mounted 0.5-in (12.7-mm) Browning M2HB heavy machine-gun, one bow-mounted 0.3-in (7.62-mm) Browning M1919A4 machine-gun, and three smoke-dischargers on each side of the hull rear
Armour: welded aluminium
Powerplant: one 224-kW (300-hp) Mitsubishi 4ZF diesel engine with 450 litres (99 Imp gal) of internal fuel
Performance: speed, road 70 km/h (43.5 mph) and water 7 km/h (4.3 mph) driven by its tracks; range, road 300 km (186 miles); fording amphibious with a kit; gradient 60%; vertical obstacle 0.7 m (27.6 in); trench 2.0 m (6.6 ft); ground clearance 0.4 m (15.75 in)

LAND WEAPONS

Variants
Type 73: this is the main Japanese APC and is an unexceptional vehicle with modest armament and short range; it was designed in the late 1960s as successor to the Type SU 60, and entered production in 1973; the embarked infantry can fire their personal weapons from inside the vehicle, and standard equipment includes infra-red driving lights and an NBC system; the type lacks amphibious capability

Type 75: ground wind-measuring variant used in conjunction with the Type 75 self-propelled rocket-launcher

Mitsubishi Type SU 60
(Japan)
Type: tracked armoured personnel carrier
Crew: 4+6
Combat weight: 11800 kg (26,014 lb)
Dimensions: length 4.85 m (15.91 ft); width 2.40 m (7.87 ft); height including machine-gun 2.31 m (7.58 ft)
Armament system: one pintle-mounted 0.5-in (12.7-mm) Browning M2HB heavy machine-gun, and one bow-mounted 0.3-in (7.62-mm) Browning M1919A4 machine-gun
Armour: welded steel
Powerplant: one 164-kW (220-hp) Mitsubishi 8 HA 21 WT turbocharged diesel engine with ? litres (? Imp gal) of internal fuel
Performance: speed, road 45 km/h (28 mph); range, road 230 km (143 miles); fording 1.0 m (3.3 ft); gradient 60%; vertical obstacle 0.6 m (23.6 in); trench 1.82 m (6 ft); ground clearance 0.4 m (15.75 in)

Variants
SU 60: designed in the late 1950s, this was the Japanese army's primary APC until the advent of the Type 73, and amongst its features are modest armament (including the same type of unusual bow machine-gun mounting as found in the Type 73) and complete lack of night-vision, NBC and amphibious equipment; there is no provision for the embarked troops to fire their personal weapons from inside the vehicle; the standard model has been converted in small numbers to produce NBC reconnaissance vehicles, dozer vehicles and a few training vehicles adapted to resemble the Soviet BMD
SV 60: mortar carrier version with an 81-mm (3.2-in) mortar and 24 bombs; the crew is five and combat weight 12100 kg (26,675 lb)
SX 60: mortar carrier version with a 107-mm (4.2-in) mortar and eight bombs; the crew is five and combat weight 12900 kg (28,439 lb)

DAF YP-408
(Netherlands)
Type: wheeled armoured personnel carrier
Crew: 2+10
Combat weight: 12000 kg (26,455 lb)
Dimensions: length 6.23 m (20.44 ft); width 2.40 m (7.87 ft); height including machine-gun 2.37 m (7.78 ft)
Armament system: one pintle-mounted 0.5-in (12.7-mm) Browning M2HB heavy machine-gun, and three smoke-dischargers on each side of the hull front
Armour: welded steel varying in thickness from 8 to 15 mm (0.315 to 0.59 in)
Powerplant: one 123-kW (165-hp) DAF DS 575 turbocharged diesel engine with 200 litres (44 Imp gal) of internal fuel, and driving a 6 x 8 layout
Performance: speed, road 80 km/h (50 mph); range, road 500 km (311 miles); fording 1.2 m (3.9 ft); gradient 70%; side slope 70%; vertical obstacle 0.7 m (27.6 in); trench 1.2 m (3.9 ft); ground clearance 0.52 m (20.4 in)

Variants
PWI-S(GR): designation of the basic APC version of the **YP-408** developed in the late 1950s, and produced in the mid- and late 1960s; the type lacks amphibious and NBC capability, but can be fitted with night-vision equipment; only the rear doors have provision for the embarked infantry to fire their personal weapons from inside the vehicle
PWI-S(PC): command vehicle similar to the PWI-S(GR) and designed for use by platoon commanders; it is fitted with additional radios and vision equipment
PWCO: command vehicle for company and battalion commanders with some of the seats removed to make room for mapboards; this model also has provision for the erection of a tented extension to the rear
PW-GWT: ambulance vehicle for two litters and four sitting wounded
PW-V: cargo model with a payload of 1500 kg (3,309 lb)
PW-MT: mortar tractor for the 120-mm (4.72-in) Brandt mortar and carrying the seven-man mortar crew plus 50 bombs
PWAT: designation of the basic vehicle modified to carry the TOW anti-tank system
PWRDR: battlefield radar surveillance vehicle with ZB 298 radar

BRAVIA Chaimite V-200
(Portugal)
Type: wheeled armoured personnel carrier
Crew: 11
Combat weight: 7300 kg (16,093 lb)
Dimensions: length 5.606 m (18.39 ft); width 2.26 m (7.41 ft); height including turret 2.26 m (7.41 ft)
Armament system: the BRAVIA-designed turret can accommodate one 0.5-in (12.7-mm) Browning M2HB heavy machine-gun and one co-axial 7.62-mm (0.3-in) machine-gun, or two 7.62-mm machine-guns, or two 5.56-mm (0.219-in) machine-guns

Armour: welded steel varying in thickness from 6.35 to 7.94 mm (0.25 to 0.313 in)
Powerplant: one 157-kW (210-hp) Model M75 petrol engine or V-6 diesel engine with 300 litres (66 Imp gal) of internal fuel, and driving a 6 x 6 layout
Performance: speed, road 99 km/h (61.5 mph) and water 7 km/h (4.3 mph) driven by its wheels; range, road 965 km (600 miles) with petrol engine or 1529 km (950 miles) with diesel engine; fording amphibious; gradient 65%; side slope 40%; vertical obstacle 0.9 m (35.4 in); ground clearance 0.41 m (16.1 in)

Variants
Chaimite V-200: designed in the 1960s for the Portuguese army, and very similar to the Cadillac Gage Commando, the V-200 is the baseline APC of this useful series, and the optional equipment includes a number of armament additions (a multi-role grenade-launcher and four 3.5-in/88.9-mm recoilless rifles) and night-vision equipment; standard equipment includes a front-mounted winch, and the embarked troops can fire their personal weapons from inside the vehicle
Chaimite V-300: upgunned model with a crew of four or five and able to carry a one-man Oerlikon GAD-AOA turret with a 20-mm KAA-001 cannon, or any of several two-man turrets armed with two machine-guns or one 20-mm cannon and one machine-gun
Chaimite V-400: upgunned model with a Mecar or GIAT 90-mm (3.54-mm) gun
Chaimite V-500: command and communications model
Chaimite V-600: mortar-carrying model with an 81- or 120-mm (3.2- or 4.72-in) mortar, crew and ammunition
Chaimite V-700: anti-tank model with HOT or Swingfire guided missiles
Chaimite V-800: ambulance model
Chaimite V-900: crash rescue model
Chaimite V-1000: riot-control model with a diversity of armament options

Sandock-Austral Ratel 20
(South Africa)
Type: wheeled infantry fighting vehicle
Crew: 4+7
Combat weight: 18500 kg (40,785 lb)
Dimensions: length, hull 7.212 m (23.66 ft); width 2.516 m (8.25 ft); height overall 2.915 m (9.56 ft)
Armament system: one 20-mm GIAT M693 cannon with 1,200 rounds, two 7.62-mm (0.3-in) machine-guns (one co-axial and one AA) with 6,000 rounds, and two smoke-dischargers on each side of the manually-powered turret; there is also a 7.62-mm (0.3-in) AA and local defence machine-gun at the hull rear

Armour: welded steel varying in thickness from 6 to 20 mm (0.24 to 0.79 in)
Powerplant: one 210-kW (282-hp) D 3256 BTXF turbocharged diesel engine with 430 litres (94.5 Imp gal) of internal fuel, and driving a 6 x 6 layout
Performance: speed, road 105 km/h (65 mph); range, road 1000 km (621 miles); fording 1.2 m (3.9 ft); gradient 60%; side slope 30%; vertical obstacle 0.35 m (13.8 in); trench 1.15 m (3.8 ft); ground clearance 0.34 m (13.4 in)

Variants
Ratel 20: infantry fighting vehicle notable for its good range, good firepower, modest maintenance requirements and excellent cross-country mobility combined with first-class anti-mine protection; as with other variants of the Ratel, the embarked troops can fire their personal weapons from inside the hull, and optional equipment includes night-vision gear and air conditioning
Ratel 60: version of the Ratel 20 with a 60-mm (2.36-in) breech-loading mortar (plus some 50 bombs) in place of the 20-mm cannon and its ammunition
Ratel 90: baseline model that entered service in 1976 as a fire-support vehicle designed to the exacting requirements of the South African army; this model is armed with a 90-mm (3.54-in) GIAT F1 rifled gun in exactly the same turret as used on the Eland armoured car, and 69 rounds of ammunition are carried; some 6,000 rounds of ammunition are carried for the three 7.62-mm machine-guns (one co-axial, one turret AA and one hull AA); the crew is 4+6, one infantryman being dropped to allow the carriage of 40 90-mm rounds in the hull; the current **Ratel 90 Mk 3** incorporates the hard lessons of South Africa's internal and external military operations, and the earlier **Ratel 90 Mk 1** and **Ratel 90 Mk 2** vehicles are being brought up to this improved standard; combat weight is 19000 kg (41,887 lb)
Ratel Command: command post model featuring a two-man turret fitted with one 0.5-in (12.7-mm) Browning M2HB heavy machine-gun with 300 rounds and a 7.62-mm AA machine-gun; there is also a rear-mounted 7.62-mm machine-gun, some 3,600 rounds of 7.62-mm ammunition being carried; the crew is 3+6, and a mass of command equipment is carried; combat weight is 18000 kg (39,683 lb)
Ratel Logistic: this is the logistic support model of the Ratel series, a much developed 8 x 8 model lengthened to 8.739 m (28.67 ft) and fitted with a 239-kW (320-hp) ADE 423 T turbocharged diesel; the combat weight is 29000 kg (63,933 lb), and the crew is three including a gunner for the 0.5-in machine-gun with 600 rounds; designed to support the Ratel IFV and FSV, the logistic model has the same cross-country mobility and can carry water, fuel, ammunition and supplies of several other types

Autocamiones BMR-600
(Spain)
Type: wheeled infantry fighting vehicle
Crew: 2+11
Combat weight: 13750 kg (30,313 lb)
Dimensions: length 6.15 m (20.18 ft); width 2.50 m (8.20 ft); height including armament 2.36 m (7.74 ft)
Armament system: one cupola-mounted 7.62-mm (0.3-in) machine-gun with 2,500 rounds
Armour: welded aluminium
Powerplant: one 228-kW (336-hp) Pegaso 9157/8 diesel engine with 300 litres (66 Imp gal) of internal fuel, and driving a 6 x 6 layout
Performance: speed, road 100 km/h (62 mph) and water 10 km/h (6.2 mph) driven by two waterjets; range, road 700 km (435 miles); fording amphibious; gradient 68%; side slope 30%; vertical obstacle 0.8 m (31.5 in); trench 1.2 m (3.9 ft); ground clearance 0.4 m (15.75 in)

Variants
BMR-600: this is the Spanish army's new infantry fighting vehicle, designed in the early 1970s and built from 1979 as the **Pegaso 3560/1**; the troops can in most models fire their weapons from inside the hull, there is a rear-mounted winch, and options include air conditioning, deletion of amphibious capability and wheel-propelled amphibious capability; the basic vehicle can also be fitted out as an ambulance, and can carry the HCT turret with twin launchers for HOT anti-tank missiles
Pegaso 3560/3: mortar carrying model with an 81-mm (3.2-in) mortar firing from the rear hull
Pegaso 3560/4: mortar tractor for the Esperanza 120-mm (4.72-in) mortar, with crew and ammunition accommodated in the hull
Pegaso 3560/5: battalion command post model
Pegaso 3564: fire-support model with a TS 90 turret carrying a 90-mm (3.54-in) GIAT CS Super (F4) rifled gun
Pegaso 3562: cavalry model, otherwise known as the VEC reconnaissance vehicle

Santa Barbara BLR
(Spain)
Type: wheeled armoured personnel carrier
Crew: 3+12
Combat weight: 11600 kg (25,573 lb)
Dimensions: length 5.65 m (18.54 ft); width 2.50 m (8.20 ft); height to hull top 1.99 m (6.53 ft)
Armament system: one cupola-mounted 7.62-mm (0.3-in) machine-gun
Armour: welded steel
Powerplant: one 127-kW (170-hp) Pegaso 9220 or 167-kW (224-hp) Pegaso 9100/41 diesel engine with 250 litres (55 Imp gal) of internal fuel, and driving a 4 x 4 layout
Performance: speed, road 86 km/h (53 mph); range, road 800 km (497 miles); fording 1.1 m (3.6 ft); gradient 75%; side slope 30%; vertical obstacle not revealed; ground clearance 0.26 m (10.25 in)

Variant
BLR: this private-venture design is a comparatively simple vehicle better suited to internal security than front-line operations

Hägglund & Söner Pansarbandvagn 302
(Sweden)
Type: tracked infantry fighting vehicle
Crew: 2+10
Combat weight: 13500 kg (29,762 lb)
Dimensions: length 5.35 m (17.55 ft); width 2.86 m (9.38 ft); height to turret top 2.50 m (8.20 ft)
Armament system: one 20-mm Hispano-Suiza cannon with 505 rounds in a manually-powered turret, and six smoke-dischargers
Armour: welded steel
Powerplant: one 209-kW (280-hp) Volvo-Penta THD 100B turbocharged diesel engine with 285 litres (63 Imp gal) of internal fuel
Performance: speed, road 66 km/h (41 mph) and water 8 km/h (5 mph) driven by its tracks; range, road 300 km (186 miles); fording amphibious; gradient 60%; vertical obstacle 0.61 m (24 in); trench 1.8 m (5.9 ft); ground clearance 0.4 m (15.75 in)

Variants
Pbv 302 Mk 1: this pioneering infantry fighting vehicle was designed in the early 1960s and produced between 1966 and 1971, its main operational disadvantages being lack of NBC protection and the inability of the embarked troops to fire their personal weapons from inside the vehicle (through there are long roof hatches for semi-exposed firing); the basic model can also be used to carry 2000 kg (4,409 lb) of cargo or as an ambulance (four litters, or six litters with the aid of a special kit)
Pbv 302 Mk 2: improved Mk 1 with a rear-mounted observation cupola for the embarked troops' commander, two Lyran launchers for illumination flares, and spaced armour over the front of the vehicle
Bplpbv 3023: fire-direction post model with extra radios and a ranging group of seven men in addition to the crew of three
Epbv 3022: observation post model with the commander's hatch replaced by an observation cupola with rangefinder
Stripbv 3021: command post model with extra radios and mapboards

MOWAG Improved Tornado
(Switzerland)
Type: tracked mechanized infantry combat vehicle
Crew: 3+7
Combat weight: 22300 kg (49,162 lb)

Dimensions: length 6.70 m (21.98 ft); width 3.15 m (10.33 ft); height to turret top 2.86 ft (9.38 ft)
Armament system: one 35-mm Oerlikon KDE cannon with 100 rounds and one 7.62-mm (0.3-in) co-axial machine-gun with 500 rounds in an electrically-powered turret, two remotely-controlled 7.62-mm (0.3-in) machine-guns (optional) at the rear of the hull, and smoke-dischargers
Armour: welded steel
Powerplant: one 390-hp (291-kW) Detroit Diesel 8V-71T turbocharged diesel engine with ? litres (? Imp gal) of internal fuel
Performance: speed, road 66 km/h (41 mph); range, road 400 km (249 miles); fording 1.3 m (4.27 ft); gradient 60%; vertical obstacle 0.85 m (33.5 in); trench 2.2 m (7.2 ft); ground clearance) .45 m (17.7 in)

Variant
Improved Tornado: this is a very capable MICV based on the company's **Tornado** prototype, developed during the 1970s from the Marder, which had also been designed by MOWAG; the embarked troops can fire their personal weapons from inside the vehicle, and standard equipment includes an NBC system and night-vision devices; the type can be fitted with a variety of one- and two-man turrets carrying 25- or 35-mm cannon; there is provision for the embarked troops to fire their personal weapons from inside the vehicle

MOWAG Piranha 8 x 8
(Switzerland)
Type: wheeled armoured personnel carrier
Crew: 15
Combat weight: 12300 kg (27,116 lb)
Dimensions: length 6.365 m (20.88 ft); width 2.50 m (8.20 ft); height without armament 1.85 m (6.07 ft)
Armament system: see Variants (below)
Armour: welded steel
Powerplant: one 300-hp (224-kW) Detroit Diesel 6V-53T turbocharged diesel engine with 300 litres (66 Imp gal) of internal fuel, and driving an 8 x 8 layout
Performance: speed, road 100 km/h (62 mph) and water 10.5 km/h (6.5 mph) driven by two propellers; range, road 780 km (485 miles); fording amphibious; gradient 70%; side slope 30%; vertical obstacle 0.5 m (19.7 in); trench not revealed; ground clearance 0.5 m (19.7 in)

Variants
Piranha 4 x 4: possessing a combat weight of 7800 kg (17,196 lb), a length of 5.32 m (17.45 ft) and a crew of 10, this 4 x 4 version is powered by a 216-hp (161-kW) 6V-53 diesel and can be fitted with turrets armed with one or two 7.62-mm (0.3-in) machine-gun(s), or with one externally-mounted 0.5-in (12.7-mm) Browning M2HB heavy machine-gun, or with one 20-mm Oerlikon KAA-001 cannon; the series was designed in the early 1970s for the export market, and deliveries began in 1976; alternative roles are anti-tank warfare with missiles, command post, cargo-carrying, ambulance, mortar carrying, reconnaissance and recovery, and standard features such as an NBC system and air conditioning can be supplemented by active or passive night-vision devices
Piranha 6 x 6: possessing a combat weight of 10500 kg (23,148 lb), a length of 5.97 m (19.59 ft) and a crew of 14, this 6 x 6 version is essentially a lengthened version of the 4 x 4 model powered by the 300-hp (224-kW) 6V-53T diesel and able to carry more infantrymen and a more advanced turret; as in the 4 x 4 model the embarked troops can fire their personal weapons from inside the hull, and the turret options are the Oerlikon GAD-AOA with a 20-mm cannon, the Oerlikon GBD-COA with a 25-mm cannon, the Oerlikon GDD-AOE and GDD-BOE with a 35-mm cannon, the British Scorpion turret with a 76-mm (3-in) L23 gun, and the Belgian CM 90 turret with a Cockerill 90-mm (3.54-in) gun; other armament options are an internally-mounted 81-mm (3.2-in) mortar and anti-tank missiles, the latter being either two ready-to-fire TOWs in a Norwegian Thune-Eureka one-man turret, or two ready-to-fire TOWs in a US TOW 2 elevating launcher; the options available on the 4 x 4 model are also available on the 6 x 6 variant, and other models are an ambulance and a battlefield surveillance model fitted with RASIT radar; derivatives of the Piranha 6 x 6 include the Canadian Cougar, Grizzly and Husky, and a number of Chilean models built by Cardoen, which has also adapted the type with the Brazilian ET-90 turret and 90-mm (3.54-in) gun
Piranha 8 x 8: this is a stretched model with superior cross-country and load-carrying capabilities; the armament options are the same as those for the 6 x 6 model with the addition of AA turrets armed with two 20- or 30-mm cannon, a towed 120-mm (4.72-in) mortar and a launcher with two banks of 15 81-mm rockets; the Piranha 8 x 8 is also the basis of the US Marine Corps' Canadian-developed LAV-25 vehicle

MOWAG Grenadier
(Switzerland)
Type: wheeled armoured personnel carrier
Crew: 1+8
Combat weight: 6100 kg (13,448 lb)
Dimensions: length 4.84 m (15.88 ft); width 2.30 m (7.55 ft); height to turret top 2.12 m (6.96 ft)
Armament system: see Variant (below)
Armour: welded steel
Powerplant: one 151-kW (202-hp) petrol engine with 180 litres (40 Imp gal) of internal fuel, and driving a 4 x 4 layout
Performance: speed, road 100 km/h (62 mph) and water 10 km/h (6.2 mph) driven by a single propeller;

range, road 550 km (342 miles); fording amphibious; vertical obstacle 0.4 m (15.75 in); ground clearance 0.25 m (9.8 in)

Variant
Grenadier: designed as a second-line APC, the Grenadier entered service in 1967, and can be fitted with a number of armaments types, including a single machine-gun, or a 20- or 25-mm cannon, or an anti-tank missile system, or an 81-mm (3.2-in) rocket-launcher system; the embarked troops have only limited provision for firing their weapons from inside the vehicle

MOWAG Roland
(Switzerland)
Type: wheeled armoured personnel carrier
Crew: 3+3
Combat weight: 4900 kg (10,802 lb)
Dimensions: length 4.73 m (15.52 ft); width 2.05 m (6.73 ft); height to turret top 2.10 m (6.89 ft)
Armament system: one 7.62-mm (0.3-in) machine-gun in a manually-powered turret
Armour: welded steel
Powerplant: one 151-kW (202-hp) petrol engine with 170 litres (37.4 Imp gal) of internal fuel, and driving a 4 x 4 layout
Performance: speed, road 110 km/h (68 mph); range 570 km (354 miles); fording 1.0 m (3.3 ft); gradient 60%; side slope 30%; vertical obstacle 0.4 m (15.75 in); ground clearance 0.42 m (16.5 in)

Variant
Roland: designed as a private venture in the 1960s, the Roland has sold well to African and Latin American countries, mainly in the internal security role; the basic vehicle can also be fitted with manual transmission, with a useful saving in weight and overall dimensions at the expense of driver fatigue; many optional items are offered, including night-vision devices, an obstacle-clearing blade and provision for the embarked troops to fire their personal weapons from inside the vehicle

BMP-2
(USSR)
Type: tracked mechanized infantry combat vehicle
Crew: 3+7
Combat weight: 14600 kg (32,187 lb)
Dimensions: length 6.71 m (22.01 ft); width 3.09 m (10.14 ft); height 2.06 m (6.76 ft)
Armament system: one 30-mm 2A42 cannon with 500 rounds, one co-axial 7.62-mm (0.3-in) PKT machine-gun with ? rounds, one launcher for ? AT-5 'Spandrel' anti-tank missiles and three smoke-dischargers on each side of the electrically-operated turret; the type can also generate smoke by injecting fuel into the exhaust system

Armour: welded steel
Powerplant: one 298-kW (400-hp) 5D20 turbocharged diesel engine with ? litres (? Imp gal) of internal fuel
Performance: speed, road 60 km/h (37 mph) and water 7 km/h (4.3 mph) driven by its tracks; range, road 500 km (311 miles); fording amphibious; gradient 60%; vertical obstacle 0.7 m (27.6 in); trench 2.0 m (6.6 ft); ground clearance not revealed

Variant
BMP-2: developed from the baseline BMP-1, the BMP-2 entered service in the late 1970s and is a considerable advance on its predecessor in that the commander is located in the turret rather than behind the driver, that a more powerful turbocharged diesel is used, and that the low-velocity 73-mm (2.87-in) gun in a one-man turret is replaced by a powerful 30-mm cannon in the two-man turret; it is also believed that the frontal arc has been uparmoured; the embarked infantry can use their personal weapons from inside the vehicle, and the anti-tank system fitted is the advanced AT-5 'Spandrel'; standard equipment includes an NBC system and night-vision devices; the 30-mm turret of the BMP-2 has also been fitted to the hull of the 2S1 (SO-122 Gvozdika) self-propelled howitzer to produce a vehicle whose precise role is as yet uncertain, but which offers greater buoyancy and thus better amphibious capability than the BMP-2, which is only marginally amphibious when fully laden; the type lacks the extensive armour protection of the BMP-2 and is probably intended for the reconnaissance role with the Warsaw Pact and export customers reluctant to wear the cost of the better defended BMP-2; the type's exact designation remains unknown

BMP-1
(USSR)
Type: tracked mechanized infantry combat vehicle
Crew: 3+8
Combat weight: 13500 kg (29,762 lb)
Dimensions: length 6.74 m (22.11 ft); width 2.94 m (9.65 ft); height over searchlight 2.15 m (7.05 ft)
Armament system: one 73-mm (2.87-in) 2A28 smooth-bore gun with 40 rounds, one co-axial 7.62-mm (0.3-in) PKT machine-gun with 2,000 rounds and one launcher for five AT-3 'Sagger' anti-tank missiles in the electrically-powered turret; the type can generate smoke by injecting fuel into the exhaust system
Armour: welded steel varying in thickness from 6 to 33 mm (0.28 to 1.3 in)
Powerplant: one 224-kW (300-hp) 5D20 diesel engine with 460 litres (101 Imp gal) of internal fuel
Performance: speed, road 80 km/h (50 mph) and water 8 km/h (5 mph) driven by its tracks; range, road 500 km (311 miles); fording amphibious; gradient 60%; side slope 30%; vertical obstacle 0.8

m (31.5 in); trench 2.2 m (7.2 ft); ground clearance 0.39 m (15.4 in)

Variants
BMP-1: developed in the early 1960s as the primary APC of the infantry units within Soviet tank divisions (and thus the counterpart of the wheeled BTR-60 developed for motorized rifle divisions), the BMP-1 was at first known in the West as the **M1967** and then as the **BMP-76PB**, and was developed via the US Army-designated **BMP M1966** interim model to the definitive **BMP M1970** with better amphibious capabilities and a small snorkel; the BMP-1 is a very useful mechanized infantry combat vehicle let down by its generally poor armament, the 73-mm low-pressure gun in the one-man turret generating so mediocre a muzzle velocity that the projectile is highly susceptible to crosswinds; the embarked infantrymen can use their personal weapons from inside the vehicle, and the BMP-1 has full NBC protection plus night-vision devices; variations on this basic theme include a model with a rear ramp in place of the two doors, a 30-mm cannon in place of the 73-mm gun, a launcher for AT-4 'Spigot' anti-tank missiles in place of the 'Sagger', and additions on the turret (such as the AGS-17 grenade-launcher or six smoke-dischargers) and hull (applique armour as proved necessary by operations in Afghanistan)
BMP-1 (Command): two command models whose proper designations remain unknown are those designated **BMP M1974** and **BMP M1978** by the US Army; the former retains armament but has a revised interior with mapboards, extra radios etc, while the latter (**BMP-1KSh**) is unarmed and has extra radios
BMP-1 (Reconnaissance): known to the US Army as the **BMP-R** or **BMP M1976**, this version has a 73-mm gun in a two-man turret; the **BMP M1976/2** improved model has a parabolic antenna above the turret
BMP-1 ARV: this armoured recovery vehicle is a Czech development with a 1500-kg (3,309-lb) capacity crane in place of the turret
PRP-3: this artillery fire-control version is known to the US Army as the **BMP M1975** and has a crew of five, a two-man turret, radar above the hull rear, and armament of one 7.62-mm machine-gun
BMP-PPO: this training model has eight roof cupolas for trainees, but no turret

BMD-1
(USSR)
Type: tracked airborne combat vehicle
Crew: 7
Combat weight: 6700 kg (14,771 lb)
Dimensions: length 5.40 m (17.72 ft); width 2.63 m (8.63 ft); height 1.97 m (6.46 ft)
Armament system: one 73-mm (2.87-in) 2A28 smooth-bore gun with 40 rounds, one co-axial 7.62-mm (0.3-in) PKT machine-gun with 2,000 rounds and one launcher rail for two AT-3 'Sagger' anti-tank missiles in an electrically-powered turret, and two bow-mounted 7.62-mm (0.3-in) PKT machine-guns; the type can generate smoke by injecting fuel into the exhaust system
Armour: welded steel varying in thickness from 7 mm (0.28 in) to 23 mm (0.91 in)
Powerplant: one 224-kW (300-hp) 5D20 diesel engine with 300 litres (66 Imp gal) of internal fuel
Performance: speed, road 70 km/h (43.5 mph) and water 10 km/h (6.2 mph) driven by two waterjets; range, road 320 km (199 miles); fording amphibious; gradient 60%; vertical obstacle 0.8 m (31.5 in); trench 1.6 m (5.25 ft); ground clearance 0.45 m (17.7 in)

Variants
BMD-1: originally known in the West as the **M1970**, the BMD-1 was designed specially for the Soviets' large airborne arm, and is a useful vehicle within its limitations of small size, comparatively light protection and the same turret as the BMP-1 with its low-pressure 73-mm (2.87-in) gun; troop accommodation is cramped even for three men, who cannot use their personal weapons from inside the vehicle, but extra firepower is provided by two flexible bow-mounted machine-guns; standard equipment includes an NBC system and night-vision devices; variants of the basic model include one without the 'Sagger' installation over the 73-mm gun replaced by an AT-4 'Spigot' launcher on the right of the turret, and the **BMD-1 M1981/1** with a 30-mm cannon in place of the 73-mm gun
BMD-2: originally designated **BMD M1979** in the West, this is a slightly lengthened version with six rather than five road wheels; the type was developed concurrently with the BMD-1 as a transport rather than assault vehicle, and has appeared in several subtypes such as the **BMD-2 M1979/1** armoured personnel carrier (two 7.62-mm bow machine-guns, and rear accommodation for two 30-mm AGS-17 grenade launchers plus their crews), and the **BMD-2KSh (BMD M1979/3)** command post model (no bow machine-guns or turret, but carrying extra radios and 'Clothes Rail' folding antenna round the superstructure); the BMD-2 has a crew of 1+9, a combat weight of 7500 kg (16,534 lb) and a length of 5.95 m (19.52 ft); power is provided by one 224-kW (300-hp) 5D20 diesel for a road speed of 60 km/h (37.3 mph) and a road range of 500 km (311 miles); the type's maximum armour thickness is 16 mm (0.63 in)
BMD Mortar: mortar-carrier model with an 82-mm (3.2-in) mortar in the troop compartment

BTR-70
(USSR)
Type: wheeled armoured personnel carrier
Crew: 2+9
Combat weight: 11500 kg (25,353 lb)

Dimensions: length 7.85 m (25.75 ft); width 2.80 m (9.19 ft); height 2.45 m (8.04 ft)
Armament system: one 14.5-mm (0.57-in) KPV heavy machine-gun with 500 rounds and one co-axial 7.62-mm (0.3-in) PKT machine-gun with 2,000 rounds in a manually-powered turret
Armour: welded steel
Powerplant: two 86-kW (115-hp) ZMZ-4905 petrol engines with about 350 litres (77 Imp gal) of internal fuel, and driving an 8 x 8 layout
Performance: speed, road 80 km/h (50 mph) and water 10 km/h (6.2 mph) driven by one waterjet; range, road 600 km (373 miles); fording amphibious; gradient 60%; vertical obstacle 0.5 m (19.7 in); trench 2.0 m (6.6 ft); ground clearance not revealed

Variants
BTR-70: designed in the 1960s as successor to the BTR-60P, the BTR-70 entered service in the late 1960s and initially received the Western designations **SPW-70** and **BTR M1970**; the embarked infantry can use their personal weapons from inside the vehicle, and standard equipment includes a front-mounted winch, a central tyre pressure regulation system, and an NBC system
BTR-70/AGS-17: version seen in Afghanistan with an AGS-17 30-mm grenade-launcher on the roof
BTR-70MS: turretless version optimized for communications
BTR-70KShM: command post version
BREM: armoured recovery vehicle with a bow-mounted crane and other items of specialized equipment
BTR-80: updated and improved development of the BTR-70, but with the two petrol engines of the earlier model replaced by one 194-kW (260-hp) V-8 diesel engine for greater reliability and range combined with reduced fire risk; the one-man turret of the BTR-70 is retained, though the 14.5-in (0.57-in) heavy machine-gun can be elevated to +60 rather than +30 degrees for improved capability against low-flying aircraft and battlefield helicopters; located on the rear of the turret is a bank of six smoke-dischargers; other improvements include better firing ports and hatches for the embarked troops

BTR-60PB
(USSR)
Type: wheeled armoured personnel carrier
Crew: 2+14
Combat weight: 10300 kg (22,707 lb)
Dimensions: length 7.56 m (24.80 ft); width 2.825 m (9.27 ft); height to turret top 2.13 m (6.99 ft)
Armament system: one 14.5-mm (0.57-in) KPV heavy machine-gun with 500 rounds and one co-axial 7.62-mm (0.3-in) PKT machine-gun with 2,000 rounds in a manually-powered turret

Armour: welded steel varying in thickness from 5 to 9 mm (0.20 to 0.35 in)
Powerplant: two 67-kW (90-hp) GAZ-49B petrol engines with 290 litres (64 Imp gal) of internal fuel, and driving an 8 x 8 layout
Performance: speed, road 80 km/h (50 mph) and water 10 km/h (6.2 mph) driven by one waterjet; range, road 500 km (311 miles); fording amphibious; gradient 60%; side slope 40%; vertical obstacle 0.4 m (15.75 in); trench 2.0 m (6.6 ft); ground clearance 0.475 m (18.7 in)

Variants
BTR-60P: this vehicle was developed in the 1950s as successor to the BTR-152 in motorized rifle divisions, and entered service in 1961; though clearly better than the non-amphibious BTR-152, the BRT-60P has an open-topped crew and troop compartment (with firing ports so that personal weapons can be used from inside the vehicle), and the armament is a single pintle-mounted 7.62-mm SGMB or PKB machine-gun, sometimes replaced by a 12.7-mm (0.5-in) DShKM heavy machine-gun
BTR-60PA: improved model introduced in 1961 and featuring an enclosed troop compartment with NBC protection; this variant can carry 2+16 men (though 2+12 is more common), has a combat weight of 9980 kg (22,002 lb)
BTR-60PB: essentially the BTR-60PA fitted with the same turret and armament as used on the BRDM-2, BTR-70 and Czech OT-64; the nominal troop capacity is seldom used, a more normal complement being 2+8; there is also a forward air control version of this model, the armament being removed from the turret and replaced by a transparent cover so that the FAC officer can undertake forward observation and call in aircraft by means of the additional radio equipment carried in the troop compartment
BTR-60PU: command post version of the BTR-60P with mapboards and extra radio equipment

BTR-50PK
(USSR)
Type: tracked armoured personnel carrier
Crew: 2+20
Combat weight: 14200 kg (31,305 lb)
Dimensions: length 7.08 m (23.23 ft); width 3.14 m (10.30 ft); height to hull top 1.97 m (6.46 ft)
Armament system: one pintle-mounted 7.62-mm (0.3-in) SGMB machine-gun with 1,250 rounds
Armour: welded steel varying in thickness from 6 to 10 mm (0.24 to 0.39 in)
Powerplant: one 179-kW (240-hp) diesel engine with 400 litres (88 Imp gal) of internal fuel
Performance: speed, road 44 km/h (27.3 mph) and water 11 km/h (6.8 mph) driven by two waterjets; range, road 400 km (249 miles); fording amphibious;

gradient 70%; side slope 40%; vertical obstacle 1.1 m (43.3 in); trench 2.8 m (9.2 ft); ground clearance 0.37 m (14.6 in)

Variants
BTR-50P: developed in the middle of the 1950s on the chassis of the PT-76 light tank, the BTR-50P was the Soviet army's standard tracked APC until the advent of the BMP-1; the type suffers from a severe tactical limitation in that the crew compartment is open-topped, the troops embarking and disembarking over the sides; standard equipment includes limited night-vision equipment; ramps at the hull rear allow a 57-mm (2.24-in) ZIS-2 anti-tank gun or 85-mm (3.35-in) D-44 field gun to be loaded onto the vehicle
BTR-50PA: derivative without loading ramps and often armed with a 14.5-mm (0.57-in) KPVT machine-gun
BTR-50PK: improved model with covered troop compartment and an NBC system
BTR-50PU: command post model with extra radios and mapboards etc in the troop compartment; there are two subvariants, the **BTR-50PU Model 1** with a single projecting bay at the rear, and the **BTR-50PU Model 2** with two projecting bays
MTK: mine-clearing model of the BTR-50PK with a launcher to fire explosive-filled tubes designed to clear paths through minefields
MTP: technical support model of the BTR-50PK with a workshop built onto the rear and also used for the front-line delivery of fuel
OT-62A: Czech APC version also known as the **TOPAS** and similar in appearance to the BTR-50PU; this model is unarmed
OT-62B: improved Czech model fitted with a small turret (on the right) housing a 7.62-mm (0.3-in) vz 59T machine-gun and accommodating on its right an 82-mm (3.2-in) T-21 recoilless rifle; this variant is also known as the **TOPAS-2A**
OT-62C: version used by the Polish army and known as the **TOPAS-2AP**; this has a centrally-mounted turret for one 14.5-mm KPVT and one 7.62-mm KPT machine-guns; the type can also carry two four-man mortar teams each with its 82-mm mortar and bombs
OT-62 Ambulance: ambulance model
OT-62 Command: this is a simple development of the OT-62 series with additional radios
WPT-TOPAS: Polish recovery vehicle development of the OT-62A with a winch, hand-operated crane and one 7.62-mm PK machine-gun in addition to personal weapons and an RPG-7 anti-tank rocket-launcher

MT-LB
(USSR)
Type: tracked multi-purpose vehicle
Crew: 2+11
Combat weight: 11900 kg (26,235 lb)
Dimensions: length 6.454 m (21.17 ft); width 2.85 m (9.35 ft); height to turret top 1.865 m (6.12 ft)
Armament system: one 7.62-mm (0.3-in) PKT machine-gun with 2,500 rounds in a manually-powered turret
Armour: welded steel varying in thickness from 7 to 14 mm (0.28 to 0.55 in)
Powerplant: one 179-kW (240-hp) YaMZ 238 V diesel engine with 450 litres (99 Imp gal) of internal fuel
Performance: speed, road 61.5 km/h (38.2 mph) and water 6 km/h (3.7 mph) driven by its tracks; range, road 500 km (311 miles); fording amphibious; gradient 60%; side slope 30%; vertical obstacle 0.7 m (27.6 in); trench 2.7 m (8.9 ft); ground clearance 0.4 m (15.75 in)

Variants
MT-LB: developed from the MT-L unarmoured carrier in the later part of the 1960s as successor to the AT-P tracked artillery tractor, the MT-LB was initially known in the West as the **M1970** and has since been seen in a number of roles including that of armoured personnel carrier in snowy or marshy conditions, ambulance, artillery tractor, cargo carrier, command vehicle, engineer vehicle (with rear-mounted plough for use in reverse) and artillery fire-control vehicle; standard equipment includes an NBC system and night-vision devices, but there is only limited capability for the embarked troops to use their personal weapons from inside the vehicle
MT-LBV: MT-LB fitted with tracks 565 mm (22.24 in) wide rather than the standard 350 mm (13.78 in) wide to reduce ground pressure (by almost half) and so facilitate operation in snowy or marshy conditions
MT-LBU: command post version with extra radios, land navigation equipment and a tented rear extension
MT-SON: artillery fire-control version with 'Pork Trough' radar mounted on the roof
MT-LB/'Big Fred': artillery- and mortar-locating version with 'Big Fred' radar mounted on the roof
MTP-LB: turretless repair version with a front-mounted A-frame, crane, winch and many items of specialized repair kit
RKhM: chemical reconnaissance vehicle with a redesigned superstructure and dispenser for lane-marking

BTR-152V1
(USSR)
Type: wheeled armoured personnel carrier
Crew: 2+17
Combat weight: 8950 kg (19,731 lb)
Dimensions: length 6.83 m (22.41 ft); width 2.32 m (7.61 ft); height 2.05 m (6.73 ft)
Armament system: one pintle-mounted 12.7-mm (0.5-in) DShKM heavy machine-gun with ? rounds, or one 7.62-mm (0.3-in) SGMB machine-gun with 1,250 rounds

Armour: welded steel varying in thickness from 4 to 13.5 mm (0.16 to 0.53 in)
Powerplant: one 82-kW (110-hp) ZIL-123 petrol engine with 300 litres (66 Imp gal) of internal fuel, and driving a 6 x 6 layout
Performance: speed, road 75 km/h (47 mph); range, road 780 km (485 miles); fording 0.8 m (2.6 ft); gradient 55%; vertical obstacle 0.6 m (23.6 in); trench 0.69 m (2.25 ft); ground clearance 0.295 m (11.6 in)

Variants
BTR-152: introduced in 1951, this was the Soviet army's first APC after World War II, and is based on the chassis and automotive system of the ZIL-151 truck; it is still in limited service but suffers the grave disadvantages on the modern battlefield of lacking NBC protection, amphibious capability, and overhead cover for the embarked troops who can, however, use their personal weapons from inside the vehicle; the vehicle's machine-gun can be supplemented by two side-mounted 7.62-mm SGMB machine-guns
BTR-152V1: improved model based on the chassis and automotive system of the ZIL-157 truck, and featuring a front-mounted winch and central tyre pressure regulation system with external air lines
BTR-152V2: model without winch
BTR-152V3: model with internal air lines, winch and infra-red driving lights
BTR-152K: much improved BTR-152V3 with overhead protection for the troop compartment
BTR-152A: AA vehicle with a manually-operated turret for two 14.5-mm (0.57-in) KPV machine-guns; variations on this theme include Egyptian vehicles with a Czech M53 mounting for four 12.7-mm DShKM machine-guns, and Palestine Liberation Organization vehicles with a Soviet ZU-23 mounting for two 23-mm cannon
BTR-152U: command post model with raised superstructure and windows

BTR-40
(USSR)
Type: wheeled armoured personnel carrier
Crew: 2+8
Combat weight: 5300 kg (11,584 lb)
Dimensions: length 5.00 m (16.40 ft); width 1.90 m (6.23 ft); height without armament 1.75 m (5.74 ft)
Armament system: one pintle-mounted 7.62-mm (0.3-in) SGMB machine-gun with 1,250 rounds
Armour: welded steel to a maximum thickness of 8 mm (0.315 in)
Powerplant: one 60-kW (80-hp) GAZ-40 petrol engine with 120 litres (26.4 Imp gal) of internal fuel, and driving a 4 x 4 layout
Performance: speed, road 80 km/h (50 mph); range, road 285 km (177 miles); fording 0.8 m (2.6 ft); gradient 60%; side slope 30%; vertical obstacle 0.47 m (18.5 in); trench 0.7 m (2.3 ft) with channels; ground clearance 0.275 m (10.8 in)

Variants
BTR-40: developed shortly after the BTR-152, the BTR-40 was based on the GAZ-63 truck and designed primarily for reconnaissance and command post duties, entering service in 1951; the type is still widely used by Soviet clients in the Middle East and Africa despite the fact that it lacks amphibious capability, overhead cover for the troop compartment, night-vision devices and even the central tyre pressure regulation system otherwise standard on Soviet front-line vehicles
BTR-40A: AA version in a manually-operated turret for two 14.5-mm (0.57-in) KPV heavy machine-guns with 2,400 rounds
BTR-40B: improved APC with overhead cover for the troop compartment
BTR-40kh: chemical reconnaissance version with provision for the fixing of marker poles
BTR-40/'Sagger': East German training version with an overhead triple launcher for AT-3 'Sagger' anti-tank missiles

AT-P
(USSR)
Type: tracked artillery tractor and armoured personnel carrier
Crew: 3+6
Combat weight: 6000 kg (13,228 lb)
Dimensions: length 4.45 m (14.60 ft); width 2.50 m (8.20 ft); height 1.83 m (6.00 ft)
Armament system: one bow-mounted 7.62-mm (0.3-in) SGMT machine-gun
Armour: welded steel to a maximum thickness of 12 mm (0.47 in)
Powerplant: one 82-kW (110-hp) ZIL-123F petrol engine with ? litres (? Imp gal) of internal fuel
Performance: speed, road 50 km/h (31 mph); range, road 500 km (311 miles); fording 0.7 m (2.3 ft); gradient 60%; side slope 30%; vertical obstacle 0.7 m (27.6 in); trench 1.22 m (4.0 ft); ground clearance 0.3 m (11.8 in)

Variants
AT-P: this vehicle was introduced in the 1950s as a tractor for artillery pieces of up to 122-mm (4.8-in) calibre; it can also carry a 1200-kg (2,645-lb) payload, but as it lacks overhead cover, NBC protection and amphibious capability has generally been replaced by the MT-LB
AT-P (Command): improved version with a commander's cupola, overhead protection and extra radio equipment
AT-P (Reconnaissance): artillery reconnaissance version with full-width rear compartment and specialized equipment

GKN Sankey MCV-80 Warrior
(UK)
Type: tracked infantry combat vehicle
Crew: 2+8
Combat weight: 52,910 lb (24000 kg)
Dimensions: length 20.80 ft (6.34 m); width 9.95 ft (3.034 m); height overall 8.97 ft (2.735 m)
Armament system: one 30-mm ROF L21 Rarden cannon with ? rounds, one co-axial McDonnell Douglas Helicopters 7.62-mm (0.3-in) EX-34 Chain Gun with ? rounds, and four smoke-dischargers on each side of the electrically-powered turret
Armour: welded aluminium hull and welded steel turret
Powerplant: one 550-hp (410-kW) Rolls-Royce CV8 TCA diesel engine with 170 Imp gal (772 litres) of internal fuel
Performance: speed, road 47 mph (75 km/h); range, road 310 miles (499 km); fording 4.25 ft (1.3 m); gradient 60%; side slope 40%; vertical obstacle 30 in (0.76 m); trench 8.25 ft (2.5 m); ground clearance 19.25 in (0.49 m)

Variants

MCV-80 Warrior: due to enter service with the British army in the late 1980s, the MCV-80 is the UK's first true mechanized infantry combat vehicle, and though featuring good armament, protection and performance, has no provision for the embarked troops to use their personal weapons from inside the vehicle, and also lacks amphibious capability; standard equipment includes an NBC system and night-vision devices; the manufacturer has proposed a whole series of derivatives, and the most significant of these are the types itemized below, for which the British army has a perceived need
MCV/C: command post vehicle with additional radios etc
MCV/CRV: combat repair vehicle armed with a 7.62-mm L7 machine-gun and fitted with an extending-jib crane
MCV/MOR: mortar-carrier version with a 7.62-mm L7 machine-gun and a rear-mounted 81-mm (3.1-in) mortar
MCV/MRV: mechanized recovery vehicle
MCV/P: platoon command vehicle armed with a 7.62-mm L7 machine-gun

GKN Sankey FV432
(UK)
Type: tracked armoured personnel carrier
Crew: 2+10
Combat weight: 33,685 lb (15280 kg)
Dimensions: length 17.23 ft (5.25 m); width 9.19 ft (2.80 m); height including machine-gun 7.50 ft (2.29 m)
Armament system: one cupola-mounted 7.62-mm (0.3-in) L7A2 machine-gun with 1,600 rounds and three smoke-dischargers on each side of the hull front, or one 7.62-mm (0.3-in) L7A2 machine-gun with 1,600 rounds and four smoke-discharger on each side of an electrically-powered turret
Armour: welded steel varying in thickness from 6 to 12 mm (0.24 to 0.47 in)
Powerplant: one 240-hp (179-kW) Rolls-Royce K60 No.4 Mk 4F multi-fuel engine with 100 Imp gal (454 litres) of internal fuel
Performance: speed, road 32.5 mph (52.3 km/h) and water 4.1 mph (6.6 km/h) driven by its tracks and supported by a flotation screen; fording 3.5 ft (1.07 m) without preparation and amphibious with preparation; gradient 60%; vertical obstacle 24 in (0.61 m); trench 6.75 ft (2.06 m); ground clearance 16 in (0.41 m)

Variants

FV432: evolved from the FV420 series of armoured vehicles which failed to enter production in the 1950s, the FV432 APC carrier appeared in the early 1960s, production beginning in 1963 of the **FV432 Mk 1** with a highly noticeable exhaust on the left of the vehicle; successive models were the **FV432 Mk 1/1** and **FV432 Mk 2** with exhausts running over the roof, and the current **FV432 Mk 2/1** with a flush rather than protruding NBC pack; the type has night-vision devices, and can be made amphibious with a simple kit; there is no provision for the embarked troops to use their personal weapons from inside the vehicle, or even to survey the terrain before disembarking; a few FV432s with the 30-mm Rarden cannon-armed turret of the FV721 Fox were produced, and these unwieldy vehicles serve with the British Army's Berlin Brigade; other variants (often without formal designation) are itemized below
FV432 Ambulance: unarmed and with provision for a crew of two plus four litters or two litters and five sitting wounded
FV432 Cargo: this model can carry a payload of 8,100 lb (3674 kg)
FV432 Carl Gustav: anti-tank model with a bar above the troop compartment for the 84-mm (3.31-in) Carl Gustav rocket-launcher; a comparable model carries the Milan anti-tank system, the launcher and missiles being disembarked for firing
FV432 Command: weighing 34,170 lb (15500 kg), this version has a crew of seven, two mapboards, extra radio equipment and a tended rear extension
FV432 Minelayer: this tows a Bar minelayer able to lay anti-tank mines at the rate of 600 per hour; a Ranger anti-personnel minelayer can be fitted on the hull roof so that combined anti-tank/anti-personnel minefields can be laid
FV432 Mortar: this carries the 81-mm (3.2-in) L16 mortar, a crew of six and 160 bombs at a combat weight of 36,155 lb (16400 kg)
FV432 Radar: this carries either the Marconi ZB 298

short-range battlefield surveillance radar or the EMI Cymbeline medium-range mortar-locating radar

FV432 Recovery: this has a rear-mounted winch of 40,340-lb (18298-kg) capacity

FV432 Royal Artillery: this is a fire-control vehicle fitted with Marconi Field Artillery Computer Equipment (FACE) and a Plessey sound ranging system; as with other Royal Artillery-operated members of the series, the armament is a single 7.62-mm Bren Gun

FV432 Royal Engineers: this tows the Giant Viper mine-clearing system

FV434: this is the maintenance carrier version operated by REME and designed for the field changing of MBT powerpacks with the air of a crane

FV438: this is an anti-tank variant with two launchers and 16 Swingfire anti-tank missiles

FV439: operated by the Royal Signals, this specialized communications model has additional equipment including parts of the Ptarmigan and Wavell systems

Alvis Stormer
(UK)
Type: tracked armoured personnel carrier
Crew: 3+8
Combat weight: 26,025 lb (11805 kg)
Dimensions: length 17.67 m (5.38 m); width 7.87 ft (2.40 m); height to top of machine-gun 7.87 ft (2.49 m)
Armament system: one cupola-mounted 7.62-mm (0.3-in) L7A2 machine-gun
Armour: welded aluminium
Powerplant: one 250-hp (186-kW) Perkins T6.3544 diesel engine with 73 Imp gal (332 litres) of internal fuel
Performance: speed, road 45 mph (72 km/h) and water 6 mph (9.6 km/h) driven by one propeller or 4 mph (6.4 km/h) driven by its tracks; range, road 330 miles (531 km); fording 3.6 ft (1.1 m) without preparation and amphibious with preparation; gradient 60%; vertical obstacle 18 in (0.46 m); trench 5.9 ft (1.8 m); ground clearance 16.75 in (0.43 m)

Variants
Stormer: this is the Alvis development of the FV4333 armoured vehicle designed in the late 1970s by the Military Vehicles and Engineering Establishment, the development being centred on the incorporation of proved components from the Scorpion CVR(T) series and the evolution of the type as the core of a large number of variants (air-defence vehicle, mortar carrier, anti-tank vehicle, missile launcher, maintenance and recovery vehicle, engineer vehicle, electronic warfare vehicle, command vehicle, logistics vehicle, ambulance, squad carrier, mobile protected gun/light assault vehicle, assault gun, low-profile 75-mm/2.95-in gun vehicle and low-profile 90-mm/3.54-in gun vehicle); of the vehicles ordered by Malaysia some have the Helio FVT900 turret armed with a 20-mm Oerlikon cannon and co-axial 7.62-mm machine-gun, and the others the Thyssen Henschel TH-1 turret with twin 7.62-mm machine-guns; the British army has ordered the type as the launch vehicle for the new Shorts Starstreak hypervelocity SAM, which is under development as the point defence AA weapon of British field formations

Alvis FV603 Saracen
(UK)
Type: 2+10
Combat weight: 22,420 lb (10170 kg)
Dimensions: length 17.17 ft (5.23 m); width 8.33 ft (2.54 m); height to turret top 8.08 ft (2.46 m)
Armament system: two 7.62-mm (0.3-in) machine-guns (one L3A3 in a manually-powered turret and one Bren Gun on the hull-rear ring mounting) with 3,000 rounds, and three smoke-dischargers on each side of the hull front
Armour: welded steel varying in thickness from 8 to 16 mm (0.315 to 0.63 in)
Powerplant: one 160-hp (119-kW) Rolls-Royce B80 Mk 6A petrol engine with 44 Imp gal (200 litres) of internal fuel, and driving a 6 x 6 layout
Performance: speed, road 45 mph (72 km/h); range, road 250 miles (402 km); fording 3.5 ft (1.07 m) without preparation and 6.5 ft (1.98 m) with preparation; gradient 42%; vertical obstacle 18 in (0.46 m); trench 5.0 ft (1.52 m); ground clearance 17 in (0.43 m)

Variants
FV603: designed in the late 1940s as part of the British army's first post-war generation of light AFVs, the Saracen APC entered production in 1952 and has since seen extensive worldwide service; the type lacks NBC protection and night-vision devices, and must be regarded as obsolete by modern standards despite the fact that the crew can fire their personal weapons from inside the vehicle, unlike the situation in later British APCs; the initial **FV603A Saracen Mk 1** vehicles had the B80 Mk 3A engine, but by the time of the **FV603B Saracen Mk 2** the B80 Mk 6A had become standard, and this engine was retained in the **FV603C Saracen Mk 3** with reverse-flow cooling for hot-climate operations; Saracens used in Northern Ireland for security operations are generally fitted with wire screens to defeat the HEAT warheads of RPG-7 anti-tank rockets, and often have water cannon or other riot-control weapons in place of the machine-guns

FV604: command post version with mapboards, extra radio equipment and a tented rear extension for a six-man crew

FV610(A): Royal Artillery command vehicle with

specialist Royal Artillery equipment and a tented rear extension for its six-man crew
FV611: ambulance model with a crew of two and accommodation for 10 sitting wounded, or three litters and two sitting wounded, or two litters and six sitting wounded

Alvis FV103 Spartan
(UK)
Type: tracked armoured personnel carrier
Crew: 3+4
Combat weight: 18,015 lb (8172 kg)
Dimensions: length 16.81 ft (5.125 m); width 7.40 ft (2.26 m); height 7.48 ft (2.28 m)
Armament system: one cupola-mounted 7.62-mm (0.3-in) L37A1 machine-gun with 3,000 rounds, and four smoke-dischargers mounted on each side of the hull front
Armour: welded aluminium
Powerplant: one 190-hp (142-kW) Jaguar J60 No.1 Mk 100B petrol engine with 85 Imp gal (386 litres) of internal fuel
Performance: speed, road 50 mph (80 km/h); range, road 300 miles (483 km); fording 3.5 ft (1.07 m) without preparation and amphibious with preparation; gradient 60%; vertical obstacle 19.75 in (0.5 m); trench 6.75 ft (2.06 m); ground clearance 14 in (0.36 m)

Variants
FV102 Striker: entering service in 1975, the Striker is the specialist anti-tank missile variant of the Scorpion CVR(T) series, and is based on the hull of the FV103; this version has a crew of three, a combat weight of 18,400 lb (8346 kg) and an armament of one cupola-mounted L37A1 machine-gun and one quintuple launcher with 10 Swingfire anti-tank missiles
FV103 Spartan: this is the APC variant of the Scorpion CVR(T) series, and entered service in 1978 for specialist transport tasks such as the carriage of Blowpipe SAM teams and Royal Engineer assault pioneer teams, and for the resupply of the FV102 Striker with Swingfire missiles; other projected variants on the Spartan theme include an AA model (mounting a turret with two 20-mm cannon or four Blowpipe-type missiles) and an anti-tank model (with a TOW launcher and nine missiles, or an HCT turret with four HOT missiles, or an MCT turret with 10 Milan missiles)
FV104 Samaritan: this is the ambulance variant of the Scorpion CVR(T) series and entered production in 1978; the hull is basically similar to that of the FV105 Sultan, and can carry a crew of two plus four litters, or two litters and three sitting wounded, or five sitting wounded; the combat weight is 19,100 lb (8664 kg)
FV105 Sultan: this is the command variant of the Scorpion CVR(T) series and has a combat weight of 19,100 lb (8664 kg) with a crew of five or six; two radios are carried, and the raised superstructure carries mapboards and other command equipment; there is a cupola-mounted 7.62-mm L37A1 machine-gun for local defence
FV106 Samson: this is the armoured recovery variant of the Scorpion CVR(T) series, and has a combat weight of 19,265 lb (8739 kg); the type entered service in 1978, and specialist equipment includes a winch, A-frame and role-dedicated tools and appliances
Streaker: this is the Alvis private-venture high-mobility load-carrier variant of the Scorpion CVR(T) series, and can carry a load of 6,615 lb (3001 kg); the type has a flat rear deck and the option of a light unarmoured or heavier armoured cab

GKN Sankey AT105-P Saxon
(UK)
Type: wheeled armoured personnel carrier
Crew: 2+8/10
Combat weight: 23,525 lb (10671 kg)
Dimensions: length 16.96 ft (5.17 m); width 8.17 ft (2.49 m); height to commander's cupola 8.62 ft (2.63 m)
Armament system: one cupola-mounted 7.62-mm (0.3-in) L7A1 machine-gun
Armour: welded steel
Powerplant: one 164-hp (122-kW) Bedford 600 diesel engine with 35 Imp gal (160 litres) of internal fuel, and driving a 4 x 4 layout
Performance: speed, road 60 mph (96 km/h); range, road 315 miles (507 km); fording 3.67 ft (1.12 m); gradient 60%; vertical obstacle 16 in (0.41 m); ground clearance 11.5 in (0.29 m)

Variants
AT105-A: ambulance model based on the AT-105P and able to carry four litters with a crew of four, or two litters and four sitting wounded with a crew of three
AT105-E: upgraded version of the AT105-P with a turret-mounted 7.62-mm machine-gun and optional riot-control equipment
AT105-P: this is the basic rifle section transport model derived from the internal-security AT104, and entered service in 1985; a large number of optional fits are possible, including night-vision devices, internal-security equipment and revised armament
AT105-Q: command version of the AT105-P with interior fitted out as a command post, and armed with one or two turret-mounted 7.62-mm machine-guns
AT105/ARWEN: internal-security version with a turret-mounted ARWEN 37V riot-control weapon, grenade-launchers and other items of internal-security equipment

AT105 Recovery: recovery version with a side-mounted winch and specialist equipment

Humber FV1611 'Pig'
(UK)
Type: wheeled armoured personnel carrier
Crew: 2+6/8
Combat weight: 12,765 lb (5790 kg)
Dimensions: length 16.16 ft (4.93 m); width 6.71 ft (2.04 m); height 6.96 ft (2.12 m)
Armament system: generally none
Armour: welded steel
Powerplant: one 120-hp (89-kW) Rolls-Royce B60 Mk 5A petrol engine with 32 Imp gal (145 litres) of internal fuel, and driving a 4 x 4 layout
Performance: speed, road 40 mph (64 mph); range, road 250 miles (402 km)

Variants
FV1609: produced as an interim vehicle pending deliveries of the FV603 Saracen, the open-topped FB1609 was produced in the late 1940s by converting FV1601A 1-ton trucks to the APC role with armoured bodies
FV1611: definitive model with troop compartment provided with overhead cover; although obsolete by front-line standards, these vehicles have a useful internal-security role in Northern Ireland as they are less overtly military than later vehicles; FV1611s have thus been fitted with a wide assortment of internal-security equipment and uparmoured to a weight of some 15,435 lb (7001 kg) to make their bodies impenetrable by high-velocity rifle rounds
FV1612: radio vehicle with crew of three
FV1613: ambulance model with a crew of two and accommodation for three litters, one litter and four sitting wounded, or eight sitting wounded

FMC M2 Bradley
(USA)
Type: tracked infantry fighting vehicle
Crew: 3+7
Combat weight: 49,800 lb (22590 kg)
Dimensions: length 21.17 ft (6.45 m); width 10.50 ft (3.20 m); height to turret roof 8.42 ft (2.565 m)
Armament system: one 25-mm M242 McDonnell Douglas Chain Gun with 900 rounds, one co-axial 7.62-mm (0.3-in) M240C machine-gun with 2,340 rounds, one retractable two-tube launcher with seven TOW missiles, and two smoke-dischargers on each side of the turret; the turret is electrically-powered, the main gun has stabilization in elevation and azimuth, and a day/night optical fire-control system is fitted in addition to the TOW sight and control unit; the type can generate smoke by injecting fuel into the exhaust system
Armour: welded aluminium, aluminium appliqué and laminate

Powerplant: one 500-hp (373-kW) Cummins VTA-903T turbocharged diesel engine with 175 US gal (662 litres) of internal fuel
Performance: speed, road 41 mph (66 km/h) and water 4.5 mph (7.2 km/h) driven by its tracks; range, road 300 miles (483 km); fording amphibious; gradient 60%; side slope 40%; vertical obstacle 36 in (0.91 m); trench 8.33 ft (2.54 m); ground clearance 17 in (0.43 m)

Variant
M2 Bradley: designed as a mechanized infantry combat vehicle for the US Army from the early 1960s this series originated with the XM701 experimental vehicle of 1965 and the Armored Infantry Fighting Vehicle of 1967 onwards; neither of these types or their derivatives found favour with the US Army, which issued its definitive specification in 1972, and this resulted in the XM723 prototype that eventually led through tortuous development and political programmes to the Fighting Vehicles System comprising the M2 Infantry Fighting Vehicle, the M3 Cavalry Fighting Vehicle, and the launch vehicle for the General-Support Rocket System (now known as the Multiple-Launch Rocket System); the first XM2 prototypes were built in 1978 and in December 1979 the type was standardized as the M2, with production deliveries beginning in 1981; the M2 and M3 are designed to scout for and support the M1 Abrams MBT on advanced-technology battlefields, and are thus designed for maximum crew protection and decisive offensive firepower; standard equipment includes night-vision devices, an NBC system and provision for the embarked troops to fire their personal weapons from inside the vehicle; offensive punch is augmented by the embarked troops' three Light Anti-tank Weapons; the basic vehicle is now being upgraded with the improved TOW 2 launcher, better 25-mm ammunition and several detail modifications, and future plans call for improved navigation and night-vision capability, and superior protection against biological/chemical agents and electromagnetic pulse effects

FMC M3 Bradley
(USA)
Type: tracked cavalry fighting vehicle
Crew: 3+2
Combat weight: 49,480 lb (22444 kg)
Dimensions: length 21.17 ft (6.45 m); width 10.50 ft (3.20 m); height to turret roof 8.42 ft (2.565 m)
Armament system: one 25-mm M242 McDonnell Douglas Chain Gun with 1,500 rounds, one co-axial 7.62-mm (0.3-in) M240C machine-gun with 4,540 rounds, one retractable two-tube launcher for 12 TOW anti-tank missiles, and two smoke-dischargers on each side of the turret; the turret is electrically-powered, the main gun is stabilized in elevation and

azimuth, and a day/night optical fire-control system is fitted in addition to the TOW sight and control unit; the type can generate smoke by injecting fuel into the exhaust system
Armour: welded aluminium and laminate
Powerplant: one 500-hp (373-kW) Cummins VTA-903T turbocharged diesel engine with 175 US gal (662 litres) of internal fuel
Performance: speed, road 41 mph (66 km/h) and water 4.5 mph (7.2 km/h) driven by its tracks; range, road 300 miles (483 km); fording amphibious; gradient 60%; side slope 40%; vertical obstacle 36 in (0.91 m); trench 8.33 ft (2.54 m); ground clearance 17 in (0.43 m)

Variant
M3 Bradley: this counterpart of the M2 in the Fighting Vehicle System is designed for reconnaissance with enhanced anti-tank capability, and carries two rather than seven embarked infantrymen; as a result more ammunition of all types can be carried, but the embarked infantry have no provision to fire their personal weapons from inside the vehicle; other modifications include lack of anti-mine applique armour, but the same improvements being made to the M2 are also being made to the M3

FMC Armored Infantry Fighting Vehicle
(USA)
Type: tracked armoured infantry fighting vehicle
Crew: 3+7
Combat weight: 30,175 lb (13687 kg)
Dimensions: length 17.25 ft (5.26 m); width 9.25 ft (2.82 m); height to turret roof 8.60 ft (2.62 m)
Armament system: one 25-mm Oerlikon KBA-BO2 cannon with 324 rounds and one co-axial 7.62-mm (0.3-in) FN MAG machine-gun with 1,940 rounds in an electro-hydraulically powered turret with optional main gun stabilization, and six smoke-dischargers on the hull front
Armour: welded aluminium and spaced laminate steel
Powerplant: one 264-hp (197-kW) Detroit Diesel 6V-53T turbocharged diesel engine with 110 US gal (416 litres) of internal fuel
Performance: speed, road 38 mph (61.2 km/h) and water 3.9 mph (6.3 km/h) driven by its tracks; range, road 305 miles (491 km); fording amphibious; gradient 60%; side slope 30%; vertical obstacle 25 in (0.635 m); trench 5.33 ft (1.625 m); ground clearance 17 in (0.43 m)

Variants
AIFV (Belgium): Belgium was the second country to adopt the AIFV, and the designation of this series is **AIFV-B**, which is operated by the Belgian forces in a number of forms alongside the closely-related M113A1-B APC series; the major differences in the Belgian series are the individual NBC air supply, suspension modelled on that of the M113A1-B, an engine compartment fire-extinguishing system, and a troop compartment heater; the three models of the AIFV-B are the **AIFV-B-25 mm** squad transport (with a turret-mounted 25-mm cannon, a crew of seven, provision for a Milan anti-tank missile launcher and two Lyran flare launchers), the **AIFB-B-.50** squad transport (with a cupola-mounted 0.5-in/12.7-mm Browning M2HB heavy machine-gun), and the **AIFV-B-CP** command post model (with a pintle-mounted 0.5-in Browning M2HB heavy machine-gun and troop compartment fitted out for command post operations with a crew of seven)
AIFV (Netherlands): the Netherlands was the first country to adopt the AIFV, which was developed by FMC in the mid-1960s on the basis of the M113 armoured personnel carrier with a power-operated turret and firing ports for the embarked infantry; this XM765 led to the Product Improved M113A1 of 1970 with a centrally-mounted turret and the commander's cupola behind it; trials confirmed the unsuitability of this arrangement, and the type was redesigned as the AIFV with side-by-side turret and commander's cupola (the former on the right and the latter on the left); the first Dutch order was placed in 1975, and deliveries of what the Dutch designate the **YPR 765 PRI** began in 1977; this is the standard AIFV, and variants are the **YPR 765 PRCO-B** command post model (crew of nine and combat weight of 30,200 lb/ 13699 kg), the **YPR 765 PRCO-C** multi-role version armed with a 0.5-in M2HB and possessing a nine-man crew at a combat weight of 27,335 lb/12399 kg (**YPR 765 PRCO-C1** for battalion command, **YPR 765 PRCO-C2** for battalion gunnery control, **YPR 765 PRCO-C3** for mortar fire-control, **YPR 765 PRCO-C4** for AA fire-control and **YPR 765 PRCO-C5** for observation post duties), the **YPT 765 PRRDR** battlefield surveillance model with ZB 298 radar, the **YPR 765 PRRDR-C** radar and command model, the **YPR 765 PRGWT** ambulance with provision for four litters, the 10-man **YPR 765 PRI/I** squad vehicle with cupola-mounted 0.5-in M2, the **YPR 765 PRMR** tractor for the 120-mm (4.72-in) Brandt mortar and 51 bombs, the **YPR 765 PRVR** 4,500-lb (2041-kg) payload cargo vehicle, the **YPR 765 PRAT** anti-tank vehicle with Emerson TOW launcher system and the **YPR 806 PRBRG** armoured recovery vehicle with roof-mounted crane
AIFV (Philippines): these are basically similar to the Dutch YPR 765 PRIs but have a 0.5-in M2HB heavy machine-gun instead of the Oerlikon cannon
Daewoo KIFV: this South Korean vehicle appears to be based closely on the AIFV, but uses British automatic transmission, aluminium armour and other components, while it is powered by a 209-kW (280-hp) MAN D-2848M diesel for maximum road and water speeds of 74 and 6 km/h (46 and 3.7 mph) at a combat weight of 12900 kg (28,439 lb); the KIFV

has a crew of 3+9, and the current armament comprises one 0.5-in (12.7-mm) Browning M2 heavy machine-gun in a turret at the right rear, one 7.62-mm (0.3-in) M60 machine-gun and a bank of six smoke-dischargers; heavier armament installations are being studied, and current variants include a mortar carrier, an armoured recovery vehicle with a roof-mounted hydraulic crane, an AA vehicle with a 20-mm Vulcan cannon, a command post vehicle and an ammunition carrier

FMC M113A1
(USA)
Type: tracked armoured personnel carrier
Crew: 2+11
Combat weight: 24,595 lb (11156 kg)
Dimensions: length 15.95 ft (4.86 m); width 8.81 ft (2.69 m); height 8.20 ft (2.50 m)
Armament system: one cupola-mounted 0.5-in (12.7-mm) Browning M2HB heavy machine-gun with 2,000 rounds
Armour: welded aluminium varying in thickness from 0.5 to 1.5 in (12.7 to 38.1 mm)
Powerplant: one 215-hp (160-kW) Detroit Diesel 6V-53 diesel engine with 95 US gal (360 litres) of internal fuel
Performance: speed, road 42 mph (67.6 km/h) and water 3.6 mph (5.8 km/h) driven by its tracks; range, road 300 miles (483 km); fording amphibious; gradient 60%; vertical obstacle 24 in (0.61 m); trench 5.5 ft (1.68 m); ground clearance 16 in (0.41 m)

Variants
M113: development of this most important Western armoured personnel carrier began in 1956 to a US Army requirement for an APC that would be amphibious and air-portable, yet well protected and possessing good cross-country performance, and capable of development into a variety of other forms; in 1960 the T113E1 trials vehicle was standardized as the 22,615-lb (10258-kg) M113 with the 209-hp (156-kW) Chrysler 75M petrol engine and 80 US gal (302 litres) of fuel for a range of 200 miles (322 km); there is no provision for the embarked troops to fire their personal weapons from inside the vehicle; US Army procurement was in the order of 5,300 vehicles and many experimental and/or trials developments have been made (in the USA and elsewhere) of the basic model and its developments
M113A1: standardized in 1963, the M113A1 has a diesel engine and increased fuel capacity for greater range with less danger of fire; US Army procurement was in the order of 12,700 vehicles
M113A2: designed as the **Product Improved M113A1**, this 25,000-lb (11341-kg) variant is being produced in the period up to 1989 by converting some 18,000 M113s and M113A1s and procuring 2,660 new-build vehicles; the main alterations are improvement of engine cooling and exhaust, plus the strengthening and improvement of the suspension and shock absorption systems to give greater ground clearance and better cross-country performance
M113A2 Recovery Vehicle: this is a simple derivative of the M113A2 with a small roof-mounted crane, a rear-mounted winch, and two or three rear-mounted spades; this model was originally known as the **M806**, and there was also an **M806A1** based on the M113A1
M113A3: from 1987 M113A1s and M113A2s will be further upgraded to M113A3 standard with developments for greater reliability and performance; these derive mainly from the use of the 275-hp (205-kW) 6V-53T turbocharged diesel with a more modern transmission system, though other improvements are armoured external fuel tanks, and anti-spall linings in the crew compartments
M48 Chaparral: this is a light AA missile-launch vehicle using a chassis derived from that of the M548 cargo carrier fitted at the rear with a turntable-mounted four-arm launcher (for a total of 12 MIM-72 SAMs derived from the widely-deployed IR-homing AIM-9 Sidewinder air-to-air missile) and the gunner's turret; once alerted by the MPQ-49 forward-area alerting radar or by a visual sighting, the gunner slews the missile mounting to follow the target until an aural bleep indicates that the missile seeker has locked-on, whereupon the missile is fired for autonomous homing; the engagement envelope includes a maximum range of 6,500 yards (5945 m) and minimum/maximum altitudes of 150/10,000 ft (46/3050 m)
M106: mortar-carrier version with a 107-mm (4.2-in) M30 mortar, mounted on a turntable in the erstwhile troop compartment, and 93 bombs; there is also a 26,445-lb (11995-kg) **M106A1** version with 88 bombs; both models have the cupola-mounted M2HB machine-gun but only 1,000 rounds
M113 Dozer: this is similar to the standard APC but has a front-mounted dozer blade; there is also an **M113A1 Dozer** version
M125: mortar-carrier version with an 81-mm (3.2-in) M29 mortar, mounted on a turntable in the erstwhile troop compartment, with 114 bombs and the same machine-gun armament as the M106; there are also **M125A1** and **M125A2** versions, the former having a combat weight of 24,825 lb (11261 kg)
M163: AA version fitted with an M168 Vulcan 20-mm cannon on a powered mounting and provided with 2,100 rounds, lead-computing sight and VPS-2 range-only radar; this model weighs 27,140 lb (12311 kg), has including turret of 8.98 ft (2.74 m) and has a four-man crew; the cannon has two rates of fire, 1,000 rounds per minute for the surface role to a range of 3,300 yards (3020 m), and 3,000 rounds per minute for the AA role to an effective range of 5,250 ft (1600 m); the limitations of the type are lack of

gun range, and restriction to clear-weather operation by lack of search and tracking radar; there are several add-on improvement packages to produce the **Autotrack Vulcan Air Defense System**, the **Product Improved Vulcan Air Defense System** and the **Improved Fire-Control System/Vulcan Air Defense System**

M548: unarmoured cargo-carrier version based on the M113A1 with a payload of 12,000 lb (5443 kg); the equivalent of the M113A2 is the **M548A1**, and currently under investigation is a stretched model with a payload of 16,000 lb (7258 kg); the M548 chassis is also the basis for the **M48** Chaparral SAM system, **Tracked Rapier** SAM system, **MSM-Fz** anti-tank minelayer for the West German army, **M752** launch vehicle for the Lance battlefield missile, and **M688** loader/transporter support vehicle for the M752

M577: five-man command post version which entered service in 1962; this variant has a higher roof over the previous troop compartment, extra radios and generating capacity, and a tented rear extension; the armament is a single 7.62-mm (0.3-in) machine-gun with 1,000 rounds, and improved models are the 25,380-lb (11513-kg) **M577A1** and the **M577A2**

M901 Improved TOW Vehicle: this anti-tank model entered service in 1979, and is an M113A2 fitted with a sighting cupola, a retractable two-tube launcher/sight assembly and 12 TOW missiles; local defence is entrusted to a single 7.62-mm machine-gun with 1,000 rounds

Lynx: this is a Canadian development of the basic M113A1 with a lower superstructure and equipment suiting it for the command and reconnaissance roles; the crew is three, and the combat weight 19,345 lb (8775 kg); armament comprises one 0.5-in M2HB heavy machine-gun with 1,000 rounds (mounted externally on a small forward turret) and one 7.62-mm machine-gun with 2,000 rounds (pintle-mounted at the rear of the roof); there are three smoke-dischargers on each side of the turret; the Dutch use a similar vehicle under the designation **M113 C & R**, though many of these vehicles have been retrofitted with an Oerlikon GBD-AOA turret carrying a 25-mm KBA-B cannon with 200 rounds; the performance of these two variants is usefully better than that of the basic M113 series, with a maximum speed of 44 mph (70.8 km/h) and a range of 325 miles (523 km)

M113 (Australia): the Australian army uses many M113 models, most of them with slight modifications to suit local conditions and requirements; the most different model is the **M113A1 FSV (Scorpion)** fire-support vehicle fitted with the turret and 76-mm (3-in) L23 gun of the British Scorpion reconnaissance vehicle; this variant is 15.96 ft (4.86 m) long, and is fitted with extra buoyancy aids; others of the M113 series are fitted with the Cadillac Gage T50 turret armed with two 7.62-mm machine-guns (5,000 rounds) or one 0.5-in and one 7.62-mm machine-guns (2,500 and 3,000 rounds respectively)

M113 (Belgium): Belgium uses the M113A1 under the designation **M113A1-B** with local modifications such as an engine compartment fire-extinguishing system, suspension to M113A2 standard, individual NBC-protected air supply to each crew member, and a different heater/air-conditioning system; the variants in Belgian service are the **M113A1-B-AMB** ambulance (crew of three and accommodation for four litters, or six sitting wounded, or two litters and three sitting wounded), **M113A1-B-ATK** anti-tank and squad transport vehicle (crew of 10, provision for an anti-tank missile launcher, and a cupola-mounted 0.5-in machine-gun plus a pintle-mounted 7.62-mm machine-gun), **M113A1-B-CP** command post vehicle (crew of five, mapboards and a tented rear extension), **M113A1-B-ENG** assault pioneer vehicle (crew of eight, demolition charges and a self-recovery winch), **M113A1-B-MIL** dedicated anti-tank vehicle (crew of five, provision for a Milan anti-tank launcher, and a cupola-mounted 0.5-in machine-gun), **M113A1-B-MTC** maintenance vehicle (crew of five, roof-mounted crane and winch), **M113A1-B-REC** recovery vehicle (crew of four, roof-mounted crane, specialist equipment and self-recovery winch) and **M113A1-B-TRG** training model (two clear canopies)

M113 (Brazil): Brazil's 700 M113 series vehicles are being re-engined with the Mercedes-Benz OM 352A diesel

M113 (Israel): Israeli M113 series vehicles have been extensively modified to remove amphibious capability, allow the carriage of up to four 0.5-in or 7.62-mm machine-guns, and permit the carriage of personal kit in hull-side baskets; other modifications are appliqué armour to resist HEAT warheads, and different armament including a 60-mm (2.36-in) Soltam mortar or TOW/Mapats anti-tank missile launcher

M113 (New Zealand): many of New Zealand's M113A1s are fitted with the same T50 turret as the Australian vehicles

M113 (Norway): Norway's variant is the **NM-135**, essentially the M113A1 fitted with a Swedish Hägglund & Söner turret containing a Rheinmetall MK 20 Rh 202 20-mm cannon and co-axial 7.62-mm machine-gun

M113 (Singapore): these vehicles have been modified locally into carriers for 81- and 120-mm mortars

M113 (Spain): these are generally similar to their US counterparts, though the mortar-carriers use Spanish weapons

M113 (Switzerland): Swiss **Schützenpanzer 63** vehicles are similar to the M113A1, though the installation of the Hägglund & Söner turret results in the designation **Schützenpanzer 63/73**; other Swiss vehicles in the series are the **Feuerleitpanzer 63** artillery command vehicle, **Geniepanzer 63** armoured bulldozer, **Kommandopanzer 63** and upgunned **Kommandopanzer 63/73** higher-echelon command vehicles, **Kranpanzer 63** repair vehicle, **Minenwerferpanzer 64** mortar-carrier, **Schützenpanzer Kommando 63** low-echelon command vehicle and **Ubermittlungspanzer 63** radio vehicle

M113 (Taiwan): the M113A1 has been used as the basis for the development of an armoured infantry fighting vehicle in Taiwan; this new vehicle combines the basic hull and systems of the M113 with a 215-hp (160-kW) Perkins 8.640 diesel engine driving a Taiwanese transmission system, additional armour (a laminate steel layer over the front and sides of the vehicle, sandwiching polyurethane foam for additional buoyancy), and a shielded 0.5-in (12.7-mm) Browning M2HB heavy machine-gun at the front of the vehicle; the result is a vehicle with the basic capabilities of the M113A2; the type is also being developed as a true AIFV with 20- or 30-mm cannon, rocket-launch vehicle with the Kung Feng IV 126-mm (4.96-in) multiple launcher, anti-tank vehicle with US or Taiwanese missiles, mortar carrier (81-, 107- and 120-mm/3.2-, 4.2- and 4.72-in types), and command post vehicle

M113 (West Germany): West Germany uses large numbers of the M113 series, most of them brought up to M113A1 standard, and substantial quantities locally modified for special tasks such as mortar location (Green Archer radar), battlefield surveillance (RATAC radar) and mortar-carrying (120-mm Tampella mortar)

FMC M59
(USA)
Type: tracked armoured personnel carrier
Crew: 2+10
Combat weight: 42,600 lb (19323 kg)
Dimensions: length 18.42 ft (5.61 m); width 10.71 ft (3.26 m); height to top of cupola 9.08 ft (2.77 m)
Armament system: one cupola-mounted 0.5-in (12.7-mm) Browning M2HB heavy machine-gun with 2,205 rounds
Armour: welded steel varying in thickness from 10 to 16 mm (0.39 to 0.63 in)
Powerplant: two 127-hp (95-kW) General Motors Model 302 petrol engines with 137 US gal (518 litres) of internal fuel
Performance: speed, road 32 mph (51.5 km/h) and water 4.3 mph (6.9 km/h) driven by its tracks; range, road 102 miles (164 km); fording amphibious; gradient 60%; vertical obstacle 18 in (0.46 m); trench 5.5 ft (1.68 m); ground clearance 18 in (0.46 m)

Variants
M59: produced as successor to the M75 and standardized in December 1953, the M59 is a massive and decidedly underpowered and under-ranged vehicle; it is amphibious only in calm water, and lacks the capability for the embarked troops to use their personal weapons from the inside of the vehicle; the shortcomings of this vehicle prompted the development of the far more satisfactory M113
M84: mortar-carrier version carrying the 107-mm (4.2-in) M30 mortar and 88 bombs

FMC/International Harvester M75
(USA)
Type: tracked armoured personnel carrier
Crew: 2+10
Dimensions: length 17.04 ft (5.19 m); width 9.33 ft (2.84 m); height to top of commander's cupola 9.10 ft (2.78 m)
Armament system: one pintle-mounted 0.5-in (12.7-mm) Browning M2HB heavy machine-gun with 1,800 rounds
Armour: welded steel varying in thickness from 0.375 to 1 in (9.5 to 25.4 mm)
Powerplant: one 295-hp (220-kW) Continental AO-895-4 petrol engine with 150 US gal (568 litres) of internal fuel
Performance: speed, road 44 mph (71 km/h); range, road 115 miles (185 km); fording 4.0 ft (1.22 m) without preparation and 6.67 ft (2.03 m) with preparation; gradient 60%; vertical obstacle 18 in (0.46 m); trench 5.5 ft (1.68 m); ground clearance 18 in (0.48 m)

Variant
M75: designed in the late 1940s on the basis of the M41 light tank, the M75 APC was not regarded as a great success because of its height and lack of performance; there is no provision for the embarked troops to fire their personal weapons from inside the vehicle, and no provision for NBC protection is available

Cadillac Gage Commando V-300
(USA)
Type: multi-role wheeled armoured vehicle
Crew: 3+9
Combat weight: 28,960 lb (13136 kg)
Dimensions: length 21.00 ft (6.40 m); width 8.33 ft (2.54 m); height to hull top 6.50 ft (1.98 m)
Armament system: see Variants (below)
Armour: welded steel
Powerplant: one 235-hp (175-kW) VT-504 diesel engine with 70 US gal (265 litres) of internal fuel, and driving a 6 x 6 layout
Performance: speed, road 57 mph (92 km/h) and water 3 mph (4.8 km/h) driven by its wheels; range, road 435 miles (700 km); fording amphibious; gradient 60%; side slope 30%; vertical obstacle 24 in (0.61 m); ground clearance 14 in (0.36 m)

Variants
Commando V-300: a logical development of the 4 x

LAND WEAPONS 69

4 Commando V-150 series, the 6 x 6 Commando V-300 offers greater capabilities plus superior cross-country performance with a basic hull and automotive system capable of accepting a number of armament options; development began in the late 1970s, and the first production vehicles were delivered to Panama in 1983; other options are an NBC system, night-vision devices, heating and air-conditioning; the current armament options are a series of Cadillac Gage one- and two-man turrets mounting a 90-mm (3.54-mm) Cockerill Mk III rifled gun, or a 76-mm (3-in) Royal Ordnance L23A1 rifled gun, or a 25-mm M242 McDonnell Douglas Chain Gun plus two 7.62-mm (0.3-in) machine-guns (one co-axial and one AA) and smoke-dischargers, or a 20-mm Oerlikon 204 GK cannon, or twin 0.5-in (12.7-mm) Browning M2HB or 7.62-mm machine-guns (one or one of each); other options under development are an anti-tank version with TOW missiles, a mortar-carrier with an 81-mm (3.2-in) mortar and 60 bombs, an ambulance, a recovery vehicle, an AA vehicle and a command vehicle

Commando V-300A: improved version designed for the fire-support role with a combat weight of 37,500 lb (17010 kg) and incorporating a three-man Cadillac Gage turret armed with the ROF Nottingham 105-mm (4.13-in) Low-Recoil Force gun; the crew is four, and powered by a 270-hp (201-kW) Detroit Diesel Allison VT-504 diesel the type has a maximum road speed of 52 mph (84 km/h) and a range of 400 miles (644 km)

Cadillac Gage Commando V-150/20-mm Turret (USA)
Type: multi-role wheeled armoured vehicle
Crew: 3+9
Combat weight: 21,800 lb (9888 kg)
Dimensions: length 18.67 ft (5.69 m); width 7.41 ft (2.26 m); height to hull top 6.50 ft (1.98 m)
Armament system: see Variants (below)
Powerplant: one 202-hp (151-kW) diesel engine with 80 US gal (303 litres) of internal fuel, driving a 4 x 4 layout
Performance: speed, road 55 mph (88.5 km/h) and water 3 mph (4.8 km/h) driven by its wheels; range, road 400 miles (644 km); fording amphibious; gradient 60%; side slope 30%; vertical obstacle 24 in (0.61 m); ground clearance 15 in (0.38 m)

Variants
Commando V-100: this original model appeared in 1963, with deliveries beginning in the following year; this 12-man model weighed 16,250 lb (7371 kg) and could attain 62 mph (100 km/h) on its 200-hp (149-kW) Chrysler 362 petrol engine
Commando V-200: stretched version of the V-100 produced in very small numbers, the Commando V-200 had a weight of 28,065 lb (12730 kg) and could attain 60 mph (96 km/h) on its 275-hp (205-kW) Chrysler 440 CID engine
Commando V-150: this model appeared in 1971 and replaced both the V-100 and V-200 series in production; the type can be used as a standard APC with front and rear pintle-mounted machine-guns (0.5-in/12.7-mm Browning M2HB heavy machine-gun and/or 7.62-mm/0.3-in medium machine-gun), and optional equipment includes an NBC system, smoke-dischargers, troop compartment firing ports, winch etc; the type can also be fitted with a number of turreted armament options as detailed below
Commando V-150/MG Turret: this carries a one-man Cadillac Gage turret fitted with two 7.62-mm machine-guns or one 0.5-in and one 7.62-mm machine-guns; this version carries a crew of 3+7, and has a weight of 19,800 lb (8981 kg) and overall height 8.33 ft (2.54 m)
Commando V-150/1-m MG Turret: this carries a Cadillac Gage 1-metre turret fitted with two 0.5-in or 7.62-mm machine-guns, or with one of each calibre; this version carries a crew of 3+7, and has a weight of 19,300 lb (8754 kg) and overall height of 8.50 ft (2.59 m)
Commando V-150/1-m Cannon Turret: this carries a one-man Cadillac Gage turret fitted with one 20-mm Oerlikon 204 GK cannon with 400 rounds and a co-axial 7.62-mm machine-gun with 3,200 rounds; this version carries a crew of 3+5, and has a weight of 20,000 lb (9072 kg) and an overall height of 9.50 ft (2.90 m)
Commando V-150/20-mm Turret: this version carries a two-man Cadillac Gage turret armed with one 20-mm Oerlikon 204 GK cannon with 400 rounds and two 7.62-mm machine-guns (one co-axial and one AA) with 3,200 rounds; this version carries a crew of 3+2, and has a weight of 21,800 lb (9888 kg) and an overall height of 8.33 ft (2.54 m)
Commando V-150/25-mm Turret: this version carries a two-man turret armed with one 25-mm M242 McDonnell Douglas Chain Gun with 630 rounds and two 7.62-mm M240C machine-guns (one co-axial and one AA) with 1,600 rounds; delivered as the **Commando V-150S** for the Light Armored Vehicle competition, this version carries a crew of 3+9, and has a weight of 23,000 lb (10433 kg) and a length of 10.14 ft (6.14 m)
Commando V-150/40-mm Turret: this version carries a one-man turret armed with a Mk 19 40-mm grenade-launcher, a co-axial 0.5-in heavy machine-gun and four smoke-dischargers mounted on each side of the turret
Commando V-150 ADV: this anti-aircraft version carries a 20-mm Vulcan rotary-barrel cannon system
Commando V-150/75-mm Turret: this version carries the two-man turret of the British Scorpion CVR(T) with its 76-mm (3-in) L23A1 gun and 41 rounds, plus two 7.62-mm machine-guns (one co-axial and one

AA) and 3,200 rounds; this version carries a crew of 3+3, and has a weight of 21,800 lb (9888 kg)
Commando V-150/90-mm Turret: this version carries a two-man turret fitted with a 90-mm (3.54-mm) Cockerill Mk III gun and 39 rounds, plus two 7.62-mm machine-guns (one co-axial and one AA) and 3,200 rounds; the crew is 3+4, and the weight 21,800 lb (9888 kg)
Commando V-150/81-mm Mortar: this version has a turntable-mounted M29 81-mm (3.2-in) mortar with 62 bombs in the erstwhile troop compartment now fitted with concertina roof-top hatches; the crew is five and the weight 19,500 lb (8845 kg)
Commando V-150/TOW: this anti-tank version is similar to the mortar-carrier version, but has a TOW launcher and seven missiles; provision is made for the location of a 7.62-mm machine-gun (2,000 rounds) on any of four positions; the crew is 4+2 and the weight 19,750 lb (8958 kg)
Commando V-150/Command: this variation on the standard model had a pod located over the roof opening, with provision for the location of a 7.62-mm machine-gun (2,000 rounds) in any of the four sides; the crew is 3+7 and the overall height 7.58 ft (2.31 m)
Commando V-150/Base Security: this is similar to the mortar-carrier version in having concertina roof hatches, and carries a crew of 3+8; a 7.62-mm machine-gun (3,000 rounds) or 40-mm grenade-launcher can be located at any of four positions, and the weight is 20,000 lb (9072 kg)
Commando V-150/ERV: this police emergency rescue vehicle has a crew of 3+9, and the roof pod has eight gun ports and provision on the roof hatch for a 7.62-mm or 0.5-in machine-gun; the weight is 20,200 lb (9162 kg)
Commando V-150/Recovery: this is the winch-fitted recovery vehicle of the series, and armament comprises a 7.62-mm or 0.5-in machine-gun plus 2,200 rounds
Vulcan-Commando: version of the Commando V-150 fitted with a 20-mm Vulcan cannon turret (the same as that fitted to the M163 derivative of the M113 APC) and three internally-controlled stabilizing jacks; the type has a crew of four and weighs 22,500 lb (10206 kg)

Cadillac Gage Commando Ranger
(USA)
Type: wheeled armoured personnel carrier
Crew: 2+6
Combat weight: 10,000 lb (4536 kg)
Dimensions: length 15.42 ft (4.70 m); width 6.62 ft (2.02 m); height 6.50 ft (1.98 m)
Armament system: see Variant (below)
Armour: welded steel
Powerplant: one 180-hp (134-kW) Dodge 360 CID petrol engine with 32 US gal (121 litres) of internal fuel, and driving a 4 x 4 layout
Performance: speed, road 70 mph (113 km/h); range, road 345 miles (556 km); fording not revealed; gradient 60%; side slope 30%; vertical obstacle 10 in (0.25 m); ground clearance 8 in (0.20 m)

Variant
Commando Ranger: this vehicle was designed in the late 1970s as a low-cost security vehicle, and deliveries to the US Air Force began in 1980; the type can be used as a standard APC (with the option for one or two turreted 7.62-mm/0.3-in or 0.5-in/12.7-mm machine-guns), or as a command, ambulance or reconnaissance vehicle

Verne Dragoon 300
(USA)
Type: multi-role wheeled armoured vehicle
Crew: 3+6
Combat weight: 28,000 lb (12701 kg)
Dimensions: length 18.33 ft (5.59 m); width 8.00 ft (2.44 m); height to turret periscopes 8.67 ft (2.64 m)
Armament system: see Variants (below)
Armour: welded steel
Powerplant: one 300-hp (224-kW) Detroit Diesel 6V-53T turbocharged diesel engine with 90 US gal (341 litres) of internal fuel, and driving a 4 x 4 layout
Performance: speed, road 72 mph (116 km/h) and water 3 mph (4.8 km/h) driven by its wheels; range, road 650 miles (1046 km); fording amphibious; gradient 60%; side slope 30%; vertical obstacle 39 in (0.99 m); ground clearance 15 in (0.38 m)

Variants
Dragoon 300 APC: this basic armoured personnel carrier type designed and produced by the Verne Corporation can also be used as a reconnaissance, command, recovery and escort vehicle with a ring- or pintle-mounting 7.62-mm (0.3-in) or 0.5-in (12.7-mm) machine-gun; as with the other variants, there is a wide range of optional equipment such as NBC protection, night-vision devices and the like, and the manufacturer has proposed a large family of associated vehicle types offering the customer great commonality of parts and operating techniques, with consequent logistical and training advantages
Dragoon 300 LWC: this light weapons carrier type can be armed with two 7.62-mm or 0.5-in machine-guns
Dragoon 300 SWC: this special weapons carrier type can be armed with an 81-mm (3.2-in) mortar, TOW anti-tank missile launcher and other such weapons
Dragoon 300 HWC: this heavy weapons carrier type can be armed with turreted weapons such as 20-, 25- and 30-mm cannon, or 76-mm (3-in) and 90-mm

(3.54-in) rifled guns together with co-axial and AA machine-guns

M2A1
(USA)
Type: halftrack armoured personnel carrier
Crew: 2+8
Combat weight: 19,600 lb (8891 kg)
Dimensions: length with winch 10.13 ft (6.14 m); width with mine racks 7.29 ft (2.22 m); height overall 8.83 ft (2.69 m)
Armament system: one pintle-mounted 0.5-in (12.7-mm) Browning M2HB heavy machine-gun with 700 rounds, and one pintle-mounted 7.62-mm (0.3-in) machine-gun with 7,750 rounds
Armour: welded and bolted steel varying in thickness from 0.25 to 0.5 in (6.4 to 12.7 mm)
Powerplant: one 128-hp (95-kW) White 160 AX petrol engine with 60 US gal (227 litres) of internal fuel
Performance: speed, road 40 mph (64 km/h); range, road 175 miles (282 km); fording 2.67 ft (0.81 m); gradient 60%; vertical obstacle 12 in (0.30 m); ground clearance 9.2 in (0.23 m)

Variants
M2: this halftrack APC and reconnaissance vehicle was produced in the late 1930s by combining the chassis and automotive system of the M2 unarmoured halftrack with the armoured body of the M3A1 scout car, and the result was one of the most important vehicle types of World War II, many of whose examples are still in widespread service to this day despite their obsolescence by first-division standards; the total complement is 10, and the armament single 7.62-mm (0.3-in) and 0.5-in (12.7-mm) machine-guns on a skate mounting round the open-topped troop compartment
M2A1: as M2 but with a ring-mounted 0.5-in machine-gun; instead of the front-mounted winch this can be fitted with a roller to help cross-country mobility, this reducing overall length to 19.56 ft (5.96 m)
M3: similar to the M2 but developed by Diamond T with a longer body, pedestal mount for 7.62-mm machine-gun, and accommodation for 13 men
M3A1: as M3 but with ring mount
M3A2: successor to the M2, M2A1, M3 and M3A1 with accommodation for a maximum of 12, and armament comprising four pintle-mounted machine-guns in the troop compartment, and a ring-mounted machine-gun in the front right-hand side of the compartment
M5: similar to the M3 built by International Harvester
M5A1: equivalent of the M3A1
M5A2: equivalent of the M3A2
M9A1: International Harvester equivalent of the M2A1
M4: mortar-carrier version of the M2 with an 81-mm (3.2-in) mortar designed to be dismounted before firing; 97 bombs carried
M4A1: mortar-carrier with the 81-mm mortar designed for onboard firing
M13: M3 fitted with an AA mounting for two 0.5-in machine-guns
M14: M5 fitted with an AA mounting for two 0.5-in machine-guns
M15: M3 fitted with a turreted AA installation of one 37-mm M1A2 gun and two 0.5-in machine-guns
M15A1: lightened version of the M15
M16: M3 fitted with an AA mounting for four 0.5-in machine-guns and 5,000 rounds
M16A1: lightened version of the M16
M17: M5 fitted with an AA mounting for four 0.5-in machine-guns and 5,000 rounds
M21: 81-mm mortar-carrier version of the M3
M2/M3/M5/M9 (Israel): Israel still uses a large number of halftracks, most of them rebuilt to a common standard with the 172-hp (128-kW) Detroit Diesel 6V-53 diesel engine; these Israeli vehicles are used for 120-mm (4.72-in) mortar-carrying, for AA defence (pedestal or turreted 20-mm cannon), for anti-tank operations (106-mm/4.17-in recoilless rifle, SS.11 missile and 90-mm/3.54-in gun) and as ambulances, command vehicles, ammunition carriers and electronic warfare vehicles

FMC LVTP7
(USA)
Type: 3+25
Combat weight: 50,350 lb (22839 kg)
Dimensions: length 26.06 ft (7.94 m); width 10.73 ft (3.27 m); height to turret roof 10.25 ft (3.12 m)
Armament system: one 0.5-in (12.7-mm) Browning M2HB heavy machine-gun with 1,000 rounds in an electro-hydraulically powered turret
Armour: welded aluminium varying in thickness from 0.26 to 1.77 in (6.7 to 45 mm)
Powerplant: one 400-hp (298-kW) Detroit Diesel 8V-53T turbocharged diesel engine with 180 US gal (681 litres) of internal fuel
Performance: speed, road 40 mph (64 km/h) and water 8.4 mph (13.5 km/h) driven by two waterjets or 4.5 mph (7.2 km/h) driven by its tracks; range, road 300 miles (483 km); endurance, water 7 hours; fording amphibious; gradient 60%; side slope 60%; vertical obstacle 36 in (0.91 m); trench 8.0 ft (2.44 m); ground clearance 16 in (0.41 m)

Variants
LVTP7: this vehicle was developed from 1964 as successor to the US Marine Corps' LVTP5 series of amphibious assault vehicles, which were troublesome to maintain and possessed generally inadequate performance; deliveries began in 1971, and since that

time several variants have been produced; the type is not fitted with an NBC system, and the amphibious role precludes the provision of firing ports in the troop compartment; the type can be fitted out as an ambulance (six litters) or cargo carrier (payload of 10,000 lb/4536 kg)

LVTP7A1: the US Marine Corps' surviving vehicles have been modernized in the period 1982–85 to this upgraded standard, now designated **LVT7A1**, with a 400-hp Cummins VT 400 turbocharged diesel engine and 171 US gal (647 litres) of fuel, smoke-generating capability, passive night-vision devices, secure radio, improve fire-extinguishing systems, better ventilation, and updated electric and hydraulic systems; this modification increases combat weight to 52,770 lb (23936 kg) and maximum road speed to 45 mph (72.4 km/h); the type is also undergoing trials with revised armament, in the form of a new electrically-powered turret fitted with a 40-mm Mk 19 grenade-launcher and 0.5-in machine-gun

LVTC7: command version with the turret replaced by a hatch, and a crew of 12 including four command staff and five radio operators in a revised troop compartment; this version has a combat weight of 44,110 lb (20008 kg), and the armament is a single pintle-mounted 7.62-mm (0.3-in) M60D machine-gun with 1,000 rounds; the type is being upgraded to **LVTC7A1** standard

LVTR7: recovery version with the same hatch and armament as the LVTC7; this version has much specialist recovery and repair equipment, a roof-mounted hydraulic crane with extending jib and a high-capacity winch; the crew is five and the combat weight 51,440 lb (23333 kg); the type is being upgraded to **LVTR7A1** standard

Ingersoll (Borg-Warner) LVTP5A1
(USA)
Type: tracked amphibious assault vehicle
Crew: 3+34
Combat weight: 82,500 lb (37422 kg)
Dimensions: length 29.67 ft (9.04 m); width 11.70 ft (3.57 m); height to turret top 9.58 ft (2.92 m)
Armament system: one 0.3-in (7.62-mm) Browning M1919A4 machine-gun with 2,000 rounds in a manually-powered turret
Armour: welded steel varying in thickness from 0.25 to 0.62 in (6.37 to 15.86 mm)
Powerplant: one 810-hp (604-kW) Continental LV-1790-1 petrol engine with 456 US gal (1726 litres) of internal fuel
Performance: speed, road 30 mph (48 km/h) and water 6.8 mph (10.9 km/h) driven by its tracks; range, road 190 miles (306 km); fording amphibious; gradient 70%; side slope 60%; vertical obstacle 36 in (0.91 m); trench 12.0 ft (3.66 m); ground clearance 18 in (0.46 m)

Variants
LVTP5A1: in the 1960s all surviving LVTP5 amphibious assault vehicles (produced between 1952 and 1957) of the US Marine Corps were upgraded to this standard with a small snorkel and other improvements, the LVTP5 clearly having inadequate freeboard when fully loaded; the type can also be fitted out as an ambulance with accommodation for 12 litters; Taiwan is assessing the feasibility of re-engining its LVTP5A1s with a diesel powerplant for better range and reduced fire risk

LVTC5: command vehicle with the troop compartment outfitted as a command post with mapboards, extra radios etc

LVTE1: engineer vehicle fitted with a front-mounted dozer blade and rocket-propelled mineclearing system

LVTH6: fire-support version with a turret-mounted M49 105-mm (4.13-in) howitzer and 100 rounds, plus a co-axial 0.3-in machine-gun with 1,000 rounds and a 0.5-in AA machine-gun with 1,050 rounds

LVTR1: recovery version with a roof-mounted crane and winch, plus many items of specialist equipment

M-980
(Yugoslavia)
Type: tracked mechanized infantry combat vehicle
Crew: 2+8
Combat weight: 13000 kg (28,660 lb)
Dimensions: length 6.40 m (21.00 ft); width 2.59 m (8.50 ft); height over missile launcher 2.50 m (8.20 ft)
Armament system: one turret-mounted 20-mm Oerlikon HS 804 cannon with 400 rounds, one co-axial 7.62-mm (0.3-in) machine-gun with 2,250 rounds, and one twin-rail launcher for four AT-3 'Sagger' anti-tank missiles
Armour: welded steel varying in thickness from 8 to 30 mm (0.315 to 1.18 in)
Powerplant: one 194-kW (260-hp) Hispano-Suiza HS 115-2 turbocharged diesel engine with ? litres (? Imp gal) of internal fuel
Performance: speed, road 60 km/h (37.3 mph) and water 7.5 km/h (4.7 mph) driven by its tracks; range, road 500 km (311 miles); fording amphibious; gradient 60%; vertical obstacle 0.8 m (31.5 in); trench 2.2 m (7.2 ft); ground clearance 0.4 m (15.75 in)

Variant
M-980: this Yugoslav MICV appeared in the early 1970s, and its powerplant and automotive system are related closely to those of the French AMX-10P; the type is fitted as standard with an NBC system, and the embarked troops can fire their personal weapons from inside the vehicle

M-60P
(Yugoslavia)
Type: tracked armoured personnel carrier
Crew: 3+10
Combat weight: 11000 kg (24,250 lb)
Dimensions: length 5.02 m (16.47 ft); width 2.77 m (9.09 ft); height including machine-gun 2.385 m (7.82 ft)
Armament system: one cupola-mounted 0.5-in (12.7-mm) Browning M2HB heavy machine-gun, and one bow-mounted 7.92-mm (0.312-in) machine-gun
Armour: welded steel varying in thickness from 10 to 25 mm (0.39 to 0.98 in)
Powerplant: one 104-kW (140-hp) FAMOS diesel engine with 150 litres (33 Imp gal) of internal fuel
Performance: speed, road 45 km/h (28 mph); range, road 400 km (249 miles); fording 1.35 m (4.4 ft); gradient 60%; side slope 40%; vertical obstacle 0.6 m (23.6 in); trench 2.0 m (6.6 ft); ground clearance 0.4 m (15.75 in)

Variants
M-60P: developed in the early 1960s, the M-60P is an unremarkable APC lacking amphibious capability and NBC protection, though there is provision for the embarked troops to fire their personal weapons from inside the vehicle; the type also exists in ambulance, command, radio and recovery forms
M-60PB: anti-tank version with two 82-mm (3.23-in) recoilless rifles on a traversing mount above the left-hand side of the roof

ARTILLERY (SELF-PROPELLED AND TOWED)

122-mm SP howitzer (China): see NORINCO Type 63 APC

Model 77 DANA 152-mm SP howitzer
(Czechoslovakia)
Type: wheeled self-propelled howitzer
Crew: probably 3
Combat weight: 23000 kg (50,705 lb)
Dimensions: length, gun forward 10.40 m (34.12 ft) and hull 8.87 m (29.10 ft); width, turret 2.97 m (9.74 ft) and hull 2.722 m (8.93 ft); height to top of machine-gun 3.525 m (11.56 ft)
Armament system: one 152-mm (6-in) 2S3 howitzer with ? rounds of ready-use ammunition, and one 12.7-mm (0.5-in) DShKM AA heavy machine-gun in a hydraulically-powered turret; direct- and indirect-fire sights are fitted
Armour: welded steel
Powerplant: one 257-kW (345-hp) Tatra T3-930-51 multi-fuel engine with ? litres (? Imp gal) of internal fuel, and driving an 8 x 8 layout
Performance: speed, road 80 km/h (50mph); range, road 1000 km (621 miles); fording 1.4 m (4.6 ft); gradient 60%; vertical obstacle not revealed; trench 2.0 m (6.6 ft); ground clearance not revealed

Variant
DANA: this very useful SP howitzer was introduced into service in 1981, and is based on the chassis and automotive system of the Tatra 815 8 x 8 high-mobility truck, which offers the same type of cross-country mobility as a tracked chassis at considerably less cost; the howitzer crew and ammunition are carried in a separate vehicle, the Soviet-designed ordnance having automatic loading powered by a hydraulic motor which also traverses the turret through the frontal arc (traverse being limited by the jacks) and provides power for elevation between −3 and +60 degrees; for firing the vehicle is stabilized by three hydraulically-actuated jacks, and the ordnance can fire a 43.5-kg (95.9-lb) HE projectile to 17410 m (19,040 yards); other ammunition types are HE RAP (to a range of some 24000 m/26,245 yards), AP HE, illuminating and smoke

Royal Ordnance SP122 SP howitzer
(UK/Egypt)
Type: tracked self-propelled howitzer
Crew: 5
Combat weight: 44,092 lb (2000 kg)
Dimensions: length 25.25 ft (7.70 m); width 9.25 ft (2.82 m); height 8.83 ft (2.96 m)
Armament system: one 122-mm (4.8-in) D-30 howitzer with 80 rounds, and one 0.5-in (12.7-mm) Browning M2HB AA heavy machine-gun with 500 rounds; direct- and indirect-fire sights are fitted
Armour: welded steel
Powerplant: one 300-hp (224-kW) Perkins TV8.540 diesel engine with ? Imp gal (? litres) of internal fuel
Performance: speed, road 34 mph (55 km/h); range, road 186 miles (300 km); fording 3.25 ft (1,0 m); gradient 60%, side slope 30%, vertical obstacle 30 in (0.76 m); trench 7.25 ft (3.3 m); ground clearance not revealed

Variant
SP122: this is the British design in an Egyptian army competition for a new SP howitzer carrying the

locally-built version of the Soviet D-30 howitzer; the other competitor is an American entry from BMY based on the chassis of the M109 series of SP weapons; the SP122 has a fixed superstructure with the ordnance projecting through the front, allowing elevation from −5 to +70 degrees, and traverse 30 degrees left and right; the basic chassis is also to be considered for a number of other vehicle types, including an ammunition carrier, an ambulance, a command post and a recovery vehicle

GIAT GCT 155-mm SP gun
(France)
Type: tracked self-propelled gun
Crew: 4
Combat weight: 42000 kg (92,593 lb)
Dimensions: length, gun forward 10.25 m (33.63 ft) and hull 6.70 m (21.98 ft); width 3.15 m (10.33 ft); height to turret top 3.25 m (10.66 ft)
Armament system: one 155-mm (6.1-in) GIAT gun with 42 rounds of ready-use ammunition, one 7.62-mm (0.3-in) or 0.5-in (12.7-mm) AA machine-gun with 2,050 or 800 rounds, and two smoke-dischargers on the front of the hydraulically-powered turret; direct- and indirect-fire sights are fitted
Armour: welded steel
Powerplant: one 537-kW (720-hp) Hispano-Suiza HS 110 supercharged multi-fuel engine with 970 litres (213 Imp gal) of internal fuel
Performance: speed, road 60 km/h (37.3 mph); range, road 450 km (280 miles); fording 2.1 m (6.9 ft); gradient 60%; side slope 30%; vertical obstacle 0.93 m (36.6 in); trench 1.9 m (6.2 ft); ground clearance 0.42 m (16.5 in)

Variant
GCT: based on the chassis of the AMX-30 MBT, the GCT entered service in 1978 initially with Saudi Arabia; the vehicle uses an automatic loading system for the main ordnance, and this makes possible a high rate of fire using the onboard ammunition supply (eight rounds per minute, with a burst capability of six rounds in 45 seconds); reloading of the ammunition supply can be undertaken while the gun is in action, four men accomplishing a complete reload in 15 minutes; optional equipment includes an NBC system and specialized artillery items; the French designation is **155 AU F1**, and the L/40 ordnance can be elevated from −4 to +66 degrees in the 360-degree traverse turret; projectile types are HE, HE RAP, anti-tank mine, smoke and illuminating; typical range is 21200 m (23,185 yards) with the 43.75-kg (96.45-lb) HE projectile, though the 42.5-kg (93.7-lb) HE RAP attains 30500 m (33,355 yards); the type can also fire the US Copperhead cannon-launched guided projectile

Creusot-Loire Mk F3 155-mm SP gun/howitzer
(France)
Type: tracked self-propelled gun/howitzer
Crew: 2
Combat weight: 17400 kg (38,360 lb)
Dimensions: length, overall 6.22 m (20.41 ft); width 2.72 m (8.92 ft); height 2.085 m (6.84 ft)
Armament system: one 155-mm (6.1-in) ATS gun/howitzer with no ready-use ammunition on a limited-traverse mounting; direct- and indirect-fire sights are fitted
Armour: welded steel varying in thickness from 10 to 20 mm (0.39 to 0.79 in)
Powerplant: one 186-kW (250-hp) SOFAM 8Gbx petrol engine or one 209-kW (280-hp) Detroit Diesel 6V-53T turbocharged diesel engine with 450 litres (99 Imp gal) of internal fuel
Performance: speed, road 60 km/h (27.3 mph) with petrol engine or 64 km/h (40 mph) with diesel engine; range, road 300 km (186 miles) with petrol engine or 400 km (249 miles) with diesel engine; fording 1.0 m (3.3 ft); gradient 40%; vertical obstacle 0.6 m (23.6 in); trench 1.5 m (4.9 ft); ground clearance 0.48 m (18.9 in)

Variant
Mk F3: this self-propelled gun is based on the shortened chassis of the AMX-13 light tank, and since its introduction in the later 1950s has been offered in petrol- or diesel-engined models; kits are available to permit the operators of petrol-engined versions to convert their guns to diesel power; on its unprotected rear-mounted position the gun can be traversed 20 degrees left and 30 degrees right between 0 and +50-degree elevation, and 16 degrees left and 30 degrees right between +50 and +67-degree elevation; the vehicle is stabilized while firing by twin spades at the rear; the eight-man gun crew and 25 rounds are carried in an accompanying VCA tracked vehicle; the L/33 ordnance fires HE, HE RAP, smoke and illuminating projectiles, the 43.75-kg (96.45-lb) HE types reaching 20050 m (21,925 yards)

Creusot-Loire Mk 61 105-mm SP howitzer
(France)
Type: tracked self-propelled howitzer
Crew: 5
Combat weight: 16500 kg (36,357 lb)
Dimensions: length, overall 5.70 m (18.70 ft); width 2.65 m (8.69 ft); height to top of cupola 2.70 m (8.86 ft)
Armament system: one 105-mm (4.13-in) Modèle 50 howitzer with 56 rounds on a manually-powered limited-traverse mounting, and one or two 7.5- or 7.62-mm (0.295- or 0.3-in) machine-guns (one for AA defence and one optional for local defence) with 2,000 rounds; direct- and indirect-fire sights are fitted

Armour: welded steel varying in thickness from 10 to 20 mm (0.39 to 0.79 in)
Powerplant: one 186-kW (250-hp) SOFAM 8Gbx petrol engine with 415 litres (91 Imp gal) of internal fuel
Performance: speed, road 60 km/h (37.3 mph); range, road 350 km (217 miles); fording 0.8 m (2.6 ft); gradient 60%; side slope 30%; vertical obstacle 0.65 m (25.6 in); trench 1.6 m (5.2 ft); ground clearance 0.275 m (10.8 in)

Variant
Mk 61: designed in the late 1940s, the Mk 61 is now obsolescent in terms of chassis (derived from that of the AMX-13 light tank) and ordnance, and was accepted for French service in 1958; the type lacks NBC protection and amphibious capability, and the ordnance can be traversed 20 degrees left and right between −4.5 and +66-degree elevation on its mounting in the rear-mounted armoured fighting compartment; the L/23 or L/30 ordnance fires HE or HEAT projectiles, the former weighing 16 kg (35.3 kg) and attaining a range of 15000 m (16,405 yards, and the latter being capable of penetrating 350 mm (13.78 in) of armour

Soltam L-33 155-mm SP gun/howitzer
(Israel)
Type: tracked self-propelled gun/howitzer
Crew: 8
Combat weight: 41500 kg (91,490 lb)
Dimensions: length, gun forward 8.46 m (27.79 ft) and hull 6.47 m (21.23 ft); width 3.50 m (11.48 ft); height 3.45 m (11.32 ft)
Armament system: one 155-mm (6.1-in) Soltam M-68 gun/howitzer with 60 rounds on a manually-powered limited-traverse mounting, and one 7.62-mm (0.3-in) AA machine-gun; direct- and indirect-fire sights are fitted
Armour: welded steel varying in thickness from 16 to 64 mm (0.63 to 2.5 in)
Powerplant: one 460-hp (343-kW) Cummins VT 8-460-Bi diesel engine with 820 litres (180 Imp gal) of internal fuel
Performance: speed, road 38 km/h (23.6 mph); range, road 260 km (162 miles); fording 0.9 m (3.0 ft); gradient 60%; side slope 30%; vertical obstacle 0.91 m (36 in); trench 2.3 m (7.5 ft); ground clearance 0.43 m (17 in)

Variant
L-33: this simple yet effective Israeli development installs an M-68 gun/howitzer in a massive fixed fighting compartment on the chassis of an obsolete M4A3E8 Sherman medium tank with horizontal-volute spring suspension rather than vertical-volute spring suspension; the ordnance can traverse a total of 60 degrees on its mounting at the front of the fighting compartment, and elevate from −3 to +52 degrees, and the type has proved successful since it entered Israeli service in 1973; the L/33 ordnance fires all NATO ammunition, including a 43.7-kg (96.3-lb) HE projectile to 20000 m (21,870 yards)

EEFA Bourges M-50 155-mm SP howitzer
(France/Israel)
Type: tracked self-propelled howitzer
Crew: 8
Combat weight: 31000 kg (68,342 lb)
Dimensions: length, overall 6.10 m (20.01 ft); width 2.98 m (9.78 ft); height 2.80 m (9.19 ft)
Armament system: one 155-mm (6.1-in) Modèle 50 howitzer with ? rounds on a manually-powered limited-traverse mounting; direct- and indirect-fire sights are fitted
Armour: welded steel
Powerplant: one 450-hp (336-kW) Ford GAA petrol engine with 168 US gal (636 litres) of internal fuel
Performance: speed, road about 45 km/h (28 mph); range, road 160 km (100 miles); fording 0.9 m (3.0 ft); gradient 60%; vertical obstacle 0.61 m (24 in); trench 2.26 m (7.4 ft); ground clearance 0.43 m (17 in)

Variant
M-50: this self-propelled howitzer was developed in France during the 1950s to meet an Israeli requirement, and combines a French Modèle 50 howitzer with an open-topped fighting compartment built onto the chassis and lower hull of an M4 Sherman medium tank (vertical-volute spring suspension type) reworked to move the engine forward; the maximum elevation of the rear-mounted ordnance is +69 degrees; ammunition types include HE, which weighs 43 kg (94.8 lb) and reaches a range of 17600 m (19,250 yards)

OTO Melara Palmaria 155-mm SP howitzer
(Italy)
Type: tracked self-propelled howitzer
Crew: 5
Combat weight: 46000 kg (101, 411 lb)
Dimensions: length, gun forward 11.474 m 37.74 ft) and hull 7.40 m (24.28 ft); width 3.386 m (11.11 ft); height without machine-gun 2.874 m (9.43 ft)
Armament system: one 155-mm (6.1-in) OTO Melara L/41 howitzer with 30 rounds and one 0.5-in (12.7-mm) or 7.62-mm (0.3-in) AA machine-gun in a hydraulically-powered turret; direct- and indirect-fire control sights fitted
Armour: welded steel and aluminium
Powerplant: one 559-kW (750-hp) MTU turbocharged multi-fuel engine with 900 litres (198 Imp gal) of internal fuel
Performance: speed, road 60 km/h (37.3 mph); range, road 400 km (249 miles); fording 1.2 m (3.9 ft)

76 LAND WEAPONS

without preparation and 4.0 m (13.1 ft) with preparation; gradient 60%; side slope 30%; vertical obstacle 1.0 m (39.3 in); trench 3.0 m (9.8 ft); ground clearance 0.4 m (15.75 in)

Variant
Palmaria: this is the chassis of the OF-40 MBT modified to take a turret accommodating an OTO Melara howitzer with an automatic loader permitting a rate of fire of one round every 15 seconds, or one round per minute for one hour, or one round every three minutes indefinitely; the turret can traverse through 360 degrees, and the ordnance elevates from −4 to +70 degrees, firing a 43.5-kg (95.9-lb) HE round to 24700 m (27,010 yards) or a 43.5-kg (95.9-lb) HE RAP round to 30000 m (32,810 yards); other projectile types are smoke and illuminating

Mitsubishi Type 75 155-mm SP howitzer
(Japan)
Type: tracked self-propelled howitzer
Crew: 6
Combat weight: 25300 kg (55,776 lb)
Dimensions: length, gun forward 7.79 m (25.56 ft) and hull 6.64 m (21.78 ft); width 3.09 m (10.14 ft); height to turret roof 2.545 m (8.35 ft)
Armament system: one 155-mm (6.1-in) NSJ howitzer with 28 rounds and one 0.5-in (12.7-mm) Browning M2HB AA heavy machine-gun with 1,000 rounds in a hydraulically-powered turret; direct- and indirect-fire sights are fitted
Armour: welded aluminium
Powerplant: one 336-kW (450-hp) Mitsubishi 6ZF diesel engine with 650 litres (143 Imp gal) of internal fuel
Performance: speed, road 47 km/h (29.2 mph); range, road 300 km (186 miles); fording 1.3 m (4.25 ft); gradient 60%; side slope 30%; vertical obstacle 0.7 m (27.6 in); trench 2.5 m (8.2 ft); ground clearance 0.4 m (15.75 in)

Variant
Type 75: this Japanese self-propelled howitzer entered service in 1978, and resembles the US M109 series in layout, with the L/30 ordnance located in a rear-mounted turret that can traverse 360 degrees and permits ordnance elevation between −5 and +70 degrees; projectile types include HE (to 19000 m/ 20,770 yards) and HE RAP (to 24000 m/26,245 yards); an NBC system is fitted

Komatsu/Japan Steel Works Type 74 105-mm SP howitzer
(Japan)
Type: tracked self-propelled howitzer
Crew: 5
Combat weight: 16500 kg (36,376 lb)
Dimensions: length 5.90 m (19.36 ft); width 2.90 m (9.51 ft); height 2.39 m (7.84 ft)
Armament system: one 105-mm (4.13-in) NSJ howitzer with ? rounds and one 0.5-in (12.7-mm) Browning M2HB AA heavy machine-gun in a hydraulically-powered turret; direct- and indirect-fire sights are fitted
Armour: welded steel
Powerplant: 224-kW (300-hp) Mitsubishi 4ZF diesel engine with ? litres (? Imp gal) of internal fuel
Performance: speed, road 50 km/h (31 mph); range, road 300 km (186 miles); fording 1.2 m (3.9 ft) without preparation and amphibious with preparation; gradient 60%; vertical obstacle 0.61 m (24 in); trench 2.0 m (6.6 ft); ground clearance 0.4 m (15.75 in)

Variant
Type 74: based on the chassis and automotive system of the Type 73 APC, the Type 74 self-propelled howitzer began to enter service in 1975; the rear-mounted turret traverses through 360 degrees; no details of ammunition types or performance have been revealed; it is thought that the type has an NBC system

Armscor G6 155-mm SP howitzer
(South Africa)
Type: wheeled self-propelled howitzer
Crew: 6
Combat weight: 46000 kg (101,411 lb)
Dimensions: length, gun forward 10.335 m (33.91 ft) and hull 9.20 m (30.18 ft); width 3.40 m (1.15 ft); height to top of turret 3.20 m (10.5 ft)
Armament system: one 155-mm (6.1-in) Armscor G5 howitzer with 47 rounds, one 0.5-in (12.7-mm) Browning M2HB AA machine-gun with 900 rounds, and four 81-mm (3.2-in) grenade-launchers on each side of the hydraulically-powered turret; direct- and indirect-fire sights are fitted
Armour: welded steel
Powerplant: one 391-kW (525-hp) diesel engine with 700 litres (154 Imp gal) of fuel, driving a 6 x 6 layout
Performance: speed, road 90 km/h (56 mph); range, road 600 km (373 miles); fording 1.0 m (3.3 ft); gradient 50%; side slope 30%; vertical obstacle 0.45 m (17.7 in); trench 1.0 m (3.3 ft)

Variant
G6: this self-propelled howitzer was inspired by a Canadian idea, but realized entirely in South Africa as a long-range weapon able to support other armoured vehicles in all types of terrain; the type entered service in the mid-1980s; the rear-mounted turret is capable of 360-degree traverse, and the L/45 ordnance can be elevated between −5 and +75 degrees, firing HE, HE BB, HE RAP, smoke and

illuminating projectiles; the HE projectile weighs 45.5 kg (100.3 lb) and reaches 30000 m (32,810 yards), while the 47-kg (103.6-lb) HE base-bleed projectile attains a prodigious 40000 m (43,745 yards)

Bofors Bandkanon 1A 155-mm SP gun
(Sweden)
Type: tracked self-propelled gun
Crew: 5
Combat weight: 53000 kg (116,843 lb)
Dimensions: length, overall 11.00 m (36.08 ft) and hull 6.55 m (21.49 ft); width 3.37 m (11.06 ft); height over machine-gun 3.85 m (12.63 ft)
Armament system: one 155-mm (6.1-in) Bofors gun with 14 rounds and one 7.62-mm (0.3-in) AA machine-gun in a manually-powered turret; a direct-fire sight is fitted
Armour: welded steel varying in thickness from 10 to 20 mm (0.39 to 0.79 in)
Powerplant: one 240-hp (179-kW) Rolls-Royce K60 diesel engine and one 300-shp (224-kW) Boeing Model 502/10MA gas turbine with 1445 litres (318 Imp gal) of internal fuel
Performance: speed, road 28 km/h (17.4 mph); range, road 230 km (143 miles); fording 1.0 m (3.3 ft); gradient 60%; side slope 30%; vertical obstacle 0.95 m (37.4 in); trench 2.0 m (6.6 ft); ground clearance 0.37 m (14.6 in)

Variant
Bandkanon 1A: entering service in 1966, this self-propelled gun is a massive and fairly slow equipment that uses many automotive components from the S-tank MBT, but has an automatic loader that permits a high rate of fire and can be recharged in a mere two minutes; the L/50 ordnance is located in a large turret at the rear of the vehicle, and can be traversed 15 degrees left and 15 degrees right above an elevation of 0 degrees; elevation limits are −3 and +40 degrees, and the 48-kg (105.8-lb) HE projectile attains 25600 m (27,995 yards)

T-34/122 122-mm SP howitzer
(Syria)
Type: tracked self-propelled howitzer
Crew: 6
Combat weight: 29000 kg (63,933 lb)
Dimensions: length overall 6.01 m (19.70 ft); width 2.99 m (9.81 ft); height 2.70 m (8.86 ft)
Armament system: one 122-mm (4.8-in) D-30 howitzer with 40 rounds mounted on a manually-powered turntable; direct- and indirect-fire sights are fitted
Armour: welded and cast steel
Powerplant: one 368-kW (494-hp) V-2 diesel engine with 560 litres (123 Imp gal) of internal fuel
Performance: speed, road 60 km/h (37.3 mph); range, road 300 km (186 miles); fording 1.3 m (4.3 ft); gradient 35%; vertical obstacle 0.7 m (27.6 in); trench 2.29 m (7.5 ft); ground clearance 0.4 m (15.75 in)

Variant
T-34/122: introduced in the early 1970s, the T-34/122 is a conversion of the elderly T-34 medium tank to mount an exposed 122-mm Soviet howitzer in place of the turret, 10 four-round ammunition boxes being attached to the hull sides; the L/40 ordnance has a practical traverse limit of 120 degrees (60 degrees left and right of the centreline to the rear), and can be elevated from −7 to +70 degrees, firing HE, HEAT and smoke projectiles

2S7 M1975 203-mm SP gun
(USSR)
Type: tracked self-propelled gun
Crew: 4
Combat weight: 40000 kg (88,183 lb)
Dimensions: length, overall 12.80 m (42.00 ft) and hull 10.50 m (34.45 ft); width 3.50 m (11.48 ft); height 3.50 m (11.48 ft)
Armament system: one 203-mm (8-in) gun with ? rounds on a limited-traverse powered mounting
Armour: welded steel
Powerplant: one 336-kW (450-hp) diesel engine with ? litres (? Imp gal) of internal fuel
Performance: not known

Variant
2S7: known in the West as the **M1975**, this is a highly impressive item of Soviet equipment, and the type is probably the largest AFV in the Warsaw Pact's current inventory: the chassis appears to combine features of the SA-12 'Gladiator' SAM launcher and T-72 MBT, and has the massive but unprotected gun mounted at its rear with a powered loading system; it is unlikely that any more than two ready-use rounds are carried on the M1975 vehicle, the bulk of the ammunition supply being carried on an accompanying vehicle; the M1975 has a large rear-mounted stabilizing spade; the weapon is deployed at front (army group) level, and the gun can be elevated to 60 degrees to deliver projectiles (conventional and nuclear) to a range of 30000 m (32,810 yards)

2S3 (M1973) 152-mm SP gun/howitzer
(USSR)
Type: tracked self-propelled gun/howitzer
Crew: 6
Combat weight: 23000 kg (50,705 lb)
Dimensions: length, gun forward 7.78 m (25.52 ft) and hull 7.14 m (23.43 ft); width 3.20 m (10.50 ft); height 2.72 m (8.92 ft)
Armament system: one 152.4-mm (6-in) modified D-20 gun/howitzer with 46 rounds and one 7.62-mm (0.3-in) AA machine-gun in a powered turret; direct- and indirect-fire sights are fitted

Armour: welded steel
Powerplant: one 388-kW (520-hp) diesel engine with 500 litres (110 Imp gal) of internal fuel
Performance: speed, road 55 km/h (34 mph); range, road 300 km (186 miles); fording 1.1 m (3.6 ft); gradient 60%; trench 2.8 m (9.2 ft); ground clearance 0.45 m (17.7 in)

Variant
2S3: originally known in the West as the **M1973**, this useful self-propelled gun/howitzer combines the well-tried D-20 ordnance in a neat turret on a chassis/automotive system combination based on that of the self-propelled launcher for the SA-4 'Ganef' SAM; the type has night-vision devices and an NBC system, but is not amphibious; the turret traverses through 360 degrees, and the L/34 ordnance can be elevated from −3 to +65 degrees; ammunition types are HE-FRAG (a 43.5-kg/95.9-lb projectile to 18500 m/20,230 yards), HE RAP (to 24000 m/26,245 yards), HEAT, AP-T, smoke, illuminating and nuclear (2-kiloton yield); it is believed that an automatic loader may be fitted

2S1 (M1974) 122-mm SP howitzer
(USSR)
Type: tracked self-propelled howitzer
Crew: 4
Combat weight: 16000 kg (35,273 lb)
Dimensions: length 7.30 m (23.95 ft); width 2.85 m (9.35 ft); height 2.40 m (7.87 ft)
Armament system: one 122-mm (4.8-in) modified D-30 howitzer with 40 rounds in an electrically-powered turret; direct- and indirect-fire sights are fitted
Armour: welded steel
Powerplant: one 179-kW (240-hp) YaMZ 238 V diesel engine with 550 litres (121 Imp gal) of internal fuel
Performance: speed, road 60 km/h (37.3 mph) and water 4.5 km/h (2.8 mph) driven by its tracks; range, road 500 km (311 miles); fording amphibious; gradient 60%; vertical obstacle 1.1 m (43.3 in); trench not revealed; ground clearance 0.46 m (18.1 in)

Variants
2S1: originally known in the West as the **M1974**, this vehicle uses a modified version of the D-30 ordnance in a new turret on a hull based on that of the MT-LB multi-role vehicle, at least in its chassis and automotive system; the type is amphibious, and has night-vision devices and an NBC system; the turret traverses 360 degrees, and the L/40 ordnance elevates from −3 to +70 degrees; the main ammunition types are HE (21.72-kg/47.9-lb projectile to 15300 m/16,730 yards), HE RAP (to 21900 m/23,950 yards), HEAT-FS (21.63-kg/47.7-lb projectile capable of penetrating 460 mm/18.1 in of armour), smoke and illuminating

M1974-1 ACRV: artillery command and reconnaissance vehicle with a large superstructure and issued at battalion level; this is armed with a 12.7-mm (0.5-in) DShK heavy machine-gun; the crew is five and the combat weight 14000 kg (30,864 lb)
M1974-2 ACRV: battery command model with a laser rangefinder
M1974-3 ACRV: battalion staff vehicle with additional communications gear
M1974 ACRV/'Big Fred': version fitted with 'Big Fred' artillery- and mortar-locating radar
M1979 MCV: mineclearing vehicle with a turret and three rockets for the discharge of explosive-filled hose

2S9 self-propelled gun/mortar
(USSR)
Type: tracked airborne assault self-propelled gun/mortar
Crew: 4
Combat weight: 9000 kg (19,841 lb)
Dimensions: length, gun forward 5.95 m (19.52 ft); width 2.70 m (8.86 ft); height 2.40 m (7.87 ft)
Armament system: one 120-mm (4.72-in) gun/mortar with 60 rounds; direct- and indirect-fire sights are fitted
Armour: welded steel varying in thickness from 7 to 16 mm (0.28 to 0.63 in)
Powerplant: one 224-kW (300-hp) Type 5D-20 diesel engine with ? litres (? Imp gal) of fuel
Performance: speed, road 60 km/h (37.3 mph) and water ? km/h (? mph) driven by two waterjets; range, road 500 km (311 miles); fording amphibious

Variant
2S9: also known in Soviet service terminology as the **SO-120**, this is an important new light support vehicle introduced in 1984 and based on the lengthened chassis of the BMD-2 tracked carrier; the vehicle is designed to provide Soviet airborne formations with powerful fire support using a turret-mounted 120-mm (4.72-in) gun/mortar of hybrid type using a clip feed mechanism to generate a rate of fire approaching 30 rounds per minute

Vickers FV433 Abbot 105-mm SP gun/howitzer
(UK)
Type: tracked self-propelled gun/howitzer
Crew: 4
Combat weight: 36,500 lb (16556 kg)
Dimensions: length, gun forward 19.16 ft (5.84 m) and hull 18.07 ft (5.709 m); width 8.67 ft (2.64 m); height without armament 8.17 ft (2.18 m)
Armament system: one turret-mounted 105-mm (4.13-in) L13A1 gun with 40 rounds, one 7.62-mm (0.3-in) L4A4 AA machine-gun, and three smoke-dischargers on each side of the powered turret; direct- and indirect-fire sights are fitted

Armour: welded steel varying in thickness from 6 to 12 mm (0.24 to 0.46 in)
Powerplant: one 240-hp (179-kW) Rolls-Royce K60 Mk 4G multi-fuel engine with 85 Imp gal (386 litres) of internal fuel
Performance: speed, road 29.5 mph (47.5 km/h) and water 3 mph (4.8 km/h) driven by its tracks; range 240 miles (386 km); fording 4.0 ft (1.22 m) without preparation and amphibious with preparation; gradient 60%; side slope 30%; vertical obstacle 24 in (0.61 m); trench 6.75 ft (2.06 m); ground clearance 16 in (0.41 m)

Variants
FV433 Abbot: this is a member of the FV430 series of AFVs, and uses many of the same automotive components as the FV432 APC with advantages in training, maintenance and spares holdings; the type began to enter service in 1964, and features night-vision devices and NBC protection; the turret provides 360-degree traverse, and the ordnance can be elevated between −5 and +70 degrees; the ammunition types are HE (35.5-lb/16.1-kg projectile to 18,600 yards/17010 m), HESH, smoke and illuminating; the maximum rate of fire is 12 rounds per minute (a rate aided by the provision of a powered rammer), and ammunition is brought up by the Alvis Stalwart 6 x 6 high-mobility truck
Value Engineered Abbot: this is the export version bought by India, and differs in having no night-vision or NBC gear, a generally reduced standard of equipment, no AA machine-gun and smoke-dischargers and a diesel-only 213-hp (159-kW) version of the K60 engine; this version weighs 35,050 lb (15899 kg) and has slightly shorter lengths

Pacific Car and Foundry M110A2 8-in SP howitzer
(USA)
Type: tracked self-propelled howitzer
Crew: 5
Combat weight: 62,500 lb (28350 kg)
Dimensions: length, gun forward 35.21 ft (10.73 m) and hull 18.75 ft (5.72 m); height to top of barrel (travelling position) 10.31 ft (3.14 m)
Armament system: one 8-in (203-mm) M201 howitzer with 2 rounds on a hydraulically-powered limited-traverse mounting; an indirect-fire sight is fitted
Armour: welded steel
Powerplant: one 405-hp (302-kW) Detroit Diesel 8V-71T turbocharged diesel engine with 260 US gal (984 litres) of internal fuel
Performance: speed, road 34 mph (54.7 km/h); range, road 325 miles (523 km); fording 3.5 ft (1.07 m); gradient 60%; side slope 30%; vertical obstacle 42 in (1.07 m); trench 6.25 ft (1.91 m); ground clearance 15.5 in (0.39 m)

Variants
M110: this initial model began to enter service in 1962, and as the type was conceived as a long-range delivery system for massive shells, no NBC system was fitted and the ordnance is mounted in an exposed position above the rear of the vehicle, only two ready-use rounds being carried; most of the 13-man crew is accommodated in the accompanying M548 support vehicle, which also carries most of the ready-use ammunition; the M2A2 ordnance has a short L/25 barrel (producing an overall length of 24.50 ft/7.47 m and a combat weight of 58,500 lb/26536 kg), and the mounting over the rear of the hull has a traverse arc of 60 degrees (30 degrees left and right) together with elevation from −2 to +65 degrees; a hydraulically-powered loading and ramming system is fitted, and ammunition types include HE (204-lb/93.53-kg projectile to 18,400 yards/16825 m), HE grenade-launching, chemical and nuclear (the M422 projectile with 0.5- or 10-kiloton W33 warhead); the standard rate of fire is one round in two minutes
M110A1: improved model introduced in 1977 with a longer-barrelled L/37 M201 ordnance for greater propellant load and improved projectile range; in this model the fuel capacity was reduced from the 300 US gal (1137 litres) of the M110, thus reducing range, though the extra weight also adversely affected speed, trench-crossing capability and ground clearance, aspects in which the M110 equalled the M107
M110A2: definitive version of the M110A1 introduced in 1978 with a double-baffle muzzle brake; this fires the same rounds as the M110 together with an HE RAP projectile (31,825 yards/29100 m) and the M753 nuclear RAP (selectable 0.5-, 1- or 2-kiloton W79-1 warhead to 31,825 yards/29100 m)
M578: this is the armoured recovery vehicle of the type, using the same chassis and automotive system but fitted with a winch and other specialized equipment

Pacific Car and Foundry M107 175-mm SP gun
(USA)
Type: tracked self-propelled gun
Crew: 5
Combat weight: 62,100 lb (28169 kg)
Dimensions: length, gun forward 36.93 ft (11.26 m) and hull 18.75 ft (5.72 m); width 10.33 ft (3.15 m); height to top of barrel (travelling position) 12.07 ft (3.68 m)
Armament system: one 175-mm (6.89-in) M113 gun with 2 rounds on a hydraulically-powered limited-traverse mounting; direct- and indirect-fire sights are fitted
Armour: welded steel
Powerplant: one 405-hp (302-kW) Detroit Diesel 8V-71T turbocharged diesel engine with 300 US gal (1137 litres) of internal fuel
Performance: speed, road 35 mph (56 km/h); range,

road 450 miles (724 km); fording 3.5 ft (1.07 m); gradient 60%; side slope 30%; vertical obstacle 42 in (1.07 m); trench 7.5 ft (2.3 m); ground clearance 18.3 in (0.47 m)

Variant
M107: this was designed as the partner for the M110 howitzer, and intended to deliver very long-range fire; one of the main features of the system was the use of identical chassis and ordnance mountings, allowing the rapid change of the M107 into the M110 and vice versa, and the type began to enter service in 1962; the mounting over the rear of the hull provides for a traverse arc of 60 degrees (30 degrees left and right), and the L/60 ordnance can be elevated from −2 to +65 degrees, diring at 147.25-lb (66.8-kg) projectile to 35,750 yards (32690 m); like the M110 the M107 is stabilized for firing by two rear-mounted spades, and normally fires at the rate of one round every two minutes, though one round per minute can be achieved for short periods thanks to the provision of a hydraulically-powered loading and ramming system

Cleveland Army Tank Plant/Bowen-McLaughlin-York M109A2 155-mm SP howitzer
(USA)
Type: tracked self-propelled howitzer
Crew: 6
Combat weight: 55,000 lb (24948 kg)
Dimensions: length, gun forward 29.92 ft (9.12 m) and hull 20.31 ft (6.19 m); width 10.17 ft (3.10 m); height including machine-gun 10.75 ft (3.28 m)
Armament system: one 155-mm (6.1-in) M126 howitzer with 36 rounds and one 0.5-in (12.7-mm) Browning M2HB AA heavy machine-gun with 500 rounds in a hydraulically-powered turret; direct- and indirect-fire sights are fitted
Armour: welded aluminium
Powerplant: one 405-hp (302-kW) Detroit Diesel 8V-71T turbocharged diesel engine with 135 US gal (511 litres) of internal fuel
Performance: speed, road 35 mph (56.3 km/h); range, road 215 miles (346 km); fording 3.25 ft (1.0 m); gradient 60%; side slope 40%; vertical obstacle 21 in (0.53 m); trench 6.0 ft (1.83 m); ground clearance 17.75 in (0.45 m)

Variants
M109: this type entered service late in 1962, and the series has since been built and developed extensively as successor to the M44 series with greater operational capabilities combined with better crew protection; the basic model weighs 44,200 lb (20049 kg) and has infra-red driving lights but no NBC system; the fording capability is 6.0 ft (1.83 m) without preparation, though the addition of nine air bags gives the type a limited amphibious capability, track-propelled water speed being 4 mph (6.4 km/h); the turret is accommodated at the rear of the vehicle and provides 360-degree traverse, while the L/20 ordnance can be elevated from −3 to +75 degrees; the vehicle is stabilized for firing by two rear-mounted spades, and the ammunition stowage amounts to 28 rounds; the ammunition types available include HE (94.6-lb/42.9-kg projectile to 16,000 yards/14630 m), HE grenade- and mine-launching, HE RAP, illuminating, smoke, chemical, nuclear (1-kiloton W48 warhead) and laser-guided Copperhead (two of these 22,000-yard/20115-m ranged cannon-launched guided projectiles generally being carried in place of other rounds); the standard rate of fire is one round per minute, though three rounds per minute can be achieved
M109A1: improved M109 with longer L/33 M185 ordnance and new charge system extending HE range to 19,800 yards (18105 m) and RAP range to 26,250 yards (24005 m); the type was produced by M109 conversion, and began to enter service in 1973; the type weighs 53,065 lb (24070 kg and has an overall length of 29.67 ft (9.04 m)
M109A2: entering service in 1979, this is the definitive M109 model based on the M109A2 but with many detail improvements over earlier types and a turret bustle for greater ammunition capacity
M109A3: designation of earlier vehicles brought up to M109A2 standard
M109A4: designation of earlier vehicles upgraded in the Howitzer Extended Life Program (HELP) with the RAM-D (Reliability, Availability, Maintainability – Durability) package, improved battlefield survivability features and the addition of an NBC protection system
M109A5: designation of M109A1 and M109A2 vehicles to be upgraded from the late 1980s onwards upgraded in the Howitzer Improvement Program (HIP) with the RAM-D package of reliability developments, plus extended range and an improved fire-control system; this last features a high degree of automation and also an inertial reference system; the nature of the ordnance is as yet uncertain, the choice still lying between the current M185 or a new ordnance with an L/39 or L/53 barrel for significantly improved range
M109AL: designation of Israeli M109A1s with stowage racks to permit the external carriage of personal kit, so boosting internal ammunition capacity
M109G: designation of West German examples of the M109 with a revised breech (boosting range to 18500 m/20,230 yards with the standard HE projectile), a new muzzle brake, tracks with features of the Leopard 1's tracks and a 7.62-mm (0.3-in) MG3 AA machine-gun; the type also has three smoke-dischargers on each side of the turret, and German fire-control equipment

M109L: designation of Italian M109s fitted with an L/39 OTO Melara howitzer able to fire the ammunition range of the FH-70 howitzer to ranges of between 24000 m (26,245 yards) with standard projectiles and 30000 m (32,810 yards) with RAPs; the ordnance is also fitted with the muzzle brake of the M109G

M109U: designation of Swiss vehicles, the original **Panzerhaubitze 66** type being the M109, which was followed by the **Panzerhaubitze 74** (M109A1) and **Panzerhaubitze 66/74** (Pzh 66 upgraded to Pzh 74 standard); these Swiss vehicles have a revised electrical system and a loading system that operates at all angles of ordnance elevation

M109 (Netherlands): the Dutch army uses a number of turretless M109s for driver training

M109 (Taiwan): under the local designation **XT-69**, Taiwan has combined the chassis and automotive system of the M109 with an exposed mounting for an ordnance derived from the US 155-mm M114 towed howitzer but fitted with a multi-baffle muzzle brake; the crew is five, and it is believed that some 25 rounds of ammunition are carried, the ordnance having a maximum range of some 15000 m (16,405 yards)

M992: field artillery ammunition support vehicle derived from the M109 with a large superstructure in place of the turret; in this superstructure is an extending conveyor system for the movement of 155-mm ammunition, stowage amounting to 93 projectiles, 99 propellant charges and 104 fuses; the type can also be used for 175- and 203-mm (6.89 and 8-in) ammunition, stowage of the latter being 98 projectiles, 53 charges and 56 fuses; it has also be proposed that the M992 be used as the basis of a whole family of self-propelled artillery support vehicles including a fire-direction centre vehicle, a command post vehicle, an ambulance, a rearmament vehicle and a maintenance assistance vehicle

Massey Harris M44A1 155-mm SP howitzer
(USA)
Type: tracked self-propelled howitzer
Crew: 5
Combat weight: 64,000 lb (29030 kg)
Dimensions: length 20.21 ft (6.16 m); width 10.62 ft (3.24 m); height over tarpaulin cover 10.20 ft (3.11 m)
Armament system: one 155-mm (6.1-in) M45 howitzer with 24 rounds on a manually-powered limited-traverse mounting, and one 0.5-in (12.7-mm) Browning M2HB AA heavy machine-gun with 900 rounds; direct- and indirect-fire sights are fitted
Armour: welded steel to a standard thickness of 0.5 in (12.7 mm)
Powerplant: one 500-hp (373-kW) Continental AOSI-895-5 petrol engine with 150 US gal (568 litres) of internal fuel
Performance: speed, road 35 mph (56.3 km/h); range, road 82 miles (132 km); fording 3.5 ft (1.07 m); gradient 60%; vertical obstacle 30 in (0.76 m); trench 6.0 ft (1.83 m); ground clearance 19 in (0.48 m)

Variant
M44A1: this weapon uses the same basic chassis and automotive system as the 105-mm (4.13-in) M52, which are very similar to those of the M41 light tank; this was the last US self-propelled artillery equipment with an open-topped fighting compartment; the original **M44** had a normally-aspirated engine, but was replaced in the mid-1950s by the M44A1 with a fuel-injected engine; the L/20 ordnance has a traverse of 60 degrees (30 degrees left and right) and can be elevated from −5 to +65 degrees; the ammunition types are HE, HE grenade-launching, smoke, illuminating and chemical, and range with the 94.6-lb (42.9-kg) HE projectile is 16,000 yards (14630 m); the normal rate of fire is one round per minute with the vehicle stabilized by two rear-mounted spades

Cadillac (General Motors) M108 105-mm SP howitzer
(USA)
Type: tracked self-propelled howitzer
Crew: 5
Combat weight: 49,500 lb (22453 kg)
Dimensions: length 20.06 ft (6.11 m); width 10.81 ft (3.295 m); height including machine-gun 10.35 ft (3.155 m)
Armament system: one 105-mm (4.13-in) M103 howitzer with 87 rounds and one 0.5-in (12.7-mm) Browning M2HB AA heavy machine-gun with 500 rounds in a manually-powered turret; direct- and indirect-fire sights are fitted
Armour: welded aluminium
Powerplant: one 405-hp (302-kW) Detroit Diesel 8V-71T turbocharged diesel engine with 135 US gal (511 litres) of internal fuel
Performance: speed, road 35 mph (56.3 km/h) and water 4 mph (6.4 km/h) driven by its tracks; range, road 240 miles (386 km); fording 6.0 ft (1.83 m) without preparation and amphibious with preparation; gradient 60%; vertical obstacle 21 in (0.53 m); trench 6.0 ft (1.83 m); ground clearance 17.75 in (0.45 m)

Variant
M108: introduced in 1964, the M108 shares a common chassis and automotive system with the M109 155-mm (6.1-in) self-propelled howitzer, and like that vehicle can be fitted with flotation bags to provide a measure of amphibious capability; the type lacks an NBC system, but has infra-red driving lights; the turret traverses through 360 degrees, and the ordnance can be elevated from −4 to +74 degrees; the ammunition types available are HE, HE grenade-launching, illuminating, smoke and chemical, and typical performance includes a range of 12,600 yards

(11520 m) with an HE round that weighs 39.9 lb (18.1 kg) complete

Detroit Arsenal M52A1 105-mm SP howitzer
(USA)
Type: tracked self-propelled howitzer
Crew: 5
Combat weight: 53,000 lb (24041 kg)
Dimensions: length 19.00 ft (5.80 m); width 10.33 ft (3.15 m); height including machine-gun 10.88 ft (3.32 m)
Armament system: one 105-mm (4.13-in) M49 howitzer with 102 rounds and one 0.5-in (12.7-mm) Browning M2HB AA heavy machine-gun with 900 rounds in a manually-powered turret; direct- and indirect-fire sights are fitted
Armour: welded steel to a standard thickness of 0.5 in (12.7 mm)
Powerplant: one 500-hp (373-kW) Continental AOSI-895-5 petrol engine with 180 US gal (681 litres) of internal fuel
Performance: speed, road 42 mph (67.6 km/h); range, road 100 miles (161 km); fording 4.0 ft (1.22 m); gradient 60%; vertical obstacle 36 in (0.91 m); trench 6.0 ft (1.83 m); ground clearance 19.33 in (0.49 m)

Variant
M52A1: this is the 105-mm (4.13-in) equivalent of the 155-mm (6.1-in) M44, introduced in 1954 and possessing the same basic characteristics as the M44 apart from its different primary armament and associated ammunition in a turret capable of 360-degree traverse; the ordnance can be elevated between −10 and +65 degrees, and the ammunition types include HE (to 12,325 yards/11270 m), HEAT, HE grenade-launching, smoke, illuminating and chemical the original normally-aspirated AOS-895-3 engine of the **M52** was soon supplanted by the fuel-injected AOSI-895-5; this version has no need of rear-mounted spades

American Locomotive/Federal Machine/Pressed Steel Car M7 Priest 105-mm SP howitzer
(USA)
Type: tracked self-propelled howitzer
Crew: 7
Combat weight: 50,634 lb (22967 kg)
Dimensions: length 19.75 ft (6.02 m); width 9.42 ft (2.87 m); height 9.58 ft (2.92 m)
Armament system: one 105-mm (4.13-in) M1A2 howitzer with 69 rounds on a manually-powered limited-traverse mounting, and one 0.5-in (12.7-mm) Browning M2HB AA heavy machine-gun with 300 rounds; direct- and indirect-fire sights are fitted
Armour: cast and welded steel varying in thickness from 0.5 to 4.5 in (12.7 to 114.3 mm)
Powerplant: one 340-hp (254-kW) Continental R-975-C1 petrol engine with 179 US gal (678 litres) of internal fuel
Performance: speed, road 26 mph (42 km/h); range, road 125 miles (201 km); fording 4.0 ft (1.22 m); gradient 60%; vertical obstacle 24 in (0.61 m); trench 7.5 ft (2.3 m); ground clearance 17.1 in (0.43 m)

Variants
M7: introduced in 1942 on the basis of the M3 medium tank with an M1A2 howitzer, the M7 series found immediate favour as a reliable and hard-hitting front-line support weapon; the howitzer is located at the front of the fighting compartment, can be traversed through 38 degrees (12 degrees left and 26 degrees right), and can be elevated from −5 to +33 degrees, firing the standard 33-lb (14.97-kg) HE projectile to a range of 12,000 yards (10975 m);
M7B1: later-production standard based on the chassis of the M4A3 medium tank with the 500-hp (373-kW) Ford GAA engine and fitted with the improved M2 howitzer
M7B2: refined version of the M7B1 with higher 'pulpit' for the AA machine-gun

CITEFA Model 77 155-mm howitzer
(Argentina)
Type: towed howitzer
Calibre: 155 mm (6.1 in)
Barrel length: 33 calibres
Muzzle brake: double baffle
Carriage: split-trail type with two road wheels and two castors; no shield
Weight: 8000 kg (17,637 lb) in travelling order
Dimensions: length 10.15 m (33.30 ft); width 2.67 m (8.76 ft); height 2.2 m (7.22 ft)
Traverse/elevation: 70 degrees total/0 to +67 degrees
Rate of fire: 1 round per minute (normal) and 4 rounds per minute (maximum)
Range: 22000 m (24,060 yards) with standard round and 25300 m (27,670 yards) with RAP
Crew: ?

Variants
Model 77: this is the ordnance of the French Mk F3 self-propelled howitzer mounted on a bottom carriage of Argentine design and manufacture; the type fires separate-loading HE, HE RAP, smoke and illuminating rounds, the HE projectile weighing 43 kg (94.8 lb) and having a muzzle velocity of 765 m (2,510 ft) per second
Model 81: updated version with Argentine-built ordnance and detail modifications

SRC International GC 45 155-mm gun/howitzer
(Belgium)
Type: towed gun/howitzer
Calibre: 155 mm (6.1-in)
Barrel length: 45 calibres
Muzzle brake: multi-baffle

Carriage: split-trail type with four road wheels and two castors; no shield
Weight: 8222 kg (18,126 lb) in travelling and firing orders
Dimensions: length, travelling with muzzle to rear 13.614 m (44.67 ft), travelling with muzzle over trails 9.144 m (30.00 ft) and firing 10.82 m (35.50 ft); width, travelling 2.692 m (8.83 ft) and firing 10.364 m (34.00 ft); height, travelling 3.28 m (10.75 ft)
Traverse/elevation: 80 degrees total/−5 to +69 degrees
Rate of fire: 2 rounds per minute (normal) and 4 rounds per minute (burst rate for 15 minutes)
Range: 30000 m (32,810 yards) with normal round and 38000 m (41,560 yards) with base-bleed round
Crew: 8

Variants
GC 45: designed in Belgium and Canada by PRB and the Space Research Corporation, the GC 45 is an advanced weapon designed to fire separate-loading ammunition (including all standard NATO 155-mm ammunition types) and fitted with an automatic breech and pneumatic rammer; the type is notable for the fact that its extended-range full-bore projectile reaches 30000 m (32,810 yards) without rocket assistance; the ERFB Mk 10 projectile weighs 45.4 kg (100 lb), contains 8.8 kg (19.4 lb) of HE and has a maximum muzzle velocity of 897 m (2,943 ft) per second
GHN 45: developed by Voerst-Alpine in Austria, this is an improved GC 45 with features designed to ease manufacture and to aid field reliability and handling; the most important modification is the optional addition of an auxiliary power unit for battlefield mobility; the APU is a 89-kW (120-hp Porsche unit with 60 litres (13.2 Imp gal) of fuel, permitting a maximum speed of 30 km/h (18.6 mph) and a range of 150 km (93 miles); the dimensions are slightly different from those of the Belgian model, and the weights with and without the APU are 12140 and 9800 kg (26,764 and 21,605 lb); the base-bleed round attains a range of 39600 m (43,305 yards)

MECAR Field Mount 90-mm anti-tank gun
(Belgium)
Type: towed anti-tank gun
Calibre: 90 mm (3.54 in)
Barrel length: 32.2 calibres
Muzzle brake: none
Carriage: three-leg (tripod) type with two road wheels; shield
Weight: 880 kg (1,940 lb) in travelling and firing orders
Dimensions: length, travelling 3.50 m (11.48 ft); width, travelling 1.25 m (4.10 ft); height, travelling 1.36 m (4.46 ft)

Traverse/elevation: 360 degrees/−10 to +12 degrees
Rate of fire: 10 rounds per minute (normal) and 18 rounds per minute (maximum)
Range: see Variant (below)
Crew: 3/4

Variant
Field Mount: this is the MECAR 90-mm (3.54-in) anti-tank gun (designed for installation in light armoured vehicles) carried on a towed carriage with twin road wheels that lift to allow the folding tripod legs to be opened out for 360-degree traverse; the fixed ammunition options are HEAT-CAN-90 anti-tank (a 2.44-kg/5.4-lb projectile with a muzzle velocity of 633 m/2,077 ft per second and the ability to defeat 375 mm/14.76 in of armour at 1000 m/1,095 yards), HE-CAN-90 anti-personnel (4-kg/8.8-lb projectile with a muzzle velocity of 338 m/1,109 ft per second and a range of 3000 m/3,280 yards), CNT-CAN-90 canister (effective range 250 m/275 yards), and smoke

M53 100-mm field and anti-tank gun
(Czechoslovakia)
Type: towed field and anti-tank gun
Calibre: 100 mm (3.94 in)
Barrel length: 67.35 calibres (including muzzle brake)
Muzzle brake: double baffle
Carriage: split-trail type with two road wheels and two castors; shield
Weight: 4280 kg (9,436 lb) in travelling order and 4210 kg (9,281 lb) in firing position
Dimensions: length, travelling 9.10 m (29.86 ft); width, travelling 2.36 m (7.74 ft); height, travelling 2.606 m (8.55 ft)
Traverse/elevation: 60 degrees total/−6 to +42 degrees
Rate of fire: 8/10 rounds per minute
Range: 21000 m (22,965 yards)
Crew: 6

Variant
M53: introduced in the early 1950s, the M53 is a dual-purpose field and anti-tank gun firing fixed ammunition, including a 16-kg (35.3-lb) APC-T projectile at 1000 m (3,281 ft) per second to penetrate 185 mm (7.28 in) of armour at 1000 m (1,095 yards), a 12.36-kg (27.25-lb) HEAT-FS projectile at 900 m (2,953 ft) per second to penetrate 380 mm (14.96 in) of armour at any range, a 5.69-kg (12.54-lb) HVAPDS-T projectile at 1415 m (4,642 ft) per second to penetrate 200 mm (7.87 in) of armour at 1000 m (1,095 yards), and a 15.59-kg (34.37-lb) FRAG-HE projectile at 900 m (2,953 ft) per second

M52/55 85-mm field and anti-tank gun
(Czechoslovakia)
Type: towed field and anti-tank gun

Calibre: 85 mm (3.35 in)
Barrel length: 59.65 calibres
Muzzle brake: double baffle
Carriage: split-trail type with two road wheels; shield
Weights: 2168 kg (4,780 lb) in travelling order and 2111 kg (4,654 lb) in firing position
Dimensions: length, travelling 7.52 m (24.67 ft); width, travelling 1.98 m (6.50 ft); height, travelling 1.515 m (4.97 ft)
Traverse/elevation: 60 degrees total/−6 to +38 degrees
Range: maximum 16160 m (17,675 yards)
Crew: 7

Variants
M52: introduced in 1952, this is a dual-purpose field and anti-tank gun, and among the fixed ammunition types available are AP-T (a 9.2-kg/20.28-lb projectile with a muzzle velocity of 820 m/2,690 ft per second to penetrate 123 mm/4.84 in of armour at 1000 m/ 1,095 yards), HEAT-FS, HVAP-T and HE
M52/55: improved model of 1955 with slightly greater weights

Tampella M-74 155-mm gun/howitzer
(Finland)
Type: towed gun/howitzer
Calibre: 155 mm (6.1 in)
Barrel length: 38.65 calibres
Muzzle brake: single baffle
Carriage: split-trail type with four road wheels; no shield
Weight: 9500 kg (20,944 lb) in travelling and firing orders
Dimensions: length, travelling 7.50 m (24.61 ft); width, travelling 2.58 m (8.46 ft); height, travelling 21.0 m (6.56 ft)
Traverse/elevation: 90 degrees total/−5 to +52 degrees
Rate of fire: 2 rounds per minute (normal) or 4 rounds per minute (short bursts)
Maximum range: 24000 m (26,245 yards) with normal round and 30000 m (32810 yards) with base-bleed round
Crew: 8

Variant
M-74: this useful weapon, currently entering service with the Finnish army under the designation **155K83**, is apparently a development of the Israeli Soltam M-68 with a longer barrel, no barrel counterweight, provision for an extra propellant charge and a number of detail modifications; the ammunition is of the separate-loading type, and the HE projectile weighs 43.6 kg (96.12 lb) with a muzzle velocity of 850 m (2,789 ft) per second

Tampella M-60 122-mm field gun
(Finland)
Type: towed field gun
Calibre: 122 mm (4.8 in)
Barrel length: 53 calibres
Muzzle brake: single baffle
Carriage: split-trail type with four road wheels; no shield
Weight: 8500 kg (18,739 lb) in travelling and firing orders
Dimensions: length, travelling 7.20 m (23.62 ft); width, travelling 2.58 m (8.46 ft); height, travelling 2.00 m (6.56 ft)
Traverse/elevation: 90 degrees total/−5 to +50 degrees
Rate of fire: 4 rounds per minute
Maximum range: 25000 m (27,340 yards)
Crew: 8

Variant
M-60: entering service in 1964, this powerful equipment has a semi-automatic breech and fires separate-loading ammunition; the 25-kg (55.1-lb) HE projectile has a muzzle velocity of 950 m (3,117 ft) per second; the four-wheel bogie can be powered hydraulically from the towing vehicle for improved cross-country mobility

Tampella M-61/37 105-mm light field howitzer
(Finland)
Type: towed light field howitzer
Calibre: 105 mm (4.13 in)
Barrel length: not known
Muzzle brake: single baffle
Carriage: split-trail type with two road wheels; shield
Weight: 1800 kg (3,968 lb) in travelling and firing orders
Dimensions: not known
Traverse/elevation: 53 degrees total/+6 to +45 degrees
Rate of fire: 7 rounds per minute
Range: 13400 m (14,655 yards)
Crew: 6

Variants
M-61/37: this is a new, Finnish-designed ordnance fitted on the carriage of the obsolete M37 field howitzer, and firing separate-loading ammunition with a muzzle velocity of 600 m (1,969 ft) per second for the 14.9-kg (32.8-lb) HE projectile
M-37/10: this is another hybrid equipment, with a Tampella ordnance on the carriage of the Soviet M10 field howitzer

GIAT TR 155-mm gun
(France)
Type: towed gun
Calibre: 155 mm (6.1 in)

Barrel length: 40 calibres
Muzzle brake: double baffle
Carriage: split-trail type with two road wheels; no shield
Weight: about 10650 kg (23,479 lb) in travelling order
Dimensions: length, travelling with barrel over trails 8.75 m (28.71 ft) and firing 10.00 m (32.81 ft); width, travelling 3.09 m (10.14 ft) and firing 8.40 m (27.56 ft); height, firing 1.65 m (5.41 ft)
Traverse/elevation: 27 degrees left and 38 degrees right/−5 to +66 degrees
Rate of fire: 2 rounds per minute (normal), or 6 rounds per minute (for 2 minutes only), or 3 rounds in 18 seconds (with cold barrel)
Maximum range: 24000 m (26,245 yards) with hollow-base projectile and 33000 m (36,090 yards) with RAP
Crew: 8

Variant
TR: entering production in 1984, the TR is a modern equipment designed for the support of motorized divisions; a hydraulically-powered rammer is standard to reduce crew fatigue, and helps the equipment to generate an high initial rate of fire with its standard French separate-loading ammunition; types include HE with a projectile weight of 43.25 kg (95.35 lb) and a muzzle velocity of 830 m (2,723 ft) per second for a range of 24000 m (26,245 yards), HE RAP with a projectile weight of 43.5-kg (95.9 lb) and a muzzle velocity of 830 m (2,723 ft) per second for a range of 33000 m (36,090 yards), anti-tank mine-launching with a projectile weight of 46 kg (101.4 lb) and a load of six 0.55-kg (1.2-lb) mines, smoke and illuminating; the TR can also fire the US M107 round and the ammunition of the FH-70 series

Modèle 50 155-mm howitzer
(France)
Type: towed howitzer
Calibre: 155 mm (6.1 in)
Barrel length: 28.45 calibres including muzzle brake
Muzzle brake: multi baffle
Carriage: split trail with four road wheels; no shield
Weights: 9000 kg (19,841 lb) in travelling order and 8100 kg (17,857 lb) in firing position
Dimensions: length, travelling 7.80 m (25.59 ft) and firing 7.15 m (23.46 ft); width, travelling 2.75 m (9.02 m) and firing 6.80 m (22.31 ft); height, travelling 2.50 m (8.20 ft) and firing 1.65 m (5.42 ft)
Traverse/elevation: 80 degrees total/−4 to +69 degrees
Rate of fire: 3/4 rounds per minute
Maximum range: 18000 m (19,685 yards) with standard ammunition or 23300 m (25,480 yards) with RAP
Crew: 11

Variant
Modèle 50: developed in the period immediately after World War II, this equipment is now verging on the obsolescent but is still in limited service; the type fires separate-loading ammunition including a 43-kg (94.8-lb) HE projectile with a muzzle velocity of 650 m (2,133 ft) per second

M 18/49
(Germany/Czechoslovakia)
Type: towed field howitzer
Calibre: 105 mm (4.13 in)
Barrel length: 28 calibres
Muzzle brake: double baffle
Carriage: split-trail type with two wheels; shield
Weight: 1750 kg (3,858 lb) in travelling and firing orders
Dimensions: length, travelling 5.86 m (19.62 ft); width, travelling 2.10 m (6.89 ft); height, travelling 1.80 m (5.91 ft)
Traverse/elevation: 60 degrees total/−5 to +42 degrees
Rate of fire: 6/8 rounds per minute
Maximum range: 12320 m (13,475 yards) with HE round
Crew: 8

Variants
M18: current designation for the weapon developed in 1928/9 by the Germans as the **leichte Feldhaubitze 18**; this is now an obsolete weapon that nonetheless remains in fairly widespread service, and weighs 1985 kg (4,376 lb); the type has no muzzle brake and fires a 14.8-kg (30.86-lb) separate-loading HE projectile at a muzzle velocity of 470 m (1,542 ft) per second to 10675 m (11,675 yards); the crew is six and maximum traverse 56 degrees
M18M: improved version originally known as the **leichte Feldhaubitze 18M** (and as the **leichte Feldhaubitze 18/40** when fitted on the tubular split-trail carriage of the PaK 40 anti-tank gun for a weight of 1800 kg/3,968 lb) with a double-baffle muzzle brake, a weight of 2040 kg (4,497 lb) and a muzzle velocity of 540 m (1,772 ft) per second to fire a 14.8-kg projectile to 12320 m (13,475 yards); the crew is 10
M 18/49: version produced after World War II in Czechoslovakia with modern wheels and possessing the same improved traverse arc as the M 18/40

Rheinmetall/OTO Melara/Vickers FH-70 155-mm field howitzer
(West Germany/Italy/UK)
Type: towed field howitzer
Calibre: 155 mm (6.1 in)
Barrel length: 38.85 calibres
Muzzle brake: single baffle
Carriage: split-trail type with auxiliary power unit, two road wheels and two castors; no shield

Weight: 9300 kg (20,503 lb) in travelling and firing orders
Dimensions: length, travelling 9.80 m (32.15 ft) and firing 12.43 m (40.78 ft); width, travelling 2.204 m (7.23 ft) and firing 9.80 m (32.15 ft); height, travelling 2.56 m (8.40 ft) and firing 2.192 m (7.19 ft)
Traverse/elevation: 56 degrees total/−5 to +70 degrees
Rate of fire: 2 rounds per minute (sustained fire), or 6 rounds per minute (normal), or 3 rounds in 13 seconds (burst), or 3 rounds in 8 seconds (with the optional 'flick' loader)
Maximum range: 24000 m (26,245 yards) with standard ammunition and 30000 m (32,810 yards) with RAP
Crew: 8

Variants
FH-70: developed jointly by West Germany, Italy and the UK in the late 1960s and entering service from 1978, the FH-70 is an advanced field howitzer firing its own special separate-loading ammunition as well as standard NATO types and the Copperhead cannon-launched guided projectile; the type uses a semi-automatic breech and loader for reduced crew fatigue and high rates of fire (with a 'flick' loader optional for higher rates still), while the addition of an 1800-cc (110-cu in) Volkswagen petrol-engined APU provides power for getting the equipment into and out of action and also gives good battlefield mobility at speeds of up to 16 km/h (10 mph), gradients of 34% and a fording depth of 0.75 m (2.5 ft); included in the ammunition types are FRAG-HE (43.5-kg/95.9-lb projectile with 11.3 kg/24.9 lb of HE), HE RAP, smoke and illuminating; the use of an eight-charge propellant system allows muzzle velocities between 213 and 827 m (700 and 2,715 ft) per second
FH-70R: improved model developed as a private venture by Rheinmetall; this version has an L/46 barrel, making possible ranges of 30000 m (32,810 yards) with the standard projectile and of 36000 m (39,370 yards) with a base-bleed projectile; Rheinmetall is also developing combustible charge containers for this weapon and other 155-mm types

Soltam Model 839P 155-mm gun/howitzer
(Israel)
Type: towed gun/howitzer
Calibre: 155 mm (6.1 in)
Barrel length: 43 calibres
Muzzle brake: single baffle
Carriage: split-trail type with auxiliary power unit, four road wheels and two castors; no shield
Weight: 10850 kg (23,920 lb) in travelling and firing orders
Dimensions: length, travelling 7.50 m (24.61 ft) and firing at 0-degree elevation 10.80 m (35.43 ft); width,
travelling 2.58 m (8.46 ft) and firing 8.00 m (26.25 ft); height, travelling 2.10 m (6.89 ft) and firing at 70-degree elevation 7.20 m (23.62 ft)
Traverse/elevation: 78 degrees total/−3 to +70 degrees
Rate of fire: 2 rounds per minute (sustained fire) or 4 rounds per minute (short periods)
Maximum range: 23500 m (25,700 yards)
Crew: 8

Variants
Model 839P: this version of the M-71 gun/howitzer entered service in 1984 and uses the same ordnance on a revised carriage fitted with a 60-kW (80-hp) Deutz diesel-engined APU; this provides a road speed of 17 km/h (10.6 mph), a range of 70 km (43.5 miles) and a gradient capability of 34%, as well as power for trail-spreading/opening, wheel lifting/lowering and firing platform lowering/lifting; a pneumatically-operated rammer eases crew fatigue and increases rate of fire, and the ordnance fires separate-loading ammunition of various types including FRAG-HE (43.5-kg/95.9-lb projectile with 8.5 kg/18.74 lb of HE) and HE RAP (4.5 kg/9.9 lb of HE)
Model 845P: entering service in 1985, this is a version of the Model 839P with a 45-calibre barrel and a range of 28500 m (31,170 yards) with a 'special projectile'
M-71: developed in the late 1960s from the M-68 gun/howitzer, the M-71 differs from its predecessor in having an L/39 barrel and a pneumatically-operated rammer operable at all angles of elevation; the type has four road wheels, weighs 9200 kg (20,282 lb) in travelling order, has useful traverse/elevation arcs of 84 degrees/−3 to +52 degrees, and can fire to a maximum range of 23500 m (25,700 yards) at the rate of 2/4 rounds per minute
M-68: introduced to service in 1970, this equipment uses virtually the same carriage as the Tampella M-60; the barrel has a length of 33 calibres, and can fire a 43.7-kg (96.3-lb) HE projectile to a range of 21000 m (22,965 yards with a muzzle velocity of 725 m (2,379 ft) per second, rate of fire being 2/4 rounds per minute; the ordnance can also fire the Tampella nine-charge ammunition range with muzzle velocities from 253 to 820 m (830 to 2,690 ft) per second; the weight is 9500 kg (20,944 lb) in travelling order, and the ordnance has traverse/elevations arcs of 90 degrees/−5 to +52 degrees

OTO Melara Modello 56 105-mm pack howitzer
(Italy)
Type: towed pack howitzer
Calibre: 105 mm (4.13 in)
Barrel length: 14.08 calibres
Muzzle brake: multi baffle
Carriage: split-trail type with two road wheels; optional shield

Weight: 1290 kg (2,844 lb) in travelling and firing orders
Dimensions: length, travelling 3.65 m (11.98 ft) and firing 4.80 m (15.75 ft); width, travelling 1.50 m (4.92 ft) and firing 2.90 m (9.51 ft); height, travelling and firing 1.93 m (6.33 ft)
Traverse/elevation: 36 degrees total/−5 to +65 degrees
Rate of fire: 3 rounds per minute (for 1 hour) or 4 rounds per minute (for 30 minutes)
Maximum range: 10575 m (11,565 yards) with standard ammunition
Crew: 7

Variant
Modello 56: this widely-used pack howitzer entered production in 1957, and amongst its attractions is the fact that it can be broken down into 11 sections (the heaviest weighing 122 kg/269 lb) for animal or light helicopter transport; in the field role the Modello 56's wheels are overslung, but for the dedicated anti-tank role they are underslung to reduce height to 1.55 m (5.09 ft), increase traverse to 56 degrees, and reduce elevation to an arc between −5 and +22 degrees; in the anti-tank role the rate of fire is 8 rounds per minute with the semi-fixed 16.7-kg (36.8-lb) HEAT round, which has a muzzle velocity of 387 m (1,270 ft) per second to penetrate 102 mm (4 in) of armour; the standard HE round has a weight of 16.7 kg (36.8 lb) and a muzzle velocity of 472 m (1,550 ft) per second

Kia Machine Tool KH179 155-mm howitzer
(South Korea)
Type: towed howitzer
Calibre: 155 mm (6.1 in)
Barrel length: 39.33 calibres
Muzzle brake: double baffle
Carriage: split-trail type with two road wheels; no shield
Weight: 6860 kg (15,123 lb) in travelling and firing orders
Dimensions: length, travelling 10.30 m (33.79 ft); width, travelling 2.438 m (8.00 ft); height, travelling 2.77 m (9.09 m)
Traverse/elevation: 23.5 degrees left and 25 degrees right/−2 to +63.5 degrees
Rate of fire: 2 rounds per minute (normal) and 4 rounds per minute (maximum)
Maximum range: 22000 m (24,060 yards) with standard HE and 30000 m (32,810 yards) with RAP
Crew: 11

Variant
KH179: just as the KH178 is a reworking of the US M101, the KH179 is an updating of the US M114 with a longer barrel and detail improvements; the type fires all standard NATO separate-loading 155-mm (6.1-in) ammunition

Kia Machine Tool KH178 105-mm light howitzer
(South Korea)
Type: towed light howitzer
Calibre: 105 mm (4.13 in)
Barrel length: 39 calibres
Muzzle brake: double baffle
Carriage: split-trail type with two road wheels; optional shield
Weight: 2697 kg (5,946 lb) in travelling and firing orders
Dimensions: length 7.60 m (24.93 ft); width 2.10 m (6.89 ft); height not revealed
Traverse/elevation: 46 degrees total/−5 to +65 degrees
Rate of fire: 3 rounds per minute (sustained fire) or 10 rounds per minute (maximum)
Maximum range: 14700 m (16,075 yards) with normal ammunition and 18000 kg (19,685 yards) with RAP
Crew: 8

Variant
KH178: this South Korean equipment, currently entering service, is derived from the US M101A2 howitzer with a longer barrel, and incorporates features of the British Light Gun and the West German Rheinmetall conversion of the M101; the weapon fires the semi-fixed ammunition types of the M101 with a muzzle velocity of 663 m (2,175 ft) per second

Armscor G5 155-mm gun/howitzer
(South Africa)
Type: towed gun/howitzer
Calibre: 155 mm (6.1 in)
Barrel length: 45 calibres
Muzzle brake: single baffle
Carriage: split-trail type with auxiliary power unit, four road wheels and two castors; no shield
Weight: 15500 kg (29,762 lb) in travelling and firing orders
Dimensions: length, travelling 9.10 m (29.86 ft) and firing 11.20 m (36.75 ft); width, travelling 2.50 m (8.20 ft) and firing 8.60 m (28.22 ft); height, travelling 2.30 m (7.55 ft)
Traverse/elevation: 84 degrees up to +15-degree elevation and 65 degrees above +15-degree elevation/−3 to +75 degrees
Rate of fire: 2 rounds per minute (sustained fire) and 3 rounds per minute (for 15 minutes)
Maximum range: 30000 m (32,810 yards) with standard ammunition, and 37500 m (41,010 yards) with HE base-bleed ammunition
Crew: 8

Variant
G5: this extremely potent piece of equipment was developed from 1975 on the basis of the GC 45, though the process altered virtually every feature of the Canadian/Belgian weapon; the type entered South African service in 1983 and has proved very successful in terms of range, accuracy and cross-country mobility; battlefield capabilities are considerably aided by the installation of a 51-kW (68-hp) Magirus-Deutz diesel-engined APU for hydraulic functions such as lowering/raising the firing platform and castors, and for opening/closing the trails; crew fatigue is reduced by the provision of a pneumatically-operated rammer; a three-charge propellant system is used for muzzle velocities between 250 and 897 m (820 and 2,943 ft) per second, and the projectile types are a 45.5-kg (100.3-lb) HE, 47-kg (103.6-lb) base-bleed HE, smoke, illuminating and white phosphorus; a cargo-carrying round is under development

Santa Barbara SB 155/39 155-mm howitzer
(Spain)
Type: towed howitzer
Calibre: 155 mm (6.1 in)
Barrel length: 39 calibres
Muzzle brake: double baffle
Carriage: split-trail type with two road wheels and two castors; no shield
Weight: 9000 kg (19,841 lb) in travelling order
Dimensions: length, travelling 8.20 m (26.90 ft) and firing 12.5 m (41.01 ft); width, travelling 2.60 m (8.53 ft); height, travelling 2.40 m (7.87 ft) and firing 1.525 m (5.00 ft)
Traverse/elevation: 60 degrees total/−5 to +70 degrees
Rate of fire: 2 rounds per minute (60 minutes), or 3 rounds per minute (30 minutes), or 1 round per minute (indefinitely), or 4 rounds per minute (bursts)
Maximum range: 24000 m (26,245 yards) with standard ammunition
Crew: 6

Variants
SB 155/39: entering service in the later 1980s, this is a conventional but useful equipment firing the full range of NATO separate-loading ammunition, and for towing the barrel is turned through 180 degrees to rest over the closed trails
SB 155/39 REMA: due to enter Spanish service in the later 1980s, this is a useful equipment fitted with 48.8-kW (65.5-hp) Citroen APU for battlefield mobility with a speed of 12 km/h (7.5 mph) and a gradient capability of 30%, and for the generation of hydraulic power for functions such as wheel lowering/raising and trail opening/closing
SB 155/39 ATP: under development for service in the late 1980s onwards, this is the self-propelled version of the family with a maximum road speed of 70 km/h (43.5 mph) and gradient/sideslope capabilities of 60/30%

Reinosa R-50/26 105-mm field howitzer
(Spain)
Type: towed field howitzer
Calibre: 105 mm (4.13 in)
Barrel length: 31.9 calibres
Muzzle brake: double baffle
Carriage: split-trail type with two road wheels; no shield
Weight: 1950 kg (4,299 lb) in travelling and firing orders
Dimensions: length, travelling 6.08 m (19.95 ft); width, travelling 2.10 m (6.89 ft); height, travelling 2.20 m (7.22 ft)
Traverse/elevation: 50 degrees total/−5 to +45 degrees
Rate of fire: 6 rounds per minute
Maximum range: 11450 m (12,520 yards)
Crew: 6

Variant
R-50/26: introduced in 1950, this Spanish-designed weapon was designed to fire its own range of semi-fixed ammunition as well as standard US ammunition; the latter is slightly lighter than the Spanish type, has less muzzle velocity and thus less range; the Spanish HE projectile weighs 15.27 kg (33.66 lb) and has a muzzle velocity of 504 m (1,654 ft) per second, and the HEAT projectile can penetrate 85 mm (3.35 in) of armour at 1000 m (1,095 yards); later versions of this weapon are designated **R-50/53**, **R-58** and **NR-61**

Bofors FH-77A 155-mm field howitzer
(Sweden)
Type: towed field howitzer
Calibre: 155 mm (6.1 in)
Barrel length: 38 calibres
Muzzle brake: pepperpot
Carriage: split-trail type with an auxiliary power unit, two road wheels and two castors; no shield
Weight: 11500 kg (25,353 lb) in travelling order
Dimensions: length, travelling 11.60 m (38.06 ft) and firing 11.20 m (36.75 ft); width, travelling 2.64 m (8.66 ft) and firing 7.18 m (23.56 ft); height, travelling 2.75 m (9.02 ft) and firing at 50-degree elevation 6.75 m (22.15 ft)
Traverse/elevation: 50 degrees at an elevation of less than +5 degrees and 60 degrees at an elevation of more than +5 degrees/−3 to +50 degrees
Rate of fire: 6 rounds every other minute (for 20 minutes), or six rounds in 20/25 seconds, or 3 rounds in 6/8 seconds
Maximum range: 22000 m (24,060 yards)
Crew: 6

Variants
FH-77A: the FH-77A entered production in 1975, and is a highly capable equipment though only at great cost and considerable size; one of the main features of the equipment is its use of a Volvo B20 auxiliary power unit for hydraulic power to drive the road wheels and castors, providing good battlefield mobility (at a maximum speed of 8 km/h; 5 mph) and competent into/out of action times; elevation and traverse are also hydraulically powered, and a powered rammer is used in conjunction with crane-supplied loading table (the crane delivering three projectiles at a time) and semi-automatic breech make possible very high burst rates of fire; the FH-77A fires particularly powerful separate-loading ammunition, the HE projectile weighing 42.2 kg (93 lb); the weapon uses a six-charge propellant system offering muzzle velocities between 310 and 774 m (1,017 and 2,539 ft) per second with projectiles that include HE, base-bleed HE, smoke and illuminating
FH-77B: export version of the FH-77A with a longer L/39 barrel, a screw rather than vertical sliding wedge breech mechanism, a fully-automated ammunition-handling system and improvements to cross-country mobility (including power take-off from the tractor to provide full 8 x 8 drive for the tractor/ordnance combination); weight is increased to 11900 kg (26,235 lb), and a bagged rather than cartridge charge propellant system is used; the standard range of the FH-77B is 24000 m (26,245 yards) rising to 30000 m (32,810 yards) with an extended-range full-bore round
Karin: version of the FH-77A with 120-mm (4.72-in) calibre and intended for the coastal role; this uses a conveyor-type loading system, a modified laying system so that moving ship targets can be followed, and remote-control facilities so that the weapon can be used in conjunction with a fire-control centre accommodated in a container-type body with radar, optronic director and computers

Bofors m/39 150-mm field howitzer
(Sweden)
Type: towed field howitzer
Calibre: 150 mm (5.9 in)
Barrel length: 24 calibres
Muzzle brake: pepperpot
Carriage: split-trail type with two road wheels; no shield
Weight: 5720 kg (12,610 lb) in travelling order
Dimensions: length, travelling with barrel forward 7.37 m (24.18 ft), travelling with barrel out of battery 6.55 m (21.49 ft) and firing 7.27 m (23.85 ft); width, travelling 2.50 m (8.20 ft) and firing 5.53 m (18.14 ft); height, travelling 2.50 m (8.20 ft)
Traverse/elevation: 45 degrees total/−5 to +66 degrees
Rate of fire: 4/6 rounds per minute
Maximum range: 14600 m (15,965 yards)
Crew: 8

Variant
m/39: now an obsolescent weapon, the m/39 uses separate-loading ammunition with a 41.5-kg (91.5-lb) HE projectile fired at a muzzle velocity of 580 m (1,903 ft) per second; the m/39 has solid wheels, the improved **m/39b** having pneumatic tyres

Bofors Tp 4140 105-mm field howitzer
(Sweden)
Type: towed field howitzer
Calibre: 105 mm (4.13 in)
Barrel length: 32 calibres
Muzzle brake: single baffle
Carriage: four-leg split-trail type with two road wheels; no shield
Weights: 3000 kg (6,614 lb) in travelling order and 2800 kg (6,173 lb) in firing position
Dimensions: length, travelling 6.80 m (32.15 ft) and firing 5.52 m (18.11 ft); width, travelling 1.81 m (5.94 ft) and firing 4.90 m (16.08 ft); height, travelling 1.85 m (6.07 ft) and firing 1.20 m (3.94 ft)
Traverse/elevation: 360 degrees/−5 to +60 degrees
Rate of fire: 25 rounds per minute
Maximum range: 15600 m (17,060 yards)
Crew: 4

Variant
Tp 4140: this Swedish equipment was developed after World War II and bears strong conceptual relationships to an inter-war Czech design; the four-trail carriage opens into a cruciform firing platform allowing 360-degree traverse, and the type fires fixed ammunition to make possible a very high rate of fire; the HE projectile weighs 15.3 kg (33.7 lb) at a muzzle velocity of 640 m (2,100 ft) per second

Federal Construction M46 105-mm field howitzer
(Switzerland)
Type: towed field howitzer
Calibre: 105 mm (4.13 in)
Barrel length: 22 calibres
Muzzle brake: multi-baffle
Carriage: split-trail type with two road wheels; shield
Weight: 1840 kg (4,056 lb) in firing position
Dimensions: not revealed
Traverse/elevation: 56 to 72 degrees total depending on elevation/−5 to +67 degrees
Rate of fire: 6/10 rounds per minute
Maximum range: 10000 m (10,935 yards)
Crew: 7

Variant
M46: designed in Sweden by Bofors and built in Switzerland up to 1953, this is an obsolescent weapon

that fires separate-loading HE ammunition, the 15.1-kg (33.3-lb) projectile being fired at a muzzle velocity of 490 m (1,080 ft) per second

Federal Construction M35 105-mm field gun
(Switzerland)
Type: towed field gun
Calibre: 105 mm (4.13 in)
Barrel length: 42 calibres
Muzzle brake: double baffle
Carriage: split-trail type with two road wheels; shield
Weight: 4245 kg (9,358 lb) in firing position
Dimensions: not revealed
Traverse/elevation: 60 degrees total/−3 to +45 degrees
Rate of fire: 6 rounds per minute
Maximum range: 17500 m (19,140 yards) with old ammunition and 21000 m (22,965 yards) with new ammunition
Crew: 9

Variant
M35: designed in Sweden by Bofors but built in Switzerland, the M35 is an obsolete equipment whose utility has been improved by the adoption of a new separate-loading ammunition type with greater muzzle velocity (800 rather than 785 m/2,625 rather than 2,575 ft per second for the 15.15-kg/33.4-lb HE projectile); other projectiles are AP and illuminating

Federal Construction M57 90-mm anti-tank gun
(Switzerland)
Type: towed anti-tank gun
Calibre: 90 mm (3.54 in)
Barrel length: 33.7 calibres
Muzzle brake: multi-baffle
Carriage: split-trail type with two road wheels; shield
Weights: 600 kg (1,323 lb) in travelling order and 570 kg (1,257 lb) in firing position
Dimensions: not revealed
Traverse/elevation: 70 degrees total at less than 11-degree elevation and 44 degrees total at more than 11-degree elevation/−15 to +23 degrees
Rate of fire: 6 rounds per minute (normal) or 20 rounds per minute (maximum)
Maximum range: 3000 m (3,280 yards) overall, or 900 m (985 yards) against a static target, or 700 m (765 yards) against a moving target
Crew: 5

Variant
M57: introduced in 1958, this light anti-tank equipment fires a 2.7-kg (5.95-lb) HEAT projectile at a muzzle velocity of 600 m (1,969 ft) per second to penetrate 250 mm (9.84 in) of armour

Federal Construction M50 90-mm anti-tank gun
(Switzerland)
Type: towed anti-tank gun
Calibre: 90 mm (3.54 in)
Barrel length: 32.2 calibres
Muzzle brake: none
Carriage: split-trail type with two road wheels; shield
Weights: 600 kg (1,323 lb) in travelling order and 556 kg (1,226 lb) in firing position
Dimensions: not revealed
Traverse/elevation: 66 degrees total at less than 11-degree elevation and 34 degrees at more than 11-degree elevation/−10 to +32 degrees
Rate of fire: 6 rounds per minute (normal) and 20 rounds per minute (maximum)
Maximum range: 3000 m (3,280 yards) overall, or 700 m (765 yards) against a static target or 500 m (550 yards) against a moving target
Crew: 5

Variant
M50: introduced in 1953, this light anti-tank equipment fires a 1.95-kg (4.3-lb) HEAT projectile at a muzzle velocity of 600 m (1,969 ft) per second to penetrate 250 mm (9.84 in) of armour

S-23 180-mm towed gun
(USSR)
Type: towed gun
Calibre: 180 mm (7.09 in)
Barrel length: 48.88 calibres
Muzzle brake: pepperpot
Carriage: split-trail type with four road wheels and a two-wheel limber; no shield
Weight: 21450 kg (47,288 lb) in firing position
Dimensions: length, travelling 10.485 m (34.40 ft); width, travelling 2.996 m (9.83 ft); height, travelling 2.621 m (8.60 ft)
Traverse/elevation: 44 degrees total/−2 to +50 degrees
Rate of fire: 1 round per 2 minutes (sustained fire) and 1 round per minute (burst fire)
Maximum range: 30400 m (33,245 yards) with normal ammunition and 43800 m (47,900 yards) with RAP
Crew: 16

Variant
S-23: originally thought in the West to be a 203-mm (8-in) weapon, the S-23 was at first known to NATO as the **M1955** and originated as a naval weapon in the early 1950s; to facilitate towed transport the ordnance can be withdrawn from battery and the tails supported on a two-wheel limber; the type fires separate-loading ammunition with variable bag charges, and the projectile types include an 84.09-kg (185.4-lb) HE type with a muzzle velocity of 790 m (2,592 ft) per second, an HE RAP type with a muzzle velocity of

850 m (2,789 ft) per second, a 97.7-kg (215.4-lb) concrete-piercing type, with a nuclear type with a yield of 0.2 kilotons

D-20 152-mm gun/howitzer
(USSR)
Type: towed gun/howitzer
Calibre: 152.4 mm (6 in)
Barrel length: 37 calibres
Muzzle brake: double baffle
Carriage: split-trail type with two road wheels and two castors; shield
Weights: 5700 kg (12,566 lb) in travelling order and 5650 kg (12,456 lb) in firing position
Dimensions: length, travelling 8.69 m (28.51 ft); width, travelling 2.32 m (7.61 ft); height, travelling 1.925 m (6.32 ft)
Traverse/elevation: 58 degrees total/−5 to +63 degrees
Rate of fire: 5/6 rounds per minute
Maximum range: 17400 m (19,030 yards) with standard ammunition and 24000 m (26,245 yards) with RAP
Crew: 10

Variants
D-20: developed after World War II for a service debut in the early 1950s, the D-20 was designed as successor to the ML-20 and was initially called the **M1955** in the West; the recoil system and carriage are identical with those of the D-74 field gun, and the type can fire separate-loading case ammunition of several types, including a 43.51-kg (95.9-lb) FRAG-HE projectile at a muzzle velocity of 655 m (2,149 ft) per second, a HE RAP projectile to 24000 m (26,245 yards), a 48.78-kg (107.5-lb) AP-T projectile at a muzzle velocity of 600 m (1,969 ft) per second to penetrate 124 mm (4.88 in) of armour at 1000 m (1,095 yards), a HEAT projectile, a concrete-piercing projectile, a smoke projectile, an illuminating projectile, a chemical projectile and a nuclear projectile with a yield of 0.2 kilotons
Type 66: designation of the Chinese-made copy of the D-20

D-1 152-mm howitzer
(USSR)
Type: towed howitzer
Calibre: 152.4 mm (6 in)
Barrel length: 25 calibres
Muzzle brake: double baffle
Carriage: split-trail type with two road wheels; shield
Weights: 3640 kg (8,025 lb) in travelling order and 2600 kg (5,732 lb) in firing position
Dimensions: length, travelling 7.558 m (24.80 ft); width, travelling 1.994 m (6.54 ft); height, travelling 1.854 m (6.08 ft)
Rate of fire: 3/4 rounds per minute

Maximum range: 12400 m (13,560 yards)
Crew: 10

Variants
D-1: developed for service in 1943 as the **M1943**, this equipment combined the recoil system and carriage of the M30 (M1938) 122-mm (4.8-in) howitzer and the ordnance of the M10 (M1938) 152-mm howitzer with a double-baffle muzzle brake; it is currently being replaced in service by the M1973 self-propelled howitzer; separate-loading case-type ammunition is used, including a 40-kg (88.2-lb) FRAG-HE projectile at a muzzle velocity of 508 m (1,667 ft) per second, a chemical projectile, a HEAT projectile, a semi-AP projectile, a smoke projectile and an illuminating projectile
Type 54: designation of the Chinese-made copy of the D-1

ML-20 152-mm gun/howitzer
(USSR)
Type: towed gun/howitzer
Calibre: 152.4 mm (6 in)
Muzzle brake: multi-baffle
Carriage: split-trail type with four road wheels and a two-wheel limber; shield
Weights: 8073 kg (17,798 lb) in travelling order and 7270 kg (16,027 lb) in firing position
Dimensions: length, travelling 7.21 m (23.65 ft); width, travelling 2.312 m (7.59 ft); height, travelling 2.26 m (7.41 ft)
Traverse/elevation: 58 degrees total/−2 to +65 degrees
Rate of fire: 4 rounds per minute
Maximum range: 17265 m (18,880 yards)
Crew: 9

Variant
ML-20: introduced in 1938 as the **M1937**, this equipment uses the carriage of the A-19 (M1931/37) gun, the ordnance being pulled back out of battery for towed transport; the ML-20 has been replaced in front-line Soviet service by the D-20 but remains in valuable service elsewhere; the ordnance fires separate-loading case-type ammunition, and amongst the projectile types are a 43.51-kg (95.9-lb) HE projectile with a muzzle velocity of 655 m (2,149 ft) per second and an AP-T projectile able to penetrate 124 mm (4.88 in) of armour at 1000 m (1,095 yards), as well as HEAT, concrete-piercing, smoke and illuminating projectiles

M1946 130-mm field gun
(USSR)
Type: towed field gun
Calibre: 130 mm (5.12 in)
Barrel length: 55 calibres
Muzzle brake: pepperpot

Carriage: split-trail type with two road wheels and a two-wheel limber; shield
Weights: 8450 kg (18,629 lb) in travelling order and 7700 kg (16,975 lb) in firing position
Dimensions: length, travelling 11.73 m (38.48 ft); width, 2.45 m (8.04 ft); height, travelling 2.55 m (8.37 ft)
Traverse/elevation: 50 degrees total/−2.5 to +45 degrees
Rate of fire: 5/6 rounds per minute
Maximum range: 27150 m (29,690 yards)
Crew: 9

Variants
M1946: introduced in the early 1950s as successor to the A-19 (M1931/37) gun, the M1946 has its ordnance pulled right back out of battery for towed transport; the type fires separate-loading case-type ammunition with projectiles such as a 33.4-kg (73.6-lb) FRAG-HE type with a muzzle velocity of 1050 m (3,445 ft) per second, and an APC-T type with the same muzzle velocity and able to penetrate 230 mm (9.06 in) of armour at 1000 m (1,095 yards), as well as HE RAP, smoke and illuminating types
M1956 (Mod): upgraded version with a longer barrel, cradle and recuperator
Type 59: this is the Chinese copy of the M1946, and has also been produced in an alternative version, the **Type 59-1** with a copy of the Soviet ordnance fitted with a scaled-up version of the 122-mm (4.8-in) D-74 gun and installed on a D-74 carriage; unlike the Soviet D-74 this latter Chinese equipment does not use a limber for towed transport; the Chinese version has a travelling weight of 5630 kg (12,411 lb), has a barrel length of 50 calibres, possesses traverse/elevation arcs of 56 degrees/−6 to +44 degrees, and fires a 33.4-kg projectile with a muzzle velocity of 930 m (3,051 ft) per second to a maximum range of 27400 m (29,965 yards) at the rate of five or six rounds per minute

SM-4-1 130-mm coastal gun
(USSR)
Type: towed coast-defence gun
Calibre: 130 mm (5.12 in)
Barrel length: 58.46 calibres
Muzzle brake:
Carriage: four-leg split-trail type with two four-wheel axles; no shield
Weights: 19000 kg (41,887 lb) in travelling order and 16000 kg (35,273 lb) in firing position
Dimensions: length, travelling 12.80 m (42.00 ft); width, travelling 2.85 m (9.35 ft); height, travelling 3.05 m (10.01 ft)
Traverse/elevation: 360 degrees/−5 to +45 degrees
Rate of fire: 5 rounds per minute
Maximum range: 29500 m (32,260 yards)
Crew: not known

Variant
SM-4-1: developed soon after World War II, this is still a powerful coast-defence equipment when used in conjunction with a radar fire-control system for engagement of moving targets in all weather conditions; in the firing position the axles are removed and the four stabilizing legs extended and staked down; the ordnance fires separate-loading ammunition, and the projectiles in service are a 33.4-kg (73.6-lb) HE type with a muzzle velocity of 1050 m (3,445 ft) per second, and a 33.6-kg (74.1-lb) APHE type with the same muzzle velocity and able to penetrate 250 mm (9.84 in) of armour at 1000 m (1,095 yards)

D-74 122-mm field gun
(USSR)
Type: towed field gun
Calibre: 121.92 mm (4.8 in)
Barrel length: 47 calibres
Muzzle brake: double baffle
Carriage: split-trail type with two road wheels and two castors: shield
Weights: 5550 kg (12,235 lb) in travelling order and 5500 kg (12,125 lb) in firing position
Dimensions: length, travelling 9.875 m (32.40 ft); width, travelling 2.35 m (7.71 ft); height, travelling 2.745 m (9.01 ft)
Traverse/elevation: 58 degrees total/−5 to +45 degrees
Rate of fire: 6/7 rounds per minute
Maximum range: 24000 m (26,245 yards)
Crew: 10

Variants
D-74: designed in the late 1940s as a possible successor to the A-19 (M1931/37) gun, the D-74 was at first known in the West as the **M1955**, and not adopted by the Soviet army, which instead accepted the 130-mm (5.12-in) M1946; the D-74 uses the same carriage as the D-20 152-mm (6-in) gun/howitzer, and was widely produced for export; the type is installed on a circular firing platform for operation, and fires separate-loading case-type ammunition; among the projectiles available are FRAG-HE (weighing 27.3 kg/60.2 lb and fired at a muzzle velocity of 885 m/2,904 ft per second), APC-T (weighing 25 kg/55.1 lb and fired at the same muzzle velocity for armour penetration of 185 mm/7.28 in at 1000 m/1,095 yards), chemical, illuminating and smoke
Type 60: designation of the Chinese-made copy of the D-74

D-30 122-mm howitzer
(USSR)
Type: towed howitzer
Calibre: 121.92 mm (4.8 in)
Barrel length: 35.5 calibres

Muzzle brake: multi baffle
Carriage: three-leg split-trail type with two road wheels; shield
Weights: 3210 kg (29,123 lb) in travelling order and 3150 kg (6,944 lb) in firing position
Dimensions: length, travelling 5.40 m (17.72 ft); width, travelling 1.95 m (6.40 ft); height, travelling 1.66 m (5.45 ft)
Traverse/elevation: 360 degrees/−7 to +70 degrees
Rate of fire: 7/8 rounds per minute
Maximum range: 15400 m (16,840 yards) with standard ammunition and 21900 m (23,950 yards) with RAP
Crew: 7

Variant
D-30: entering service in the early 1960s as successor to the M-30, this equipment offers useful advantages such as 360-degree traverse (on its three-leg firing platform) and greater range; the weapon is towed by its muzzle and fires separate-loading case-type ammunition; among the projectile types available are FRAG-HE (weighing 21.76 kg/47.97 lb and fired at a muzzle velocity of 690 m/2,264 ft per second), HEAT-FS (weighing 21.63 lb/47.7 lb and fired at a muzzle velocity of 740 m/2,428 ft per second for an armour penetration of 460 mm/18.1 in at any range), HE RAP, chemical, illuminating and smoke

M-30 122-mm howitzer
(USSR)
Type: towed howitzer
Calibre: 121.92 mm (4.8 in)
Barrel length: 22.7 calibres
Muzzle brake: none
Carriage: split-trail type with two road wheels; shield
Weight: 2450 kg (5,401 lb) in travelling and firing orders
Dimensions: length, travelling 5.90 m (19.36 ft); width, travelling 1.975 m (6.48 ft); height, travelling 1.82 m (5.97 ft)
Traverse/elevation: 49 degrees total/−3 to +63.5 degrees
Rate of fire: 5/6 rounds per minute
Maximum range: 11800 m (12,905 yards)
Crew: 8

Variants
M-30: introduced in 1939 as the **M1938**, the M-20 remained the Soviet army's basic divisional howitzer until the advent of the D-30; the equipment uses the same carriage as the D-1 (M1943) 152-mm (6-in) howitzer, and fires separate-loading case-type ammunition; the projectile types include FRAG-HE (weighing 21.76 kg/47.97 lb and fired at a muzzle velocity of 515 m/1,690 ft per second), HEAT (weighing 13.3 kg/29.3 lb and fired at 570 m/1,870 ft per second to penetrate 200 mm/7.87 in of armour at 1000 m/1,095 yards), chemical, illuminating and smoke
Type 54: designation of the Chinese-made copy of the M-30

A-19 122-mm gun
(USSR)
Type: towed gun
Calibre: 121.92 mm (4.8 in)
Barrel length: 46 calibres
Muzzle brake: none
Carriage: split-trail type with two road wheels and a two-wheel limber; shield
Weights: 8050 kg (17,747 lb) in travelling order and 7250 kg (15,983 lb) in firing position
Traverse/elevation: 58 degrees total/−2 to +65 degrees
Rate of fire: 5/6 rounds per minute
Maximum range: 20800 m (22,745 yards)
Crew: 8

Variant
A-19: introduced in the late 1930s as the **M1931/37** updated version of the M1931, the A-19 was designed as a corps-level weapon and has been replaced in front-line Soviet service by the M1946 and D-74 guns, though it continues in widespread service with Soviet clients; the ordnance fires separate-loading case-type ammunition, and the projectile types include FRAG-HE (weighing 25 kg/55.1 lb and fired at a muzzle velocity of 800 m/2,625 ft per second), AP-T (weighing the same and having the same muzzle velocity to penetrate 160 mm/6.3 in of armour at 1000 m/1,095 yards), concrete-piercing and smoke

T-12 100-mm anti-tank gun
(USSR)
Type: towed anti-tank gun
Calibre: 100 mm (3.94 in)
Barrel length: 84.8 calibres
Muzzle brake: pepperpot
Carriage: split-trail type with two road wheels; shield
Weight: 3000 kg (6,614 lb) in travelling and firing orders
Dimensions: length, travelling 9.162 m (30.06 ft); width, travelling 1.70 m (5.58 ft); height, travelling 1.448 m (4.75 ft)
Traverse/elevation: 27 degrees total/−10 to +20 degrees
Rate of fire: 10 rounds per minute
Maximum range: 8500 m (9,295 yards)
Crew: 6

Variants
T-12: designed as successor to the 85-mm (3.35-in) D-48 anti-tank gun, the T-12 was introduced in the mid-1960s and uses fixed ammunition of two types, namely APFSDS with a projectile weighing 5.5 kg

(12.13 lb) fired at a muzzle velocity of 1500 m (4,921 ft) per second to penetrate 406 mm (16 in) of armour at 500 m (550 yards), and HEAT with a projectile weighing 9.5 kg (20.94 lb) fired at a muzzle velocity of 990 m (3,248 yards) per second to penetrate 400 mm (15.75 in) of armour at any range up to 1200 m (1,310 yards)
T-12A: modified version with larger wheels, a weight of 3100 kg (6,834 lb) and an overall length of 9.64 m (31.63 ft)

BS-3 100-mm field and anti-tank gun
(USSR)
Type: towed field and anti-tank gun
Calibre: 100 mm (3.94 in)
Barrel length: 60.7 calibres
Muzzle brake: double baffle
Carriage: split-trail type with four road wheels; shield
Weight: 3650 kg (8,047 lb) in travelling and firing orders
Dimensions: length, travelling 9.37 m (30.74 ft); width, travelling 2.15 m (17.05 ft); height, travelling 1.50 m (4.92 ft)
Traverse/elevation: 58 degrees total/−5 to +45 degrees
Rate of fire: 8/10 rounds per minute
Maximum range: 21000 m (22,965 yards)
Crew: 6

Variants
BS-3: introduced in 1944 as the **M1944**, this powerful weapon was derived from a naval ordnance and fires fixed ammunition; among the projectile types are FRAG-HE (weighing 15.6 kg/34.4 lb and fired at a muzzle velocity of 900 m/2,953 ft per second), APC-T (weighing 16 kg/35.3 lb and fired at a muzzle velocity of 1000 m/3,281 ft per second to penetrate 185 mm/7.28 in of armour at 1000 m/1,095 yards), HVAPDS-T (weighing 5.69 kg/12.54 lb and fired at a muzzle velocity 1415 m/4,642 ft per second to penetrate 200+ mm/7.87+ in of armour at 1000 m/1,095 yards) and HEAT-FS (weighing 12.36 kg/27.25 lb fired at a muzzle velocity of 900 m/2,953 ft per second to penetrate 380 mm/14.96 in of armour at any range up to 1000 m/1,095 yards)
Type 59: provisional designation of the Chinese copy of the BS-3 being adopted as successor to the 85-mm (3.35-in) Type 56 as an anti-tank and counter-battery weapon; the equipment's weight is estimated at 3450 kg (7,606 lb), and the AP-T projectile's penetration at 157 mm (6.18 in) of armour at 1000 m (1,095 yards); maximum range with the 15.6-kg (34.4-lb) FRAG-HE projectile is believed to be 20000 m (21,870 yards)

SD-44 85-mm field and anti-tank gun
(USSR)
Type: towed field and anti-tank gun
Calibre: 85 mm (3.35 in)
Barrel length: 53 calibres
Muzzle brake: double baffle
Carriage: split-trail type with an auxiliary power unit, two road wheels and one castor; shield
Weight: 2250 kg (4,960 lb) in travelling and firing orders
Dimensions: length, travelling 8.22 m (26.97 ft); width, travelling 1.78 m (5.84 ft); height, travelling 1.42 m (4.66 ft)
Traverse/elevation: 54 degrees total/−7 to +35 degrees
Rate of fire: 10 rounds per minute (sustained fire) and 15 rounds per minute (burst fire)
Maximum range: 15650 m (17,115 yards)
Crew: 7

Variant
SD-44: this equipment is essentially the D-44 divisional gun fitted with a 10.4-kW (14-hp) M-72 petrol engine to provide airborne formations with a relatively powerful ordnance possessing limited self-mobility at a maximum road speed of 25 km/h (15.5 mph); the ordnance fires fixed ammunition, and the projectile types include FRAG-HE (weighing 9.6 kg/ 21.2 lb and fired at a muzzle velocity of 792 m/2,598 ft per second), AP-T (weighing 9.2 kg/20.3 lb and fired at the same muzzle velocity to penetrate 125 mm /4.92 in of armour at 1000 m/1,095 yards), HVAP-T (weighing 5.06 kg/11.16 lb and fired at 1030 m/3,379 ft per second to penetrate 180 mm/7.09 in of armour at 1000m/1,095 yards) and HEAT-FS (weighing 7.34 kg/16.2 lb and fired at a muzzle velocity of 840 m/2,756 ft per second to penetrate 300 mm/11.8 in of armour at any range)

D-48 85-mm anti-tank gun
(USSR)
Type: towed anti-tank gun
Calibre: 85 mm (3.35 in)
Barrel length: 74 calibres
Muzzle brake: pepperpot
Carriage: split-trail type with two road wheels and one castor; shield
Weight: 2350 kg (5,181 lb) in travelling and firing orders
Dimensions: (estimated) length, travelling 8.717 m (28.60 ft); width, travelling 1.585 m (5.20 ft); height, travelling 1.89 m (6.20 ft)
Traverse/elevation: 54 degrees total/−6 to +35 degrees
Rate of fire: 8 rounds per minute (sustained fire) and 15 rounds per minute (burst fire)
Maximum range: 18970 m (20,745 yards)
Crew: 6

Variant
D-48: this was originally thought in the West to be a

100-mm (3.94-in) weapon when introduced in the mid-1950s, and it does indeed use a 100-mm type of fixed round, though necked down in this application to fire an 85-mm (3.35-in) projectile, either a 9.7-kg (221.4-lb) APHE type with a muzzle velocity of 1000 m (3,281 ft) per second to penetrate 190 mm (7.48 in) of armour at 1000 m (1,095 yards) and a 9.3-kg (20.5-lb) HVAP type with a muzzle velocity of some 1200 m (3,937 ft) per second to penetrate 240 mm (9.45 in) of armour at the same range; the type began to leave Soviet front-line service in the later 1960s

D-44 85-mm field and anti-tank gun
(USSR)
Type: towed field and anti-tank gun
Calibre: 85 mm (3.35 in)
Barrel length: 55.2 calibres including muzzle brake
Muzzle brake: double baffle
Carriage: split-trail type with two road wheels and one castor; shield
Weights: 1725 kg (3,803 lb) in travelling order and 1703 kg (3,754 lb) in firing position
Dimensions: length, travelling 8.34 m (27.36 ft); width, travelling 1.78 m (5.84 ft); height, travelling 1.42 m (4.66 ft)
Traverse/elevation: 54 degrees total/−7 to +35 degrees
Rate of fire: 15 rounds per minute (sustained fire) and 20 rounds per minute (burst fire)
Maximum range: 16560 m (18,110 yards)
Crew: 8

Variants
D-44: introduced in the late 1940s as successor to the ZIS-3 (M1942) 76.2-mm (3-in) divisional gun, the D-44 is no longer in first-line Soviet service but remains a widely encountered piece of equipment elsewhere; the same ordnance is used in the SD-44, and the D-44 thus fires the same ammunition types
Type 56: this is the Chinese-made copy of the D-44 with only very slight modifications, and the anti-armour performance is generally comparable to that of the Soviet equipment

M1966 76-mm mountain howitzer
(USSR)
Type: towed mountain howitzer
Calibre: 76.2 mm (3 in)
Muzzle brake: none
Barrel length: not known
Carriage: split-trail type with two road wheels; shield
Weight: 780 kg (1,720 lb) in travelling order
Dimensions: length, travelling 4.80 m (15.75 ft); width, travelling 1.50 m (8.20 ft); height, travelling not known
Traverse/elevation: 50 degrees total/−5 to +65 degrees
Rate of fire: 15 rounds per minute

Maximum range: about 11000 m (12,030 yards)
Crew: 6

Variant
M1966: otherwise known as the **M1969** in the Western provisional terminology, this mountain howitzer appeared in the second half of the 1960s; the type can fire all the ammunition types available to the ZIS-3 76.2-mm (3-in) divisional gun, and with the HEAT-FS round the M1966 can penetrate 280 mm (11.02 in) of armour; it is believed that the weapon can be broken down in components small enough for animal or light helicopter transport

ZIS-3 76-mm field gun
(USSR)
Type: towed field gun
Calibre: 76.2 mm (3 in)
Barrel length: 45 calibres including muzzle brake
Muzzle brake: double baffle
Carriage: split-trail type with two road wheels; shield
Weight: 1116 kg (2,460 lb) in travelling and firing orders
Dimensions: length, travelling 6.095 m (20.00 ft); width, travelling 1.645 m (5.40 ft); height, travelling 1.375 m (4.51 ft)
Traverse/elevation: 54 degrees total/−5 to +37 degrees
Rate of fire: 15 rounds per minute (sustained fire) and 20 rounds per minute (burst fire)
Maximum range: 13290 m (14,535 yards)
Crew: 7

Variants
ZIS-3: although no longer in first-line service with any major element of the Warsaw Pact armies, this **M1942** weapon is still widely used elsewhere in the world despite its age; the ordnance fires fixed ammunition, including FRAG-HE, AP-T, HVAP-T, HEAT and smoke
Type 54: designation of the Chinese-made copy of the ZIS-3

Ch-26 57-mm anti-tank gun
(USSR)
Type: towed anti-tank gun
Calibre: 57 mm (2.24 in)
Barrel length: 71.4 calibres including muzzle brake
Muzzle brake: double baffle
Carriage: split-trail type with an auxiliary power unit and three road wheels; shield
Weight: 1250 kg (2,756 lb) in travelling and firing orders
Dimensions: length, travelling 6.112 m (20.05 ft); width, travelling 1.80 m (5.91 ft); height, travelling 1.22 m (4.00 ft)
Traverse/elevation: 56 degrees total/−4 to +15 degrees

Rate of fire: 12 rounds per minute
Maximum range: 6700 m (7,325 yards)
Crew: 5

Variant
Ch-26: designed for airborne formations, the Ch-26 remains in second-line service with a few countries, and is powered by a 10.4-kW (14-hp) petrol engine which provides a maximum road speed of 40 km/h (25 mph); the ordnance fires fixed ammunition whose projectile types include FRAG-HE (weighing 3.75 kg/8.27 lb and fired with a muzzle velocity of 695 m/2,280 ft per second), AP-T (weighing 3.14 kg/6.92 lb and fired with a muzzle velocity of 980 m/3,215 ft per second to penetrate 106 mm/4.17 in of armour at 500 m/550 yards), and HVAP-T (weighing 1.76 kg/3.88 lb and fired at a muzzle velocity of 1255 m/4,117 ft per second to penetrate 140 mm/5.51 in of armour at 500 m/550 yards)

ZIS-2 57-mm anti-tank gun
(USSR)
Type: towed anti-tank gun
Calibre: 57 mm (2.24 in)
Barrel length: 73 calibres
Muzzle brake: none
Carriage: split-trail type with two road wheels; shield
Weight: 1150 kg (2,535 lb) in travelling and firing orders
Dimensions: length, travelling 6.795 m (22.29 ft); width, travelling 1.70 m (5.58 ft); height, travelling 1.37 m (4.49 ft)
Traverse/elevation: 56 degrees total/−5 to +25 degrees
Rate of fire: 25 rounds per minute
Maximum range: 8400 m (9,185 yards)
Crew: 7

Variants
ZIS-2: introduced in 1943, this **M1943** equipment is obsolescent but remains in widespread service with Soviet clients; it fires the same ammunition types as the Ch-26 and has comparable anti-armour performance
Type 55: designation of the Chinese-made copy of the ZIS-2

5.5-in Medium Gun
(UK)
Type: towed gun
Calibre: 5.5 in (139.7 mm)
Barrel length: 29.9 calibres
Muzzle brake: none
Carriage: split-trail type with two road wheels; no shield
Weight: 12,900 lb (5851 kg) in travelling and firing orders
Dimensions: length, travelling 24.67 ft (7.52 m); width, travelling 8.33 ft (2.54 m); height, travelling 8.58 ft (2.62 m)
Traverse/elevation: 60 degrees total/−5 to +45 degrees
Rate of fire: 2 rounds per minute
Maximum range: 16,200 yards (14815 m) with 100-lb (45.36-kg) HE projectile and 18,000 yards (16460 m) with 80-lb (36.29-kg) HE projectile
Crew: 10

Variant
5.5-in Medium Gun: introduced in 1942 and now obsolescent, this is a moderately powerful equipment that fires separate-loading ammunition with two types of HE projectile, with smoke and illuminating projectiles also available; in South Africa the equipment is designated the **G2**

L118A1 105-mm Light Gun
(UK)
Type: towed field gun
Calibre: 105 mm (4.13 in)
Barrel length: 29.2 calibres
Muzzle brake: double baffle
Carriage: split trail type with two road wheels; no shield
Weight: 4,100 lb (1860 kg) in travelling and firing orders
Dimensions: length, travelling with gun forward 20.75 ft (6.32 m), travelling with barrel over trails 16.00 ft (4.88 m) and in firing position with barrel horizontal 23.00 ft (7.01 m); width, travelling and in firing position 5.83 ft (1.78 m); height, travelling with barrel forward 8.63 ft (2.63 m)
Traverse/elevation: 11 degrees total on carriage or 360 degrees on firing platform/−5.5 to +70 degrees
Rate of fire: 3 rounds per minute (sustained fire), or 6 rounds per minute (for 3 minutes) or 8 rounds per minute (for 1 minute)
Maximum range: 18,800 yards (17190 m)
Crew: 6

Variants
L118A1: introduced in 1974, the L118A1 is the basic electrically-fired British version of the Light Gun and is a thoroughly modern piece of equipment designed to fulfil the pack howitzer and gun roles while being airportable under medium-lift helicopters; the ordnance fires semi-fixed ammunition with a seven-charge propellant system offering ranges between 2,750 and 18,800 yards (2515 and 17190 m); the projectile types available are 35.5-lb (16.1-kg) HE, 23.1-lb (10.49-kg) HESH, illuminating and three types of smoke
L119A1: version developed for the US M1 ammunition series with different rifling, a single-baffle

muzzle brake, percussion firing and a maximum range of 12,000 yards (10975 m)
L127A1: version developed for Switzerland with different rifling, a double-baffle muzzle brake and percussion firing

25-pdr Mk 3 field gun
(UK)
Type: towed field gun
Calibre: 87.6 mm (3.45 in)
Barrel length: 28.3 calibres
Muzzle brake: double baffle
Carriage: box-type with two road wheels; shield
Weight: 3,970 lb (1801 kg) in travelling and firing positions
Dimensions: length, travelling 26.00 ft (7.92 m); width, travelling 6.96 ft (2.12 m); height, travelling 5.41 ft (1.65 m)
Traverse/elevation: 8 degrees total on carriage and 360 degrees on firing platform/−5 to +40 degrees
Rate of fire: 5/10 rounds per minute
Maximum range: 13,400 yards (12255 m)
Crew: 6

Variant
25-pdr Mk 3: a classic weapon introduced in 1939 as the **25-pdr Mk 1** (or **18/25-pdr**) with an 18-pdr ordnance on a 25-pdr carriage and entering definitive service in 1940 as the **25-pdr Mk 2** before being developed into the 25-pdr Mk 3 of 1942 with a muzzle brake to make feasible the firing of armour-piercing rounds at a high muzzle velocity; now obsolete, the type still serves in some numbers with smaller armies and fires separate-loading ammunition, including a 25-lb (11.34-kg) HE projectile with a muzzle velocity of 1,700 ft (518 m) per second, a 20.3-lb (9.2-kg) AP projectile with a muzzle velocity of 2,000 ft (609 m) per second and able to penetrate 2.75 in (70 mm) of armour at 400 yards (365 m), an illuminating projectile and a smoke projectile

17-pdr anti-tank gun
(UK)
Type: towed anti-tank gun
Calibre: 76.2 mm (3 in)
Barrel length: 58.3 calibres
Muzzle brake: double baffle
Carriage: split-trail type with two road wheels; shield
Weights: 6,700 lb (3039 kg) in travelling order and 6,445 lb (2923 kg) in firing position
Dimensions: length, travelling 24.75 ft (7.54 m); width, travelling 7.30 ft (2.225 m); height, travelling 5.50 m (1.68 m)
Traverse/elevation: 60 degrees total/−6 to +16.5 degrees
Rate of fire: 10 rounds per minute
Maximum range: 10,000 yards (9145 m)
Crew: 6

Variant
17-pdr: introduced in 1943, this obsolete equipment is still in service with smaller armies, and fires fixed ammunition; the 15.4-lb (6.98-kg) HE projectile has a muzzle velocity of 2,875 ft (876 m) per second, the 17-lb (7.7-kg) AP projectile has a muzzle velocity of 2,900 ft (884 m) per second and can penetrate 4.3 in (109 mm) of armour at 1,000 yards (915 m), the 17-lb APC projectile has the same muzzle velocity and can penetrate 4.65 in (118 mm) of armour at 1,000 yards (915 m), and the 7.6-lb (3.4-kg) APDS projectile has a muzzle velocity of 3,945 ft (1203 m) per second and can penetrate 9.1 in (231 mm) of armour at 1,000 yards (915 m); the maximum effective anti-tank range is 1,650 yards (1510 m)

M115 8-in howitzer
(USA)
Type: towed howitzer
Calibre: 8 in (203.2 mm)
Barrel length: 25 calibres
Muzzle brake: none
Carriage: split-trail type with eight road wheels and a two-wheel limber; no shield
Weights: 32,000 lb (14515 kg) in travelling order and 29,700 lb (13472 kg) in firing position
Dimensions: length, travelling 36.00 ft (10.97 m); width, travelling 9.33 ft (2.84 m) and in firing position 22.50 ft (6.86 m); height, travelling 9.00 ft (2.74 m)
Traverse/elevation: 60 degrees total/−2 to +65 degrees
Rate of fire: 1 round per 2 minutes (sustained fire) or 1 round per minute (burst fire)
Maximum range: 18,375 yards (16800 m)
Crew: 14

Variant
M115: introduced in 1940 as the M1 and redesignated M115 after World War II, this is the largest-calibre towed equipment in current service; the ordnance fires separate-loading ammunition with a seven-charge propellant system; the projectile types include HE (weighing 204 lb/92.53 kg and fired at a muzzle velocity of 1,925 ft/587 m per second to the weapon's maximum range), HE grenade-launching (weighing 200 lb/90.72 kg and carrying 104 M43 grenades, fired at the same muzzle velocity to the maximum range), HE grenade-launching (weighing 206.5 lb/93.66 kg and carrying 195 M42 grenades, fired at a muzzle velocity of 1,950 ft/594 m per second to a range of 17,500 yards/16000 m), chemical (weighing 204 lb and fired at a muzzle velocity of 1,925 ft per second to the maximum range) and nuclear (M422 projectile with a W33 warhead of 0.5- or 10-kiloton yield, or M753 projectile with a W79-1 warhead of 0.5-, 1- or 2-kiloton yield)

M198 155-mm howitzer
(USA)
Type: towed howitzer
Calibre: 155 mm (6.1 in)
Barrel length: 39.3 calibres
Muzzle brake: double baffle
Carriage: split-trail type with two road wheels; no shield
Weight: 15,790 lb (7162 kg) in travelling and firing orders
Dimensions: length, travelling with barrel forward 40.67 ft (12.40 m), travelling with barrel over trails 23.25 ft (7.09 m) and in firing position 37.08 ft (11.30 m); width, travelling 9.17 ft (2.79 m) and in firing position 28.00 ft (8.53 m); height, travelling with barrel forward 9.92 ft (3.02 m), travelling with barrel over trails 6.96 ft (2.12 m) and in firing position with barrel horizontal 5.92 ft (1.80 m)
Traverse/elevation: 45 degrees total/−5 to +72 degrees
Rate of fire: 4 rounds per minute
Maximum range: 19,850 yards (18150 m) with HE grenade-launching projectile and 32,800 yards (29995 m) with HE RAP
Crew: 11

Variant
M198: intended as successor to the M114 series, the M198 towed howitzer began to enter US service in 1978; the type is designed to fire standard NATO separate-loading ammunition, and projectile types that can be used are two types of HE mine-launchers (each weighing 102.5 lb/42.5 kg and carrying 36 anti-personnel mines), two HE grenade-launchers (a 102.57-lb/46.53-kg type with 88 dual-purpose grenades and a 95-lb/43.09-kg type with 60 anti-personnel grenades), an HE RAP (weighing 96 lb/43.54 kg and fired at a muzzle velocity of 2,710 ft/826 m per second), two HE mine-launchers (each weighing 103 lb/46.72 kg with nine anti-tank mines), the Copperhead laser-guided anti-tank projectile, three chemical/gas types, two illuminating types, two smoke types and one nuclear type (M454 with a 0.1-kiloton W48 warhead, to be replaced by the W82 enhanced-radiation 'neutron' warhead)

M59 155-mm gun
(USA)
Type: towed gun
Calibre: 155 mm (6.1 in)
Barrel length: 45 calibres
Muzzle brake: none
Carriage: split-trail type with eight road wheels and a two-wheel limber; no shield
Weights: 30,600 lb (13880 kg) in travelling order and 27,780 lb (12601 kg) in firing position
Dimensions: length, travelling 36.17 ft (11.02 m); width, travelling 8.25 ft (2.51 m); height, travelling 8.92 ft (2.72 m)
Traverse/elevation: 60 degrees total/−2 to +63 degrees
Rate of fire: 1 round per minute (sustained fire), or 2 rounds per minute (short periods) or 4 rounds per minute (burst fire)
Maximum range: 24,000 yards (21945 m)
Crew: 14

Variant
M59: introduced as the M1 series in 1938 and redesignated M59 after World War II, this 'Long Tom' is a cumbersome weapon but still very accurate; the M59 fires separate-loading ammunition, the three basic projectiles being a 95.7-lb (43.4-kg) HE type with a muzzle velocity of 2,745 ft (837 m) per second, a 99-lb (44.9-kg) AP type with a muzzle velocity of 2,800 ft (854 m) per second and able to penetrate 3 in (76 mm) of armour, and a smoke type

M114A1 howitzer
(USA)
Type: towed howitzer
Calibre: 155 mm (6.1 in)
Barrel length: 20 calibres
Muzzle brake: none
Carriage: split-trail type with two road wheels; shield
Weights: 12,785 lb (5799 kg) in travelling order and 12,700 lb (5761 kg) in firing position
Dimensions: length, travelling 24.00 ft (7.32 m); width, travelling 8.00 ft (2.44 m); height, travelling 5.92 ft (1.80 m)
Traverse/elevation: 24 degrees left and 25 degrees right/−2 to +63 degrees
Rate of fire: 40 rounds per hour (sustained fire), or 16 rounds in the first 10 minutes, or 8 rounds in the first four minutes, or 2 rounds in the first 30 seconds
Maximum range: 16,000 yards (14630 m)
Crew: 11

Variants
M114: introduced in 1941 as the M1 series, this type was redesignated M114 after World War II, and though obsolescent in its basic form it is still in very widespread service firing much of the standard NATO separate-loading ammunition; typical projectiles are the 94.6-lb (42.91-kg) HE type with a muzzle velocity of 1,850 ft (564 m) per second, the 95-lb (43.1-kg) HE grenade-launcher type with 60 anti-personnel grenades and the same muzzle velocity, three chemical/gas types, two illuminating types, two smoke types and the nuclear M454 with the 0.1-kiloton W48 warhead (to be replaced by the W82 enhanced-radiation 'neutron' warhead)
M114A1: version of the M114 on the M1A2 rather than M1A1 carriage with a screw- rather than rack/pinion-operated firing jack

M114A2: designation of US M114A1s fitted with the ordnance of the M198

M65: designation of the copy of the M114A1 made in Yugoslavia, and differing only in detail from the US original

M139: updated version developed by RDM (ex-Space Research Corporation) in the Netherlands to fire a new range of ammunition including an extended-range full-bore type; the ordnance has an L/39 barrel with a multi-baffle muzzle brake (increasing overall length to 10.00 m/32.81 ft) and a weight of 7500 kg (16,534 lb); range with the standard M107 HE projectile is 18100 m (19,975 yards), rising to 24600 m (26,900 yards) with the standard ERFB projectile and to 30400 m (33,245 yards) with the ERFB base-bleed projectile; the designation M139 covers new-build equipments, and the designation **M114/39** is allocated to conversions from M114 standard, both types having all but identical ballistic performances

FH155(L): updated version proposed by Rheinmetall in West Germany, with an improved shield, a new barrel and breech, and a double-baffle muzzle brake

T65: designation of the M114A1 made under licence in Taiwan, and generally similar to the US original apart from a weight of 5000 kg (11,023 lb) and a range of 15000 m (16,405 yards)

M102 105-mm howitzer
(USA)
Type: towed howitzer
Calibre: 105 mm (4.13 in)
Barrel length: 32 calibres
Muzzle brake: none
Carriage: box-type with two road wheels and a longitudinal roller for traverse; no shield
Weight: 3,300 lb (1497 kg) in travelling and firing orders
Dimensions: length, travelling 17.00 ft (5.18 m); width, travelling 6.44 ft (1.96 m); height, travelling 5.23 ft (1.59 m)
Traverse/elevation: 360 degrees on the baseplate/−5 to +75 degrees
Rate of fire: 10 rounds per minute
Maximum range: 12,575 yards (11500 m) with standard ammunition and 16,500 yards (15090 m) with HE RAP
Crew: 8

Variant
M102: designed as a lightweight successor to the M101 for service with airborne and air-mobile formations, the M102 entered the inventory in 1966 and is notable for its compact design and aluminium box trail; the type fires the same semi-fixed ammunition as the M101, but its longer barrel allows a slightly higher muzzle velocity and thus enhanced range

M101A1 105-mm howitzer
(USA)
Type: towed field howitzer
Calibre: 105 mm (4.13 in)
Barrel length: 22.5 calibres
Muzzle brake: none
Carriage: split-trail type with two road wheels; shield
Weight: 4,980 lb (2259 kg) in travelling and firing orders
Dimensions: length, travelling 19.67 ft (6.00 m); width, travelling 7.08 ft (2.16 m) and in firing position 12.00 ft (3.66 m); height, travelling 5.16 ft (1.57 m)
Traverse/elevation: 46 degrees total/−5 to +66 degrees
Rate of fire: 3 rounds per minute (sustained fire) and 10 rounds per minute (burst fire)
Maximum range: 12,325 yards (11270 m) with standard ammunition and 15,965 yards (14600 m) with HE RAP
Crew: 8

Variants
M101: introduced in 1940, the M2 series was redesignated M101 after World War II; though it is an obsolescent equipment by modern standards it is still in widespread service; the ordnance fires a wide range of NATO standard semi-fixed ammunition, the projectiles including one HE, two HE grenade-launching, one HE RAP, one HEP, one anti-personnel, one chemical, two smoke and one illuminating types; these have complete round weights varying from 33.45 to 46.4 lb (15.17 to 21.06 kg), and muzzle velocities ranging from 1,400 to 1,800 ft (427 to 549 m) per second

M101A1: improved version using the M2A2 rather than M2A1 carriage

FH105(L): improved version of the M101A1 developed in West Germany by Rheinmetall; this has an L/35.5 barrel and a single-baffle muzzle brake, the effect being an increase in the muzzle velocity of the standard HE round from 1,550 to 1,970 ft (472 to 600 m) per second and a lengthening of the range from 12,325 to 15,420 yards (11270 to 14100 m) at the expense of an additional 280 kg (617 lb) of weight and a weapon difficult to tow because it is unbalanced

105-mm Howitzer C1: Canadian licence-built version of the M101A1 with an improved ordnance and breech

M116 75-mm pack howitzer
(USA)
Type: towed pack howitzer
Calibre: 75 mm (2.95 in)
Barrel length: 15.9 calibres
Muzzle brake: none
Carriage: box-type with two road wheels; shield
Weight: 1,440 lb (653 kg) in travelling and firing orders

Dimensions: length, travelling 12.00 ft (3.66 m); width, travelling 3.91 ft (1.19 m); height, travelling 3.08 m (0.94 m)
Traverse/elevation: 6 degrees total/−5 to +45 degrees
Rate of fire: 3 rounds per minute (sustained fire) and 6 rounds per minute (burst fire)
Maximum range: 9,600 yards (8780 m)
Crew: 5

Variant
M116: introduced in 1927 as the M1 series, this equipment became the M116 after World War II, and was designed for pack animal transport (eight mule-transportable loads), paradropping (nine loads), or towing by a light vehicle; the type fires either fixed ammunition (HEAT-T with a 15.65-lb/7.1-kg projectile) or semi-fixed ammunition (HE with a 18.25-lb/8.27-kg projectile, and smoke)

M56 105-mm howitzer
(Yugoslavia)
Type: towed howitzer
Calibre: 155 mm (6.1 in)
Barrel length: 27.9 calibres
Muzzle brake: multi-baffle
Carriage: split-trail type with two road wheels; shield
Weights: 2100 kg (4,630 lb) in travelling order and 2060 kg (4,541 lb) in firing position
Dimensions: length, travelling 6.17 m (20.24 ft); width, travelling 2.15 m (7.05 ft); height, travelling 1.56 m (5.12 ft)
Traverse/elevation: 52 degrees total/−12 to +68 degrees
Rate of fire: 16 rounds per minute
Maximum range: 13000 m (14,215 yards)
Crew: 7

Variant
M56: this Yugoslav equipment was introduced in the second half of the 1950s and is based loosely on the German leFH 18/40 but firing US M1 ammunition; the ordnance fires semi-fixed ammunition such as HE, AP-T, HESH-T, illuminating and smoke

M48(B-1) 76-mm mountain gun
(Yugoslavia)
Calibre: 76.2 mm (3 in)
Barrel length: 15.45 calibres
Muzzle brake: multi-baffle
Carriage: folding split-trail type with two road wheels; shield
Weights: 720 kg (1,587 lb) in travelling order and 705 kg (1,554 lb) in firing position
Dimensions: length, travelling 2.42 m (7.94 ft); width, travelling 1.46 m (4.79 ft); height, travelling 1.22 m (4.00 ft)
Traverse/elevation: 50 degrees total/−15 to +45 degrees
Rate of fire: 25 rounds per minute
Maximum range: 8750 m (9,570 yards)
Crew: 6

Variant
M48: designed after World War II to meet the special requirements of Yugoslav mountain units, this equipment entered service in the late 1940s and has proved moderately successful; there are four subtypes, namely the **M48(B-1)** basic model that can be broken down into eight mule loads, the **M48(B-1A1-1)** that cannot be broken down for pack transport, the **M48(B-1A2)** with alloy wheels and solid tyres, and the **M48B-2** updated version; the ordnance fires semi-fixed HE, HEAT and smoke rounds

AA SYSTEMS (SELF-PROPELLED AND TOWED)

Steyr 4K tracked self-propelled AA gun mountings (Austria): see Steyr 4K series tracked armoured personnel carriers

Type 63
(China)
Type: tracked self-propelled AA gun mounting
Crew: 6
Combat weight: 32000 kg (70,547 lb)
Dimensions: length, guns forward 6.432 m (21.10 ft) and hull 7.53 m (24.70 ft); width 3,27 m (10.73 ft); height to top of turret 2.995 m (9.83 ft)
Armament system: two 37-mm cannon with ? rounds in an electrically-powered turret
Armour: cast and welded steel
Powerplant: one 373-kW (500-hp) V-2-34 diesel engine with 590 litres (130 Imp gal) of internal fuel
Performance: speed, road 55 km/h (34 mph); range, road 300 km (186 miles); fording 1.32 m (4.33 ft) without preparation; gradient 60%; vertical obstacle 0.73 m (29 in); trench 2.5 m (8.2 ft); ground clearance 0.41 m (16 in)

Variant
Type 63: this is a Chinese-designed equipment based on the chassis of the obsolete T-34 medium tank; the tank turret has been replaced by a tall slab-sided turret with an open top, this mounting the two 37-

mm cannon and a ready-use supply of FRAG-T and AP-T ammunition loaded in five-round clips (and replenished from boxes attached to the outside of the hull); the turret can traverse through 360 degrees, and the cannon can be elevated in an arc from −5 to +85 degrees; fire-control is of the basic optical type, and the cannon have a practical rate of fire (per barrel) of 80 rounds per minute; the projectiles have a muzzle velocity of 880 m (2,887 ft) per second, the effective slant range being 3000 m (3,280 yards) and armour-penetration capability 46 mm (1.81 in) at 500 m (550 yards); the cannon have a horizontal range of 9500 m (10,390 yards)

M53/59
(Czechoslovakia)
Type: wheeled self-propelled AA gun mounting
Crew: 5
Combat weight: 10300 kg (22,707 lb)
Dimensions: length 6.92 m (22.70 ft); width 2.35 m (7.71 ft); height including magazines 2.95 m (9.68 ft)
Armament system: two 30-mm M53 cannon with 800 rounds on an electrically-powered turntable mounting
Armour: welded steel to a maximum thickness of 10 mm (0.39 in)
Powerplant: one 82-kW (110-hp) Tatra T 212-2 diesel engine with 120 litres (26 Imp gal) of fuel, driving a 6 x 6 layout
Performance: speed, road 60 km/h (37.3 mph); range, road 500 km (311 miles); fording 0.8 m (2.6 ft); gradient 60%; side slope 30%; vertical obstacle 0.46 m (18.1 in); trench 0.69 m (2.25 ft); ground clearance 0.4 m (15.75 in)

Variants
M53/59: introduced in the late 1950s, this mounting combines the chassis and automotive system of the Praga V3S truck with an armoured body and turntable-mounted armament; traverse is 360 degrees and elevation from −2 to +85 degrees, and ammunition feed comprises a 50-round magazine over each weapon (fed by 10-round clips of HEI and API ammunition) for cyclic and practical rates of 500 and 150 rounds per minute; the horizontal range is 9700 m (10,610 yards) and the effective AA range 3000 m (9,845 ft); the system is essentially a clear-weather type, and it suffers the disadvantages of lacking NBC protection and night-vision devices
M53/70: later version with improved fire control

GIAT Shahine, SICA tracked self-propelled SAM launcher (France): see GIAT AMX-30 MBT

GIAT AMX-30R tracked self-propelled SAM launcher (France): see GIAT AMX-30 MBT

AMX-30SA tracked self-propelled AA gun system (France/Saudi Arabia): see GIAT AMX-30 MBT

GIAT AMX-13 DCA
(France)
Type: tracked self-propelled AA gun system
Crew: 3
Combat weight: 17200 kg (37,919 lb)
Dimensions: length 5.40 m (17.72 ft); width 2.50 m (8.20 ft); height with radar erected 3.80 m (12.47 ft)
Armament system: two 30-mm Hispano-Suiza HSS-831A cannon with 600 rounds and two smoke-dischargers on each side of the hydraulically-powered turret
Armour: welded steel varying in thickness from 10 to 20 mm (0.39 to 0.79 in)
Powerplant: one 186-kW (250-hp) SOFAM 8Gxb petrol engine with 415 litres (91 Imp gal) of fuel
Performance: speed, road 60 km/h (37.3 mph); range, road 300 km (186 miles); fording 0.6 m (2.0 ft); gradient 60%; side slope 30%; vertical obstacle 0.65 m (25.6 in); trench 1.7 m (5.6 ft); ground clearance 0.43 m (16.9 in)

Variant
AMX-13 DCA: this is the AMX-13 light tank chassis (in fact identical with that of the Mk 61 self-propelled howitzer) fitted with the SAMM S 401A turret carrying two 30-mm cannon, Oeil Noir 1 search and ranging radar, and a fire-control system; the turret traverses through 360 degrees, and the guns can be elevated from −5 to +85 degrees; ammunition types are HEI, HEI-T, SAPHEI-T, TP-T and TP, the cyclic rate 600 rounds per gun, and the maximum effective AA range 3500 m (11,485 ft); deliveries began in 1968

GIAT AMX-10 RAA wheeled self-propelled AA gun mounting (France): see GIAT AMX-10 RC wheeled armoured reconnaissance vehicle

Panhard ERC wheeled self-propelled AA gun mountings (France): see Panhard ERC 90 F4 Sagaie wheeled armoured car

Panhard AML wheeled self-propelled AA gun mountings (France): see Panhard AML H 90 light wheeled armoured car

Renault VAB wheeled self-propelled SAM launcher and AA gun mounting (France): see Renault VAB TT wheeled armoured personnel carrier

Panhard M3/VDA wheeled self-propelled AA gun mounting (France): see Panhard M3 wheeled armoured personnel carrier

Thomson-CSF/Matra Crotale P4R
(France)
Type: wheeled self-propelled AA missile system
Crew: 2 each in the radar and SAM vehicles
Combat weight: radar vehicle 12500 kg (27,557 lb) and SAM vehicle 14800 kg (32,628 lb)

Dimensions: length 6.22 m (20.41 ft); width 2.65 m (8.69 ft); height 2.04 m (6.69 ft)
Armament system: (SAM vehicle only) four ready-to-fire R.440 SAMs mounted two on each side of the electrically-powered turret/radar unit
Armour: welded steel varying in thickness from 3 to 5 mm (0.12 in to 0.2 in)
Powerplant: one petrol engine
Performance: speed, road 70 km/h (43.5 mph); range, road 500 km (311 miles); fording 0.68 m (2.25 ft); gradient 40%; vertical obstacle 0.3 m (11.8 in); ground clearance 0.45 m (17.7 in)

Variants
Cactus: initial version of this all-weather low-altitude mobile air-defence system, whose development was funded primarily by South Africa
Crotale P4R: standard version in two halves; the radar vehicle has a Thomson-CSF Mirador IV pulse-Doppler radar able to detect targets at 18-km (11.2-mile) range, and its computer system can handle 30 targets while tracking 12 simultaneously and passing information to the SAM vehicle, which has 17-km (10.6-km) range monopulse radar for the tracking of one target and the guiding of two SAMs
Crotale 4000: upgraded version of the Crotale P4R

Krauss-Maffei/Contraves 5PZF-B2 Gepard
(West Germany/Switzerland)
Type: tracked self-propelled AA gun system
Crew: 3
Combat weight: 47300 kg (104,277 lb)
Dimensions: length, guns forward 7.73 m (25.36 ft) and hull 6.85 m (22.47 ft); width 3.37 m (11.06 ft); height with radar erected 4.03 m (13.22 ft)
Armament system: two 35-mm Oerlikon KDA cannon with 660 rounds and four smoke-dischargers on each side of the hydraulically-powered turret
Armour: welded steel
Powerplant: one 619-kW (830-hp) MTU MB 838 Ca M500 multi-fuel engine with 985 litres (217 Imp gal) of fuel
Performance: speed, road 65 km/h (40.4 mph); range, road 550 km (342 miles); fording 2.5 m (8.2 ft); gradient 60%; vertical obstacle 1.15 m (45.3 in); trench 3.0 m (9.8 ft); ground clearance 0.44 m (17.3 in)

Variants
5PZF-B2 Gepard: entering service in 1976, this extremely capable all-weather low-level air-defence system is based on the hull and automotive system of the Leopard 1 MBT (though slightly lengthened and carrying thinner armour) fitted with an Oerlikon turret with twin 35-mm cannon, a Contraves fire-control system and Siemens radars; these last are an MPDR-12 pulse-Doppler search radar with a range of 15 km (9.3 miles) and a similarly-ranged Siemens-Albis tracking radar, whose inputs are fed into the fire-control system to produce a fire solution for the two cannon, which have a cyclic rate of 550 rounds per minute per gun and are provided with HEI, HEI-T, SAPHEI-T, APDS-T, TP and TP-T ammunition; normal target engagement range is in the order of 2000 to 3000 m (2,185 to 3,280 yards)
5PZF-B2L: improved version with a Siemens laser rangefinder
5PZF-C CA 1: version for the Dutch army with Hollandse Signaalapparaten pulse-Doppler integrated search and tracking radars; this revised radar fit reduces overall height to 3.70 m (12.14 ft) on what the Dutch army designates the **PRTL**

Euromissile Roland/Marder
(West Germany/France)
Type: tracked self-propelled AA missile system
Crew: 3
Combat weight: 32500 kg (71,649 lb)
Dimensions: length 6.915 m (22.69 ft); width 3.24 m (10.63 ft); height with radar folded 2.92 m (9.58 ft)
Armament system: two launchers with 10 Roland 2 SAMs mounted on each side of the electro-hydraulically powered turret/radar unit, and one 7.62-mm (0.3-in) MG3 machine-gun for local defence
Armour: welded steel
Powerplant: one 447-kW (600-hp) MTU MB 833 Ea 500 diesel engine with 652 litres (143 Imp gal) of fuel
Performance: speed, road 70 km/h (43.5 mph); range, road 520 km (323 miles); fording 1.5 m (4.9 ft) with preparation; gradient 60%; vertical obstacle 1.15 m (45.3 in); trench 2.5 m (8.2 ft); ground clearance 0.44 m (17.3 in)

Variants
Roland/Marder: entering service in 1978, this mobile battlefield SAM system is designed for the engagement of low-level targets ranging from stationary helicopters to supersonic aircraft, and is based on the hull of the Marder MICV with a mounting above the rear hull for two Roland 2 missile launchers and the radars; these later are a Siemens pulse-Doppler search radar, which has a search range of 18 km (11.2 miles), and a Siemens monopulse tracking radar to which any targets detected by the search radar are then allocated; there is also an optical sight for daylight use
Roland/Leopard: currently under development, this system is based on the hull of the Leopard 1 MBT

Hägglund & Söner Lvrbv 701
(Sweden)
Type: tracked self-propelled AA missile system
Crew: 4
Combat weight: 9700 kg (21,384 lb)
Dimensions: length 4.81 m (15.78 ft); width 2.54 m

(8.33 ft); height with missile system raised 2.98 m (9.78 ft)
Armament system: one extendible launcher with ? RBS 70 Rayrider
Armour: welded steel
Powerplant: one 136-hp (101-kW) Ford Model 2658E petrol engine with 260 litres (57 Imp gal) of fuel
Performance: speed, road 41 km/h (25.5 mph); range, road 350 km (217 miles); fording 0.9 m (3.0 ft); gradient 60%; vertical obstacle not revealed; trench 1.5 m (4.9 ft); ground clearance 0.33 m (13 in)

Variants
Lvrbv 701: entering service with the Swedish army in the mid-1980s, this is a conversion of obsolete Ikv 102 and Ikv 103 infantry cannon vehicles to low-altitude air-defence vehicles; the infantry cannon hulls have been rebuilt, with the gun aperture plated over, the lengthening of the crew compartment, the replacement of the powerplant and transmission and the upgrading of communications; the elevating SAM launcher (surmounted by an identification friend or foe device) rises through a two-piece hatch for firing
PS 701/R: radar-warning version fitted with Ericsson Giraffe radar on an elevating arm

BRDM-2/SA-9 'Gaskin' wheeled self-propelled SAM launcher (USSR):
see BRDM-2 wheeled scout car

ZSU-57-2
(USSR)
Type: tracked self-propelled AA gun mounting
Crew: 6
Combat weight: 28100 kg (61,949 lb)
Dimensions: length, guns forward 8.48 m (27.82 ft); width 3.27 m (10.73 ft); height 2.75 m (9.02 ft)
Armament system: two 57-mm S-68 cannon with 316 rounds in an electro-hydraulically powered turret
Armour: welded steel varying in thickness from 10.6 to 15 mm (0.42 to 0.59 in)
Powerplant: one 388-kW (520-hp) V-54 diesel engine with 812 litres (179 Imp gal) of internal fuel and 400 litres (88 Imp gal) of auxiliary fuel
Performance: speed, road 50 km/h (31 mph); range, road 420 km (261 miles) with standard fuel and 595 km (370 miles) with auxiliary fuel; fording 1.4 m (4.6 ft); gradient 60%; side slope 30%; vertical obstacle 0.8 m (31.5 in); trench 2.7 m (8.9 ft); ground clearance 0.425 m (16.75 in)

Variant
ZSU-57-2: now obsolescent, this was at the time of its introduction in the mid-1950s an extremely potent air-defence weapon with a twin 57-mm cannon turret on a hull based on that of the T-54 but carrying thinner armour; its limitations are comparatively slow rates of traverse and elevation, and restriction to clear-weather operations because of its lack of radar; the guns each have a cyclic/practical rate of 120/70 rounds per minute, and fire FRAG-T or APC-T rounds to a maximum horizontal range of 12000 m (13,125 yards) and a maximum AA range of 4000 m (13,125 ft)

ZSU-23-4
(USSR)
Type: tracked self-propelled AA gun system
Crew: 4
Combat weight: 19000 kg (41,887 lb)
Dimensions: length 6.54 m (21.46 ft); width 2.95 m (9.68 ft); height without radar 2.25 m (7.38 ft)
Armament system: four 23-mm AZP-23 cannon with 2,000 rounds in an electrically-powered turret
Armour: welded steel varying in thickness from 10 to 15 mm (0.39 to 0.59 in)
Powerplant: 209-kW (280-hp) V-6R diesel engine with 250 litres (55 Imp gal) of fuel
Performance: speed, road 44 km/h (27.3 mph); range, road 260 km (162 miles); fording 1.07 m (3.5 ft); gradient 60%; side slope 30%; vertical obstacle 1.1 m (43.3 in); trench 2.8 m (9.2 ft); ground clearance 0.4 m (15.75 in)

Variant
ZSU-23-4: this is a highly capable AA gun system based on components of the PT-76 light amphibious tank, and the hull is all but identical with that of the SA-6 'Gainful' SAM system; the type entered service in the mid-1960s, and is designed for the short-range defence of divisions and regiments against low-level air threats; the radar used in the ZSU-23-4 is the 'Gun Dish' type, which is mounted above the rear of the turret; this radar can detect possible targets at a range of 20 km (12.4 miles) and track them at 8 km (5 miles); the four water-cooled cannon each have a cyclic rate between 800 and 1,000 rounds per minute, and fire HEI-T and API-T ammunition to a maximum effective range of 2500 m (8,200 ft) in the AA and surface roles; the turret can traverse through 360 degrees, and the guns can be elevated from −4 to +85 degrees, high traverse and elevation rates allowing the ZSU-23-4 to engage fast-moving and fast-crossing targets even at short ranges; there are a number of subvariants of the basic ZSU-23-4, most of them featuring alterations to the cooling and stowage facilities; standard equipment includes an NBC system and night-vision devices
ZSU-23-4M: latest version with ammunition boxes attached to the turret sides, a digital fire-control computer, and improved 'Gun Dish' radar that can be linked to off-set radars and fire-control systems

SA-4 'Ganef'
(USSR)
Type: tracked self-propelled AA missile system
Crew: 5
Combat weight: 25000 kg (55,115 lb)
Dimensions: length, with missiles 9.46 m (31.04 ft) and hull 7.80 m (25.59 ft); width, vehicle 3.20 m (10.50 ft); height with missiles 4.472 m (14.67 ft)
Armament system: one twin-rail launcher for two 'Ganef' SAMs
Armour: welded steel
Powerplant: one 447-kW (600-hp) diesel engine with 500 litres (110 Imp gal) of fuel
Performance: speed, road 50 km/h (31 mph); range, road 300 km (186 miles); fording not revealed; gradient 60%; vertical obstacle 1.0 m (39.4 in); trench 3.2 m (10.5 ft); ground clearance 0.44 m (17.3 in)

Variants
SA-4 'Ganef': this is a medium-range missile system designed for the medium- and high-altitude air defence of Soviet fronts (army groups) and armies within the integrated Soviet air-defence network for mobile forces; the system was introduced in the early 1960s, and though it lacks night-vision devices and amphibious capability, it is fitted with an NBC system; target detection at ranges of more than 100 km (62 miles) is tasked to separate regimental-level 'Long Track' surveillance and 'Thin Skin' height-finding radars, information being passed to the 'Ganef' battery's 'Pat Hand' fire-control and missile-guidance radar, which assumes control once the target is in range; this controls the training and elevation of the launcher, and the firing and guidance of the missile until the latter's semi-active radar terminal homing system takes over; the engagement envelope includes minimum/maximum ranges of 9.3/72 km (5.8/44.7 miles) and minimum/maximum altitudes of 1100/24000 m (3,610/78,740 ft)
SA-4/'Pat Hand': derivative of the 'Ganef' launcher with 'Pat Hand' fire-control and missile-guidance radar
GMZ: minelayer using the same hull

SA-6 'Gainful'
(USSR)
Type: tracked self-propelled AA missile system
Crew: 3
Combat weight: 14000 kg (30,864 lb)
Dimensions: length including missiles 7.389 m (24.24 ft) and hull 6.79 m (22.28 ft); width 3.18 m (10.43 ft); height including missiles 3.45 m (11.32 ft)
Armament system: one three-rail launcher for three 'Gainful' SAMs
Armour: welded steel varying in thickness from 10 to 15 mm (0.39 to 0.59 in)
Powerplant: one 209-kW (280-hp) V-6R diesel engine with 250 litres (55 Imp gal) of fuel
Performance: speed, road 44 km/h (27.3 mph); range, road 250 km (162 miles); fording 1.0 m (3.3 ft); gradient 60%; side slope 30%; vertical obstacle 1.1 m (43.3 in); trench 2.8 m (9.2 ft); ground clearance 0.4 m (15.75 in)

Variants
SA-6 'Gainful': based on the same chassis as the ZSU-23-4 self-propelled AA gun system, the SA-6 'Gainful' is designed for the medium-range protection of armies and divisions from low- and medium-altitude air threats, and entered service in the mid-1960s; the type has NBC protection and night-vision devices, but lacks amphibious capability; target detection at ranges of more than 100 km (62 miles) is undertaken by regimental-level 'Long Track' surveillance and 'Thin Skin' height-finding radars, data being passed to the SA-6 battery's 60/90-km (37/56-mile) range 'Straight Flush' target-tracking, fire-control and missile-guidance radar for the preparation of a fire-problem solution leading to missile launch and guidance to the point at which the missile's semi-active radar terminal homing system takes over; the engagement envelope with the standard SA-6A missile has minimum/maximum ranges of 4/30 km (2.5/18.7 miles) and minimum/maximum altitudes of 100/18000 m (330/59,055 ft)
SA-6/'Straight Flush': version of the SA-6 hull mounting the 'Straight Flush' target-tracking and missile-guidance radar
SA-11 'Gadfly': entering service in the mid-1980s, the SA-11 system is essentially similar to the SA-6 apart from its use of a new missile, offering an engagement envelope with minimum/maximum ranges of 3/30 km (1.86/18.6 miles) and minimum/maximum altitudes of 30/15000 m (100/49,215 ft)
SA-11/'Clamshell': version of the SA-6/SA-11 hull with the radar associated with the SA-11 system, notably the 'Clamshell' 3D surveillance radar and 'Flap Lid' tracking and guidance radar

SA-8A 'Gecko'
(USSR)
Type: wheeled self-propelled AA missile system
Crew: 4
Combat weight: 9000 kg (19,841 lb)
Dimensions: length 9.14 m (30.00 ft); width 2.90 m (9.51 ft); height with radar folded 4.2 m (13.78 ft)
Armament system: one four-rail launcher for four 'Gecko' SAMs
Armour: none
Powerplant: one 130-kw (174-hp) diesel engine with ? litres (? Imp gal) of fuel
Performance: speed, road 65 km/h (40.4 mph); range, road not known; fording amphibious driven by a waterjet; gradient not known; vertical obstacle not known; trench not known; ground clearance not known

Variants
SA-8A 'Gecko': entering service in 1974, the 'Gecko' system is designed for the short-range air-defence of divisions against low- and medium-altitude air threats; the system is based on a wheeled hull with 6 x 6 drive and has NBC protection plus night-vision devices; operating with the support of regimental-level 'Long Track' surveillance and 'Thin Skin' height-finding radars, the radar system of the 'Gecko' is designated 'Land Role' by NATO, and comprises a 30-km (18.6-mile) range surveillance radar, a target-tracking radar and two missile-guidance radars, each of the last being topped by a low-light-level TV for alternative use in ECM conditions; the engagement envelope for the system includes minimum/maximum ranges of 1600/12000 m (1,750/13,125 yards) and minimum/maximum altitudes of 50/13000 m (50/ 42,650 ft)
SA-8B 'Gecko': upgraded version introduced in the late 1970s with three launch canisters rather than two launch rails on each side of the surveillance radar; these canisters carry a lengthened and more capable missile with consequent improvement of the engagement envelope

SA-13 'Gopher'
(USSR)
Type: tracked self-propelled AA missile system
Crew: 3/4
Combat weight: 12500 kg (27,557 lb)
Dimensions: length 6.60 m (21.65 ft); width 2.90 m (9.51 ft); height in travelling position 2.30 m (7.55 ft)
Armament system: four container-launchers for up to 20 'Gopher' SAMs
Armour: welded steel to a maximum thickness of 14 mm (0.55 in)
Powerplant: one 179-kW (240-hp) YaMZ 238 V diesel engine with ? litres (? Imp gal) of fuel
Performance: speed, road 55 km/h (34 mph); range, road 450 km (280 miles); fording amphibious; gradient 60%; side slope 30%; vertical obstacle 0.7 m (27.6 in); trench 2.7 m (8.9 ft); ground clearance 0.4 m (15.75 in)

Variants
SA-13 'Gopher': being introduced from the mid-1980s as a more mobile replacement for the SA-9 system with less limited missiles, the SA-13 system is based on the hull of the MT-LB multi-role vehicle and is designed for the short-range defence of divisions against low-level air threats; the IR-homing missiles are carried in two pairs, one of each side of the elevating launcher with its centrally-mounted tracking radar; the engagement envelope includes minimum/ maximum ranges of 500/1000 m (550/10000 yards) and minimum/maximum altitudes of 10/10000 m (33/ 32,810 ft), though the effective ceiling is thought to be 3200 m (10,500 ft); there are two subvariants of the system, designated **SA-13 TELAR-1** and **SA-13 TELAR-2** by NATO, the former having four 'Hat Box' passive radar-detection antennae and the latter none; the SA-13 system is sometimes seen with 'Gaskin' missiles

BTR-152A wheeled self-propelled AA gun mounting
(USSR): see BTR-152V1 wheeled armoured personnel carrier

BTR-40A wheeled self-propelled AA gun mounting
(USSR): see BTR-40 wheeled armoured personnel carrier

BAe Dynamics Tracked Rapier
(UK)
Type: tracked self-propelled AA missile system
Crew: 3
Combat weight: 30,885 lb (14009 kg)
Dimensions: length 21.00 ft (6.40 m); width 9.19 ft (2.80 m); height with tracker raised 9.12 ft (2.78 m)
Armament system: one launch system with two four-round container-launchers for eight Rapier SAMs, plus two six-tube (vehicle front) and two four-tube (vehicle rear) smoke-dischargers
Armour: welded aluminium to a maximum thickness of 1 in (25.4 mm)
Powerplant: one 210-hp (157-kW) Detroit Diesel 6V-53 diesel engine with 87.5 Imp gal (386 litres) of fuel
Performance: speed, road 30 mph (48 km/h) and water 3.5 mph (5.6 km/h) driven by its tracks; range, road 185 miles (298 km); fording amphibious with screen; gradient 60%; side slope 30%; vertical obstacle 24 in (0.61 m); trench 5.5 ft (1.68 m); ground clearance 16 in (0.41 m)

Variant
Tracked Rapier: introduced to service in 1983, this is the high-mobility version of the Rapier SAM system based on the hull of the US M548 cargo carrier, and is designed for the short-range defence of armoured and mechanized divisions against low-altitude air threats; apart from the eight Rapiers, the rear-mounted launcher contains the pulse-Doppler surveillance radar and mounts the elevating antenna for the missile guidance system; this latter uses a microwave link for automatic command-to-line-of-sight guidance using a cab-mounted optical (TV) tracker; Blindfire radar is an optional trailer-mounted extra for use in adverse weather conditions; the engagement envelope includes minimum/maximum ranges of 900/ 7,100+ yards (825/6490+ m) and minimum/ maximum altitudes of 0/10,000 ft (0/3050 m); the initial **Tracked Rapier Mk 1A** was produced in limited numbers as it is capable only of clear-weather operations, while the main production version is the **Tracked Rapier Mk 1B** with TOTE (Tracker, Optical, Thermally Enhanced) for adverse-weather and night

capability; Mk 1A equipments are to be upgraded to Mk 1B standard in due course; the type is also under development for possible US Army purchase with a revised launcher mounting a pair of external 0.5-in (12.7-mm) M2HB heavy machine-guns, and a number of electronic improvements such as capability for on-the-move surveillance and thermal sights for the commander and SAM operator

FMC M48 Chaparral tracked self-propelled SAM launcher (USA): see FMC M113A1 tracked armoured personnel carrier

FMC M163 tracked self-propelled AA gun mounting (USA): see FMC M113A1 tracked armoured personnel carrier

Cadillac Gage Vulcan-Commando wheeled self-propelled AA gun mounting (USA): see Cadillac Gage Commando V-150 wheeled armoured personnel carrier

Cadillac (General Motors) M42A1
(USA)
Type: tracked self-propelled AA gun mounting
Crew: 6
Combat weight: 49,500 lb (22453 kg)
Dimensions: length, guns forward 20.85 ft (6.36 m) and hull 19.09 ft (5.82 m); width 10.58 ft (3.23 m); height 9.33 ft (2.84 m)
Armament system: two 40-mm M2A1 cannon with 480 rounds on a hydraulically-powered turntable mounting, and one 7.62-mm (0.3-in) AA machine-gun with 1,750 rounds
Armour: welded steel varying in thickness from 0.35 to 1.2 in (9.05 to 31 mm)
Powerplant: one 500-hp (373-kW) Continental AOSI-895-5 petrol engine with 140 US gal (530 litres) of fuel
Performance: speed, road 45 mph (72.4 km/h); range, road 100 miles (161 km); fording 3.33 ft (1.02 m); gradient 60%; side slope 30%; vertical obstacle 28 in (0.71 m); trench 6.0 ft (1.83 m); ground clearance 17.25 in (0.44 m)

Variant
M42A1: based on the hull of the M24 light tank, the M42 with normally-aspirated AOS-895-3 engine was introduced in 1952 but was replaced from 1956 by the fuel-injected M42A1, to which standard most M42s were improved; the type was designed for the low-level defence of armoured and mechanized divisions, and is now thoroughly obsolete as it lacks armament protection, radar search and control, an NBC system and other modern essentials; the armament turntable ensures a traverse of 360 degrees, and the guns can be elevated from −5 to +85 degrees; the practical rate of each barrel is 120 rounds per minute, and the maximum AA range is 16,400 ft (5000 m); the ammunition types available are AP-T, HE-T and TP-T

BOV/20-mm
(Yugoslavia)
Type: wheeled self-propelled AA gun mounting
Crew: 4
Combat weight: 800 kg (17,637 lb)
Dimensions: length 6.00 m (19.69 ft); width 2.00 m (6.56 ft); height 3.00 m (9.84 ft)
Armament system: three 20-mm Hispano-Suiza HSS-666 cannon with ? rounds and ? smoke-dischargers mounted on each side of the powered turret
Armour: welded steel
Powerplant: one 104-kW (140-hp) engine with ? litres (? Imp gal) of fuel, driving a 4 x 4 layout
Performance: speed, road 100 km/h (62 mph); range, road 500 km (311 miles); fording amphibious; gradient 60%; side slope 30%; vertical obstacle not known; ground clearance not known

Variant
BOV/20-mm: this is a simple wheeled AA mounting with three 20-mm cannon on the open-topped turret together with a neat electro-optical tracking system; the turret can be traversed through 360 degrees, and the guns elevated from −5 to +80 degrees; each barrel has a practical rate of 150 rounds per minute, and ammunition feed is from 50-round drums; the effective AA range is 1500 m (4,920 ft), and the cannon fire AP, API, API-T, HE, HEI and HEI-T ammunition

M53 30-mm AA gun mounting
(Czechoslovakia)
Type: towed twin 30-mm AA gun mounting
Calibre: 30 mm
Barrel length: 81 calibres overall
Carriage: four-wheel platform with outriggers; no shield
Weights: 2100 kg (4,630 lb) in travelling order and 1750 kg (3,858 lb) in firing position
Dimensions: length, travelling 7.587 m (24.89 ft); width, travelling 1.758 m (5.77 ft); height, travelling 1.575 m (5.17 ft)
Traverse/elevation: 360 degrees/−10 to +85 degrees
Rate of fire (per barrel): 450/500 rounds per minute (cyclic) and 100 rounds per minute (practical)
Horizontal range: 9700 m (10,610 yards) maximum
Slant range: 3000 m (3,280 yards) effective
Crew: 4

Variant
M53: this weapon entered service in the late 1950s, and offers considerable range advantages over Soviet towed 23-mm equipments; the M53 suffers from lack of radar fire-control, and is thus limited to clear-

weather operation; ammunition is fed by 10-round clips and the two basic ammunition types are HEI (with a 0.45-kg/0.99-lb projectile) and AP-T (with a 0.54-kg/1.2-lb projectile able to pierce 55 mm/2.16 of armour at 500 m/550 yards); both types are fired at a muzzle velocity of 1000 m (3,281 ft) per second

M53 12.7-mm AA gun mounting
(Czechoslovakia)
Type: towed quadruple AA gun mounting
Calibre: 12.7 mm (0.5 in)
Barrel length: 125 calibres overall
Carriage: two-wheel platform with outriggers; no shield
Weights: 2830 kg (6,239 lb) in travelling order and 628 kg (1,384 lb) in firing position
Dimensions: length, travelling 2.90 m (9.51 ft); width, travelling 1.57 m (5.15 ft); height, travelling 1.78 m (5.84 ft)
Traverse/elevation: 360 degrees/−7 to +90 degrees
Rate of fire (per barrel): 550/600 rounds per minute (cyclic) and 80 rounds per minute (practical)
Horizontal range: 6500 m (7,110 yards) maximum
Slant range: 5600 m (6,125 yards) maximum and 1000 m (1,095 yards) effective
Crew: 6

Variant
M53: introduced in the mid-1950s, this equipment combines four Soviet DShK heavy machine-guns with a Czech mounting and carriage; the machine-guns are fed from 50-round belts, and fire a 49.5-g (1.75-oz) API projectile at a muzzle velocity of 840 m (2,756 ft) per second, sufficient to penetrate 20 mm (0.79 in) of armour at 500 m (550 yards)

GIAT 76T2 Cerbère 20-mm AA gun mounting
(France)
Type: towed twin AA gun mounting
Calibre: 20 mm
Barrel length: 103.25 calibres overall
Carriage: two-wheel platform with outriggers; shield
Weights: 2019 kg (4,451 lb) in travelling order without ammunition and 1513 kg (3,336 lb) in firing position with ammunition
Dimensions: length, travelling 5.05 m (16.57 ft) and in firing position 4.02 m (13.19 ft); width, travelling 2.39 m (7.84 ft) and in firing position 2.06 m (6.76 ft); height, travelling 2.075 m (6.81 ft) and in firing position 2.06 m (6.76 ft)
Traverse/elevation: 360 degrees/−3.5 to +81.5 degrees powered or −5 to +83 degrees manually
Rate of fire (per barrel): 900 rounds per minute (cyclic)
Horizontal range: not revealed
Slant range: 2000 m (2,185 yards) effective
Crew: 3

Variant
76T2 Cerbère: this is a French development of the West German MK 20 Rh 202 mounting with the original cannon replaced by two French GIAT M693 (F2) weapons; the type can be operated in the powered or manual modes, and there is a selectable dual-feed mechanism, ammunition supply amounting to 270 rounds per barrel; the Cerbère can be used with its on-mounting sight, or in conjunction with a radar fire-control system, or with a helmet-mounted target-indicator system

GIAT 53T2 Tarasque 20-mm AA gun mounting
(France)
Type: towed single 20-mm AA gun mounting
Calibre: 20 mm
Barrel length: 103.25 calibres overall
Carriage: two-wheel platform with outriggers; shield
Weights: 840 kg (1,852 kg) in travelling order with ammunition and 660 kg (1,455 lb) in firing position with ammunition
Dimensions: length, travelling not revealed; width, travelling 1.90 m (6.23 ft); height, travelling not revealed
Traverse/elevation: 360 degrees/−8 to +83 degrees
Rate of fire: 740 rounds per minute (cyclic) and 200 rounds per minute (practical)
Horizontal range: 6000 m (6,560 yards) maximum
Slant range: 2000 m (2,185 yards) effective
Crew: 3

Variant
53T2 Tarasque: introduced in 1982, the Tarasque is a light AA mounting (with secondary anti-AFV and anti-personnel capabilities) designed for rapid cross-country movement; the M693 (F2) weapon is hydraulically powered, and is dual-fed by belts for 100 HEI and 40 APDS rounds, the former having a muzzle velocity of 1050 m (3,445 ft) per second and the latter of 1293 m (4,242 ft) per second; the APDS projectile can penetrate 20 mm (0.79 in) of armour at 1000 m (1,095 yards)

Kuka Arrow (West Germany): see HA1 Artemis (Greece)

Rheinmetall MK 20 Rh 202 20-mm AA gun mounting
(West Germany)
Type: towed twin 20-mm AA gun mounting
Calibre: 20 mm
Barrel length: 130.5 calibres overall
Carriage: two-wheel platform with outriggers; shield
Weights: 2160 kg (4,762 lb) in travelling order without ammunition and 1640 kg (3,616 lb) in firing position with ammunition
Dimensions: length, travelling 5.035 m (16.52 ft) and in firing position 4.05 m (13.29 ft); width, travelling 2.36 m (7.74 ft) and in firing position 2.30 m (7.55

ft); height, travelling 2.075 m (6.81 ft) and in firing position 1.67 m (5.48 ft)
Traverse/elevation: 360 degrees/−3.5 to +81.6 degrees powered and −5.5 to +83.5 degrees manually
Rate of fire (per barrel): 1,000 rounds per minute (cyclic)
Horizontal range: 3000 m (3,280 yards) maximum
Slant range: 2000 m (2,185 yards) effective
Crew: 3/4

Variant
MK 20 Rh 202: this twin-barrel light AA equipment was produced to a West German specification, but has also been one of the most successful export weapons of its type as the designers found an excellent combination of accuracy (using a computerized optical sight), ammunition supply and rates of traverse and elevation using an onboard power supply (a small petrol engine driving a hydraulic system); each barrel has its own 270-round ammunition box, and there are another 10 rounds in the feed mechanism, the ammunition types being APDS-T, API.T, HEI and HEI-T fired at muzzle velocities between 1045 and 1150 m (3,428 and 3,773 ft) per second; it is planned that West German equipments be upgunned with the 25-mm Mauser Model E cannon

HAI Artemis 30 30-mm AA gun system
(Greece)
Type: towed twin 30-mm AA gun system
Calibre: 30 mm
Barrel length: 86.25 calibres overall
Carriage: four-wheel platform with outriggers; shield
Weight: 7100 kg (15,653 lb) in travelling and firing orders
Dimensions: not revealed
Traverse/elevation: 360 degrees/−5 to +85 degrees
Rate of fire (per barrel): 800 rounds per minute (cyclic)
Horizontal range: 5000 m (5,470 yards) effective
Slant range: 3500 m (3,830 yards) effective
Crew: not revealed

Variant
Artemis 30: though co-ordinated and built in Greece by Hellenic Arms Industries, this impressive system is in reality a co-operative venture, the guns and carriage being West German and the fire-control system Swedish; the on-carriage optronic fire-control system allows useful performance in the standard powered and emergency unpowered modes, though the equipment is designed for battery use under the remote control of a computerized fire-control centre using a Siemens target acquisition and tracking radar; the circular tanks in the central pedestal hold 250 rounds for each barrel, the ammunition being of the type developed in the USA for the powerful GAU-8 Avenger tank-busting aircraft cannon, but made under licence in West Germany for the Mauser Model F (MK 30) cannon used in this and other applications
Kuka Arrow: this is West German equipment similar to the Artemis, produced to a Thai requirement, and featuring two 30-mm Mauser MK 30 Model F cannon controlled by Contraves Skyguard radar

RAMTA TCM Mk 3 20-, 23- or 25-mm AA gun mounting
(Israel)
Type: towed twin 20-, 23- or 25-mm AA gun mounting
Calibre: 20 mm, 23 mm or 25 mm
Barrel length: dependent on specific cannon type
Carriage: two-wheel platform with outriggers; shield
Weight: (with 20-mm HS-404 cannon) 1350 kg (2,976 lb) in towed configuration
Dimensions: not revealed
Traverse/elevation: 360 degrees/−10 to +90 degrees mechanically or −6 to +85 degrees electrically
Rate of fire (per barrel): (with HS-404 cannon) 700 rounds per minute (cyclic) and 150 rounds per minute (practical)
Horizontal range: (with HS-404 cannon) 5700 m (6,235 yards) maximum
Slant range: (with HS-404 cannon) 1200 m (1,315 yards) effective
Crew: 3

Variant
TCM Mk 3: this equipment entered service in 1984, and is essentially a product-improved TCM-20 with more modern assemblies (for greater reliability and less maintenance) plus a more advanced sight, the option of a night sight and control by a computerized fire-control system with laser ranger; the same type of electric drive is used, and the accommodation of the weapons on special adaptors means that most types of 20-, 23- and 25-mm cannon can be installed; the mounting can also be fitted on the back of light armoured vehicles of the halftrack type still so favoured by the Israelis

RAMTA TCM-20 20-mm AA gun mounting
(Israel)
Type: towed twin 20-mm AA gun mounting
Calibre: 20 mm
Barrel length:
Carriage: two-wheel platform with outriggers; shield
Weight: 1350 kg (2,976 lb) in travelling order with two loaded magazines
Dimensions: length, travelling 3.27 m (10.73 ft); width, travelling 1.70 m (5.58 ft); height, travelling 1.63 m (5.35 m)
Traverse/elevation: 360 degrees/−10 to +90 degrees
Rate of fire (per barrel): 700 rounds per minute (cyclic) and 150 rounds per minute (practical)
Horizontal range: 5700 m (6,235 yards) maximum

Slant range: 1200 m (1,315 yards) effective
Crew: 3

Variant
TCM-20: developed in Israel during the late 1960s, the TCM-20 is in essence an updated version of the US M55 mounting armed with two 20-mm HS-404 cannon rather than four 0.5-in (12.7-mm) Browning M2HB heavy machine-guns; the mounting is traversed and the weapon elevated by electric motors powered by two 12-volt batteries, the charge of the batteries being topped up by a small petrol engine; the type has built up an enviable combat record against low-flying aircraft, helicopters and light armoured vehicles, and can also be installed on the back of vehicles such as the M2 and M3 halftracks still widely employed by the Israelis; the two HS-404 cannon have been modified to use HS-804 ammunition (fed from 60-round drums) including APHE-T, APDS-T and HE-T; the type is optically controlled, but is often used with a radar warning system

Breda 40L70 40-mm AA gun mounting
(Italy)
Type: towed twin 40-mm AA gun mounting
Calibre: 40 mm
Barrel length: 70 calibres
Carriage: four-wheel platform with outriggers; glas-fibre gunhouse
Weight: 11000 kg (24,250 lb) in firing position with ammunition
Dimensions: length, travelling 8.05 m (26.41 ft); width, travelling 3.20 m (10.50 ft); height, travelling about 3.65 m (11.98 ft)
Traverse/elevation: 360 degrees/−13 to +85 degrees
Rate of fire (per barrel): 300 rounds per minute (cyclic)
Horizontal range: 12500 m (13,670 yards) maximum
Slant range: 4000 m (4,375 yards) effective
Crew: 0

Variant
Breda 40L70: this is a land mounting developed from the successful Breda 40L70 Compact naval installation, and is designed for fully automatic operation with an external power source and control from a radar fire-control system; the circular magazine in the bottom of the firing platform holds 444 rounds, which are loaded in 4-round clips and fed automatically to the breeches; the standard range of Bofors 40-mm ammunition is used

Kongsberg FK 20-2 20-mm AA gun mounting
(Norway)
Type: towed single 20-mm AA mounting
Calibre: 20 mm
Barrel length: 130.5 calibres overall
Carriage: two-wheel platform with tripod legs; shield
Weights: 660 kg (1,455 lb) in travelling order and 440 kg (970 lb) in firing position
Dimensions: length, travelling 4.00 m (13.12 ft); width, travelling 1.86 m (6.10 ft); height, travelling 2.20 m (7.22 ft)
Traverse/elevation: 360 degrees/−8 to +83 degrees
Rate of fire: 1,000 rounds per minute (cyclic)
Horizontal range: 7000 m (7,655 yards) maximum
Slant range: 1500 m (1,640 yards) effective
Crew: 3

Variant
FK 20-2: designed to meet a West German and Norwegian requirement, this neat manually-operated light mounting is intended for operation in the surface-to-air and surface-to-surface roles with APDS-T, API-T, HEI and HEI-T ammunition carried in two 75-round side magazines plus 10 rounds in the feed mechanism and fired from the MK 20 Rh 202 cannon at muzzle velocities between 1045 and 1150 m (3,428 and 3,773 ft) per second

Bofors 40L70 Type B 40-mm AA gun mounting
(Sweden)
Type: towed single 40-mm towed AA gun mounting
Calibre: 40 mm
Barrel length: 70 calibres
Carriage: four-wheel platform with outriggers; shield
Weight: 5150 kg (11,354 lb) in travelling order
Dimensions: length, travelling 7.29 m (23.92 ft); width, travelling 2.23 m (7.30 ft); height, travelling 2.35 m (7.71 ft)
Traverse/elevation: 360 degrees/−4 to +90 degrees
Rate of fire: 300 rounds per minute (cyclic)
Horizontal range: 12500 m (13,670 yards) maximum
Slant range: 4000 m (4,375 yards) effective
Crew: 4/6

Variants
40L70 Type A: introduced after World War II, this remains one of the most powerful light weapons in the world (in both the surface-to-air and surface-to-surface roles) thanks to its high rate of fire (resulting from the use of an automatic breech and the ramming of each round during the run-out), good range and excellent ammunition; the Type A is the original model without on-carriage power, and weighs 4800 kg (10,582 lb) in travelling order; ammunition types are APC-T, HCHE, HE-T and PFHE fired at muzzle velocities between 1005 and 1030 m (3,297 and 3,379 ft) per second, and the ammunition is fed in 4-round clips into an optional overhead stay holding 26 rounds and from two 48-round racks at the rear of the carriage
40L70 Type B: improved model with on-carriage power generator, but otherwise similar to the Type

A and operated in local control by a gunner on the left of the ordnance

40L70 BOFI Fair Weather: development of the Type B with Bofors Optronic Fire-control Instrument and proximity-fused ammunition; this model weighs 5500 kg (12,125 lb) in travelling order, and the BOFI equipment uses a laser rangefinder, a day/night image-intensifying sight and a fire-control computer to generate aiming and firing instructions for the gunner, with early warning provided optionally by an off-carriage radar system

40L70 BOFI All Weather: development of the 40L70 BOFI Fair Weather system with the addition of a pulse-Doppler radar for automatic target acquisition and tracking capability; this model has a travelling weight of 5700 kg (12,566 lb)

Breda 40L70: this is the 40L70 built under licence in Italy, where Breda has developed an optional automatic feeding device, which takes ammunition in groups of three from a magazine pre-loaded with 144 rounds in 4-round clips; the travelling weight of the Breda equipment is 5300 kg (11,684 lb) and there are some other detail differences in dimensions

Bofors m/36 40L60 40-mm AA gun mounting
(Sweden)
Type: towed single 40-mm AA gun mounting
Calibre: 40 mm
Barrel length: 56 or 60 calibres
Carriage: four-wheel platform with outriggers; no shield
Weight: 2400 kg (5,291 lb) in travelling and firing orders
Dimensions: length, travelling 6.38 m (20.93 ft); width, travelling 1.72 m (5.64 ft); height, travelling 2.00 m (6.56 ft)
Traverse/elevation: 360 degrees/−5 to +90 degrees
Rate of fire: 120 rounds per minute (cyclic)
Horizontal range: 4750 m (5,195 yards) maximum
Slant range: 2500 m (2,735 yards) effective
Crew: 4/6

Variants
40L60: introduced in 1936 as the **m/36**, this manually-operated equipment is still in widespread service as it uses a powerful round and is still effective in regions of less advanced technology; it is found on three carriages, the m/38 for a weight of 2150 kg (4,740 lb), the m/39 as described above, and the m/49e for a weight of 2050 kg (4,519 lb); the ordnance fires AP or HE projectiles at a muzzle velocity of 850 m (2,788 ft) per second
Automatic Anti-Aircraft Gun Mk 1: designation of the British licence-made version with a different carriage, a travelling weight of 5,045 lb (2288 kg) and a number of detail modifications
Automatic Anti-aircraft Gun M1: designation of the US licence-made version with a different carriage, a travelling weight of 5,855 lb (2656 kg) and a number of detail modifications

Oerlikon GDF-002 35-mm AA gun mounting
(Switzerland)
Type: towed twin 35-mm AA gun mounting
Calibre: 35 mm
Barrel length: 90 calibres
Carriage: four-wheel platform with outriggers; shield
Weight: 6700 kg (14,771 lb) in travelling order with ammunition
Dimensions: length, travelling 7.80 m (25.59 ft) and in firing position 8.83 m (28.97 ft); width, travelling 2.26 m (7.41 m) and in firing position 4.49 m (14.73 ft); height, travelling 2.60 m (8.53 ft) and in firing position 1.72 m (5.64 ft)
Traverse/elevation: 360 degrees/−5 to +92 degrees
Rate of fire (per barrel): 550 rounds per minute (cyclic)
Horizontal range: not revealed
Slant range: 4000 m (4,375 yards) effective
Crew: 3

Variants
GDF-001: introduced in the early 1960s as the **2 ZLA 353 MK**, this is the heavyweight of the Oerlikon AA series, and an exceptionally potent weapon capable of effective use in the surface-to-air and surface-to-surface roles with its HEI, HEI-T and SAPHEI-T ammunition fired at a muzzle velocity of 1175 m (3,855 ft) per second from the two KDB (formerly 353 MK) cannon, which are fed automatically from two 56-round containers replenished in 7-round clips from the two 63-round reserve containers carried on the carriage; the mounting has three operating modes, namely remote electric control from the Super Fledermaus or Skyguard fire-control radar, local electric control with the Xaba optical sight, and local manual control with handwheels
GDF-002: updated version available from 1980 with Ferranti Type GSA Mk 3 sights and digital data transmission; the type is also available with optional packages such as camouflage, automatic reloaders, a gunner's cab, and integrated power source, and a Minisight Gun King incorporating a laser rangefinder

Oerlikon GBF-BOB Diana 25-mm AA gun system
(Switzerland)
Type: towed twin 25-mm AA gun system
Calibre: 25 mm
Barrel length: 92 calibres
Carriage: two-wheel platform with outriggers; gunner's cab
Weight: 2100 kg (4,630 lb) in firing position with ammunition
Dimensions: length, travelling 4.295 m (14.09 ft); width, travelling 2.10 m (6.89 ft); height, travelling 2.13 m (6.99 ft)

Traverse/elevation: 360 degrees/−5 to +85 degrees
Rate of fire (per barrel): 800 rounds per minute (cyclic)
Horizontal range: 6000 m (6,560 yards) maximum
Slant range: 2500 m (2,735 yards) effective
Crew: 4/5

Variants
GBF-BOB Diana: introduced in the mid-1980s, this is an advanced AA weapon with secondary capability against light AFVs, and provides a capability between Oerlikon's twin 20-mm and twin 35-mm weapons; the carriage has its own electrical and hydraulic power, but the system can also be operated manually in emergencies; the cannon are two Oerlikon KBB weapons with dual-feed capability for the 250 ready-use APDS-T and HEI rounds, which have muzzle velocities of 1355 and 1160 m (4,446 and 3,806 ft) per second respectively; the system can be used in conjunction with the Skyguard radar fire-control or Alerter surveillance radar systems, but is designed principally for autonomous operation with its own Contraves Gun King optical fire-control system incorporating a laser rangefinder and digital computer
GBF-AOA Diana: this version has twin Oerlikon KBA cannon, each with a cyclic rate of 570 rounds per minute and firing APDS-T, HEI and SAPHEI-T ammunition; this equipment has a lower muzzle velocity than the GBF-BOB and is fitted with an Officine Galileo P75 optical sight

Oerlikon GBI-AO1 25-mm AA gun mounting
(Switzerland)
Type: towed single 25-mm AA gun mounting
Calibre: 25 mm
Barrel length: 87.3 calibres
Carriage: two-wheel platform with tripod legs; no shield
Weights: 666 kg (1,468 lb) in travelling order with ammunition and 506 kg (1,116 lb) in firing position with ammunition
Dimensions: length, travelling 4.72 m (15.49 ft); width, travelling 1.80 m (5.91 ft); height, travelling 1.65 m (5.41 ft)
Traverse/elevation: 360 degrees/−10 to +70 degrees
Rate of fire: 570 rounds per minute (cyclic) and 170 rounds per minute (practical)
Horizontal range: 6000 m (6,560 yards) maximum
Slant range: 2500 m (2,735 yards) effective
Crew: 3

Variant
GBI-AO1: though designed primarily for AA use, the manually-operated GBI-AO1 mounting can be used against battlefield targets such as light AFVs, the Oerlikon KBA-C cannon being able to fire APDS-T projectiles at a muzzle velocity of 1335 m (4,380 ft) per second, plus HEI, HEI-T, SAPHEI and SAPHEI-T projectiles at 1100 m (3,609 ft) per second; these rounds are accommodated in two 40-round containers (one on each side of the weapon), a dual-feed mechanism allowing selection of round type

Oerlikon GAI-DO1 20-mm AA gun mounting
(Switzerland)
Type: towed twin 20-mm AA gun mounting
Calibre: 20 mm
Barrel length: 95 calibres including flash suppressor
Carriage: two-wheel platform with outriggers; shield
Weights: 1800 kg (3,968 lb) in travelling order with ammunition and 1330 kg (2,932 lb) in firing position with ammunition
Dimensions: length, travelling 4.59 m (15.06 ft); width, travelling 1.86 m (6.10 ft); height, travelling 2.34 m (7.68 ft)
Traverse/elevation: 360 degrees/−3 to +81 degrees
Rate of fire (per barrel): 1,000 rounds per minute (cyclic)
Horizontal range: 5700 m (6,235 yards) maximum
Slant range: 1500 m (1,640 yards) effective
Crew: 5

Variant
GAI-DO1: designed in the mid-1970s as the **HS-666A** and available from 1978, this equipment has successfully bridged the tactical gap between Oerlikon's single-barrel 20-mm weapons and more capable equipments such as the twin 35-mm GDF-002; the equipment is hydraulically-powered, with reversion to manual operation in emergencies, and though intended primarily for AA use can also be deployed for a number of battlefield roles; the two Oerlikon KAD-B cannon can fire AP-T, HEI, HEI-T, SAPHEI and SAPHEI-T ammunition (the first at a muzzle velocity of 1150 m/3,773 ft per second and the other four at a muzzle velocity of 1100 m/3,609 ft per second) fed from a 120-round magazine on the side of each weapon; the GAI-DO1 can also be tied into a fire-control system and early warning radar

Oerlikon GAI-CO4 20-mm AA gun mounting
(Switzerland)
Type: towed single 20-mm AA gun mounting
Calibre: 20 mm
Barrel length: 95 calibres including flash suppressor
Carriage: two-wheel platform with tripod legs; no shield
Weight: 589 kg (1,299 lb) in travelling order with ammunition and 435 kg (959 lb) in firing position with ammunition
Dimensions: length, in firing position 3.87 m (12.70 ft); width, in firing position 1.70 m (5.58 ft); height, in firing position 1.45 m (4.76 ft)
Traverse/elevation: 360 degrees/−7 to +83 degrees
Rate of fire: 1,050 rounds per minute (cyclic)

Horizontal range: 5700 m (6,235 yards) maximum
Slant range: 1500 m (1,640 yards) effective
Crew: 3

Variants
GAI-CO1: developed as the **HS-639-B3.1** and fitted with an Oerlikon KAD-B13-3 (formerly HS-820-SL7 A3-3) cannon, this manually-operated equipment has weights (in travelling order and in firing position respectively, each with ammunition) of 534 and 370 kg (1,177 and 816 lb); the type fires the same ammunition as the GAI-DO1, and is single-fed by a 75-round magazine on the right of the weapon, so limiting the type's utility in dual-role operations
GAI-CO3: developed as the **HS-639-B4.1** and fitted with an Oerlikon KAD-AO1 (formerly HS-820 SAA1) cannon, this equipment fires the same ammunition as the GAI-DO1, though rounds are fed to the cannon from an overhead drum containing 50 rounds; the type has weights (in travelling order and in firing position respectively, complete with ammunition) of 510 and 432 kg (1,124 and 952 lb)
GAI-CO4: developed as the **HS-639-B5** and fitted with an Oerlikon KAD-B14 (formerly HS-820-SL7 A4) cannon, this is an improved version of the GAI-CO1 with a dual-feed mechanism and two 75-round magazines for greater capability in the surface-to-air and secondary surface-to-surface roles

Oerlikon GAI-BO1 20-mm AA gun mounting
(Switzerland)
Type: towed single 20-mm AA gun mounting
Calibre: 20 mm
Barrel length: 120 calibres
Carriage: two-wheel platform with cruciform legs; no shield
Weights: 547 kg (1,206 lb) in travelling order and 405 kg (893 lb) in firing position
Dimensions: length, travelling 3.85 m (12.63 ft); width, travelling 1.55 m (5.09 ft); height, travelling 2.50 m (8.20 ft)
Traverse/elevation: 360 degrees/−5 to +85 degrees)
Rate of fire: 1,000 rounds per minute (cyclic)
Horizontal range: 5700 m (6,235 yards) maximum
Slant range: 1500 m (1,640 yards) effective
Crew: 3

Variant
GAI-BO1: the lightweight of the Oerlikon AA family, this equipment was designed as the **10 ILa/5TG** and is fitted with an Oerlikon KAB-001 cannon; the equipment is manually operated, and the cannon fires the same type of ammunition as the GAI-DO1, fed from a 50- or a 20-round drum, or from a 8-round box

KS-30 130-mm AA gun mounting
(USSR)
Type: towed 130-mm AA gun mounting
Calibre: 130 mm (5.12 in)
Barrel length: 64.7 calibres overall
Muzzle brake: none
Carriage: eight-wheel platform with outriggers; no shield
Weight: 29500 kg (65,035 lb) in travelling order and 24900 kg (54,894 lb) in firing position
Dimensions: length, travelling 11.52 m (37.80 ft); width, travelling 3.03 m (9.95 ft); height, travelling 3.05 m (10.00 ft)
Traverse/elevation: 360 degrees/−5 to +80 degrees
Rate of fire: 10/12 rounds per minute
Horizontal range: 27000 m (29,530 yards) maximum
Slant range: 13700 m (14,985 yards) effective
Crew: 15/20

Variant
KS-30: introduced in the early 1950s as the initial Soviet response to the threat of high-flying US strategic bombers, the KS-30 is an obsolete weapon that can be used in the AA (33.4-kg/73.6-lb HE projectile) and anti-tank (33.4-kg/73.6-lb APHE projectile) roles; the ammunition is of the separate-loading type and fired at a muzzle velocity of 970 m (3,182 ft) per second with the aid of a semi-automatic breech, an automatic fuse setter and a power rammer; the type has on-carriage sights, but is generally operated in conjunction with 'Fire Wheel' radar and the appropriate director

KS-19 100-mm AA gun mounting
(USSR)
Type: towed 100-mm AA gun mounting
Calibre: 100 mm (3.94 in)
Barrel length: 57.4 calibres overall
Muzzle brake: multi-baffle
Carriage: four-wheel platform with outriggers; no shield
Weight: 9550 kg (21,054 lb) in travelling order
Dimensions: length, travelling 9.45 m (31.00 ft); width, travelling 2.35 m (7.71 ft); height, travelling 2.20 m (7.22 ft)
Traverse/elevation: 360 degrees/−3 to +85 degrees
Rate of fire: 15 rounds per minute
Horizontal range: 21000 m (22,965 yards) maximum
Slant range: 13700 m (14,985 yards) with proximity-fused ammunition
Crew: 15

Variants
KS-19: introduced in the late 1940s as successor to the USSR's standard 85-mm (3.3.35-in) AA guns, the KS-19 is obsolescent but still in widespread service with Soviet clients; the type has a semi-automatic breech, an automatic fuse setter and a power rammer

for the fixed ammunition, and is designed for dual-role operation against aircraft (HE, FRAG-HE and FRAG projectiles fired at a muzzle velocity of 900 m/2,953 ft per second) and tanks (AP-T and APC-T projectiles fired at muzzle velocities of 1000 m/3,281 ft and 900 m per second respectively); the weapon has on-carriage fire-control, but is generally used with 'Fire Can' or 'Whiff' radars and appropriate directors
Type 59: designation of the Chinese-made KS-19

KS-18 85-mm AA gun mounting
(USSR)
Type: towed 85-mm AA gun mounting
Calibre: 85 mm (3.35 in)
Barrel length: 67.6 calibres overall
Muzzle brake: T-shaped
Carriage: four-wheel platform with outriggers; shield
Weight: 500 kg (11,023 lb) in travelling and firing orders
Dimensions: length, travelling 8.20 m (26.90 ft); width, travelling 2.15 m (7.05 ft); height, travelling 2.25 m (7.38 ft)
Traverse/elevation: 360 degrees/−3 to +82 degrees
Rate of fire: 15/20 rounds per minute
Horizontal range: 18000 m (19,685 yards) maximum
Slant range: 10000 m (10,935 yards)
Crew: 7

Variants
KS-12: introduced in 1939 as the **M1939** with a 55.2-calibre barrel, this equipment is still in limited service with Soviet clients; it fires fixed FRAG anti-aircraft ammunition at a muzzle velocity of 792 m (2,598 ft) per second or HVAP anti-tank ammunition at 1030 m (3,379 ft) per second, and a good rate of fire is ensured by the use of a semi-automatic breech; this weapon has a weight of 4300 kg (9,480 lb) in travelling and firing orders, and a travelling length of 7.05 m (23.13 ft); the type has a multi-baffle muzzle brake
KS-18: much developed model introduced as the **M1944** and generally used in conjunction with 'Fire Can' radar and appropriate director; the longer barrel of this ordnance increases standard muzzle velocity to 900 m (2,953 ft) per second, with consequent improvements in effective and maximum ranges
Type 56: designation of the Chinese-made KS-12

S-60 57-mm AA gun mounting
(USSR)
Type: towed single 57-mm AA gun mounting
Calibre: 57 mm
Barrel length: 77 calibres overall
Muzzle brake: pepperpot
Carriage: four-wheel platform with outriggers; no shield
Weights: 4660 kg (10,273 lb) in travelling order and 4500 kg (9,921 lb) in firing position
Dimensions: length, travelling 8.50 m (27.89 ft); width, travelling 2.05 m (6.74 ft); height, travelling 2.37 m (7.78 ft)
Traverse/elevation: 360 degrees/−4 to +85 degrees
Rate of fire: 110 rounds per minute (cyclic) and 70 rounds per minute (practical)
Horizontal range: 12000 m (13,125 yards) maximum
Slant range: 4000 m (4,375 yards) with on-carriage control and 6000 m (6,560 yards) with off-carriage control
Crew: 7

Variants
S-60: introduced in the late 1940s as a heavy tactical AA weapon to replace the 37-mm M1939, this equipment fires FRAG-T and APC-T ammunition (of the fixed type and fed in 4-round clips) at a muzzle velocity of 1000 m (3,281 ft) per second; the weapon can be used with its on-carriage fire-control system, but is far more capable when used with 'Fire Can' or 'Flap Wheel' radar and appropriate director; night-vision sights have also been seen on the type, which can be operated manually or with servo-assistance
Type 59: designation of the Chinese-built S-60

M1939 37-mm AA gun mounting
(USSR)
Type: towed single 37-mm AA gun mounting
Calibre: 37 mm
Barrel length: 73.75 calibres overall
Carriage: four-wheel platform with outriggers; optional shield
Weight: 2100 kg (4,630 lb) in firing position without shield
Dimensions: length, travelling 6.04 m (19.80 ft); width, travelling without shield 1.94 m (6.35 ft); height, travelling without shield 2.11 m (6.91 ft)
Traverse/elevation: 360 degrees/−5 to +85 degrees
Rate of fire: 170 rounds per minute (cyclic) and 80 rounds per minute (practical)
Horizontal range: 9500 m (10,390 yards) maximum
Slant range: 3000 m (3,280 yards) effective
Crew: 8

Variants
M1939: a clear-weather system now thoroughly obsolete, the manually-operated M1939 is still widely used by Soviet clients, and fires fixed ammunition (fed in 5-round clips) of the FRAG-T and AP-T types at a muzzle velocity of 880 m (2,887 ft) per second
Type 55: designation of the Chinese-made M1939
Type 63: twin-barrel version of the M1939 produced in China largely for export

ZU-23 23-mm AA gun mounting
(USSR)
Type: towed twin 23-mm AA gun mounting
Calibre: 23 mm
Barrel length: 87.4 calibres overall

Carriage: two-wheel triangular platform with screw jacks; no shield
Weight: 950 kg (2,094 lb) in travelling and firing orders with ammunition
Dimensions: length, travelling 4.57 m (14.99 ft); width, travelling 1.83 m (6.00 ft); height, travelling 1.87 m (6.14 ft)
Traverse/elevation: 360 degrees/−10 to +90 degrees
Rate of fire (per barrel): 800/1,000 rounds per minute (cyclic) and 200 rounds per minute (practical)
Horizontal range: 7000 m (7,655 yards) maximum
Slant range: 2500 m (2,735 yards) effective
Crew: 5

Variant
ZU-23: introduced in the mid-1960s as successor to the ZPU series, the ZU-23 is the most important towed AA equipment in the Soviet armoury, and has been widely exported to client states, many of whom have installed the equipment on vehicles; the equipment is manually-operated and designed only for clear-weather operations with two types of ammunition (API-T and HEI-T) fired at a muzzle velocity of 970 m (3,182 ft) per second and delivered from boxes (one on the side of each barrel) containing 50 belted rounds

ZPU-4 14.5-mm AA gun mounting
(USSR)
Type: towed quadruple 14.5-mm AA gun mounting
Calibre: 14.5 mm (0.57 in)
Barrel length: 93 calibres overall
Carriage: four-wheel platform with outriggers; no shield
Weight: 1810 kg (3,990 lb) in travelling and firing orders
Dimensions: length, travelling 4.53 m (14.86 ft); width, travelling 1.72 m (5.64 ft); height, travelling 2.13 m (6.99 ft)
Traverse/elevation: 360 degrees/−10 to +90 degrees
Rate of fire (per barrel): 600 rounds per minute (cyclic) and 150 rounds per minute (practical)
Horizontal range: 8000 m (8,740 yards) maximum
Slant range: 1400 m (1,530 yards) effective
Crew: 5

Variants
ZPU-1: single-barrel version on a two-wheel carriage, introduced soon after World War II and based on the KPV heavy machine-gun with 150 rounds of belted ammunition (API, API-T and I-T) in a side-mounted magazine; like the later ZPUs this is a clear-weather weapon and has a weight of 413 kg (190 lb) in travelling and firing orders
ZPU-2: introduced from 1949, this is the twin-barrel weapon of the series and was produced in early and late models, the former having travelling and firing weights of 994 and 639 kg (2,191 and 1,409 lb) respectively, and the latter of 649 and 621 kg (1,431 and 1,369 lb) respectively
ZPU-4: introduced in 1949, this is the four-barrel model of the ZPU series
Type 56: designation of the Chinese-made ZPU-4
Type 58: designation of the Chinese-made ZPU-2

M118 90-mm AA gun mounting
(USA)
Type: towed 90-mm AA gun mounting
Calibre: 90 mm (3.54 in)
Barrel length: 50 calibres overall
Muzzle brake: none
Carriage: four-wheel platform with outriggers; shield
Weight: 32,300 lb (14651 kg) in travelling order
Dimensions: length, travelling 29.50 ft (8.99 m); width, travelling 8.60 ft (2.62 m); height, travelling 10.08 ft (3.07 m)
Traverse/elevation: 360 degrees/−10 to +80 degrees
Rate of fire: 28 rounds per minute (burst fire) and 23 rounds per minute (sustained fire)
Horizontal range: 20,750 yards (18975 m) maximum
Slant range: 9,300 yards (8505 m) effective
Crew: 10/12

Variants
M117: introduced in 1940 as the M1 and redesignated after World War II, this was in its time a capable AA weapon weighing 19,015 lb (8625 kg) in travelling order and 14,650 lb (6645 kg) in firing position; it was deleted from US service in the 1960s but still serves with a number of other countries
M118: this weapon was developed from the M1 during World War II as the M2, with the object of equipping the US field forces with a multi-role (AA, anti-tank and field) gun offering significantly better AA and tactical capabilities than those of the M1; the type was redesignated after World War II, and is still used by some American allies; the weapon fires three types of projectile, namely APHE at a muzzle velocity of 2,800 ft (853 m) per second, HE at a muzzle velocity of 2,700 ft (823 m) per second and HVAP-T at a muzzle velocity of 3,355 ft (1023 m) per second, the fixed ammunition being loaded by a power rammer; the type is generally used with the M33 radar fire-control system

M167 Vulcan 20-mm AA gun system
(USA)
Type: towed sextuple-barrel 20-mm AA gun system
Calibre: 20 mm
Barrel length: 76.2 calibres overall
Carriage: two-wheel platform with outriggers; no shield
Weights: 3,500 lb (1588 kg) in travelling order and 3,450 lb (1565 kg) in firing position

Dimensions: length, travelling 16.09 ft (4.91 m); width, travelling 6.50 ft (1.98 m); height, travelling 6.69 ft (2.04 m)
Traverse/elevation: 360 degrees/−5 to +80 degrees
Rate of fire: selectable 1,000 (surface-to-surface) or 3,000 (surface-to-air) rounds per minute (cyclic)
Horizontal range: 6,500 yards (5,945 m) maximum
Slant range: 1,300 yards (1190 m) effective
Crew: 1

Variants
M167: this is the towed version of the M163 self-propelled AA mounting, and though it features a capable fire-control system (with range-only radar and a lead-computing sight) and the formidable Vulcan six-barrel cannon plus 500 rounds of ammunition (AP-T, HEI and HEI-T all fired at a muzzle velocity of 3,380 ft/1030 m per second), the equipment is limited by the need for external power and its lack of all-weather capability; the type has been improved in reliability and cross-country mobility by the addition of an extra wheel on each side
M167A1: improved M167
Basic Vulcan: export version of the M167 with a range-updating computer instead of the range radar and other modifications

M55 0.5-in AA gun mounting
(USA)
Type: towed quadruple 0.5-in AA gun mounting
Calibre: 0.5 in (12.7 mm)
Barrel length: 90 calibres overall
Carriage: two-wheel platform; shield
Weight: 2,950 lb (1338 kg) in travelling order
Dimensions: length, travelling 9.48 ft (2.89 m); width, travelling 6.86 ft (2.09 m); height, travelling 5.27 ft (1.61 m)
Traverse/elevation: 360 degrees/−10 to +90 degrees
Rate of fire (per barrel): 500 rounds per minute (cyclic) and 150 rounds per minute (practical)
Horizontal range: 1,650 yards (1510 m) effective
Slant range: 1,300 yards (1190 m) effective
Crew: 4

Variants
M55: designed in World War II and now obsolete though still in widespread service, the M55 is electrically operated in azimuth and elevation, the two 6-volt batteries for the task being charged by a small petrol engine; each barrel is fed by a 210-round belt, and the ammunition types that can be used are AP, API, API-T, ball and incendiary; the mount can also be installed on the back of light armoured vehicles, halftracks and the like
M55 (Modernized): evolved as a kit in Brazil, this type has modern 12-volt electrics and other improvements

20/3 mm M55 A2 20-mm AA gun mounting
(Yugoslavia)
Type: towed triple 20-mm AA gun mounting
Calibre: 20 mm
Barrel length: 97.8 calibres overall
Carriage: two-wheel platform with tripod legs; no shield
Weight: 1100 kg (2,425 lb) in travelling and firing orders with ammunition
Dimensions: length, travelling 4.30 m (14.11 ft); width, travelling 1.27 m (4.17 ft); height, travelling 1.47 m (4.82 ft)
Traverse/elevation: 360 degrees/−5 to +83 degrees
Rate of fire (per barrel): 700 rounds per minute (cyclic)
Horizontal range: 5500 m (6,015 yards) maximum
Slant range: 2000 m (2,185 yards) effective
Crew: 6

Variants
M55 A2: introduced in 1955, this is the basic model of a three-equipment series, and uses three licence-built HS-804 cannon on a variation of the HS 630-3 carriage; the type is manually operated, and the weapons fire HEI and HEI-T ammunition at a muzzle velocity of 850 m (2,789 ft) per second, or API and API-T ammunition at 840 m (2,756 ft) per second; each barrel is fed from a 60-round drum
M55 A3 B1: this is the version of the M55 A2 with a 6-kW (8-hp) Wankel engine to provide hydraulic power for traverse and elevation; the weight of this variant is 1236 kg (2,725 lb)
M55 A4 B1: this combines the triple-barrel armament of the previous M55 variants with a derivative of the HS-666A (Oerlikon GAI-DO1) carriage revised to accommodate a Wankel engine for powered traverse and elevation; this model weighs 1350 kg (2,976 lb) and has a licence-built Officine Galileo P56 computerized sight

20/1 mm M75 20-mm AA gun mounting
(Yugoslavia)
Type: towed single 20-mm AA gun mounting
Calibre: 20 mm
Barrel length: 97.8 calibres overall
Carriage: two-wheel platform with tripod legs; no shield
Weight: 260 kg (573 lb) in travelling and firing orders with ammunition
Dimensions: length, travelling not known; width, travelling 1.51 m (4.95 ft); height, travelling not known
Traverse/elevation: 360 degrees/−10 to +83 degrees
Rate of fire: 700 rounds per minute (cyclic)
Horizontal range: 5500 m (6,015 yards) maximum
Slant range: 2000 m (2,185 yards) effective
Crew: 4/6

116 LAND WEAPONS

Variant
M75: this is a lightweight manually-operated mounting for a single HS-804 cannon; the type fires the same ammunition as the M55 series, using either a 60-round drum (HEI and HEI-T rounds) or a 10-round box (AP-T, API and API-T rounds)

MULTIPLE ROCKET SYSTEMS

CITEFA SLAM-Pampero multiple rocket system
(Argentina)
Type: 15-tube multiple rocket system
Rocket dimensions: diameter 105 mm (4.13 in); length not revealed
Rocket weights: whole round 30 kg (66.14 lb); warhead 11 kg (24,25 lb) HE
Rocket range: 12000 m (13,125 yards)
Reload time: not revealed
Launcher traverse/elevation: not revealed
Crew: not revealed
Mounting: trailer or vehicle rear

Variant
SLAM-Pampero: little is known of this comparatively new Argentine system, which is powered by a solid-propellant rocket and stabilized by four wrap-round fins; a salvo is believed to saturate an area of 300 by 200 m (330 by 220 yards)

CITEFA SAPBA-1 multiple rocket system
(Argentina)
Type: 32-tube multiple rocket system
Rocket dimensions: diameter 127 mm (5 in); length 2.228 m (7.31 ft)
Rocket weights: whole round 54 kg (119.05 lb); warhead 18 kg (39.7 lb) HE
Rocket range: 20000 m (21,870 yards)
Reload time: 5 minutes for all four 9-tube pods
Launcher traverse/elevation: 90 degrees left and right/ 0 to +60 degrees
Crew: 5
Mounting: vehicle rear (generally Fiat 697 6 x 6 truck)

Variant
SAPBA-1: this is a capable area-saturation multiple rocket system whose projectiles can be fired singly or in a ripple at 0.5-second intervals; the rocket is powered by a solid-propellant rocket motor, stabilized by four fins, and has either a FRAG-HE warhead with a lethal area of 1000 m (10,765 sq ft) or a proximity-fused anti-personnel type that disperses 4,500 steel pellets over an area of 6000 m^2 (64,585 sq ft)

AVIBRAS FGT108-RA1 multiple rocket system
(Brazil)
Type: 16-tube multiple rocket system
Rocket dimensions: diameter 108 mm (4.25 in); length 0.967 m (3.17 ft)
Rocket weights: whole round 17 kg (37.5 lb); warhead 3 kg (6.6 lb) HE
Rocket range: 7000 m (7,655 yards)
Reload time: not revealed
Launcher traverse/elevation: 12 degrees left and right/ −1 to +50 degrees
Crew: 4
Mounting: two-wheel trailer weighing 802 kg (1,768 lb) loaded

Variant
FGT108-RA1: this neat little system is fired at a maximum rate of two rounds per second, and is normally grouped in batteries of four launchers each with 64 reloads (four 16-round containers); the rocket is powered by a solid-propellant rocket motor and spin-stabilized by six canted rocket nozzles

AVIBRAS SBAT-70 multiple rocket system
(Brazil)
Type: 32-tube multiple rocket system
Rocket dimensions: diameter 70 mm (2.76 in); length not revealed
Rocket weights: whole round 9 kg (19.8 lb); warhead 4 kg (8.8 lb) HEAT, FRAG-HE, flechette or smoke
Rocket range: 7500 m (8,200 yards)
Reload time: not revealed
Launcher traverse/elevation: 12 degrees left and right/ 0 to +50 degrees
Crew: 4
Mounting: two-wheel trailer

Variant
SBAT-70: this rocket is powered by a solid-propellant rocket motor and stabilized by four folding fins, and the trailer has a loaded weight of some 1000 kg (2,205 lb)

AVIBRAS SBAT-127 multiple rocket system
(Brazil)
Type: 12-rail multiple rocket system

Rocket dimensions: diameter 127 mm (5 in); length not revealed
Rocket weights: whole round 48 or 61 kg (105.8 or 134.5 lb) depending on warhead; warhead 22 or 35 kg (48.5 or 77.2 lb) HE
Rocket range: 14000 or 12500 m (15,310 or 13,670 yards) depending on warhead weight
Reload time: not revealed
Launcher traverse/elevation: not revealed
Crew: not revealed
Mounting: trailer or vehicle rear

Variant
SBAT-127: developed from the 127-mm (5-in) aircraft rocket, the SBAT-127 is powered by a solid-propellant rocket motor and stabilized by fins

AVIBRAS ASTROS SS-30 multiple rocket system
(Brazil)
Type: 32-tube multiple rocket system
Rocket dimensions: diameter 127 mm (5 in); length 3.10 m (10.17 ft)
Rocket weights: whole round 66 kg (145.5 lb); warhead 20 kg (44 lb) HE
Rocket range: 30000 m (32,810 yards)
Reload time: not revealed
Launcher traverse/elevation: not revealed
Crew: not revealed
Mounting: ASTROS II launcher on rear of TECTRAN 6 x 4 vehicle

Variant
SS-30: designed for launch from an armoured cross-country vehicle, this rocket has a solid-propellant motor and wrap-round rear fins, and the system is designed to operate in conjunction with the Contraves Fieldguard fire-control system for maximum accuracy; all three vehicles in the ASTROS II system (rocket launcher, resupply type and fire-control type) use the same TECTRAN 6 x 4 chassis

AVIBRAS ASTROS SS-40 multiple rocket system
(Brazil)
Type: 16-tube multiple rocket system
Rocket dimensions: diameter 180 mm (7.09 in); length 3.98 m (13.06 ft)
Rocket weights: whole round 146 kg (321.9 lb); warhead 54 kg (119 lb) HE or cluster munition
Rocket range: 35000 m (38,275 yards)
Reload time: not revealed
Launcher traverse/elevation: not revealed
Crew: not revealed
Mounting: ASTROS launcher on rear of TECTRAN 6 x 4 vehicle

Variant
SS-40: designed to use another version of the TECTRAN-mounted launcher and the same Contraves Fieldguard fire-control system, the SS-40 rocket has a solid-propellant motor and wrap-round fins, but its warhead is a cluster-munition type designed to carry dual-purpose anti-tank/anti-personnel bomblets; the size of the area sown with bomblets depends on the fused burst altitude of the rocket

AVRIBRAS ASTROS SS-60 multiple rocket system
(Brazil)
Type: 4-tube multiple rocket system
Rocket dimensions: diameter 300 mm (11.81 in); length 5.20 m (17.06 ft)
Rocket weights: whole round 517 kg (1,140 lb); warhead 160 kg (352.7 lb) cluster munition
Rocket range: 60000 m (65,615 yards)
Reload time: not revealed
Launcher traverse/elevation: not revealed
Crew: not revealed
Mounting: ASTROS launcher on rear of TECTRAN 6 x 4 vehicle

Variant
SS-60: largest of the SS series, the SS-60 has a solid-propellant rocket and wrap-round fins, and is designed to deliver a substantial load of dual-purpose anti-tank/anti-personnel bomblets over considerable range; like the other rocket-launchers in the series it uses the Contraves Fieldguard fire-control system; the size of the area sown with bomblets depends on the fused burst altitude of the rocket

320-mm multiple rocket system
(China)
Type: 10-rail multiple rocket system
Rocket dimensions: diameter 320 mm (12.6 in); length not revealed
Rocket weights: not revealed
Rocket range: not revealed
Reload time: not revealed
Launcher traverse/elevation: not revealed
Crew: 6
Mounting: rear of CA-30A 6 x 6 truck

Variant
320-mm MRS: this Chinese system, whose proper designation remains unknown, is designed for the laying of anti-tank minefields, each rocket having a solid-propellant motor and fixed cruciform fins, and holding five mines; the rockets are mounted in an over-and-under arrangement on a launcher frame based on that of the Soviet BM-13 equipment

Type 70 multiple rocket system
(China)
Type: 19-tube multiple rocket system

Rocket dimensions: diameter 130 mm (5.12 in); length 1.05 m (3.45 ft)
Rocket weights: whole round 32.8 kg (72.3 lb); warhead 14.7 kg (32.4 lb) FRAG-HE
Rocket range: 10370 m (11,340 yards)
Reload time: not revealed
Launcher traverse/elevation: 360 degrees/not revealed
Crew: 6
Mounting: rear of Type 531 APC

Variants
Type 63: this system is used by the Chinese as replacement for the Soviet BM-13-16, and the solid-propellant fin-stabilized rockets are thought to have a performance comparable to that of the Soviet 140-mm (5.51-in) types; the launch tubes are disposed as a row of 10 over a row of nine; this model of the 130-mm (5.12-in) type is based on the rear of an NJ-230 2.5-ton 4 x 4 truck
Type 70: this has the same launcher and rockets, but in this application located above the hull of the Type 531 APC

Type 81 multiple rocket system
(China)
Type: 12-tube multiple rocket system
Rocket dimensions: diameter 107 mm (4.21 in); length 0.837 m (2.75 ft)
Rocket weights: whole round 18.8 kg (41.45 lb); warhead 8.33 kg (18.36 lb) FRAG-HE or 7.54 kg (16.62 lb) incendiary
Rocket range: 8500 m (9,295 yards) with lighter warhead or 7900 m (8,640 yards) with heavier warhead
Reload time: 3 minutes
Launcher traverse/elevation: 360 degrees/−3 to +57 degrees
Crew: 4
Mounting: rear of 4 x 4 truck

Variants
Type 63: designed in the late 1950s, this system has three four-round horizontal rows of launch tubes, and the combination can also be installed on the back of light vehicles; the rocket is spin-stabilized and powered by a solid-propellant motor; this model has a crew of 5 and 32-degree traverse
Type 81: this is the truck-mounted variant for greater tactical mobility; 12 reload rockets are carried on the truck

M51 multiple rocket system
(Czechoslovakia)
Type: 32-tube multiple rocket system
Rocket dimensions: diameter 130 mm (5.12 in); length 0.80 m (2.62 ft)
Rocket weights: whole round 24.2 kg (53.4 lb); warhead not revealed

Rocket range: 8200 m (8,970 yards)
Reload time: 2 minutes
Launcher traverse/elevation: 240 degrees/0 to +50 degrees
Crew: 6
Mounting: rear of Praga V3S or ZIL-151/157 6 x 6 trucks (Czech and most export models) or Steyr 680 M3 6 x 6 truck (Austrian model)

Variant
M51: developed in the 1950s, this system uses a solid-propellant spin-stabilized rocket, reload rounds being carried in the launcher base structure

SAKR-18 multiple rocket system
(Egypt)
Type: 21-, 30- and 40-tube multiple rocket system
Rocket dimensions: diameter 122 mm (4.8 in); length 3.25 m (10.66 ft)
Rocket weights: whole round 67 kg (147.7 lb); warhead 21 kg (46.3 lb) cluster munition
Rocket range: 18000 m (19.685 yards)
Reload time: not revealed
Launcher traverse/elevation: not revealed
Crew: not revealed
Mounting: light AFV or truck

Variant
SAKR-18: available in three types of launcher configuration, this system is based on a simple rocket designed to carry a payload of 21 anti-tank bomblets or 28 anti-personnel bomblets for area saturation purposes

SAKR-30 multiple rocket system
(Egypt)
Type: 40-tube multiple rocket system
Rocket dimensions: diameter 122 mm (4.8 in); length 2.50, 3.10 or 3.16 m (8.20, 10.17 or 10.37 ft)
Rocket weights: whole round 56.5, 61.5 or 63 kg (124.6, 135.6 or 138.9 lb); warhead 17.5 kg (38.6 lb) FRAG-HE, 23 kg (50.7 lb) cluster munition or 24.5 kg (54 lb) cluster munition
Rocket range: 30000 m (32,810 yards)
Reload time: not revealed, but probably 10 minutes
Launcher traverse/elevation: 120 degrees left and right/0 to +50 degrees
Crew: 6
Mounting: rear of Ural-375D 6 x 6 truck

Variant
SAKR-30: this is essentially the Soviet BM-21 launcher system fitted with a rocket designed and produced in Egypt, and using an advanced solid-propellant motor for very good range with all three rocket types; the system also uses (in its two larger forms) an indigenously produced submunition warhead with 28 anti-tank or 35 anti-personnel bomblets (medium round)

or five anti-tank mines (largest round), the purpose again being the area-saturation of defended territory

VAP multiple rocket system
(Egypt)
Type: 12-frame multiple rocket system
Rocket dimensions: diameter 80 mm (3.15 in); length 1.50 m (4.92 ft)
Rocket weights: whole round 12 kg (26.5 lb); warhead not revealed
Rocket range: 8000 m (8,750 yards)
Reload time: not revealed
Launcher traverse/elevation: 360 degrees/not revealed
Crew: 2/3
Mounting: light truck or AFV

Variant
VAP: this lightweight Egyptian rocket is powered by a solid-propellant motor and fin-stabilized, and designed to carry a FRAG-HE or illuminating warhead; the pedestal-mounted system can be installed in most light trucks and AFVs, thus providing the Egyptian armed forces with a potent weapon of great tactical flexibility

LARS-2 multiple rocket system
(West Germany)
Type: 36-tube multiple rocket system
Rocket dimensions: diameter 110 mm (4.33 in); length 2.26 m (7.41 ft)
Rocket weights: whole round 35.15 kg (77.5 lb); warhead 17.2 kg (37.9 lb)
Rocket range: 14000 m (15,310 yards)
Reload time: 15 minutes
Launcher traverse/elevation: 95 degrees left and right/ −9 to +55 degrees
Crew: 5
Mounting: rear of MAN 7-ton 6 x 6 truck

Variant
LARS-2: developed in the mid- to late 1960s for the West German army as the LARS-1, the Light Artillery Rocket System has in the 1980s been upgraded to LARS-2 standard with improved rockets, a new fire-control system and a MAN rather than Magirus-Deutz truck for improved cross-country capability; the solid-propellant rockets are fin-stabilized, and are fired singly or in ripples at 0.5-second intervals, a full salvo being launched in 17.5 seconds; the LARS-2 is used in conjunction with the Contraves Fieldguard fire-control system, and a diversity of rocket types is available, the most important being the dispenser (with five AT2 hollow-charge anti-tank mines) and the radar target (used to calibrate the system before full fire is opened); less important types are FRAG-HE, smoke and a dispenser type with eight AT1 anti-tank mines; the weight of the entire truck-mounted system is 17480 kg (38,536 lb)

IMI MAR-290 multiple rocket system
(Israel)
Type: 4-tube multiple rocket system
Rocket dimensions: diameter 290 mm (11.4 in); length 5.45 m (17.88 ft)
Rocket weights: whole round 600 kg (1,323 lb); warhead 320 kg (705.5 lb)
Rocket range: 25000 m (27,340 yards)
Reload time: 10 minutes
Launcher traverse/elevation: 360 degrees/0 to +60 degrees
Crew: 4
Mounting: chassis of Centurion MBT

Variant
MAR-290: this system was developed by Israel Military Industries during the 1960s, and entered service on the chassis of Sherman medium tanks modified to accept a quadruple frame rack for this exceptionally powerful solid-propellant fin-stabilized rocket; in the early 1980s the system was developed onto the chassis of the Centurion with a four-tube launcher fitted onto the turret ring for 360-degree traverse; in this guise the whole system weighs 50800 kg (111,993 lb); the reloading process is controlled by a single man making extensive use of hydraulics to manoeuvre the heavy rockets from the reload vehicle into the launch tube

IMI BM-24 multiple rocket system
(Israel)
Type: 12-rack multiple rocket system
Rocket dimensions: diameter 240 mm (945 in); length 1.29 m (4.23 ft)
Rocket weights: whole round 110.5 kg (243.6 lb); warhead 48.3 kg (106.5 lb)
Rocket range: 10700 m (11,700 yards)
Reload time: 4 minutes
Launcher traverse/elevation: 140 degrees/0 to 65 degrees
Crew: 6
Mounting: rear of ZIL-157 6 x 6 truck

Variant
IMI BM-24: Israel has captured from various Arab combatants in recent years sufficient Soviet BM-24 equipments to take the type into her own inventory and to undertake manufacture of an Israeli development of the Soviet rocket; this is a solid-propellant type with spin stabilization, and on detonation the rocket breaks into some 12,000 steel fragments that cover an area of 12000 m^2 (14,352 sq yards); a salvo blankets an area of 125000 m^2 (14,950 sq yards)

IMI LAR-160 multiple rocket system
(Israel)
Type: 36-tube multiple rocket system

Rocket dimensions: diameter 160 mm (6.3 in); length 3.311 m (10.86 ft)
Rocket weights: whole round 110 kg (242.5 lb); warhead 50 kg (110 lb) cluster munition
Rocket range: 30000 m (32,810 yards)
Reload time: not revealed
Launcher traverse/elevation: 360 degrees/0 to 54 degrees
Crew: not revealed
Mounting: chassis of AMX-13 light tank

Variant
LAR-160: this is a highly advanced multiple rocket system of Israeli design and manufacture, based on a solid-propellant rocket with wrap-round stabilizing fins, and currently deployed on the chassis of the obsolete AMX-13 light tank (two 18-tube pods, for a combat weight of 19200 kg/43,328 lb) and M548 cargo carrier (two 13-tube pods, for a combat weight of 12800 kg/28,219 lb); it could also be fitted on the M809 6 x 6 truck (two 13-tube pods, for a combat weight of 14170 kg/31,239 lb) or on the M47 tank chassis (two 25-tube pods, for a combat weight of 45000 kg/99,206 lb); the LARS-160 battery comprises six launcher systems and one Contraves Fieldguard or Westinghouse Quickfire radar fire-control system; the submunition payload of the rocket comprises 187 M42 bomblets, a FASCAM scatterable mine unit, a chemical or biological agent type, Skeet anti-tank weapons, a 155-mm (6.1-in) howitzer projectile, or an illuminating type

SNIA FIROS 6 multiple rocket system
(Italy)
Type: 48-tube multiple rocket system
Rocket dimensions: diameter 51 mm (2 in); length 1.05 m (3.44 ft)
Rocket weights: whole round 4.8 kg (10.6 lb); warhead 2.2 kg (4.4 lb) HEI, FRAG-HE, anti-tank/anti-personnel, illuminating or smoke
Rocket range: 6550 m (7,165 yards)
Reload time: 5 minutes
Launcher traverse/elevation: 360 degrees/−5 to +45 degrees
Crew: 2/3
Mounting: rear of light wheeled vehicle

Variant
FIROS 6: this system was designed by SNIA as an area-saturation equipment to be fitted on the rear of any lightweight 4 x 4 vehicle or on an APC, the type firing a developed version of the solid-propellant folding-fin aircraft rocket; the rockets are ripple-fired at the rate of 10 per second; a typical installation is the 480-kg (1,058-lb) loaded launcher assembly on a Fiat 1107 truck for a combat weight of 2670 kg (5,886 lb) with 48 reload rockets

SNIA FIROS 25 multiple rocket system
(Italy)
Type: 40-tube multiple rocket launcher
Rocket dimensions: diameter 122 mm (4.8 in); length 3.174 m (10.41 ft) with cluster munition warhead or 2.575 m (8.45 ft) with conventional warhead
Rocket weights: whole round 63 kg (138.9 lb) with cluster munition warhead or 52.5 kg (115.75 lb) with conventional warhead; cluster munition warhead 27.5 kg (60.6 lb) or conventional warhead 17 kg (37.5 lb)
Rocket range: 22000 m (24,060 yards) with cluster munition warhead or 25000 m (27,340 yards) with conventional warhead
Reload time: not revealed
Launcher traverse/elevation: 105 degrees left and right/0 to +60 degrees
Crew: 3/4
Mounting: rear of most 6 x 6 trucks

Variant
FIROS 25: essentially an enlarged development of the FIROS 6 system, this equipment fires solid-propellant rockets carrying either a conventional warhead (HE, HE-FRAG, or smoke) or a cluster munition warhead (seven anti-tank mines, 66 anti-personnel mines, or 84 dual-role anti-tank/anti-personnel bomblets); the launchers are grouped in batteries of six with a fire-control vehicle and six reload vehicles, the whole system being designed for the rapid saturation of large areas

Type 67 multiple rocket system
(Japan)
Type: two-rail multiple rocket system
Rocket dimensions: diameter 307 mm (12.09 in); length 4.50 m (14.76 ft)
Rocket weights: whole round 573 kg (1,263 lb); warhead not revealed
Rocket range: 28000 m (30,620 yards)
Reload time: not revealed
Launcher traverse/elevation: not revealed
Crew: 2
Mounting: rear of Hino 4-ton 6 x 6 truck

Variant
Type 67: now obsolescent system, the Type 67 launcher was designed in the mid-1960s and fires the Type 68 rocket, which has fixed cruciform fins and a solid-propellant motor, and carries a FRAG-HE warhead

Type 75 multiple rocket system
(Japan)
Type: 30-frame multiple rocket system
Rocket dimensions: diameter 131.5 mm (5.18 in); length 1.856 m (6.09 ft)
Rocket weights: whole round 43 kg (94.8 lb); warhead 15 kg (33 lb) HE

Rocket range: 15000 m (16,405 yards)
Reload time: not revealed
Launcher traverse/elevation: 50 degrees left and right/ 0 to +50 degrees
Crew: 3
Mounting: armoured launch vehicle based on the Type 73 APC

Variant
Type 75: developed in the first half of the 1970s, this equipment is in Japanese service only, and can fire its rockets individually or in a 12-second ripple salvo; the rocket is fin-stabilized and powered by a solid-propellant motor

130-mm multiple rocket system
(South Korea)
Type: 36-tube multiple rocket system
Rocket dimensions: diameter 130 mm (5.12 in); length 2.40 m (7.87 ft)
Rocket weights: whole round 55 kg (121.25 lb); warhead 21 kg (46.3 lb) HE
Rocket range: 23000 m (25,155 yards)
Reload time: not revealed
Launcher traverse/elevation: 120 degrees left and right/0 to +55 degrees
Crew: 3
Mounting: rear of KM809A1 6 x 6 truck

Variant
130-mm MRS: designed for South Korea's extremes of terrain and climate, this useful multiple rocket system can be installed on most 5-ton 6 x 6 trucks, and the rockets can be fired individually or in a ripple of two rockets per second

WP-8 multiple rocket system
(Poland)
Type: 8-tube multiple rocket system
Rocket dimensions: diameter 140 mm (5.51 in); length 1.092 m (3.58 ft)
Rocket weights: whole round 39.6 kg (81.35 lb); warhead 18.8 kg (41.4 lb) HE
Rocket range: 9800 m (10,715 yards)
Reload time: 2 minutes
Launcher traverse/elevation: 14 degrees left and right/ −12 to +50 degrees
Crew: 5
Mounting: two-wheel trailer

Variant
WP-8: this Polish equipment is designed for the use of airborne forces, and uses the same solid-propellant rocket as the Soviet RPU-14 and BM-14 launchers; the loaded weight of the trailer is 688 kg (1,517 lb), and the type can be towed by a UAZ-469 4 x 4 light truck

Armscor Valkyri multiple rocket system
(South Africa)
Type: 24-tube multiple rocket system
Rocket dimensions: diameter 127 mm (5 in); length 2.68 m (8.8 ft)
Rocket weights: not revealed
Rocket range: 22000 m (24,060 yards)
Reload time: 10 minutes
Launcher traverse/elevation: 52 degrees left and right/ 50-degree arc
Crew: 2
Mounting: rear of SAMIL 4 x 4 truck

Variant
Valkyri: designed in direct response to the threat of Soviet-supplied BM-21 equipments in the hands of African 'freedom fighters', the Valkyri is a highly capable piece of kit designed for area saturation in bush conditions, the solid-propellant rocket being fin-stabilized to deliver a load of 3,500 steel balls to the target area, where the lethal area is 1500 m^2 (1,795 sq yards); the rockets can be fired individually or in ripples at the rate of one rocket per second, range being variable between 8000 and 22000 m (8,750 and 24,060 yards) by the use of spoiler rings on the rockets to slow them; each launcher is accompanied by a vehicle with 72 reload rockets

R-6/B-2 multiple rocket system
(Spain)
Type: 32-frame multiple rocket system
Rocket dimensions: diameter 108 mm (4.25 in); length 0.935 m (3.07 ft)
Rocket weights: whole round 19.4 kg (42.8 lb); warhead 8.8 kg (19.4 lb HE, smoke or incendiary
Rocket range: 10000 m (10,935 yards)
Reload time: not revealed
Launcher traverse/elevation: 45 degrees left and right/ not revealed
Crew: 2/3
Mounting: rear of 1-ton 4 x 4 truck

Variant
R-6: now obsolete, this spin-stabilized solid-propellant rocket is fired from a B-2 launcher containing four horizontal rows of eight frames; the rockets can be fired individually or in a ripple at 1-second intervals

Santa Barbara Teruel multiple rocket system
(Spain)
Type: 40-tube multiple rocket system
Rocket dimensions: diameter 140 mm (5.51 in); length 2.044 m (6.71 ft) with HE warhead and 2.14 m (7.02 ft) with cluster munition warhead
Rocket weights: whole round 56 kg (123.5 lb) with HE warhead and 59 kg (130 lb) with cluster munition

warhead; warhead 18.6 kg (41 lb) HE or 21 kg (46.3 lb) cluster munition
Rocket range: 18200 m (19,905 yards)
Reload time: 5 minutes
Launcher traverse/elevation: 120 degrees left and right/0 to +55 degrees
Crew: 2/3
Mounting: rear of Pegaso 3050 6 x 6 truck

Variant
Teruel: the launcher comprises two sub-assemblies each containing five horizontal rows each of four launcher tubes; the rockets have folding fins for stabilization, and are powered by solid-propellant motors; launcher batteries have six launchers, and each launcher truck is accompanied by a vehicle with 80 reload rockets; the cluster munition warhead can contain 28 impact-fused anti-tank grenades, or six anti-tank mines or 42 anti-personnel grenades; the Teruel is designed to supplant the multitude of older rockets and launchers still in Spanish service

E-3/L-21 multiple rocket system
(Spain)
Type: 21-frame multiple rocket system
Rocket dimensions: diameter 216 mm (8.50 in); length 1.406 m (4.61 ft)
Rocket weights: whole round 101 kg (222.7 lb); warhead 37.5 kg (123 lb) HE, smoke or incendiary
Rocket range: 14500 m (15,855 yards)
Reload time: 14 minutes
Launcher traverse/elevation: 45 degrees left and right/not revealed
Crew: not revealed
Mounting: rear of Banieros Pante III 6 x 6 truck

Variant
E-3: the E-3 is an obsolete rocket has a solid-propellant motor and spin stabilization, the latter imparted by helical rails in the L-21 launch frame

D-3/L-10 multiple rocket system
(Spain)
Type: 10-frame multiple rocket system
Rocket dimensions: diameter 300 mm (11.81 in); length 1.898 m (6.23 ft)
Rocket weights: whole round 247.5 kg (545.6 lb); warhead 89.4 kg (197 lb) HE, smoke or incendiary
Rocket range: 17700 m (19,355 yards)
Reload time: not revealed
Launcher traverse/elevation: 45 degrees left and right/not revealed
Crew: not revealed
Mounting: rear of Banieros Pante III 6 x 6 truck

Variant
D-3: now a thoroughly obsolete system, the spin-stabilized solid-propellant D-3 rocket is an enlarged version of the E-3, and like that equipment fired individually at the rate of one round every two seconds, in this instance from an L-10 launcher containing two horizontal rows of five frames; the system is reloaded mechanically as a result of the rocket weight

G-3/L-8 multiple rocket system
(Spain)
Type: 8-frame multiple rocket system
Rocket dimensions: diameter 381 mm (15 in); length 2.657 m (8.72 ft)
Rocket weights: whole round 527.5 kg (1,162.9 lb); warhead 217 kg (478.4 lb) HE
Rocket range: 23500 m (25,700 yards)
Reload time: not revealed
Launcher traverse/elevation: 22.5 degrees left and right/not revealed
Crew: not revealed
Mounting: rear of 6 x 6 truck

Variant
G-3: a large, cumbersome and obsolete equipment, the G-3 is the largest of the Spanish rockets, and like its smaller brethren is a solid-propellant type with spin stabilization for launch from the L-8 launcher which is only slightly longer than the rocket

Oerlikon RWK-014 multiple rocket system
(Switzerland)
Type: 30-tube multiple rocket system
Rocket dimensions: diameter 81 mm (3.12 in); length 1.80 m (5.91 ft)
Rocket weights: whole round 15.7 or 19.6 kg (34.6 or 43.2 lb) depending on warhead: warhead 7 or 11 kg (15.4 or 24.25 lb) FRAG-HE
Rocket range: 10400 m (11,375 yards)
Reload time: not revealed
Launcher traverse/elevation: 360 degrees/−10 to +50 degrees
Crew:
Mounting: most light AFVs

Variant
RWK-014: this Swiss launcher fires folding fin-stabilized solid-propellant SNORA rockets developed jointly with SNIA in Italy, and a salvo of 30 rockets is fired in 3 seconds; the loaded launcher weighs some 1410 kg (3,108 lb)

Kung Feng III and IV multiple rocket systems
(Taiwan)
Type: 40-tube multiple rocket system
Rocket dimensions: diameter 126 mm (4.96 in); length about 1.60 m (5.25 ft)
Rocket weights: whole round about 25 kg (55.1 lb); warhead not revealed
Rocket range: 9000 m (9,845 yards)

Reload time: not revealed
Launcher traverse/elevation: 0 degrees/not revealed
Crew: not revealed
Mounting: two-wheel trailer or M113 APC

Variants
Kung Feng III: this is the trailer-mounted version towed by a 0.75-ton truck, the launcher comprising two banks of launchers, each containing five horizontal rows of four tubes; the 40 rockets are ripple-fired in 16 seconds
Keng Feng IV: this is the vehicle-mounted version, generally carried on the back of an M113 APC but also used on LVTP7 amphibious APCs

Kung Feng VI multiple rocket system
(Taiwan)
Type: 45-tube multiple rocket system
Rocket dimensions: diameter 117 mm (4.61 in); length 1.80 m (5.91 ft)
Rocket weights: whole round 60 kg (132.3 lb); warhead not revealed
Rocket range: 1500 m (16,405 yards)
Reload time: 15 minutes
Launcher traverse/elevation: 90 degrees left and right/ not revealed
Crew: not revealed
Mounting: rear of M52 6 x 6 truck

Variant
Kung Feng VI: this truck-mounted system comprises five horizontal rows of nine tubes, and the rockets are ripple-fired in 22 seconds

BM-25 multiple rocket system
(USSR)
Type: 6-frame multiple rocket system
Rocket dimensions: diameter 250 mm (9.84 in); length 5.822 m (19.10 ft)
Rocket weights: whole round 455 kg (1,003 lb); warhead not revealed
Rocket range: at least 30000 m (32,810 yards)
Reload time: 10/20 minutes
Launcher traverse/elevation: 3 degrees left and right/ 0 to +55 degrees
Crew: 8/12
Mounting: rear of KrAZ-214 6 x 6 truck

Variant
BM-25: introduced in the 1950s, this is the largest rocket to have entered Soviet service since World War II, and is now obsolete; the type is powered by a storable liquid-propellant rocket, and is stabilized by four fixed fins which engage in helical grooves in the launcher for imparted spin; the warhead is of the HE type; the whole equipment weighs 18145 kg (40,002 lb)

BM-24 multiple rocket system
(USSR)
Type: 12-frame multiple rocket system
Rocket dimensions: diameter 240 mm (9.45 in); length 1.18 m (3.87 ft)
Rocket weights: whole round 112.5 kg (248 lb); warhead 46.9 kg (103.4 lb) HE
Rocket range: 11000 m (12,030 yards)
Reload time: 3/4 minutes
Launcher traverse/elevation: 70 degrees left and right/ 0 to +65 degrees
Crew: 6
Mounting: rear of ZIL-157 6 x 6 truck

Variants
BM-24: this equipment was introduced to Soviet service in the early 1950s, and was one of the most important support weapons available to motorized rifle divisions until the advent of the BM-21; the weight of the complete equipment is 9200 kg (20,282 lb); the rocket is powered by a solid-propellant motor and is spin-stabilized
BM-24T: version with tube rather than frame launchers, and mounted on the AT-S armoured and tracked artillery tractor

BM-27 multiple rocket system
(USSR)
Type: 16-tube multiple rocket system
Rocket dimensions: diameter 220 mm (8.66 in); length 4.80 m (15.75 ft)
Rocket weights: whole round 360 kg (793.7 lb); warhead not revealed
Rocket range: 40000 m (43,745 yards)
Reload time: 15/20 minutes
Launcher traverse/elevation: 360 degrees/+15 to +55 degrees
Crew: 6
Mounting: rear of ZIL-135 8 x 8 truck

Variant
BM-27: introduced in 1977, this modern and highly capable cross-country equipment comprises a three-bank launcher (four, six and six tubes) firing solid-propellant rockets in support of Soviet manoeuvre formations; the rocket can deliver HE, chemical and cluster munition warheads, the last including FRAG-HE, incendiary and mine types; the whole equipment weighs 22750 kg (50,154 kg), and is supported in the field by another ZIL-135 carrying 16 reload rockets, a reloading crane and a power rammer

BMD-20 multiple rocket system
(USSR)
Type: 4-frame multiple rocket system
Rocket dimensions: diameter 200 mm (7.87 in); length 3.11 m (10.20 ft)

Rocket weights: whole round 194 kg (427.7 lb); warhead not revealed
Rocket range: 20000 m (21,870 yards)
Reload time: 6/10 minutes
Launcher traverse/elevation: 10 degrees left and right/+9 to +60 degrees
Crew: 6
Mounting: rear of ZIL-157 6 x 6 truck

Variant
BMD-20: introduced into Soviet service in the first half of the 1950s, this equipment is obsolete but still in service with some Soviet clients; the rocket is fin-stabilized (with spin imparted by guide rails in the frames) and powered by a solid-propellant motor; the whole equipment weighs 8700 kg (19,180 lb)

BM-14-17 multiple rocket system
(USSR)
Type: 17-tube multiple rocket system
Rocket dimensions: diameter 140 mm (5.51 in); length 1.092 m (3.58 ft)
Rocket weights: complete round 39.6 kg (87.3 lb); warhead 18.8 kg (41.45 lb) HE
Rocket range: 9800 m (10,715 yards)
Reload time: 2 minutes
Launcher traverse/elevation: 90 degrees left and right/0 to +50 degrees
Crew: 7
Mounting: rear of GAZ-53A 4 x 4 truck

Variants
BM-14-17: this equipment was introduced to Soviet service in the mid- and late 1950s as the complement to the BM-14-16 equipment in motorized rifle and tank divisions; the whole equipment weighs 5323 kg (11,735 lb), and the launcher comprises two horizontal rows, with nine tubes over eight; the rocket is spin-stabilized and powered by a solid-propellant motor
BM-14-16: introduced in the early 1950s, this equipment has a combat weight of 6432 kg (14,180 lb) including the ZIL-131 6 x 6 truck; the rocket is the same M-14-OF fired by the BM-14-17, and the type can carry smoke or chemical loads as alternatives to the HE payload; this equipment has a crew of seven, and the launcher traverse/elevation arcs are 105 degrees left and right/0 to +45 degrees
RPU-14: this is a towed equipment using the M-14-OF rocket, with four rows of four tubes on a two-wheel trailer; the equipment was designed for the use of Soviet airborne formations

BM-13-16 multiple rocket system
(USSR)
Type: 8-rail multiple rocket system
Rocket dimensions: diameter 132 mm (5.20 in); length 1.743 m (5.72 ft)

Rocket weights: whole round 42.5 kg (93.7 lb); warhead 18.5 kg (40.8 lb) HE
Rocket range: 9000 m (9,845 yards)
Reload time: 5/10 minutes
Launcher traverse/elevation: 10 degrees left and right/+15 to +45 degrees
Crew: 6
Mounting: rear of ZIL-151 6 x 6 truck

Variant
BM-13-16: introduced in 1940, and one of the decisive weapons of World War II on the Eastern Front, this 'Katyusha' system is still in service with some Soviet client states; the equipment has a combat weight of 6430 kg (14,175 lb), and the eight launcher rails each carry two rockets (one above and the other below)

BM-21 multiple rocket system
(USSR)
Type: 40-tube multiple rocket system
Rocket dimensions: diameter 122 mm (4.8 in); length 1.905 m (6.25 ft) or 3.23 m (10.60 ft)
Rocket weights: whole round 45.8 or 77 kg (101 or 169.75 lb); warhead 19.4 kg (42.8 lb)
Rocket range: 11000 m (12,030 yards) with short rocket or 20380 m (22,290 yards) with long rocket
Reload time: 10 minutes
Launcher traverse/elevation: 120 degrees left and right/0 to +50 degrees
Crew: 6
Mounting: rear of Ural-375D 6 x 6 truck

Variants
BM-21: introduced in the early 1960s, this equipment is the standard area support equipment of Soviet motorized rifle and tank divisions (being deployed at the rate of one battery of six equipments to the divisional artillery regiment), though the BM-27 is in the process of supplanting it; the equipment has a combat weight of 11500 kg (25,353 kg), and the solid-propellant rocket has wrap-round fins for stabilization after initial spin has been imparted by helical grooves in the launch tube; the rocket comes in two types for medium- or long-range use, and three types of warhead are available (FRAG-HE, chemical and smoke); among the chemical loads for the warhead is the blood agent hydrogen cyanide; China also uses a version of this equipment on a Chinese-designed 6 x 6 truck, and India has developed the indigenous **LRAR** using the BM-21 launcher and rocket
M1972: introduced in the early 1970s as a companion to the BM-21 with improved cross-country capability, this Czech version uses the standard launcher and rocket mounted on the rear of a Tatra 813 8 x 8 truck together with a reload pack of 40 rockets located between the launcher and the armoured cab; the combat weight of the whole equipment is 33700 kg

(74,295 lb), and the type is known to bear the Czech designation **RM-70**

M1975: this Soviet equipment is a 12-tube version of the BM-21 mounted on a light truck, and is designed as the replacement for the BM-14-17 and RPU-14 with Soviet airborne formations

M1976: this Soviet equipment is an improved version of the basic BM-21 mounted on a ZIL-131 3.5-ton 6 x 6 truck

BM-21/Bucegi: this is a Romanian equipment with a 21-tube launcher mounted on a Bucegi SR-114 4 x 4 truck; the proper designation is unknown

BM-21/Isuzu: this is a 30-tube Palestine Liberation Organization development, with two banks (each of three five-tube horizontal rows) mounted on the back of a Japanese Isuzu 2.5-ton 6 x 6 truck, and used operationally against the Israelis during the Lebanon fighting of 1982

Vought Multiple-Launch Rocket System
(USA)
Type: 12-tube multiple rocket system
Rocket dimensions: diameter 8.94 in (227 mm); length 13.00 ft (3.96 m)
Rocket weights: whole round 680 lb (308.4 kg); warhead not revealed
Rocket range: more than 33,000 yards (30175 m), but see Variant (below)
Reload time: 10 minutes
Launcher traverse/elevation: 360 degrees/not revealed
Crew: 3
Mounting: adapted version of the M2 Bradley IFV

Variant
MLRS: this system is a key ingredient of the NATO powers' plans for the defeat of any Warsaw Pact incursion into Western Europe, the type offering the possibility of generating swiftly but at long-range large areas impassable to AFVs and men; the system is based on an existing high-mobility chassis, which carries at its rear a frame arrangement into which the accompanying Heavy Expanded Mobility Tactical Truck can fit two of its four pre-loaded six-tube rocket pods, another four loaded pods being accommodated on the HEMTT-drawn Heavy Expanded Mobility Ammunition Trailer; the rocket is a folding fin type with solid-propellant motor, and can carry any of four singularly advanced warheads currently under development or in service; these are a cluster munition with 644 M77 shaped-charge FRAG-HE dual-purpose anti-tank/anti-personnel bomblets (fired to 33,000 yards/30175 m), a cluster munition with 28 AT-2 parachute-retarded anti-tank mines (fired to a range of 43,750 yards/40005 m), a cluster munition with six active-radar terminally-guided free-fall anti-tank shaped charge bomblets (fired to a range of 45,925 yards/41995 m) and a chemical warhead with 92 lb (41.7 kg) of binary nerve gas agent; other warheads are under investigation for future development; the combat weight of the MRLS is 55,535 lb (25191 kg)

M77 Oganj multiple rocket system
(Yugoslavia)
Type: 32-tube multiple rocket system
Rocket dimensions: diameter 128 mm (5.04 in); length 2.60 m (8.53 ft)
Rocket weights: whole round 65 kg (143.3 lb); warhead 20 kg (44.1 lb) FRAG-HE
Rocket range: 20000 m (21,870 yards)
Reload time: 2 minutes
Launcher traverse/elevation: 360 degrees/0 to +50 degrees
Crew: not revealed
Mounting: rear of FAP 2020BS 6 x 4 truck

Variant
M77 Oganj: developed for service in the mid-1970s, this is a useful Yugoslav equipment with a combat weight of about 13000 kg (28,660 lb) complete with a reload pack of 32 rockets, which are of the solid-propellant type; the rockets can be fired individually or in a ripple salvo of 18 seconds

M63 Plaman multiple rocket system
(Yugoslavia)
Type: 32-tube multiple rocket system
Rocket dimensions: diameter 128 mm (5.04 in); length 0.814 m (2.67 ft)
Rocket weights: whole round 23.1 kg (50.9 lb); warhead 7.55 kg (16.6 lb) HE
Rocket range: 8600 m (9,405 yards)
Reload time: 5 minutes
Launcher traverse/elevation: 15 degrees left and right/ 0 to +48 degrees
Crew: 3/5
Mounting: two-wheel trailer

Variant
M63 Plaman: this equipment uses a spin-stabilized solid-propellant rocket, and the rockets can be fired individually or in a ripple with intervals of 0.2, 0.4 or 0.6 seconds; the same rocket is also used in three export models (one, eight or 16 tubes)

UNGUIDED ANTI-TANK WEAPONS

FM Model 1968 anti-tank weapon
(Argentina)
Type: four-man anti-tank recoilless gun
Calibre: 105 mm (4.13 in)
Length: launcher, overall 4.02 m (13.19 ft)
Weights: launcher 459.4 kg (1,012.8 lb); round 15.6 kg (34.4 lb) for HE type and 11.1 kg (24.5 lb) for hollow-charge HE type
Warhead: HE or shaped-charge HE
Range: 9200 m (10,060 yards) maximum and 1200 m (1,315 yards) effective

Variant
Model 68: this is an unexceptional anti-tank recoilless gun mounted on a two-wheel trailer for towing, and aimed with the aid of a spotting rifle; the gun can be elevated between −7 and +40 degrees, and the armour penetration of the shaped-charge warhead is 200 mm (7.87 in)

MECAR RL-83 Blindicide anti-tank weapon
(Belgium)
Type: two-man anti-tank rocket launcher
Calibre: 83 mm (3.27 in)
Lengths: launcher 0.62 m (2.03 ft) for carriage and 1.70 m (5.58 ft) ready for firing; rocket 0.57 m (1.87 ft)
Weights: launcher 8.4 kg (18.52 lb); rocket 2.4 kg (5.29 lb)
Warhead: shaped-charge HE
Range: 900 m (985 yards)

Variants
RL-83 Blindicide: this is a widely used weapon notable for its range and the fact that it can fire HEAT, anti-personnel, smoke, illuminating and incendiary warheads; the armour penetration is 300 mm (11.81 in)
M-50: this is the Swiss development of the RL-83 with a range of 300 m (330 yards)
M-58: lightened version of the M-50
M-75: improved M-58 firing a rocket with fold-out fins

MECAR MPA 75 anti-tank weapon
(Belgium)
Type: one-man two-tube anti-tank and anti-fortification rocket launcher
Calibre: 75 mm (2.95 in)
Length: launcher 0.51 m (1.67 ft)
Weights: total with rockets 2.6 kg (5.7 lb); rocket 0.95 kg (2.09 lb)
Warhead: shaped-charge FRAG-HE
Range: 100 m (109 yards) in the anti-tank role and 300 m (328 yards) in the anti-personnel role

Variant
MPA 75: a light dual-role type carrying two small rockets in a disposable launcher, the MPA 75 is designed for use against personnel and heavier targets, the penetration figures being 270 mm (10.63 in) against armour and 600 mm (23.6 in) against concrete

M59A anti-tank weapon
(Czechoslovakia)
Type: four-man anti-tank recoilless gun
Calibre: 82 mm (3.2 in)
Length: launcher overall 4.597 m (15.08 ft)
Weights: launcher 385 kg (848.8 lb); round 6 kg (13.2 kg)
Warhead: HE and shaped-charge HE
Range: 6657 m (7,280 yards) maximum and 1200 m (1,315 yards) effective

Variant
M59: this Czech recoilless gun is generally towed on a two-wheel trailer but can also be mounted on light vehicles; the type is ranged with the aid of a ranging rifle, and the armour penetration is 250 mm (9.84 in)
M59A: improved version with a radially-finned firing chamber for more rapid heat dissipation

M-55 anti-tank weapon
(Finland)
Type: one-man anti-tank rocket launcher
Calibre: 55 mm (2.16 in)
Lengths: 0.94 m (3.08 ft) unloaded and 1.24 m (4.07 ft) loaded; rocket not revealed
Weights: launcher 8.5 kg (18.74 lb); rocket 2.5 kg (5.51 lb)
Warhead: shaped-charge HE
Range: 200 m (220 yards)

Variant
M-55: this light anti-tank launcher can fire between three and five rounds per minute, and the rocket warhead can penetrate 200 m (7.87 in) of armour

Thomson-Brandt SARPAC anti-tank weapon
(France)
Type: one-man anti-tank rocket launcher
Calibre: 68 mm (2.68 in)
Lengths: launcher 0.734 m (2.41 ft) for carriage and 0.997 m (3.27 ft) ready for firing
Weights: launcher 1.9 kg (4.19 lb) unloaded; rocket 1.09 kg (2.4 lb)

Warhead: shaped-charge HE
Range: 200 m (220 yards)

Variant
SARPAC: this weapon was designed as a one-shot type, but has since been developed with a launcher able to fire some 20 rounds; the ROCHAR anti-tank round can penetrate 300 mm (11.8 in) of armour, and the other two rocket types are the ROCAP anti-personnel and ROCLAIR illuminating types, which have ranges up to 700 m (765 yards)

GIAT ACL-APX anti-tank weapon
(France)
Type: one-man anti-tank rocket launcher
Calibre: 80 mm (3.15 in)
Lengths: launcher 1.40 m (459 ft); rocket 0.53 m (1.74 ft)
Weights: launcher 8.8 kg (18.96 lb); rocket 3.6 kg (7.94 lb)
Warhead: shaped-charge HE
Range: 580 m (635 yards)

Variant
ACL-APX: this simple multi-shot system fires a HEAT round which can penetrate 270 mm (10.63 in) of armour; the launcher can also fire an anti-personnel rocket

Saint Etienne LRAC 89-STRIM anti-tank weapon
(France)
Type: one-man anti-tank rocket launcher
Calibre: 88.9 mm (3.50 in)
Lengths: launcher 1.168 m (3.83 ft) for carriage and 1.60 m (5.25 ft) for firing; rocket 0.60 m (1.97 ft)
Weights: launcher 5.4 kg (11.9 lb) for unloaded and 8.6 kg (18.96 lb) complete for firing; rocket 2.20 kg (4.85 lb)
Warhead: 0.565-kg (1.25-lb) shaped-charge HE
Range: 500 m (550 yards) effective

Variant
LRAC 89-STRIM: this equipment entered service in 1969, and is designed primarily for infantry use at range up to 400 m (440 yards) and incidence angles of less than 75 degrees; the warhead is capable of penetrating 480 mm (18.9 in) of armour at 0 degrees

Nobel PZF 44 Lanze anti-tank weapon
(West Germany)
Type: one-man anti-tank rocket launcher
Calibre: 44 mm (1.73 in)
Lengths: launcher 0.888 m (2.91 ft) for carriage and 1.162 m (3.81 ft) ready for firing; rocket 0.55 m (1.80 ft)
Weights: 7.8 kg (17.2 lb) for carriage and 10.3 kg (22.7 lb) ready for firing; rocket 1.5 kg (3.31 lb)

Warhead: shaped-charge HE
Range: 400 mm (440 yards)

Variant
PZF44: this lightweight shoulder-fired weapon can penetrate 370 mm (14.57 in) of armour

IMI B-300 anti-tank weapon
(Israel)
Type: one-man anti-tank rocket launcher
Calibre: 82 mm (3.23 in)
Lengths: launcher 0.755 m (2.498 ft) and loaded weapon 1.35 m (4.43 ft)
Weights: launcher 3.5 kg (7.71 lb) unloaded and 8 kg (17.64 lb); rocket 3 kg (6.61 lb)
Warhead: not revealed
Range: 400 m (440 yards)

Variants
B-300: designed for the engagement of light AFVs and strongpoints, the B-300 equipment comprises a launcher and a backpack with three rockets in their containers, which clip onto the rear of the launcher as required; the shaped-charge HE warhead can penetrate 400 mm (15.75 in) of armour
SMAW: the B-300 has been accepted for service with the US Marine Corps as the Shoulder-Launched Multi-purpose Assault Weapon; this incorporates a number of improvements, including a flash-producing spotting rifle

Instalza M-65 anti-tank weapon
(Spain)
Type: one-man anti-tank rocket launcher
Calibre: 88.9 mm (3.50 in)
Lengths: launcher 0.83 m (2.72 ft) for carriage and 1.60 m (5.25 ft) ready for firing; rocket not revealed
Weights: launcher 5.4 kg (11.9 lb); rocket 2.0 to 3.1 kg (4.41 to 6.83 lb) according to type
Warhead: shaped-charge HE
Range: 450 m (490 yards)

Variant
M-65: this is a typical 'bazooka' weapon, and can fire any of three rocket types with good anti-armour and anti-concrete penetration

FFV Miniman anti-tank weapon
(Sweden)
Type: one-man anti-tank rocket launcher
Calibre: 74 mm (2.91 in)
Lengths: launcher 0.90 m (2.95 ft) loaded; rocket 0.325 m (1.07 ft)
Weights: launcher 2.9 kg (6.4 lb) loaded; rocket 0.88 kg (1.94 lb)
Warhead: shaped-charge HE
Range: 250 m (275 yards)

Variant
Miniman: a one-shot throwaway weapon, the Miniman can penetrate 300 mm (11.8 in) of armour

FFV Carl Gustav M2-550 anti-tank weapon
(Sweden)
Type: one-man anti-tank rocket launcher
Calibre: 84 mm (3.31 in)
Lengths: launcher 1.13 m (3.71 ft); rocket not revealed
Weights: launcher 16.1 kg (35.5 lb); rocket 3.2 kg (7.05 lb)
Warhead: shaped-charge HE
Range: 700 m (765 yards)

Variants
Carl Gustav M2: this is the original version of this widely used equipment, capable of firing a 2.5-kg (5.51-lb) HEAT rocket to penetrate 2228 mm (8.98 in) of armour at 400 m (440 yards) against a moving target and 500 m (550 yards) against a stationary target; other rocket types are HE (to 1000 m/1,095 yards) and smoke (to 1300 m/1,420 yards) at a rate of six rounds per minute
Carl Gustav M2-550: improved version with a rocket-assisted round for greater range
Carl Gustav M2-597: upgraded model with a more powerful rocket and fin-stabilization; the rocket weighs 7 kg (15.4 lb) and can penetrate 900 mm (35.4 in) of armour at ranges up to 300 m (330 yards)
Carl Gustav M3: lightweight version of the basic M2 series, the loaded weight being only 8 kg (17.6 lb)

Bofors PV-1110 anti-tank weapon
(Sweden)
Type: three-man anti-tank recoilless gun
Calibre: 90 mm (3.54 in)
Length: barrel 3.70 m (12.14 ft)
Weights: launcher and trailer 260 kg (573 lb); round 3.1 kg (6.83 lb)
Warhead: shaped-charge HE
Range: 900 m (984 yards) effective

Variant
PV-1110: this neat recoilless weapon is either towed on a light two-wheel trailer or pedestal-mounted on a light vehicle; the weapon can be fitted at 6 rounds per minute, and the armour penetration is 380 mm (14.96 in); the barrel can be elevated between −10 and +15 degrees, and total traverse is between 75 and 115 degrees according to the elevation

RPG-2 anti-tank weapon
(USSR)
Type: one-man anti-tank rocket launcher
Calibre: 40 mm (1.57 in) for launcher and 82 mm (3.2 in) for rocket
Lengths: launcher 1.494 m (4.90 ft); rocket not revealed
Weights: launcher 2.83 kg (6.24 lb); rocket 1.84 kg (4.06 lb)
Warhead: shaped-charge HE
Range: 150 m (165 yards)

Variants
RPG-2: this weapon is now obsolete, but still found in some numbers all over the Middle and Far East; the launcher can deliver between four and six rounds per minute, and the HEAT warhead can penetrate 180 mm (7.09 in) of armour
Type 56: Chinese-built model with a warhead able to penetrate 250 mm (9.84 in) of armour

RPG-7 anti-tank weapon
(USSR)
Type: one-man anti-tank rocket launcher
Calibre: 40 mm (1.57 in) for launcher and 85 mm (3.35 in) for rocket
Lengths: launcher 0.99 m (3.25 ft); rocket 0.65 m (2.13 ft)
Weights: launcher 7 kg (15.4 lb); rocket 2.25 kg (4.96 lb)
Warhead: shaped-charge HE
Range: 500 m (545 yards)

Variants
RPG-7: this is the world's most numerous anti-tank rocket launcher, and was widely used by soviet and Soviet bloc forces before being passed to 'freedom fighters' and 'liberation armies'; the HEAT warhead can penetrate 320 mm (12.6 in) of armour
RPV-7V: improved model with smaller overall dimensions but a heavier rocket
RPV-7D: version of the RPV-7V for airborne forces, designed to fold as a means of reducing transport length
RPG-16: considerably updated version whose rocket is far less susceptible to crosswind effects than those of predecessor launchers
Type 69: Chinese-built version of the RPG-7

RPG-18 anti-tank weapon
(USSR)
Type: one-man anti-tank weapon
Calibre: 70 mm (2.76 in) for launcher and 64 mm (2.52 in) for rocket
Lengths: launcher 0.705 m (27.76 in); closed and 1.00 m (39.37 in) open; rocket not known
Weights: complete round 4.0 kg (8.82 lb); rocket 2.5 kg (5.51 lb)
Warhead: shaped-charge HE
Range: 350 m (382 yards) maximum and 200 m (218 yards) effective

Variants
RPG-18: this is a simple one-shot rocket launcher bvery similar to the US M72 LAW; at the maximum effective range the 65-mm (2.56-in) diameter warhead can penetrate 250 mm (9.84 in) of armour

SPG-9 anti-tank weapon
(USSR)
Type: two- or three-man anti-tank recoilless gun
Calibre: 73 mm (2.87 in)
Length: launcher, overall 2.11 m (6.92 ft)
Weights: launcher with tripod 59.5 kg (131.2 lb); round not known
Warhead: HEAT
Range: 1300 m (1,420 yards)

Variant
SPG-9: this is an effective anti-tank weapon firing a fin-stabilized rocket-boosted round for an armour penetration capability of 390+ mm (15.35+ in)

B-10 anti-tank weapon
(USSR)
Type: two- or three-man anti-tank recoilless gun
Calibre: 82 mm (3.2 in)
Length: launcher, overall 1.677 m (5.50 ft); barrel 1.659 m (5.44 ft)
Weights: launcher in travelling order 87.6 kg (193.1 lb); round 4.5 kg (9.92 lb) for HE and 3.6 kg (7.94 lb) for HEAT
Warhead: HE or HEAT
Range: 4500 m (4,920 yards) maximum for HE round, and 400 m (440 yards) effective for HEAT round

Variants
B-10: now only in limited service in areas other than the Middle East, the B-10 is a fairly substantial recoilless gun fitted with a shield and fired from a light tripod; the weapon can be manhandled on a two-wheel dolly; the armour penetration of the HEAT round is 240 mm (9.45 in), and the rate os fire is some 6 or 7 rounds per minute
Type 65: designation of the Chinese-made B-10

B-11 anti-tank weapon
(USSR)
Type: three-man anti-tank recoilless gun
Calibre: 107 mm (4.2 in)
Length: launcher, overall 3.314 m (10.87 ft); barrel 2.718 m (8.92 ft)
Weights: launcher 305 kg (672 lb); round 13.6 kg (30 lb) for HE and 9 kg (19.8 lb) for HEAT
Warhead: HE or HEAT
Range: 6650 m (7,275 yards) maximum with HE and 450 m (495 yards) effective with HEAT

Variant
B-11: this obsolescent is fired from a light tripod and can be transported on a two-wheel dolly, and the armour penetration is in the order of 380 mm (14.96 in) with the fin-stabilized HEAT projectile; the rate of fire is 6 rounds per minute

Hunting LAW 80 anti-tank weapon
(UK)
Type: one-man anti-tank rocket launcher
Calibre: 94 mm (3.70 in)
Lengths: launcher 1.0 m (3.28 ft) for carriage and 1.50 m (4.92 ft) for firing; rocket not revealed
Weights: 9.5 kg (20.9 lb) complete for firing; rocket 4 kg (8.82 lb) in case
Warhead: shaped-charge HE
Range: 500 m (550 yards)

Variant
LAW 80: designed to maintain a capability against the latest MBTs, this lightweight shoulder-launched weapon is designed to penetrate 600 mm (23.6 in) of armour; the weapon is aimed with the aid of a small spotting weapon firing a round with flash-producing warhead

M18A1 anti-tank weapon
(USA)
Type: two-man anti-tank rocket launcher
Calibre: 57 mm (2.24 in)
Length: launcher 5.15 ft (1.57 ft)
Weight: 44.5 lb (20.2 kg)
Warhead: shaped-charge HE
Range: 500 yards (460 m)

Variants
M18A1: though obsolescent and now a large weapon for its capabilities, this 'bazooka' still has useful performance in less severe tactical situations
Type 36: designation of the Chinese-made copy of the M18A1

M20 75-mm anti-tank weapon
(USA)
Type: two- or three-man anti-tank recoilless gun
Calibre: 75 mm (2.95 in)
Length: 6.83 ft (2.083 m)
Weights: launcher and tripod mount 148 lb (67 kg); round 14.4 lb (6.53 kg)
Warhead: HE, HEP and HEAT
Range: 7,175 yards (6560 m) maximum with HEP and 3,500 yards (3200 m) maximum with HEAT

Variants
M20: an obsolescent weapon still in extensive worldwide service, the M20 can be used on either of two tripod mounts
Type 52: designation of the Chinese-made M20
Type 56: designation of the Chinese-built version of

the M20 able to fire only specially-developed Chinese ammunition

M20 3.5-in anti-tank weapon
(USA)
Type: two-man anti-tank rocket launcher
Calibre: 3.5 in (88.9 mm)
Lengths: launcher 5.08 ft (1,549 m); rocket 3.5 ft (1.07 m)
Weights: launcher 12.13 lb (5.5 kg); rocket 8.9 lb (4.04 kg)
Warhead: HEAT
Range: 1,300 yards (1190 m) maximum and 120 yards (110 m) effective

Variants
M20: now obsolete, this important weapon is still in widespread service in less advanced countries
Type 51: designation of the Chinese-made M20

M27 anti-tank weapon
(USA)
Type: five-man anti-tank recoilless gun
Calibre: 105 mm (4.13 in)
Length: launcher, overall 11.19 ft (3.41 m); barrel 8.92 ft (2.718 m)
Weight: launcher without mounting 365 lb (166 kg)
Warhead: HE, WP, HEAT, HEP and HEP-T
Range: 9,250 yards (8450 m) maximum and 1,100 yards (1005 m) effective

Variants
M27: this is an obsolete and fairly massive weapon designed for operations on a tripod or alternatively on a vehicle mounting; traverse arcs of 60 or 80 degrees are possible (depending on the mounting) and the elevation arc is −13 to +50 degrees
M27A1: designation of the M27 fitted with a forcing cone in the firing chamber

Watervliet Arsenal M40A1 anti-tank weapon
(USA)
Type: two-man anti-tank recoilless gun
Calibre: 106 mm (4.17 in)
Lengths: launcher, overall 11.15 ft (3.40 m); barrel 9.33 ft (2.845 m)
Weights: gun with spotting rifle but no mounting 287 lb (130 kg); round 35.9 lb (16.3 kg) for HEAT and 37.9 lb (17.2 kg) for HEP-T
Warhead: HEAT and HEP-T
Range: 8,425 yards (7705 m) maximum and 1,200 yards (1100 m) effective

Variant
M40A1: a powerful but obsolescent weapon fitted with a 0.5-in (12.7-mm) spotting rifle, the M40A1 can be located on either of two tripod mounts (weighing 194 lb/88 kg and 117 lb/53 kg) on a light vehicle; the mounts provide for 360-degree traverse, and elevation arcs of between −17 and +65 degrees or −17 and +50 degrees

US Army Munitions Command M72A2 anti-tank weapon
(USA)
Type: one-man anti-tank rocket launcher
Lengths: launcher 2.15 ft (0.655 m) for carriage and 2.92 ft (0.893 m) ready for firing; rocket 1.67 ft (0.508 m)
Weights: launcher 3.0 lb (1.36 kg); rocket 2.2 lb (1.0 kg)
Warhead: shaped-charge HE
Range: 325 yards (300 m)

Variant
M72A2: this is an obsolescent weapon of the single-shot fire-and-discard type, able to penetrate 12 in (305 mm) of armour

Table 1.1 Machine-guns

designation (country)	calibre	weight (with mounting)	length overall	muzzle velocity	rate of fire (cyclic)	feed
FN MAG (Belgium)	7.62 mm (0.3 in)	20.6 kg (45.4 lb) on tripod	1.26 m (49.61 in)	840 m (2,756 ft) per second	600–1,000 rounds per minute	50-round belt
FN Minimi/M249 (Belgium/USA)	5.56 mm (0.219 in)	6.5 kg (14.33 lb) on bipod	1.05 m (41.34 in)	915 m (3,002 ft) per second	750–1,000 rounds per minute	100- or 200-round belt, or 30-round box
vz 59 (Czechoslovakia)	7.62 mm (0.3 in)	19.24 kg (42.42 lb) on tripod with heavy barrel	1.215 m (47.84 in)	830 m (2,723 ft) per second	700–800 rounds per minute	50- or 250-round belt
AA 52 (France)	7.5 mm (0.295 in)	11.37 kg (25.07 lb) on bipod with heavy barrel	1.245 m (49.02 in)	840 m (2,756 ft) per second	700 rounds per minute	50-round belt

LAND WEAPONS

MG3 (West Germany)	7.62 mm (0.3 in)	11.05 kg (24.36 lb) on bipod	1.225 m (48.23 in)	820 m (2,690 ft) per second	700–1,300 rounds per minute	50-round belt
HK 21A1 (West Germany)	7.62 mm (0.3 in)	8.3 kg (18.3 lb) on bipod	1.03 m (40.55 in)	800 m (2,625 ft) per second	900 rounds per minute	100-round belt
MG Model 62 (Japan)	7.62 mm (0.3 in)	10.7 kg (23.6 lb) on bipod	1.20 m (47.24 in)	855 m (2,805 ft) per second	550 rounds per minute	belt
Ultimax 100 (Singapore)	5.56 mm (0.219 in)	4.7 kg (10.36 lb) on bipod	1.03 m (40.55 in)	990 m (3,248 ft) per second	400–600 rounds per minute	20- or 30-round box, or 100-round drum
DPM/Type 53 (USSR/China)	7.62 mm (0.3 in)	10.0 kg (22.05 lb) on bipod	1.27 m (50.0 in)	840 m (2,756 ft) per second	600 rounds per minute	49-round drum
RP-46/Type 58 (USSR/China)	7.62 mm (0.3 in)	13.0 kg (28.66 lb) on bipod	1.283 m (50.51 in)	840 m (2,756 ft) per second	600 rounds per minute	250-round belt
RPD/Type 56 (USSR/China)	7.62 mm (0.3 in)	7.1 kg (15.65 lb) on bipod	1.036 in (40.79 in)	700 m (2,297 ft) per second	700 rounds per minute	250-round belt
SGM/Type 57 (USSR/China)	7.62 mm (0.3 in)	40.7 kg (89.7 lb) on tripod	1.12 m (44.09 in)	800 m (2,625 ft) per second	650 rounds per minute	250-round belt
RPK/Type 74 (USSR/China)	7.62 mm (0.3 in)	5.0 kg (11.02 lb) on bipod	1.035 m (40.75 in)	732 m (2,402 ft) per second	660 rounds per minute	30- or 40-round box, or 75-round drum
PRK-74 (USSR)	5.45 mm (0.2146 in)	5.0 kg (11.02 lb) on bipod	1.035 m (40.75 in)	732 m (2,402 ft) per second	660 rounds per minute	30- or 40-round box, or 75-round drum
PKS/Type 80 (USSR/China)	7.62 mm (0.3 in)	16.5 kg (36.4 lb) on tripod	1.16 m (45.67 in)	825 m (2,707 ft) per second	690–720 rounds per minute	100-, 200- or 250-round belt
DShK-38/46/Type 54 (USSR/China)	12.7 mm (0.5 in)	35.7 kg (78.7 lb) without mounting	1.588 in (62.52 in)	845 m (2,772 ft) per second	550-600 rounds per minute	50-round belt
KPV (USSR)	14.5 mm (0.57 in)	49.1 kg (108.2 lb) without mounting	2.006 in (78.97 in)	1000 m (3,281 ft) per second	600 rounds per minute	belt
L4A4 Bren (UK)	7.62 mm (0.3 in)	21.0 lb (9.53 kg) on bipod	44.6 in (1.133 m)	2,700 ft (823 m) per second	500 rounds per minute	30-round box
L7A2 (UK)	7.62 mm (0.3 in)	25.0 lb (10.9 kg) on bipod	48.5 in (1.232 m)	2,750 ft (838 m) per second	750–1,000 rounds per minute	50-round box or 100-round belt
LSW (UK)	5.56 mm (0.219 in)	12.345 lb (5.6 kg) on bipod	35.43 in (0.90 in)	3,182 ft (970 m) per second	700–800 rounds per minute	30-round box
M1919A4 (USA)	0.3 in (7.62 mm)	45.0 lb (20.41 kg) on tripod	41.0 in (1.041 m)	2,800 ft (854 m) per second	400–500 rounds per minute	250-round belt
M60 (USA)	7.62 mm (0.3 in)	23.17 lb (10.51 kg) on bipod	43.5 in (1.105 m)	2,805 ft (855 m) per second	550 rounds per minute	50-round belt

132 LAND WEAPONS

| M2HB (USA) | 0.5 in (12.7 mm) | 128 lb (58.06 kg) on tripod | 65.1 in (1.654 m) | 2,900 ft (844 m) per second | 450–575 rounds per minute | 110-round belt |

Table 1.2 Mortars

designation (country)	type	calibre	weight (maximum)	round weight (HE)	rate of fire	range
Thomson-Brandt 60-mm Commando (France)	light mortar	60 mm	7.7 kg (16.98 lb)	1.7 kg (3.75 lb)	12–20 rounds per minute	100 to 1050 m (110 to 1,150 yards)
Thomson-Brandt 60-mm Light (France)	light mortar	60 mm	14.8 kg (32.63 lb)	1.65 kg (3.64 lb)		100 to 2000 m (110 to 2,185 yards)
Thomson-Brandt MO-81C (France)	medium mortar	81 mm	39.4 kg (86.7 lb)	3.3 kg (7.28 lb)	12–15 rounds per minute	120 to 4100 m (130 to 4,485 yards)
Thomson-Brandt MO-120-60 (France)	medium/heavy mortar	120 mm	94 kg (207.2 lb)	13 kg (28.66 lb)	8–15 rounds per minute	600 to 6610 m (655 to 7,230 yards)
Thomson-Brandt MO-120-LT (France)	wheeled heavy mortar	120 mm	203 kg (447.5 lb)	13 kg (28.66 lb)	8–12 rounds per minute	600 to 6560 m (655 to 7,175 yards)
Thomson-Brandt MO-120-RT-61 (France)	wheeled heavy rifled mortar	120 mm	582 kg (1,283 lb)	18.7 kg (41.23 lb)	10–12 rounds per minute	1100 to 13000 m (1,200 to 14,215 yards)
IMI 52-mm (Israel)	light mortar	52 mm	7.9 kg (17.4 lb)	1.02 kg (2.25 lb)	20–35 rounds per minute	130 to 420 m (140 to 460 yards)
Soltam 60-mm (Israel)	light mortar	60.75 mm	14.5 kg (32 lb)	1.59 kg (3.5 lb)		150 to 2555 m (155 to 2,795 yards)
Soltam M64 (Israel)	medium mortar	81 mm	43 kg (94.8 lb)	4.0 kg (8.82 lb)	12.15 rounds per minute	150 to 4600 m (165 to 5,030 yards)
Soltam 120-mm (Israel)	wheeled medium/heavy mortar	120 mm	218 kg (480.6 lb)	12.9 kg (28.4 lb)	5–10 rounds per minute	400 to 6200 m (435 to 6,780 yards)
Soltam M65 (Israel)	wheeled medium/heavy mortar	120 mm	365 kg (804.7 lb)	16.7 kg (36.8 lb)	5–10 rounds per minute	400 to 6200 m (435 to 6,780 yards)
Soltam M66 (Israel)	wheeled heavy mortar	160 mm	1700 kg (3,748 lb)	40 kg (88.2 lb)	5–8 rounds per minute	up to 9300 m (10,170 yards)
Esperanza Model L (Spain)	light mortar	60 mm	12.0 kg (26.5 lb)	1.43 kg (3.15 lb)	30 rounds per minute	up to 1975 m (2,160 yards)
Esperanza Model L-N (Spain)	medium mortar	81 mm	46.0 kg (101.4 lb)	4.13 kg (9.1 lb)	10–15 rounds per minute	up to 6200 m (5,685 yards)
Esperanza 105-mm Model L (Spain)	wheeled medium mortar	105 mm	214 kg (461.7 lb)	9.4 kg (20.7 lb)	10–12 rounds per minute	up to 6000 m (6,560 yards)
Esperanza 120-mm Model L (Spain)	wheeled heavy mortar	120 mm	542 kg (1,195 lb)	16.745 kg (36.9 lb)	8–12 rounds per minute	up to 5940 m (6,495 yards)
Bofors Lyran (Sweden)	illuminating mortar	71 mm	17.0 kg (37.5 lb)	1.17 kg (2.58 lb)		400 to 800 m (435 to 875 yards)

M37/Type 53 (USSR/China)	medium mortar	82 mm	56.0 kg (123.5 lb)	3.05 kg (6.72 lb)	15–25 rounds per minute	100 to 3000 m (110 to 3,280 yards)
M43/Type 55 (USSR/China)	wheeled heavy mortar	120 mm	500 kg (1,102 lb)	15.4 kg (33.95 lb)	12–15 rounds per minute	460 to 5700 m (500 to 6,235 yards)
M43 160-mm (USSR)	wheeled breech-loading heavy mortar	160 mm	1270 kg (2,800 lb)	40.8 kg (133.9 lb)	3 rounds per minute	630 to 5150 m (690 to 5,630 yards)
M-160 (USSR)	wheeled breech-loading heavy mortar	160 mm	1470 kg (3,241 lb)	41.5 kg (91.5 lb)	2–3 rounds per minute	750 to 8040 m (820 to 8,795 yards)
M-240 (USSR)	wheeled breech-loading heavy mortar	250 mm	3900 kg (8,598 lb)	100 kg (220.5 lb)	1 round per minute	1500 to 9700 m (1,640 to 10,610 yards)
51-mm Mortar (UK)	light mortar	51.25 mm	13.84 lb (6.28 kg)	2.03 lb (0.92 kg)		55 to 875 yards (50 to 800 m)
3-in Mortar Mk 5 (UK)	medium mortar	81.5 mm	132 lb (59.9 kg)	9.75 lb (4.42 kg)	15 rounds per minute	500 to 2m800 yards (460 to 2560 m)
L16/M252 (UK/USA)	medium mortar	81.4 mm	83.45 lb (37.85 kg)	9.26 lb (4.2 kg)		200 to 6,180 yards (180 to 5650 m)
M224 (USA)	light mortar	60 mm	46.0 lb (20.9 kg)			up to 3,825 yards (3005 m)
M1 (USA)	medium mortar	81 mm	132 lb (59.88 kg)		18–30 rounds per minute	up to 3,300 yards (3015 m)
M29A1 (USA)	medium mortar	81 mm	115 lb (52.2 kg)		12–30 rounds per minute	up to 5,180 yards (4735 m)
M2A1 (USA)	medium/heavy mortar	107 mm	330 lb (149.7 kg)	25.5 lb (11.56 kg)	5 rounds per minute	600 to 4,400 yards (550 to 4025 m)
M30 (USA)	rifled medium/heavy mortar	107 mm	675 lb (306 kg)	27.0 lb (12.25 kg)	3–9 rounds per minute	1,000 to 6,200 yards (915 to 5670 m)

SECTION 2

WARSHIPS

SUBMARINES

Nuclear-powered ballistic missile type

'Benjamin Franklin' class SSBN
(USA)
Type: nuclear-powered ballistic missile submarine
Displacement: 6,650 tons light, 7,250 tons surfaced and 8,250 tons dived
Dimensions: length 425.0 ft (129.5 m); beam 33.0 ft (10.1 m); draught 31.5 ft (9.6 m)
Gun armament: none
Missile armament: 16 vertical launch tubes for 16 submarine-launched ballistic missiles (UGM-96A Trident I C4 in SSBN640, 641, 643, 655, 657 and 658, and UGM-73A Poseidon C3 in the others)
Torpedo armament: four 21-in (533-mm) Mk 65 tubes (all bow) for 12 Mk 48 dual-role wire-guided torpedoes
Anti-submarine armament: ? UUM-44A SUBROC tube-launched missiles, and torpedoes (see above)
Electronics: one BPS-11A or BPS-15 surface-search radar, one BQR-7 passive sonar, one BQR-15 towed-array sonar, one BQR-19 sonar, one BQR-21 passive detection and tracking sonar, one BQS-4 sonar, Mk 84 SLBM fire-control system, three Mk 2 SINS, one Mk 113 underwater weapons fire-control system, one WSC-3 satellite communications transceiver, and one satellite navigation receiver
Propulsion: one Westinghouse S5W pressurized water-cooled reactor supplying steam to two sets of geared turbines delivering 15,000 shp (11185 kW) to one shaft
Performance: maximum speed 20 kts surfaced and about 25 kts dived; diving depth 1,150 ft (350 m) operational and 1,525 ft (465 m) maximum
Complement: 13+130

Class
1. USA

Name	No.	Builder	Laid down	Commissioned
Benjamin Franklin	SSBN640	GD (EB Div)	May 1963	Oct 1965
Simon Bolivar	SSBN641	Newport News	Apr 1963	Oct 1965
Kamehameha	SSBN642	Mare Island NY	May 1963	Dec 1965
George Bancroft	SSBN643	GD (EB Div)	Aug 1963	Jan 1966
Lewis & Clark	SSBN644	Newport News	Jul 1963	Dec 1965
James K. Polk	SSBN645	GD (EB Div)	Nov 1963	Apr 1966
George C. Marshall	SSBN654	Newport News	Mar 1964	Apr 1966
Henry L. Stimson	SSBN655	GD (EB Div)	Apr 1964	Aug 1966
George Washington Carver	SSBN656	GD (EB Div)	Aug 1964	Jun 1966
Francis Scott Key	SSBN657	GD (EB Div)	Dec 1964	Dec 1966
Mariano G. Vallejo	SSBN658	Mare Island NY	Jul 1964	Dec 1966
Will Rogers	SSBN659	GD (EB Div)	Mar 1966	Apr 1967

Note
This was the partner class (built with quieter machinery) to the 'Lafayette' type, the two succeeding the pioneer 'George Washington' and 'Ethan Allen' classes as launch platforms for the Polaris SLBM; the boats were subsequently modified to carry the more capable Poseidon missile, and some are now the main element of the US Navy's Trident I force pending the delivery of more 'Ohio' class SSBNs

'Delta I' class SSBN
(USSR)
Type: nuclear-powered ballistic missile submarine
Displacement: 8,750 tons surfaced and 10,000 tons dived
Dimensions: length 136.0 m (446.1 ft); beam 11.6 m (38.0 ft); draught 10.0 m (32.8 ft)
Gun armament: none
Missile armament: 12 vertical launch tubes for 12 SS-N-8 submarine-launched ballistic missiles
Torpedo armament: six 533-mm (21-in) tubes (all bow), two of them with 406-mm (16-in) liners, for a typical load of 12 533-mm (21-in) anti-ship torpedoes or eight 533-mm (21-in) anti-ship and six 406-mm (16-in) anti-submarine torpedoes
Anti-submarine armament: torpedoes (see above)
Electronics: one 'Snoop Tray' surface-search and navigation radar, one low-frequency bow sonar, one medium-frequency torpedo fire-control sonar, one 'Brick Group' ESM system, one 'Pert Spring' satellite navigation system, and extensive communications systems

Propulsion: two pressurized water-cooled reactors supplying steam to two sets of geared turbines delivering 22400 kW (30,040 shp) to two shafts
Performance: maximum speed 20 kts surfaced and 26 kts dived; diving depth 400 m (1,315 ft) operational and 600 m (1,970 ft) maximum
Complement: 130

Class
1. USSR

18 boats

Note
The 'Delta I' class is essentially an updated and enlarged 'Yankee' class and, built between 1970 and 1977 at Severdvinsk and Komsomolsk, was for its time the world's largest submarine

'Delta II' class SSBN
(USSR)
Type: nuclear-powered ballistic missile submarine
Displacement: 9,750 tons surfaced and 11,000 tons dived
Dimensions: length 152.7 m (500.9 ft); beam 11.8 m (38.7 ft); draught 10.2 m (33.5 ft)
Gun armament: none
Missile armament: 16 vertical launch tubes for 16 SS-N-8 submarine-launched ballistic missiles
Torpedo armament: six 533-mm (21-in) tubes (all bow), two of them with 406-mm (16-in) liners, for a typical load of 12 533-mm (21-in) anti-ship torpedoes or eight 533-mm (21-in) anti-ship and six 406-mm (16-in) anti-submarine torpedoes
Anti-submarine armament: torpedoes (see above)
Electronics: one 'Snoop Tray' surface-search and navigation radar, one low-frequency bow sonar, one medium-frequency torpedo fire-control sonar, one 'Brick Group' ESM system, one 'Pert Spring' satellite navigation system, and extensive communications systems
Propulsion: two pressurized water-cooled reactors supplying steam to two sets of geared turbines delivering 22400 kW (30,040 shp) to two shafts
Performance: maximum speed 20 kts surfaced and 25 kts dived; diving depth 400 m (1,315 ft) operational and 600 m (1,970 ft) maximum
Complement: 140

Class
1. USSR

4 boats

Note
Produced in the mid-1970s at Komsolmolsk, the 'Delta II' class was an interim development (with a missile section lengthened by 16.2 m/53.2 ft to accommodate an additional four missile tubes) pending the arrival of the 'Delta III' class

'Delta III' class SSBN
(USSR)
Type: nuclear-powered ballistic missile submarine
Displacement: 9,750 tons surfaced and 11,000 tons dived
Dimensions: length 150.0 m (492.0 ft); beam 12.0 m (39.4 ft); draught 10.2 m (33.4 ft)
Gun armament: none
Missile armament: 16 vertical launch tubes for 16 SS-N-18 submarine-launched ballistic missiles
Torpedo armament: six 533-mm (21-in) tubes (all bow), two of them with 406-mm (16-in) liners, for a typical load of 12 533-mm (21-in) anti-ship torpedoes or eight 533-mm (21-in) anti-ship and six 406-mm (16-in) anti-submarine torpedoes
Anti-submarine armament: torpedoes (see above)
Electronics: one 'Snoop Tray' surface-search and navigation radar, one low-frequency bow sonar, one medium-frequency torpedo fire-control sonar, one 'Brick Group' ESM system, one 'Pert Spring' satellite navigation system, and extensive communications systems
Propulsion: two pressurized water-cooled reactors supplying steam to two sets of geared turbines delivering 22400 kW (30,040 shp) to two shafts
Performance: maximum speed 20 kts surfaced and 24 kts dived; diving depth 400 m (1,315 ft) operational and 600 m (1,970 ft) maximum
Complement: 140

Class
1. USSR

14 boats

Note
Introduced in 1976, the 'Delta III' class was essentially the 'Delta II' revised to accommodate the longer and more capable SS-N-18 SLBM; these boats form the main strength of the Soviet SSBN force

'Delta IV' class SSBN
(USSR)
Type: nuclear-powered ballistic missile submarine
Displacement: 10,100 tons surfaced and 11,600 tons dived
Dimensions: length 157.5 m (516.7 ft); beam 12.0 m (39.4 ft); draught: 8.7 m (28.5 ft)
Gun armament: none
Missile armament: 16 vertical launch tubes for 16 SS-N-23 submarine-launched ballistic missiles
Torpedo armament: six 533-mm (21-in) tubes (all bow), two of them with 406-mm (16-in) liners, for a typical load of 12 533-mm (21-in) anti-ship torpedoes or eight 533-mm (21-in) anti-ship and six 406-mm (16-in) anti-submarine torpedoes
Anti-submarine armament: torpedoes (see above)

Electronics: one 'Snoop Tray' surface-search radar, one low-frequency bow sonar, one medium-frequency torpedo fire-control sonar, one 'Brick Group' ESM system, or 'Pert Spring' satellite navigation system, and one 'Cod Eye' navigation system
Propulsion: two pressurized water-cooled reactors supplying steam to four sets of geared steam turbines delivering ? kW (? hp) to two shafts
Performance: maximum speed 20 kts surfaced and 23.5 kts dived; diving depth 400 m (1,315 ft) operational and 600 m (1,970 ft) maximum
Complement: 150

Class
1. USSR
2+? boats building or planned

Note
Introduced in 1985, this is a much-improved development of the 'Delta III' class with more advanced systems and a primary missile armament of new SS-N-23 SLBMs; the type makes an interesting contrast with the massive 'Typhoon' class SSBNs, offering good capability with far smaller infra-red, acoustic and magnetic signatures, so allowing it to operate in areas other than the Soviet 'SSBN sanctuaries' close to the USSR's coasts; the type's hydroplanes have been relocated from the sail to the hull, a clear indication that the Soviets intend the type for under-ice operations

'Hotel II' class SSBN
(USSR)
Type: nuclear-powered ballistic missile submarine
Displacement: 4,500 tons surfaced and 5,500 tons dived
Dimensions: length 115.2 m (377.2 ft); beam 9.1 m (29.8 ft); draught 7.6 m (25.0 ft)
Gun armament: none
Missile armament: three vertical launch tubes for three SS-N-5 submarine-launched ballistic missiles
Torpedo armament: six 533-mm (21-in) tubes (all bow) for a maximum of 16 533-mm (21-in) anti-ship/anti-harbour conventional or nuclear torpedoes
Anti-submarine armament: two 406-mm (16-in) tubes (both stern) for four 406-mm (16-in) anti-submarine torpedoes
Electronics: one 'Snoop Tray' surface-search and navigation radar, one 'Herkules' sonar, and one 'Feniks' sonar
Propulsion: two pressurized water-cooled reactors supplying steam to two sets of geared turbines delivering 22400 kW (30,040 shp) to two shafts
Performance: maximum speed: 20 kts surfaced and 26 kts dived
Complement: 90

Class
1. USSR
5 boats

Note
Built at Severdvinsk between 1958 and the mid-1960s, the 'Hotel' series parallels the 'Golf' class but with nuclear rather than diesel-electric propulsion; the type was planned as an SSN class to work in concert with the 'November' class on attacks on US bases and major forces with nuclear-armed torpedoes, but was then recast as an interim SSBN with continued capability against bases and harbours with torpedoes

'Hotel III' class SSBN
(USSR)
Type: nuclear-powered ballistic missile submarine
Displacement: 5,000 tons surfaced and 6,000 tons dived
Dimensions: length 115.2 m (377.2 ft); beam 9.1 m (29.8 ft); draught 7.6 m (25.0 ft)
Gun armament: none
Missile armament: six vertical launch tubes for six SS-N-8 submarine-launched ballistic missiles
Torpedo armament: six 533-mm (21-in) tubes (all bow) for a maximum of 16 533-mm (21-in) anti-ship/anti-harbour conventional or nuclear torpedoes
Anti-submarine armament: two 406-mm (16-in) tubes (both stern) for four 406-mm (16-in) anti-submarine torpedoes
Electronics: one 'Snoop Tray' surface-search and navigation radar, one 'Herkules' sonar, and one 'Feniks' sonar
Propulsion: two pressurized water-cooled reactors supplying steam to two sets of geared turbines delivering 22400 kW (30,040 shp) to two shafts
Performance: maximum speed 20 kts surfaced and 26 kts dived
Complement: 90

Class
1. USSR
1 boat

Note
This is essentially an upgraded 'Hotel II' type with a revised missile armament

'Lafayette' class SSBN
(USA)
Type: nuclear-powered ballistic missile submarine
Displacement: 6,650 tons light, 7,250 tons surfaced and 8,250 tons dived
Dimensions: length 425.0 ft (129.5 m); beam 33.0 ft (10.1 m); draught 31.5 ft (9.6 m)

Gun armament: none
Missile armament: 16 vertical launch tubes for 16 submarine-launched ballistic missiles (UGM-96A Trident I C4 in SSBN627, 629, 630 and 632, and UGM-73A Poseidon C3 in the others)
Torpedo armament: four 21-in (533-mm) Mk 65 tubes (all bow) for 12 Mk 48 dual-role wire-guided torpedoes
Anti-submarine armament: ? UUM-44A SUBROC tube-launched missiles, and torpedoes (see above)
Electronics: one BPS-11A or BPS-15 surface-search radar, one BQR-7 passive sonar, one BQR-15 towed-array sonar, one BQR-19 sonar, one BQR-21 passive detection and tracking sonar, one BQS-4 active sonar, Mk 84 SLBM fire-control system, three Mk 2 SINS, one Mk 113 underwater weapons fire-control system, one WSC-3 satellite communications transceiver, and one satellite navigation receiver
Propulsion: one Westinghouse S5W pressurized water-cooled reactor supplying steam to two sets of geared turbines delivering 15,000 shp (11185 kW) to one shaft
Performance: maximum speed 20 kts surfaced and about 25 kts dived; diving depth 1,150 ft (350 m) operational and 1,525 ft (465 m) maximum
Complement: 13+130

Class
1. USA

Name	No.	Builder	Laid down	Commissioned
Lafayette	SSBN616	GD (EB Div)	Jan 1961	Apr 1963
Alexander Hamilton	SSBN617	GD (EB Div)	Jun 1961	Jun 1963
Andrew Jackson	SSBN619	Mare Island NY	Apr 1961	Jul 1963
John Adams	SSBN620	Portsmouth NY	May 1961	May 1964
James Monroe	SSBN622	Newport News	Jul 1961	Dec 1963
Woodrow Wilson	SSBN624	Mare Island NY	Sep 1961	Dec 1963
Henry Clay	SSBN625	Newport News	Oct 1961	Feb 1964
Daniel Webster	SSBN626	GD (EB Div)	Dec 1961	Apr 1964
James Madison	SSBN627	Newport News	Mar 1962	Jul 1964
Tecumseh	SSBN628	GD (EB Div)	Jun 1962	May 1964
Daniel Boone	SSBN629	Mare Island NY	Feb 1962	Apr 1964
John C. Calhoun	SSBN630	Newport News	Jun 1962	Sep 1964
Ulysses S. Grant	SSBN631	GD (EB Div)	Aug 1962	Jul 1964
Von Steuben	SSBN632	Newport News	Sep 1962	Sep 1964

Note
Built to succeed the 'George Washington' and 'Ethan Allen' class SSBNs, the 'Lafayette' class was first armed with the Polaris, then the Poseidon and finally (in some cases only) with the Trident I SLBM

'L'Inflexible' class SSBN
(France)
Type: nuclear-powered ballistic missile submarine
Displacement: 8,080 tons surfaced and 8,920 tons dived
Dimensions: length 128.7 m (422.2 ft); beam 10.6 m (34.7 ft); draught 10.0 m (32.8 ft)
Gun armament: none
Missile armament: 16 vertical launch tubes for 16 MSBS M-4 submarine-launched ballistic missiles, and four SM.39 Exocet tube-launched anti-ship missiles
Torpedo armament: four 533-mm (21-in) tubes (all bow) for 14 L5 and F17 torpedoes
Anti-submarine armament: anti-submarine torpedoes (see above)
Electronics: one Calypso surface-search and navigation radar, one passive ESM system, one missile fire-control system, one DLT D3 torpedo fire-control system, one INS, one DUUX 2 passive ranging sonar, and one DSUX 21 multi-role sonar
Propulsion: one pressurized water-cooled reactor supplying steam to two sets of geared turbines powering two turbo-alternators/one electric motor delivering 11925 kW (15,990 shp) to one shaft
Performance: maximum speed 18 kts surfaced and 25 kts dived; diving depth 250 m (820 ft) operational and 330 m (1,085 ft) maximum
Complement: 15+120

Class
1. France

Name	No.	Builder	Laid down	Commissioned
L'Inflexible		Cherbourg ND	Mar 1980	Apr 1985

Note
This singleton design is an intermediate stage between the 'Le Redoutable' class and the class of 15,000-tonners planned for the 1990s, and is thus a valuable trials platform (as well as being an operational boat) for the latest systems and weapons

'Le Redoutable' class SSBN
(France)
Type: nuclear-powered ballistic missile submarine
Displacement: 8,045 tons surfaced and 8,940 tons dived
Dimensions: length 128.7 m (422.1 ft); beam 10.6 m (34.8 ft); draught 10.0 m (32.8 ft)
Gun armament: none
Missile armament: 16 vertical launch tubes for 16 MSBS M-20 (being replaced by M-4) submarine-launched ballistic missiles, and ? SM.39 Exocet tube-launched anti-ship missiles
Torpedo armament: four 533-mm (21-in) tubes (all bow) for 18 L5 and F17 torpedoes
Anti-submarine armament: anti-submarine torpedoes (see above)
Electronics: one Calypso surface-search and navigation radar, one passive ESM system, one missile fire-control system, one DLT D3 torpedo fire-control system, one INS, one DSUX 2 multi-role sonar, and one DUUX 2 passive ranging sonar
Propulsion: one pressurized water-cooled reactor supplying steam to two sets of geared turbines powering two turbo-alternators/one electric motor delivering 11925 kW (15,990 shp) to one shaft; there is also an auxiliary powerplant comprising two diesels delivering 975 kW (1,307 hp)
Performance: maximum speed 20+ kts surfaced and 25 kts dived; diving depth 250 m (820 ft) operational and 330 m (1,085 ft) maximum
Complement: 15+120

Class
1. France

Name	No.	Builder	Laid down	Commissioned
Le Foudroyant	S610	Cherbourg ND	Dec 1969	Jun 1974
Le Redoutable	S611	Cherbourg ND	Mar 1964	Dec 1971
Le Terrible	S612	Cherbourg ND	Jun 1967	Dec 1973
L'Indomptable	S613	Cherbourg ND	Dec 1971	Dec 1976
Le Tonnant	S614	Cherbourg ND	Oct 1974	May 1980

Note
This was the first class of French SSBNs, *Le Redoutable* preceding the other boats by a considerable margin so that she could serve as a trials platform for the various advanced systems and M-1 first-generation SLBM; all the boats are to carry the sonar suite of the later 'L'Inflexible' class

'Ohio' class SSBN
(USA)
Type: nuclear-powered ballistic missile submarine
Displacement: 16,765 tons surfaced and 18,750 tons dived
Dimensions: length 560.0 ft (170.7 m); beam 42.0 ft (12.8 m); draught 36.5 ft (11.1 m)
Gun armament: none
Missile armament: 24 vertical launch tubes for 24 UGM-96 A Trident I C4 submarine-launched ballistic missiles
Torpedo armament: four 21-in (533-mm) Mk 68 tubes (all amidships) for ? Mk 48 dual-role wire-guided torpedoes
Anti-submarine armament: anti-submarine torpedoes (see above)
Electronics: one BPS-15A surface-search radar, one WLR-8(V) ESM system, one BQQ-6 sonar, one

Class
1. USA

Name	No.	Builder	Laid down	Commissioned
Ohio	SSBN726	GD (EB Div)	Apr 1976	Nov 1981
Michigan	SSBN727	GD (EB Div)	Apr 1977	Sep 1982
Florida	SSBN728	GD (EB Div)	Jun 1977	Jul 1982
Georgia	SSBN729	GD (EB Div)	Apr 1979	Feb 1984
Henry M. Jackson	SSBN730	GD (EB Div)	Jan 1981	Oct 1984
Alabama	SSBN731	GD (EB Div)	Apr 1981	May 1985
Alaska	SSBN732	GD (EB Div)	Jan 1985	Jan 1986
Nevada	SSBN733	GD (EB Div)	Aug 1983	Aug 1986
Tennessee	SSBN734	GD (ED Div)		

Note
After the Soviet 'Typhoon' class the world's largest submarine type, the 'Ohio' class was designed from the early 1970s to carry the Trident I SLBM, with the capability to accept the considerably larger and heavier (but also more potent and accurate) Trident II when this becomes available in the 1990s; the whole 'Ohio' programme has been beset by severe technical and budgetary problems, but the boats are now proving themselves admirably capable in their designed role; from SSBN734 onwards the boats are to be built for the Trident II, and earlier boats will be retrofitted with the type during refits; the class is planned to total 24 SSBNs

BQS-13 active/passive search sonar, one BQ5-15 sonar, one BQR-19 sonar, one BQS-23 towed-array sonar, one Mk 98 SLBM fire-control system, one Mk 118 torpedo fire-control system, two Mk 2 SINS, and one WSC-3 satellite communications transceiver
Propulsion: one General Electric S8G pressurized water-cooled reactor supplying steam to two sets of geared turbines delivering 60,000 shp (44740 kW) to one shaft
Performance: maximum speed 28 kts surfaced and 30 kts dived; diving depth 985 ft (300 m) operational and 1,640 ft (500 m) maximum
Complement: 15+145

'Resolution' class SSBN
(UK)
Type: nuclear-powered ballistic missile submarine
Displacement: 7,500 tons surfaced and 8,400 tons dived
Dimensions: length 425.0 ft (129.5 m); beam 33.0 ft (10.1 m); draught 30.0 ft (9.1 m)
Gun armament: none
Missile armament: 16 vertical launch tubes for 16 UGM-27C Polaris A3TK submarine-launched ballistic missiles
Torpedo armament: six 21-in (533-mm) tubes (all bow) for Mk 24 Tigerfish dual-role wire-guided and Mk 8 anti-ship torpedoes
Anti-submarine armament: anti-submarine torpedoes (see above)
Electronics: one Type 1003 surface-search and navigation radar, one Type 2001 long-range active/passive sonar, one Type 2007 long-range passive sonar, one Type 2023 retractable towed-array sonar, one ESM system, and INS
Propulsion: one Rolls-Royce pressurized water-cooled reactor supplying steam to one set of English Electric geared turbines delivering 15,000 shp (11185 kW) to one shaft
Performance: maximum speed 20 kts surfaced and 25 kts dived; diving depth 1,150 ft (350 m) operational and 1,525 ft (465 m) maximum
Complement: 13+130

Class
1. UK

Name	No.	Builder	Laid down	Commissioned
Resolution	S22	Vickers	Feb 1964	Oct 1967
Repulse	S23	Vickers	Mar 1965	Sep 1968
Renown	S26	Cammell Laird	Jun 1964	Nov 1968
Revenge	S27	Cammell Laird	May 1965	Dec 1969

Note
The boats were developed from the early 1960s to carry the Polaris SLBM, and are similar in concept and basic design to the US 'Lafayette' and 'Benjamin Franklin' classes, though the SLBM system has been developed along British lines with the adoption of the Polaris A3TK Chevaline missile/warhead package; the class is to be phased out when the UK's Trident II SSBN class becomes operational

'Typhoon' class SSBN
(USSR)
Type: nuclear-powered ballistic missile submarine
Displacement: about 26,000 tons surfaced and 30,000 tons dived
Dimensions: length 183.0 m (600.4 ft); beam 22.9 m (75.1 ft); draught 15.0 m (49.2 ft)
Gun armament: none
Missile armament: 20 vertical launch tubes for 20 SS-N-20 submarine-launched ballistic missiles
Torpedo armament: six 533-mm (21-in) tubes (all bow), two of them with 406-mm (16-in) liners, for a typical load of 16 533-mm (21-in) anti-ship torpedoes or 10 533-mm (21-in) anti-ship and eight 406-mm (16-in) anti-submarine torpedoes, and two 650-mm (25.6-in) tubes (both bow) for six Type 65 torpedoes, SS-N-15 anti-submarine missiles and SS-N-16 anti-submarine missiles
Anti-submarine armament: torpedoes and missiles (see above)
Electronics: one 'Snoop Pair' surface-search and navigation radar, one 'Rim Hat' ESM system, one low-frequency bow sonar, one medium-frequency torpedo fire-control sonar, one 'Shot Gun' VHF communications antenna, one 'Park Lamp' VLF/LF communications antenna, one 'Pert Spring' satellite navigation system, and other communications and navigation systems
Propulsion: two pressurized water-cooled reactors supplying steam to two sets of geared turbines delivering 60000 kW (80,460 shp) to two shafts
Performance: maximum speed 20 kts surfaced and 24 kts dived; diving depth 400 m (1,315 ft) operational and 600 m (1,970 ft) maximum
Complement: about 150

Class
1. USSR

4 boats plus ? building and planned

Note
The 'Typhoon' class is currently the world's largest

submarine, and entered production at Severodvinsk in 1975; it is likely that the type comprises two modified 'Delta III' hulls combined side-by-side (with a smaller and slightly higher control-section hull between them) in an outer casing; the result is a massive submarine with 20 launch tubes for the new SS-N-20 SLBM, with the useful advantages of separate powerplant arrangements in each of the two main hulls; the relocation of the hydroplanes from the sail to the hull indicates that the type is designed for under-ice operations

'Xia' class SSBN
(China)
Type: nuclear-powered ballistic missile submarine
Displacement: about 8,000 tons dived
Dimensions: length 120.0 m (393.7 ft); beam 10.0 m (32.8 ft); draught 8.0 m (26.25 ft)
Gun armament: none
Missile armament: 14 vertical launch tubes for 14 JL-1 (CSS-N-4) submarine-launched ballistic missiles
Torpedo armament: probably four 533-mm (21-in) tubes (all bow) for ? 533-mm (21-in) dual-role torpedoes
Anti-submarine armament: torpedoes (see above)
Electronics: one surface-search and navigation radar, several sonar systems, and communications and navigation systems
Propulsion: two pressurized water-cooled reactors supplying steam to two sets of geared turbines delivering ? kW (? shp) to two shafts
Performance: maximum speed 18 kts surfaced and 22 kts dived; diving depth about 300 m (985 ft) operational
Complement: about 95

Class
1. China

2 boats plus 4 (?) building or planned

Note
Though these boats are limited by comparison with Western and Soviet SSBNs, their appearance marks an important milestone for the growing Chinese navy

'Yankee I' and 'Yankee II' class SSBNs
(USSR)
Type: nuclear-powered ballistic missile submarine
Displacement: 7,700 tons surfaced and 9,300 tons dived
Dimensions: length 129.5 m (4,214.6 ft); beam 11.6 m (38.0 ft); draught 7.8 m (25.6 ft)
Gun armament: none
Missile armament: ('Yankee I' class) 16 vertical launch tubes for 16 SS-N-6 submarine-launched ballistic missiles, or ('Yankee II' class) 12 vertical launch tubes for 12 SS-N-17 submarine-launched ballistic missiles
Torpedo armament: six 533-mm (21-in) tubes (all bow), two of them with 406-mm (16-in) liners, for a typical load of 12 533-mm (21-in) anti-ship or eight 533-mm (21-in) anti-ship and six 406-mm (16-in) anti-submarine torpedoes
Anti-submarine armament: 406-mm (16-in) torpedoes (see above)
Electronics: one 'Snoop Tray' surface-search and navigation radar, one 'Brick Group' ESM system, one low-frequency bow sonar, one medium-frequency torpedo fire-control sonar, and extensive communications and navigation systems
Propulsion: two pressurized water-cooled reactors supplying steam to two sets of geared turbines delivering 22400 kW (30,040 shp) to two shafts
Performance: maximum speed 20 kts surfaced and 27 kts dived; diving depth 400 m (1,315 ft) operational and 600 m (1,970 ft) maximum
Complement: 130

Class
1. USSR

19 'Yankee I' class boats
1 'Yankee II' class boat

Note
The 'Yankee I' class was the first modern SSBN design to be built by the Soviets, and was based on plans of the US 'Benjamin Franklin' and 'Lafayette' classes secured by Soviet intelligence; total production between 1967 and 1974 was 34 boats built at Severodvinsk and Komsomolsk; several boats have been deactivated (some becoming SSNs) in accordance with the SALT treaties, and one has been revised as the sole 'Yankee II' class boat with launch tubes for 12 SS-N-17 missiles

Nuclear-powered cruise missile type

'Charlie I' class SSGN
(USSR)
Type: nuclear-powered cruise missile submarine
Displacement: 4,000 tons surfaced and 5,000 tons dived
Dimensions: length 93.9 m (308.0 ft); beam 9.9 m (32.5 ft); draught 7.5 m (24.6 ft)
Gun armament: none
Missile armament: eight launch tubes for eight SS-N-7 submarine-launched anti-ship missiles
Torpedo armament: six 533-mm (21-in) tubes (all bow), two of them with 406-mm (16-in) liners, for a typical load of four 533-mm (21-in) HE dual-role and two 533-mm (21-in) nuclear dual-role torpedoes

Anti-submarine armament: two SS-N-15 missiles launched through two dedicated 618-mm (24.33-in) tubes, 533-mm (21-in) torpedoes (see above), and four 406-mm (16-in) torpedoes
Mines: up to 24 AMD-1000 mines instead of torpedoes
Electronics: one 'Snoop Tray' surface-search and navigation radar, one low-frequency bow sonar, one medium-frequency underwater weapons fire-control sonar, one combined 'Brick Spit' and 'Brick Pulp' intercept and threat-warning ESM system, and extensive communications and navigation equipment
Propulsion: one pressurized water-cooled reactor supplying steam to two sets of geared turbines delivering 15000 kW (20,115 shp) to one shaft
Performance: maximum speed 20 kts surfaced and 27 kts dived; diving depth 400 m (1,325 ft) operational and 600 m (1,970 ft) maximum
Complement: 90

Class
1. USSR

10 boats

Note
Built between 1965 and 1972 at Gorki, the 'Charlie I' class was designed to counter the US Navy's overwhelming superiority in surface attack capability (the US carrier battle groups) and is thus armed with a mix of conventional/nuclear torpedoes and anti-ship missiles, the latter accommodated in two four-tube banks in the bows for submerged launch

'Charlie II' and 'Charlie III' class SSGNs
(USSR)
Type: nuclear-powered cruise missile submarine
Displacement: 4,400 tons surfaced and 5,500 tons dived
Dimensions: length 102.9 m (337.5 ft); beam 9.9 m (32.5 ft); draught 7.8 m (25.6 ft)
Gun armament: none
Missile armament: eight launch tubes for ('Charlie II' class) eight SS-N-9 'Siren' submarine-launched anti-ship missiles or ('Charlie III' class) eight SS-N-22 submarine-launched anti-ship missiles
Torpedo armament: six 533-mm (21-in) tubes (all bow), two of them with 406-mm (16-in) liners, for a typical load of four 533-mm (21-in) HE dual-role and two 533-mm (21-in) nuclear dual-role torpedoes
Anti-submarine armament: two SS-N-15 or SS-N-16 missiles launched through two dedicated 618-mm (24.33-in) tubes, 533-mm (21-in) torpedoes (see above), and six 406-mm (16-in) torpedoes
Mines: up to 24 AMD-1000 mines instead of torpedoes
Electronics: one 'Snoop Tray' surface-search and navigation radar, one low-frequency bow sonar, one medium-frequency underwater weapons fire-control sonar, one combined 'Brick Spit' and 'Brick Pulp' intercept and threat-warning ESM system, and extensive communications and navigation systems
Propulsion: one pressurized water-cooled reactor supplying steam to two sets of geared turbines delivering 15000 kW (20,115 shp) to two shafts
Performance: maximum speed 20 kts surfaced and 26 kts dived; diving depth 400 m (1,325 ft) operational and 600 m (1,970 ft) maximum
Complement: 90

Class
1. USSR

4 'Charlie II' class boats
5 'Charlie III' class boats

Note
Built at Gorki between 1972 and 1979 as successor the 'Charlie I' class, the 'Charlie II' type is a revised design reflecting the growing importance of anti-submarine warfare with a 9-m (29.5-ft) hull extension to provide for the launch and control systems associated with the SS-N-15 and SS-N-16 underwater-launched missiles; the class also carries the longer-range SS-N-9 in place of the SS-N-7; built from 1979 at Gorki, the 'Charlie III' is based on the 'Charlie II, but carries SS-N-22 anti-ship missiles

'Echo II' class SSGN
(USSR)
Type: nuclear-powered cruise missile submarine
Displacement: 4,800 tons surfaced and 5,800 tons dived
Dimensions: length 117.3 m (384.7 ft); beam 9.2 m (30.2 ft); draught 7.8 m (25.5 ft)
Gun armament: none
Missile armament: eight launch tubes for eight SS-N-3A 'Shaddock' or SS-N-12 'Sandbox' surface-launched anti-ship missiles
Torpedo armament: six 533-mm (21-in) tubes (all bow) for a typical load of 16 HE dual-role and four nuclear dual-role torpedoes
Anti-submarine armament: two 406-mm (16-in) tubes (both stern) for two torpedoes, and 533-mm (21-in) torpedoes (see above)
Mines: up to 40 AMD-1000 mines instead of 533-mm (21-in) torpedoes
Electronics: one 'Snoop Slab' surface-search and navigation radar, one 'Front Piece' and one 'Front Door' radars for the SS-N-3 system, one 'Herkules' sonar, one 'Feniks' sonar, one 'Stop Light' ECM system, and extensive communications and navigation systems
Propulsion: one pressurized water-cooled reactor supplying steam to two sets of geared turbines delivering 22400 kW (30,040 shp) to two shafts
Performance: maximum speed 20 kts surfaced and 25

kts dived; diving depth 300 m (985 ft) operational and 500 m (1,640 ft) maximum
Complement: 100

Class
1. USSR

28 boats

Note
Derived from the 'Echo I' class SSN, and built at Severodvinsk and Komsomolsk between 1962 and 1967, the 'Echo II' class has a primary armament of eight surface-launched anti-ship missiles (SS-N-3A 'Shaddock' or, in updated boats, SS-N-12 'Sandbox'), reflecting the Soviet realization in the 1960s of the very real threat posed by US carrier battle groups; the need to surface for firing (so that the external missile launchers can be elevated to some 25/30 degrees and the front of the sail hinged through 180 degrees to expose the 'Front' series missile-control radar) is a severe tactical limitation despite the range of the missiles and the use of a Tupolev Tu-95 'Bear-D' mid-course update aircraft

'Oscar' class SSGN
(USSR)
Type: nuclear-powered cruise missile submarine
Displacement: 10,000 tons surfaced and 15,000 tons dived
Dimensions: length 150.0 m (491.2 ft); beam 18.3 m (60.0 ft); draught 11.0 m (36.1 ft)
Gun armament: none
Missile armament: 24 vertical launch tubes for 24 SS-N-19 surface-launched anti-ship missiles
Torpedo armament: six 533-mm (21-in) tubes (all bow), two of them with 406-mm (16-in) liners, for a typical load of 10 533-mm HE dual-role and four 533-mm (21-in) nuclear dual-role torpedoes
Anti-submarine armament: two SS-N-15 and two SS-N-16 missiles launched through two dedicated 618-mm (24.33-in) tubes, 10 406-mm (16-in) torpedoes, and 533-mm (21-in) torpedoes (see above)
Mines: up to 48 AMD-1000 mines instead of 533-mm (21-in) torpedoes
Electronics: one 'Snoop' series surface-search and navigation radar, one low-frequency bow sonar, one medium-frequency underwater weapons fire-control sonar, one combined 'Brick Spit' and 'Brick Pulp' intercept and threat-warning system, and extensive communications and navigation systems
Propulsion: two pressurized water-cooled reactors supplying steam to two sets of geared turbines delivering about 44750 kW (60,010 shp) to two shafts
Performance: maximum speed 20 kts surfaced and 35 kts dived; diving depth 500 m (1,640 ft) operational and 830 m (2,725 ft) maximum
Complement: about 130

Class
1. USSR

3 boats plus ? building or planned

Note
Building at Severodvinsk from 1980, the 'Oscar' class marks the beginning of a new stage in Soviet anti-ship capabilities through the adoption of a much improved propulsion arrangement and two 12-tube banks of launch tubes for the surface-launched SS-N-19 long-range anti-ship missile; the class is designed to deal with US and NATO surface-vessel threats to Soviet battle groups, generally though centred on major units such as helicopter/aircraft carriers or the 'Kirov' class battle-cruisers

'Papa' class SSGN
(USSR)
Type: nuclear-powered cruise missile submarine
Displacement: about 6,100 tons surfaced and 7,000 tons dived
Dimensions: length 150.0 m (492.1 ft); beam 11.5 m (37.7 ft); draught 7.6 m (25.0 ft)
Gun armament: none
Missile armament: 10 launch tubes for 10 SS-N-9 'Siren' submarine-launched anti-ship missiles
Torpedo armament: six 533-mm (21-in) tubes (all bow), two of them with 406-mm (16-in) liners, for a typical load of four 533-mm (21-in) HE dual-role and two 533-mm (21-in) nuclear dual-role torpedoes
Anti-submarine armament: two SS-N-15 or SS-N-16 missiles launched through two dedicated 618-mm (24.33-in) tubes, six 406-mm (16-in) torpedoes, and 533-mm (21-in) torpedoes (see above)
Mines: up to 24 AMD-1000 mines instead of 533-mm (21-in) torpedoes
Electronics: one 'Snoop Tray' surface-search and navigation radar, one low-frequency bow sonar, one medium-frequency underwater weapons fire-control sonar, one combined 'Brick Spit' and 'Brick Pulp' intercept and threat-warning system, and extensive communications and navigation systems
Propulsion: one pressurized water-cooled reactor supplying steam to two sets of geared turbines delivering 29850 kW (40,030 shp) to one shaft
Performance: maximum speed 20 kts surfaced and 40 kts dived; diving depth 400 m (1,325 ft) operational and 600 m (1,970 ft) maximum
Complement: 90

Class
1. USSR

1 boat

Note
Built at Severodvinsk and launched in 1970, the sole 'Papa' class unit was the prototype for the 'Oscar' class

Ballistic missile type

'Wuhan' class SSB'
(USSR/China)
Type: ballistic missile submarine
Displacement: 2,350 tons surfaced and 2,850 tons dived
Dimensions: length 98.0 m (321.4 ft); beam 8.5 m (27.9 ft); draught 6.4 m (21.0 ft)
Gun armament: none
Missile armament: three vertical launch tubes for three HY-1 (CCS-N-2) submarine-launched ballistic missiles
Torpedo armament: six 533-mm (21-in) tubes (all bow) for 12 torpedoes
Anti-submarine armament: torpedoes (see above)
Electronics: one surface-search and navigation radar, various sonar systems, and extensive communications and navigation systems
Propulsion: diesel-electric arrangement, with three diesels delivering 4500 kW (6,035 bhp) and three electric motors delivering 9000 kW (12,070 hp) to three shafts
Performance: maximum speed 17 kts surfaced and 14 kts dived; range 42000 km (26,100 miles) at cruising speed
Complement: 12+74

Class
1. China

1 boat

Note
This is essentially a development type, supplied by the USSR as a 'Golf' class boat and later adapted by the Chinese at the beginning of their SLBM programme

'Golf II' class SSB
(USSR)
Type: ballistic missile submarine
Displacement: 2,600 tons surfaced and 3,000 tons dived
Dimensions: length 98.0 m (321.4 ft); beam 8.5 m (27.9 ft); draught 6.4 m (21.0 ft)
Gun armament: none
Missile armament: three vertical launch tubes for three SS-N-5 'Serb' submarine-launched ballistic missiles
Torpedo armament: six 533-mm (21-in) tubes (all bow) for a typical load of 12 dual-role torpedoes
Anti-submarine armament: torpedoes (see above)
Electronics: one 'Snoop Tray' surface-search and navigation radar, one 'Herkules' sonar, one 'Feniks' sonar, and extensive communications and navigation systems
Propulsion: diesel-electric arrangement, with three diesels delivering 4500 kW (6,035 hp) and three electric motors delivering 9000 kW (12,070 hp) to three shafts
Performance: maximum speed 17 kts surfaced and 14 kts dived; range 42000 km (26,100 miles) at cruising speed
Complement: 12+75

Class
1. USSR

13 boats

Note
Built at Severodvinsk during the early 1960s, the 'Golf II' class was developed from the out-of-service 'Golf I' class with three SS-N-5 rather than three SS-N-4 SLBMs; the type is now allocated to the theatre nuclear role in Europe

Cruise missile type

'Juliett' class SSN
(USSR)
Type: cruise missile submarine
Displacement: 3,000 tons surfaced and 3,8000 tons surfaced and 3,800 tons dived
Dimensions: length 86.7 m (284.4 ft); beam 10.1 m (33.1 ft); draught 7.0 m (22.9 ft)
Gun armament: none
Missile armament: four launch tubes for four SS-N-3A 'Shaddock' surface-launched anti-ship missiles
Torpedo armament: six 533-mm (21-in) tubes (all bow) for a typical load of 18 dual-role torpedoes
Anti-submarine armament: torpedoes (see above)
Electronics: one 'Snoop Slab' surface-search and navigation radar, one 'Front Piece' and one 'Front Door' radar for the SS-N-3 system, one 'Herkules' sonar, one 'Feniks' sonar, one 'Stop Light' ESM system, and extensive communications and navigation systems
Propulsion: diesel-electric arrangement, with three diesels delivering 6000 kW (8,045 bhp) and three electric motors delivering 4500 kW (6,035 hp) to two shafts
Performance: maximum speed 19 kts surfaced and 14 kts dived; range 28000 km (17,400 miles) at surfaced cruising speed

Complement: 79
Class
1. USSR

15 boats

Note
Built from 1961 in Leningrad, the 'Juliett' class was designed as a conventionally-powered interim type of missile-armed attack submarine pending the development of more potent nuclear-powered types; the primary armament is four SS-N-3A 'Shaddock' missiles carried in elevating launchers, one pair forward of the sail and the other pair aft of it

Nuclear-powered attack type

'Akula' class SSN
(USSR)
Type: nuclear-powered attack submarine
Displacement: 8,000 tons dived
Dimensions: length 107.0 m (351.0 ft); beam 11.2 m (36.75 ft); draught 7.5 m (24.6 ft)
Gun armament: none
Missile armament: ? SS-N-21 anti-ship cruise plus SS-N-15 and SS-N-16 anti-submarine missiles launched through the two 650-mm (25.6-in) torpedo tubes
Torpedo armament: four 533-mm (21-in) and two 650-mm (25.6-in) tubes (all bow) for cruise and anti-ship missiles, plus ? dual-role HE and/or nuclear torpedoes
Anti-submarine armament: torpedoes (see above)
Electronics: one surface-search radar, active and passive sonars, various fire-control system, one ESM system, one inertial navigation system, and extensive communication and navigation systems
Propulsion: one pressurized water-cooled nuclear reactor delivering ? kW (? shp) to ? shafts
Performance: not revealed
Complement: not revealed

Class
1. USSR

2 boats plus ? more building or planned

Note
This is clearly an exceptionally capable design blending SSN and SSGN capabilities in a single fairly large hull

'Alpha' class SSN
(USSR)
Type: nuclear-powered attack submarine
Displacement: 2,800 tons surfaced and 3,680 tons dived
Dimensions: length 81.0 m (265.75 ft); beam 9.5 m (31.2 ft); draught 8.0 m (26.25 ft)
Gun armament: none
Missile armament: none
Torpedo armament: six 533-mm (21-in) tubes (all bow), two of them with 406-mm (21-in) liners, for a typical load of eight 533-mm (21-in) HE dual-role and two 533-mm (21-in) nuclear dual-role torpedoes
Anti-submarine armament: two SS-N-15 missiles launched from two dedicated 618-mm (24.33-in) tubes, 10 406-mm (16-in) torpedoes, and 533-mm (21-in) torpedoes (see above)
Electronics: one 'Snoop Tray' surface-search and navigation radar, one low-frequency passive bow sonar, one medium-frequency active/passive underwater weapons fire-control sonar, one 'Brick Spit/Brick Pulp' combined ESM system, inertial navigation system, and extensive communications and navigation systems
Propulsion: one liquid-metal-cooled reactor supplying steam to two sets of geared turbines delivering 30000 kW (40,230 shp) to one shaft; the type also possesses an auxiliary propulsion plant comprising one diesel-electric arrangement delivering power to two auxiliary shafts
Performance: maximum speed 20 kts surfaced and 45 kts dived; diving depth 600 m (1,970 ft) operational and 1000 m (3,280 ft) maximum
Complement: 60

Class
1. USSR

6 boats plus ? building and planned

Note
The 'Alpha' class design underwent a very difficult and protracted genesis, but has matured as an advanced type with performance (speed and diving depth) unmatched by Western boats

'Churchill' and 'Valiant' class SSNs
(UK)
Type: nuclear-powered attack submarine
Displacement: 4,000 tons light, 4,400 tons standard and 4,900 tons dived
Dimensions: length 285.0 ft (86.9 m); beam 33.2 ft (10.1 m); draught 27.0 ft (8.2 m)
Gun armament: none
Missile armament: ? UGM-84A Sub-Harpoon tube-launched anti-ship missiles
Torpedo armament: six 21-in (533-mm) tubes (all bow) for 26 Mk 24 Tigerfish dual-role wire-guided and Mk 8 anti-ship torpedoes
Anti-submarine armament: torpedoes (see above)
Mines: up to 64 Mk 5 Stonefish or Mk 6 Sea Urchin mines instead of torpedoes
Electronics: one Type 1006 surface-search and navi-

gation radar, one Type 2001 active/passive chin sonar, one Type 2024 towed-array sonar, one Type 2007 long-range passive sonar, one Type 2019 active/passive range and intercept sonar, one Type 197 passive ranging sonar, one DCB torpedo fire-control system, one ESM system, one INS, and extensive communications and navigation systems
Propulsion: one Rolls-Royce pressurized water-cooled reactor supplying steam to two sets of English Electric geared turbines delivering 20,000 shp (14915 kW) to one shaft
Performance: maximum speed 20 kts surfaced and 29 kts dived; diving depth 985 ft (300 m) operational and 1,640 ft (500 m) maximum
Complement: 13+90

Class
1. UK ('Churchill' class)

Name	No.	Builder	Laid down	Commissioned
Churchill	S46	Vickers	Jun 1967	Jul 1970
Conqueror	S48	Cammell Laird	Dec 1967	Nov 1971
Courageous	S50	Vickers	May 1968	Oct 1971

2. UK ('Valiant' class)

Name	No.	Builder	Laid down	Commissioned
Valiant	S102	Vickers	Jan 1962	Jul 1966
Warspite	S103	Vickers	Dec 1963	Oct 1971

Note
Developed from the 'Dreadnought' class with a British rather than American reactor, these two very similar designs are notable for the quietness of the submerged operation, the later 'Churchill' class being less noisy still than the 'Valiant' class; the designed role was anti-submarine warfare, but as new and still quieter boats enter service the five boats are being shifted to the anti-ship role with UGM-84 Sub-Harpoon missiles and torpedoes

'Echo I' class SSN
(USSR)
Type: nuclear-powered attack submarine
Displacement: 4,300 tons surfaced and 5,200 tons dived
Dimensions: length 114.0 m (373.9 ft); beam 9.1 m (29.8 ft); draught 7.3 m (23.9 ft)
Gun armament: none
Missile armament: none (provision for six SS-N-3 'Shaddock' anti-ship missiles having been removed)
Torpedo armament: six 533-mm (21-in) tubes (all bow) for a typical load of 16 HE dual-role and four nuclear dual-role torpedoes
Anti-submarine armament: two 406-mm (16-in) tubes (both stern) for two torpedoes
Electronics: one 'Snoop Tray' surface-search and navigation radar, one 'Herkules' sonar, one 'Feniks' sonar, one 'Stop Light' ECM system, and extensive communications and navigation systems
Propulsion: one pressurized water-cooled reactor supplying steam to two sets of geared turbines delivering 22400 kW (30,040 shp) to two shafts
Performance: maximum speed 20 kts surfaced and 28 kts dived; diving depth 300 m (985 ft) operational and 500 m (1,640 ft) maximum
Complement: 12+80

Class
1. USSR

3 boats

Note
Built at Komsomolsk between 1960 and 1962 as cruise missile-armed boats parallel to the conventionally-powered 'Juliett' class SSGs, the 'Echo I' class was converted to SSN status between 1969 and 1974, the four SS-N-3A 'Shaddock' launchers being removed and new role-specialized equipment being fitted to bring the boats to a standard approximating that of the 'November' class; the class has clear design affinities to the 'Hotel' and 'November' classes, and is now obsolescent because of its high underwater noise and other tactical limitations

'Ethan Allen' class SSN
(USA)
Type: nuclear-powered attack and special forces insertion submarine
Displacement: 6,955 tons standard and 7,900 tons dived
Dimensions: length 410.5 ft (125.2 m); beam 33.0 m (10.1 m); draught 30.0 ft (9.1 m)
Gun armament: none
Missile armament: none (the launch tubes for 16 UGM-27 Polaris submarine-launched ballistic missiles have been filled with cement and sealed, the associated fire-control system and one SINS being removed at the same time, and some of the missile tube bases later being removed during the boats' conversion to the 'amphibious transport' role)
Torpedo armament: four 21-in (533-mm) Mk 65 tubes

(all bow) for eight Mk 48 dual-role wire-guided torpedoes
Anti-submarine armament: torpedoes (see above)
Electronics: one BPS-15 surface-search and navigation sonar, one BQS-4 multi-function sonar, one BQR-15 sonar, one Mk 112 torpedo fire-control system, one ESM system, and extensive communications and navigation systems including one WSC-3 satellite communications transceiver
Propulsion: one Westinghouse S5W pressurized water-cooled reactor supplying steam to two sets of General Electric geared turbines delivering 15,000 shp (11185 kW) to one shaft
Performance: maximum speed 20 kts surfaced and 25 kts dived
Complement: 13+124, plus berthing, equipment stowage and airlock equipment for an unspecified number of special forces troopers delivered for clandestine missions

Class
1. USA

Name	No.	Builder	Laid down	Commissioned
Sam Houston	SSN609	Newport News	Dec 1959	Mar 1962
John Marshall	SSN611	Newport News	Apr 1960	May 1962

Note
These are SSBNs of an obsolete type modified as training boats for the US SSN force, and as delivery vehicles for clandestine operations

'Glenard P. Lipscomb' class SSN
(USA)
Type: nuclear-powered attack submarine
Displacement: 5,815 tons standard and 6,480 tons dived
Dimensions: length 365.0 ft (111.3 m); beam 31.7 ft (9.7 m); draught 31.0 ft (9.5 m)
Gun armament: none
Missile armament: four UGM-84A Sub-Harpoon tube-launched anti-ship missiles
Torpedo armament: four 21-in (533-mm) Mk 63 tubes (all amidships) for 15 Mk 48 dual-role wire-guided torpedoes
Anti-submarine armament: four UUM-44A SUBROC tube-launched missiles, and torpedoes (see above)
Electronics: one BPS-15 surface-search and navigation radar, one BQQ-2/5 attack sonar system, one BQS-14 under-ice sonar, one towed-array sonar, one Mk 117 underwater weapons fire-control system, one Mk 2 SINS, and extensive communications and navigation equipment including one WSC-3 satellite communications transceiver
Propulsion: one Westinghouse S5Wa pressurized water-cooled reactor supplying steam to one General Electric turbo-electric drive delivering ? hp (? kW) to one shaft
Performance: maximum speed 18 kts surfaced and 24 kts dived; diving depth 1,315 ft (400 m) operational and 1,970 ft (600 m) maximum
Complement: 13+115

Class
1. USA

Name	No.	Builder	Laid down	Commissioned
Glenard P. Lipscomb	SSN685	GD (EB Div)	Jun 1971	Dec 1974

Note
Though produced as an experimental boat with turbo-electric drive (offering reduced underwater noise but slower speed because of the weight and bulk of the drive system), this singleton unit is fully operational

'Han' class SSN
(China)
Type: nuclear-powered attack submarine
Displacement: 5,000 tons surfaced and 5,700 tons dived
Dimensions: (estimated) length 100.0 m (328.1 ft); beam 11.0 m (36.1 ft); draught 6.5 m (21.3 ft)
Gun armament: none
Missile armament: none
Torpedo armament: six 533-mm (21-in) tubes (all bow) for ? torpedoes
Anti-submarine armament: torpedoes (see above)
Mines: between 24 and 28 depending on type, carried instead of torpedoes
Electronics: not known, but certainly including one surface-search and navigation radar, several sonar systems, and extensive communications and navigation equipment
Propulsion: one pressurized water-cooled reactor supplying steam to two sets of geared turbines delivering ? kW (? shp) to one shaft
Performance: maximum speed 18 kts surfaced and 22 kts dived; diving depth 300 m (985 ft) operational
Complement: about 75

Class
1. China

3 boats with another 2 building and ? planned

Note
Little is known of this Chinese class, which marks the emergence of the Chinese navy into a blue-water force; it is also possible that these are in fact SSGNs with a primary armament of six Fl-1 or FL-7 anti-ship cruise missiles

'Los Angeles' class SSN
(USA)
Type: nuclear-powered attack submarine
Displacement: 6,200 tons standard and 6,900 tons dived
Dimensions: length 360.0 ft (109.7 m); beam 33.0 ft (10.1 m); draught 32.3 ft (9.9 m)
Gun armament: none
Missile armament: four UGM-84A Sub-Harpoon tube-launched anti-ship missiles, and 12 BGM-109 Tomahawk submarine-launched anti-ship cruise missiles (tube-launched weapons in SSBN712 to SSN718, and vertically launched from two six-tube Vertical Launch System banks in the remaining boats)
Torpedo armament: four 21-in (533-mm) Mk 67 tubes (all amidships) for 26 weapons including Mk 48 dual-role wire-guided torpedoes
Anti-submarine armament: four tube-launched UUM-44A SUBROC missiles, and torpedoes (see above)
Mines: up to 78 Mk 57, Mk 60 and Mk 67 mines instead of other tube-launched weapons
Electronics: one BPS-15A surface-search and navigation radar, one BQQ-5 multi-function attack sonar with BQR-7 passive and BQS-13 active/passive search elements, one BQS-15 close-range sonar, one BQR-15 towed-array passive sonar, one Mk 113 or (from SSN 712 onwards) on Mk 117 underwater weapons fire-control system with UYK-7 tactical computers, one ESM system, and extensive communications and navigation systems including one WSC-3 satellite communications transceiver
Propulsion: one General Electric S6G pressurized water-cooled reactor supplying steam to two sets of geared turbines delivering about 35,000 shp (26100 kW) to one shaft
Performance: maximum speed 18 kts surfaced and 31 kts dived; diving depth 1,475 ft (450 m) operational and 2,460 ft (750 m) maximum
Complement: 13+120

Class
1. USA

Name	No.	Builder	Laid down	Commissioned
Los Angeles	SSN688	Newport News	Jan 1972	Nov 1976
Baton Rouge	SSN689	Newport News	Nov 1972	Jun 1977
Philadelphia	SSN690	GD (EB Div)	Aug 1972	Jun 1977
Memphis	SSN691	Newport News	Jun 1973	Dec 1977
Omaha	SSN692	GD (EB Div)	Jan 1973	Mar 1978
Cincinatti	SSN693	Newport News	Apr 1974	Jun 1978
Groton	SSN694	GD (EB Div)	Aug 1973	Jul 1978
Birmingham	SSN695	Newport News	Apr 1975	Dec 1978
New York City	SSN696	GD (EB Div)	Dec 1973	Mar 1979
Indianapolis	SSN697	GD (EB Div)	Oct 1974	Jan 1980
Bremerton	SSN698	GD (EB Div)	May 1976	Mar 1981
Jacksonville	SSN699	GD (EB Div)	Feb 1976	May 1981
Dallas	SSN700	GD (EB Div)	Oct 1976	Jul 1981
La Jolla	SSN701	GD (EB Div)	Oct 1976	Oct 1981
Phoenix	SSN702	GD (EB Div)	Jul 1977	Dec 1981
Boston	SSN703	GD (EB Div)	Aug 1978	Jan 1982
Baltimore	SSN704	GD (EB Div)	May 1979	Jul 1982
City of Corpus Christi	SSN705	GD (EB Div)	Sep 1979	Jan 1983
Albuquerque	SSN706	GD (EB Div)	Dec 1979	May 1983
Portsmouth	SSN707	GD (EB Div)	May 1980	Oct 1983
Minneapolis-St Paul	SSN708	GD (EB Div)	Jan 1981	Mar 1984
Hyman G. Rickover	SSN709	GD (EB Div)	Jul 1981	Jul 1984
Augusta	SSN710	GD (EB Div)	Mar 1982	Jan 1985
San Francisco	SSN711	Newport News	May 1977	Apr 1981
Atlanta	SSN712	Newport News	Aug 1978	Feb 1982
Houston	SSN713	Newport News	Jan 1979	Sep 1982
Norfolk	SSN714	Newport News	Aug 1979	May 1983
Buffalo	SSN715	Newport News	Jan 1980	Nov 1983
Salt Lake City	SSN716	Newport News	Aug 1980	May 1984
Olympia	SSN717	Newport News	Mar 1981	Nov 1984
Honolulu	SSN718	Newport News	Nov 1981	Jul 1985

Providence	SSN719	GD (EB Div)	Oct 1982	Jul 1985
Pittsburgh	SSN720	GD (EB Div)	Apr 1983	Nov 1985
Chicago	SSN721	Newport News	Jan 1983	Dec 1985
	SSN722	Newport News	Jul 1983	1986
	SSN723	Newport News	Jan 1984	1986
Louisville	SSN724	GD (EB Div)	Sep 1984	Nov 1986
	SSN725	GD (EB Div)	Apr 1985	
Newport News	SSN750	Newport News	Mar 1984	
San Juan	SSN751	GD (EB Div)	Sep 1985	
	SSN752	GD (EB Div)	Dec 1985	
	SSN753	Newport News	Mar 1985	
	SSN754	GD (EB Div)	Apr 1986	
	SSN755	GD (EB Div)	Sep 1986	
	SSN756	Newport News		
	SSN757	Newport News		
	SSN758	GD (EB Div)		
	SSN759	Newport News		

Note
The 'Los Angeles' class is the world's largest and the Western world's most important SSN design, combining the speed capability of the 'Skipjack' class with the sensors and weapons of the 'Permit' and 'Sturgeon' classes in a hull with double the installed power of previous classes; even so the boats are significantly slower and shallower-diving than the new generation of Soviet SSNs epitomized by the 'Alpha' class; current boats carry 12 tube-launched Tomahawk cruise missiles (at considerable sacrifice to anti-submarine weapon stowage), but from the 33rd boat onwards provision is to be made for 12 vertical launch tubes for Tomahawks in the flooding space between the front of the pressure hull and the sonar array in the bows; the good underwater speed of the class allows the boats to be used for the escort of fast carrier battle groups and other surface strike forces, and from the 34th boat the hydroplanes are relocated from the sail to the hull

'Mike' class SSN
(USSR)
Type: nuclear-powered attack submarine
Displacement: 9,700 tons dived
Dimensions: length 110.0 m (360.9 ft); beam 12.0 m (39.4 ft); draught 9.0 m (29.5 ft)
Gun armament: none
Missile armament: ? SS-N-21 anti-ship cruise plus SS-N-15 and SS-N-16 anti-submarine missiles launched through the two 650-mm (25.6-in) torpedo tubes
Torpedo armament: two 533-mm (21-in) and two 650-mm (25.6-in) tubes (all bow) for ? dual-role HE and/or nuclear torpedoes
Anti-submarine armament: torpedoes (see above)
Electronics: one surface-search radar, various sonar electronic support measures and fire-control systems, and extensive communication and navigation systems
Propulsion: two reactors (probably of the liquid metal-cooled type) supplying steam to one set of geared turbines delivering 44750 kW (60,010 shp) to one shaft
Performance: speed 35+ kts dived
Complement: about 95

Class
1. USSR

2 boats plus ? building or planned

Note
Like the 'Akula' class the 'Mike' class combines SSN and SSGN capability

'Narwhal' class SSN
(USA)
Type: nuclear-powered attack submarine
Displacement: 4,450 tons standard and 5,350 tons dived
Dimensions: length 314.0 ft (95.8 m); beam 38.0 ft (11.6 m); draught 27.0 ft (8.2 m)
Gun armament: none
Missile armament: four UGM-84A Sub-Harpoon tube-launched anti-ship missiles
Torpedo armament: four 21-in (533-mm) Mk 63 tubes (all amidships) for 15 Mk 48 dual-role wire-guided torpedoes
Anti-submarine armament: four tube-launched UUM-44A SUBROC missiles, and torpedoes (see above)
Mines: up to 46 Mk 57, Mk 60 and Mk 67 mines instead of torpedoes
Electronics: one BPS-15 surface-search and navigation radar, one BQQ-5 attack sonar with BQS-6 active and BQR-7 passive elements, one BQS-8 under-ice sonar, one Mk 117 underwater weapons fire-control system, one ESM system, and extensive communications and navigation systems including one WSC-3 satellite communications transceiver
Propulsion: one General Electric S5G pressurized water-cooled reactor supplying steam to two sets of geared turbines delivering 17,000 shp (12675 kW) to one shaft
Performance: maximum speed 18 kts surfaced and 26 kts dived; diving depth 1,315 ft (400 m) operational and 1,970 ft (600 m) maximum
Complement: 12+95

152 WARSHIPS

Class
1. USA

Name	No.	Builder	Laid down	Commissioned
Narwhal	SSN671	GD (EB Div)	Jan 1966	Jul 1969

Note
Constructed like the USS *Glenard P. Lipscomb* as an experimental boat (in this instance with a natural convection rather than pumped circulation reactor design), this singleton unit is fully operational and in most respects similar to the 'Sturgeon' class

'November' class SSN
(USSR)
Type: nuclear-powered attack submarine
Displacement: 4,200 tons surfaced and 5,000 tons dived
Dimensions: length 109.7 m (359.8 ft); beam 9.1 m (29.8 ft); draught 6.7 m (21.9 ft)
Gun armament: none
Missile armament: none
Torpedo armament: eight 533-mm (21-in) tubes (all bow) for 18 HE dual-role and six nuclear dual-role torpedoes
Anti-submarine armament: two 406-mm (16-in) tubes (both stern) for two torpedoes, and 533-mm (21-in) torpedoes (see above)
Mines: up to 48 AMD-1000 mines instead of 533-mm (21-in) torpedoes
Electronics: one 'Snoop Tray' surface-search and navigation radar, one 'Herkules' sonar, one 'Feniks' sonar, one 'Stop Light' ECM system, and extensive communications and navigation systems
Propulsion: one pressurized water-cooled reactor supplying steam to two sets of geared turbines delivering 22370 kW (30,000 shp) to two shafts
Performance: maximum speed 20 kts surfaced and 30 kts dived; diving depth 300 m (985 ft) operational and 500 m (1,640 ft) maximum
Complement: 86

Class
1. USSR
12 boats

Note
Built at Severodvinsk between 1958 and 196?, the 'November' class was the Soviets' first SSN type, and has since developed an uncertain reputation for mechanical unreliability and poor radiation shielding; the boats are particularly noisy, but their prime role remains the engagement of US carrier battle groups with nuclear-armed torpedoes

'Permit' class SSN (USA): see 'Thresher' class SSN

'Rubis' class SSN
(France)
Type: nuclear-powered attack submarine
Displacement: 2,385 tons surfaced and 2,670 tons dived
Dimensions: length 72.1 m (236.5 ft); beam 7.6 m (25.0 ft); draught 6.4 m (21.0 ft)
Gun armament: none
Missile armament: four SM.39 Exocet tube-launched anti-ship missiles
Torpedo armament: four 533-mm (21-in) tubes (all bow) for 14 F17 wire-guided and L5 torpedoes
Anti-submarine armament: torpedoes (see above)
Mines: up to 28 instead of torpedoes
Electronics: one DRUA 23 surface-search and navigation radar, one DSUV 22 passive sonar, one DUUA 2B active sonar, one DUUX 2 passive detection and ranging sonar, one TUUM 1 underwater telephone, one ESM system, one underwater weapons fire-control system, and extensive communications and navigation systems
Propulsion: one CAS pressurized water-cooled reactor supplying steam to two turbo-alternators/one electric motor delivering ? kW (? hp) to one shaft
Performance: maximum speed 18 kts surfaced and 25 kts dived; diving depth 300 m (985 ft) operational and 500 m (1,640 ft) maximum
Complement: 9+35+22

Class
1. France

Name	No.	Builder	Laid down	Commissioned
Rubis	S601	Cherbourg ND	Dec 1976	Feb 1983
Saphir	S602	Cherbourg ND	Sep 1979	Mar 1984
Casabianca	S603	Cherbourg ND	Sep 1981	Jul 1987
	S604	Cherbourg ND		
	S605	Cherbourg ND		
	S606	Cherbourg ND	Oct 1984	

Note
Otherwise known as the **'SNA72' class**, the 'Rubis' class is the world's smallest SSN design, and is essentially a nuclear-

powered version of the 'Agosta' class patrol submarine with sail- rather than hull-mounted hydroplanes for maximum underwater manoeuvrability

'Sierra' class SSN
(USSR)
Type: nuclear-powered attack submarine
Displacement: 8,000 tons dived
Dimensions: length 110.0 m (360.9 ft); beam 11.0 m (36.1 ft); draught 7.4 m (24.3 ft)
Gun armament: none
Missile armament: ? SS-N-21 anti-ship cruise plus SS-N-15 and SS-N-16 anti-submarine missiles fired through the two 650-mm (25.6-in) torpedo tubes
Torpedo armament: four 533-mm (21-in) and two 650-mm (25.6-in) tubes (all bow) for ? dual-role HE and/or nuclear torpedoes
Anti-submarine armament: torpedoes (see above)
Electronics: one surface-search radar, various sonar, electronic support measures and fire-control systems, and extensive communication and navigation systems
Propulsion: two pressurized water-cooled reactors supplying steam to one set of geared turbines delivering 30000 kW (40,230 shp) to one shaft
Performance: speed 32 kts dived
Complement: about 85

Class
1. USSR

2 boats plus ? more building or planned

Note
This is another new Soviet class combining SSN and SSGN capabilities in a single hull

'Skate' class SSN
(USA)
Type: nuclear-powered attack submarine
Displacement: 2,570 tons standard and 2,860 tons dived
Dimensions: length 267.7 ft (81.5 m); beam 25.0 ft (7.6 m); draught 22.0 ft (6.7 m)
Gun armament: none
Missile armament: none
Torpedo armament: eight 21-in (533-mm) tubes (six Mk 56 bow and two Mk 57 stern) for ? Mk 48 dual-role wire-guided torpedoes
Anti-submarine armament: torpedoes (see above)
Electronics: one BPS-12 surface-search and navigation radar, one BQS-4 active/passive sonar, one Mk 101 torpedo fire-control system, and extensive communications and navigation systems
Propulsion: one pressurized water-cooled reactor (Westinghouse S3W in SSN578 and SSN583, and Westinghouse S4W in SSN579 and SSN584) supplying steam to two sets of Westinghouse geared turbines delivering 6,600 shp (4920 kW) to two shafts
Performance: maximum speed 20+ kts surfaced and 25+ kts dived
Complement: 11+76

Class
1. USA

Name	No.	Builder	Laid down	Commissioned
Skate	SSN578	GD (EB Div)	Jul 1955	Dec 1957
Swordfish	SSN579	Portsmouth NY	Jan 1956	Sep 1958
Sargo	SSN583	Mare Island NY	Feb 1956	Oct 1958

Note
These were the US Navy's first series-produced SSNs, and are in essence conventional boats fitted with a simplified version of the nuclear powerplant developed for the USS *Nautilus*; the boats are no longer considered first-line assets

'Skipjack' class SSN
(USA)
Type: nuclear-powered attack submarine
Displacement: 3,075 tons standard and 3,515 tons dived
Displacement: length 251.7 ft (76.7 m); beam 31.5 ft (9.6 m); draught 29.4 ft (8.9 m)
Gun armament: none
Missile armament: none
Torpedo armament: six 21-in (533-mm) tubes (all bow) for 24 Mk 48 dual-role wire-guided torpedoes
Anti-submarine armament: torpedoes (see above)
Mines: up to 48 Mk 57 mines instead of torpedoes
Electronics: one BPS-12 surface-search and navigation radar, one BQS-4 active/passive sonar, one Mk 101 torpedo fire-control system, and extensive communications and navigation systems
Propulsion: one Westinghouse S5W pressurized water-cooled reactor supplying steam to two sets of geared turbines (Westinghouse in SSN585 and General Electric in the others) delivering 15,000 shp (11185 kW) to one shaft
Performance: maximum speed 18 kts surfaced and 30 kts dived; diving depth 985 ft (300 m) operational and 1,640 ft (500 m) maximum
Complement: 12+108

154 WARSHIPS

Class
1. USA

Name	No.	Builder	Laid down	Commissioned
Skipjack	SSN585	GD (EB Div)	May 1956	Apr 1959
Scamp	SSN588	Mare Island NY	Jan 1959	Jun 1961
Sculpin	SSN590	Ingalls	Feb 1958	Jun 1961
Shark	SSN591	Newport News	Feb 1958	Feb 1961

Note
Though a small class and somewhat elderly, the 'Skipjack' class is still highly valuable for its high speed, though it is likely to be relegated to a training and target role in the near future as the 'Los Angeles' class approaches full strength

'SNA 72' class SSN: (France): see 'Rubis' class SSN

'Sturgeon' class SSN
(USA)
Type: nuclear-powered attack submarine
Displacement: 3,640 tons standard and 4,650 tons dived
Dimensions: length 292.2 ft (89.0 m); beam 31.7 ft (9.5 m); draught 29.5 ft (9.0 m)
Gun armament: none
Missile armament: four UGM-84A Sub-Harpoon tube-launched anti-ship missiles in SSN638, 639, 646, 652, 660, 662, 665, 667/670, 679, 684, 686 and 687, and provision for BGM-109 Tomahawk tube-launched anti-ship cruise missiles in all but SSN637, 647, 650, 651, 653, 661, 664, 666, 674/678, 680 and 683
Torpedo armament: four 21-in (533-mm) Mk 63 tubes (all amidships) for 15 Mk 48 dual-role wire-guided torpedoes
Anti-submarine armament: four UUM-44A SUBROC tube-launched missiles in boats fitted with the Mk 113 underwater weapons fire-control system, and torpedoes (see above)
Mines: up to 46 Mk 57, Mk 60 or Mk 67 mines instead of torpedoes
Electronics: one BPS-15 surface-search and navigation radar, one BQQ-2 attack sonar with BQS-6 active and BQR-7 passive elements, one Mk 113 (SSN637, 650, 653, 661, 672, 675/678, 683 and 687) or Mk 117 (other boats) underwater weapons fire-control system, one BQS-8 under-ice sonar, one BQS-12 (SSN637/639, 646/653 and 660/664) or BQS-13 (other boats) active/passive search sonar, one ESM system, and extensive communications and navigation systems including one WSC-3 satellite communications transceiver
Propulsion: one Westinghouse S5W pressurized water-cooled reactor supplying steam to two sets of geared turbines delivering 15,000 shp (11185 kW) to one shaft
Performance: maximum speed 20+ kts surfaced and 30+ kts dived; diving depth 1,325 ft (400 m) operational and 1,970 ft (600 m) maximum
Complement: 13+115

Class
1. USA

Name	No.	Builder	Laid down	Commissioned
Sturgeon	SSN637	(GD (Quincy)	Aug 1963	Mar 1967
Whale	SSN638	GD (Quincy)	May 1964	Oct 1968
Tautog	SSN639	Ingalls	Jan 1964	Aug 1968
Grayling	SSN646	Portsmouth NY	May 1964	Oct 1969
Pogy	SSN647	Ingalls	May 1964	May 1971
Aspro	SSN648	Ingalls	Nov 1964	Feb 1969
Sunfish	SSN649	GD (Quincy)	Jan 1965	Mar 1969
Pargo	SSN650	GD (Quincy)	Jun 1964	Jan 1968
Queenfish	SSN651	Newport News	May 1964	Dec 1966
Puffer	SSN652	Ingalls	Feb 1965	Aug 1969
Ray	SSN653	Newport News	Apr 1965	Apr 1967
Sand Lance	SSN660	Portsmouth NY	Jan 1965	Sep 1971
Lapon	SSN661	Newport News	Jul 1965	Dec 1967
Gurnard	SSN662	Mare Island NY	Dec 1964	Dec 1968
Hammerhead	SSN663	Newport News	Nov 1965	Jun 1968
Sea Devil	SSN664	Newport News	Apr 1966	Jan 1969
Guitarro	SSN665	Mare Island NY	Dec 1965	Sep 1972
Hawkbill	SSN666	Mare Island NY	Sep 1966	Feb 1971
Bergall	SSN667	GD (Quincy)	Apr 1966	Jun 1969

Name	No.	Builder	Laid down	Commissioned
Spadefish	SSN668	Newport News	Dec 1966	Aug 1969
Seahorse	SSN669	GD (Quincy)	Aug 1966	Sep 1969
Finback	SSN670	Newport News	Jun 1967	Feb 1970
Pintado	SSN672	Mare Island NY	Oct 1967	Sep 1971
Flying Fish	SSN673	GD (Quincy)	Jun 1967	Apr 1970
Trepang	SSN674	GD (Quincy)	Oct 1967	Aug 1970
Bluefish	SSN675	GD (Quincy)	Mar 1968	Jan 1971
Billfish	SSN676	GD (Quincy)	Sep 1968	Mar 1971
Drum	SSN677	Mare Island NY	Aug 1968	Apr 1972
Archerfish	SSN678	GD (Quincy)	Jun 1969	Dec 1971
Silversides	SSN679	GD (Quincy)	Oct 1969	May 1972
William H. Bates	SSN680	Ingalls	Aug 1969	May 1973
Batfish	SSN681	GD (Quincy)	Feb 1970	Sep 1972
Tunny	SSN682	Ingalls	May 1970	Jan 1974
Parche	SSN683	Ingalls	Dec 1970	Aug 1974
Cavalla	SSN684	GD (Quincy)	Jun 1970	Feb 1973
L. Mendel Rivers	SSN686	Newport News	Jun 1971	Feb 1975
Richard B. Russell	SSN687	Newport News	Oct 1971	Aug 1975

Note
This classic SSN design is basically an enlarged version of the 'Thresher/Permit' class design with additional features for quiet running and improved electronics, the last nine boats being lengthened to allow the incorporation of additional electronics; like the 'Permit' class boats the 'Sturgeon' class units are being fitted with the BQQ-5 sonar, Mk 117 fire-control system and more modern missile armament; nine of the boats are allocated to the 'Holy Stone' programme of quiet reconnaissance of potentially hostile shores

'Swiftsure' class SSN
(UK)
Type: nuclear-powered attack submarine
Displacement: 4,200 tons standard and 4,500 tons dived
Dimensions: length 272.0 ft (82.9 m); beam 32.3 ft (9.8 m); draught 27.0 ft (8.2 m)
Gun armament: none
Missile armament: five UGM-84A Sub-Harpoon tube-launched anti-ship missiles
Torpedo armament: five 21-in (533-mm) tubes (all bow) for 20 Mk 24 Tigerfish dual-role wire-guided and Mk 8 anti-ship torpedoes
Anti-submarine armament: torpedoes (see above)
Mines: up to 50 Mk 5 Stonefish or Mk 6 Sea Urchin mines instead of torpedoes
Electronics: one Type 1006 surface-search and navigation radar, one Type 2001 long-range active/passive chin sonar, one Type 2024 towed-array sonar, one Type 2019 long-range passive intercept and range sonar, one Type 2007 long-range passive sonar, one Type 197 passive range sonar, one DCB underwater weapons fire-control system, one ESM system, and extensive communications and navigation systems
Propulsion: one Rolls-Royce pressurized water-cooled reactor supplying steam to two sets of General Electric geared turbines delivering 15,000 shp (11185 kW) to one shaft
Performance: maximum speed 20 kts surfaced and 30+ kts dived; diving depth 1,325 ft (400 m) operational and 1,970 ft (600 m) maximum
Complement: 12+85

Class
1. UK

Name	No.	Builder	Laid down	Commissioned
Swiftsure	S126	Vickers	Jun 1969	Apr 1973
Sovereign	S108	Vickers	Sep 1970	Jul 1974
Superb	S109	Vickers	Mar 1972	Nov 1976
Sceptre	S104	Vickers	Feb 1974	Feb 1978
Spartan	S105	Vickers	Apr 1976	Sep 1979
Splendid	S106	Vickers	Nov 1977	Mar 1981

Note
Introduced in the early 1970s, the 'Swiftsure' class of SSN obviously has affinities with the earlier 'Valiant' and 'Churchill' classes, but has a shorter, fuller hull to maximize internal volume and increase strength for deeper diving; the type is used for the anti-submarine screening of surface task forces, but also possesses a powerful anti-ship capability; the loss of one tube and seven torpedoes compared with previous classes is balanced by a very much faster reload time; sonar capability is to be enhanced during refits as the Type 2001 is replaced by the Type 2020

'Thresher' or 'Permit' class SSN
(USA)
Type: nuclear-powered attack submarine
Displacement: 3,750 tons standard (except SSN613-615 3,800 tons standard) and 4,470 tons dived (except SSN613-615 4,245 tons dived)
Dimensions: length 278.5 ft (84.9 m) or (SSN605) 297.4 ft (90.7 m) or (SSN613/615) 292.2 ft (89.1 m); beam 31.7 ft (9.6 m); draught 28.4 ft (8.7 m)
Gun armament: none
Missile armament: four UGM-84A Sub-Harpoon tube-launched anti-ship missiles
Torpedo armament: four 21-in (533-mm) Mk 63 tubes (all amidships) for 14 Mk 48 dual-role wire-guided torpedoes
Anti-submarine armament: four UUM-44A SUBROC tube-launched missiles in boats fitted with the Mk 113 underwater weapons fire-control system, and torpedoes (see above)
Mines: up to 46 Mk 57, Mk 60 and Mk 67 mines instead of torpedoes

Electronics: one BPS-11 surface-search and navigation radar, one BQQ-2 attack sonar with BQS-6 active and BQR-7 passive elements, one Mk 113 (SSN604-606 and SSN612-615) or Mk 117 (other boats) underwater weapons fire-control system, one ESM system, and extensive communications and navigation systems including one WSC-3 satellite communications transceiver
Propulsion: one Westinghouse S5W pressurized water-cooled reactor supplying steam to two sets of geared turbines delivering 15,000 shp (11185 kW) to one shaft, except SSN605 which has an ungeared contra-rotating turbine driving co-axial contra-rotating propellers
Performance: maximum speed 18 kts surfaced and 27 kts dived; diving depth 1,325 ft (400 m) operational and 1,970 ft (600 m) maximum
Complement: 13+115 (more in boats used for 'Holy Stone' reconnaissance work)

Class
1. USA

Name	No.	Builder	Laid down	Commissioned
Permit	SSN594	Mare Island NY	Jul 1959	May 1962
Plunger	SSN595	Mare Island NY	Mar 1960	Nov 1962
Barb	SSN596	Ingalls	Nov 1959	Aug 1963
Pollack	SSN603	New York SB	Mar 1960	May 1964
Haddo	SSN604	New York SB	Sep 1960	Dec 1964
Jack	SSN605	Portsmouth NY	Sep 1960	Dec 1964
Tinosa	SSN606	Portsmouth NY	Nov 1959	Oct 1964
Dace	SSN607	Ingalls	Jun 1960	Apr 1964
Guardfish	SSN612	New York SB	Feb 1961	Dec 1966
Flasher	SSN613	GD (EB Div)	Apr 1961	Jul 1966
Greenling	SSN614	GD (EB Div)	Aug 1961	Nov 1967
Gato	SSN615	GD (EB Div)	Dec 1961	Jan 1968
Haddock	SSN621	Ingalls	Apr 1961	Dec 1967

Note
This was the US Navy's first deep-diving SSN with sonar in the bows (for optimum search and detection) and the torpedo tubes angled outwards amidships; mid-life updates are being used to replace the original BQQ-2 sonar suite and Mk 113 analog fire-control system with the later BQQ-5 sonar suite (with facility for a clip-on towed array) and Mk 117 digital fire-control system; other improvements are Sub-Harpoon and Tomahawk missiles, and provision for the ASW-SOW when this becomes available; the class was renamed the 'Permit' class after the loss of the lead boat, USS *Thresher*, in a diving accident in 1963

'Trafalgar' class SSN
(UK)
Type: nuclear-power attack submarine
Displacement: 4,700 tons surfaced and 5,200 tons dived
Dimensions: length 280.1 ft (85.4 m); beam 32.1 ft (9.8 m); draught 26.9 ft (8.2 m)
Gun armament: none
Missile armament: five UGM-84A Sub-Harpoon tube-launched anti-ship missiles
Torpedo armament: five 21-in (533-mm) tubes (all bow) for 20 Mk 24 Tigerfish dual-role wire-guided and Mk 8 anti-ship torpedoes

Anti-submarine armament: torpedoes (see above)
Mines: up to 50 Mk 5 Stonefish or Mk 6 Sea Urchin mines instead of torpedoes
Electronics: one Type 1006 surface-search and navigation radar, one Type 2020 multi-function sonar, one Type 2024 towed-array sonar, one Type 2007 long-range passive sonar, one Type 2019 active/passive ranging and intercept sonar, one Type 197 passive ranging sonar, one DCB underwater weapons fire-control system, one ESM system, and extensive communications and navigation systems
Propulsion: one Rolls-Royce pressurized water-cooled reactor supplying steam to two sets of General

Electric geared turbines delivering 15,000 shp (11185 kW) to one shaft
Performance: maximum speed 20 kts surfaced and 32 kts dived; diving depth 1,315 ft (400 m) operational and 1,970 ft (600 m) maximum
Complement: 12+85

Class
1. UK

Name	No.	Builder	Laid down	Commissioned
Trafalgar	S107	Vickers	1978	May 1983
Turbulent	S87	Vickers	1979	Apr 1984
Tireless	S88	Vickers	1981	Oct 1985
Torbay	S90	Vickers	Dec 1982	Feb 1987
Trenchant	S91	Vickers	Apr 1984	
Talent	S92	Vickers	1985	
Triumph	S93	Vickers		

Note
This is in reality an improved 'Swiftsure' class design with features to reduce radiated noise; these include anechoic tilings, pump-jet propulsion and a quieter reactor system

'Victor I' class SSN
(USSR)
Type: nuclear-powered attack submarine
Displacement: 4,200 tons surfaced and 5,200 tons dived
Dimensions: length 93.9 m (307.7 ft); beam 10.0 m (32.8 ft); draught 7.3 m (23.9 ft)
Gun armament: none
Missile armament: none
Torpedo armament: six 533-mm (21-in) tubes (all bow), two of them with 406-mm (16-in) liners, for a typical load of eight 533-mm (21-in) HE dual-role and two 533-mm (21-in) nuclear dual-role torpedoes
Anti-submarine armament: two SS-N-15 tube-launched missiles, 10 406-mm (16-in) torpedoes, and 533-mm (21-in) torpedoes (see above)
Mines: up to 36 AMD-1000 mines instead of 533-mm (21-in) torpedoes
Electronics: one 'Snoop Tray' surface-search and navigation radar, one low-frequency bow sonar, one medium-frequency underwater weapons fire-control sonar, one combined 'Brick Spit' and one 'Brick Pulp' intercept and threat-warning system, and extensive communications and navigation systems
Propulsion: one pressurized water-cooled reactor supplying steam to two sets of geared turbines delivering 22370 kW (30,000 shp) to one main and two auxiliary shafts
Performance: maximum speed 20 kts surfaced and 32 kts dived; diving depth 400 m (1,325 ft) operational and 600 m (1,970 ft) maximum
Complement: 90

Class
1. USSR
14 boats

Note
Built between 1965 and 1974 at the Admiralty Yard in Leningrad, the 'Victor I' class was developed with the 'Charlie' and 'Yankee' classes as the Soviets' second generation of nuclear-powered attack submarines, and was the first Soviet SSN with a teardrop hull for high underwater speed combined with quieter operation

'Victor II' class SSN
(USSR)
Type: nuclear-powered attack submarine
Displacement: 4,800 tons surfaced and 5,800 tons dived
Dimensions: length 100.0 m (328.1 ft); beam 10.0 m (32.8 ft); draught 7.3 m (23.9 ft)
Gun armament: none
Missile armament: none
Torpedo armament: six 533-mm (21-in) tubes (all bow), two of them with 406-mm (16-in) liners, for a typical load of eight 533-mm (21-in) HE dual-role and two 533-mm (21-in) nuclear dual-role torpedoes
Anti-submarine armament: two SS-N-15 and two SS-N-16 missiles launched from dedicated 618-mm (24.33-in) tubes, 10 406-mm (16-in) torpedoes, and 533-mm (21-in) torpedoes (see above)
Mines: up to 36 AMD-1000 mines instead of torpedoes
Electronics: one 'Snoop Tray' surface-search and navigation radar, one low-frequency bow sonar, one medium-frequency underwater weapons fire-control sonar, one combined 'Brick Spit' and one 'Brick Pulp' intercept and threat-warning ESM system, and extensive communications and navigation systems
Propulsion: one pressurized water-cooled reactor supplying steam to two sets of geared turbines delivering 22370 kW (30,000 shp) to one main and two auxiliary shafts
Performance: maximum speed 20 kts surfaced and 31 kts dived; diving depth 400 m (1,325 ft) operational and 600 m (1,970 ft) maximum
Complement: 90

Class
1. USSR

7 boats

Note
Built at Gorki and the Admiralty Yard, Leningrad between 1971 and 1972, the 'Victor II' class is an interim design based on the 'Victor I' with the hull lengthened by 6.1 m (20 ft) forward of the sail for the launch and control systems for the new SS-N-15 and SS-N-16 underwater-launched anti-submarine missiles

'Victor III' class SSN
(USSR)
Type: nuclear-powered attack submarine
Displacement: 5,000 tons surfaced and 6,000 tons dived
Dimensions: length 104.0 m (341.2 ft); beam 10.0 m (32.8 ft); draught 7.3 m (23.9 ft)
Gun armament: none
Missile armament: none
Torpedo armament: six 533-mm (21-in) tubes (all bow), two of them with 406-mm (16-in) liners, for a typical load of eight 533-mm (21-in) HE dual-role and two 533-mm (21-in) nuclear dual-role torpedoes
Anti-submarine armament: two SS-N-15 and two SS-N-16 missiles launched from dedicated 618-mm (24.33-in) tubes, 10 406-mm (16-in) torpedoes, and 533-mm (21-in) torpedoes (see above)
Mines: up to 36 AMD-1000 mines instead of torpedoes
Electronics: one 'Snoop Tray' surface-search and navigation radar, one low-frequency bow sonar, one low-frequency towed-array sonar, one medium-frequency underwater weapons fire-control sonar, one combined 'Brick Spit' and one 'Brick Pulp' passive intercept and threat-warning ESM system, and extensive communications and navigation systems
Propulsion: one pressurized water-cooled reactor supplying steam to two sets of geared turbines delivering 22370 kW (30,000 shp) to one main and two auxiliary shafts
Performance: maximum speed 20 kts surfaced and 30 kts dived; diving depth 400 m (1,325 ft) operational and 600 m (1,970 ft) maximum
Complement: 90

Class
1. USSR

20 boats

Note
Definitive version of the 'Victor' design, the 'Victor III' class was built at the Admiralty Yard, Leningrad and at Komsomolsk between 1974 and 1984; the type has the capabilities of the 'Victor II' class, but is lengthened by 3 m (9.85 ft) forward of the fin and features a bulbous fairing at the top of the fin for a retractable towed-array sonar

Attack type

'Agosta' class SS
(France)
Type: attack submarine
Displacement: 1,480 tons surfaced and 1,760 tons dived
Dimensions: length 67.2 m (221.7 ft); beam 6.8 m (22.3 ft); draught 5.4 m (17.7 ft)
Gun armament: none
Missile armament: ? SM.39 Exocet tube-launched anti-ship missiles
Torpedo armament: four 550-mm (21.65-in) tubes (all bow) with 533-mm (21-in) liners for 23 F17 and E18 dual-role torpedoes
Anti-submarine armament: torpedoes (see above)
Mines: up to 46 TSM3510 (MCC23) mines instead of torpedoes
Electronics: one DRUA 23 surface-search and navigation radar, one DSUV 2H passive sonar, one DUUA 1D active sonar, one DUUA 2A active search and attack sonar, one DUUX 2A passive ranging sonar, one action information and underwater weapons fire-control system, ARUR and ARUD ESM systems, and extensive communications and navigation equipment
Propulsion: diesel-electric arrangement, with two SEMT-Pielstick A16 PA4 185 main diesels delivering 3430 kW (4,600 bhp), one cruising diesel delivering 23 kW (31 bhp) and one electric motor delivering 3500 kW (4,695 hp) to one shaft
Performance: maximum speed 12 kts surfaced and 20 kts dived; diving depth 300 m (985 ft) operational and 500 m (1,640 ft) maximum; range 15750 km (9,785 miles) at 9 kts snorting and 400 km (249 miles) at 3.5 kts dived; endurance 45 days
Complement: 7+45

Class
1. France

Name	No.	Builder	Laid down	Commissioned
Agosta	S620	Cherbourg ND	Nov 1972	Jul 1977
Bévéziers	S621	Cherbourg ND	May 1973	Sep 1977
La Praya	S622	Cherbourg ND	1974	Mar 1978
Ouessant	S623	Cherbourg ND	1974	Jul 1978

2. Pakistan

Name	No.	Builder	Laid down	Commissioned
Hashmat	S135	Dubigeon	Sep 1976	Feb 1979
Hurmat	S136	Dubigeon		Feb 1980

(These Pakistani boats carry at least four UGM-84A Sub-Harpoon missiles instead of Exocets, but are otherwise similar to the French submarines.)

3. Spain ('S70' class)

Name	No.	Builder	Laid down	Commissioned
Galerna	S71	Bazan	Sep 1977	Jun 1982
Siroco	S72	Bazan	1978	May 1983
Mistral	S73	Bazan	May 1980	Jun 1985
Tramontana	S74	Bazan	Dec 1981	1985

(The Spanish boats have a different electronics fit compared with the French units, with DRUA 33C radar, two DUAA-21 active sonars, one DSUV 2H passive sonar, and one DUUX 2A [first pair] or DUUX-5 [second pair] ranging sonar; the boats are to be fitted from UGM-84A Sub-Harpoon rather than Exocet missiles.)

Note
A successful design that has generated a considerable degree of export success, the 'Agosta' class has a rapid-reload system for its four bow tubes, and can launch its torpedoes quietly at all depths down to the designed maximum

'Asashio' class SS
(Japan)
Type: attack submarine
Displacement: 1,650 tons standard
Dimensions: length 88.0 m (288.7 ft); beam 8.2 m (26.9 ft); draught 4.9 m (16.2 ft)
Gun armament: none
Missile armament: none
Torpedo armament: six 21-in (533-mm) tubes (all bow) for 16 torpedoes
Anti-submarine armament: two 12.75-in (324-mm) tubes (both stern) for four swim-out Mk 46 torpedoes
Electronics: one ZPS-3 surface-search and navigation radar, one JQS-3A sonar, one JQQ-2A sonar, one UQS-1 sonar, one torpedo fire-control system, and extensive communications and navigation systems
Propulsion: diesel-electric arrangement, with two Kawasaki/MAN diesels delivering 2160 kW (2,895 shp) and two electric motors delivering 4700 kW (6,305 hp) to two shafts
Performance: maximum speed 14 kts surfaced and 18 kts dived
Complement: 80

Class
1. Japan

Name	No.	Builder	Laid down	Commissioned
Arashio	SS565	Mitsubishi	Jul 1967	Jul 1969

Note
This is an obsolescent design

'Barbel' class SS
(USA)
Type: attack submarine
Displacement: 1,740 tons standard, 2,145 tons surfaced and 2,895 tons dived
Dimensions: length 219.5 ft (66.9 m); beam 29.0 ft (8.8 m); draught 28.0 ft (8.5 m)
Gun armament: none
Missile armament: none
Torpedo armament: six 21-in (533-mm) Mk 58 tubes (all bow) for? Mk 48 dual-role wire-guided torpedoes
Anti-submarine armament: torpedoes (see above)
Electronics: one BPS-12 surface-search and navigation radar, one BQS-4 active sonar, one Mk 101 torpedo fire-control system, and extensive communication and navigation systems
Propulsion: diesel-electric arrangement, with three Fairbanks-Morse diesels delivering 4,800 bhp (3580 kW) and two General Electric electric motors delivering 3,150 hp (2350 kW) to one shaft
Performance: maximum speed 15 kts surfaced and 25 kts dived
Complement: 8+74

160 WARSHIPS

Class
1. USA

Name	No.	Builder	Laid down	Commissioned
Barbel	SS580	Portsmouth NY	May 1956	Jan 1959
Bluefish	SS581	Ingalls	Apr 1957	Oct 1959
Bonefish	SS582	New York SB	Jun 1957	Jul 1959

Note
This was the last diesel-electric submarine class developed for the US Navy, and is of advanced teardrop design

'Casma' class SS (Peru): see 'Type 209/1' class SS

'Daphné' class SS
(France)
Type: attack submarine
Displacement: 860 tons surfaced and 1,040 tons dived
Dimensions: length 57.75 m (189.5 ft); beam 6.76 m (22.2 ft); draught 4.62 m (15.2 ft)
Gun armament: none
Missile armament: none
Torpedo armament: 12 550-mm (21.65-in) tubes (eight bow and four stern) for 12 L3, E14 and E15 torpedoes
Anti-submarine armament: anti-submarine torpedoes (included in the total above)
Mines: up to 24 TSM3510 (MCC23) mines instead of torpedoes
Electronics: one DRUA 31F surface-search and navigation radar, one DSUV 2 passive search sonar, one DUUA 1 active sonar, one DUUA 2A active search and attack sonar, one DUUX 2 passive ranging sonar, one action information and torpedo fire-control system, and extensive communications and navigation systems
Propulsion: diesel-electric arrangement, with two SEMT-Pielstick diesels delivering 1820 kW (2,440 bhp) and two electric motors delivering 1940 kW (2,600 hp) to two shafts
Performance: maximum speed 13.5 kts surfaced and 16 kts dived; diving depth 300 m (985 ft) operational and 575 m (1,885 ft) maximum; range 18500 km (11,495 miles) at 7 kts surfaced or 5560 km (3,455 miles) at 7 kts snorting
Complement: 6+39

Class
1. France

Name	No.	Builder	Laid down	Commissioned
Daphné	S641	Dubigeon	Mar 1958	Jun 1964
Diane	S642	Dubigeon	Jul 1958	Jun 1964
Doris	S643	Cherbourg ND	Sep 1958	Aug 1964
Flore	S645	Cherbourg ND	Sep 1958	May 1964
Galatée	S646	Cherbourg ND	Sep 1958	Jul 1964
Junon	S648	Cherbourg ND	Jul 1961	Feb 1966
Venus	S649	Cherbourg ND	Aug 1961	Jan 1966
Psyché	S650	Brest ND	May 1965	Jul 1969
Sirène	S651	Brest ND	May 1965	Mar 1970

2. Pakistan

Name	No.	Builder	Laid down	Commissioned
Hangor	S131	Brest Arsenal	Dec 1967	Jan 1970
Shushuk	S132	CN Ciotat	Dec 1967	Jan 1970
Mangro	S133	CN Ciotat	Jul 1968	Aug 1970
Ghazi	S134	Dubigeon	May 1967	Oct 1969

(These boats differ only marginally from the French norm.)

3. Portugal

Name	No.	Builder	Laid down	Commissioned
Albacora	S163	Dubigeon	Sep 1965	Oct 1967
Barracuda	S164	Dubigeon	Oct 1965	May 1968
Delfin	S166	Dubigeon	May 1967	Oct 1969

(These boats differ only marginally from the French norm.)

4. South Africa

Name	No.	Builder	Laid down	Commissioned
Maria van Riebeeck	S97	Dubigeon	Mar 1968	Jun 1970
Emily Hobhouse	S98	Dubigeon	Nov 1968	Jan 1971

Johanna van der Merwe	S99	Dubigeon	Apr 1969	Jul 1971

(These boats differ only marginally from the French norm.)

5. Spain ('S60' class)

Name	No.	Builder	Laid down	Commissioned
Delfin	S61	Bazan	Aug 1968	May 1973
Tonina	S62	Bazan	1969	Jul 1973
Marsopa	S63	Bazan	Mar 1971	Apr 1975
Narval	S64	Bazan	1971	Nov 1975

(These boats differ from the French standard only in their radar fit, comprising DRUA 31 or DRUA 33 with ECM equipment.)

Note
Designed in the mid-1950s as deep-diving ocean-going submarines with modest speed, the units of the 'Daphné' class has proved highly successful, but the design is obsolescent by the latest standards; the armament is located externally in fixed tubes, so doing away with the need for a torpedo room and reload rounds

'Delfinen' class SS
(Denmark)
Type: attack submarine
Displacement: 595 tons surfaced and 645 tons dived
Dimensions: length 54.0 m (177.2 ft); beam 4.7 m (15.4 ft); draught 4.2 m (13.8 ft)
Gun armament: none
Missile armament: none
Torpedo armament: four 533-mm (21-in) tubes (all bow) for four torpedoes
Anti-submarine armament: torpedoes (see above)

Electronics: one surface-search and navigation radar, one active sonar, one passive sonar, one torpedo fire-control system, and extensive communications and navigation systems
Propulsion: diesel-electric arrangement, with two Burmeister & Wain diesels delivering 900 (1,205 bhp) and two electric motors delivering 895 kW (1,200 hp) to two shafts
Performance: maximum speed 16 kts surfaced and dived; range 7400 km (4,600 miles) at 8 kts surfaced
Complement: 33

Class
1. Denmark

Name	No.	Builder	Laid down	Commissioned
Delfinen	S326	Royal Dockyard	Jun 1954	Sep 1958
Spaekhuggeren	S327	Royal Dockyard	Nov 1954	Jun 1959
Springeren	S328	Royal Dockyard	Jan 1961	Oct 1964

Note
These boats are now decidedly elderly and are to be replaced as soon as finance permits

'Dolfijn' and 'Potvis' class SSs
(Netherlands)
Type: attack submarine
Displacement: 1,495 tons surfaced and 1,110 tons dived
Dimensions: length 79.5 m (260.9 ft); beam 7.8 m (25.8 ft); draught 5.0 m (16.4 ft)
Gun armament: none
Missile armament: none
Torpedo armament: eight 21-in (533-mm) tubes (four bow and four stern) for Mk 48 dual-role wire-guided torpedoes

Anti-submarine armament: torpedoes (see above)
Electronics: one Type 1001 surface-search and navigation radar, active and passive sonar systems, one WM-8 torpedo fire-control system, and extensive communications and navigation systems
Propulsion: diesel-electric arrangement, with two MAN diesels delivering 2300 kW (3,085 shp) and two electric motors delivering 3120 kW (4,185 hp) to two shafts
Performance: maximum speed 14.5 kts surfaced and 17 kts dived; diving depth 300 m (985 ft) operational
Complement: 67

Class
1. Netherlands ('Dolfijn' class)

Name	No.	Builder	Laid down	Commissioned
Zeehond	S809	Rotterdamse DD	Dec 1954	Mar 1961

2. Netherlands ('Potvis' class)

Name	No.	Builder	Laid down	Commissioned
Potvis	S804	Wilton	Sep 1962	Nov 1965
Tonijn	S805	Wilton	Nov 1962	Feb 1966

Note
Based on the exterior lines of the US 'Guppy' modification series, these two very similar classes are unusual in having triple pressure hulls, a long upper one containing the crew spaces and the two shorter side-by-side lower ones the batteries and engine spaces

'Draken' class SS
(Sweden)
Type: attack submarine
Displacement: 835 tons surfaced and 1,110 tons dived
Dimensions: length 69.0 m (226.4 ft); beam 5.1 m (16.7 ft); draught 4.6 m (15.1 ft)
Gun armament: none
Missile armament: none
Torpedo armament: four 533-mm (21-in) tubes (all bow) for 12 torpedoes
Anti-submarine armament: torpedoes (see above)
Mines: up to 24 instead of torpedoes
Electronics: one surface-search and navigation radar, various sonar systems, one torpedo fire-control system, and communications and navigation systems
Propulsion: diesel-electric arrangement, with two SEMT-Pielstick diesels and one electric motor delivering 1250 kW (1,675 hp) to one shaft
Performance: maximum speed 17 kts surfaced and 20 kts dived
Complement: 36

Class
1. Sweden

Name	No.	Builder	Laid down	Commissioned
Delfinen	Del	Karlskrona	1959	Jun 1962
Nordkaparen	Nor	Kockums	1959	Apr 1962
Springaren	Spr	Kockums	1960	Nov 1962
Vargen	Vgn	Kockums	1958	Nov 1961

Note
These are the survivors of a 12-strong class derived from the German Type XXI of World War II, the latter six (including these boats) having one rather than two propellers

'Foxtrot' class SS
(USSR)
Type: attack submarine
Displacement: 1,950 tons surfaced and 2,500 tons dived
Dimensions: length 91.5 m (300.1 ft); beam 8.0 m (26.2 ft); draught 6.1 m (20.0 ft)
Gun armament: none
Missile armament: none
Torpedo armament: six 533-mm (21-in) tubes (all bow) for a typical load of 22 torpedoes
Anti-submarine armament: four 406-mm (16-in) tubes (all stern) for torpedoes carried in the total above, and 533-mm (21-in) torpedoes (see above)
Mines: up to 44 AMD-100 mines instead of 533-mm (21-in) torpedoes
Electronics: one 'Snoop Tray' surface-search and navigation radar, one 'Herkules' sonar, one 'Feniks' sonar, one torpedo fire-control system, one 'Stop Light' ECM system, and communications and navigation systems
Propulsion: diesel-electric arrangement, with three diesels delivering 4500 kW (6,035 bhp) and three electric motors delivering 4500 kW (6,035 hp) to three shafts
Performance: maximum speed 18 kts surfaced and 16 kts dived; diving depth 300 m (985 ft) operational and 500 m (1,640 ft) maximum; range 37000 km (22,990 miles) at low cruising speed; endurance 70 days
Complement: 8+67

Class
1. Cuba
3 boats
2. India

Name	No.
Kursurja	S20
Karanj	S21
Kanderi	S22
Kalvari	S23
Vela	S40
Vagir	S41
Vagli	S42
Vagsheer	S43

3. Libya

Name	No.
Al Badr	311
Al Ahad	313
Al Fateh	312
Al Matrega	314
plus two others	

4. USSR

45 boats

Note
Built at the Sudomekh Yard in Leningrad in two major groups for the Soviet navy (45 boats between 1958 and 1968, and 17 boats between 1971 and 1974) plus more for export, the 'Foxtrot' class is a capable design providing strong evidence of the USSR's continued commitment to quiet-running diesel-electric patrol submarines for coastal operations even at a time of hectic building in the nuclear-powered submarine programme

'Glavkos' class (Greece): see 'Type 209/0' class and 'Type 209/1' class

'Guppy IA' class SS
(USA)
Type: attack submarine
Displacement: 1,870 tons surfaced and 2,440 tons dived
Dimensions: length 308.0 ft (93.8 m); beam 27.0 ft (8.2 m); draught 17.0 ft (5.2 m)
Gun armament: none
Missile armament: none
Torpedo armament: 10 21-in (533-mm) tubes (six bow and four stern) for 24 torpedoes
Anti-submarine armament: none
Electronics: one surface-search and navigation radar, active and passive sonar, one torpedo fire-control system, and communications and navigation systems
Propulsion: diesel-electric arrangement, with three Fairbanks-Morse 38D diesels delivering 4,800 bhp (3580 kW) and two Elliot Motor electric motors delivering 5,400 hp (4925 kW) to two shafts
Performance: maximum speed 17 kts surfaced and 15 kts dived; diving depth 400 ft (120 m) operational
Complement: 84

Class
1. Peru

Name	No.	Builder	Laid down	Commissioned
Pacocha	S48	Portsmouth NY	Dec 1943	Jun 1944
La Pedrera	S49	Portsmouth NY	Feb 1944	Jul 1944

Note
Developed in the early stages of the 'Guppy' programme from World War II patrol submarines, these boats are at best suited for training

'Guppy II' class SS
(USA)
Type: attack submarine
Displacement: 1,870 tons surfaced and 2,420 tons dived
Dimensions: length 307.5 ft (93.6 m); beam 27.2 ft (8.3 m); draught 18.0 ft (5.5 m)
Gun armament: none
Missile armament: none
Torpedo armament: 10 21-in (533-mm) tubes (six bow and four stern) for 24 torpedoes
Anti-submarine armament: none
Electronics: one surface-search and navigation radar, active and passive sonar, one torpedo fire-control system, and communications and navigation equipment
Propulsion: diesel-electric arrangement, with three Fairbanks-Morse 38D diesels delivering 4,800 bhp (3580 kW) and two Elliot Motor electric motors delivering 5,400 hp (4025 kW) to two shafts
Performance: maximum speed 18 kts surfaced and 15 kts dived; diving depth 400 ft (120 m) operational
Complement: 7+68

Class
1. Brazil

Name	No.	Builder	Laid down	Commissioned
Guanabara	S10	Electric Boat	Jun 1944	Apr 1946
Bahia	S12	Portsmouth NY	Nov 1944	Jun 1945
Ceara	S14	Boston NY	Feb 1944	Mar 1946

2. Taiwan

Name	No.	Builder	Laid down	Commissioned
Hai Shih	736	Portsmouth NY	Jul 1944	Mar 1945

| Hai Pao | 794 | Federal SB | Aug 1943 | Apr 1946 |

(The Taiwanese boats differ from the norm in having DUUG 1B sonar plus WLR-1 and WLR-3 passive ESM suites.)

3. Venezuela

Name	No.	Builder	Laid down	Commissioned
Picua	S22	Boston NY	Feb 1944	Feb 1951

Note
Though slightly more capable than the 'Guppy I' boats, these too are second-line boats at best

'Guppy IIA' class SS
(USA/Turkey)
Type: attack submarine
Displacement: 1,850 tons surfaced and 2,440 tons dived
Dimensions: length 306.0 ft (93.2 m); beam 27.0 ft (8.2 m); draught 17.0 ft (5.2 m)
Gun armament: none
Missile armament: none
Torpedo armament: 10 21-in (533-mm) tubes (six bow and four stern) for 24 torpedoes
Anti-submarine armament: none
Mines: up to 40 instead of torpedoes

Electronics: one BPS-9 surface-search and navigation radar, one BQR-2 passive sonar, one BQS-2 active sonar, one GQR-3 passive tracking sonar, one torpedo fire-control system, and communications and navigation systems
Propulsion: diesel-electric arrangement, with three General Motors diesels delivering 4,800 shp (3580 kW) and two electric motors delivering 5,400 hp (4025 kW) to two shafts
Performance: maximum speed 17 kts surfaced and 15 kts dived; diving depth 400 ft (120 m) operational; range 13,800 miles (22210 km) at 10 kts surfaced
Complement: 82

Class
1. Greece

Name	No.	Builder	Laid down	Commissioned
Papanikolis	S114	Manitowoc SB	Jul 1943	Apr 1944

2. Turkey

Name	No.	Builder	Laid down	Commissioned
Burakreis	S335	Portsmouth NY	Nov 1943	Jun 1944
Muratkreis	S336	Portsmouth NY	Sep 1943	Apr 1944
Orucreis	S337	Portsmouth NY	Jul 1943	Feb 1944
Ulucalireis	S338	Portsmouth NY	Apr 1944	Oct 1944
Cerbe	S340	Portsmouth NY	Dec 1943	Nov 1944
Preveze	S345	Electric Boat	Feb 1944	Apr 1945
Birinci Inonu	S346	Portsmouth NY	Mar 1944	Aug 1944

Note
Though best described as obsolete, these boats may retain a slight combat value in the shallow waters of the Black Sea and Aegean

'Guppy III' class SS
(USA/Turkey)
Type: attack submarine
Displacement: 1,975 tons surfaced and 2,450 tons dived
Dimensions: length 326.5 ft (99.4 m); beam 27.0 ft (8.2 m); draught 17.0 ft (5.2 m)
Gun armament: none
Missile armament: none
Torpedo armament: 10 21-in (533-mm) tubes (six bow and four stern) for 24 torpedoes
Anti-submarine armament: none
Mines: up to 40 instead of torpedoes

Electronics: one SS-2 surface-search and navigation radar, one BQR-2 passive sonar, one BQS-2 active sonar, one BQG-4 passive fire-control sonar, one Mk 106 torpedo fire-control system, and communications and navigation equipment
Propulsion: diesel-electric arrangement, with four General Motors 278A diesels delivering 6,400 bhp (4770 kW) and two General Electric electric motors delivering 5,400 hp (4025 kW) to two shafts
Performance: maximum speed 17.5 kts surfaced and 15 kts dived; diving depth 400 ft (120 m)
Complement: 8+78

Class
1. Brazil

Name	No.	Builder	Laid down	Commissioned
Goiaz	S15	Cramps SB	Aug 1943	Jan 1946
Amazonas	S16	Electric Boat	Jun 1944	Jun 1946

2. Greece

Name	No.	Builder	Laid down	Commissioned
Katsonis	S115	Portsmouth NY	Mar 1945	Jan 1946

3. Turkey

Name	No.	Builder	Laid down	Commissioned
Canakkale	S341	Electric Boat	Apr 1944	Aug 1945
Ikinci Inonu	S333	Electric Boat	Apr 1944	Nov 1945

Note
These boats lack more than nominal combat value, but are useful trainers

'Hai Lung' class SS (Taiwan): see 'Zwaardvis' class SS

'Heroj' class SS
(Yugoslavia)
Type: attack submarine
Displacement: 1,070 tons standard, 1,170 tons surfaced and 1,350 tons dived
Dimensions: length 64.0 m (210.0 ft); beam 7.2 m (23.6 ft); draught 5.0 m (16.4 ft)
Gun armament: none
Missile armament: none
Torpedo armament: six 533-mm (21-in) tubes (all bow) for ? Swedish dual-role torpedoes
Anti-submarine armament: torpedoes (see above)
Mines: up to ? instead of torpedoes
Electronics: one 'Snoop' series surface-search and navigation radar, Soviet active sonar system, Krupp-Atlas PRS 3 passive sonar, Swedish torpedo fire-control system, and communications and navigation systems
Propulsion: diesel-electric arrangement, with two diesels and two electric motors delivering 1800 kW (2,415 hp) to one shaft
Performance: maximum speed 16 kts surfaced and 10 kts dived; range 11175 km (6,945 miles) at 10 kts surfaced
Complement: 55

Class
1. Yugoslavia

Name	No.	Builder	Laid down	Commissioned
Heroj	821	Uljanik SY	1964	1968
Junak	822	S & DE Factory	1965	1969
Uskok	823	Uljanik SY	1966	1970

Note
These small boats are well suited to Adriatic operations

'Kilo' class SS
(USSR)
Type: attack submarine
Displacement: 2,500 tons surfaced and 3,200 tons dived
Dimensions: length 70.0 m (229.7 ft); beam 9.0 m (29.5 ft); draught 7.0 m (23.0 ft)
Gun armament: none
Missile armament: none
Torpedo armament: six 533-mm (21-in) tubes (all bow) for a typical load of 18 dual-role torpedoes
Anti-submarine armament: torpedoes (see above)
Mines: up to 36 AMD-1000 mines instead of torpedoes
Electronics: one 'Snoop Tray' surface-search and navigation radar, one low-frequency bow sonar, one medium-frequency torpedo fire-control sonar, one action information system, one 'Brick Pulp' ESM system, and communications and navigation systems
Propulsion: diesel-electric arrangement, with four diesels and two electric motors delivering ? kW (? hp) to one shaft
Performance: maximum speed 14 kts surfaced and 24 kts dived; diving depth 450 m (1,475 ft) operational and 650 m (2,135 ft) maximum
Complement: about 55

Class
1. India

Name
Sindhughosh and five more on order

2. USSR

11 boats plus ? building and planned

Note
Designed to replace the 'Whiskey' class in the USSR's Pacific Fleet, the 'Kilo' class is of thoroughly modern medium-range design that entered production at Komsomolsk during the late 1970s, the first unit being launched in 1980; the type is notable for its advanced hull shape (offering good underwater speed with considerable diving depth) and suite of new sonar equipments

'Ming' class SS
(China)
Type: attack submarine
Displacement: about 1,600 tons surfaced and 2,000 tons dived
Dimensions: (estimated) length 79.0 m (259.2 ft); beam 7.3 m (240.0 ft); draught 6.5 m (21.3 ft)
Gun armament: none
Missile armament: none
Torpedo armament: probably eight 533-mm (21-in) tubes (all bow) for 14 dual-role torpedoes
Anti-submarine armament: torpedoes (see above)
Mines: up to 28 instead of torpedoes
Electronics: one surface-search and navigation radar, active and passive sonar, one torpedo fire-control system, and communications and navigation systems
Propulsion: diesel-electric arrangement, with two diesels and two electric motors delivering ? kW (? hp) to two shafts
Performance: maximum speed 16 kts surfaced and 14 kts dived; diving depth 250 m (820 ft) operational; range 15750 km (9,785 miles) at 10 kts surfaced
Complement: 10+50

Class
1. China

3 boats plus? more building and planned

Note
Little is known of this design, which is nevertheless clear evidence that the Chinese are extending the scope of their navy from purely defensive coastal operations

'Näcken' class SS
(Sweden)
Type: attack submarine
Displacement: 1,030 tons surfaced and 1,125 tons dived
Dimensions: length 49.5 m (162.4 ft); beam 5.6 m (18.4 ft); draught 5.6 m (18.4 ft)
Gun armament: none
Missile armament: none
Torpedo armament: six 533-mm (21-in) tubes (all bow) for 10 Tp 61 dual-role wire-guided torpedoes
Anti-submarine armament: two 400-mm (15.75-in) tubes (both bow) for two Tp 42 wire-guided torpedoes, and 533-mm (21-in) torpedoes (see above)
Mines: up to 20 mines instead of 533-mm (21-in) torpedoes
Electronics: one surface-search and navigation radar, active and passive sonars, one AI-FCS torpedo fire-control system, and communications and navigation equipment
Propulsion: diesel-electric arrangement, with two MTU 16V652 diesels delivering 1570 kW (2,105 bhp) and one Jeumont-Schneider electric motor delivering 1100 kW (1,475 hp) to one shaft
Performance: maximum speed 20 kts surfaced and 25 kts dived
Complement: 19

Class
1. Sweden

Name	No.	Builder	Laid down	Commissioned
Näcken	Näk	Kockums	Nov 1972	Apr 1980
Najad	Nad	Kockums	Sep 1973	Jun 1981
Neptun	Nep	Kockums	Mar 1974	Dec 1980

Note
This is a capable design of compact size but good performance, based on a high degree of automation and a small crew

'Narhvalen' class SS
(Denmark)
Type: attack submarine
Displacement: 420 tons surfaced and 450 tons dived
Dimensions: length 44.3 m (145.3 ft); beam 4.6 m (150.0 ft); draught 4.2 m (13.8 ft)
Gun armament: none
Missile armament: none
Torpedo armament: eight 533-mm (21-in) tubes (all bow) for eight Tp 61 dual-role wire-guided torpedoes
Anti-submarine armament: ? Tp 41 torpedoes, and dual-role torpedoes (see above)
Mines: up to 16 instead of torpedoes
Electronics: one surface-search and navigation radar,

Krupp-Atlas CSU 3-2 active and PRS 3-4 passive sonars, one torpedo fire-control system, and communications and navigation systems
Propulsion: diesel-electric arrangement, with two MTU diesels delivering 1120 kW (1,500 bhp) and one electric motor delivering 1120 kW (1,500 hp) to one shaft
Performance: maximum speed 12 kts surfaced and 17 kts dived
Complement: 22

Class
1. Denmark

Name	No.	Builder	Laid down	Commissioned
Narhvalen	S320	Royal Dockyard	Feb 1965	Feb 1970
Nordkaparen	S321	Royal Dockyard	Mar 1966	Dec 1970

Note
This useful type is based on the West German 'Type 205' and Norwegian 'Type 207' class designs

'Oberon' and 'Porpoise' class SSs
(UK)
Type: attack submarine
Displacement: 2,030 tons surfaced and 2,410 tons dived
Dimensions: length 295.2 ft (90.0 m); beam 26.5 ft (8.1 m); draught 18.0 ft (5.5 m)
Gun armament: none
Missile armament: none
Torpedo armament: six 21-in (533-mm) tubes (all bow) for 20 Mk 24 Tigerfish dual-role wire-guided and Mk 8 anti-ship torpedoes
Anti-submarine armament: two 21-in (533-mm) tubes (both stern) for two short torpedoes, and torpedoes (see above)
Mines: up to 50 Mk 5 Stonefish and Mk 6 Sea Urchin mines instead of 533-mm (21-in) long torpedoes
Electronics: one Type 1002 surface-search and navigation radar, one Type 2051 Triton (updated boats) or Type 187 attack sonar, one Type 2007 long-range passive sonar, one Type 186 sonar, one DCH torpedo fire-control system, one action information system, one ESM system, and communications and navigation systems
Propulsion: diesel-electric arrangement, with two Admiralty Standard Range diesels delivering 7,360 bhp (5490 kW) and two electric motors delivering 6,000 hp (4475 kW) to two shafts
Performance: maximum speed 12 kts surfaced and 17 kts dived; diving depth 650 ft (200 m) operational and 1,115 ft (340 m) maximum; range 10,350 miles (16655 km) at surfaced cruising speed
Complement: 7+62

Class
1. Australia

Name	No.	Builder	Laid down	Commissioned
Oxley	S57	Scotts SB	Jul 1964	Apr 1967
Otway	S59	Scotts SB	Jun 1965	Apr 1968
Onslow	S60	Scotts SB	Dec 1967	Dec 1969
Orion	S61	Scotts SB	Oct 1972	Jun 1977
Otama	S62	Scotts SB	May 1973	Apr 1978
Ovens	S70	Scotts SB	Jun 1966	Apr 1969

(These Australian boats have been upgraded with Sperry Micropuffs passive ranging sonar, Krupp CSU3-41 attack sonar, Singer Librascope SFCS-RAN-Mk 1 fire-control system, Mk 48 wire-guided torpedoes, and UGM-84A Sub-Harpoon submarine-launched anti-ship missiles.)

2. Brazil

Name	No.	Builder	Laid down	Commissioned
Humaita	S20	Vickers	Nov 1970	Jun 1973
Tonelero	S21	Vickers	Nov 1971	Dec 1977
Riachuelo	S22	Vickers	May 1973	Mar 1977

(The Brazilian boats are similar to the British submarines except that they have Vickers TIOS-B fire-control systems and are being updated with Mk 24 Mod 1 Tigerfish torpedoes.)

3. Canada

Name	No.	Builder	Laid down	Commissioned
Ojibwa	72	Chatham DY	Sep 1962	Sep 1965
Onondaga	73	Chatham DY	Jun 1964	Jun 1967
Okanagan	74	Chatham DY	Mar 1965	Jun 1968

(The Canadian boats are similar to the British boats, but have better air-conditioning and a high level of Canadian equipment; the Submarine Operational Update Program has added Sperry Micropuffs passive ranging sonar, Singer Librascope fire-control and provision for at least four UGM-84A Sub-Harpoon submarine-launched anti-ship missiles; the main torpedo armament comprises Mk 37C anti-submarine weapons, being replaced in the SOUP modernization by Mk 48 Mod 4 wire-guided torpedoes.)

4. Chile

Name	No.	Builder	Laid down	Commissioned
O'Brien	22	Scott-Lithgow	Jan 1971	Apr 1976
Hyatt	23	Scott-Lithgow	Jan 1972	Sep 1976

(The Chilean boats are similar to the British standard, but carry West German SUT torpedoes.)

5. UK ('Oberon' class)

Name	No.	Builder	Laid down	Commissioned
Odin	S10	Cammell Laird	Apr 1959	May 1962
Orpheus	S11	Vickers	Apr 1959	Nov 1960
Olympus	S12	Vickers	Mar 1960	Jul 1962
Osiris	S13	Vickers	Jan 1962	Jan 1964
Onslaught	S14	Chatham DY	Apr 1959	Aug 1962
Otter	S15	Scotts SB	Jan 1960	Aug 1962
Oracle	S16	Cammell Laird	Apr 1960	Feb 1963
Ocelot	S17	Chatham DY	Nov 1960	Jan 1964
Otus	S18	Scotts SB	May 1961	Oct 1963
Opossum	S19	Cammell Laird	Dec 1961	Jun 1964
Opportune	S20	Scotts SB	Oct 1962	Dec 1964
Onyx	S21	Cammell Laird	Nov 1964	Nov 1967

6. UK ('Porpoise' class)

Name	No.	Builder	Laid down	Commissioned
Sealion	S07	Cammell Laird	Jun 1958	Jul 1961

Note
One of the most successful conventional submarine designs since World War II, the 'Oberon' class was developed from the 'Porpoise' class with improved soundproofing and other quietening features, and is still a capable though elderly coastal-waters type

'Porpoise' class SS (UK): see 'Oberon' class SS

'Potvis' class SS (Netherlands): see 'Dolfijn' class SS

'Romeo' class SS
(USSR)
Type: attack submarine
Displacement: 1,400 tons surfaced and 1,800 tons dived
Dimensions: length 76.8 m (251.9 ft); beam 7.3 m (23.9 ft); draught 5.5 m (18.0 ft)
Gun armament: none
Missile armament: none
Torpedo armament: eight 533-mm (21-in) tubes (six bow and two stern) for 18 dual-role torpedoes
Anti-submarine armament: torpedoes (see above)
Mines: up to 36 AMD-1000 mines instead of torpedoes
Electronics: one 'Snoop Plate' surface-search and navigation radar, one 'Herkules' sonar, one 'Feniks' sonar, one torpedo fire-control system, one 'Stop Light' ECM system, and communications and navigation systems
Propulsion: diesel-electric arrangement, with two diesels delivering 3000 kW (4,025 bhp) and two electric motors delivering 3000 kW (4,025 hp) to two shafts
Performance: maximum speed 17 kts surfaced and 14 kts dived; diving depth 300 m (985 ft) operational and 500 m (1,640 ft) maximum; range 29650 km (18,425 miles) at 10 kts surfaced; endurance 45 days
Complement: 54

Class
1. Algeria

2 boats
2. Bulgaria

Name	No.
Pobeda	11
Slava	12

3. China

94 boats

4. Egypt

10 boats

5. North Korea

15 boats
6. Syria

2 boats

7. USSR

5 boats

Note
Developed from the 'Whiskey' class design and placed in production at Gorki in 1958, the 'Romeo' class coincided with the development of new nuclear-powered types and was thus built only in limited numbers in the USSR, most production being undertaken by the nascent Chinese submarine industry

'S60' class (France): see 'Daphné' class

'S70' class (Spain): see 'Agosta' class

'Salta' class (Argentina): see 'Type 209/1' class

'Sauro' class SS
(Italy)
Type: attack submarine
Displacement: 1,455 tons surfaced and 1,630 tons dived
Dimensions: length 63.9 m (210.0 ft); beam 6.8 m (22.5 ft); draught 5.7 m (18.9 ft)
Gun armament: none
Missile armament: none
Torpedo armament: six 533-mm (21-in) tubes (all bow) for 12 A184 dual-role wire-guided torpedoes
Anti-submarine armament: torpedoes (see above)
Mines: up to 24 instead of torpedoes
Electronics: one BPS 704 surface-search and navigation radar, one IPD 70/S active/passive sonar, one MD100 Mk 1 passive ranging sonar, one SACTIS action-data system, one torpedo fire-control system, one ESM system, and extensive communications and navigation systems
Propulsion: diesel-electric arrangement, with three Fiat-built General Motors A210 16NM diesels delivering 2400 kW (3,220 bhp) and one electric motor delivering 2720 kW (3,650 hp) to one shaft
Performance: maximum speed 11 kts surfaced, 12 kts snorting, and 20 kts dived; diving depth 250 m (820 ft) operational and 410 m (1,345 ft) maximum; range 13000 km (6,080 miles) at surfaced cruising speed, 23150 km (14,385 miles) at 4 kts snorting, and 750 km (465 miles) at 4 kts or 38 km (24 miles) at 20 kts dived
Complement: 41

Class
1. Italy (**'Tipo 1081'** or **'Sauro Batch 1/2' class**)

Name	No.	Builder	Laid down	Commissioned
Nazario Sauro	S518	Italcantieri	Jul 1974	Mar 1980
Fecia di Cossato	S519	Italcantieri	Nov 1975	Mar 1980
Leonardo da Vinci	S520	Italcantieri	Jun 1978	Oct 1981
Guglielmo Marconi	S521	Italcantieri	Oct 1979	Sep 1982

2. Italy (**'Sauro Batch 3/4' class**)

Name	No.	Builder	Laid down	Commissioned
Salvatore Pelose	S522	Italcantieri		
Prini	S523	Italcantieri		
	S524	Italcantieri		
	S525	Italcantieri		

(Based on the original 'Sauro' class, this closely related subtype has a submerged displacement of 2,000 tons as well as enhanced weapons [A184 wire-guided torpedoes and UGM-84A Sub-Harpoon tube-launched anti-ship missiles] and sensors, the two boats of the 'Sauro Batch 3' subvariant being designed for the IPD 70/S fire-control system, and those of the 'Sauro Batch 4' subvariant for the SEPA Mk 3 fire-control system; the boats are powered by Fincantieri A210.16HM diesel generators, and the use of HY80 steel in the hull provides an operational diving depth of 300 m/985 ft.)

'Sava' class SS
(Yugoslavia)
Type: attack submarine
Displacement: 500 tons standard and 965 tons dived
Dimensions: length 65.8 m (215.8 ft); beam 7.0 m (22.9 ft); draught 5.5 m (18.0 ft)
Gun armament: none
Missile armament: none
Torpedo armament: six 533-mm (21-in) tubes (all bow) for 10 dual-role torpedoes
Anti-submarine armament: torpedoes (see above)
Mines: up to 20 instead of torpedoes
Electronics: one surface-search and navigation radar, Soviet active sonar, Krupp-Atlas PRS 3 passive sonar, one torpedo fire-control system, and communications and navigation systems
Propulsion: diesel-electric arrangement, with two diesels and two electric motors delivering 1800 kW (2,415 hp) to one shaft
Performance: maximum speed 16 kts dived; diving depth 300 m (985 ft) operational; endurance 28 days
Complement: 35

Class				
1. Yugoslavia				
Name	No.	Builder	Laid down	Commissioned
Sava	831	S & DE Factory	1975	1978
Drava	832	S & DE Factory		1979

Note
This neat local design uses a high proportion of imported equipment

'Sjöormen' class SS
(Sweden)
Type: attack submarine
Displacement: 1,125 tons surfaced and 1,400 tons dived
Dimensions: length 51.0 m (167.3 ft); beam 6.1 m (20.0 ft); draught 5.8 m (19.0 ft)
Gun armament: none
Missile armament: none
Torpedo armament: four 533-mm (21-in) tubes (all bow) for eight Tp 61 dual-role wire-guided torpedoes
Anti-submarine armament: two 400-mm (15.75-in) tubes (both bow) for four Tp 42 wire-guided torpedoes, and 533-mm (21-in) torpedoes (see above)
Mines: up to 16 instead of 533-mm (21-in) torpedoes
Electronics: one Terma surface-search and navigation radar, one Krupp-Atlas CSU 3 sonar, one Ericsson AI-FCS torpedo fire-control system, one action information system, one ESM system, and extensive communications and navigation systems
Propulsion: diesel-electric arrangement, with four Hedemora-Pielstick 12 PA4 diesels and one Asea electric motor delivering 1640 kW (2,200 hp) to one shaft
Performance: maximum speed 15 kts surfaced and 20 kts dived; diving depth 150 m (485 ft) operational and 250 m (820 ft) maximum; endurance 21 days
Complement: 18

Class				
1. Sweden				
Name	No.	Builder	Laid down	Commissioned
Sjöormen	Sor	Kockums	1965	Jul 1968
Sjölejonet	Sle	Kockums	1966	Dec 1968
Sjöhunden	Shu	Kockums	1966	Jun 1969
Sjöbjörnen	Sbj	Karlskrona	1967	Feb 1969
Sjöhasten	Shä	Karlskrona	1966	Sep 1969

Note
Specially designed for Sweden's coastal waters and carrying an armament of potent wire-guided torpedoes, the 'Sjöormen' class is notable for its quietness and underwater manoeuvrability

'Sutjeska' class SS
(Yugoslavia)
Type: attack submarine
Displacement: 820 tons surfaced and 945 tons dived
Dimensions: length 60.0 m (196.8 ft); beam 6.8 m (22.3 ft); draught 4.9 m (16.1 ft)
Gun armament: none
Missile armament: none
Torpedo armament: six 533-mm (21-in) tubes (four bow and two stern) for eight dual-role torpedoes
Anti-submarine armament: torpedoes (see above)
Mines: up to 12 mines instead of torpedoes
Electronics: one surface-search and navigation radar, Soviets sonars, one torpedo fire-control system, and communications and navigation systems
Propulsion: diesel-electric arrangement, with two Sulzer diesels and two electric motors delivering 1340 kW (1,795 hp) to two shafts
Performance: maximum speed 14 kts surfaced and 9 kts dived; range 8900 km (5,530 miles) at 8 kts surfaced
Complement: 38

Class				
1. Yugoslavia				
Name	No.	Builder	Laid down	Commissioned
Sutjeska	811	Uljanik SY	1957	Sep 1960
Neretva	812	Uljanik SY	1957	1962

Note
This Yugoslav design combines American engines with Soviet sensors and weapons

'Tango' class SS
(USSR)
Type: attack submarine
Displacement: 3,000 tons surfaced and 3,700 tons dived
Dimensions: length 92.0 m (301.8 ft); beam 9.0 m (29.5 ft); draught 7.0 m (23.0 ft)
Gun armament: none
Missile armament: none
Torpedo armament: eight 533-mm (21-in) tubes (six bow and two stern) for 18 dual-role torpedoes
Anti-submarine armament: two SS-N-15 missiles, and torpedoes (see above)
Mines: up to 36 AMD-1000 mines instead of torpedoes
Electronics: one 'Snoop Tray' surface-search and navigation radar, one low-frequency bow sonar, one medium-frequency underwater weapons fire-control sonar, one action information system, one 'Brick Pulp (Modified)' ESM system, and extensive communications and navigation systems
Propulsion: diesel-electric arrangement, with three diesels and three electric motors delivering 4500 kW (6,035 hp) to three shafts
Performance: maximum speed 15 kts surfaced and 16 kts dived; diving depth 300 m (985 ft) operational and 500 m (1,640 ft) maximum
Complement: 62

Class
1. USSR
20 boats

Note
This class was produced in small numbers at Gorki from 1972 as an interim type in the role of replacement for the long-range 'Foxtrot' boats serving with the Black Sea and Northern Fleets

'Toti' class SS
(Italy)
Type: attack submarine
Displacement: 525 tons surfaced and 585 tons dived
Dimensions: length 46.2 m (151.5 ft); beam 4.7 m (15.4 ft); draught 4.0 m (13.1 ft)
Gun armament: none
Missile armament: none
Torpedo armament: four 533-mm (21-in) tubes (all bow) for six A184 dual-role wire-guided torpedoes
Anti-submarine armament: torpedoes (see above)
Mines: up to 12 instead of torpedoes
Electronics: one 3RM 20/SMG surface-search and navigation radar, one IPD 44 active sonar, one MD 64 passive sonar, one Velox M5 passive ranging sonar, one torpedo fire-control system, one ESM system, and extensive communications and navigation systems
Propulsion: diesel-electric arrangement, with two Fiat MB 820 N/I diesels delivering 1640 kW (2,200 bhp) and one electric motor delivering 1640 kW (2,200 hp) to one shaft
Performance: maximum speed 14 kts surfaced and 15 kts dived; diving depth 180 m (590 ft) operational and 300 m (985 ft) maximum; range 5550 km (3,450 miles) at 5 kts surfaced
Complement: 4+22

Class
1. Italy

Name	No.	Builder	Laid down	Commissioned
Attilio Bagnolino	S505	Italcantieri	Apr 1965	Jun 1968
Enrico Toti	S506	Italcantieri	Apr 1965	Jan 1968
Enrico Dandolo	S513	Italcantieri	Mar 1967	Sep 1968
Lazzaro Mocenigo	S514	Italcantieri	Jun 1967	Jan 1969

Note
The boats were designed in the early 1960s as Italy's first indigenous submarines since World War II, and are optimized for anti-submarine warfare in the difficult conditions of the Mediterranean

'TR 1700' class SS
(West Germany/Argentina)
Type: attack submarine
Displacement: 2,100 tons surfaced and 2,300 tons dived
Dimensions: length 65.0 m (213.2 ft); beam 7.3 m (23.9 ft); draught 6.5 m (21.3 ft)
Gun armament: none
Missile armament: none
Torpedo armament: six 533-mm (21-in) tubes (all bow) for 22 SST4 dual-role wire-guided torpedoes
Anti-submarine armament: torpedoes (see above)
Mines: up to 44 instead of torpedoes
Electronics: one surface-search and navigation radar, active and passive sonars, one passive ranging sonar, one Sinbads torpedo fire-control system, one ESM system, and extensive communications and navigation systems
Propulsion: diesel-electric arrangement, with four diesel generators and one electric motor delivering 6600 kW (8,850 hp) to one shaft
Performance: maximum speed 13 kts surfaced, 15 kts

snorting and 25 kts dived; diving depth 300 m (985 ft) operational and 500 m (1,640 ft) maximum; range 27800 km (17,275 miles) at 5 kts surfaced, and 200 km (125 miles) at 15 kts or 37 km (23 miles) at 25 kts dived
Complement: 26

Class
1. Argentina

Name	No.	Builder	Laid down	Commissioned
Santa Cruz	S33	Thyssen	Dec 1980	Oct 1984
San Juan	S34	Thyssen	Mar 1982	Nov 1985
Santiago de Estero	S35	Tandanor	Oct 1983	
		Tandanor		
		Tandanor		
		Tandanor		

Note
This is an ocean patrol type with extensive torpedo reload capacity

'Type 205' class SS
(West Germany)
Type: attack submarine
Displacement: 420 tons surfaced and 450 tons dived
Dimensions: length 43.9 m (144.0 ft); beam 4.6 m (15.1 ft); draught 4.3 m (14.1 ft)
Gun armament: none
Missile armament: none
Torpedo armament: eight 533-mm (21-in) tubes (all bow) for eight SST4 dual-role wire-guided torpedoes
Anti-submarine armament: torpedoes (see above)
Mines: up to 16 instead of torpedoes
Electronics: one Calypso surface-search and navigation radar, one active sonar, one passive sonar, one WM-7 torpedo fire-control system, one ESM system, and extensive communications and navigation systems
Propulsion: diesel-electric arrangement, with two MTU 820Db diesels delivering 1570 kW (2,105 bhp) and one Siemens electric motor delivering 1125 kW (1,510 hp) to one shaft
Performance: maximum speed 10 kts surfaced and 17 kts dived; diving depth 150 m (490 ft) operational
Complement: 4+18

Class
1. West Germany

Name	No.	Builder	Laid down	Commissioned
U1	S180	Howaldtswerke	Feb 1965	Jun 1967
U2	S181	Howaldtswerke	Sep 1964	Oct 1966
U9	S188	Howaldtswerke	Dec 1964	Apr 1967
U10	S189	Howaldtswerke	Jul 1965	Nov 1967
U11	S190	Howaldtswerke	Apr 1966	Jun 1968
U12	S191	Howaldtswerke	Sep 1966	Jan 1969

Note
Designed for coastal operations and built in anti-magnetic steel, these boats now have limited operational utility and are used mainly for training

'Type 206' class SS
(West Germany)
Type: attack submarine
Displacement: 450 tons surfaced and 500 tons dived
Dimensions: length 48.6 m (159.4 ft); beam 4.6 m (15.1 ft); draught 4.5 m (14.8 ft)
Gun armament: none
Missile armament: none
Torpedo armament: eight 533-mm (21-in) tubes (all bow) for eight SST4 dual-role wire-guided torpedoes
Anti-submarine armament: torpedoes (see above)
Mines: up to 16 mines instead of torpedoes, plus another 24 in two external strap-on containers
Electronics: one Calypso surface-search and navigation radar, one active sonar, one passive panoramic sonar, one WM-8/8 torpedo fire-control system, one ESM system, and extensive communications and navigation systems
Propulsion: diesel-electric arrangement, with two MTU diesels delivering 900 kW (1,205 bhp) and one electric motor delivering 1350 kW (1,810 hp) to one shaft
Performance: maximum speed 10 kts surfaced and 17 kts dived; diving depth 150 m (490 ft) operational and 250 m (820 ft) maximum; range 8350 km (5,190 miles) at 5 kts surfaced
Complement: 4+18

Class
1. West Germany

Name	No.	Builder	Laid down	Commissioned
U 13	S192	Howaldtswerke	Nov 1969	Apr 1973
U 14	S193	Rheinstahl	Sep 1970	Apr 1973
U 15	S194	Howaldtswerke	May 1970	Apr 1974
U 16	S195	Rheinstahl	Apr 1970	Nov 1973
U 17	S196	Howaldtswerke	Oct 1970	Nov 1973
U 18	S197	Rheinstahl	Jul 1971	Dec 1973
U 19	S198	Howaldtswerke	Jan 1971	Nov 1973
U 20	S199	Rheinstahl	Feb 1972	May 1974
U 21	S170	Howaldtswerke	Apr 1971	Aug 1974
U 22	S171	Rheinstahl	May 1972	Jul 1974
U 23	S172	Howaldtswerke	Aug 1972	May 1975
U 24	S173	Rheinstahl	Jul 1972	Oct 1974
U 25	S174	Howaldtswerke	Oct 1971	Jun 1974
U 26	S175	Rheinstahl	Nov 1972	Mar 1975
U 27	S176	Howaldtswerke	Jan 1972	Oct 1974
U 28	S177	Rheinstahl	Jan 1972	Dec 1974
U 29	S178	Howaldtswerke	Feb 1972	Nov 1974
U 30	S179	Rheinstahl	Apr 1973	Mar 1975

(Twelve of these boats are to be upgraded to **'Type 206A'** class standard with the Krupp-Atlas SLW 83 combat information system [based on the CSU 83/DBQS-21D active sonar] and the DM2A3 or Seal 3 dual-role heavyweight torpedo version of the SST4.)

Note
These boats are intended solely for coastal operations

'Type 207' class SS

(West Germany/Norway)
Type: attack submarine
Displacement: 370 tons surfaced and 435 tons dived
Dimensions: length 45.4 m (149.0 ft); beam 4.6 m (15.1 ft); draught 4.3 m (14.1 ft)
Gun armament: none
Missile armament: none
Torpedo armament: eight 533-mm (21-in) tubes (all bow) for eight Tp 61 anti-ship wire-guided and Mk 37C anti-submarine dual-role wire-guided torpedoes
Anti-submarine armament: torpedoes (see above)
Mines: up to 16 mines instead of torpedoes
Electronics: one surface-search and navigation radar, active and passive sonars, one WM-8/10 (being replaced by MSI-70U) action information and torpedo fire-control system, one ESM system, and extensive communications and navigation systems
Propulsion: diesel-electric arrangement, with two MTU 820 diesels and one electric motor delivering 1275 kW (1,710 hp) to one shaft
Performance: maximum speed 10 kts surfaced and 17 kts dived
Complement: 5+13

Class
1. Norway

Name	No.	Builder	Laid down	Commissioned
Kinn	S316	Rheinstahl	1962	May 1965
Utsira	S301	Rheinstahl	1963	Jul 1965
Stadt	S307	Rheinstahl	1962	Sep 1965
*		Rheinstahl	1962	Dec 1965
*		Rheinstahl	1962	Feb 1966
Sklinna	S305	Rheinstahl	1963	May 1966
Skolpen	S306	Rheinstahl	1963	Aug 1966
*		Rheinstahl	1963	Nov 1966
Stord	S308	Rheinstahl	1964	Feb 1967
Svenner	S309	Rheinstahl	1965	Jun 1967
Kaura	S315	Rheinstahl	1961	Feb 1965
Kya	S317	Rheinstahl	1961	Jun 1964
Kobben	S318	Rheinstahl	1961	Aug 1964
Kunna	S319	Rheinstahl	1961	Oct 1964

*boats for Denmark

Note
This is a stronger-hulled development of the West German 'Type 205' class; Norway is modernizing six boats for service

into the 1990s, and three more are being upgraded for sale to Denmark as replacements for the obsolete 'Delfinen' class submarines; the boats for Denmark were *Utstein* (S5302), *Utvaer* (S303) and *Stadt* (S307), and the current *Kinn* and *Stadt* were *Ula* (S300) and *Utstein* (S302)

'Type 209/0' or 'Type 1100' class SS
(West Germany/Greece)
Type: attack submarine
Displacement: 990 tons surfaced and 1,290 tons dived
Dimensions: length 54.3 m (178.1 ft); beam 6.2 m (20.3 ft); draught 5.5 m (18.0 ft)
Gun armament: none
Missile armament: none
Torpedo armament: eight 533-mm (21-in) tubes (all bow) for 14 SST4 dual-role wire-guided torpedoes
Anti-submarine armament: torpedoes (see above)
Mines: up to 28 mines instead of torpedoes
Electronics: one surface-search and navigation radar, one Krupp-Atlas CSU 3-2 active sonar, one Krupp-Atlas PRS 3-4 passive sonar, one passive ranging sonar, one torpedo fire-control system, one ESM system, and extensive communications and navigation systems
Propulsion: diesel-electric arrangement, with four MTU diesels delivering 7040 kW (9,440 bhp) and one Siemens electric motor delivering 3700 kW (4,960 hp) to one shaft
Performance: maximum speed 10 kts surfaced and 22 kts dived; diving depth 300 m (985 ft) operational and 500 m (1,640 ft) maximum; endurance 50 days
Complement: 31

Class
1. Greece ('**Glavkos**' class)

Name	No.	Builder	Laid down	Commissioned
Glavkos	S110	Howaldtswerke	Sep 1968	Nov 1971
Nereus	S111	Howaldtswerke	Jan 1969	Feb 1972
Triton	S112	Howaldtswerke	Jun 1969	Nov 1972
Proteus	S113	Howaldtswerke	Oct 1969	Aug 1972

'Type 209/1' or 'Type 1200' class SS
(West Germany/Greece)
Type: attack submarine
Displacement: 1,185 tons surfaced and 1,285 tons dived
Dimensions: length 55.9 m (183.4 ft); beam 6.3 m (20.5 ft); draught 5.5 m (17.9 ft)
Gun armament: none
Missile armament: none
Torpedo armament: eight 533-mm (21-in) tubes (all bow) for 14 SST4 dual-role wire-guided torpedoes
Anti-submarine armament: torpedoes (see above)
Mines: up to 28 mines instead of torpedoes
Electronics: one surface-search and navigation radar, one Krupp-Atlas CSU 3-4 active/passive sonar, one torpedo fire-control system, one ESM system, and extensive communications and navigation systems
Propulsion: diesel-electric arrangement, with four MTU 12V493 TY60 diesels delivering 1800 kW (2,415 bhp) and one Siemens electric motor delivering 3725 kW (4,995 hp) to one shaft
Performance: maximum speed 10 kts surfaced and 22 kts dived; diving depth 300 m (985 ft) operational and 500 m (1,640 ft) maximum
Complement: 32

Class
1. Argentina ('**Salta**' class)

Name	No.	Builder	Laid down	Commissioned
Salta	S31	Howaldtswerke	Apr 1970	Mar 1974
San Luis	S32	Howaldtswerke	Oct 1970	May 1974

2. Colombia

Name	No.	Builder	Laid down	Commissioned
Pijao	SS28	Howaldtswerke	Apr 1972	Apr 1975
Tayrona	SS29	Howaldtswerke	May 1972	Jul 1975

(The Colombian boats are similar to the Greek units, but have separate Krupp-Atlas CSU 3-2 active and PRS 3-4 passive sonars, the WM-8/24 combat system and SUT torpedoes.)

3. Greece ('**Glavkos**' class)

Name	No.	Builder	Laid down	Commissioned
Posydon	S116	Howaldtswerke	Jan 1976	Mar 1979
Amphitrite	S117	Howaldtswerke	Apr 1976	Jul 1979
Okeanos	S118	Howaldtswerke	Oct 1976	Nov 1979
Pontos	S119	Howaldtswerke	Jan 1977	Apr 1980

4. Peru ('**Casma**' class)

Name	No.	Builder	Laid down	Commissioned
Casma	S31	Howaldtswerke	Jul 1977	Dec 1980
Antofagasta	S32	Howaldtswerke	Oct 1977	Feb 1981
Pisagua	S33	Howaldtswerke	Aug 1978	Jul 1983
Chipana	S34	Howaldtswerke	Nov 1978	Sep 1982
Islay	S45	Howaldtswerke	May 1971	Aug 1974
Arica	S46	Howaldtswerke	Nov 1971	Jan 1975

(Similar to the Greek boats, these Peruvian boats fall into two subclasses, S45 and S45 having the WM-8/24 combat system with separate CSU 3-2 active and PRS 3-4 passive sonars, and the four later boats having the Sinbads combat system and integrated CSU 3-4 active/passive sonar; the boats carry Mk 37C and SST-4 torpedoes, though a recent purchase has been the Italian A184 torpedo to be interfaced with the SEPA Mk 3 combat system, which can also handle the SST-4 torpedo.)

5. Turkey

Name	No.	Builder	Laid down	Commissioned
Atilay	S347	Howaldtswerke	Dec 1972	Jul 1975
Saldiray	S348	Howaldtswerke	Jan 1973	Jan 1977
Batiray	S349	Howaldtswerke	Jun 1975	Jul 1978
Yildiray	S350	Golcuk NY	May 1976	Jul 1981
Koganay	S351	Golcuk NY	Mar 1980	Aug 1984
Titiray	S352	Golcuk NY		

(Turkey plans to build 12 of this class, the last six to a slightly enlarged design; the combat system is either the WM-8/24 [S347 and S348] or Sinbads [other boats] interfaced with Krupp-Atlas CSU 3-4 active/passive sonar.)

'Type 209/2' or 'Type 1300' class SS
(West Germany/Ecuador)
Type:
Displacement: 1,285 tons surfaced and 1,390 tons dived
Dimensions: length 59.5 m (195.1 ft); beam 6.3 m (20.5 ft); beam 5.4 m (17.9 ft)
Gun armament: none
Missile armament: none
Torpedo armament: eight 533-mm (21-in) tubes (all bow) for 16 SUT dual-role wire-guided torpedoes
Anti-submarine armament: torpedoes (see above)
Mines: up to 32 mines instead of torpedoes

Electronics: one surface-search and navigation radar, one Krupp-Atlas CSU 3-4 active/passive sonar, one WM-8/24 torpedo fire-control system, one ESM system, and extensive communications and navigation systems
Propulsion: diesel-electric arrangement, with four MTU diesels and one electric motor delivering 3725 kW (4,995 hp) to one shaft
Performance: maximum speed 10 kts surfaced and 21.5 kts dived; diving depth 300 m (985 ft) operational and 500 m (1,640 ft) operational
Complement: 33

Class
1. Ecuador

Name	No.	Builder	Laid down	Commissioned
Shyri	S11	Howaldtswerke	Aug 1974	Nov 1977
Huancavilca	S12	Howaldtswerke	Jan 1975	Mar 1978

2. Indonesia

Name	No.	Builder	Laid down	Commissioned
Cakra	401	Howaldtswerke	Nov 1977	Mar 1981
Nanggala	402	Howaldtswerke	Mar 1978	Jul 1981
	403	Howaldtswerke		
	404	Howaldtswerke		

(The Indonesian boats have the Sinbads combat system with separate Krupp-Atlas CSU 3-2 active and PRS 3-4 passive sonars.)

3. Venezuela

Name	No.	Builder	Laid down	Commissioned
Sabalo	S31	Howaldtswerke	May 1973	Aug 1976
Caribe	S32	Howaldtswerke	Aug 1973	Mar 1977

(The Venezuelan boats fire the SST-4 heavyweight torpedo with the aid of the WM-8/24 combat system and associated Krupp-Atlas CSU 3-4 active/passive sonar.)

'Type 209/3' or 'Type 1400' class SS
(West Germany/Chile)
Type: attack submarine
Displacement: 1,260 tons surfaced and 1,390 tons dived
Dimensions: length 56.4 m (185.0 ft); beam 6.1 m (20.0 ft); draught 5.0 m (16.4 ft)
Gun armament: none
Missile armament: none
Torpedo armament: eight 533-mm (21-in) tubes (all bow) for 16 SST4 dual-role wire-guided torpedoes
Anti-submarine armament: torpedoes (see above)
Mines: up to 32 instead of torpedoes
Electronics: one surface-search and navigation radar, Krupp-Atlas CSU 3-4 active/passive sonar, one underwater weapons fire-control system, and one Porpoise ESM system
Propulsion: diesel-electric arrangement, with four MTU-Siemens diesel generators and one Siemens electric motor delivering ? kW (? hp) to one shaft
Performance: speed 12 kts surfaced and 20 kts dived
Complement: 36

Class
1. Brazil

Name	No.	Builder	Laid down	Commissioned
Tupi		Howaldtswerke	Mar 1985	
Timbiri		Ast Ilhas	Aug 1985	
Tamoio		Ast Ilhas		

(These Brazilian boats are close to the Chilean units in displacement, dimensions, propulsion and performance, but the armament fit is optimized for the Mk 24 Mod 1 Tigerfish wire-guided torpedo with Ferranti KAFS fire-control; the boats will also have the Marconi Manta ESM system.)

2. Chile

Name	No.	Builder	Laid down	Commissioned
Thompson	20	Howaldtswerke	Nov 1980	Sep 1984
Simpson	21	Howaldtswerke	Feb 1982	Jan 1985

'Type 209/4' or 'Type 1500' class SS
(West Germany/India)
Type: attack submarine
Displacement: 1,660 tons surfaced and 1,850 tons dived
Dimensions: length 64.4 m (211.3 ft); beam 6.5 m (21.3 ft); draught 6.0 m (19.7 ft)
Gun armament: none
Missile armament: none
Torpedo armament: eight 533-mm (21-in) tubes (all bow) for 14 dual-role torpedoes
Anti-submarine armament: torpedoes (see above)
Electronics: one surface-search and navigation radar, sonar, fire-control and electronic support measures systems, and extensive communications and navigation systems
Propulsion: diesel-electric arrangement, with four MTU-Siemens diesel generators and one Siemens electric motor delivering ? kW (? hp) to one shaft
Performance: speed 11 kts surfaced and 22 kts dived; range 14825 km (9,210 miles) surfaced
Complement: 40

Class
1. India

Name	No.	Builder	Laid down	Commissioned
Shishumar		Howaldtswerke	1982	1986
		Howaldtswerke	Dec 1982	
		Mazagon	Jun 1984	
		Mazagon		
		Mazagon		
		Mazagon		

(The 'Type 1500' is derived closely from the basic 'Type 209' series, but has an integral rescue system.)

Note
The 'Type 209' class was designed for the export market as an ocean-going type, but its short length and handiness have also attracted customers with a coastal requirement; the type's versatility is attested by its development in five subvariants to meet a range of customer requirements

'Type 210' or 'Ula' class SS
(West Germany/Norway)
Type: attack submarine
Displacement: 940 tons standard and 1,300 tons dived
Dimensions: length 59.0 m (193.6 ft); beam 5.4 m (17.7 ft); draught 4.6 m (15.1 ft)
Gun armament: none
Missile armament: none

Torpedo armament: eight 533-mm (21-in) tubes (all bow) for 14 DM2A3 dual-role wire-guided torpedoes
Anti-submarine armament: torpedoes (see above)
Electronics: one surface-search and navigation radar, Krupp-Atlas CSU-83 (DBQS-21F) sonar, Kongsberg MSI-70U fire-control system, one electronic support measures system, and extensive communications and navigation systems
Propulsion: diesel-electric arrangement, with two MTU 16V SB83 diesels and one Siemens electric motor delivering 1880 kW (2,520 hp) to one shaft
Performance: speed 11 kts surfaced and 23 kts dived
Complement: 18

Class
1. Norway

Name	No.	Builder	Laid down	Commissioned
Ula	S300	Thyssen Nordseewerke	Jan 1987	
Utsira	S301	Thyssen Nordseewerke		
Utstein	S302	Thyssen Nordseewerke		
Utvaer	S303	Thyssen Nordseewerke		
Uthaung	S304	Thyssen Nordseewerke		
Uredd	S305	Thyssen Nordseewerke		

Note
This is a highly capable coastal class designed to supplant the 'Type 207' class

'Type 540' class SS
(West Germany/Israel)
Type: attack submarine
Displacement: 420 tons surfaced and 600 tons dived
Dimensions: length 45.0 m (146.7 ft); beam 4.7 m (15.4 ft); draught 3.7 m (12.0 ft)
Gun armament: none
Missile armament: possibly one SLAM launcher for Blowpipe surface-to-air missiles
Torpedo armament: eight 533-mm (21-in) tubes (all bow) for 10 Mk 37C and NT-37E dual-role wire-guided torpedoes
Anti-submarine armament: torpedoes (see above)
Mines: up to 20 mines instead of torpedoes
Electronics: one surface-search and navigation radar, active and passive sonars, one torpedo fire-control system, one ESM system, and extensive communications and navigation systems
Propulsion: diesel-electric arrangement, with two MTU 12V493 TY60 diesels and one electric motor delivering 1350 kW (1,810 hp) to one shaft
Performance: maximum speed 11 kts surfaced and 17 kts dived; diving depth 300 m (985 ft) operational and 500 m (1,640 ft) maximum
Complement: 22

Class
1. Israel

Name	No.	Builder	Laid down	Commissioned
Gal		Vickers	1973	Jan 1977
Tanin		Vickers	1974	1977
Rahav		Vickers	1974	Dec 1977

Note
These boats are basically similar to the West German 'Type 206' class, but built in the UK to avoid political complications

'Type 1100' class SS (West Germany): see 'Type 209/0' class

'Type 1200' class SS (West Germany): see 'Type 209/1' class

'Type 1300' class SS (West Germany): see 'Type 209/2' class

'Type 1400' class SS (West Germany): see 'Type 209/3' class

'Type 1500' class SS (West Germany): see Type 209/4' class

'Type 2400' class (UK): see 'Upholder' class

'Ula' class SS (Norway): see 'Type 219' class SS

'Upholder' class SS
(UK)
Type: attack submarine
Displacement: 2,160 tons surfaced and 2,400 tons dived
Dimensions: length 230.6 ft (70.3 m); beam 25.0 ft (7.6 m); draught 17.7 ft (5.5 m)
Gun armament: none
Missile armament: at least four UGM-84A Sub-Harpoon tube-launched anti-ship missiles

Torpedo armament: six 21-in (533-mm) tubes (all bow) for 18 Mk 24 Tigerfish and Spearfish dual-role wire-guided torpedoes
Anti-submarine armament: torpedoes (see above)
Mines: up to 44 Mk 5 Stonefish and Mk 6 Sea Urchin mines instead of torpedoes
Electronics: one Type 1006 surface-search and navigation radar, one Type 2040 passive sonar, one Type 2019 active/passive range and intercept sonar, one Type 2026 towed-array sonar, one DCC underwater weapons fire-control system, one Racal ESM system, one Lion action information system, and extensive communications and navigation systems
Propulsion: diesel-electric arrangement, with two Paxman Valenta 16 RPA 200S diesels and one GEC electric motor delivering 5,400 hp (4030 kW) to one shaft
Performance: maximum speed 12 kts surfaced, 10+ kts snorting and 20 kts dived; diving depth 985 ft (300 m) operational and 1,640 ft (500 m) maximum; range 9,200+ miles (14805+ km) at surfaced cruising speed
Complement: 7+13+24

Class
1. UK

Name	No.	Builder	Laid down	Commissioned
Upholder	S40	Vickers		
Unseen	S41	Cammell Laird		
Ursula	S42	Cammell Laird		
Unicorn	S43			
and eight more planned				

Note
This class is designed to replace the 'Oberon' class, and great emphasis has been placed on automation and standardization; features are very low radiated noise, an automated weapon-handling system, and a quick recharge facility for the batteries

'Uzushio' class SS
(Japan)
Type: attack submarine
Displacement: 1,900 tons surfaced and 2,430 tons dived
Dimensions: length 72.0 m (236.2 ft); beam 9.9 m (32.5 ft); draught 7.5 m (24.6 ft)
Gun armament: none
Missile armament: none
Torpedo armament: six 533-mm (21-in) tubes (all amidships) for ? Mk 48 dual-role wire-guided torpedoes
Anti-submarine armament: torpedoes (see above)
Mines: can be carried instead of torpedoes
Electronics: one surface-search and navigation radar, various active and passive sonar systems, one torpedo fire-control system, one ESM system, and extensive communications and navigation systems
Propulsion: diesel-electric arrangement, with two Kawasaki/MAN V8V24/30 diesels delivering 2550 kW (3,420 bhp) and one electric motor delivering 5375 kW (7,210 hp) to one shaft
Performance: maximum speed 12 kts surfaced and 20 kts dived; diving depth 200 m (655 ft) operational
Complement: 10+70

Class
1. Japan

Name	No.	Builder	Laid down	Commissioned
Uzushio	SS566	Kawasaki	Sep 1968	Jan 1971
Makishio	SS567	Mitsubishi	Jun 1969	Feb 1972
Isoshio	SS568	Kawasaki	Jul 1970	Nov 1972
Narushio	SS569	Mitsubishi	May 1971	Sep 1973
Kuroshio	SS570	Kawasaki	Jul 1972	Nov 1974
Takashio	SS571	Mitsubishi	Jul 1973	Jan 1976
Yaeshio	SS572	Kawasaki	Apr 1975	Mar 1978

Note
Designed for operations in Japan's coastal waters, the 'Uzushio' class has a teardrop hull for high speed and quiet running

'Västergötland' class SS
(Sweden)
Type: attack submarine
Displacement: 1,070 tons surfaced and 1,140 tons dived
Dimensions: length 48.5 m (159.1 ft); beam 6.1 m (20.0 ft); draught 6.1 m (20.0 m)
Gun armament: none
Missile armament: none
Torpedo armament: four 533-mm (21-in) tubes (all bow) for ? Tp 61 dual-role wire-guided torpedoes
Anti-submarine armament: three 400-mm (15.75-in)

tubes (all stern) for ? Tp 42 wire-guided torpedoes, and 533-mm (21-in) torpedoes (see above)
Mines: up to 16 mines instead of 533-mm (21-in) torpedoes
Electronics: one Terma surface-search and navigation radar, one Krupp-Atlas CSU 83 bow sonar, one torpedo fire-control system, one action information system, one ESM system, and extensive communications and navigation systems
Propulsion: diesel-electric arrangement, with two Hedemora VRA/1546 diesels and one Jeumont-Schneider electric motor delivering 1610 kW (2,160 hp) to one shaft
Performance: maximum speed 17 kts surfaced and 20 kts dived; diving depth 300 m (985 ft) operational and 500 m (1,640 ft) maximum
Complement: 21

Class
1. Sweden

Name	No.	Builder	Laid down	Commissioned
Västërgötland	Vgd	Karlskrona	1983	
Halsingland	Hgd	Karlskrona	Jan 1984	
Södermanland	Sod	Karlskrona		
Östergötland	Ogd	Karlskrona		

Note
This is a very advanced design based on the 'Näcken' class with better armament and the latest in integrated sonar/fire-control systems

'Whiskey' class SS
(USSR)
Type: attack submarine
Displacement: 1,080 tons surfaced and 1,350 tons dived
Dimensions: length 76.0 m (249.3 ft); beam 6.5 m (21.3 ft); draught 4.9 m (16.1 ft)
Gun armament: none
Missile armament: none
Torpedo armament: four 533-mm (21-in) tubes (all bow) for 14 dual-role torpedoes
Anti-submarine armament: two 406-mm (16-in) tubes (both stern) for two torpedoes, and 533-mm (21-in) torpedoes (see above)
Mines: up to 28 AMD-1000 mines instead of 533-mm (21-in) torpedoes
Electronics: one 'Snoop Plate' surface-search and navigation radar, one 'Tamir' sonar, one torpedo fire-control system, one 'Stop Light' ECM system, and communications and navigation systems
Propulsion: diesel-electric arrangement, with two diesels delivering 2985 kW (4,000 bhp) and two electric motors delivering 2000 kW (2,680 hp) to two shafts
Performance: maximum speed 18 kts surfaced and 14 kts dived; diving depth 300 m (985 ft) operational and 500 m (1,640 ft) maximum; range 24000 km (14,915 miles) at 8 kts surfaced
Complement: 54

Class
1. Albania

3 boats

2. China

16 boats

3. Cuba

1 boat

4. Egypt

6 boats

5. North Korea

4 boats

6. Poland

4 boats

7. USSR

50 'Whiskey' class boats

Note
Built in large numbers at a number of yards between 1950 and 1957, these elderly boats are notable for their great strength and first-class range/endurance, but are now being phased out of front-line service by most operators

'Yuushio' class SS
(Japan)
Type: attack submarine
Displacement: 2,200 tons surfaced and 2,730 tons dived
Dimensions: length 76.0 m (249.3 ft); beam 9.9 m (32.5 ft); draught 7.5 m (24.6 ft)
Gun armament: none
Missile armament: at least four UGM-84A Sub-Harpoon tube-launched anti-ship missiles (SS577 onwards)

Torpedo armament: six 533-mm (21-in) tubes (all amidships) for 18 Mk 37C anti-submarine and Mk 48 dual-role wire-guided torpedoes
Anti-submarine armament: torpedoes (see above)
Mines: up to 44 mines instead of torpedoes
Electronics: one ZPS-4 surface-search and navigation radar, one ZQQ-4 (BQS-4) bow sonar, one SQS-36 (J) fin sonar, one underwater weapons fire-control system, one action information system, one ESM system, and communications and navigation systems
Propulsion: diesel-electric arrangement, with two Kawasaki/MAN diesels delivering 2900 kW (3,890 bhp) and one electric motor delivering 5375 kW (7,210 hp) to one shaft
Performance: maximum speed 12 kts surfaced and 20 kts dived; diving depth 300 m (985 ft) operational and 500 m (1,640 ft) maximum
Complement: 75

Class
1. Japan

Name	No.	Builder	Laid down	Commissioned
Yuushio	SS573	Mitsubishi	Dec 1976	Feb 1980
Mochishio	SS574	Kawasaki	May 1978	Mar 1981
Setoshio	SS575	Mitsubishi	Apr 1979	Mar 1982
Okishio	SS576	Kawasaki	Apr 1980	Mar 1983
Nadashio	SS577	Mitsubishi	Apr 1981	Mar 1984
Hamashio	SS578	Kawasaki	Apr 1982	1985
Akishio	SS579	Mitsubishi	Apr 1983	
Takeshio	SS580	Kawasaki	Apr 1984	
	SS581	Mitsubishi	Apr 1985	
	SS582	Kawasaki	Apr 1986	

Note
This is an improved 'Uzushio' class design with greater diving depth; an upgraded **'Yuushio (Improved)' class** is also in its early stages, this slightly larger 2,400-ton type having later electronics and provision for UGM-84 Sub-Harpoon anti-ship missiles

'Zeeuleeuw' class SS
(Netherlands)
Type: attack submarine
Displacement: 2,450 tons surfaced and 2,800 tons dived
Dimensions: length 67.7 m (222.1 ft); beam 8.4 m (27.6 ft); draught 7.0 m (23.0 ft)
Gun armament: none
Missile armament: at least four UGM-84A Sub-Harpoon tube-launched anti-ship missiles
Torpedo armament: six 533-mm (21-in) tubes (all bow) for 20 Mk 48 dual-role wire-guided torpedoes
Anti-submarine armament: torpedoes (see above)
Mines: up to 40 mines instead of torpedoes
Electronics: one Type 1001 surface-search and navigation radar, one Octopus active/passive sonar, one Type 2026 towed-array sonar, one Gipsy III underwater weapons fire-control system, one SEWACO VIII action information system, one ESM system, and extensive communications and navigation systems
Propulsion: diesel-electric arrangement, with three SEMT-Pielstick 12 PA4-V200 diesels delivering 5150 kW (6,905 hp) and one Holec electric motor delivering 4100 kW (5,500 hp) to one shaft
Performance: maximum speed 12 kts surfaced and 20 kts dived; diving depth 450 m (1,475 ft) operational and 650 m (2,135 ft) maximum; range 16000 km (9,940 miles) at 9 kts snorting
Complement: 49

Class
1. Netherlands

Name	No.	Builder	Laid down	Commissioned
Walrus	S802	Rotterdamse DD	Oct 1979	
Zeeleeuw	S803	Rotterdamse DD	Sep 1981	1987
Dolfijn	S804	Rotterdamse DD	Jun 1986	
Bruinvis	S805	Rotterdamse DD		
	S806	Rotterdamse DD		
	S807	Rotterdamse DD		

Note
Designated the 'Walrus' class until the disastrous fire on the lead boat in 1986, this design is based on the 'Zwaardvis' class, but has greater diving depth and manoeuvrability combined with later electronics and a smaller crew; the last results from a high degree of automation

'Zwaardvis' class SS
(Netherlands)
Type: attack submarine
Displacement: 2,350 tons surfaced and 2,640 tons dived
Dimensions: length 66.2 m (217.2 ft); beam 10.3 m (33.8 ft); draught 7.1 m (23.3 ft)
Gun armament: none
Missile armament: at least four UGM-84A Sub-Harpoon tube-launched anti-ship missiles
Torpedo armament: six 533-mm (231-in) tubes (all bow) for 20 Mk 37C anti-submarine and Mk 48 dual-role wire-guided torpedoes
Anti-submarine armament: torpedoes (see above)
Mines: up to 40 mines instead of torpedoes
Electronics: one Type 1001 surface-search and navigation radar, one Krupp-Atlas low-frequency active bow sonar, one medium-frequency sonar, one WM-8/7 underwater weapons fire-control system, one SEWACO VIII action information system, one ESM system, and extensive communications and navigation systems
Propulsion: diesel-electric arrangement, with three diesels delivering 3150 kW (4,225 bhp) and one electric motor delivering 3800 kW (5,100 hp) to one shaft
Performance: maximum speed 13 kts surfaced and 20 kts dived; diving depth 300 m (985 ft) operational and 500 m (1,640 ft) maximum
Complement: 8+60

Class
1. Netherlands

Name	No.	Builder	Laid down	Commissioned
Zwaardvis	S806	Rotterdamse DD	Jul 1966	Aug 1972
Tijgerhaai	S807	Rotterdamse DD	Jul 1966	Oct 1972

2. Taiwan ('**Zwaardvis Mk 2**' or '**Hai Lung' class**)
2 boats on order
(These Taiwanese boats have a submerged displacement of 2,650 tons and a diesel-electric propulsion system with Bronswerk O-RUB 215 diesels, and the armament is thought to comprise Mk 37C torpedoes used in conjunction with the Sinbads-M fire-control system and associated SIASS-Z sonar.)

Note
This useful design was based on the US 'Barbel' class, but with additional noice-reduction features such as a sprung false deck for the machinery

AIRCRAFT-CARRIERS

'Clemenceau' class CV
(France)
Type: multi-role aircraft-carrier
Displacement: 27,310 tons standard and 32,780 tons full load
Dimensions: length 265.0 m (869.4 ft); beam 31.72 m (104.1 ft); draught 8.6 m (28.2 ft); flightdeck length 257.0 m (843.2 ft) and width 29.5 m (96.8 ft)
Gun armament: eight 100-mm (3.9-in) L/55 DP in single mountings
Missile armament: two Naval Crotale launchers for ? Matra R.440 SAMs
Torpedo armament: none
Anti-submarine armament: aircraft and helicopters (see below)
Aircraft: 40 fixed-wing (20 Dassault-Breguet Super Etendards, 10 Vought F-8 Crusaders and 10 Dassault-Breguet Alizés) and four rotary-wing (two Aérospatiale Super Frelons and two Aérospatiale Alouette IIIs)
Armour: flightdeck, hull sides and bulkheads (magazine and engine room areas) and island
Electronics: one DRBV 50 surveillance radar, one DRBV 23B air-search radar, one DRBV 20C air-warning radar, two DRBI 10 height-finging radars, two DRBC 31 fire-control radars, two DRBC 32 fire-control radars, one Decca 1226 navigation radar, one NRBA landing radar, one SQS-503 sonar, one SENIT 2 action information system, and several ESM systems
Propulsion: six boilers supplying steam to two sets of Parsons geared turbines delivering 126,000 shp (93960 kW) to two shafts
Performance: maximum speed 32 kts; range 13900 km (8,635 miles) at 18 kts or 6500 km (4,040 miles) at 32 kts
Complement: 64+476+798

Class
1. France

Name	No.	Builder	Laid down	Commissioned
Clemenceau	R98	Brest ND	Nov 1955	Nov 1961
Foch	R99	Ch. Atlantique	Feb 1957	Jul 1963

Note
France is determined to maintain a multi-role aircraft-carrier strength, and these two ships are being kept up to date and fully operational pending the introduction of France's two planned nuclear-powered aircraft-carriers in the 1990s, allowing the *Clemenceau* be paid off in 1992 and her sister in 1998 if all goes well; the ships each have two elevators and two catapults

'Colossus' class CVLS
(UK/Brazil)
Type: light anti-submarine aircraft-carrier
Displacement: 15,890 tons standard, 17,500 tons normal and 19,890 tons full load
Dimensions: length 695.0 ft (211.8 m); beam 80.0 ft (24.4 m); draught 24.5 ft (7.5 m); flightdeck length 690.0 ft (210.3 m) and width 121.0 ft (37.0 m)
Gun armament: 10 40-mm Bofors L/60 AA in two Mk 2 quadruple and one Mk 1 twin mounting
Missile armament: none
Anti-submarine armament: aircraft and helicopters (see below)
Aircraft: generally seven fixed-wing (Grumman S-2E Trackers) and eight rotary-wing (Sikorsky SH-3D Sea Kings), though a complement of helicopters (three types) is sometimes carried
Electronics: one SPS-40B air-search radar, one SPS-4 surface-search radar, two SPS-8B combat-control radars, one Raytheon 1402 navigation radar, two SPG-34 40-mm fire-control radars, two Mk 63 gun fire-control systems, and one Mk 51 gun fire-control system
Propulsion: four boilers supplying steam to two sets of Parsons geared turbines delivering 40,000 shp (29830 kW) to two shafts
Performance: maximum speed 24 kts; range 13,800 miles (22210 km) at 14 kts or 7,140 miles (11490 km) at 23 kts
Complement: 1,000 plus an air group of about 300

Class
1. Argentina

Name	*No.*	*Builder*	*Laid down*	*Commissioned*
Veinticinco de Mayo	V2	Cammell Laird	Dec 1942	Jan 1945

(This Argentine ship differs considerably in appearance and equipment from the Brazilian aircraft-carrier, and normally operates with 18 fixed-wing [Dassault-Breguet Super Etendard, McDonnell Douglas A-4 Skyhawk and Grumman S-2 Tracker] and four rotary-wing [Sikorsky SH-3 Sea King and Aérospatiale Alouette III] aircraft; the electronic fit comprises LW-01 and LW-02 surveillance radars, one DA-02 surface-search radar, one ZW-01 surface-warning and navigation radar, and one CAAIS action information system; and the armament comprises nine 40-mm Bofors L/70 AA in single mountings.)

2. Brazil

Name	*No.*	*Builder*	*Laid down*	*Commissioned*
Minas Gerais	A11	Swan Hunter	Nov 1942	Jan 1945

Note
Though the two hulls were built as sisters, their development in South American hands has caused a divergence, the Argentine ship being a small multi-role type and the Brazilian unit an anti-submarine type with different aircraft and sensors

'Enterprise' class CVN
(USA)
Type: nuclear-powered multi-role aircraft-carrier
Displacement: 75,700 tons standard and 89,600 tons full load
Dimensions: length 1,123 ft (342.5 m); beam 133.0 ft (40.5 m); draught 35.8 ft (10.9 m); flightdeck length 1,123.0 ft (342.5 m) and width 248.3 ft (75.7 m)
Gun armament: three 20-mm Phalanx Mk 16 CIWS mountings
Missile armament: three Mk 29 octuple launchers for 24 RIM-7 NATO Sea Sparrow SAMs
Torpedo armament: none
Anti-submarine armament: aircraft and helicopters (see below)
Aircraft: about 90 in a multi-role carrier air wing
Electronics: one SPS-48C 3D radar, one SPS-49 air-search radar, one SPS-10 surface-search radar, one SPS-65 threat-warning radar, one SPS-58 low-level threat-warning radar, three SPN-series aircraft landing radars, three Mk 91 SAM fire-control radars, Naval Tactical Data System, one SLQ-29(V)3 ESM system, URN-26 TACAN, four Mk 36 Super RBOC launchers, and one OE-82 satellite communications system
Propulsion: eight Westinghouse A2W pressurized water-cooled reactors supplying steam to four sets of Westinghouse geared turbines delivering 280,000 shp (208795 kW) to four shafts
Performance: maximum speed 35 kts; range 460,000 miles (741245 km) at 20 kts
Complement: 165+2,993 plus an air group of about 300+2,325

Class				
1. USA				
Name	*No.*	*Builder*	*Laid down*	*Commissioned*
Enterprise	CVN65	Newport News	Feb 1958	Nov 1961

Note
This was the world's first nuclear-powered aircraft-carrier, and is in essence a 'Forrestal' class carrier enlarged and modified for the massive nuclear powerplant; as with other advanced US carriers she has four aircraft elevators and four catapults, and has been extensively upgraded in electronics and aircraft to serve into the 21st century

'Forrestal' class CV
(USA)

Type: multi-role aircraft-carrier
Displacement: 59,650 tons (CV59) or 60,000 tons (others) standard and 78,000 tons full load
Dimensions: length (CV59/61) 1,039.0 ft (316.7 m) or (CV62) 1,046.5 ft (319.0 m); beam 129.5 ft (39.5 m); draught 37.0 ft (11.3 m); flightdeck length (CV59/61) 1,039.0 ft (316.7 m) or (CV62) 1,046.5 ft (319.0 m), and width (CV59/61) 238.0 ft (72.5 m) or (CV62) 238.5 ft (72.7 m)
Gun armament: three 20-mm Phalanx Mk 16 CIWS mountings (CV60 only)
Missile armament: two Mk 25 (CV59/60) or two Mk 29 (CV61/62) octuple launchers for 16 RIM-7 NATO Sea Sparrow SAMs, to be replaced in all ships by three Mk 29 launchers with 24 SAMs
Anti-submarine armament: aircraft and helicopters (see below)
Aircraft: about 90 in a multi-role carrier air wing
Electronics: one SPS-48 3D radar, one SPS-49 air-search radar or (CV61) one SPS-43A air-search radar, one SPS-10 surface-search radar, one SPS-58 low-level threat-warning radar, one LN-66 navigation radar, two Mk 115 (CV59/60) or Mk 91 (CV61/62) SAM fire-control radars (to be modified to three as the triple Mk 29 launcher installation is completed), Naval Tactical Data System, one SLQ-29(V)3 ESM system, three Mk 36 Super RBOC launchers, URN-20 TACAN, and one OE-82 satellite communications system
Propulsion: eight boilers supplying steam to four sets of Westinghouse geared turbines delivering 280,000 shp (208795 kW) to four shafts
Performance: maximum speed 33 kts (CV59) or 34 kts (CV60/62); range 13,800 miles (22210 km) at 20 kts or 4,600 miles (7400 km) at 30 kts
Complement: about 145+2,800 plus an air group of about 300+2,200

Class				
1. USA (4)				
Name	*No.*	*Builder*	*Laid down*	*Commissioned*
Forrestal	CV59	Newport News	Jul 1952	Oct 1955
Saratoga	CV60	New York NY	Dec 1952	Apr 1956
Ranger	CV61	Newport News	Aug 1954	Aug 1957
Independence	CV62	New York NY	Jul 1955	Jan 1959

Note
These were the first aircraft-carriers to be designed for jet aircraft, and are much-upgraded and capable ships with four elevators and four catapults; they are slated to remain in service up to 2000 or slightly later

'Garibaldi' class CVLS
(Italy)

Type: light anti-submarine aircraft-carrier
Displacement: 10,100 tons standard and 13,370 tons full load
Dimensions: length 180.0 m (590.4 ft); beam 30.4 m (99.7 ft); draught 6.7 m (22.0 ft); flightdeck length 174.0 m (570.2 ft) and width 30.4 m (99.7 ft)
Gun armament: six 40-mm Breda L/70 AA in three Dardo twin mountings
Missile armament: two twin Teseo container-launchers for 10 Otomat Mk 2 anti-ship missiles, and two Albatros octuple launchers for 16 Aspide SAMs
Anti-submarine armament: two triple ILAS 3 tube mountings for 324-mm (12.75-in) A 244S or 12.75-in (324-mm) Mk 46 torpedoes, and helicopter-launched weapons (see below)
Aircraft: up to 18 Agusta-Sikorsky ASH-3 Sea King helicopters
Electronics: one RAN 10S air-search radar, one SPS 702 combined air- and surface-search radar, three RTN 20X radars used in conjunction with the Dardo fire-control system, two RTN 30X radars used in conjunction with two Argo NA30 Albatros fire-control systems, one SPN 703 navigation radar, one DE 1160 hull-mounted sonar, several ESM systems, TACAN, two Breda SCLAR-D chaff launchers, and one IPN 10 action information system
Propulsion: COGAG (COmbined Gas turbine And Gas turbine) arrangement, with four General Elec-

tric/Fiat LM2500 gas turbines delivering 59655 kW (80,000 shp) to two shafts
Performance: maximum speed 30 kts; range 13000 km (8,080 miles) at 20 kts
Complement: 550 plus an air group of 230

Class
1. Italy

Name	No.	Builder	Laid down	Commissioned
Giuseppe Garibaldi	C551	Italcantieri	Jun 1981	Jul 1985

Note
Similar in basic concept to the British 'Invincible' class, the singleton 'Garibaldi' class light carrier may be equipped also with STOVL Sea Harrier aircraft, and has two aircraft lifts

'Hermes' class CV
(UK/India)
Type: aircraft-carrier
Displacement: 23,900 tons standard and 28,700 tons full load
Dimensions: length 744.3 ft (226.9 m); beam 90.0 ft (27.4 m); draught 28.5 ft (8.7 m); flightdeck length 744.3 ft (226.9 m) and width 160.0 ft (48.8 m)
Gun armament: none
Missile armament: two quadruple launchers for about 40 Sea Cat SAMs
Torpedo armament: none
Anti-submarine armament: helicopter-launched weapons (see below)
Aircraft: normally six Sea Harrier FRS.Mk 51 STOVL fighters and nine Sea King Mk 42 helicopters
Armour: belt 1/2 in (25.4/50.8 mm) over machinery spaces and magazines; deck 0.75 in (19 mm)
Electronics: one Type 965 air-search radar, one Type 993 surface-search radar, one Type 1006 navigation radar, two GWS 22 SAM-control systems, one TACAN, one Type 184 hull-mounted sonar, several active/passive ECM systems, and two Corvus chaff launchers
Propulsion: four boilers supplying steam to two sets of Parsons geared turbines delivering 76,000 shp (56675 kW) to two shafts
Performance: maximum speed 28 kts
Complement: 1,350 including air group

Class
1. India

Name	No.	Builder	Laid down	Commissioned
Virat		Vickers	June 1944	Nov 1959

Note
(Originally the British *Hermes*, this ship was sold to India in 1985 and is fitted with a 7.5-degree 'ski jump' for the launching of Sea Harrier aircraft.)

'Independence' class CVL
(USA/Spain)
Type: light aircraft-carrier
Displacement: 13,000 tons standard and 16,415 tons full load
Dimensions: length 623.0 ft (189.9 m); beam 71.5 ft (21.8 m); draught 26.0 ft (7.9 m); flightdeck length 545.0 ft (166.1 m) and width 108.0 ft (32.9 m)
Gun armament: 26 40-mm Bofors L/60 AA in two Mk 2 quadruple and nine Mk 1 twin mountings
Missile armament: none
Torpedo armament: none
Anti-submarine armament: none
Aircraft: seven fixed-wing and 20 rotary-wing
Electronics: one SPS-6 air-search radar, one SPS-40 air-search radar, one SPS-10 surface-search radar, one SPS-8 height-finding radar, two navigation radars, two Mk 29 radars used in conjunction with two Mk 57 gun fire-control systems, two Mk 28 radars used in conjunction with two Mk 63 gun fire-control systems, URN-20 TACAN, and one WLR-1 ESM system
Propulsion: four boilers supplying steam to four sets of General Electric geared turbines delivering 100,000 shp (74570 kW) to four shafts
Performance: maximum speed 24 kts; range 8,300 miles (13355 km) at 15 kts
Complement: 1,112 excluding air group

Class
1. Spain

Name	No.	Builder	Laid down	Commissioned
Dedalo	R01	New York SB	Aug 1942	Jul 1943

Note
Now decidedly long in the tooth, the *Dedalo* operates a mix of STOVL aircraft, anti-submarine helicopters, electronic warfare helicopters and utility helicopters as required

'Invincible' class CVL
(UK)
Type: light aircraft-carrier
Displacement: 16,000 tons standard and 19,500 tons full load
Dimensions: length 677.0 ft (206.6 m); beam 90.0 ft (27.5 m); draught 24.0 ft (7.3 m); flightdeck length 550.0 ft (167.8 m) and width 105.0 ft (31.9 m)
Gun armament: two (three in R09) 20-mm Phalanx CIWS mountings, and two 20-mm AA in single mountings (to be replaced by three 25-mm Goalkeeper CIWS mountings)
Missile armament: one twin launcher for 30+ Sea Dart SAMs
Torpedo armament: none
Anti-submarine armament: helicopter-launched weapons (see below)
Aircraft: eight fixed-wing (BAe Sea Harrier FRS.Mk 1) and 12 rotary-wing (Westland Sea King HAS.Mk 5 and AEW.Mk 2), though five Sea Harriers and nine Sea Kings is a more common complement
Electronics: one Type 996 3D radar, one Type 1022 air-search radar, one Type 992R air-search radar, two Type 1006 navigation radars, two Type 909 radars used in conjunction with two GWS 30 SAM fire-control systems, one ADAWS 5 action information system, one Type 2016 hull-mounted sonar, one UAA-1 Abbey Hill ESM system, and two Corvus chaff launchers
Propulsion: COGAG (COmbined Gas turbine And Gas turbine) arrangement, with four Rolls-Royce Olympus TM3B gas turbines delivering 112,000 shp (83520 kW) to two shafts
Complement: 131+265+605 plus an air group of 320

Class
1. UK

Name	No.	Builder	Laid down	Commissioned
Invincible	R05	Vickers	Jul 1973	Jul 1980
Illustrious	R06	Swan Hunter	Oct 1976	Jul 1982
Ark Royal	R09	Swan Hunter	Dec 1978	Jul 1985

Note
These are small mixed helicopter/STOVL carriers, but have proved capable anti-submarine types with secondary multi-role capability

'Jeanne d'Arc' class CVLS
(France)
Type: light anti-submarine aircraft-carrier/assault ship
Displacement: 10,000 tons standard and 12,365 tons full load
Dimensions: length 182.0 m (597.1 ft); beam 24.0 m (78.7 ft); drught 7.3 m (24.0 ft); flightdeck length 62.0 m (203.4 ft) and width 21.0 m (68.9 ft)
Gun armament: four 100-mm (3.9-in) L/55 DP in single mountings
Missile armament: six container-launchers for six MM.38 Exocet anti-ship missiles
Torpedo armament: none
Anti-submarine armament: helicopter-launched weapons (see below)
Aircraft: up to eight Westland Lynx Mk 2 helicopters, or when used as a commando carrier 14 Aérospatiale SA 321 Super Frelon helicopters
Capacity: up to 700 troops and associated equipment
Electronics: one DRBV 22D air-search radar, one DRBV 50 combined air- and surface-search radar, one DRBI 10 height-finding radar, one DRBN 32 navigation radar, three DRBC 32A fire-control radars, one SQ5-503 hull-mounted sonar, URN-6 TACAN, one SENIT 2 action information system, and two Syllex chaff launchers
Propulsion: four boilers supplying steam to two sets of Rateau-Bretagne geared turbines delivering 30000 kW (40,230 shp) to two shafts
Performance: maximum speed 26.5 kts; range 11125 km (6,915 miles) at 15 kts
Complement: 30+183+404

Class
1. France

Name	No.	Builder	Laid down	Commissioned
Jeanne d'Arc	R97	Brest ND	Jul 1960	Jun 1964

Note
Designed for anti-submarine, assault and training roles, this single-ship class can carry a diverse assortment of aircraft and has an extensive array of modern electronics

'John F. Kennedy' class CV
(USA)
Type: multi-role aircraft-carrier
Displacement: 61,000 tons standard and 82,000 tons full load
Dimensions: length 1,047.5 ft (319.5 m); beam 130.0

ft (39.6 m); draught 35.9 ft (10.9 m); flightdeck length 1,047.5 ft (319.5 m) and width 367.5 ft (81.6 m)
Gun armament: one 20-mm Phalanx Mk 16 CIWS mounting
Missile armament: three Mk 29 octuple launchers for 24 RIM-7 NATO Sea Sparrow SAMs
Torpedo armament: none
Anti-submarine armament: aircraft and helicopters (see below)
Aircraft: about 90 in a multi-role carrier air wing
Electronics: one SPS-48C 3D radar, one SPS-49 long-range air-search radar, one SPS-65 low-level threat-warning radar, one SPS-10F surface-search radar, one SPN-10 navigation radar, two SPN-40 landing radars, three Mk 91 SAM fire-control systems, Naval Tactical Data System, one SLQ-29(V)3 ESM system, four Mk 36 Super RBOC chaff launchers, and one OE-82 satellite communications antenna
Propulsion: eight boilers supplying steam to four sets of Westinghouse geared turbines delivering 280,000 shp (208795 kW) to four shafts
Performance: maximum speed 30+ kts; range 13,800 miles (22210 km) at 20 kts
Complement: 142+2,811 plus an air group of about 300+2,200

Class
1. USA

Name	No.	Builder	Laid down	Commissioned
John F. Kennedy	CV67	Newport News	Oct 1964	Sep 1968

Note
This singleton unit is essentially an upgraded version of the 'Kitty Hawk' class with the underwater protection of the US nuclear-powered aircraft-carriers; the ship carries four aircraft elevators and four catapults, and is intended to remain operational until well into the next century

'Kiev' class CV
(USSR)
Type: multi-role hybrid aircraft-carrier/guided-missile cruiser
Displacement: 36,000 tons standard and 42,000 tons full load
Dimensions: length 275.0 m (902.0 ft); beam 41.2 m (135.0 ft); draught 10.0 m (32.8 ft); flightdeck length 189.0 m (620.0 ft) and width 50.0 m (164.0 ft) including sponsons
Gun armament: four 76-mm (3-in) L/60 DP in two twin mountings, and eight 30-mm AK-630 CIWS mountings
Missile armament: four twin launchers for 24 SS-N-12 'Sandbox' anti-ship missiles, two twin launchers for 72 SA-N-3 'Goblet SAMs, and (*Kiev* and *Minsk*) two twin launchers for about 36 SA-N-4 'Gecko' SAMs or (*Novorossiysk* and *Baku*) two groups of four octuple vertical launchers for 64 SA-N-9 SAMS
Torpedo armament: none
Anti-submarine armament: one twin SUW-N-1 launcher for 20 FRAS-1/SS-N-14 'Silex' missiles, two RBU 6000 12-barrel rocket-launchers, two quintuple 533-mm (21-in) mountings for anti-submarine torpedoes, and helicopter-launched weapons (see below)
Aircraft: normally 32, comprising 13 fixed-wing (12 Yakovlev Yak-38 'Forger-As' and one Yak-38 'Forger-B') and 19 rotary-wing (16 Kamov Ka-27 'Helix-As' and three Ka-27 'Helix-Bs')
Electronics: one 'Top Sail' 3D radar and one 'Top Steer' (*Kiev* and *Minsk*) or one 'Top Mesh/Top Plate' (*Novorossiysk* and *Baku*) 3D radar, one 'Trap Door' SSM fire-control radar, two 'Head Light' SA-N-3 fire-control radars, two 'Pop Group' SA-N-4 (*Kiev* and *Minsk*) or one/two 'Cross Sword' (to be fitted in *Novorossiysk* and *Baku*) fire-control radars, two 'Owl Screech' main armament fire-control radars, four 'Bass Tilt' CIWS fire-control radars, four 'Tin Man' optronic trackers (*Novorossiysk* only), one 'Top Knot' aircraft control radar, one 'Don Kay' navigation radar, two 'Palm Frond' navigation radars, two 'Don-2' navigation radars, one low-frequency hull-mounted sonar, one medium-frequency variable-depth sonar, extensive communications and navigation systems, and an extremely extensive electronic warfare suite including eight 'Side Globe', four 'Top Hat-A', four 'Top Hat-B', two 'Rum Tub', two 'Bell Clout' and two 'Cage Pot' antennae/housings
Propulsion: ? boilers supplying steam to four sets of geared turbines delivering 149150 kW (200,015 shp) to four shafts
Performance: maximum speed 32 kts; range 24100 km (14,975 miles) at 18 kts, or 7400 km (4,600 miles) at 30 kts
Complement: 2,500 including air group

Class
1. USSR

Name	Builder	Laid down	Commissioned
Kiev	Nikolayev South	Sep 1970	May 1975
Minsk	Nikolayev South	Dec 1972	Feb 1978

Novorossiysk	Nikolayev South	Sep 1975	Aug 1982
Baku	Nikolayev South	Dec 1978	1985

Note
These are very useful hybrid ships combining the attribute of the helicopter/STOVL carrier and guided-missile cruiser with the command and communication facilities for a major 'out-of-area' deployments; the ships are also providing the Soviets with much useful experience towards the development of a dedicated fixed-wing carrier force

'Kitty Hawk' class CV
(USA)
Type: multi-role aircraft-carrier
Displacement: 60,100 tons (CV63/64) or 60,300 tons (CV66) standard and 80,800 tons (CV63/64) or 78,500 tons (CV66) full load
Dimensions: length (CV63) 1,062.5 ft (324.0 m), (CV64) 1,072.5 ft (327.1 m) and (CV66) 1,047.5 ft (319.5 m); beam (CV63/64) 129.5 ft (39.5 m) and (CV66) 130.0 ft (39.6 m); draught 36.0 ft (11.0 m); flightdeck length (CV63) 1,062.5 ft (324.0 m), (CV64) 1,072.5 ft (327.1 m) and (CV66) 1,047.5 ft (319.5 m), and width (CV63/64) 250.0 ft (76.2 m) and (CV66) 266.0 ft (81.1 m)
Gun armament: three 20-mm Phalanx CIWS mountings
Missile armament: three Mk 29 octuple launchers for 24 RIM-7 NATO Sea Sparrow SAMs
Torpedo armament: none
Anti-submarine armament: aircraft and helicopters (see below)
Aircraft: about 90 in a multi-role carrier air wing
Electronics: one SPS-48C 3D radar, one SPS-49 long-range air-search radar, one surface-search radar (SPS-10B in CV63/64 and SPS-10F in CV66), one SPS-30 height-finding radar, one SPN-42 navigation radar, three Mk 91 SAM fire-control systems, Naval Tactical Data System, one SLQ-29(V)3 ESM system, URN-20 TACAN, four Mk 36 Super RBOC launchers, and one OE-82 satellite communications system
Propulsion: eight boilers supplying steam to four sets of Westinghouse geared turbines delivering 280,000 shp (208795 kW) to four shafts
Performance: maximum speed 30+ kts; range 13,800 miles (22210 km) at 20 kts or 4,600 miles (7400 km) at 30 kts
Complement: 145+2,775 plus an air group of about 300+2,200

Class
1. USA

Name	No.	Builder	Laid down	Commissioned
Kitty Hawk	CV63	New York SB	Dec 1956	Apr 1961
Constellation	CV64	New York NY	Sep 1957	Oct 1961
America	CV66	Newport News	Jan 1961	Jan 1965

Note
These three ships (of which the USS *America* may be considered a one-ship subclass) are of an improved 'Forrestal' class design, and carry four elevators and four catapults; they are to serve into the 21st century

'Majestic' class CVL
(UK/India)
Type: light aircraft-carrier
Displacement: 16,000 tons standard and 19,500 tons full load
Dimensions: length 700.0 ft (213.4 m); beam 80.0 ft (24.4 m); draught 24.0 ft (7.3 m); flightdeck length 700.0 ft (213.4 m) and width 128.0 ft (39.0 m)
Gun armament: nine 40-mm Bofors L/60 AA in single mountings
Missile armament: none
Torpedo armament: none
Anti-submarine armament: aircraft and helicopters (see below)
Aircraft: 12 fixed-wing (eight BAe Sea Harrier FRS.Mk 51s and four Breguet Alizés) and about 10 rotary-wing (Westland Sea Kings and Aérospatiale Alouette IIIs)
Electronics: one LW-05 air-search radar, ZW-06 surface-search radar, one LW-10 tactical radar, one LW-11 tactical radar, one Type 963 aircraft landing radar, and various communications systems
Propulsion: four boilers supplying steam to two sets of Parsons geared turbines delivering 40,000 shp (29830 kW) to two shafts
Performance: maximum speed 24.5 kts; range 13,800 miles (22210 km) at 14 kts or 7,150 miles (11505 km) at 23 kts
Complement: 1,075

Class
1. India

Name	No.	Builder	Laid down	Commissioned
Vikrant	R11	Vickers-Armstrong	Oct 1943	Mar 1961

Note
Though technically obsolescent, the *Vikrant* has been upgraded and represents a significant force in local naval strengths as a result of her range and modern aircraft; she is to be supplemented by another elderly carrier, ex-HMS *Hermes*, after the latter has been extensively refitted

'Midway' class CV
(USA)
Type: multi-role aircraft-carrier
Displacement: 51,000 tons (CV41) or 52,500 tons (CV43) standard and 62,200 tons (CV41) or 62,200 tons (CV43) full load
Dimensions: length 979.0 ft (298.4 m); beam 121.0 ft (36.9 m); draught 35.3 ft (10.8 m); flightdeck length 979.0 ft (298.4 m) and width 238.0 ft (72.5 m)
Gun armament: three 20-mm Phalanx Mk 16 CIWS mountings
Missile armament: two Mk 25 octuple launchers for 16 RIM-7 Sea Sparrow SAMs (CV41 only)
Torpedo armament: none
Anti-submarine armament: aircraft and helicopters (see below)
Aircraft: about 75 in a multi-role carrier air wing
Electronics: one SPS-48C 3D air-search radar (CV41 only), one SPS-65V low-level threat-detection radar (CV41 only), one air-search radar (SPS-49 in CV41 and SPS-30 in CV43), one SPS-43C long-range air-search radar (CV-43 only), one surface-search radar (SPS-10 in CV43 and SPS-10F in CV41), one LN-66 navigation radar, one SPN-35A aircraft landing radar (CV41 only), one SPN-42 aircraft landing radar (CV41 only), one SPN-44 aircraft landing radar (CV41 only) one SPN-43A aircraft landing radar (CV43 only), two Mk 115 SAM fire-control systems (CV41 only), URN-20 TACAN, Naval Tactical Data System, one SLQ-29 ESM system, four Mk 36 Super RBOC chaff launchers, and one OE-82 satellite communications antenna
Propulsion: 12 boilers supplying steam to four sets of Westinghouse geared turbines delivering 212,000 shp (158090 kW) to four shafts
Performance: maximum speed 32 kts; range 17,275 miles (27800 km) at 15 kts
Complement: 125+2,385 plus an air group of about 225+1,725

Class
1. USA

Name	No.	Builder	Laid down	Commissioned
Midway	CV41	Newport News	Oct 1943	Sep 1945
Coral Sea	CV43	Newport News	Jul 1944	Oct 1947

Note
Though two of the oldest carriers afloat, the ships of the 'Midway' class have been extensively upgraded over the years, but are generally relegated to secondary areas as their short flightdecks prevent the embarkation of the latest aircraft types (notably the Grumman F-14 and Lockheed S-3); both ships have three deck-edge lifts, but while CV43 has three catapults CV41 has only two; both units are to be replaced in the 1990s

'Moskva' class CVS
(USSR)
Type: anti-submarine helicopter-carrier
Displacement: 16,500 tons standard and 20,000 tons full load
Dimensions: length 190.5 m (624.8 ft); beam 23.0 m (75.4 ft); draught 11.0 m (36.1 ft); flightdeck length 81.0 m (265.7 ft) and width 34.0 m (111.5 ft)
Gun armament: four 57-mm L/70 AA in two twin mountings
Missile armament: two twin launchers for 48 SA-N-3 'Goblet' SAMs
Torpedo armament: none
Anti-submarine armament: one SUW-N-1 twin launcher for 20 FRAS-1 missiles, two RBU 6000 12-barrel rocket-launchers, and helicopter-launched weapons (see below)
Aircraft: 18 Kamov Ka-25 'Hormone-A' or Ka-27 'Helix-A' helicopters
Electronics: one 'Top Sail' 3D radar, one 'Head Net-C' 3D radar, two 'Head Light' SAM-control radars, two 'Muff Cob' AA gun-control radars, three 'Don-2' navigation radars, one low-frequency hull-mounted sonar, one medium-frequency variable-depth sonar, two twin chaff launchers, and a very extensive ESM system including eight 'Side Globe' and eight 'Bell' series antennae/housings
Propulsion: four boilers supplying steam to two sets of geared turbines delivering 74570 kW (100,000 shp) to two shafts
Performance: maximum speed 30 kts; range 16675 km (10,360 miles) at 18 kts or 5200 km (3,230 miles) at 30 kts
Complement: 840 excluding air group

Class			
1. USSR			
Name	*Builder*	*Laid down*	*Commissioned*
Moskva	Nikolayev South	1962	May 1967
Leningrad	Nikolayev South	1964	1968

Note
Hybrid designs featuring elements of missile cruiser and helicopter carrier design, the two 'Moskva' class ships have provided the Soviet navy with anti-submarine and command facilities for out-of-area deployments, and have generated useful data about the operation of aircraft from large warships

'Nimitz' class CVN
(USA)
Type: nuclear-powered multi-role aircraft-carrier
Displacement: 72,700 tons light, 81,600 tons standard and 91,485 tons (CVN68/70) or 96,350 tons (CVN71/73) full load
Dimensions: length 1,092.0 ft (332.9 m); beam 234.0 ft (40.8 m); draught 37.0 ft (11.2 m); flightdeck length 1,092.0 ft (332.9 m), and width (CVN68/70) 252.0 ft (76.8 m) or (CVN71/73) 257.0 ft (78.4 m)
Gun armament: three (CVN68/69) or four (CVN70/73) 20-mm Phalanx Mk 16 CIWS mountings
Missile armament: three Mk 25 (CVN68/69) or Mk 29 (CVN70/73) octuple launchers for 24 RIM-7 Sea Sparrow SAMs
Torpedo armament: none
Anti-submarine armament: aircraft and helicopters (see below)
Aircraft: 90 or more in a multi-role carrier air wing
Electronics: one 3D air-search radar (SPS-48B in CVN68/71 and SPS-48E in CVN72/73), one SPS-40 long-range air-search radar, one SPS-10F surface-search radar, one LN-66 navigation radar, two SPN-42 aircraft landing radars, one SPN-43 aircraft landing radar, one SPN-44 aircraft landing radar, three Mk 115 (CVN68/69) or four Mk 91 (CVN70/73) SAM fire-control systems, Naval Tactical Data System, URN-20 TACAN, one SLQ-29(V)3 ESM system, Mk 36 Super RBOC chaff launchers, and one OE-82 satellite communications antenna
Propulsion: two Westinghouse A2W pressurized water-cooled reactors supplying steam to four sets of geared turbines delivering 260,000 shp (193880 kW) to four shafts
Performance: maximum speed 30+ kts
Complement: about 145+2,900 plus an air group of 300+2,325

Class				
1. USA				
Name	*No.*	*Builder*	*Laid down*	*Commissioned*
Nimitz	CVN68	Newport News	Jun 1968	May 1975
Dwight D. Eisenhower	CVN69	Newport News	Aug 1970	Oct 1977
Carl Vinson	CVN70	Newport News	Oct 1975	Feb 1982
Theodore Roosevelt	CVN71	Newport News	Oct 1981	Oct 1986
Abraham Lincoln	CVN72	Newport News	Nov 1984	
George Washington	CVN73	Newport News	Sep 1986	

Note
These are the world's largest warships, and are highly capable offensive/defensive platforms carrying the latest aircraft and electronics; the use of a more modern (and more compact) reactor arrangement than that of the 'Enterprise' design allows greater volume for aircraft weapons and fuel, the limiting factor in modern aircraft-carrier operations; the 'Nimitz' class ships each carry sufficient ordnance and fuel for 16 days of continuous operations, whereas the 'Enterprise' class carries sufficient for only 12 days of operations; from the fourth ship Kevlar armour is built into the design and the hull design has additional protective features, and the proposed class total is eight highly capable ships; each ship has four deck-edge lifts and four catapults

'Principe de Asturias' class CVL
(Spain)
Type: light aircraft-carrier
Displacement: 14,700 tons full load
Dimensions: length 195.1 m (639.9 ft); beam 24.7 m (81.0 ft); draught 9.1 m (29.8 ft); flightdeck length 175.0 m (574.0 ft) and width 32.0 m (105.0 ft)
Gun armament: four 20-mm Meroka CIWS mountings
Missile armament: none
Torpedo armament: none
Anti-submarine armament: none
Aircraft: 17 fixed-wing and rotary-wing
Electronics: one SPS-52D 3D air-search radar, one SPS-55 surface-search radar, one SPN-35A aircraft landing radar, four Meroka fire-control radars, one action information system, URN-22 TACAN and one ESM system

Propulsion: COGAG (COmbined Gas turbine And Gas turbine) arrangement, with two General Electric LM2500 gas turbines delivering 46,000 shp (34300 kW) to two shafts
Performance: maximum speed 26 kts; range 13,900 km (8635 km) at 20 kts
Complement: 793 including air group

Class
1. Spain

Name	No.	Builder	Laid down	Commissioned
Principe de Asturias	R11	Bazan	Oct 1979	1986

Note
Designed to replace the *Dedalo*, this single ship is based on the US Navy's abortive Sea Control Ship and has a ski jump flightdeck, facilities for helicopters and V/STOL aircraft, two aircraft elevators and modern electronics for multi-role and command deployment

BATTLESHIPS AND BATTLE-CRUISERS

'Iowa' class BB
(USA)
Type: battleship
Displacement: 48,425 tons standard and 57,500 tons full load
Dimensions: length 887.2 ft (270.4 m); beam 108.2 ft (33.0 m); draught 38.0 ft (11.6 m)
Gun armament: nine 16-in (406-mm) L/50 in three Mk 7 triple mountings, 12 5-in (127-mm) L/38 DP in six Mk 28 twin mountings, and four 20-mm Phalanx Mk 16 CIWS mountings
Missile armament: eight quadruple Armored Box Launchers Mk 143 for 32 BGM-109 Tomahawk anti-ship and land attack cruise missiles, and four quadruple Mk 141 container-launchers for 16 RGM-84A Harpoon anti-ship missiles
Torpedo armament: none
Anti-submarine armament: none
Aircraft: provision for three or four helicopters on a platform aft
Armour: 13.5/1.62-in (343/41-mm) belt, 6-in (152-mm) decks, 17/7.25-in (432/184-mm) turrets, 17.3-in (439-mm) barbettes, and 17.5/7.25-in (445/184-mm) conning tower
Electronics: one SPS-49 long-range air-search radar, one SPS-10F surface-search radar, one LN-66 navigation radar, four Mk 25 radars used in conjunction with four Mk 37 gun fire-control systems, two Mk 13 radars used in conjunction with two Mk 38 gun directors, one Mk 27 radar used in conjunction with one Mk 40 gun director, two Mk 51 gun directors, six Mk 56 gun directors, four Mk 37 gun directors, two Mk 34 radars used in conjunction with two Mk 63 gun directors, one SLQ-31(V)3 ESM system, one Combat Engagement Center, eight Mk 36 Super RBOC chaff launchers, and two WSC-3 satellite communications transceivers
Propulsion: eight boilers supplying steam to four sets of geared turbines (General Electric in BB61 and 63, Westinghouse in BB62 and 64) delivering 212,000 shp (158090 kW) to four shafts
Performance: maximum speed 33 kts; range 17,275 miles (27800 km) at 17 kts
Complement: 65+1,450

Class
1. USA

Name	No.	Builder	Laid down	Commissioned
Iowa	BB61	New York NY	Jun 1940	Feb 1943 (Apr 1984)
New Jersey	BB62	Philadelphia NY	Sep 1940	May 1943 (Dec 1982)
Missouri	BB63	New York NY	Jan 1941	Jun 1944 (1986)
Wisconsin	BB64	Philadelphia NY	Jan 1941	Apr 1944

Note
Though these four ships had been mothballed since shortly after World War II (apart from brief excursions by all four ships in the Korean War and by the USS *New Jersey* in the Vietnam War), the development of the Soviet navy's surface force in the second half of the 1970s resulted in their reappraisal, which resulted in positive appreciation of the battleships' high speed, great protection, ability to accommodate modern missiles and a few helicopters while still retaining a massive shore-bombardment capability, and suitability for electronic enhancement as command ships for surface battle groups; the *New Jersey* and *Iowa* have been brought up to the described standard, and the other two ships are due for recommissioning later in the decade

'Kirov' class CBN
(USSR)
Type: nuclear-powered guided-missile battle-cruiser
Displacement: 22,000 tons standard and 28,000 tons full load
Dimensions: length 250.0 m (820.2 ft); beam 28.5 m (93.5 ft); draught 10.0 m (32.8 ft)
Gun armament: two 100-mm (3.93-in) DP in single mountings or (*Frunze*) two 130-mm (5.12-in) in one twin mounting, and eight 30-mm AK-630 CIWS mountings
Missile armament: 20 vertical launch tubes for 20 SS-N-19 anti-ship missiles, 12 vertical launch tubes for 96 SA-N-6 SAMs, two twin launchers for 40 SA-N-4 'Gecko' SAMs, and (*Frunze* only) three groups (one of eight and two of four) octuple verticle launchers for 64 SA-N-9 SAMs
Torpedo armament: two 533-mm (21-in) quintuple mountings for dual-role torpedoes
Anti-submarine armament: one (except *Frunze*) twin launcher for 16 SS-N-14 'Silex' missiles, one RBU 6000 12-barrel rocket-launcher, two RBU 1000 six-barrel rocket-launchers, and helicopter-launched weapons (see below)
Aircraft: three to five Kamov Ka-27 'Helix' helicopters in a hangar aft
Electronics: one 'Top Pair' 3D radar, one 'Top Steer' 3D radar, two 'Eye Bowl' SS-N-14 control radars (not in *Frunze*), two 'Top Dome' SA-N-6 control radars, two 'Pop Group' SA-N-4 control radars, one 'Kite Screech' main armament gun-control radar, four 'Bass Tilt' AA gun-control radars, three 'Palm Frond' navigation radars, one low-frequency bow sonar, one low-frequency variable-depth sonar, two twin chaff launchers, and an extremely extensive ESM system including eight 'Side Globe', four 'Rum Tub' and two 'Round House' antennae/housings
Propulsion: two pressurized water-cooled reactors supplying oil-fired superheated steam to ? sets of geared turbines delivering 119300 kW (160,000 shp) to ? shafts
Performance: maximum speed 35 kts
Complement: 900

Class 1. USSR

Name	Builder	Laid down	Commissioned
Kirov	Baltic Yard 189	Jun 1973	Sep 1980
Frunze	Baltic Yard 189	Jan 1978	Nov 1983
	Baltic Yard 189	1982	

Note
These are exceptionally powerful missile cruisers of new concept and size, powered by a novel hybrid nuclear/oil-fired propulsion system, fitted out with extensive electronics and command facilities as escort to the USSR's new aircraft-carriers or as leaders of detached surface battle groups, and sporting an assortment of weapons suited to the whole gamut of surface-to-surface, surface-to-air and surface-to-underwater tasks

GUIDED-MISSILE CRUISERS

'Andrea Doria' class CG
(Italy)
Type: anti-submarine and anti-aircraft guided-missile escort cruiser
Displacement: 5,000 tons standard and 6,500 tons full load
Dimensions: length 149.3 m (489.8 ft); beam 17.2 m (56.4 ft); draught 5.0 m (16.4 ft)
Gun armament: eight (six in *Caio Duilio*) 76-mm (3-in) OTO Melara L/62 DP in single mountings
Missile armament: one twin launcher for 40 RIM-67A Standard SM-2 ER SAMs
Torpedo armament: none
Anti-submarine armament: two Mk 32 triple mountings for 12.75-in (324-mm) Mk 46 torpedoes, and helicopter-launched weapons (see below)
Aircraft: four Agusta-Bell AB.212ASW helicopters in a hangar aft
Electronics: one SPS-52 3D radar, one SPS-40 long-range air-search radar, two SPG-55C radars used in conjunction with the Mk 76 SAM fire-control system, four (three in *Caio Duilio*) RTN 10X radars used in conjunction with four (three in *Caio Duilio*) Argo NA9 gun fire-control systems, one SPQ 2 navigation radar, one SQS-23 long-range hull-mounted active sonar, SCLAR chaff launchers, one ESM system, and one action information system
Propulsion: four boilers supplying steam to two sets of geared turbines (CNR in *Andrea Doria* and Ansaldo in *Caio Duilio*) delivering 44750 kW (60,010 shp) to two shafts
Performance: maximum speed 31 kts; range 9250 km (5,750 miles) at 17 kts
Complement: 45+425

Class 1. Italy				
Name	*No.*	*Builder*	*Laid down*	*Commissioned*
Andrea Doria	C553	CNR	May 1958	Feb 1964
Caio Duilio	C554	Navalmeccanica	May 1958	Nov 1964

Note
Designed as escort cruisers for Mediterranean operations, these are in essence small helicopter carriers (though the helicopter facility is somewhat cramped) for anti-submarine and anti-aircraft use

'Bainbridge' class CGN
(USA)
Type: nuclear-powered anti-aircraft and anti-ship guided-missile escort cruiser
Displacement: 7,700 tons standard and 8,580 tons full load
Dimensions: length 565.0 ft (172.3 m); beam 58.0 ft (17.7 m); draught 29.0 ft (8.8 m)
Gun armament: two 20-mm Phalanx CIWS CIWS mountings, and two 20-mm AA in one Mk 67 twin mounting
Missile armament: two Mk 141 quadruple container-launchers for eight RGM-84A Harpoon anti-ship missiles, and two Mk 10 twin launchers for 80 RIM-67B Standard SM-2 ER SAMs
Torpedo armament: none
Anti-submarine armament: one Mk 16 octuple launcher for eight RUR-5A ASROC missiles, and two Mk 32 triple mountings for 12.75-in (324-mm) Mk 46 torpedoes
Aircraft: none
Electronics: one SPS-39 3D radar, one SPS-37 air-search radar, one SPS-10F surface-search radar, four SPG-55B radars used in conjunction with four Mk 76 SAM fire-control systems, one Mk 11 weapon direction system (being replaced by one Mk 14 weapon direction system), one SQQ-23A long-range hull-mounted active sonar used in conjunction with the Mk 111 underwater weapons fire-control system, Naval Tactical Data System, one SLQ-32(V)3 ESM system, four Mk 36 Super RBOC chaff launchers, URN-20 TACAN, OE-82 satellite communications system, SSR-1 satellite communications receiver, and WSC-1 satellite communications transceiver
Propulsion: two General Electric D2G pressurized water-cooled reactors supplying steam to two sets of geared turbines delivering 60,000 shp (44740 kW) to two shafts
Performance: maximum speed: 38 kts
Complement: 32+509, plus flag accommodation of 6+12

Class 1. USA				
Name	*No.*	*Builder*	*Laid down*	*Commissioned*
Bainbridge	CGN25	Bethlehem Steel	May 1959	Oct 1962

Note
This is a nuclear-powered derivative of the 'Leahy' class

'Belknap' class CG
(USA)
Type: anti-aircraft, anti-submarine and anti-ship guided-missile escort cruiser
Displacement: 6,570 tons standard and 7,930 tons full load
Dimensions: length 547.0 ft (166.7 m); beam 54.8 ft (16.7 m); draught 28.8 ft (8.8 m)
Gun armament: one 5-in (127-mm) L/54 DP in a Mk 42 mounting, and two 20-mm Phalanx Mk 16 CIWS mountings
Missile armament: two Mk 141 quadruple container-launchers for eight RGM-84A Harpoon anti-ship missiles (except in *Jouett* and *Horne*), one Mk 10 twin launcher for 40 RIM-67B Standard SM-2 ER SAMs, and (to be fitted) BGM-109 Tomahawk cruise missiles
Torpedo armament: none
Anti-submarine armament: two Mk 32 triple mountings for 12.75-in (324-mm) Mk 46 torpedoes, up to 20 RUR-5A ASROC missiles fired from the Mk 10 launcher, and helicopter-launched weapons (see below)
Aircraft: one Kaman SH-2F Seasprite helicopter in a hangar aft
Electronics: one SPS-48C 3D radar, one SPS-49 (CG26/28 and CG30) or SPS-40 (others) air-search radar, one SPS-10F surface-search radar, two SPG-55B radars used in conjunction with two Mk 76 SAM fire-control systems, one Mk 7 (CG27), Mk 11 (CG31/34) or Mk 14 (others) weapon direction system, one SPG-53F radar used in conjunction with the Mk 68 gun fire-control system, one SQS-53A (CG26) or SQS-26 (others) bow-mounted sonar used in conjunction with one Mk 116 (CG26) or Mk 114 (others) underwater weapons fire-control system, Naval Tactical Data System, one SLQ-32(V)3 ESM

system, four Mk 36 Super RBOC chaff launchers, URN-20 TACAN, OE-82 satellite communications system, SRR-1 satellite communications receiver, and WSC-3 satellite communications transceiver
Propulsion: four boilers supplying steam to two sets of geared turbines delivering 85,000 shp (63385 kW) to two shafts

Performance: maximum speed 32.5 kts; range 8,175 miles (13155 km) at 20 kts
Complement: 25+450 including squadron staff, or (CG26) 520 including flag staff

Class
1. USA

Name	No.	Builder	Laid down	Commissioned
Belknap	CG26	Bath Iron Works	Feb 1962	Nov 1964
Josephus Daniels	CG27	Bath Iron Works	Apr 1962	May 1965
Wainwright	CG28	Bath Iron Works	Jul 1962	Jan 1966
Jouett	CG29	Puget Sound NY	Sep 1962	Dec 1966
Horne	CG30	San Francisco NY	Dec 1962	Apr 1967
Sterett	CG31	Puget Sound NY	Sep 1962	Apr 1967
William H. Standley	CG32	Bath Iron Works	Jul 1963	Jul 1966
Fox	CG33	Todd Shipyard	Jan 1963	May 1966
Biddle	CG34	Bath Iron Works	Dec 1963	Jan 1967

Note
These ships were produced after a tortuous design period and have since been used for a host of trials purposes, but are designed mainly for the anti-aircraft and anti-submarine escort of aircraft-carriers, though they are limited by being single-ended missile ships and carrying only one helicopter; a useful surface-to-surface capability has been added

'California' class CGN
(USA)
Type: nuclear-powered anti-submarine, anti-ship and anti-aircraft guided-missile escort cruiser
Displacement: 9,560 tons standard and 11,105 tons full load
Dimensions: length 596.0 ft (181.7 m); beam 61.0 ft (18.6 m); draught 31.5 ft (9.6 m)
Gun armament: two 5-in (127-mm) L/54 DP in Mk 45 single mountings, and two 20-mm Phalanx Mk 16 CIWS mountings
Missile armament: two quadruple Armored Box Launchers Mk 143 for eight BGM-109 Tomahawk anti-ship and land attack cruise missiles, two Mk 141 quadruple container-launchers for eight RGM-84A Harpoon anti-ship missiles, and two Mk 13 single launchers for 80 RIM-66C Standard SM-2 MR SAMs
Torpedo armament: none
Anti-submarine armament: one Mk 16 octuple launcher for eight RUR-5A ASROC missiles, and four Mk 32 single mountings for 12.75-in (324-mm) Mk 46 torpedoes

Aircraft: none
Electronics: one SPS-48C 3D radar, one SPS-40B air-search radar, one SPS-55 surface-search radar, four SPG-51D radars used in conjunction with two Mk 74 SAM fire-control systems, one SPG-60 search-and-track radar and one SPQ-9A track-while-scan radar used in conjunction with one Mk 86 gun fire-control system, one Mk 13 weapon direction system, one SQS-26CX bow-mounted sonar used in conjunction with one Mk 114 underwater weapons fire-control system, Naval Tactical Data System, URN-20 TACAN, T Mk 6 Fanfare torpedo decoy system, four Mk 36 Super RBOC chaff launchers, one SLQ-32(V)3 ESM system, one OE-82 satellite communications system, one SRR-1 satellite communications receiver, and one WSC-3 satellite communications transceiver
Propulsion: two General Electric D2G pressurized water-cooled reactors supplying steam to two sets of geared turbines delivering 60,000 shp (44740 kW) to two shafts
Performance: maximum speed 30+ kts
Complement: 30+519

Class
1. USA

Name	No.	Builder	Laid down	Commissioned
California	CGN36	Newport News	Jan 1970	Feb 1974
South Carolina	CGN37	Newport News	Dec 1970	Jan 1975

Note
This was originally to have been a five-ship class, but was curtailed so that extra finance could be allocated to the succeeding and more capable 'Virginia' class; the class is notable for its nuclear propulsion system, whose cores have three times the life of those in the preceding 'Bainbridge' class; the type suffers from lack of useful helicopter facility, and though the design is double-ended, the launchers are of the single-rail type so limiting the simultaneous-engagement capability to two

aircraft, a poor situation in saturation air attacks against carrier battle groups; the surface-to-surface missile capability is good

'Colbert' class CG
(France)

Type: anti-aircraft and anti-ship guided-missile escort cruiser
Displacement: 8,500 tons standard and 11,300 tons full load
Dimensions: length 180.8 m (593.2 ft); beam 20.2 m (66.1 ft); draught 7.7 m (25.2 ft)
Gun armament: two 100-mm (3.9-in) L/55 DP in single mountings, and 12 57-mm Bofors L/70 AA in six twin mountings
Missile armament: four container-launchers for four MM.38 Exocet anti-ship missiles, and one twin launcher for 48 Masurca SAMs
Torpedo armament: none
Anti-submarine armament: none
Aircraft: provision for a helicopter on a platform aft
Armour: 50/80-mm (2/3.15-in) belt, and 50-mm (2-in) deck
Electronics: one DRBV 50 combined air- and surface-search radar, one DRBV 23C air-search radar, one DRBV 20 air-warning radar, one DRBI 10D height-finding radar, two DRBR 51 SAM-control radars, one DRBC 32C 100-mm gun-control radar, two DRBC 31 57-mm gun-control radars, one RM416 navigation radar, one SENIT 1 action information system, URN-20 TACAN, one ESM system, and two Syllex chaff launchers
Propulsion: four boilers supplying steam to two sets of CEM/Parsons geared turbines delivering 64130 kW (86,000 shp) to two shafts
Performance: maximum speed 31.5 kts; range 7400 km (4,600 miles) at 25 kts
Complement: 24+190+346

Class
1. France

Name	No.	Builder	Laid down	Commissioned
Colbert	C611	Brest ND	Dec 1953	May 1959

Note
Developed from the pre-World War II 'De Grasse' class, the 'Colbert' class is designed for area-defence of surface forces, and is extensively fitted for her role as flagship of the French Mediterranean fleet

'Kara' class CG
(USSR)

Type: anti-submarine and anti-aircraft guided-missile cruiser
Displacement: 8,200 tons standard and 10,000 tons full load
Dimensions: length 173.2 m (568.0 ft); beam 18.0 m (59.0 ft); draught 6.7 m (22.0 ft)
Gun armament: four 76-mm (3-in) L/60 DP in two twin mountings, and four 30-mm AK-630 CIWS mountings
Missile armament: two twin launchers for 72 SA-N-3 'Goblet' SAMs, and two twin launchers for 36 SA-N-4 'Gecko' SAMs; *Azov* has only one SA-N-3 and one SA-N-4 system
Torpedo armament: two quintuple 533-mm (21-in) mountings for 10 dual-role torpedoes
Anti-submarine armament: two quadruple container-launchers for eight SS-N-14 'Silex' anti-submarine missiles, torpedoes (see above), two RBU 1000 six-barrel rocket-launchers, and helicopter-launched weapons (see below)
Aircraft: one Kamov Ka-25 'Hormone-A' helicopter in a hangar aft
Electronics: one 'Top Sail' 3D radar, two 'Don Kay' navigation and surface-search radars, one 'Don 2' navigation radar, two 'Head Light' SA-N-3 missile-control radars (one 'Head Light' and one 'Top Dome' in *Azov*), two 'Pop Group' SA-N-4 missile-control radars, two 'Owl Screech' 76-mm gun-control radars, two 'Bass Tilt' 30-mm gun-control radars, hull-mounted sonar(s), one medium-frequency variable-depth sonar, one BAT-1 torpedo countermeasures system, one 'High Pole-A' IFF, one 'High Pole-B' IFF, two chaff launchers, and an extremely comprehensive ESM suite including two 'Tee Plinth', eight 'Side Globe', two 'Bell Slam', one 'Bell Clout' and one 'Bell Tap' housings
Propulsion: COGAG (COmbined Gas turbine And Gas turbine) arrangement, with two gas turbines delivering 18000 kW (24,140 shp) and four gas turbines delivering 75000 kW (100,575 shp) to two shafts
Performance: maximum speed 33 kts; range 14825 km (9,210 miles) at 18 kts or 3700 km (2,300 miles) at 32 kts
Complement: 30+510

Class
1. USSR

Name	Builder	Laid down	Commissioned
Nikolayev	Nikolayev North	1969	1971
Ochakov	Nikolayev North	1970	1973
Kerch	Nikolayev North	1971	1974
Azov	Nikolayev North	1972	1975
Petropavlovsk	Nikolayev North	1973	1976
Tashkent	Nikolayev North	1974	1977
Tallinn	Nikolayev North	1975	1979

Note
Built at Nikolayev North between 1969 and 1979, this class is essentially a refined version of the 'Kresta II' class with COGAG propulsion, more advanced weapons and optimization for the anti-submarine role though still retaining a useful air-defence capability; the ships also have good command facilities, and are thus well suited to lead anti-submarine task groups

'Krasina' class CG
(USSR)
Type: anti-ship and anti-aircraft guided-missile cruiser
Displacement: 11,000 tons standard and 13,000 tons full load
Dimensions: length 187.0 m (613.5 ft); beam 22.3 m (73.2 ft); draught 7.6 m (25.0 ft)
Gun armament: two 130-mm (5.1-in) L/70 in one twin mounting, and six 30-mm AK-630 CIWS mountings
Missile armament: eight twin container-launchers for 16 SS-N-12 anti-ship missiles, eight vertical launch tubes for about 84 SA-N-6 SAMs, and two twin launchers for 40 SA-N-4 'Gecko' SAMs
Torpedo armament: two quadruple or quintuple mountings for 533-mm (21-in) dual-role torpedoes
Anti-submarine armament: two 12-barrel RBU 6000 rocket-launchers, and torpedoes (see above)
Aircraft: one Kamov Ka-25 'Hormone-B' or Ka-27 'Helix-B' missile-targeting helicopter
Electronics: one 'Top Pair' 3D radar, one 'Top Steer' 3D radar, three 'Palm Frond' surface-search and navigation radars, one 'Trap Door' SS-N-12 control radar, one 'Top Dome' SA-N-6 control radar, two 'Pop Group' SA-N-4 control radars, one 'Kite Screech' 130-mm gun-control radar, three 'Bass Tilt' 30-mm gun-control radars, two 'Tee Plinth' IR surveillance systems, one low-frequency hull sonar, one medium-frequency variable-depth sonar, 'High Pole-B' IFF, two chaff launchers, and an extremely extensive ESM system including eight 'Side Globe', four 'Rum Tub' and ? 'Bell' series antennae/housings
Propulsion: COGAG (COmbined Gas turbine And Gas turbine) arrangement, with four gas turbines delivering about 130000 kW (174,330 shp) to two shafts
Performance: maximum speed 34 kts
Complement: about 600

Class
1. USSR

Name	Builder	Laid down	Commissioned
Slava	61 Kommuna Yard, Nikolayev	1976	1982
	61 Kommuna Yard, Nikolayev	1978	1986
	61 Kommuna Yard, Nikolayev	1979	1987

Note
Built at Nikolayev North in succession to the 'Kara' class, the 'Krasina' class has the same basic escort role as its predecessor type, but features more advanced weapons (though centred on anti-ship missiles) and electronics on a development of the same basic hull; it is anticipated that the class will total eight units

'Kresta I' class CG
(USSR)
Type: anti-ship and anti-aircraft guided-misile cruiser
Displacement: 6,140 tons standard and 7,500 tons full load
Dimensions: length 155.5 m (510.0 ft); beam 17.0 m (55.7 ft); draught 6.0 m (19.7 ft)
Gun armament: four 57-mm L/80 DP in two twin mountings, and (*Drozd* only) four 30-mm AK-630 CIWS mountings
Missile armament: two twin container-launchers for four SS-N-3B 'Shaddock' anti-ship missiles, and two twin launchers for 32 or 44 SA-N-1 'Goa' SAMs
Torpedo armament: two quintuple 533-mm (21-in) mountings for 10 dual-role torpedoes
Anti-submarine armament: torpedoes (see above),

two RBU 6000 12-barrel rocket-launchers, and two RBU 1000 six-barrel rocket-launchers
Aircraft: one Kamov Ka-25 'Hormone-B' missile-targeting helicopter in a hangar aft
Electronics: one 'Head Net-C' 3D radar, one 'Big Net' air-search radar, two 'Plinth Net' surface-search radars, one 'Scoop Pair' SSM-control radar, two 'Peel Group' SAM-control radars, two 'Muff Cob' 57-mm gun-control radars, two 'Bass Tilt' CIWS-control radars (*Drozd* only), two 'Don 2' navigation radars, one 'High Pole-B' IFF, one medium-frequency hull-mounted sonar, and a comprehensive ESM suite including eight 'Side Globe' housings
Propulsion: four boilers supplying steam to two sets of geared turbines delivering 75000 kW (101,575 shp) to two shafts
Performance: maximum speed 34 kts; range 13000 km (8,080 miles) at 18 kts or 3700 km (2,300 miles) at 32 kts
Complement: 375

Class
1. USSR

Name	Builder	Laid down	Commissioned
Admiral Zozulya	Zhdanov Yard	Sep 1964	1967
Vladivostok	Zhdanov Yard	1965	1968
Vitse-Admiral Drozd	Zhdanov Yard	1965	1968
Sevastopol	Zhdanov Yard	1966	1969

Note
Built in Leningrad, the 'Kresta I' class is an interim design between the anti-ship 'Kynda' and anti-submarine 'Kresta II' classes, with features from both roles but optimized for anti-ship warfare; the ships were the first Soviet warships with full helicopter accommodation, in this instance a 'Hormone-B' missile guidance type

'Kresta II' class CG
(USSR)
Type: anti-submarine and anti-aircraft guided-missile cruiser
Displacement: 6,000 tons standard and 7,800 tons full load
Dimensions: length 158.5 m (520.0 ft); beam 16.9 m (55.45 ft); draught 6.0 m (19.7 ft)
Gun armament: four 57-mm L/80 DP in two twin mountings, and four 30-mm AK-630 CIWS mountings
Missile armament: two twin launchers for 48 SA-N-3 'Goblet' SAMs
Torpedo armament: two 533-mm (21-in) quintuple mountings for 10 dual-role torpedoes
Anti-submarine armament: two quadruple container-launchers for eight SS-N-14 'Silex' anti-submarine missiles, torpedoes (see above), two RBU 6000 12-barrel rocket-launchers, two RBU 1000 six-barrel rocket-launchers, and helicopter-launched weapons (see below)
Aircraft: one Kamov Ka-25 'Hormone-A' helicopter in a hangar aft
Electronics: one 'Top Sail' 3D radar, one 'Head Net-C' 3D radar, two 'Don Kay' surface-search and navigation radars, two 'Head Light' SAM-control radars, two 'Muff Cob' 57-mm gun-control radars, two 'Bass Tilt' CIWS-control radars, one hull-mounted sonar, and a comprehensive ESM suite including eight 'Side Globe' housings
Propulsion: four boilers supplying steam to two sets of geared turbines delivering 75000 kW (100,575 shp) to two shafts
Performance: maximum speed 34 kts; range 13000 km (8,080 miles) at 18 kts or 3250 km (2,020 miles) at 32 kts
Complement: 380/400

Class
1. USSR

Name	Builder	Laid down	Commissioned
Kronshtadt	Zhdanov Yard	1966	1969
Admiral Isakov	Zhdanov Yard	1967	1970
Admiral Nakhimov	Zhdanov Yard	1968	1971
Admiral Makarov	Zhdanov Yard	1969	1972
Marshal Voroshilov	Zhdanov Yard	1970	1973
Admiral Oktabrsky	Zhdanov Yard	1970	1973
Admiral Isachenko	Zhdanov Yard	1971	1974
Marshal Timoshenko	Zhdanov Yard	1972	1975
Vasily Chapaev	Zhdanov Yard	1973	1976
Admiral Yumaschev	Zhdanov Yard	1974	1977

Note
Following the 'Kresta I' class on the Leningrad slips, the 'Kresta II' class ships are fully optimized for the anti-submarine role (with specialized missiles and a 'Hormone-A' ASW helicopter instead of a 'Hormone-B' targeting helicopter) and carry a more advanced surface-to-air missile type

'Kynda' class CG
(USSR)

Type: anti-ship guided-missile cruiser
Displacement: 4,400 tons standard and 5,700 tons full load
Dimensions: length 142.0 m (465.8 ft); beam 15.8 m (51.8 ft); draught 5.3 m (17.4 ft)
Gun armament: four 76-mm (3-in) L/60 DP in two twin mountings, and (in *Varyag* only) four 30-mm AK-630 CIWS mountings
Missile armament: two quadruple container-launchers for 16 SS-N-3B 'Shaddock' anti-ship missiles, and one twin launcher for 22 SA-N-1 'Goa' SAMs
Torpedo armament: two 533-mm (21-in) triple mountings for six dual-role torpedoes
Anti-submarine armament: two 12-barrel RBU 6000 rocket-launchers, torpedoes (see above), and helicopter-launched weapons (see below)
Aircraft: provision for one Kamov Ka-25 'Hormone-A' helicopter on a platform aft
Electronics: two 'Head Net-A' air-search radars, or (in *Fokin* only) one 'Head Net-A' air-search radar and one 'Head Net-C' 3D radar, or (in *Varyag* only) two 'Head Net-C' 3D radars, two 'Plinth Net' surface-search radars (not in *Golovko*), two 'Scoop Pair' SSM-control radars, one 'Peel Group' SAM-control radar, two 'Owl Screech' 76-mm gun-control radars, two 'Bass Tilt' CIWS-control radars (in *Varyag* only), two 'Don-2' navigation radars, one 'High Pole-B' IFF, one high-frequency hull-mounted sonar, and various electronic warfare systems
Propulsion: four boilers supplying steam to two sets of geared turbines delivering 75000 kW (100,575 shp) to two shafts
Performance: speed 34 kts; range 11125 mk (6,915 miles) at 15 kts or 2775 km (1,725 miles) at 34 kts
Complement: 390

Class
1. USSR

Name	Builder	Laid down	Commissioned
Grozny	Zhdanov Yard	1959	Jun 1962
Admiral Fokin	Zhdanov Yard	1960	Aug 1963
Admiral Golovko	Zhdanov Yard	1961	Jul 1964
Varyag	Zhdanov Yard	1962	Feb 1965

Note
Built at Leningrad between 1959 and 1965 as dedicated anti-ship cruisers with long-range missiles, these vessels introduced the standard Soviet pattern of enclosed pyramid mast for the support of extensive electronic antennae; the class was curtailed at four units as the Soviets began to appreciate the growing importance of anti-submarine rather than anti-ship warfare in the 1960s

'Leahy' class CG
(USA)

Type: anti-aircraft, anti-submarine and anti-ship guided-missile escort cruiser
Displacement: 5,670 tons standard and 7,800 tons full load
Dimensions: length 533.0 ft (162.5 m); beam 55.0 ft (16.8 m); draught 25.0 ft (7.6 m)
Gun armament: two 20-mm Phalanx Mk 16 CIWS mountings
Missile armament: two Mk 141 quadruple container-launchers for eight RGM-84A Harpoon anti-ship missiles, and two Mk 10 twin launchers for 80 RIM-67B Standard SM-2 ER SAMs
Torpedo armament: none
Anti-submarine armament: one Mk 16 octuple launcher for eight RUR-5A ASROC missiles, two Mk 32 triple mountings for 12.75-in (324-mm) Mk 46 torpedoes, and helicopter-launched weapons (see below)
Aircraft: provision for a Kaman SH-2F Seasprite helicopter on a platform aft
Electronics: one SPS-48C 3D radar, one SPS-49 air-search radar, one SPS-10F surface-search radar, four SPG-55B radars used in conjunction with four Mk 76 SAM fire-control systems, one Mk 14 weapon direction system, one SQQ-23B bow-mounted sonar used in conjunction with one Mk 114 underwater weapons fire-control system, Naval Tactical Data System, OE-82 satellite communications system, SRR-1 satellite communications receiver, WSC-3 satellite communications transceiver, URN-20 TACAN, four Mk 36 Super RBOC chaff launchers, and one SLQ-32(V)3 ESM system
Propulsion: four boilers supplying steam to two sets of geared turbines delivering 85,000 shp (63385 kW) to two shafts
Performance: speed 33 kts; range 9200 miles (14,805 km) at 20 kts
Complement: 27+397 plus flag accommodation of 6+18

Class
1. USA

Name	No.	Builder	Laid down	Commissioned
Leahy	CG16	Bath Iron Works	Dec 1959	Aug 1962

198 WARSHIPS

Harry E. Yarnell	CG17	Bath Iron Works	May 1960	Feb 1963
Worden	CG18	Bath Iron Works	Sep 1960	Aug 1963
Dale	CG19	New York SB	Sep 1960	Nov 1963
Richmond K. Turner	CG20	New York SB	Jan 1961	Jun 1964
Gridley	CG21	Puget Sound Bridge	Jul 1960	May 1963
England	CG22	Todd SB	Oct 1960	Dec 1963
Halsey	CG23	San Francisco NY	Aug 1960	Jul 1963
Reeves	CG24	Puget Sound NY	Jul 1960	May 1964

Note
This was the first US cruiser class optimized for the carrier-escort anti-aircraft role with a medium/long-range missile armament, the two twin launchers in a double-ended arrangement allowing the simultaneous engagement of four targets; the ships have been extensively upgraded in electronics and weapons, the latter being developed from the original Terrier to the Standard SM-1 and then the Standard SM-2 series; recent developments have upgraded the close-in air-defence arrangements and added a potent anti-ship missile armament

'Long Beach' class CGN
(USA)
Type: nuclear-powered anti-aircraft, anti-submarine and anti-ship guided-missile escort cruiser
Displacement: 14,200 tons standard and 17,350 tons full load
Dimensions: length 721.2 ft (219.9 m); beam 73.2 ft (22.3 m); draught 29.7 ft (9.1 m)
Gun armament: two 5-in (127-mm) L/38 DP in Mk 30 single mountings, and two 20-mm Phalanx Mk 16 CIWS mountings
Missile armament: two quadruple Armored Box Launchers Mk 143 for eight BGM-109 Tomahawk cruise missiles, four Mk 141 quadruple container-launchers for 16 RGM-84A Harpoon anti-ship missiles, and two Mk 10 twin launchers for 120 RIM-67B Standard SM-2 ER SAMs
Torpedo armament: none
Anti-submarine armament: one Mk 16 octuple launcher for 20 RUR-5A ASROC missiles, and two Mk 32 triple mountings for 12.75-in (324-mm) Mk 46 torpedoes
Aircraft: provision for a helicopter on a platform aft
Electronics: one SPS-48C 3D radar, one SPS-49B air-search radar, one SPS-65 low-level threat-warning radar, one SPS-10 surface-search radar, four SPG-55B radars used in conjunction with four Mk 76 SAM fire-control systems, two SPG-49B radars used in conjunction with two Mk 56 gun fire-control systems, two Mk 35 radars used in conjunction with two Mk 56 gun fire-control systems, one Mk 14 weapon direction system, one SQQ-23B hull-mounted sonar used in conjunction with one Mk 111 underwater weapons fire-control system, URN-20 TACAN, one SLQ-32(V)3 ESM system, two Mk 36 Super RBOC chaff launchers, one OE-82 satellite communications system, one SRR-1 satellite communications receiver, and one WSC-3 satellite communications transceiver
Propulsion: two Westinghouse C1W pressurized water-cooled reactors supplying steam to two sets of General Electric geared turbines delivering 80,000 shp (59655 kW) to two shafts
Performance: maximum speed 36 kts
Complement: 64+857

Class
1. USA

| Name | No. | Builder | Laid down | Commissioned |
| Long Beach | CGN9 | Bethlehem | Dec 1957 | Sep 1961 |

Note
This single-ship class was the world's first nuclear-powered surface combatant type, and was designed for the anti-aircraft escort of carrier battle groups with two twin Talos and two twin Terrier launchers for long- and medium-range missiles respectively; the ship has been greatly updated in electronics and weapons, and is currently configured as a single-ended ship with two twin SAM launchers plus an ASROC launcher and a powerful anti-ship missile armament

'Sverdlov' class CG
(USSR)
Type: guided-missile command cruiser
Displacement: 16,000 tons standard and 17,500 tons full load
Dimensions: length 210.0 m (689.0 ft); beam 22.0 m (72.2 ft); draught 7.5 m (24.5 ft)
Gun armament: nine (*Dzerzhinsky* and *Zhdanov*) or six (*Admiral Senyavin*) 152-mm (6-in) L/50 in three or two triple mountings, 12 100-mm (3.9-mm) L/60 DP in six twin mountings, 16 37-mm AA in eight twin mountings, and 16 (*Admiral Senyavin*) or eight (*Zhdanov*) 30-mm AA in eight or four twin mountings
Missile armament: one twin launcher for (*Dzerzhinsky*) 30 SA-N-2 'Guideline' or (*Admiral Senyavin* and *Zhdanov*) 18 SA-N-4 'Gecko' SAMs
Torpedo armament: none
Anti-submarine armament: none
Aircraft: none

Mines: up to 200 (not in *Admiral Senyavin* and *Zhdanov*)
Armour: 100/125-mm (3.93/4.92-in) belt, 40/50-mm (1.6/2-in) ends, 25/76-mm (1/3-in) decks, 125-mm (4.92-in) turrets and 150-mm (5.9-in) conning tower
Electronics: (*Dzerzhinsky*) one 'Big Net' air-search radar, one 'Slim Net' air-search radar, one 'Low Sieve' surface-search radar, one 'Fan Song-E' SAM-control radar, one 'Top Bow' and two 'Sun Visor' 152-mm gun-control radars, six 'Egg Cup' 100-mm gun-control radars, two 'Sun Visor' 100-mm gun-control radars, one 'Neptun' navigation radar, and various ECM systems
Eletronics: (*Admiral Zenyavin* and *Zhdanov*) one 'Top Trough' air-search radar, one 'Slim Net' air-search radar, one 'Low Sieve' surface-search radar, one 'Pop Group' SAM-control radar, one 'Sun Visor' and two 'Top Bow' 152-mm gun-control radars, six 'Egg Cup' 100-mm gun-control radars, four (*Admiral Senyavin*) or two (*Zhdanov*) 30-mm fire-control radars, one 'Neptun' navigation radar, and various ECM systems
Propulsion: six boilers supplying steam to two sets of geared turbines delivering 82000 kW (109,965 shp) to two shafts
Performance: maximum speed 32 kts; range 16000 km (9,945 miles) at 18 kts
Complement: about 70+940

Class
1. USSR

Name	Builder	Laid down	Commissioned
Dzerzhinsky	Nosenko Yard, Nikolayev	May 1949	Nov 1952
Zhdanov	Nosenko Yard, Nikolayev	1950	1954
Admiral Senyavin	Amur Yard, Komsomolsk	1952	1957

Note
These three ships are the most extensively developed of the 14-strong 'Sverdlov' class gun cruisers, all possessing a limited SAM armament; the *Dzerzhinsky* was the trials ship for the unsuccessful SA-N-2 naval version of the 'Guideline' SAM, and had by the late 1970s been placed in reserve with the Black Sea Fleet; the *Admiral Senyavin* and *Zhdanov* were converted in 1971 and 1972 into command cruisers with a limited SAM armament and extensive command facilities, *Admiral Senyavin* now serving with the Pacific Fleet and the *Zhdanov* with the Black Sea Fleet; the *Admiral Nakhimov* was converted as trials ship for the SS-N-1 'Scrubber' anti-ship missile, but was then scrapped in 1961; the other nine surviving units of this class are the *Admiral Lazarev, Admiral Ushakov, Aleksandr Nevsky, Aleksandr Suvarov, Dmitri Pozharsky, Mickail Kutuzov, Murmansk, Oktyabrskaya Revolutsiya* and *Sverdlov*, of which three (one in reserve) serve with the Pacific Fleet, two with the Black Sea Fleet, two with the Baltic Fleet and two with the Northern Fleet as gunfire ships in support of amphibious operations; the ships differ in detail, but each has a full-load displacement of 20,000 tons, a primary armament of 12 152-mm (6-in) guns in four triple turrets, a secondary armament of 12 100-mm (3.94-in) guns in six twin turrets, and an AA armament of 37-mm guns (32 such weapons in 16 twin mountings except in the *Admiral Ushakov, Aleksandr Suvarov* and *Oktyabrskaya Revolutsiya* which have 28 such weapons in 14 twin mountings plus 16 30-mm AA in eight twin mountings); the last unit of the class was *Ordzhonikidze*, transferred to Indonesia in 1962 as the *Irian* but scrapped in 1972 for lack of spares

'Ticonderoga' class CG
(USA)
Type: anti-aircraft, anti-submarine and anti-ship guided-missile escort cruiser
Displacement: 9,350 tons full load
Dimensions: length 563.3 ft (171.8 m); beam 55.0 ft (16.8 m); draught 31.0 ft (9.5 m)
Gun armament: two 5-in (127-mm) L/54 DP in Mk 45 single mountings, and two 20-mm Phalanx Mk 16 CIWS mountings
Missile armament: two Mk 141 quadruple container-launchers for 16 RGM-84A Harpoon anti-ship missiles, and two Mk 26 twin launchers for up to 68 RIM-67B Standard SM-2 ER SAMs; from CG52 onwards the ships will each have two 61-cell Mk 41 Vertical Launch Systems for 12 BGM-109 Tomahawk cruise missiles, plus ASROC(VL) and Standard missiles in a total of 122 missiles
Torpedo armament: none
Anti-submarine armament: two Mk 32 triple mountings for 12.75-in (324-mm) Mk 46 torpedoes, up to 20 RUR-5A ASROC missiles as part of the total missile strength of the Mk 26 launchers (see above), and helicopter-launched weapons (see below)
Aircraft: two Sikorsky SH-60B Seahawk helicopters in a hangar aft
Electronics: two SPY-1A or (from CG59 onwards) SPY-1B 3D phased-array pairs of long-range search, target-tracking and missile-control radars used in conjunction with the AEGIS Weapon-Control System Mk 7 and four SPG-62 SAM-control radars, one SPS-49(V) air-search radar, one SPS-55 surface-search radar, four Mk 99 missile directors, one SP-9A track-while-scan radar used in conjunction with the Mk 86 gun fire-control system, one SQS-53A (SQS-53B from CG54 onwards) hull-mounted long-

range active sonar and one SQR-19 tactical towed-array sonar used in conjunction with the Mk 116 underwater weapons fire-control system, one OE-82 satellite communications system, four SRR-1 satellite communications receivers, two WSC-3 satellite communications transceivers, four Mk 36 Super RBOC chaff launchers, and one SLQ-32(V)3 ESM system

Propulsion: COGAG (COmbined Gas turbine And Gas turbine) arrangement, with four General Electric LM2500 gas turbines delivering 80,000 shp (59655 kW) to two shafts
Performance: speed 30+ kts; range 6,900 miles (11105 km) at 20 kts
Complement: 33+342

Class
1. USA

Name	No.	Builder	Laid down	Commissioned
Ticonderoga	CG47	Ingalls SB	Jan 1980	Jan 1983
Yorktown	CG48	Ingalls SB	Oct 1981	Jul 1984
Vincennes	CG49	Ingalls SB	Oct 1982	Jul 1985
Valley Forge	CG50	Ingalls SB	Apr 1983	Jan 1986
Thomas S. Gates	CG51	Bath Iron Works	Nov 1983	Jan 1987
Bunker Hill	CG52	Ingalls SB	Jan 1984	
Mobile Bay	CG53	Ingalls SB	Jun 1984	Feb 1987
Antietau	CG54	Ingalls SB	Nov 1984	
Leyte Gulf	CG55	Ingalls SB	Apr 1985	
San Jacinto	CG56	Ingalls SB		
Lake Champlain	CG57	Ingalls SB		
	CG58	Bath Iron Works		
Princeton	CG59	Ingalls SB		
	CG60	Bath Iron Works		
	CG61	Bath Iron Works		
	CG62	Ingalls SB		
	CG63	Ingalls SB		
	CG64	Ingalls SB		
	CG65	Ingalls SB		

'Truxtun' class CGN
(USA)
Type: nuclear-powered anti-aircraft, anti-submarine and anti-ship guided-missile escort cruiser
Displacement: 8,200 tons standard and 9,125 tons full load
Dimensions: length 564.0 ft (171.9 m); beam 58.0 ft (17.7 m); draught 31.0 ft (9.4 m)
Gun armament: one 5-in (127-mm) L/54 DP in a Mk 42 mounting, and two 20-mm Phalanx Mk 16 CIWS mountings
Missile armament: two Mk 141 quadruple container-launchers for eight RGM-84A Harpoon anti-ship missiles, and one Mk 10 twin launcher for up to 60 RIM-67B Standard SM-2 ER SAMs
Torpedo armament: none
Anti-submarine armament: four Mk 32 single mountings for 12.75-in (324-mm) Mk 46 torpedoes, up to 20 RUR-5A ASROC missiles included in the Mk 10 launcher total (see above), and helicopter-launched weapons (see below)
Aircraft: one Kaman SH-2F Seasprite helicopter in a hangar aft

Electronics: one SPS-48 3D radar, one SPS-40 air-search radar, one SPS-10 surface-search radar, one SPG-53F radar used in conjunction with the Mk 86 gun fire-control system, two SPG-55B radars used in conjunction with two Mk 76 SAM fire-control systems, one Mk 14 weapon direction system, one SQS-26 bow-mounted 'bottom-bounce' sonar used in conjunction with the Mk 114 underwater weapons fire-control system, Naval Tactical Data System, URN-20 TACAN, OE-82 satellite communications system, SRR-1 satellite communications receiver, WSC-3 satellite communications transceiver, four Mk 36 Super RBOC chaff launchers, and one SLQ-32(V)3 ESM system
Propulsion: two General Electric D2G pressurized water-cooled reactors supplying steam to two sets of geared turbines delivering 60,000 shp (44740 kW) to two shafts
Performance: speed 38 kts
Complement: 31+503, plus flag accommodation for 6+12

Class
1. USA

Name	No.	Builder	Laid down	Commissioned
Truxtun	CGN35	New York SB	Jun 1963	May 1967

Note
This is the nuclear-powered version of the otherwise similar 'Belknap' class

'Type 82' class CG
(UK)
Type: anti-aircraft and anti-submarine guided-missile escort cruiser
Displacement: 6,100 tons standard and 7,100 tons full load
Dimensions: length 507.0 ft (154.5 m); beam 55.0 ft (16.8 m); draught 23.0 ft (7.0 m)
Gun armament: one 4.5-in (114-mm) L/55 DP in a Mk 8 single mounting, and two 20-mm AA in Mk 7 single mountings
Missile armament: one twin launcher for 40 Sea Dart SAMs
Torpedo armament: none
Anti-submarine armament: one launcher for 40 Ikara missiles, one Limbo Mk 10 mortar, and helicopter-launched weapons (see below)
Aircraft: one Westland Wasp HAS.Mk 1 helicopter on a platform aft
Electronics: one Type 965 air-search radar, one Type 992Q surface-search and target indication radar, two Type 909 radars used in conjunction with two GWS 30 SAM fire-control systems, one Type 1006 navigation radar, one Type 162 hull-mounted side-looking classification sonar, one Type 184 hull-mounted medium-range panoramic search and attack sonar and one Type 170 hull-mounted short-range search and attack sonar used in conjunction with the GWS 40 Ikara fire-control system, one Type 182 torpedo decoy system, one ADAWS 2 action-information system, one SCOT satellite communications antenna, two Corvus chass launchers, and one UAA-1 Abbey Hill ESM system
Propulsion: COSAG (Combined Steam and Gas turbine) arrangement, with two boilers supplying steam to two sets of ASR turbines delivering 30,000 shp (22370 kW), and two Rolls-Royce Olympus TM1A gas turbines delivering 30,000 shp (22370 kW) to two shafts
Performance: speed 28 kts; range 5,750 miles (9255 km) at 18 kts
Complement: 29+378

Class
1. UK

Name	No.	Builder	Laid down	Commissioned
Bristol	D23	Swan Hunter	Nov 1967	Mar 1973

Note
Classified by the British as a destroyer, the single 'Type 82' class ship has the capabilities of a cruiser, and was to have been the lead ship of a class of four carrier escorts optimized for area defence against aircraft and submarines; the ship has the space and electronic capabilities to serve as a command vessel

'Virginia' class CGN
(USA)
Type: nuclear-powered anti-aircraft, anti-ship and anti-submarine guided-missile escort cruiser
Displacement: 8,625 tons standard and 11,000 tons full load
Dimensions: length 585.0 ft (178.4 m); beam 63.0 ft (19.2 m); draught 29.5 ft (9.0 m)
Gun armament: two 5-in (127-mm) L/54 DP in Mk 45 single mountings, and two 20-mm Phalanx Mk 16 CIWS mountings
Missile armament: three quadruple Armored Box Launchers Mk 143 for 12 BGM-109 Tomahawk cruise missiles, two Mk 141 quadruple container-launchers for eight RGM-84A Harpoon anti-ship missiles, and two Mk 26 twin launchers for up to 68 RIM-67B Standard SM-2 ER SAMs
Torpedo armament: none
Anti-submarine armament: two Mk 32 triple mountings for 12.75-mm (3243-mm) Mk 46 torpedoes, up to 20 RUR-5A ASROC missiles in the Mk 26 total (see above), and helicopter weapons (see below)
Aircraft: provision for two helicopters being deleted to provide space for the Tomahawk ABLs (see above)
Electronics: one SPS-48C 3D radar, one SPS-40B air-search radar, one SPS-55 surface-search radar, two SPG-51D radars used in conjunction with one Mk 74 SAM fire-control system, one SPG-60D search-and-tracking radar and one SPQ-9A track-while-scan radar used in conjunction with one Mk 86 gun fire-control system, one Mk 14 weapon direction system, one SQS-53A bow-mounted sonar used in conjunction with one Mk 116 underwater weapons fire-control system, one T Mk 6 Fanfare torpedo decoy system, Naval Tactical Data System, OE-82 satellite communications system, SRR-1 satellite communications receiver, WSC-3 satellite communications transceiver, URN-20 TACAN, four Mk 36 Super RBOC chaff launchers, and one SLQ-32(V)3 ESM system
Propulsion: two General Electric D2G pressurized water-cooled reactors supplying steam to two sets of geared turbines delivering 100,000 shp (74570 kW) to two shafts
Performance: speed 40 kts
Complement: 30+502

Class
1. USA

Name	No.	Builder	Laid down	Commissioned
Name	No.	Builder	Laid down	Commissioned
Virginia	CGN38	Newport News	Aug 1972	Sep 1976
Texas	CGN39	Newport News	Aug 1973	Sep 1977
Mississippi	CGN40	Newport News	Feb 1975	Aug 1978
Arkansas	CGN41	Newport News	Feb 1975	Oct 1980

Note
These are essentially improved 'California' class ships, optimized for the area defence of nuclear-powered carriers with a double-ended design with two twin launchers for Standard SAMs, plus a useful anti-submarine and anti-ship missile capability

'Vittorio Veneto' class CG
(Italy)
Type: anti-submarine, anti-aircraft and anti-ship guided-missile escort cruiser
Displacement: 7,500 tons standard and 8,850 tons full load
Dimensions: length 179.6 m (589.0 ft); beam 19.4 m (63.6 ft); draught 6.0 m (19.7 ft)
Gun armament: eight 76-mm (3-in) OTO Melara L/62 DP in single mountings, and six 40-mm Bofors L/70 AA in three Breda Compact twin mountings
Missile armament: one Mk 20 Aster twin launcher for about 40 RIM-67A Standard SM-2 ER SAMs and 20 RUR-5A ASROC missiles, and four Teseo container-launchers for four Otomat anti-ship missiles
Torpedo armament: none
Anti-submarine armament: RUR-5A ASROC missiles (see above), two Mk 32 triple mountings for 12.75-mm (324-mm) Mk 46 torpedoes, and helicopter-launched weapons (see below)

Aircraft: up to nine Agusta-Bell AB.212ASW helicopters on a flight deck aft
Electronics: one SPS-52C 3D radar, one SPS-40 air-search radar, one SPS 70 surface-search radar, two SPG-55C radars used in conjunction with the Aster SAM fire-control system, four RTN 10X radars used in conjunction with four Argo NA9 76-mm gun fire-control systems, two RTN 16X radars used in conjunction with two Dardo AA gun fire-control systems, one 3RM 7 navigation radar, one SQS-23 hull-mounted long-range sonar, URN-20 TACAN, two SCLAR chaff launchers, and one UAA-1 Abbey Hill ESM system
Propulsion: four boilers supplying steam to two sets of Tosi geared turbines delivering 54500 kW (73,095 ship) to two shafts
Performance: speed 32 kts; range 9250 km (5,750 miles)
Complement: 50+500

Class
1. Italy

Name	No.	Builder	Laid down	Commissioned
Vittorio Veneto	C550	Italcantieri	Jun 1965	Jul 1969

Note
Based on the 'Andrea Doria' class, the design of the *Vittorio Veneto* was recast radically when it was realized that the earlier ships' helicopter facilities were far too small, and the result is an anti-submarine and anti-aircraft dual-role ship with impressive helicopter capabilities

GUIDED-MISSILE DESTROYERS

'Almirante' class DDG
(Chile)
Type: guided-missile destroyer
Displacement: 2,730 tons standard and 3,300 tons full load
Dimensions: length 402.0 ft (122.5 m); beam 43.0 ft (13.1 m); draught 13.3 ft (4.0 m)
Gun armament: four 4-in (102-mm) L/60 in Mk(N)R single mountings, and four 40-mm Bofors L/70 AA in single mountings
Missile armament: four container-launchers for four MM.38 Exocet anti-ship missiles, and two quadruple launchers for 16 Sea Cat SAMs
Torpedo armament: none
Anti-submarine armament: two Mk 32 triple mountings for 12.75-in (324-mm) Mk 44 torpedoes, and two

Squid depth-charge mortars
Aircraft: none
Electronics: one AWS 1 air-search and target-indication radar, and other systems
Propulsion: two boilers supplying steam to two sets of Parsons/Pametrada geared turbines delivering 54,000 shp (40270 kW) to two shafts
Performance: speed 34.5 kts; range 6,900 miles (11100 km)
Complement: 17+249

Class
1. Chile

Name	No.	Builder	Laid down	Commissioned
Almirante Riveros	18	Vickers-Armstrong	Apr 1957	Dec 1960
Almirante Williams	19	Vickers-Armstrong	Jun 1956	Mar 1960

Note
These ships are obsolete by all but local standards, despite a capable anti-ship missile armament

'Amatsukaze' class DDG
(Japan)
Type: anti-aircraft and anti-submarine guided-missile destroyer
Displacement: 3,050 tons standard and 4,000 tons full load
Dimensions: length 131.0 m (429.8 ft); beam 13.4 m (44.0 ft); draught 4.2 m (13.8 ft)
Gun armament: four 3-in (76-mm) L/50 DP in two Mk 33 twin mountings
Missile armament: one Mk 13 single launcher for 40 RIM-66B Standard SM-1 MR SAMs
Torpedo armament: none
Anti-submarine armament: one Mk 16 octuple launcher for eight RUR-5A ASROC missiles, two Mk 32 triple mountings for 12.75-in (324-mm) Mk 46 torpedoes, and two Hedgehog Mk 15 mortars
Aircraft: none
Electronics: one SPS-52 3D radar, one SPS-29 air-search radar, one OPS-17 surface-search radar, two SPG-51C radars used in conjunction with two Mk 73 SAM fire-control systems, one GFCS-2 gun fire-control system, one SQS-23 hull-mounted long-range sonar used in conjunction with the Mk 114 underwater weapons fire-control system, and various ESM systems including two chaff launchers, one NOLR-6 ECM system and one OLT-1 jammer
Propulsion: two boilers supplying steam to two sets of Ishikawajima/General Electric geared turbines delivering 44750 kW (60,010 shp) to two shafts
Performance: speed 33 kts; range 12975 km (8,065 miles) at 18 kts
Complement: 290

Class
1. Japan

Name	No.	Builder	Laid down	Commissioned
Amatsukaze	DD163	Mitsubishi	Nov 1962	Feb 1965

Note
This is a dual-capable anti-submarine and anti-aircraft destroyer of obsolescent type, though still useful in Japan's defence-oriented strategy

'Anshan' class DDG
(China)
Type: anti-ship guided-missile destroyer
Displacement: 1,855 tons standard and 2,040 tons full load
Dimensions: length 112.8 m (370.0 ft); beam 10.2 m (33.5 ft); draught 3.8 m (12.5 ft)
Gun armament: four 130-mm (5.1-in) L/50 DP in single mountings, and eight 37-mm AA in four twin mountings
Missile armament: two twin container-launchers for four HY-2 (CSS-N-1) anti-ship missiles
Torpedo armament: none
Anti-submarine armament: two BMB-1 mortars with 24 depth charges
Mines: up to 60
Aircraft: none
Electronics: one 'Cross Bird' air-search radar, one 'Ball End' surface-search radar, one 'Square Tie' missile-control radar, one 'Post Lamp' gun fire-control radar, one 'Mina' optical fire-control system, 'Yard Rake' IFF, 'Ski Pole' IFF, and hull-mounted sonar
Propulsion: three boilers supplying steam to two sets of Tosi geared turbines delivering 35800 kW (48,010 shp) to two shafts
Performance: speed 32 kts; range 4800 km (2,980 miles) at 19 kts
Complement: 246

204 WARSHIPS

Class
1. China

Name	No.	Builder	Laid down	Commissioned
Anshan	101	Nikolayev-Dalzavod	1935	1940
Zhangzhun	102	Nikolayev-Dalzavod	1936	1941
Jilin	103	Nikolayev-Komsomolsk	1936	1941
Fuzhun	104	Nikolayev-Komsomolsk	1935	1941

Note
These old ships were built as 'Gordy' class gun destroyers, but have been rebuilt and upgraded in Chinese service

'Arleigh Burke' class DDG
(USA)
Type: anti-aircraft, anti-ship and anti-submarine guided-missile escort destroyer
Displacement: 8,200 tons standard and 8,500 tons full load
Dimensions: length 466.0 ft (142.1 m); beam 59.0 ft (18.0 m); draught 27.0 ft (8.2 m)
Gun armament: one 5-in (127-mm) L/54 DP in a Mk 45 single mounting, and two 20-mm Phalanx Mk 16 CIWS mountings
Missile armament: two Mk 141 quadruple container-launchers for eight RGM-84A Harpoon anti-ship missiles, and two Vertical Launch Systems (one 32-cell and one 64-cell) for 96 RUR-5A ASROC, Standard SM-2 SAM and BGM-109 Tomahawk cruise missiles
Torpedo armament: none
Anti-submarine armament: ASROC missiles in the VLS (see above), two Mk 32 triple mountings for 12.75-in (324-mm) Mk 46 or Mk 50 Barracuda torpedoes, and helicopter-launched weapons (see below)
Aircraft: one Sikorsky SH-60B Seahawk helicopter on a platform aft
Electronics: one SPY-1D phased-array air-search radar used in conjunction with the AEGIS Weapon-Control System Mk 7, one SPS-67 surface-search radar, three SPG-62 radars used in conjunction with three Mk 99 SAM fire-control systems, one Seafire gun fire-control system, one SQS-53C bow-mounted sonar and one SQR-19 towed-array sonar used in conjunction with the Mk 116 underwater weapons fire-control system, Automatic Data Action System, TACAN, satellite navigation system, two Mk 36 Super RBOC chaff launchers, and one SLQ-32(V)2 ESM system
Propulsion: COGAG (COmbined Gas turbine And Gas turbine) arrangement, with four General Electric LM2500 gas turbines delivering 100,000 shp (74570 kW) to two shafts
Performance: speed 32 kts; range 5,750 miles (9250 km) at 20 kts
Complement: 23+302

Class
1. USA

Name	No.	Builder	Laid down	Commissioned
Arleigh Burke	DDG51	Bath Iron Works	Jul 1986	

Note
This is a very impressive design of advanced guided-missile destroyer intended as partner for the 'Ticonderoga' class CGs in replacement of the 'Coontz' class DDGs and 'Leahy/Belknap' class CGs; the type features an 'austere' version of the AEGIS radar/SAM-control system, and some 29 units are currently planned

'Audace' class DDG
(Italy)
Type: anti-submarine and anti-aircraft guided-missile escort destroyer
Displacement: 3,950 tons standard and 4,560 tons full load
Dimensions: length 136.6 m (448.0 ft); beam 14.2 m (46.6 ft); draught 4.6 m (15.1 ft)
Gun armament: two 127-mm (5-in) OTO Melara Compact L/54 DP in single mountings, and four 76-mm (3-in) OTO Melara Compact L/62 DP in single mountings
Missile armament: one Mk 13 single launcher for 40 RIM-66B Standard SM-1 MR SAMs
Torpedo armament: none
Anti-submarine armament: two ILAS 3 triple mountings for 12 12.75-in (324-mm) Mk 46 or 324-mm (12.75-in) A 244S torpedoes, two 533-mm (21-in) twin mountings for 12 A184 torpedoes, and helicopter-launched weapons (see below)
Aircraft: two Agusta-Bell AB.212ASW or one Agusta-Sikorsky ASH-3 helicopter in a hangar aft
Electronics: one SPS-52 3D radar, one RAN 20S air-search radar, one SPQ 2 surface-search radar, two

SPG-51 radars used in conjunction with two Mk 74 SAM fire-control systems, three RTN 10X radars used in conjunction with three Argo NA10 gun fire-control systems, one CWE 610 hull-mounted sonar, two SCLAR chaff launchers, and various ESM systems

Propulsion: four boilers supplying steam to two sets of geared turbines delivering 54500 kW (73,085 shp) to two shafts
Performance: speed 34 kts; range 5560 km (3,455 miles)
Complement: 30+350

Class
1. Italy

Name	No.	Builder	Laid down	Commissioned
Ardito	D550	Italcantieri	Jul 1968	Dec 1973
Audace	D551	CNR, Riva Trigosa	Apr 1968	Nov 1972

Note
Developed from the 'Impavido' class, the two 'Audaces' have greater seaworthiness and habitability combined with anti-ship, anti-submarine and anti-aircraft sensors and armament; the Otomat Mk 2 anti-ship missile systems is to be retrofitted, giving the ships powerful capability against other surface vessels

'Audace (Improved)' class DDG
(Italy)
Type: guided-missile destroyer
Displacement: about 5,000 tons
Dimensions: length 137.0 m (449.5 ft); beam not revealed; draught not revealed
Gun armament: four 76-mm (3-in) OTO Melara Compact L/72 DP in single mountings, and 12 40-mm Bofors L/70 AA in six Breda Compact twin mountings
Missile armament: four container-launchers for four Otomat Mk 2 anti-ship missiles, one Mk 13 single launcher for ? RIM-67 Standard SM-2 SAMs, and two Albatros octuple launchers for ? Aspide SAMs
Torpedo armament: none
Anti-submarine armament: two ILAS 3 triple mountings for 12 12.75-in (324-mm) Mk 46 or 324-mm (12.75-in) A 244S torpedoes, plus helicopter-launched weapons (see below)
Aircraft: two Agusta-Sikorsky ASH-3D/H helicopters in a hangar aft

Electronics: one SPS-52C 3D radar, one surface-search radar, two SPG-51D radars used in conjunction with two Mk 74 SAM fire-control systems, two RTN 30X radars used in conjunction with the Albatros SAM system, two RAN 20S 76-mm fire-control radars, three RTX 20X radars used in conjunction with three Dardo 40-mm fire-control systems, one DE 1164 bow-mounted sonar and one variable-depth sonar used in conjunction with an underwater weapons fire-control system, one action information system, one electronic support measures system, and two SCLAR chaff launchers
Propulsion: CODOG (COmbined Diesel Or Gas turbine) arrangement, with two Fiat/General Electric LM2500 gas turbines delivering 55,000 shp (41,015 kW) and two GMT B230, 20DVM diesels delivering 11,000 hp (8200 kW) to two shafts
Performance: speed 30+ kts
Complement: not revealed

Class
1. Italy

Name	No.	Builder	Laid down	Commissioned
Animoso		CNR, Riva Trigosa		
Ardimentoso		CNR, Riva Trigosa		

'Boxer' class DDG (UK): see 'Broadsword' class DDG

'Broadsword' or 'Type 22' class DDG (UK)
Type: anti-aircraft, anti-ship and anti-submarine guided-missile escort destroyer
Displacement: (F88/F91) 3,500 tons standard and 4,400 tons full load, or (F92/F101) 4,100 tons standard and 4,600 tons full load
Dimensions: length (F88/F91) 430.0 ft (131.2 m) or (F92/F101) 471.0 ft (143.6 m); beam 48.5 ft (14.8 m); draught 19.9 ft (6.0 m)

Gun armament: two 40-mm Bofors L/60 AA in single mountings, and two 20-mm AA in single mountings, or (from F98 onwards) one 4.5-in (114-mm) L/55 DP in a single mounting and two CIWS mountings
Missile armament: four container-launchers for four MM.38 Exocet anti-ship missiles or (F98 onwards) two quadruple container-launchers for eight RGM-84A Harpoon anti-ship missiles, and two sextuple launchers for 60 Sea Wolf SAMs
Torpedo armament: none
Anti-submarine armament: helicopter-launched weapons (see below), and (from F90 onwards) two

STWS triple mountings for 12.75-in (324-mm) Mk 46 or Stingray torpedoes
Aircraft: two Westland Lynx HAS.Mk 2 or (F94 onwards) two EH.101 or Westland Sea King helicopters in a hangar aft
Electronics: one Type 967 air-search radar, one Type 968 air-search radar, one Type 1006 navigation radar, two Type 910 radars used in conjunction with the GWS 25 SAM fire-control system, one GWS 50 Exocet fire-control system, one Type 2008 sonar, one Type 2016 sonar, one Type 182 torpedo decoy system, one Type 2031(Z) towed-array sonar (from F92 onwards), one SCOT satellite communications antenna, one CAAIS action information system, two Mk 36 Super RBOC chaff launchers, and one UAA-1 Abbey Hill ESM system

Propulsion: COGOG (COmbined Gas turbine Or Gas turbine) arrangement, with two Rolls-Royce Olympus TM3B gas turbines delivering 56,000 bhp (41760 kW) and two Rolls-Royce Tyne RM1A gas turbines delivering 8,500 bhp (6340 kW), or (from F94 onwards) COGAG (COmbined Gas turbine And Gas turbine) arrangement, with two Rolls-Royce Spey SM1A gas turbines delivering 18,770 shp (13995 kW) and two Rolls-Royce Tyne RM1A gas turbines delivering 8,500 bhp (6340 kW), in each arrangement to two shafts
Performance: speed 30 kts on Olympus engines or 18 kts on Tyne engines; range 5,200 miles (8370 km) at 18 kts on Tyne engines
Complement: 18+205, with a maximum possible of 290

Class
1. UK ('**Broadsword Batch 1**' class)

Name	No.	Builder	Laid down	Commissioned
Broadsword	F88	Yarrow	Feb 1975	May 1979
Battleaxe	F89	Yarrow	Feb 1976	Mar 1980
Brilliant	F90	Yarrow	Mar 1977	May 1981
Brazen	F91	Yarrow	Aug 1978	Jul 1982

2. UK ('**Broadsword Batch 2**' or '**Boxer**' class)

Name	No.	Builder	Laid down	Commissioned
Boxer	F92	Yarrow	Nov 1979	Jan 1984
Beaver	F93	Yarrow	Jun 1980	Dec 1984
Brave	F94	Yarrow	May 1982	Jul 1986
London	F95	Yarrow	Feb 1983	
Sheffield	F96	Swan Hunter	Mar 1984	
Coventry	F97	Swan Hunter	Mar 1984	

3. UK ('**Broadsword Batch 3**' or '**Cornwall**' class)

Name	No.	Builder	Laid down	Commissioned
Cornwall	F98	Yarrow	Dec 1983	
Cumberland	F99	Yarrow	Dec 1983	
Campbeltown	F100	Cammell Laird	Dec 1985	
Chatham	F101	Swan Hunter	May 1986	

Note
This class was designed as successor to the 'Leander' class, and though rated as frigates the ships are in reality destroyers optimized for anti-submarine warfare in North Atlantic waters; F88 to F91 are '**Broadsword Batch 1**' ships, which proved too short to take the new Type 2031(Z) towed-array sonar, so the next six ships (F92 to F97) were the lengthened '**Broadsword Batch 2**' or '**Boxer**' **class** with this sonar and (from F94 onwards) a revised propulsion arrangement and provision for larger helicopters; the lessons of the Falklands campaign in 1982 dictated the succeeding '**Broadsword Batch 3**' or '**Cornwall**' **class** (F98 to F101) with improved gun, point-defence and anti-ship missile armament for its planned four units

'C65' class DDG
(France)
Type: anti-submarine and anti-ship guided-missile escort destroyer
Displacement: 3,500 tons standard and 3,900 tons full load
Dimensions: length 127.0 m (416.7 ft); beam 13.4 m (44.0 ft); draught 5.8 m (18.9 ft)
Gun armament: two 100-mm (3.9-in) L/55 DP in single mountings
Missile armament: four container-launchers for four MM.38 Exocet anti-ship missiles
Torpedo armament: none
Anti-submarine armament: one single launcher for 13 Malafon missiles, one 305-mm (12-in) quadruple mortar, and two single launchers for 533-mm (21-in) L5 torpedoes
Aircraft: none
Electronics: one DRBV 13 surveillance radar, one DRBV 22A air-search radar, one DRBC 32B gun fire-control radar, one DRBN 32 navigation radar, one DUBV 23 hull-mounted active search and attack

sonar, one DUBV 43 variable-depth sonar, one SENIT 3 action information system, and two Syllex chaff launchers
Propulsion: two boilers supplying steam to two sets of Rateau geared turbines delivering 21350 kW (28,635 shp) to two shafts
Performance: speed 27 kts; range 9250 km (5,750 miles) at 18 kts
Complement: 15+89+125

Class
1. France

Name	No.	Builder	Laid down	Commissioned
Aconit	D609	Lorient ND	Jan 1966	Mar 1973

Note
This is the obsolescent predecessor to the 'F67' class

'C70/AA' class DDG
(France)
Type: anti-aircraft and anti-ship guided-missile escort destroyer
Displacement: 4,000 tons standard and 4,480 tons full load
Dimensions: length 139.0 m (456.0 ft); beam 14.0 m (45.9 ft); draught 5.7 m (18.7 ft)
Gun armament: one 100-mm (3.9-in) L/55 DP in a single mounting, and two 20-mm AA in single mountings
Missile armament: four twin container-launchers for eight MM.40 Exocet anti-ship missiles, one Mk 13 single launcher for 40 RIM-66 Standard SM-1 MR SAMs, and two Sadral point-defence systems with ? Mistral SAMs
Torpedo armament: none
Anti-submarine armament: two single launchers for 10 533-mm (21-in) L5 torpedoes
Aircraft: one Aérospatiale SA365F Dauphin helicopter on a platform aft
Electronics: one DRBJ 11B 3D radar, one DRBV 26 surveillance radar, one DRBC 32D fire-control radar, two SPG-51C SAM fire-control radars, two RM1229 navigation radars, one Vega weapon direction system, one Panda fire-control director, one DUBA 25A hull-mounted sonar, one EBTF towed-array sonar, one ARBR 17 radar detector, one ARBB 33 jammer, one Vampir IR surveillance system, one SENIT 6 action information system, two Dagaie chaff launchers, and two Sagaie chaff launchers
Propulsion: four SEMT-Pielstick 18 PA6 BTC diesels delivering 31800 kW (42,645 shp) to two shafts
Performance: speed 30 kts; range 15200 km (9,445 miles) at 17 kts or 9250 km (5,750 miles) at 24 kts
Complement: 251

Class
1. France

Name	No.	Builder	Laid down	Commissioned
Cassard	D614	Lorient ND	Sep 1982	1988
Jean Bart	D617	Lorient ND	May 1984	1989
Jean Bart	D616	Lorient ND	May 1986	
Cheralier Paul	D617	Lorient ND	May 1986	

Note
This is a capable anti-aircraft destroyer type with a totally revised armament and propulsion system compared with the 'C70/ASW' type

'C70/ASW' class DDG
(France)
Type: anti-submarine and anti-ship guided-missile escort destroyer
Displacement: 3,830 tons standard and 4,350 tons full load
Dimensions: length 139.0 m (456.0 ft); beam 14.0 m (45.9 ft); draught 5.7 m (18.7 ft)
Gun armament: one 100-mm (3.9-in) L/55 DP in a single mounting, and two 20-mm AA in single mountings
Missile armament: four container-launchers for four MM.38 Exocet (MM.40 Exocet from D642 onwards) anti-ship missiles, and one Naval Crotale octuple launcher for 26 Matra R.440 SAMs
Torpedo armament: none
Anti-submarine armament: two single launchers for 10 533-mm (21-in) L5 torpedoes, and helicopter-launched weapons (see below)
Aircraft: two Westland Lynx Mk 2 helicopters in a hangar aft
Electronics: one DRBV 51C surveillance radar, one DRBV 26 air-search radar, one DRBC 32D fire-control radar, one SPG-51C SAM-control radar, two RM1226 navigation radars, one DUBV 23D hull-mounted active search and attack sonar, one DUBV 43B variable-depth sonar (D640/D643) or DSBV 61 towed-array sonar (D644/D646), one Panda optical

fire-control director, one Vega system, one SENIT 4 action information system, one ARBB 32B jammer, one ABRB 16 radar detector and two Syllex chaff launchers (D640/D642) or one ABRB 17 radar detector and two Dagaie chaff launchers (D643/D646 and to be retrofitted in earlier ships)
Propulsion: CODOG (COmbined Diesel Or Gas turbine) arrangement, with two Rolls-Royce Olympus TM3B gas turbines delivering 52,000 bhp (38775 kW) and two SEMT-Pielstick 16 PA6 CV280 diesels delivering 7750 kW (10,395 bhp) to two shafts
Performance: speed 30 kts on gas turbines or 21 kts on diesels; range 15750 km (9,785 miles) at 17 kts on diesels
Complement: 15+90+111

Class
1. France

Name	No.	Builder	Laid down	Commissioned
Georges Leygues	D640	Brest ND	Sep 1974	Dec 1979
Dupleix	D641	Brest ND	Oct 1975	Jun 1981
Montcalm	D642	Brest ND	Dec 1975	May 1982
Jean de Vienne	D643	Brest ND	Oct 1979	May 1984
Primauguet	D644	Brest ND	Nov 1981	1986
La Motte-Piquet	D645	Brest ND	Feb 1982	1988
	D646	Brest ND	Feb 1984	1990

'Charles F. Adams' class DDG
(USA)
Type: anti-aircraft, anti-ship and anti-submarine guided-missile escort destroyer
Displacement: 3,370 tons standard and 4,500 tons full load
Dimensions: length 437.0 ft (133.2 m); beam 47.0 ft (14.3 m); draught 22.0 ft (6.7 m)
Gun armament: two 5-in (127-mm) L/54 DP in Mk 42 single mountings
Missile armament: one Mk 11 twin launcher (DDG2/DDG14) for 42 RGM-84A Harpoon anti-ship missiles and RIM-66B Standard SM-1 MR SAMs, or one Mk 13 single launcher (DDG15/DDG24) for 40 RGM-84A Harpoon anti-ship missiles and RIM-66B Standard SM-1 MR SAMs
Torpedo armament: none
Anti-submarine armament: one Mk 16 octuple launcher for eight or (some ships only) 12 RUR-5A ASROC missiles, and two Mk 32 triple mountings for 12.75-in (324-mm) Mk 46 torpedoes
Aircraft: none
Electronics: one SPS-39A (being replaced by SPS-52) 3D radar, one SPS-37 (DDG2/DDG14) or SPS-40D (DDG14/DDG24) air-search radar, one SPS-65 low-level air-search radar (DDG19, 20 and 22 only), one SPS-10 surface-search radar, two SPG-51C radars used in conjunction with two Mk 74 SAM fire-control systems, one SPG-53A radar used in conjunction with one Mk 68 gun fire-control system, one Mk 4 (being replaced by Mk 13) weapon direction system, one SQS-23A long-range active sonar (hull-mounted in DDG2/DDG19 and bow-mounted in DDG20/DDG24) used in conjunction with the Mk 111 (DDG2/DDG15) or Mk 114 (DDG16/DDG24) underwater weapons fire-control system, one T Mk 6 Fanfare torpedo decoy system, one OE-82 satellite communications antenna, one SRR-1 satellite communications receiver, two WSC-3 satellite communications transceivers, URN-20 TACAN, one WLR-6 ECM system, one ULQ-6B ECM system and two Mk 36 Super RBOC chaff launchers; of the ships of this class DDG17, 19, 20 and 22/24 are being electronically upgraded with the Anti-Ship Missile Defense programme, which includes SLQ-31 and SLQ-32 ESM systems, SPS-40C or SPS-40D radar instead of the SPS-40 as part of the SYS-1 Integrated Automatic Detection and Tracking System, Mk 86 fire-control system with SPG-60 and SPQ-9 radars for joint control of the gun and improved RIM-66C Standard SM-2 MR SAM armament, and SQQ-23A sonar
Propulsion: four boilers supplying steam to two sets of General Electric or Westinghouse geared turbines delivering 70,000 shp (52200 kW) to two shafts
Performance: speed 31.5 kts; range 6,900 miles (11105 km) at 14 kts or 1,840 miles (2960 km) at 30 kts
Complement: 21+340

Class
1. Australia ('**Perth**' class)

Name	No.	Builder	Laid down	Commissioned
Perth	D38	Defoe SB	Sep 1962	Jul 1965
Hobart	D39	Defoe SB	Oct 1962	Dec 1965
Brisbane	D41	Defoe SB	Feb 1965	Dec 1967

(The Australian ships are similar to the American norm, but have the Ikara anti-submarine missile [two single launchers and 32 missiles per ship] with UQC 1D and UQN 1 sonars.)

2. USA

Name	No.	Builder	Laid down	Commissioned
Charles F. Adams	DDG2	Bath Iron Works	Jun 1958	Sep 1960
John King	DDG3	Bath Iron Works	Aug 1958	Feb 1961
Lawrence	DDG4	New York SB	Oct 1958	Jan 1962
Claude V. Ricketss	DDG5	New York SB	May 1959	May 1962
Barney	DDG6	New York SB	May 1959	Aug 1962
Henry B. Wilson	DDG7	Defoe SB	Feb 1958	Dec 1960
Lynde McCormick	DDG8	Defoe SB	Apr 1958	Jun 1961
Towers	DDG9	Todd Pacific	Apr 1958	Jun 1961
Sampson	DDG10	Bath Iron Works	Mar 1959	Jun 1961
Sellers	DDG11	Bath Iron Works	Aug 1959	Oct 1961
Robison	DDG12	Defoe SB	Apr 1959	Dec 1961
Hoel	DDG13	Defoe SB	Jun 1959	Jun 1962
Buchanan	DDG14	Todd Pacific	Apr 1959	Feb 1962
Berkeley	DDG15	New York SB	Jun 1960	Dec 1962
Joseph Strauss	DDG16	New York SB	Dec 1960	Apr 1963
Conyngham	DDG17	New York SB	May 1961	Jul 1963
Semmes	DDG18	Avondale Marine	Aug 1960	Dec 1962
Tattnall	DDG19	Avondale Marine	Nov 1960	Apr 1963
Goldsborough	DDG20	Puget Sound Bridge	Jan 1961	Nov 1963
Cochrane	DDG21	Puget Sound Bridge	Jul 1961	Mar 1964
Benjamin Stoddert	DDG22	Puget Sound Bridge	Jun 1962	Sep 1964
Richard E. Byrd	DDG23	Todd Pacific	Apr 1961	Mar 1964
Waddell	DDG24	Todd Pacific	Feb 1962	Aug 1964

'Charles F. Adams (Modified)' or 'Type 103B' class DDG
(USA/West Germany)
Type: anti-aircraft, anti-submarine and anti-ship guided-missile destroyer
Displacement: 3,370 tons standard and 4,500 tons full load
Dimensions: length 437.0 ft (133.2 m); beam 47.0 ft (14.3 m); draught 222.0 ft (6.7 m)
Gun armament: one 5-in (127-mm) L/54 DP in Mk 42 single mountings (but see below, Missile armament)
Missile armament: two quadruple container-launchers for eight RGM-84A Harpoon anti-ship missiles, one Mk 13 single launcher for 40 RGM-84A Harpoon anti-ship missiles and RIM-67A Standard SM-1 ER SAMs, and (as a retrofit) two EX-31 24-round container-launchers for 48 RAM-ASDM short-range SAMs
Torpedo armament: none
Anti-submarine armament: one Mk 16 octuple launcher for eight RUR-5A ASROC missiles, and two Mk 32 triple mountings for 12.75-in (324-mm) Mk 46 torpedoes
Aircraft: none
Electronics: one SPS-52 3D radar, one SPS-40 air-search radar, one SPS-10 surface-search radar, one SPG-60 radar and one SPQ-9 radar used in conjunction with the Mk 86 SAM fire-control system, one Mk 68 gun fire-control system, one SQS-23 hull-mounted long-range active sonar, URN-22 TACAN, one SATIR 1 action information system, and various ESM systems
Propulsion: four boilers supplying steam to two sets of geared turbines delivering 70,000 shp (52200 kW) to two shafts
Performance: speed 31.5 kts; range 5,200 miles (8370 km) at 20 kts
Complement: 19+319

Class
1. West Germany

Name	No.	Builder	Laid down	Commissioned
Lütjens	D185	Bath Iron Works	Mar 1966	Mar 1969
Mölders	D186	Bath Iron Works	Apr 1966	Sep 1969
Rommel	D187	Bath Iron Works	Aug 1967	May 1970

Note
These are very seaworthy multi-role ships with a 'mack' (combined mast and stack) layout

'Coontz' class DDG (USA): see 'Farragut' class DDG

'Cornwall' class DDG: (UK): see 'Broadsword' class DDG

'County' class DDG
(UK)
Type: anti-aircraft and anti-ship guided-missile destroyer

Displacement: 6,200 tons standard and 6,800 tons full load
Dimensions: length 520.5 ft (158.7 m); beam 54.0 ft (16.5 m); draught 20.5 ft (6.3 m)
Gun armament: two 4.5-in (114-mm) L/45 DP in one Mk 6 twin mounting, and two 20-mm AA in single mountings
Missile armament: four container-launchers for four MM.38 Exocet anti-ship missiles, one twin launcher for 36 Sea Slug Mk 2 SAMs, and two quadruple launchers for 32 Sea Cat SAMs
Torpedo armament: none
Anti-submarine armament: two STWS triple mountings for 12.75-in (324-mm) Mk 46 or Stingray torpedoes, and helicopter-launched weapons (see below)
Aircraft: one Westland Lynx HAS.Mk 2 helicopter in a hangar aft
Electronics: one Type 992Q surveillance radar, one Type 965M air-search radar, one Type 978M height-finding radar, one Type 901 Sea Slug-control radar, two Type 904 radars used in conjunction with two GWS 22 Sea Cat fire-control systems, one Type 903 radar used in conjunction with the MRS3 gun fire-control system, one Type 978 or Type 1006 navigation radar, one Type 176 hull-mounted panoramic sonar, one Type 177 hull-mounted medium-range panoramic sonar, one Type 192 hull-mounted sonar, one Type 182 torpedo decoy system, one ADAWS 1 action information system, two Corvus chaff launchers, and extensive ESM systems
Propulsion: COSAG (COmbined Steam And Gas turbine) arrangement, with two boilers supplying steam to two sets of AEI geared turbines delivering 30,000 shp (22370 kW) and four G.6 gas turbines delivering 30,000 shp (22370 kW) to two shafts
Performance: speed 30 kts; range 4,000 miles (6440 km) at 28 kts
Complement: 34f438

Class
1. Chile

Name	No.	Builder	Laid down	Commissioned
Prat	11	Swan Hunter	Mar 1966	Mar 1970
Almirante Cochrane	12	Fairfield SB	Jan 1966	Jul 1970
Aguirre	13	Vickers	Sep 1962	Oct 1966

(These Chilean ships may be refitted with Franch weapons and electronics.)

2. Pakistan

Name	No.	Builder	Laid down	Commissioned
Babur	C84	Swan Hunter	Feb 1960	Nov 1963

(The Sea Slug SAM system has been removed from this Pakistani vessel, which can now operate a Sea King helicopter with provision for anti-ship missiles in addition to its conventional anti-submarine weapons and sensors.)

3. UK

Name	No.	Builder	Laid down	Commissioned
Fife	D20	Fairfield SB	Jun 1962	Jun 1966

Note
These large destroyers are nearly as capable as cruisers, but the SAM system is now obsolescent

'Daring' class DDG
(UK/Peru)
Type: anti-ship guided-missile destroyer
Displacement: 2,800 tons standard and 3,600 tons full load
Dimensions: length 390.0 ft (118.9 m); beam 43.0 ft (13.1 m); draught 18.0 ft (5.5 m)
Gun armament: four 4.5-in (114-mm) L/45 DP in two Mk 6 twin mountings, and four 40-mm Bofors L/60 AA in two Mk 5 twin mountings
Missile armament: eight (DD73) or six (DD74) container-launchers for eight or six MM.38 Exocet anti-ship missiles
Torpedo armament: none
Anti-submarine armament: one Squid mortar
Aircraft: provision for one light helicopter on a platform aft
Electronics: one AWS 1 air-search radar, one combined air- and surface-search radar, one Decca navigation radar, and one hull-mounted sonar
Propulsion: two boilers supplying steam to two sets of General Electric geared turbines delivering 54,000 shp (40270 kW) to two shafts
Performance: speed 32 kts; range 3,450 miles (5550 km) at 20 kts
Complement: 297

Class
1. Chile

Name	No.	Builder	Laid down	Commissioned
Palacios	DD73	Yarrow	Apr 1947	Mar 1954

Ferre	DD74	Yarrow	Sep 1946	Apr 1953

Note
The ships are obsolete by all but local standards

'Exeter' class DDG: (UK): see 'Sheffield' class DDG

'F67' class DDG
(France)
Type: anti-submarine and anti-ship guided-missile escort destroyer
Displacement: 4,580 tons standard and 5,745 tons full load
Dimensions: length 152.75 m (501.1 ft); beam 15.3 m (50.2 ft); draught 5.7 m (18.7 ft)
Gun armament: two 100-mm (3.9-in) L/55 DP in single mountings, and two 20-mm AA in single mountings
Missile armament: four container-launchers for four MM.38 Exocet anti-ship missiles, and one Naval Crotale octuple launcher for 26 Matra R.440 SAMs
Torpedo armament: none
Anti-submarine armament: one launcher for 13 Malafon missiles, and two single launchers for 10 533-mm (21-in) L5 torpedoes
Aircraft: none
Electronics: one DRBV 51 combined air- and surface-search radar, one DRBV 26 air-search radar, one DRBC 32D gun fire-control radar used in conjunction with the Vega fire-control system, two RM1226 navigation radars, one DUBV 23 hull-mounted search and attack sonar, one DUBV 43 variable-depth sonar, one SENIT 3 action information system, and two Dagaie chaff launchers
Propulsion: four boilers supplying steam to two sets of Rateau geared turbines delivering 40500 kW (54,320 shp) to two shafts
Performance: speed 32 kts; range 9250 km (5,750 miles) at 18 kts or 3500 km (2,175 miles) at 30 kts
Complement: 17+113+162

Class
1. France

Name	No.	Builder	Laid down	Commissioned
Tourville	D610	Lorient ND	Mar 1970	Jun 1974
Duguay-Trouin	D611	Lorient ND	Feb 1971	Sep 1975
De Grasse	D612	Lorient ND	Jun 1972	Oct 1977

Note
Designed for a primary anti-submarine role, these ships also possess good short-range anti-aircraft capability for self-defence

'Farragut' class DDG
(USA)
Type: anti-aircraft, anti-ship and anti-submarine guided-missile escort destroyer
Displacement: 4,700 tons standard and 5,800 tons full load
Dimensions: length 512.5 ft (156.3 m); beam 52.5 ft (16.0 m); draught 23.4 ft (7.1 m)
Gun armament: one 5-in (127-mm) L/54 DP in a Mk 42 single mounting
Missile armament: two Mk 141 quadruple container-launchers for eight RGM-84A Harpoon anti-ship missiles, and one Mk 10 twin launcher for 40 RIM-67B Standard SM-2 ER SAMs
Torpedo armament: none
Anti-submarine armament: one Mk 16 octuple launcher for eight RUR-5A ASROC missiles, two Mk 32 triple mountings for 12.75-in (324-mm) Mk 46 torpedoes, and (sometimes) helicopter-launched weapons (see below)
Aircraft: provision for a Kaman SH-2F Seasprite helicopter on a platform aft
Electronics: one SPS-48 3D radar, one SPS-29E (DDG37 and DDG38) or SPS-49 (DDG39/DDG46) long-range air-search radar, one SPS-10B surface-search radar, two SPG-55B radars used in conjunction with two Mk 76 SAM fire-control systems, one SPG-53 radar used in conjunction with the Mk 68 gun fire-control system, one Mk 11 (being replaced by Mk 14) weapon direction system, one SQQ-23 hull-mounted long-range sonar used in conjunction with the Mk 111 underwater weapons fire-control system, one T Mk 6 Fanfare torpedo decoy system, Naval Tactical Data System, URN-20 TACAN, OE-82 satellite communications system, SRR-1 satellite communications receiver, WSC-3 satellite communications transceiver, two Mk 36 Super RBOC chaff launchers, and various ESM systems
Propulsion: four boilers supplying steam to two sets of geared turbines delivering 85,000 shp (63385 kW) to two shafts
Performance: speed 33 kts; range 5,750 miles (9255 km) at 20 kts
Complement: 25+378, plus flag accommodation of 7+12

Class
1. USA

Name	No.	Builder	Laid down	Commissioned
Farragut	DDG37	Bethlehem Steel	Jun 1957	Dec 1960
Luce	DDG38	Bethlehem Steel	Oct 1957	May 1961
MacDonough	DDG39	Bethlehem Steel	Apr 1958	Nov 1961
Coontz	DDG40	Puget Sound NY	Mar 1957	Jul 1960
King	DDG41	Puget Sound NY	Mar 1957	Nov 1960
Mahan	DDG42	San Francisco NY	Jul 1957	Aug 1960
Dahlgren	DDG43	Philadelphia NY	Mar 1958	Apr 1961
William V. Pratt	DDG44	Philadelphia NY	Mar 1958	Nov 1961
Dewey	DDG45	Bath Iron Works	Aug 1957	Dec 1959
Preble	DDG46	Bath Iron Works	Dec 1957	May 1960

Note
Also known as the **'Coontz' class**, these ships are capable anti-submarine and long-range anti-aircraft destroyers with good anti-ship missile armament and advanced electronics

'Halifax' class DDG
(Canada)
Type: anti-submarine and anti-ship guided-missile escort destroyer
Displacement: 4,255 tons full load
Dimensions: length 437.9 ft (133.5 m); beam 53.8 ft (16.4 m); draught 15.1 ft (4.6 m)
Gun armament: one 57-mm Bofors L/70 DP in a single mounting, and one 20-mm Phalanx Mk 16 CIWS mounting
Missile armament: two Mk 141 quadruple container-launchers for eight RGM-84 Harpoon anti-ship missiles, and two octuple vertical-launch systems for ? RIM-7 Sea Sparrow SAMs
Torpedo armament: none
Anti-submarine armament: two Mk 32 twin mountings for 12.75-in (324-mm) Mk 46 torpedoes, and helicopter-launched weapons (see below)
Aircraft: two Sikorsky CH-124A Sea King helicopters in a hangar aft
Electronics: one CMR 1820 air-search radar, one Type 1031 surface-search radar, one Raytheon 1629C navigation radar, one WM-25 fire-control system, one SQS-505 hull-mounted sonar and one SQR-19 towed-array sonar used in conjunction with an underwater weapons fire-control system, four Mk 36 Super RBOC chaff launchers, and one CANEWS ECM system
Propulsion: two LM2500 gas turbines delivering 50,000 shp (37285 kW) to two shafts
Performance: speed 29+ kts; range 6,550 miles (10540 km) at 15 kts
Complement: 225

Class
1. Canada

Name	No.	Builder	Laid down	Commissioned
Halifax	DD330	St John SB	Apr 1986	
Vancouver	DD331	Marine Industries		
Ville de Quebec	DD332	St John SB		
Toronto	DD333	Marine Industries		
Regina	DD334	St John SB		
Calgary	DD335	Marine Industries		

'Halland' class DDG
(Sweden)
Type: anti-ship and anti-submarine guided-missile destroyer
Displacement: 2,800 tons standard and 3,400 tons full load
Dimensions: length 121.0 m (397.2 ft); beam 12.6 m (41.3 ft); draught 5.5 m (18.0 ft)
Gun armament: four 120-mm (4.7-in) L/50 DP in two twin mountings, two 57-mm Bofors L/50 AA in a twin mounting, and six Bofors 40-mm L/48 AA in single mountings
Missile armament: one Mk 20 twin launcher for RB 08A anti-ship missiles
Torpedo armament: one 533-mm (21-in) quintuple and one 533-mm (21-in) triple mounting for dual-role torpedoes
Anti-submarine armament: two 375-mm (14.76-in) Bofors four-barrel rocket-launchers, and torpedoes (see above)
Aircraft: none
Electronics: one 9 LV 200 combined air- and surface-search radar, one WM-20 fire-control radar, one hull-mounted search sonar, and one hull-mounted attack sonar

Propulsion: two boilers supplying steam to two sets of De Laval geared turbines delivering 43250 kW (58,000 shp) to two shafts

Performance: speed 35 kts; range 5500 km (3,415 miles) at 20 kts
Complement: 18+272

Class
1. Sweden

Name	No.	Builder	Laid down	Commissioned
Halland	J18	Gotaverken	1951	Jun 1955

Note
This ship is now totally obsolete by Western European standards

'Hamburg' or 'Type 101A' class DDG
(West Germany)
Type: guided-missile destroyer
Displacement: 3,340 tons standard and 4,680 tons full load
Dimensions: length 133.8 m (439.0 ft); beam 13.4 m (44.0 ft); draught 6.2 m (20.3 ft)
Gun armament: three 100-mm (3.9-in) L/55 DP in single mountings, and eight 40-mm Breda AA in four twin mountings
Missile armament: two twin container-launchers for four MM.38 Exocet anti-ship missiles
Torpedo armament: none
Anti-submarine armament: two 375-mm (14.76-in) Bofors four-barrel rocket-launchers, one 533-mm (21-in) quadruple mounting for ? torpedoes, and two depth-charge throwers
Aircraft: none
Mines: between 60 and 80 depending on type
Electronics: one LW-02 air-search radar, one DA-02 surface-search and target indication radar, one Decca navigation radar, two WM-45 main armament fire-control radars, two WM-45 AA armament fire-control radars, one ELAC 1BV hull-mounted sonar used in conjunction with a Hollandse Signaalapparaten underwater weapons fire-control system, one SCLAR chaff launcher, and various ESM systems
Propulsion: four boilers supplying steam to two sets of Wahodag geared turbines delivering 50700 kW (67,990 shp) to two shafts
Performance: speed 34 kts; range 11125 km (6,915 miles) at 13 kts or 1700 km (1,055 km) at 34 kts
Complement: 19+249

Class
1. West Germany

Name	No.	Builder	Laid down	Commissioned
Hamburg	D181	H.C. Stülcken	Jan 1959	Mar 1964
Schleswig-Holstein	D182	H.C. Stülcken	Aug 1959	Oct 1964
Bayern	D183	H.C. Stülcken	Sep 1960	Jul 1965
Hessen	D184	H.C. Stülcken	Feb 1961	Oct 1968

Note
These ships are obsolescent by West European standards, being lacking in modern electronics and anti-aircraft armament

'Haruna' class DDG
(Japan)
Type: anti-submarine and anti-ship guided-missile destroyer
Displacement: 4,700 tons standard and 6,300 tons full load
Dimensions: length 153.0 m (502.0 ft); beam 17.5 m (57.4 ft); draught 5.1 m (16.7 ft)
Gun armament: two 5-in (127-mm) L/54 DP in Mk 42 single mountings, and two 20-mm Phalanx Mk 16 CIWS mountings
Missile armament: two Mk 141 quadruple container-launchers for eight RGM-84A Harpoon anti-ship missiles, and one octuple launcher for eight RIM-7 Sea Sparrow SAMs
Torpedo armament: none
Anti-submarine armament: one Mk 16 octuple launcher for 16 RUR-5A ASROC missiles, two Mk 32 triple mountings for 12.75-in (324-mm) Mk 46 torpedoes, and helicopter-launched weapons (see below)
Aircraft: three Sikorsky SH-3 Sea King helicopters in a hangar aft
Electronics: one SPS-52B 3D radar, one OPS-17 surface-search radar, two Type 72 gun fire-control radars, one WM-25 SAM-control radar, URN-20 TACAN, one OQS-3 hull-mounted sonar, one SQS-3S(J) variable-depth sonar, and one comprehensive ESM system
Propulsion: ? boilers supplying steam to two sets of General Electric/Ishikawajima geared turbines delivering 52200 kW (70,000 shp) to two shafts
Performance: speed 32 kts
Complement: 364

214 WARSHIPS

Class
1. Japan

Name	No.	Builder	Laid down	Commissioned
Haruna	DD141	Mitsubishi	Mar 1970	Feb 1973
Hiei	DD142	Ishikawajima	Mar 1972	Nov 1974

Note
Predecessors to the 'Shirane' class units, the 'Haruna' class destroyers can each operate three large Sea King anti-submarine helicopters as part of a powerful anti-submarine armament now complemented by adequate anti-ship and close air-defence armament

'Hatakaze' class DDG
(Japan)
Type: anti-aircraft, anti-submarine and anti-ship guided-missile destroyer
Displacement: 4,600 tons standard
Dimensions: length 150.0 m (492.1 ft); beam 16.4 m (53.8 ft); draught 4.7 m (15.4 ft)
Gun armament: two 5-in (127-mm) L/54 DP in Mk 42 single mountings, and two 20-mm Phalanx Mk 16 CIWS mountings
Missile armament: two Mk 141 quadruple container-launchers for eight RGM-84A Harpoon anti-ship missiles, and one Mk 13 single launcher for 40 RIM-66 Standard SM-1 MR SAMs
Torpedo armament: none
Anti-submarine armament: one Mk 16 octuple launcher for 16 RUR-5A ASROC missiles, and two Type 68 triple mountings for 12.75-in (324-mm) Mk 46 torpedoes
Aircraft: provision for one Sikorsky SH-3 Sea King helicopter on a platform aft
Electronics: one OPS-12 3D radar, one OPS-11 air-search radar, one OPS-28 surface-search radar, two SPG-51 radars used in conjunction with two Mk 73 SAM fire-control systems, one GFCS-2 gun fire-control system, one OQS-4 hull-mounted sonar used in conjunction with the Mk 114 underwater weapons fire-control system, one action information system, one NOLQ-1 ECM system, one OLT-3 ESM system, and four Mk 36 Super RBOC chaff launchers
Propulsion: COGAG (COmbined Gas turbine And Gas turbine) arrangement, with two Rolls-Royce Olympus TM3B gas turbines and two Rolls-Royce Spey SM1A gas turbines delivering a total of 90,200 shp (67260 kW) to two shafts
Performance: speed 32 kts
Complement: 260

Class
1. Japan

Name	No.	Builder	Laid down	Commissioned
Hatakaze	DD171	Mitsubishi	1983	Mar 1986
	DD172	Mitsubishi	1984	1988

Note
This is a useful and balanced design optimized for anti-aircraft warfare but possessing capable anti-ship and anti-submarine armament

'Hatsuyuki' class DDG
(Japan)
Type: anti-submarine and anti-ship guided-missile destroyer
Displacement: 2,950 tons standard and 3,700 tons full load
Dimensions: length 128.0 m (419.9 ft); beam 13.7 m (44.9 ft); draught 4.3 m (14.1 ft)
Gun armament: one 76-mm (3-in) OTO Melara Compact L/62 DP in a single mounting, and two 20-mm Phalanx Mk 16 CIWS mountings
Missile armament: two Mk 141 quadruple container-launchers for eight RGM-84A Harpoon anti-ship missiles, and one Mk 29 octuple launcher for eight RIM-7 NATO Sea Sparrow SAMs
Torpedo armament: none
Anti-submarine armament: one Mk 16 octuple launcher for 16 RUR-5A ASROC missiles, two Type 68 triple mountings for 12.75-in (324-mm) Mk 46 torpedoes, and helicopter-launched weapons (see below)
Aircraft: one Sikorsky SH-3 Sea King helicopter in a hangar aft
Electronics: one OPS-14B air-search radar, one OPS-18 surface-search radar, one FCS2-12A missile-control system, one GFCS2-21 gun-control system, one OQS-4 hull-mounted sonar, one NOLR-6C passive ECM system, one OLT-3 ESM system, and two Mk 36 Super RBOC chaff launchers
Propulsion: COGOG (COmbined Gas turbine Or Gas turbine) arrangement, with two Rolls-Royce Olympus TM3B gas turbines delivering 45,000 shp (33555 kW) and two Rolls-Royce Tyne RM1C gas turbines delivering 19,680 shp (7965 kW) to two shafts
Performance: speed 30 kts
Complement: 190

Class
1. Japan

Name	No.	Builder	Laid down	Commissioned
Hatsuyuki	DD122	Sumitomo	Mar 1979	Mar 1982
Shirayuki	DD123	Hitachi	Dec 1979	Feb 1983
Mineyuki	DD124	Mitsubishi	May 1981	Mar 1984
Sawayuki	DD125	Ishikawajima	Apr 1981	Feb 1984
Hamayuki	DD126	Mitsui	Apr 1981	Jan 1984
Isoyuki	DD127	Ishikawajima	Apr 1982	Jan 1985
Harayuki	DD128	Sumitomo	Apr 1982	Mar 1985
Yamayuki	DD129	Hitachi	Oct 1983	1986
Matsuyuki	DD130	Ishikawajima	Oct 1983	1986
	DD131	Mitsui	Jan 1984	
	DD132	Sumitomo	Dec 1983	
	DD133	Mitsubishi	May 1984	

Note
These are comparatively small destroyers with a neat balance of anti-submarine, anti-ship and anti-aircraft armament

'Hatsuyuki (Improved)' class DDG
(Japan)
Type: anti-submarine and anti-ship guided-missile destroyer
Displacement: 3,400 tons standard
Dimensions: length 136.5 m (447.8 ft); beam 14.6 m (47.9 ft); draught 4.45 m (14.6 ft)
Gun armament: one 76-mm (3-in) OTO Melara Compact in a single mounting, and two 20-mm Phalanx Mk 16 CIWS mountings
Missile armament: two Mk 141 quadruple container-launchers for eight RGM-84A Harpoon anti-ship missiles, and one Mk 29 octuple launcher for eight RIM-7 NATO Sea Sparrow SAMs
Torpedo armament: none
Anti-submarine armament: one Mk 16 octuple launcher for 16 RUR-5A ASROC missiles, two Type 68 triple mountings for 12.75-in (324-mm) Mk 46 torpedoes, and helicopter-launched weapons (see below)
Aircraft: one Sikorsky SH-3 Sea King or SH-60B Seahawk helicopter in a hangar aft
Electronics: one OPS-14C air-search radar, one OPS-28 surface-search radar, one GFCS-2-22 gun fire-control system, one MFCS-2-12E SAM fire-control system, one OQS-4A hull-mounted sonar, one variable-depth sonar, one SQR-19 towed-array sonar, one OLR-9C passive ECM system, one NOLR-6C passive ECM system, and two Mk 36 Super RBOC chaff launchers
Propulsion: COGAG (COmbined Gas turbine And Gas turbine) arrangement, with four Rolls-Royce Spey SM1A gas turbines delivering 75,000 shp (55925 kW) to two shafts
Performance: speed 30 kts
Complement: 230

Class
1. Japan

Name	No.	Builder	Laid down	Commissioned
Asagiri	DD151	Ishikawajima-Harima	Mar 1985	
	DD152			
	DD153			
	DD153			
	DD154			
	DD155			
	DD156			
	DD157			
	DD158			

Note
This is a much developed version of the basic 'Hatsuyuki' design made possible by the adoption of quadruple Spey gas turbines as the main propulsion system; this has permitted optimum arrangement of the machinery spaces and also increased power so that higher speeds can be attained; the opportunity has also been made to revise the disposition of the missile armament and to upgrade the electronics fit

'Impavido' class DDG
(Italy)
Type: anti-aircraft guided-missile destroyer
Displacement: 3,200 tons standard and 3,990 tons full load
Dimensions: length 131.3 m (429.5 ft); beam 13.6 m (44.7 ft); 4.5 m (14.8 ft)
Gun armament: length 131.3 m (429.5 ft); beam 13.6 m (44.7 ft); 4.5 m (14.8 ft)
Gun armament: two 5-in (127-mm) L/38 DP in one Mk 38 twin mounting, and four 76-mm (3-in) OTO Melara L/62 DP in single mountings
Missile armament: one Mk 13 single launcher for 40 RIM-66 Standard SM-1 MR SAMs
Torpedo armament: none
Anti-submarine armament: two Mk 32 triple mountings for 12.75-in (324-mm) Mk 46 torpedoes
Aircraft: provision for one light helicopter on a platform aft
Electronics: one SPS-52B 3D radar, one SPS-12 air-search radar, one SPQ 2 surface-search radar, two SPG-51B radars used in conjunction with two Mk 73 SAM fire-control systems, three RTN 10X radars used in conjunction with three Argo NA10 gun fire-control systems, one SQS-23 hull-mounted sonar, two SCLAR chaff launchers, and one ESM system
Propulsion: four boilers supplying steam to two sets of Tosi geared turbines delivering 52200 kW (70,000 shp) to two shafts
Performance: speed 33 kts; range 6100 km (3,790 miles) at 20 kts or 2775 km (1,725 miles) at 30 kts
Complement: 23+317

Class
1. Italy

Name	No.	Builder	Laid down	Commissioned
Impavido	D570	CNR, Riva Trigoso	Jun 1957	Nov 1963
Intrepido	D571	Italcantieri	May 1959	Jul 1964

Note
A capable design, the 'Impavido' class is hampered by lack of habitability and inadequate seaworthiness

'Kanin' class DDG
(USSR)
Type: anti-aircraft and anti-submarine guided-missile escort destroyer
Displacement: 3,700 tons standard and 4,700 tons full load
Dimensions: length 139.0 m (455.9 ft); beam 14.7 m (48.2 ft); draught 5.0 m (16.4 ft)
Gun armament: eight 57-mm L/70 AA in two quadruple mountings, and eight 30-mm AA in four twin mountings
Missile armament: one twin launcher for 22 SA-N-1 'Goa' SAMs
Torpedo armament: two 533-mm (21-in) quintuple mountings for 10 dual-role torpedoes
Anti-submarine armament: three RBU 6000 12-barrel rocket-launchers, and torpedoes (see above)
Aircraft: provision for one Kamov Ka-25 'Hormone' helicopter on a platform aft
Electronics: one 'Head Net-C' 3D radar, two 'Don Kay' navigation radars, one 'Peel Group' SAM-control radar, one 'Hawk Screech' 57-mm gun-control radar, two 'Drum Tilt' 30-mm gun-control radars, one hull-mounted sonar, and various ESM systems including four 'Top Hat' and two 'Bell Squat' housings
Propulsion: four boilers supplying steam to two sets of geared turbines delivering 60000 kW (80,470 shp) to two shafts
Performance: speed 34 kts; range 8350 km (5,190 miles) at 16 kts or 2050 km (1,275 miles) at 33 kts
Complement: 350

Class
1. USSR

Name
Boyky
Derzky
Gnevny
Gordy
Gremyashchy
Uporny
Zhguchy
Zorky

Note
Built between 1958 and 1960 as missile-armed 'Krupny' class anti-ship destroyers these obsolescent ships are designated as anti-submarine destroyers but lack much capability in this direction, having been converted (five ships in Leningrad in 1968/72 and the other three ships at Far Eastern yards in 1974/1978) as SAM-equipped escorts for anti-submarine forces

'Kashin' class DDG
(USSR)
Type: anti-aircraft guided-missile escort destroyer
Displacement: 3,750 tons standard and 4,500 tons full load
Dimensions: length 143.3 m (470.7 ft); beam 15.8 m (51.8 ft); draught 4.7 m (15.4 ft)
Gun armament: four 76-mm (3-in) L/60 DP in two twin mountings
Missile armament: two twin launchers for 32 SA-N-1 'Goa' SAMs

Torpedo armament: one 533-mm (21-in) quintuple mounting for five dual-role torpedoes
Anti-submarine armament: two RBU 6000 12-barrel rocket-launchers, two RBU 1000 six-barrel rocket-launchers, torpedoes (see above), and helicopter-launched weapons (see below)
Aircraft: one Kamov Ka-25 'Hormone-A' helicopter on a platform aft
Electronics: one 'Head Net-C' 3D radar and one 'Big Net' air-search radar or two 'Head Net-A' air-search radars, two 'Peel Group' SAM-control radars, two 'Owl Screech' gun-control radars, two 'Don' navigation radars, one hull-mounted high-frequency sonar, 'High Pole-B' IFF, and two 'Watch Dog' ECM systems
Propulsion: four gas turbines delivering 70100 kW (94,020 shp) to two shafts
Performance: speed 38 kts; range 8350 km (5,190 miles) at 18 kts or 2600 km (1,615 miles) at 36 kts
Complement: 280

Class
1. USSR

Name	Builder
Komsomolets Ukrainy	Nikolayev North
Krazny-Kavaz	Nikolayev North
Krazny-Krim	Nikolayev North
Obraztsovy	Zhdanov Yard
Odarenny	Zhdanov Yard
Provorny	Nikolayev North
Reshitelny	Nikolayev North
Skory	Nikolayev North
Smetlivy	Nikolayev North
Soobrazitelny	Nikolayev North
Sposobny	Nikolayev North
Steregushchy	Zhdanov Yard
Strogy	Nikolayev North

Note
Built between 1961 and 1972 as anti-submarine destroyers, these are in fact anti-aircraft destroyers designed for the screening of anti-submarine task groups; this was the world's first major warship class with gas turbine propulsion; *Provorny* was used as trials ship for the SA-N-7 SAM system (aft), and otherwise has two twin 76-mm (3-in) gun mountings, two RBU 6000 rocket-launchers and two quintuple tube mountings; her radar fit includes one 'Top Steer' 3D radar and six 'Front Dome' SAM-control radars

'Kashin (Modified)' class DDG
(USSR)
Type: anti-aircraft and anti-ship guided-missile escort destroyer
Displacement: 3,950 tons standard and 4,750 tons full load
Dimensions: length 146.5 m (480.6 ft); beam 15.8 m (51.8 ft); draught 4.7 m (15.4 ft)

Gun armament: four 76-mm (3-in) L/60 DP in two twin mountings, and four 30-mm AK-630 CIWS mountings
Missile armament: four container-launchers for four SS-N-2C 'Styx' anti-ship missiles, and two twin launchers for 32 SA-N-1 'Goa' SAMs
Torpedo armament: one 533-mm (21-in) quintuple mounting for five dual-role torpedoes
Anti-submarine armament: two RBU 6000 12-barrel rocket-launchers, torpedoes (see above), and helicopter-launched weapons (see below)
Aircraft: provision for one Kamov Ka-25 'Hormone-A' helicopter on a platform aft
Electronics: either one 'Head Net-C' 3D radar and one 'Big Net' air-search radar or two 'Head Net-A' air-search radars, two 'Peel Group' SAM-control radars, two 'Owl Screech' 76-mm gun-control radars, two 'Bass Tilt' CIWS-control radars, two 'Don Kay' or 'Don 2' navigation radars, one hull-mounted medium-frequency sonar, one variable-depth sonar, 'High Pole-B' IFF, two 'Bell Shroud' and two 'Bell Squat' ECM systems, and four chaff launchers
Propulsion: four gas turbines delivering 70100 kW (94,020 shp) to two shafts
Performance: speed 37 kts; range 8350 km (5,190 miles) at 18 kts
Complement: 300

Class
1. USSR

Name	Builder
Ognevoy	Zhdanov Yard
Sderzhanny	Nikolayev North
Slavny	Zhdanov Yard
Smely	Nikolayev North
Smyshlenny	Nikolayev North
Stroyny	Nikolayev North

Note
These have greater anti-ship capability and variable-depth sonar, the conversion programme being undertaken between 1972 and 1980

'Kashin II' class DDG
(USSR)
Type: anti-aircraft and anti-ship guided-missile destroyer
Displacement: 3,950 tons standard and 4,950 tons full load
Dimensions: length 146.5 m (480.5 ft); beam 15.8 m (51.8 ft); draught 4.8 m (15.7 ft)
Gun armament: two 76-mm (3-in) L/60 DP in a twin mounting, eight 30-mm AA in four twin mountings, and ('Kashin II [Mod]' class) two 30-mm AK-630 CIWS mountings
Missile armament: four container-launchers for four SS-N-2C 'Styx' anti-ship missiles, and two twin launchers for 44 SA-N-1 'Goa' SAMs

Torpedo armament: one 533-mm (21-in) quintuple mounting for five dual-role torpedoes
Anti-submarine armament: two RBU 6000 12-barrel rocket-launchers, torpedoes (see above), and helicopter-launched weapons (see below)
Aircraft: provision for one Kamov Ka-25 'Hormone-A' or ('Kashin II [Mod]' class) Ka-27 'Helix-A' helicopter on a platform aft
Electronics: one 'Head Net-C' 3D radar, one 'Big Net' air-search radar, two 'Peel Group' SAM-control radars, one 'Owl Screech' 76-mm gun-control radar, two 'Drum Tilt' AA gun-control radars, two 'Bass Tilt' CIWS-control radars ('Kashin II [Mod] class'), two 'Don Kay' navigation radars, one hull-mounted medium-frequency sonar, one variable-depth sonar, two 'High Pole-B' IFF, four chaff launchers, and ESM systems including two 'Watch Dog', two 'Top Hat-A' and two 'Top Hat-B'
Propulsion: four gas turbines delivering 71,575 kW (95,995 shp) to two shafts
Performance: speed 35 kts; range 8350 km (5,190 miles) at 18 kts or 1675 km (1,040 miles) at 35 kts
Complement: 320

Class
1. India

Name	*No.*	*Builder*	*Commissioned*
Rajput	D51	Nikolayev	1980
Rana	D52	Nikolayev	1982
Ranjit	D53	Nikolayev	1983
Ranvir*	D54	Nikolayev	1986
*	D55	Nikolayev	
*	D56	Nikolayev	

*'Kashin II (Mod)' class

Note
An export variant developed with good anti-ship and anti-aircraft capabilities, plus full provision for an anti-submarine helicopter; it is likely that the total Indian requirement is six units

'Kidd' class DDG
(USA)
Type: anti-aircraft, anti-submarine and anti-ship guided-missile escort destroyer
Displacement: 6,210 tons light and 8,300 tons full load
Dimensions: length 563.3 ft (171.8 m); beam 55.0 ft (16.8 m); draught 30.0 ft (9.1 m)
Gun armament: two 5-in (127-mm) L/54 DP in Mk 45 single mountings, and two 20-mm Phalanx Mk 16 CIWS mountings
Missile armament: two Mk 141 quadruple container-launchers for eight RGM-84A Harpoon anti-ship missiles, and two Mk 26 twin launchers for 52 RIM-67A Standard SM-2 ER SAMs
Torpedo armament: none
Anti-submarine armament: two Mk 32 triple launchers for 12.75-in (324-mm) Mk 46 torpedoes, 16 RUR-5A ASROC missiles fired from the Mk 26 launcher, and helicopter-launched weapons (see below)
Aircraft: two Kaman SH-2F Seasprite or Sikorsky SH-60B Seahawk helicopters in a hangar amidships
Electronics: one SPS-48C 3D radar, one SPS-55 surface-search and navigation radar, two SPG-51D radars used in conjunction with two Mk 74 SAM fire-control systems, one SPG-60 search-and-track radar and one SPQ-9A track-while-scan radar used in conjunction with two Mk 86 gun fire-control systems, one SQS-53A hull-mounted sonar and one SQR-19 towed-array sonar used in conjunction with one Mk 116 underwater weapons fire-control system, Automatic Data Action System, two Mk 36 Super RBOC chaff launchers, and one SLQ-32(V)2 ESM system
Propulsion: four General Electric LM2500 gas turbines delivering 80,000 shp (59655 kW) to two shafts
Performance: speed 33 kts; range 9,200 miles (14805 km) at 17 kts or 3,800 miles (6115 km) at 30 kts
Complement: 20+318

Class
1. USA

Name	*No.*	*Builder*	*Laid down*	*Commissioned*
Kidd	DDG993	Ingalls SB	Jun 1978	May 1981
Callaghan	DDG994	Ingalls SB	Oct 1978	Aug 1981
Scott	DDG995	Ingalls SB	Feb 1979	Oct 1981
Chandler	DDG996	Ingalls SB	May 1979	Mar 1982

Note
These are the world's most powerful multi-role destroyers (really cruisers), and were built to an Iranian order cancelled after the overthrow of the Shah; the units are notable for the excellent blend of anti-aircraft, anti-submarine and anti-ship capabilities

'Kildin (Modified)' class DDG
(USSR)
Type: anti-ship guided-missile destroyer
Displacement: 3,000 tons standard and 3,600 tons full load
Dimensions: length 126.6 m (414.9 ft); beam 13.0 m (42.6 ft); draught 4.9 m (16.1 ft)
Gun armament: four 76-mm (3-in) L/60 DP in two twin mountings, and 16 57-mm L/70 AA in four quadruple mountings
Missile armament: four container-launchers for four SS-N-2C 'Styx' anti-ship missiles
Torpedo armament: two 533-mm (21-in) quintuple mountings for 10 dual-role torpedoes
Anti-submarine armament: two RBU 2500 16-barrel rocket-launchers, and torpedoes (see above)
Aircraft: none
Electronics: one 'Head Net-C' 3D radar or (in *Bedovy* only) one 'Strut Pair' air-search radar, one 'Owl Screech' 76-mm gun-control radar, two 'Hawk Screech' 57-mm gun-control radars, one 'Don-2' navigation radar, one hull-mounted sonar, one 'Square Head' IFF, one 'High Pole-A' IFF, and two 'Watch Dog' ECM housings
Propulsion: four boilers supplying steam to two sets of geared turbines delivering 53700 kW (72,025 shp) to two shafts
Performance: speed 35 kts; range 8350 km (5,190 miles) at 16 kts or 1850 km (1,150 miles) at 34 kts
Complement: 300

Class
1. USSR

Name	Builder
Bedovy	Nikolayev
Neulovimy	Leningrad
Prozorlivy	Nikolayev

Note
Built in the late 1950s, these are obsolescent anti-ship destroyers with very limited anti-submarine and anti-aircraft capabilities; a less capable companion is the **'Kildin' class** *Neuderzhimy* armed with a single launcher for nine obsolete SS-N-1 'Scrubber' anti-ship missiles in place of the after 130-mm (5.1-in) turret of the 'Kotlin' class gun destroyer on which the type is based

'Luda' or 'Type 051' class DDG
(China)
Type: anti-ship guided-missile destroyer
Displacement: 3,250 tons standard and 3,900 tons full load
Dimensions: length 131.0 m (429.8 ft); beam 13.7 m (45.0 ft); draught 4.6 m (15.0 ft)
Gun armament: four 130-mm (5.1-in) L/50 in two twin mountings, eight 37-mm AA in four twin mountings, and eight 25-mm AA in four twin mountings
Missile armament: two triple container-launchers for six HY-2 (CSS-N-1) anti-ship missiles
Torpedo armament: none
Anti-submarine armament: two FQF-2500 12-barrel rocket-launchers, four BMB-series depth-charge throwers, and two depth-charge racks
Mines: up to 80
Aircraft: none
Electronics: one 'Bean Sticks' air-search radar, one 'Square Tie' missile-control radar, 'Wasp Head' and 'Post Lamp' (nos 105/109) or 'Sun Visor-B' and 'Rice Lamp' (others) gun-control radars, one 'Fin Curve' navigation radar, one hull-mounted medium-frequency sonar, and one 'Square Head' IFF
Propulsion: four boilers supplying steam to two sets of geared turbines delivering 53700 kW (72,025 shp) to two shafts
Performance: speed 36 kts; range 5925 km (3,680 miles) at 18 kts
Complement: 340

Class
1. China

Name	No.	Builder	Commissioned
	105	Luda	1971
	106	Luda	1972
	107	Luda	1972
	108	Luda	1973
	109	Luda	1974
	110	Luda	
	111	Luda	
	131	Guangzhou	1974
	132	Guangzhou	1974
	133	Luda	1979
	161	Guangzhou	1980
	162		1980
	163		1980
	164		1981
	165		

Note
Not comparable with Western or Soviet destroyers, the 'Luda' class is derived in part from the 'Kotlin' class, and is expected to number 16 units in all; plans to modernize the class with Western electronics and SAMs have been postponed for financial reasons, but the existence of this class even its present form provides China with powerful long-range capability by local standards

'Manchester' class DDG (UK): see 'Sheffield' class DDG

'Meko 360H2' class DDG
(West Germany/Nigeria)
Type: anti-ship and anti-submarine guided-missile destroyer
Displacement: 3,630 tons full load
Dimensions: length 125.6 m (412.0 ft); beam 15.0 m (49.2 ft); draught 4.3 m (14.1 ft)
Gun armament: one 127-mm (5-in) OTO Melara Compact L/54 DP in a single mounting, and eight 40-mm Bofors L/70 AA in four Breda twin mountings
Missile armament: one quadruple and four single container-launchers for eight Otomat anti-ship missiles, and one octuple Albatros launcher for eight Aspide SAMs
Torpedo armament: none
Anti-submarine armament: two STWS triple mountings for 324-mm (12.75-in) A 244S torpedoes, and helicopter-launched weapons (see below)
Aircraft: one Westland Lynx helicopter in a hangar aft
Electronics: one AWS 5 surface-search radar, one WM-25 and one STIR surface-search and fire-control radars, one RM1226 navigation radar, one KAE 80 hull-mounted sonar, and two SCLAR chaff launchers
Propulsion: CODOG (COmbined Diesel Or Gas turbine) arrangement, with two MTU 20V956 TB92 diesels delivering 7500 kW (10,060 shp) and two Rolls-Royce Olympus TM3B gas turbines delivering 56,000 hp (41760 kW) to two shafts
Performance: speed 30.5 kts on gas turbines; range 12000 km (7,455 miles) at cruising speed
Complement: 200, with a maximum of 237 possible

Class
1. Argentina

Name	No.	Builder	Laid down	Commissioned
Almirante Brown	D10	Blohm und Voss	Sep 1980	Feb 1982
La Argentina	D11	Blohm und Voss	Mar 1981	May 1983
Heroina	D12	Blohm und Voss	Aug 1981	Nov 1983
Sarandi	D13	Blohm und Voss	Mar 1982	Apr 1984

(The Argentine ships differ from the Nigerian unit in several important respects: the SSM armament is eight MM.40 Exocets in two quadruple container-launchers, the anti-submarine fit is two ILAS 3 triple mountings for 324-mm/12.75-in A 244S torpedoes, the electronics are mainly Dutch [DA-08A air- and surface-search, and ZW-06 navigation radars] with French processing equipment plus German ECM and sonar, and the COGOG propulsion matches two Olympus turbines with two Tyne turbines for respective maximum speeds of 30.5 and 20.5 kts.)

2. Nigeria

Name	No.	Builder	Laid down	Commissioned
Aradu	F89	Blohm und Voss	Dec 1978	Sep 1981

Note
These were the world's first large warships built to a modular design, with consequent advantages when the ships receive new weapons or undergo major refits

'Muntenia' class DDG
(Romania)
Type: anti-ship guided-missile destroyer
Displacement: 4,500 tons standard
Dimensions: length 145.0 m (475.7 ft); beam about 17.0 m (55.8 ft); draught not known
Gun armament: two 76-mm (3-in) L/59 DP in a twin mounting, and eight 30-mm AA in four AK-230 twin mountings
Missile armament: four twin container-launchers for eight SS-N-2C 'Styx' anti-ship missiles, and provision for SAM systems
Torpedo armament: two twin 533-mm (21-in) mountings for ? dual-role torpedoes
Anti-submarine armament: torpedoes (see above)
Aircraft: one IAR-316B Alouette III helicopter in a hangar aft
Electronics: one 'Strut Curve' surveillance radar, one 'Owl Screech' gun fire-control radar, two 'Tilt' series AA fire-control radars, sonar and other systems
Propulsion: COGOG or CODOG (COmbined Gas turbine Or Gas turbine, or COmbined Diesel Or Gas turbine) arrangement, delivering ? kW (? hp) to two shafts
Performance: speed 28+ kts
Complement: ?

Class
1. Romania

Name	No.	Builder	Laid down	Commissioned
Muntenia		Mangalia		Aug 1985

Note
Though rated a cruiser by the Romanians, this is a conventional missile destroyer optimized for anti-ship operations, and few details are yet clear about the type

'Niteroi' or 'Vosper Thornycroft Mk 10' class DDG
(UK/Brazil)
Type: anti-ship and anti-submarine guided-missile escort destroyer
Displacement: 3,200 tons standard and 3,800 tons full load
Dimensions: length 424.0 ft (129.2 m); beam 44.2 ft (13.5 m); draught 18.2 ft (5.5 m)
Gun armament: one (F40, 41, 44 and 45) or two (F42 and 43) 4.5-in (114-mm) L/55 DP in Mk 8 single mountings, and three 40-mm Bofors L/70 AA in single mountings
Missile armament: two twin container-launchers for four MM.38 Exocet anti-ship missiles (F42 and 43 only), and two triple launchers for 60 Sea Cat SAMs
Torpedo armament: none
Anti-submarine armament: one 375-mm (14.76-in) Bofors two-barrel rocket-launcher, two STWS triple mountings for six 12.75-in (324-mm) Mk 46 torpedoes, one depth-charge rail, one Branik launcher (except F42 and 43) for 10 Ikara missiles, and helicopter-launched weapons (see below)
Aircraft: one Westland Lynx helicopter in a hangar aft
Electronics: one AWS 2 air-search radar, one ZW-06 surface-search radar, two RTN 10X weapon-control and tracking radars, one CAAIS action information system, one EDO 610E hull-mounted sonar and (F40, 41, 44 and 45) one EDO 700E variable-depth sonar, and one RDL 2/3 ESM system
Propulsion: CODOG (COmbined Diesel Or Gas turbine) arrangement, with four MTU diesels delivering 11760 kW (15,775 shp) and two Rolls-Royce Olympus TM3B gas turbines delivering 56,000 shp (41760 kW) to two shafts
Performance: speed 30 kts on gas turbines or 22 kts on diesels; range 9825 km (6,105 miles) at 16 kts on two diesels, or 7775 km (4,830 miles) at 19 kts on four diesels, or 2400 km (1,490 miles) at 28 kts on gas turbines
Complement: 21+179

Class
1. Brazil

Name	No.	Builder	Laid down	Commissioned
Niteroi	F40	Vosper Thornycroft	Jun 1972	Nov 1976
Defensora	F41	Vosper Thornycroft	Dec 1972	Mar 1977
Constituicao	F42	Vosper Thornycroft	Mar 1974	Mar 1978
Liberal	F43	Vosper Thornycroft	May 1975	Nov 1978
Independencia	F44	Navyard	Jun 1972	Sep 1979
Uniao	F45	Navyard	Jun 1972	Sep 1980

Note
These are well-automated multi-role ships with good weapons and sensors; F42 and F43 are optimized for general-purpose use, and the others for the anti-submarine role; Brazil has also built a training ship, the *Brasil*, to the **'Niteroi (Modified)' class** design; this has one 76-mm (3-in) gun, two Pielstick diesels for a speed of 18 kts, a crew of 26+69+120, and accommodation for 200 cadets

'SAM Kotlin' class DDG
(USSR)
Type: anti-aircraft guided-missile escort destroyer
Displacement: 2,850 tons standard and 3,600 tons full load
Dimensions: length 127.5 m (418.2 ft); beam 12.9 m (42.3 ft); draught 4.6 m (15.1 ft)
Gun armament: two 130-mm (5.1-in) L/58 DP in a twin mounting, four or (in *Bravy*) 12 45-mm AA in quadruple mountings, and (in *Nesokrushimy*, *Skrytny* and *Soznatelny* only) eight 30-mm AA in four twin mountings
Missile armament: one twin launcher for 20 SA-N-1 'Goa' SAMs
Torpedo armament: one 533-mm (21-in) quintuple mounting for five dual-role torpedoes
Anti-submarine armament: two RBU 6000 12-barrel rocket-launchers or (in *Bravy* and *Skromny* only) two RBU 2500 16-barrel rocket-launchers, and torpedoes (see above)
Aircraft: none
Electronics: one 'Head Net-C' 3D radar or (in *Bravy* only) one 'Head Net-A' air-search radar, one 'Peel Group' SAM-control radar, one 'Sun Visor' 130-mm

gun-control radar, one 'Egg Cup' 130-mm gun-control radar, one 'Hawk Screech' 45-mm gun-control radar, two 'Drum Tilt' 30-mm gun-control radars (in *Nesokrushimy*, *Skrytny* and *Soznatelny* only), one or two 'Don Kay' or 'Don 2' navigation radars, one hull-mounted high-frequency sonar, 'High Pole-B' IFF, and two 'Watch Dog' ECM housings
Propulsion: four boilers supplying steam to two sets of geared turbines delivering 53700 kW (72,025 shp) to two shafts
Performance: speed 36 kts; range 8350 km (5,190 miles) at 15 kts or 2000 km (1,245 miles) at 35 kts
Complement: 300

Class
1. Poland

Name	No.
Warszawa	275

2. USSR

Name
Bravy
Nakhodchivy
Nastochivy
Nesokrushimy
Skromny
Skrytny
Soznatelny
Vozbuzhdenny

Note
Built in the late 1950s as gun destroyers, these are obsolescent anti-aircraft destroyers designed for the screening of less important surface forces, a role to which they were converted in 1961 (*Bravy*) and between 1967 and 1972 (others)

'Sheffield' or 'Type 42' class DDG
(UK)
Type: anti-aircraft and anti-submarine guided-missile escort destroyer
Displacement: (D86/92 and D108) 3,850 tons standard and 4,350 tons full load, or (D95/98) 4,775 tons standard and 5,350 tons full load
Dimensions: length (D86/92 and D108) 412.0 ft (125.6 m) or (D95/98 463.0 ft (141.1 m); beam (D86/92 and D108) 47.0 ft (14.3 m) or (D95/98 49.0 ft (14.9 m); draught 19.0 ft (5.8 m)
Gun armament: one 4.5-in (114-mm) L/55 DP in a Mk 8 single mounting, four 30-mm AA in two twin mountings, and four 20-mm AA in single mountings
Missile armament: one twin launcher for 24 (D86/92 and D108) or 40 (D95/98) Sea Dart SAMs
Torpedo armament: none
Anti-submarine armament: two triple mountings for 12.75-in (324-mm) Mk 46 or Stingray torpedoes, and helicopter-launched weapons (see below)
Aircraft: one Westland Lynx HAS.Mk 2 helicopter in a hangar aft
Electronics: one Type 1022 air-search radar, one Type 992Q or Type 992R surveillance and target-indication radar, two Type 909 radars used in conjunction with the GWS 30 SAM fire-control system, one Type 1006 navigation radar, one Type 184M hull-mounted medium-range panoramic search and attack sonar or (later ships) one Type 2016 hull-mounted passive sonar, one Type 170B hull-mounted short-range search and attack sonar, one Type 162M side-looking classification sonar, one Type 182 torpedo decoy system, one ADAWS 4 (D86/92) or ADAWS 8 (others) action information system, one Type 675 ECM system, one UAA-1 Abbey Hill ESM system, two SCOT satellite communications antennae, and two Corvus or Shield chaff launchers
Propulsion: COGOG (COmbined Gas turbine Or Gas turbine) arrangement, with two Rolls-Royce Olympus TM3B gas turbines delivering 56,000 shp (41760 kW) and two Rolls-Royce Tyne RM1A gas turbines delivering 8,500 shp (6340 kW) to two shafts
Performance: speed 29 kts on Olympus turbines or 18 kts on Tyne turbines; range 4,600 miles (7400 km) at 18 kts or 750 miles (1205 km) at 29 kts
Complement: 24+229, with a maximum of 312 possible

Class
1. Argentina

Name	No.	Builder	Laid down	Commissioned
Hercules	D1	Vickers	Jun 1971	Jul 1976
Santisima Trinidad	D2	AFNE Rio Santiago	Oct 1971	Jul 1981

(These Argentine units have suffered severe maintenance problems since the Falklands campaign of 1982 as no British spares have been obtainable; the ships are basically 'Type 42 Batch 1' class vessels, but have four single MM.38 Exocet anti-ship missile container/launchers, and two triple ILAS 3 tube mountings for A 244S anti-submarine torpedoes.)

2. UK ('**Type 42 Batch 1**' or '**Sheffield**' class)

Name	No.	Builder	Laid down	Commissioned
Birmingham	D86	Cammell Laird	Mar 1972	Dec 1976
Newcastle	D87	Swan Hunter	Feb 1973	Mar 1978
Glasgow	D88	Swan Hunter	Apr 1974	May 1979
Cardiff	D108	Vickers	Nov 1972	Sep 1979

3. UK ('Type 42 Batch 2' or 'Exeter' class)

Name	No.	Builder	Laid down	Commissioned
Exeter	D89	Swan Hunter	Jul 1976	Sep 1980
Southampton	D90	Vosper Thornycroft	Oct 1976	Oct 1981
Nottingham	D91	Vosper Thornycroft	Feb 1978	Apr 1982
Liverpool	D92	Cammell Laird	Jul 1978	Jul 1982

4. UK ('Type 42 Batch 3' or 'Manchester' class)

Name	No.	Builder	Laid down	Commissioned
Manchester	D95	Vickers	May 1978	Dec 1982
Gloucester	D96	Vosper Thornycroft	Oct 1979	May 1985
Edinburgh	D97	Cammell Laird	Sep 1980	1985
York	D98	Swan Hunter	Jan 1980	1985

'Shirane' class DDG
(Japan)
Type: anti-submarine and anti-ship guided-missile destroyer
Displacement: 5,200 tons standard and 6,800 tons full load
Dimensions: length 159.0 m (521.5 ft); beam 17.5 m (57.5 ft); draught 5.3 m (17.5 ft)
Gun armament: two 5-in (127-mm) L/54 DP in Mk 42 single mountings, and two 20-mm Phalanx Mk 16 CIWS mountings
Missile armament: two Mk 141 quadruple container-launchers for eight RGM-84A Harpoon anti-ship missiles, and one octuple launcher for eight RIM-7 Sea Sparrow SAMs
Torpedo armament: none
Anti-submarine armament: one Mk 16 octuple launcher for 16 RUR-5A ASROC missiles, two Type 68 triple mountings for 12.75-in (324-mm) Type 46 torpedoes, and helicopter-launched weapons (see below)
Aircraft: three Sikorsky SH-3 Sea King helicopters in a hangar amidships
Electronics: one OPS-12 3D radar, one OPS-28 surface-search radar, one OPS-22 navigation radar, two Type 72 gun-control systems, one WM-25 missile-control radar, one OQS-101 hull-mounted sonar, one SQS-35(J) variable-depth sonar, one SQR-18 passive towed-array sonar, URN-20A TACAN, and one ESM system
Propulsion: ? boilers supplying steam to two sets of geared turbines delivering 52200 kW (70,010 shp) to two shafts
Performance: speed 32 kts
Complement: 350

Class
1. Japan

Name	No.	Builder	Laid down	Commissioned
Shirane	DD143	Ishikawajima	Feb 1977	Mar 1980
Kurama	DD144	Ishikawajima	Feb 1978	Mar 1981

Note
These are improved 'Haruna' class destroyers with reduced noise generation

'Sovremenny' class DDG
(USSR)
Type: anti-ship guided-missile destroyer
Displacement: 6,200 tons standard and 7,800 tons full load
Dimensions: length 156.0 m (511.8 ft); beam 17.3 m (56.8 ft); draught 6.5 m (21.3 ft)
Gun armament: four 130-mm (5.1-in) L/60 DP in two twin mountings, and four 30-mm AK-630 CIWS mountings
Missile armament: two quadruple container-launchers for eight SS-N-22 anti-ship missiles, and two single launchers for 48 SA-N-7 SAMs
Torpedo armament: two 533-mm (21-in) twin mountings for dual-role torpedoes
Anti-submarine armament: two six-barrel RBU 1000 rocket-launchers, torpedoes (see above), and helicopter-launched weapons (see below)
Aircraft: one Kamov Ka-25 'Hormone-A' or Ka-27 'Helix-A' helicopter in a hangar amidships
Mines: fitted with rails for 30 to 50 mines
Electronics: one 'Top Pair' or (*Osmotritelny* onwards) one 'Head Net/Top Plate' 3D radar, one 'Band Stand' SSM-control radar, six 'Front Dome' SAM-control radars, one 'Kite Screech' 130-mm gun-control radar, two 'Bass Tilt' CIWS-control radars, three 'Palm Frond' surface-search and navigation radars, two medium-frequency hull-mounted sonars, and very extensive ESM systems including two 'Bell Shroud' and two 'Bell Squat' housings plus two chaff launchers
Propulsion: ? boilers supplying steam to two sets of turbo-pressurized turbines delivering 82000 kW (109,965 shp) to two shafts
Performance: speed 34 kts; range 12000 km (7,455 miles) at 20 kts or 4450 km (2,765 miles) at 32 kts
Complement: about 350

224 WARSHIPS

Class
1. USSR

Name	Builder	Laid down	Commissioned
Sovremenny	Zhdanov Yard	1976	Aug 1980
Otchyanny	Zhdanov Yard	1977	May 1982
Otlichnny	Zhdanov Yard	1978	May 1983
Osmotritelny	Zhdanov Yard	1979	Jun 1984
	Zhdanov Yard	1980	1985
	Zhdanov Yard	1981	1986
	Zhdanov Yard	1982	1987
	Zhdanov Yard	1983	1988

Note:
This is an extremely capable multi-role type with little short of cruiser capability in its primary surface strike task; it is expected that some 10 or 12 of the class will be built

'Spruance' class DDG
(USA)
Type: anti-submarine and anti-ship guided-missile escort destroyer
Displacement: 5,830 tons light and 7,810 tons full load
Dimensions: length 563.2 ft (171.7 m); beam 55.1 ft (16.8 m); draught 29.0 ft (8.8 m)
Gun armament: two 5-in (127-mm) L/54 DP in Mk 45 single mountings, and two 20-mm Phalanx Mk 126 CIWS mountings
Missile armament: two (none in *Foster* and *Kinkaid*) Mk 141 quadruple container-launchers for eight RGM-84A Harpoon anti-ship missiles, one Mk 41 Vertical Launch System for up to 61 BGM-109 Tomahawk cruise missiles and RUR-5A ASROC anti-submarines missiles (to be fitted in all ships in replacement of the Mk 16 ASROC launcher currently fitted), and one Mk 29 octuple launcher for 24 RIM-7 Sea Sparrow SAMs
Torpedo armament: none
Anti-submarine armament: one Mk 16 octuple launcher for 24 RUR-5A ASROC missiles (to be replaced, see above), two Mk 32 triple mountings for 14 12.75-in (324-mm) Mk 46 torpedoes, and helicopter-launched weapons (see below)

Aircraft: one Sikorsky SH-3 Sea King or two Kaman SH-2F Seasprite helicopters in a hangar aft (to be replaced by one Sikorsky SH-60B Seahawk helicopter in a hangar aft)
Electronics: one SPS-40B air-search radar, one SPS-55 surface-search radar, one SPG-60 search-and-track radar and one SPQ-9A track-while-scan radar used in conjunction with the Mk 91 SAM and Mk 86 gun fire-control systems, one SQS-53 (to be modernized to SQS-53C) bow-mounted 'bottom-bounce' sonar and one SQR-19 towed-array sonar used in conjunction with the Mk 116 underwater weapons fire-control system, one T Mk 6 Fanfare torpedo decoy system, Naval Tactical Data System, URN-20 TACAN, one OE-82 satellite communications system, one SRR-1 satellite communications receiver, three WSC-3 satellite communications transceivers, two Mk 36 Super RBOC chaff launchers, and one SLQ-32(V)2 ESM system
Propulsion: four General Electric LM2500 gas turbines delivering 80,000 shp (59655 kW) to two shafts
Performance: speed 33 kts; range 6,900 miles (11105 km) at 20 kts
Complement: 18+306

Class
1. USA

Name	No.	Builder	Laid down	Commissioned
Spruance	DDG963	Ingalls SB	Nov 1972	Sep 1975
Paul F. Foster	DDG964	Ingalls SB	Feb 1973	Feb 1976
Kinkaid	DDG965	Ingalls SB	Apr 1973	Jul 1976
Hewitt	DDG966	Ingalls SB	Jul 1973	Sep 1976
Elliott	DDG967	Ingalls SB	Oct 1973	Jan 1976
Arthur W. Radford	DDG968	Ingalls SB	Jan 1974	Apr 1977
Peterson	DDG969	Ingalls SB	Apr 1974	Jul 1977
Caron	DDG970	Ingalls SB	Jul 1974	Oct 1977
David R. Ray	DDG971	Ingalls SB	Sep 1974	Nov 1977
Oldendorf	DDG972	Ingalls SB	Dec 1974	Mar 1978
John Young	DDG973	Ingalls SB	Feb 1975	May 1978
Comte de Grasse	DDG974	Ingalls SB	Apr 1975	Aug 1978
O'Brien	DDG975	Ingalls SB	May 1975	Dec 1977
Merrill	DDG976	Ingalls SB	Jun 1975	Mar 1978
Briscoe	DDG977	Ingalls SB	Jul 1975	Jun 1978

Name	No.	Builder	Laid down	Commissioned
Stump	DDG978	Ingalls SB	Aug 1975	Aug 1978
Conolly	DDG979	Ingalls SB	Sep 1975	Oct 1978
Moosbrugger	DDG980	Ingalls SB	Nov 1975	Dec 1978
John Hancock	DDG981	Ingalls SB	Jan 1976	Mar 1979
Nicholson	DDG982	Ingalls SB	Feb 1976	May 1979
John Rodgers	DDG983	Ingalls SB	Aug 1976	Jul 1979
Leftwich	DDG984	Ingalls SB	Nov 1976	Aug 1979
Cushing	DDG985	Ingalls SB	Dec 1976	Sep 1979
Harry W. Hill	DDG986	Ingalls SB	Jan 1977	Nov 1979
O'Bannon	DDG987	Ingalls SB	Feb 1977	Dec 1979
Thorn	DDG988	Ingalls SB	Aug 1977	Feb 1980
Deyo	DDG989	Ingalls SB	Oct 1977	Mar 1980
Ingersoll	DDG990	Ingalls SB	Dec 1977	Apr 1980
Fife	DDG991	Ingalls SB	Mar 1978	May 1980
Fletcher	DDG992	Ingalls SB	Apr 1978	Jul 1980
Hayler	DDG997	Ingalls SB	Oct 1980	Mar 1983

'Suffren' class DDG

(France)
Type: anti-aircraft, anti-submarine and anti-ship guided-missile escort destroyer
Displacement: 5,090 tons standard and 6,090 tons full load
Dimensions: length 157.6 m (517.1 ft); beam 15.5 m (51.0 ft); draught 6.1 m (20.0 ft)
Gun armament: two 100-mm (3.9-in) L/55 DP in single mountings, and four 20-mm AA in single mountings
Missile armament: four container-launchers for four MM.38 Exocet anti-ship missiles, and one twin launcher for 48 Masurca SAMs
Torpedo armament: none
Anti-submarine armament: one single launcher for 13 Malafon missiles, and four single launchers for 533-mm (21-in) L5 torpedoes
Aircraft: none
Electronics: one DRBI 23 3D radar, one DRBV 50 combined air- and surface-search radar, two DRBR 51 SAM-control radars, one SRBC 32A gun-control radar, one DRBN 32 navigation radar, one DUBV 23 hull-mounted active search and attack sonar, one DUBV 43 variable-depth sonar, SENIT 1 action information system, TACAN, two Dagaie chaff launchers, and one ESM system
Propulsion: four boilers supplying steam to two sets of Rateau geared turbines delivering 54000 kW (72,425 shp) to two shafts
Performance: speed 34 kts; range 9500 km (5,905 miles) at 18 kts or 4450 km (2,765 miles) at 29 kts
Complement: 23+332

Class
1. France

Name	No.	Builder	Laid down	Commissioned
Suffren	D602	Lorient ND	Dec 1962	Jul 1967
Duquesne	D603	Brest ND	Nov 1964	Apr 1970

Note
These are area-defence destroyers designed to protect French aircraft-carriers from air and submarine attack, and were the first French ships designed from scratch as missile carriers, in this instance with stabilizers to provide a stable launch platform

'T47' class DDG

(France)
Type: anti-aircraft guided-missile escort destroyer
Displacement: 2,750 tons standard and 3,740 tons full load
Dimensions: length 128.5 m (421.6 ft); beam 13.0 m (42.5 ft); draught 6.3 m (20.7 ft)
Gun armament: six 57-mm L/60 DP in three twin mountings
Missile armament: one Mk 13 single launcher for 40 RIM-24B Tartar and RIM-66B Standard SM-1 MR SAMs
Torpedo armament: none
Anti-submarine armament: one 375-mm (14.76-in) six-barrel rocket-launcher, and two 550-mm (21.7-in) triple mountings for K2 and L3 torpedoes
Aircraft: none
Electronics: one SPS-39A/B 3D radar, one DRBV 22 air-search radar, one DRBV 31 surface-search and navigation radar, two SPG-51B SAM-control radars, one DUBA 1 hull-mounted sonar, one DUBV 24 hull-mounted panoramic search and attack sonar, SENIT 2 action information system, and one ESM system

Propulsion: four boilers supplying steam to two sets of geared turbines delivering 47000 kW (63,035 shp) to two shafts
Performance: speed 32 kts; range 9250 km (5,750 miles) at 18 kts or 2225 km (1,385 miles) at 32 kts
Complement: 17+83+177

Class
1. France

Name	No.	Builder	Laid down	Commissioned
Dupetit Thouars	D625	Brest ND	Mar 1952	Sep 1956
De Chayla	D630	Brest ND	Jul 1953	Jun 1957

Note
These are obsolescent area-defence anti-aircraft destroyers of single-ended capability and limited radar capacity

'Tachikaze' class DDG
(Japan)
Type: anti-aircraft, anti-submarine and anti-ship guided-missile escort destroyer
Displacement: 3,850 tons (DD168 and 169) or 3,900 tons (DD170) standard and 4,800 tons full load
Dimensions: length (DD168 and 169) 135.0 m (442.9 ft) or (DD170) 143.0 m (469.2 ft); beam 14.3 m (46.9 ft); draught 4.6 m (15.0 ft)
Gun armament: two 5-in (127-mm) L/54 DP in Mk 42 single mountings, and one 20-mm Phalanx Mk 16 CIWS mounting
Missile armament: two Mk 141 quadruple container-launchers for eight RGM-84A Harpoon anti-ship missiles (DD170 only), and one Mk 13 single launcher for 40 RIM-66B Standard SM-1 MR SAMs
Torpedo armament: none
Anti-submarine armament: one Mk 16 octuple launcher for 16 RUR-5A ASROC missiles, and two Mk 32 triple mountings for 12.75-in (324-mm) Mk 46 torpedoes
Aircraft: none
Electronics: one SPS-52B 3D radar, one OPS-11B or (in DD170 only) OPS-28 air-search radar, one OPS-17 surface-search radar, two SPG-51C radars used in conjunction with two Mk 73 SAM fire-control systems, two GFCS-1 (GFCS-2 in DD170) gun fire-control systems, one OQS-3 hull-mounted sonar used in conjunction with the Mk 114 underwater weapons fire-control system, one NOLQ-1 (NOLR-5 in DD168) ECM system, and one OLT-3 ESM system
Propulsion: ? boilers supplying steam to two sets of Westinghouse geared turbines delivering 60,000 shp (44740 kW) to two shafts
Performance: speed 32 kts
Complement: 260 (DD168), 250 (DD169) and 270 (DD170)

Class
1. Japan

Name	No.	Builder	Laid down	Commissioned
Tachikaze	DD168	Mitsubishi	Jun 1973	Mar 1976
Asakaze	DD169	Mitsubishi	May 1976	Mar 1979
Sawakaze	DD170	Mitsubishi	Sep 1979	Mar 1983

Note
These are medium-range air-defence destroyers developed from the 'Takatsuki' class, and also possess a useful anti-submarine capability

'Takatsuki' class DDG
(Japan)
Type: anti-submarine and anti-ship guided-missile destroyer
Displacement: 3,050 tons (DD164/165) or 3,100 tons (DD166/167) standard and 4,500 tons full load
Dimensions: length 136.0 m (446.2 ft); beam 13.4 m (44.0 ft); draught 4.4 m (14.5 ft)
Gun armament: one (DD164/165) or two (DD166/167) 5-in (127-mm) L/54 DP in Mk 42 single mountings, and two 20-mm Phalanx Mk 16 CIWS mountings
Missile armament: two Mk 141 quadruple container-launchers for eight RGM-84A Harpoon anti-ship missiles (DD164/165 only), and one octuple launcher for eight RIM-7 Sea Sparrow SAMs
Torpedo armament: none
Anti-submarine armament: one Mk 16 octuple launcher for 16 RUR-5A ASROC missiles, one 375-mm (14.76-in) Bofors four-barrel rocket-launcher, two Type 68 triple mountings for 12.75-in (324-mm) Mk 46 torpedoes, and (DD166/167 only) helicopter-launched weapons (see below)
Aircraft: (DD166/167 only) provision for one helicopter on a platform aft
Electronics: one OPS-11B air-search radar, one OPS-17 surface-search radar, two Mk 56 fire-control radars, one SQS-23 (DD165/166) or OQS-3 (DD166/167) hull-mounted sonar, one SQR-18 towed-array sonar (DD164/165 only), and one NOLQ-1 or NOLR-1 ESM system

Propulsion: two boilers supplying steam to two sets of Westinghouse geared turbines delivering 60,000 shp (44740 kW) to two shafts

Performance: speed 32 kts; range 12975 km (8,065 miles) at 20 kts
Complement: 275

Class
1. Japan

Name	No.	Builder	Laid down	Commissioned
Takatsuki	DD164	Ishikawajima	Oct 1964	Mar 1967
Kikuzuki	DD165	Mitsubishi	Mar 1966	Mar 1968
Mochizuki	DD166	Ishikawajima	Nov 1966	Mar 1969
Nagatsuki	DD167	Mitsubishi	Mar 1968	Feb 1970

Note
These anti-submarine destroyers have been upgraded with strong anti-ship armament and an improved anti-aircraft capability

'Tromp' class DDG
(Netherlands)
Type: anti-submarine, anti-aircraft and anti-ship guided-missile destroyer
Displacement: 3,900 tons standard and 4,580 tons full load
Dimensions: length 138.2 m (453.3 ft); beam 14.8 m (48.6 ft); draught 4.6 m (15.1 ft)
Gun armament: two 120-mm (4.7-in) L/50 DP in a twin mounting
Missile armament: two Mk 141 quadruple container-launchers for eight RGM-84A Harpoon anti-ship missiles, one Mk 13 single-launcher for 40 RIM-66B Standard SM-1 MR SAMs, and one Mk 29 octuple launcher for 60 RIM-7 NATO Sea Sparrow SAMs
Torpedo armament: none
Anti-submarine armament: two Mk 32 triple mountings for 12.75-in (324-mm) Mk 46 torpedoes, and helicopter-launched weapons (see below)
Aircraft: one Westland SH-14B Lynx helicopter in a hangar aft

Electronics: one 3D-MITR 3D radar, two ZW-05 surface-search radars, two SPG-51C Standard-control radars, two WM-25 gun and Sea Sparrow fire-control radars, two Decca navigation radars, one Type 162 hull-mounted side-looking classification sonar and one CWE-610 hull-mounted sonar used in conjunction with an underwater weapons fire-control system, one SEWACO I action information system, one Daisy data-handling system, two Corvus chaff launchers, and one Sphinx ESM system
Propulsion: COGOG (COmbined Gas turbine Or Gas turbine) arrangement, with two Rolls-Royce Olympus TM3B gas turbines delivering 50,000 hp (37285 kW) and two Rolls-Royce Tyne RM1A gas turbines delivering 8,000 hp (5965 kW) to two shafts
Performance: speed 30 kts; range 9250 km (5,750 miles) at 18 kts
Complement: 34+267

Class
1. Netherlands

Name	No.	Builder	Laid down	Commissioned
Tromp	F801	Koninklijke Maatschappij	Sep 1971	Oct 1975
De Ruyter	F802	Koninklijke Maatschappij	Dec 1971	Jun 1976

Note
These are well-balanced multi-role destroyers with advanced electronics and good anti-ship, anti-aircraft and anti-submarine armament

'Type 42' class DDG (UK): see 'Sheffield' class DDG

'Type 101A' class DDG (West Germany): see 'Hamburg' class DDG

'Type 103A' class DDG (West Germany): see 'Charles F. Adams (Modified)' class DDG

'Udaloy' class DDG
(USSR)
Type: anti-submarine and anti-aircraft guided-missile escort destroyer
Displacement: 6,700 tons standard and 8,200 tons full load
Dimensions: length 162.0 m (531.5 ft); beam 19.3 m (63.3 ft); draught 6.2 m (20.3 ft)

Gun armament: two 100-mm (3.9-in) L/56 DP in single mountings, and four 30-mm AK-630 CIWS mountings
Missile armament: two quadruple container-launchers for eight SS-N-14 'Silex' anti-submarine missiles, and four silo groups each containing two octuple vertical launchers for 64 SA-N-9 SAMs
Torpedo armament: two 533-mm (21-in) quadruple mountings for eight dual-role torpedoes
Anti-submarine armament: two RBU 6000 12-barrel rocket-launchers, SS-N-14 missiles (see above), torpedoes (see above), and helicopter-launched weapons (see below)
Aircraft: two Kamov Ka-27 'Helix-A' helicopters in twin hangars aft
Mines: fitted with rails for ? mines
Electronics: two 'Strut Pair' air-search or (*Vasilevsky* onwards) 'Top Mesh/Top Plate' 3D radars, two 'Eye Bowl' SS-N-14 control radars, two 'Cross Sword' SA-N-9 control radars (third ship onwards), one 'Kite Screech' 100-mm gun-control radar, two 'Bass Tilt' CIWS-control radars, three 'Palm Frond' surface-search and navigation radars, two 'High Pole-B' IFF, one bow-mounted low-frequency active sonar, one variable-depth sonar, and very extensive ESM systems including two 'Bell Shroud' and two 'Bell Squat' housings plus two chaff launchers
Propulsion: COGOG (COmbined Gas turbine Or Gas turbine) arrangement, with four gas turbines delivering about 82000 kW (109,965 shp) to two shafts
Performance: speed 35 kts; range 11125 km (6,915 miles) at 20 kts or 3700 km (2,300 miles) at 32 kts
Complement: about 30

Class
1. USSR

Name	Builder	Laid down	Commissioned
Udaloy	Yantar	1978	Nov 1980
Vitse-Admiral Kulakov	Zhdanov Yard	1978	Sep 1981
Marshal Vasilevsky	Zhdanov Yard	1979	Jul 1983
Admiral Zakharov	Yantar	1979	Oct 1983
Admiral Spiridonov	Yantar	1981	Sep 1984
	Zhdanov Yard	1981	1985
	Yantar	1983	1986
	Zhdanov Yard	1981	1986
	Yantar	1984	1988

Note
This is a powerful and well-balanced design for the escort of major warships, and some 12 of the class are expected

DESTROYERS

'Akizuki' class DD
(Japan)
Type: destroyer
Displacement: 2,300 tons standard and 2,890 tons full load
Dimensions: length 118.0 m (387.1 ft); beam 12.0 m (39.4 ft); draught 4.0 m (13.1 ft)
Gun armament: three 5-in (127-mm) L/54 in Mk 39 single mountings, and four 3-in (76-mm) L/50 DP in two Mk 33 twin mountings
Missile armament: none
Torpedo armament: one Type 65 quadruple mounting for 21-in (533-mm) torpedoes
Anti-submarine armament: two Type 68 triple mountings for 12.75-in (324-mm) Mk 46 torpedoes, and one 375-mm (14.76-in) Bofors four-barrel rocket-launcher
Aircraft: none
Electronics: one OPS-1 air-search radar, one OPS-15 surface-search radar, one Mk 34 gun-control radar used in conjunction with Mk 57 and Mk 63 fire-control systems, one SQS-29 hull-mounted sonar, and one OQA-1 variable-depth sonar
Propulsion: two boilers supplying steam to two sets of geared turbines delivering 33,500 kW (44,925 shp) to two shafts
Performance: speed 32 kts
Complement: 330

Class
1. Japan

Name	No.	Builder	Laid down	Commissioned
Akizuki	DD161	Mitsubishi	Jul 1958	Feb 1960
Teruzuki	DD162	Mitsubishi	Aug 1958	Feb 1960

Note
Now thoroughly obsolete, these two units can be used effectively only as convoy escort command ships

'Allen M. Sumner' class DD
(USA/Brazil)
Type: destroyer
Displacement: 2,200 tons standard and 3,320 tons full load
Dimensions: length 376.5 ft (114.8 m); beam 40.9 ft (12.5 m); draught 19.0 ft (5.8 m)
Gun armament: six 5-in (127-mm) L/38 DP in three Mk 38 twin mountings
Missile armament: one triple launcher for ? Sea Cat SAMs
Torpedo armament: none
Anti-submarine armament: two Mk 32 triple mountings for 12.75-in (324-mm) Mk 44 torpedoes, two Hedgehog mortars, and depth charges
Aircraft: none
Electronics: one SPS-6 air-search radar, one SPS-10 surface-search radar, one Mk 35 radar used in conjunction with the Mk 56 gun fire-control system, and one SQS-31 hull-mounted active sonar used in conjunction with the Mk 105 underwater weapons fire-control system
Propulsion: four boilers supplying steam to two sets of geared turbines delivering 60,000 shp (44740 kW) to two shafts
Performance: speed 34 kts; range 5,300 miles (8530 km) at 15 kts or 1,140 miles (1835 km) at 31 kts
Complement: 274

Class
1. Brazil

Name	No.	Builder	Laid down	Commissioned
Mato Grosso	D34	Federal SB	1944	Nov 1944

2. Taiwan

Name	No.	Builder	Laid down	Commissioned
Po Yang	928	Bath Iron Works	1943	Jun 1944
Yuen Yang	944	Federal SB	1943	Jun 1944
Huei Yang	972	Federal SB	1943	May 1944
Heng Yang	976	Bethlehem Steel	1943	Jun 1944
Hsiang Yang	986	Bethlehem Steel	1943	Apr 1944
Hua Yang	988	Bethlehem Steel	1944	Mar 1945

(The Taiwanese ships are similar to the Brazilian vessel, but three of them carry Hsiung Feng [Gabriel] anti-ship missiles [one triple launcher in 988 and two twin launchers in 944 and 986] plus the associated RTN 10X radar. A later standard of upgrading has been seen in another unit, and this may become standard for the Taiwanese units with a main gun armament of two 5-in/127-mm guns in a twin mounting, one 76-mm/3-in OTO Melara Compact and two single 40-mm Bofors L/70 AA, a missile armament of five Hsiung Feng anti-ship missiles and one Sea Chaparral SAM launcher, an aircraft fit of one Hughes 500 helicopter in a telescoping hangar aft, and a fire-control suite comprising one Honeywell H930 weapon-control system, two RCA HR-76 directors and a Kollmorgen optronic director.)

'Allen M. Sumner (FRAM II)' class DD
(USA/Brazil)
Type: destroyer
Displacement: 2,200 tons standard and 3,320 tons full load
Dimensions: length 376.5 ft (114.8 m); beam 40.9 ft (12.5 m); draught 19.0 ft (5.8 m)
Gun armament: six 5-in (127-mm) L/38 DP in three Mk 38 twin mountings
Missile armament: none
Torpedo armament: none
Anti-submarine armament: two Mk 32 triple mountings for 12.75-in (324-mm) Mk 44 torpedoes, two Hedgehog mortars, and helicopter-launched weapons (see below)
Aircraft: provision for one Westland Wasp HAS.Mk 1 light helicopter on a platform aft
Electronics: one SPS-40 air-search radar, one SPS-10 surface-search radar, one Mk 25 radar used in conjunction with the Mk 37 gun fire-control system, and one SQS-40 hull-mounted active sonar and one SQA-10 variable-depth sonar used in conjunction with the Mk 105 underwater weapons fire-control system
Propulsion: four boilers supplying steam to two sets of geared turbines delivering 60,000 shp (44740 kW) to two shafts
Performance: speed 34 kts; range 5,300 miles (8530 km) at 15 kts or 1,140 miles (1835 km) at 31 kts
Complement: 274

Class
1. Brazil

Name	No.	Builder	Laid down	Commissioned
Sergipe	D35	Bethlehem Steel	1944	Feb 1945

Name	No.	Builder	Laid down	Commissioned
Alagoas	D36	Bethlehem Steel	1944	Jun 1946
Rio Grande do Norte	D37	Bethlehem Steel	1943	Mar 1945
Espirito Santo	D38	Bethlehem Steel	1943	Jul 1944

2. Chile

Name	No.	Builder	Laid down	Commissioned
Ministro Zenteno	16	Todd Pacific	1944	Dec 1944
Ministro Portales	17	Federal SB	1943	May 1944

(The Chilean ships are similar to the Brazilian units, but have two 40-mm AA in single mountings, and a different helicopter.)

3. Greece

Name	No.	Builder	Laid down	Commissioned
Miaoulis	D211	Federal SB	1943	Mar 1944

(This ship has an Aérospatiale Alouette III helicopter.)

4. Iran

Name	No.	Builder	Laid down	Commissioned
Babr	61	Todd Pacific	1943	Oct 1944
Palang	62	Todd Pacific	1944	Jan 1945

(The Iranian vessels are laid up for lack of spares, but are fitted to carry eight Standard missiles on four launchers.)

5. South Korea

Name	No.	Builder	Laid down	Commissioned
Dae Gu	DD917	Bath Iron Works	1944	Sep 1944
Inchon	DD918	Federal SB	1943	Mar 1944

(The South Korean ships each have a helicopter in a hangar aft, and carry one 20-mm Phalanx Mk 16 CIWS mounting.)

6. Taiwan

Name	No.	Builder	Laid down	Commissioned
Lo Yang	949	Bethlehem Steel	1943	May 1944
Nan Yang	954	Bethlehem Steel	1944	Oct 1945

(These Taiwanese units are generally similar to the same country's 'Allen M. Sumner' class destroyers, but carry a Hughes Model 500 helicopter on an Israeli-supplied flightdeck carried in place of Y turret.)

7. Turkey

Name	No.	Builder	Laid down	Commissioned
Zafer	D356	Federal SB	1944	Mar 1945

(The Turkish ship is similar to the Greek unit, but has one OTO Melara Compact 76-mm/3-in L/62 DP gun for anti-aircraft and anti-FAC defence.)

'Ayanami' class DD

(Japan)
Type: destroyer
Displacement: 1,700 tons standard and 2,500 tons full load
Dimensions: length 109.0 m (357.6 ft); beam 10.7 m (35.1 ft); draught 3.7 m (12.0 ft)
Gun armament: six 3-in (76-mm) L/50 DP in three Mk 33 twin mountings
Missile armament: none
Torpedo armament: one 21-in (533-mm) quadruple mounting for four torpedoes
Anti-submarine armament: two Mk 32 triple mountings (DD112) for 12.75-in (324-mm) Mk 46 torpedoes or two Mk 4 single mountings (DD110/111) for 12.75-in (324-mm) Mk 32 torpedoes, and two Hedgehog Mk 15 mortars
Aircraft: none
Electronics: one OPS-1 or OPS-2 air-search radar, one OPS-15 or OPS-16 surface-search radar, one Mk 34 radar used in conjunction with the Mk 57 or 63 gun fire-control systems, one OQS-12 hull-mounted sonar, and (DD110) one OQA-1 variable-depth sonar
Propulsion: two boilers supplying steam to two sets of Mitsubishi/Escher-Wyss geared turbines delivering 26100 kW (35,000 shp) to two shafts
Performance: speed 32 kts; range 11,100 km (6900 miles) at 18 kts
Complement: 230

Class
1. Japan

Name	No.	Builder	Laid down	Commissioned
Takanami	DD110	Mitsui	Nov 1958	Jan 1960

| Oonami | DD111 | Ishikawajima | Mar 1959 | Aug 1960 |
| Makinami | DD112 | Iino | Mar 1959 | Oct 1960 |

Note
As with other gun destroyer classes, this Japanese type is little improved from the concepts of World War II and is now of little use other than for convoy work

'Carpenter (FRAM I)' class DD
(USA/Turkey)
Type: destroyer
Displacement: 2,425 tons standard and 3,540 tons full load
Dimensions: length 390.5 ft (119.0 m); beam 41.0 ft (12.5 m); draught 20.9 ft (6.4 m)
Gun armament: two 5-in (127-mm) L/38 DP in a Mk 38 twin mounting, and two 35-mm AA in a twin mounting
Missile armament: none
Torpedo armament: none
Anti-submarine armament: one Mk 16 octuple launcher for eight RUR-5A ASROC missiles, and two Mk 32 triple mountings for 12.75-in (324-mm) Mk 46 torpedoes
Aircraft: none
Electronics: one SPS-40 air-search radar, one SPS-10 surface-search radar, one SPG-35 radar used in conjunction with the Mk 56 gun fire-control system, one SQS-23 hull-mounted long-range active sonar used in conjunction with the Mk 114 underwater weapons fire-control system, and one Mk 1 target designation system
Propulsion: four boilers supplying steam to two sets of General Electric geared turbines delivering 60,000 shp (44740 kW) to two shafts
Performance: speed 33 kts
Complement: 15+260

Class
1. Turkey

Name	No.	Builder	Laid down	Commissioned
Alcitepe	D346	Bath Iron Works	Oct 1945	Nov 1949
Anittepe	D348	Consolidated Steel	Jul 1945	Dec 1949

Note
The provision of an ASROC launcher provides these obsolescent ships with a limited but useful shallow-water anti-submarine capability

'D20' class DD (Spain): see 'Fletcher' class DD

'D60' class DD (Spain): see 'Gearing (FRAM I)' class DD

'Fletcher' class DD
(USA/Greece)
Type: destroyer
Displacement: 2,050 tons standard and 3,050 tons full load
Dimensions: length 376.5 ft (114.7 m); beam 39.5 ft (12.0 m); draught 18.0 ft (5.5 m)
Gun armament: four 5-in (127-mm) L/38 DP in Mk 30 single mountings, six 3-in (76-mm) L/55 DP in three Mk 33 twin mountings, and 10 40-mm Bofors AA in two single and two quadruple mountings
Missile armament: none
Torpedo armament: one 21-in (533-mm) quintuple mounting for five torpedoes (in D06, 16, 56 and 85)
Anti-submarine armament: one Hedgehog mortar, depth-charge racks, and side-launching racks for 12.75-in (324-mm) Mk 44 or Mk 46 torpedoes, or (D16 and 85 only) two Mk 32 triple mountings for 12.75-in (324-mm) Mk 46 torpedoes
Aircraft: none
Electronics: one SPS-6 air-search radar, one SPS-10 surface-search radar, one Mk 25 radar used in conjunction with the Mk 37 main armament fire-control system, one Mk 35 gun-control radar used in conjunction with the one Mk 56 and two Mk 63 secondary armament fire-control systems, two Mk 51 AA armament fire-control systems, and one SQS-39 or SQS-43 hull-mounted sonar
Propulsion: four boilers supplying steam to two sets of General Electric geared turbines delivering 60,000 shp (44740 kW) to two shafts
Performance: speed 32 kts; range 6,900 miles (11105 km) at 15 kts or 1,450 miles (2335 km) at 32 kts
Complement: 250

Class
1. Brazil

Name	No.	Builder	Laid down	Commissioned
Piaui	D31	Federal SB	Mar 1943	Sep 1943

Name	No.	Builder	Laid down	Commissioned
Santa Catarina	D32	Bethlehem Steel	May 1943	Feb 1944
Maranhao	D33	Puget Sound NY	Aug 1943	Feb 1945

(These Brazilian ships are similar to the Greek vessels; the anti-submarine fit is two Mk 32 triple mountings for 12.75-in/324-mm Mk 46 torpedoes, and the sonar equipment comprises SQS-4, SQS-29 or SQS-34 sets.)

2. Greece

Name	No.	Builder	Laid down	Commissioned
Aspis	D06	Boston NY	Apr 1942	Jun 1943
Velos	D16	Boston NY	Feb 1941	May 1943
Kimon	D42	Bath Iron Works	Aug 1942	Mar 1943
Lonchi	D56	Boston NY	Apr 1942	Jul 1943
Nearchos	D65	Federal SB	Jun 1942	Dec 1942
Sfendoni	D85	Consolidated Steel	May 1941	Oct 1942

3. Mexico

Name	No.	Builder	Laid down	Commissioned
Cuitlahuac	E02	Consolidated Steel	Jul 1941	Feb 1943

(This ship lacks the 3-in/76-mm armament and has no anti-submarine capability.)

4. South Korea

Name	No.	Builder	Laid down	Commissioned
Chung Mu	DD911	Bath Iron Works	Oct 1942	May 1943
Pusan	DD913	Federal SB	Mar 1943	Sep 1943

(These ships have no 3-in/76-mm armament, and are fitted with SQS-20 hull-mounted sonar.)

5. Spain ('D20' class)

Name	No.	Builder	Laid down	Commissioned
Lepanto	D21	Gulf SB	Jun 1941	Jun 1943
Almirante Ferrandiz	D22	Gulf SB	Jun 1941	Sep 1943
Alcala Galiano	D24	Todd Pacific	Jun 1943	Jun 1944
Jorge Juan	D25	Federal SB	May 1943	Dec 1943

(The Spanish 'D20' class falls into two main subgroups, D21/22 having five 5-in/127-mm, six 40-mm and six 20-mm guns while D23/25 have four 5-in/127-mm and six 3-in/76-mm guns plus a triple 21-in/533-mm torpedo tube mounting; all the ships have two Hedgehog mortars and two Mk 32 triple mountings for 12.75-in/324-mm Mk 46 torpedoes, and there are detail alterations in the electronics fit.)

6. Taiwan

Name	No.	Builder	Laid down	Commissioned
Kun Yang	934	Bethlehem Steel	Dec 1942	Dec 1943
Ching Yang	947	Bethlehem Steel	Jan 1942	Apr 1943
Kwei Yang	956	Bethlehem Steel	Nov 1942	Dec 1943
An Yang	997	Bethlehem Steel	Jul 1942	May 1943

(The Taiwanese ships differ considerably from the Greek vessels, having a secondary armament of four 3-in/76-mm guns, and [except 947] an AA fit of two quadruple 40-mm and six single 20-mm guns plus one sextuple launcher for RIM-72B/C Sea Chaparral SAMs; 934 is fitted for minelaying and retains the quintuple torpedo mounting, but the other have instead two Mk 32 triple mountings for 12.75-in/324-mm Mk 46 torpedoes; there are differences in the sonar fit, and all the Taiwanese ships carry BLR-1 or SLR-2 passive detection systems; the average complement is 270.)

'Friesland' class DD

(Netherlands/Peru)
Type: destroyer
Displacement: 2,495 tons standard and 3,070 tons full load
Dimensions: length 116.0 m (380.5 ft); beam 11.7 m (38.5 ft); draught 5.2 m (17.0 ft)
Gun armament: four 120-mm (4.7-in) Bofors L/50 DP in two twin mountings, and four 40-mm Bofors L/70 AA in single mountings
Missile armament: none
Torpedo armament: none
Anti-submarine armament: two 375-mm (14.76-in) Bofors four-barrel rocket-launchers, and two depth-charge racks
Aircraft: none
Electronics: one LW-03 air-search radar, one DA-05 combined air- and surface-search radar, one WM-45 gun-control radar, one CWE 10-N hull-mounted sonar, and one PAE 1-N hull-mounted sonar
Propulsion: four boilers supplying steam to two sets of Werkspoor geared turbines delivering 44750 kW (60,020 shp) to two shafts
Performance: speed 36+ kts; range 7400 km (4,600 miles) at 18 kts
Complement: 284

Class
1. Peru

Name	No.	Builder	Laid down	Commissioned
Bolognesi	DD70	Dok en Werfmaatschappij	Oct 1953	Oct 1957
Castilla	DD71	Koninklijke Maatschappij	Feb 1954	Oct 1957
Guise	DD72	Nederlandse Dok	Jan 1954	Aug 1957
Capitan Quinones	DD76	Koninklijke Maatschappij	Nov 1953	Oct 1956
Villar	DD77	Nederlandse Dok	Mar 1956	Aug 1958
Galvez	DD78	Nederlandse Dok	Feb 1952	Sep 1956
Diez Canseco	DD79	Rotterdamse Droogdok	Jan 1954	Feb 1957

Note
This is one of the largest gun destroyer classes still in service, but has by modern standards been rendered obsolete (even by local standards) through the purchase of missile-armed frigates by neighbouring countries

'Gearing (FRAM I)' class DD
(USA/Greece)
Type: destroyer
Displacement: 2,425 tons standard and 3,500 tons full load
Dimensions: length 390.5 ft (119.0 m); beam 41.2 ft (12.6 m); draught 19.0 ft (5.8 m)
Gun armament: four 5-in (127-mm) L/38 DP in two Mk 38 twin mountings, one 76-mm (3-in) OTO Melara Compact L/62 DP in a single mounting, and one 40-mm Bofors L/60 AA in a single mounting
Missile armament: none
Torpedo armament: none
Anti-submarine armament: one Mk 16 octuple launcher for eight RUR-5A ASROC missiles, and two Mk 32 triple mountings for 12.75-in (324-mm) Mk 44 or Mk 46 torpedoes
Aircraft: none
Electronics: one SPS-37 or SPS-40 air-search radar, one SPS-10 surface-search radar, one Mk 25 or Mk 28 radar used in conjunction with the Mk 37 main armament fire-control system, one TRN 10X radar used in conjunction with the Argo NA10 76-mm gun fire-control system, and one SQS-23 hull-mounted long-range active sonar used in conjunction with the Mk 114 underwater weapons fire-control system
Propulsion: four boilers supplying steam to two sets of Westinghouse geared turbines delivering 60,000 shp (44740 kW) to two shafts
Performance: speed 32.5 kts; range 5,525 miles (8890 km) at 15 kts
Complement: 16+253

Class
1. Brazil

Name	No.	Builder	Laid down	Commissioned
Marcilio Dias	D25	Consolidated Steel	May 1944	Mar 1945
Mariz e. Barros	D26	Consolidated Steel	Dec 1944	Oct 1945

(The Brazilian ships are similar to the Greek units, but lack the 76-mm/3-in gun and have instead provision for a Westland Wasp light helicopter.)

2. Ecuador

Name	No.	Builder	Laid down	Commissioned
Presidente E. Alfaro	DD01	Consolidated Steel	Apr 1945	May 1946

(This unit is comparable to the Brazilian ships, but lacks the Mk 16 ASROC launcher.)

3. Greece

Name	No.	Builder	Laid down	Commissioned
Kanaris	D212	Consolidated Steel	Jan 1945	Sep 1945
Kountouriotis	D213	Bethlehem Steel	May 1945	Mar 1946
Sachtouris	D214	Bethlehem Steel	Mar 1945	Jan 1946
Tompazis	D215	Todd Pacific	Oct 1944	May 1945
Apostolis	D216	Bath Iron Works	Dec 1944	Jun 1945
Kriezis	D217	Consolidated Steel	Apr 1945	Feb 1946

4. Mexico

Name	No.	Builder	Laid down	Commissioned
Quetzalcoatl	E03	Bethlehem Steel	Aug 1944	Apr 1945
Netzahualcoyotl	E04	Bethlehem Steel	Sep 1944	May 1945

(The ships lack the Mk 16 ASROC launcher and the 76-mm/3-in gun.)

5. Pakistan

Name	No.	Builder	Laid down	Commissioned
Alamgir	D160	Bethlehem Steel	Nov 1944	Dec 1946
Shah Jahan	D164	Bethlehem Steel	Oct 1944	Jun 1945
Tariq	D165	Federal SB	Mar 1945	Jan 1946
Taimur	D166	Todd Pacific	Jun 1945	Mar 1949
Tughril	D167	Todd Pacific	Oct 1944	Aug 1945
Tippu Sultan	D168	Bethlehem Steel	May 1945	Apr 1946

(These ships are comparable with the Greek navy's units, but lack the 76-mm/3-in gun and have provision instead for a light helicopter.)

6. South Korea

Name	No.	Builder	Laid down	Commissioned
Taejon	DD919	Consolidated Steel	Apr 1945	Apr 1946
Kwang Ju	DD921	Bath Iron Works	Jul 1945	May 1946
Kang Won	DD922	Federal SB	Oct 1944	Sep 1945
Kyong Ki	DD923	Consolidated Steel	Oct 1944	Jul 1945
Jeon Ju	DD925	Consolidated Steel	Jun 1944	Mar 1945

(The South Korean ships are closer to the original American pattern of this class, with six 5-in/127-mm guns in three twin mountings, one twin 40-mm Bofors mounting, no Mk 16 ASROC launcher and provision for an Aérospatiale Alouette III helicopter for longer-range anti-submarine work.)

7. Spain ('D60' class)

Name	No.	Builder	Laid down	Commissioned
Churruca	D61	Federal SB	1944	Jun 1945
Gravina	D62	Consolidated Steel	1944	Jul 1945
Mendez Nunez	D63	Consolidated Steel	1945	Nov 1945
Langara	D64	Consolidated Steel	1944	May 1945
Blas de Lezo	D65	Bath Iron Works	1945	Nov 1945

(These ships are close to the Greek standard, but have a Hughes 500 helicopter instead of the 76-mm/3-in gun; D65 has no Mk 16 ASROC launcher.)

8. Taiwan

Name	No.	Builder	Laid down	Commissioned
Kai Yang	915	Todd Pacific	1945	Oct 1945
Chien Yang	921	Todd Pacific	1945	Feb 1946
Te Yang	925	Bath Iron Works	1945	Jul 1945
Shen Yang	932	Bath Iron Works	1945	Sep 1945
Liao Yang	938	Bath Iron Works	1944	May 1945
Dang Yang	966	Bethlehem Steel	1945	Mar 1947
Han Yang	978	Bath Iron Works	1944	May 1945
Lao Yang	981	Todd Pacific	1945	Jun 1946
Lai Yang		Bethlehem Steel	1945	Jun 1946
Chao Yang	912	Federal SB	1945	Jul 1946
		Consolidated Steel	May 1945	Oct 1946

(The Taiwanese units lack the 76-mm/3-in gun, but carry three Hsiung Feng [Gabriel] anti-ship missiles and extra 40-mm Bofors guns; 915 and 966 have no Mk 16 ASROC launcher.)

9. Turkey

Name	No.	Builder	Laid down	Commissioned
Yucatepe	D345	Todd Pacific	Sep 1944	Jun 1945
Savastepe	D348	Consolidated Steel	Jan 1945	Dec 1945
Kilic Ali Pasha	D349	Consolidated Steel	Jun 1945	Oct 1946
Piyale Pasha	D350	Bath Iron Works	Apr 1945	Nov 1945
M. Fevzi Cakmak	D351	Bethlehem Steel	1944	Sep 1946
Gayret	D352	Todd Pacific	1945	Jul 1946
Adatepe	D353	Bethlehem Steel	1945	Jun 1946

(The Turkish ships are very similar to the Greek units.)

Note
Though obsolete by modern standards, and now becoming increasingly difficult to maintain in fully serviceable condition, these units generally retain a local importance through the carriage of ASROC launchers, light anti-submarine helicopters and anti-FAC armament

'Gearing (FRAM II)' class DD
(USA/South Korea)
Type: destroyer
Displacement: 2,425 tons standard and 3,500 tons full load
Dimensions: length 390.5 ft (119.0 m); beam 41.2 ft (12.6 m); draught 19.0 ft (5.8 m)
Gun armament: six 5-in (127-mm) L/38 DP in three Mk 38 twin mountings, two 40-mm Bofors L/60 AA in a twin mounting, and one 20-mm Phalanx Mk 16 CIWS mounting (DD915) or two 30-mm AA in one Emerlec twin mounting (DD916)
Missile armament: one Mk 141 quadruple container-launcher for four RGM-84A Harpoon anti-ship missiles
Torpedo armament: none
Anti-submarine armament: two Mk 32 triple mountings for 12.75-in (324-mm) Mk 46 torpedoes, and two Mk 11 Hedgehog mortars
Aircraft: one Aérospatiale Alouette III helicopter on a platform aft
Electronics: one SPS-40 air-search radar, one SPS-10 surface-search radar, one Mk 25 or Mk 28 radar used in conjunction with the Mk 37 gun fire-control system, and one SQS-29 hull-mounted active sonar
Propulsion: four boilers supplying steam to two sets of General Electric geared turbines delivering 60,000 shp (44740 kW) to two shafts
Performance: speed 32.5 kts; range 6,675 miles (10740 km) at 15 kts
Complement: 280

Class
1. Greece

Name	No.	Builder	Laid down	Commissioned
Themistocles	D210	Bath Iron Works	May 1944	Dec 1944

(This ship is similar to the South Korean units, and has provision for an Aérospatiale Alouette III light helicopter.)

2. South Korea

Name	No.	Builder	Laid down	Commissioned
Chung Buk	DD915	Bath Iron Works	Jun 1944	Jan 1945
Jeong Buk	DD916	Bath Iron Works	Sep 1944	Apr 1945

3. Taiwan

Name	No.	Builder	Laid down	Commissioned
Fu Yang	963	Bath Iron Works	1945	Aug 1945

(This unit is similar to the South Korean ships but carries three Hsiung Feng [Gabriel] anti-ship missiles and an AA armament of eight 40-mm Bofors guns.)

4. Turkey

Name	No.	Builder	Laid down	Commissioned
Kocatepe	D354	Bethlehem Steel	1944	Jun 1945

(This ship lacks missile armament, guns totalling four 5-in/127-mm in two mountings, two 40-mm in on twin mounting, two 40-mm in single mountings and two 35-mm in a twin mounting.)

Note
Despite an increasing degree of serviceability problems, the various components of this elderly class retain a

measure of first-line capability through the addition of more modern sensor and armament, typically anti-ship missiles and an enhanced anti-aircraft capability

'Halland (Modified)' class DD
(Sweden/Colombia)
Type: destroyer
Displacement: 2,650 tons standard and 3,300 tons full load
Dimensions: length 121.1 m (397.2 ft); beam 12.4 m (40.7 ft); draught 4.7 m (15.4 ft)
Gun armament: six 120-mm (4.7-in) Bofors L/50 DP in three twin mountings, and four 40-mm Bofors AA in single mountings
Missile armament: none
Torpedo armament: one 533-mm (21-in) quadruple mounting for four torpedoes
Anti-submarine armament: one 375-mm (14.76-in) Bofors four-barrel rocket-launcher
Aircraft: none
Electronics: one LW-03 air-search radar, one DA-02 surface-search radar, and four WM-20 fire-control radars
Propulsion: two boilers supplying steam to two sets of De Laval geared turbines delivering 55,000 shp (41015 kW) to two shafts
Performance: speed 25 kts (D05) or 32 kts (D06)
Complement: 21+227

Class
1. Colombia

Name	No.	Builder	Laid down	Commissioned
Veinte de Julio	D05	Kockums	Oct 1955	Jun 1958
Siete de Agosto	D06	Gotaverken	Nov 1955	Oct 1958

Note
These units are obsolete by local standards, the armament having been developed no at all from the basic gun type of the 1940s and early 1950s

'Holland' class DD
(Netherlands/Peru)
Type: destroyer
Displacement: 2,215 tons standard and 2,765 tons full load
Dimensions: length 113.1 m (371.1 ft); beam 11.4 m (37.5 ft); draught 5.1 m (16.8 ft)
Gun armament: four 120-mm (4.7-in) Bofors L/50 DP in two twin mountings, and one 40-mm Bofors AA in a single mounting
Missile armament: none
Torpedo armament: none
Anti-submarine armament: two 375-mm (14.76-in) Bofors four-barrel rocket-launchers, and two depth-charge racks
Aircraft: none
Electronics: one LW-03 air-search radar, one DA-02 surface-search radar, one ZW-01 surface-search radar, three WM-45 fire-control radars, and one hull-mounted sonar
Propulsion: four boilers supplying steam to two sets of Parsons/Werkspoor geared turbines delivering 33500 kW (44,925 shp) to two shafts
Performance: speed 32 kts; range 7400 km (4,600 miles) at 18 kts
Complement: 247

Class
1. Peru

Name	No.	Builder	Laid down	Commissioned
Garcia y Garcia	DD75	Rotterdamse Droogdok	Apr 1950	Dec 1954

Note
This ship is thoroughly obsolete

'Iroquois' or 'Tribal' class DD
(Canada)
Type: destroyer
Displacement: 3,550 tons standard and 4,700 tons full load
Dimensions: length 426.0 ft (129.8 m); beam 50.0 ft (15.2 m); draught 15.5 ft (4.7 m)
Gun armament: one 5-in (127-mm) OTO Melara Compact L/54 DP in a single mounting
Missile armament: two quadruple launchers for 32 RIM-7 Canadian Sea Sparrow SAMs
Torpedo armament: none
Anti-submarine armament: two Mk 32 triple mountings for 12.75-in (324-mm) Mk 46 torpedoes, one Mk NG 10 Limbo mortar, and helicopter-launched weapons (see below)
Aircraft: two Sikorsky CH-124 Sea King helicopters in a hangar amidships

Electronics: one SPS-501 long-range air-warning radar with LW-03 antenna, one SPQ 2D surface-search radar, two WM-22 fire-control radars, one SQS-505 hull-mounted medium-range search and attack sonar, one SQS-505 variable-depth sonar, one SQS-501 classification sonar, one SLQ-25 Nixie torpedo decoy system, one CCS 280 action information system, and URN-20 TACAN
Propulsion: COGOG (COmbined Gas turbine Or Gas turbine) arrangement, with two Pratt & Whitney FT4A2 gas turbines delivering 50,000 shp (37285 kW) and two Pratt & Whitney FT12AH3 gas turbines delivering 7,400 shp (5520 kW) to two shafts
Performance: speed 29+ kts on main engines or 18 kts on cruising engines; range 5,200 miles (8370 km) at 20 kts
Complement: 20+225, plus an air unit of 7+33

Class
1. Canada

Name	No.	Builder	Laid down	Commissioned
Iroquois	280	Marine Industries	Jan 1969	Jul 1972
Huron	281	Marine Industries	Jan 1969	Dec 1972
Athabaskan	282	Davies SB	Jun 1969	Nov 1972
Algonguin	283	Davies SB	Sep 1969	Sep 1973

Note
Though not fitted with any major offensive/defensive missile armament, these four units are still capable anti-submarine vessels as a result of their size, sensors and advanced anti-submarine helicopters; between 1986 and 1991 the ships are to be taken in hand under the TRUMP (TRibal Update and Modernization Project) for development into dual-role area-defence and anti-submarine DDGs; this is an extensive programme involving the strengthening of the hull to permit operations under all Atlantic conditions, a host of engineering improvements (including the replacement of the FT12AH3 gas turbines by two Detroit Diesel Allison 570KF gas turbine cruising engines), the replacement of the Sea Sparrow systems by a Mk 41 Vertical Launch System for Standard SM-2 SAMs, the addition of a 76-mm (3-in) OTO Melara Super Rapid gun and one Phalanx Mk 16 CIWS mounting, the modernization of the electronic suite (with LW-08 long-range radar, DA-08 medium-range radar, STIR 1.8 fire-control radar, one CANEWS electronic warfare system, one Shield decoy system and a SHINPADS command and control system) and a new torpedo-handling system for the retained anti-submarine capability (based on the same sensors and helicopters)

'Kotlin' class DD
(USSR)
Type: destroyer
Displacement: 2,850 tons standard and 3,600 tons full load
Dimensions: length 127.5 m (418.2 ft); beam 12.9 m (42.3 ft); draught 4.6 m (15.1 ft)
Gun armament: four 130-mm (5.10-in) L/58 DP in two twin mountings, and 16 45-mm L/85 AA in four quadruple mountings
Missile armament: none
Torpedo armament: two 533-mm (21-in) quintuple mountings for 10 dual-4ole torpedoes
Anti-submarine armament: two RBU 2500 16-barrel rocket-launchers, torpedoes (see above), and (not in *Svetly*) six depth-charge throwers
Mines: 80
Aircraft: (in *Svetly* only) one Kamov Ka-25 'Hormone' helicopter on a platform aft
Electronics: one 'Slim Net' surface-search radar, one 'Sun Visor' 130-mm gun-control radar, two 'Egg Cup' 130-mm gun-control radars, two 'Hawk Screech' 45-mm gun-control radars, one Post Lamp or 'Top Bow' fire-control radar, two 'Don' navigation radars, one hull-mounted sonar, 'Square Head' IFF, and 'High Pole-A' IFF
Propulsion: four boilers supplying steam to two sets of geared turbines delivering 53700 kW (72,020 shp) to two shafts
Performance: speed 36 kts; range 8350 km (5,190 miles) at 15 kts or 2050 km (1,275 miles) at 35 kts
Complement: 285

Class
1. USSR

Name
Dalnevostochny Komsomolets
Spleshny
Spokoiny
Svetly
Vesky
Vliyatelny

Note
Built in the mid-1950s at Nikolayev and Leningrad, these are obsolete gun destroyers used mainly for patrol and training

'Kotlin (Modified)' class DD
(USSR)
Type: destroyer
Displacement: 2,850 tons standard and 3,600 tons full load
Dimensions: length 127.5 m (418.2 ft); beam 12.9 m (42.3 ft); draught 4.6 m (15.1 ft)
Gun armament: four 130-mm (5.1-in) L/58 DP in two twin mountings, eight 45-mm L/85 AA in two quadruple mountings, and four or eight 25-mm AA in two or four twin mountings
Missile armament: none

Torpedo armament: one 533-mm (21-in) quintuple mounting for 10 dual-role torpedoes
Anti-submarine armament: two RBU 2500 16-barrel rocket-launchers, two RBU 600 six-barrel rocket-launchers, torpedoes (see above), and six depth-charge throwers
Mines: 80
Aircraft: none
Electronics: one 'Slim Net' surface-search radar, one 'Sun Visor' 130-mm gun-control radar, two 'Egg Cup' 130-mm gun-control radars, two 'Hawk Screech' 45-mm gun-control radars, one 'Post Lamp' or 'Top Bow' fire-control radar, two 'Don' navigation radars, one hull-mounted sonar, 'Square Head' IFF, and 'High Pole-A' IFF
Propulsion: four boilers supplying steam to two sets of geared turbines delivering 53700 kW (72,020 shp) to two shafts
Performance: speed 36 kts; range 8350 km (5,190 miles) at 15 kts or 2050 km (1,275 miles) at 35 kts
Complement: 285

Class
1. USSR

Name
Blagorodny
Blestyashchy
Burlivy
Byvaly
Moskovsky Komsomolets
Naporisty
Plamenny
Svedushchy
Vdokhnovenny
Vozmushchenny
Vyderzhanny
Vyzyvayushchy

Note
Built at Leningrad and Nikolayev in the mid- and late 1950s, these gun destroyers were upgraded in anti-submarine capability during the early 1960, but are at best obsolescent and suited only for patrol and training duties

'Minegumo' class DD
(Japan)
Type: destroyer
Displacement: 2,050 tons standard and 2,150 tons full load
Dimensions: length 114.9 m (377.0 ft); beam 11.8 m (38.7 ft); draught 4.0 m (13.1 ft)
Gun armament: two 3-in (76-mm) L/50 DP in a Mk 33 twin mounting, and one 76-mm (3-in) OTO Melara Compact L/62 DP in a single mountings
Missile armament: none
Torpedo armament: none
Anti-submarine armament: one Mk 16 octuple launcher for eight RUR-5A ASROC missiles, two Type 68 triple mountings for 12.75-in (324-mm) Mk 46 torpedoes, and one 375-mm (14.76-in) Bofors four-barrel rocket-launcher
Aircraft: none
Electronics: one OPS-11 air-search radar, one OPS-17 surface-search radar, one Mk 35 gun-control radar used in conjunction with the Type 2 gun fire-control system, one OQS-3 hull-mounted sonar and one SQS-35(J) variable-depth sonar used in conjunction with the Mk 114 underwater weapons fire-control system
Propulsion: six Mitsubishi diesels delivering 19800 kW (26,555 bhp) to two shafts
Performance: speed 27 kts; range 12975 km (8,065 miles) at 20 kts
Complement: 210

Class
1. Japan

Name	No.	Builder	Laid down	Commissioned
Minegumo	DD116	Mitsui	Mar 1967	Aug 1968
Natsugumo	DD117	Uraga	Jun 1967	Apr 1969
Murakumo	DD118	Maizuru	Oct 1968	Aug 1970

Note
These ships are obsolescent, and suitable only for anti-submarine duties in lesser areas

'Murasame' class DD
(Japan)
Type: destroyer
Displacement: 1,800 tons standard and 2,500 tons full load
Dimensions: length 108.0 m (354.3 ft); beam 11.0 m (36.0 ft); draught 3.7 m (12.2 ft)
Gun armament: three 5-in (127-mm) L/54 DP in Mk 39 single mountings, and four 3-in (76-mm) L/50 DP in two Mk 33 twin mountings
Missile armament: none
Torpedo armament: none
Anti-submarine armament: two Mk 4 launchers (in DD108) or two Type 68 triple mountings (in DD107 and DD109) for 12.75-in (324-mm) Mk 46 torpedoes, one Hedgehog mortar, and (in DD108 only) one Mk 1 Y-gun depth charge projector and one depth charge rack
Aircraft: none
Electronics: one OPS-1 air-search radar, one OPS-15 surface-search radar, two Mk 34 radars used in conjunction with Mk 57 and Mk 63 gun fire-control

systems, one SQS-29 hull-mounted active sonar, and (in DD109 only) one OQA-1 variable-depth sonar
Propulsion: two boilers supplying steam to two sets of geared turbines delivering 22375 kW (30,010 shp) to two shafts

Performance: speed 30 kts; range 11125 km (6,915 miles) at 18 kts
Complement: 250

Class
1. Japan

Name	No.	Builder	Laid down	Commissioned
Murasame	DD107	Mitsubishi	Dec 1957	Feb 1959
Yudachi	DD108	Ishikawajima	Dec 1957	Mar 1959
Harusame	DD109	Uraga	Jun 1958	Dec 1959

Note
Another obsolescent Japanese destroyer type, this class was based on the 'Ayanami' class, and has limited anti-submarine capability in shallow waters with its Mk 44/46 torpedoes

'Skory' and 'Skory (Modified)' class DDs
(USSR)
Type: destroyer
Displacement: 2,240 tons standard and 3,080 tons full load
Dimensions: length 120.5 m (395.2 ft); beam 11.9 m (38.9 ft); draught 4.6 m (15.1 ft)
Gun armament: four 130-mm (5.1-in) L/50 DP in two twin mountings, two 85-mm (3.5-in) L/55 DP in a twin mounting ('Skory' class), five 57-mm AA in single mountings ('Skory [Modified]' class), eight 37-mm AA in four twin mountings or seven 37-mm AA in seven single mountings ('Skory' class), and four 25-mm AA in two twin mountings (some 'Skory' class units)
Missile armament: none
Torpedo armament: two ('Skory' class) or one ('Skory [Modified]' class) 533-mm (21-in) mountings for 10 or five torpedoes
Anti-submarine armament: four depth-charge throwers ('Skory' class) or two RBU 2500 16-barrel rocket-launchers ('Skory [Modified]' class)
Mines: 80
Electronics ('Skory' class): one 'Cross Bird' or 'Knife Rest' air-search radar, one 'High Sieve' surface-search radar, one 'Top Bow' or 'Half Bow' 130-mm gun-control radar, one 'Post Lamp' target-designation radar, one or two 'Don-2' navigation radars, one hull-mounted high-frequency sonar, one 'Square Head' IFF, one 'High Pole-A' IFF, and two 'Watch Dog' ECM housings
Electronics ('Skory [Modified]' class): one 'Slim Net' surface-search radar, one 'Top Bow' 130-mm gun-control radar, two 'Hawk Screech' AA gun-control radars, one or two 'Don-2' navigation radars, one hull-mounted high-frequency sonar, one 'Square Head' IFF, one 'High Pole-A' IFF, and two 'Watch Dog' ECM housings
Propulsion: four boilers supplying steam to two sets of geared turbines delivering 44750 kW (60,020 shp) to two shafts

Performance: speed 33 kts; range 7225 km (4,490 miles) at 13 kts or 1700 km (1,055 miles) at 32 kts
Complement: 280

Class
1. Egypt

Name	No.
6 October*	666
Al Zaffer	822
Damiet*	844
Suez	888

2. USSR

Name
Besstrashny
*Bessumny**
*Bezboyaznenny**
Bezuderzhny
Bezuprechny
Buyny
*Ognenny**
*Ostorozhny**
Ostry
*Otchetlivy**
Ozhivlenny
Serdity
Serezny
Solidny
Sovershenny
Statny
Stepenny
Stoyky
Stremitelny
Surovy
Svobodny
Vdumchivy
Vnimatelny
*Volny**
Vrazumitelny
* 'Skory (Modified)' class

Note
Built at Komsomolsk, Leningrad, Nikolayev and Severodvinsk to the extent of 72 units between 1949 and 1954 on the basis of the pre-World War II 'Ognevoy' class, the

'Skory' class marked a high point in Soviet gun destroyer design and production, but is now totally obsolete for anything but second-line and training duties; the 'Skory (Modified)' class was produced in the early 1960s by conversion, but adds little to anti-submarine capability

'T47' class DD
(France)
Type: destroyer
Displacement: 2,750 tons standard and 3,900 tons full load
Dimensions: 132.5 m (434.6 ft); beam 12.7 m (41.7 ft); draught 5.9 m (19.4 ft)
Gun armament: two 100-mm (3.9-in) L/55 DP in single mountings, and two 20-mm AA in single mountings
Missile armament: none
Torpedo armament: none
Anti-submarine armament: one launcher for 13 Malafon missiles, one 375-mm (14.76-in) six-barrel rocket-launcher, and two 550-mm (21.7-in) triple mountings for K2 or L3 torpedoes
Aircraft: none
Electronics: one DRBV 22A air-search radar, one DRBV 50 combined air- and surface-search radar, two DRBV 32A 100-mm gun-control radars, one DRBN 32 navigation radar, one DUBV 23 hull-mounted active search and attack sonar, one DUBV 43 variable-depth sonar, and ESM systems
Propulsion: four boilers supplying steam to two sets of geared turbines delivering 47000 kW (63,035 shp) to two shafts
Performance: speed 32 kts; range 9250 km (5,750 miles) at 18 kts
Complement: 15+104+151

Class
1. France

Name	No.	Builder	Laid down	Commissioned
Maille Brézé	D627	Lorient ND	Oct 1953	May 1957
Vauquelin	D628	Lorient ND	Mar 1953	Nov 1956
D'Estrées	D629	Brest ND	May 1953	Mar 1957
Guépratte	D632	A.C. Bretagne	Aug 1953	Jun 1957

Note
Though built as gun destroyers, these ships were given a useful anti-submarine capability in a conversion programme running from 1968 to 1971

'Tribal' class DD (Canada): see 'Iroquois' class DD

'Yamagumo' class DD
(Japan)
Type: destroyer
Displacement: 2,050 tons standard and 2,150 tons full load
Dimensions: length 114.9 m (377.0 ft); beam 11.8 m (38.7 ft); draught 4.0 m (13.1 ft)
Gun armament: four 3-in (76-mm) L/50 DP in two Mk 33 twin mountings
Missile armament: none
Torpedo armament: none
Anti-submarine armament: one Mk 16 octuple launcher for eight RUR-5A ASROC missiles, two Type 68 triple mountings for 12.75-in (324-mm) Mk 46 torpedoes, and one 375-mm (14.76-in) Bofors four-barrel rocket-launcher
Aircraft: none
Electronics: one OPS-11 air-search radar, one OPS-17 surface-search radar, two Mk 35 radars used in conjunction with Mk 56 and Mk 63 gun fire-control systems, one SQS-23 hull-mounted sonar (DD113/115) or one OQS-3 hull-mounted sonar (DD119/121), one SQS-35(J) variable-depth sonar (DD113/114 and DD120/121), one SLR-1 and one SLR-2 ECM systems, and one NOLR-6 ESM system
Propulsion: six Mitsubishi diesels delivering 19800 kW (26,555 bhp) to two shafts
Performance: speed 27 kts; range 12975 km (8,065 miles) at 20 kts
Complement: 210

Class
1. Japan

Name	No.	Builder	Laid down	Commissioned
Yamagumo	DD113	Mitsui	Mar 1964	Jan 1966
Makigumo	DD114	Uraga	Jun 1964	Mar 1966
Asagumo	DD115	Maizuru	Jun 1965	Aug 1967
Aokumo	DD119	Sumitomo	Oct 1970	Nov 1972
Akigumo	DD120	Sumitomo	Jul 1972	Jul 1974
Yugumo	DD121	Sumitomo	Feb 1976	Mar 1978

Note
These Japanese destroyers are characterized by their multi-diesel powerplant (and thus good range) and an effective anti-submarine capability through the use of ASROC missiles

GUIDED-MISSILE FRIGATES

'A69' or 'D'Estienne d'Orves' class FFG
(France)
Type: coastal anti-submarine and anti-ship guided-missile frigate
Displacement: 950 tons standard and 1,170 tons (1,250 tons in later ships) full load
Dimensions: length 80.0 m (262.5 ft); beam 10.3 m (33.8 ft); draught 3.0 m (9.8 ft)
Gun armament: one 100-mm (3.9-in) L/55 DP in a single mounting, and two 20-mm AA in single mountings
Missile armament: two (four in F792 onwards) container-launchers for two MM.38 Exocet (four MM.40 Exocet in F792 onwards) anti-ship missiles
Torpedo armament: none

Anti-submarine armament: one 375-mm (14.76-in) six-barrel rocket-launcher, and four single tubes for 550-mm (21.7-in) L3 or 533-mm (21-in) L5 torpedoes
Aircraft: none
Electronics: one DRBV 51A combined air- and surface-search radar, one DRBC 32E fire-control radar, one DRBN 32 navigation radar, one DUBA 25 hull-mounted attack sonar, one Vega weapon direction system, two Dagaie chaff launchers, and one ARBR 16 ESM system
Propulsion: two SEMT-Pielstick 12PC2V diesels delivering 8950 kW (12,000 bhp) to two shafts
Performance: speed 24 kts; range 8340 km (5,185 miles) at 15 kts
Complement: 5+29+45

Class
1. Argentina

Name	No.	Builder	Laid down	Commissioned
Drummond	P1	Lorient ND	Mar 1976	Nov 1978
Guerrico	P2	Lorient ND	Oct 1976	Nov 1978
Granville	P3	Lorient ND	1978	Jun 1981

(These ships differ from the French standard by having additional AA armament [two 40-mm Bofors], different anti-submarine armament [two Mk 32 triple mountings for 12.75-in/324-mm Mk 46 torpedoes] and a revised electronic suite with Diodon sonar and improved ESM capability.)

2. France

Name	No.	Builder	Laid down	Commissioned
D'Estienne d'Orves	F781	Lorient ND	Sep 1972	Sep 1976
Amyot d'Indville	F782	Lorient ND	Sep 1973	Oct 1976
*Drogou**	F783	Lorient ND	Oct 1973	Sep 1976
Détroyat	F784	Lorient ND	Dec 1974	May 1977
Jean Moulin	F785	Lorient ND	Jan 1975	May 1977
*Quarter Maitre Anquetil**	F786	Lorient ND	Aug 1975	Feb 1978
*Commandant de Pimodan**	F787	Lorient ND	Sep 1975	May 1978
2er Maitre Le Bihan	F788	Lorient ND	Nov 1976	Jul 1979
Lieutenant Le Henaff	F789	Lorient ND	Mar 1977	Feb 1980
Lieutenant Lavallée	F790	Lorient ND	Nov 1977	Aug 1980
Commandant l'Herminier	F791	Lorient ND	May 1979	Dec 1981
*1er Maitre l'Her**	F792	Lorient ND	Jul 1979	Dec 1981
*Commandant Blaison**	F793	Lorient ND	Nov 1979	Apr 1982
*Enseigne Jacoubet**	F794	Lorient ND	Jun 1980	Oct 1982
*Commandant Ducuing**	F795	Lorient ND	Oct 1980	Mar 1983
*Commandant Birot**	F796	Lorient ND	Mar 1981	Mar 1984
*Commandant Bouan**	F797	Lorient ND	Oct 1981	Nov 1984

* ships fitted with Exocet

'Amazon' or 'Type 21' class FFG
(UK)
Type: anti-submarine and anti-ship guided-missile frigate
Displacement: 2,750 tons standard and 3,250+ tons full load
Dimensions: length 384.0 ft (117.0 m); beam 41.7 ft (12.7 m); draught 19.5 ft (5.9 m)
Gun armament: one 4.5-in (114-mm) L/55 DP in a Mk 8 single mounting, and four 20-mm AA in single mountings
Missile armament: four container-launchers for four MM.38 Exocet anti-ship missiles, and one quadruple launcher for 24 Sea Cat SAMs (to be replaced by GWS 25 Sea Wolf)
Torpedo armament: none
Anti-submarine armament: two STWS triple mountings for 12.75-in (324-mm) Stingray and Mk 46 torpedoes, and helicopter-launched weapons (see below)
Aircraft: one Westland Lynx HAS.Mk 2 helicopter in a hangar aft
Electronics: one Type 992Q surveillance and target-indication radar, one Type 978 navigation radar, two GWS 24 SAM-control systems, two RTN 10X gun-control radars, one Type 184M hull-mounted medium-range search and attack sonar, one Type 162M hull-mounted side-looking classification sonar, one Type 182 torpedo decoy system, one CAAIS action information system, one SCOT satellite communications antenna, two Corvus chaff launchers, and one UAA-1 Abbey Hill ESM system
Propulsion: COGOG (COmbined Gas turbine Or Gas turbine) arrangement, with two Rolls-Royce Olympus TM3B gas turbines delivering 56,000 shp (41760 kW), and two Rolls-Royce Tyne RM1A gas turbines delivering 8,500 shp (6340 kW) to two shafts
Performance: speed 30 kts on Olympus gas turbines or 18 kts of Tyne gas turbines; range 4,600 miles (7400 km) at 17 kts or 1,380 miles (2220 km) at 30 kts
Complement: 13+162, with a maximum of 192 possible

Class
1. UK

Name	No.	Builder	Laid down	Commissioned
Amazon	F169	Vosper Thornycroft	Nov 1969	May 1974
Active	F171	Vosper Thornycroft	Jul 1971	Jun 1977
Ambuscade	F172	Yarrow	Sep 1971	Sep 1975
Arrow	F173	Yarrow	Sep 1972	Jul 1976
Alacrity	F174	Yarrow	Mar 1973	Jul 1977
Avenger	F175	Yarrow	Oct 1974	Jul 1978

Note
Designed by a private consortium as replacement for the 'Type 41' and 'Type 61' ('Leopard' and 'Salisbury') class frigates, the type has good seaworthiness and habitability, but lacks the potential for 'stretch' so necessary in modern warship design and thus will not be extensively updated in the ships' mid-life refits and so relegated to secondary duties

'Baleares' class FFG (Spain): see 'Knox' class FFG

'Bremen' or 'Type 122' class FFG
(West Germany)
Type: anti-ship and anti-submarine guided-missile frigate
Displacement: 2,900 tons standard and 3,750 tons full load
Dimensions: length 130.5 m (428.1 ft); beam 14.4 m (48.5 ft); draught 6.0 m (19.7 ft)
Gun armament: one 76-mm (3-in) OTO Melara Compact L/62 DP in a single mounting
Missile armament: two Mk 141 quadruple container-launchers for eight RGM-84A Harpoon anti-ship missiles, one Mk 29 octuple launcher for 24 RIM-7 NATO Sea Sparrow SAMs, and two EX-31 21-round container-launchers for 42 RIM-116 RAM SAMs
Torpedo armament: none
Anti-submarine armament: four Mk 32 single mountings for 12.75-in (324-mm) Mk 46 torpedoes, and helicopter-launched weapons (see below)
Aircraft: two Westland Lynx HAS.Mk 88 helicopters in a hangar aft
Electronics: one DA-08 air-search radar, one ZW-06 surface-search radar, one STIR surveillance and target-indication radar, one WM-25 fire-control radar, one 3RM 20 navigation radar, one SATIR action information system, one DSQS-21 BZ hull-mounted active-passive sonar, one SLQ-25 Nixie torpedo decoy system, four Mk 36 Super RBOC chaff launchers, and one FL1800S ESM system
Propulsion: CODOG (COmbined Diesel Or Gas turbine) arrangement, with two MTU 20V956 TB92 diesels delivering 7750 kW (10,395 shp) and two General Electric LM2500 gas turbines delivering 51,600 shp (38480 kW) to two shafts
Performance: speed 32 kts on gas turbines or 20 kts on diesels; range 7400 km (4,600 miles) at 18 kts
Complement: 203, with a maximum of 225 possible

Class
1. West Germany

Name	No.	Builder	Laid down	Commissioned
Bremen	F207	Bremer Vulkan	Jul 1979	Apr 1982
Niedersachsen	F208	AG Weser	Nov 1970	Oct 1982
Rheinland-Pfalz	F209	Blohm und Voss	Jun 1980	Apr 1983
Emden	F210	Thyssen	Jun 1980	Oct 1983
Köln	F211	Blohm und Voss	Dec 1980	Oct 1984
Karlsruhe	F212	Howaldtswerke	May 1981	Oct 1984

Note
This is the West German CODOG version of the Dutch 'Kortenaer' class gas turbine frigate design; advanced features include full NBC protection and a Prairie/Masker bubble system for the hull and propellers to reduce the level of radiated noise; each of the class is to receive two RAM point-defence missile installations on the hangar roof when the system becomes operational

'Brooke' class FFG
(USA)
Type: anti-submarine and anti-aircraft guided-missile escort frigate
Displacement: 2,645 tons standard and 3,425 tons full load
Dimensions: length 414.5 ft (126.3 m); beam 44.2 ft (13.5 m); draught 15.0 ft (4.6 m)
Gun armament: one 5-in (127-mm) L/38 DP in a Mk 30 single mounting
Missile armament: one Mk 22 single launcher for 16 RIM-66B Standard SM-1 MR SAMs
Torpedo armament: none
Anti-submarine armament: one Mk 16 octuple launcher for eight (FFG1/3) or 16 (FFG4/6) RUR-5A ASROC missiles, two Mk 32 triple mountings for 12.75-in (324-mm) Mk 46 torpedoes, and helicopter-launched weapons (see below)
Aircraft: one Kaman SH-2F Seasprite helicopter in a hangar aft
Electronics: one SPS-52B 3D radar, one SPS-10F surface-search radar, one SPG-51C radar used in conjunction with the Mk 74 SAM fire-control system, one SPG-35 radar used in conjunction with the Mk 56 gun fire-control system, one LN-66 navigation radar, one SQS-26 bow-mounted 'bottom-bounce' sonar used in conjunction with the Mk 114 underwater weapons fire-control system, one Mk 4 weapon direction system, OE-82 satellite communications antenna, SRR-1 satellite communications receiver, WSC-3 satellite communications transceiver, one SRN-15 TACAN, two Mk 36 Super RBOC chaff launchers, and one SLQ-32(V)2 ESM system
Propulsion: two boilers supplying steam to one set of Westinghous (FFG1/3) or General Electric (FFG4/6) geared turbines delivering 35,000 shp (26100 kW) to one shaft
Performance: speed 27.2 kts; range 4,600 miles (7400 km) at 20 kts
Complement: 17+231

Class
1. USA

Name	No.	Builder	Laid down	Commissioned
Brooke	FFG1	Lockheed SB	Dec 1962	Mar 1966
Ramsey	FFG2	Lockheed SB	Feb 1963	Jun 1967
Schofield	FFG3	Lockheed SB	Apr 1963	May 1968
Talbot	FFG4	Bath Iron Works	May 1964	Apr 1967
Richard L. Page	FFG5	Bath Iron Works	Jan 1965	Aug 1967
Julius A. Furer	FFG6	Bath Iron Works	Jul 1965	Nov 1967

Note
Based on the same hull and machinery as the other class, this is the anti-aircraft counterpart of the anti-submarine 'Garcia' class; the type is designed for the support of smaller task forces, for which a single-ended layout is deemed sufficient; the types also possess a useful anti-submarine ASROC/helicopter capability

'Chengdu' class FFG
(China)
Type: anti-ship guided-missile frigate
Displacement: 1,250 tons standard and 1,450 tons full load
Dimensions: length 91.5 m (300.1 ft); beam 10.1 m (33.1 ft); draught 3.2 m (10.5 ft)
Gun armament: three 100-mm (3.9-in) L/50 DP in single mountings, four 37-mm AA in two twin mountings, and four 14.5-mm (0.57-in) AA in two twin mountings
Missile armament: one twin container-launcher for two HY-2 (CSS-N-1) anti-ship missiles
Torpedo armament: none

Anti-submarine armament: two MBU 6000 rocket-launchers, and four BMB-2 depth-charge throwers
Mines: 60
Aircraft: none
Electronics: one 'Slim Net' surface-search radar, one 'Sun Visor' 100-mm gun-control radar, one 'Square Tie' missile-control radar, one 'Neptune' navigation radar, one 'High Pole-A' IFF, and one 'Watch Dog' ECM system

Propulsion: two boilers supplying steam to two sets of geared turbines delivering 14915 kW (20,000 shp) to two shafts
Performance: speed 28 kts; range 3700 km (2,300 miles) at 10 kts
Complement: 175

Class
1. China

Name	No.	Builder	Laid down	Commissioned
Guiyang	505	Hutong		1958
Kunming	506	Guangzhou		1959
Chengdu	507	Hutong	1955	1958
Guilin	508	Guangzhou	1955	1959

Note
These are four Soviet 'Riga' class frigates assembled in China and updated with anti-ship missiles and other more modern features

'D'Estienne d'Orves' class FFG (France): see 'A69' class FFG

'Commandant Rivière' class FFG
(France)
Type: anti-ship and anti-submarine guided-missile escort frigate
Displacement: 1,750 tons standard and 2,250 tons full load
Dimensions: length 103.0 m (337.9 ft); beam 11.5 m (37.7 ft); draught 4.3 m (14.1 ft)
Gun armament: two 100-mm (3.9-in) L/55 DP in single mountings, and two 30-mm AA in single mountings
Missile armament: four container-launchers for four MM.38 Exocet anti-ship missiles (not in *Balny*)
Torpedo armament: none
Anti-submarine armament: two triple mountings for 550-mm (21.7-in) K2 and L3 torpedoes, and one 305-mm (12-in) quadruple mortar
Aircraft: provision for one light helicopter on a platform aft
Electronics: one DRBV 22A air-search radar, one DRBV 50 combined air- and surface-search radar, one DRBC 32C missile-control radar, one DRBC 32A gun-control radar, one DRBN 32 navigation radar, one DUBA 3 hull-mounted sonar, one SQS-17 hull-mounted sonar, and two Dagaie chaff launchers
Propulsion: four SEMT-Pielstick diesels delivering 11920 kW (15,990 bhp) to two shafts, except *Balny* which has a CODAG (COmbined Diesel And Gas turbine) arrangement with two diesels and one gas turbine
Performance: speed 25 kts; range 13900 km (8,635 miles) at 15 kts
Complement: 10+61+96

Class
1. France

Name	No.	Builder	Laid down	Commissioned
Victor Schoelcher	F725	Lorient ND	Oct 1957	Oct 1962
Commandant Bory	F726	Lorient ND	Mar 1958	Mar 1964
Admiral Charner	F727	Lorient ND	Nov 1958	Dec 1962
Doudart de Lagrée	F728	Lorient ND	Mar 1960	May 1963
Balny	F729	Lorient ND	Mar 1960	Mar 1962
Commandant Rivière	F733	Lorient ND	Apr 1957	Dec 1962
Commandant Bourdais	F740	Lorient ND	Apr 1959	Mar 1963
Protet	F748	Lorient ND	Sep 1961	May 1964
Enseigne Henry	F749	Lorient ND	Sep 1962	Jan 1965

Note
Nicely designed escort frigates for all climatic regions, the ships of this class have flag accommodation and a balanced weapons fit

'Descubierta' class FFG
(Spain)
Type: anti-ship and anti-submarine guided-missile escort frigate
Displacement: 1,275 tons standard and 1,480 tons full load
Dimensions: length 88.8 m (291.3 ft); beam 10.4 m (34.0 ft); draught 3.8 m (12.5 ft)
Gun armament: one 76-mm (3-in) OTO Melara Compact L/62 DP in a single mounting, one 40-mm Bofors L/70 AA in a Breda single mounting, and one 20-mm Meroka CIWS mounting
Missile armament: two Mk 141 quadruple container-launchers for eight RGM-84A Harpoon anti-ship missiles, and one Albatros octuple launcher for 24 Aspide SAMs
Torpedo armament: none
Anti-submarine armament: two Mk 32 triple mountings for 12.75-in (324-mm) Mk 46 torpedoes, and one 375-mm (14.76-in) Bofors twin-barrel rocket-launcher
Aircraft: none
Electronics: one DA-05 air-search radar, one DA-02 surface-search and target-indication radar, one ZW-06 navigation radar, one WM-22/41 or WM-25 fire-control radar, one optronic fire-control system, one Raytheon 1160B hull-mounted sonar, one Raytheon 1167 variable-depth sonar, one SEWACO action information system, two chaff launchers, and one Beta ESM system
Propulsion: four MTU/Bazan 16V956 TB91 diesels delivering 13400 kW (17,970 bhp) to two shafts
Performance: speed 25.5 kts; range 7400 km (4,600 miles) at 18 kts
Complement: 116

Class
1. Egypt

Name	No.	Builder	Laid down	Commissioned
El Suez		Bazan	Oct 1979	1983
Abu Qir	946	Bazan	Dec 1979	Sep 1984

(These Egyptian ships are similar to the Spanish units but have two 40-mm AA and no Meroka CIWS mounting, and the missile fit has yet to be decided.)

2. Morocco

Name	No.	Builder	Laid down	Commissioned
Colonel Errhamani	51	Bazan	Mar 1979	1982

(This Moroccan unit differs from the Spanish norm in having two 40-mm AA in single mountings, no Meroka CIWS, four MM.38 Exocet rather than Harpoon anti-ship missiles, a revised Hollandse Signaalapparaten electronic fit, and a less powerful MTU/Bazan powerplant delivering 12000 kW [16,095 bhp] for a speed of 26 kts; the complement is 100.)

3. Spain

Name	No.	Builder	Laid down	Commissioned
Descubierta	F31	Bazan	Nov 1974	Nov 1978
Diana	F32	Bazan	Jul 1975	Jun 1979
Infanta Elena	F33	Bazan	Jan 1976	Apr 1980
Infanta Cristina	F34	Bazan	Sep 1976	Nov 1980
Cazadora	F35	Bazan	Dec 1977	Jul 1981
Vencedora	F36	Bazan	May 1978	Jan 1982

Note
This class is based on (but extensively upgraded from) the Portuguese 'Joao Coutinho' class frigate, with diesel propulsion for greater reliability and range, and revised armament and sensor fits offering a nice balance between anti-ship, anti-submarine and anti-aircraft capabilities

'Duke' or 'Type 23' class FFG
(UK)
Type: anti-submarine, and anti-ship and anti-aircraft guided-missile frigate
Displacement: 3,000 tons standard and 3,700 tons full load
Dimensions: length 436.2 ft (133.0 m); beam 49.2 ft (15.0 m); draught 14.1 ft (4.3 m)
Gun armament: one 4.5-in (114-mm) L/55 DP in a Mk 8 single mounting, and two 30-mm Goalkeeper CIWS mountings
Missile armament: two Mk 141 quadruple container-launchers for eight RGM-84 Harpoon anti-ship missiles, and one vertical launch system for 32 Sea Wolf SAMs
Torpedo armament: none
Anti-submarine armament: two STWS triple mountings for 12.75-in (324-mm) Stingray torpedoes, and helicopter-launched weapons (see below)
Aircraft: one or two Westland Lynx HAS.Mk 2 or EHI EH.101 helicopters in a hangar aft
Electronics: one Type 996 combined air- and surface-search radar, one Type 1007 navigation radar, two Type 911 radars used in conjunction with the GWS 25 SAM fire-control system, one Type 2050 bow-mounted sonar, one Type 2031 towed-array sonar,

one action information system, one UAA-1 Abbey Hill ESM system, and four Sea Gnat chaff launchers
Propulsion: CODAG (COmbined Diesel And Gas turbine) arrangement, with four Paxman Valenta 12 RPA 200 CZ diesels delivering 7,000 hp (5220 kW) and two Rolls-Royce Spey SM1A gas turbines delivering 34,000 shp (25355 kW) to two shafts
Performance: speed 28 kts; range 9,000 miles (14485 km) at 15 kts
Complement: 143, with a maximum possible of 177

Class
1. UK

Name	No.	Builder	Laid down	Commissioned
Norfolk	F230	Yarrow	1985	
	F231	Swan Hunter	1985	

'E-71' class FFG (Belgium): see 'Wielingen' class FFG

'Espora' class FFG (Argentina): see 'Meko 140A16' class FFG

'F70' class FFG (Spain): see 'Knox' class FFG

'F80' class FFG (Spain): see 'Oliver Hazard Perry' class FFG

'F2000' or 'Madina' class FFG
(France/Saudi Arabia)
Type: anti-ship and anti-submarine guided-missile frigate
Displacement: 2,250 tons standard and 2,610 tons full load
Dimensions: length 115.0 m (377.3 ft); beam 12.5 m (41.0 ft); draught 4.7 m (15.3 ft)
Gun armament: one 100-mm (3.9-in) Creusot-Loire Compact L/55 DP in a single mounting, and four 40-mm Bofors L/70 AA in two twin Dardo mountings
Missile armament: two quadruple container-launchers for eight Otomat Mk 2 anti-ship missiles, and one Naval Crotale octuple launcher for 26 Matra R.440 SAMs
Torpedo armament: none
Anti-submarine armament: four single mountings for 533-mm (21-in) F17 wire-guided torpedoes, and two Mk 32 single mountings for 12.75-in (324-mm) Mk 46 torpedoes
Aircraft: one Aérospatiale SA 365F Dauphin 2 helicopter in a hangar aft
Electronics: one DRBV 15 combined air- and surface-search radar, one DRBC 32E SAM-control radar, one Castor II fire-control radar, three Naja optronic fire-control systems, two TM1226 navigation radars, one Sylosat navigation system, one Diodon TSM2630 hull-mounted sonar, one Sorel variable-depth sonar, one SENIT VI action information system, two Dagaie chaff launchers, and one DR4000 ESM system
Propulsion: CODAD (COmbined Diesel And Diesel) arrangement, with four SEMT-Pielstick 16PA6 BTC diesels delivering 24240 kW (32,510 shp) to two shafts
Performance: speed 30 kts; range 12000 km (7,455 miles) at 18 kts
Complement: 15+164

Class
1. Saudi Arabia

Name	No.	Builder	Laid down	Commissioned
Madina	F702	Lorient ND	Oct 1981	Dec 1984
Hofouf	F704	CNIM	Jun 1982	Nov 1985
Abha	F706	CNIM	Dec 1982	1986
Taif	F708	CNIM	Jun 1983	Feb 1987

Note
This is a very ambitious design produced for a technologically-immature navy, and combines state-of-the-art weapons and sensors in a nicely arranged multi-role package; the predominant feature of the balanced design is the choice of ship-launched long-range Otomat Mk 2 anti-ship missiles and helicopter-launched short-range AS.15TT anti-ship missiles

'Fatahillah' class FFG
(Netherlands/Indonesia)
Type: anti-submarine and anti-ship guided-missile frigate
Displacement: 1,200 tons standard and 1,450 tons full load
Dimensions: length 84.0 m (275.6 ft); beam 11.1 m (36.4 ft); draught 3.3 m (10.7 ft)
Gun armament: one 120-mm (4.7-in) Bofors L/46 DP in a single mounting, one (in 361 and 362) or two (in 363) 40-mm Bofors AA in single mountings, and two 20-mm AA in single mountings
Missile armament: four container-launchers for four MM.38 Exocet anti-ship missiles
Torpedo armament: none
Anti-submarine armament: one 375-mm (14.76-in)

Bofors two-barrel rocket-launcher, and two Mk 32 triple mountings for 12.75-in (324-mm) Mk 46 or 324-mm (12.75-in) A 244S torpedoes (in 361 and 362 only) or helicopter-launched weapons (in 363 only, see below)
Aircraft: one Nurtanio-MBB BO105 helicopter in a hangar aft (in 363 only)
Electronics: one DA-05 combined air- and surface-search radar, one WM-28 fire-control radar, one Lirod optronic fire-control system, one AC1229 navigation radar, one PHS-32 hull-mounted sonar, one T Mk 6 Fanfare torpedo decoy system, one Daisy action information system, two Corvus chaff launchers, and one Susie I ESM system
Propulsion: CODOG (COmbined Diesel Or Gas turbine) arrangement, with two MTU diesels delivering 6000 kW (8,045 shp) and one Rolls-Royce Olympus TM3B gas turbine delivering 22,360 shp (16675 kW) to two shafts
Performance: speed 30 kts; range 7875 km (4,895 miles) at 18 kts
Complement: 89

Class
1. Indonesia

Name	No.	Builder	Laid down	Commissioned
Fatahillah	361	Wilton-Fijenoord	Jan 1977	Jul 1979
Malahayati	362	Wilton-Fijenoord	Jul 1977	Aug 1980
Nala	363	Wilton-Fijenoord	Jan 1978	1980

Note
A balanced anti-ship/anti-submarine design with advanced electronic and optronic sensors and fire-control equipment, the 'Fatahillah' class design is also notable for its full NBC protection

'Godavari' class FFG
(India)
Type: anti-ship and anti-submarine guided-missile escort frigate
Displacement: 3,850 tons full load
Dimensions: length 414.6 ft (126.4 m); beam 47.6 ft (14.5 m); draught 29.5 ft (9.0 m)
Gun armament: one 76-mm (3-in) OTO Melara Compact L/62 DP in a single mounting, and two 30-mm AK-630 CIWS mountings
Missile armament: two container-launchers for two SS-N-2C 'Styx' anti-ship missiles, and one launcher for SA-N-4 'Gecko' SAMs
Torpedo armament: none
Anti-submarine armament: two ILAS 3 triple mountings for 324-mm (12.75-in) A244S torpedoes, and helicopter-launched weapons (see below)
Aircraft: two Westland Sea King Mk 42/42A helicopters in a hangar aft
Electronics: one 'Head Net-C' 3D radar, one LW-05 air-search radar, one 'Pop Group' SAM-control radar, two 'Drum Tilt' CIWS-control radars, one Type 978 navigation radar, one Type 184 hull-mounted panoramic search and attack sonar, and one Fathom Oceanology variable-depth sonar
Propulsion: two boilers supplying steam to two sets of geared turbines delivering 30,000 shp (22370 kW) to two shafts
Performance: speed 27 kts; range 5,200 miles (8370 km) at 12 kts
Complement: 250

Class
1. India

Name	No.	Builder	Laid down	Commissioned
Godavari	F51	Magazon Docks	1978	1983
Ganga	F52	Magazon Docks	1981	Dec 1985
Gomati	F53	Magazon Docks	1980	1984

Note
This class is an Indian development of the British 'Leander' class hull with two large anti-submarine helicopters, British, Dutch and Soviet electronics, Soviet missiles, Italian and Swedish anti-submarine armament, and Italian and Soviet guns

'Groningen' class FFG
(Netherlands)
Type: anti-ship and anti-submarine guided-missile frigate
Displacement: 2,650 tons standard and 3,050 tons full load
Dimensions: length 122.0 m (400.25 ft); beam 14.4 m (47.25 ft); draught 6.0 m (19.7 ft)
Gun armament: one 76-mm (3-in) OTO Melara Compact L/62 DP in a single mounting, one 25-mm Goalkeeper CIWS mounting, and two 20-mm AA in single mountings
Missile armament: two Mk 141 quadruple container-launchers for eight RGM-84 Harpoon anti-ship missiles, and one vertical launch system for ? SAMs
Torpedo armament: none

Anti-submarine armament: two Mk 32 twin mountings for 12.75-in (324-mm) Mk 46 torpedoes, and helicopter-launched weapons (see below)
Aircraft: one Westland Lynx Mk 81 helicopter in a hangar aft
Electronics: one SMART 3D radar, one LW-08 air- and surface-search radar, one ZW-06 surface-search and target-designation radar, two STIR fire-control radars, one PH-36 hull-mounted sonar and either one variable-depth or towed-array sonar used in conjunction with an underwater weapons fire-control system, one torpedo decoy system, one SEWACO VII action information system, two Mk 36 RBOC chaff launchers, and one Ramses ESM system
Propulsion: CODOG (COmbined Diesel Or Gas turbine) arrangement, with two Werkspoor SW280 diesels delivering 6000 kW (8,045 hp) and one Rolls-Royce Spey SM1A gas turbine delivering 36,000 shp (26845 kW) to two shafts
Performance: speed 30 kts on gas turbine or 22 kts on diesels; range 9250 km (5,750 miles) at 18 kts
Complement: 16+121

Class
1. Netherlands

Name	No.	Builder	Laid down	Commissioned
Karel Doorman	F827	Koninklijke Maatschappij	1985	
Van Speijk	F828	Koninklijke Maatschappij		
Willem van den Zwann	F829	Koninklijke Maatschappij		
Tjerk Hiddes	F830	Koninklijke Maatschappij		
Van Amstel	F831	Koninklijke Maatschappij		
Abraham van der Hulst	F832	Koninklijke Maatschappij		
Van Nes	F833	Koninklijke Maatschappij		
Van Galen	F834	Koninklijke Maatschappij		

Note
This was originally designated the 'M' class, and is a mide-sized design intended to replace the 'Van Speijk' class currently being transferred to Indonesia

'Inhauma' or 'V-28' class FFG
(Brazil)
Type: anti-submarine and anti-ship guided-missile frigate
Displacement: 1,600 tons standard and 1,900 tons full load
Dimensions: length 95.8 m (314.2 ft); beam 11.4 m (37.4 ft); draught 3.7 m (12.1 ft)
Gun armament: one 4.5-in (114-mm) L/55 DP in a Mk 8 single mounting, and two 40-mm Bofors L/70 AA in single mountings
Missile armament: two twin container-launchers for four MM.38 Exocet anti-ship missiles
Torpedo armament: none
Anti-submarine armament: two Mk 32 triple mountings for 12.75-in (324-mm) Mk 46 torpedoes, and helicopter-launched weapons (see below)
Aircraft: one Westland Lynx Mk 21 helicopter in a hangar aft
Electronics: one AWS 4 air-search radar, one navigation radar, one RTN 10X radar used in conjunction with the WSA-420 fire-control system, one CAAIS 450 action information system, one hull-mounted medium-frequency sonar, one data-link system, one Shield chaff/decoy launcher, and one ESM system
Propulsion: CODOG (COmbined Diesel Or Gas turbine) arrangement, with two MTU 16V956 TB91 diesels delivering 2570 kW (3,450 shp) and one General Electric LM2500 gas turbine delivering 27,500 shp (20505 kW) to two shafts
Performance: speed 29 kts or gas turbine or 15 kts on diesels; range 7400 km (4,600 miles) at cruising speed
Complement: 120

Class
1. Brazil

Name	No.	Builder	Laid down	Commissioned
Jaceguary	V28	Arsenal de Marinha	Sep 1983	1987
Inhauma	V29	Arsenal de Marinha	Sep 1984	1988
Frontin	V30	Verolme	May 1986	
Julio de Noronha	V31	Verolme	Aug 1986	

Note
This is a capable patrol design using the best of imported systems and weapons

'Ishikari' class FFG
(Japan)
Type: anti-ship and anti-submarine guided-missile frigate
Displacement: 1,290 tons
Dimensions: length 85.0 m (278.8 ft); beam 10.6 m (34.7 ft); draught 5.9 m (19.2 ft)
Gun armament: one 76-mm (3-in) OTO Melara Compact L/62 DP in a single mounting
Missile armament: two Mk 141 quadruple container-launchers for eight RGM-84A Harpoon anti-ship missiles
Torpedo armament: none
Anti-submarine armament: two Type 68 triple mountings for 12.75-in (324-mm) Mk 46 torpedoes, and one 375-mm (14.76-in) Bofors four-barrel rocket-launcher
Aircraft: none
Electronics: one OPS-28 surface-search radar, one OPS-19 navigation radar, sonar and other systems
Propulsion: CODOG (COmbined Diesel Or Gas turbine) arrangement, with one 6 DRV diesel delivering 3500 kW (4,700 shp) and one Rolls-Royce Olympus TM3B gas turbine delivering 22,500 shp (16780 kW) to two shafts
Performance: speed 25 kts
Complement: 90

Class
1. Japan

Name	No.	Builder	Laid down	Commissioned
Ishikari	DE226	Mitsui	May 1979	Mar 1981

Note
Though fully operational, the sole 'Ishikari' class frigate may be regarded as a prototype for new Japanese frigate construction, being the first new design of this type since the mid-1960s and the now-obsolescent non-missile 'Chikugo' class

'Jacob van Heemskerck' class FFG
(Netherlands)
Type: anti-aircraft and anti-ship guided-missile frigate
Displacement: 3,000 tons standard and 3,750 tons full load
Dimensions: length 130.5 m (428.1 ft); beam 14.6 m (47.9 ft); draught 4.3 m (14.1 ft)
Gun armament: one 30-mm Goalkeeper CIWS mounting
Missile armament: two Mk 141 quadruple container-launchers for eight RGM-84A Harpoon anti-ship missiles, one Mk 13 single launcher for 40 RIM-66B Standard SM-1 MR SAMs, and one Mk 29 octuple launcher for 24 RIM-7 NATO Sea Sparrow SAMs
Torpedo armament: none
Anti-submarine armament: two Mk 32 twin mountings for 12.75-in (324-mm) Mk 46 torpedoes
Aircraft: none
Electronics: one SMART 3D radar, one DA-05 surface-search radar, three STIR fire-control radars, one ZW-06 navigation radar, one PHS-36 bow-mounted sonar, one SEWACO II action information system, one Daisy data-handling system, two Corvus chaff launchers, and various EW systems including one Ramses ECM system
Propulsion: COGOG (COmbined Gas turbine Or Gas turbine) arrangement, with two Rolls-Royce Olympus TM3B gas turbines delivering 50,000 shp (37285 kW) and two Rolls-Royce Tyne RM1C gas turbines delivering 8,000 shp (5965 kW) to two shafts
Performance: speed 30 kts; range 8700 km (5,405 miles) at 16 kts
Complement: 176

Class
1. Netherlands

Name	No.	Builder	Laid down	Commissioned
Jacob van Heemskerck	F812	Koninklijke Maatschappij	Jan 1981	Jan 1986
Witte de With	F813	Koninklijke Maatschappij	Dec 1981	1986

Note
This is the air-defence counterpart of the anti-submarine 'Kortenaer' class frigate, the helicopter facilities of the 'Kortenaers' being replaced by a single-ended missile installation; the ships are also fitted to serve as flagships

'Jiangdong' or 'Type 053J' class FFG
(China)
Type: anti-aircraft guided-missile escort frigate
Displacement: 1,570 tons standard and 2,000 tons full load
Dimensions: length 103.2 m (338.5 ft); beam 10.2 m (38.5 ft); draught 3.1 m (10.2 ft)
Gun armament: four 100-mm (3.9-in) L/56 DP in two twin mountings, and eight 37-mm AA in four twin mountings

Missile armament: two twin launchers for ? CSA-N-? SAMs
Torpedo armament: none
Anti-submarine armament: two RBU 1200 five-barrel rocket-launchers, and four BMB-2 depth-charge throwers
Aircraft: none
Electronics: one air-search radar, one surface-search radar, two missile-control radars, one 'Sun Visor' gun-control radar, one 'Fin Curve' navigation radar, one hull-mounted sonar, 'Yard Rake' IFF, 'Ski Pole' IFF, two 'Square Head' IFF, and one 'Jug Pair' ESM system
Propulsion: two diesels delivering 17900 kW (24,000 shp) to two shafts
Performance: speed 28 kts; range 7400 km (4,600 miles) at 15 kts
Complement: 180

Class
1. China

Name	No.	Builder	Laid down	Commissioned
Zhongdong	531	Shanghai	1971	1977
	532	Shanghai	1972	1978

Note
To number perhaps five units built at the Hutong yard in Shanghai, the 'Jiangdong' class has undergone an extremely lengthy development and building programme as China's first SAM-equipped warship design

'Jianghu I', 'Jianghu II' and 'Jianghu III' class FFGs
(China)
Type: anti-ship guided-missile frigate
Displacement: 1,570 tons standard and 2,000 tons full load
Dimensions: length 103.2 m (338.5 ft); beam 10.2 m (38.5 ft); draught 3.1 m (10.2 ft)
Gun armament: ('Jianghu I' and 'Jianghu II' classes) two 100-mm (3.9-in) L/56 DP in two single mountings or ('Jianghu III' class) four 100-mm (3.9-in) L/56 DP in two twin mountings, and 12 37-mm AA in six twin mountings
Missile armament: two twin container-launchers for four FL-1 (CSS-N-1) anti-ship missiles
Torpedo armament: none
Anti-submarine armament: two or four RBU 1200 five-barrel rocket-launchers, four BMB-2 depth-charge throwers, and two depth-charge racks
Mines: 60

Aircraft: none
Electronics: one MX 902 air-search radar, one 'Eye Shield' surface-search radar, one 'Square tie' ('Jianghu I' class) or 'Won Wok' ('Jianghu II' class) fire-control radar, one 'Fin Curve' navigation radar, one optronic fire-control director ('Jianghu II'), one Pegas-2M hull-mounted high-frequency sonar, one 'High Pole' or two 'Square Head' IFF, and one 'Watch Dog' ESM system
Propulsion: ('Jianghu I' and 'Jianghu III' classes) two diesels delivering 17900 kW (24,005 shp) to two shafts, or ('Jianghu II' class) CODOG (Combined Diesel Or Gas turbine) arrangement, with ? diesels and ? gas turbines delivering ? kW (? shp) to two shafts
Performance: speed ('Jianghu I' and 'Jianghu III' classes) 28 kts or ('Jianghu II' class) 30 kts; range 7400 km (4,600 miles) at 15 kts
Complement: 180

Class
1. China

Name	No.	Builder	Laid down	Commissioned
	510*	Shanghai	1974	1976
	511	Shanghai	1974	1976
	512*	Shanghai	1974	1976
	513	Shanghai	1975	1977
	514	Shanghai	1976	1978
	515	Shanghai	1976	1978
	516	Shanghai	1977	1979
	517	Shanghai	1978	1980
	518*	Shanghai	1979	1981
	519	Shanghai	1980	1982
	520	Shanghai	1981	1983
	525	Shanghai		
	527	Shanghai		
	533	Shanghai		
	534	Shanghai		
	535***	Shanghai		

	538	Shanghai
	544*	Shanghai
	551	Shanghai
	552	Shanghai

*'Jianghu II' class
**'Jianghu IV' class
***'Jianghu V' class

2. Egypt

Name	No.	Builder	Laid down	Commissioned
Najim az Zafir	951	Shanghai		Nov 1984
		Shanghai		Apr 1985

(These are in essence 'Jianghu III' class ships, but each has four 57-mm L/70 AA in two twin mountings rather than the Chinese 100-mm/3.9-in fit.)

Note
The basic 'Jianghu' class design is a development of the 'Jiangdong' class with greater emphasis on anti-ship warfare, such a capability being offered by the missile installation (the FL-1 or HY-2 version of the Soviet SS-N-2 'Styx' in early units, and the C-801 smaller but no less capable missile in later units); it is believed that the type is to number some 27–30 units in all, built at the Hutung and Tungmang yards in Shanghai; the 'Jianghu I' subclass (three or four ships) has a diesel powerplant with a rounded stack; the 'Jianghu II' subclass units have what is apparently an indigenously CODOG (COmbined Diesel Or Gas turbine) powerplant; the 'Jianghu III' subclass (some 18 ships) is the definitive early model with increased gun armament; the 'Jianghu IV' subclass has improved anti-submarine capability on a lengthened 100-mm (360.8-ft) hull with a flightdeck and hangar aft for one Harbin Z-9 helicopter; the other armament is a twin container-launcher for two FL-1 anti-ship missile, two 100-m (3.9-in) L/56 DP guns in a twin mounting, and eight 37-mm L/63 AA in twin mountings controlled by 'Rice Lamp' radars; the 'Jianghu V' class has the same lengthened hull as the 'Jianghu IV' subclass for a standard displacement of 1,675 tons; the primary armament comprises four container-launchers for four of the newer C-801 anti-ship missile type, and the other weapons are four 100-m (3.9-in) L/56 DP in twin mountings, and eight 37-mm L/63 AA in twin mountings controlled by 'Rice Lamp' radars; the propulsion is two 8950-kW (12,005-bhp) licence-built Pielstick diesels for a maximum speed of 28 kts

'Knox' class FFG
(USA)
Type: anti-submarine and anti-ship guided-missile escort frigate
Displacement: 3,010 tons standard and 3,875 tons full load
Dimensions: length 438.0 ft (133.5 m); beam 46.8 ft (14.3 m); draught 24.8 ft (7.8 m)
Gun armament: one 5-in (127-mm) L/54 DP in a Mk 42 single mounting, and one 20-mm Phalanx Mk 16 CIWS mounting
Missile armament: up to eight RGM-84A Harpoon anti-ship missiles instead of RUR-5A ASROC missiles (see below), and (in FF1052/1069 and FF1071/77) one Mk 25 octuple launcher for eight RIM-7 Sea Sparrow SAMs
Torpedo armament: none
Anti-submarine armament: one Mk 16 octuple launcher for up to 16 RUR-5A ASROC missiles, two Mk 32 twin mountings for 24 12.75-in (324-mm) Mk 46 torpedoes, and helicopter-launched weapons (see below)
Aircraft: one Kaman SH-2F Seasprite helicopter in a hangar aft
Electronics: one SPS-40 air-search radar, one SPS-10 surface-search radar, one SPS-58 threat-warning radar (some ships), one LN-66 navigation radar, one SPG-53 radar used in conjunction with the Mk 68 gun fire-control system, one Mk 115 SAM fire-control system, one Mk 1 weapon direction system, one SQS-26CX bow-mounted sonar and (in FF1052, 1056, 1063/1071 and 1073/1076) one SQS-35A variable-depth sonar (being replaced by SQR-18A towed-array sonar) used in conjunction with the Mk 114 underwater weapons fire-control system, OE-82 satellite communications system, SRR-1 satellite communications receiver, WSC-3 satellite communications transceiver, and one SLQ-32(V)2 ESM system
Propulsion: two boilers supplying steam to one set of Westinghouse geared turbines delivering 35,000 shp (26100 kW) to one shaft
Performance: speed 27 kts; range 5,200 miles (8370 km) at 20 kts
Complement: 17+270

Class
1. Spain ('**Baleares**' or '**F70**' class)

Name	No.	Builder	Laid down	Commissioned
Baleares	F71	Bazan	Oct 1968	Sep 1973
Andalucia	F72	Bazan	Jul 1969	May 1974
Cataluna	F73	Bazan	Aug 1970	Jan 1975

Name				
Asturias	F74	Bazan	Mar 1971	Dec 1975
Extremadura	F75	Bazan	Nov 1971	Nov 1976

(This is the Spanish version of the 'Knox' class design, and has a Mk 22 single launcher for 13 RIM-66B Standard SM-1 MR SAMs and three RGM-84A Harpoon anti-ship missiles, and its anti-submarine armament is augmented by two single launchers for 19 Mk 37 19-in (485-mm) torpedoes in addition to the 22 Mk 46 torpedoes carried for the Mk 32 tube mountings; the type also carries a revised electronic fit with an Elsag ECM suite.)

2. USA

Name	No.	Builder	Laid down	Commissioned
Knox	FF1052	Todd Pacific	Oct 1965	Apr 1969
Roark	FF1053	Todd Pacific	Feb 1966	Nov 1969
Gray	FF1054	Todd Pacific	Nov 1966	Apr 1970
Hepburn	FF1055	Todd Pacific	Jun 1966	Jul 1969
Connole	FF1056	Avondale	Mar 1967	Aug 1969
Rathburne	FF1057	Lockheed SB	Jan 1968	May 1970
Meyerkord	FF1058	Todd Pacific	Sep 1966	Nov 1969
W.S. Sims	FF1059	Avondale	Apr 1967	Jan 1970
Lang	FF1060	Todd Pacific	Mar 1967	Mar 1970
Patterson	FF1061	Avondale	Oct 1967	Mar 1970
Whipple	FF1062	Todd Pacific	Apr 1967	Aug 1970
Reasoner	FF1063	Lockheed SB	Jan 1969	Jul 1971
Lockwood	FF1064	Todd Pacific	Nov 1967	Dec 1970
Stein	FF1065	Lockheed SB	Jun 1970	Jan 1972
Marvin Shields	FF1066	Todd Pacific	Apr 1968	Apr 1971
Francis Hammond	FF1067	Todd Pacific	Jul 1967	Jul 1970
Vreeland	FF1068	Avondale	Mar 1968	Jun 1970
Bagley	FF1069	Lockheed SB	Sep 1970	May 1972
Downes	FF1070	Todd Pacific	Sep 1968	Aug 1971
Badger	FF1071	Todd Pacific	Feb 1968	Dec 1970
Blakely	FF1072	Avondale	Jun 1968	Jul 1970
Robert E. Peary	FF1073	Lockheed SB	Dec 1970	Sep 1972
Harold E. Holt	FF1074	Todd Pacific	May 1968	Mar 1971
Trippe	FF1075	Avondale	Jul 1968	Sep 1970
Fanning	FF1076	Todd Pacific	Dec 1968	Jul 1971
Ouellet	FF1077	Avondale	Jan 1969	Dec 1970
Joseph Hewes	FF1078	Avondale	May 1969	Apr 1971
Bowen	FF1079	Avondale	Jul 1969	May 1971
Paul	FF1080	Avondale	Sep 1969	Aug 1971
Aylwin	FF1081	Avondale	Nov 1969	Sep 1971
Elmer Montgomery	FF1082	Avondale	Jan 1970	Oct 1971
Cook	FF1083	Avondale	Mar 1970	Dec 1971
McCandless	FF1084	Avondale	Jun 1970	Mar 1972
Donald B. Beary	FF1085	Avondale	Jul 1970	Jul 1972
Brewton	FF1086	Avondale	Oct 1970	Jul 1972
Kirk	FF1087	Avondale	Dec 1970	Sep 1972
Barbey	FF1088	Avondale	Feb 1971	Nov 1972
Jesse L. Brown	FF1089	Avondale	Apr 1971	Feb 1973
Ainsworth	FF1090	Avondale	Jun 1971	Mar 1973
Miller	FF1091	Avondale	Aug 1971	Jun 1973
Thomas C. Hart	FF1092	Avondale	Oct 1971	Jul 1973
Capodanno	FF1093	Avondale	Oct 1971	Nov 1973
Pharris	FF1094	Avondale	Feb 1972	Jan 1974
Truett	FF1095	Avondale	Apr 1972	Jun 1974
Valdez	FF1096	Avondale	Jun 1972	Jul 1974
Moinester	FF1097	Avondale	Aug 1972	Nov 1974

Note

This was the largest single warship class produced in the West between World War II and the advent of the 'Oliver Hazard Perry' class FFG, and was developed to provide the US Navy with an ocean-going second-line anti-submarine force; the class was modelled on the preceding 'Garcia/Brooke' type, though the hull was enlarged as a non-pressure-fired boiler type was used; the type has proved successful, but has been criticized for lack of manoeuvrability (resulting from the use of a single propeller) and lack of anti-ship capability, though this latter has been remedied by the use of the left-hand pair of ASROC cells for Harpoon anti-ship missiles; the class is fitted with the Prairie/Masker bubble system to reduce underwater radiated noise; the seaworthiness of the class has been improved by the addition of higher bulwarks and strakes forward, and electronic capabilities enhanced by the installation of SQR-18A towed-array sonar (in 18 units as a replacement for the SQS-35A variable-depth sonar) and the upgrading of the SLQ-32(V)1 ESM system to SLQ-32(V)2 standard

'Koni' class FFG
(USSR)
Type: guided-missile patrol frigate
Displacement: 1,700 tons standard and 2,000 tons full load
Dimensions: length 95.0 m (311.6 ft); beam 12.8 m (42.0 ft); draught 4.2 m (13.7 ft)
Gun armament: four 76-mm (3-in) L/60 DP in two twin mountings, and four 30-mm AA in two twin mountings
Missile armament: one twin launcher for 18 SA-N-4 'Gecko' SAMs, and (Libyan and Yugoslav vessels only) four container-launchers for four 'Styx' anti-ship missiles (SS-N-2C on Libyan ship and SS-N-2B on Yugoslav ships)
Torpedo armament: none
Anti-submarine armament: two (one in Libyan variant) RBU 6000 12-barrel rocket-launchers, racks for 24 depth charges, and (Libyan variant only) for 406-mm (16-in) tubes for anti-submarine torpedoes
Mines: 20/30 depending on type
Aircraft: none
Electronics: one 'Strut Curve' air-search radar, one 'Plank Shave' SS-N-2 targeting radar (Libyan variant only), one 'Pop Group' SAM-control radar, one 'Hawk Screech' 76-mm gun-control radar, one 'Drum Tilt' 30-mm gun-control radar, one hull-mounted medium-frequency sonar, one 'High Pole-B' IFF, and two 'Watch Dog' ECM systems
Propulsion: CODAG (COmbined Diesel And Gas turbine) arrangement, with two diesels delivering 9000 kW (12,070 shp) and one gas turbine delivering 13425 kW (18,000 shp) to three shafts
Performance: speed 28 kts on gas turbine or 22 kts on diesels; range 3700 km (2,300 miles) at 14 kts
Complement: 110

Class
1. Algeria

Name	No.	Commissioned
Murat Reis	901	Dec 1980
Rais Kellich	902	
Rais Korfo	903	

2. Cuba

Name	No.	Commissioned
Mariel	350	Sep 1981
	356	

3. East Germany

Name	No.	Commissioned
Rostock	141	Jul 1978
Berlin	142	May 1979
	143	Jan 1986

4. Libya

Name	No.	Commissioned
El Hani	212	Jun 1986

5. USSR

Name	No.	Commissioned
Timofey Ulyantsev		1976

6. Yugoslavia

Name	No.	Commissioned
Split	R31	Feb 1980
Koper	R32	1982

Note
These light general-purpose frigates are built at the Black Sea yard of Zelenodolsk purely for export, the USSR retaining one unit for the training of crews

'Kortenaer' class FFG
(Netherlands)
Type: anti-submarine and anti-ship guided-missile frigate
Displacement: 3,050 tons standard and 3,785 tons full load
Dimensions: length 130.5 m (428.1 ft); beam 14.4 m (47.2 ft); draught 66.2 m (20.3 ft)
Gun armament: one 76-mm (3-in) OTO Melara Compact L/62 DP in a single mounting, and one 25-mm Goalkeeper CIWS mounting
Missile armament: two Mk 141 quadruple container-launchers for eight RGM-84A Harpoon anti-ship missiles, and one Mk 29 octuple launcher for 24 RIM-7 NATO Sea Sparrow SAMs
Torpedo armament: none
Anti-submarine armament: two Mk 32 twin mountings for 12.75-in (324-mm) Mk 46 torpedoes, and helicopter-launched weapons (see below)
Aircraft: two Westland SH-14B Lynx helicopters in a hangar aft
Electronics: one LW-08 long-range air-search radar, one ZW-06 surface-search and navigation radar, one DR-05 surface-search and target-acquisition radar, one STIR surveillance and target-indicator radar, one WM-25 fire-control system, one SQS-505 bow-mounted medium-range search and attack sonar, one SEWACO II action information system, one Daisy data-handling system, two Corvus chaff launchers, and one Sphinx (F807/811) or Ramses ESM system
Propulsion: COGOG (COmbined Gas turbine Or Gas turbine) arrangement, with two Rolls-Royce Olympus TM3B gas turbines delivering 51,600 shp (38480 kW) and two Rolls-Royce Tyne RM1C gas turbines delivering 9,800 shp (7310 kW) to two shafts
Performance: speed 30 kts; range 8700 km (5,405 miles) on Tyne gas turbines at 16 kts
Complement: 176, with accommodation for 200 possible

Class
1. Greece

Name	No.	Builder	Laid down	Commissioned
Elli	F450	Koninklijke Maatschappij	Jul 1977	Oct 1981
Limnos	F451	Koninklijke Maatschappij	Jun 1978	1982

(The Greek ships are very similar to the Dutch units, but have Agusta-Bell AB.212ASW helicopters, and a fire-control suite reduced to one STIR and one WM-25 radars.)

2. Netherlands

Name	No.	Builder	Laid down	Commissioned
Kortenaer	F807	Koninklijke Maatschappij	Apr 1975	Oct 1978
Callenburgh	F808	Koninklijke Maatschappij	Jun 1975	Jul 1979
Van Kinsbergen	F809	Koninklijke Maatschappij	Sep 1975	Apr 1980
Banckert	F810	Koninklijke Maatschappij	Feb 1978	Oct 1980
Piet Heyn	F811	Koninklijke Maatschappij	Apr 1977	Apr 1981
Abraham Crijnssen	F816	Koninklijke Maatschappij	Oct 1978	Jan 1983
Philips Van Almonde	F823	Dok en Werfmaatschappij	Oct 1977	Dec 1981
Bloys Van Treslong	F824	Dok en Werfmaatschappij	Apr 1978	Nov 1982
Jan Van Brakel	F825	Koninklijke Maatschappij	Nov 1979	Apr 1983
Pieter Florisz	F826	Koninklijke Maatschappij	Jan 1981	Oct 1983

Note
Designed for anti-submarine operations in the Atlantic, the 'Kortenaer' class has matured into an excellent anti-submarine and anti-ship type

'Krivak I', 'Krivak II' and 'Krivak III' class FFGs (USSR)

Type: guided-missile patrol frigate
Displacement: 3,000 tons standard and 3,700 tons ('Krivak I' class) or 3,800 tons ('Krivak II' and 'Krivak III' classes) full load
Dimensions: length 123.5 m (405.2 ft); beam 14.0 m (45.9 ft); draught 4.7 m (15.4 ft)
Gun armament: four 76-mm (3-in) L/60 DP in two twin mountings ('Krivak I' class), or two 100-mm (3.9-in) L/56 DP in single mountings ('Krivak II' class) or one 100-mm (3.9-in) L/56 DP in a single mounting and two 30-mm AK-630 CIWS mountings ('Krivak III' class)
Missile armament: two (one in 'Krivak III' class) twin launchers for 36 (18 in 'Krivak III' class) SA-N-4 'Gecko' SAMs
Torpedo armament: none
Anti-submarine armament: one quadruple container-launcher for four SS-N-14 'Silex' missiles (not in 'Krivak III' class), two RBU 6000 12-barrel rocket-launchers, and two quadruple mountings for eight 533-mm (21-in) torpedoes
Mines: 30/40 depending on type
Aircraft: one Kamov Ka-27 'Helix-A' helicopter in a hangar aft ('Krivak III' class only)
Electronics: one 'Head Net-C' 3D radar, two 'Eye Bowl' SS-N-14 control radars (not in 'Krivak III' class), two (one in 'Krivak III' class) 'Pop Group' SAM-control radars, one 'Owl Screech' ('Krivak I' class) or 'Kite Screech' ('Krivak II' and 'Krivak III' classes) gun fire-control radar, one 'Bass Tilt' CIWS-control radar ('Krivak III' class only), one 'Don Kay', 'Palm Frond-A/B' or 'Don 2' navigation radar, one hull-mounted medium-frequency sonar, one medium-frequency variable-depth sonar, one 'High Pole-B' or ('Krivak III' class) 'Salt Pot' IFF, four chaff launchers, two 'Bell Shroud' ECM systems, and two 'Bell Squat ECM' systems
Propulsion: COGAG (COmbined Gas turbine And Gas turbine) arrangement, with two gas turbines delivering 10440 kW (14,000 shp) and two gas turbines delivering 41000 kW (54,980 shp) to two shafts
Performance: speed 32 kts; range 7400 km (4,600 miles) at 15 kts or 2775 km (1,725 miles) at 32 kts
Complement: 180

Class
1. USSR ('Krivak I' class)

Name	Builder
Bditelny	Zhdanov
Bezukoroznenny	Kamysch-Burun
Bezzavetny	Kamysch-Burun
Bodry	Zhdanov
Deyatelny	Kamysch-Burun
Doblestny	Kamysch-Burun
Dostoyny	Kamysch-Burun
Druzhny	Zhdanov
Ladny	Kamysch-Burun
Leningradsky Komsomolets	Kaliningrad
Letuchy	Kaliningrad
Poryvisty	Kamysch-Burun
Pylky	Kaliningrad
Razumny	Zhdanov
Razyashchy	Zhdanov
Retivy	Kaliningrad
Silny	Zhdanov
Storozhevoy	Zhdanov
Svirepy	Zhdanov
Zadorny	Kaliningrad
Zharky	Kaliningrad

2. USSR ('Krivak II' class)

Name	Builder
Bessmenny	Kaliningrad
Gorelivy	Kaliningrad
Gromky	Kaliningrad
Grozyashchy	Kaliningrad
Neukrotimy	Kaliningrad
Pytlivy	Kaliningrad
Razytelny	Kaliningrad
Revnostny	Kaliningrad
Rezky	Kaliningrad
Rezvy	Kaliningrad
Ryanny	Kaliningrad

3. (USSR ('Krivak III' class)

Name	Builder
Menzhinsky	Kamysch-Burun

Note
The 'Krivak I' class units were built between 1970 and 1982 as anti-submarine frigates, the revised 'Krivak II' class units following between 1976 and 1981; both types were re-rated as patrol frigates in the late 1976s, experience having shown their inadequate size and endurance for blue water ASW; the 'Krivak III' appeared in mid-1984 and remedies some of the earlier classes' limitations; the type is in series production as the USSR's primary missle frigate type for the 1990s, with perhaps 25 or 30 of the class planned

'Leander Batch 2' class FFG
(UK)
Type: anti-submarine and anti-ship guided-missile frigate
Displacement: 2,450 tons standard and 3,200 tons full load
Dimensions: length 372.0 ft (113.4 m); beam 41.0 ft (12.5 m); draught 19.0 ft (5.8 m)
Gun armament: two 40-mm Bofors L/60 AA in single mountings
Missile armament: four container-launchers for four MM.38 Exocet anti-ship missiles, and three quadruple launchers for Sea Cat SAMs
Torpedo armament: none
Anti-submarine armament: two Mk 32 triple mountings for 12.75-in (324-mm) Mk 46 or Stingray torpedoes, and helicopter-launched weapons (see below)
Aircraft: one Westland Lynx HAS.Mk 2/3 or Westland Wasp HAS.Mk 1 helicopter in a hangar aft
Electronics: one Type 965 air-search radar, one Type 994 combined air- and surface-search radar, two Type 903 radars used in conjunction with the MRS 3/GWS 22 gun/SAM fire-control system, one Type 1006 navigation radar, one Type 184 hull-mounted medium-range panoramic search and attack sonar, and two Corvus chaff launchers
Propulsion: two boilers supplying steam to two sets of White/English Electric geared turbines delivering 30,000 shp (22370 kW) to two shafts
Performance: speed 28 kts; range 4,600 miles (7400 km) at 15 kts
Complement: 20+203

Class
1. New Zealand

Name	No.	Builder	Laid down	Commissioned
Waikato	F55	Harland & Wolff	Jan 1964	Sep 1966
Southland	F104	Yarrow	Dec 1959	Sep 1963

(*Waikato* has a twin 4.5-in/114-mm mounting, no Bofors guns, 12 Sea Cat SAMs for one quadruple launcher and a hangar large enough to take a Lynx helicopter, whereas the *Southland* has two quadruple SAM launchers, no 4.5-in/114-mm guns, two Bofors guns and one launcher for Ikara anti-submarine missiles.)

2. UK

Name	No.	Builder	Laid down	Commissioned or converted
Cleopatra*	F28	Devonport Dockyard	Jun 1963	Nov 1975
Sirius*	F40	Portsmouth Dockyard	Aug 1963	Oct 1977
Phoebe*	F42	Alexander Stephen	Jun 1963	Apr 1977
Minerva*	F45	Vickers	Jul 1973	Mar 1979

Danae	F47	Devonport Dockyard	Dec 1964	Dec 1980
Juno**	F52	Thornycroft	Jul 1964	Jul 1967
Argonaut	F56	Hawthorn Leslie	Nov 1964	Mar 1980
Penelope	F127	Vickers	Mar 1961	Mar 1981

* 'Leander Batch 2 TA' class
** training ship

Note
Developed in the late 1950s and early 1960s as the Royal Navy's standard general-purpose frigate, the 'Leander' class has since been extensively developed into specialized classes; of the narrow-beam ships a number were converted with Exocet anti-ship missiles, and these **'Leander Batch 2' class** ships in turn fall into two major subclasses, namely the basic 'Leander Batch 2' class when fitted with the armament and sensors described above, and **'Leander Batch 2 TA' class** when fitted with only two Sea Cat launchers and Type 2031(I) towed-array sonar in addition to the Type 184 equipment

'Leander Batch 3' or 'Broad-Beam Leander' class FFG
(UK)
Type: anti-ship, anti-submarine and anti-aircraft guided-missile frigate
Displacement: 2,500 tons standard and 2,960 tons full load
Dimensions: length 372.0 ft (113.4 ft); beam 43.0 ft (13.1 m); draught 18.0 ft (5.5 m)
Gun armament: two 4.5-in (114-mm) L/45 DP in one Mk 6 twin mounting and two 20-mm AA in single mountings (unconverted ships), or two 40-mm Bofors AA in single mountings (converted ships)
Missile armament: (unconverted ships) one quadruple launcher for Sea Cat SAMs, or (converted ships) four container-launchers for four MM.38 Exocet anti-ship missiles, and one sextuple launcher for 30 Sea Wolf SAMs
Torpedo armament: none
Anti-submarine armament: one Limbo three-barrel mortar (unconverted ships), or two STWS triple mountings for 12.75-in (324-mm) Mk 46 or Stingray torpedoes, plus (both types) helicopter-launched weapons (see below)
Aircraft: (unconverted ships) one Westland Wasp HAS.Mk 1 or (converted ships) one Westland Lynx HAS.Mk 2/3 helicopter in a hangar aft
Electronics: one Type 965 air-search radar and one Type 994 combined air- and surface-search radar (unconverted ships) or one Type 967 air-search and one Type 968 surface-search radars (converted ships), one Type 910 radar used in conjunction with the GWS 25 SAM fire-control system, one Type 177 hull-mounted medium-range panoramic search and attack sonar and one Type 170B hull-mounted short-range search and attack sonar (unconverted ships) or one Type 2016 hull-mounted sonar (converted ships), one CAAIS action information system one SCOT satellite communications antenna, two Corvus chaff launchers, and one UAA-1 Abbey Hill ESM system
Propulsion: two boilers supplying steam to two sets of White/English Electric geared turbines delivering 30,000 shp (22370 kW) to two shafts
Performance: speed 28 kts; range 4,600 miles (7400 km) at 15 kts
Complement: 19+241

Class
1. Chile

Name	No.	Builder	Laid down	Commissioned
Condell	06	Yarrow	Jun 1971	Dec 1973
Almirante Lynch	07	Yarrow	Dec 1971	May 1974

(These Chilean ships have the twin 4.5-in/114-mm mounting, one quadruple launcher with 16 Sea Cat SAMs, two Mk 32 triple mountings for 12.75-in/324-mm Mk 46 torpedoes, and no variable-depth sonar.)

2. New Zealand

Name	No.	Builder	Laid down	Commissioned
Canterbury	F421	Yarrow	Apr 1969	Oct 1971
Wellington	F69	Vickers	Oct 1966	Oct 1969

(These two New Zealand ships are fairly similar to the British ships, the standard armament fit being two 4.5-in/114-mm guns, one quadruple SAM launcher and two Mk 32 triple mountings for 12.75-in/324-mm Mk 46 torpedoes.)

3. UK

Name	No.	Builder	Laid down	Commissioned or converted
Achilles	F12	Yarrow	Dec 1967	Jul 1970
Diomede	F16	Yarrow	Jan 1968	Apr 1971

*Andromeda**	F57	Portsmouth Dockyard	May 1966	Dec 1980
*Hermione**	F58	Alexander Stephen	Dec 1965	Jan 1983
*Jupiter**	F60	Yarrow	Oct 1966	Jul 1983
Apollo	F70	Yarrow	May 1969	May 1972
*Scylla**	F71	Devonport Dockyard	May 1967	Apr 1983
Ariadne	F72	Yarrow	Nov 1969	Feb 1973
*Charybdis**	F75	Harland & Wolff	Jan 1967	Jun 1982

* converted ships

Note
This is the most potent of the 'Leander' class conversions, providing good anti-ship, anti-submarine and anti-aircraft capabilities

'Lupo' class FFG
(Italy)
Type: anti-submarine and anti-ship guided-missile escort frigate
Displacement: 2,210 tons standard and 2,525 tons full load
Dimensions: length 113.2 m (371.3 ft); beam 11.3 m (37.1 ft); draught 3.7 m (12.1 ft)
Gun armament: one 127-mm (5-in) OTO Melara Compact L/54 DP in a single mounting, and four 40-mm Bofors L/70 AA in two Dardo twin mountings
Missile armament: eight container-launchers for eight Otomat Mk 2 anti-ship missiles, and one Albatros octuple launcher for 24 Aspide SAMs
Torpedo armament: none
Anti-submarine armament: two Mk 32 triple mountings for 12.75-in (324-mm) Mk 46 torpedoes, and helicopter-launched weapons (see below)
Aircraft: one Agusta-Bell AB.212ASW helicopter in a hangar aft

Electronics: one RAN 10S air-search radar, one RAN 11/LX combined air- and surface-search radar, one SPQ 2F surface-search radar, one 3RM 20 navigation radar, onne RTN 10X gun-control radar, two RTN 20X Dardo-control radars, one Raytheon 1160B hull-mounted sonar, one IPN 10 action information system, two SCLAR chaff launchers, and one ESM system
Propulsion: CODOG (COmbined Diesel Or Gas turbine) arrangement, with two General Motors diesels delivering 7,800 shp (5815 kW) and two Fiat/General Electric LM2500 gas turbines delivering 50,000 shp (37285 kW) to two shafts
Performance: speed 35 kts on gas turbines or 21 kts on diesels; range 8000 km (4,970 miles) at 16 kts on diesels
Complement: 16+169

Class
1. Iraq

Name	No.	Builder	Laid down	Commissioned
Hittin		CNR, Ancona	Mar 1982	Mar 1985
Thi Qar		CNR, Ancona	Sep 1982	1986
Al Yarmook		CNR, Ancona	Mar 1983	1986
Al Qadisyaa		CNR, Ancona	Sep 1983	Feb 1986

(The Iraqi ships are very similar to the Italian ships other than having a fixed rather than telescopic hangar, which restricts the SAM armament to the eight missiles in the launcher, no reloads being carried.)

2. Italy

Name	No.	Builder	Laid down	Commissioned
Lupo	F564	CNR, Ancona	Oct 1974	Sep 1977
Sagittario	F565	CNR, Ancona	Feb 1976	Nov 1978
Perseo	F566	CNR, Ancona	Feb 1977	Mar 1980
Orsa	F567	CNR, Ancona	Aug 1977	Mar 1980

3. Peru

Name	No.	Builder	Laid down	Commissioned
Meliton Carvajal	F51	CNR, Ancona	Aug 1974	Feb 1979
Manuel Villavicencio	F52	CNR, Ancona	Oct 1976	Jun 1979
Montero	F53	SIMAC, Peru	Oct 1978	Jan 1984
Mariateque	F54	SIMAC, Peru	1979	Feb 1984

(The Peruvian ships are similar to the Italian units, having a telescopic hangar and SAM reloads, though the latter are manually rather than automatically reloaded into the octuple launcher, and the 40-mm mountings carried higher than in the Italian ships.)

258 WARSHIPS

4. Venezuela

Name	No.	Builder	Laid down	Commissioned
Mariscal Sucre	F21	CNR, Ancona	Nov 1976	May 1980
Almirante Brion	F22	CNR, Ancona	Jun 1977	Mar 1981
General Urdaneta	F23	CNR, Ancona	Jan 1978	Aug 1981
General Soublette	F24	CNR, Ancona	Aug 1978	Dec 1981
General Salom	F25	CNR, Ancona	Nov 1978	Feb 1982
Jose Felix Ribas	F26	CNR, Ancona	Aug 1979	Oct 1982

(The Venezuelan ships are similar to the Italian units, but have a fixed hangar, making it impossible to carry reloads rounds for the SAM launcher, and carry SQS-29 hull-mounted sonar matched to the 324-mm/12.75-in A 244S torpedoes fired from the two ILAS 3 triple mountings.)

Note
The basic 'Lupo' class frigate has proved popular and successful in service, though slightly limited in capability and 'strech' by the restrictions of a comparatively small hull design

'Madina' class FFG (Saudi Arabia): see 'F2000' class

'Maestrale' class FFG
(Italy)
Type: anti-submarine and anti-ship guided-missile frigate
Displacement: 3,040 tons standard and 3,200 tons full load
Dimensions: length 122.7 m (405.0 ft); beam 12.9 m (42.5 ft); draught 8.4 m (27.4 ft)
Gun armament: one 127-mm (5-in) OTO Melara Compact L/54 DP in a single mounting, and four 40-mm Bofors L/70 AA in two Dardo twin mountings
Missile armament: four container-launchers for four Otomat Mk 2 anti-ship missiles, and one Albatros octuple launcher for 24 Aspide SAMs
Torpedo armament: two single 533-mm (21-in) mountings for A 184 wire-guided dual-role torpedoes
Anti-submarine armament: two ILAS 3 triple mountings for 324-mm (12.75-in) A244S torpedoes, and helicopter-launched weapons (see below)
Aircraft: two Agusta-Bell AB.212ASW helicopters in a hangar aft

Electronics: one RAN 10S (MM/SPS 774) air-search radar, one SPQ 2F (MM/SPS 702) surface-search radar, one MM/SPN 703 navigation radar, one RTN 30X radar used in conjunction with the Albatros SAM fire-control system, one RTN 10X radar used in conjunction with the Argo NA10 gun fire-control system, two RTN 20X radars used in conjunction with two Dardo fire-control systems, one DE 1160 hull-mounted sonar, one DE 1164 variable-depth sonar, one IPN 10 Action information system, one SADOC 2 data-handling system two SCLAR chaff launchers, and one ESM system
Propulsion: CODOG (COmbined Diesel Or Gas turbine) arrangement, with two General Motors 230.50 DVM diesels delivering 11,000 shp (8200 kW) and two Fiat/General Electric LM2500 gas turbines delivering 50,000 shp (37285 kW) to two shafts
Performance: speed 32 kts on gas turbines or 21 kts on diesels; range 11125 km (6,915 miles) at 16 kts
Complement: 24+208

Class
1. Italy

Name	No.	Builder	Laid down	Commissioned
Maestrale	F570	CNR, Riva Trigoso	Mar 1978	Feb 1982
Grecale	F571	CNR, Muggiano	Mar 1979	Sep 1982
Libeccio	F572	CNR, Riva Trigoso	Aug 1979	Sep 1982
Scirocco	F573	CNR, Riva Trigoso	Feb 1980	Jan 1983
Aliseo	F574	CNR, Riva Trogoso	Aug 1980	Jun 1983
Euro	F575	CNR, Riva Trigoso	Apr 1981	Dec 1983
Espero	F576	CNR, Riva Trigoso	1981	1984
Zeffiro	F577	CNR, Riva Trigoso	1982	1984

Note
The 'Maestrale' class is a development of the 'Lupo' design optimized for Italian requirements and sacrificing speed for better seakeeping and superior anti-submarine capability

'Meko 140A16' or 'Espora' class FFG
(West Germany/Argentina)
Type: anti-submarine and anti-ship guided-missile frigate

Displacement: 1,470 tons standard and 1,700 tons full load
Dimensions: length 91.2 m (299.1 ft); beam 12.2 m (40.0 ft); draught 3.3 m (10.8 ft)

Gun armament: one 76-mm (3-in) OTO Melara Compact L/62 DP in a single mounting, and four 40-mm Bofors L/70 AA in two Dardo twin mountings
Missile armament: four container-launchers for four MM.40 Exocet anti-ship missiles
Torpedo armament: none
Anti-submarine armament: two ILAS 3 triple mountings for 12 324-mm (12.75-in) A 244S torpedoes, and helicopter-launched weapons (see below)
Aircraft: one Westland Lynx helicopter on a platform amidships (first three ships) or in a telescopic hangar (last three ships)
Electronics: one DA-05/2 air-search radar, one Decca surface-search radar, one TM1226 navigation radar, one WM-28 fire-control system, two Lirod optronic fire-control systems, one Krupp ASO-4 hull-mounted sonar, one Graseby G1738 torpedo decoy system, one Daisy data-handling system, two Dagie chaff launchers, one RDC-2ABC ESM system, and one RCM-2 ESM system
Propulsion: two SEMT-Pielstick 16PC2-5V400 diesels delivering 16850 kW (22,600 shp) to two shafts
Performance: speed 27 kts; range 7400 km (4,600 miles) at 18 kts
Complement: 100

Class
1. Argentina

Name	No.	Builder	Laid down	Commissioned
Espora	P4	AFNE Rio Santiago	Oct 1980	1984
Rosales	P5	AFNE Rio Santiago	Jul 1981	Nov 1985
Spiro	P6	AFNE Rio Santiago	Jan 1982	
Parker	P7	AFNE Rio Santiago	Jun 1983	
Robinson	P8	AFNE Rio Santiago	Dec 1983	
Gomez Roca	P9	AFNE Rio Santiago	1984	

Note
Based on the design of the Portuguese 'Joao Coutinho' class but featuring modular design and construction for ease of modification and updating, the 'Meko 140A16' class is designed for coastal and short-range anti-ship and anti-submarine use in support of the 'Meko 360' class destroyers

'Meko 200' class FFG
(West Germany/Turkey)
Type: anti-ship and anti-sumarine guided-missile frigate
Displacement: 2,000 tons standard and 2,400 tons full load
Dimensions: length 110.5 m (362.5 ft); beam 14.2 m (46.6 ft); draught 4.0 m (13.1 ft)
Gun armament: one 5-in (127-mm) L/54 DP in a Mk 45 single mounting, and one 25-mm Seaguard CIWS mounting
Missile armament: two Mk 141 quadruple container-launchers for eight RGM-84 Harpoon anti-ship missiles, and one Albatros octuple launcher for ? Aspide SAMs
Torpedo armament: none
Anti-submarine armament: two Mk 32 triple mountings for 12.75-in (324-mm) Mk 46 torpedoes, and helicopter-launched weapons (see below)
Aircraft: one Agusta-Bell AB.212ASW helicopter in a hangar aft
Electronics: one SPS-49 air-search radar, one Dolphin surface-search radar, one STIR target-designation and tracking radar and one Albis tracking radar used in conjunction with one WM-28 fire-control system, one Krupp-Atlas hull-mounted sonar, one SEWACO action information system, and two Hycor chaff launchers
Propulsion: four MTU 20V1163 TB93 diesels delivering 15200 kW (20,385 hp) to two shafts
Performance: speed 28 kts; range 7400 km (4,600 miles) at 20 kts
Complement: 179

Class
1. Turkey

Name	No.	Builder	Laid down	Commissioned
Yavuz	F240	Blohm und Voss	May 1985	
Turgut	F241	Howaldtswerke	Sep 1985	
Fatih	F242	Golcuk NY	Nov 1985	
Yildirim	F243	Golcuk NY	Jun 1986	

Note
Building and operating this useful class will greatly expand the industrial and operational capacity of Turkey

'Niels Juel' class FFG
(Denmark)
Type: anti-ship guided-missile patrol frigate
Displacement: 1,320 tons full load
Dimensions: length 84.0 m (275.5 ft); beam 10.3 m (33.8 ft); draught 3.1 m (10.2 ft)
Gun armament: one 76-mm (3-in) OTO Melara Compact L/62 DP in a single mounting
Missile armament: two Mk 141 quadruple container-launchers for eight RGM-84A Harpoon anti-ship missiles, one Mk 29 octuple launcher for eight RIM-7 NATO Sea Sparrow SAMs, and (to be fitted) one launcher for RAM SAMs
Torpedo armament: none
Anti-submarine armament: four Mk 32 single mountings for 12.75-in (324-mm) Mk 46 torpedoes
Mines: have minelaying capability
Aircraft: none
Electronics: one AWS 5 tactical radar, one Philips 9 GR 600 surface-search radar, one Skanter Mk 009 navigation radar, two RTN 10X radars used in conjunction with the Philips 9 LV 200 Mk 2 action information and fire-control system, one PMS 26 hull-mounted sonar, and two chaff launchers
Propulsion: CODOG (COmbined Diesel Or Gas turbine) arrangement, with one MTU 20V956 TB92 diesel delivering 3350 kW (4,495 shp) and one General Electric LM2500 gas turbine delivering 18,400 shp (13720 kW) to two shafts
Performance: speed 28 kts on gas turbine or 20 kts on diesel; range 4625 km (2,875 miles) at 18 kts
Complement: 90

Class
1. Denmark

Name	No.	Builder	Laid down	Commissioned
Niels Juel	F354	Aalborg Vaerft	Oct 1976	Aug 1980
Olfert Fischer	F355	Aalborg Vaeft	Dec 1978	Oct 1981
Peter Tordenskiold	F356	Aalborg Vaerft	Dec 1979	Apr 1982

Note
These ships are tailored neatly into the Danish concept of seaward defence, and possess good anti-ship (with missiles and mines) and adequate anti-aircraft self-protection capabilities at the expense of anti-submarine power

'Nilgiri' class FFG
(India)
Type: anti-ship and anti-submarine guided-missile frigate
Displacement: 2,450 tons standard and 2,960 tons full load
Dimensions: length 372.0 ft (113.4 m); beam 43.0 ft (13.1 m); draught 18.0 ft (5.5 m)
Gun armament: two 4.5-in (114-mm) L/45 DP in a Mk 6 twin mounting, and two 20-mm AA in single mountings
Missile armament: two container-launchers for two SS-N-2B 'Styx' anti-ship missiles (*Vindhyagiri* and *Taragiri* only), and one (*Nilgiri* and *Himgiri*) or two (others) quadruple launchers for 16 or 32 Sea Cat SAMs
Torpedo armament: none
Anti-submarine armament: one Limbo three-barrel mortar (not *Vindhyagiri* and *Taragiri*), or (*Vindhyagiri* and *Taragiri* only) two ILAS 3 triple mountings for A 244S torpedoes and one 375-mm (14.76-in) Bofors four-barrel rocket-launcher, and (all) helicopter-launched weapons (see below)
Aircraft: one Aérospatiale Alouette III (first four) or Westland Sea King (last two) helicopter in a hangar aft
Electronics: one Type 965 (*Nilgiri* only) or LW-05 air-search radar, one Type 993 surface-search radar (*Nilgiri* only), one navigation radar, one Type 904 fire-control radar (*Nilgiri* only), one GWS 22 (*Nilgiri* and *Himgiri* only) or two WM-44 SAM fire-control systems, one Type 184 hull-mounted medium-range sonar, one Type 199 (first four ships) or Thomson-CSF (other two ships) variable-depth sonar, and one Type 667/Type UA-8 ECM/ESM system
Propulsion: two boilers supplying steam to two sets of geared turbines delivering 30,000 shp (22370 kW) to two shafts; it is believed that *Vindhyagiri* and *Taragiri* have more power
Performance: speed 27 kts or (*Vindhyagiri* and *Taragiri* only) 28 kts; range 5,180 miles (8335 km) at 12 kts
Complement: 17+250

Class
1. India

Name	No.	Builder	Laid down	Commissioned
Nilgiri	F33	Mazagon Docks	Oct 1966	Nov 1972
Himgiri	F34	Mazagon Docks	1967	Nov 1974
Udaygiri	F35	Mazagon Docks	Jan 1973	Feb 1977
Dunagiri	F36	Mazagon Docks	Sep 1970	Feb 1976
Vindhyagiri	F38	Mazagon Docks	1975	Jul 1981
Taragiri	F41	Mazagon Docks	1974	May 1980

Note
These ships are essentially to the 'Broad-Beam Leander' class design with significant modifications to the after half for improved anti-submarine helicopter capability

'Oliver Hazard Perry' class FFG
(USA)
Type: anti-aircraft, anti-submarine and anti-ship guided-missile escort frigate
Displacement: 2,770 tons standard and 3,660 tons full load
Dimensions: length 445.0 ft (135.6 m) or (FFG8, 36/43 and 45/61) 453.0 ft (138.1 m); beam 45.0 ft (13.7 m); draught 24.5 ft (5.7 m)
Gun armament: one 76-mm (3-in) OTO Melara Compact L/62 DP in a Mk 75 single mounting, and one 20-mm Phalanx Mk 16 CIWS mounting
Missile armament: one Mk 13 single launcher for four RGM-84A Harpoon anti-ship missiles and 36 RIM-66B Standard SM-1 MR SAMs
Torpedo armament: none
Anti-submarine armament: two Mk 32 triple mountings for 24 12.75-in (324-mm) Mk 46 torpedoes, and helicopter-launched weapons
Aircraft: two Kaman SH-2F Seasprite helicopters in a hangar aft (being replaced from 1986 in FFG36 onwards by Sikorsky SH-60B Seahawk helicopters in a modification requiring the lengthening of the ships to 453.0 ft/138.1 m by an 8-ft/2.4-m increase of the rearward slope of the transom)
Electronics: one SPS-49 long-range air-search radar, one SPS-55 surface-search and navigation radar, one STIR-modified SPG-60 surveillance and target-indication radar, one Mk 92 (WM-28) fire-control system, one Mk 13 weapon direction system, one SQS-56 hull-mounted medium-range sonar, one SQR-19 towed-array sonar (FFG36/43 and FFG45/61), one T Mk 6 Fanfare torpedo decoy system, URN-25 TACAN, two Mk 36 Super RBOC chaff launchers, two OE-82 satellite communications systems, one SRR-1 satellite communications receiver, one WSC-3 satellite communications transceiver, and one SLQ-32(V)2 ESM system
Propulsion: two General Electric LM2500 gas turbines delivering 41,000 shp (30575 kW) to one shaft
Performance: speed 29 kts; range 5,200 miles (8370 km) at 20 kts
Complement: 13+180

Class
1. Australia

Name	No.	Builder	Laid down	Commissioned
Adelaide	F01	Todd Pacific	Jan 1977	Nov 1980
Canberra	F02	Todd Pacific	Mar 1978	Mar 1981
Sydney	F03	Todd Pacific	Jan 1980	Jan 1983
Darwin	F04	Todd Pacific	Jul 1981	Jul 1984
	F05	Williamstown DY	Jul 1985	
	F06	Williamstown DY	Aug 1986	

(The Australian ships, eventually to number 10, are in all vital respects similar to the US units.)

2. Spain ('**F80**' class)

Name	No.	Builder	Laid down	Commissioned
Santa Maria	F81	Bazan	1983	1985
Victoria	F82	Bazan	1983	1986
Numancia	F83	Bazan	1984	1987

(The Spanish ships have been much delayed as a result of the programme for the light carrier *Principe de Asturias*, but the ships will be similar to the US units other than having SPS-49 air-search radar, SPS-55 surface-search radar, Mk 92 search and tracking radar, STIR target-illuminating radar, RAN 12L/X missile-detecting radar, Raytheon navigation radar, SQR-19 towed-array sonar, Elettronica warfare systems, four Mk 36 RBOC chaff launchers, and a 20-mm Meroka CIWS mounting instead of the 20-mm Phalanx; it is expected that the eventual class total will be six.)

3. USA

Name	No.	Builder	Laid down	Commissioned
Oliver Hazard Perry	FFG7	Bath Iron Works	Jun 1975	Dec 1977
McInerney	FFG8	Bath Iron Works	Nov 1977	Nov 1979
Wadsworth	FFG9	Todd Pacific	Jul 1977	Feb 1980
Duncan	FFG10	Todd Pacific	Apr 1977	May 1980
Clark	FFG11	Bath Iron Works	Jul 1978	May 1980
George Philio	FFG12	Todd Pacific	Dec 1977	Oct 1980
Samuel Eliot Morison	FFG13	Bath Iron Works	Dec 1978	Oct 1980
Sides	FFG14	Todd Pacific	Aug 1978	May 1981
Estocin	FFG15	Bath Iron Works	Apr 1978	Jan 1981

Clifton Sprague	FFG16	Bath Iron Works	Sep 1979	Mar 1981
John A. Moore	FFG19	Todd Pacific	Dec 1978	Nov 1981
Antrim	FFG20	Todd Pacific	Jun 1978	Sep 1981
Flatley	FFG21	Bath Iron Works	Nov 1979	Jun 1981
Fahrion	FFG22	Todd Pacific	Dec 1978	Jan 1982
Lewis B. Puller	FFG23	Todd Pacific	May 1979	Apr 1982
Jack Williams	FFG24	Bath Iron Works	Feb 1980	Sep 1981
Copeland	FFG25	Todd Pacific	Oct 1979	Aug 1982
Gallery	FFG26	Bath Iron Works	May 1980	Dec 1981
Mahlon S. Tisdale	FFG27	Todd Pacific	Mar 1980	Nov 1982
Boone	FFG28	Todd Pacific	Mar 1970	May 1982
Stephen W. Groves	FFG29	Bath Iron Works	Sep 1980	Apr 1982
Reid	FFG30	Todd Pacific	Oct 1980	Feb 1983
Stark	FFG31	Todd Pacific	Aug 1979	Oct 1982
John L. Hall	FFG32	Bath Iron Works	Jan 1981	Jun 1982
Jarrett	FFG33	Todd Pacific	Feb 1981	Jul 1983
Aubrey Fitch	FFG34	Bath Iron Works	Apr 1981	Oct 1982
Underwood	FFG36	Bath Iron Works	Jul 1981	Jan 1983
Crommelin	FFG37	Todd Pacific	May 1980	Jun 1983
Curts	FFG38	Todd Pacific	Jul 1981	Oct 1983
Doyle	FFG39	Bath Iron Works	Oct 1981	May 1983
Halyburton	FFG40	Todd Pacific	Sep 1980	Jan 1984
McClusky	FFG41	Todd Pacific	Oct 1981	Dec 1983
Klakring	FFG42	Bath Iron Works	Mar 1982	Aug 1983
Thach	FFG43	Todd Pacific	Mar 1982	Mar 1984
Dewert	FFG45	Bath Iron Works	Jun 1982	Nov 1983
Rentz	FFG46	Todd Pacific	Sep 1982	Jun 1984
Nicholas	FFG47	Bath Iron Works	Sep 1982	Mar 1984
Vandergrift	FFG48	Todd Pacific	Oct 1981	Nov 1984
Robert G. Bradley	FFG49	Bath Iron Works	Dec 1982	Aug 1984
Jesse L. Taylor	FFG50	Bath Iron Works	Feb 1983	Nov 1984
Gary	FFG51	Todd Pacific	Dec 1982	Nov 1984
Carr	FFG52	Todd Pacific	Mar 1982	Aug 1985
Hawes	FFG53	Bath Iron Works	Aug 1983	Feb 1985
Ford	FFG54	Todd Pacific	Jul 1983	Jun 1985
Elrod	FFG55	Bath Iron Works	Nov 1983	Jul 1985
Simpson	FFG56	Bath Iron Works	Feb 1984	Aug 1985
Reuben James	FFG57	Todd Pacific	Nov 1983	Feb 1986
Samuel B. Roberts	FFG58	Todd Pacific	May 1984	Apr 1986
Kauffman	FFG59	Todd Pacific	Apr 1985	Oct 1986
Rodney M. Davies	FFG60	Todd Pacific	Feb 1985	Mar 1987
Ingraham	FFG61	Todd Pacific	Dec 1986	Nov 1988

Note
This is one of the US Navy's most important classes of small combatants

'Oslo' class FFG
(Norway)
Type: anti-ship and anti-submarine guided-missile coastal frigate
Displacement: 1,450 tons standard and 1,745 tons full load
Dimensions: length 96.6 m (317.0 ft); beam 11.2 m (36.7 ft); draught 5.3 m (17.4 ft)
Gun armament: four 3-in (76-mm) L/50 DP in two Mk 33 twin mountings
Missile armament: six container-launchers for six Penguin Mk II anti-ship missiles, and one Mk 29 octuple launcher for 24 RIM-7 NATO Sea Sparrow SAMs
Torpedo armament: none
Anti-submarine armament: two Mk 32 triple mountings for 12.75-in (324-mm) Mk 46 torpedoes, and one six-barrel rocket-launcher
Aircraft: none
Electronics: one DRBV 22 air-search radar, one TM1226 navigation radar, one WM-22 fire-control system, one SQS-36 hull-mounted medium-range sonar, and one Terne III Mk 3 hull-mounted sonar
Propulsion: two boilers supplying steam to one set of Ljungstrom/De Laval geared turbines delivering 20,000 shp (14915 kW) to one shaft
Performance: speed 25 kts; range 8350 km (5,190 miles) at 15 kts
Complement: 11+140

Class
1. Norway

Name	No.	Builder	Laid down	Commissioned
Oslo	F300	Marinens Hovedwerft	1963	Jan 1966
Bergen	F301	Marinens Hovedwerft	1964	Jun 1967
Trondheim	F302	Marinens Hovedwerft	1963	Jun 1966
Stavanger	F304	Marinens Hovedwerft	1965	Dec 1967
Narvik	F305	Marinens Hovedwerft	1964	Nov 1966

Note
The design of this Norwegian class is based on that of the US 'Dealey' class destroyer escort with a higher bow and a sensor/weapon fit suited to the coastal protection role off Norway's long and storm-ridden shores

'Peder Skram' class FFG
(Denmark)
Type: anti-ship and anti-submarine guided-missile frigate
Displacement: 2,030 tons standard and 2,720 tons full load
Dimensions: length 112.6 m (396.6 ft); beam 12.0 m (39.5 ft); draught 3.6 m (11.8 ft)
Gun armament: two 5-in (127-mm) L.38 DP in a Mk 38 twin mounting, and four 40-mm Bofors L/60 AA in single mountings
Missile armament: two Mk 141 quadruple container-launchers for eight RGM-84A Harpoon anti-ship missiles, and one quadruple launcher for 16 RIM-7 NATO Sea Sparrow SAMs
Torpedo armament: two 533-mm (21-in) twin mountings for wire-guided dual-role torpedoes
Anti-submarine armament: torpedoes (see above), and depth charges
Aircraft: none
Electronics: two CWS-3 combined air- and surface-search radars, one NWS-1 tactical radar, one NWS-2 navigation radar, three CGS-1 fire-control systems, and one PMS 26 hull-mounted search and attack sonar
Propulsion: CODOG (COmbined Diesel Or Gas turbine) arrangement, with two General Motors 16-567D diesels delivering 4,800 shp (3580 kW) and two Pratt & Whitney GG4A-3 gas turbines delivering 44,000 shp (32810 kW) to two shafts
Performance: speed 32.5 kts on gas turbines or 16.5 kts on diesels
Complement: 115

Class
1. Denmark

Name	No.	Builder	Laid down	Commissioned
Peder Skram	F352	Helsingors J&M	Sep 1964	May 1966
Herluf Trolle	F353	Helsingors J&M	Dec 1964	Apr 1967

Note
These Danish frigates feature good anti-ship and anti-submarine capability plus adequate anti-aircraft self-protection, and feature a high degree of automation to reduce manning requirements

'Saam' class FFG (Saudi Arabia): see 'Vosper Thornycroft Mk 5' class FFG

'Type 053H' class FFG (China): see 'Jianghu' class

'Type 053J' class FFG (China): see 'Jiangdong' class

'Type 12' class FFG (UK): see 'Whitby' class FFG

'Type 21' class FFG (UK): see 'Amazon' class FFG

'Type 23' class FFG (UK): see 'Duke' class

'Type 122' class FFG (West Germany): see 'Bremen' class FFG

'Type FS1500' class FFG
(West Germany/Colombia)
Type: anti-ship and anti-submarine guided-missile frigate
Displacement: 1,500 tons standard and 1,800 tons full load
Dimensions: length 95.3 m (312.7 ft); beam 11.3 m (37.1 ft); draught 3.3 m (10.8 ft)
Gun armament: one 76-mm (3-in) OTO Melara Compact L/62 DP in a single mounting, and two 40-mm Bofors L/70 AA in a Breda twin mounting
Missile armament: two quadruple container-launchers for eight MM.40 Exocet anti-ship missiles, and one Albatros octuple launcher for eight Aspide SAMs
Torpedo armament: none
Anti-submarine armament: two Mk 32 triple mountings for 12.75-in (324-mm) Mk 46 torpedoes, and helicopter-launched weapons (see below)

Aircraft: one MBB BO105 helicopter in a hangar aft
Electronics: one Sea Tiger combined air- and surface-search radar, one Canopus optronic director used in conjunction with a Castor fire-control radar system, one Krupp ASO 4-2 hull-mounted sonar, two Dagaie chaff launchers, one Phoenix ESM system, and one Scimitar ECM system
Propulsion: four MTU 20V1163 TB82 diesels delivering 17900 kW (24,005 bhp) to two shafts
Performance: speed 26.5 kts on four diesels and 18 kts on two diesels; range 10500 km (6,525 miles) at 18 kts
Complement: 92

Class
1. Colombia

Name	No.	Builder	Laid down	Commissioned
Almirante Padilla	51	Howaldtswerke	Mar 1981	Oct 1983
Caldas	52	Howaldtswerke	Jun 1981	Feb 1984
Antioquia	53	Howaldtswerke	Jun 1981	Apr 1984
Independente	54	Howaldtswerke	Jun 1981	Jul 1984

2. Malaysia

Name	No.	Builder	Laid down	Commissioned
Kasturi	F25	Howaldtswerke	Jan 1983	Aug 1984
Lekir	F26	Howaldtswerke	Jan 1983	Aug 1984

(The Malaysian ships differ quite considerably from the Colombian vessels in a number of important respects; at 97.3 m/319.2 ft and 3.5 m/11.5 ft the length and draught are slightly altered; the gun armament is one 100-mm/3.9-in L/55 Creusot-Loire DP, one 57-mm Bofors L/70 DP and four 30-mm AA in two Emerlec twin mountings; the missile armament comprises two MM.38 Exocets in two container-launchers; the anti-submarine armament comprises one 375-mm/14.76-in twin-barrel rocket-launcher, and there is only provision for a helicopter; the electronics fit is different, comprising one LW-08 combined air- and surface-search radar, one RM626 navigation radar, one WM-22 fire-control system, one Lirod optronic director, one ASO 84-5 sonar, two Dagaie chaff launchers, one RAPIDS ESM system, and one Scimitar ECM system; the complement is 124.)

'Ulsan' class FFG
(South Korea)
Type: anti-ship guided-missile frigate
Displacement: 1,600 tons standard and 1,940 tons full load
Dimensions: length 102.0 m (334.6 ft); beam 11.3 m (37.1 ft); draught 3.4 m (11.2 ft)
Gun armament: two 76-mm (3-in) OTO Melara Compact L/62 DP in single mountings, and eight 30-mm AA in four Emerlec twin mountings
Missile armament: two Mk 141 quadruple container-launchers for eight RGM-84A Harpoon anti-ship missiles
Torpedo armament: none
Anti-submarine armament: two Mk 32 triple mountings for 12.75-in (324-mm) Mk 46 torpedoes, and depth charges
Aircraft: none
Electronics: one DA-05 combined air- and surface-search radar, one ZW-06 surface-search and navigation radar, two WM-28 fire-control systems, one PAS 32 hull-mounted sonar, four chaff launchers, and various EW systems
Propulsion: CODOG (COmbined Diesel Or Gas turbine) arrangement, with two MTU 16V538 diesels delivering 5070 kW (6,800 shp) and two General Electric LM2500 gas turbines delivering 53,600 shp (39970 kW) to two shafts
Performance: speed 35 kts on gas turbines or 18 kts on diesels; range 7400 km (4,600 miles) at 18 kts on diesels
Complement: 125

Class
1. South Korea

Name	No.	Builder	Laid down	Commissioned
Ulsan	FF951	Hyundai Shipyards		Jan 1981
Seoul	FF952	Korea Shipbuilding		Mar 1985
	FF953	Hyundai Shipyards		
	FF954	Korea Shipbuilding		

Note
Despite their small displacement, these are potent ships with modern weapons and sensors tailored to tactical conditions around the Korean peninsula

'V-28' class FFG (Brazil): see 'Inhauma' class FFG

'Van Speijk' class FFG
(Netherlands)
Type: anti-submarine and anti-ship guided-missile frigate
Displacement: 2,255 tons standard and 2,835 tons full load
Dimensions: length 113.4 m (372.0 ft); beam 12.5 m (41.0 ft); draught 5.5 m (18.0 ft)
Gun armament: one 76-mm (3-in) OTO Melara Compact L/62 DP in a single mounting
Missile armament: two Mk 141 quadruple container-launchers for eight RGM-84A Harpoon anti-ship missiles, and two quadruple launchers for 32 Sea Cat SAMs
Torpedo armament: none
Anti-submarine armament: two Mk 32 triple mountings for 12.75-in (324-mm) Mk 46 torpedoes, and helicopter-launched weapons (see below)
Aircraft: one Westland SH-14B Lynx helicopter in a hangar aft
Electronics: one LW-03 air-search radar, one DA-05 combined air- and surface search and target-indication radar, one TM1229C navigation radar, one WM-45 gun fire-control system, one WM-44 SAM fire-control system, one CWE 610 hull-mounted sonar, one SQR-18A towed-array sonar, one Type 170B hull-mounted short-range search and attack sonar, one Type 162 hull-mounted side-looking classification sonar, one SEWACO II action information system, one Daisy data-handling system, and various EW systems
Propulsion: two boilers supplying steam to two sets of Werkspoor/English Electric geared turbines delivering 22370 kW (30,000 shp) to two shafts
Performance: speed 30 kts; range 8350 km (5,190 miles) at 12 kts
Complement: 175

Class
1. Indonesia

Name	No.	Builder	Laid down	Commissioned
		Nederlandse Dok	Jun 1964	Aug 1967
		Nederlandse Dok	Oct 1963	Feb 1967

(These two Indonesian units were originally the Dutch *Tjerk Hiddes* and *Van Speijk* respectively, and were bought early in 1986 for delivery late in the same year; another two units are following.)

2. Netherlands

Name	No.	Builder	Laid down	Commissioned
Isaac Sweers	F814	Nederlandse Dok	May 1965	May 1968
Evertsen	F815	Koninklijke Maatschappij	Jul 1965	Dec 1967

Note
Essentially 'Leander Batch 2' class ships, these Dutch-built frigates have a particularly potent anti-ship capability without detriment to the anti-aircraft and anti-submarine fits

'Vosper Thornycroft Mk 5' class FFG
(UK/Iran)
Type: anti-ship and anti-submarine guided-missile frigate
Displacement: 1,110 tons standard and 1,400 tons full load
Dimensions: length 310.0 ft (94.4 m); beam 36.3 ft (11.1 m); draught 14.0 ft (4.3 m)
Gun armament: one 4.5-in (114-mm) L/55 DP in a Mk 8 single mounting, and two 35-mm AA in a twin mounting
Missile armament: one quintuple container-launcher for five Sea Killer anti-ship missiles, and one Albatros quadruple launcher for Aspide SAMs
Torpedo armament: none
Anti-submarine armament: one Limbo Mk 10 mortar
Aircraft: none
Electronics: one AWS 1 air-search radar, two Sea Hunter surface-search and target-acquisition radars, one Type 170 hull-mounted short-range search and attack sonar, one Type 174 hull-mounted sonar, and one RDL 1 ESM system
Propulsion: CODOG (COmbined Diesel Or Gas turbine) arrangement, with two Paxman Ventura diesels delivering 3,800 shp (2835 kW) and two Rolls-Royce Olympus TM3B gas turbines delivering 46,000 shp (34300 kW) to two shafts
Performance: speed 39 kts; range 3,700 miles (5955 km) at 17 kts
Complement: 125

Class
1. Iran ('Saam' class)

Name	No.	Builder	Laid down	Commissioned
Saam	71	Vosper Thornycroft	May 1967	May 1971

Zaal	72	Vickers	Mar 1968	Mar 1971
Rostam	73	Vickers	Dec 1967	Jun 1972
Faramarz	74	Vosper Thornycroft	Jul 1968	Feb 1972

Note
The serviceability of these nicely-balanced frigates is now seriously in doubt as a result of the protracted Gulf War

'Vosper Thornycroft Mk 7' class FFG
(UK/Libya)
Type: anti-ship guided-missile frigate
Displacement: 1,325 tons standard and 1,625 tons full load
Dimensions: length 330.0 ft (100.6 m); beam 36.0 ft (11.0 m); draught 11.2 ft (3.4 m)
Gun armament: one 4.5-in (114-mm) L/55 DP in a Mk 8 single mounting, two 40-mm Bofors L/70 AA in single mountings, and two 35-mm AA in a twin mounting
Missile armament: four container-launchers for four Otomat anti-ship missiles, and one Albatros quadruple launcher for Aspide SAMs
Torpedo armament: none
Anti-submarine armament: two ILAS 3 triple mountings for 324-mm (12.75-in) A 244S torpedoes
Aircraft: none
Electronics: one RAN-series air-search radar, one RAN-series surface-search radar, one RTN-series radar used in conjunction with the Argo NA-series fire-control system, and one Diodon hull-mounted sonar
Propulsion: CODOG (Combined Diesel Or Gas turbine) arrangement, with two Paxman Ventura diesels delivering 3,500 shp (2610 kW) and two Rolls-Royce Olympus TM2A gas turbines delivering 46,400 shp (34600 kW) to two shafts
Performance: speed 37.5 kts on gas turbines; range 6,550 miles (10540 km) at 17 kts
Complement: not known

Class
1. Libya

Name	No.	Builder	Laid down	Commissioned
Dat Assawari	F01	Vosper Thornycroft	Sep 1968	Feb 1973

Note
This is an enlarged Vosper Mk 5 frigate with mainly Italian weapons and sensors

'Whitby' or 'Type 12' class FFG
(UK/India)
Type: anti-ship and anti-submarine guided-missile frigate
Displacement: 2,145 tons standard and 2,545 tons (F40) or 2,555 tons (F43) full load
Dimensions: length 369.8 ft (112.7 m); beam 41.0 ft (12.5 m); draught 17.8 ft (5.4 m)
Gun armament: four 40-mm Bofors AA in one twin and two single mountings
Missile armament: two container-launchers for two SS-N-2B 'Styx' anti-ship missiles
Torpedo armament: none
Anti-submarine armament: two Limbo Mk 10 mortars
Aircraft: none
Electronics: one Type 277 tactical radar, one Type 993 tactical radar, one 'Square Tie' missile-control radar, one fire-control system, one Type 162 hull-mounted side-looking classification sonar, one Type 170 hull-mounted short-range search and attack sonar, and one Type 174 hull-mounted sonar
Propulsion: two boilers supplying steam to two sets of English Electric geared turbines delivering 30,000 shp (22370 kW) to two shafts
Performance: speed 30 kts; range 5,200 miles (8370 km) at 12 kts
Complement: 11+220

Class
1. India

Name	No.	Builder	Laid down	Commissioned
Talwar	F40	Cammell Laird	1957	1960
Trishul	F42	Harland & Wolff	1957	1960

Note
Developed in India from the British original with Soviet anti-ship missiles and sensors, this class is generally obsolete

'Wielingen' or 'E-71' class FFG
(Belgium)
Type: anti-ship and anti-submarine guided-missile escort frigate
Displacement: 1,880 tons standard and 2,285 tons full load
Dimensions: length 106.4 m (349.0 ft); beam 12.3 m (40.3 ft); draught 4.6 m (18.4 ft)
Gun armament: one 100-mm (3.9-in) L/55 DP in a single mounting, and one 30-mm Goalkeeper CIWS mounting
Missile armament: four container-launchers for four MM.38 Exocet anti-ship missiles, and one Mk 29 octuple launcher for eight RIM-7 NATO Sea Sparrow SAMS
Torpedo armament: none
Anti-submarine armament: one 375-mm (14.76-in) six-barrel rocket-launcher, and two single 533-mm (21-in) mountings for 10 L5 torpedoes
Aircraft: none
Electronics: one DA-05 combined air- and surface-search radar, one TM1645/9X navigation radar, one WM-25 fire-control system, one Panda optical director, one SEWACO IV action information system, one SQS-505A hull-mounted medium-range search and attack sonar, one SLQ-25 Nixie torpedo decoy system, two Mk 36 Super RBOC chaff launchers, and one ELCOS 1 ESM system
Propulsion: CODOG (COmbined Diesel Or Gas turbine) arrangement, with two Cockerill CO-240 diesels delivering 4475 kW (6,000 shp) and one Rolls-Royce Olympus TM3B gas turbine delivering 28,000 shp (20880 kW) to two shafts
Performance: speed 29 kts on gas turbine or 20 kts on diesels; range 8350 km (5,190 miles) at 18 kts on diesels
Complement: 15+145

Class
1. Belgium

Name	No.	Builder	Laid down	Commissioned
Wielingen	F910	Boelwerf	Mar 1974	Jan 1978
Westdiep	F911	Cockerill	Sep 1974	Jan 1978
Wandelaar	F912	Boelwerf	Mar 1975	Oct 1978
Westhinder	F913	Cockerill	Dec 1975	Oct 1978

Note
Of simple design with only moderate power, these are useful escort frigates with a good balance of weapons and sensors from several NATO countries

'Yubari' class FFG
(Japan)
Type: anti-ship and anti-submarine guided-missile coastal frigate
Displacement: 1,470 tons standard and 1,690 tons full load
Dimensions: length 91.0 m (298.6 ft); beam 10.8 m (35.4 ft); draught 3.5 m (11.5 ft)
Gun armament: one 76-mm (3-in) OTO Melara Compact L/62 DP in a single mounting, and one 20-mm Phalanx Mk 16 CIWS mounting
Missile armament: two Mk 141 quadruple container-launchers for eight RGM-84A Harpoon anti-ship missiles
Torpedo armament: none
Anti-submarine armament: one 375-mm (14.76-in) Bofors four-barrel rocket-launcher, and two Type 68 triple mountings for 12.75-in (324-mm) Mk 46 torpedoes
Aircraft: none
Electronics: one OPS-28 surface-search radar, one OPS-19 navigation radar, one Mk 1 gun fire-control system, one OQS-1 hull-mounted sonar, two Mk 36 Super RBOC chaff launchers, one OLT-3 ECM system, and one NOLQ-6 ESM system
Propulsiion: CODOG (COmbined Diesel Or Gas turbine) arrangement, with one Mitsubishi 6 DRV diesel delivering 3500 kW (4,695 shp) and one Rolls-Royce Olympus TM3B gas turbine delivering 22,500 shp (16780 kW) to two shafts
Performance: speed 26 kts
Complement: 98

Class
1. Japan

Name	No.	Builder	Laid down	Commissioned
Yubari	DE227	Sumitomo	Feb 1981	Mar 1983
Yubetsu	DE228	Hitachi	Feb 1982	Mar 1984

Note
The 'Yubari' class was developed as an enlarged version of the semi-experimental *Ishikari*, with weapons and sensors optimized for coastal anti-ship and anti-submarine operations under cover of Japanese air strength; a class of three **'Yubari (Improved)' class** FFGs is to be built in the near future

FRIGATES

'Almirante Pereira da Silva' class FF
(Portugal)
Type: frigate
Displacement: 1,450 tons standard and 1,950 tons full load
Dimensions: length 95.9 m (314.6 ft); beam 11.3 m (36.9 ft); draught 5.3 m (17.4 ft)
Gun armament: four 3-in (76-mm) L/50 DP in two Mk 33 twin mountings
Missile armament: none
Torpedo armament: none
Anti-submarine armament: two 375-mm (14.76-in) Bofors four-barrel rocket-launchers, and two Mk 32 triple mountings for 12.75-in (324-mm) Mk 44/46 torpedoes

Aircraft: none
Electronics: one MLA 1b air-search radar, one Type 978 tactical radar, one RM316P navigation radar, two SPG-34 radars used in conjunction with two Mk 63 gun fire-control systems, one SQS-32A hull-mounted long-range search sonar, one SQA-10 variable-depth sonar, one DUBA 3A hull-mounted attack sonar, and several EW systems
Propulsion: two boilers supplying steam to two sets of De Laval geared turbines delivering 20,000 shp (14915 kW) to two shafts
Performance: speed 27 kts; range 5975 km (3,715 miles) at 15 kts
Complement: 12+154

Class
1. Portugal

Name	No.	Builder	Laid down	Commissioned
Almirante Pereira da Silva	F472	Estaleiros Navais	Jun 1962	Dec 1966
Almirante Gago Coutinho	F473	Estaleiros Navais	Dec 1963	Nov 1967
Almirante Magalhaes Correa	F474	Estaleiros Navais	Aug 1965	Nov 1968

Note
This simple class was developed from the US 'Dealey' class destroyer escort, and is now obsolescent and suffering serviceability problems

'Alpino' class FF
(Italy)
Type: frigate
Displacement: 2,000 tons standard and 2,700 tons full load
Dimensions: length 113.3 m (371.7 ft); beam 13.3 m (43.6 ft); draught 3.9 m (12.7 ft)
Gun armament: six 76-mm (3-in) OTO Melara L/62 DP in single mountings
Missile armament: none
Torpedo armament: none
Anti-submarine armament: two Mk 32 triple mountings for 12.75-in (324-mm) Mk 46 torpedoes, one Mk 113 depth-charge thrower, and helicopter-launched weapons (see below)
Aircraft: two Agusta-Bell AB.212ASW helicopters in a hangar aft

Electronics: one SPS-12 air-search radar, one SPQ 2 surface-search radar, three RTN 10X radars used in conjunction with three Orion series gun fire-control systems, one SQS-43 hull-mounted sonar, one SQA-10 variable-depth sonar, and one SCLAR chaff launcher
Propulsion: CODAG (COmbined Diesel And Gas turbine) arrangement, with four Tosi diesels delivering 12520 kW (16,790 shp) and two Tosi/Metrovick gas turbines delivering 15,000 shp (11185 kW) to two shafts
Performance: speed 28 kts on diesels and gas turbines or 20 kts on diesels; range 6500 km (4,040 miles) at 18 kts
Complement: 13+150

Class
1. Italy

Name	No.	Builder	Laid down	Commissioned
Alpino	F580	CNR, Riva Trigoso	Feb 1963	Jan 1968
Carabiniere	F581	CNR, Riva Trigoso	Jan 1965	Apr 1968

Note
Though obsolescent, these ships maintain an anti-submarine utility with their medium-size helicopters

'Annapolis' class FF
(Canada)
Type: frigate
Displacement: 2,400 tons standard and 3,000 tons full load
Dimensions: length 371.0 ft (113.1 m); beam 42.0 ft (12.8 m); draught 14.4 ft (4.4 m)
Gun armament: two 3-in (76-mm) L/60 DP in a Mk 33 twin mounting
Missile armament: one octuple launcher for 24 RIM-7 Canadian Sea Sparrow SAMs
Torpedo armament: none
Anti-submarine armament: two Mk 32 triple mountings for 12.75-in (324-mm) Mk 46 torpedoes, one Limbo Mk 10 mortar, and helicopter-launched weapons (see below)
Aircraft: one Sikorsky CH-124 Sea King helicopter in a hangar amidships
Electronics: one SPS-12 air-search radar, one SPS-10 surface-search radar, one Sperry Mk II navigation radar, one SPG-48 radar used in conjunction with the Mk 69 gun fire-control system, one Type 501 hull-mounted bottom-classification sonar, one Type 502 hull-mounted Limbo-control sonar, one Type 503 hull-mounted search sonar, one Type 504 variable-depth sonar, one SQS-10/11 hull-mounted sonar, one SLQ-25 Nixie torpedo decoy system, URN-20 TACAN, one CCS-280 action information system, and one ESM system
Propulsion: two boilers supplying steam to two sets of English Electric geared turbines delivering 30,000 shp (22370 kW) to two shafts
Performance: speed 28 kts; range 5,475 miles (8810 km) at 14 kts
Complement: 11+199

Class
1. Canada

Name	No.	Builder	Laid down	Commissioned
Annapolis	265	Halifax Shipyards	Jul 1960	Dec 1964
Nipigon	266	Marine Industries	Apr 1960	May 1964

Note
The viability of this class depends primarily on its large anti-submarine helicopter

'Baptista de Andrade' class FF
(Portugal)
Type: frigate
Displacement: 1,250 tons standard and 1,380 tons full load
Dimensions: length 84.6 m (277.5 ft); beam 10.3 m (33.8 ft); draught 3.6 m (11.8 ft)
Gun armament: one 100-mm (3.9-in) L/55 DP in a single mounting, and two 40-mm Bofors L/70 AA in single mountings
Missile armament: provision for the wartime shipping of two single container-launchers for two MM.38 Exocet anti-ship missiles
Torpedo armament: none
Anti-submarine armament: two Mk 32 triple mountings for 12.75-in (324-mm) Mk 44/46 torpedoes
Aircraft: none
Electronics: one AWS 2 air-search radar, one TM626 navigation radar, one Pollux radar used in conjunction with the Panda fire-control system, and one Diodon hull-mounted sonar
Propulsion: two SEMT-Pielstick PA6V-280 diesels delivering 8170 kW (10,960 shp) to two shafts
Performance: speed 23.5 kts; range 1100 km (6,835 miles) at 18 kts
Complement: 107

Class
1. Portugal

Name	No.	Builder	Laid down	Commissioned
Bapiste de Andrade	F486	Bazan	1972	Nov 1974
Joao Roby	F487	Bazan	1972	Mar 1975
Afonso Cerqueira	F488	Bazan	1973	Jun 1975
Oliveira E. Carmo	F489	Bazan	1972	Feb 1975

Note
These ships are now suitable only for patrol and training

'Bergamini' class FF
(Italy)
Type: frigate
Displacement: 1,650 tons full load
Dimensions: length 94.0 m (308.4 ft); beam 11.4 m (37.4 ft); draught 3.2 m (10.5 ft)
Gun armament: two 76-mm (3-in) OTO Melara L/62 DP in single mountings
Missile armament: none
Torpedo armament: none
Anti-submarine armament: two Mk 32 triple mountings for 12.75-in (324-mm) Mk 46 torpedoes, one Mk

113 depth-charge thrower, and helicopter-launched weapons (see below)
Aircraft: one Agusta-Bell AB.204B helicopter in a hangar amidships
Electronics: one SPS-12 air-search radar, one SPQ 2 surface-search and navigation radar, one Orion 3 radar used in conjunction with the OG3 gun fire-control system, one SQS-40 hull-mounted sonar, and one SPR-A ESM system
Propulsion: four Fiat CB/LR diesels delivering 11920 kW (15,985 shp) to two shafts
Performance: speed 24 kts; range 5500 km (3,420 miles) at 18 kts
Complement: 13+150

Class
1. Italy

Name	No.	Builder	Laid down	Commissioned
Virginio Fasan	F594	Navalmeccanica	Mar 1960	Oct 1962
Carlo Margottini	F585	Navalmeccanica	May 1957	May 1962

Note
These ships are now wholly obsolete by advanced standards in the Mediterranean

'Berk' class FF
(Turkey)
Type: frigate
Displacement: 1,450 tons standard and 1,950 tons full load
Dimensions: length 95.2 m (312.2 ft); beam 11.8 m (37.8 ft); draught 4.4 m (14.4 ft)
Gun armament: two 3-in (76-mm) L/50 DP in a Mk 33 twin mounting
Missile armament: none
Torpedo armament: none
Anti-submarine armament: two Mk 32 triple mountings for 12.75-in (324-mm) Mk 46 torpedoes, two Hedgehog Mk 11 mortars, and one depth-charge rack
Aircraft: provision for a light helicopter on a platform aft
Electronics: one SPS-40 air-search radar, one SPS-10 surface-search radar, two SPG-34 radars used in conjunction with two Mk 63 gun fire-control systems, and one SQS-11 hull-mounted sonar
Propulsion: four Fiat/Tosi 3-016-RSS diesels delivering 17,900 kW (24,000 shp) to one shaft
Performance: speed 25 kts
Complement: not known

Class
1. Turkey

Name	No.	Builder	Laid down	Commissioned
Berk	D358	Golcuk Navy Yard	Mar 1967	Jul 1972
Peyk	D359	Golcuk Navy Yard	Jun 1968	Jul 1975

Note
Based on the US 'Claud Jones' class escort, these frigates lack modern weapons and sensors, but gave the Turks a start into major warship construction

'Bronstein' class FF
(USA)
Type: frigate
Displacement: 2,360 tons standard and 2,650 tons full load
Dimensions: length 371.5 ft (113.2 m); beam 40.5 ft (12.3 m); draught 23.0 ft (7.0 m)
Gun armament: two 3-in (76-mm) L/50 DP in a Mk 33 twin mounting
Missile armament: none
Torpedo armament: none
Anti-submarine armament: one Mk 16 octuple launcher for eight RUR-5A ASROC missiles, two Mk 32 triple mountings for 12.75-in (324-mm) Mk 46 torpedoes, and helicopter-launched weapons (see below)
Aircraft: provision for one Kaman SH-2F Seasprite helicopter on a platform aft
Electronics: one SPS-40 air-search radar, one SPS-10 surface-search radar, one SPG-35 radar used in conjunction with the Mk 56 gun fire-control system, one Mk 1 weapon direction system, one SQS-26 bow-mounted 'bottom-bounce' sonar and one SQR-15 towed-array sonar used in conjunction with the Mk 114 underwater weapons fire-control system, OE-82 satellite communications system, and SRR-1 satellite communications receiver
Propulsion: two boilers supplying steam to one set of De Laval geared turbines delivering 20,000 shp (14915 kW) to one shaft
Performance: speed 26 kts; range 3,685 miles (5930 km) at 20 kts
Complement: 15+202

Class				
1. USA				
Name	No.	Builder	Laid down	Commissioned
Bronstein	FF1037	Avondale	May 1961	Jun 1963
McCloy	FF1038	Avondale	Sep 1961	Oct 1963

Note
Essentially development ships for the new generation of US frigates of the 1960s, these two units are operational but lack the speed and defensive armament for first-line use

'Cannon' class FF
(USA/Greece)
Type: frigate
Displacement: 1,240 tons standard and 1,900 tons full load
Dimensions: length 306.0 ft (93.3 m); beam 36.7 ft (11.2 m); draught 14.0 ft (4.3 m)
Gun armament: three 3-in (76-mm) L/50 DP in Mk 22 single mountings, four 40-mm Bofors L/60 AA in two Mk 1 twin mountings, and 14 20-mm AA in seven twin mountings
Missile armament: none
Torpedo armament: none
Anti-submarine armament: two Mk 32 single mountings for 12.75-in (324-mm) Mk 44/46 torpedoes, one Hedgehog mortar, eight depth-charge throwers, and one depth-charge rack
Aircraft: none
Electronics: one surface-search and navigation radar, one Mk 26 radar used in conjunction with the Mk 52 gun fire-control system, and hull-mounted sonar
Propulsion: diesel-electric arrangement, with four General Motors diesels powering two electric motors delivering 6,000 shp (4475 kW) to two shafts
Performance: speed 19.25 kts; range 10,350 miles (16655 km) at 12 kts
Complement: 220

Class				
1. Greece				
Name	No.	Builder	Laid down	Commissioned
Aetos	D01	Tampa SB	Mar 1943	May 1944
Ierax	D31	Tampa SB	Apr 1943	Jul 1944
Leon	D54	Federal SB	Feb 1943	Aug 1943
Panthir	D67	Federal SB	Sep 1943	Jan 1944

2. Philippines

Name	No.	Builder	Laid down	Commissioned
Datu Siratuna	PF77	Federal SB	Nov 1942	Jul 1943
Rajah Humabon	PF78	Norfolk NY	Jan 1943	Aug 1943

(The Filipino ships have SPS-5 surface-search and SPS-6C air-search radars, six rather than four 40-mm Bofors and six rather than 14 20-mm AA guns, no anti-submarine torpedoes, and a complement of about 165.)

3. Thailand

Name	No.	Builder	Laid down	Commissioned
Pin Klao	3	Western Pipe	1943	Sep 1943

(The Thai ship has three 3-in/76-mm and four 40-mm guns, two Mk 32 triple mountings for 12.75-in/324-mm Mk 46 torpedoes, and SPS-5 surface-search radar.)

4. Uruguay

Name	No.	Builder	Laid down	Commissioned
Uruguay	1	Federal SB	Dec 1942	Jul 1943
Artigas	2	Federal SB	Aug 1943	Dec 1943

(The Uruguayan ship has three 3-in/76-mm and two 40-mm guns, no anti-submarine torpedoes, one SPS-6 air-search and one SPS-10 surface-search radar, and a complement of 160.)

Note
Thoroughly obsolete, these ships are suitable only for patrol and training

'Charles Lawrence' and 'Crosley' class FFs
(USA/Taiwan)
Type: frigate
Displacement: 1,400 tons standard and 2,130 tons full load
Dimensions: length 306.0 ft (93.3 m); beam 37.0 ft (11.3 m); draught 12.6 ft (3.8 m)
Gun armament: two 5-in (127-mm) L/38 DP in Mk 30 single mountings, six 40-mm Bofors L/60 AA in three twin mountings, and four or eight 20-mm AA

in four single or twin mountings
Missile armament: none
Torpedo armament: none
Anti-submarine armament: two Mk 32 triple mountings for 12.75-in (324-mm) Mk 44/46 torpedoes, two Hedgehog mortars, and depth charges
Aircraft: none
Electronics: one SPS-5 surface-search radar, one Decca 707 navigation radar, one Mk 26 radar used in conjunction with the Mk 37 gun fire-control system, and one hull-mounted sonar
Propulsion: two boilers supplying steam to a General Electric turbo-electric drive delivering 12,000 shp (8950 kW) to two shafts
Performance: speed 23.6 kts; range 5,750 miles (9255 km) at 15 kts
Complement: 200

Class
1. Ecuador

Name	No.	Builder	Laid down	Commissioned
Moran Valverde	DD02	Philadelphia NY	1943	Sep 1943

(This Ecuadorean ship is an ex-'Charles Lawrence' class frigate/high-speed transport, and still serves in the latter role with a capacity for 162 troops that can be ferried in two LCUs; the armament is one 5-in/127-mm and four 40-mm AA guns, and the electronics include SPS-6 air-search and SPS-10 surface-search radars; the ship has a small helicopter platform aft.)

2. Mexico

Name	No.	Builder	Laid down	Commissioned
Tehuantepec	B05	Consolidated Steel	1943	Apr 1945
Usumacinta	B06	Consolidated Steel	1943	May 1945
Coahuila	B07	Bethlehem Steel	1944	Mar 1945
Chihuahua	B08	Norfolk NY	1943	Oct 1943

(These Mexican ships are similar to the Ecuadorean unit but has six 40-mm AA guns an SC combined air- and surface-search radar; the first three are ex-'Crosley' class ships, and the last an ex-'Charles Lawrence' class vessel.)

3. South Korea

Name	No.	Builder	Laid down	Commissioned
Kyong Nam	DE882	Defoe SB	1944	May 1945
Ah San	DE823	Bethlehem Steel	1943	Jun 1945
Ung Po	DE825	Bethlehem Steel	1943	Jun 1945
Kyong Puk	DE826	Charleston NY	1943	Jan 1945
Jon Nam	DE827	Charleston NY	1943	Mar 1944
Chi Ju	DE828	Charleston NY	1943	Apr 1945

(These South Korean ships comprise two ex-'Charles Lawrence' class [DE826/827] and four ex-'Crosley' class units, all still used in the high-speed transport role with 160 troops and four LVCP landing craft apiece; the armament is one 5-in/127-mm and six 40-mm AA guns.)

4. Taiwan

Name	No.	Builder	Laid down	Commissioned
Tien Shan	615	Defoe SB	1944	Jun 1945
Lu Shan	821	Defoe SB	1942	Aug 1943
Yu Shan	826	Charleston NY	1943	Nov 1944
Wen Shan	834	Bethlehem Steel	1942	Jul 1943
Fu Shan	835	Charleston NY	1943	Jul 1944
Chung Shan	845	Bethlehem Steel	1943	Sep 1943
Hua Shan	854	Defoe SB	1944	Apr 1945
Tai Shan	878	Charleston NY	1943	Jan 1945
Shou Shan	893	Bethlehem Steel	1944	Oct 1944

(These Taiwanese ships comprise three ex-'Charles Lawrence' class [821, 834 and 845] and six ex-'Crosley' class units, and have vestigial fast-transport capability for 160 troops with one LVCP landing craft.)

Note
Designed as high-speed raider transports, these elderly ships are now suitable only for patrol and training even in remoter areas

'Chikugo' class FF
(Japan)
Type: frigate
Displacement: 1,470/1,500 tons standard and 1,700 tons full load
Dimensions: length 93.1 m (305.5 ft); beam 10.8 m (35.5 ft); draught 3.5 m (11.5 ft)
Gun armament: two 3-in (76-mm) L/50 DP in a Mk 33 twin mounting, and two 40-mm Bofors AA in a Mk 1 twin mounting

Missile armament: none
Torpedo armament: none
Anti-submarine armament: one Mk 16 octuple launcher for eight RUR-5A ASROC missiles, and two Mk 32 triple mountings for 12.75-in (324-mm) Mk 46 torpedoes
Aircraft: none
Electronics: one OPS-14 air-search radar, one OPS-29 surface-search radar, one OPS-19 navigation radar, one Mk 33 radar used in conjunction with the GFCS-1 3-in gun fire-control system, one Mk 51 40-mm gun fire-control system, one OQS-3 hull-mounted sonar, one SQS-35J variable-depth sonar, and one NORL-5 ESM system
Propulsion: four diesels delivering 11930 kW (16,000 shp) to two shafts
Performance: speed 25 kts
Complement: 165

Class
1. Japan

Name	No.	Builder	Laid down	Commissioned
Chikugo	DE215	Mitsui	Dec 1968	Jul 1970
Ayase	DE216	Ishikawajima	Dec 1969	May 1971
Mikuma	DE217	Mitsui	Mar 1970	Aug 1971
Tokachi	DE218	Mitsui	Dec 1970	May 1972
Iwase	DE219	Mitsui	Aug 1971	Dec 1972
Chitose	DE220	Hitachi	Oct 1971	Aug 1973
Niyodo	DE221	Mitsui	Sep 1972	Feb 1974
Teshio	DE222	Hitachi	Jul 1973	Jan 1975
Yoshino	DE223	Mitsui	Sep 1973	Feb 1975
Kumano	DE224	Hitachi	May 1974	Nov 1975
Noshiro	DE225	Mitsui	Jan 1976	Aug 1977

Note
Designed for coastal operations, the ships of this class are notably quiet

'Claud Jones' class FF
(USA/Indonesia)
Type: frigate
Displacement: 1,450 tons standard and 1,750 tons full load
Dimensions: length 310.0 ft (94.5 m); beam 37.0 ft (11.3 m); draught 18.0 ft (5.5 m)
Gun armament: (341 and 342) one 3-in (76-mm) L/50 DP in a Mk 34 single mounting, two 37-mm AA in a twin mounting, and two 25-mm AA in a twin mounting; or (343 and 344) two 3-in (76-mm) L/50 DP in Mk 34 single mountings, and two 25-mm AA in a twin mounting
Missile armament: none
Torpedo armament: none
Anti-submarine armament: two Mk 32 triple mountings for 12.75-in (324-mm) Mk 44 torpedoes, and two Hedgehog mortars
Aircraft: none
Electronics: one SPS-6 air-search radar, one SPS-10 surface-search radar, one Decca navigation radar, one Mk 70 3-in gun fire-control system, and one SQS-29/32 series hull-mounted sonar
Propulsion: four Fairbanks-Morse 38D81/8 diesels delivering 9,200 shp (6860 kW) to one shaft
Performance: speed 22 kts
Complement: 165

Class
1. Indonesia

Name	No.	Builder	Laid down	Commissioned
Samadikun	341	Avondale	Oct 1957	May 1959
Martadinata	342	American SB	Oct 1958	Nov 1959
Monginsidi	343	Avondale	Jun 1957	Feb 1959
Ngurah Rai	344	American SB	Nov 1958	Mar 1960

Note
These are simple escorts now useful only for patrol and training

'Comandante Joao Belo' class FF
(Portugal)
Type: frigate
Displacement: 1,750 tons standard and 2,250 tons full load
Dimensions: length 103.7 m (340.3 ft); beam 11.7 m (38.4 ft); draught 4.8 m (15.7 ft)
Gun armament: three 100-mm (3.9-in) L/55 DP in single mountings, and two 40-mm Bofors L/70 AA in single mountings
Missile armament: none
Torpedo armament: none

Anti-submarine armament: one 305-mm (12-in) four-barrel mortar, and two 533-mm (21-in) triple mountings for six torpedoes
Aircraft: provision for a light helicopter being added
Electronics: one DRBV 22A air-search radar, one DRBV 50 combined air- and surface-search radar, one RM316 navigation radar, one DRBC 31D gun-control radar, one SQS-17A hull-mounted search sonar, and one DUBA 3A hull-mounted attack sonar
Propulsion: four SEMT-Pielstick diesels delivering 11920 kW (15,985 shp) to two shafts
Performance: speed 25 kts; range 13900 km (8,635 miles) at 15 kts
Complement: 14+186

Class
1. Portugal

Name	No.	Builder	Laid down	Commissioned
Comandante Joao Belo	F480	A & C de Nantes	Sep 1965	Jul 1967
Comandante Hermenegildo Capelo	F481	A & C de Nantes	May 1966	Apr 1968
Comandante Roberto Ivens	F482	A & C de Nantes	Dec 1966	Nov 1968
Comandante Sacadura Cabral	F483	A & C de Nantes	Aug 1967	Jul 1969

Note
These useful but lightly armed general-purpose frigates are based on the French 'Commandant Rivière' class

'Crosley' class FF (USA): see 'Charles Lawrence' class FF

'Garcia' class FF
(USA)
Type: frigate
Displacement: 2,620 tons standard and 3,405 tons full load
Dimensions: length 414.5 ft (126.3 m); beam 44.2 ft (13.5 m); draught 24.0 ft (7.3 m)
Gun armament: two 5-in (127-mm) L/38 DP in Mk 30 single mountings
Missile armament: up to four RGM-84A Harpoon anti-ship missiles instead of RUR-5A ASROC missiles (see below)
Torpedo armament: none
Anti-submarine armament: one Mk 16 octuple launcher for eight (FF1040/1045) or 16 (FF1047/1051) RUR-5A ASROC missiles, and two Mk 32 triple mountings for 12.75-in (324-mm) Mk 46 torpedoes
Aircraft: one Kaman SH-2F Seasprite helicopter in a hangar aft (not in FF1048 and 1050)
Electronics: one SPS-40 air-search radar, one SPS-10 surface-search radar, one LN-66 navigation radar, one SPG-35 radar used in conjunction with the Mk 56 gun fire-control system, one Mk 1 weapon direction system, one SQS-26AXR (FF1040/1045) or SQS-26BR (FF1047/1051) bow-mounted 'bottom-bounce' sonar and (FF1048 and 1050 only) one BQR-15 towed-array sonar used in conjunction with the Mk 114 underwater weapons fire-control system, one T Mk 6 Fanfare torpedo decoy system, Naval Tactical Data System (FF1047 and 1049 only), OE-82 satellite communications antenna, SRR-1 satellite communications receiver, WSC-3 satellite communications transceiver, SRN-15 TACAN, one WLR-1 ECM system, one WLR-3 ECM system, and one ULQ-6 ECM system
Propulsion: two boilers supplying steam to two sets of Westinghouse or General Electric geared turbines delivering 35,000 shp (26100 kW) to two shafts
Performance: speed 27.5 kts; range 4,600 miles (7400 km) at 20 kts
Complement: about 18+260

Class
1. USA

Name	No.	Builder	Laid down	Commissioned
Garcia	FF1040	Bethlehem Steel	Oct 1962	Dec 1964
Bradley	FF1041	Bethlehem Steel	Jan 1963	May 1965
Edward McDonnell	FF1043	Avondale	Apr 1963	Feb 1965
Brumby	FF1044	Avondale	Aug 1963	Aug 1945
Davidson	FF1045	Avondale	Sep 1963	Dec 1965
Voge	FF1047	Defoe SB	Nov 1963	Nov 1966
Sample	FF1048	Lockheed SB	Jul 1963	Mar 1968
Koelsch	FF1049	Defoe SB	Feb 1964	Jun 1967
Albert David	FF1050	Lockheed SB	Apr 1963	Oct 1968
O'Callaghan	FF1051	Defoe SB	Feb 1964	Jul 1968

Note
These are comparatively simple escort vessels optimized for anti-submarine work with ASROC missiles and (in most units) a LAMPS I helicopter

'Glover' class FF

(USA)
Type: frigate
Displacement: 2,645 tons standard and 3,425 tons full load
Dimensions: length 414.5 ft (126.3 m); beam 44.2 ft (13.5 m); draught 24.0 ft (7.3 m)
Gun armament: one 5-in (127-mm) L/38 DP in a Mk 30 single mounting
Missile armament: none
Torpedo armament: none
Anti-submarine armament: one Mk 16 octuple launcher for eight RUR-5A ASROC missiles, two Mk 32 triple mountings for 12.75-in (324-mm) Mk 46 torpedoes, and helicopter-launched weapons (see below)
Aircraft: provision for one Kaman SH-2F Seasprite helicopter on a platform aft
Electronics: one SPS-40 air-search radar, one SPS-10 surface-search radar, one LN-66 navigation radar, one SPG-35 radar used in conjunction with the Mk 56 gun fire-control system, one Mk 1 weapon designation system, one SQS-26AXR bow-mounted 'bottom-bounce' sonar and one SQS-35 variable-depth sonar used in conjunction with the Mk 114 underwater weapons fire-control system, OE-82 satellite communications system, and SRR-1 satellite communications receiver
Propulsion: two boilers supplying steam to a set of Westinghouse geared turbines delivering 35,000 shp (26100 kW) to one shaft
Performance: speed 27 kts; range 4,600 miles (7400 km) at 20 kts
Complement: 18+252

Class
1. USA

Name	No.	Builder	Laid down	Commissioned
Glover	FF1098	Bath Iron Works	Jul 1963	Nov 1965

Note
Designed and built as an experimental type, this ship has limited armament and performance, but features advanced sensors

'Isuzu' class FF

(Japan)
Type: frigate
Displacement: 1,490 tons standard and 1,700 tons full load
Dimensions: length 94.0 m (308.3 ft); beam 10.4 m (34.2 ft); draught 3.5 m (11.5 ft)
Gun armament: four 3-in (76-mm) L/50 DP in two Mk 33 twin mountings
Missile armament: none
Torpedo armament: one 533-mm (21-in) quadruple mounting for four torpedoes
Anti-submarine armament: one 375-mm (14.76-in) Bofors four-barrel rocket-launcher, two Type 68 triple mountings for 12.75-in (324-mm) Mk 46 torpedoes, one Mk 1 Y-gun depth-charge thrower, and (DE211 and 214) one depth-charge rack
Electronics: one OPS-1 air-search radar, one OPS-16 surface-search radar, one ORD-1 navigation radar, one SPG-34 radar used in conjunction with the Mk 63 gun fire-control system, one SQS-29 hull-mounted sonar and (DE212/213 only) one OQA-1 variable-depth sonar used in conjunction with the Mk 105 underwater weapons fire-control system, and one BLR-1 ECM system
Propulsion: four Mitsubishi or Mitsui diesels delivering 11920 kW (15,985 shp) to two shafts
Performance: speed 25 kts
Complement: 180

Class
1. Japan

Name	No.	Builder	Laid down	Commissioned
Isuzu	DE211	Mitsui	Apr 1960	Jul 1961
Mogami	DE212	Mitsubishi	Aug 1960	Oct 1961
Kitakami	DE213	Ishikawajima	Jun 1962	Feb 1964
Ooi	DE214	Maizuru	Jun 1962	Jan 1964

Note
This class is now obsolete and suitable only for patrol and training

'Jiangnan' class FF

(China)
Type: frigate
Displacement: 1,150 tons standard and 1,500 tons full load
Dimensions: length 90.8 m (297.8 ft); beam 10.0 m (32.8 ft); draught 3.9 m (12.8 ft)
Gun armament: three 100-mm (3.9-in) L/56 DP in single mountings, eight 37-mm AA in four twin mountings, and four 12.7-mm (0.5-in) AA in two twin mountings
Missile armament: none

Torpedo armament: none
Anti-submarine armament: two RBU 1200 five-barrel rocket-launchers, four BMB-2 depth-charge throwers, and two depth-charge racks
Mines: 60
Aircraft: none
Electronics: one 'Ball End' surface-search radar, one 'Wok Won' gun-control radar, one 'Fin Curve' navigation radar, one hull-mounted high-frequency sonar, and one 'High Pole-B' IFF
Propulsion: two diesels delivering 17900 kW (24,000 shp) to two shafts
Performance: speed 28 kts; range 5200 km (3,230 miles) at 10 kts
Complement: about 180

Class
1. China

Name	No.	Builder	Laid down	Commissioned
	501	Tunglung, Guangzhou	1965	1967
	502	Tunglung, Guangzhou	1966	1968
	503	Tunglung, Guangzhou	1966	1968
	504	Tunglung, Guangzhou	1967	1969
	509	Jiangnan, Shanghai	1965	1967

Note
These are in essence Chinese-developed enlargements of the 'Chengdu' or 'Riga' class general-purpose frigate

'Joao Coutinho' class FF
(Portugal)
Type: frigate
Displacement: 1,205 tons standard and 1,380 tons full load
Dimensions: length 84.6 m (277.5 ft); beam 10.3 m (33.8 ft); draught 3.6 m (11.8 ft)
Gun armament: two 3-in (76-mm) L/50 DP in a Mk 34 twin mounting, and two 40-mm Bofors L/70 AA in a twin mounting
Missile armament: none
Torpedo armament: none
Anti-submarine armament: one Hedgehog mortar, two depth-charge throwers, and two depth-charge racks
Aircraft: none
Electronics: one MLA 1b air-search radar, one TM626 navigation radar, one SPG-34 radar used in conjunction with the Mk 63 gun fire-control system, one Mk 51 AA fire-control system, and one QCU-2 hull-mounted sonar
Propulsion: two SEMT-Pielstick PA6V-280 diesels delivering 8170 kW (10,960 shp) to two shafts
Performance: speed 24.5 kts; range 10925 km (6,790 miles) at 18 kts
Complement: 9+91

Class
1. Portugal

Name	No.	Builder	Laid down	Commissioned
Antonio Enes	F471	Bazan	Apr 1968	Jun 1971
Joao Coutinho	F475	Blohm und Voss	Sep 1968	Mar 1970
Jacinto Candido	F476	Blohm und Voss	Apr 1968	Jun 1970
General Pereira d'Eca	F477	Blohm und Voss	Oct 1968	Oct 1970
Augusto de Castilho	F484	Bazan	Aug 1968	Nov 1970
Honorio Barreto	F485	Bazan	Jul 1968	Apr 1971

Note
Lacking modern armament and sensors, these general-purpose frigates are suited mostly to patrol tasks

'Köln' or 'Type 120' class FF
(West Germany)
Type: frigate
Displacement: 2,100 tons standard and 2,700 tons full load
Dimensions: length 109.9 m (360.5 ft); beam 11.0 m (36.1 ft); draught 5.1 m (16.7 ft)
Gun armament: two 100-mm (3.9-in) L/55 DP in single mountings, and six 40-mm Bofors L/60 AA in two single and two twin mountings
Missile armament: none
Torpedo armament: none
Anti-submarine armament: two 375-mm (14.76-in) Bofors four-barrel rocket-launchers, four single 533-mm (21-in) mountings for ? torpedoes, and two depth-charge throwers
Mines: 80
Aircraft: none
Electronics: one ZW-series surface-search radar, one DA-02 target designation radar, two WM-45 100-mm gun fire-control systems, two WM-45 40-mm gun fire-control systems, and one PAE/CWE hull-mounted sonar used in conjunction with a Hollandse Signaalapparaten underwater weapons fire-control system

Propulsion: CODAG (COmbined Diesel And Gas turbine) arrangement, with four MAN diesels delivering 8940 kW (11,990 shp) and two Brown-Boveri gas turbines delivering 17900 kW (24,005 shp) to two shafts
Performance: speed 28 kts on diesels and gas turbines or 18 kts on diesels; range 1675 km (1,040 miles) at 28 kts
Complement: 17+193

Class
1. Turkey

Name	No.	Builder	Laid down	Commissioned
Gelibolu	D360	H.C. Stülcken	Dec 1958	Dec 1962
Gemlik	D361	H.C. Stülcken	Apr 1958	Oct 1961

(These Turkish ships are in all essential respects identical with the West German units.)

2. West Germany

Name	No.	Builder	Laid down	Commiss
Augsburg	F222	H.C. Stülcken	Oct 1958	Apr 1962
Lübeck	F224	H.C. Stülcken	Oct 1959	Jun 1963
Braunschweig	F225	H.C. Stülcken	Jul 1960	Jun 1964

Note
Though heavily armed, these general-purpose frigates are obsolete by the advanced standards prevailing in the NATO northern sector

'Leander Batch 1' class FF
(UK)
Type: frigate
Displacement: 2,450 tons standard and 2,860 tons full load
Dimensions: length 372.0 ft (113.4 m); beam 41.0 ft (12.5 m); draught 14.8 ft (4.5 m)
Gun armament: two 40-mm Bofors L/60 AA in single mountings
Missile armament: two quadruple launchers for Sea Cat SAMs
Torpedo armament: none
Anti-submarine armament: one launcher for ? Ikara missiles, one Limbo Mk 10 mortar, and helicopter-launched weapons (see below)
Aircraft: one Westland Lynx HAS.Mk 2 helicopter in a hangar aft
Electronics: one Type 994 combined air- and surface-search radar, one Type 1006 navigation radar, two Type 903 radars used in conjunction with the GWS 22 SAM fire-control system, one Type 199 variable-depth sonar, one Type 184 hull-mounted medium-range panoramic search and attack sonar and one Type 170 hull-mounted short-range search and attack sonar used in conjunction with the GWS 40 Ikara fire-control system, one CAAIS action information system, and two Corvus chaff launchers
Propulsion: two boilers supplying steam to two sets of White/English Electric geared turbines delivering 30,000 shp (22370 kW) to two shafts
Performance: speed 28 kts; range 4,600 miles (7400 km) at 15 kts
Complement: 19+238

Class
1. UK

Name	No.	Builder	Laid down	Conversion completed
Aurora	F10	John Brown	Jun 1961	Mar 1976
Euryalus	F15	Scotts Engineering	Nov 1961	Mar 1976
Galatea	F18	Swan Hunter	Dec 1961	Sep 1974
Arethusa	F38	J. Samuel White	Sep 1962	Apr 1977
Dido	F104	Yarrow	Dec 1959	Oct 1978
Leander	F109	Harland & Wolff	Apr 1959	Dec 1972

Note
Built as general-purpose frigates, these units were converted into anti-submarine escorts with an Ikara mounting in place of the 4.5-in (114-mm) gun mounting, anti-aircraft strength being boosted by a second SAM launcher

'Leopard' or 'Type 41' class FF
(UK/India)
Type: frigate
Displacement: 2,250 tons standard and 2,515 tons full load
Dimensions: length 339.8 ft (103.6 m); beam 40.0 ft (12.2 m); draught 16.0 ft (4.9 m)
Gun armament: two 4.5-in (114-mm) L/45 DP in a Mk 6 twin mounting, and two 40-mm Bofors L/60 AA in single mountings

Missile armament: none
Torpedo armament: none
Anti-submarine armament: one Squid three-barrel mortar
Aircraft: none
Electronics: one Type 960 air-search radar, one Type 293 surface-search radar, one Type 275 gun-control radar, one Type 978 navigation radar, one hull-mounted sonar, and one ECM system
Propulsion: eight ASR diesels delivering 14,400 shp (10740 kW) to two shafts
Performance: speed 24 kts; range 8,625 miles (13880 km) at 16 kts
Complement: 210

Class
1. Bangladesh

Name	No.	Builder	Laid down	Commissioned
Abu Bakr	F15	John Brown	Aug 1953	Mar 1957
Ali Haider	F17	William Denny	Nov 1953	Dec 1959

(This Bangladeshi ship is very similar in all essential respects to the Indian units, but has one rather than two 40-mm AA guns, and a complement of 15+220.)

2. India

Name	No.	Builder	Laid down	Commissioned
Brahmaputra	F31	John Brown	1956	Mar 1958
Beas	F37	Vickers	1957	May 1960
Betwa	F39	Vickers	1957	Dec 1960

Note
These ships are obsolete for anything but patrol and training

'Mackenzie' class FF
(Canada)
Type: frigate
Displacement: 2,380 tons standard and 2,880 tons full load
Dimensions: length 366.0 ft (111.6 m); beam 42.0 ft (12.8 m); draught 13.5 ft (4.1 m)
Gun armament: four 3-in (76-mm) L/50 DP in one Mk 6 and one Mk 33 twin mountings
Missile armament: none
Torpedo armament: none
Anti-submarine armament: two Limbo Mk 10 mortars, and two single launchers for Mk 43 torpedoes
Aircraft: none
Electronics: one SPS-12 air-search radar, one SPS-10 surface-search radar, one SPG-48 radar used in conjunction with the Mk 69 gun fire-control system, one SQS-501 bottom-classification sonar, one SQS-502 Limbo-control sonar, one SQS-505 hull-mounted search sonar, one SQS-10 hull-mounted sonar, one SLQ-25 Nixie torpedo decoy system, and ECM systems
Propulsion: two boilers supplying steam to two sets of English Electric geared turbines delivering 30,000 shp (22370 kW) to two shafts
Performance: speed 28 kts; range 3,150 miles (5070 km) at 14 kts
Complement: 11+199

Class
1. Canada

Name	No.	Builder	Laid down	Commissioned
Mackenzie	261	Vickers	Dec 1958	Oct 1962
Saskatchewan	262	Victoria Machinery	Jul 1959	Feb 1963
Yukon	263	Burrard Dry Dock	Oct 1959	May 1963
Qu'Appelle	264	Davie SB	Jan 1960	Sep 1963

Note
These are totally obsolete escorts

'Makut Rajakumarn' class FF
(UK/Thailand)
Type: frigate
Displacement: 1,650 tons standard and 1,900 tons full load
Dimensions: length 320.0 ft (97.6 m); beam 36.0 ft (11.0 m); draught 18.1 ft (5.5 m)
Gun armament: one 4.5-in (114-mm) L/55 DP in a Mk 8 single mounting, and two 40-mm Bofors L/60 AA in single mountings
Missile armament: one quadruple launcher for Sea Cat SAMs
Torpedo armament: none
Anti-submarine armament: one Limbo Mk 10 mortar, two depth-charge throwers, and one depth-charge rack

Aircraft: none
Electronics: one LW-04 air-search radar, one DA-05 combined air- and surface-search radar, one Type 626 navigation radar, one WM-22 gun fire-control system, one WM-44 SAM fire-control system, one Type 170B hull-mounted short-range search and attack sonar, one Type 162 hull-mounted side-looking classification sonar, one MS 27 hull-mounted lightweight search and attack sonar, and one Hollandse Signaalapparaten action information system
Propulsion: CODOG (COmbined Diesel Or Gas turbine) arrangement, with one Crossley/Pielstick 12PC2V diesel delivering 4475 kW (6,000 shp) and one Rolls-Royce Olympus TM3B gas turbine delivering 23,125 shp (17245 kW) to two shafts
Performance: speed 26 kts on gas turbine or 18 kts on diesel; range 5,750 miles (9250 km) at 18 kts on diesel or 1,380 miles (2220 km) at 26 kts on gas turbine
Complement: 16+124

Class
1. Thailand

Name	No.	Builder	Laid down	Commissioned
Makut Rajakumarn	7	Yarrow	Jan 1970	May 1973

Note
This ship provides good patrol and training capabilities for a developing navy, and its advanced features include modern sensors/electronics and a comparatively high degree of automation

'Mirka I' and 'Mirka II' class FFs
(USSR)
Type: frigate
Displacement: 950 tons standard and 1,150 tons full load
Dimensions: length 82.4 m (270.3 ft); beam 9.1 m (29.9 ft); draught 3.0 m (9.8 ft)
Gun armament: four 76-mm (3-in) L/60 DP in two twin mountings
Missile armament: none
Torpedo armament: none
Anti-submarine armament: four ('Mirka I' class) or two ('Mirka II' class) RBU 6000 12-barrel rocket-launchers, and one ('Mirka I' class) or two ('Mirka II' class) quintuple mountings for five or 10 406-mm (16.in) torpedoes
Aircraft: none
Electronics: one 'Slim Net' or (in some 'Mirka II' class ships) 'Strut Curve' air-search radar, one 'Hawk Screech' gun-control radar, one 'Don-2' navigation radar, one hull-mounted medium-frequency sonar, one high-frequency variable-depth sonar (in most ships), two 'Square Head' IFF, two 'High Pole-B' IFF, and two 'Watch Dog' ECM housing
Propulsion: CODAG (COmbined Diesel And Gas turbine) arrangement, with two diesels delivering 9000 kW (12,070 shp) and two gas turbines delivering 22400 kW (30,045 shp) to two shafts
Performance: speed 36 kts; range 4650 km (2,890 miles) at 20 kts
Complement: 98

Class
1. USSR

9 'Mirka I' class ships
9 'Mirka II' class ships

Note
Built at Kaliningrad between 1964 and 1965 ('Mirka I') and 1965 and 1966 ('Mirka II'), these two classes were derived from the 'Petya' class with greater optimization for the coastal anti-submarine role; the ships have since been relegated to the patrol and training roles

'Najin' class FF
(North Korea)
Type: frigate
Displacement: 1,500 tons
Dimensions: length 100.0 m (328.8 ft); beam 10.0 m (32.8 ft); draught 2.7 m (8.9 ft)
Gun armament: two 100-mm (3.9-in) L/56 DP in single mountings, four 57-mm AA in two twin mountings, four 25-mm AA in two twin mountings, and eight 14.5-mm (0.57-in) AA in four twin mountings
Missile armament: none
Torpedo armament: one 533-mm (21-in) triple mounting
Anti-submarine armament: two RBU 1200 five-barrel rocket-launchers, two depth-charge throwers, and two depth-charge racks
Mines: about 30
Aircraft: none
Electronics: one 'Skin Head' surface-search radar, one 'Pot Head' surface-search radar, one 'Ski Pole' IFF, one hull-mounted sonar, and one variable-depth sonar
Propulsion: two diesels delivering 11200 kW (15,020 shp) to two shafts
Performance: speed 26 kts; range 7400 km (4,600 miles) at 14 kts
Complement: 180

Class
1. North Korea

Name	No.	Builder	Laid down	Commissioned
	3025		1971	1973
	3026			1975
	3027			1976
			1976	1979

Note
Virtually nothing is known of these North Korean ships, which may be receiving a twin SS-N-2A 'Styx' anti-ship missile launcher

'Obuma' class FF
(Netherlands/Nigeria)
Type: frigate
Displacement: 1,725 tons standard and 2,000 tons full load
Dimensions: length 109.8 m (360.2 ft); beam 11.3 m (37.0 ft); draught 3.5 m (11.5 ft)
Gun armament: two 4-in (102-mm) L/45 DP in a Mk 19 twin mounting, and four 40-mm Bofors AA in single mountings
Missile armament: none
Torpedo armament: none
Anti-submarine armament: one Squid three-barrel mortar
Aircraft: none
Electronics: one AWS 4 surface-search radar, and one Type 293 tactical radar
Propulsion: four MAN diesels delivering 11920 kW (15,985 shp) to two shafts
Performance: speed 26 kts; range 6,500 km (4,040 miles) at 15 kts
Complement: 216

Class
1. Nigeria

Name	No.	Builder	Laid down	Commissioned
Obuma	F87	Wilton-Fijenoord	Apr 1964	Sep 1965

Note
The ship provides a useful patrol capability at the same time as offering an emergent navy training in modern ship design features and operation

'Petya I', 'Petya I (Modified)', 'Petya II' and 'Petya II (Modified)' class FFs
(USSR)
Type: frigate
Displacement: 950 tons standard and 1,150 tons or ('Petya II' classes) 1,160 tons full load
Displacement: length 81.8 m (268.4 ft) or ('Petya II' classes) 82.5 m (270.7 ft); beam 9.1 m (29.9 ft); draught 2.9 m (9.5 ft)
Gun armament: four 76-mm (3-in) L/60 DP in two twin mountings
Missile armament: none
Torpedo armament: none
Anti-submarine armament: four ('Petya I' class) or two ('Petya I [Modified]' class) RBU 2500 16-barrel rocket-launchers, or ('Petya II' classes) two RBU 6000 12-barrel rocket-launchers, one ('Petya I' classes) or two ('Petya II' classes) 406-mm (16-in) quintuple mountings for five or 10 torpedoes, and two or ('Petya I [Modified]' class) one racks for 24 (or 12) depth charges
Mines: 20/30 depending on type (not in 'Petya I [Modified]' class)
Aircraft: none
Electronics: one 'Slim Net' ('Petya I' classes) or 'Strut Curve' ('Petya II' classes) air-search radar, one 'Hawk Screech' gun-control radar, one 'Don-2' or ('Petya I' class) 'Neptun' navigation radar, two 'Square Head' and one 'High Pole-B' IFF ('Petya I' classes) or one 'High Pole-B' IFF ('Petya II' classes), one hull-mounted high-frequency sonar, one high-frequency dipping sonar, one medium-frequency variable-depth sonar ('Petya I [Modified]' and 'Petya II [Modified]' classes only), and two 'Watch Dog' ECM housings
Propulsion: CODAG (COmbined Diesel And Gas turbine) arrangement, with one diesel delivering 4,475 kW (6,000 shp) and two gas turbines delivering 22,375 kW (30,010 shp) to two shafts
Performance: speed 35 kts; range 7400 km (4,600 miles) at 20 kts or 925 km (575 miles) at 35 kts
Complement: 98

Class
1. Ethiopia

2 'Petya II' class ships

2. India

Name	No.	Builder
Arnala	P68	Kaliningrad

Androth	P69	Kaliningrad
Anjadip	P73	Kaliningrad
Andaman	P74	Kaliningrad
Amini	P75	Kaliningrad
Kamorta	P77	Kaliningrad
Kadmath	P78	Kaliningrad
Kiltan	P79	Kaliningrad
Kavaratti	P80	Kaliningrad
Katchal	P81	Kaliningrad
Kanjar	P82	Kaliningrad
Amindivi	P83	Kaliningrad

(These Indian ships are all of the 'Petya II' class.)

3. Syria

2 'Petya I' class ships

4. USSR

7 'Petya I' class ships
11 'Petya I (Modified)' class ships
23 'Petya II' class ships
1 'Petya II (Modified)' class ship

5. Vietnam

4 'Petya I' class ships
2 'Petya II' class ships

Note
The 'Petya I' class (18 ships) was built at Kaliningrad and Komsomolsk between 1961 and 1964, and was succeeded at both yards by the 'Petya II' class (27 ships) between 1964 and 1969, the later class having more torpedo tubes and better anti-submarine rocket-launchers; both types were used for coastal anti-submarine operations, and many units were converted as trials ships as later designs replaced the 'Petyas' in first-line service; most of the ships are now used for patrol and training

'PF103' class FF
(USA/Iran)
Type: frigate
Displacement: 900 tons standard and 1,135 tons full load
Dimensions: length 275.0 ft (83.8 m); beam 33.0 ft (10.1 m); draught 10.0 ft (3.0 m)
Gun armament: two 3-in (76-mm) L/50 DP in Mk 34 single mountings, two 40-mm Bofors AA in single mountings, and two 23-mm AA in a twin mounting
Missile armament: none
Torpedo armament: none
Anti-submarine armament: two Mk 32 triple mountings for 12.75-in (324-mm) Mk 44 torpedoes, and two Mk 9 racks for 24 depth charges
Aircraft: none
Electronics: one SPS-6C air-search radar, one SPS-53E navigation radar, one SPG-34 radar used in conjunction with the Mk 63 gun fire-control system, and one Mk 51 40-mm gun fire-control system
Propulsion: two Fairbanks-Morse diesels delivering 6,000 shp (4475 kW) to two shafts
Performance: speed 20 kts
Complement: 150

Class
1. Iran

Name	No.	Builder	Laid down	Commissioned
Bayandor	81	Levingstone SB	Aug 1962	May 1964
Naghdi	82	Levingstone SB	Sep 1962	Jul 1964
Milanian	83	Levingstone SB	May 1967	Feb 1969
Kahnamuie	84	Levingstone SB	Jun 1967	Feb 1969

2. Thailand

Name	No.	Builder	Laid down	Commissioned
Tapi	5	American SB	Apr 1970	Nov 1971
Khirirat	6	Norfolk SB	Feb 1972	Aug 1974

(The Thai ships differ from the Iranian units in having an armament of one 76-mm/3-in Oto Melara Compact L/62 DP and one 40-mm Bofors L/70 AA guns with associated WM-25 fire-control system, and additional anti-submarine armament in the form of a Hedgehog mortar and SQS-17A hull-mounted sonar.)

Note
These are simple escort and patrol frigates of limited operational value

'President' class FF
(UK/South Africa)
Type: frigate
Displacement: 2,380 tons standard and 2,800 tons full load
Dimensions: length 370.0 ft (112.8 m); beam 41.1 ft (12.5 m); draught 17.3 ft (5.3 m)
Gun armament: two 4.5-in (114-mm) L/50 DP in a Mk 6 twin mounting, and two 40-mm Bofors L/70 AA in a twin mounting
Missile armament: none
Torpedo armament: none
Anti-submarine armament: two Mk 32 triple mountings for 12.75-in (324-mm) Mk 44 torpedoes, one Limbo Mk 10 mortar, and helicopter-launched weapons (see below)
Aircraft: one Westland Wasp helicopter on a platform aft
Electronics: one Jupiter air-search radar, one Type 293 combined air- and surface-search radar, one Argo

NA9C fire-control system, one Type 177 hull-mounted medium-range panoramic search and attack sonar, and one Type 174 hull-mounted sonar
Propulsion: two boilers supplying steam to two sets of geared turbines delivering 30,000 shp (22370 kW) to two shafts
Performance: speed 29 kts; range 5,200 miles (8370 km) at 12 kts
Complement: 13+190

Class
1. South Africa

Name	No.	Builder	Laid down	Commissioned
President Pretorius	F145	Yarrow	Nov 1960	Mar 1964
President Steyn	F147	Alexander Stephen	May 1960	Apr 1963

Note
These are 'Rothesay' class general-purpose frigates in origin, and are of limited operational capability by all but local standards

'Rahmat' class FF
(UK/Malaysia)
Type: frigate
Displacement: 1,250 tons standard and 1,600 tons full load
Dimensions: length 308.0 ft (93.9 m); beam 34.1 ft (10.4 m); draught 14.8 ft (4.5 m)
Gun armament: one 4.5-in (114-mm) L/45 DP in a Mk 8 single mounting, and two 40-mm Bofors AA in single mountings
Missile armament: one quadruple launcher for Sea Cat SAMs
Torpedo armament: none
Anti-submarine armament: one Limbo Mk 10 mortar
Aircraft: provision for a light helicopter on a hatch over the Limbo well
Electronics: one LW-02 air-search radar, one WM-22 gun-control system, and one WM-44 SAM-control system
Propulsion: CODOG (COmbined Diesel Or Gas turbine) arrangement, with one Crossley/Pielstick diesel delivering 2870 kW (3,850 shp) and one Rolls-Royce Olympus TM1B gas turbine delivering 19,500 shp (14545 kW) to two shafts
Performance: speed 26 kts on gas turbine or 16 kts on diesel; range 6,900 miles (11105 km) at 16 kts
Complement: 140

Class
1. Malaysia

Name	No.	Builder	Laid down	Commissioned
Rahmat	F24	Yarrow	Feb 1966	Mar 1971

Note
This is a very useful general-purpose frigate by local standards, and its high level of automation provides valuable saving in technically trained crew

'Restigouche (Improved)' class FF
(Canada)
Type: frigate
Displacement: 2,390 tons standard and 2,900 tons full load
Dimensions: length 371.0 ft (113.1 m); beam 42.0 ft (12.8 m); draught 14.1 ft (4.3 m)
Gun armament: two 3-in (76-mm) L/70 in a Mk 6 twin mounting
Missile armament: none
Torpedo armament: none
Anti-submarine armament: one Mk 16 octuple launcher for eight RUR-5A ASROC missiles, and one Limbo Mk 10 mortar
Aircraft: none
Electronics: one SPS-12 air-search radar, one SPS-10 surface-search radar, one Sperry Mk II navigation radar, one SPG-48 radar used in conjunction with the Mk 69 fire-control system, one SQS-501 hull-mounted bottom-classification sonar, one SQS-505 hull-mounted medium-range search and attack sonar, one SQS-505 variable-depth sonar, and one SLQ-25 Nixie torpedo decoy system
Propulsion: two boilers supplying steam to two sets of English Electric geared turbines delivering 30,000 shp (22370 kW) to two shafts
Performance: speed 28 kts; range 5,475 miles (8810 km) at 14 kts
Complement: 13+201

Class
1. Canada

Name	No.	Builder	Laid down	Commissioned
Gatineau	236	Davie SB	Apr 1953	Feb 1959

Restigouche	257	Vickers	Jul 1953	Jun 1958
Kootenay	258	Burrard Dry Dock	Aug 1952	Mar 1959
Terra Nova	259	Victoria Machinery	Nov 1952	Jun 1959

Note
Though elderly, these units have a limited anti-submarine value by virtue of their ASROC launchers

'Riga' class FF
(USSR)
Type: frigate
Displacement: 1,260 tons standard and 1,510 tons full load
Dimensions: length 91.5 m (300.1 ft); beam 10.1 m (33.1 ft); draught 3.2 m (10.5 ft)
Gun armament: three 100-mm (3.9-in) L/50 DP in single mountings, four 37-mm AA in two twin mountings, and (some ships) four 25-mm AA in two twin mountings
Missile armament: none
Torpedo armament: one twin or triple mounting for two or three 533-mm (21-in) torpedoes
Anti-submarine armament: two RBU 2500 16-barrel rocket-launchers, and racks for 24 depth charges
Mines: 50
Aircraft: none
Electronics: one 'Slim Net' air-search radar, one 'Sun Visor-B' 100-mm gun-control radar, one 'Wasp Head' AA gun-control radar, one 'Don-2' or 'Neptun' navigation radar, one hull-mounted high-frequency sonar, one dipping sonar (some ships only), two 'Square Head' IFF, one 'High Pole' IFF, and two 'Watch Dog' ECM housings

Propulsion: two boilers supplying steam to two sets of geared turbines delivering 14900 kW (19,985 shp) to two shafts
Performance: speed 28 kts; range 370 km (2,300 miles) at 15 kts or 1300 km (810 miles) at 27 kts
Complement: 175

Class
1. Bulgaria

Name	No.
Druzki	15
Smeli	16

2. Indonesia

Name	No.
Jos Sudarso	351
Lambung Mangkurat	357

3. USSR

34 ships

Note
Some 64 'Riga' class limited anti-ship/anti-submarine frigates were built between 1955 and 1958 at Kaliningrad, Komsomolsk and Nikolayev for coastal duties round the USSR, and several examples were exported; the class is now elderly, but still perform usefully in the simple patrol role

'River' class FF
(Australia)
Type: frigate
Displacement: 2,100 tons standard and 2,700 tons full load
Dimensions: length 370.0 ft (112.8 m); beam 41.0 ft (12.5 m); draught 17.3 ft (5.3 m)
Gun armament: two 4.5-in (114-mm) L/45 DP in a Mk 6 twin mounting
Missile armament: one quadruple launcher for Sea Cat SAMs
Torpedo armament: none
Anti-submarine armament: one launcher for Ikara missiles, and two Mk 32 triple mountings for 12.75-in (324-mm) Mk 46 torpedoes
Aircraft: provision for a light helicopter on a platform aft

Electronics: one LW-02 air-search radar, one WM-22 surface-search and fire-control radar, one SPS-55 (D45/49) or Type 978 navigation radar, one Type 162 hull-mounted side-looking classification sonar, one Type 177 hull-mounted medium-range panoramic search and attack sonar, one Type 170 hull-mounted short-range search and attack sonar, one Type 185 hull-mounted sonar, and (after modernization) one Mulloka hull-mounted sonar
Propulsion: two boilers supplying steam to two sets of geared turbines delivering 30,000 shp (22370 kW) to two shafts
Performance: speed 30 kts; range 3,900 miles (6275 km) at 12 kts
Complement: 13+234

Class
1. Australia

Name	No.	Builder	Laid down	Commissioned
Parramatta	D46	Cockatoo Island DY	Jan 1957	Jul 1961
Stuart	D48	Cockatoo Island DY	Mar 1959	Jun 1963
Derwent	D49	HMA Naval Dockyard	Jun 1958	Apr 1964

| Swan | D50 | HMA Naval Dockyard | Aug 1965 | Jan 1970 |
| Torrens | D53 | Cockatoo Island DY | Aug 1965 | Jan 1971 |

Note
Though built in Australia, these are in essence 'Leander Batch 1' class general-purpose frigates optimized with local weapons and sensors for the anti-submarine role

'Rothesay' or 'Type 12 (Modified)' class FF
(UK)
Type: frigate
Displacement: 2,380 tons standard and 2,800 tons full load
Dimensions: length 370.0 ft (112.8 m); beam 41.0 ft (12.5 m); draught 17.3 ft (5.3 m)
Gun armament: two 4.5-in (114-mm) L/45 DP in a Mk 6 twin mounting, and up to four 20-mm AA in single mountings
Missile armament: one quadruple launcher for Sea Cat SAMs
Torpedo armament: none
Anti-submarine armament: one Limbo Mk 10 mortar, and helicopter-launched weapons (see below)

Aircraft: one Westland Wasp HAS.Mk 1 helicopter in a hangar aft
Electronics: one Type 994 surface-search radar, one Type 978 navigation radar, one MRS 3 gun fire-control system, one GWS 20 SAM fire-control system, one Type 174 hull-mounted sonar, one Type 170 hull mounted short-range search and attack sonar, and one Type 162 hull-mounted side-looking classification sonar
Propulsion: two boilers supplying steam to two sets of ASR geared turbines delivering 30,000 shp (22370 kW) to two shafts
Performance: speed 30 kts; range 5,200 miles (8370 km) at 12 kts
Complement: 15+220

Class
1. New Zealand

| Name | No. | Builder | Laid down | Commissioned |
| Otago | F111 | John I. Thornycroft | 1957 | Jun 1960 |

(This New Zealand vessel is similar to the British units, but has an anti-submarine armament of two Mk 32 triple mountings for 12.75-in [324-mm] Mk 46 torpedoes, and an electronic fit that includes Type 993 and Type 277 search radars, Type 275 fire-control radar, and Type 1006 navigation radar; the complement is 13+229.)

2. UK

Name	No.	Builder	Laid down	Commissioned
Yarmouth	F101	John Brown	Nov 1957	Mar 1960
Rothesay	F107	Yarrow	Nov 1956	Apr 1960
Plymouth	F126	Devonport Dockyard	Jul 1958	May 1961

Note
These ships are obsolete for all but patrol and limited anti-submarine service

'Salisbury' or 'Type 61' class FF
(UK/Bangladesh)
Type: frigate
Displacement: 2,170 tons standard and 2,410 tons full load
Dimensions: length 339.8 ft (103.6 m); beam 40.0 ft (12.2 m); draught 15.5 ft (4.7 m)
Gun armament: two 4.5-in (114-mm) L/45 DP in a Mk 6 twin mounting, and two 40-mm Bofors L/70 AA in Mk 9 single mountings
Missile armament: none
Torpedo armament: none
Anti-submarine armament: one Squid mortar
Aircraft: none

Electronics: one Type 965 long-range air-search radar, one Type 993 combined air- and surface-search radar, one Type 278M height-finding radar, one Type 986 air-warning radar, one Type 978 navigation radar, one Type 275 radar used in conjunction with the Mk 6M gun fire-control system, one Type 174 hull-mounted sonar, and one Type 164B hull-mounted sonar
Propulsion: eight ASR1 diesels delivering 14,400 shp (10740 kW) to two shafts
Performance: speed 24 kts; range 8,650 miles (13920 km) at 16 kts
Complement: 14+223

Class
1. Bangladesh

| Name | No. | Builder | Laid down | Commissioned |
| Umar Farooq | F16 | Hawthorn Leslie | Aug 1953 | Apr 1958 |

Note
Obsolete by Western standards, this type still has patrol and general-purpose capability in the Bay of Bengal

'Tribal' or 'Type 81' class FF
(UK/Indonesia)
Type: frigate
Displacement: 2,300 tons standard and 2,700 tons full load
Dimensions: length 360.0 ft (109.7 m); beam 42.5 ft (13.0 m); draught 18.0 ft (5.5 m)
Gun armament: two 4.5-in (114-mm) L/55 DP in Mk 5 single mountings, and two 20-mm AA in single mountings
Missile armament: two quadruple launchers for 32 Sea Cat SAMs
Torpedo armament: none
Anti-submarine armament: one Limbo mortar, and helicopter-launched weapons (see below)
Aircraft: one Westland Wasp HAS.Mk 1 helicopter on a platform aft
Electronics: one Type 965 air-search radar, one Type 993 air-search radar, one Type 978 navigation radar, one GWS 22 SAM fire-control system, one Type 177 medium-range panoramic search sonar, one Type 170B short-range search and attack sonar, one Type 162 side-looking classification sonar, and two Corvus chaff launchers
Propulsion: COSAG (COmbined Steam And Gas turbine) arrangement, with one boiler supplying steam to one Metrovicj geared turbine delivering 12,500 shp (9,320 kW) and one AEI G.6 gas turbine delivering 7,500 shp (5595 kW) to one shaft
Performance: speed 25 on steam and gas turbines, or 17 kts on gas turbine; range 5,200 miles (8370 km) at 12 kts
Complement: 13+240

Class
1. Indonesia

Name	No.	Builder	Laid down	Commissioned
Martha Kristina Tiyahadu	331	Alex Stephen	Dec 1960	Apr 1964
Wilhelmus Zakarias Johannes	332	Thornycroft	Nov 1958	Feb 1963
Hasanuddin	333	Devonport DY	Oct 1959	Feb 1962

Note
Built as part of the interim seven-strong 'Type 81' class of gun frigates for the Royal Navy, these ships were sold to Indonesia in 1985 as training ships, though they retain the full combat capability of the RN standard, plus air-conditioning, fin-stabilization and gas turbine propulsion for cold starts and emergency speed

'Type 12 (Modified)' class FF (UK): see 'Rothesay' class FF

'Type 41' class FF (UK): see 'Leopard' class FF

'Type 41/61' class FF
(UK/Malaysia)
Type: frigate
Displacement: 2,300 tons standard and 2,520 tons full load
Dimensions: length 304.0 ft (92.7 m); beam 40.0 ft (12.2 m); draught 16.0 ft (4.9 m)
Gun armament: one 100-mm (3.9-in) L/55 DP in a single mounting, and two 40-mm Bofors AA in single mountings
Missile armament: none
Torpedo armament: none
Anti-submarine armament: one Limbo Mk 10 mortar
Aircraft: none
Electronics: one AWS 1 air-search radar, one Type 170 hull-mounted short-range search and attack sonar, and one Type 176 hull-mounted sonar
Propulsion: eight ASR1 diesels delivering 14,400 shp (10740 kW) to two shafts
Performance: speed 24 kts; range 5,525 miles (8890 km) at 15 kts
Complement: 210

Class
1. Malaysia

Name	No.	Builder	Laid down	Commissioned
Hang Tuah	F76	Yarrow	1965	May 1973

Note
Though rated a frigate, this hybrid 'Leopard' and 'Salisbury' class ship is in reality a capable training ship with secondary patrol capability

'Type 61' class FF (UK): see 'Rothesay' class FF

'Type 81' class FF (UK): see 'Tribal' class FF

'Type 120' class FF (West Germany): see 'Köln' class FF

CORVETTES

'Assad' class FFL
(Italy/Libya)
Type: guided-missile corvette
Displacement: 670 tons full load
Dimensions: length 61.7 m (202.4 ft); beam 9.3 m (30.5 ft); draught 2.2 m (7.6 ft)
Gun armament: one 76-mm (3-in) OTO Melara Compact L/62 DP in a single mounting, and two 35-mm AA in a twin mounting
Missile armament: four container-launchers for four Otomat anti-ship missiles
Torpedo armament: none
Anti-submarine armament: two ILAS 3 triple mountings for 324-mm (12.75-in) A 244S torpedoes
Mines: up to 16
Aircraft: none
Electronics: one RAN 11L/X combined air- and surface-search radar, one TM1226 navigation radar, one RTN 10X radar used in conjunction with the Argo NA10 gun fire-control system, one Diodon hull-mounted search and attack sonar, and one IPN 10 action information system
Propulsion: four MTU 16V956 TB91 diesels delivering 13420 kW (18,000 shp) to four shafts
Performance: speed 34 kts; range 8150 km (5,065 miles) at 14 kts
Complement: 58

Class
1. Iraq

Name	No.	Builder	Laid down	Commissioned
*Hussa el Hussair**	F210	CNR, Muggiano	Jan 1982	1984
Abdullah Ibn Abi Serk	F211	CNR, Breda, Mestre	Mar 1982	1984
*Tariq Ibn Zyiad**	F212	CNR, Muggiano	May 1982	1984
Kalid Ibn al Walid	F213	CNR, Breda, Mestre	Jun 1982	1984
Saab Ibn Abi Wakkas	F214	CNR, Breda, Marghera	Sep 1982	1985
Salah al Din al Ayubi	F215	CNR, Breda, Marghera	Sep 1982	1985

* fitted with helicopter

(The Iraqi vessels are believed to be similar in all essential respects to the Libyan units, but have a missile armament of six Otomat anti-ship missiles and one Albatros quadruple launcher for eight Aspide SAMs; two of the ships are fitted to carry a light helicopter in a telescopic hangar aft, and these two Muggiano-built ships do not have the twin 40-mm mounting; the basic complement is 51 without aircrew.)

2. Libya

Name	No.	Builder	Laid down	Commissioned
Assad el Tadjer	412	CNR, Muggiano	May 1976	Sep 1979
Assad el Touggour	413	CNR, Muggiano	May 1976	Feb 1980
Assad al Khali	414	CNR, Muggiano	Oct 1977	Sep 1980
Assad al Hudud	415	CNR, Muggiano	May 1978	Sep 1980

'Badr' class FFL
(USA/Saudi Arabia)
Type: guided-missile corvette
Displacement: 732 tons standard and 815 tons full load
Dimensions: length 245.0 ft (74.7 m); beam 31.5 ft (9.6 m); draught 14.6 ft (4.5 m)
Gun armament: one 76-mm (3-in) OTO Melara Compact L/62 DP in a single mounting, and one 20-mm Phalanx Mk 16 CIWS mounting
Missile armament: two Mk 141 quadruple container-launchers for eight RGM-84A Harpoon anti-ship missiles
Torpedo armament: none
Anti-submarine armament: two Mk 32 triple mountings for 12.75-in (324-mm) Mk 46 torpedoes
Aircraft: none
Electronics: one SPS-40B air-search radar, one SPS-55 surface-search and navigation radar, one Mk 92 fire-control system, one SQS-56 hull-mounted medium-range sonar, and one SLQ-32 ESM system
Propulsion: CODOG (COmbined Diesel Or Gas turbine) arrangement, with two MTU diesels delivering 3000 kW (4,025 shp) and one General Electric LM2500 gas turbine delivering 23,000 shp (17150 kW) to two shafts
Performance: speed 30 kts on gas turbine or 16 on diesels
Complement: 7+51

Class
1. Saudi Arabia

Name	No.	Builder	Laid down	Commissioned
Badr	612	Tacoma BB	Oct 1979	Nov 1980
Al Yarmook	614	Tacoma BB	Jan 1980	May 1981
Hitteen	616	Tacoma BB	May 1980	Oct 1981
Tabuk	618	Tacoma BB	Sep 1980	Jan 1982

Note
These are highly capable anti-ship vessels with good anti-submarine capability

'Esmeraldas' class FFL
(Italy/Ecuador)
Type: guided-missile corvette
Displacement: 620 tons standard and 685 tons full load
Dimensions: length 62.3 m (204.4 ft); beam 9.3 m (30.5 ft); draught 2.5 m (8.2 ft)
Gun armament: one 76-mm (3-in) OTO Melara Compact L/62 DP in a single mounting, and four 40-mm Bofors L/70 AA in two Dardo twin mountings
Missile armament: six single container-launchers for six MM.40 Exocet anti-ship missiles, and one Albatros quadruple launcher for four Aspide SAMs
Torpedo armament: none
Anti-submarine armament: two ILAS 3 triple mountings for 324-mm (12.75-in) A 244S torpedoes
Aircraft: provision for a light helicopter on a platform amidships
Electronics: one RAN 10S combined air- and surface-search radar, one RTN 10X radar used in conjunction with the Argo NA10 gun fire-control system, one RTN 20X radar used in conjunction with the Dardo fire-control system, one 3RM 20 navigation radar, one Diodon hull-mounted search and attack sonar, one IPN 20 action information system, one SCLAR chaff launcher, and one Gamma ESM system
Propulsion: four MTU MA20V956 TB92 diesels delivering 18200 kW (24,405 shp) to four shafts
Performance: speed 37 kts; range 7900 km (4,910 miles) at 14 kts
Complement: 51

Class
1. Ecuador

Name	No.	Builder	Laid down	Commissioned
Esmeraldas	CM11	CNR, Muggiano	Sep 1979	Aug 1982
Manabi	CM12	CNR, Ancona	Feb 1980	Jan 1983
Los Rios	CM13	CNR, Muggiano	Dec 1979	Nov 1982
El Oro	CM14	CNR, Ancona	Mar 1980	Feb 1983
Galapagos	CM15	CNR, Muggiano	Dec 1980	Nov 1983
Loja	CM16	CNR, Ancona	Mar 1981	Feb 1984

Note
These are extremely capable ships developed from the basic 'Assad' class desgin, and with the capabilities of a modern guided-missile frigate

'Grisha I' and 'Grisha III' class FFLs
(USSR)
Type: guided-missile corvette
Displacement: 950 tons standard and 1,200 tons full load
Dimensions: length 72.0 m (236.2 ft); beam 10.0 m (32.8 ft); draught 3.7 m (12.1 ft)
Gun armament: two 57-mm L/80 DP in one twin mounting, and ('Grisha III' class only) one 30-mm AK-630 CIWS mounting
Missile armament: one twin launcher for 18 SA-N-4 'Gecko' SAMs
Torpedo armament: two twin mountings for 533-mm (21-in) dual-role torpedoes
Anti-submarine armament: two RBU 6000 12-barrel rocket-launchers, torpedoes (see above), and rails for 12 depth charges
Mines: 20/30 depending on type
Aircraft: none
Electronics: one 'Strut Curve' air-search radar, one 'Pop Group' SAM-control radar, one 'Muff Cob' ('Grisha I' class) or 'Bass Tilt' ('Grisha III' class) gun-control radar, one 'Don-2' navigation radar, one hull-mounted medium-frequency sonar, one high-frequency dipping sonar, one 'High Pole-A' IFF, and two 'Watch Dog' ECM systems
Propulsion: CODAG (COmbined Diesel And Gas turbine) arrangement, with four diesels delivering 11920 kW (15,985 shp) and one gas turbine delivering 17900 kW (24,000 shp) to two shafts
Performance: speed 36 kts; range 3700 km (2,300 miles) at 20 kts or 925 km (575 miles) at 30 kts
Complement: 80

Class
1. USSR

16 'Grisha I' class ships
34 'Grisha III' class ships plus ? more building and planned

Note
Built between 1968 and 1974, the 'Grisha I' class succeeded the 'Mirka' and 'Petya' classes as the USSR's primary anti-submarine corvette type for coastal and escort deployment; the improved 'Grisha III' class followed in 1975, and the 'Grisha II' class is a maritime patrol model operated by the KGB with a second twin 57-mm mounting in place of the SA-N-4 'Gecko' self-protection SAM launcher

'HDP 1000' class FFL
(South Korea)
Type: corvette
Displacement: 1,400 tons full load
Dimensions: length 81.5 m (267.4 ft); beam 9.8 m (32.2 ft); draught 3.0 m (9.8 ft)
Gun armament: one 76-mm (3-in) OTO Melara Compact L/62 DP in a single mounting, four 30-mm AA in two Emerlec twin mountings, four 20-mm AA in one twin and two single mountings, and two 7.62-mm (0.3-in) machine-guns
Missile armament: fitted for but not with four RGM-84 Harpoon anti-ship missiles
Torpedo armament: none
Anti-submarine armament: none
Aircraft: none
Electronics: two surface-search and navigation radars, one WM-28 fire-control system, one action information system, and one ESM system
Propulsion: two diesels delivering 9,600 bhp (7160 kW) to two shafts
Performance: speed 22 kts; range 12975 km (8,065 miles) at 18 kts
Complement: 69

Class
1. South Korea

Name	No.	Builder	Laid down	Commissioned
Ma-San Ho		Korea SEC		1982
		Hyundai		1983
		Hyundai		1983
		Hyundai		1983
An-Yang Ho		Hyundai		1983

Note
This is a useful coastal patrol type with considerable capacity for upgrading in the event of hostilities

'Minerva' class FFL
(Italy)
Type: corvette
Displacement: 1,285 tons full load
Dimensions: length 87.0 m (285.4 ft); beam 10.3 m (33.8 ft); draught 5.5 m (18.0 ft)
Gun armament: one 76-mm (3-in) OTO Melara Compact L/62 DP in a single mounting, and four 40-mm Bofors L/70 AA in two Dardo twin mountings
Missile armament: one Albatros octuple launcher for ? Aspide SAMs, and provision for the retrofitting of Otomat anti-ship missiles
Torpedo armament: none
Anti-submarine armament: two ILAS 3 triple mountings for 324-mm (12.75-in) A 244S torpedoes
Aircraft: none
Electronics: one MM/SPS 774 search radar, one MM/SPN 728 navigation radar, one RTN 20X radar used in conjunction with the Albatros system, two RTN 10X radars used in conjunction with two Dardo systems, one hull-mounted sonar and one DE 1167 variable-depth sonar used in conjunction with an underwater weapons fire-control system, one Mini-SADOC action information system, two SCLAR chaff launchers, two Barricade chaff launchers, and one ESM system
Propulsion: two General Motors 230.20 DVM diesels delivering 11,600 hp (8650 kW) to two shafts
Performance: speed 25 kts; range 4050 km (2,515 miles) at 18 kts
Complement: 9+112

Class
1. Italy

Name	No.	Builder	Laid down	Commissioned
Minerva		CNR	1983	
Urania		CNR	1983	
Danaide		CNR	1984	
Sfinge		CNR	1984	
Chimera		CNR		
Driade		CNR		

| Fenice | CNR |
| Sibilla | CNR |

Note
This is an effective design intended to replace the survivors of Italy's obsolete frigate classes in short-range operations

'Nanuchka I', 'Nanuchka II' and 'Nanuchka III' class FFLs
(USSR)
Type: guided-missile corvette
Displacement: 780 tons standard and 900 tons full load
Dimensions: length 59.3 m (194.6 ft); beam 12.6 m (41.3 ft); draught 2.4 m (7.9 ft)
Gun armament: two 57-mm L/70 AA in a twin mounting, or ('Nanuchka III' class only) one 76-mm (3-in) L/59 DP in a single mounting and one 30-mm AK-630 CIWS mounting
Missile armament: two triple container-launchers for six SS-N-9 anti-ship missiles or ('Nanuchka II' class only) two twin container-launchers for four SS-N-2B 'Styx' anti-ship missiles, and one twin launcher for 18 SA-N-4 'Gecko' SAMs
Torpedo armament: none
Anti-submarine armament: none
Aircraft: none
Electronics: one 'Band Stand' combined air- and surface-search radar, one 'Peel Pair' and one 'Spar Stump' surface-search and navigation radars, two 'Fish Bowl' SSM-control radars (not in 'Nanuchka II' class units), one 'Pop Group' SAM-control radar, one 'Muff Cob' main armament gun-control radar or ('Nanuchka II' class only) one 'Bass Tilt' CIWS-control radar, one 'Square Head' IFF, 'High Pole-B' IFF, two chaff launchers, and two passive ECM systems including one 'Bell Tap'
Propulsion: six diesels delivering 17900 kW (24,010 shp) to three shafts
Performance: speed 34 kts; range 8350 km (5,190 miles) at 15 kts or 2400 km (1,490 miles) at 33 kts
Complement: 70

Class
1. Algeria

Name	No.	Commissioned
Ras Hamidou	801	Jul 1980
Salah Reis	802	Feb 1981
Reis Ali	803	May 1982
	804	Jan 1983

(These Algerian ships are standard 'Nanuchka II' class units.)

2. India

Name	No.	Commissioned
Vijay Durg	K71	Mar 1977
Sindhu Durg	K72	
Hos Durg	K73	

(These Indian ships are standard 'Nanuchka II' class units, and six ships are planned.)

3. Libya

Name	No.	Commissioned
Ean Mara	416	Oct 1981
Ean el Gazala	417	Feb 1983
Ean Zara	418	Feb 1984
	419	1985

(These Libyan units are standard 'Nanuchka II' class units.)

4. USSR

17 'Nanuchka I' class ships with more building
8 'Nanuchka III' class ships with more building

Note
Built at the Petrovsky yard in Leningrad from 1969 to 1974 ('Nanuchka I and II') and at Petrovsky and on the Pacific from 1977 ('Nanuchka III'), this basic type offers excellent coastal anti-ship capability and performance on a small hull

'Parchim' class FFL
(East Germany)
Type: corvette
Displacement: 960 tons standard and 1,200 tons full load
Dimensions: length 72.5 m (237.9 ft); beam 9.4 m (30.8 ft); draught 3.5 m (11.5 ft)
Gun armament: two 57-mm L/80 AA in a twin mounting, and two 30-mm AA in a twin mounting
Missile armament: two quadruple launchers for 32 SA-N-5 'Grail' SAMs
Torpedo armament: none
Anti-submarine armament: two RBU 6000 12-barrel rocket-launchers, four single mountings for 406-mm (16-in) torpedoes, and 24 depth charges
Mines: 20/30 depending on type
Aircraft: none
Electronics: one 'Strut Curve' air-search radar, one TSR333 navigation radar, one 'Muff Cob' gun-control radar, one hull-mounted medium-frequency sonar, one high-frequency dipping sonar (some ships only), one 'High Pole-B' IFF, one 'Cross Loop-B' IFF, two chaff launchers, and two 'Watch Dog' ECM housings
Propulsion: two diesels delivering 9000 kW (12,070 shp) to two shafts
Performance: speed 25 kts
Complement: 60

Class
1. East Germany

Name	Builder	Commissioned
Wismar	Peenewerft	Apr 1981
Parchim	Peenewerft	Sep 1982
Bad Doberan	Peenewerft	1981
Buetzow	Peenewerft	1981

Perleberg	Peenewerft	1981
Luebz	Peenewerft	Feb 1982
Teterow	Peenewerft	1982
Purna	Peenewerft	1982
Anklam	Peenewerft	1983
Waren	Peenewerft	1983
Gustrow	Peenewerft	1983
Ribnitz-Darmgarten	Peenewerft	1983
Ludwichslust	Peenewerft	1983
	Peenewerft	1984
Gabebusch	Peenewerft	1985
Bergen	Peenewerft	1985

Note
This is a coastal anti-submarine design based on the 'Grisha' class; a total of 18 ships is anticipated.

'Pauk' class FFL
(USSR)
Type: corvette
Displacement: 580 tons full load
Dimensions: length 58.0 m (190.3 ft); beam 10.5 m (34.4 ft); draught 2.5 m (8.2 ft)
Gun armament: one 76-mm (3-in) L/59 DP in a single mounting, and one 30-mm AK-630 CIWS mounting
Missile armament: one quadruple launcher for eight SA-N-5 'Grail' SAMs
Torpedo armament: none
Anti-submarine armament: two RBU 1200 five-barrel rocket-launchers, four single mountings for four 406-mm (16-in) torpedoes, and two depth-charge racks
Aircraft: none
Electronics: one 'Plank Shave' combined air- and surface-search radar, one 'Spin Trough' navigation radar, one 'Bass Tilt' gun-control radar, and one medium-frequency dipping sonar, 'Square Head' IFF, and 'High Pole-B' IFF
Propulsion: two diesels delivering 8950 kW (12,005 shp) to two shafts
Performance: speed 34 kts
Complement: about 80

Class
1. USSR

20 ships with ? more building or planned

Note
Replacing the 'Poti' class, this coastal anti-submarine class is derived from the 'Tarantul' class

'Poti' class FFL
(USSR)
Type: corvette
Displacement: 500 tons standard and 580 tons full load
Dimensions: length 60.0 m (196.8 ft); beam 8.0 m (26.2 ft); draught 2.8 m (9.2 ft)
Gun armament: two 57-mm L/80 DP in a twin mounting

Missile armament: none
Torpedo armament: none
Anti-submarine armament: two RBU 6000 12-barrel rocket-launchers, and four mountings for 406-mm (16-in) torpedoes
Electronics: one 'Strut Curve' air-search radar, one 'Spin Trough' navigation radar, one 'Muff Cob' gun-control radar, one hull-mounted high-frequency sonar, and one 'High Pole-B' IFF
Propulsion: CODAG (COmbined Diesel And Gas turbine) arrangement, with two M503A diesels delivering 6000 kW (8,045 shp) and two gas turbines delivering 22400 kW (30,045 shp) to two shafts
Performance: speed 35 kts; range 1125 km (6,915 miles) at 10 kts or 925 km (575 miles) at 34 kts
Complement: 80

Class
1. Bulgaria
3 ships
2. Romania
3 ships
3. USSR
58 ships

Note
This is an obsolescent coastal anti-submarine and patrol type built between 1961 and 1968

'Ratanakosin' class FFL
(USA/Thailand)
Type: guided-missile corvette
Displacement: 890 tons full load
Dimensions: length 252.0 ft (76.8 m); beam 31.0 ft (9.4 m); draught 8.0 ft (2.4 m)
Gun armament: one 76-mm (3-in) OTO Melara Compact L/62 DP in a single mounting, two 40-mm Bofors L/70 AA in a Breda twin mounting, and two 20-mm AA in single mountings
Missile armament: two Mk 141 quadruple container-launchers for eight RGM-84 Harpoon anti-ship missiles, and one Albatros octuple launcher for ? Aspide SAMs
Torpedo armament: none
Anti-submarine armament: two Mk 32 triple mountings for 12.75-in (324-mm) Mk 46 torpedoes
Aircraft: none
Electronics: one LW-series air-search radar, one WM-25 fire-control system, one Krupp Atlas hull-mounted sonar, one action information system, one Dagaie chaff launcher, and one Elettronica ESM system
Propulsion: four MTU diesels delivering ? kW (? hp) to two shafts
Performance: speed 30 kts
Complement: 95

Class				
1. Thailand				
Name	*No.*	*Builder*	*Laid down*	*Commissioned*
Ratanakosin	1	Tacoma	Feb 1984	1986
Sukhu Thai	2	Tacoma	Mar 1984	1987
	3	Ital-Thai		

Note
This is a good coastal design with well-balanced multi-role capability in weapons and sensors

'Saar 5' class FFL
(Israel)
Type: guided-missile corvette
Displacement: 850 tons
Dimensions: length 77.2 m (253.2 ft); beam 8.8 m (28.9 ft); draught 4.2 m (13.8 ft)
Gun armament: two 76-mm (3-in) OTO Melara Compact L/62 DP in single mountings, and six 30-mm AA in three TCM-30 twin mountings
Missile armament: four container-launchers for four anti-ship missiles (probably a mix of RGM-84A Harpoon and Gabriel Mk II missiles)
Torpedo armament: none
Anti-submarine armament: one 375-mm (14.76-in) Bofors three-barrel rocket-launcher, two Mk 32 triple mountings for 12.75-in (324-mm) Mk 46 torpedoes, and helicopter-launched weapons (see below)
Aircraft: one light helicopter in a hangar amidships
Electronics: an extensive radar, sonar and ESM fit
Propulsion: CODAG (COmbined Diesel And Gas turbine) arrangement, with two MTU diesels delivering 3000 kW (4,025 shp) and one General Electric LM2500 gas turbine delivering 24,000 shp (17895 kW) to two shafts
Performance: speed 42 kts on gas turbine or 25 kts on diesels; range 8350 km (5,190 miles) at cruising speed on diesels
Complement: 45

Class				
1. Israel				
Name	*No.*	*Builder*	*Laid down*	*Commissioned*
		Israel Shipyards, Haifa		

Note
Developed in Israel, this Eastern Mediterranean type offers exceptional general-purpose and anti-ship capabilities on a high-performance hull, and a total of perhaps six units is planned

'Sleipner' class FFL
(Norway)
Type: corvette
Displacement: 600 tons standard and 780 tons full load
Dimensions: length 69.0 m (227.8 ft); beam 8.0 m (26.2 ft); draught 2.4 m (8.2 ft)
Gun armament: one 3-in (76-mm) L/50 DP in a Mk 34 single mounting, and one 40-mm Bofors AA in a single mounting
Missile armament: none
Torpedo armament: none
Anti-submarine armament: two Mk 32 triple mountings for 12.75-in (324-mm) Mk 46 torpedoes, and one Terne six-barrel rocket-launcher
Aircraft: none
Electronics: one TM1229 surface-search and navigation radar, one Type 202 navigation radar, one TVT-300 gun fire-control system, one SQS-36 hull-mounted medium-range sonar, and one Terne III Mk 3 hull-mounted sonar
Propulsion: four MTU diesels delivering 6700 kW (8,985 shp) to two shafts
Performance: speed 20+ kts
Complement: 62

Class				
1. Norway				
Name	*No.*	*Builder*	*Laid down*	*Commissioned*
Sleipner	F310	Nylands Verksted	1963	Apr 1965
Aeger	F311	Akers	1964	Mar 1967

Note
These are obsolescent coastal anti-submarine ships

'Tarantul I' and 'Tarantul II' class FFLs
(USSR)
Type: guided-missile corvette
Displacement: 480 tons standard and 580 tons full load
Dimensions: length 56.0 m (183.7 ft); beam 9.5 m (31.2 ft); draught 2.5 m (8.2 ft)
Gun armament: one 76-mm (3-in) DP in a single mounting, and two 30-mm AK-630 CIWS mountings
Missile armament: two twin container-launchers for four SS-N-2C 'Styx' anti-ship missiles ('Tarantul I' class) or two twin container-launchers for four SS-N-22 anti-ship missiles ('Tarantul II' class), and one quadruple launcher for 16 SA-N-5 'Grail' SAMs
Torpedo armament: none
Anti-submarine armament: none
Aircraft: none
Electronics: one 'Band Stand' combined air- and surface-search radar ('Tarantul II' class only), one 'Plank Shave' surveillance radar, one 'Light Bulb' target-data system, one 'Bass Tilt' gun-control radar, one 'Spin Trough' navigation radar, one 'Square Head' IFF, one 'High Pole-B' IFF, two chaff launchers, and four passive ECM systems
Propulsion: CODOG (COmbined Diesel Or Gas turbine) arrangement, with two diesels delivering 3000 kW (4,025 shp) and two gas turbines delivering 18000 kW (24,140 shp) to three shafts
Performance: speed 40 kts
Complement: 50

Class
1. East Germany

4 'Tarantul I' class ships with ? more building or planned

2. India

5 'Tarantul I' class ships building

3. Poland

3 'Tarantul I' class ships with 5 more building or planned

4. USSR

2 'Tarantul II' class ships with more building
16 'Tarantul II' class ships with ? more building

Note
Built at the Petrovsky yard in Leningrad since 1978, the 'Tarantul' class appears to have a role similar to that of the 'Nanuchka' class, though its capabilities are less

'Thetis' or 'Type 420' class FFL
(West Germany)
Type: corvette
Displacement:
Dimensions: length 69.7 m (228.7 ft); beam 8.5 m (27.0 ft); draught 4.2 m (14.0 ft)
Gun armament: two 40-mm Bofors L/70 AA in a twin mounting
Missile armament: none
Torpedo armament: none
Anti-submarine armament: one 375-mm (14.76-in) Bofors four-barrel rocket-launcher, and one quadruple mounting for 533-mm (21-in) torpedoes
Aircraft: none
Electronics: one Kelvin Hughes 14/9 radar, one TRS-N radar, and one ELAC 1BV hull-mounted sonal
Propulsion: two MAN diesels delivering 5100 kW (6,840 shp) to two shafts
Performance: speed 19.5 kts
Complement: 70

Class
1. West Germany

Name	No.	Builder	Laid down	Commissioned
Thetis	P6052	Rolandwerft		Jul 1961
Hermes	P6053	Rolandwerft		Dec 1961
Najade	P6054	Rolandwerft		May 1962
Triton	P6055	Rolandwerft		Nov 1962
Theseus	P6056	Rolandwerft		Aug 1963

Note
These vessels are obsolete

'Turunmaa' class FFL
(Finland)
Type: corvette
Displacement: 660 tons standard and 770 tons full load
Dimensions: length 74.1 m (243.1 ft); beam 7.8 m (25.6 ft); draught 2.4 m (7.9 ft)
Gun armament: one 120-mm (4.7-in) Bofors L/46 DP in a single mounting, two 40-mm Bofors AA in single mountings, and two 23-mm AA in a twin mounting
Missile armament: none
Torpedo armament: none
Anti-submarine armament: two RBU-series rocket-launchers, and two depth-charge racks
Aircraft: none
Electronics: one WM-22 navigation and fire-control radar, and one navigation radar
Propulsion: CODOG (COmbined Diesel Or Gas turbine) arrangement, with three MTU diesels delivering 2235 kW (3,000 shp) and one Rolls-Royce Olympus TM1A gas turbine delivering 15,000 shp (11185 kW) to three shafts

Performance: speed 35 kts on gas turbine or 17 kts on diesels
Complement: 70

Class				
1. Finland				
Name	*No.*	*Builder*	*Laid down*	*Commissioned*
Turunmaa	03	Wärtsilä	Mar 1967	Aug 1968
Karjala	04	Wärtsilä	Mar 1967	Oct 1968

Note
These flotilla leaders are no longer up to advanced West European standards

'Type 420' class FFL (West Germany): see 'Thetis' class FFL

'Vosper Mk 1' class FFL
(UK/Ghana)
Type: corvette
Displacement: 440 tons standard and 500 tons full load
Dimensions: length 177.0 ft (53.9 m); beam 28.5 ft (8.7 m); draught 13.0 ft (4.0 m)
Gun armament: one 4-in (102-mm) L/40 DP in a Mk 23 single mounting, and one 40-mm Bofors AA in a single mounting
Missile armament: none
Torpedo armament: none
Anti-submarine armament: one Squid three-barrel mortar
Aircraft: none
Electronics: one AWS 1 air-search radar, one Decca 45 navigation radar, and one Type 1674 hull-mounted sonar
Propulsion: two MTU diesels delivering 5300 kW (7,110 shp) to two shafts
Performance: speed 20 kts; range 3,350 miles (5390 km) at 14 kts
Complement: 6+48

Class				
1. Ghana				
Name	*No.*	*Builder*	*Laid down*	*Commissioned*
Kromantse	F17	Vosper		Jul 1964
Keta	F18	Vickers		May 1965
2. Libya				
Name	*No.*	*Builder*	*Laid down*	*Commissioned*
Tobruk	C01	Vosper		Apr 1966

(This Libyan ship is similar to the two Ghanaian unit, but has two rather than one 40-mm guns, two Paxman Ventura diesels for a speed of 18 kts and a complement of 5+58.)

Note
Though rated as corvettes, these are in reality training vessels

'Vosper Thornycroft Mk 3' class FFL
(UK/Nigeria)
Type: corvette
Displacement: 500 tons standard and 650 tons full load
Dimensions: length 202.0 ft (61.6 m); beam 31.0 ft (9.5 ft); draught 11.3 ft (3.5 m)
Gun armament: two 4-in (102-mm) L/45 DP in a Mk 19 twin mounting, two 40-mm Bofors AA in single mountings, and two 20-mm AA in single mountings
Missile armament: none
Torpedo armament: none
Anti-submarine armament: none
Aircraft: none
Electronics: one AWS 1 air-search radar, one TM626 navigation radar, one WM-22 fire-control system, and one MS 22 hull-mounted sonar
Propulsion: two MAN V24/30-B diesels delivering 5960 kW (7,995 shp) to two shafts
Performance: speed 22 kts; range 3,450 miles (5550 km) at 14 kts
Complement: 8+59

Class				
1. Nigeria				
Name	*No.*	*Builder*	*Laid down*	*Commissioned*
Dorina	F81	Vosper Thornycroft	Jan 1970	Jun 1972
Otobo	F82	Vosper Thornycroft	Sep 1970	Nov 1972

Note
These are of greater patrol and training value than of real combat capability

'Vosper Thornycroft Mk 9' class FFL
(UK/Nigeria)
Type: corvette
Displacement: 850 tons full load
Dimensions: length 226.0 ft (69.0 m); beam 31.5 ft (9.6 m); draught 9.8 ft (3.0 m)
Gun armament: one 76-mm (3-in) OTO Melara Compact L/62 DP in a single mounting, one 40-mm Bofors L/70 AA in a single mounting, and two 20-mm AA in single mountings
Missile armament: one triple launcher for 12 Sea Cat SAMS
Torpedo armament: none
Anti-submarine armament: one 375-mm (14.76-in) Bofors two-barrel rocket-launcher
Aircraft: none
Electronics: one AWS 2 surface-search radar, one TM1226 navigation radar, one WM-24 fire-control system, and one PMS 26 hull-mounted sonar
Propulsion: two MTU 20V956 TB92 diesels delivering 13120 kW (17,595 shp) to two shafts
Performance: speed 27 kts; range 2,550 miles (4105 km) at 14 kts
Complement: 90

Class
1. Nigeria

Name	No.	Builder	Laid down	Commissioned
Erin'mi	F83	Vosper Thornycroft	Oct 1975	Jan 1980
Enyimiri	F84	Vosper Thornycroft	Feb 1977	Jul 1980

Note
By local standards these are modestly capable coastal and offshore vessels

FAST ATTACK CRAFT

'Al Mansur' class FAC(M) (Oman): see 'Brooke Marine 37.5-m' class FAC(M)

'Al Siddiq' class FAC(M)
(USA/Saudi Arabia)
Type: fast attack craft (missile)
Displacement: 384 tons full load
Dimensions: length 190.5 ft (58.1 m); beam 26.5 ft (8.1 m); draught 11.0 ft (3.4 m)
Gun armament: one 76-mm (3-in) OTO Melara Compact L/62 DP in a single mounting, one 20-mm Phalanx Mk 16 CIWS mounting, two 20-mm AA in single mountings, and one 81-mm (3.2-in) mortar
Missile armament: two twin container-launchers for four RGM-84A Harpoon anti-ship missiles
Torpedo armament: none
Anti-submarine armament: none
Electronics: one SPS-40B air-search radar, one SPS-55 surface-search radar, and one Mk 92 fire-control system
Propulsion: CODOG (COmbined Diesel Or Gas turbine) arrangement, with two MTU diesels delivering 2980 kW (3,995 shp) and one General Electric LM2500 gas turbine delivering 23,000 bhp (17150 kW) to two shafts
Performance: speed 38 kts on gas turbine or 15.5 kts on diesels
Complement: 5+33

Class
1. Saudi Arabia

Name	No.	Builder	Commissioned
Al Siddiq	511	Peterson Builders	Dec 1980
Al Farouq	513	Peterson Builders	Jun 1981
Abdul Aziz	515	Peterson Builders	Aug 1981
Faisal	517	Peterson Builders	Nov 1981
Kahlid	519	Peterson Builders	Jan 1982
Amyr	521	Peterson Builders	Jun 1982
Tariq	523	Peterson Builders	Aug 1982
Oqbah	525	Peterson Builders	Oct 1982
Abu Obaidah	527	Peterson Builders	Dec 1982

'Al Wafi' class FAC(G) (Oman): see 'Brooke Marine 37.5-m' class FAC(M)

'Alia' class FAC(M) (Israel): see 'Saar 4.5' class FAC(M)

'Asheville' class FAC(G) (USA): see 'PSMM Mk 5' class FAC(M)

'Brooke Marine 32.6-m' class FAC(M)
(UK/Kenya)
Type: fast attack craft (missile)
Displacement: 120 tons standard and 145 tons full load
Dimensions: length 107.0 ft (32.6 m); beam 20.0 ft (6.1 m); draught 5.6 ft (1.7 m)

Gun armament: two 40-mm Bofors L/60 AA in single mountings
Missile armament: four container-launchers for four Gabriel anti-ship missiles
Torpedo armament: none
Anti-submarine armament: none
Electronics: surface-search and navigation radar
Propulsion: two Paxman Valenta diesels delivering 5,400 bhp (4025 kW) to two shafts
Performance: speed 25.5 kts; range 2,875 miles (4625 km) at 12 kts
Complement: 3+18

Class
1. Kenya

Name	No.	Builder	Commissioned
Madaraka	P3121	Brooke Marine	Jun 1975
Jamhuri	P3122	Brooke Marine	Jun 1975
Harambe	P3123	Brooke Marine	Aug 1975

'Brooke Marine 37.5-m' class FAC(M)
(UK/Oman)
Type: fast attack craft (missile)
Displacement: 166 tons standard and 185 tons full load
Dimensions: length 123.0 ft (37.5 m); beam 22.5 ft (6.9 m); draught 6.0 ft (2.2 m)
Gun armament: two 40-mm Bofors L/70 AA in a Breda twin mounting
Missile armament: two single container-launchers for two MM.38 Exocet anti-ship missiles
Torpedo armament: none
Anti-submarine armament: none
Electronics: one TM1229 surface-search and navigation radar, one Laurence-Scott optical director, and one Sea Archer fire-control system
Propulsion: two Paxman Ventura 16RP200 diesels delivering 4,800 bhp (3580 kW) to two shafts
Performance: speed 25 kts; range 3,800 miles (6115 km) at 15 kts
Complement: 4+24

Class
1. Morocco ('Kebir' class)

Name	No.	Builder	Commissioned
El Yadekh	341	Brooke Marine	1982
El Mourakeb	342	Brooke Marine	1983
	343	Mers-el-Kebir	1985
	344	Mers-el Kebir	1985
	345	Mers-el-Kebir	
	346	Mers-el-Kebir	

(These Algerian craft are based on the standard 'Brooke 37.5-m' hull, but have a full-load displacement of 200 tons and carry an armament of one 76-mm/3-in OTO Melara Compact DP and one 20-mm AA guns; they are each powered by two MTU 12V538 TB92 diesels whose 4500 kW/6,035 bhp provides a speed of 29 kts; the complement is 3+24.)

2. Oman ('Al Mansur' class)

Name	No.	Builder	Commissioned
Al Mansur	B2	Brooke Marine	Mar 1973

3. Oman ('Al Wafi class)

Name	No.	Builder	Commissioned
Al Wafi	B4	Brooke Marine	Mar 1977
Al Fulk	B5	Brooke Marine	Mar 1977
Al Mujahid	B6	Brooke Marine	Jul 1977
Al Jabbar	B7	Brooke Marine	Oct 1977

(The 'Al Wafi' class is a variant of the 'Al Mansur' class optimised for the fast attack craft [gun] role with a main armament of one 76-mm/3-in Oto Melara Compact L/62 DP and a secondary armament of one 20-mm AA; the displacements are 135 tons standard and 155 tons full load, and the complement is 3+24.)

'Constitucion' class FAC(M/G)
(UK/Venezuela)
Type: fast attack craft (missile/gun)
Displacement: 170 tons
Dimensions: length 121.0 ft (36.9 m); beam 23.3 ft (7.1 m); draught 6.0 ft (1.8 m)
Gun armament: one 76-mm (3-in) OTO Melara Compact L/62 DP in a single mounting (P11, 13 and 15 only), and one 40-mm Bofors L/70 AA in a single mounting
Missile armament: two container-launchers for two Otomat anti-ship missiles (P12, 14 and 16 only)
Torpedo armament: none
Anti-submarine armament: none
Electronics: one SPQ 2D surface-search radar, and one RTN 10X radar used in conjunction with the Argo NA10 fire-control system
Propulsion: two MTU diesels delivering 5370 kW (7,205 shp) to two shafts
Performance: speed 31 kts; range 2500 km (1,555 miles) at 16 kts
Complement: 3+14

Class
1. Venezuela

Name	No.	Builder	Commissioned
Constitucion	P11	Vosper Thornycroft	Aug 1974
Federacion	P12	Vosper Thornycroft	Mar 1975
Independencia	P13	Vosper Thornycroft	Sep 1974
Libertad	P14	Vosper Thornycroft	Jun 1975
Patria	P15	Vosper Thornycroft	Jan 1975
Victoria	P16	Vosper Thornycroft	Sep 1975

'Cormoran' class FAC(M)
(Spain/Egypt)
Type: fast attack craft (missile)
Displacement: 357 tons full load
Dimensions: length 52.0 m (170.7 ft); beam 7.5 m (24.8 ft); draught 1.9 m (6.2 ft)
Gun armament: one 76-mm (3-in) OTO Melara Compact L/62 DP in a single mounting, and one 40-

mm Bofors L/70 AA in a single mounting
Missile armament: four container-launchers for four MM.38 Exocet anti-ship missiles
Torpedo armament: none
Anti-submarine armament: none
Electronics: one ZW-06 navigation radar, one WM-28 fire-control system, and one optical fire-control system
Propulsion: three MTU/Bazan diesels delivering 10050 kW (13,475 shp) to three shafts
Performance: speed 32 kts; range 3700 km (2,300 miles) at 15.5 kts
Complement: 31

Class
1. Egypt

Name	*No.*	*Builder*	*Commissioned*
		Bazan	
		Bazan	
		Bazan	
		Bazan	
		Bazan	
		Bazan	

2. Somalia

Name	*No.*	*Builder*	*Commissioned*
		Bazan	

3. Sudan

Name	*No.*	*Builder*	*Commissioned*
		Bazan	

'Dvora' class FAC(M)
(Israel)
Type: fast attack craft (missile)
Displacement: 47 tons
Dimensions: length 21.6 m (70.8 ft); beam 5.5 m (18.0 ft); draught 1.0 m (3.3 ft)
Gun armament: one 20-mm AA in a single mounting, and two 0.5-in (12.7-mm) machine-guns
Missile armament: two container-launchers for two Gabriel anti-ship missiles
Torpedo armament: none
Anti-submarine armament: none
Electronics: one Decca 926 surface-search and navigation radar
Propulsion: two MTU 12V331 TC81 diesels delivering 2030 kW (2,725 shp) to two shafts
Performance: speed 36 kts; range 1300 km (810 miles) at 27 kts
Complement: 10

Class
1. Israel
2 craft
2. Taiwan (**'Tzu Chiang class**)
50 craft plus ? more building or planned

'Flagstaff 2' FAH(M)
(USA/Israel)
Type: fast attack hydrofoil (missile)
Displacement: 91.5 tons
Dimensions: length 84.0 ft (25.6 m); beam 21.3 ft (6.5 m); draught 5.0 ft (1.6 m)
Gun armament: four 30-mm AA in two twin mountings, or one 76-mm (3-in) OTO Melara Compact L/62 DP in a single mounting and two 30-mm AA in a twin mounting
Missile armament: four container-launchers for four Gabriel or RGM-84A Harpoon anti-ship missiles
Torpedo armament: none
Anti-submarine armament: none
Electronics: surface-search and navigation radars
Propulsion: two diesels for hullborne operation, and one Allison gas turbine for foilborne operation
Performance: speed 52 kts; range 1850 km (1,150 miles)
Complement: not known

Class
1. Israel

Name	*No.*	*Builder*	*Commissioned*
Shimrit	M161	Lantana	1982
Livnit	M162	Lantana	
Snapirit	M163	Lantana	

'Hauk' class FAC(M)
(Norway)
Type: fast attack craft (missile)
Displacement: 120 tons standard and 155 tons full load
Dimensions: length 36.5 m (119.7 ft); beam 6.2 m (20.3 ft); draught 3.6 m (11.8 ft)
Gun armament: one 40-mm Bofors AA in a single mounting, and one 20-mm AA in a single mounting
Missile armament: six container-launchers for six Penguin Mk II anti-ship missiles
Torpedo armament: two mountings for two 533-mm (21-in) torpedoes
Anti-submarine armament: none
Electronics: one MSI-805 surface-search radar and fire-control system
Propulsion: two MTU 16V538 TB91 diesels delivering 5370 kW (7,205 shp) to two shafts
Performance: speed 34 kts; range 815 km (505 miles) at 34 kts
Complement: 22

Class
1. Norway

Name	*No.*	*Builder*	*Commissioned*
Hauk	P986	Bergens Mek	Aug 1977
Orn	P987	Bergens Mek	Jan 1979
Terne	P988	Bergens Mek	Mar 1979
Tjeld	P989	Bergens Mek	May 1979
Skarv	P990	Bergens Mek	Jul 1979

Name	No.	Builder	
Teist	P991	Bergens Mek	Sep 1979
Jo	P992	Bergens Mek	Nov 1979
Lom	P993	Bergens Mek	Jan 1980
Stegg	P994	Bergens Mek	Mar 1980
Falk	P995	Bergens Mek	Apr 1980
Ravn	P996	Westamarin	May 1980
Gribb	P997	Westamarin	Jul 1980
Geir	P998	Westamarin	Sep 1980
Erle	P999	Westamarin	Dec 1980

'Hegu' class FAC(G) (China): see 'Komar' class FAC(M)

'Helsinki' class FAC(M)
(Finland)
Type: fast attack craft missile
Displacement: 280 tons standard and 300 tons full load
Dimensions: length 45.0 m (147.6 ft); beam 8.9 m (29.2 ft); draught 11.8 ft)
Gun armament: one 57-mm Bofors L/70 DP in a single mounting, and two 23-mm AA in a twin mounting
Missile armament: eight container-launchers for eight RBS 15F anti-ship missiles
Torpedo armament: none
Anti-submarine armament: none
Electronics: one 9 LV 225 surface-search and fire-control radar
Propulsion: three MTU 16V538 TB92 diesels delivering 8200 kW (10,995 shp) to three shafts
Performance: speed 30 kts
Complement: 30

Class
1. Finland

Name	No.	Builder	Commissioned
Helsinki	60	Wärtsilä	Sep 1981
Turku	61	Wärtsilä	Jun 1985
Oulu	62	Wärtsilä	Oct 1985
	63	Wärtsilä	Jun 1986

plus another eight units planned

'Huangfen' class FAC(M) (China): see 'Osa' class FAC

'Huchuan' class FAH(T)
(China)
Type: fast attack hydrofoil (torpedo)
Displacement: 39 tons standard and 45 tons full load
Dimensions: length 21.8 m (71.5 ft); beam 5.0 m (16.4 ft) over foils; draught 1.0 m (3.3 ft) hullborne
Gun armament: four 14.5-mm (0.57-in) machine-guns in two twin mountings
Missile armament: none
Torpedo armament: two mountings for two 533-mm (21-in) torpedoes
Anti-submarine armament: none
Electronics: one 'Skin Head' surface-search radar
Propulsion: three M50F diesels delivering 2700 kW (3,620 shp) to three shafts
Performance: speed 55 kts foilborne; range 925 km (575 miles) at cruising speed
Complement: 4+11/15

Class
1. Albania
32 craft
2. China
130 craft
3. Pakistan
4 craft
4. Romania
19 craft
5. Tanzania
4 craft
6. Zaire
4 craft

'Hugin' class FAC(M)
(Sweden)
Type: fast attack craft (missile)
Displacement: 120 tons standard and 150 tons full load
Dimensions: length 36.6 m (120.0 ft); beam 6.3 m (20.7 ft); draught 1.7 m (5.6 ft)
Gun armament: one 57-mm Bofors L/70 DP in a single mounting
Missile armament: six container-launchers for six Penguin Mk II anti-ship missiles (to be replaced by six RBS 15M anti-ship missiles)
Torpedo armament: none
Anti-submarine armament: none
Mines: fitted for minelaying
Electronics: one 9 LV 200 Mk 2 surface-search and fire-control system, one Skanter Mk 009 navigation radar, and one SQ 3D/SF hull-mounted sonar
Propulsion: two MTU 20V672 TY90 diesels delivering 4560 kW (6,115 shp) to two shafts
Performance: speed 32 kts
Complement: 3+19

Class
1. Sweden

Name	No.	Builder	Commissioned
Jagaren	P150	Bergens Mek	Nov 1972
Hugin	P151	Bergens Mek	Jul 1978
Munin	P152	Bergens Mek	Jul 1978
Magne	P153	Bergens Mek	Oct 1978
Mode	P154	Westamarin	Jan 1979
Vale	P155	Westamarin	Apr 1979

Vidar	P156	Westamarin	Aug 1979
Mjolner	P157	Westamarin	Oct 1979
Mysing	P158	Westamarin	Feb 1980
Kaparen	P159	Bergens Mek	Aug 1980
Vaktaren	P160	Bergens Mek	Sep 1980
Snapphanen	P161	Bergens Mek	Jan 1980
Spejaren	P162	Bergens Mek	Mar 1980
Styrbjorn	P163	Bergens Mek	Jun 1980
Starkodder	P164	Bergens Mek	Aug 1981
Tordon	P165	Bergens Mek	Oct 1981
Tirfing	P166	Bergens Mek	Jan 1982

'Jaguar' class FAC(T)
(west Germany)
Type: fast attack craft (torpedo)
Displacement: 160 tons standard and 190 tons full load
Dimensions: length 42.5 m (139.4 ft); beam 7.2 m (23.4 ft); draught 2.4 m (7.9 ft)
Gun armament: two 40-mm Bofors L/70 AA in single mountings
Missile armament: none
Torpedo armament: four mountings for four 533-mm (21-in) torpedoes
Anti-submarine armament: none
Mines: up to four if two torpedo tubes are removed
Electronics: navigation radar
Propulsion: four MTU diesels delivering 8960 kW (12,020 shp) to four shafts
Performance: speed 42 kts
Complement: 39

Class
1. Greece

Name	*No.*	*Builder*	*Commissioned*
Hesperos	P50	Lürssen	1958
Kentauros	P52	Lürssen	1958
Kyklon	P53	Lürssen	1958
Lelaps	P54	Lürssen	1958
Skorpios	P55	Lürssen	1958
Tyfon	P56	Lürssen	1958

2. Indonesia

Name	*No.*	*Builder*	*Commissioned*
Beruang	652	Lürssen	1959
Harimau	654	Lürssen	1960

3. Saudi Arabia

Name	*No.*	*Builder*	*Commissioned*
Dammam		Lürssen	1969
Khabar		Lürssen	1969
Maccah		Lürssen	1969

4. Turkey

Name	*No.*	*Builder*	*Commissioned*
Tufan	P331	Lürssen	1962
Mizrak	P333	Lürssen	1962
Kalkan	P335	Lürssen	1959
Karayel	P336	Lürssen	1962

5. Turkey ('Kartal' class)

Name	*No.*	*Builder*	*Commissioned*
Denizkusu	P321	Lürssen	1967
Atmaca	P322	Lürssen	1967
Sahin	P323	Lürssen	1967
Kartal	P324	Lürssen	1967
Meltem	P325	Lürssen	1968
Pelikan	P326	Lürssen	1968
Albatros	P327	Lürssen	1968
Simsek	P328	Lürssen	1968
Kasirga	P329	Lürssen	1967

(The 'Kartal' class is a variant of the 'Jaguar' class produced for the Turkish navy's FAC[M] role; the type has a length of 42.8 m/140.5 ft, a beam of 7.1 m/23.5 ft and a draught of 2.2 m/7.2 ft; the armament is fairly impressive, comprising two single 40-mm Bofors L/70 AA guns, four container-launchers for four Penguin Mk II anti-ship missiles, and four mountings for four 533-mm/21-in torpedoes; the primary electronic system is a WM-28 fire-control system.)

'Jerong' class FAC(G) (Malaysia): see 'Lürssen TNC-45' class FAC(M)

'Kaman' class FAC(M) (Iran): see 'La Combattante II' class FAC(M)

'Kartal' class FAC(M) (Turkey): see 'Jaguar' class FAC(T)

'Komar' class FAC(M)
(USSR)
Type: fast attack craft (missile)
Displacement: 68 tons standard and 75 tons full load
Dimensions: length 26.8 m (87.9 ft); beam 6.2 m (20.3 ft); draught 1.5 m (4.9 ft)
Gun armament: two 25-mm AA in a twin mounting
Missile armament: two container-launchers for two SS-N-2A 'Styx' anti-ship missiles
Torpedo armament: none
Anti-submarine armament: none
Electronics: one 'Square Tie' surface-search radar, one 'Dead Duck' IFF, and one 'High Pole-A' IFF
Propulsion: four M50F diesels delivering 3600 kW (4,830 shp) to four shafts
Performance: speed 40 kts; range 740 km (460 miles) at 30 kts
Complement: 4+19

Class
1. Algeria

6 craft

2. Bangladesh

4 'Hegu' class craft

3. China

2 craft, plus
95 locally-built **'Hegu' class** developments with four 25-mm AA in two twin mountings

4. Cuba
14 craft

5. Egypt
10 craft (including 6 'Hegu' class)

6. North Korea
10 craft, plus 10 locally-built **'Sohang' class** developments

7. Syria
6 craft

'La Combattante II' class FAC(M)
(France/Greece)
Type: fast attack craft (missile)
Displacement: 234 tons standard and 275 tons full load
Dimensions: length 47.0 m (154.2 ft); beam 7.1 m (23.3 ft); draught 2.5 m (8.2 ft)
Gun armament: four 35-mm AA in two twin mountings
Missile armament: four container-launchers for four MM.38 Exocet anti-ship missiles
Torpedo armament: two mountings for two 533-mm (21-in) torpedoes
Anti-submarine armament: none
Electronics: one Triton surface-search and navigation radar, and one Pollux missile-control radar
Propulsion: four MTU 16V538 diesels delivering 8940 kW (11,990 shp) to four shafts
Performance: speed 36.5 kts; range 3700 km (2,300 miles) at 15 kts or 1575 km (980 miles) at 25 kts
Complement: 4+36

Class
1. Greece

Name	No.	Builder	Commissioned
Anthipoploiarhos Anninos	P14	CMN	Jun 1972
Ipoploiarhos Arliotis	P15	CMN	Apr 1972
Ipoploiarhos Konidis	P16	CMN	Jul 1972
Ipoploiarhos Batsis	P17	CMN	Dec 1971

2. Iran (**'Kaman' class**)

Name	No.	Builder	Commissioned
Kaman	P221	CMN	Aug 1977
Zoubin	P222	CMN	Sep 1977
Khadang	P223	CMN	Mar 1978
Peykan	P224	CMN	Mar 1978
Joshan	P225	CMN	Mar 1978
Falakhon	P226	CMN	Mar 1978
Shamshir	P227	CMN	Mar 1978
Gorz	P228	CMN	Aug 1978
Gardouneh	P229	CMN	Sep 1978
Khanjar	P230	CMN	Aug 1981
Heyzeh	P231	CMN	Aug 1981
Tabarzin	P232	CMN	Aug 1981

(These Iranian craft are similar in hull and machinery to the Greek units, but have a different armament and electronic fit; the armament comprises one 76-mm/3-in OTO Melara Compact L/62 DP in a single mounting, one 40-mm Bofors L/70 AA in a single mounting and four container-launchers for four RGM-84A Harpoon anti-ship missiles used in conjunction with a WM-28 fire-control system; the Iranians lack Harpoon missiles because of a US embargo on the delivery of weapons to the country; the propulsion comprises four MTU 16V538 TB91 diesels delivering 10740 kW/14,405 shp, but as the full laden displacement is 275 tons the maximum speed is 34.5 kts; a complement of 31 is carried.)

3. Libya (**'La Combattante IIG' class**)

Name	No.	Builder	Commissioned
Sharara	518	CMN	Feb 1982
Shehab	522	CMN	Apr 1982
Wahg	524	CMN	May 1982
Waheed	526	CMN	Jul 1982
Shouaiai	528	CMN	Sep 1982
Shoula	532	CMN	Oct 1982
Shafak	534	CMN	Dec 1982
Bark	536	CMN	Mar 1982
Rad	538	CMN	May 1983
Laheeb	542	CMN	Jul 1983

(These Libyan craft are to an enlarged 'La Combattante II' design with a full-load displacement of 311 tons on a length of 49.4 m/162.1 ft, a beam of 7.1 m/23.3 ft and a draught of 2.0 m/6.6 ft; the armament comprises one 76-mm/3-in OTO Melara Compact L/62 DP in a single mounting and two 40-mm Bofors L/70 AA in a Breda twin mounting plus four container-launchers for four Otomat anti-ship missiles; the electronics fit is one Triton surface-search radar, one Castor radar used in conjunction with the Vega fire-control system, and one Panda optronic fire-control system; four MTU 20V538 TB91 diesels deliver 13420 kW/18,000 shp to four shafts for a speed of 39 kts and a range of 2960 km/1,840 miles at 15 kts; the complement is 27.)

4. Malaysia ('**Perdana' class**)

Name	No.	Builder	Commissioned
Perdana	P3501	CMN	Dec 1972
Serang	P3502	CMN	Jan 1973
Ganas	P3503	CMN	Feb 1973
Ganyang	P3504	CMN	Mar 1973

(These Malaysian craft are close to the norm in displacement and dimensions; the armament comprises one 57-mm Bofors and one 40-mm Bofors AA, each in single mountings, plus two container-launchers for two MM.38 Exocet anti-ship missiles used in conjunction with the Triton surface-search radar and the Pollux radar with the Vega fire-control system; four MTU diesels deliver 10400 kW/13,950 shp to four shafts for a speed of 36.5 kts and a range of 1500 km/930 miles at 25 kts; the complement is 5+30.)

5. West Germany ('**Type 148' class**)

Name	No.	Builder	Commissioned
Tiger	P6141	CMN	Oct 1972
Iltis	P6142	CMN	Jan 1973
Luchs	P6143	CMN	Apr 1973
Marder	P6144	CMN	Jun 1973
Leopard	P6145	CMN	Aug 1973
Fuchs	P6146	Lürssen/CMN	Oct 1973
Jaguar	P6147	CMN	Nov 1973
Lowe	P6148	Lürssen/CMN	Jan 1974
Wolf	P6149	CMN	Feb 1974
Panther	P6150	Lürssen/CMN	Mar 1974
Haher	P6151	CMN	Jun 1974
Storch	P6152	Lürssen/CMN	Jul 1974
Pelikan	P6153	CMN	Sep 1974
Elster	P6154	Lürssen/CMN	Nov 1974
Alk	P6155	CMN	Jan 1975
Dommel	P6156	Lürssen/CMN	Feb 1975
Weihe	P6157	CMN	Apr 1975
Pinguin	P6158	Lürssen/CMN	May 1975
Reiher	P6159	CMN	Jun 1975
Kranich	P6160	Lürssen/CMN	Aug 1975

(These West German craft are derivatives of the standard 'La Combattante II' class design, with standard and full-load displacements of 235 and 265 tons on dimensions that include a length of 47.0 m/154.2 ft, a beam of 7.6 m/24.9 ft and a draught of 2.5 m/8.2 ft; the armament consists of one 76-mm/3-in OTO Melara Compact L/62 DP and one 40-mm Bofors L/70 AA, each in a single mounting, plus four container-launchers for four MM.38 Exocet anti-ship missiles; the electronics include one Triton combined air- and surface-search radar, one Pollux radar used in conjunction with the Vega fire-control system, and one 3RM 20 navigation radar; the powerplant comprises four MTU 872 diesels delivering 8940 kW/11,990 shp to four shafts for a speed of 35+ kts and a range of 1125 km/700 miles at 30 kts; the complement is 4+26.)

'La Combattante III' class FAC(M)
(France/Greece)
Type: fast attack craft (missile)
Displacement: 395 tons standard and 425 tons full load (P20/23) or 330 tons standard and 430 tons full load (P24/29)
Dimensions: length 56.0 m (183.7 ft); beam 7.9 m (25.9 ft); draught 2.5 m (8.2 ft)
Gun armament: two 76-mm (3-in) OTO Melara Compact L/62 DP in single mountings, and four 30-mm AA in two Emerlec twin mountings
Missile armament: four container-launchers for four MM.38 Exocet anti-ship missiles (P20/23) or six container-launchers for six Penguin Mk II anti-ship missiles (P24/29)
Torpedo armament: two mountings for two 533-mm (21-in) torpedoes (P20/23 only)
Anti-submarine armament: none
Electronics: one Triton surface-search radar, one Castor and one Pollux radar used in conjunction with the Vega fire-control system, and two Panda optronic fire-control systems
Propulsion: (P20/23) four MTU 20V538 TB91 diesels delivering 13420 kW (18,000 shp), or (P24/29) four MTU diesels delivering 11200 kW (15,020 shp), in each case to four shafts
Performance: speed 35.7 kts; (P20/23) or 32.5 kts (P24/29); range 3700 km (2,300 miles) at 15 kts or 1300 km (810 miles) at 32.6 kts
Complement: 5+37

Class
1. Greece

Name	No.	Builder	Commissioned
Antiploiarhos Laskos	P20	CMN	Apr 1977
Plotarhis Blessas	P21	CMN	Jul 1977
Ipoploiarhos Mikonios	P22	CMN	Feb 1978
Ipoploiarhos Troupakis	P23	CMN	Nov 1977
Simeoforos Kavaloudis	P24	Hellenic Shipyards	Jul 1980
Anthipoploiarhos Kostakos	P25	Hellenic Shipyards	Sep 1980
Ipoploiarhos Deyiannis	P26	Hellenic Shipyards	Dec 1980
Simeoforos Xenos	P27	Hellenic Shipyards	Mar 1981
Simeoforos Simitzopoulos	P28	Hellenic Shipyards	Jun 1981
Simeoforos Starakis	P29	Hellenic Shipyards	Oct 1981

2. Nigeria ('La Combattante IIIB' class)

Name	No.	Builder	Commissioned
Siri	P181	CMN	Feb 1981
Ayam	P182	CMN	Jun 1981
Ekun	P183	CMN	Sep 1981

(These Nigerian craft have standard and full-load displacements of 395 and 430 tons on a length of 56.2 m/184.4 ft, a beam of 7.6 m/24.9 ft and a draught of 2.1 m/7.0 ft; the armament comprises one 76-mm/3-in OTO Melara Compact L/62 DP in a single mounting, two 430-mm Bofors L/70 AA in a Breda twin mounting, four 30-mm AA in two Emerlec twin mountings and four container-launchers for four MM.38 Exocet anti-ship missiles; four MTU 16V956 TB92 diesels deliver 14920 kW/20,010 shp to four shafts for a speed of 41 kts; the complement is 42.)

3. Qatar ('La Combattante IIIM' class)

Name	No.	Builder	Commissioned
Damsah	Q01	CMN	Nov 1982
Al Ghariyah	Q02	CMN	Feb 1983
Rbigah	Q03	CMN	May 1983

(These Qatari craft are similar to the Tunisian units in all essential respects.)

4. Tunisia ('La Combattante IIIM' class)

Name	No.	Builder	Commissioned
La Galité	501	CMN	Oct 1984
Tunis	502	CMN	Nov 1984
Carthage	503	CMN	Dec 1984

(Though similar to the Greek units, these Tunisian craft have a beam enlarged to 8.2 m/26.9 ft for standard and full-load displacements of 385 and 430 tons; the armament comprises one 76-mm/3-in OTO Melara Compact L/62 DP in a single mounting, two 40-mm Bofors L/70 AA in a Breda twin mounting, two 30-mm AA in twin mountings and eight container-launchers for eight MM.40 Exocet anti-ship missiles; amongst the electronic fit are a Sylosat satellite navigation system, two Naja optronic fire-control systems, and one Dagaie chaff launcher; the propulsion system comprises five MTU 20V538 TB93 diesels delivering 14400 kW/19,315 shp to four shafts for a speed of 38.5 kts; the complement is 42.)

'Lazaga' class FAC(M) (Spain): see 'Lürssen FPB-57' class FAC(M)

'Lazaga (Modified)' class FAC(M) (Morocco): see 'Lürssen FPB-57' class FAC(M)

'Lürssen FPB-57' class FAC(M)
(West Germany/Spain)
Type: fast attack craft (missile)
Displacement: 275 tons standard and 400 tons full load
Dimensions: length 58.1 m (190.6 ft); beam 7.6 m (24.9 ft); draught 2.6 m (8.5 ft)
Gun armament: one 76-mm (3-in) OTO Melara Compact L/62 DP in a single mounting, one 40-mm Bofors L/70 AA in a single mounting, and two 20-mm AA in single mountings
Missile armament: one quadruple container-launcher for four RGM-84A Harpoon anti-ship missiles
Torpedo armament: none
Anti-submarine armament: provision for two Mk 32 triple mountings for 12.75-in (324-mm) Mk 46 torpedoes, and depth-charge racks
Electronics: one WM-22/41 surface-search and fire-control system, one Raytheon TM 1620/6X navigation radar, one HSM Mk 22 optical fire-control system, one ELAC hull-mounted sonar, and one ECM system
Propulsion: two MTU/Bazan MA15 TB91 diesels delivering 6000 kW (8,045 shp) to two shafts
Performance: speed 30 kts; range 11300 km (7,020 miles) at 17 kts
Complement: 4+30

Class
1. Ghana

Name	No.	Builder	Commissioned
Achimota	P28	Lürssen	1980
Yogaga	P29	Lürssen	1980

(These Ghanaian craft are of the original 'FPB-57' class of fast attack craft [gun], and have a full-load displacement of 376 tons; the length is 58.1 m/190.6 ft, the beam 7.6 m/24.9 ft and the draught 2.8 m/9.2 ft; the armament comprises one 76-mm/3-in OTO Melara Compact L/62 DP and one 40-mm Bofors L/70 AA, each in a single mounting, and the electronics include surface-search and navigation radars; three MTU diesels deliver 8505 kW/11,405 shp to three shafts for a speed of 38 kts; the complement is 60.)

2. Kuwait

Name	No.	Builder	Commissioned
Istiqlal	P5701	Lürssen	Nov 1982
Sabhan	P5703	Lürssen	Mar 1983

(The Kuwaiti craft are standard 'FPB-57' class boats with a full-load displacement of 410 tons, the same dimensions as the Spanish 'Lazaga' class units and the same armament as the Moroccan 'Lazaga [Modified]' class; the electronics fit includes one Decca 1226 radar, one 9 LV 228 fire-control system, a Lynx optical fire-control system and two Dagaie chaff launchers; the powerplant comprises four MTU 16V538 TB92 diesels delivering 11640 kW/15,610 shp for a maximum speed of 41 kts; the complement is 55.)

3. Morocco ('Lazaga [Modified]' class)

Name	No.	Builder	Commissioned
El Khatabi	304	Bazan	Jul 1981
Commandant Azouggargh	305	Bazan	Aug 1982
Commandant Boutouba	306	Bazan	Nov 1981
Commandant El Harty	307	Bazan	Feb 1982

(These Moroccan craft are derived from the Spanish vessels, and have a full-load displacement of 410 tons; the dimensions are a length of 57.4 m/188.3 ft, a beam of 7.6 m/24.9 m and a draught of 2.7 m/8.9 ft; the gun armament comprises one 76-mm/3-in OTO Melara Compact L/62 DP and one 40-mm Bofors L/70 AA, each in single mountings, and the missile fit comprises four container-launchers for four MM.40 Exocet anti-ship missiles; there is no torpedo or anti-submarine armament; the electronics include one ZW-06 surface-search radar and one WM-20 fire-control system; the speed is 36 kts, and the complement amounts to 41; another six units are on order.)

3. Nigeria

Name	No.	Builder	Commissioned
Ekpe	P178	Lürssen	Aug 1981
Damisa	P179	Lürssen	Aug 1981
Agu	P180	Lürssen	Aug 1981

(These Nigerian craft are based on the standard 'FPB-57' hull with a full-load displacement of 410 tons; the armament comprises one 76-mm/3-in OTO Melara Compact L/62 DP in a single mounting, two 40-mm Bofors L/70 AA in a Breda twin mounting, four 30-mm AA in two Emerlec twin mountings and four container-launchers for four MM.40 Exocet anti-ship missiles used with the WM-29 fire-control system; the powerplant is four MTU 16V956 TB92 diesels delivering 15000 kW/20,120 shp to two shafts for a speed of 42 kts and a range of 2400 km/1,490 miles at 30 kts; the complement is 40.)

4. Singapore

Name	No.	Builder	Commissioned
		Singapore SB	
		Singapore SB	
		Singapore SB	

(These Singapore units are of the fast attack craft [gun] type, with a full-load displacement of 410 tons and the standard dimensions; the armament consists of one 76-mm/3-in OTO Melara Compact L/62 DP in a single mounting and two 40-mm Bofors L/70 AA in a Breda twin mounting; the powerplant comprises four MTU diesels delivering 11940 kW/16,015 shp to two shafts for a speed of 38 kts; the complement is 40.)

5. Spain ('Lazaga' class)

Name	No.	Builder	Commissioned
Lazaga	PC01	Lürssen	Jul 1975
Alsedo	PC02	Bazan	Feb 1977
Cadarso	PC03	Bazan	Jul 1976
Villamil	PC04	Bazan	Apr 1977
Bonifaz	PC05	Bazan	Jul 1977
Recalde	PC06	Bazan	Dec 1977

6. Turkey

Name	No.	Builder	Commissioned
Dogan	P340	Lürssen	Jun 1977
Marti	P341	Taskizak NY	Jul 1978
Tayfun	P342	Taskizak NY	Jul 1979
Volkan	P343	Taskizak NY	Jul 1980
Ruzgar	P344	Taskizak NY	1983
Poyraz	P345	Tazkizak NY	1984

(These Turkish units are to the standard 'FPB-57' class specification, and have an armament of one 76-mm/3-in OTO Melara Compact L/62 DP in a single mounting, two 35-mm AA in a twin mounting, and two quadruple container-launchers for eight RGM-84A Harpoon anti-ship missiles used in conjunction with the WM-28 fire-control system; the four MTU diesels deliver 13400 kW/17,970 shp to two shafts for a speed of 38 kts, and the complement is 39.)

7. West Germany ('Type 143A' class)

Name	No.	Builder	Commissioned
Gepard	P6121	AEG/Lürssen	Dec 1982
Puma	P6122	AEG/Lürssen	Feb 1983
Hermelin	P6123	AEG/Kroger	May 1983
Nerz	P6124	AEG/Lürssen	Jul 1983
Zobel	P6125	AEG/Lürssen	Sep 1983
Frettchen	P6126	AEG/Lürssen	Dec 1983
Dachs	P6127	AEG/Lürssen	Mar 1984
Ozelot	P6128	AEG/Luørssen	May 1984
Wiesel	P6129	AEG/Kroger	Jul 1984
Hyane	P6130	AEG/Kroger	Nov 1984

(These West German craft are to a modified 'FPB-57' class design, with standard and full-load displacements of 295 and 390 tons, and dimensions that include a length of 57.5 m/188.6 ft, a beam of 7.6 m/24.9 ft and a draught of 2.5 m/8.2 ft; the armament comprises one 76-mm/3-in OTO Melara Compact L/62 DP in a single mounting, four container-launchers for four MM.38 Exocet anti-ship missiles and one launcher for RIM-116 RAM SAMs; the electronics suite includes one WM-27 fire-control system, one 3RM 20 navigation radar and one AGIS command/control system; four MTU 16V956 TB91 diesels deliver 13400 kW/17,970 shp to four shafts for a speed of 40 kts; the complement is 4+30.)

8. West Germany ('Type 143B' class)

Name	No.	Builder	Commissioned
Albatros	P6111	Lürssen	Nov 1976
Falke	P6112	Lürssen	Apr 1976
Geier	P6113	Lürssen	Jun 1976
Bussard	P6114	Lürssen	Aug 1976
Sperber	P6115	Kroger	Sep 1976
Greif	P6116	Lürssen	Nov 1976
Kondor	P6117	Kroger	Dec 1976
Seeadler	P6118	Lürssen	Mar 1977
Habicht	P6119	Kroger	Dec 1977
Kormoran	P6120	Lürssen	Jul 1977

(These West German craft were originally of the 'Type 143' class with an armament of two 76-mm/3-in OTO Melara Compact L/62 DP in single mountings, four container-launchers for four MM.38 Exocet anti-ship missiles, and two mountings for two 533-mm/21-in wire-guided torpedoes; the after 76-mm mounting has been removed for installation on 'Type 143A' class craft, and a launcher for the RIM-116 SAM is to be installed in what has become the 'Type 143B' class; the displacement, dimensions and electronics are the same as those for the 'Type 143A' class, but the powerplant comprises four MTU diesels delivering 11920 kW/15,985 shp to four shafts for a speed of 35+ kts; the complement is 40.)

'Lürssen TNC-45' class FAC(M)
(West Germany/United Arab Emirates)
Type: fast attack craft (missile)
Displacement: 234 tons standard and 269 tons full load
Dimensions: length 44.9 m (147.3 ft); beam 7.0 m (23.0 ft); draught 2.3 m (7.5 ft)
Gun armament: one 76-mm (3-in) OTO Melara Compact L/62 DP in a single mounting, and two 40-mm Bofors L/70 AA in a Breda twin mounting
Missile armament: four container-launchers for four MM.40 Exocet anti-ship missiles
Torpedo armament: none
Anti-submarine armament: none
Electronics: one 9 LV 200 Mk 2 surface-search and fire-control system, one Decca navigation radar, one Panda optical fire-control system, one Cutlass ECM system, and one Dagaie chaff launcher
Propulsion: four MTU diesels delivering 10740 kW (14,405 shp) to four shafts
Performance: speed 40 kts
Complement: 40

Class
1. Argentina

Name	No.	Builder	Commissioned
Intrepida	P85	Lürssen	Jul 1974
Indomita	P86	Lürssen	Dec 1974

(These Argentine craft were built as fast attack craft [gun] but have recently been converted into fast attack craft [missile], with an armament of one 76-mm/3-in OTO Melara Compact L/62 DP and two 40-mm Bofors L/70 AA, all in single mountings, complemented by two mountings for two 533-mm/21-in torpedoes and Otomat anti-ship missiles; the craft are slightly beamier than the norm, at 7.4 m/24.3 ft, and have a full-load displacement of 268 tons; a speed of 40 kts is possible on the 8940 kW/11,990 shp provided to four shafts by four MTU diesels; the complement is 35, and the electronics include a WM-20 fire-control system.)

2. Bahrain

Name	No.	Builder	Commissioned
Ahmed el Fateh	B485	Lürssen	1983
Al Jabery	B486	Lürssen	1984

(These Bahraini craft are standard FAC[M] units with an armament and propulsion layout/performance identical with those of the UAE craft; the electronics fit includes a WM-28 fire-control system, one Panda optical fire-control system and one Dagaie chaff launcher.)

3. Ecuador

Name	No.	Builder	Commissioned
Quito	LM31	Lürssen	Jul 1976
Guayaquil	LM32	Lürssen	Dec 1977
Cuenca	LM33	Lürssen	Jul 1977

(These Ecuadorean craft approximate closely to the norm, but the secondary armament comprises two 35-mm AA in a twin mounting, and the electronic fit is one Triton search radar and one Pollux radar used in conjunction with the Vega fire-control system; the complement is 35.)

4. Ghana

Name	No.	Builder	Commissioned
Dzata	P28	Lürssen	Jul 1980
Sebo	P29	Lürssen	Jul 1980

(These Ghanaian craft are fast attack craft [gun] with an armament of two 40-mm Bofors L/70 AA in single mountings; the two MTU diesels deliver 5370 kW/7,205 shp, and the range is 3300 km/2,050 miles at 16 kts; the complement is 55.)

5. Israel ('**Saar 2' class**)

Name	No.	Builder	Commissioned
Mivtach	311	CMN	1968
Miznag	312	CMN	1968
Mifgav	313	CMN	1968
Eilath	314	CMN	1968
Haifa	315	CMN	1968
Akko	316	CMN	1968

(These Israeli craft were built in France to avoid political problems with West Germany, and are exceptionally well equipped; the dimensions are normal, and the standard and full-load displacements are 220 and 250 tons; the armament comprises between one and three 40-mm Bofors L/70 AA in Breda single mountings, between two and eight container-launchers for Gabriel anti-ship missiles [two trainable triple container-launchers being fitted on the after ring mountings if only one 40-mm gun is carried], and four Mk 32 mountings for 12.75-in/324-mm Mk 46 anti-submarine torpedoes if no triple container-launcher units are shipped; the sensor fit includes one Neptune surface-search radar, one RTN 10X fire-control radar, one EDO 780 variable-depth sonar, and a number of Israeli-developed ECM/ESM systems; the four MTU diesels deliver 10060 kW/13,490 shp for a speed of 40+ kts; the complement is between 35 and 40.)

6. Israel ('**Saar 3' class**)

Name	No.	Builder	Commissioned
Saar	331	CMN	1969
Soufa	332	CMN	1969
Gaash	333	CMN	1969
Herev	341	CMN	1969
Hanit	342	CMN	1969
Hetz	343	CMN	1969

(These Israeli craft are very similar to the 'Saar 2' class units, but have no anti-submarine provision and a more powerful gun armament, centred on a forward-located 76-mm/3-in OTO Melara Compact L/62 DP with the option of two 40-mm Bofors L/70 AA or two triple container-launchers for Gabriel anti-ship missiles aft.)

7. Kuwait

Name	No.	Builder	Commissioned
Al Boom	P4501	Lürssen	1983
Al Betteel	P4503	Lürssen	1983
Al Sanbouk	P4505	Lürssen	1983
Al Saadi	P4507	Lürssen	1983
Al Ahmadi	P4509	Lürssen	1983
Al Abdali	P4511	Lürssen	1983

(These Kuwaiti craft have the standard displacement and dimensions, and are armed in the same way as the UAE craft; the electronic fit is one 9 LV 228 fire-control system, one Decca 1226 navigation radar, one Lynx optical fire-control system, and two Dagaie chaff launchers; the propulsion system comprises four MTU 16V956 TB92 diesels delivering 11600 kW/15,560 shp to four shafts for a speed of 41 kts; the complement is 55.)

8. Malaysia ('Jerong' class)

Name	No.	Builder	Commissioned
Jerong	P3505	Hong-Leong/Lürssen	Mar 1976
Todak	P3506	Hong-Leong/Lürssen	Jun 1976
Paus	P3507	Hong Leong/Lürssen	Aug 1976
Yu	P3508	Hong-Leong/Lürssen	Nov 1976
Baung	P3509	Hong-Leong/Lürssen	Jan 1977
Pari	P3510	Hong-Leong/Lürssen	Mar 1977

(These Malaysian craft are fast attack craft [gun] with an armament of one 57-mm Bofors L/70 gun and one 40-mm Bofors L/70 AA gun in single mountings; the three MTU diesels deliver 7380 kW/9,900 shp for a speed of 32 kts; the complement is 41.)

9. Singapore

Name	No.	Builder	Commissioned
Sea Wolf	P76	Lürssen	1972
Sea Lion	P77	Lürssen	1972
Sea Dragon	P78	Singapore SB	1974
Sea Tiger	P79	Singapore SB	1974
Sea Hawk	P80	Singapore SB	1975
Sea Scorpion	P81	Singapore SB	1975

(These Singapore craft approximate to the norm in displacement, dimensions, propulsion and performance, but the armament comprises single 57-mm Bofors and 40-mm Bofors guns plus a missile fit of two single and one triple container-launchers for five Gabriel anti-ship missiles supported by a WM-28 fire-control system.)

10. Thailand

Name	No.	Builder	Commissioned
Prabparapak		Singapore SB	Jul 1976
Hanhak Sattru		Singapore SB	Nov 1976
Suphairin		Singapore SB	Feb 1977

(These Thai craft are essentially similar to the Singapore units.)

11. United Arab Emirates

Name	No.	Builder	Commissioned
Ban Yas	4501	Lürssen	Nov 1980
Marban	4502	Lürssen	Nov 1980
Rodqm	4503	Lürssen	Jul 1981
Shaheen	4504	Lürssen	Jul 1981
Sagar	4505	Lürssen	Sep 1981
Tarif	4506	Lürssen	Sep 1981

'Matka' class FAH(M)
(USSR)
Type: fast attack hydrofoil (missile)
Displacement: 200 tons standard and 230 tons full load
Dimensions: length 39.9 m (130.9 ft); beam 8.0 m (26.2 ft); draught 1.8 m (5.9 ft)
Gun armament: one 76-mm (3-in) DP in a single mounting, and one 30-mm AK-630 CIWS mounting
Missile armament: two container-launchers for two SS-N-2C 'Styx' anti-ship missiles
Torpedo armament: none
Anti-submarine armament: none
Electronics: one 'Plank Shave' surface-search and SSM-control radar, one 'Cheese Cake' navigation radar, one 'Bass Tilt' gun-control radar, one 'Square Head' IFF, one 'High Pole-B' IFF, and two chaff launchers

Propulsion: three M504 diesels delivering 11185 kW (15,000 shp) to three shafts
Performance: speed 42 kts; range 2775 km (1,725 miles) at 14 kts hullborne or 1110 km (690 miles) at 35 kts foilborne
Complement: 30

Class
1. USSR

17 craft

'Minister' class FAC(M) (South Africa): see 'Saar 4' class FAC(M)

'Mol' class FAC(T) (USSR): see 'Shershen' class FAC(T)

'MV 400' class FAC(G)
(Italy/Thailand)
Type: fast attack craft (gun)
Displacement: 450 tons full load
Dimensions: length 60.4 m (198.2 ft); beam 8.1 m (26.6 ft); draught 4.5 m (14.8 ft)
Gun armament: two 76-mm (3-in) OTO Melara Compact L/62 DP in single mountings, and two 40-mm Bofors L/70 AA in a Breda twin mounting
Missile armament: provision for the retrofitting of anti-ship missiles
Torpedo armament: none
Anti-submarine armament: none
Electronics: one WM-22/61 tactical radar and fire-control system, one 3RM 20 navigation radar, one LIROD optronic fire-control system, and four Hycor Mk 135 chaff launchers
Propulsion: three MTU 20V538 TB92 diesels delivering 11190 kW (15,010 shp) to three shafts
Performance: speed 30 kts; range 4600 km (2,860 miles) at 18 kts
Complement: 45

Class
1. Thailand

Name	No.	Builder	Commissioned
Chon Buri	1	CN Breda	Dec 1982
Songkhla	2	CN Breda	Jan 1983
Phuket	3	CN Breda	May 1983

'October' class FAC(M)
(Egypt)
Type: fast attack craft (missile)
Displacement: length 25.5 m (84.0 ft); beam 6.1 m (20.0 ft); draught 1.3 m (5.0 ft)
Gun armament: four 30-mm AA in two Laurence-Scott twin mountings
Missile armament: two container-launchers for two Otomat anti-ship missiles
Torpedo armament: none
Anti-submarine armament: none
Electronics: one S810 surface-search radar, one Sapphire missile-control radar, and one Matilda ESM system
Propulsion: four CRM 18WD/S2 diesels delivering 4020 kW (3,590 shp) to four shafts
Performance: speed 40 kts; range 750 km (465 miles) at 30 kts
Complement: 20

Class
1. Egypt

6 craft

'Osa I' and 'Osa II' class FAC(M)
(USSR)
Type: fast attack craft (missile)
Displacement: 165 tons standard and 210 tons full load
Dimensions: length 39.0 m (127.9 ft); beam 7.8 m (25.6 ft); draught 1.8 m (5.9 ft)
Gun armament: four 30-mm AA in two twin mountings
Missile armament: four container-launchers for four SS-N-2A 'Styx' ('Osa I' class) or SS-N-2C 'Styx' ('Osa II' class) anti-ship missiles, and (some craft) one quadruple launcher for four SA-N-5 'Grail' SAMs
Torpedo armament: none
Anti-submarine armament: none
Electronics: one 'Square Tie' surface-search radar, one 'Bass Tilt' gun-control radar, two 'Square Head' IFF, and one 'High Pole-B' IFF
Propulsion: three M503A ('Osa I' class) or M504 ('Osa II' class) diesels delivering 8940 kW (11,990 shp) or 11190 kW (15,010 shp) to three shafts
Performance: speed 38 kts ('Osa I' class) or 40 kts ('Osa II' class); range 1500 km (930 miles) at 30 kts or 925 km (575 miles) at 35 kts
Complement: 5+25

Class
1. Algeria

3 'Osa I' class craft
9 'Osa II' class craft

2. Benin

2 'Osa I' class craft

3. Bulgaria

3 'Osa I' class craft
1 'Osa II' class craft

4. China

115 'Osa' and locally-built **'Huangfen' class** craft

5. Cuba

5 'Osa I' class craft
13 'Osa II' class craft

6. East Germany
15 'Osa I' class craft

7. Egypt
8 'Osa I' class craft

8. Ethiopia
3 'Osa II' class craft

9. Finland
4 **'Tuima' class** craft

10. India
8 'Osa I' class craft
8 'Osa II' class craft

11. Iraq
4 'Osa I' class craft
8 'Osa II' class craft

12. Libya
12 'Osa II' class craft

13. North Korea
8 'Osa I' class craft
8 locally-built **'Soju' class** craft

14. Poland
13 'Osa I' class craft

15. Romania
5 'Osa I' class craft

16. Somalia
2 'Osa II' class craft

17. South Yemen
6 'Osa II' class craft

18. Syria
6 'Osa I' class craft
6 'Osa II' class craft

19. USSR
70 'Osa I' class craft
50 'Osa II' class craft

20. Vietnam
8 'Osa II' class craft

21. Yugoslavia
10 'Osa I' class craft

'P400' class FAC(M)
(France)
Type: fast attack craft (missile)
Displacement: 380 tons standard and 445 tons full load
Dimensions: length 54.5 m (178.8 ft); beam 7.7 m (25.3 ft); draught 2.5 m (8.3 ft)
Gun armament: one 40-mm Bofors L/60 AA in a single mounting, and one 20-mm AA in a single mounting
Missile armament: two container-launchers for two MM.38 Exocet anti-ship missiles (not generally fitted in peace), and (to replace 20-mm gun) one Sadral launcher for Mistral SAMs
Torpedo armament: none
Anti-submarine armament: none
Electronics: surface-search radar, and navigation radar
Propulsion: two SEMT-Pielstick 16PA4-200 VGDS diesels delivering 5960 kW (7,995 shp) to two shafts
Performance: speed 24 kts; range 7400 km (4,600 miles) at 15 kts
Complement: 3+21, plus berthing and stores provision for the accommodation of 20 troops in the overseas transport role

Class
1. France

Name	No.	Builder	Commissioned
L'Audacieuse	P682	CMN	1985
La Boudeuse	P683	CMN	1985
La Capricieuse	P684	CMN	1985
La Fougeuse	P685	CMN	1985
La Glorieuse	P686	CMN	1985
La Gracieuse	P687	CMN	1985
La Moqueuse	P688	CMN	1986
La Railleuse	P689	CMN	1986
La Rieuse	P690	CMN	1986
La Tapageuse	P691	CMN	1986

2. Gabon
2 craft

'Patra' class FAC(M)
(France)
Type: fast attack craft (missile)
Displacement: 115 tons standard and 148 tons full load
Dimensions: length 47.7 m (156.5 ft); beam 5.9 m (19.4 ft); draught 1.6 m (5.2 ft)
Gun armament: one 40-mm Bofors L/60 AA in a single mounting, and one 12.7-mm (0.5-in) machine-gun
Missile armament: six SS.12M anti-ship missiles
Torpedo armament: none
Anti-submarine armament: none
Electronics: one surface-search and navigation radar
Propulsion: two AGO 195 V12 CZSHR diesels delivering 3730 kW (5,000 shp) to two shafts
Performance: speed 26 kts; range 3250 km (2,020 miles) at 10 kts or 1400 km (870 miles) at 20 kts
Complement: 1+17

Class
1. France

Name	No.	Builder	Commissioned
Trident	P670	Auroux	Dec 1976
Glaive	P671	Auroux	Apr 1977
Epée	P672	CMN	Oct 1976
Pertuisane	P673	CMN	Jan 1977

2. Ivory Coast

Name	No.	Builder	Commissioned
L'Ardent		Auroux	Oct 1978
L'Intrepide		Auroux	Oct 1978

(The Ivory Coast units are similar to the French craft apart from their missile armament, which comprises four container-launchers for four MM.40 Exocet anti-ship missiles.)

'Pegasus' class FAH(M)
(USA)
Type: fast attack hydrofoil (missile)
Displacement: 198 tons light and 242 tons full load
Dimensions: length 131.9 ft (40.5 m) with foils extended and 145.3 ft (44.3 m) with foils retracted; beam 47.5 ft (14.5 m) with foils extended and 28.2 ft (8.6 m) with foils retracted; draught 23.2 ft (7.1 m) with foils extended and 6.2 ft (1.9 m) with foils retracted
Gun armament: one 76-mm (3-in) OTO Melara Compact L/62 DP in a Mk 75 single mounting
Missile armament: two Mk 141 quadruple container-launchers for eight RGM-84A Harpoon anti-ship missiles
Torpedo armament: none
Anti-submarine armament: none
Electronics: one SPS-63 surface-search and navigation radar, one WM-28 (PHM-1) or Mk 92 (others) fire-control system, one OE-82 satellite communication system, one SRR-1 satellite communications receiver, one WSC-3 satellite communications transceiver, two Mk 34 Super RBOC chaff launchers, and one SLR-20 ECM system
Propulsion: two MTU 8V331 TC81 diesels delivering 1220 kW (1,635 shp) to two waterjets for hullborne operation, and one General Electric LM2500 gas turbine delivering 16,765 shp (12500 kW) to two waterjets for foilborne operation
Performance: speed 48 kts foilborne or 12 kts hullborne; range 1,950 miles (3140 km) at 9 kts or 800 miles (1285 km) at 40 kts
Complement: 4+19

Class
1. USA

Name	No.	Builder	Commissioned
Pegasus	PHM1	Boeing	Jul 1977
Hercules	PHM2	Boeing	Jul 1982
Taurus	PHM3	Boeing	Oct 1981
Aquila	PHM4	Boeing	Dec 1981
Aries	PHM5	Boeing	Apr 1982
Gemini	PHM6	Boeing	Jun 1982

'Perdana' class FAC(M) (Malaysia): see 'La Combattante II' class

'PR 72' class FAC(M)
(France/Peru)
Type: fast attack craft (missile)
Displacement: 470 tons standard and 560 tons full load
Dimensions: length 64.0 m (210.0 ft); beam 8.35 m (27.4 ft); draught 1.6 m (5.2 ft)
Gun armament: one 76-mm (3-in) OTO Melara Compact L/62 DP in a single mounting, two 40-mm Bofors L/70 AA in a Breda twin mounting, and two 20-mm AA in single mountings
Missile armament: four container-launchers for four MM.38 Exocet anti-ship missiles
Torpedo armament: none
Anti-submarine armament: none
Electronics: one Triton surface-search radar, one Castor II radar used in conjunction with the Vega fire-control system, one TM1226 navigation radar, and one Panda optical fire-control system
Propulsion: four SACM/AGO 240 V16 diesels delivering 16400 kW (21,995 shp) to four shafts
Performance: speed 34 kts; range 4625 km (2,875 miles) at 16 kts
Complement: 36

Class
1. Morocco ('**PR 72M**' class)

Name	No.	Builder	Commissioned
Okba	33	SFCN	Dec 1976
Triki	34	SFCN	Jul 1977

(These Moroccan craft are smaller than the Peruvian units, with standard and full-load displacements of 375 and 445 tons on dimensions that include a length of 57.5 m/188.8 ft, a beam of 7.6 m/24.9 ft and a draught of 2.1 m/7.1 ft; the armament of these fast attack craft [gun] comprises one 76-mm/3-in OTO Melara Compact L/62 DP and one 40-mm Bofors L/70 AA, each in a single mounting, and there is provision for the installation of four container-launchers for four MM.38 Exocet anti-ship missiles; the powerplant comprises four AGO diesels delivering 8240 kW/11,050 shp for a speed of 28 kts; the complement is 5+48.)

2. Peru ('**PR 72P**' class)

Name	No.	Builder	Commissioned
Velarde	P21	SFCN	Jul 1980
Santillana	P22	SFCN	Jul 1980
De Los Heros	P23	SFCN	Nov 1980
Herrera	P24	SFCN	Feb 1981
Larrea	P25	SFCN	Jun 1981
Sanchez Carrion	P26	SFCN	Sep 1981

3. Senegambia ('**PR 72S**' class)

Name	No.	Builder	Commissioned
Njambuur		SFCN	1982

(This Senegambian unit is similar to the Moroccan unit, but is a fast attack craft [gun] with a length of 58.7 m/192.5 ft and a draught of 2.2 m/7.2 ft for standard and full-load displacements of 375 and 451 tons; the armament comprises two 76-mm/3-in OTO Melara Compact L/62 DP and two 20-mm AA, all in single mountings, and the electronic fit is less capable than that of the Moroccan craft; the four AGO V16 RVR diesels deliver 9540 kW/12,795 shp) to four shafts for a speed of 29 kts; the complement is 39.)

'Province' class FAC(M)
(UK/Oman)
Type: fast attack craft (missile)
Displacement: 420 tons full load
Dimensions: length 186.0 ft (56.7 m); beam 26.9 ft (8.2 m); draught 8.9 ft (2.7 m)
Gun armament: one 76-mm (3-in) OTO Melara Compact in a single mounting, and two 40-mm Breda L/70 AA in a twin mounting

Missile armament: two triple (B10) or quadruple (B11 and B12) container-launchers for six or eight MM.40 Exocet anti-ship missiles
Torpedo armament: none
Anti-submarine armament: none
Electronics: one AWS 4 (B10) or TM1226C (B11 and B12) surface-search and tactical radar used in conjunction with a Sea Archer (B10) or Philips 307 (B11 and B12) fire-control system, one optronic fire-control system, and other systems
Propulsion: four Paxman Valenta diesels delivering 18,200 hp (13570 kW) to four shafts
Performance: speed 40 kts
Complement: 40, plus provision for 19 passengers

Class
1. Kenya

Name	No.	Builder	Commissioned
		Vosper Thornycroft	
		Vosper Thornycroft	

(These Kenyan boats were ordered late in 1984, and probably have a primary armament of eight Gabriel anti-ship missiles supported by a 76-mm/3-in OTO Melara Compact gun, plus a comprehensive range of radar and other electronic systems.)

2. Oman

Name	No.	Builder	Commissioned
Dhofar	B10	Vosper Thornycroft	Aug 1982
Al Sharqiyah	B11	Vosper Thornycroft	Nov 1983
Al Bat'nah	B12	Vosper Thornycroft	Jan 1984

'PSMM Mk 5' class FAC(M)
(USA/South Korea)
Type: fast attack craft (missile)
Displacement: 240 tons standard and 270 tons full load
Dimensions: length 165.0 ft (50.3 m); beam 24.0 ft (7.3 m); draught 9.5 ft (2.9 m)
Gun armament: one 3-in (76-mm) L/50 DP in a Mk 34 single mounting, one 40-mm Bofors AA in a single mounting, and two 0.5-in (12.7-mm) machine-guns
Missile armament: four Standard launchers for eight RIM-66 anti-ship missiles (PGM352, 353 and 355) or two twin container-launchers for four RGM-84A Harpoon anti-ship missiles (others)
Torpedo armament: none
Anti-submarine armament: none
Electronics: surface-search and navigation radar, and two RBOC chaff launchers
Propulsion: six Avco Lycoming TF35 gas turbines delivering 16,800 shp (12525 kW) to two shafts
Performance: speed 40+ kts; range 2,750 miles (4425 km) at 18 kts
Complement: 5+27

Class
1. Indonesia

Name	No.	Builder	Commissioned
Mandau	621	Korea-Tacoma	Oct 1979
Rencong	622	Korea-Tacoma	Oct 1979
Badik	623	Korea-Tacoma	1980
Kris	624	Korea-Tacoma	1980
	625	Korea-Tacoma	
	626	Korea-Tacoma	
	627	Korea-Tacoma	
	628	Korea-Tacoma	

(These Indonesian craft differ somewhat from the South Korean units, having a full-load displacement of 290 tons on a hull with a length of 53.6 m/175.8 ft, a beam of 8.0 m/26.2 ft and a draught of 1.6 m/5.2 ft; the armament comprises one 57-mm Bofors L/70 DP and one 40-mm Bofors L/70 AA, each in a single mounting, plus four container-launchers for four MM.38 Exocet anti-ship missiles used in conjunction with a WM-25 fire-control system; the powerplant is a CODOG arrangement with two MTU diesels and one General Electric LM2500 gas turbine for speeds of 41 kts [gas turbine] or 17 kts [diesels]; the complement is 32.)

2. Philippines

Name	No.	Builder	Commissioned
		Korea-Tacoma	
		Korea-Tacoma	
		Korea-Tacoma	

(These craft are in all major respects similar to the Indonesian units.)

3. South Korea

Name	No.	Builder	Commissioned
Paek Ku 52	PGM352	Tacoma BB	Mar 1975
Paek Ku 53	PGM353	Tacoma BB	Mar 1975
Paek Ku 55	PGM355	Tacoma BB	Feb 1976
Paek Ku 56	PGM356	Tacoma BB	Feb 1976
Paek Ku 57	PGM357	Korea-Tacoma	1977
Paek Ku 58	PGM358	Korea-Tacoma	1977
Paek Ku 59	PGM359	Korea-Tacoma	1977
Paek Ku 61	PGM361	Korea-Tacoma	1978

4. Taiwan

Name	No.	Builder	Commissioned
Lung Chiang	PGG581	Tacoma BB	1979
	PGG582	China SB	1979
Suikiang	PGG583	China SB	
	PGG584	China SB	
	PGG585	China SB	
	PGG586	China SB	
	PGG587	China SB	
	PGG588	China SB	

(These Taiwanese craft are similar to the South Korean units, but their dimensions include a length of 164.5 ft/50.2 m, a beam of 23.9 ft/7.3 m and a draught of 7.5 ft/2.3 m; the gun armament comprises one 76-mm/3-in OTO Melara Compact L/62 DP in a single mounting and two 35-mm AA in an Emerlec twin mounting, and the missile fit is two container-launchers for two Hsiung Feng [Gabriel] anti-ship missiles used in conjunction with the RAN 11L/X radar and associated Argo NA10 fire-control system and IPN 10 action information system or, in the last six craft, with the RCA R76 C5 fire-control system; the CODAG powerplant has three Avco Lycoming TF35 gas turbines delivering 15,000 shp/11185 kW and three diesels delivering 2,800 shp/2145 kW to three shafts for speeds of 40 or 20 kts respectively; the complement is 5+29.)

Note
The 'PSSM Mk 5' class design is the Patrol Ship Multi-Mission Mk 5 derivative of the **'Asheville' class** FAC(G); 17 of this patrol combatant type were built for the US Navy between 1966 and 1971 for coastal patrol and blockade, but were not extensively used; the calss has a full-load displacement of 245 tons on dimensions that includes a length of 164.5 ft (50.1 m), a beam of 23.8 ft (7.3 m) and a draught of 9.5 ft (2.9 m) for a speed of 38 kts on the CODOG (COmbined Diesel Or Gas turbine) arrangement of two Cummins diesels delivering 3,500 shp (2611 kW) and one General Electric gas turbine delivering 13,300 shp (9922 kW) to two shafts; the standard armament is one 3-in (76-mm) gun in a Mk 76 single mounting, one 40-mm Bofors AA in a Mk 3 single mounting, and two 0.5-in (12.7-mm) machine-guns in a twin mounting; the craft are now out of US service and either exported (one to South Korea, and two each to Colombia, Taiwan and Turkey) or awaiting transfer.

'Ramadan' class FAC(M)

(UK/Egypt)
Type: fast attack craft (missile)
Displacement: 312 tons full load
Dimensions: length 170.6 ft (52.0 m); beam 25.0 ft (7.6 m); draught 6.6 ft (2.0 m)
Gun armament: one 76-mm (3-in) OTO Melara Compact L/62 DP in a single mounting, and two 40-mm Bofors L/70 AA in a Breda twin mounting
Missile armament: four container-launchers for four Otomat Mk 1 anti-ship missiles
Torpedo armament: none
Anti-submarine armament: none
Electronics: one S820 surface-search radar, two ST802 radars, one Sapphire gun fire-control system, one CAAIS action information system, and one Cutlass ECM system
Propulsion: four MTU 20V538 diesels delivering 12800 kW (17,170 shp) to four shafts
Performance: speed 40 kts; range 3700 km (2,300 miles) at 16 kts
Complement: 40

Class
1. Egypt

Name	No.	Builder	Commissioned
Ramadan	561	Vosper Thornycroft	1981
Khyber	562	Vosper Thornycroft	1982
El Kadesseya	563	Vosper Thornycroft	1982
El Yarmouk	564	Vosper Thornycroft	1982
Hettein	565	Vosper Thornycroft	1982
Badr	566	Vosper Thornycroft	1982

'Ratcharit' class FAC(M)

(Italy/Thailand)
Type: fast attack craft (missile)
Displacement: 235 tons standard and 270 tons full load
Dimensions: length 49.8 m (163.4 ft); beam 7.5 m (24.6 ft); draught 2.3 m (7.5 ft)
Gun armament: one 76-mm (3-in) OTO Melara Compact L/62 DP in a single mounting, and one 40-mm Bofors L/70 AA in a single mounting
Missile armament: four container-launchers for four MM.38 Exocet anti-ship missiles
Torpedo armament: none
Anti-submarine armament: none
Electronics: one navigation radar, and one WM-20 series fire-control system
Propulsion: three MTU 20V538 TB92 diesels delivering 10050 kW (13,480 shp) to three shafts
Performance: speed 37 kts; range 3700 km (2,300 miles) at 15 kts
Complement: 45

Class
1. Thailand

Name	No.	Builder	Commissioned
Ratcharit	4	CNR, Breda	Aug 1979
Witthayakhom	5	CNR, Breda	Nov 1979
Udomdet	6	CNR, Breda	Feb 1980

'Reshef' class FAC(M) (Israel): see 'Saar 4' class FAC

'Saar 2' class FAC(M) (Israel): see 'Lürssen TNC-45' class FAC(M)

'Saar 3' class FAC(M) (Israel): see 'Lürssen TNC-45' class FAC(M)

'Saar 4' or 'Reshef' class FAC(M)

(Israel)
Type: fast attack (missile)
Displacement: 415 tons standard and 450 tons full load
Dimensions: length 58.0 m (190.6 ft); beam 7.6 m (24.9 ft); draught 2.4 m (8.0 ft)
Gun armament: two 76-mm (3-in) OTO Melara Compact L/62 DP in single mountings, and two 20-mm AA in single mountings
Missile armament: four container-launchers for four RGM-84A Harpoon anti-ship missiles, and five container-launchers for five Gabriel Mk II anti-ship missiles
Torpedo armament: none
Anti-submarine armament: none
Aircraft: one light helicopter on a platform aft (*Tarshish* only, which thus has only one 76-mm gun)
Electronics: one Neptune surface-search radar, one RTN 10X radar used in conjunction with the Argo NA10 fire-control system, one ELAC hull-mounted sonar (not in *Yaffo*), four large and 72 small chaff launchers, and one MN-53 ESM system
Propulsion: four MTU diesels delivering 10440 kW (14,000 shp) to four shafts
Performance: speed 32 kts; range 7400 km (4,600 miles) at 17.5 kts or 3075 km (1,910 miles) at 30 kts
Complement: 45

Class
1. Chile

Name	No.	Builder	Commissioned
Casma		Haifa Shipyard	Mar 1974
Chipana		Haifa Shipyard	Oct 1973

(These Chilean craft are similar to the Israeli norm, but have a missile armament of six Gabriel anti-ship missiles and a less sophisticated sensor fit.)

2. Israel

Name	No.	Builder	Commissioned
Reshef		Haifa Shipyard	Apr 1973
Kidon		Haifa Shipyard	Sep 1974
Tarshish		Haifa Shipyard	Mar 1975
Yaffo		Haifa Shipyard	Apr 1975
Nitzhon		Haifa Shipyard	Mar 1979

Komemiut	Haifa Shipyard	Oct 1979
Atsmout	Haifa Shipyard	Nov 1979
Moledet	Haifa Shipyard	1980

3. South Africa ('**Minister**' class)

Name	*No.*	*Builder*	*Commissioned*
Jan Smuts	P1561	Haifa Shipyard	Sep 1977
P. W. Botha	P1562	Haifa Shipyard	Dec 1977
Frederic Creswell	P1563	Haifa Shipyard	May 1978
Jim Fouche	P1564	Sandock Austral	Dec 1978
Frans Erasmus	P1565	Sandock Austral	Jul 1979
Oswald Pirow	P1566	Sandock Austral	Mar 1980
Hendrik Mentz	P1567	Sandock Austral	Sep 1982
Kobie Coetzee	P1568	Sandock Austral	Mar 1983

(These South African craft are a variation on the Israeli norm, with a full-load displacement of 430 tons on dimensions that include a length of 62.2 m/204.0 ft, a beam of 7.8 m/25.6 ft and a draught of 2.4 m/8.0 ft; the armament comprises the standard two 76-mm/3-in OTO Melara Compact L/62 DP in single mountings and six container-launchers for six Skorpioen [Gabriel Mk II] anti-ship missiles; the propulsion and performance are similar to those of the Israeli craft, and the complement is 47.)

4. Taiwan

Name	*No.*	*Builder*	*Commissioned*
Sui Kiang		Sandock Austral	1985

(This is essentially a 'Minister' class unit armed with six Hsiung Feng [Gabriel II] anti-ship missiles.)

'Saar 4.5' or 'Alia' class FAC(M)
(Israel)
Type: fast attack craft (missile)
Displacement: 488 tons standard
Dimensions: length 61.7 m (202.4 ft); beam 7.6 m (24.9 ft); draught 2.5 m (8.2 ft)
Gun armament: one 76-mm (3-in) OTO Melara Compact L/62 DP in a single mounting (not *Alia* and *Geoula*), two 30-mm AA in a TCM-30 twin mounting, two 20-mm AA in single mountings, and four 0.5-in (12.7-mm) machine-guns
Missile armament: four container-launchers for four RGM-84A Harpoon anti-ship missiles, four (*Alia* and *Geoula*) or eight (others) container-launchers for four or eight Gabriel Mk II anti-ship missiles, and one launcher for Barak SAMs
Torpedo armament: none
Anti-submarine armament: helicopter-launched weapons (see below)
Aircraft: one Bell Model 205 helicopter in a hangar amidships (*Alia* and *Geoula* only)
Electronics: one Neptune surface-search radar, one RTN 10X radar used in conjunction with the Argo NA10 fire-control system, one ELAC hull-mounted sonar, chaff launchers, and one MN-53 ESM system
Propulsion: four MTU diesels delivering 10440 kW (14,000 shp) to four shafts
Performance: speed 31 kts
Complement: 53

Class
1. Israel

Name	*No.*	*Builder*	*Commissioned*
Alia		Haifa Shipyard	1980
Geoula		Haifa Shipyard	1981
Romach		Haifa Shipyard	1981
Keshet		Haifa Shipyard	1982

'Shanghai II, III and IV' class FAC(G)s
(China)
Type: fast attack craft (gun)
Displacement: 120 tons standard and 155 tons full load
Dimensions: length 39.0 m (128.0 ft); beam 5.5 m (18.0 ft); draught 1.7 m (5.5 ft)
Gun armament: four 37-mm L/63 AA in two twin mountings and four 25-mm AA in two twin mountings ('Shanghai II' class) or two 57-mm L/70 AA in a twin mounting and one 25-mm AA in a single mounting ('Shanghai III and IV' classes); some craft have two 75-mm (2.95-in) Type 56 recoilless rifles in a twin mounting
Missile armament: none
Torpedo armament: none
Anti-submarine armament: two throwers for eight depth charges
Mines: up to 10
Electronics: one 'Skin Head' or 'Pot Head' surface-search radar, one hull-mounted sonar, and (some craft) one variable-depth sonar
Propulsion: two M50F and two Type 3D-12 diesels delivering 2385 kW (3,200 shp) to four shafts
Performance: speed 30 kts; range 1500 km (930 miles) at 17 kts
Complement: 8+30

Class
1. Albania

6 'Shanghai II' class craft

2. Bangladesh

8 'Shanghai II' class craft

3. Cameroun

2 'Shanghai II' class craft

4. China

310 'Shanghai II, III and IV' class craft

5. Congo

3 'Shanghai II' class craft

6. Egypt

4 'Shanghai II' class craft

7. Guinea

6 'Shanghai II' class craft

8. North Korea

8 'Shanghai II' class craft

9. Pakistan

12 'Shanghai II' class craft

10. Romania

20 'Shanghai II' class craft

11. Sierra Leone

3 'Shanghai II' class craft

12. Sri Lanka

7 'Shanghai II' class craft

13. Tanzania

7 'Shanghai II' class craft

14. Tunisia

2 'Shanghai II' class craft

15. Vietnam

8+ 'Shanghai II' class craft

16. Zaire

4 'Shanghai II' class craft

'Shershen' and 'Mol' class FAC(T)s
(USSR)
Type: fast attack craft (torpedo)
Displacement: ('Shershen' class) 145 tons standard and 175 tons full load, or ('Mol' class) 160 tons standard and 200 tons full load
Dimensions: ('Shershen' class) length 34.7 m (113.8 ft); beam 6.7 m (222.0 ft); draught 1.5 m (4.9 ft)
Dimensions: ('Mol' class) length 39.0 m (127.9 ft); beam 8.1 m (26.6 ft); draught 1.8 m (5.9 ft)
Gun armament: four 30-mm AA in two twin mountings
Missile armament: none
Torpedo armament: four mountings for four 533-mm (21-in) torpedoes
Anti-submarine armament: two racks for 12 depth charges
Electronics: one 'Pot Drum' surface-search radar, one 'Drum Tilt' gun-control radar, one 'Square Head' IFF, and one 'High Pole-A' IFF
Propulsion: ('Shershen' class) three M503A diesels delivering 8940 kW (11,990 shp) to three shafts, or ('Mol' class) three M504 diesels delivering 11190 kW (15,010 shp) to three shafts
Performance: speed ('Shershen' class) 47 kts or ('Mol' class) 36 kts; range 1500 km (930 miles) at 30 kts
Complement: ('Shershen' class) 23 or ('Mol' class) 25

Class
1. Angola

4 'Shershen' class craft

2. Bulgaria

6 'Shershen' class craft

3. Cape Verde

2 'Shershen' class craft with torpedo armament

4. Congo

1 'Shershen' class craft without torpedo armament

5. East Germany

18 'Shershen' class craft

6. Egypt

6 'Shershen' class craft

7. Ethiopia

2 'Mol' class craft

8. Guinea

2 'Shershen' class craft

9. Guinea-Bissau

2 'Shershen' class craft

10. North Korea

4 'Shershen' class craft

11. Somalia

4 'Mol' class craft

12. Sri Lanka

1 'Mol' class craft

13. USSR

30 'Shershen' class craft

14. Vietnam

16 'Shershen' class craft

15. Yugoslavia

15 'Shershen' class craft

'Snögg' class FAC(M/T)
(Norway)
Type: fast attack craft (missile and torpedo)
Displacement: 100 tons standard and 125 tons full load
Dimensions: length 36.5 m (119.8 ft); beam 6.2 m (20.5 ft); draught 1.3 m (5.0 ft)
Gun armament: one 40-mm Bofors L/70 AA in a single mounting
Missile armament: four container-launchers for four Penguin anti-ship missiles
Torpedo armament: four mountings for four 533-mm (21-in) Tp 61 torpedoes
Anti-submarine armament: none
Electronics: one Philips surface-search radar, one TM626 navigation radar, one TORCI torpedo fire-control systems, and (being fitted) on MSI-805 fore-control system with one TVT-300 optronic tracker and a laser rangefinder
Propulsion: two MTU 16V538 TB92 diesels delivering 5370 kW (7,200 shp) to two shafts
Performance: speed 32 kts; range 1000 km (620 miles) at 32 kts
Complement:

Class
1. Norway

Name	No.	Builder	Commissioned
Snögg	P980	Batservice	1970
Rapp	P981	Batservice	1970
Snar	P982	Batservice	1970
Rask	P983	Batservice	1971
Kvikk	P984	Batservice	1971
Kjapp	P985	Batservice	1971

'Sohang' class FAC(M) (North Korea): see 'Komar' class FAC(M)

'Soju' class FAC(M) (North Korea): see 'Osa' class FAC

'Sparviero' class FAH(M)
(Italy)
Type: fast attack hydrofoil (missile)
Displacement: 62.5 tons full load
Dimensions: length 24.6 m (80.7 ft) foilborne and 23.0 m (75.4 ft) hullborne; beam 12.1 m (39.7 ft) foilborne and 7.0 (22.9 ft) hullborne; draught 4.4 m (14.4 ft) hullborne
Gun armament: one 76-mm (3-in) OTO Melara Compact L/62 DP in a single mounting
Missile armament: two container-launchers for two Otomat Mk 1 (P420) or Otomat Mk 2 (others) anti-ship missiles
Torpedo armament: none
Anti-submarine armament: none
Electronics: one 3RM 7 surface-search and navigation radar, and one RTN 10X radar used in conjunction with the Argo NA10 fire-control system
Propulsion: CODOG (COmbined Diesel Or Gas turbine) arrangement, with one diesel delivering 120 kW (160 shp) to one retractable propeller for hullborne operation, and one Rolls-Royce Proteus gas turbine delivering 4,500 shp (3355 kW) to one waterjet for foilborne operation
Performance: speed 50 kts foilborne or 8 kts hullborne; range 2225 km (1,385 miles) at 8 kts or 740 km (460 miles) at 45 kts
Complement: 2+8

Class
1. Italy

Name	No.	Builder	Commissioned
Sparviero	P420	Alinavi	Jul 1974
Nibbio	P421	CNR, Muggiano	Mar 1982
Falcone	P422	CNR, Muggiano	Mar 1982
Astore	P423	CNR, Muggiano	Feb 1983
Grifone	P424	CNR, Muggiano	Feb 1983
Gheppio	P425	CNR, Muggiano	Jan 1983
Condor	P426	CNR, Muggiano	Jan 1984

'Spica-M' class FAC(M) (Malaysia): see 'Spica II' class FAC(M/T)

'Spica I' class FAC(M/T)
(Sweden)
Type: fast attack craft (missile and torpedo)
Displacement: 185 tons standard and 215 tons full load
Dimensions: length 42.7 m (140.1 ft); beam 7.1 m (23.3 ft); draught 2.6 m (8.5 ft)
Gun armament: one 57-mm Bofors L/70 DP in a single mounting
Missile armament: four or eight container-launchers for four or eight RBS 15F anti-ship missiles if two or four torpedo-tube mountings are landed
Torpedo armament: six mountings for six 533-mm (21-in) Tp 61 torpedoes
Anti-submarine armament: none
Electronics: one Skanter Mk 009 surface-search radar, and one WM-22 fire-control system
Propulsion: three Rolls-Royce Proteus gas turbines delivering 12,720 shp (9485 kW) to three shafts
Performance: speed 40 kts
Complement: 6+34

Class
1. Sweden

Name	No.	Builder	Commissioned
Spica	T121	Götaverken	1966
Sirius	T122	Go4taverken	1966

Capella	T123	Götaverken	1966
Castor	T124	Karlskrona Varvet	1967
Vega	T125	Karlskrona Varvet	1967
Virgo	T126	Karlskrona Varvet	1967

'Spica II' class FAC(M/T)
(Sweden)
Type: fast attack craft (missile and torpedo)
Displacement: 190 tons standard and 215 tons full load
Dimensions: length 43.6 m (143.0 ft); beam 7.1 m (23.3 ft); draught 2.4 m (7.4 ft)
Gun armament: one 57-mm Bofors L/70 DP in a single mounting
Missile armament: four container-launchers for four RBS 15M anti-ship missiles
Torpedo armament: six mountings for six 533-mm (21-in) Tp 61 torpedoes
Anti-submarine armament: none
Electronics: one Sea Giraffe combined air- and surface-search radar, one 9 LV 200 fire-control system, and one action information system
Propulsion: three Rolls-Royce Proteus gas turbines delivering 12,900 shp (9620 kW) to three shafts
Performance: speed 40.5 kts
Complement: 27

Class
1. Malaysia (**'Spica-M' class**)

Name	*No.*	*Builder*	*Commissioned*
Handalan	P3511	Karlskrona Varvet	Aug 1979
Perkasa	P5312	Karlskrona Varvet	Aug 1979
Pendekar	P5313	Karlskrona Varvet	Aug 1979
Gempita	P5314	Karlskrona Varvet	Aug 1979

(These Malaysian craft are essentially diesel-engined versions of the Swedish 'Spica II' class craft, with three MTU 16V538 TB91 diesels delivering 8055 kW/10,805 shp to three shafts for a speed of 34.5 kts and a range of 3425 km/2130 miles at 14 kts; the full-load displacement is 240 tons; the armament comprises one 57-mm Bofors L/70 DP and one 40-mm Bofors L/70 AA guns, each in a single mounting, and four container-launchers for four MM.38 Exocet anti-ship missiles; the electronic suite includes one 6 GR 600 surface-search radar, one Decca navigation radar, one 9 LV 200 fire-control system and an optronic AA fire-control system.)

2. Sweden

Name	*No.*	*Builder*	*Commissioned*
Norrköping	T131	Karlskrona Varvet	May 1973
Nynäshamn	T132	Karlskrona Varvet	Sep 1973
Norrtälje	T133	Karlskrona Varvet	Feb 1974
Varberg	T134	Karlskrona Varvet	Jun 1974
Västeras	T135	Karlskrona Varvet	Oct 1974
Västervik	T136	Karlskrona Varvet	Jan 1975
Umea	T137	Karlskrona Varvet	May 1975
Pitea	T138	Karlskrona Varvet	Sep 1975
Lulea	T139	Karlskrona Varvet	Nov 1975
Halmstad	T140	Karlskrona Varvet	Apr 1976
Stromstad	T141	Karlskrona Varvet	Sep 1976
Ystad	T142	Karlskrona Varvet	Jan 1976

'Spica III' or 'Stockholm' class FAC(M)
(Sweden)
Type: fast attack craft (missile)
Displacement: 315 tons full load
Dimensions: length 58.0 m (190.3 ft); beam 7.5 m (24.6 ft); draught 2.0 m (6.6 ft)
Gun armament: one 57-mm Bofors L/70 DP in a single mounting, and one 40-mm Bofors L/70 AA in a single mounting
Missile armament: six container-launchers for six RBS 15M anti-ship missiles
Torpedo armament: two mountings for two 533-mm (21-in) Tp 61 torpedoes
Anti-submarine armament: two mountings for two 400-mm (15.75-in) Tp 42 torpedoes, and depth charges
Electronics: one Sea Giraffe air- and surface-search radar, one 9 LV 200 fire-control system, one TSM 2642 variable-depth sonar, chaff launchers, and one EWS 905 ESM system
Propulsion: CODAG (COmbined Diesel And Gas turbine) arrangement, with two MTU 16V396 TB93 diesels delivering 3130 kW (4,200 shp) and one Allison 570KF gas turbine delivering 6,000 shp (4475 kW) to three shafts
Performance: speed 32 kts on all engines or 20 kts on diesels; range 1850 km (1,150 miles) at 20 kts
Complement: 30

Class
1. Sweden

Name	No.	Builder	Commissioned
Karlskrona	K11	Karlskrona Varvet	Mar 1985
Malmö	K12	Karlskrona Varvet	Oct 1985

'Stenka' class FAC(G/T)
(USSR)
Type: fast attack craft (gun/torpedo)
Displacement: 170 tons standard and 210 tons full load
Dimensions: length 39.0 m (127.9 ft); beam 7.8 m (25.6 ft); draught 1.8 m (5.9 ft)
Gun armament: four 30-mm AA in two twin mountings
Missile armament: none
Torpedo armament: none
Anti-submarine armament: four mountings for four 406-mm (16-in) torpedoes, and two depth-charge racks
Electronics: one 'Pot Drum' surface-search radar, one 'Muff Cob' gun-control radar, one variable-depth sonar, two 'Square Head' IFF, and one 'High Pole' IFF
Propulsion: three M503A diesels delivering 8940 kW (11,990 shp) to three shafts
Performance: speed 36 kts
Complement: 30

Class
1. USSR

90 craft

'Stockholm' class FAC(M): see 'Spica III' class FAC(M)

'Storm' class FAC(M)
(Norway)
Type: fast attack craft (missile)
Displacement: 100 tons standard and 125 tons full load
Dimensions: length 36.5 m (119.8 ft); beam 6.2 m (20.5 ft); draught 1.5 m (4.9 ft)
Gun armament: one 76-mm (3-in) Bofors L/50 DP in a single mounting, and one 40-mm Bofors L/60 AA in a single mounting
Missile armament: four or six container-launchers for four or six Penguin Mk II anti-ship missiles
Torpedo armament: none
Anti-submarine armament: none
Electronics: one TM1226 surface-search and navigation radar, one WM-26 fire-control radar, and (being fitted) one MSI-805 fire-control system with one TVT-300 optronic tracker and a laser rangefinder
Propulsion: two MTU MB872A diesels delivering 5370 kW (7,200 shp) to two shafts
Performance: speed 32 kts
Complement: 4+21

Class
1. Norway

Name	No.	Builder	Commissioned
Storm	P960	Bergens Mek	1968
Blink	P961	Bergens Mek	1985
Glimt	P962	Bergens Mek	1966
Skjold	P963	Westermoen	1966
Trygg	P964	Bergens Mek	1966
Kjekk	P965	Bergens Mek	1966
Djerv	P966	Westermoen	1966
Skudd	P967	Bergens Mek	1966
Arg	P968	Bergens Mek	1966
Steil	P969	Westermoen	1967
Brann	P970	Bergens Mek	1967
Tross	P971	Bergens Mek	1967
Hvass	P972	Westermoen	1967
Traust	P973	Bergens Mek	1967
Brott	P974	Bergens Mek	1967
Odd	P975	Westermoen	1967
Brask	P977	Bergens Mek	1967
Rokk	P978	Westermoen	1968
Gnist	P979	Bergens Mek	1968

'Tuima' class FAC(M) (Finland): see 'Osa' class FAC

'Turya' class FAH(T)
(USSR)
Type: fast attack hydrofoil (torpedo)
Displacement: 190 tons standard and 250 tons full load
Dimensions: length 39.3 m (128.7 ft); beam 7.8 m (25.6 ft); draught 1.8 m (5.9 ft)
Gun armament: two 57-mm L/70 AA in a twin mounting, and two 25-mm AA in a twin mounting
Missile armament: none
Torpedo armament: four mountings for four 533-mm (21-in) dual-role torpedoes
Anti-submarine armament: none
Electronics: one 'Pot Drum' surface-search radar, one 'Muff Cob' gun-control radar, one variable-depth sonar, one 'Square Head' IFF, and one 'High Pole' IFF
Propulsion: three M504 diesels delivering 11190 kW (15,010 shp) to three shafts
Performance: speed 45 kts
Complement: 30

Class	
1. Cuba	6 craft
2. USSR	30 craft

'Type 143' class FAC(M) (West Germany): see 'Lürssen FPB-57' class FAC(M)

'Type 148' class FAC(M) (West Germany): see 'La Combattante II' class FAC(M)

'Tzu Chiang' class FAC(M) (Taiwan): see 'Dvora' class FAC(M)

'Willemoes' class FAC(M)
(Denmark)
Type: fast attack craft (missile)
Displacement: 260 tons full load
Dimensions: length 46.0 m (150.9 ft); beam 7.4 m (24.0 ft); draught 2.5 m (8.2 ft)
Gun armament: one 76-mm OTO Melara Compact L/62 DP in a single mounting
Missile armament: two Mk 141 quadruple container-launchers for eight RGM-84A Harpoon anti-ship missiles
Torpedo armament: two mountings for two 533-mm (21-in) torpedoes
Anti-submarine armament: none
Electronics: one Skanter Mk 009 surface-search radar, one NWS-3 navigation radar, and one 9 LV 200 fire-control system
Propulsion: CODOG (COmbined Diesel Or Gas turbine) arrangement, with two General Motors 6V-71 diesels delivering 1,600 shp (1195 kW) and three Rolls-Royce Proteus gas turbines delivering 12,750 shp (9510 kW) to three shafts
Performance: speed 38 kts on gas turbines or 12 kts on diesels
Complement: 6+19

Class
1. Denmark

Name	No.	Builder	Commissioned
Bille	P540	Frederikshavn Vaerft	Oct 1976
Bredal	P541	Frederikshavn Vaerft	Jan 1977
Hammer	P542	Frederikshavn Vaerft	Apr 1977
Huitfeld	P543	Frederikshavn Vaerft	Jun 1977
Krieger	P544	Frederikshavn Vaerft	Sep 1977
Norby	P545	Frederikshavn Vaerft	Nov 1977
Rodsteen	P546	Frederikshavn Vaerft	Feb 1978
Sehested	P547	Frederikshavn Vaerft	May 1978
Suenson	P548	Frederikshavn Vaerft	Aug 1978
Willemoes	P549	Frederikshavn Vaerft	Jun 1976

AMPHIBIOUS WARFARE VESSELS

'Alligator' class LST
(USSR)
Type: landing ship tank
Displacement: 3,400 tons standard and 4,500 tons full load
Dimensions: length 114.0 m (374.0 ft); beam 15.5 m (50.9 ft); draught 4.5 m (14.8 ft)
Gun armament: two 57-mm L/80 AA in a twin mounting, two rocket-launchers, and ('Alligator Type 4' class) four 25-mm AA in two twin mountings
Missile armament: two quadruple launchers for ? SA-N-5 'Grail' SAMs
Torpedo armament: none
Anti-submarine armament: none
Aircraft: none
Capacity: 1,700 tons handled with the aid of two 5-ton cranes ('Alligator Type 1' class) or one 15-ton crane ('Alligator Types 2, 3 and 4' class); ramps are built into the bow and stern for roll-on/roll-off operations with up to 50 tanks; the Naval Infantry complement is about 120, though 250 can be accommodated for short voyages
Electronics: one 'Don-2' or 'Spin Trough' navigation radar, one 'Muff Cob' gun-control radar, and one 'High Pole-B' IFF
Propulsion: two diesels delivering 6710 kW (9,000 shp) to two shafts
Performance: speed 18 kts; range 20000 km (12,425 miles) at 15 kts
Complement: 100

Class
1. USSR

4 'Alligator Type 1' class units
2 'Alligator Type 2' class units
6 'Alligator Type 3' class units
2 'Alligator Type 4' class units

Note
This class was built at Kaliningrad between 1964 and 1977, and provides the Soviet navy with a useful roll-on/roll-off medium-lift capability; the first two subclasses are optimized for transport and the latter two for over-the-beach assault, each ship being able to carry the equipment of a Naval Infantry battalion landing team

'Anchorage' class LSD
(USA)
Type: landing ship dock
Displacement: 8,600 tons light and 13,600 tons full load
Dimensions: length 553.3 ft (168.8 m); beam 84.0 ft (25.6 m); draught 20.0 ft (6.1 m)
Gun armament: six 3-in (76-mm) L/50 DP in three Mk 33 twin mountings, and two 20-mm Phalanx Mk 16 CIWS mountings
Missile armament: none
Torpedo armament: none
Anti-submarine armament: none
Aircraft: provision for one helicopter on a platform aft
Capacity: the docking well in the stern measures 430 ft (131.1 ft) in length and 50 ft (15.2 ft) in width, and can accommodate three LCUs, or 29 LCM(6)s or many LVTs; there is deck space for one LCM, and davits can take one LCPL and one LCVP; two 50-ton cranes are provided for the handling of freight; the troop accommodation is 28+348
Electronics: one SPS-40 air-search radar, one SPS-10 surface-search radar, OE-82 satellite communications system, SRR-1 satellite communications receiver, WSC-3 satellite communications transceiver, and one Mk 36 Super RBOC chaff launcher
Propulsion: two boilers supplying steam to two sets of De Laval geared turbines delivering 24,000 shp (17895 kW) to two shafts
Performance: speed 22 kts
Complement: 21+376

Class
1. USA

Name	No.	Builder	Laid down	Commissioned
Anchorage	LSD36	Ingalls SB	Mar 1967	Mar 1969
Portland	LSD37	General Dynamics	Sep 1967	Oct 1970
Pensacola	LSD38	General Dynamics	Mar 1969	Mar 1971
Mount Vernon	LSD39	General Dynamics	Jan 1970	May 1972
Fort Fisher	LSD40	General Dynamics	Jul 1970	Dec 1972

Note
This is virtually a repeat of the 'Thomaston' class with a tripod rather than pole mast as the major distinguishing feature; the removable helicopter platform covers virtually the whole docking well, which is larger than that of the 'Thomaston' class

'Austin' class LPD
(USA)
Type: amphibious transport dock
Displacement: 10,000 tons light and 15,900 tons (LPD4/6) or 16,550 tons (LPD7/10) or 16,900 tons (AGF11 and LPD12/13) or 17,000 tons (LPD14/15) full load
Dimensions: length 570.0 ft (173.8 m); beam 100.0 ft (30.5 m); draught 23.0 ft (7.0 m)
Gun armament: two 3-in (76-mm) L/50 DP in a Mk 33 twin mounting, and two 20-mm Phalanx Mk 16 CIWS mountings
Missile armament: none
Torpedo armament: none
Anti-submarine armament: none
Aircraft: up to six Boeing Vertol CH-46 Sea Knight helicopters
Capacity: the docking well measures 395 ft (120.4 ft) in length and 50 ft (15.2 m) in width, and can accommodate one LCU and three LCM(6)s, or nine LCM(6)s, or four LCM(8)s, or 28 LVTPs; freight is handled by two forklifts, six cranes and one elevator; the troop accommodation is 930 in LPD4/6 and LPD14/15 or 840 in LPD7/10, AGF11 and LPD12/13
Electronics: one SPS-40C air-search radar, one SPS-10F surface-search radar, URN-20 TACAN, OE-82 satellite communications system, SRR-1 satellite communications receiver, WSC-3 satellite communications transceiver, and one Mk 36 Super RBOC chaff launcher
Propulsion: two boilers supplying steam to two sets of De Laval geared turbines delivering 24,000 shp (17895 kW) to two shafts
Performance: speed 21 kts; range 8,865 miles (14,265 km) at 20 kts
Complement: 27+446 plus flag accommodation for about 90 in LPD7/10, AGF11 and LPD12/13

Class
1. USA

Name	No.	Builder	Laid down	Commissioned
Austin	LPD4	New York NY	Feb 1963	Feb 1965
Ogden	LPD5	New York NY	Feb 1963	Jun 1965
Duluth	LPD6	New York NY	Dec 1963	Dec 1965
Cleveland	LPD7	Ingalls SB	Nov 1964	Apr 1967
Dubuque	LPD8	Ingalls SB	Jan 1965	Sep 1967
Denver	LPD9	Lockheed SB	Feb 1964	Oct 1968
Juneau	LPD10	Lockheed SB	Jan 1965	Jul 1969
Coronado	AGF11	Lockheed SB	May 1965	May 1970
Shreveport	LPD12	Lockheed SB	Dec 1965	Dec 1970
Nashville	LPD13	Lockheed SB	Mar 1966	Feb 1970
Trenton	LPD14	Lockheed SB	Aug 1966	Mar 1971
Ponce	LPD15	Lockheed SB	Oct 1966	Jul 1971

Note
This is an enlarged version of the 'Raleigh' class design, with the same size of docking well but lengthened by 12 m (39.4 ft) just forward of this to provide additional vehicle accommodation; the helicopter platform above the docking well is not removable, and is provided (except in LPD4) with a telescoping hangar to accommodate one of the six Boeing Vertol CH-46 helicopters that can be carried; *Coronado* is used as a flagship

'Batral' class LSM
(France)
Type: landing ship medium
Displacement: 750 tons standard and 1,410 tons full load
Dimensions: length 80.0 ft (262.4 ft); beam 13.0 m (42.6 ft); draught 3.0 m (9.8 ft)
Gun armament: two 40-mm Bofors AA in single mountings, one 81-mm (3.2-in) mortar, and two 12.7-mm (0.5-in) machine-guns
Missile armament: none
Torpedo armament: none
Anti-submarine armament: none
Aircraft: provision for one helicopter on a platform aft
Capacity: up to 380 tons, including troops and 12 vehicles, offloaded with the aid of a 10-ton derrick into one LCVP and one LCPS; troop accommodation is provided for 5+133 (first pair) or 188 (second pair)
Electronics: one Decca navigation radar, and two hull-mounted sonars
Propulsion: two SACM diesels delivering 2680 kW (3,595 shp) to two shafts
Performance: speed 16 kts; range 8350 km (5,190 miles) at 13 kts
Complement: 4+35

Class
1. Chile

Name	No.	Builder	Laid down	Commissioned
Maipo	91	Asmar	1980	Aug 1982
Rancagua	92	Asmar	1980	1983
Chacabuco	93	Asmar	1982	1984

2. France

Name	No.	Builder	Laid down	Commissioned
Champlain	L9030	Brest ND		Oct 1974
Francis Garnier	L9031	Brest ND		Jun 1974
Dumont d'Urville	L9032	CMN		1983
Jacques Cartier	L9033	CMN		1983
	L9034	AFO		
	L9035	AFO		

3. Gabon

Name	No.	Builder	Laid down	Commissioned
President el Hadji Omar Bongo		AFO		Oct 1984

4. Ivory Coast

Name	No.	Builder	Laid down	Commissioned
L'Elephant		Dubigeon	1975	Feb 1977

5. Morocco

Name	No.	Builder	Laid down	Commissioned
Daoud Ben Aicha		Dubigeon		May 1977
Ahmed Es Sakali		Dubigeon		Sep 1977
Abou Abdallah el Ayachi		Dubigeon		Mar 1978

6. Panama

Name	No.	Builder	Laid down	Commissioned
		Chantiers de la Manche		
		Chantiers de la Manche		

Note
This is a useful light amphibious type that has proved commercially attractive in the export market

'BDC' class LST
(France)
Type: landing ship tank
Displacement: 1,400 tons standard and 4,225 tons full load
Dimensions: length 102.1 m (335.0 ft); beam 15.5 m (50.9 ft); draught 3.2 (10.5 ft)
Gun armament: two 120-mm (4.7-in) mortars in a twin mounting, three 40-mm Bofors AA in single mountings, and one 20-mm AA in a single mounting
Missile armament: none
Torpedo armament: none
Anti-submarine armament: none
Aircraft: two Aérospatiale Alouette III helicopters in a hangar amidships (L9007 and 9009 only)
Capacity: up to 1,800 tons; troop accommodation is provided for a standard complement of 170, or 335 in austere conditions or 807 in emergency conditions, these men being moved ashore with the aid of four LCVPs
Electronics: navigation radar
Propulsion: two SWMT-Pielstick 16PA1 diesels delivering 1490 kW (2,000 shp) to two shafts
Performance: speed 11 kts; range 34250 km (21,285 miles) at 10 kts
Complement: 6+69

Class
1. France

Name	No.	Builder	Laid down	Commissioned
Argens	L9003	Chantiers de Bretagne		Jun 1960
Bidassoa	L9004	Chantiers de la Seine		Oct 1961
Trieux	L9007	Chantiers de Bretagne		Mar 1960
Dives	L9008	Chantiers de la Seine		Apr 1961
Blavet	L9009	Chantiers de Bretagne		Jan 1961

Note
The design of this class is based on that of the US 'LST1' type

'Blue Ridge' class LCC
(USA)
Type: amphibious force command ship
Displacement: 19,290 tons full load
Dimensions: length 620.0 ft (189.0 m); beam 82.0 ft (25.0 m); draught 29.0 ft (8.8 m)
Gun armament: four 3-in (76-mm) L/50 DP in two Mk 33 twin mountings, and two 20-mm Phalanx Mk 16 CIWS mountings
Missile armament: two Mk 25 octuple launchers for RIM-7 Sea Sparrow SAMs
Torpedo armament: none
Anti-submarine armament: none
Aircraft: provision for two utility helicopters on a platform aft
Capacity: three LCPs, two LCVPs and one personnel launch
Electronics: one SPS-48 3D radar, one SPS-40 air-search radar, one SPS-10 surface-search radar, two Mk 115 SAM fire-control systems, two Mk 56 gun fire-control systems, URN-20 TACAN, OE-82 satellite communications system, SRR-1 satellite communications receiver, WSC-3 satellite communications transceiver, Naval Tactical Data System, Link 11 and Link 14 automatic data-transmission systems, one Mk 36 Super RBOC chaff launcher, and one ESM system
Propulsion: two boilers supplying steam to one set of General Electric geared turbines delivering 22,000 shp (16405 kW) to one shaft
Performance: speed 23 kts; range 15,000 miles (24140 km) at 16 kts
Complement: 50+775, plus command staff (see below)

Class				
1. USA				
Name	*No.*	*Builder*	*Laid down*	*Commissioned*
Blue Ridge	LCC19	Philadelphia NY	Feb 1967	Nov 1970
Mount Whitney	LCC20	Newport News	Jan 1969	Jan 1971

Note
Based on the 'Iwo Jima' class design with the major hangar spaces adapted as offices, operations rooms and accommodation for the command staff (a maximum of 200+500), the two units of the 'Blue Ridge' class are the world's only custom-designed command ships, originally schemed as integrated air/land/sea amphibious control vessels but in fact used as fleet command vessels

'Fearless' class LSD
(UK)
Type: amphibious assault ship
Displacement: 11,060 tons standard and 12,120 tons full load
Dimensions: length 520.0 ft (158.5 m); beam 80.0 ft (24.4 m); draught 20.5 ft (6.2 m)
Gun armament: two 40-mm Bofors L/70 AA in single mountings
Missile armament: four quadruple launchers for Sea Cat SAMs
Torpedo armament: none
Anti-submarine armament: none
Aircraft: five Westland Wessex HU.Mk 5 helicopters on a platform aft
Capacity: the docking well can accommodate four LCM(9)s, and the capacity of these craft can be supplemented by four davit-borne LCVPs; a typical load is 15 MBTs, seven 3-ton trucks and 20 Land Rovers; the normal troop complement is 400, but 700 troops can be carried under austere conditions
Electronics: one Type 994 combined air- and surface-search radar, one Type 978 navigation radar, one CAAIS action-information system, one ESM system, and two Corvus chaff launchers
Propulsion: two boilers supplying steam to two sets of English Electric geared turbines delivering 22,000 shp (16405 kW) to two shafts
Performance: speed 21 kts; range 5,750 miles (9250 km) at 20 kts
Complement: 580

Class				
1. UK				
Name	*No.*	*Builder*	*Laid down*	*Commissioned*
Fearless	L10	Harland & Wolff	Jul 1962	Nov 1965
Intrepid	L11	John Brown	Dec 1962	Mar 1967

Note
Designed for the assault transport of an amphibious group, each of these ships has command accommodation for an integrated brigade staff for the control of air/land/sea operations

'Frosch I' and 'Frosch II' class LSTs
(East Germany)
Type: landing ship tank
Displacement: 1,950 tons ('Frosch I' class) or 2,000 tons ('Frosch II' class) standard and 4,000 tons full load
Dimensions: length 91.0 m (298.4 ft); beam 11.0 m (36.1 ft); draught 2.8 m (9.2 ft)
Gun armament: four 57-mm L/70 AA in two twin mountings, four 30-mm ('Frosch I' class) or 25-mm ('Frosch II' class) AA in two twin mountings, and ('Frosch I' class only) two 40-tube BM-21 rocket-launchers
Missile armament: none
Torpedo armament: none
Anti-submarine armament: none
Aircraft: none
Capacity: up to 800 tons including 12 tanks or 16 lighter AFVs; the troop accommodation is normally 60, though 260 can be carried for short periods
Electronics: one 'Strut Curve' air-search radar, one 'Square Head' surface-search radar, one TSR 333 navigation radar, two 'Muff Cob' 57-mm gun-control radars, two 'Drum Tilt' 30-mm gun-control radars, one 'Square Head' IFF, one 'High Pole' IFF, and several ECM systems
Propulsion: two diesels delivering ? kW (? shp) to two shafts
Performance: speed 16 kts
Complement: 40

Class
1. East Germany
12 'Frosch I' class ships
2 'Frosch II' class ships

Note
Though based on the Soviet 'Ropucha' class, the 'Frosch I' type has considerably heavier armament and was built

between 1975 and 1979 at Wolgast by the Peenewerft yard for assault landings covered by the rocket-launchers and guns; the type can also be used as a stern minelayer; built in 1980, the two 'Frosch II' class units are thought to be assault cargo transports

'Ivan Rogov' class LPD
(USSR)
Type: amphibious transport dock
Displacement: 14,000 tons full load
Dimensions: length 159.0 m (521.6 ft); beam 24.5 m (80.2 ft); draught 6.5 m (21.2 ft)
Gun armament: two 76-mm (3-in) L/60 DP in a twin mounting, four 30-mm AK-630 CIWS mountings, and one 40-tube BM-21 rocket-launcher
Missile armament: one twin launcher for ? SA-N-4 'Gecko' SAMs
Torpedo armament: none
Anti-submarine armament: none
Aircraft: up to five Kamov Ka-25 'Hormone-C' or Ka-27 'Helix-C' helicopters on a platform forward and in a hangar aft
Capacity: the docking well measures some 79.0 m (259.2 ft) in length and 13.0 m (42.6 ft) in width, and can accommodate two 'Lebed' class hovercraft and one 'Ondatra' class LCM, or three 'Gus' class hovercraft; the normal troop accommodation is for a Naval Infantry battalion of 522 with 40 tanks, though the tank complement is halved when maximum landing craft/hovercraft are carried
Electronics: one 'Head Net-C' 3D radar, one 'Owl Screech' main armament gun-control radar, two 'Bass Tilt' CIWS-control radars, one 'Pop Group' SAM-control radar, two 'Don Kay' navigation radars, one 'High Pole' IFF, two 'Bell Shroud' ESM systems, two 'Bell Squat' ESM systems, and two chaff launchers
Propulsion: two gas turbines delivering 33550 kW (45,000 shp) to two shafts
Performance: speed 26 kts; range 18500 km (11,500 miles) at 12 kts
Complement: 400

Class
1. USSR

Name	Builder	Laid down	Commissioned
Ivan Rogov	Kaliningrad		1978
Aleksandr Nikolayev	Kaliningrad	1979	

Note
These are the largest amphibious warfare vessels yet built by the Soviets, and are each designed to carry a reinforced Naval Infantry battalion landing team with all its equipment plus 10 PT-76 light amphibious tanks, or the tank battalion of a Naval Infantry regiment; it is thought that four of the class will be produced, one for each Soviet fleet

'Iwo Jima' class LPH
(USA)
Type: amphibious assault ship
Displacement: 17,000 tons light and 18,000 tons (LPH2, 3 and 7) or 18,300 tons (LPH9/10) or 17,705 tons (LPH11) or 17,515 tons (LPH12) full load
Dimensions: length 592.0 ft (180.5 m); beam 84.0 ft (25.6 m); draught 26.0 ft (7.9 m)
Gun armament: four 3-in (76-mm) L/50 DP in two Mk 33 twin mountings, and two 20-mm Phalanx Mk 16 CIWS mountings
Missile armament: two Mk 25 octuple launchers for RIM-7 Sea Sparrow SAMs (deleted as the Phalanx systems are fitted)
Torpedo armament: none
Anti-submarine armament: none
Aircraft: 19 Boeing Vertol CH-46 Sea Knight or 11 Sikorsky CH-53 Sea Stallion helicopters (or a mixture of the two), or a mix of 24 CH-46, CH-53 Bell AH-1 and Bell UH-1 helicopters, and (with a reduced helicopter complement) four BAe/McDonnell Douglas AV-8 Harrier aircraft
Capacity: accommodation is provided for a US Marine Corps battalion landing team of 144+1,602 plus its equipment, artillery and vehicles; the accommodation amounts to 4,300 sq ft (400 m²) for vehicles and 37,400 sq ft (3475 m²) for palletized stores plus bulk storage for 6,500 US gal (24605 litres) of vehicle fuel and 405,000 US gal (1533090 litres) of turbine fuel
Electronics: one SPS-40 air-search radar, one SPS-10 surface-search radar, one SPN-120 navigation radar, one SPN-10 or SPN-43 aircraft recovery radar, two Mk 63 gun fire-control systems, two Mk 115 SAM fire-control systems, URN-20 TACAN, OE-82 satellite communications system, SRR-1 satellite communications receiver, WSC-3 satellite communications transceiver, various ECM and ESM systems, and one Mk 36 Super RBOC chaff launcher
Propulsion: two boilers supplying steam to one set of geared turbines delivering 22,000 shp (16405 kW) to one shaft
Performance: speed 23 kts; range 11,500 miles (18505 km) at 20 kts
Complement: 47+635

Class				
1. USA				
Name	No.	Builder	Laid down	Commissioned
Iwo Jima	LPH2	Puget Sound NY	Apr 1959	Aug 1961
Okinawa	LPH3	Puget Sound NY	Apr 1960	Apr 1962
Guadalcanal	LPH7	Philadelphia NY	Sep 1961	Jul 1963
Guam	LPH9	Philadelphia NY	Nov 1962	Jan 1965
Tripoli	LPH10	Ingalls SB	Jun 1964	Aug 1966
New Orleans	LPH11	Philadelphia NY	Mar 1966	Nov 1968
Inchon	LPH12	Ingalls SB	Apr 1968	Jun 1970

Note
Based on the modified design of the World War II type of escort carrier, these ships are each intended for the transport and helicopter-assault landing of a reinforced battalion landing team of the US Marine Corps; the types can now operate AV-8 Harrier STOVL close-support aircraft

'LSM 1' class LSM
(USA)
Type: landing ship medium
Displacement: 1,095 tons full load
Dimensions: length 203.5 ft (62.0 m); beam 34.6 ft (10.5 m); draught 8.5 ft (2.6 m)
Gun armament: (typical) two 40-mm Bofors AA in a twin mounting, and a varying number of 20-mm AA in single mounting
Missile armament: none
Torpedo armament: none
Anti-submarine armament: none
Aircraft: none
Capacity: up to 900 tons, though 740 tons is the maximum for beaching at a draught of 3.4 ft (1.0 m)
Electronics: one surface-search and navigation radar
Propulsion: two Fairbanks-Morse or General Motors diesels delivering 2,800 shp (2090 kW) to two shafts
Performance: speed 12 kts
Complement: about 60

Class	
1. China	
14 ships	
2. Dominican Republic	
1 ship	
3. Ecuador	
2 ships	
4. Greece	
5 ships	
5. Israel	
3 ships	
6. Paraguay	
1 ship	
7. Peru	
2 ships	
8. Philippines	
4 ships	
9. South Korea	
8 ships	
10. Thailand	
3 ships	
11. Turkey	
5 ships	
12. Venezuela	
1 ship	
13. Vietnam	
? ships	

Note
These elderly vessels are of limited combat value except in low-intensity theatres, but have considerable value for local transport and the like

'LST 1-1152' class LST
(USA)
Type: landing ship tank
Displacement: 1,655 tons standard and 3,640 tons full load
Dimensions: length 328.0 ft (100.0 m); beam 50.0 ft (15.2 m); draught 14.0 ft (4.3 m)
Gun armament: (typical) up to 10 40-mm Bofors AA in two twin and six single mountings, or two 3-in (76-mm) L/50 DP in a Mk 33 twin mounting and six 40-mm Bofors AA
Missile armament: none
Torpedo armament: none
Anti-submarine armament: none
Aircraft: none
Capacity: up to 1,875 tons, though 446 tons is the maximum for beaching at the designed draught
Electronics: one surface-search and navigation radar
Propulsion: two General Motors diesels delivering 1,700 shp (1270 kW) to two shafts

Performance: speed 11.6 kts; range 6,900 miles (11105 km) at 9 kts
Complement: 100/125

Class	
1. Brazil	
1 ship	
2. Chile	
2 ships	
3. China	
15+ ships	
4. Ecuador	
1 ship	
5. Greece	
5 ships	
6. Indonesia	
9 ships	
7. Malaysia	
2 ships	
8. Mexico	
2 ships	
9. Peru	
3 ships	
10. Philippines	
24 ships	
11. Singapore	
6 ships	
12. South Korea	
8 ships	
13. Taiwan	
21 ships	
14. Thailand	
4 ships	
15. Turkey	
2 ships	
16. Vietnam	
3 ships	

Note
These elderly vessels are of limited combat value except in low-intensity theatres, but are much used for local transport and communications

'Newport' class LST

(USA)
Type: landing ship tank
Displacement: 8,450 tons full load
Dimensions: length 522.3 ft (159.2 m); beam 69.5 ft (21.2 m); draught 17.5 ft (5.3 m)
Gun armament: four 3-in (76-mm) L/50 DP in two Mk 33 twin mountings, and two 20-mm Phalanx Mk 16 CIWS mountings
Missile armament: none
Torpedo armament: none
Anti-submarine armament: none
Aircraft: provision for one helicopter on a platform aft
Capacity: 500 tons of vehicles on a floor area of 19,000 sq ft (1765 m), the vehicles exiting over a 75-ton capacity 112.0-ft (34.1-m) derrick-supported bow ramp or, if amphibious, through a stern gate; the parking area is sufficient for 25 LVTPs and 17 2.5-ton trucks, or for 21 MBTs and 17 2.5-ton trucks; each ship can also carry four sections of pontoon causeway; the troop accommodation is 20+400
Electronics: one SPS-10 surface-search radar, one LN-66 navigation radar, OE-82 satellite communications system, WSC-3 satellite communications transceiver, and one Mk 36 Super RBOC chaff launcher
Propulsion: six General Motors or Alco diesels delivering 16,000 shp (11930 kW) to two shafts
Performance: speed 20 kts; range 2875 miles (4625 km) at cruising speed
Complement: 14+211

Class
1. USA

Name	*No.*	*Builder*	*Laid down*	*Commissioned*
Newport	LST1179	Philadelphia NY	Nov 1966	Jun 1969
Manitowoc	LST1180	Philadelphia NY	Feb 1967	Jan 1970
Sumter	LST1181	Philadelphia NY	Nov 1967	Jun 1970
Fresno	LST1182	National Steel & SB	Dec 1967	Nov 1969
Peoria	LST1183	National Steel & SB	Feb 1968	Feb 1970
Frederick	LST1184	National Steel & SB	Apr 1968	Apr 1970
Schenectady	LST1185	National Steel & SB	Aug 1968	Jun 1970
Cayuga	LST1186	National Steel & SB	Sep 1968	Aug 1970
Tuscaloosa	LST1187	National Steel & SB	Nov 1968	Oct 1970
Saginaw	LST1188	National Steel & SB	May 1969	Jan 1971
San Bernardino	LST1189	National Steel & SB	Jul 1969	Mar 1971

Boulder	LST1190	National Steel & SB	Sep 1969	Jun 1971
Racine	LST1191	National Steel & SB	Dec 1969	Jul 1971
Spartanburg County	LST1192	National Steel & SB	Feb 1970	Sep 1971
Fairfax County	LST1193	National Steel & SB	Mar 1970	Oct 1971
La Moure County	LST1194	National Steel & SB	May 1970	Dec 1971
Barbour County	LST1195	National Steel & SB	Aug 1970	Feb 1972
Harlan County	LST1196	National Steel & SB	Nov 1970	Apr 1972
Barnstable County	LST1197	National Steel & SB	Dec 1979	May 1972
Bristol County	LST1198	National Steel & SB	Feb 1971	Aug 1972

Note
This class was the final expression of US experience with LSTs in World War II, with a pointed bow and dropping ramp to ensure a 20-kt cruising speed; the type has bow and stern gates for through loading and unloading

'Ouragan' class LSD
(France)
Type: landing ship dock
Displacement: 5,800 tons light and 8,500 tons full load
Dimensions: length 149.0 m (488.9 ft); beam 23.0 m (75.4 ft); draught 5.4 m (17.7 ft)
Gun armament: two 120-mm (4.7-in) mortars, and four 40-mm Bofors AA in single mountings
Missile armament: none
Torpedo armament: none
Anti-submarine armament: none
Aircraft: the main helicopter platform can accommodate three Aérospatiale Super Frelon or 10 Aérospatiale Alouette III helicopters, while the removable platform aft can accommodate one Super Frelon or three Alouette III helicopters
Capacity: the docking well measures 120.0 m (393.7 ft) in length and 14.0 m (45.9 ft) in width, and can accommodate two EDIC landing craft each carrying 11 tanks, or 18 LCM(6)s loaded for the direct-assault role, in which case troop accommodation is 343; in the logistics role as freight load of 1,500 tons can be carried, typical loads being one 400-ton craft, or 12 50-ton barges, or 120 AMX-13 light tanks, or 84 amphibious tractors, or 340 jeeps, or 18 Super Frelons, or 80 Alouette IIIs; there are two 35-ton cranes for the handling of freight
Electronics: one DRBN 32 navigation radar, and (*Ouragan* only) one SQS-17 hull-mounted sonar
Propulsion: two SEMT-Pielstick diesels delivering 6410 kW (8,595 shp) to two shafts
Performance: speed 17 kts; range 16675 km (10,360 miles) at 15 kts
Complement: 238

Class
1. France

Name	No.	Builder	Laid down	Commissioned
Ouragan	L9021	Brest ND	Jun 1962	Jun 1965
Orage	L9022	Brest ND	Jun 1966	Apr 1968

Note
Designed for logistic transport as well as amphibious assault, the two 'Ouragans' are now elderly as due for replacement

'Polnochny' class LSM
(USSR)
Type: landing ship medium
Displacement: ('Polnochny Type A' class) 780 tons standard and 800 tons full load; ('Polnochny Type B' class) 790 tons standard and 850 tons full load; ('Polnochny Type C' class) 700 tons standard and 1,150 tons full load
Dimensions: ('Polnochny Type A' class) length 73.0 m (239.5 ft); beam 8.5 m (27.9 ft); draught 1.8 m (5.8 ft); ('Polnochny Type B' class) length 76.0 m (249.3 ft); beam 8.5 m (27.9 ft); draught 1.8 m (5.8 ft); ('Polnochny Type C' class) length 82.0 m (269.0 ft); beam 10.0 m (32.8 ft)' draught 1.8 m (5.8 ft)
Gun armament: ('Polnochny Type A' class) two 140-mm (5.5-in) rocket-launchers and two 14.5-mm (0.57-in) machine-guns, or two 30-mm AA, or two 25-mm AA; ('Polnochny Type B' class) two 140-mm (5.5-in) rocket-launchers and two or four 30-mm AA in twin mountings; ('Polnochny Type C' class) two 140-mm (5.5-in) rocket-launchers and four 30-mm AA in two twin mountings
Missile armament: (some ships only) four quadruple launchers for 32 SA-N-5 'Grail' SAMs
Torpedo armament: none
Anti-submarine armament: none
Aircraft: none
Capacity: up to 350 tons including six tanks
Electronics: ('Polnochny Type A' class) one 'Don-2' or 'Spin Trough' navigation radar and (ships with 30-mm guns) one 'Drum Tilt' AA-control radar; ('Polnochny Type B' class) one 'Don-2' or 'Spin Trough' navigation radar and one 'Drum Tilt' AA-control radar; ('Polnochny Type C' class) one 'Don-2' navigation radar and one 'Drum Tilt' AA-control radar
Propulsion: ('Polnochny Type A and B' classes) two

diesels delivering 2980 kW (3,995 shp) to two shafts; ('Polnochny Type C' class) two diesels delivering 3730 kW (5,005 shp) to two shafts
Performance: speed 18 kts ('Polnochny Type A and B' classes) or 20 kts ('Polnochny Type C' class)
Complement: 40

Class
1. Algeria

1 'Polnochny Type A' class ship

2. Angola

3 'Polnochny Type B' class ships

3. Cuba

3 'Polnochny Type B' class ships

4. Egypt

3 'Polnochny Type A' class ships

5. Ethiopia

1 'Polnochny Type C' class ship

6. India

2 'Polnochny Type A' class ships

4+ 'Polnochny Type C' class ships

7. Iraq

4 'Polnochny Type C' class ships

8. Libya

3 'Polnochny Type C' class ships

9. Poland

23 'Polnochny' class ships

10. Somalia

1 'Polnochny Type A' class ship

11. South Yemen

1 'Polnochny Type B' class ship

12. USSR

50 'Polnochny' class ships

13. Vietnam

3 'Polnochny Type B' class ships

Note
These ships were all built at the Polnochny yard at Gdansk in Poland between 1961 and 1973

'Raleigh' class LPD
(USA)
Type: amphibious transport dock
Displacement: 8,040 tons light and 13,600 tons full load
Dimensions: length 521.8 ft (159.1 m); beam 100.0 ft (30.5 m); draught 22.0 ft (6.7 m)
Gun armament: six 3-in (76-mm) L/50 DP in three Mk 33 twin mountings, and two 20-mm Phalanx Mk 16 CIWS mountings
Missile armament: none
Torpedo armament: none
Anti-submarine armament: none
Aircraft: six Boeing Vertol CH-46 Sea Knight helicopters on a platform aft
Capacity: the docking well measures 168.0 ft (51.2 m) in length and 50.0 ft (15.2 m) in width, and can accommodate one LCU and three LCM(6)s, or four LCM(8)s or 20 LVTs; further ship-to-shore capability is provided by two LCM(6)s or four LCPLs launched by crane; the troop capacity is 143+996
Electronics: one SPS-40 air-search radar, one SPS-10 surface-search radar, one Mk 56 gun fire-control system, two Mk 51 gun fire-control systems, OE-02 satellite communications system, SRR-1 satellite communications receiver, WSC-3 satellite communications transceiver, URN-20 TACAN, and one Mk 36 Super RBOC chaff launcher
Propulsion: two boilers supplying steam to two sets of De Laval geared turbines delivering 24,000 shp (17895 kW) to two shafts
Performance: speed 21 kts; range 11,050 miles (17,785 km) at 16 kts
Complement: 26+400

Class
1. USA

Name	No.	Builder	Laid down	Commissioned
Raleigh	LPD1	New York NY	Jun 1960	Sep 1962
Vancouver	LPD2	New York NY	Nov 1960	May 1963
La Salle	AGF3	New York NY	Apr 1962	Feb 1964

Note
These are to a concept developed from that of the LSD, with increased troop/vehicle accommodation at the expense of well deck area; the USS *La Salle* is used as a command ship in the Indian Ocean

'Ropucha' class LST
(USSR)
Type: landing ship tank
Displacement: 3,450 tons standard and 4,400 tons full load
Dimensions: length 110.0 m (360.9 ft); beam 14.5 m (47.6 ft); draught 3.6 m (11.5 ft)
Gun armament: four 57-mm AA in two twin mountings
Missile armament: (some ships only) four quadruple

launchers for 32 SA-N-5 'Grail' SAMs
Torpedo armament: none
Anti-submarine armament: none
Aircraft: none
Capacity: 230 troops plus 24 tanks, or 450 tons of stores
Electronics: one 'Strut Curve' air-search radar, one 'Muff Cob' gun-control radar, one 'Don-2' navigation radar, and one 'High Pole-B' IFF
Propulsion: four diesels delivering 7460 kW (10,005 shp) to two shafts
Performance: speed 17 kts
Complement: 95

Class
1. South Yemen

1 ship

2. USSR

12 ships 19 ships plus ? more building and planned

Note
Built at the Polnochny yard at Gdansk in Poland since 1975, these LSTs provide the USSR with a useful AFV lift capability in a roll-on/roll-off design

'Tarawa' class LHA
(USA)
Type: amphibious assault ship
Displacement: 39,300 tons full load
Dimensions: length 820.0 ft (249.9 m); beam 106.6 ft (32.5 m); draught 26.0 ft (7.9 m)
Gun armament: three 5-in (127-mm) L/54 DP in Mk 45 single mountings, six 20-mm AA in Mk 67 single mountings, and two 20-mm Phalanx Mk 16 CIWS mountings
Missile armament: none
Torpedo armament: none
Anti-submarine armament: none
Aircraft: up to 19 Sikorsky CH-53 Sea Stallion or 26 Boeing Vertol CH-46 Sea Knight helicopters, and (by a reduction of embarked helicopter strength) a varying number of BAe/McDonnell Douglas AV-8 Harrier aircraft
Capacity: apart from two full-length hangar decks under the flightdeck, there is a docking well measuring 268.0 ft (81.7 m) in length and 78.0 ft (23.8 m) in width, and this is able to accommodate four LCUs; other landing capacity is provided by six LCM(6)s; vehicle accommodation amounts to 33,730 sq ft (3135 m^2), and palletized stores to a volume of 116,900 cu ft (3310 m^3) can also be carried; liquid storage is provided for 10,000 US gal (37855 litres) of vehicle fuel and 400,000 US gal (1514160 litres) of turbine fuel; the troop accommodation can take one US Marine Corps reinforced battalion of 172+1,731 men
Electronics: one SPS-52B 3D radar, one SPS-40B air-search radar, one SPS-10F surface-search radar, one SPN-35 navigation radar, one SPG-60 search-and-track radar and one SPQ-9A track-while-scan radar used in conjunction with four Mk 86 gun fire-control systems, URN-22 TACAN, Integrated Tactical Amphibious Warfare Date System, OE-82 satellite communications system, SRR-1 satellite communications receiver, WSC-3 satellite communications transceiver, SLQ-32(V) ESM systems, and two Mk 36 Super RBOC chaff launchers
Propulsion: two boilers supplying steam to two sets of Westinghouse geared turbines delivering 140,000 shp (104400 kW) to two shafts
Performance: speed 24 kts; range 11,500 miles (18505 km) at 20 kts
Complement: 57+880

Class
1. USA

Name	No.	Builder	Laid down	Commissioned
Tarawa	LHA1	Ingalls SB	Nov 1971	May 1976
Saipan	LHA2	Ingalls SB	Jul 1972	Oct 1977
Belleau Wood	LHA3	Ingalls SB	Mar 1973	Sep 1978
Nassau	LHA4	Ingalls SB	Aug 1973	Jul 1979
Peleliu	LHA5	Ingalls SB	Nov 1976	May 1980

Note
These are the world's largest amphibious warfare ships, and combine in a single hull the capabilities of the LPH, LPD, amphibious command ship and amphibious cargo ship; each ship carries a reinforced battalion landing team of the US Marine Corps, and has facilities such as a 300-bed hospital

'TCD 90' class LSD
(France)
Type: landing ship dock
Displacement: 11,000 tons full load
Dimensions: length 160.0 m (524.9 ft); beam 22.0 m (72.2 ft); draught not known
Gun armament: six 40-mm Bofors AA in one twin and four single mountings
Missile armament: two sextuple Sadral launchers for ? Mistral SAMs
Torpedo armament: none
Anti-submarine armament: none
Aircraft: provision for four Aérospatiale AS 332 Super Puma helicopters in a hangar amidships

328 WARSHIPS

Capacity: the docking well measures 120.0 m (393.7 ft) in length and 14.0 m (45.9 ft) in width, and can accommodate 10 CTM landing craft medium, or one EDIC landing craft tank and four CTMs, plus two LCVPs; there is a removable vehicle deck, and when this is used the stores area increases from 970 m² (10,440 sq ft) to 1360 m² (14,640 sq ft); the troop accommodation is 470

Electronics: one DRBV 51 air-search radar, three surface-search and navigation radars, and extensive communications equipment
Propulsion: two SEMT-Pielstick 16 PC diesels delivering 14900 kW (19,980 shp) to two shafts
Performance: speed 20 kts
Complement: 210

Class
1. France

Name	No.	Builder	Laid down	Commissioned
		Brest ND	1985	

Note
These ships are being produced as replacements for the 'Ouragan' class, and three are planned

'Thomaston' class LSD
(USA)
Type: landing ship dock
Displacement: 6,880 tons light and 12,000 tons full load
Dimensions: length 510.0 ft (155.5 m); beam 84.0 ft (25.6 m); draught 19.0 ft (5.8 m)
Gun armament: six 3-in (76-mm) L/50 DP in three Mk 33 twin mountings
Missile armament: none
Torpedo armament: none
Anti-submarine armament: none
Aircraft: provision for helicopters on a platform over the docking well
Capacity: the docking well measures 391.0 ft (119.2 m) in length and 48.0 ft (14.6 m) in width, and can accommodate 21 LCM(6)s, or three LCUs and six LCM(6)s, or 50 LVTs with another 30 housed on the mezzanine and upper decks; the troop accommodation is 340
Electronics: one SPS-6 air-search radar, one SPS-10 surface-search radar, OE-82 satellite communications system, SRR-1 satellite communications receiver, WSC-3 satellite communications transceiver, and one Mk 36 Super RBOC chaff launcher
Propulsion: two boilers supplying steam to two sets of General Electric geared turbines delivering 24,000 shp (17895 kW) to two shafts
Performance: speed 22.5 kts; range 11,500+ miles (18505+ km) at cruising speed
Complement: 21+379

Class
1. USA

Name	No.	Builder	Laid down	Commissioned
Thomaston	LSD28	Ingalls SB	Mar 1953	Sep 1954
Plymouth Rock	LSD29	Ingalls SB	May 1953	Nov 1954
Fort Snelling	LSD30	Ingalls SB	Aug 1953	Jan 1955
Point Defiance	LSD31	Ingalls SB	Nov 1953	Mar 1955
Spiegel Grove	LSD32	Ingalls SB	Sep 1954	Jun 1956
Alamo	LSD33	Ingalls SB	Oct 1954	Aug 1956
Hermitage	LSD34	Ingalls SB	Apr 1955	Dec 1956
Monticello	LSD35	Ingalls SB	Jun 1955	Mar 1957

Note
These were the USA's first post-World War II LSDs, and incorporated lessons from the Korean War

'Wasp' class LHD
(USA)
Type: amphibious assault ship
Displacement: 40,500 tons full load
Dimensions: length 843.0 ft (256.9 m); beam 105.0 ft (32.0 m); draught 26.0 ft (7.9 m)
Gun armament: three 20-mm Phalanx Mk 16 CIWS mountings
Missile armament: two Mk 29 octuple launchers for 16 RIM-7 Sea Sparrow SAMs
Torpedo armament: none
Anti-submarine armament: none
Aircraft: Bell UH-1N Huey and AH-1T SeaCobra, Boeing Vertol CH-46 Sea Knight, and Sikorsky CH-53 Sea Stallion/Super Stallion helicopters, and BAe McDonnell Douglas AV-8B Harrier II aircraft; helicopter capacity is 38 CH-46s in the hangar and more on deck
Capacity: the docking well can accommodate air-cushion vehicles, LCUs, LCMs and LVTs; the vehicle

accommodation extends to 22,000 sq ft (2044 m^2), and that for stores to 100,900 cu ft (2857 m^3); troop accommodation is provided for 1,870 men
Electronics: one SPS-52C 3D radar, one SPS-49 air-search radar, one SPS-64 surface-search radar, one SPS-67 surface-search radar, two SPS-23 target-acquisition radars, TACAN, Integrated Tactical Amphibious Warfare System, one SLQ-25 Nixie torpedo decoy system, OE-82 satellite communications system, SRR-1 satellite communications receiver, WSC-3 satellite communications transceiver, one SLQ-32(V)3 ESM system, and two Mk 36 Super RBOC chaff launchers
Propulsion: two boilers supplying steam to two sets of Westinghouse geared turbines delivering 70,000 shp (52200 kW) to two shafts
Performance: speed 20 kts
Complement: 98+982

Class
1. USA

Name	No.	Builder	Laid down	Commissioned
Wasp	LHD1	Ingalls SB	Aug 1985	
Essex	LHD2			
	LHD3			
	LHD4			
	LHD5			

Note
This new class resembles the LHA but incorporates features of the LPD for greater operational capability with more helicopters than the LHA plus a useful landing craft/hovercraft capacity; the type is slated to replace the 'Iwo Jima' class, and 11 or 12 units are planned

'Whidbey Island' class LSD
(USA)
Type: landing ship dock
Displacement: 11,125 tons light and 15,725 tons full load
Dimensions: length 609.6 ft (185.9 m); beam 84.0 ft (25.6 m); draught 20.5 ft (6.3 m)
Gun armament: two 20-mm Phalanx Mk 16 CIWS mountings
Missile armament: none
Torpedo armament: none
Anti-submarine armament: none
Aircraft: two Sikorsky CH-53 Sea Stallion helicopters or BAe/McDonnell Douglas AV-8 Harrier aircraft on a flightdeck aft
Capacity: the docking well measures 440.0 ft (134.1 m) in length and 50.0 ft (15.2 m) in width, and can accommodate four assault air-cushion vehicles, three LCUs, 21 LCM(6)s or 64 LVTPs; including four loaded air-cushion vehicles in the dock, the vehicle area amounts to 12,500 sq ft (1161.25 m^2), and up to 5,000 cu ft (141.6 m^3) of palletized freight can also be carried; the troop capacity is 338
Electronics: one SPS-49V air-search radar, one SPS-67V surface-search radar, one SPS-64 navigation radar, OE-82 satellite communications antenna, SRR-1 satellite communications receiver, WSC-3 satellite communications transceiver, one SLQ-32(V) ESM system, and one Mk 36 Super RBOC chaff launcher
Propulsion: four Colt-Pielstick 16PC2V-V400 diesels delivering 41,600 shp (31020 kW) to two shafts
Performance: speed 20+ kts
Complement: 19+337

Class
1. USA

Name	No.	Builder	Laid down	Commissioned
Whidbey Island	LSD41	Lockheed SB	Aug 1981	Feb 1985
Germantown	LSD42	Lockheed SB	Aug 1982	Dec 1985
	LSD43	Lockheed SB	Jun 1983	
	LSD44	Avondale		
	LSD45	Avondale		
	LSD46	Avondale		

Note
This class (planned as a total of eight ships) is designed as replacement for the 'Thomaston' class, and is in essence an improved 'Anchorage' class design; there are also plans for another eight units to a **'Whidbey Island (Modified)' class** design with two rather than four LCAC hovercraft to permit the carriage of more vehicles and stores

Table 2.1 Mine warfare vessels

class (country)	number in service or planned	full-load displacement	length	speed	armament	equipment	crew
'MHCAT' inshore mine hunter (Australia)	6 Australia	170 tons	101.75 ft (31.0 m)	10 kts	2 x 7.62-mm (0.3-in) machine-guns	2 x PAP 104 destructor sleds 1 x DSQS-11H minehunting sonar	13
'Tripartite' minehunter (Belgium, France & Netherlands)	15 Belgium 10 France 15 Netherlands	544 tons	51.60 m (169.3 ft)	15 kts	1 x 20-mm AA 2 x 0.5-in (12.7-mm) AA	1 x mechanical sweep 2 x PAP 104 destructor sleds 1 x DUBM 21A minehunting sonar 1 x DUBM 41B minehunting sonar	29–48
'Circé' minehunter (France)	5 France	510 tons	50.90 m (167.0 ft)	15 kts	1 x 20-mm AA	2 x PAP 104 destructor sleds 1 x DUBM 20A minehunting sonar	48
'Kondor II' ocean minesweeper (East Germany)	24 East Germany	400 tons	55.00 m (180.4 ft)	21 kts	2 or 3 x triple 25-mm AA	1 x mechanical sweep ? x sonars	40
'Type 331' minehunter (West Germany)	12 West Germany	463 tons	47.10 m (154.5 ft)	17 kts	1 x 40-mm AA	2 x PAP 104 destructor sleds 1 x Type 195M or DSQS-11 minehunting sonar	46
'Type 351' drone minesweeper control craft (West Germany)	6 West Germany	488 tons	47.10 m (154.5 ft)	16.5 kts	1 x 40-mm AA	1 x mechanical sweep 3 x F-1 Troika drone craft 1 x DSQS-11 minehunting sonar	43
'Lerici' mine countermeasures craft (Italy)	10 Italy 4 Malaysia 2 Nigeria	502 tons	50.00 m (164.0 ft)	15 kts	1 x 20-mm AA	1 x MIN-14 destructor sled 1 x SQQ-14 minehunting sonar	40
'Hatsushima' coastal minesweeper (Japan)	15 Japan	440 tons	55.0 m (180.4 ft)	14 kts	1 x 40-mm AA	1 x mechanical sweep	45
'Takami' coastal minesweeper (Japan)	19 Japan	490 tons	46.0 m (150.9 ft)	14 kts	1 x 20-mm AA	1 x mechanical sweep	43
'Dokkum' coastal minesweeper and minehunter (Netherlands)	8 Netherlands	453 tons	45.7 m (149.8 ft)	16 kts	2 x 40-mm AA	1 x mechanical sweep 1 x Type 193 minehunting sonar	27–36

Table 2.1 Mine warfare vessels–*continued*

'Krogulec' coastal minesweeper (Poland)	12 Poland	500 tons	58.0 m (190.3 ft)	16 kts	3 x twin 25-mm AA	1 x mechanical sweep ? x sonars	?
'Landsort' mine countermeasures craft (Sweden)	6 Sweden	340 tons	47.5 m (155.8 ft)	14 kts	1 x 40-mm AA	1 x catamaran sweeper 1 x Thomson-CSF 2022 minehunting sonar	24
'Arkö' coastal minesweeper (Sweden)	9 Sweden	300 tons	44.4 m (145.6 ft)	14.5 kts	1 x 40-mm AA	1 x mechanical sweep	25
'Hunt' mine countermeasures craft (UK)	8+ UK	725 tons	197.0 ft (60.0 m)	17 kts	1 x 40-mm AA	2 x PAP 104 destructor sleds	45
'River' ocean minesweeper (UK)	12 UK		156.0 ft (47.5 m)	14 kts	(1 x 40-mm AA) 2 x 7.62-mm (0.3-in) machine-guns	1 x Mk 9 EDATS towed sweep ? x sonars	30
'Ton' coastal minesweeper and minehunter (UK)	6 Argentina 2 Eire 6 South Africa 4 Turkey 24 UK 4 Yugoslavia	440 tons	153.0 ft (46.3 m)	15 kts	1 x 40-mm AA	1 x mechanical sweep 1 x Type 193 minehunting sonar	29–38
'Natya' ocean minesweeper (USSR)	6 India 6 Libya 35 USSR	765 tons	61.0 m (200.1 ft)	18 kts	2 x quadruple SAMs 2 x twin 30-mm AA 2 x twin 25-mm AA 2 x ASW rocket-launchers 20 mines	1 x mechanical sweep 1 x Tamir minehunting sonar	60
'T43' ocean minesweeper (USSR)	1 Albania 1 Algeria 2 Bulgaria 23 China 6 Egypt 2 Indonesia 2 Iraq 11 Poland 1 Syria 35 USSR	580 tons	58.0 m (190.2 ft)	14 kts	1 x twin 37-mm AA 1 x twin 25-mm AA 2 x depth-charge throwers 20 mines	1 x mechanical sweep ? x sonar	65
'Yurka' ocean minesweeper (USSR)	4 Egypt 45 USSR 1 Vietnam	540 tons	52.0 m (170.6 ft)	19 kts	2 x quadruple SAMs 2 x twin 30-mm AA 20 mines	1 x mechanical sweep 1 x Tamir minehunting sonar	50

'Sonya' coastal minesweeper (USSR)	1 Bulgaria 2 Cuba 50 USSR	450 tons	48.0 m (148.5 ft)	15 kts	1 x quadruple SAM 1 x twin 30-mm AA 1 x twin 25-mm AA	1 x mechanical sweep 1 x minehunting sonar	43
'Vanya' coastal minesweeper (USSR)	4 Bulgaria 2 Syria 70 USSR	245 tons	40.0 m (131.2 ft)	16 kts	1 x twin 30-mm AA 16 mines	1 x mechanical sweep 1 x minehunting sonar	30
'Avenger' mine countermeasures craft (USA)	21 USA	1,240 tons	224.0 ft (68.3 m)	14 kts	2 x 0.5-in (12.7-mm) machine-gune	1 x MNS destructor sled 1 x mechanical sweep 1 x SQQ-30 minehunting sonar	72
'Aggressive' ocean minesweeper (USA)	3 Belgium 5 France 4 Italy 4 Spain 20 USA	735 tons	172.0 ft (52.4 m)	14 kts	1 x 40-mm AA	1 x mechanical sweep or destructor sled 1 x SQQ-14 minehunting sonar	73
'Adjutant' coastal minesweeper (USA)	6 Belgium 6 Denmark 3 Fiji 5 France 5 Greece 2 Iran 16 Italy 9 Norway 5 Pakistan 2 Singapore 8 South Korea 8 Spain 14 Taiwan 4 Thailand 2 Tunisia 12 Turkey 1 Uruguay	390 tons	144.0 ft (43.9 m)	13.5 kts	1 x 40-mm AA	1 x mechanical sweep 1 x minehunting sonar	40

Table 2.2 Naval guns

designation (country)	role	calibre	mounting	weight of mounting	projectile weight	muzzle velocity	rate of fire	range
130-mm (China)	DP	130 mm (5.12 in)	twin-barrel		33.4 kg (73.6 lb)	945 m (3,100 ft) per second	15 rounds per minute	29300 m (32,045 yards)
37-mm (China)	AA	37 mm	single- or twin-barrel		1.416 kg (3.12 lb)	875 m (2,871 ft) per second	160/180 rounds per minute per barrel	8500 m (9,295 yards)
M1968-II (France)	DP	100 mm (3.94 in)	single-barrel	22 tonnes	13.5 kg (29.76 lb)	870 m (2,854 ft) per second	60/80 rounds per minute	17000 m (18,590 yards)
Creusot-Loire 100-mm Compact (France)	DP	100 mm (3.94 in)	single-barrel	17 tonnes	13.5 kg (29.76 lb)	870 m (2,854 ft) per second	90 rounds per minute	17000 m (18,590 yards)

Table 2.2 Naval guns–*continued*

Name	Type	Calibre	Barrels	Weight	Shell	Muzzle velocity	Rate of fire	Range
TCM-30 (Israel)	AA	30 mm	twin-barrel		0.36 kg (0.79 lb)	1080 m (3,543 ft) per second	1,300 rounds per minute	3000 m (3,280 yards)
OTO Melara 127 Compact (Italy)	DP	127 mm (5 in)	single-barrel	34 tonnes		807 m (2,648 ft) per second	45 rounds per minute	15000 m (16,405 yards)
OTO Melara 76 Compact (Italy)	DP	76 mm (3 in)	single-barrel	7.35 tonnes	6.3 kg (13.9 lb)	925 m (3,035 ft) per second	85/100 rounds per minute	8000 m (8,750 yards)
OTO Melara 76 Super Rapid (Italy)	DP	76 mm (3 in)	single-barrel		6.3 kg (13.9 lb)	925 m (3,035 ft) per second	100+ rounds per minute	8000 m (8,750 yards)
Breda 40 Compact Type 70 (Italy)	AA	40 mm	twin-barrel	7.3 tonnes with 736 rounds or 6.3 tonnes with 444 rounds	0.88 kg (1.94 lb)	1000 m (3,281 ft) per second	300 rounds per minute per barrel	4000 m (4,375 yards)
Breda 40 Type 106 (Italy)	AA	40 mm	twin-barrel	6.61 tonnes	0.88 kg (1.94 lb)	1000 m (3,281 ft) per second	300 rounds per minute per barrel	4000 m (4,375 yards)
Breda 40 Type 564 (Italy)	AA	40 mm	single-barrel	3.4 tonnes	0.88 kg (1.94 lb)	1000 m (3,281 ft) per second	300 rounds per minute	4000 m (4,375 yards)
Breda 30 (Italy)	AA	30 mm	twin-barrel				800 rounds per minute per barrel	3000 m (3,280 yards)
Goalkeeper (Netherlands/USA)	point defence AA	30 mm	seven-barrel	6.73 tonnes		3,350 ft (1021 m) per second	4,200 rounds per minute	3000 m (3,280 yards)
Meroka (Spain)	point defence AA	20 mm	12-barrel	4.5 tonnes			3,600 rounds per minute	2000+ m (2,185 yards)
Bofors 120-mm (Sweden)	DP	120 mm (4.72 in)	twin-barrel	28.5 tonnes	23.5 kg (51.8 lb)	850 m (2,789 ft) per second	40 rounds per minute per barrel	20500 m (23,515 yards)
Bofors SAK Mk 2 (Sweden)	DP	57 mm	single-barrel	6 tons	2.8 kg (6.17 lb)	1025 m (3,363 ft) per second	220 rounds per minute	17000 m (18,590 yards)
Bofors SAK Mk 1 (Sweden)	DP	57 mm	single-barrel	6 tons	2.4 kg (5.3 lb)	1025 m (3,363 ft) per second	200 rounds per minute	17000 m (18,590 yards)
Bofors 40-mm L/70 (Sweden)	AA	40 mm	single-barrel	1.7 to 3.8 tons	0.96 kg (2.12 lb)	1000 m (3,281 ft) per second	300 rounds per minute	4000 m (4,375 yards)
Bofors 40-mm L/60 (Sweden)	AA	40 mm	single-barrel	1.2 to 2.5 tonnes	0.9 kg (1.98 lb)	830 m (2,723 ft) per second	120 rounds per minute	3000 m (3,280 yards)
Oerlikon GDM-A (Switzerland)	AA	35 mm	twin-barrel	6.52 tonnes	0.55 kg (1.21 lb)	1175 m (3,855 ft) per second	550 rounds per minute per barrel	3500 m (3,830 yards)
Oerlikon GCM-A (Switzerland)	AA	30 mm	twin-barrel	1.29 to 2.56 tonnes	0.36 kg (0.79 lb)	1080 m (3,543 ft) per second	650 rounds per minute per barrel	3000 m (3,280 yards)

Table 2.2 Naval guns–*continued*

Name	Role	Calibre	Barrels	Weight	Shell weight	Muzzle velocity	Rate of fire	Range
Oerlikon GBM-A (Switzerland)	AA	25 mm	single-barrel	0.6 tonne	0.18 kg (0.4 lb)	1100 m (3,609 ft) per second	570 rounds per minute	2000 m (2,185 yards)
Seaguard (Switzerland, Italy & UK)	point defence AA	25 mm	four-barrel	5.7 tonnes	0.625 kg (1.38 lb)	1355 m (4,446 ft) per second	850 rounds per minute per barrel	3500 m (3,850 yards)
Oerlikon GAM-B (Switzerland)	AA	20 mm	single-barrel	0.5 tonne	0.125 kg (0.28 lb)	1050 m (3,444 ft) per second	600 rounds per minute	2000 m (2,185 yards)
152-mm (USSR)	shove bombardment	152 mm (6 in)	triple-barrel		50 kg (110.2 lb)	915 m (3,002 ft) per second	10 rounds per minute per barrel	27000 m (29,530 yards)
130-mm L/70 (USSR)	DP	130 mm (5.12 in)	twin-barrel			950 m (3,117 ft) per second	65 rounds per minute per barrel	28000 m (30,620 yards)
130-mm L/80 (USSR)	DP	130 mm (5.12 in)	twin-barrel		27 kg (59.5 lb)	875 m (2,871 ft) per second	10 rounds per minute per barrel	28000 m (30,620 yards)
100-mm (USSR)	DP	100 mm (3.94 in)	twin-barrel	35 tonnes	16 kg (35.3 lb)	900 m (2,953 ft) per second	20 rounds per minute per barrel	18000 m (19,685 yards)
100-mm (USSR)	DP	100 mm (3.94 in)	single-barrel			900 m (2,953 ft) per second	80 rounds per minute	15000 m (16,405 yards)
76-mm (USSR)	DP	76 mm (3 in)	twin-barrel		16 kg (35.3 lb)	900 m (2,953 ft) per second	45 rounds per minute per barrel	15000 m (16,405 yards)
76-mm (USSR)	DP	76 mm (3 in)	single-barrel		16 kg (35.3 lb)	900 m (2,953 ft) per second	120 rounds per minute	16000 m (17,500 yards)
57-mm L/70 (USSR)	AA	57 mm	one-, two- or four-barrel		2.85 kg (6.28 lb)	1000 m (3,281 ft) per second	120 rounds per minute per barrel	4500 m (4,920 yards)
57-mm L/80 (USSR)	AA	57 mm	twin-barrel		2.85 kg (6.28 lb)	1000 m (3,281 ft) per second	120 rounds per minute per barrel	6000 m (6,560 yards)
37-mm (USSR)	AA	37 mm	single- or twin-barrel		0.73 kg (1.61 lb)	880 m (2,887 ft) per second	160 rounds per minute per barrel	4000 m (4,375 yards)
AK-630 (USSR)	point-defence AA	30 mm	six-barrel		0.54 kg (1.2 lb)	1000 m (3,281 ft) per second	3,000 rounds per minute	3000 m (3,280 yards)
AK-230 (USSR)	AA	30 mm	two-barrel		0.54 kg (1.2 lb)	1000 m (3,281 ft) per second	1,050 rounds per minute	2500 m (2,735 yards)
Vickers Mk 8 (UK)	DP	4.5 in (114.3 mm)	single-barrel		46.3 lb (21 kg)	2,854 ft (870 m) per second	25 rounds per minute	24,000 yards (21945 m)
Vickers Mk 6 (UK)	DP	4.5 in (114.3 mm)	twin-barrel	45 tonnes	55 lb (24.9 kg)	2,790 ft (850 m) per second	20 rounds per minute per barrel	20,250 yards (18515 m)

Table 2.2 Naval guns–*continued*

LS30R (UK)	AA	30 mm	single-barrel	1,764 lb (800 kg)	0.79 lb (0.357 kg)	3,543 to 3,937 ft (1080 to 1200 m) per second	40 rounds per minute	4,400 yards (4025 m)
Turret Mk 7 (USA)	shore bombardment	16 in (406 mm)	triple-barrel	1735 tonnes	1,900 lb (862 kg) HE	2,500 ft (762 m) per second	2 rounds per minute per barrel	41,530 yards (37975 m)
Mounting Mk 45 (USA)	DP	5 in (127 mm)	single-barrel	21.68 tonnes	70 lb (31.8 kg)		20 rounds per minute	26,000 yards (23775 m)
Mounting Mk 42 (USA)	DP	5 in (127 mm)	single-barrel	66.86 tonnes	70 lb (31.8 kg)	2,657 ft (810 m) per second	40 rounds per minute	26,000 yards (23775 m)
Mounting Mk 33 (USA)	DP	3 in (76 mm)	twin-barrel	14.5 tonnes	13 lb (5.9 kg)	2,705 ft (824 m) per second	25 rounds per minute per barrel	14,000 yards (12800 m)
Emerlec-30 (USA)	AA	30 mm	twin-barrel	1.9 tonnes	0.36 kg (0.79 lb)	1080 m (3,543 ft) per second	600 rounds per minute per barrel	3,300 yards (3020 m)
Phalanx Mk 16 (USA)	point-defence AA	20 mm	six-barrel	6.09 tonnes	about 2.2 lb (1.0 kg)	3,600 ft (1097 m) per minute	3,000 rounds per minute	1,650 yards (1510 m)

Table 2.3 Surface skimmers

designation (country)	role	weight	length	speed	payload
SR.N4 Mk3 (UK)	transport	300 tons	185.0 ft (56.4 m)	65 kts	280 passengers and 37 vehicles
SR.N3 Mk 8 (UK)	transport	16.7 tons	60.0 ft (18.3 m)	50 kts	55 troops
BH7 Mk 5A (UK)	transport	55 tons	78.33 ft (23.9 m)	58 kts	170 troops or 60 troops and 3 vehicles
'Aist' (USSR)	transport	270 tons	47.8 m (156.8 ft)	80 kts	220 troops, or 2 MBTs, or 4 light tanks or 60 tons of freight
'Gus' (USSR)	transport	26.7 tons	21.33 m (70.0 ft)	60 kts	25 troops or freight
'Lebed' (USSR)	transport	85 tons	24.8 m (81.4 ft)	70 kts	120 troops, or 2 light tanks or 40 tons of freight
'Tsaplya' (USSR)	transport	100 tons	24.0 m (78.75 ft)	65 kts	160 troops, or 80 troops and 1 light tank, or 25 tons of freight
'Utenok' (USSR)	transport	80 tons	26.3 m (86.3 ft)	65 kts	1 MBT
Bell LACV-30 (USA)	transport	51.34 tons	76.5 ft (23.32 m)	33.5 kts	30 tons of freight
Bell LCAC (USA)	transport	170 tons	88.0 ft (26.82 m)	50 kts	1 MBT or 75 tons of freight

Table 2.4 Torpedoes

designation (country)	role	length	diameter	weight	warhead	speed	range	guidance
E14 (France)	submarine-launched anti-ship	4.191 m (13.75 ft)	550 mm (21.65 in)	900 kg (1,984 lb)	200 kg (441 lb) HE	25 kts	5500 m (6,015 yards)	passive acoustic
E15 (France)	submarine-launched anti-ship	6.00 m (19.69 ft)	550 mm (21.65 in)	1350 kg (2,976 lb)	300 kg (661 lb) HE	25 kts	12000 m (13,125 yards)	passive acoustic
F17 (France)	submarine-launched anti-ship	5.914 m (19.40 ft)	533 mm (21 in)	1410 kg (3,108 lb)	250 kg (551 lb) HE	35 kts	18000 m (19,685 yards)	wire and/or passive acoustic
L3 (France)	ship- and submarine-launched anti-submarine	4.30 m (14.11 ft)	550 mm (21.65 in)	910 kg (2,005 lb)	200 kg (441 lb) HE	25 kts	5500 m (6,015 yards)	active acoustic
L4 (France)	air-launched anti-submarine	3.13 m (10.27 ft)	533 mm (21 in)	540 kg (1,190 lb)	104 kg (229 lb) HE	30 kts	5500 m (6,015 yards)	active acoustic
L5 (France)	multi-role	4.40 m (14.44 ft)	533 mm (21 in)	L5 Mod 1 1000 kg (2,205 lb); L5 Mod 3 1300 kg (2,866 lb); L5 Mod 4 920 kg (2,028 lb); L5 Mod 4P 930 kg (2,050 lb)	150 kg (331 lb) HE	35 kts	9250 m (10,115 yards)	active/passive acoustic
SUT (West Germany)	ship- and submarine-launched anti-ship and anti-submarine	6.62 m (21.72 ft) with wire casket	533 mm (21 in)	1414 kg (3,117 lb)	260 kg (573 lb) HE	23 or 35 kts	28000 or 12000 m (30,620 or 13,125 yards)	wire and/or active/passive acoustic
Seal (West Germany)	ship- and submarine-launched anti-ship	6.55 m (21.49 ft) with wire casket	533 mm (21 in)	1370 kg (3,020 lb)	260 kg (573 lb) HE	23 or 35 kts	28000 or 12000 m (30,520 or 13,125 yards)	wire and/or active/passive acoustic
SST4 (West Germany)	ship- and submarine-launched anti-ship	6.50 m (21.33 ft) with wire casket	533 mm (21 in)	1414 kg (3,117 lb)	260 kg (573 lb) HE	23 or 35 kts	28000 or 12000 m (30,620 or 13,125 yards)	wire and/or active/passive acoustic
Seeschlange (West Germany)	ship- and submarine-launched anti-submarine	4.62 m (15.16 ft) with wire casket	533 mm (21 in)		100 kg (220 lb) HE			wire and/or active/passive acoustic

Table 2.4 Torpedoes–*continued*

Name	Type	Length	Diameter	Weight	Warhead	Speed	Range	Guidance
A184 (Italy)	ship- and submarine-launched anti-ship and anti-submarine	6.00 m (19.69 ft)	533 mm (21 in)	1265 kg (2,789 lb)	250 kg (551 lb) HE	24 or 36 kts	25000 or 10000 m (27,340 or 10,935 yards)	wire and/or active/passive acoustic
A 244S (Italy)	air- and ship-launched anti-submarine	2.70 m (8.86 ft)	324 mm (12.75 in)	221 kg (487 lb)	34 kg (75 lb) HE	30 kts	6500 m (7,110 yards)	active/passive acoustic
Tp 422 (Sweden)	air-launched anti-submarine	2.60 m (8.53 ft) with wire casket	400 mm (15.75 in)	298 kg (657 lb)	45 kg (99 lb) HE	15 or 25 kts	20000 or 10000 m (21,870 or 10,935 yards)	wire and passive acoustic
Tp 423 (Sweden)	ship- and submarine-launched anti-ship and anti-submarine	2.60 m (8.53 ft) with wire casket	400 mm (15.75 in)	298 kg (657 lb)	45 kg (99 lb) HE	15 or 25 kts	20000 or 1000 m (21,879 or 10,935 yards)	wire and passive acoustic
Tp 432 (Sweden)	air-launched anti-submarine	2.60 m (8.53 ft) with wire casket	400 mm (15.75 in)	280/350 kg (617/772 lb)	45 kg (99 lb) HE	15, 25 or 35 kts	30000, 20000 or 10000 m (32,810, 21,870 or 10,935 yards)	wire and passive acoustic
Tp 613 (Sweden)	ship-, submarine- and shore-launched anti-ship	7.025 m (23.05 ft)	533 mm (21 in)	1850 kg (4,078 lb)	250 kg (551 lb)	45 kts	30000 m (32,810 yards)	wire and acoustic
Type 65 (USSR)	submarine-launched anti-ship	9.15 m (30.02 ft)	650 mm (25.6 in)		15-kiloton nuclear	30 or 50 kts	100 or 50 km (62 or 31 miles)	wake-homing
533-mm type (USSR)	ship- and submarine-launched anti-ship and anti-submarine	about 8.25 m (27.07 ft)	533 mm (21 in)		HE or 15-kiloton nuclear			acoustic
450-mm type (USSR)	air-launched anti-submarine		450 mm (17.72 in)		HE			acoustic
406-mm type (USSR)	air- and submarine-launched anti-submarine		406 mm (16 in)		HE			acoustic
Mk 8 (UK)	submarine-launched anti-ship	22.00 ft (6.71 m)	21 in (533 mm)	3,355 lb (1522 kg)	750 lb (340 kg) HE	45 kts	7,000 yards (6400 m)	gyroscopic
Mk 24 Tigerfish (UK)	submarine-launched anti-ship and anti-submarine	21.21 ft (6.464 m)	21 in (533 mm)	3,410 lb (1547 kg)	295 lb (134 kg) HE	24 or 35 kts	23,000 or 14,250 yards (21030 or 13030 m)	wire and active/passive acoustic

338 WARSHIPS

Table 2.4 Torpedoes–*continued*

Spearfish (UK)	submarine-launched anti-submarine and anti-ship	27.9 ft (8.5 m)	21 in (533 mm)	4,400 lb (1996 kg)	550 lb (249 kg) HE	24 or 60 kts	28.75 or 17.75 miles (46.25 or 28.6 km)	wire and active/passive acoustic
Stingray (UK)	air- and ship-launched anti-submarine	8.52 ft (2.60 m)	12.75 in (324 mm)	585 lb (265 kg)	88 lb (40 kg) HE	45 kts	12,000 yards (10975 m)	active/passive acoustic
NT-37E Mod 2 (USA)	submarine-launched anti-submarine and anti-ship	14.78 ft (4.506 m)	19 in (484.6 mm)	1,650 lb (748 kg)	330 lb (150 kg) HE	22.4 or 33.6 kts	23,750 yards (21715 m)	wire and active/passive acoustic
NT-37E Mod 3 (USA)	submarine-launched anti-submarine and anti-ship	12.62 ft (3.946 in)	19 in (484.6 mm)	1,412 lb (640 kg)	330 lb (150 kg) HE	22.4 or 33.6 kts	20,000 yards (18290 m)	gyroscopic and active/passive acoustic
Mk 44 Mod 1 (USA)	air- and ship-launched anti-submarine	8.44 ft (2.57 m)	12.75 in (324 mm)	433 lb (196 kg)	73 lb (33 kg) HE	30 kts	6,000 yards (5485 m)	active acoustic
Mk 46 Mod 5 (USA)	air- and ship-launched anti-submarine	8.50 ft (2.59 m)	12.75 in (324 mm)	508 lb (230 kg)	95 lb (43 kg) HE	40/45 kts	12,000 yards (10975 m)	active/passive acoustic
Mk 50 Barracuda (USA)	air- and ship-launched anti-submarine	9.50 ft (2.90 m)	12.75 in (324 mm)	800 lb (363 kg)	100 lb (45 kg) HE	55+ kts	15,000 yards (13715 m)	active/passive acoustic
Mk 48 Mods 1 and 3 (USA)	submarine-launched anti-ship and anti-submarine	19.17 ft (5.84 m)	21 in (533 mm)	3,480 lb (1579 kg)	650 lb (295 kg) HE	48 kts	35,000 yards (32005 m)	wire and/or active/passive acoustic
Mk 48 Mod 4 (USA)	submarine-launched anti-ship and anti-submarine	19.17 ft (5.84 m)	21 in (533 mm)	3,480 lb (1579 kg)	650 lb (295 kg) HE	55 kts	30,800 yards (28165 m)	wire and/or active/passive acoustic
Mk 48 Mod 5 (USA)	submarine-launched anti-ship and anti-submarine	19.17 ft (5.84 m)	21 in (533 mm)	3,480 lb (1579 kg)	650 lb (295 kg) HE	60 kts	41,800 yards (38220 m)	wire and/or active/passive acoustic

SECTION 3

AIRCRAFT

Aeritalia F-104S Starfighter
(Italy)
Type: multi-purpose interceptor and attack fighter
Accommodation: pilot seated on a Martin-Baker IQ7A ejector seat
Armament (fixed): (optional in place of missile-guidance package) one General Electric M61A1 Vulcan rotary-barrel 20-mm cannon
Armament (disposable): this is carried on two wingtip, four underwing and three underfuselage hardpoints, up to a maximum weight of 3400 kg (7,495 lb); for the interception role the normal armament is two AIM-7 Sparrow or Aspide air-to-air missiles under the wings and two AIM-9 Sidewinder air-to-air missiles at the wingtips; for the ground-attack role the weapons that can be accommodated are a launcher for 19 2.75-in (69.85-mm) rockets, a launcher for five 5-in (127-mm) rockets), 250-lb (113-kg) bombs, 500-lb (227-kg) bombs, 1,000-lb (454-kg) bombs, and other weapons (napalm tanks, fire bombs, cluster bombs and a 750-lb/340-kg demolition bomb)
Electronics and operational equipment: communication and navigation equipment, plus R21G/H multi-role radar for air-interception, air-to-surface mapping and terrain avoidance, Litton LN-3-2A inertial navigation system, a bombing computer, an air-data computer and an optical gunsight
Powerplant and fuel system: one 8120-kg (17,900-lb) afterburning thrust Fiat-built General Electric J79-GE-19 turbojet, and a total internal fuel capacity of 3390 litres (746 Imp gal) in five fuselage bag tanks, plus provision for two 740-litre (163-Imp gal) underwing and two 645-litre (142-Imp gal) wingtip drop-tanks
Performance: maximum speed 2330 km/h (1,450 mph) or Mach 2.2 at 11000m (36,090 ft) and 1465 km/h (910 mph) or Mach 1.2 at sea level; cruising speed 980 km/h (609 mph) at 11000 m (36,090 ft); service ceiling 17680 m (58,005 ft); combat radius 1245 km (774 mile) radius with maximum fuel; ferry range 2920 km (1,815 miles) with maximum internal and external fuel
Weights: empty 6760 kg (14,903 lb); normal take-off 9840 kg (21,693 lb); maximum take-off 14060 kg (30,996 lb)
Dimensions: span 6.68 m (21 ft 11 in) without tiptanks; length 16.69 m (54 ft 9 in); height 4.11 m (13 ft 6 in); wing area 18.22 m² (196.1 sq ft)

Variant
F-104S: Italian development of the basic F-104 concept but derived more specifically from the F-104G multi-role fighter with particular emphasis on optimized air-to-air capability through the use of the AIM-7 Sparrow medium-range air-to-air missile and appropriate avionics; production totalled 245 aircraft (205 for the Italian air force and 40 for the Turkish air force), the first machine flying in December 1968 and production ending in 1976 to complete world production of the Starfighter family

Aeritalia G91R/1
(Italy)
Type: tactical close support and reconnaissance fighter
Accommodation: pilot seated on a Martin-Baker ejector seat
Armament (fixed): four 0.5-in (12.7-mm) Colt-Browning M3 machine-guns with 300 rounds per gun
Armament (disposable): a variety of stores can be accommodated on four underwing hardpoints to a maximum of 680 kg (1,500 lb); the inner pair can each take one 227-kg (500-lb) bomb, or one AS.20 or AS.30 air-to-surface missile, or one cluster of six 5-in (127-mm) rockets, or a honeycomb pack of 31 2.75-in (70-mm) FFAR rockets, or a pod containing one 0.5-in (12.7-mm) machine-gun with 250 rounds; the outer pair can each take one 250-lb (113-kg) bomb, or honeycomb pack of 19 2.75-in (70-mm) FFAR rockets, or a pod containing one 0.5-in (12.7-mm) machine-gun with 250 rounds
Electronics and operational equipment: communication and navigation equipment, the latter including Bendix DRA-12A Doppler radar, and a SFOM sight
Powerplant and fuel system: one 2268-kg (5,000-lb) thrust Fiat-built Rolls-Royce (Bristol) Orpheus Mk 803 turbojet, and a total internal fuel capacity of ? litres (? Imp gal) plus provision for drop tanks
Performance: maximum speed 1085 km/h (674 mph) at 1500 m (4,920 ft); cruising speed 650 km/h (403 mph); initial climb rate 1830 m (6,005 ft) per minute; service ceiling 13100 m (42,980 ft); combat radius 320 km (199 miles) on a typical mission; ferry range 1850 km (1,150 miles)
Weights: empty 3100 kg (6,835 lb); normal take-off 5440 kg (11,995 lb); maximum 5500 kg (12,125 lb)
Dimensions: span 8.56 m (28 ft 1 in); length 10.30 m (33 ft 9.25 m); height 4.00 m (13 ft 1.25 in); wing area 16.42 m² (176.74 sq ft)

Variants
G91R/1: initial service version resulting from a 1954 NATO requirement for a light attack fighter for European service; the prototype first flew in August 1956, and proved fully able to meet the requirements of the specification, but in the event only West Germany and Italy adopted the type, which is now obsolescent and being phased out of front-line service
G91R/1A: updated G91R/1 introduced in 1959 with the improved navigation package of the G91R/3
G91R/1B: strengthened version of the G91R/1A with reinforced structure and landing gear, and incorporating a number of equipment changes dictated by operational experience
G91R/3: version of the G91R/3 built in West

Germany, and incorporating more capable navigational equipment (including Doppler radar and a position-and-homing indicator) plus a revised gun armament of two DEFA 552 30-mm cannon
G91R/4: version of the G91R/3 with the armament of the G91R/1 and detail equipment modifications
G91T/1: Italian-built operational trainer version of the G91R series, with a gun armament of two 0.5-in (12.7-mm) machine-guns and provision for external stores on two underwing hardpoints
G91T/2: West German-built operational trainer version of the G91R very similar to the G91T/1

Aeritalia G91Y
(Italy)
Type: tactical close support and reconnaissance fighter
Accommodation: pilot seated on a Martin-Baker ejector seat
Armament (fixed): two 30-mm DEFA 552 cannon with 125 rounds per gun
Armament (disposable): up to 1814 kg (4,000 lb) can be accommodated on four underwing hardpoints, typical loads being four 1,000-lb (454-kg) bombs, or four 750-lb (340-kg) napalm tanks, or four Aerea pods each containing 28 2-in (51-mm) rockets, or four pods each with four 5-in (127-mm) Zuni rockets, or four pods each with seven 2-in (51-mm) rockets, or a mixture of these loads; two AIM-9 Sidewinder missiles can also be carried
Electronics and operational equipment: communication and navigation equipment, plus a Smiths head-up display and a Ferranti ISIS sight
Powerplant and fuel system: two 4,080-lb (1850-kg) afterburning thrust General Electric J85-GE-13A turbojets, and a total internal fuel capacity of ? litres (? Imp gal) plus provision for drop tanks
Performance: maximum speed 1140 km/h (708 mph) or Mach 0.93 at sea level, and 1040 km/h (646 mph) or Mach 0.95 at 9145 m (30,000 ft); cruising speed 800 km/h (497 mph) or Mach 0.75 at 10670 m (35,000 ft); initial climb rate 5180 m (17,000 ft) per minute; service ceiling 12500 m (41,010 ft); combat radius 385 km (240 miles) on a lo-lo sortie with a 1320 kg (2,910 lb) load; ferry range 3500 km (2,175 miles)
Weights: empty 3682 kg (8,117 lb); normal take-off 7800 kg (17,196 lb); maximum take-off 8700 kg (19,180 lb)
Dimensions: span 9.01 m (29 ft 6.5 in); length 11.67 m (38 ft 3.5 in); height 4.43 m (14 ft 6.5 in); wing area 18.13 m² (195.15 sq ft)

Variant
G91Y: sole production version of the much-improved twin-engined version of the G91 family, with the twin-engine powerplant offering some 63 per cent more power (for an increase in empty weight of only 18 per cent) as well as the tactical advantages of twin-engined reliability and greater payload and/or greater fuel; the type was based on the slightly lengthened airframe developed for the G91T series, and apart from more modern avionics, the G91Y offered such advantages as a pressurized and air-conditioned cockpit and a zero/zero ejector seat; the first G91Y flew in December 1966, and production totalled 77 aircraft

Aeritalia G222
(Italy)
Type: general-purpose transport
Accommodation: crew of three on the flightdeck, and up to 53 troops, or 42 paratroops, or 9000 kg (19,841 lb) of freight in the hold
Armament (fixed): none
Armament (disposable): none
Electronics and operational equipment: communication and navigation equipment
Powerplant and fuel system: two 2535-kW (3,400-shp) Fiat-built General Electric T64-GE-P4D turboprops, and a total internal fuel capacity of 12000 litres (2,638 Imp gal) in four integral wing tanks
Performance: maximum speed 540 km/h (336 mph) at 4575 m (15,000 ft); cruising speed 439 km/h (273 mph) at 6000 m (19,685 ft); initial climb rate 520 m (1,705 ft) per minute; service ceiling 7620 m (25,000 ft); range 1370 km (851 miles) with maximum payload, or 4635 km (2,880 miles) for ferrying with maximum fuel
Weights: empty 15400 kg (33,950 lb); maximum take-off 28000 kg (61,728 lb)
Dimensions: span 28.70 m (94 ft 2 in); length 22.70 m (74 ft 5.5 in); height 9.80 m (32 ft 1.75 in); wing area 82.00 m² (882.6 sq ft)

Variants
G222: though originating in a NATO requirement for a V/STOL tactical transport aircraft, the design was recast in the late 1960s as a trim twin-engined transport of conventional though attractive type geared to the requirements of the Italian air force; the prototype was first flown in July 1970, and production for the Italian air force amounted to 50 aircraft in several versions; small numbers of this basic model were also exported; the type can also be fitted out for firefighting operations, and the manufacturer has proposed airborne early warning, inflight-refuelling and anti-submarine warfare versions
G222RM: navaid calibration version
G222VS: electronic warfare version developed for the Italian air force in the electronic and signals intelligence roles; specialist equipment carried by this model includes signal-processing and data-recording gear
G222L: version produced for Libya with European avionics and 4,860-shp (3624-kW) Rolls-Royce RTy.20 Mk 801 turboprops in place of the US engines

used in other model, the greater power offering higher payload or, alternatively, better performance in 'hot-and-high' conditions with the standard payload; the type has a maximum take-off weight of 29000 kg (63,933 lb); the G222L first flew under the designation G222T in May 1980

Aeritalia/Aermacchi/EMBRAER AMX
(Italy/Brazil)
Type: interdiction, close air support and reconnaissance aircraft
Accommodation: pilot seated on a Martin-Baker IT10LY zero/zero ejector seat
Armament (fixed): one 20-mm M61A1 Vulcan six-barrel rotary canon with 350 rounds (Italian aircraft) or two 30-mm DEFA 553 cannon with ? rounds per gun (Brazilian aircraft)
Armament (disposable): this is carried on one double underfuselage hardpoint (rated at 907 kg/2,000 lb), on four underwing hardpoints (the inner pair rated at 907 kg/2,000 lb each and the outer pair at 454 kg/1,000 lb each) and on two wingtip missile rails up to a maximum of 3800 kg (7,377 lb) of free-fall, retarded and special-purpose bombs, rocket pods, air-to-surface missiles, anti-ship missiles and air-to-air missiles
Electronics and operational equipment: communication and navigation equipment, plus Fiar Pointer ranging radar, head-up display, head-down display, stores management and delivery system, inertial navigation system, GEC/Aeritalia air-data and central computers, comprehensive electronic countermeasures used in conjunction with an Elettronica radar-warning receiver, and laser ranger and marked-target seeker optionally replaceable by a sensor pallet carried instead of the standard centreline reconnaissance pod
Powerplant and fuel system: one 5000-kg (11,023-lb) thrust Piaggio-built Rolls-Royce Spey Mk 807 non-afterburning turbofan, and a total internal fuel capacity of ? litres (? Imp gal) in a segmented fuselage and two integral wing tanks, plus provision for two 1000-litre (220-Imp gal) and two 500-litre (110-Imp gal) drop tanks; provision for inflight refuelling
Performance: maximum speed 1163 km/h (723 mph) or Mach 0.95 at sea level with maximum payload; combat radius 370 km (230 miles) on a lo-lo-lo mission with 2722-kg (6,000-lb) payload; ferry range 2965 km (1,842 miles)
Weights: empty 6000 kg (13,228 lb); maximum take-off 11500 kg (25,353 lb)
Dimensions: span without wingtip AAMs 8.874 m (29 ft 1.4 in); length 13.575 m (44 ft 6.5 in); height 4.58 m (15 ft 0.2 in); wing area 21.00 m² (226.04 sq ft)

Variant
AMX: originated in Italy but now a joint Italo-Brazilian project (the work split being 70 and 30 per cent respectively for anticipated production of 187 and 79 aircraft), the AMX was designed as a lightweight fighter-bomber with STOL performance from semi-prepared strips, and was planned as replacement for the Aeritalia G91 series in operational partnership with the Panavia Tornado; keynotes of the design are basic simplicity and low cost (resulting in transonic rather than supersonic performance) with utmost versatility of weapon capability, though great attention has been paid to 'stretch' potential, especially in the avionics (a planned anti-ship single-seater with Thomson-CSF Agave radar will later offer the option of more advanced radar such as the Ferranti Sea Vixen, Ericsson PS-05, Fiar Grifo and Fiar/Westinghouse Altair); in the reconnaissance role the type can carry any one of three centreline-mounted pods, one carrying three cameras for tactical reconnaissance, the second carrying a long focal-length side-looking camera for operation reconnaissance, and the third carrying high-altitude mapping equipment; the first protype flew in 1985, and the type is due to enter service in 1987 or early 1988; plans for further development envisage a two-seater for the operational conversion and electronic warfare roles

Aermacchi M.B.326L
(Italy)
Type: basic and advanced trainer with secondary light attack capability
Accommodation: pupil and instructor seated in tandem on Martin-Baker Mk 06A ejector seats
Armament (fixed): none
Armament (disposable): this is carried on six underwing hardpoints, the inner four each rated at 454 kg (1,000 lb) and the outer pair each at 340 kg (750 lb), up to a maximum weight of 1814 kg (4,000 lb); a very wide assortment of stores can be carried, typical loads consisting of two LAU-3/A launchers each with 19 2.75-in (69.85-mm) rockets and two launchers each with eight Oerlikon-Bührle SURA 81-mm (3.19-in) rockets; or two pods each with a single 12.7 mm (0.5-in) Browning M3 machine-gun and four launchers each with six SURA 81-mm (3.19-in) rockets; or one pod with one 7.62-mm (0.3-in) Minigun, one pod with one 12.7 mm (0.5-in) machine-gun, two Matra 122 launchers each with 68-mm (2.68-in) rockets and two launchers each with six SURA 81-mm (3.19-in) rockets; or two 500-lb (227-kg) bombs and eight 5-in (127-mm) HVAR rockets; or two AS.12 air-to-surface missiles; or two Matra SA-10 packs each with a 30-mm Aden cannon and 150 rounds; or one reconnaissance pod and one 12.7-mm gun pod
Electronics and operational equipment: communication and navigation equipment, plus Doppler radar, SFOM Type 83 fixed or Ferranti LFS 5/102A gyro gunsight, and provision for a bombing computer and laser rangefinder

Powerplant and fuel system: one 4,000-lb (1814-kg) thrust Rolls-Royce (Bristol) Viper Mk 362-43 turbojet, and a total internal fuel capacity of 1660 litres (366 Imp gal) in three fuselage and two non-jettisonable wingtip tanks, plus provision for two drop-tanks of up to 340-litre (75-Imp gal) capacity each
Performance: maximum speed 890 km/h (593 mph) at 1525 m (5,000 ft) clean, and 685 km/h (426 mph) at 9145 m (30,000 ft) with external stores; initial climb rate 1980 m (6,500 ft) per minute; service ceiling about 14325 m (47,000 ft); combat radius 268 km (167 miles) on a lo-lo sortie with internal fuel and 1280-kg (2,822-lb) weapon load; ferry range 2130 km (1,323 miles)
Weights: empty 2964 kg (6,534 lb); normal take-off 4211 kg (9,285 lb); maximum take-off 5897 kg (13,000 lb)
Dimensions: span 10.85 m (35 ft 7 in) over tiptanks; length 10.673 m (35 ft 0.25 in); height 3.72 m (12 ft 2 in); wing area 19.35 m² (208.3 sq ft)

Variants
M.B.326: initial version of this highly successful two-seat trainer and light attack aircraft series. the design was started in 1954, and the first prototype flew in December 1957 on the power of a 1,750-lb (794-kg) Bristol Viper 8 turbojet; the type was adopted by the Italian air force from 1962 with the 2,500-lb (1134-kg) Viper 11, and has no provision for armament
M.B.326B: armed trainer for Tunisia derived from the proposed M.B.326A with the Viper 11 and six underwing hardpoints for a total of 907 kg (2,000 lb) of disposable stores
M.B.326E: hybrid type for Italy with the fuselage and powerplant of the M.B.326 combined with the wings and armament provision of the M.B.326GB
M.B.326F: armed trainer for Ghana, basically similar to the M.B.326B
M.B.326GB: upgraded version based on the M.B.326G unarmed trainer with the 3,410-lb (1547-kg) Viper 20 Mk 540 turbojet and the ability to carry 1814 kg (4,000 lb) of disposable stores under the wings; under the local designation **EMBRAER EMB-326 (AT-26 Xavante)**, the basically similar **M.B.326GC** is built under licence in Brazil and generally flown as a single-seater
M.B.325H: armed two-seat trainer for Australia with the Viper 11 engine
M.B.326K: much improved single-seat operational training and light attack model based on the airframe of the M.B.326GB but fitted with the 4,000-lb (1814-kg) Viper 20 Mk 540 engine (in the prototype, which first flew in August 1970) or Viper 632-43 (in production aircraft); the volume of the erstwhile rear cockpit is used for avionics, fuel and inbuilt armament (two 30-mm DEFA 553 cannon with 125 rounds per gun), and the type can be provided with a limited air-to-air capability by the carriage of two Matra 550 Magic AAMs; a wide assortment of underwing armament can be lifted, typical weapons being older-generation air-to-surface missiles (AS.11 and AS.12), free-fall bombs, gun pods etc; the type was built in South Africa with the Viper Mk 540 engine as the **Atlas Impala Mk 2**
M.B.326L: armed two-seater based on the M.B.326K with a second cockpit instead of the inbuilt cannon armament
M.B.326M: Viper 11-engined unarmed two-seater for South Africa; the type was also built in that country as the armed **Atlas Impala Mk 1**

Aermacchi M.B.339A
(Italy)
Type: basic and advanced trainer with secondary attack capability
Accommodation: pupil and instructor seated in tandem on Martin-Baker IT10F ejector seats
Armament (fixed): none
Armament (disposable): this is carried on six underwing hardpoints, the inner four each rated at 454 kg (1,000 lb) and the outer pair each at 340 kg (750 lb), up to a maximum weight of 1814 kg (4,004 lb); an exceptionally wide variety of stores can be accommodated, including two Macchi gun pods each containing a DEFA 553 cannon with 120 rounds or an AN/M-3 0.5-in (12.7-mm) machine-gun with 350 rounds, or six SUU-11/A pods each with a 7.62-mm (0.3-in) Minigun and 1,500 rounds, or two Matra 550 Magic or AIM-9 Sidewinder air-to-air missiles, or four 1,000-lb (454-kg) bombs, or six 750-lb (340-kg) bombs, or six 500-lb (227-kg) bombs, or six Aerea AL-25-50 launchers for 25 50-mm (1.97-in) rockets, or six Aerea AL-18-50 launchers for 18 50-mm (1.97-in) rockets, or six Aerea AL-12-80 launchers for 12 81-mm (3.19-in) rockets, or six Matra 155 or F2 launchers each for seven 68-mm (2.68-in) rockets, or six LAU-32/G launchers for seven 2.75-in (69.85-mm) rockets, or six LAU-68/A launchers for 19 2.75-in (69.85-in) rockets, or four LAU-10/A launchers for four 5-in (127-mm) rockets
Electronics and operational equipment: communication and navigation equipment, plus Elettronica ECM pod, or one photo-reconnaissance pod with four 70-mm Vinten cameras, a radar-wirning receiver, and a reflector gunsight (Aeritalia, Saab or Thomson-CSF)
Powerplant and fuel system: one 1814-kg (4,000-lb) thrust Piaggio-built Rolls-Royce Viper Mk 632-43 turbojet, and a total internal fuel capacity of 1413 litres (311 Imp gal) in two fuselage and two non-jettisonable wingtip tanks, plus provision for two 325-litre (71.5-Imp gal) drop-tanks
Performance: maximum speed 817 km/h (508 mph) or Mach 0.75 at 9145 m (30,000 ft), and 898 km/h (558 mph) or Mach 0.73 at sea level; initial climb rate

2010 m (6,595 ft) per minute; service ceiling 14630 m (48,000 ft); combat radius 270 km (168 miles) on a lo-lo-lo sortie with six 500-lb (227-kg) bombs; ferry range 2110 km (1,310 miles) with two drop-tanks
Weights: empty 3135 kg (6,911 lb); normal take-off 4400 kg (9700 lb); maximum take-off 5895 kg (12,996 lb) with maximum external stores
Dimensions: span 10.858 m (35 ft 7.5 in); length 10.972 m (36 ft 10 in); height 3.90 m (12 ft 9.5 in); wing area 19.30 m² (207.74 sq ft)

Variants
M.B.339A: first flown in prototype form in August 1976, the M.B.339A is a much-improved conceptual development of the M.B.326 with more advanced aerodynamics and structure for improved performance and handling using the same basic engine as the M.B.326 series; apart from its better handling characteristics, the chief operational advantage of the M.B.339A is the considerably better rear seating arrangement for the instructor, who is now raised above the level of the front-seat pupil for good forward vision; a slightly modified version is the **M.B.339PAN** for the Italian air force aerobatic team, with smoke-generating equipment but no tiptanks
M.B.339B: version of the M.B.339A designed for the dual training and light attack roles
M.B.339C: introduced in December 1985, this is the attack-optimized two-seat version of the series with digital avionics compatible with the latest weapons, and fitted with GEC AD-660 Doppler navigation and AD-620K integrated tactical navigation system linked with a Litton LR80 inertial platform, two Kaiser Sabre head-up displays, a Logic sores-management system, a Fiar/Ericsson laser ranger, Aeritalia CRT displays and provision for items such as the Elettronica ELT/156 radar-warning receiver, ELT/555 jammer pod and Tracor ALE-40 chaff/flare dispenser; the weapon options are similar to those of the Veltro 2 (which is to be offered with avionics improvements of the M.B.339C) plus provision for AGM-65 Maverick air-to-surface missiles and Marte II anti-ship missiles; the type is powered by the 4,450-lb (2019-kg) Viper Mk 680 turbojet, and the capacity of the tip tanks is increased to 510 litres (112 Imp gal) each for a total fuel capacity of 1773 litres (390 Imp gal)

Aermacchi M.B.339K Veltro 2
(Italy)
Type: light attack aircraft
Accommodation: pilot seated on a Martin-Baker IT10F ejector seat
Armament (fixed): two 30-mm DEFA 553 cannon with 125 rounds per gun
Armament (disposable): this is carried on six underwing hardpoints, the inner four each rated at 454 kg (1,000 lb) and the outer pair each at 340 kg (750 lb), up to a maximum weight of 1815 kg (4,001 lb); an exceptionally wide variety of stores can be accommodated, including two Macchi gun pods each containing a DEFA 553 cannon with 120 rounds or an AN/M-3 0.5-in (12.7-mm) machine-gun with 350 rounds, or six SUU-11/A pods each with a 7.62-mm (0.3-in) Minigun and 1,500 rounds, or two Matra 550 Magic or AIM-9 Sidewinder air-to-air missiles, or four 1,000-lb (454-kg) bombs, or six 750-lb (340-kg) bombs, or six 500-lb (227-kg) bombs, or six Aerea AL-25-50 launchers for 25 50-mm (1.97-in) rockets, or six Aerea AL-18-50 launchers for 18 50-mm (1.97-in) rockets, or six Aerea AL-12-80 launchers for 12 81-mm (3.19-in) rockets, or six Matra 155 or F2 launchers each for seven 68-mm (2.68-in) rockets, or six LAU-32/G launchers for seven 2.75-in (69.85-mm) rockets, or six LAU-68/A launchers for 19 2.75-in (69.85-in) rockets, or four LAU-10/A launchers for four 5-in (127-mm) rockets
Electronics and operational equipment: communication and navigation equipment, plus a Saab-Scania RGS2 gyro gunsight and options (head-up display, ECM etc) to customer requirements
Powerplant and fuel system: one 1814-kg (4,000-lb) thrust Piaggio-built Rolls-Royce Viper Mk 632-43 turbojet, and a total internal fuel capacity of 1660 litres (365 Imp gal) in a three-cell fuselage tank and two non-jettisonable tiptanks, or 2030 litres (446.5 Imp gal) with larger tiptanks, plus provision for two 325-litre (71.5-Imp gal) drop-tanks
Performance: maximum speed 900 km/h (559 mph) or Mach 0.73 at sea level; initial climb rate 2286 m (7,500 ft) per minute; service ceiling 13565 m (44,500 ft); combat radius 648 km (403 miles) on a hi-lo-hi sortie with a 1088-kg (2,400-lb) load, or 376 km (234 miles) on lo-lo-lo sortie with the same load
Weights: empty 3175 kg (7,000 lb); normal take-off 5980 kg (10,980 lb); maximum take-off 6150 kg (13,558 lb)
Dimensions: span 10.858 m (35 ft 7.5 in) over standard tiptanks, or 11.045 m (35 ft 2.75 in) over circular-section tiptanks; length 10.792 m (35 ft 5 in); height 3.90 m (12 ft 9.5 in); wing area 19.30 m² (207.74 sq ft)

Variant
M.B.339K Veltro 2: this is the single-seat light attack aircraft derivative of the M.B.339A analogous to the M.B.326K; the rear cockpit is again used for fuel, avionics (including an inertial navigation system and a head-up display) and two 30-mm DEFA 553 cannon; the model is powered by the 4,450-lb (2019-kg) Viper 680 turbojet

Aero L-29 Delfin 'Maya'
(Czechoslovakia)
Type: two-seat basic and advanced jet trainer with limited secondary attack capability

Accommodation: pupil and instructor seated in tandem on ejector seats
Armament (fixed): none
Armament (disposable): two 7.62 mm (0.3-in) machine-gun pods, or two 100-kg (220-lb) bombs, or eight unguided rockets, or two drop-tanks, mounted on one pylon beneath each wing
Electronics and operational equipment: communications and navigation equipment
Powerplant and fuel system: one 890-kg (1,960-lb) thrust Motorlet M701 turbojet, and a total internal fuel capacity of 1050 litres (231 Imp gal) of fuel in two fuselage tanks, plus provision for two 150-litre (33-Imp gal) underwing drop-tanks
Performance: maximum speed 655 km/h (407 mph) at 5000 m (16,405 ft); initial climb rate 840 m (2,755 ft) per minute; service ceiling 11000 m (36,090 ft); range 640 km (397 miles) on internal fuel, or 895 km (555 miles) with auxiliary fuel
Weights: empty 2280 kg (5,027 lb); normal take-off 3280 kg (7,231 lb); maximum take-off 3540 kg (7,804 lb)
Dimensions: span 10.29 m (33 ft 9 in); length 10.81 m (35 ft 5.5 in); height 3.13 m (10 ft 3 in); wing area 19.78 m² (212.88 sq ft)

Variant
L-29 Delfin: designed to meet Warsaw Pact requirements for a simple yet capable basic and advanced jet trainer, the aircraft designated **'Maya'** by NATO first flew in April 1959, and was subsequently built in very large numbers for the Warsaw Pact countries and for export; among the type's advantages are great strength, viceless handling characteristics, and the ability to operated from semi-prepared, sandy and even waterlogged airstrips; the **L-29A Delfin Akrobat** was a single-seat aerobatic model produced in very small numbers

Aero L-39ZA Albatros
(Czechoslovakia)
Type: light attack aircraft
Accommodation: two in tandem on ejector seats
Armament (fixed): a pod containing one two-barrel GSh-23L 23-mm cannon and between 150 and 180 rounds may be attached under the fuselage
Armament (disposable): a maximum of 1100 kg (2,425 lb) of stores can be accommodated on four underwing hardpoints, the inner pair each rated at 500 kg (1,102 lb) and the outer pair each at 250 kg (551 lb); typical loads include 500-, 250- and 100-kg (1,102-, 551- and 220-lb) bombs, UB-16 rocket-launcher pods each with 16 55-mm (2.17-in) unguided rockets, S-130 130-mm (5.12-in) unguided rockets, AA-2 or AA-8 air-to-air missiles, a five-camera reconnaissance pod, and 350-litre (77-Imp gal) drop tanks
Electronics and operational equipment: communications and navigation equipment, plus an ASP-3-NMO-39 sight
Powerplant and fuel system: one 1720-kg (3,972-lb) thrust Walter Titan turbofan (licence-built Ivchenko AI-25-TL) or a 1900-kg (4,188-lb) thrust updated version of the same engine, and 1255 litres (276 Imp gal) of internal fuel in five fuselage bag tanks and two non-jettisonable tiptanks, plus additional fuel instead of the second cockpit and provision for two 350-litre (77-Imp gal) drop-tanks
Performance: maximum speed with stores 630 km/h (391 mph) at 6000 m (19,685 ft); initial climb rate 960 m (3,160 ft) per minute; service ceiling 9000 m (29,530 ft); range 780 km (485 miles) with stores, and 1600 km (994 miles) for ferrying
Weights: empty 3330 kg (7,341 lb); normal take-off 4570 kg (10,075 lb); maximum take-off 5270 kg (11,618 lb)
Dimensions: span 9.46 m (31 ft 0.5 in); length 12.32 m (40 ft 5 in); height 4.72 m (15 ft 5.5 in); wing area 18.80 m² (202.36 sq ft)

Variants
L-39C: designed as successor to the L-29 in much the same way as the Aermacchi M.B.339 was developed from the basis of the M.B.326, the L-39 was the initial production version of this important tandem-seat basic and advanced trainer, and first flew in prototype form in November 1968; apart from a host of aerodynamic and other changes (including modular construction for ease of replacement, and features for simple maintenance access), the L-39 introduced turbofan power for economy combined with performance' the type is also being developed in an improved 'second-generation' version for service in the later 1980s; this variant has generally upgraded and modernized airframe and powerplant, plus considerably improved avionics
L-39V: target-tug version of the L-39
L-39ZA: development of the L-39ZO for ground attack and reconnaissance, featuring reinforced landing gear and provision for an underfuselage gun pod
L-39ZO: weapons training version of the L-39 with four hardpoints under the reinforced wings
L-39MS: usefully uprated variant introduced in 1987 with a 2400-kg (5,291-lb) turbofan for better performance though not increased weapons load; the type also has an improved airframe and revised avionics, the latter including cathode ray tube cockpit displays

Aérospatiale CM.170-1 Magister
(France)
Type: multi-role trainer and light attack aircraft
Accommodation: pupil and instructor in tandem
Armament (fixed): two 7.5- or 7.62-mm (0.295- or 0.3-in) machine-guns in the nose with 200 rounds per gun

Armament (disposable): this is carried on two underwing hardpoints, and can be made up of two Matra 181 launchers each with 18 37-mm (1.46-in) rockets, or two Matra F2 launchers each with six 68-mm (2.68-in) rockets, or four 25-kg (55-lb) air-to-surface rockets, or eight 88-mm (3.46-in) air-to-surface rockets, or two 50-kg (110-lb) bombs, or two AS.11 air-to-surface missiles
Electronics and operational equipment: communication and navigation equipment, plus a gyro gunsight
Powerplant and fuel system: two 400-kg (882-lb) thrust Turboméca Marboré IIA turbojets and a total internal fuel capacity of 730 litres (161 Imp gal) in two fuselage tanks and 250 litres (55 Imp gal) in two non-jettisonable wingtip tanks
Performance: maximum speed 715 km/h (414 mph) at 9145 m (30,000 ft), and 650 km/h (403 mph) at sea level; initial climb rate 1020 m (3,345 ft) per minute; service ceiling 11000 m (36,090 ft); range 925 km (575 miles)
Weights: empty 2150 kg (4,740 lb); normal take-off 3100 kg (6,834 lb); maximum take-off 3200 kg (7,055 lb)
Dimensions: span 12.15 m (39 ft 10 in) over tiptanks; length 10.06 m (33 ft 0 in); height 2.80 m (9 ft 2 in); wing area 17.30 m² (186.1 sq ft)

Variants
CM.170-1 Magister: designed as the world's first jet trainer by Fouga (later Potez and later still Aérospatiale), the CM.170 flew in prototype form during July 1952 on the power of two 400-kg (882-lb) Turboméca Marboré II turbojets, and entered service with a number of detail improvements; the type was built to the extent of 761 aircraft in France, Finland, West Germany and Israel
CM.170-2 Super Magister: uprated version with two 480-kg (1,058-lb) Marboré VI turbojets; production amounted to 137 aircraft
CM.175 Zéphyr: designation of 32 navalized aircraft for the French navy, basically to CM.170–1 standard but fitted with arrester hooks
IAI Improved Fouga: designation of the enhanced model available from Israel Aircraft Industries as rebuilt aircraft or as an updating kit for local modification; in this programme the airframe is strengthened, the avionics are brought up to modern standards, Marboré VI engines are fitted, the inbuilt armament is standardized as two 0.3-in (7.62-mm) Browning machine-guns with 180 rounds per gun, and four underwing hardpoints are fitted for the carriage of rocket-launcher pods, Minigun pods, free-fall bombs, reconnaissance pods etc; the type has a maximum take-off weight of 3400 kg (7,496 lb)
IAI AMIT Fouga: serving with the Israeli air force as the **Tzukit** (thrush), this is the Advanced Multi-mission Improved Trainer version of the Magister incorporating all the Improved Fouga modifications plus extra avionics to improve reliability and operational capability

Aérospatiale SE 313B Alouette II
(France)
Type: light utility helicopter
Accommodation: pilot and passenger at the front of the cabin, and three passengers, or two litters and one attendant, or freight in the rear of the cabin
Armament (fixed): none
Armament (disposable): a wide variety of external stores can be carried to operator requirements, typical fits being magazine- or belt-fed machine-guns, 20-mm cannon, rocket-launcher pods or AS.12 air-to-surface missiles
Electronics and operational equipment: communication and navigation equipment
Powerplant and fuel system: one 395-kW (530-shp) Turboméca Artouste IIC6 turboshaft derated to 269 kW (360 shp), and a total internal fuel capacity of 580 litres (128 Imp gal) in a fuselage tank
Performance: maximum speed 185 km/h (115 mph) at sea level; cruising speed 165 km/h (102 mph) at sea level; initial climb rate 252 m (825 ft) per minute; service ceiling 2150 m (7,050 ft); range 300 km (186 miles) with a 390-kg (860-lb) payload, or 565 km (351 miles) with maximum fuel
Weights: empty 895 kg (1,973 lb); maximum take-off 1600 kg (3,527 lb)
Dimensions: main rotor diameter 10.20 m (33 ft 5.6 in); length (rotor folded) 9.66 m (31 ft 8.3 in); height 2.75 m (9ft 0.25 in); main rotor disc area 81.71 m² (879.6 sq ft)

Variant
SE 313B Alouette II: this Alouette (lark) was the world's first turboshaft-powered helicopter to enter production (by Sud-Aviation, later Aérospatiale), such a powerplant having been pioneered by the SE 3130 Alouette II which first flew in March 1955 on the power of a 269-kW (360-shp) Turboméca Artouste I; the Alouette II served with some 22 air arms, and production amounted to 923 helicopters for civil and military operators; apart from what was for the time good performance, the Alouette II was notable for its ability to carry a 600-kg (1,323-lb) slung load; the type is now obsolescent except in the communications role

Aérospatiale SA 318C Alouette II Astazou
(France)
Type: light utility helicopter
Accommodation: pilot and passenger seated side-by-side at the front of the cabin, and up to three passengers, or two litters and one attendant, or freight in the cabin rear
Armament (fixed): none
Armament (disposable): a wide variety of external

stores can be carried to operator requirements, typical fits being magazine- or belt-fed machine-guns, 20-mm cannon, rocket-launcher pods or two AS.11 air-to-surface missiles
Electronics and operational equipment: communication and navigation equipment
Powerplant and fuel system: one 395-kW (530-shp) Turboméca Astazou IIA turboshaft derated to 269 kW (360 shp), and a total internal fuel capacity of 580 litres (128 Imp gal) in a fuselage tank
Performance: maximum speed 206 km/h (127 mph) at sea level; cruising speed 180 km/h (112 mph) at sea level; initial climb rate 400 m (1,312 ft) per minute; service ceiling 3300 m (10,825 ft); range 100 km (62 miles) with 600-kg (1,323-lb) payload, and 720 km (447 miles) with maximum fuel
Weights: empty 890 kg (1,961 lb); maximum take-off 1650 kg (3,630 lb)
Dimensions: main rotor diameter 10.20m (33 ft 5.6 in); length (fuselage with tail rotor turning) 9.75 m (31 ft 11.75 in); height 2.75 m (9 ft 0.25 in); main rotor disc area 81.71 m² (879.6 sq ft)

Variant
SA 318C Alouette II Astazou: first flown in January 1961, this is a much improved version of the SE 313B with the fuel-efficient Astazou IIA turboshaft, which also offered greater reliability than the Artouste as it was down-rated from 395 kW (530 shp) to 269 kW (360 shp) in this application; production lasted into 1975, some 350 aircraft being built; of the Alouette II series some 963 went to military operators

Aérospatiale SA 315B Lama
(France)
Type: general-purpose helicopter
Accommodation: pilot and passenger side-by-side on the front seat, with three passengers behind them
Armament (fixed): none
Armament (disposable): none
Electronics and operational equipment: communication and navigation equipment
Powerplant and fuel system: one 649-kW (870-shp) Turboméca Artouste IIIB turboshaft derated to 410 kW (510 shp), and a total internal fuel capacity of 575 litres (126 Imp gal) in a centre fuselage tank
Performance: cruising speed 192 km/h (119 mph); initial climb rate 330 m (1,080 ft) per minute; service ceiling 5400 m (17,715 ft); range 515 km (320 miles)
Weights: empty 1021 kg (2,251 lb); normal take-off 1950 kg (4,300 lb); maximum take-off 2300 kg (5,070 lb) with an externally-slung load of 1135 kg (2,502 lb)
Dimensions: main rotor diameter 11.0 m (36 ft 1.75 in); length (fuselage) 10.24 m (33 ft 7.25 in); height 3.09 m (10 ft 1.25 in); main rotor disc area 95.38 m² (1,026.7 sq ft)

Variant
SA 315B Lama: developed to meet an Indian requirement of the late 1960s for a general-purpose helicopter with good 'hot-and-high' performance, the SA 315B first flew in March 1969 and is in essence the airframe of the Alouette II with the dynamic system of the Alouette III; the type began to enter service in 1970, and has since been built under licence in India as the **HAL Cheetah** and in Brazil as the **Helibras HB 315B Gaviao**

Aérospatiale SA 316B Alouette III
(France)
Type: light utility helicopter
Accommodation: pilot and two passengers side-by-side at the front of the cabin, and up to four passengers, or two litters and two attendants, or freight at the cabin rear
Armament (fixed): none
Armament (disposable): a wide variety of armament fits are possible to operator requirements, typical fits being (in the cabin) one 7.62-mm (0.3-in) AA52 machine-gun with 1,000 rounds, or one MG 151/20 20-mm cannon with 480 rounds, or one M621 20-mm cannon with 350 rounds, and (on external rails) four launchers each with a variable number of 68-mm (2.68-in) rockets, or four AS.11 air-to-surface missiles, or four AS.12 air-to-surface missiles, or two Mk 46 homing torpedoes
Electronics and operational equipment: communication and navigation equipment, plus APX-Bézu 260 gyro-stabilized sight for use in conjunction with the missile fit
Powerplant and fuel system: one 649-kW (870-shp) Turboméca Artouste IIIB turboshaft derated to 425 kW (570 shp), and a total internal fuel capacity of 575 litres (126 Imp gal) in one fuselage tank
Performance: maximum speed 210 km/h (130 mph) at sea level; cruising speed 185 km/h) 115 mph) at sea level; initial climb rate 260 m (850 ft) per minute; service ceiling 3200 m (10,500 ft); range 540 km (335 miles)
Weights: empty 1122 kg (2,474 lb); maximum take-off 2200 kg (4,850 lb)
Dimensions: main rotor diameter 11.02 m (36 ft 1.75 in); length (fuselage with rotor folded) 10.03 m (32 ft 10.75 in); height (to rotor head) 3.00 m (9ft 10 in); main rotor disc area 95.38 m² (1,026.7 sq ft)

Variants
SA 316A Alouette III: developed under the designation SE 3160, the Alouette III was a basic upgrading and updating of the Alouette II, featuring greater power (matched by an upgraded transmission system), a fully-covered pod-and-boom fuselage, a large cabin and other improvements to offer higher payload (including a slung load of 750 kg/1,653 lb) and performance together with the reliability and alti-

tude performance as the Alouette II; the prototype first flew in February 1959, and production followed in 1961

SA 316B Alouette III: introduced in 1968, this improved model has strengthened transmission and improved landing gear, a combination allowing a further increase in payload; the type was built under licence in India as the **HAL Chetak** and in Switzerland by FFA

SA 316C Alouette III: limited-production version powered by the Artouste IIID turboshaft

ICA-Brasov IAR-316B Alouette III: designation of the basic type licence-built in Romania; a developed version of the type is the **IAR-317 Airfox**, locally evolved on the basis of the Alouette III/Artouste IIIB combination as a light attack helicopter with a tandem two-seat forward fuselage with solid nose; the prototype first flew in 1984 and the type is being developed with two fixed 20-mm cannon and pylons for the carriage of 600 kg (1,323 lb) of external stores such as unguided rockets or IR-homing air-to-air missiles; the type has a standard range of 525 km (326 miles), this being boosted to 180 km (503 miles) with two external tanks; the type's service ceiling is 6200 m (20,340 ft)

Aérospatiale SA 319B Alouette III Astazou
(France)

Type: light utility helicopter
Accommodation: pilot and two passengers side-by-side at the front of the cabin, and up to four passengers, or two litters and two attendants, or freight in the cabin rear
Armament (fixed): none
Armament (disposable): a wide variety of armament fits are possible to operator requirements, typical fits being (in the cabin) one 7.62-mm (0.3-in) AA52 machine-gun with 1,000 rounds, or one MG 151/20 20-mm cannon with 480 rounds, or one M621 20-mm cannon with 350 rounds, and (on external rails) four launchers each with a variable number of 68-mm (2.68-in) rockets, or four AS.11 air-to-surface missiles, or four AS.12 air-to-surface missiles, or two Mk 46 homing torpedoes
Electronics and operational equipment: communication and navigation equipment, plus alternative provision for an APX-Bézu 260 gyro-stabilized sight (battlefield model), or Omera ORB 31 search radar (anti-ship model), or ORB 31 radar and magnetic anomaly detection (MAD) gear (anti-submarine model)
Powerplant and fuel system: one 649-kW (870-shp) Turboméca Astazou XIV turboshaft derated to 447-kW (600 shp), and a total internal fuel capacity of 575 litres (126 Imp gal) in one fuselage tank
Performance: maximum speed 220 km/h (136 mph) at sea level; cruising speed 197 km/h (122 mph) at sea level; initial climb rate 270 m (885 ft) per minute;

range 605 km (375 miles) with six passengers
Weights: empty 1108 kg (2,442 lb); maximum take-off 2250 kg (4,960 lb)
Dimensions: main rotor diameter 11.02 m (36 ft 1.75 in); length (fuselage with rotor folded) 10.03 m (32 ft 10.75 in); height (to rotor head) 3.00 m (9 ft 10 in); main rotor disc area 95.38 m² (1,026.7 sq ft)

Variants
SA 319B Alouette III Astazou: just as the SA 318C was developed as an Astazou-powered SE 313B, the SA 319B was produced and first flown in 1967 as a more economical and more reliable machine as the SA 319B with the Astazou XIV down-rated from 649 kW (870 shp) to 447 kW (600 shp) in this application; total production of the Alouette III amounted to 1,453, the greater portion of them for military service in a number of roles

Atlas Alpha-XH1: this is a South African gunship development combining the dynamic system of the Alouette III Astazou with a few fuselage and an advanced weapon system; the crew of two is seated in tandem in upward staggered separate cockpits, and the gun armament comprises a GA1 20-mm cannon (with 1,000 rounds) in a servo-actuated mounting under the fuselage; it is likely that this prototype will be developed into a production model with stub wings for the carriage of rocket-launchers and/or anti-tank missiles for additional capability in the battlefield role

Aérospatiale SA 321G Super Frelon
(France)

Type: anti-submarine helicopter
Accommodation: crew of two on the flightdeck, plus mission crew of three and up to 27 passengers in the cabin
Armament (fixed): none
Armament (disposable): this is carried on the sides of the fuselage outside the cabin, and can comprise four Mk 46 homing torpedoes, or two AM.39 Exocet anti-ship missiles, or eight 250-kg (551-lb) mines, or Mk 49, Mk 52 or Mk 54 depth charges
Electronics and operational equipment: communication and navigation equipment, plus a self-contained navigation system with Doppler radar, Sylphe panoramic search radar in outrigger floats and AQS-13 dunking sonar (ASW role), or Omera-Segid Héraclès ORB 31D or ORB 32 designation radar (anti-ship role); other fits can include minesweeping gear and a 275-kg (606-lb) capacity rescue winch
Powerplant and fuel system: three 1170-kW (1,570-shp) Turboméca Turmo IIIC6 turboshafts, and a total internal fuel capacity of 3975 litres (874 Imp gal) in flexible tanks under the floor of the central fuselage, plus provision for four 500-litre (110-Imp gal) auxiliary tanks (two internal and two external)
Performance: maximum speed 248 km/h (154 mph) at sea level cruising speed 248 km/h (154 mph) at sea

level; initial climb rate 300 m (985 ft) per minute; service ceiling 3100 m (10,170 ft); range 815 km (506 miles) on normal fuel at sea level, or 1020 km (633 miles) at sea level with a 3500-kg (7,716-lb) payload
Weights: empty 6863 kg (15,130 lb); maximum take-off 13000 kg (28,660 lb)
Dimensions: main rotor diameter 18.90 m (62 ft 0 in); length (fuselage) 19.40 m (63 ft 7.75 in); height (blades and tail folded) 5.20 m (17 ft 0.75 in); main rotor disc area 280.55 m² (3,019 sq ft)

Variants
SA 321G Super Frelon: developed as the production version of the SA 3210 Super Frelon (itself a development of the SA 3200 Frelon, or hornet) designed with the aid of Sikorsky in the USA, the SA 321G first flew in November 1965 and provides the French navy with a powerful amphibious anti-submarine helicopter for the clearing of the approaches of the approaches to the bases for French nuclear submarines
SA 321H Super Frelon: simplified land-based transport helicopter with three Turmo IIIE turboshafts, and a payload of 30 troops, or 15 litters, or a load of 5000 kg (11,023 lb) carried internally or externally
SA 321K Super Frelon: non-amphibious transport and assault version developed for Israel on the basis of the SA 321J commercial model; delivered from 1967 with Turmo engines, the helicopters have since been re-engined with General Electric T58 turboshafts
SA 321L Super Frelon: modified version of the SA 321K sold to China, Libya and South Africa
Harbin Z-8: Chinese licence-built SA 321J with Chinese engines

Aérospatiale SA 330L Puma
(France)
Type: medium-lift transport helicopter
Accommodation: crew of one or two on the flight-deck, with provision for a third crew member if required, and up to 20 troops, or six litters and six seated casualties, or freight in the cabin, or a slung load of 3200 kg (7,055 lb)
Armament (fixed): none
Armament (disposable): a wide variety of armament can be carried, including 20-mm cannon, 7.62-mm (0.3-in) machine-guns, rockets and missiles
Electronics and operational equipment: communication and navigation equipment, plus optional radar (Bendix RDR 1400 or RCA Primus 40 or 50) for search-and-rescue role
Powerplant and fuel system: two 1175-kW (1,575-shp) Turboméca Turmo IVC turboshafts, and a total internal fuel capacity of 1544 litres (340 Imp gal) in one auxiliary and four main tanks under the cabin floor, plus provision for 2600 litres (572 Imp gal) of ferry fuel in four internal and two external tanks

Performance: cruising speed 258 km/h (160 mph); initial climb rate 366 m (1,200 ft) per minute; service ceiling 4800 m (15,750 ft); range 550 km (341 miles)
Weights: empty 3615 kg (7,970 lb); maximum take-off 7400 kg (16,315 lb)
Dimensions: main rotor diameter 15.00 m (49 ft 2.5 in); length (fuselage) 14.06 m (46 ft 1.5 in); height 5.14 m (16 ft 10.5 in); main rotor disc area 176.7 m² (1,902.2 sq ft)

Variants
SA 330B Puma: designed from the early 1960s in response to a French army requirement for an all-weather day/night tactical transport helicopter, the Puma first flew in prototype from during April 1965, and the SA 330B initial production version (powered by two Turmo IIIC4 turboshafts) began to enter service in March 1969, proving highly successful in its intended role and paving the way for substantial export orders; the French are currently evaluating the possibility of a battlefield surveillance derivative of the SA 330B providing 360-degree radar coverage with a retractable ventral antenna
SA 330C Puma: export version with Turmo IVC turboshafts
SA 330E Puma: version of the SA 330B for the UK, which designates the type **Puma HC. Mk 1**
SA 330H Puma: uprated version of the SA 330C
SA 330L Puma: improved export model with Turmo IVC turboshafts, inlet deicing and rotor blades of composite structure
ICA-Brasov IAR-330 Puma: version built under licence in Romania

Aérospatiale AS 332B Super Puma
(France)
Type: medium-lift multi-role helicopter
Accommodation: crew of one or two on the flight-deck, with provision for a third crew member if necessary, and up to 21 troops, or six litters and seven seated casualties or freight in the cabin, or a slung load of 4500 kg (9,921 lb)
Armament (fixed): none
Armament (disposable): typically this can comprise one M621 20-mm cannon with 720 rounds, or two 7.62-mm (0.3 in) machine-guns (NF1 or FN MAG) with 200 rounds per gun, or two rocket pods, each containing 19 2.75-in (69.85-mm) rockets, or 22 68-mm (2.68-in) rockets, or 36 68-mm (2.68-in) rockets
Electronics and operational equipment: communication and navigation equipment, including a Decca or Nadir self-contained navigation system with a navigation computer, roller-map display etc
Powerplant and fuel system: two Turboméca Makila 1A turboshafts each rated at 1327 kW (1,789 shp) for contingencies and at 1145 kW (1,535 shp) for continuous running, and a total internal fuel capacity of 1560 litres (343 Imp gal) in flexible underfloor cells

plus provision for 2600 litres (572 Imp gal) of ferry fuel in four internal and two external auxiliary tanks
Performance: cruising speed 280 km/h (173 mph) at sea level; initial climb rate 528 m (1,732 ft) per minute; range 635 km (394 miles) with standard tankage, and 1720 km (1,068 miles) with full auxiliary tankage
Weights: empty 4100 kg (9,039 lb); normal take-off 8350 kg (18,410 lb); maximum take-off 9350 kg (20,615 lb)
Dimensions: main rotor diameter 15.58 m (51 ft 1.5 in); length (fuselage) 14.76 m (48 ft 5 in); height 4.92 m (16 ft 1.75 in); main rotor disc area 190.6 m² (2,052.1 sq ft)

Variants
AS 332B Super Puma: the Super Puma was designed from 1974 as a successor to the basic SA 330 Puma with Makila turboshafts for greater payload (21 rather than 20 troops) and performance, and with features to reduce cabin noise, maintenance requirements and operational vulnerability; the AS 331 prototype was first flown in September 1977 to test the revised dynamic system (new powerplant and upgraded transmission), and the first AS 332 flew in September 1978; deliveries began in November 1981, and the Super Puma has proved attractive for its low-maintenance but damage-resistant composite rotor blades, self-sealing fuel tanks, duplicated electric and hydraulic lines, damage-resistant transmission and other tactically desirable features
AS 332F Super Puma: naval version with a folding tail and suitable for the search-and-rescue role (with Bendix RDR 1400 or RCA Primus 40/50 radar), anti-submarine role (with HS 12 dunking sonar and two Mk 46 torpedoes) or anti-ship role (with ORB 3214 radar and two AM.39 Exocet, or six AS.15TT, or one Exocet and three AS.15TT anti-ship missiles)
AS 332M Super Puma: upgraded version with the cabin lengthened by 0.76 m (2.5 ft) to make possible the carriage of 25 troops
NAS-332 Super Puma: designation of the series built under licence in Indonesia

Aérospatiale SA 341F Gazelle
(France)
Type: light utility helicopter
Accommodation: crew of one or two side-by-side at the front of the cabin, and up to three passengers on a bench seat behind them, or a slung load of 700 kg (1,540 lb)
Armament (fixed): none
Armament (disposable): this is carried on pylons projecting one on each side of the fuselage, and can comprise two machine-gun pods, each with a single 7.62-mm (0.3-in) AA52 or FN MAG machine-gun (alternatively, a Minitat turret with one 7.62-mm/0.3-in Minigun can be fitted under the fuselage), or two twin AS.11 air-to-surface missiles, or two AS.12 air-to-surface missiles, or four TOW anti-tank missiles, or two pods each with seven 2.75-in (69.85-mm) rockets, or two pods each with 10 68-mm (2.68-in) rockets, or two pods each with 12 68-mm (2.68-in) rockets
Electronics and operational equipment: communication and navigation equipment, and (AS.11 or AS.12 installation) an APX-Bézu 334 gyro-stabilized sight or (TOW installation) and XM26 sight
Powerplant and fuel system: one 440-kW (590-shp) Turboméca Astazou IIIC turboshaft, and a total internal fuel capacity of 445 litres (98 Imp gal) in one fuselage tank, plus provision for 290 litres (64 Imp gal) of auxiliary fuel in two ferry tanks (one under the baggage compartment floor and one in the cabin)
Performance: maximum speed 264 km/h (164 mph) at sea level; cruising speed 233 km/h (144 mph) at sea level; initial climb rate 540 m (1,770 ft) per minute; service ceiling 5000 m (16,405 ft); range 670 km (416 miles) with maximum fuel, or 360 km (223 miles) with one pilot and 500-kg (1,102-lb) payload
Weights: empty 920 kg (2,028 lb); maximum take-off 1800 kg (3,968 lb)
Dimensions: main rotor diameter 10.50 m (34 ft 5.5 in); length (fuselage) 9.53 m (31 ft 3.2 in); height 3.18 m (10 ft 5.25 in); main rotor disc area 86.6 m² (932.1 sq ft)

Variants
SA 341B Gazelle: this important light helicopter originated in a French army requirement of the early 1960s for a light observation helicopter (successor to the Alouette II series) characterized by good manoeuvrability and considerable speed; the type was developed initially as the SA 340 with a conventional tail rotor (first flown in April 1967), though this was replaced in later prototypes with the 'fenestron' shrouded rotor; the initial production model was the SA 341B with an Astazou IIIA turboshaft, and this model was adopted for British army service as the **Gazelle AH.Mk 1** with the Astazou IIIN turboshaft
SA 341C Gazelle: naval version of the SA 341B for the Royal Navy, which accepted the type as the **Gazelle HT.Mk 2**
SA 341D Gazelle: trainer version of the SA 341B for the RAF, which accepted the type as the **Gazelle HT.Mk 3**
SA 341E Gazelle: communications version for the RAF, which accepted the type as the **Gazelle HCC.Mk 4**
SA 341F Gazelle: initial version for the French army with the Astazou IIIC turboshaft; the **SA 341F/Canon Gazelle** is an interim gunship conversion with one 20-mm M621 cannon fixed on the starboard side of the cabin; there is also a reconnaissance version with a simplified version of the APX397 Athos magnifying sight used on the SA 342M

SA 341H Gazelle: export version of the SA 341B/F type with the Astazou IIIB turboshaft, and built under licence in Yugoslavia by Soko with a possible armament of AT-3 'Sagger' wire-guided anti-tank missiles

SA 342K Gazelle: the SA 342 designation is used for the uprated model powered by the Astazou XIV turboshaft for greater performance and payload, especially under 'hot-and-high' operating conditions; the SA 342K series has the 649-kW (870-shp) Astazou XIVH with momentum-separation shrouds over the inlet

SA 342L Gazelle: military version of the uprated SA 342J civil helicopter with an improved 'fenestron' tail; rotor and a number of detail modifications allowing the type to operate at a maximum take-off weight of 1900 kg (4,189 lb) with a wide assortment of military loads

SA 342M Gazelle: dedicated anti-tank helicopter for the French army, and powered by the 640-kW (858-shp) Astazou XIVM turboshaft; the type is fitted with the SFIM APX397 gyro-stabilized sight for the guidance of four or six HOT and anti-tank missiles in two twin or triple installations, though the type can also be armed with two 7.62-mm (0.3-in) machine-guns or one 20-mm cannon; deliveries began in June 1980, and other modifications are the inclusion of Doppler radar plus self-contained navigation system, an autopilot, an infra-red sight, and an exhaust deflector for reduced vulnerability to ground-launched IR-homing missiles

Aérospatiale AS 350B Ecureuil
(France)
Type: general-purpose helicopter
Accommodation: pilot and up to five passengers in the cabin, or a 750-kg (1,653-lb) slung load
Armament (fixed): none
Armament (disposable): rocket pods or machine-gun pods
Electronics and operational equipment: communication and navigation equipment
Powerplant and fuel system: one 468-kW (641-shp) Turboméca Arriel turboshaft, and a total internal fuel capacity of 530 litres (117 Imp gal) in a fuselage tank
Performance: maximum cruising speed 232 km/h (144 mph); initial climb rate 480 m (1,575 ft) per minute; service ceiling 4575 m (15,000 ft); range with maximum fuel 700 km (435 miles)
Weights: empty 1065 kg (92,348 lb); maximum take-off 2100 kg (4,630 lb) with a slung load
Dimensions: main rotor diameter 10.69 m (35 ft 0.75 in); length (fuselage) 10.91 m (35 ft 9.5 in); height 3.15 m (10 ft 4 in); main rotor disc area 89.75 m² (966.1 sq ft)

Variants
AS 350B Ecureuil: designed in the early 1970s as a fuel-economical successor to the Alouette II on the civil market, and first flown in June 1974, the Ecureuil (squirrel) is one of the manufacturer's most successful light helicopters and has also sold moderately well on the military market, largely as a communications machine; the type is built in Brazil as the **Helibras HB 350 Esquilo**

AS 350C Astar: version for the American market with the Avco Lycoming LTS101 turboshaft

AS 350D Astar: uprated helicopter for the American market with a more powerful version of the LTS101

AS 350L Ecureuil: dedicated military version of the Ecureuil series with outrigger pylons in a removable through-cabin beam for the carriage of light weapons, a TOW anti-tank missile installation also being available; other modifications are a strengthened floor, upgraded landing gear and sliding rather than hinged doors

AS 351 Ecureuil: version under development as an uprated single-engine type with the 560-kW (751-shp) Turboméca TM333 turboshaft

AS 355E Ecureuil 2/Twinstar: much developed version (for the world and the American markets respectively) first flown in September 1979 with two 425-shp (317-kW) Allison 250-C20F turboshafts for great operational reliability; aircraft for the French services have two 330-kW (443-shp) Turboméca TM319 turboshafts

AS 355F Ecureuil 2/Twinstar: uprated version with greater payload and wider-chord main rotor blades; the type can lift a slung payload of 1045 kg (2,300 lb), and typical roles in military service are pilot training and casevac

AS 355M Ecureuil 2: twin-engine equivalent of the AS 350L Ecureuil with the same armament provisions and comparable modifications for improved operational reliability

Aérospatiale SA 365F Dauphin 2
(France)
Type: anti-ship helicopter
Accommodation: crew of two seated side-by-side in the front of the cabin, and up to eight passengers in the cabin
Armament (fixed): none
Armament (disposable): this is carried on two struts (one projecting from each side of the cabin), and comprises a pair of AS.15TT anti-ship missiles on each strut
Electronics and operational equipment: communication and navigation equipment, plus Thomson-CSF Agrion 15 radar on a roll-stabilized pivot mount under the nose for use in conjunction with the AS.15TT missiles
Powerplant and fuel system: two 529-kW (710-shp) Turboméca Arriel 520M turboshafts, and a total internal fuel capacity of 1140 litres (250 Imp gal) in four underfloor tanks and one centre-fuselage tank,

plus provision for 475 litres (104 Imp gal) in an optional ferry tank in the cabin
Performance: maximum speed 252 km/h (156 mph) at sea level; cruising speed 259 km/h (161 mph) at optimum altitude; range 900 km (559 miles) with maximum fuel at sea level
Weights: empty 2141 kg (4,720 lb); maximum take-off 3900 kg (8,598 lb)
Dimensions: main rotor diameter 11.93 m (39 ft 1.75 in); length (fuselage) 12.11 m (39 ft 8.75 in); height 3.99 m (13 ft 1 in); main rotor disc area 111.78 m² (1,203.2 sq ft)

Variants
SA 360 Dauphin: developed in the late 1960s and early 1970s as successor to the Alouette III series, the SA 360 first flew in prototype form in June 1972 is a trim helicopter with a 783-kW (1,050-shp) Turboméca Astazou XVIIIA turboshaft, a 'fenestron' tail rotor and accommodation for a crew of two plus up to 12 passengers, or four litters and a medical attendant, or 2.5 m³ (88.2 cu ft) of freight; only limited sales were made to the military, largely in the communications role, and though an SA 361H anti-tank/assault transport was developed as a private venture, Aérospatiale soon appreciated that both military and civil applications would be better served by a twin-engine powerplant
SA 365C Dauphin 2: initial twin-engine version, flown in prototype form during January 1975 with two Turboméca Arriel turboshafts and produced later in the decade with 492-kW (660-shp) Arriel 1A turboshafts or, in the improved SA 365N Dauphin 2 version, with 529-kW (710-shp) Arriel 1C turboshafts; the SA 365N introduced a large degree of composite construction as well as retractable landing gear
SA 365F Dauphin 2: versatile naval development of the SA 365N intended primarily for the anti-ship role with Agrion 15 search radar, optional Crouzet magnetic anomaly detection with a towed 'bird', an armament of two or four AS.15TT short-range anti-ship missiles, and the avionics for mid-course targeting update of ship-launched Otomat long-range anti-ship missiles; the type is also available in search-and-rescue configuration with ORB 32 search radar, a rescue winch, an automatic navigation system and a hover/transition coupler; the manufacturer is also proposing a more advanced anti-submarine capability in a derivative with HS 12 dunking sonar, magnetic anomaly detection and an armament of two light-weight homing torpedoes
SA 365M Panther: first flown in February 1984, the Panther is a dedicated military version of the Dauphin 2 series with two 634-kW (850-shp) Turboméca TM333-1M turboshafts for a maximum take-off weight of 4100 kg (9,039 lb), strengthened cabin floor and landing gear, sliding rather than hinged doors, crash-resistant fuel tanks and the ability to carry eight or 10 troops in the assault role, or two quadruple launch units for HOT anti-tank missiles in the offensive role; the HOT missiles are targeted with the aid of a Viviane day/night unit for a SFIM gyro-stabilized platform holding a TRT Hector infra-red camera and SAT deviation-measuring equipment; other elements of the advanced avionics suite are Lacroix chaff dispenser and a Sherloc radar-warning receiver; alternative armament fits on the weapons beam are 20-mm cannon pods or two pods each containing 22 68-mm (2.68-in) unguided rockets
SA 366G Dauphin 2: development of the SA 365N for the US Coast Guard's short-range recovery role with two 618-shp (461-kW) Avco Lycoming LTS101-750A-1 turboshafts and much US-made equipment; the type has a crew of three, and can detect people in the water with the aid of forward-looking infra-red equipment before hauling them to safety with a rescue winch; the prototype first flew in July 1980, and the improved **SA 366G-1** has been accepted for the USCG with the designation **HH-65A Dolphin**
Harbin Z-9 Zaitun: designation of the SA 365M built under licence in China

Agusta A 109A Mk II
(Italy)
Type: general-purpose helicopter
Accommodation: crew of one or two on the flight-deck, and up to seven troops, or two litters and two attendants, or 630 kg (1,389 lb) of freight in the cabin
Armament (fixed): none
Armament (disposable): this is carried on two pylons, one on each side of the cabin, and can be made up of an extremely wide variety of stores depending on the mission; Agusta has made great weapon-qualification efforts to boost sales.
Electronics and operational equipment: communication and navigation equipment, plus specialized avionics
Powerplant and fuel system: two 420-shp (313-kW) Allison 250-C20B turboshafts, derated to 346 shp (258 kW) each for normal twin-engine output, and a total internal fuel capacity of 560 litres (123 Imp gal) in two tanks located in the lower rear fuselage, plus provision for auxiliary tanks containing 138 kg (304 lb) or 198 kg (436 lb)
Performance: cruising speed 272 km/h (160 mph); initial climb rate 503 m (1,650 ft) per minute; service ceiling 4570 m (15,000 ft); range 555 km (345 miles) with maximum fuel
Weights: empty between 1560 and 1790 kg (3,439 and 3,946 lb) depending on role; normal take-off 2330 kg (5,137 lb) as a medevac helicopter; maximum take-off 2600 kg (5,732 lb) as an anti-tank or ECM helicopter
Dimensions: main rotor diameter 11.00 m (36 ft 1 in); length (fuselage) 10.706 m (35 ft 1.5 in); height (over fin) 3.30 m (10 ft 10 in); main rotor disc area 95.03 m² (1,022.9 sq ft)

Variants
A 109A: this was the first production model of this important twin-turboshaft multi-role helicopter which, among other things, has compact lines and retractable landing gear; the type first flew in August 1971 and has since sold well on the civil market, with a few passing to the military in the utility and communications roles; the type was qualified for an extremely wide assortment of weapons, but no armed model have been sold

A 109A Mk II: uprated development of the A 109A with strengthened transmission to permit the use of 740 shp (552 kW) under continuous-cruise conditions; again the type has been qualified with a multitude of weapons options, but again no sales of overtly armed helicopters have been made

A 109K: variant developed with an eye to the African and Middle Eastern nations, and thus optimized for good performance under 'hot-and-high' conditions with two 539-kW (723-shp) Turboméca Arriel 1K turboshafts driving a strengthened transmission; the type also has a slightly longer nose to make possible the installation of additional avionics should particular customers so require

Agusta A 129 Mangusta
(Italy)
Type: anti-tank helicopter
Accommodation: co-pilot/gunner and pilot seated in tandem
Armament (fixed): none
Armament (disposable): this is carried on four hardpoints under the stub wings, the inner pair each rated at 300 kg (661 lb) and the outer pair each at 200 kg (441 lb); the primary armament is four, six or eight BGM-71 TOW anti-tank missiles (carried in two-, three- or four-round pods) on the outer hardpoints, leaving the inner stations for two pods each with a single 12.7-mm (0.5-in) machine-gune, or two pods each for seven 2.7-in (69.85-mm) rockets; alternatively, rocket-launcher pods can be carried on all four hardpoints, or two seven- or 19-tube launchers can be matched with two 12.7-mm (0.5-in) machine-gun pods
Electronics and operational equipment: communication and navigation equipment, plus the sight associated with the TOW installation, comprising an undernose fitting with a laser rangefinder and forward-looking infra-red, and a self-protection system with Elettronica/E-Systems radar-warning receiver, ITT radar jammer, Sanders infra-red jammer, Perkin-Elmer radar and laser warning receiver and an engine infra-red diffuser
Powerplant and fuel system: two 815-shp (608-kW) Rolls-Royce Gem 2-2 turboshafts, and a total internal fuel capacity of ? litres (? Imp gal) in two separate but cross-fed fuselage fuel systems
Performance: maximum speed 270 km/h (168 mph) at sea level; cruising speed 250 km/h (155 mph) at 2000 m (6,560 ft); initial climb rate 600 m (1,968 ft) per minute; endurance 2.5 hours on an anti-tank mission
Weights: empty 2530 kg (5,578 lb); maximum take-off 3655 kg (8,058 lb)
Dimensions: main rotor diameter 11.90 m (39 ft 0.5 in); length (fuselage) 12.275 m (40 ft 3.25 in); height (rotor head) 3.35 m (11 ft 0 in); main rotor disc area 111.2 m^2 (1,196.95 sq ft)

Variants
A 129 Mangusta: this is the first tandem-seat anti-tank helicopter developed in Europe; though derived conceptually from the A 109 series, the Mangusta (mongoose) is a totally new machine offering good performance, powerful anti-tank armament and associated sights, and a small fuselage profile featuring vertically stagged tandem seating for the gunner (front cockpit) and pilot (rear cockpit); the type first flew in September 1983, and is scheduled to enter service in the second half of the 1980s; the type offers full day and night offensive capability, and the Integrated Multiplex System monitors the helicopter and all its systems via two computers, so leaving the crew to devote their attentions to the mission

A 129 Tonal: this is the proposed Mk 2 co-production version to be developed and produced with Westland in the UK and with other European nations (possibly Belgium, the Netherlands and Spain); the type will have advanced avionics and weapons, and possibly only a single engine in the form of the Rolls-Royce/Turboméca RTM322 turboshaft

Agusta (Bell) AB.204AS
(Italy/USA)
Type: anti-submarine helicopter
Accommodation: pilot and co-pilot on the flightdeck, and two mission crew or up to eight passengers or freight in the cabin
Armament (fixed): none
Armament (disposable): two Mk 44 torpedoes (anti-submarine role) or two AS.12 air-to-surface missiles (anti-ship role)
Electronics and operational equipment: communication and navigation equipment, plus APN-195 search radar, Bendix AQS-13B dunking sonar, automatic stabilization system, and hover-coupling equipment
Powerplant and fuel system: one 1,290-shp (962-kW) General Electric T58-GE-3 turboshaft, and a total internal fuel capacity of 705 litres (155 Imp gal) in fuselage tanks
Performance: cruising speed 167 km/h (104 mph) at sea level; combat radius 110-km (68-mile) for a 1.66-hour sonar search patrol
Weights: empty 2940 kg (6,481 lb); maximum take-off 4310 kg (9,501 lb)

Dimensions: main rotor diameter 14.63 m (48 ft 0 in); length (fuselage) 12.67 m (41 ft 7 in); height 3.84 m (12 ft 7.25 in); main rotor disc area 168.1 m² (1,809.56 sq ft)

Variants
AB.204B: built between 1961 and 1974 for a number of armed forces, this is the Italian licence-built version of the Bell Model 204B utility helicopter; the type can be fitted with a number of light armament installations, and is found with main rotors of 44-ft (13.41-m) or 48-ft (14.63-m) diameter; the type is also found with the Avco Lycoming T53, General Electric T58 or Rolls-Royce Gnome H.1200 turboshaft
AB.204AS: dedicated naval version developed in Italy, and able to handle anti-submarine and/or light anti-ship tasks from comparatively small naval platforms

Agusta (Bell) AB.205 (Italy): see Bell UH-1H Iroquois

Agusta (Bell) AB.206 (Italy): see Bell Model 206L-1 LongRanger II

Agusta (Bell) AB.212ASW
(Italy/USA)
Type: anti-submarine and anti-shipping strike helicopter
Accommodation: crew of two on the flightdeck, and one or two systems operators in the cabin; alternative cabin loads are seven passengers, or four litters and an attendant
Armament (fixed): provision for an Emerson gun turret
Armament (disposable): this is carried on the sides of the fuselage, and can comprise two Sea Killer or AS.12 anti-ship missiles, or for the anti-submarine role two Mk 46 homing torpedoes or depth charges
Electronics and operational equipment: communication and navigation equipment, plus Canadian Marconi APN-208(V)2 Doppler radar, Canadian Marconi CMA-708B/ASW computer, Bendix AQS-13B dunking sonar and SMA/APS search radar, and (with AS.12 missile fit) XM58 gyro-stabilized sight; provision for TG-2 data-link for mid-course guidance of ship-launched Otomat missiles
Powerplant and fuel system: one 1,875-shp (1398-kW) Pratt & Whitney Canada PT6T-6 Turbo Twin Pac turboshaft, and a total internal fuel capacity of 815 litres (179 Imp gal) in five fuselage fuel cells, plus provision for one internal or two external auxiliary fuel tanks
Performance: maximum speed 195 km/h (121 mph) at sea level; cruising speed 185 km/h (115 mph) with armament; initial climb rate 396 m (1,300 ft) per minute; range 665 km (413 miles) with auxiliary tanks, or an anti-ship strike 615 km (312 miles)
Weights: empty 3420 kg (7,540 lb); maximum take-off 5070 kg (11,177 lb) for an ASW mission with two Mk 46 torpedoes
Dimensions: main rotor diameter 14.63 m (48 ft 0 in); length (fuselage and turning tail rotor) 14.00 m (45 ft 11.25 in); height (rotor head) 3.92 m (12 ft 10.25 in); main rotor disc area 168.1 m² (1,809.56 sq ft)

Variants
AB.212: this is the Italian licence-built version of the Bell Model 212/UH-1N series, with only marginal differences from the American original, and sold in southern Europe, North Africa and the Middle East
AB.212ASW: advanced anti-submarine and limited anti-ship model developed by Italy for use from larger naval platforms and from shore bases; the airframe is strengthened in comparison with that of the AB.212, but the avionics and powerplant are different so that the weapons can be used to full effect; the type can carry a slung load of 2268 kg (5,000 lb), and has proved popular with southern European, Middle Eastern and South American navies for its compact dimensions but very useful avionics/weapon combination; Italian helicopters are being upgraded with the Marte Mk II system for air-launched Sea Killer anti-ship missiles

Agusta (Bell) AB.412 Grifone
(Italy/USA)
Type: multi-role helicopter
Accommodation: pilot and up to 14 troops, or six litters and two attendants, or freight in the cabin
Armament (fixed): one 25-mm Hughes Chain Gun cannon
Armament (disposable): this is carried on two fuselage-side hardpoints, and can consist of four launchers each with seven 2.75-in (69.85-mm) rockets, or four air-to-surface missiles, or four air-to-air missiles, or two 12.7-mm (0.5-in) machine-gun pods, or other stores to suit operator requirements
Electronics and operational equipment: communication and navigation equipment, plus a self-contained navigation system permitting 'nap-of-the-earth' flight at night and under adverse weather conditions; also to be fitted at cutomer discretion are a variety of extra sensors, ECM etc
Powerplant and fuel system: one 1,800-shp (1342-kW) Pratt & Whitney Canada PT6T-3B-1 Turbo Twin Pac coupled turboshaft flat-rated to 1,308 shp (975 kW) for take-off and to 1,130 shp (843 kW) for continuous running, and a total internal fuel capacity of 821 litres (181 Imp gal) in five fuselage fuel cells, plus provision for 610 litres (134 Imp gal) in auxiliary tanks
Performance: cruising speed 230 km/h (143 mph) at sea level; initial climb rate 442 m (1,450 ft) per minute; service ceiling 4330 m (14,205 ft); range 370

km (230 miles) with maximum payload, or 420 km (261 miles) with maximum fuel
Weights: empty 2823 kg (6,223 lb); maximum take-off 5262 kg (11,600 lb)
Dimensions: main rotor diameter 14.02 m (46 ft 0 in); length (fuselage) 12.92 m (42 ft 4.75 in); height 4.32 m (14 ft 2.25 in); main rotor disc area 154.4 m² (1,662.0 sq ft)

Variants
AB.412: this is the Italian licence-built copy of the Bell Model 412, a development of the Model 212 with a four-blade main rotor; the type is intended primarily for civil applications, but can also be used in the utility military role
AB.412 Grifone: dedicated military version of the AB.412, intended as a multi-role aircraft but with particular capabilities in the assault transport and hard-target attack roles, in the former carrying 15 troops and in the latter using missiles such as the TOW anti-tank type and the Sea Skua anti-ship type

Agusta (Sikorsky) ASH-3H
(Italy/USA)
Type: multi-role helicopter
Accommodation: crew of two on the flightdeck, and cabin accommodation for two ASW systems operators, or 31 troops, or 15 litters and one attendant, or 25 survivors
Armament (fixed): none
Armament (disposable): this is carried externally, and comprises (ASW role) four Mk 46 homing torpedoes or depth charges, or (anti-ship role) two Sea Killer Mk 2 or AM.39 Exocet or AGM-84 Harpoon missiles, or alternatively four AS.12 air-to-surface missiles
Electronics and operational equipment: communication and navigation equipment plus Doppler radar, Bendix AQS-13B dunking sonar and SMA/APS radar, the last specially adapted for successful operation in an atmosphere of dense electromagnetic radiation so as to provide targeting and guidance data for long-range missiles
Powerplant and fuel system: two 1,500-shp (1119-kW) General Electric T58-GE-100 turboshafts, and a total internal fuel capacity of 3180 litres (700 Imp gal) in underfloor bag tanks, plus provision for one internal auxiliary fuel tank
Performance: cruising speed 222 km/h (138 mph); initial climb rate 670 m (2,200 ft) per minute; service ceiling 3720 m (12,200 ft); range 580 km (360 miles) with 31 troops, or 1165 km (724 miles) with maximum standard fuel
Weight: maximum take-off 9525 kg (21,000 lb)
Dimensions: main rotor diameter 18.90 m (62 ft 0 in); length (fuselage) 16.69 m (54 ft 9 in); height 5.13 m (16 ft 10 in); main rotor disc area 280.5 m² (3,019.1 sq ft)

Variants
ASH-3D: this is the Italian licence-built version of the Sikorsky SH-3D Sea King anti-submarine helicopter, and differs only slightly from the original in items such as airframe strengthening, a revised tailplane, slightly different armament and avionics, and an uprated powerplant
ASH-3D/TS: VIP transport version of the Italian series, otherwise designated **AS-61A-4** and **AS-61VIP**
ASH-3H: licence-built version of the SH-3H Sea King with role optimization for anti-submarine and anti-ship warfare; the type can also carry a freight load of 2720 kg (5,996 lb) carried internally or of 3630 kg (8,003 lb) carried externally as a slung load; other tasks undertaken by the ASH-3H are anti-ship missile defence, electronic warfare and tactical trooping

AIDC AT-TC-3
(Taiwan)
Type: advanced trainer and light attack aircraft
Accommodation: pupil and instructor seated in tandem on ejector seats
Armament (fixed): two 0.5-in (12.7-mm) Colt-Browning machine-guns each with ? rounds (optional in lower-fuselage bay)
Armament (disposable): this is carried in a lower-fuselage bay, five hardpoints (one under the fuselage and four under the wings) and two wingtip rails for air-to-air missiles, up to a maximum weight of 5,511 lb (2500 kg); typical stores are air-to-air and air-to-surface missiles, rocket-launcher pods, bombs, gun pods and other stores
Electronics and operational equipment: communication and navigation equipment, plus a gunsight in the front cockpit
Powerplant and fuel system: two 3,500-lb (1588-kg) thrust Garrett TFE731-2-2L turbofans, and a total internal fuel capacity of 1180 kg (2,601 lb) in fuselage and wing tanks
Performance: maximum speed 560 mph (901 km/h) at sea level and 562 mph (904 km/h) or Mach 0.85 at 36,000 ft (10970 m); initial climb rate 8,000 ft (2440 m) per minute; service ceiling 48,000 ft (14625 m); range 1,265 miles (2035 km)
Weights: empty 8,500 lb (3855 kg); normal take-off 11,500 lb (5216 kg); maximum take-off 16,500 lb (7485 kg)
Dimensions: span 34 ft 3.75 in (10.46 m); length including nose probe 42 ft 4 in (12.90 m); height 14 ft 3.75 in (4.36 m); wing area 236.05 sq ft (21.93 m²)

Variant
AT-TC-3: developed from 1975 under the initial designation XAT-3 as an advanced trainer with secondary light attack capability; the prototype flew in September 1980 and the type began to enter service in 1984; keynotes of the design are simplicity and ease of maintenance in the airframe and powerplant,

advanced avionics, and compact dimensions for great manoeuvrability

AIDC T-CH-1
(Taiwan)
Type: trainer and light attack aircraft
Accommodation: pupil and instructor seated in tandem
Armament (fixed): none
Armament (disposable): this is carried on two underwing hardpoints, and can consist of light bombs, rocket-launcher pods and machine-gun pods
Electronics and operational equipment: communication and navigation equipment
Powerplant and fuel system: one 1,450-ehp (1081-kW) Avco Lycoming T53-L-701 turboshaft, and a total internal fuel capacity of 963 litres (212 Imp gal) in one fuselage and four wing tanks
Performance: maximum speed 590 km/h (367 mph) at 4570 m (15,000 ft); cruising speed 410 km/h (255 mph) at 4570 m (15,000 ft); initial climb rate 1035 m (3,395 ft) per minute; service ceiling 9755 m (32,000 ft); range 2010 km (1,250 miles)
Weights: empty 2608 kg (5,750 lb); normal take-off 3402 kg (7,500 lb); maximum take-off 5057 kg (11,150 lb)
Dimensions: span 12.19 m (40 ft 0 in); length 10.26 m (33 ft 8 in); height 3.66 m (12 ft 0 in); wing area 25.18 m² (271.0 sq ft)

Variant
T-CH-1: though claimed as an indigenous Taiwanese aircraft from the start of the design process in 1970, the T-CH-1 seems to be little more than the North American T-28 Trojan piston-engined trainer revised for a turboprop powerplant; the provision of light armament capability suits the type to the weapon training and light attack/counter-insurgency roles, and the type began to enter service in late 1976 after evaluation of the unarmed T-CH-1A and armed T-CH-1B prototypes had confirmed the advantages of the latter

Antonov An-2 'Colt'
(USSR)
Type: STOL utility transport
Accommodation: crew of two on the flightdeck, and up to 12 passengers or 1300 kg (2,866 lb) of freight in the cabin
Armament (fixed): none
Armament (disposable): none
Electronics and operational equipment: communication and navigation equipment
Powerplant and fuel system: one 746-kW (1,000-hp) Shvetsov ASh-62R radial piston engine, and a total internal fuel capacity of 1200 litres (264 Imp gal) in six upper-wing tanks
Performance: maximum speed 255 km/h (158 mph) at 1750 m (5,740 ft); cruising speed 190 km/h (118 mph); initial climb rate 210 m (690 ft) per minute; service ceiling 4400 m (14,435 ft); range 900 km (559 miles) with a 500-kg (1,102-lb) payload
Weights: empty 3450 kg (7,606 lb); maximum take-off 5500 kg (12,125 lb)
Dimensions: span 18.18 m (59 ft 8.5 in); length 12.74 m (41 ft 9.5 in); height 6.10 m (20 ft 0 in); wing area 71.60 m² (770.7 sq ft)

Variants
An-2 'Colt': though apparently an anachronism is these days of advanced-technology aircraft, the piston-engined, fixed-gear An-2 biplane is still a very worthy type in the STOL utility transport role for which it was designed, especially in those less-advanced areas of the world where the communist creed predominates; the type first flew in prototype form in August 1947, and more than 5,000 examples had been produced in the USSR by 1960, when continued production was switched to Poland's WSK-PZL Mielec, which produced another 10,000 by the end of 1984 as production began finally to tail off; the type has been produced in a number of subvariants over the years, the more notable types optimized for the aerial survey, air ambulance, general-purpose transport and parachuting roles
Shijiazhung Y-5: designation of the An-2 built in China since 1957 to the extent of 1,500 or more aircraft; the Chinese have also developed a turbo-prop-powered model

Antonov An-12BP 'Cub-A'
(USSR)
Type: medium transport
Accommodation: crew of five on the flightdeck and in glazed nose, plus a rear gunner, and up to 100 paratroops or a 20000-kg (44,092-lb) payload in the hold
Armament (fixed): two NR-23 23-mm cannon in the rear turret
Armament (disposable): none
Electronics and operational equipment: communication and navigation equipment, plus weather and navigation radar
Powerplant and fuel system: four 2983-ekW (4,000-ehp) Ivchenko AI-20K turboprops, plus a total internal fuel capacity of 18100 litres (3,981 Imp gal) in 22 wing tanks
Performance: maximum speed 775 km/h (482 mph); cruising speed 670 km/h (416 mph); initial climb rate 600 m (1,970 ft) per minute; service ceiling 10200 m (33,465 ft); range 3600 km (2,237 miles) with maximum payload, or 5700 km (3,542 miles) with maximum fuel
Weights: empty 28000 kg (61,728 lb); normal take-off 55100 kg (121,473 lb); maximum take-off 61000 kg (134,480 lb)

Dimensions: span 38.00 m (124 ft 8.1 in); length 33.10 m (108 ft 7.25 in); height 10.53 m (34 ft 6.5 in); wing area 121.7 m² (1,310.0 sq ft)

Variants

An-12BP 'Cub': designed to the mid-1950s as a civil and military transport based on the An-10 airliner, the An-10 may be regarded as the USSR's equivalent to the Lockheed C-130 Hercules used by the USAF and many other Western air arms; the An-12 was first flown in prototype form during 1958, and production amounted to some 900 aircraft all featuring the high wing and upswept tail designed to provide and unobstructed and easily loaded hold in a fuselage supported for rough-field operations on main landing gear units retracting into fuselage blister fairings; the primary roles of the type in air force service are tactical transport and airborne movement

An-12 'Cub-A': electronic intelligence variant with revised interior and small exterior modifications such as blade antennae in the area of the flight deck and small pressurized compartment

An-12 'Cub-B': electronic intelligence derivative of the standard aircraft for service with the Soviet naval air arm; these aircraft (perhaps 10 in all) sport four blister fairings under the forward and centre fuselage, and also feature a number of blade antennae; the specific role of the type is the collection of radiation intelligence about Western warships, the hold being provided with receivers, analyzers, recording and (probably) data-link equipment for realtime communication of findings to ship and shore bases

An-12 'Cub-C': electronic countermeasures variant of the standard aircraft with a solid ogival tailcone in place of the tail turret; the type carries several tonnes of electronic equipment including palletized jammers (operating in at least five and possibly 10 wavebands) and chaff dispensers

An-12 'Cub-D': electronic intelligence and electronic countermeasures variant of the standard aircraft with two large blister fairings extending side-by-side from the nose to the main landing gear fairings; the type also possesses active jammers and decoy systems

An-12 'Cub-?': type fitted with a palletized command system in the hold, providing Soviet forces with excellent battlefield command capabilities in areas where conventional radio equipment cannot function adequately because of terrain limitations

Hanzhong Y-8: designation of the An-12BP produced in China largely for tactical transport, though roles comparable to those of the Soviet aircraft are almost certainly planned as production steps up; the type has also been developed for maritime patrol with the H-6 (Tu-16 'Badger'), Litton APS-504 search and inertial navigation of western origins

Antonov An-14 'Clod'
(USSR)
Type: STOL utility transport
Accommodation: pilot and one passenger on the flightdeck, and up to seven passengers or 1590 kg (3,505 lb) of freight in the cabin
Armament (fixed): none
Armament (disposable): none
Electronics and operational equipment: communication and navigation equipment
Powerplant and fuel system: two 224-kW (300-hp) Ivchenko AI-4RF radial piston engines, and a total internal fuel capacity of 383 litres (84 Imp gal) in four wing tanks
Performance: maximum speed 220 km/h (137 mph) at 1000 m (3,280 ft); cruising speed 180 km/h (112 mph) at 2000 m (6,560 ft); initial climb rate 305 m (1,000 ft) per minute; service ceiling 5200 m (17,060 ft); range 650 km (404 miles) with maximum payload, and 800 km (497 miles) with maximum fuel
Weights: empty 2000 kg (4,409 lb); normal take-off 3450 kg (7,606 lb); maximum take-off 3630 kg (8,003 lb)
Dimensions: span 21.99 m (72 ft 1.75 in); length 11.44 m (37 ft 6.5 in); height 4.63 m (15 ft 2.5 in); wing area 39.72 m² (422.8 sq ft)

Variants

An-14 'Clod': designed as a light transport for civil and military applications in the mid-1950s, the An-14 first flew in prototype form during March 1958; production was undertaken between 1965 and 1975, some 300 aircraft being produced

An-28 'Cash': improved version first flown in September 1969 as the An-14M with aerodynamic refinements and two 604-kW (810-shp) Isotov TVD-850 turboprops; production began only during 1984 at the Mielec facility of WSK-PZL in Poland, the powerplant being two 716-kW (960-shp) Glushenkov TVD-10B turboprops; this model can carry a maximum of 21 passengers or a freight load of 1550 kg (3,417 lb)

Antonov An-22 'Cock'
(USSR)
Type: heavy freight transport
Accommodation: crew of five or six on the flightdeck and in the glazed nose, and up to 29 passengers plus an 80000-kg (176,367-lb) payload in the hold
Armament (fixed): none
Armament (disposable): none
Electronics and operational equipment: communication and navigation equipment, plus weather and navigation radar
Powerplant and fuel system: four 11185-ekW (15,000-ehp) Kuznetsov NK-12MA turboprops, and a total internal fuel capacity of 55800 litres (12,275 Imp gal) in integral wing tanks
Performance: maximum speed 740 km/h (460 mph); cruising speed 520 km/h (323 mph); service ceiling 7500 m (24,605 ft); range 5000 km (3,107 miles) with

maximum payload, and 10950 km (6,804 miles) with maximum fuel and a 45000-kg (99,206-lb) payload
Weights: empty 114000 kg (251,323 lb); maximum take-off 250000 kg (551,146 lb)
Dimensions: span 64.40 m (211 ft 4 in); length 57.92 m (190 ft 0 in); height 12.53 m (41 ft 1.5 in); wing area 345.0 m² (3,713.7 sq ft)

Variant
An-22 'Cock': developed primarily for resources-exploitation tasks in Siberia, the An-22 was in its time the world's largest aircraft; the type first flew in February 1965, and is characterized by its large unobstructed hold with access via a rear ramp, and by its multi-wheel landing gear to provide good soft-field performance in association with the 'blown' double-slotted flaps in the slipstream of the propellers of the four great turboprop engines; production ended in 1974 after some 100 aircraft had been made, Aeroflot and the Soviet air force transport force each receiving about equal numbers

Antonov An-24RT 'Coke'
(USSR)
Type: utility freight transport
Accommodation: crew of five including a freight-handler, and up to 5700 kg (12,566 lb) of freight in the cabin
Armament (fixed): none
Armament (disposable): none
Electronics and operational equipment: communication and navigation equipment
Powerplant and fuel system: two 1902-ekW (2,550-ehp) Ivchenko AI-24A turboprops and one 900-kg (1,985-lb) thrust RU 19-300 auxiliary turbojet in the starboard nacelle, and a total internal fuel capacity of 5240 litres (1,153 Imp gal) in four centre-section bag tanks and integral wing tanks
Performance: cruising speed 450 km/h (280 mph) at 6000 m (19,685 ft); initial climb rate 205 m (675 ft) per minute; service ceiling 9000 m (29,530 ft); range 640 km (397 miles) with maximum payload, and 3000 km (1864 miles) with maximum fuel
Weights: empty about 14725 kg (32,463 lb); maximum take-off 21800 kg (48,060 lb)
Dimensions: span 29.20 m (95 ft 9.5 in); length 23.53 m (77 ft 2.5 in); height 8.32 m (27 ft 3.5 in); wing area 74.98 m² (807.1 sq ft)

Variants
An-24V Series I 'Coke': first flown in December 1959 as the precursor of an important twin-turboprop transport family to supplant the Ilyushin Il–14 series, the An-24V Series I has accommodation for 28 to 40 passengers and entered service in October 1962 with 1902-ekW (2,550-ehp) Ivchenko AI-24 turboprops
An-24V Series II 'Coke': improved passenger version with accommodation for 52 passengers and introduced in 1967 with water-injected AI-24A turboprops plus a TG-16 gas turbine auxiliary power unit in the starboard nacelle for independent starting
An-24RV 'Coke': development of the An-24V Series II with a 900-kg (1,984-lb) thrust Tumansky RU-16-300 auxiliary turbojet (instead of the APU and designed to provide extra take-off thrust as well as APU capability) and two 2103-ekW (2,820-ehp) AI-24T turboprops
An-24T 'Coke': dedicated freighter version of the An-24V Series II with a freight door in the underside of the rear fuselage, and a freight-handling system in the cabin for a total payload of 4610 kg (10,163 lb)
An-24RT 'Coke': freighter version of the An-24RV with the same provisions as the An-24T
Xian Y-7: designation of the An-24 built in China and powered by WJ-5A-21 (AI-24A) turboprops

Antonov An-26B 'Curl'
(USSR)
Type: short-range transport
Accommodation: crew of five on the flightdeck, and up to 40 troops, or 24 litters and one attendant, or 5500 kg (12,125 lb) of freight in the cabin
Armament (fixed): none
Armament (disposable): none
Electronics and operational equipment: communication and navigation equipment, plus weather/navigation radar
Powerplant and fuel system: two 2103-ekW (2,820-ehp) Ivchenko AI-24VT turboprops and one 800-kg (1,765-lb) thrust RU 19A-300 auxiliary turbojet in the starboard nacelle, and a total internal fuel capacity of 7100 litres (1,562 Imp gal) in integral and bag tanks
Performance: cruising speed 440 km/h (273 mph) at 6000 m (19,685 ft); initial climb rate 480 m (1,575 ft) per minute; service ceiling 7500 m (24,605 ft); range 1100 km (684 miles) with maximum payload, and 2550 km (1,584 miles) with maximum fuel
Weights: empty 15020 kg (33,113 lb); normal take-off 23000 kg (50,706 lb); maximum take-off 24000 kg (52,911 lb)
Dimensions: span 29.20 m (95 ft 9.5 in); length 23.80 m (78 ft 1 in); height 8.575 m (28 ft 1.5 in); wing area 74.98 m² (807.1 sq ft)

Variants
An-26 'Curl': first seen by the West in 1969, the An-26 is clearly a pure-freighter derivative of the An-24 series, but has been radically redesigned around the rear fuselage to permit the incorporation a full-width door/ramp in its underside; the type also introduces a fully pressurised hold; the rear ramp can be swung under the fuselage to provide clear access to the hold from the beds of trucks, and to permit simple airdropping of embarked loads; other features are a 2000-kg (4,409-lb) capacity winch on a track running along the length of the hold, a floor-mounted conveyor belt,

and tip-up seats for a maximum of 40 paratroops; the type can also be outfitted for the carriage of 24 litters
An-26B 'Curl': introduced in 1981, this is an improved version with revised handling arrangements so that two men can deal with three pallets weighing 5500 kg (12,125 lb)
An-30 'Clank': first flown in 1974, the An-30 is the dedicated aerial survey member of the An-24/An-26 family based generally on the fuselage of the An-24 with the flying surfaces and powerplant An-26; the main modification is a revised forward fuselage with a raised flightdeck leaving the extensively glazed nose for the navigator; there are five camera ports (two of them oblique), and the fuselage has a darkroom and film storage facilities

Antonov An-32 'Cline'
(USSR)
Type: medium transport
Accommodation: crew of five on the flightdeck, and up to 39 troops, or 30 paratroops, or 24 litters and one attendant, or 6000 kg (13,228 lb) of freight in the cabin
Armament (fixed): none
Armament (disposable): none
Electronics and operational equipment: communication and navigation equipment, plus weather radar
Powerplant and fuel system: two 3863-ekW (5,180-ehp) Ivchenko AI-20DM turboprops, and a total internal fuel capacity of 5500 kg (12,125 lb) in integral wing tanks
Performance: cruising speed 510 km/h (317 mph) at 8000 m (26,245 ft); service ceiling 9500 m (31,168 ft); range 800 km (497 miles) with maximum payload, and 2200 km (1,367 miles) with maximum fuel
Weights: maximum take-off 26000 kg (57,319 lb)
Dimensions: span 29.20 m (95 ft 9.5 in); length 23.80 m (78 ft 1 in); height 8.575 m (28 ft 1.5 in); wing area 74.98 m² (807.1 sq ft)

Variant
An-32 'Cline': first flown in 1977, the An-32 is another derivative of the An-24 and An-26 series, but in this instance optimized for maximum performance under 'hot-and-high' conditions; to this end the wings and Ivchenko-designed powerplant are extensively modified, the former having automatic leading-edge slats and triple-slotted trailing-edge flaps, and the latter being a choice being the 3128-ekW (4,195-ehp) AI-20M turboprop for moderate-temperature areas or the 3863-ekW (5,180-ehp) AI-20DM for high-temperature areas; these engines drive 4.7-m (15.4-ft) diameter propellers rather than the 3.9-m (12.8-ft) diameter units of the earlier aircraft, and to provide adequate tip clearance the engines are raised to positions above the wings; the type can operate from airfields as high as 4500 m (14,765 ft), and is operated by the Indian air force under the name **Sutlej**

Antonov An-72 'Coaler'
(USSR)
Type: STOL light transport
Accommodation: crew of three on the flightdeck, and up to 32 passengers, or 24 litters and one attendant, or 10000 kg (22,046 lb) of freight (reduced to 3500 kg/7,717 lb for STOL operations) in the hold
Armament (fixed): none
Armament (disposable): none
Electronics and operational equipment: communication and navigation equipment, plus weather/navigation radar and Doppler navigation
Powerplant and fuel system: two 6500-kg (14,330-lb) thrust Lotarev D-36 turbofans operating as part of an upper-surface blowing system configuration for STOL capability, and a total internal fuel capacity of ? litres (? Imp gal) in integral wing tanks
Performance: maximum speed 760 km/h (472 mph); cruising speed 720 km/h (447 mph); service ceiling 10000 m (32,810 ft); range 1000 km (621 miles) with maximum payload, and 3800 km (2,361 miles) with maximum fuel
Weights: normal take-off 26500 kg (58,422 lb) from a 1000-m (1,095-yard) runway; maximum take-off 30500 kg (67,240 lb) from a 1200-m (1,310-yard) runway
Dimensions: span 25.83 m (8 ft 9 in); length 26.576 m (87 ft 2.25 in); height 8.235 m (27 ft 0.25 in); wing area about 90.0 m² (968.8 sq ft)

Variants
An-72 'Coaler': this advanced type was designed as the STOL counterpart to the An-24 and An-28 series, and first flew in prototype form in 1977; the concept used for the STOL performance is upper-surface blowing, the exhaust of the two turbofans blowing over the upper surface of the wings and flaps to produce high lift and controllability at low airspeeds; to keep the tailplane out of this disturbed airflow a T-tail is fitted on a rear fuselage derived conceptually from that of the An-26; production has been slow to get under way
An-74: production derivative of the An-72 with stretched fuselage (for a maximum payload of 10000 kg/ 22,046 lb carried over a range of 1150 km/715 miles) and revised wing; the type can also be used for artic operations when fitted with combined wheel/ski landing gear to complement its extensive deicing capability and advanced all-weather avionics

Antonov An-124 'Condor'
(USSR)
Type: strategic heavy-lift transport
Accommodation: crew of six on the flightdeck (with a full relief crew accommodated behind it), up to 88 passengers in a cabin above the hold and aft of the wing, and a maximum freight load of 150000 kg (330,693 lb)

Armament (fixed): none
Armament (disposable): none
Electronics and operational equipment: communication and navigation equipment including radar, all-weather avionics and triple inertial navigation systems
Powerplant and fuel system: four 23425-kg (51,642-lb) Lotarev D-18T non-afterburning turbofans, and a total internal fuel capacity of 220000 kg (485,009 lb) in integral wing tanks
Performance: maximum cruising speed 865 km/h (537 mph) at 12000 m (39,370 ft); cruising speed 800 km/h (497 mph); range with maximum payload 4500 km (2,796 miles); ferry range with maximum fuel 16000 km (9,942 miles)
Weights: maximum take-off 405000 kg (892,872 lb)
Dimensions: span 73.30 m (240 ft 5.8 in); length 69.50 m (228 ft 0.2 in); height 22.00 m (72 ft 2.1 in); wing area ? m² (? sq ft)

Variant
An-124 'Condor': designed in the 1970s as successor to the same bureau's An-22 strategic airlifter, the An-124 is sized to be able to lift a whole SS-20 missile system in a single load; the type entered service in 1986 as is a truly prodigious machine of typical transport configuration with a hold 36.0 m (118.1 ft) long, 6.4 m (21 ft) wide and 4.4 m (14.4 ft) high; the hold ceiling has two longitudinally-travelling gantries each rated at 10000 kg (22,046 lb) and fitted with two 5000-kg (11,023-lb) capacity transverse-moving winches; like the comparable Lockheed C-5 Galaxy, the An-124 is designed for through loading with a rear ramp and an upward-hinged nose

Atlas Alpha-XH1 (South Africa): see Aérospatiale SA 319B Alouette III Astazou

Atlas Cheetah (South Africa): see Dassault-Breguet Mirage IIIE

Atlas Impala (South Africa): see Aermacchi M.B.326

Beech C-12A Huron
(USA)
Type: light utility transport
Accommodation: crew of one or two on the flightdeck, and up to 13 passengers in the cabin
Armament (fixed): none
Armament (disposable): none
Electronics and operational equipment: communication and navigation equipment
Powerplant and fuel system: two 750-shp (559-kW) Pratt & Whitney Canada PT6A-38 turboprops, and a total internal fuel capacity of 386 US gal (1461 litres) in wing tanks, plus provision for gal (598 litres) of auxiliary fuel
Performance: maximum speed 303 mph (488 km/h) at 15,000 ft (4570 m); cruising speed 272 mph (438 km/h) at 30,000 ft (9145 m); service ceiling 29,200 ft (8900 m); range 1,825 miles (2937 km)
Weights: empty 7,800 lb (3538 kg); maximum take-off 12,500 lb (5670 kg)
Dimensions: span 54 ft 6 in (16.61 m); length 43 ft 9 in (13.34 m); height 15 ft 0 in (4.57 m); wing area 303.0 sq ft (28.15 m²)

Variants
C-12A Huron: this was the first version of the **Beech Super King Air 200** light transport to be ordered by the US forces; the Model 200 first flew in October 1972 as a much-improved King Air 100 with a T-tail, greater span and fuel capacity, more powerful engines and a higher cabin pressurization differential, all combining to increase range and cruising altitude; the C-12A is a utility transport for the US Air Force and US Army, the examples operated by the latter having been retrofitted with 850-shp (634-kW) Pratt & Whitney Canada PT6A-41 turboprops
UC-12B: version for the US Navy and Marine Corps with PT6A-41s, high-flotation landing gear and a cargo door
C-12C Huron: US Army version delivered with PT6A-41 turboprops
C-12D Huron: upgraded C-12C for the US Army with a cargo door and tip tanks, the latter increasing span to 55 ft 6 in (16.92 m)
RC-12D Huron: special-mission version of the C-12D designed for battlefield electronic surveillance in the South Korean and European theatres with the advanced USD-9(V)2 Guardrail V signals intelligence and direction-finding equipment including signal-processing gear, a secure data-link, ALQ-136 and ALQ-162 defensive electronic countermeasures pods at the wingtips (increasing span to 57 ft 10 in/17.63 m) and antennae above/below the wings; the type also features the ALQ-156 missile detection system; the first 13 aircraft has PT6A-41 turboprops and the last six more efficient PT6A-42 turboprops
UC-12D: version of the C-12D for the US Air Force and Air National Guard
C-12F: operational support aircraft for the US Air Force, based on the **Super King Air B200** and powered by PT6A-42 turboprops and able to carry a 2,500-lb (1134-kg) payload over a range of 2,400 miles (3862 km) at maximum and cruising speeds of 336 and 330 mph (541 and 531 km/h) respectively
RU-21J: despite its designation this was the first model for the US services, the type being accepted by the US Army in 1974 for the battlefield electronic reconnaissance role in the 'Cefly Lancer' programme; the aircraft have a maximum take-off weight of 15,000 lb (6804 kg) and carry a mass of electronic equipment similar to (but not as advanced as) that of the RC-12D
Super King Air 200T: special high-altitude photographic and weather reconnaissance aircraft for

France with Wild RC-10 cameras, Doppler navigation and optional tip tanks
Maritime Patrol B200T: coastal patrol aircraft based on the Super King Air B200 with high-flotation landing gear and optional tip tanks; there is a choice of two search radars (each with its antenna in a ventral radome for 360-degree coverage), forward-looking infra-red, low-light-level TV, advanced navigation systems and other modern systems; the type has a patrol range of 2,060 miles (3315 km)

Beech T-34C-1
(USA)
Type: two-seat trainer, weapons trainer, forward air control and counter-insurgency aircraft
Accommodation: two seated in tandem
Armament (fixed): none
Armament (disposable): this is carried on four underwing hardpoints, the inner pair rated at 600 lb (272 kg) each and the outer pair at 300 lb (136 kg) each, to a maximum weight of 1,200 lb (544 kg); weapons that can be carried include Mk 81 250-lb (113-kg) bombs, BLU-10/B incendiary bombs, SUU-11 pods for 7.62-mm (0.3-in) Miniguns and LAU-32 or LAU-59 rocket pods for 2.75-in (69.85-mm) rockets
Electronics and operational equipment: communication and navigation equipment, plus a Chicago Aerial CA-513 reflector sight
Powerplant and fuel system: one 715-shp (533-kW) Pratt & Whitney of Canada PT6A-25 turboprop, and a total internal fuel capacity of 129 US gal (488 litres) in two wing tanks
Performance: maximum speed 221 mph (355 km/h); initial climb rate 880 ft (268 m) per minute; range on a forward air control mission with 2.6 hours over the target 115 miles (185 km) or on an attack mission with four stores 345 miles (555 km)
Weights: empty 2,630 lb (1193 kmg); maximum take-off 5,500 lb (2495 kg)
Dimensions: span 33 ft 4 in (10.16 m); length 28 ft 8.5 in (8.75 m); height 9 ft 11 in (3.02 m); wing area 179.9 sq ft (16.71 m²)

Variants
T-34A: pure trainer version developed from the Model 35 Bonanza as the Beech Model 45 in response to a USAF requirement; the type first flew in December 1948 and was adopted as the USAF's primary trainer in 1953; the type is powered by a 225-hp (168-kW) Continental O-470-13 flat-six piston engine, has a maximum take-off weight of 2,950 lb (1315 kg) and possesses a maximum speed of 189 mph (304 km/h)
T-34B Mentor: US Navy equivalent of the T-34A
T-34C Turbine Mentor: much improved turbine-engined derivative produced in the early 1970s in response to a US Navy requirement for an updated model with considerably improved performance but greater reliability and fuel economy; the YT-34C prototype first flew in September 1973, and some 335 production aircraft were delivered between 1975 and 1984
T-34C-1: designation of the company-funded weapons trainer, forward air control and light attack version; this has four underwing hardpoints for a wide diversity of armament, and useful sales were made to a number of countries mainly in Africa and South America

Beech U-21A Ute
(USA)
Type: utility transport
Accommodation: crew of two on the flightdeck, and cabin accommodation for 10 troops, or six staff, or three litters and three sitting wounded plus an attendant, or 3,000 lb (1361 kg) of freight
Armament (fixed): none
Armament (disposable): none
Electronics and operational equipment: communication and navigation equipment
Powerplant and fuel system: two 550-shp (410-kW) Pratt & Whitney Canada PT6A-20 turboprops, and a total internal fuel capacity of 384 US gal (1454 litres) in nacelle and tip tanks
Performance: maximum speed 249 mph (401 km/h) at 11,000 ft (3355 ml); maximum cruising speed 245 mph (394 km/h) at 10,000 ft (3050 m); initial climb rate 2,000 ft (609 m) per minute; service ceiling 25,500 ft (7770 m); range with maximum payload and reserves 1,167 miles (1878 km); range with maximum fuel and reserves 1,676 miles (2697 km)
Weights: empty 5,464 lb (2478 kg); maximum take-off 9,650 lb (4377 kg)
Dimensions: span 45 ft 10.5 in (13.98 m); length 35 ft 6 in (10.82 m); height 14 ft 2.5 in (4.33 m); wing area 279.7 sq ft (25.98 m²)

Variants
VC-6A: VIP transport version of the **King Air A90** for the USAF with PT6A-20 turboprops
VC-6B: improved VIP transport version of the **King Air C90** with more capable pressurization; the basic type was also developed as the civil/military **King Air E90** series with 680-shp (507-kW) PT6A-28 turboprops flat-rated at 550 shp (410 kW)
T-44A: advanced trainer for the US Navy based on the King Air E90 but with features of the King Air C90, and with 750-shp (559-kW) PT6A-34B turboprops flat-rated at 550 shp (410 kW)
U-21A Ute: this is the designation of unpressurized military utility transport developed as the **King Air 65-90** from the King Air 90 series of pressurized civil light transports
EU/RU-21: this highly classified series serves mainly with the US Army Security Agency for specialized electronic reconnaissance with advanced electronic

sensors, sophisticated navigation and communication equipment, and many other modifications; known variants within this important series are the **EU-21A** electronic reconnaissance, **RU-21A** electronic reconnaissance/interception (with the ALQ-38 'Left Jab' direction-finding system), and **RU-21D** revised electronic reconnaissance/interception aircraft with the same powerplant as the U-21A, the **RU-21B** and **RU-21C** electronic reconnaissance/interception aircraft with 620-shp (462-kW) PT6A-29 turboprops, and the **RU-21E** and **RU-21H** upgraded and modernized electronic reconnaissance/interception aircraft with T74-CP-700 turboprops
U-21F: USAF version of the civil King Air A100 with 680-shp (507-kW) PT6A-28 turboprops and full pressurization
U-21G: updated version of the U-21A

Bell AH-1G HueyCobra
(USA)
Type: close-support and attack helicopter
Accommodation: co-pilot/weapons officer and pilot seated in tandem
Armament (fixed): one M28 chin turret able to house two 7.62-mm (0.3-in) GAU-2B/A Miniguns with 4,000 rounds per gun, or two M129 40-mm (1.57-in) grenade-launchers each with 300 rounds, or one Minigun and one M129 grenade-launcher
Armament (disposable): this is carried on four hardpoints under the stub wings, and can comprise four M159 launchers each with 19 2.75-in (69.85-mm) rockets, or four M157 launchers each with seven 2.75-in (69.85-mm) rockets, or two M18E1 7.62-mm (0.3-in) Minigun pods, or one M35 armament system with one M61A1 Vulcan rotary-barrel cannon (on the port inner hardpoint) with 1,000 rounds; the weapons officer is primarily responsible for the turreted armament and the pilot for the underwing armament, although each man can control both type of armament
Electronics and operational equipment: communication and navigation equipment, plus secure voice link, M73 rocket sight and M130 chaff-dispenser
Powerplant and fuel system: one 1,400-shp (1044-kW) Avco Lycoming T53-L-13 turboshaft derated to 1,100 shp (820 kW) for continuous running, and a total internal fuel capacity of 268 US gal (1014 litres) in one fuselage tank
Performance: maximum speed 172 mph (277 km/h) at sea level; initial climb rate 1,230 ft (375 m) per minute; service ceiling 11,400 ft (3475 m); range 357 miles (574 km)
Weights: empty 6,075 lb (2,756 kg); normal take-off 9,407 lb (4266 kg); maximum take-off 9,500 lb (4309 kg)
Dimensions: main rotor diameter 44 ft 0 in (13.41 m); length fuselage) 44 ft 7 in (13.59 m); height 13 ft 6.25 in (4.12 m); main rotor disc area 1,520.5 sq ft (141.26 m²)

Variants
AH-1G HueyCobra: first flown in September 1965 as the Bell Model 209, this important attack helicopter was evolved in response to urgent requests from the US Army in Vietnam for a helicopter gunship able to escort troop-carrying helicopters and to provide fire support for the landed troops; the result was the dynamic system of the UH-1C mated to a new fuselage of remarkable slimness (38 in/0.97 m) tailored to the width of the tandem-seated crew and supporting stub wings for disposable armament; gun armament was located in a trim turret under the nose; the type was rushed into production as the AH-1G and proved highly successful in the close support and attack roles; a few were converted as **TH-1G** trainers
AH-1Q HueyCobra: developed in the early 1970s, this was an interim anti-tank helicopter converted from AH-1G standard with provision for eight TOW anti-tank missiles (four under each outer underwing hardpoint) controlled with the aid of an M65 system for the weapons officer in the front seat
AH-1R HueyCobra: uprated AH-1G with the 1,800-shp (1342-kW) T53-L-703 turboshaft and without provision for TOW missiles

Bell AH-1S HueyCobra
(USA)
Type: anti-tank and close support helicopter
Accommodation: co-pilot/weapons officer and pilot seated in tandem
Armament (fixed): one General Electric Universal Turret in the chin position and able to accommodate weapons of 20- or 30-mm calibre (M197 three-barrel 20-mm cannot fitted in US aircraft)
Armament (disposable): this is carried on four hardpoints under the stub wings, and comprises four BGM-71 TOW anti-tank missiles (carried as two vertically-stacked twin launchers) under each outer hardpoint, and two LAU-68A/A or LAU-6A/A launchers for seven or 19 2.75-in (69.85-mm) rockets of five possible types, under the inner hardpoints
Electronics and operational equipment: communication and navigation equipment, plus a Hughes Aircraft M65 sight for the weapons officer, Kaiser head-up display for the pilot, Teledyne digital fire-control computer, Baldwin Electronics XM138 underwing stores management subsystem, Hughes laser rangefinder, Rockwell airborne laser tracker, Hughes Laser-Augmented Airborne TOW (LAAT) stabilized sight, Marconi air-data subsystem, Sander ALQ-144 infra-red jammer, APR-39 radar-warning receiver, ALQ-136 radar jammer, M130 chaff-dispenser, secure voice communications, weapons officer's and pilot's helmet sights and (possibly as a retrofit) the Hughes Aircraft FACTS (FLIR-Augmented Cobra TOW Sight)
Powerplant and fuel system: one 1,800-shp (1342-kW) Avco Lycoming T53-L-703 turboshaft, and a

total internal fuel capacity of 306 US gal (1158 litres) in one fuselage tank
Performance: maximum speed 141 mph (227 km/h); initial climb rate 1,620 ft (494 m) per minute; service ceiling 12,200 ft (3720 m); range 315 miles (507 km)
Weights: empty 6,598 lb (2993 kg); normal take-off 9,975 lb (4524 kg); maximum take-off 10,000 lb (4535 kg)
Dimensions: main rotor diameter 44 ft 0 in (13.41 m); length fuselage) 44 ft 7 in (13.59 m); height 13 ft 6.25 in (4.12 m); main rotor disc area 1,520.5 sq ft (141.26 m²)

Variants
Modified AH-1S HueyCobra: produced in the mid-1970s, this is a reworking of 315 AH-1Gs and the 92 AH-1Qs to improved anti-tank configuration with the powerplant of the AH-1R, eight TOW anti-tank missiles, better defensive capabilities and improved fire-control subsystems
Production AH-1S HueyCobra: this designation covers 100 new-build helicopters to the same basic AH-1S standard but fitted with flat-plate canopies for reduced glint, a better cockpit layout, better instrumentation for nap-of-the-earth operations and (from the 67th aircraft) composite-construction rotor blades
Up-gun AH-1S HueyCobra: this designation applies to 98 new-build aircraft identical to the Production AH-1S apart from their provision for a General Electric undernose turret (with provision for 20- and 30-mm aircraft cannon) and the M138 underwing stores subsystem
Modernized AH-1S HueyCobra: definitive production version of the single-engine HueyCobra series, with all the features of the Production and Up-gun AH-1S models plus Doppler navigation, a laser rangefinder and tracker, a ballistic computer, a pilot's head-up display, secure communications and an infrared jammer

Bell AH-1T Improved SeaCobra
(USA)
Type: close support and attack helicopter
Accommodation: co-pilot/weapons officer and pilot seated in tandem
Armament (fixed): one General Electric chin turret housing an M197 20-mm cannon with 750 rounds
Armament (disposable): this is carried on four hardpoints under the stub wings, and can comprise four LAU-61/A, LAU-68/A, LAU-68A/A, LAU-68B/A or LAU-69/A launchers for 2.75-in (69.85-mm) rockets, or two CBU-55B fuel/air explosive devices, or two M118 grenade-dispensers, or two M18E1 7.62-mm (0.3-in) Minigun pods; alternative provisions is made for the carriage of TOW or (from 1985) Hellfire anti-tank missiles
Electronics and operational equipment: communication and navigation equipment, plus mission equipment similar to that of the AH-1G
Powerplant and fuel system: one 2,050-shp (1528-kW) Pratt & Whitney Canada T400-WV-402 coupled turboshaft, and a total internal fuel capacity of 306 US gal (1158 litres) in two fuselage tanks
Performance: maximum speed 172 mph (277 km/h) at sea level; initial climb rate 1,785 ft (544 m) per minute; service ceiling 7,400 ft (2255 m); range 260 miles (418 km)
Weights: empty 8,608 lb (3904 kg); maximum take-off 14,000 lb (6350 kg)
Dimensions: main rotor diameter 48 ft 0 in (14.68 m); length (fuselage) 48 ft 2 in (14.68 m); height 14 ft 2 in (4.32 m); main rotor disc area 1,809.5 sq ft (168.1 m²)

Variants
AH-1J SeaCobra: first flown in October 1969, the AH-1J was produced for the US Marine Corps as a derivative of the AH-1G with a twin-engine powerplant, in this instance the 1,800-shp (1342–kW) Pratt & Whitney Canada T400-CP-400 coupled turboshaft flat-rated at 1,100 shp (820 kW); maximum take-off weight is 10,000 lb (4536 kg), so despite power comparable to that of the identically-dimensioned AH-1G performance is slightly degraded to a sea-level maximum speed of 207 mph (333 km/h); the maximum weapons load is 2,200 lb (998 kg)
AH-1T Improved SeaCobra: developed from the AH-1J but incorporating features of the Bell Model 309 KingCobra and Bell Model 214, the first AH-1T flew in May 1976 and is notable for its increased length (for additional fuel capacity) and upgraded transmission to handle the full 1,970 shp (1469 kW) of the T400-WV-402 coupled-turboshaft powerplant; the type can carry TOW or Hellfire anti-tank missiles on the outboard underwing hardpoints
AH-1W SuperCobra: designation of the version entering service in 1987 with considerably improved capabilities resulting from the installation of two General Electric T700-GE-401 turboshafts for a total of 3,250 shp (2424 kW) driving through a new combining gearbox; combined with new subsystems (including a Kaiser head-up display for the pilot, and an avionics suite that includes the APR-39(V)1 radar detector, APR-44(V)1 radar-warning receiver, ALE-39 chaff/flare dispenser and ALQ-144 infra-red jammer) and the latest weaponry, this provides the US Marine Corps with a highly advanced high-performance close support and attack helicopter ideally suited to support of beach-head operations from forward airstrips or from assault ships lying just offshore; the primary armament comprises eight BGM-71 TOW or AGM-114 Hellfire anti-tank missiles, supported by a pair of AIM-9L Sidewinder air-to-air missiles for battlefield self-defence

Bell UH-1B/E Iroquois
(USA)
Type: utility tactical helicopter
Accommodation: pilot, and up to eight troops, or three litters, two seated casualties and one attendant, or 3,000 lb (1361 kg) of freight in the cabin
Armament (fixed): none
Armament (disposable): this can comprise two 7.62-mm (0.3-in) M60 machine-guns in the cabin doors, or four 7.62-mm (0.3-in) machine-guns or two launchers each with 24 2.75-in (69.85-mm) rockets mounted on the fuselage sides
Electronics and operational equipment: communication and navigation equipment
Powerplant and fuel system: one 1,100-shp (820-kW) Avco Lycoming T53-L-11 turboshaft, and a total internal fuel capacity of 220 US gal (832 litres) in fuselage tanks
Performance: maximum speed 161 mph (259 km/h); cruising speed 138 mph (222 km/h) at sea level; initial climb rate 1,850 ft (563 m) per minute; service ceiling 21,000 ft (6400 m); range 285 miles (459 km)
Weights: empty 5,055 lb (2293 kg); maximum take-off 9,500 lb (4309 kg)
Dimensions: main rotor diameter 44 ft 0 in (13.41 m); length (fuselage) 38 ft 5 in (11.70 m); height 12 ft 7.25 in (3.84 m); main rotor disc area 1,520.5 sq ft (141.26 m²)

Variants
UH-1A Iroquois: generally known as the **'Huey'** rather than by its official name, this was the initial production version of the **Bell Model 204** utility helicopter developed as the US Army's first turbine-engined aircraft under the designation XH-40; the first prototype flew in October 1956; the production version was initially designated HU-1A and began to enter service in June 1959, the type being powered by a 770-shp (574-kW) T53-L-1A turboshaft and having accommodation for a crew of two plus six passengers or two litters; the type could be armed with two 7.62-mm (0.3-in) flexible machine-guns and two launchers each containing 16 2.75-in (69.85-mm) folding-fin rockets
UH-1B Iroquois: updated version with the 960-shp (716-kW) T53-L-5 or later with the 1,100-shp (820-kW) T53-L-11 turboshaft and increased accommodation UH-1C Iroquois: 1965 model based closely on the UH-1B but having a 'door-hinge' main rotor with wider-chord blades for improved speed and greater manoeuvrability
UH-1E: version of the UH-1B for the US Marine Corps with a rotor brake, special avionics and a rescue winch
UH-1F: utility version of the UH-1B for the USAF with the 1,100-shp (820-kW) General Electric T58-GE-3 turboshaft driving a 48-ft (14.63-m) diameter main rotor for a payload of 10 passengers or 4,000 lb (1814 kg) of freight in the missile site-support role; the type was also produced as the **TH-1F** trainer for the USAF
HH-1K: search-and-rescue variant of the UH-1E for the US Navy, but fitted with the 1,400-shp (1044-kW) T53-L-13 turboshaft
UH-1L: utility version of the UH-1E for the US Navy, but fitted with a T53-L-13 derated to 1,100 shp (820 shp); the **TH-1L** was a trainer version of this model
Model 204B: developed as a civil machine with the 1,100-shp (820-kW) Avco Lycoming T5311A turboshaft and a 48-ft (14.63-m) diameter main rotor, this version could carry 10 passengers and useful export sales were made to overseas air arms
Fuji (Bell) UH-1B: developed as the Fuji (Bell) Model 204B-2, this was the Japanese licence-built version of the Model 204B with the 1,100-shp (820-kW) Kawasaki T53-A-11A turboshaft plus a tractor rather than pusher tail rotor

Bell UH-1H Iroquois
(USA)
Type: utility tactical helicopter
Accommodation: pilot, and up to 14 troops, or six litters and one attendant, or 4,000 lb (1814 kg) of freight in the cabin
Armament (fixed): none
Armament (disposable): provision for machine-guns and rocket-launcher pods as required
Electronics and operational equipment: communication and navigation equipment
Powerplant and fuel system: one 1,400-shp (1044-kW) Avco Lycoming T53-L-13 turboshaft, and a total internal fuel capacity of 220 US gal (832 litres) in fuselage fuel cells, plus provision for 300 US gal (1136 litres) of auxiliary fuel in two internal tanks
Performance: maximum speed 127 mph (204 km/h); cruising speed 127 mph (204 km/h); initial climb rate 1,600 ft (488 m) per minute; service ceiling 12,600 ft (3840 m); range 320 miles (515 km)
Weights: empty 4,667 lb (2116 kg); maximum take-off 9,500 lb (4309 kg)

Variants
UH-1D Iroquois: developed with the company designation Model 205 as a version of the Model 204 series with a stretched fuselage for greater payload, this variant first flew in August 1961 and was powered by the 1,100-shp (820-kW) T53-L-11 turboshaft for a payload of one pilot plus 12 troops, or six litters and an attendant, or 4,000 lb (1818 kg) of freight; the type also introduced greater fuel capacity and provision for auxiliary fuel; the type was also built under licence in West Germany by Dornier
UH-1H Iroquois: uprated version of the UH-1D with the 1,400-shp (1044-kW) T53-L-13 turboshaft; the type was also built under licence in Taiwan by AIDC;

in the early 1980s a few UH-1Hs were converted into **EH-1H** battlefield electronic countermeasures helicopters (each fitted with the ALQ-151 'Quick Fix IA/B' communications interception, direction-fining and jamming equipment, the APR-39(V)2 radar-warning receiver, the M130 chaff and flare dispenser, and the ALQ-144 infra-red countermeasures equipment); another battlefield development of the UH-1H is the **UH-1H SOTAS** battlefield targeting helicopter fitted with a General Dynamics radar (with moving target indication facility) and a ventral antenna; some 220 of the 2,000 UH-1Hs have also been converted into **UH-1V** casevac helicopters during the 1980s
HH-1H: on-base rescue variant for the USAF with special equipment for rescue of downed aircrew
CH-118: variant of the UH-1H for the Canadian Armed forces, and originally designated CUH-1H
Model 205A-1: civil version of the basic type with a 1,400-shp (1044-kW) Avco Lycoming T5313B turboshaft derated to 1,250 shp (932 kW) and able to carry a slung load of 5,000 lb (2268 kg); many have been sold to export military customers
Agusta (Bell) AB.205: Italian licence-built version of the UH-1D/H type differing only in small details from the US pattern
Fuji (Bell) UH-1H: Japanese licence-built version of the UH-1H with a 1,400-shp (1044-kW) Kawasaki T53-K-13B turboshaft and a tractor rather than pusher tail rotor

Bell UH-1N Iroquois
(USA)
Type: utility helicopter
Accommodation: pilot, and up to 14 troops, or 4,000 lb (1814 kg) of freight in the cabin
Armament (fixed): none
Armament (disposable): machine-guns can be fitted in the cabin doors
Electronics and operational equipment: communication and navigation equipment
Powerplant and fuel system: one 1,800-shp (1342-kW) Pratt & Whitney Canada T400-CP-400 coupled turboshaft flat-rated to 1,290 shp (962 kW) for take-off and to 1,130 shp (842 kW) for continuous running, and a total internal fuel capacity of 215 US gal (814 litres) in five fuselage fuel cells, plus provision for 180 US gal (681 litres) of auxiliary fuel
Performance: cruising speed 142 mph (230 km/h) at sea level; initial climb rate 1,320 ft (402 m) per minute; service ceiling 15,000 ft (4570 m); range 248 miles (400 km)
Weights: empty 6,169 lb (2798 kg); normal take-off 10,000 lb (4536 kg); maximum take-off 10,500 lb (4762 kg)
Dimensions: main rotor diameter 48 ft 2.25 in (14.69 m) with tracking tips; length (fuselage) 42 ft 4.75 in (12.92 m); height 14 ft 10.25 in (4.53 m); main rotor disc area 1,871.9 sq ft (173.9 m²)

Variants
UH-1N Iroquois: developed from 1968 at the instigation of the Canadians, who needed a version of the highly successful UH-1 with a twin-engine powerplant for extra reliability and safety in remote areas, the Model 212 first flew in 1969 as a combination of the UH-1H's airframe with the Pratt & Whitney Canada PT6T Turbo Twin Pac coupled turboshaft powerplant offering good performance added to single-engine flight capability; the production model features the PT6T-3B (military designation T400-CP-400) and substantial sales were made to the US forces and to overseas air arms
CH-135: designation of the UH-1N for the Canadian Armed Forces
Model 212: designation of the civil model, which won useful orders for civil and third-world military operators

Bell Model 412 (USA): see Agusta (Bell) AB.412 Grifone

Bell OH-58C Kiowa
(USA)
Type: light observation helicopter
Accommodation: pilot and co-pilot or one passenger at the front of the cabin, with provision behind them for two passengers or freight
Armament (fixed): M27 armament kit centred on a 7.62-mm (0.3-in) Minigun
Armament (disposable): none
Electronics and operational equipment: communication and navigation equipment
Powerplant and fuel system: one 420-shp (313-kW) Allison T63-A-720 turboshaft, and a total internal fuel capacity of 73 US gal (276 litres) in underfloor tankage
Performance: maximum speed 134 mph (216 km/h) at 5,000 ft (1525 m); cruising speed 117 mph (188 km/h); initial climb rate 1,780 ft (543 m) per minute; service ceiling 18,900 ft (5760 m); range 305 miles (490 km) on an armed scout mission
Weights: empty 1,585 lb (719 kg); normal take-off 2,435 lb (1105 kg); maximum take-off 3,200 lb (1451 kg)
Dimensions: main rotor diameter 35 ft 4 in (10.77 m); length (fuselage) 32 ft 7 in (9.93 m); height 9 ft 6.5 in (2.91 m); main rotor disc area 980.5 sq ft (91.1 m²)

Variants
OH-58A Kiowa: developed during the early 1960s as the Model 206 in response to a US Army requirement for a light observation helicopter, the OH-4 prototype first flew in December 1962, but failed to find initial selection against the Hughes OH-6; however, in 1967 the LOH competition was reopened because of difficulties with the production programme and cost of

the Hughes helicopter, and the improved Model 206A this time prevailed, entering production as the OH-58A with the 317-shp (237-kW) Allison T63-A-700 turboshaft, some 2,200 examples being produced up to 1974
OH-58B: designation of 12 OH-58 helicopter for Austria
OH-58C Kiowa: improved development of the OH-58A with flat-plate canopy for reduced glint, an uprated powerplant for better performance under 'hot-and-high' conditions, and infra-red reduction package to reduce the chances of destruction by IR-homing missiles
OH-58D Kiowa: developed as the Model 406, this is a radically upgraded Kiowa produced in the early 1980s as part of the US Army Helicopter Improvement Program; some 580 OH-58As are being converted to the new standard, which includes the 650-shp (485-kW) Allison 250-C30R turboshaft for a maximum speed of 147 mph (237 km/h) and a range of 345 miles (556 km) with 105 US gal (398 litres) of fuel and a maximum take-off weight of 4,500 lb (2041 kg); more importantly, the OH-58D introduces a cockpit control and display subsystem and a mast-mounted sight above the 35-ft (10.67-m) four-blade main rotor; this sight includes a TV camera and forward-looking infra-red unit for observation under battlefield conditions as the helicopter hovers behind cover; the type can also designate targets for laser-homing missiles, and has an air-to-air self-defence capability through the carriage of two Stinger missiles
CH-136: designation of helicopters similar to the OH-58A for the Canadian Armed Forces, and originally designated COH-58A
CAC Model 206B-1: designation of OH-58As built under licence in Australia by Commonwealth Aircraft Corporation

Bell Model 206L-1 LongRanger II
(USA)
Type: general-purpose helicopter
Accommodation: crew of two at the front of the cabin, and to their rear five passengers, or two litters and two seated casualties/attendants, or freight
Armament (fixed): none
Armament (disposable): none
Electronics and operational equipment: communication and navigation equipment
Powerplant and fuel system: one 500-shp (373-kW) Allison 250-C28B turboshaft, and a total internal fuel capacity of 98 US gal (371 litres) in a fuselage tank
Performance: cruising speed 134 mph (215 km/h) at 5,000 ft (1525 m); initial climb rate 1,520 ft (463 m) per minute; service ceiling 19,500 ft (5945 m); range 385 miles (620 km)
Weights: empty 2,155 lb (978 kg); maximum take-off 4,150 lb (1882 kg)
Dimensions: main rotor diameter 37 ft 0 in (11.28 m); length (fuselage) 33 ft 0 in (10.13 m); height (to rotor head) 10 ft 0 in (3.05 m); main rotor disc area 1,075.2 sq ft (99.89 m^2)

Variants
Model 206A JetRanger: this is the civil helicopter derived from Bell's unsuccessful Model 206 entry in the US Army's LOH competition of 1962, and proved very successful, many examples going to overseas air arms; the type is powered by the 317-shp (237-kW) Allison 250-C18A turboshaft driving a 33.33-ft (10.16-m) diameter main rotor; the type was adopted as a primary training helicopter by the US Navy with the designation **TH-57A SeaRanger**
Model 206B JetRanger II: simple improvement of the Model 206A with the 400-shp (298-kW) Allison 250-C20 turboshaft; the type was adopted by the US Navy as the TH-57B SeaRanger primary trainer
Model 206B JetRanger III: further upgraded model with the 420-shp (313–kW) Allison 250-C20B turboshaft for better performance under 'hot-and-high' conditions; the type was adopted by the US Navy as the **TH-57C SeaRanger** instrument trainer
Model 206L LongRanger: stretched version of the JetRanger II with a 37-ft (11.28-m) diameter main rotor and 420-shp (313-kW) Allison 250-C20B turboshaft
Model 206L-1 LongRanger II: upgraded version of the Model 206L LongRanger
Model 206L-3 LongRanger III: further upgraded version with the 650-shp (485-kW) Allison 250-C30P flat-rated at 557 shp (415 kW) for better performance especially under 'hot-and-high' conditions
Model 206L TexasRanger: Bell-developed armed version of the series with the 500-shp (373-kW) Allison 250-C28B turboshaft and provision for advanced sights plus an assortment of disposable ordnance including up to four TOW anti-tank missiles
Agusta (Bell) AB.206: under this overall designation Bell's Italian licensee has produced JetRanger and LongRanger helicopters basically similar to the US originals; one development separate from the US mainstream has been the **AB.206A-1** military model with several armament options, and the anti-submarine **HKP 6** derivative for Sweden with longer landing gear legs to permit the carriage of torpedoes or depth charges under the fuselage

Bell Model 214ST
(USA)
Accommodation: pilot and co-pilot on the flightdeck, and up to 18 passengers or freight in the cabin
Armament (fixed): none
Armament (disposable): none
Electronics and operational equipment: communication and navigation equipment
Powerplant and fuel system: two 1,625-shp (1212-kW) General Electric CT7-sA turboshafts, and a total

internal fuel capacity of 435 US gal (1647 litres) in seven fuselage fuel cells, plus provision for auxiliary fuel
Performance: cruising speed 159 mph (256 km/h) at 4,000 ft (1220 m); initial climb rate 1,850 ft (564 m) per minute; service ceiling (one engine) 7,000 ft (2135 m); range 500 miles (805 km) with standard fuel, and more than 635 miles (1022 km) with standard and auxiliary fuel
Weights: maximum take-off 17,500 lb (7938 kg)
Dimensions: main rotor diameter 52 ft 0 in (15.85 m); length (fuselage) 50 ft 0 in (15.24 m); height 15 ft 10.5 in (4.84 m); main rotor disc area 2,124.0 sq ft (197.32 m²)

Variants
Model 214A Isfahan: resulting from a 1972 Iranian requirement for a high-performance medium-lift helicopter, this model was developed from the up-engined Model 214 Huey Plus development of the UH-1H; the type is powered by the 2,930-shp (2185-kW) Avco Lycoming LTC4B-8D turboshaft and has a maximum take-off weight of 16,000 lb (7257 kg)
Model 214C: search-and-rescue version of the Model 214A for the Iranian air force
Model 214ST: twin-engine utility version with excellent performance and payload, and attractive to many overseas air arms

Bell/Boeing V-22 Osprey
(USA)
Type: tilt-rotor V/STOL multi-role aircraft
Accommodation: crew of two on the flightdeck, and accommodation for 24 troops or freight in the cabin
Armament (fixed): one 0.5-in (12.7-mm) multi-barrel machine-gun in the nose
Armament (disposable): none
Electronics and operational equipment: not yet determined
Powerplant and fuel system: two 4,855-shp (3620-kW) General Electric T64-GE-717 turboshafts, and a total internal fuel capacity of ? US gal (? litres) in integral wing tanks
Performance: maximum cruising speed 300 mph (483 km/h); range with pilot and 24 troops 460 miles (740 km) at 3,000 ft (915 m)
Weights: maximum for short take-off 55,000 lb (24948 kg); maximum for vertical take-off 43,800 lb (19867 kg)
Dimensions: rotor diameter, each 38 ft 0 in (11.58 m); width overall 84 ft 6 in (25.76 m); length 56 ft 10 in (17.32 m); height, rotors in take-off position 20 ft 2 in (6.15 m); rotor disc area, total 2,268.24 sq ft (210.72 m²)

Variant
V-22 Osprey: now under development as a V/STOL assault transport for the US Marine corps, and under consideration for a host of other applications, this fascinating tilt-rotor hybrid rotary/fixed-wing aircraft offers an excellent combination of payload/range and VTOL capability, the concept having been validated by the success of the Bell XV-15 prototype, which first flew in May 1977; the type is planned to enter service in 1991 after a first flight in 1987, and among the specified requirements for the type are a hovering ceiling out of ground effect of 3,000 ft (915 m) with an external load of 8,300 lb (3765 kg), and a ferry range of 2,420 miles (3895 km)

Beriev Be-12 (M-12) 'Mail'
(USSR)
Type: maritime reconnaissance amphibian
Accommodation: crew of three or four on the flightdeck, and mission crew of undetermined size in the nose compartment and fuselage
Armament (fixed): none
Armament (disposable): this is believed to comprise depth charges and/or homing torpedoes in a fuselage bay, and a mixture of bombs, missiles and rockets on two small and two large underwing hardpoints, up to a maximum weight of some 5000 kg (11,023 lb)
Electronics and operational equipment: communication and navigation equipment, plus search radar in a nose 'thimble', magnetic anomaly detection (MAD) gear in the tail 'sting' and sonobuoys in the fuselage bay, as well as onboard computation and analysis equipment
Powerplant and fuel system: two 3124-ekW (4,190-ehp) Ivchenko AI-20D turboprops, and a total internal fuel capacity of ? litres (? Imp gal)
Performance: maximum speed 608 km/h (378 mph); cruising speed 320 km/h (199 mph); initial climb rate 912 m (2,990 ft) per minute; service ceiling 11280 m (37,000 ft); range 4000 km (2,485 miles)
Weights: empty about 18000 kg (39,680 lb); maximum take-off 29450 kg (64,925 lb)
Dimensions: span 29.71 m (97 ft 5.75 in); length 30.17 m (99 ft 0 in) with MAD 'sting' or 30.95 m (101 ft 6.5 in) with later nose radar; height 7.00 m (22 ft 11.5 in); wing area 105.0 m² (1,130.2 sq ft)

Variant
Be-12 'Mail': one of the few amphibians still in service, the Be-12 (also known as the M-12) first flew in the late 1950s, and has proved itself admirably suited to the requirements of anti-submarine warfare and maritime reconnaissance in the particular conditions of the USSR

Boeing B-52G Stratofortress
(USA)
Type: strategic heavy bomber and missile carrier
Accommodation: crew of six (pilot, co-pilot, navigator, radar operator, ECM officer and gunner
Armament (fixed): four 0.5-in (12.7-mm) Colt-

Browning M3 machine-guns in a rear barbette remotely-controlled by an ASG-15 radar fire-control system
Armament (disposable): see Variants (below)
Electronics and operational equipment: communication and navigation equipment, and a constantly updated electronics suite including Phase VI electronic countermeasures; ASQ-151 Electro-optical Viewing System (EVS) comprising two steerable chin turrets, the unit on the port side containing a Westinghouse AVQ-22 low light level TV (LLLTV) and the unit on the starboard side containing a Hughes Aircraft AAQ-6 forward-looking infra-red (FLIR) sensor; ALR-20A panoramic receiver; Motorola ALQ-122 Smart Noise Operation Equipment (SNOE) countermeasures system; Dalmo Victor ALR-46(V)4 digital radar-warning receiver; Westinghouse ALQ-153 pulse-Doppler tail warning receiver; ITT Avionics ALQ-172 jammers; Northrop ALQ-155(V) advanced ECM system; eight ALE-24 wing-mounted launchers with 1,125 chaff packages; 12 ALE-20 wing-mounted launchers with 192 flares; AFSATCOM satellite communication system; and Boeing-integrated Offensive Avionics System (OAS) Phase I including Teledyne Ryan Doppler navigation, Honeywell ASN-131 inertial navigation, IBM/Raytheon ASQ-38 analog bombing and navigation system with an IBM digital data processor, Lear Siegler attitude heading and reference system, and Tercom (terrain comparison) guidance
Powerplant and fuel system: eight 13,750-lb (6237-kg) thrust Pratt & Whitney J57-P-43WB turbojets, and a total internal fuel capacity of 46,000 US gal (174130 litres) in integral wing tanks, plus provision for two 700-US gal (2650-litre) drop-tanks under the outer wing panels; provision for inflight-refuelling
Performance: maximum speed 595 mph (957 km/h) or Mach 0.9 at high altitude, or 420 mph (676 km/h) at low level; cruising speed 509 mph (819 km/h) or Mach 0.77 at high altitude; service ceiling 55,000 ft (16765 m); range more than 7,500 miles (12070 km) with maximum internal fuel but without inflight-refuelling
Weights: maximum take-off more than 488,000 lb (221357 kg)
Dimensions: span 185 ft 0 in (56.39 m); length 160 ft 10.9 in (49.05 m); height 40 ft 8 in (12.40 m); wing area 4,000 sq ft (371.6 m²)

Variants
B-52G Stratofortress: designed as a high-speed high-altitude strategic bomber with genuine intercontinental range, the B-52 has been the manned bomber mainstay of the US Strategic Air Command since it entered service in its initial B-52B production form during 1955; since that time the type has undergone enormous development (mainly in terms of its avionics and a strengthened airframe for the low-level role adopted during 1962), and the definitive turbojet-powered model was the B-52G, which first flew in October 1952 with integral wing tankage for greater range, an improved pressurized compartment for the six crew, shorter vertical tail surfaces, a remotely-controlled tail turret, and provision for Quail decoy missiles and Hound Dog stand-off nuclear missiles; some 151 of the 193 B-52Gs survive, 90 of them fitted for the stand-off nuclear role with a primary armament of 12 AGM-86B cruise missiles under the wings (six rounds on each of two hardpoints) and the other 61 for the maritime surveillance/support role with a primary armament of AGM-84 Harpoon and GBU-15(V) air-launched anti-ship missiles; the maximum bombload is 50,000 lb (22680 kg)
B-52H Stratofortress: ultimate development of the B-52 series, this variant first flew in March 1961 as the designated carrier for the Douglas AGM-87 Skybolt stand-off nuclear missile, which was later cancelled; the B-52H was internally redesigned for the low-level role, though more obvious external changes were the use of eight 17,000-lb (7711-kg) Pratt & Whitney TF33-P-3 turbofan engines (for a range of 10,000 miles/16093 km with maximum fuel) and the adoption of an M61A1 20-mm rotary cannon in the tail position; other changes were much enhanced avionics (including provision for terrain-avoidance radar) and the latest electronic countermeasures equipment; 90 of the 102 B-52Hs remain in service with a primary armament of 20 AGM-69A SRAM stand-off nuclear missiles, coupled with the ASQ-151 EVS and introduction of Phase 6 avionics and the Offensive Avionics System; some aircraft are earmarked for conventional support of the Rapid-DeploymentTask Force (US Readiness Command) with GP and cluster bombs, the GBU-15 EO-guided glide bombs, the 'Popeye' 3,000-lb (1361-lb) IR- or radar-guided stand-off bomb, and the AGM-130 rocket-assisted glide bomb; as the Rockwell B-1B enters service the B-52Hs are being withdrawn for conversion into cruise missile aircraft, carrying 12 missiles under the wings (as in the B-52G) with provision for another eight in a modified bomb bay

Boeing KC-135A Stratotanker
(USA)
Type: inflight-refuelling tanker with secondary transport capability
Accommodation: crew of five (three pilots, radar navigator and boom operator), and up to 160 troops, or 83,000 lb (37650 kg) of freight in the cabin and hold
Armament (fixed): none
Armament (disposable): none
Electronics and operational equipment: communication and navigation equipment, and weather radar
Powerplant and fuel system: four 13,750-lb (6237-kg) thrust Pratt & Whitney J57-P-59W turbojets, and a

total internal fuel capacity of 31,200 US gal (118105 litres) in integral wing tanks and bladder tanks in the centre section, under the cabin floor and in the tail; all of this fuel is available for transfer; provision for inflight-refuelling

Performance: maximum speed 585 mph (941 km/h) at 30,000 ft (9145 m); cruising speed 532 mph (856 km/h) at 35,000 ft (10670 m); initial climb rate 1,290 ft (393 m) per minute; service ceiling about 50,000 ft (15240 m); radius 3,450 miles (5552 km) to offload 24,000 lb (10886 kg) of fuel, or 1,150 miles (1850 km) to offload 120,000 lb (54432 kg) of fuel

Weights: empty 106,306 lb (48220 kg); normal take-off 301,600 lb (136806 kg); maximum take-off 316,000 lb (143338 kg)

Dimensions: span 130 ft 10 in (39.88 m); length 134 ft 6 in (40.99 m); height 41 ft 8 in (12.69 m); wing area 2,433.0 sq ft (226.03 m²)

Variants

KC-135A Stratotanker: this is the baseline inflight-refuelling tanker for the US Strategic Air Command, though the aircraft are frequently used in support of Tactical Air Command assets at home and overseas; the type was the initial production development of Boeing privately-funded Model 367-80 prototype that also paved the way for the great Model 707 airliner; fitted with the standard Boeing-developed Flying Boom inflight-refuelling system, the initial KC-135A flew in August 1956, and the first of 732 production aircraft was delivered a mere three months later; the type has spawned an enormous number of derivatives in the tanker and other roles, and conversions from standard KC-135As included the **JC-135A** and **JKC-135A** special-duties aircraft (eventually redesignated **NC-135A** and **NKC-135A** in keeping with other test and research aircraft) and the seven **VC-135A** staff transports, of which two retained tanker capability; the 56 **KC-135Q** aircraft are KC-135As converted to deal with the special JP-7 fuel used only by the Lockheed SR-71A 'Blackbird' strategic reconnaissance aircraft

C-135A Stratolifter: designation of the initial transport version of the series, which was not produced in quantity as the Lockheed C-130 Hercules and Lockheed C-141 StarLifter proved more capable and flexible because of their optimized military transport design; production amounted to 15 aircraft (as well as three **C-135A Interim** aircraft adapted from KC-135As) powered by the J57-P-59W turbojet, and these began to enter service in 1961 with accommodation for 89,000 lb (40370 kg) of freight, or 126 passengers, or 44 litters plus 54 sitting wounded

KC-135B Stratotanker: designation of a limited-production development (17 aircraft) of the KC-135A with 18,000-lb (8165-kg) Pratt & Whitney TF33-P-5 turbofans, additional fuel capacity, an inflight-refuelling receptacle above the flightdeck and provision for service as airborne command posts, in which role all 17 production aircraft began to serve as soon as they entered service

C-135B Stratolifter: version of the C-135A with TF33-P-5 turbofans and a slightly enlarged tail (a feature also of later KC-135As); the 30 aircraft were rarely used in the transport role, five becoming **VC-135B** staff transports and 11 being turned into **WC-135B** weather reconnaissance aircraft (of which three were reconverted as **C-135C** transports)

KC-135E Stratotanker: under this designation a larger number of KC-135As operated by the Air Force Reserve and Air National Guard are being upgraded so that they can remain in useful service into the next century; a primary change is the reskinning of the undersurfaces of the wings (a programme being undertaken for all KC-135s), while other features are re-engining with Pratt & Whitney JT3D-3B turbofans (removed from surplus Model 707 airliners), engine mountings and nacelles, together with their tails; at the same time new brakes and anti-skid units are being installed

C-135F: designation of 12 KC-135A aircraft for France, designed for support of the Dassault Mirage IV intermediate-range bomber force; the 11 survivors are being re-engined (for greater economy and range) with the General Electric/SNECMA CFM56 turbofan under the revised designation **C-135FR**

KC-135R Stratotanker: radical updating of the KC-135A along the lines of the KC-135E programme but far more extensive in scope; the selected powerplant is the 22,000-lb (9979-kg) CFM International F108-CF-100 turbofan, the military version of the CFM56-2B-1 civil engine, and other features of the programme are strengthened landing gear with anti-skid units, an improved autopilot, an updated cockpit, an enlarged tailplane, a fuel capacity of 203,288 lb (92210 kg) and an auxiliary power unit for self-start capability; the result is an aircraft able to transfer 150 per cent more fuel than the KC-135A at a radius of 2,875 miles (4627 km) yet able to operate from considerably shorter runways; this last is of considerable importance, for the take-off run of a fully-laden KC-135A is so great that only the largest military (and in emergencies civil) airfields can be used, which has at times proved a tactical limitation to the type's effective employment

Boeing EC-135A
(USA)

Type: communications relay aircraft
Accommodation: crew of three or four on the flightdeck, plus a variable mission crew in the cabin
Armament (fixed): none
Armament (disposable): none
Electronics and operational equipment: communication and navigation equipment, plus classified mission electronics

Powerplant and fuel system: four 13,750-lb (6237-kg) thrust Pratt & Whitney J57-P-59W turbojets, and a total internal fuel capacity of 23,855 US gal (90299 litres) in seven integral wing tanks; provision for inflight-refuelling
Performance: maximum speed 585 mph (941 km/h) at 30,000 ft (9145 m); cruising speed 430 mph (853 km/h); initial climb rate 2,000 ft (610 m) per minute; service ceiling 50,000 ft (15240 m)
Weights: empty 98,466 lb (46633 kg); maximum take-off 297,000 lb (134717 kg)
Dimensions: span 130 ft 10 in (39.88 m); length 136 ft 3 in (41.43 m); height 41 ft 8 in (12.70 m); wing area 2,433.0 sq ft (226.03 m²)

Variants
EC-135A: confirmation of the C/KC-135's operating economics and performance in the late 1950s soon showed that the type was ideally suited to conversion in the airborne command post and communications relay roles with the cabin outfitted with a large quantity of advanced communications equipment; the availability of such aircraft would add enormously to the strategic flexibility of the Strategic Air Command, by offering the possibility of a virtually indestructible command system operating in high orbits well clear of potentially destructible ground facilities; the EC-135A designation covers six aircraft thus converted from KC-135A tankers
EC-135B ARIA: designation of at least four C-135Bs converted for the Advanced Range Instrumentation Aircraft role with a steerable antenna in a large nose fairing
EC-135C: designation of the original 17 airborne command post aircraft for SAC's Post-Attack Command Control System
EC-135G: designation of three KC-135As converted as airborne control centres for ICBM launch purposes, but also able to double as communications relay aircraft
EC-135H: designation of five airborne national command post aircraft converted from one VC-135A and four KC-135As
EC-135J: four upgraded airborne national command post aircraft converted from EC-135Cs
EC-135K: two Tactical Air Command airborne command post aircraft converted from KC-135As
EC-135L: five radio relay aircraft converted from KC-135As
EC-135N: eight satellite-tracking aircraft converted from C-135As
EC-135P: two Pacific Air Forces airborne command post aircraft converted from KC-135As

Boeing RC-135C
(USA)
Type: electronic reconnaissance aircraft
Accommodation: crew of three or four on the flight-deck, plus a variable mission crew in the cabin
Armament (fixed): none
Armament (disposable): none
Electronics and operational equipment: communication and navigation equipment, plus classified mission equipment in the cabin
Powerplant and fuel system: four 18,000-lb (8165-kg) thrust Pratt & Whitney TF33-P-9 turbofans, and a total internal fuel capacity of 23,855 US gal (90299 litres) in seven integral wing tanks; provision for inflight-refuelling
Performance: maximum speed 616 mph (991 km/h) at 25,000 ft (7620 m); cruising speed 560 mph (901 km/h); service ceiling 40,600 ft (12375 m); operational radius 2,675 miles (4305 km); ferry range 5,655 miles (9100 km)
Weights: empty 102,300 lb (46403 kg); maximum take-off 275,500 lb (124965 kg)
Dimensions: span 120 ft 10 in (39.88 m); length 128 ft 7.3 in (39.20 m); height 41 ft 8 in (12.70 m); wing area 2,433.0 sq ft (226.03 m²)

Variants
RC-135A: given the payload and range of the C/KC-135A series, it was natural that the type should be developed for the arcane science of electronic intelligence, and the first such machine was the RC-135A phot-mapping and electronic reconnaissance aircraft, based on the C-135A and using the same J57-P-59W engines
RC-135B: 10 electronic reconnaissance aircraft based on the C-135B but with Pratt & Whitney TF33-P-9 turbofans, and lacking any type of standardization in the electronic fit
RC-135C: conversions of RC-135B aircraft with side-looking airborne radar, an undernose radome and ventral camera installation
RC-135D: four electronic reconnaissance conversions from KC-135A (one) and C-135A (three) standard with side-looking airborne radar, a thimble nose radome and other systems
RC-135E: one C-135B conversion similar to the RC-135C but with a wide glassfibre radome round the forward fuselage
RC-135M: at least six C/VC-135B conversions for electronic reconnaissance in the 'Rivet Card' and 'Rivet Quick' programmes and fitted with thimble noses, teardrop fairings on the fuselage sides forward of the tail and twin-lobe ventral antennae
RC-135S: several C-135B conversions with numerous blister fairings and a dipole aerial as part of the 'Rivet Ball' programme
RC-135T: one C-135B conversion for electronic surveillance in support of Strategic Air Command operations
RC-135U: two or three aircraft with side-looking airborne radar and an undernose radome, plus electronic intelligence systems, and associated with the

'Combat Pink' and 'Combat Scent' programmes

RC-135V: seven or more conversions with side-looking airborne radar, thimble noses and extensive antenna arrays under their fuselages for the electronic intelligence role

RC-135W: several variations on the RC-135V theme with side-looking airborne radar and other electronic intelligence equipment

Boeing E-3A Sentry
(USA)
Type: airborne warning and control system (AWACS) aircraft
Accommodation: crew of four on the flightdeck, and a mission crew of 13 in the cabin
Armament (fixed): none
Armament (disposable): none, but see under Variants
Electronics and operational equipment: communication and navigation equipment, plus Westinghouse APY-1 surveillance radar with its antenna (measuring about 24 ft/7.32 m by 5 ft/1.52 m) in a rotodome (measuring 30 ft/9.14 m by 6 ft/1.83 m) turning at 6 rpm when the radar is in use but at 0.25 rpm when it is not, the antenna being backed by IFF/TADIL C antenna (AIL APX-103 system) for the location and identification of any aircraft within a radius of 230 miles (370 km); IBM 4 Pi CC-1 high-speed computer capable of 740,000 operations per second and with main and mass memory capacities of 114,688 and 802,816 words respectively; Hazeltine multi-purpose consoles and auxiliary display units (numbering nine and two respectively); two Delco ASN-119 inertial platforms updated by Northrop ARN-120 Omega; Teledyne Ryan APN-213 Doppler navigation; and a mass of communications equipment by Collins, Electronic Communications, E-Systems and Hughes
Powerplant and fuel system: four 21,000-lb (9526-kg) thrust Pratt & Whitney TF33-PW-100/100A turbofans, and a total internal fuel capacity of 23,985 US gal (90800 litres) in integral wing tanks; provision for inflight-refuelling
Performance: maximum speed 530 mph (853 km/h) at high altitude; service ceiling over 29,000 ft (8850 m); patrol endurance 6 hours at a radius of 1,000 miles (1609 km)
Weights: empty 172,000 lb (78019 kg); maximum take-off 325,000 lb (147420 kg)
Dimensions: span 145 ft 9 in (44.42 m); length 152 ft 11 in (46.61 m); height 41 ft 9 in (12.73 m); wing area 3,050.0 sq ft (283.35 m²)

Variants
E-3A Sentry: one of the most expensive but important aircraft in the current military inventory, the Sentry is a highly capable airborne warning and control system aircraft designed for three-dimensional surveillance of massive volumes of air and the direction of air operations within that volume; the E-3A is based on the airframe of the Model 707-300B, and two EC-137D prototypes were delivered for operational evaluation of the Westinghouse APY-1 and Hughes APY-2 radars in competition as the Sentry's primary sensor; the Westinghouse radar was judged superior, and this type was selected for installation in the 30-ft (9.14-m) overfuselage radome of the 34 E-3A production aircraft, which began to enter service in March 1977; the last aircraft was delivered in 1985; the first 24 aircraft have only and overland capability and the designation **Core E-3A** (pulse-Doppler radar, CC-1 computer, nine situation display consoles, two auxiliary display units and 13 communication links), while the last 10 have the designation **Standard E-3A** with additional overwater sensor capability, faster-working CC-2 computer with a 665,360-word memory, secure voice communications facility, and Joint Tactical Information Distribution System; the same Standard E-3A type (though with provision for self-defence air-to-air missiles) was ordered for the multi-national NATO early warning force, and these 18 aircraft were delivered between January 1982 and April 1985; another five aircraft have been ordered by Saudi Arabia under the semi-official designation **E-3A/Saudi**, and these have slightly less capable electronics (no JTIDS, commercial communication links and reduced electronic counter-countermeasures facility) and a powerplant comprising four 22,000-lb (9979-kg) CFM56-A2-2 turbofans
E-3B Sentry: designation of the standard to which the Core E-3As are being raised with the CC-2 computer, JTIDS, improved electronic countermeasures capability, 'Have Quick' communications system, and limited overwater sensor capability
E-3C Sentry: designation of the standard to which the Standard E-3As are being raised with five extra situation display consoles, additional UHF radio gear and the 'Have Quick' communications system; the type can also carry small underwing pylons fro self-defence AIM-9 Sidewinder air-to-air missiles
E-6A: designation of a planned 15 US Navy aircraft based on the airframe of the E-3 but powered by four 22,000-lb (9979-kg) CFM56-A2-2 turbofans and incorporating special electronics to provide a relay communications service between submerged missile-armed nuclear submarines and command post aircraft; this is the TACAMO (Take Charge and And Move Out) concept, and the E-6A is to replace Lockheed EC-130Q aircraft in the role with the ARC-182 VLF radio system transmitting through a 4.9-mile (7.9-km) trailing aerial; ALR-66(V) electronic support measures are fitted in the right wingtip

Boeing E-4B
(USA)
Type: advanced airborne command post (AABNCP) aircraft

Accommodation: two crews of three or four on the flightdeck, and about 90 battle staff and communications staff in the cabin
Armament (fixed): none
Armament (disposable): none
Electronics and operational equipment: communications and navigation equipment, and a mass of classified communications, ECM and data-analysis systems provided by Boeing, E-Systems, Electrospace Systems, Collins Radio, RCA, Burroughs and Special Systems Group; the super high frequency (SHF) satellite communication system is distinguishable by the dorsal hump above the upper deck, and the Collins low frequency/very low frequency communication system is characterized by a trailing wire antenna some 5 miles (8 km) in length; the main cabin is divided into six main stations (the National Command Authorities' work area, conference room, briefing room, battle staff work area, communications centre and rest area; the upper deck accommodates the flightdeck, navigation centre and flightcrew rest area; and the lower deck accommodates the communications equipment, power supply distribution system, spares and onboard maintenance facilities
Powerplant and fuel system: four 52,500-lb (23814-kg) thrust General Electric CF6-50E turbofans, and a total internal fuel capacity of 331,565 lb (150395 kg) in integral wing tanks; provision for inflight-refuelling
Performance: range classified, but 12-hour unrefuelled endurance is standard, and a mission endurance of 72 hours is possible
Weight: maximum take-off 797,000 lb (361520 kg)
Dimensions: span 195 ft 8 in (59.64 m); length 231 ft 4 in (70.51 m); height 63 ft 5 in (19.33 m); wing area 5,500 sq ft (511 m²)

Variant
E-4B: based on the airframe and powerplant of the Model 747-200B airliner, the E-4 was schemed in the early 1970s as replacement for the EC-135 series in the airborne national command post, and the first three such aircraft entered service in 1974 and 1975 as E-4As with equipment stripped from EC-135s and updated by E-Systems; a fourth aircraft was built to a more advanced E-4B standard, and the three E-4As have since been upgraded to the more capable standard with an extremely wide-ranging assortment of voice, teletype and data communication links

Boeing Model 707-300C
(USA)
Type: long-range transport
Accommodation: crew of three or four on the flightdeck, and up to 215 passengers or 141,100 lb (64002 kg) of freight in the cabin
Armament (fixed): none
Armament (disposable): none
Electronics and operational equipment: communication and navigation equipment
Powerplant and fuel system: four 19,000-lb (8618-kg) thrust Pratt & Whitney JT3D-7 turbofans, and a total internal fuel capacity of 23,855 US gal (90299 litres) in seven integral wing tanks
Performance: maximum speed 627 mph (1009 km/h) at high altitude; cruising speed 605 mph (973 km/h) at 25,000 ft (7620 m); initial climb rate 4,000 ft (1219 m) per minute; service ceiling 39,000 ft (11885 m); range 4,300 miles (6920 km) with maximum payload, and 7,475 miles (12030 km) with maximum fuel
Weights: empty 138,610 lb (62872 kg); maximum take-off 333,600 lb (151315 kg)
Dimensions: span 145 ft 9 in (44.42 m); length 152 ft 11 in (46.61 m); height 42 ft 5 in (12.93 m); wing area 3,050.0 sq ft (283.4 m²)

Variants
VC-137B: designation of three Model 707-153s procured as 13,500-lb (6123-kg (Pratt & Whitney JT3C-6 turbojet-engined VC-137A VIP transports and redesignated after they had been re-engined with 18,000-lb (8165-kg) Pratt & Whitney TF33-P-5 turbofans
VC-137C: designation of two presidential-use VIP transports based on the civil Model 707-300B Intercontinental
KE-3A: designation of convertible cargo/inflight-refuelling aircraft being converted from ex-airline Model 707-300s for Saudi Arabia; these aircraft are powered by CFM56-A2-2 turbofans, and are provided with three hose-and-drogue units (one ventral and two underwing) for the transfer of 123,000 lb (55792 kg) of fuel at a radius of 1,000 miles (1609 km) when the optional extra fuel tankage (for 5,000 US gal/18927 litres) is fitted in the lower cargo hold; the cabin is fitted with facilities for interchangeable freighting, passenger, mixed freight and passenger, and VIP roles; similar aircraft for other countries are designated **KC-707**
C-18: designation of the Model 707 variant being planned as a battlefield surveillance and targeting aircraft with the JSTARS (Joint Surveillance and Target Attack Reconnaissance System) equipment in a joint US Air Force and US Army programme
E-18B: designation of six aircraft being converted from ex-airline Model 707-300s as Advanced Range Instrumentation Aircraft with special electronics and a large steerable nose antenna, and intended mainly for missile and spacecraft tracking in parts of the world where the USA lacks ground-based tracking facilities
Model 707-300: this was the turbofan-engined civil model developed in Model 707-300B Intercontinental, Model 707-300C Convertible and Model 707-300C Freighter versions. several examples serve with air arms in a number of guises, including (in the case

374 AIRCRAFT

of Israel) electronic reconnaissance and intelligence-gathering; a few of the earlier Model 707-100 and Model 707-200 series aircraft (often re-engined with turbofans) also used by lesser air arms

Boeing Vertol CH-46D Sea Knight
(USA)
Type: medium assault transport helicopter
Accommodation: crew of three, and up to 25 troops or freight in the hold
Armament (fixed): none
Armament (disposable): none
Electronics and operational equipment: communication and navigation equipment
Powerplant and fuel system: two 1,400-shp (1044-kW) General Electric T58-GE-10 turboshafts, and a total internal fuel capacity of 350 US gal (1323 litres) in sponson tanks
Performance: maximum speed 166 mph (267 km/h); cruising speed 165 mph (266 km/h) at sea level; initial climb rate 1,715 ft (523 m) per minute; service ceiling 14,000 ft (5750 m); range 238 miles (383 km) with a 4,550-lb (2064-kg) payload
Weights: empty 13,067 lb (5927 kg); maximum take-off 23,000 lb (10433 kg)
Dimensions: rotor diameter (each) 51 ft 0 in (15.54 m); length (rotors turning) 84 ft 4 in (25.70 m); height (to top of rear rotor hub) 16 ft 8.5 in (5.09 m); rotor disc area (total) 4,085.6 sq ft (379.56 m²)

Variants
CH-46A Sea Knight: this standard US Marine Corps assault transport type originated as the Model 107M military derivative of the Model 107-II commercial development of the underpowered and generally unsuccessful Model 107, which first flew in April 1958 and was evaluated as the YHC-1A for the US Army; the powerplant comprises two 1,250-shp (932-kW) General Electric T58-GE-8B turboshafts, and the type began to enter service in 1965 with a payload of 25 troops, or 15 litters and two attendants, or 4,000 lb (1814 kg) of freight carried over a range of 115 miles (185 km)
HH-46A Sea Knight: US Navy vertical replenishment (vertrep) equivalent of the CH-46A
CH-46D Sea Knight: much upgraded development of the CH-46A with 1,400-shp (1044-kW) T58-GE-10 turboshafts and cambered rotor blades; the type can carry a slung load of 10,000 lb (4536 kg)
UH-46D Sea Knight: US Navy vertrep equivalent of the CH-46D; the **HH-46D** is a search-and-rescue derivative of the type
CH-46E Sea Knight: designation of CH-46A and CH-46D helicopters brought up to an improved operational standard with 1,870-shp (1394-kW) T58-GE-16 turboshafts, crash-attenuating crew seats, a crash-resistant fuel system and glassfibre rotor blades
CH-46F Sea Knight: improved version of the CH-46D with updated avionics and equipment
CH-113 Labrador: utility helicopter derived from the CH-46A and in service with the Canadian air arm since 1963
CH-113B Voyageur: Canadian army equivalent of the CH-113

Boeing Vertol CH-47D Chinook
(USA)
Type: medium transport helicopter
Accommodation: crew of two or three on the flight-deck, and up to 44 troops, or 24 litters and two attendants, or 14,322 lb (6496 kg) of freight in the cabin
Armament (fixed): none
Armament (disposable): none
Electronics and operational equipment: communication and navigation equipment
Powerplant and fuel system: two Avco Lycoming T55-L-712 turboshafts, each with an emergency rating of 4,500 shp (3356 kW) and a standard rating of 3,750 shp (2796 kW), and a total internal fuel capacity of 1,030 US gal (3899 litres) in fuselage side sponsons
Performance: maximum speed 185 mph (298 km/h) at sea level; initial climb rate 1,490 ft (454 m) per minute; service ceiling (one engine out) 14,000 ft (4270 m); radius 115 miles (185 km) with an 18,000-lb (8164-kg) load
Weights: empty 23,903 lb (10475 kg); maximum take-off 53,000 lb (24267 kg)
Dimensions: rotor diameter (each) 60 ft 0 in (18.29 m); length (fuselage) 51 ft 0 in (15.54 m); height (to top of rear rotor head) 18 ft 7.8 in (5.68 m); rotor disc area (total) 5,654.9 sq ft (525.34 m²)

Variants
CH-47A Chinook: initial production version of the Model 114/YHC-1B prototype, which first flew in September 1961 for service from late 1962; the type is essentially a scaled-up Model 107 with 2,200-shp (1641-kW) Avco Lycoming T55-L-5 turboshafts, later replaced by 2,650-shp (1976-kW) T55-L-7 turboshafts, driving larger-diameter rotors on a lengthened fuselage with quadricycle rather than tricycle landing gear but still fitted with a rear ramp for ease of loading/unloading; the extra power and size provides a payload of 44 troops or an external load of 16,000 lb (7257 kg)
CH-47B Chinook: upgraded model with 2,850-shp (2125-kW) T55-L-7C turboshafts, modified rotor blades and a number of detail improvements; service entry took place in 1967
CH-47C Chinook: much improved model with 3,750-shp (2796-kW) T55-L-11A turboshafts, strengthened transmission and an increase in internal fuel capacity to 1,040 US gal (3937 litres); service entry took place in 1968, and all CH-47As and CH-47Bs were upgraded to the same basic standard; from the late

1970s survivors have been fitted with glassfibre rotor blades

CH-47D Chinook: updated standard with 13 major improvements such more powerful engines, further strengthened transmission, composite-construction rotor blades, crash-resistant features and more advanced avionics; the type has a three-point external attachment system to cater for loads as high as 22,783 lb (10334 kg), and the first CH-47C upgraded to this standard was delivered in May 1982

CH-147: version for the Canadian Armed Forces with T55-L-11C turboshafts and avionics optimized for Canadian conditions

MH-47E Chinook: special forces version of the CH-47D with inflight-refuelling capability, terrain-following radar and other advanced features

Chinook HC.Mk 1: version for the UK and based on the CH-147 but with T55-L-11E turboshafts and a maximum external load capability of 28,000 lb (12700 kg)

Model 414: designation of military Chinooks for export

Meridionali CH-47C: designation of the Italian licence-made CH-47C and differing little from the US standard; the type has been sold in southern Europe, North Africa and the middle East

Breda Nardi NH-300C (Italy): see Hughes (McDonnell Douglas Helicopters) Model 300C

Breda Nardi NH-500MD (Italy): see Hughes (McDonnell Douglas Helicopters) OH-6A Cayuse

British Aerospace HS 748 Series 2B Military Transport
(UK)
Type: utility transport
Accommodation: crew of two on the flightdeck, and up to 58 troops, or 48 paratroops, or 24 litters and nine attendants, or 12,975 lb (5885 kg) of freight in the cabin
Armament (fixed): none
Armament (disposable): none
Electronics and operational equipment: communication and navigation equipment
Powerplant and fuel system: two 2,280-ehp (1700-kW) Rolls-Royce Dart RDa.7 Mk 536-2 turboprops, and a total internal fuel capacity of 1,440 Imp gal (6550 litres) in two integral wing tanks
Performance: cruising speed 280 mph (451 km/h); initial climb rate 1,420 ft (433 m) per minute; service ceiling 25,000 ft (7620 m); range 1,475 miles (2374 km) with a 14,027-lb (6363-kg) payload and maximum fuel, or 905 miles (1456 km) with maximum payload
Weights: empty 25,730 lb (11671 kg); normal take-off 46,500 lb (21092 kg); maximum take-off 51,000 lb (23133 kg)
Dimensions: span 102 ft 5.5 in (31.23 m); length 67 ft 0 in (20.42 m); height 24 ft 10 in (7.57 m); wing area 828.9 sq ft (77.0 m²)

Variants
Andover C.Mk 1: designation of the Avro (later HS) 780 derivative of the Avro (later HS) 748M military transport for the Royal Air Force in the STOL tactical transport (freighting, trooping and paratrooping) role; this requirement led to the strengthening of the floor in a fuselage lengthened to 77 ft 11 in (23.75 m) and fitted with an upswept tail unit for the fitting of a rear loading/unloading ramp, the installation of 3,245-ehp (2420-ekW) Dart RDa.12 Mk 201C turboprops on a 98-ft (29.87-m) wing, the use of large-diameter low-pressure tyres, and the incorporation of a hydraulic system in the landing gear to align the ramp with the tailgates of trucks; the type entered service in July 1966, and seven were later modified as navaid calibration aircraft with the designations **Andover E.Mk 3** (four aircraft) and **Andover E.Mk 3A** (three aircraft)
Andover CC.Mk 2: passenger transport version of the HS 748 Series 2
HS 748M: military transport version of the HS 748 Series 1 civil transport, later known as the HS 757 and assembled by HAL in India from British components
HS 748 Series 2A Military Transport: development of the Series 2 civil transport with 2,280-ehp (1700-ekW) Dart RDa.7 Mk 534-2 or Mk 532-2 turboprops, a strengthened floor and an upswept tail with ventral ramp
HS 748 Series 2B Military Transport: improved version with uprated powerplant and a 4-ft (1.22-m) increase in span for better performance in 'hot-and-high' conditions
HS 748 Coastguarder: version available for the maritime patrol, search-and-rescue and limited anti-ship/submarine roles with fuel capacity increased to 2,210 Imp gal (10047 litres) for greater operating range; the type has Litton APS-504(V)3 or MEL Marec II search radar, high-quality navigation systems, optional features (such as magnetic anomaly detection, electronic countermeasures, electronic support measures etc) and weapons capability

British Aerospace Buccaneer S.Mk 2B
(UK)
Type: low-level strike and attack aircraft
Accommodation: pilot and systems operator seated in tandem on Martin-Baker ejector seats
Armament (fixed): none
Armament (disposable): this is carried on the revolving door of the weapons bay and on four underwing hardpoints, each rated at 3,000 lb (1361 kg), up to a maximum weight of 16,000 lb (7258 kg); a typical stores load on the revolving door is four Mk 10 1,000-lb (454-kg) bombs; and each underwing hardpoint can accommodate one Mk 10 or Mk N1 1,000-lb (454-kg) bomb, or two 500-lb (227-kg) or 540-lb (245-kg) bombs, or one launcher with 18 68-mm (2.68-in) rockets, or one launcher with 36 2-in (50.8 mm)

rockets, or three 1,000-lb (454-kg) bombs on triple ejector units, or six 500-lb (227-kg) bombs on multiple ejector units; an alternative underwing load is three AJ.168 Martel air-to-surface missiles and one Martel systems pod, or four Sea Eagle anti-ship missiles; provision for tactical nuclear bombs
Electronics and operational equipment: communication and navigation equipment, plus Decca Doppler radar, central air-data computer, moving-map display, and Ferranti Blue Parrot search and fire-control radar with terrain-warning capability and strike-sighting and computing system; there is provision in the weapons bay for a reconnaissance pack of one F97 night and six F95 day cameras, or infra-red linescan equipment
Powerplant and fuel system: two 11,100-lb (5035-kg) thrust Rolls-Royce Spey RB.168-1A Mk 101 turbofans, and a total internal fuel capacity of 1,560 Imp gal (7092 litres) in eight integral tanks in the upper fuselage, plus provision for a 425-Imp gal (1932-litre) auxiliary tank on the revolving weapons-bay door, and a 440-Imp gal (2,000-litre) auxiliary tank in the weapons bay and/or two 250- or 430-Imp gal (1136- or 1955-litre) drop tanks; provision for an inflight-refuelling probe and a Mk 20B or Mk 20C refuelling pod for inflight-refuelling of other Buccaneers
Performance: maximum speed 645 mph (1038 km/h) or Mach 0.85 at 200 ft (61 m); initial climb rate 7,000 ft (2134 m) per minute; service ceiling more than 40,000 ft (12190 m); combat radius 1,150 miles (1851 km) on a hi-lo-hi sortie
Weights: empty about 30,000 lb (13608 kg); normal take-off 56,000 lb (25402 kg); maximum take-off 62,000 lb (28123 kg)
Dimensions: span 44 ft 0 in (13.41 m); length 63 ft 5 in (19.33 m); height 16 ft 3 in (4.95 m); wing area 514.7 sq ft (47.82 m²)

Variants
Buccaneer S.Mk 2A: designed as the Blackburn B-103 (sometimes known as NA.39) low-level carrier-borne attack aircraft, the Buccaneer first flew in April 1968 and entered service as the Buccaneer S.Mk 1 with 7,100-lb (3221-kg) de Havilland Gyron Junior Mk 101 turbojets, but was soon developed into the considerably more capable S.Mk 2 type with Rolls-Royce Spey turbofans for service from 1965; with the rundown of the Royal Navy's carrier force the Buccaneers were reallocated to the Royal Air Force, the S.Mk 2 becoming the S.Mk 2A without provision for Martel air-to-surface missiles; the type can now carry the powerful Sea Eagle anti-ship missile
Buccaneer S.Mk 2B: designation of Buccaneers built for the RAF with a bulged bomb-bay door for the carriage of additional fuel, and provision for the TV-homing Martel missile; the type can now carry the powerful Sea Eagle anti-ship missile; surviving Buccaneers are being upgraded with improved avionics such as revised Blue Parrot radar, a Ferranti FIN1063 inertial navigation system and other comparatively low-cost developments, financial restrictions having caused the cancellation of a full-scale refit with the ARI.18228 radar-warning receiver, and the Westinghouse ALQ-101-10 and Marconi Sky Shadow electronic countermeasures systems
Buccaneer S.Mk 50: derivative of the S.Mk 2 for South Africa, basically similar to the British aircraft apart from the retractable installation of two retractable 8,000-lb (3629-kg) Bristol Siddeley BS.605 rocket engines to boost performance at take-off under 'hot-and-high' conditions

British Aerospace Canberra B(I).Mk 8
(UK)
Type: light bomber/intruder
Accommodation: pilot (on a Martin-Baker ejector seat) and navigator
Armament (fixed): four Aden 20-mm cannon with 500 rounds per gun in a ventral pack carried optionally in the rear half of the bomb bay
Armament (disposable): this can be made up of some 8,000 lb (3629 kg) of stores carried in the bomb bay and on two underwing hardpoints, the latter each rated at 1,000 lb (454 kg); the bomb bay can accommodate six 1,000-lb (454-kg) bombs on two Avro triple carriers (only three bombs on one triple carrier if the ventral gunpack is carried), or one 4,000-lb (1814-kg) and two 1,000-lb (454-kg) bombs, or eight 500-lb (227-kg) bombs; the underwing hardpoints can each accept one 1,000-lb (454-kg) or two 500-lb (227-kg) bombs, or one AS.30 air-to-surface missile, one 7.62-mm (0.3-in) machine-gun pod, or one launcher with 37 2-in (50.8-mm) rockets; other stores options include 100-lb (45-kg) bombs of free-fall or retarded type
Electronics and operational equipment: communication and navigation equipment
Powerplant and fuel system: two 7,500-lb (3402-kg) thrust Rolls-Royce Avon RA.7 Mk 109 turbojets, and a total internal fuel capacity of 2,277 Imp gal (10351 litres) in three fuselage fuel cells and two integral wing leading-edge tanks, plus provision for two 250-Imp gal (1137-litre) tiptanks and one 300-Imp gal (1364-litre) tank in the forward part of the bomb bay, or one 650-Imp gal (2955-litre) ferry tank
Performance: maximum speed 541 mph (871 km/h) at 40,000 ft (12190 m) and 510 mph (821 km/h) at sea level; initial climb rate 3,600 ft (1097 m) per minute; service ceiling 48,000 ft (14630 m); range 805 miles (1295 km) with maximum bombload
Weights: empty 27,950 lb (12678 kg); normal take-off 43,000 lb (19505 kg); maximum take-off 54,950 lb (24925 kg)
Dimensions: span 63 ft 11.5 in (19.49 m) without tiptanks; length 65 ft 6 in (19.96 m); height 15 ft 7 in (4.75 m); wing area 960.0 sq ft (89.19 m²)

Variants
Canberra B.Mk 2: initial production version of the Canberra light bomber, first flown in May 1949; this version has a powerplant of two 6,500-lb (2948-kg) Rolls-Royce Avon RA.7 Mk 101 turbojets and an armament of six 1,000-lb (907-kg) bombs carried internally, and began to enter Royal Air Force service in 1951; export versions were the **Canberra B.Mk 20**, **Canberra B.Mk 52**, **Canberra B.Mk 62** and **Canberra B.Mk 82** for Australia, Ethiopia, Argentina and Venezuela
Canberra T.Mk 4: baseline dual-control trainer version also produced as the **Canberra T.Mk 13**, **Canberra T. Mk 21**, **Canberra T.Mk 64** and **Canberra T.Mk 84** for New Zealand, Australia, Argentina and Venezuela
Canberra B.Mk 6: definitive light bomber version powered by 7,500-lb (3402-kg) Avon Mk 109 turbojets and carrying eight 1,000-lb (454-kg) bombs (two of them under the wings, and replaceable by AS.30 missiles, rocket-launcher pods etc)
Canberra PR.Mk 7: developed photographic reconnaissance model bearing the same relationship to the Canberra PR.Mk 3 as the B.Mk 6 to the B.Mk 2; the main export version was the **Canberra PR.Mk 57** for India, the **Canberra PR.Mk 83** being a version of the Canberra PR.Mk 3 for Venezuela
Canberra B(I).Mk 8: definitive nocturnal intruder version with a fighter-type pilot's cockpit offset to port and the navigator relocated to a glazed nose position; export versions were the **Canberra B(I).Mk 12**, **Canberra B(I).Mk 56**, **Canberra B(I).Mk 58**, **Canberra B(I).Mk 68** and **Canberra B(I).Mk 82** and New Zealand/South Africa, Peru (refurbished rather than new-build aircraft), India, Peru (new-build aircraft) and Venezuela
Canberra PR.Mk 9: definitive photographic reconnaissance version with 10,500-lb (4763-kg) Avon Mk 206 turbojets, a service ceiling of about 60,000 ft (18290 m) and a range greater than 3,650 miles (5875 km) at a maximum take-off weight of some 56,250 lb (25514 kg); this type is still in useful service, and has a span of 67 ft 10 in (20.68 m) and extended-chord inner wings for an area of 1,045.0 sq ft (97.08 m^2); in the mid-1980s five aircraft were being refurbished and upgraded for visual and electronic reconnaissance in British colours
Canberra T.Mk 17: electronic countermeasures trainer version produced by converting Canberra B.Mk 2s with radar and communications jammers in the nose, chaff dispensers in the bomb bay, and radar warning receivers in the nose and tail
Canberra T.Mk 19: electronically-silent target version produced as conversions of Canberra T.Mk 11 night-fighter trainers
Canberra T.Mk 22: conversion of Canberra PR.Mk 7s with Blue Parrot radar and other systems for the training of Fleet Air Arm crews

British Aerospace Harrier GR.Mk 3
(UK)
Type: V/STOL close-support and reconnaissance aircraft
Accommodation: pilot seated on a Martin-Baker Mk 9D ejector seat
Armament (fixed): (optional) two Aden 30-mm cannon with 100 rounds per gun carried in place of the underfuselage strakes
Armament (disposable): this is carried on one underfuselage and four underwing hardpoints, the underfuselage and inner pair of underwing hardpoints each rated at 2,000 lb (907 kg) and the outer pair of underwing hardpoints each at 650 lb (295 kg), up to a maximum cleared weight of 5,000 lb (2268 kg) though trials have been conducted with up to 8,000 lb (3269 kg) of weapons; weapons that can be carried include Mk 83 1,000 lb (454-kg) free-fall or retarded bombs, Mk 82 500-lb (227-kg) free-fall or retarded bombs, Mk 81 250-lb (113-kg) free-fall or retarded bombs, 'Paveway' laser-guided bombs, 'Rockeye' cluster bombs, BL755 cluster bombs, Matra 155 series launchers each with 16 68-mm (2.68-in) rockets, LAU-10/A launchers each with four 5-in (127-mm) rockets, LAU-68/A launchers each with seven 2.75-in (69.85-mm) rockets, LAU-69/A launchers each with 19 2.75-in (69.85-mm) rockets, Mk 77 firebombs, AJ.168 Martel or AGM-84 Harpoon anti-ship missiles, or AIM-9L Sidewinder air-to-air missiles
Electronics and operational equipment: communication and navigation equipment, plus Ferranti FE 541 inertial navigation and attack system, Smiths head-up display, Smiths air-data computer, Marconi ARI.18223 radar-warning receiver, Ferranti Type 106 laser-ranger and marked-target seeker, Tracor ALE-40 chaff/flare dispensers, and Marconi Sky Shadow ECM
Powerplant and fuel system: one 21,500-lb (9752-kg) thrust Rolls-Royce (Bristol) Pegasus Mk 103 vectored-thrust turbofan, and a total internal fuel capacity of 630 Imp gal (2865 litres) in five fuselage and two wing integral tanks, plus provision for two 100-Imp gal (455-litre) drop-tanks or two 330-Imp gal (1500-litre) ferry tanks; provision for inflight-refuelling
Performance: maximum speed over 737 mph (1186 km/h) or Mach 0.97 at low altitude; climb to 40,000 ft (12190 m) in 2 minutes 22 seconds after VTO; service ceiling more than 50,000 ft (15240 m); combat radius 414 miles (667 km) with 3,000-lb (1361-kg) payload after VTO; ferry range 3,445 miles (5560 km) with one inflight-refuelling
Weights: empty 13,535 lb (6139 kg); maximum take-off more than 25,200 lb (11431 kg)
Dimensions: span 25 ft 3 in (7.70 m), or 29 ft 8 in (9.04 m) with low-drag ferry tips; length 46 ft 10 in (14.27 m); height 11 ft 4 in (3.45 m); wing area 201.1 sq ft (18.68 m^2), or 216.0 sq ft (20.07 m^2) with ferry tips

Variants
Harrier GR.Mk 3: designation of all current Harrier V/STOL single-seat close-support aircraft, a few being built as such but most being produced by upgrading surviving Harrier GR.Mk 1A aircraft with a laser ranger and marked-target seeker in a chisel nose increasing length by 1.25 ft (0.38 m), with a radar-warning receiver system, and with a Pegasus Mk 103 engine instead of the 20,500-lb (9299-kg) Pegasus Mk 102; the type first flew as the Hawker Siddeley P.1127 in October 1960 and was developed as the Kestrel pre-production type before emerging as the Harrier for a first flight in August 1966 and service entry in April 1969
Harrier T.Mk 4: designation of Harrier T.Mk 2A trainers upgraded with Pegasus Mk 103 turbofans and with the operation equipment of the Harrier GR.Mk 3 other than the radar-warning receiver system
Harrier T.Mk 4A: aircraft built with Pegasus Mk 103 turbofans
Harrier T.Mk 4N: trainer for the Fleet Air Arm without the laser ranger and marked-target seeker
Harrier T.Mk 60: trainer for the Indian navy with the electronics suite of the Sea Harrier FRS.Mk 51 apart from the 'Blue Fox' radar
AV-8A: designation of aircraft for the US Marine Corps, basically similar to the Harrier GR.Mk 3 but lacking the laser ranger and marked-target seeker, the radar-warning receiver and the inertial navigation system; the type has US equipment (including the Stencel SIII-S3 ejector seat) and provision for US weapons (including AIM-9 Sidewinder air-to-air missiles; the designation **AV-8S Matador** covers the version for Spain
TAV-8A: version of the Harrier T.Mk 4/4A for the US Marine Corps but without the laser ranger and marked-target seeker; the type is fitted, however, with UHF radio for use in the forward air control role; the designation **TAV-8S Matador** covers the version for Spain
AV-8C: designation of 47 AV-8A aircraft modified and life-extended by McDonnell Douglas with many of the systems improvements of the McDonnell Douglas AV-8B Harrier II, such as the lift-improvement devices, the ALE-39 chaff/flare dispenser system, onboard oxygen generation, secure voice communication equipment and an ALR-45 radar-warning system with receivers in the wingtips and tail

British Aerospace Harrier GR. Mk 5: see McDonnell Douglas/British Aerospace AV-8B Harrier II

British Aerospace Hawk T.Mk 1
(UK)
Type: basic and advanced trainer with secondary air-defence and ground-attack roles
Accommodation: pupil and instructor seated in tandem on Martin-Baker Mk 10B rocket-assisted ejector seats
Armament (fixed): optional Aden 30-mm cannon in a pod with 130 rounds under the fuselage
Armament (disposable): this is carried on two standard underwing hardpoints, each rated at 1,000 lb (454 kg), and on one optional underfuselage hardpoint (instead of the cannon pod) and two optional underwing hardpoints, all rated at 1,000 lb (454 kg) each, up to a maximum ordnance load of 5,660 lb (2567 kg) though trials have been conducted with stores up to a weight of 6,800 lb (3084 kg); demonstrated loads have included four Matra 155 launchers each with 18 68-mm (2.68-in) rockets, four launchers each with 19 2.75-in (69.85-mm) rockets, four packs each with nine 81-mm (3.19-in) rockets, four LAU-10/A launchers each with four 5-in (127-mm) rockets, four gun pods each with two 7.62-mm (0.3-in) machine-guns, five Mk 83 1,000-lb (454-kg) free-fall or retarded bombs, nine Mk 82 500-lb (227-kg) free-fall or retarded bombs, nine Mk 81 250-lb (113-kg) free-fall or retarded bombs, nine 50-Imp gal (227-kg) napalm tanks, and two AIM-9L Sidewinder air-to-air missiles for the air-defence role
Electronics and operational equipment: communication and navigation equipment, plus Ferranti F195 weapon sight and provision for a centreline reconnaissance pod
Powerplant and fuel system: one 5,200-lb (2359-kg) thrust Rolls-Royce/Turboméca Adour Mk 151 turbofan, plus a total internal fuel capacity of 375 Imp gal (1704 litres) in one fuselage bag tank and one integral wing tank, plus provision for two 100- or 130-Imp gal (455- or 592-litre) drop-tanks
Performance: maximum speed 645 mph (1038 km/h) or Mach 0.88 at 11,000 ft (3355 m); initial climb rate 9,300 ft (2835 m) per minute; service ceiling 50,000 ft (15240 m); combat radius 345 miles (556 km) with 5,600-lb (2540-kg) weapon load; range 1,920 miles (3090 km) with maximum internal fuel and two drop-tanks
Weights: empty 8,040 lb (3647 kg); normal take-off 12,284 lb (5572 kg) with weapons as a trainer; maximum take-off 18,890 lb (8569 kg)
Dimensions: span 30 ft 9.75 in (9.39 m); length 36 ft 7.75 in (11.17 m) excluding probe; height 13 ft 1.25 in (3.99 m); wing area 179.6 sq ft (16.69 m^2)

Variants
Hawk T.Mk 1: designed as the Hawker Siddeley P.1182 to succeed the Hawker Siddeley (Folland) Gnat and Hawker Hunter flying and weapons trainers, the Hawk first flew in August 1972 and has matured into a highly capable multi-role trainer and light attack aircraft characterized by perfect handling for the carriage and delivery of a diverse weapons load
Hawk T.Mk 1A: designation of 72 Hawk T.Mk 1s modified as secondary air-defence aircraft with

provision for AIM-9L Sidewinder air-to-air missiles and BL755 cluster bombs

Hawk Mk 51: designation of 50 aircraft for Finland, identical with the Hawk T.Mk 1 apart from the Saab RGS2 sight

Hawk Mk 52: designation of 12 aircraft for Kenya to Hawk T.Mk 1 standard

Hawk Mk 53: designation of 20 aircraft for Indonesia in the training and light attack roles

Hawk Mk 60: uprated model with the 5,700-lb (2586-kg) Adour Mk 861 turbofan; variants of this model have been ordered by Abu Dhabi, Dubai, Kuwait, Saudi Arabia and Zimbabwe

Hawk Mk 100: also known as the Enhanced Ground-Attack Hawk, this two-seat version is under development in the mid-1980s with Hands On Throttle And Stick (HOTAS) controls and a MIL 1553B databus to allow full integration of modern weapons with the new inertial nav/attack, head-up display, head-down display, weapon-aiming computer, stores management, flare/chaff dispenser and radar-warning systems; the type will also be able to carry a laser ranger and forward-looking infra-red as options

Hawk Mk 200: dedicated light attack single-seater under development by BAe as a private venture based on the Hawk Mk 60 but fitted with the uprated Adour Mk 871 turbofan, a new forward fuselage and advanced avionics such as a Smiths head-up display, Singer-Kearfott inertial navigation system and Ferranti laser ranger and marked-target seeker for highly accurate first-pass attacks with a wide assortment of disposable stores (including AIM-120 AMRAAM air-to-air missiles and Sea Eagle anti-ship missiles in aircraft fitted with nose rader rather than ther laser ranger and marked-target seeker) up to a weight of 6,800 lb (3085 kg) on one underfuselage and four underwing hardpoints; the additional fuselage volume allows the fitting of single or twin 25-mm Aden cannon (or comparable weapons to customer requirements); the type first flew in May 1986, and proposed production variants are the **Hawk Mk 200-60** using the nav/attack system of the Hawk Mk 60 for visual operations, and the **Hawk Mk 200-100** all-weather suite of the Hawk Mk 100

T-45A Goshawk: designation of the navalized trainer to be produced jointly by BAe and McDonnell Douglas for the US Navy as replacement for the Rockwell T-2 Buckeye and McDonnell Douglas TA-4 Skyhawk; the type is based on the Hawk T.Mk 1 but has full carrier capability with strengthened landing gear (including long-stroke main units and a twin-wheel nose unit stressed for catapult take-offs), a US Navy cockpit and the new Martin-Baker NACES ejector seats; the type is powered by the Adour Mk 861-49 turbofan derated to 5,450-lb (2472-kg) thrust for economy and longevity, and has a maximum take-off weight of 12,450 lb (5647 kg)

British Aerospace Hunter FGA.Mk 9
(UK)

Type: fighter and weapons-training aircraft

Accommodation: pilot seated on a Martin-Baker Mk 3H ejector seat

Armament (fixed): four Aden 30-mm cannon with 135 rounds per gun in a detachable ventral pack

Armament (disposable): this is carried on four underwing hardpoints, the inner pair each rated at 1,000 lb (454 kg) and the outer pair each at 500 lb (227 kg); typical underwing stores are 1,000-lb (454-kg) bombs, 500-lb (227-kg) bombs, clusters of 3-in (76.2-mm) rockets, napalm tanks, and launchers for a wide variety of rockets ranging in size from 2 in (50.8 mm) to 2.75 in (69.85 mm)

Electronics and operational equipment: communication and navigation equipment, plus ranging radar

Powerplant and fuel system: one 10,150-lb (4604-kg) thrust Rolls-Royce Avon RA.28 Mk 207 turbojet, and a total internal fuel capacity of 392 Imp gal (1,782 litres) in two fuselage and two wing tanks, plus provision for 100-, 230- or 330-Imp gal (456-, 1046- or 1500-litre) drop-tanks

Performance: maximum speed 710 mph (1,144 km/h) or Mach 0.93 at sea level, and 620 mph (978 km/h) or Mach 0.94 at high altitude; initial climb rate 8,000 ft (2440 m) per minute; service ceiling 51,500 ft (15700 m); range 1,840 miles (2965 km) with two 230-Imp gal (1046-litre) drop-tanks

Weights: empty 13,010 lb (5901 kg); normal take-off 18,000 lb (8165 kg); maximum take-off 24,600 lb (11158 kg) with two 100-Imp gal (456-litre) and two 230-Imp gal (1046-litre) drop-tanks

Dimensions: span 33 ft 8 in (10.25 m); length 45 ft 10.5 in (13.98 m); height 13 ft 2 in (4.02 m); wing area 349.0 sq ft (32.42 m^2)

Variants

Hunter F.Mk 6: the Hunter was developed as the Hawker P.1067 and first flew in prototype form during July 1951, entering service with the Royal Air Force as the Hunter F.Mk 1 in July 1954; the Hunter F.Mk 6 was the definitive fighter model and appeared in 1956 with the Avon Mk 200 series turbojet for more power and thus better performance than earlier models; variants of this basic theme are the **Hunter Mk 56** for India and **Hunter Mk 58** for Switzerland, the latter being particularly potent with their Saab weapon systems, AIM-9 Sidewinder air-to-air missiles and AGM-65 Maverick air-to-surface missiles

Hunter T.Mk 7: side-by-side conversion trainer based on the Hunter F.Mk 4 fighter but with the Avon Mk 122 turbojet; export versions generally had the Avon Mk 200 series turbojet, and included the **Hunter T.Mk 66**, **Hunter T.Mk 72**, **Hunter T.Mk 75**, **Hunter T.Mk 77** and **Hunter T.Mk 79** for India, Chile, Singapore, Abu Dhabi and Qatar

Hunter T.Mk 8: naval trainer version derived from

the Hunter F.Mk 4 with specialized equipment and arrester hooks

Hunter FGA.Mk 9: dedicated fighter/ground-attack model introduced in 1959 with strengthened landing gear, a braking parachute and greater provision for external stores; export versions included the **Hunter FGA.Mk 57**, **Hunter FGA.Mk 70**, **Hunter FGA.Mk 71**, **Hunter FGA.Mk 74**, **Hunter FGA.Mk 76** and **Hunter FGA.Mk 78** for Kuwait, Lebanon, Chile, Singapore, Abu Dhabi and Qatar

Hunter FR.Mk 10: tactical reconnaissance derivative of the Hunter FGA.Mk 9 with a fan of three cameras instead of the nose-mounted ranging radar; the **Hunter FR.Mk 74A** was the export version for Singapore

Hunter GA.Mk 11: version of the Hunter FGA.Mk 9 for the Fleet Air Arm, based on the Hunter F.Mk 4 and used for weapons training

Hunter PR.Mk 11: version for the Fleet Air Arm, based on the Hunter F.Mk 4 and used for photographic reconnaissance training

British Aerospace Jet Provost (UK): see British Aerospace Strikemaster Mk 88

British Aerospace Lightning F.Mk 6 (UK)

Type: interceptor fighter

Accommodation: pilot seated on a Martin-Baker ejector seat

Armament (fixed): two Aden 30-mm cannon with 120 rounds per gun in the front portion of the ventral fuel tank

Armament (disposable): this is carried on two lateral projections of an interchangeable pack in the lower front fuselage, and can comprise two Red Top or Firestreak air-to-air missiles; an infrequent alternative is a pair of retractable honeycomb packs each with 22 2-in (50.8-mm) rockets

Electronics and operational equipment: communication and navigation equipment, plus Ferranti AI-23S Airpass interception radar and, as an alternative to the missile package, a day reconnaissance pack (with five Vinten 360 cameras) or a night reconnaissance pack (with cameras and infra-red linescan equipment)

Powerplant and fuel system: two 16,360-lb (7420-kg) afterburning thrust Rolls-Royce Avon RB.146 Mk 301 turbojets, and a total internal fuel capacity of ? Imp gal (? litres) plus provision for a ventral tank of 610-Imp gal (2773-litre) capacity; provision for inflight-refuelling

Performance: maximum speed 1,320 mph (2112 km/h) or Mach 2 at 36,000 ft (10970 m); climb to 40,000 ft (12190 m) in 2 minutes 30 seconds; service ceiling 55,000 ft (16765 m); range 800 miles (1287 km) on internal fuel

Weights: empty 28,041 lb (12717 kg); maximum take-off 42,000 lb (19047 kg)

Dimensions: span 34 ft 10 in (10.62 m); length 55 ft 3 in (16.84 m) including probe; height 19 ft 7 in (5.97 m); wing area 458.5 sq ft (42.97 m²)

Variants

Lightning F.Mk 3: the English Electric Lightning was schemed as a Mach 2 research aircraft and then turned into a potent interception type with sparkling climb and speed but only moderate armament and poor range; the Lightning F.Mk 1 first flew in October 1958, and the Lightning F.Mk 3 followed in June 1962 with a square-topped fin, Avon Mk 301 afterburning turbojets, AI-23B radar, no inbuilt cannon, two Red Top air-to-air missiles, and provision for large overwing ferry tanks; this basic type was also modified as the **Lightning F.Mk 3A** with reduced leading-edge sweep outboard of the main landing gear hinges

Lightning T.Mk 4: unarmed side-by-side two-seat operational trainer derived from the Lightning F.Mk 1A fighter

Lightning T.Mk 5: armed side by-side two-seat operational trainer based on the Lightning F.Mk 3

Lightning F.Mk 6: ultimate development of the Royal Air Force's single-seat type, with revised wings and a large ventral tank with provision for cannon armament in its forward portion

Lightning F.Mk 52: designation of Lightning F.Mk 2s transferred to Saudi Arabia

Lightning F.Mk 53: designation of the multi-role derivative of the Lightning F.Mk 6 built for Saudi Arabia and incorporating two underwing hardpoints each able to carry two 1,000-lb (454-kg) free-fall 'slick' or retarded bombs, or two 1,000-lb (454-kg) fire bombs, or one Matra 155 launcher with 18 68-mm (2.68-in) rockets, or one machine-gun pod; there is also provision for two overwing hardpoints, each able to carry one 1,000-lb (454-kg) bomb or one drop tank

Lightning T.Mk 54: designation of Lightning T.Mk 4s transferred to Saudi Arabia

Lightning T.Mk 55: designation of the side-by-side two-seat operational conversion trainer for Saudi Arabia and based on the Lightning F.Mk 53

British Aerospace Nimrod MR.Mk 2 (UK)

Type: maritime patrol aircraft

Accommodation: crew of three on the flightdeck, and a normal mission crew of nine in the cabin (or 45 troops if the aircraft is used in its secondary transport role)

Armament (fixed): none

Armament (disposable): this is carried in a lower-fuselage bay and on two underwing hardpoints; the fuselage bay is 48 ft 6 in (14.78 m) long and can accommodate six lateral rows of anti-submarine weapons up to a weight of 13,500 lb (6124 kg) including up to nine Mk 46 or Stingray torpedoes and a variety of bombs; the underwing hardpoints can lift AIM-9L Sidewinder air-to-air missiles for self-defence, or Sea Eagle or AGM-84 Harpoon anti-ship missiles

Electronics and operational equipment: communication and navigation equipment, plus separate processing systems for tactical navigation, acoustic signals and radar; the central tactical system is produced by Marconi with a 920 ATC computer at its core and having input from a Ferranti inertial navigation system, the Loral ARI.18240/1 and Thomson-CSF electronic support measures systems, the Emerson ASQ-10A magnetic anomaly detection system, the acoustics system and the radar system; the acoustics system is based on the Marconi AQS-901 processing and display system with twin 920 ATC computers and compatible with many types of active-passive sonobuoys such as the American SSQ-41 and SSQ-53, the Australian BARRA, the British Ultra X17255 command active multi-beam type and the Canadian TANDEM; and the radar system is based on EMI Search Water long-range search radar and a Ferranti FM 1600D digital computer, the whole system being able to track several targets simultaneously in conditions of clutter and electronic countermeasures

Powerplant and fuel system: four 12,140-lb (5507-kg) thrust Rolls-Royce Spey RB.168-20 Mk 250 turbofans, and a total internal fuel capacity of 10,730 Imp gal (48780 litres) in fuselage keel, integral wing and leading-edge pod tanks, plus provision for 1,890 Imp gal (8592 litres) in six optional weapons-bay tanks; provision for inflight-refuelling

Performance: maximum speed 575 mph (926 km/h); cruising speed 547 mph (880 km/h) and patrol speed 230 mph (370 km/h); service ceiling 42,000 ft (12800 m); ferry range 5,755 miles (9265 km); mission endurance 12 hours

Weights: empty 86,000 lb (39010 kg); normal take-off 177,500 lb (80514 kg); maximum take-off 192,000 lb (87091 kg)

Dimensions: span 114 ft 10 in (35.00 m); length 126 ft 9 in (38.63 m); height 29 ft 8.5 in (9.08 m); wing area 2,121.0 sq ft (197.0 m²)

Variants

Nimrod R.Mk 1: based on the Nimrod MR.Mk 1 maritime patrol and anti-submarine aircraft, itself based on the airframe of the de Havilland Comet 4C airliner and developed from 1964 by Avro and then Hawker Siddeley, the Nimrod R.Mk 1 was produced in small numbers (just three aircraft) for the dedicated electronic intelligence role; the three aircraft were handed over to the Royal Air Force in July 1971 for commissioning in May 1974 after an extensive fitting-out programme with Loral electronic support measures in wingtip pods, and with other British and American electronic intelligence systems in the fuselage; little is known of the aircraft, but it is thought that they have excellent range through the installation of auxiliary fuel tanks in the volume reserved for weapons in the MR series; the overall length is 119 ft 9 in (36.50 m)

Nimrod MR.Mk 2: improved maritime patrol and anti-submarine aircraft produced by converting 31 of the 46 Nimrod MR.Mk 1 aircraft; the first mark had been delivered from 1969, but the pace of Soviet submarine development is such that by the mid-1970s the Nimrod MR.Mk 1s were beginning to show signs of obsolescence in their sensor and data-processing capabilities; this led to the Nimrod MR.Mk 2 upgrade programme, the much improved aircraft being redelivered from 1979 with a new tactical system, search radar (Search Water in place of the original EMI ASV-21D), acoustic processing system, displays and consoles, inertial navigation and Loral ARI.18240/1 wingtip electronic support measures pods to complement the Thomson-CSF unit podded on top of the fin; developments in the 1980s have added inflight-refuelling capability on 16 aircraft, plus the ability to carry AGM-84 Harpoon anti-ship missiles and up to four AIM-9 Sidewinder self-protection air-to-air missiles

British Aerospace Sea Harrier FRS.Mk 1
(UK)

Type: carrierborne V/STOL fighter, reconnaissance and strike aircraft

Accommodation: pilot seated on a Martin-Baker Mk 10H ejector seat

Armament (fixed): provision for two Aden 30-mm cannon with 100 rounds per gun in pods carried as replacement for the underfuselage strakes

Armament (disposable): this is carried on one underfuselage and four underwing hardpoints, the underfuselage and inner pair of underwing hardpoints each rated at 2,000 lb (907 kg) and the outer pair of underwing hardpoints each at 650 lb (295 kg), up to a maximum cleared weight of 5,000 lb (2268 kg) though trials have been flown with up to 8,000 lb (3629 kg) of external ordnance; weapons that can be carried are Mk 83 1,000-lb (454-kg) free-fall or retarded bombs, Mk 82 500-lb (227-kg) free-fall or retarded bombs, Mk 81 250-lb (113-kg) free-fall or retarded bombs, Mk 77 fire bombs, BL755 cluster bombs, 'Rockeye' cluster bombs, 'Paveway' laser-guided bombs, Matra 155 launchers each with 19 68-mm (2.68-in) rockets, AIM-9L Sidewinder air-to-air missiles, AJ.168 Martel air-to-surface missiles, AGM-84 Harpoon anti-ship missiles, Sea Eagle anti-ship missiles and the WE-177 tactical nuclear weapon

Electronics and operational equipment: communication and navigation equipment, plus Ferranti Blue Fox pulse-Doppler multi-mode radar (with TV-raster daylight-viewing tube so that flight information can be provided to the pilot on the same screen), Smiths head-up display, Smiths digital weapons-aiming computer, Decca 72 Doppler navigation, Marconi

ARI.18223 radar-warning receiver, and Ferranti FE 541 inertial navigation system
Powerplant and fuel system: one 21,500-lb (9752-kg) thrust Rolls-Royce (Bristol) Pegasus 11 Mk 104 vectored-thrust turbofan, and a total internal fuel capacity of 630 Imp gal (2865 litres) in five fuselage and two wing integral tanks, plus provision for two 100 Imp gal (456-litre) or 330-Imp gal (1500-litre) drop-tanks; provision for inflight-refuelling
Performance: maximum speed 735 mph (1183 km/h) or Mach 0.97 at sea level; cruising speed 530 mph (853 km/h) at 36,000 ft (10975 m); climb to 40,000 ft (12190 m) in 2 minutes 20 seconds; service ceiling more than 50,000 ft (15240 m); combat radius 460 miles (740 km) on a high-altitude interception, or 288 miles (463 km) on an attack sortie
Weights: empty 13,100 lb (5942 kg); maximum take-off 26,190 lb (11880 kg)
Dimensions: span 25 ft 3 in (7.70 m); length 47 ft 7 in (14.50 m); height 12 ft 2 in (3.71 m); wing area 201.1 sq ft (18.68 m²)

Variants
Sea Harrier FRS.Mk 1: this highly capable carrier-borne V/STOL fighter, reconnaissance and strike derivative of the land-based Harrier proved itself a formidable weapon in the Falklands war of 1982, and was developed to give the Royal Navy's three 'Invincible' class light carriers a potent fixed-wing air strength; development began in 1975, and the type began to enter Fleet Air Arm service in 1981; compared with the Harrier, the Sea Harrier has a revised structure less susceptible to salt-water corrosion, naval equipment, and a totally new forward fuselage; this last seats the pilot higher under a bubble canopy for air-combat visibility, has a new nav/attack system and displays, and a Blue Fox multi-mode radar
Sea Harrier FRS.Mk 2: mid-life update version of the Sea Harrier FRS.Mk 1, designed to provide the type with the ability to engage multiple beyond-visual-horizon targets (even those at low level) with four AIM-120 AMRAAM missiles; to this end Ferranti Blue Vixen coherent pulse-Doppler track-while-scan radar replaces the original Blue Fox radar to improve acquisition and look-down capabilities; other improvements are the installation of the Guardian radar-warning receiver system, the Joint Tactical Information and Distribution System for secure voice and data links, two additional underwing hardpoints, wingtip stations for two AIM-9 Sidewinder air-to-air missiles, a digital databus to make possible the integration of more modern weapons, wing and cockpit improvements, larger drop tanks, and 25-mm Aden cannon in place of the elderly 30-mm weapons
Sea Harrier FRS.Mk 51: designation of Sea Harrier FRS.Mk 1 aircraft for the Indian navy with revised systems and provision for Matra Magic rather than AIM-9 Sidewinder air-to-air missiles

British Aerospace Shackleton AEW.Mk 2
(UK)
Type: airborne early warning aircraft
Accommodation: crew of four on the flightdeck, and a tactical crew of six in the cabin
Armament (fixed): none
Armament (disposable): none
Electronics and operational equipment: communication and navigation equipment, plus APS-20F(1) search radar, 'Orange Harvest' broadband electronic support measures equipment, APX-17 IFF and selective interrogator, and processing/display equipment
Powerplant and fuel system: four 2,455-hp (1831-kW) Rolls-Royce Griffon 57A piston engines, and a total fuel capacity of 5,248 Imp gal (19285 litres) in wing tanks
Performance: maximum sped 273 mph (439 km/h); initial climb rate 850 ft (259 m) per minute; service ceiling 23,000 ft (7010 m); range 3,050 miles (4908 km)
Weights: empty 57,000 lb (25855 kg); maximum take-off 98,000 lb (44452 kg)
Dimensions: span 120 ft 0 in (36.58 m); length 87 ft 4 in (26.62 m); height 16 ft 9 in (5.10 m); wing area 1,421.0 sq ft (132.0 m²)

Variant
Shackleton AEW.Mk 2: now the only variant of the Shakleton maritime patrol, anti-submarine and airborne early warning family in service (and retained in service only because of cancellation of the Nimrod AEW.Mk 3 programme), the Shackleton AEW.Mk 2 is only one step removed from total obsolescence; the Shackleton MR series entered service in 1951, and the Shackleton AEW.Mk 2 was produced between 1971 and 1974 by conversion of surplus Shackleton MR.Mk 2s with World War II-vintage APS-20 radar stripped from Fairey Gannet AEW aircraft surplus to Fleet Air Arm requirements after the demise of its major carriers; with the cancellation of the Nimrod AEW.Mk 3 programme, the Shackleton now has to remain in service until the advent of the RAF's Boeing Sentry AWACS aircraft in the early 1990s

British Aerospace Strikemaster Mk 88
(UK)
Type: light attack and counter-insurgency aircraft
Accommodation: pilot and co-pilot/pupil seated side-by-side on Martin-Baker Mk PB4 ejector seats
Armament (fixed): two 7.62-mm (0.3-in) FN machine-guns with 550 rounds per gun in the lower edges of the inlets
Armament (disposable): this is carried on four underwing hardpoints, each rated at 1,000 lb (454 kg), up

to a maximum weight of 3,000 lb (1361 kg) with reduced fuel or 2,650 lb (1202 kg) with maximum fuel; typical weapons are 245-kg (540-lb) free-fall or retarded bombs, Mk 83 1,000-lb (454-kg) free-fall or retarded bombs, Mk 82 500-lb (227-kg) free-fall or retarded bombs, Mk 81 250-lb (113-kg) free-fall or retarded bombs, Matra 155 launchers each with 18 68-mm (2.68-in) rockets, LAU-68/A launchers each with seven 2.75-in (69.85-mm) rockets, banks each of four 81-mm (3.19-in) SURA rockets, napalm tanks, 7.62-mm (0.3-in) machine-gun pods, 20-mm cannon pods and several other ordnance items
Electronics and operational equipment: communication and navigation equipment, plus SFOM optical sight, or GM2L reflector sight or Ferranti LFS Type 5 gyro sight
Powerplant and fuel system: one 3,140-lb (1424-kg) thrust Rolls-Royce (Bristol) Viper Mk 535 turbojet, and a total internal fuel capacity of 270 Imp gal (1227 litres) in six wing bag tanks and two integral wing tanks, plus 96 Imp gal (436 litres) in non-jettisonable wingtip tanks and provision for two 50-Imp gal (227-litre) and two 75-Imp gal (341-litre) drop-tanks
Performance: maximum speed 480 mph (772 km/h) at 18,000 ft (5585 m) and 450 mph (724 km/h) at sea-level; initial climb rate 5,250 ft (1600 m) per minute; service ceiling 40,000 ft (12190 m); range 145 mile (233-km) lo-lo-lo radius with 3,000-lb (1361-kg) weapon load, or 1,382 miles (2224 km) with maximum fuel
Weights: empty 6,195 lb (2810 kg); normal take-off 10,600 lb (4808 kg) for armament training; maximum take-off 11,500 lb (5215 kg)
Dimensions: span 36 ft 10 in (11.23 m) over tiptanks; length 33 ft 8.5 in (10.27 m); height 10 ft 11.5 in (3.34 m); wing area 213.7 sq ft (19.85 m²)

Variants
Jet Provost T.Mk 3: the Hunting Jet Provost first flew in prototype form in June 1954 and continues to serve the Royal Air Force well during the later 1980s; the Jet Provost T.Mk 3 was introduced in 1958 with the 1,750-lb (794-kg) Viper 8 turbojet, ejector seats, a clear rather than framed canopy, and tip tanks; the installation of better navigational equipment resulted in the **Jet Provost T.Mk 3A**
Jet Provost T.Mk 4: improved version of 1960 with the 2,500-lb (1134-kg) Viper 11
Jet Provost T.Mk 5: definitive trainer version with full pressurization, a slightly bulged canopy and the 2,500-lb (1134-kg) Viper Mk 201; the installation of better navigational equipment resulted in the **Jet Provost T.Mk 5A**
Jet Provost T.Mk 50: designation of trainers for Sri Lanka and Sudan and strengthened trainer/light attack aircraft for Iraq and Venezuela
Strikemaster Mk 80: designation of the light attack aircraft derived from the Jet Provost with better ejector seats, provision for a wide assortment of armament, considerable structural strengthening (especially of the wings and landing gear), fuel relocated to integral and bag tanks in the wings and fixed tip tanks; the type first flew in October 1967 and has been used successfully in combat, where its great strength has proved a primary asset; variations within the type have been only small, and operators include Saudi Arabai (**Strikemaster Mk 80** and improved **Strikemaster Mk 80A**), South Yemen (**Strikemaster Mk 81**), Oman (**Strikemaster Mk 82** and improved **Strikemaster Mk 82A**), Kuwait (**Strikemaster Mk 83**), Singapore (**Strikemaster Mk 84**), Kenya (**Strikemaster Mk 87**), New Zealand (**Strikemaster Mk 88**) and Ecuador (**Strikemaster Mk 89**)
Strikemaster Mk 90: final version, delivered in 1984 to Sudan with a number of modern features

British Aerospace VC10 K.Mk 3
(UK)
Type: inflight-refuelling tanker
Accommodation: crew of four on the flightdeck, and provision for 17 passengers in the cabin
Armament (fixed): none
Armament (disposable): none
Electronics and operational equipment: communication and navigation equipment, plus weather radar, and refuelling equipment comprising two Flight Refuelling Ltd Mk 32/2800 pods under the wings and one Flight Refuelling Ltd Mk 17B hose-and-drogue unit under the rear fuselage
Powerplant and fuel system: four 21,800-lb (9888-kg) thrust Rolls-Royce Conway Mk 550B turbofans, and a total internal fuel capacity of 19,365 Imp gal (88032 litres) in integral wing tanks, plus transfer fuel in five cylindrical fuselage tanks; provision for inflight-refuelling
Performance: no data available
Weights: no data available
Dimensions: span 146 ft 2 in (44.55 m); length 171 ft 8 in (52.32 m) excluding inflight-refuelling probe; height 39 ft 6 in (12.04 m); wing area 2,932.0 sq ft (272.38 m²)

Variants
VC10 C.Mk 1: original RAF version based closely on the fuselage of the VC10 civil airliner combined with other components from the Super VC10, and bought in small numbers for strategic and VIP transport; the type can carry 150 passengers in the cabin, alternative layouts catering for 78 litters plus attendants, or for 61 sitting wounded, or for 54,000 lb (24494 kg) of freight; the type has a maximum take-off weight of 323,000 lb (146513 kg), and a range of 4,720 miles (7596 km) with maximum payload
VC10 K.Mk 2: designation of five ex-British Airways VC10s converted to an airframe, powerplant and avionics standard approximating that of the VC10

C.Mk 1 but finished as a three-point hose-and-drogue inflight-refuelling tanker with 112,000 lb (50802 kg) of transfer fuel
VC10 K.Mk 3: designation of four ex-East African Airways Super VC10s modified like the VC10, K.Mk 2s but with 190,400 lb (86364 kg) of transfer fuel

British Aerospace Victor K.Mk 2
(UK)
Type: inflight-refuelling tanker
Accommodation: crew of four or five on the flightdeck
Armament (fixed): none
Armament (disposable): none
Electronics and operational equipment: communication and navigation equipment, plus ARI.18228 radar-warning receiver, and inflight-refuelling gear comprising two Flight Refuelling Ltd Mk 20B pods under the wings and one Flight Refuelling Ltd Mk 17B hose-and-drogue unit under the rear fuselage
Powerplant and fuel system: four 20,600-lb (9344-kg) thrust Rolls-Royce Conway Mk 201 turbofans, and a total fuel capacity (standard and transfer) of 15,875 Imp gal (72168 litres) in 10 wing, seven fuselage, two bomb-bay and two non-jettisonable under-wing tanks; provision for inflight-refuelling
Performance: maximum speed over 600 mph (966 km/h) at 40,000 ft (12190 m); service ceiling over 50,000 ft (15240 m); range 4,600 miles (7403 km)
Weights: empty 114,500 lb (51936 kg); maximum take-off 238,000 lb (107955 kg)
Dimensions: span 113 ft 0 in (34.44 m); length 114 ft 11 in (35.03 m); height 28 ft 1.5 in (8.57 m); wing area 2,200.0 sq ft (204.38 m²)

Variant
Victor K.Mk 2: now the only variant of the UK's three strategic V-bombers still in service, the Victor K.Mk 2 was produced urgently in the early 1970s by the conversion of 'surplus' Victor B.Mk 2 bombers when the Vickers Valiant tanker fleet developed irremediable structural problems; no two conversions are identical, but a 'typical' Victor K.Mk 2 has some 100,000 lb (45360 kg) of transfer fuel

Britten-Norman BN-2B Defender
(UK)
Type: multi-role military aircraft
Accommodation: pilot and up to nine passengers, or eight parachutists and a despatcher, or three litters and two attendants, or more than 2,250 lb (1021 kg) of freight
Armament (fixed): none
Armament (disposable): this is carried on four optional underwing hardpoints, the inner pair each rated at 700 lb (318 kg) and the outer pair each at 450 lb (204 kg); weapons that can be accommodated are 500-lb (227-kg) bombs, 250-lb (113-kg) bombs, anti-personnel grenades, packs of 81-mm (3.19-in) SURA rockets, Matra 155 launchers each with 18 68-mm (2.68-in) rockets, Matra 122 launchers each with seven 68-mm (2.68-in) rockets, pods each with two 7.62-mm (0.3-in) machine-guns, and AS.11 air-to-surface wire-guided missiles
Electronics and operational equipment: communication and navigation equipment, plus weather radar and other sensors as specified by individual operators
Powerplant and fuel system: two 300-hp (224-kW) Avco Lycoming IO-540-K1B5 flat-six piston engines, and a total internal fuel capacity of 114 Imp gal (518 litres) in two integral wing tanks, plus provision for two 24.5 Imp gal (111-litre) bolt-on ferry tiptanks and two 50-Imp gal (227-litre) drop-tanks
Performance: maximum speed 174 mph (280 km/h) at sea level; cruising speed 159 mph (255 km/h) at 10,000 ft (3050 m); initial climb rate 1,300 ft (396 m) per minute; service ceiling 17,000 ft (5180 m); range 375 miles (603 km) with maximum internal and external payload, or 1,723 miles (2772 km) with maximum internal and external fuel
Weights: empty 4,020 lb (1824 kg); maximum take-off 6,600 lb (2993 kg)
Dimensions: span 49 ft 0 in (14.94 m), or 53 ft 0 in (16.5 m) with extended tips; length 39 ft 5.75 in (12.02 m) with optional nose extension; height 13 ft 8.75 in (4.18 m); wing area 325.0 sq ft (30.19 m²), or 337.0 sq ft (31.31 m²) with extended tips

Variants
BN-2 Islander: initial civil model first flown in June 1965, and subsequently adopted by a number of air arms for the utility and tactical transport roles
BN-2A Islander: improved model also bought by some air arms
BN-2B Islander: definitive standard transport model with increased weights and other operational improvements
BN-2B Islander/CASTOR: under this designation the British army is developing its Corps Airborne STand-Off Radar system with the Ferranti CASTOR I radar (its radome in a platypus-type nose radome) for battlefield and rear-area surveillance; the definitive system may be switched to converted BAe Canberra airframes (using the CASTOR C derivative of the Thorn-EMI Search Water maritime surveillance radar) for improved performance over the battlefield, and thus less vulnerability to enemy countermeasures
BN-2B AEW Defender: proposed airborne early warning version with Thorn-EMI Skymaster surveillance radar in a bulbous nose radome for overland and overwater detection of targets at long range
BN-2B Defender: capable multi-role military derivative capable of use in roles as different as long-range patrol to forward air control via logistical transport, trooping, casevac and paratrooping

BN-2B Maritime Defender: developed maritime version of the Defender with an enlarged nose for Bendix RDR 1400 search radar, increasing length to 36 ft 3.75 in (11.07 m)
BN-2T Turbine Islander: turbine-engined model of the 1980s with two 400-shp (298-kW) Allison 250-B17C turboprops flat-rated at 320 shp (239 kW) for better performance with a 1,610-lb (730-kg) payload
BN-2B Turbine Defender: combination of Defender capabilities with Turbine Islander performance
BN-2A Mk III Trislander: stretched version with accommodation for 17 passengers and powered by three 260-hp (194-kW) Avco Lycoming O-540-E4C5 piston engines
BN-2A Mk III Trislander M: military version of the Trislander able to lift 17 troops over a range of 750 miles (1207 km) or 16 paratroops over a comparable range; the payload is 3,700 lb (1678 kg), and the type can be fitted with nose-mounted radar for the maritime patrol role

Canadair CL-41G-5 Tebuan
(Canada)
Type: light attack aircraft
Accommodation: pilot and co-pilot seated side-by-side on ejector seats
Armament (fixed): none
Armament (disposable): this is carried on six underwing hardpoints for a maximum of 4,000 lb (1814 kg) of stores including 7.62-mm (0.3-in) Minigun pods, 500-lb (227-kg) bombs, 250-lb (113-kg) bombs, various types of rocket-launcher pods, Sidewinder air-to-air missiles and other weapons
Electronics and operational equipment: communication and navigation equipment
Powerplant and fuel system: one 2,950-lb (1338-kg) thrust General Electric J85-J4 turbojet, and a total internal fuel capacity of 250 Imp gal (1135 litres) in a five-cell fuselage tank, plus provision for two 40-Imp gal (182-litre) drop-tanks
Performance: maximum speed 470 mph (755 km/h) at 28,500 ft (8685 m); initial climb rate 4,250 ft (1295 m) per minute; service ceiling 42,200 ft (12865 m); range 1,380 miles (2220 km) with external fuel
Weights: empty 5,296 lb (2402 kg); normal take-off 10,000 lb (4536 kg); maximum take-off 11,288 lb (5120 kg)
Dimensions: span 36 ft 6 in (11.13 m); length 32 ft 0 in (9.75 m); height 9 ft 3 in (2.81 m); wing area 220.0 sq ft (20.44 m²)

Variants
CL-41A: this trim side-by-side two-seat pilot trainer was developed in Canada and first flew in January 1960 before entering service with the Canadian air arm as the **CT-114 Tutor** in 1963; the type was powered by two 2,850-lb (1293-kg) J85-CAN-40 turbojets, though surviving aircraft have received 2,950-lb (1338-kg) engines in an updating programme launched in 1976 to extend service life, improve avionics and fit provision for external fuel; the type has a maximum take-off weight of 7,787 lb (3532 kg) and a maximum speed of 495 mph (797 km/h)
CL-41G-5 Tebuan: designation of the variant produced for Malaysia in the advanced training and light attack roles with strengthened airframe and landing gear, zero/zero ejector seats and six underwing hardpoints; the extra weight but only marginally more power means performance generally slightly inferior to that of the CL-41A

Canadair CL-215
(Canada)
Type: multi-role amphibian
Accommodation: crew of two on the flightdeck, plus up to 19 passengers in the cabin, alternative payloads being 12,000 lb (5443 kg) of water or 6,260 lb (2839 kg) of freight
Armament (fixed): none
Armament (disposable): none
Electronics and operational equipment: communication and navigation equipment
Powerplant and fuel system: two 2,100-hp (1566-kW) Pratt & Whitney R-2800-83AM radial piston engines, and a total internal fuel capacity of 1,300 Imp gal (5910 litres) in 16 flexible cells forming two tanks between the wing spars
Performance: cruising speed 181 mph (291 km/h) at 10,000 ft (3050 m); initial climb rate 1,000 ft (305 m) per minute; range 1,405 miles (2260 km) with a 3,500-lb (1587-kg) payload
Weights: empty 27,750 lb (12587 kg); normal take-off 37,700 lb (17100 kg) from water; maximum take-off 43,500 lb (19731 kg) from land
Dimensions: span 93 ft 10 in (28.60 m); length 65 ft 0.5 in (19.82 m); height 29 ft 3 in (8.92 m) on land; wing area 1,080.0 sq ft (100.33 m²)

Variant
CL-215: designed as a multi-role of modest performance but great reliability and durability, the CL-215 amphibian first flew in October 1967 and has achieved a small but steady trickle of sales to civil and military operators, mainly in the fire-fighting and search-and-rescue roles; Spain and Thailand use the type for maritime patrol with RCA AVQ-21 or Bendix RDR 1400 radar in the nose, and the continued success of the type is attested by the fact that the manufacturer is planning a PW120 turboprop-powered version now that high-octane fuel for piston engines is becoming scarcer and more expensive
CL-215T: updated model featuring two Pratt & Whitney Canada PW100 turboprops for improved overall performance, and under development in the late 1980s for the carriage of an additional 1500 kg (3,307 lb) of payload

CASA C-101EB Aviojet
(Spain)

Type: basic and advanced trainer with light attack capability

Accommodation: pupil and instructor seated in tandem on Martin-Baker E10C ejector seats

Armament (fixed): this is accommodated in a quick-change pack under the rear cockpit, and comprises either a DEFA 30-mm cannon or two 12.7-mm (0.5-in) FN-Browning machine-guns

Armament (disposable): this is carried on six underwing hardpoints, the inner pair each rated at 500 kg (1,102 lb), the centre pair each at 375 kg (827 lb) and the outer pair each at 250 kg (551 lb), up to a maximum weight of 1500 kg (3,307 lb); typical underwing loads are two BR-500 500-kg (1,102-lb) bombs, or four BR-375 375-kg (827-lb) bombs, or six BR-250 250-kg (551-lb) bombs, or four BLU-27 napalm tanks, or four LAU-10/A launchers each with four 5-in (127-mm) rockets, or six LAU-68/A launchers each with 19 2.75-in (69.85-mm) rockets, or six Matra 155 launchers each with 18 68-mm (2.68-in) rockets, or two AIM-9 Sidewinder air-to-air missiles

Electronics and operational equipment: communication and navigation equipment, plus alternative pods (photographic reconnaissance, or ECM or laser designation) in place of the gun package in the fuselage

Powerplant and fuel system: one 3,500-lb (1588-kg) thrust Garrett TFE331-2-2J turbofan, and a total internal fuel capacity of 2335 litres (514 Imp gal) in one fuselage and three integral wing tanks

Performance: maximum speed 690 km/h (429 mph) at sea level, and 795 km/h (494 mph) or Mach 0.71 at 7620 m (25,000 ft); cruising speed 655 km/h (407 mph) at 9145 m (30,000 ft); initial climb rate 1152 m (3,780 ft) per minute; service ceiling 12200 m (40,025 ft); combat radius 380 km (236 miles) on a lo-lo-hi sortie with four 250-kg (551-lb) bombs and a 30-mm cannon; ferry range 3615 km (2,246 miles)

Weights: empty 3350 kg (7,385 lb); normal take-off 4850 kg (10,692 lb); maximum take-off 5600 kg (12,345 lb)

Dimensions: span 10.60 m (34 ft 9.3 in); length 12.25 m (40 ft 2.25 in); height 4.25 m (13 ft 11.25 in); wing area 20.00 m² (215.3 sq ft)

Variants

C-101EB Aviojet: designed in Spain with technical assistance from Northrop and MBB in the USA and West Germany respectively, the Aviojet is a neat tandem-seat advanced trainer that packs a useful light attack capability; the type features great strength, good handling characteristics, uncomplicated aerodynamics and structure and, for maximum fuel economy, a single high-bypass-ratio turbofan; a unusual feature is the provision in all versions of a lower-fuselage bay for the accommodation of armament, reconnaissance equipment, electronic countermeasures equipment or a laser designator; the type first flew in June 1977 and was soon ordered for the Spanish air force, with which it entered service as the E.25 Mirlo

C-101BB Aviojet: this is the armed export version powered by the 3,700-lb (1678-kg) TFE731-3-1J turbofan and sold to Honduras and Chile, the latter producing the type under licence as well as importing a few knock-down kits; Chile is qualifying its A-36 armed versions for the maritime attack role with a pair of BAe Sea Eagle missiles

C-101CC Aviojet: improved version also being produced by Chile, and powered by the 4,700-lb (2132-kg) TFE731-5-1J turbofan for additional performance and an increase in weapons load to 2250 kg (4,460 lb); the type has a maximum speed of 834 km/h (518 mph) at 4500 m (14,765 ft), a radius of 370 km (230 miles) on a lo-lo-lo close air support mission with weapons and a 50-minute loiter over the target, and a maximum take-off weight of 6300 kg (13,889 lb)

C-101DD Aviojet: yet further enhanced model first flown in May 1985 (and due to enter production in 1987) with the TFE731-5-1J turbofan, Doppler radar and a head-up display

CASA C-212 Series 200 Aviocar
(Spain)

Type: STOL utility transport

Accommodation: crew of two on the flightdeck, and up to 24 troops, or 23 paratroops, or 12 litters and four attendants, or 2770 kg (6,107 lb) of freight in the cabin

Armament (fixed): none

Armament (disposable): none

Electronics and operational equipment: communication and navigation equipment

Powerplant and fuel system: two 900-shp (671-kW) Garrett TPE331-10-501C turboprops, and a total internal fuel capacity of 2040 litres (449 Imp gal) in four integral wing tanks

Performance: maximum speed 375 km/h (233 mph); cruising speed 365 km/h (227 mph) at 3050 m (10,000 ft); initial climb rate 474 m (1,555 ft) per minute; service ceiling 8535 m (28,000 ft); range 408 km (253 miles) with maximum payload, or 1760 km (1,094 miles) with maximum fuel

Weights: empty 4115 kg (9,072 lb); maximum take-off 7450 kg (16,424 lb)

Dimensions: span 19.00 m (62 ft 4 in); length 15.16 m (49 ft 9 in); height 6.30 m (20 ft 8 in); wing area 40.00 m² (430.56 sq ft)

Variants

C-212A Aviocar: designed to replace the miscellany of military transports serving with the Spanish air arm in the 1960s, the Aviocar is a simple aircraft of modest performance, but require little maintenance and has good STOL performance with a small but nonetheless useful 2000-kg (4,409-lb) load in an unobstructed hold 5.0 m (16.4 ft) long, 2.0 m (6.6 ft) wide and 1.7 m (5.6 ft) high, accessed by a rear ramp; the type first flew in March 1971 and began to enter service in 1973 with 750-shp (560-kW) Garrett TPE331-5 turboprops, the freight capacity often being exchanged for accommodation for 15 paratroops and an instructor, or 12 litters and three sitting wounded plus attendants, or 19 passengers; the type was also made in Indonesia as the **Nurtanio NC-212**, and the basically similar **C-212 Series 100** was also bought by some military operators

C-212AV Aviocar: VIP version of the C-212A

C-212B Aviocar: designation of six pre-production C-212As converted into photo-survey aircraft with Wild RC-10 cameras and a darkroom in the hold

C-212C Aviocar: designation of civil C-212A Series 100 aircraft in military service

C-212D Aviocar: designation of the last two of the eight pre-production aircraft after modification as navigation trainers; some production aircraft followed to the same standard

C-212 Series 200 Aviocar: introduced in 1979, this is a stretched version with more powerful engines, and the hold increased in length to 6.5 m (21.33 ft) for a payload of 2770 kg (6,107 lb) of freight, or 24 troops, or 23 paratroops, or 28 passengers; four aircraft optimized for the electronic intelligence role are in service with the United Arab Emirates, these machines having automatic signals interception, classification and localizationcapability; two similar aircraft in Portuguese service also possess a jamming capability

C-212 Series 300 Aviocar: version for electronic warfare, coastal patrol etc with customer-specified role equipment; the type has greater span and maximum take-off weight, and the maritime model is offered with APS-128 radar with 240-degree scan for the patrol role and 360-degree scan for the anti-submarine role

CASA/Nurtanio (Airtech) CN-235

(Spain/Indonesia)

Type: utility transport aircraft
Accommodation: crew of two on the flightdeck plus 38 passengers or 3575 kg (7,881 lb) of freight in the cabin
Armament (fixed): none
Armament (disposable): none
Electronics and operational equipment: communication and navigation equipment
Powerplant and fuel system: two 1,700-shp (1268-kW) General Electric CT7-7 turboprops, and a total internal fuel capacity of 4000 kg (8,818 lb)
Performance: cruising speed 454 km/h (282 mph); initial climb rate 540 m (1,772 ft) per minute; service ceiling 8780 m (28800 ft); range with maximum payload 1670 km (1,038 miles)

Weights: empty 7950 kg (17,526 lb); maximum take-off 13000 kg (28,860 lb)
Dimensions: span 25.80 m (84 ft 7.75 in); length 21.30 m (59 ft 10.5 in); height 7.90 m (25 ft 11 in); wing area 60.00 m² (645.8 sq ft)

Variant
CN-235: first flown in November 1983, this light utility transport was designed jointly by Spanish and Indonesian interests, and offers civil and military operators a useful transport capability at modest cost and with limited maintenance requirements

Cessna A-37B Dragonfly

(USA)

Type: light attack aircraft
Accommodation: pilot and co-pilot seated side-by-side on Weber ejector seats
Armament (fixed): one 7.62-mm (0.3-in) General Electric GAU-2B/A Minigun in the forward fuselage
Armament (disposable): this is carried on eight underwing hardpoints, the innermost four each rated at 870 lb (394 kg), the intermediate pair each at 600 lb (272 kg) and the outer pair each at 500 lb (227 kg); among the ordnance which can be carried is the Mk 82 500-lb (227-kg) free-fall or retarded bomb, the Mk 81 250-lb (113-kg) free-fall or retarded bomb, the BLU-1B/C 750-lb (340-kg) napalm bomb, the BLU-32/B 500-lb (227-kg) napalm bomb, the M117 750-lb (340-kg) retarded demolition bomb, the SUU-11/A 7.62-mm (0.3-in) Minigun pod, the SUU-25/A flare-launcher, the SUU-20 bomb/rocket pack, the LAU-3/A launcher with 19 2.75-in (69.85-mm) rockets, the LAU-32/A launcher with seven 2.75-in (69.85-mm) rockets, the LAU-59/A launcher with seven 2.75-in (69.85-mm) rockets, the CBU-12/A bomb dispenser, the CBU-14/A bomb dispenser, the CBU-19/A canister-bomb dispenser, the CBU-22/A bomb dispenser, the CBU-24/B bomb dispenser, and the CBU-25/A bomb dispenser
Electronics and operational equipment: communication and navigation equipment, plus Chicago Aerial Industries CA-503 sight
Powerplant and fuel system: two 2,850-lb (1293-kg) General Electric J85-GE-17A turbojets, and a total internal fuel capacity of 507 US gal (1920 litres) in one fuselage, two integral wing and two wingtip tanks, plus provision for four 100-US gal (378-litre) drop-tanks; provision for inflight-refuelling
Performance: maximum speed 507 mph (816 km/h) at 16,000 ft (4875 m); cruising speed 489 mph (787 km/h) at 25,000 ft (7620 m); initial climb rate 6,990 ft (2130 m) per minute; service ceiling 41,765 ft (12730 m); range 460 miles (740 km) with 4,100 lb (1860 kg) of ordnance, or 1,012 miles (1628 km) with maximum fuel
Weights: empty 6,211 lb (2817 kg); maximum take-off 14,000 lb (6350 kg)

Dimensions: span 35 ft 10.5 in (10.93 m) over tiptanks; length 28 ft 3.25 in (6.62 m) excluding probe; height 8 ft 10.5 in (2.70 m); wing area 183.9 sq ft (17.09 m^2)

Variants

T-37B: this is the US Air Force's standard side-by-side two-seat primary trainer, the production version of the Cessna Model 318 design having flown in October 1952 as the T-37A on the power of two imported Turboméca Marboré turbojets; the type entered production with a licence-built version of the Marboré, the 920-lb (417-kg) Teledyne CAE J69-T-9 and began to enter service in 1957; all survivors were later upgraded to T-37B standard to complement some 447 new-build aircraft with more powerful engines and additional fuel (in tip tanks) plus revised navigation and communication equipment; the USAF is now investigating means to extent the T-37B's life until (or indeed if) the cancelled Northrop T-46 trainer is reinstated or replaced by another type

T-37C: export version of the T-37B with provision for reconnaissance gear and/or light armament, the latter carried on two underwing hardpoints and comprising two 250-lb (113-kg) bombs, two rocket-launcher pods or (somewhat rarely) four AIM-9 Sidewinder air-to-air missiles; this increases maximum take-off weight to 7,500 lb (3402 kg)

A-37A Dragonfly: designation of the initial light attack and counter-insurgency version of the T-37 trainer; the requirement for such an aircraft was made clear by events in the late 1950s and early 1960s, and the YAT-37D prototype (converted from a T-37B) was flown October 1963 at a maximum take-off weight of 14,000 lb (6350 kg), more than twice that of the standard T-37B; the need for such aircraft in the Vietnam War then led to the conversion of 39 T-37Bs as 'production' A-37As with 2,400-lb (1089-kg) J85-GE-5 turbojets, beefed-up structure, increased internal fuel capacity and eight underwing hardpoints

A-37B Dragonfly: definitive production version of the A-37 type based on the A-37A but re-engineered for more powerful engines and greater underwing loads, and fitted with many operational and aerodynamic refinements, stronger landing gear with hydraulic rather than electrical actuation, flak curtains, an inflight-refuelling probe and night/adverse-weather avionics

OA-37B Dragonfly: conversions of A-37Bs to forward air control aircraft

T-48A: provisional designation of proposed T-37B with considerable updating such as a pressurized cockpit, structural strengthening, a revised tail and two 1,330-lb (603-kg) thrust Garrett F109-GA-100 turbofans

Cessna O-2A
(USA)

Type: forward air control aircraft
Accommodation: pilot and observer seated side-by-side at the front of the cabin, and provision for two passengers in the rear of the cabin
Armament (fixed): none
Armament (disposable): this is carried on four underwing hardpoints, and can comprise items such as the SUU-11/A 7.62-mm (0.3-in) Minigun pod, LAU series launchers for 2.75-in (69.85-mm) rockets and SUU-21 or SUU-25 series flare pods
Electronics and operational equipment: communication and navigation equipment
Powerplant and fuel system: two 210-hp (157-kW) Continental IO-360-C/D flat-six piston engines; and a total internal fuel capacity of 92 US gal (348 litres) in two outer-wing tanks, plus provision for two 28-US gal (106-litre) optional tanks in each inner wing panel
Performance: maximum speed 199 mph (320 km/h) at sea-level; cruising speed 144 mph (232 km/h) at 10,000 ft (3050 m); initial climb rate 1,180 ft (360 m) per minute; service ceiling 19,300 ft (5885 m); range 1,060 miles (1706 km)
Weights: empty 2,848 lb (1292 kg); maximum take-off 5,400 lb (2449 kg)
Dimensions: span 38 ft 2 in (11.63 m); length 29 ft 9 in (9.07 m); height 9 ft 4 in (2.84 m); wing area 202.5 sq ft (18.81 m^2)

Variants

O-2A: designation of the 'off-the-shelf' forward air control aircraft procured by the US Air Force from 1966 on the basis of the Model 337 Super Skymaster civil aircraft; the type was fitted with revised avionics and provision four underwing hardpoints

O-2B: psychological warfare version of the O-2A with 600-watt amplifiers and directional speakers

Model 337: designation of the unusual push-pull twin-engine civil aircraft also bought by several air arms in the light transport and utility roles; the type entered service in 1965 as the updated, retractable landing gear version of the Model 336 Skymaster, which first flew in February 1961

Brico O-2: 1983 version with a single 650-shp (485-kW) Allison 250-C30 turboshaft driving a shrouded pusher propeller for increased efficiency and reduced nose, and also leaving the nose clear for a forward-looking infra-red sensor and a GAU-2B/A 7.62-mm (0.3-in) Minigun; the status of the project is not clear

Reims FTMA Milirole: introduced in May 1970, this is a upgraded military model produced by Cessna's French licensee in a number of subvariants, but all featuring STOL performance, four underwing hardpoints and a crew of two plus four passengers or two litters

Summit Sentry O2-337: introduced in 1980, this is a

military version of the T337 pressurized civil aircraft powered by two 225-hp (186-kW) Continental TSIO-470 engines and fitted with four underwing hardpoints each rated at 350 lb (159 kg)

Commonwealth Aircraft Corporation Model 206B-1 (Australia): see Bell OH-58C Kiowa

Convair F-106A Delta Dart
(USA)
Type: interceptor fighter
Accommodation: pilot only, seated on an ejector seat
Armament (fixed): one General Electric M61A1 Vulcan 20-mm rotary-barrel cannon
Armament (disposable): this is carried in three internal weapons bays, and comprises one Douglas AIR-2A Genie or AIR-2B Super Genie nuclear-warhead air-to-air rocket plus two or four Hughes AIM-4F or AIM-4G Super Falcon air-to-air missiles
Electronics and operational equipment: communication and navigation equipment, plus Hughes MA-1 interception and fire-control system tied in by datalink to the SAGE (Semi-Automatic Ground Environment) system within the NORAD (North American Air Defense) network
Powerplant and fuel system: one 24,500-lb (11113-kg) afterburning thrust Pratt & Whitney J75-P-17 turbojet, and a total internal fuel capacity of 1,514 US gal (5731 litres), plus provision for two underwing drop tanks
Performance: maximum speed 1,525 mph (2455 km/h) or Mach 2.3 at 36,000 ft (10970 m); initial climb rate about 30,000 ft (9145 m) per minute; service ceiling 57,000 ft (17325 m); range 1,150 miles (1850 km)
Weights: empty 23,646 lb (10726 kg); maximum take-off 38,700 lb (17554 kg)
Dimensions: span 38 ft 3.5 in (11.67 m); length 70 ft 8.75 in (21.56 m); height 20 ft 3.25 in (6.18 m); wing area 631.3 sq ft (58.65 m²)

Variants
F-106A Delta Dart: derived in concept from the F-102 Delta Dagger, the much-refined F-106A first flew in December 1956 and immediately displayed performance far superior to that of the F-102; other significant advances were a new and fully integrated weapons system (radar, fire control and missiles); deliveries began in July 1959 and the type still survives in small numbers as an interceptor in US airspace, the intervening gears having been marked by considerable development of the type's avionics
F-106B Delta Dart: tandem two-seat combat trainer version of the F-106A

Dassault-Breguet Alizé
(France)
Type: carrierborne anti-submarine aircraft
Accommodation: pilot seated in the cockpit, and two sensor-system operators in the fuselage
Armament (fixed): none
Armament (disposable): this is carried in a lower-fuselage weapons bay and on eight underwing hardpoints; the weapons bay can accept one 500-kg (1,102-lb) homing torpedo or three 160-kg (353-lb) depth charges; the two hardpoints under the inner wing panels can each take one 160- or 175-kg (353- or 386-lb) depth charge; and the six hardpoints under the outer wing panels can accept six 5-in (127-mm) rockets or two AS.12 anti-ship missiles
Electronics and operational equipment: communication and navigation equipment, plus Thomson-CSF DRAA 2B (Indian aircraft) or Thomson-CSF Iguane (French aircraft) search radar with its antenna in a retractable 'dustbin' radome in the lower fuselage, and sonobuoys carried in the forward portion of the main landing gear fairings; ECM equipment is being fitted to the 28 French Alizés with Iguane radar
Powerplant and fuel system: one 1,975-shp (1473-kW) Rolls-Royce Dart RDa.7 Mk 21 turboprop, and a total internal fuel capacity of ? litres (? Imp gal) in wing and fuselage fuel tanks
Performance: maximum speed 520 km/h (323 mph) at 3000 m (9,845 ft); cruising speed 370 km/h (230 mph); initial climb rate 420 m (1,380 ft) per minute; service ceiling over 6250 m (20,505 ft); range 2500 km (1,553 miles) with standard fuel, and 2870 km (1,783 miles) for ferrying with auxiliary fuel
Weights: empty 5700 kg (12,566 lb); maximum take-off 8200 kg (18,078 lb)
Dimensions: span 15.60 m (51 ft 2 in); length 13.86 m (45 ft 6 in); height 5.00 m (16 ft 4.75 in); wing area 36.00 m² (387.51 sq ft)

Variant
Alizé: developed as the Breguet Br.1050, the Alizé (tradewind) is a small carrierborne anti-submarine aircraft that first flew in October 1956; the type was based on an abortive attack aircraft, with a turboprop in the nose for range and a turbojet in the rear fuselage for speed; removal of the turbojet made room for the DRAA 2A search radar and its retractable radome, so paving the way for the Alizé; the type began to enter service in May 1959 and remains in French and Indian service, the French aircraft having been updated for continued utility in the most modern conditions

Dassault-Breguet Atlantique 2
(France)
Type: long-range maritime patrol and anti-submarine aircraft
Accommodation: crew of 12, comprising an observer in the nose, flight crew of three in the cockpit, a mission crew of six in the tactical compartment, and two observers

Armament (fixed): none
Armament (disposable): this is carried in a lower-fuselage stores bay and on four underwing hardpoints, the inner pair rated at 1000 kg (2,205 lb) each and the outer pair at 750 kg (1,653 lb) each, up to a maximum of 3500 kg (7,716 lb); the stores bay is fitted with three 'bridges' (trusses), and the two forward bridges can each take up to four Mk 46 torpedoes, or one L4 torpedo, or four depth charges, or three air/sea rescue containers, or three 250-kg (551-lb) mines; the aft bridge can lift four depth charges, or three 250-kg (551-lb) mines, or three air/sea rescue containers; the central bridge alone can accept a Mk 101 depth charge; and the two forward bridges can be adapted to take one or two AM.39 Exocet anti-ship misiles; the normal underwing load is one air-to-surface missile on each hardpoint
Electronics and operational equipment: communication and navigation equipment, plus Thomson-CSF Iguane (a pulse-compression equipment with a high-resolution side-looking synthetic-aperture mode for target identification) search radar in a retractable radome forward of the stores bay, SAT/TRT forward-looking infra-red (FLIR) sensor in trainable nose turret, Crouzet magnetic anomaly detector (MAD) in the tail 'sting', Thomson-CSF ARAR 13 radar-warning receiver in fintop and wingtip pods, 78 sonobuoys with a rear-fuselage launcher and a Thomson-CSF Sadang processor (able to deal with inputs from four active and 16 passive sonobuoys), ESD/Decca Doppler navigation, twin Sagem Uliss 53 inertial navigation platforms, Crouzet air-data computer, and Agiflite and Omera cameras
Powerplant and fuel system: two 5,665-shp (4225-kW) Rolls-Royce Tyne RTy.20 Mk 21 turboprops, and a total internal fuel capacity of 23120 litres (5,085 Imp gal) in four integral wing tanks
Performance: maximum speed 645 km/h (400 mph) at optimum altitude, and 592 km/h (368 mph) at sea level; cruising speed 315 km/h (195 mph) between sea level and 1525 m (5,000 ft); initial climb rate 610 m (2,000 ft) per minute; service ceiling 9145 m (30,000 ft); range 9075 km (5,640 miles); patrol endurance 5 hours at a radius of 1850 km (1,150 miles)
Weights: empty 25300 kg (55,776 lb); normal take-off 43900 kg (96,781 lb); maximum take-off 46200 kg (101,852 lb)
Dimensions: span 37.42 m (122 ft 9.25 in) with tip pods; length 33.63 m (110 ft 4 in); height 10.89 m (35 ft 8.75 in); wing area 120.34 m² (1,295.3 sq ft)

Variants
Atlantic 1: developed as the Breguet Br.1150 to meet a 1958 NATO requirement for a long-range maritime patrol and anti-submarine aircraft, the Atlantic was judged the best of 27 designs originating from seven NATO member countries; the type did not enter NATO-wide service, instead securing initial orders from France, West Germany, Italy and the Netherlands for 89 aircraft; the airframe was built by the five-country SEBCAT consortium, and the 6,220-ehp (4638-ekW) Rolls-Royce Tyne 21 turboprops by a four-nation consortium; the high aspect ratio of the wings combined with good fuel capacity and economical engines to provide moderate speed but great range, and the 'double-bubble' fuselage combines a capacious unpressurized weapons bay with a pressurized upper lobe for the flight crew of three, the mission crew of seven and, at the rear, two observers; the sensors (search radar, magnetic anomaly detector, electronic support measures, sonobuoys etc) are mainly of Thomson-CSF manufacture, and are computer-integrated on the Plotac display and control panels in the tactical compartment; the type has a maximum take-off weight of 43500 kg (95,900 lb), a span of 36.30 m (119.09 ft) and a range of 7970 km (4,950 miles) with 10 per cent reserves; considerable updating has been achieved within the limitations of the 1950s-vintage electronics, and five of the West German navy's 15 survivors have been extensively converted as electronic intelligence aircraft with a signals intelligence suite installed by Vought and based on a Loral electronic support measures suite with podded wingtip antennae
Atlantique 2: considerably updated version for the French navy (resulting in reversion to the French spelling of the name) developed under the designation Atlantic Nouvelle Génération; the same consortia have been revived for the construction of airframe and powerplant, and the airframe has been considerably developed to reduce the possibility of corrosion and fatigue problems; it is in the electronics that the greatest development has taken place, however, the whole suite being replaced by the latest French equipments (and expanded with the introduction of a forward-looking infra-red sensor) as detailed in the specification; the first aircraft flew in May 1981 and deliveries of 42 aircraft to the Aéronavale are scheduled to begin in 1989

Dassault-Breguet Etendard IV (France): see Dassault-Breguet Super Etendard

Dassault-Breguet Mirage IIIE
(France)
Type: fighter-bomber and strike aircraft
Accommodation: pilot seated on a Martin-Baker RM4 ejector seat
Armament (fixed): two DEFA 552A 30-mm cannon with 125 rounds per gun
Armament (disposable): this is carried on one underfuselage hardpoint, rated at 1180 kg (2,600 lb), and four underwing hardpoints, the inner pair rated at 840 kg (1,852 lb) each and the outer pair at 168 kg (370 lb) each, to a maximum of 4000 kg (8,818 lb); the underfuselage hardpoint can carry one AN 52 15-kiloton nuclear weapon, or AS.30 air-to-surface

missile, or one Matra 530 air-to-air missile, or four 125-kg (276-lb) bombs, or four 250-kg (551-lb) bombs, or one 400-kg (882-lb) bomb, or four Durandal runway-destroying bombs, or one ECM or reconnaissance pod, or one 1300-litre (286-Imp gal) drop-tank; the inner underwing hardpoints can each carry four 125- or 250-kg (276- or 551-lb) bombs, or two 400-kg (882-lb) bombs, or three Durandal bombs, or one pod with six 100-mm (3.94-in) or 18 68-mm (2.68-in) rockets, or one CEM 1 multi-store pod, or one drop tank (625-, 1300- or 1700-litre/137-, 286- or 374-Imp gal), or one JL-100R pod with 250 litres (55 Imp gal) of fuel and 18 68-mm (2.68-in) rockets, or one gun pod; and the outer underwing hardpoints can each carry one Matra 550 Magic air-to-air missile, or one pod with six 68-mm (2.68-in) rockets, or one ECM pod

Electronics and operational equipment: communication and navigation equipment, plus CSF Cyrano II fire-control radar (air-to-air and air-to-surface), CSF 97 sighting system, Marconi Doppler radar, Thomson-CSF Type BF radar-warning receiver and other specialized avionics

Powerplant and fuel system: one 6200-kg (13,670-lb) afterburning thrust SNECMA Atar 09C turbojet plus one optional (and jettisonable) 1500-kg (3,307-lb) thrust SEPR 844 rocket, and a total internal fuel capacity of 3000 litres (660 Imp gal) plus up to 3900 litres (858 Imp gal) of auxiliary fuel in drop-tanks

Performance: maximum speed clean 2350 km/h (1,460 mph) or Mach 2.2 at 12000 m (39,370 ft), and 1390 km/h (863 mph) or Mach 1.13 at sea level; cruising speed 955 km/h (593 mph) or Mach 0.9 at 11000 m (36,090 ft); climb to 11000 m (36,090 ft) in 3 minutes; service ceiling 17000 m (55,775 ft) without rocket motor; combat radius 1200 km (745 miles) radius on a lightly-armed ground-attack mission

Weights: empty 7050 kg (15,540 lb); normal take-off 9600 kg (21,165 lb); maximum take-off 13700 kg (30,200 lb)

Dimensions: span 8.22 m (26 ft 11.5 in); length 15.03 m (49 ft 3.5 in); height 4.50 m (14 ft 9 in); wing area 35.00 m² (376.75 sq ft)

Variants
Mirage IIIB: this operational conversion trainer of the Mirage III series was produced on the basis of the Mirage IIIA pre-production single-seater, which first flew in May 1958 with a 6000-kg (13,228-lb) SNECMA Atar 09B afterburning turbojet as a development of the Mirage III prototype of 1956 powered by the 4500-kg (9,921-lb) Atar 101G turbojet; the Mirage IIIA carried Cyrano Ibis radar, and though the Mirage IIIB has provision for armament it lacks radar
Mirage IIIC: initial production single-seater of the Mirage III series, based on the Mirage IIIA and optimized for the all-weather interception role; the first Mirage IIIC flew in October 1960 and the type was exported primarily to Israel and South Africa under the designations **Mirage IIICJ** and **Mirage IIICZ** respectively
Mirage IIID: two-seat operational conversion trainer development of the Mirage IIIO but with more advanced attack equipment; the type was built in Australia and France, the latter exporting several subvariants including the uprated **Mirage IIID2Z** for South Africa with the Atar 09K-50 turbojet
Mirage IIIE: single-seat version optimized for the long-range intruder and fighter-bomber roles with the Atar 09C turbojet, a fuselage lengthened by 0.3 m (11.8 in), Doppler and TACAN navigation, Cyrano II radar and other improvements; the type was widely exported in a number of subvariants, and the **Mirage IIIBE** is the two-seat trainer version of the variant
Mirage IIIO: version of the Mirage IIIE developed for licence production in Australia
Mirage IIIR: reconnaissance version of the Mirage IIIE with five Omera 31 cameras instead of nose-mounted radar, increasing length to 15.50 m (50.85 ft); the type first flew in November 1961, and was also exported in its standard version and as the uprated **Mirage IIIR2Z** for South Africa with the Atar 09K-50 turbojet
Mirage IIIRD: improved Mirage IIIR with modern Doppler navigation, automatic cameras, a SAT Cyclope infra-red reconnaissance package in a ventral fairing and two 1700-litre (374-Imp gal) underwing drop tanks
Mirage IIIS: much revised version of the Mirage IIIE for Switzerland with Hughes TARAN fire control for Hughes Falcon air-to-air missiles
Mirage 3-50: current designation of the type available in the later 1980s with fly-by-wire controls, canard foreplanes, Atar 09K-50 turbojet and other items of advanced technology
Atlas Cheetah: designation of South African Mirage IIIBZ, Mirage IIIDZ or Mirage IIID2Z aircraft rebuilt to the same basic standard as the IAI Kfir-TC2 (Elta EL-2001 or dual-role Elta EL/M-2001B radar in a drooped nose, an integrated nav/attack system with a laser designator, a chaff/flare dispenser underfuselage strakes, dogtoothed leading edges, fixed canard foreplanes, inertial navigation systems etc) for the air-to-air and air-to-surface roles, but using the 7200-kg (15,873-lb) Atar 09K-50 turbojet instead of the Israeli aircraft's General Electric J79 turbojet and thus possessing slightly inferior performance; it is likely that a single-seat version equivalent to the Kfir-C2 will be produced by converting Mirage IIICZ and Mirage IIIEZ aircraft

Dassault-Breguet Mirage IVP
(France)
Type: theatre missile-delivery aircraft
Accommodation: pilot and systems operator seated

in tandem on Martin-Baker ejector seats
Armament (fixed): none
Armament (disposable): in the strategic role, the Mirage IVP normally carries one ASMP stand-off missile under the fuselage; in the tactical role, the Mirage IVP can carry 16 450-kg (992-lb) free-fall bombs or four AS.37 Martel air-to-surface missiles on two underfuselage and two underwing hardpoints
Electronics and operational equipment: communication and navigation equipment, plus Thomson-CSF Arcana attack radar, Marconi Doppler navigation and Electronique Marcel Dassault central computer
Powerplant and fuel system: two 7000-kg (15,432-lb) afterburning thrust SNECMA Atar 09K-50 turbojets, and a total internal fuel capacity of 14000 litres (3,080 Imp gal) in integral wing and fuselage tanks, plus provision for two 2500-litre (550-Imp gal) drop-tanks; provision for inflight-refuelling
Performance: maximum speed 2340 km/h (1,454 mph) or Mach 2.2 at 13125 m (40,060 ft) for dashes, or 1965 km/h (1,221 mph) or Mach 1.7 at 20000 m (65,615 ft) for sustained flight; climb to 11000 m (36,090 ft) in 4.25 minutes; service ceiling 20000 m (65,615 ft); combat radius 1240 km (771 miles) without inflight-refuelling; ferry range 4000 km (2,486 miles)
Weights: empty 14500 kg (31,967 lb); normal take-off 31600 kg (69,665 lb); maximum take-off 33475 kg (73,799 lb)
Dimensions: span 11.85 m (38 ft 10.5 in); length 23.50 m (77 ft 1 in); height 5.40 m (17 ft 8.5 in); wing area 78.0 m² (839.6 sq ft)

Variants
Mirage IVA: first flown in June 1959 and immediately apparent in aerodynamic terms as a scaled-up Mirage III, the Mirage IVA began to enter service in 1964 as the bomber component of France's nuclear deterrent forces, each aircraft being equipped to carry one 60-kilotron AN 22 free-fall nuclear weapon on intermediate-range sorties; the type has a maximum take-off weight of 33475 kg (73,790 lb) on the power of two Atar 09K-50 afterburning turbojets, and of the 62 aircraft built 12 were converted for the strategic reconnaissance role at high and low levels
Mirage IVP: designation of 18 Mirage IVA aircraft converted in the mid- and late 1980s to carry the 150-kiloton ASMP short-range stand-off missile; in this role the Mirage IVP is optimized for low-level penetration with the aid of new Thomson-CSF Arcana Doppler radar and nav/attack equipment

Dassault-Breguet Mirage 5
(France)
Type: ground-attack aircraft and interceptor
Accommodation: pilot seated on a clear-weather Martin-Baker RM4 ejector seat
Armament (fixed): two 30-mm DEFA 552A cannon with 125 rounds per gun
Armament (disposable): this is carried on one underfuselage hardpoint, rated at 1180 kg (2,600 lb), and six underwing hardpoints, the inner tandem pairs rated at 1680 kg (3,704 lb) each and the outer pair at 168 kg (370 lb) each, to a maximum of more than 4000 kg (8,818 lb) with the use of multiple launchers; the disposition of loads is similar to that on the Mirage IIIE, and weapons which can be carried are the AN 52 nuclear bomb, AS.30 air-to-surface missile, Matra 550 Magic or AIM-9 Sidewinder air-to-air missiles, Durandal runway-destroying bomb, 125-, 250- and 500-kg (276-, 551- and 882-lb) high-explosive bombs, pods containing six 100-mm (3.94-in) or six 68-mm (2.68-in) rockets, CEM 1 multi-store pod, JL-100R pod with 18 68-mm (2.68-in) rockets and 250 litres (55 Imp gal) of fuel, ECM pod, reconnaissance pod, and 500-, 1300- and 1700-litre (110-, 286- and 374-Imp gal) drop-tanks
Electronics and operational equipment: communication and navigation equipment, including an inertial navigation system and navigation system, plus either an Aida II radar and CSF LT 102 or TAV 34 laser-ranging equipment, or an Agave or Cyrano IV M3 multi-mode radar, and Thomson-CSF Type BF radar-warning receiver
Powerplant and fuel system: one 6200-kg (13,670-lb) afterburning thrust SNECMA Atar 09C turbojet and a total internal fuel capacity of 3470 litres (763 Imp gal) plus up to 3900 litres (858 Imp gal) of auxiliary fuel in drop-tanks (normally 1000 litres/220 Imp gal for ground attack and maximum external fuel for interception)
Performance: maximum speed clean 2350 km/h (1,460 mph) or Mach 2.2 at 12000 m (39,370 ft), and 1390 km/h (863 mph) or Mach 1.13 at sea level; cruising speed 955 km/h (593 mph) or Mach 0.9 at 11000 m (36,090 ft); climb to 11000 m (36,090 ft) in 3 minutes; service ceiling 17000 m (55,775 ft); combat radius 1300 km (808 miles) on a hi-lo-hi sortie or 650 km (404 miles) on a lo-lo-lo sortie with 907-kg (2,000-lb) bombload; ferry range 4000 km (2,485 miles)
Weights: empty 6600 kg (14,550 lb); normal take-off 9600 kg (21,165 lb); maximum take-off 13700 kg (30,200 lb)
Dimensions: span 8.22 m (26 ft 11.5 in); length 15.55 m (51 ft 0.25 in); height 4.50 m (14 ft 9 in); wing area 35.00 m² (376.75 sq ft)

Variants
Mirage 5A: this version of the Mirage IIIE resulted from an Israeli requirement for a less sophisticated and thus cheaper aircraft optimized for the ground-attack role in clear-weather conditions, eliminating search radar and other equipment to permit the carriage of more fuel and weapons; the first Mirage 5 was flown in May 1967, and though the original Israeli order was embargoed by the French govern-

ment, the type soon scored useful sales in the Middle East, Africa and South America; the original aircraft entered service with the French air force under the designation **Mirage 5F**, and other useful European sales were made to Belgium; during the 1970s the advent of microminiaturization in electronics led to the development of much smaller and lighter avionics, and much of the all-weather capability deleted from the original Mirage 5 series was subsequently restored, providing customers with the Mirage III's capabilities and the Mirage 5's payload/range

Mirage 5D: two-seat operational conversion trainer version of the Mirage 5

Mirage 5R: reconnaissance version of the Mirage 5 with a fan of five cameras in the nose

Mirage 5-50: designation of the current production standard with the uprated 7200-kg (15873-lb) Atar 09K-50 turbojet, Thomson-CSF Agave or Cyrano IVM search radar and advanced equipment such as a head-up display for the accurate delivery of the type's considerable weapons load (more than 4000 kg/8,818 lb within a maximum take-off weight of 13700 kg/30,200 lb); the type is also available in the same training and reconnaissance variants as the basic Mirage 5 series

Dassault-Breguet Mirage F1C

(France)

Type: multi-role fighter and attack aircraft

Accommodation: pilot seated on a Martin-Baker F1RM4 ejector seat

Armament (fixed): two 30-mm DEFA 553 cannon with 135 rounds per gun

Armament (disposable): this is carried on one underfuselage, four underwing and two wingtip hardpoints (the last suitable only for Matra 550 Magic or AIM-9 Sidewinder air-to-air missiles), up to a maximum weight of 4000 kg (8,818 lb); the underfuselage hardpoint can accommodate four 125-, 250- or 400-kg (276-, 551- or 882-lb) bombs, or one 1000-kg (2,205-lb) bomb, or four Durandal runway-destroying bombs, or one 30-mm cannon pod, or two Belouga cluster bombs, or one laser designator pod, or one AS.37 or ARMAT anti-radiation missile, or one AS.30 air-to-surface missile, or one of two types of photo-reconnaissance pod, or one Elint pod, or one SLAR pod, or one active ECM pod; the underwing hardpoints can accommodate two Super 530 air-to-air missiles, or 10 125- or 250-kg (276- or 551-lb) bombs, or four 400-kg (882-lb) bombs, or two 1000-kg (2,205-lb) bombs, or two CEM 1 18-tube rocket-launcher/bomb pods, or four 18- or 36-tube rocket-launchers, or two 30-mm cannon pods, or four Durandal runway-destroying bombs, or 12 BAT 120 anti-vehicle or BAP 100 anti-runway bombs, or two Belouga cluster bombs, or two AS.30L air-to-surface missiles, or two active/passive ECM pods, or six 600-litre (132-Imp gal) napalm tanks

Electronics and operational equipment: communication and navigation equipment, plus Thomson-CSF Cyrano IVM multi-mode fire-control radar, Sagem Uliss 47 inertial navigation system, CSF head-up display, Doppler navigation, laser-rangefinder, Thomson-CSF Type BF radar-warning receiver and terrain-avoidance radar

Powerplant and fuel system: one 7200-kg (15,873-lb) afterburning thrust SNECMA Atar 09K-50 turbojet, and a total internal fuel capacity of 4280 litres (941 Imp gal) plus three 1200-litre (264-Imp gal) or one 2200-litre (484-Imp gal) external tank; provision for an inflight-refuelling probe

Performance: maximum speed 2350 km/h (1,460 kph) or Mach 2.2 at 12000 m (39,370 ft), and 1470 km/h (913 mph) or Mach 1.2 at sea level; initial climb rate 12780 m (41,930 ft) per minute; service ceiling 20000 m (65,615 ft); combat radius 600 km (373 miles) on lo-lo-lo sortie with six 250-kg (551-lb) bombs, or 425 km (264 miles) on a hi-lo-hi sortie with 14 250-kg (551-lb) bombs; ferry range 3300 km (2,051 miles)

Weights: empty 7400 kg (16,314 lb); normal take-off 10900 kg (24,030 lb); maximum take-off 16200 kg (35,714 lb)

Dimensions: span 8.40 m (27 ft 6.75 in); length 15.00 m (49 ft 2.5 in); height 4.50 m (14 ft 9 in); wing area 25.00 m² (269.1 sq ft)

Variants

Mirage F1A: bearing the same relationship to the Mirage F1C as the Mirage 5 to the Mirage IIIE, this is the clear-weather ground-attack version of the Mirage F1C designed for South Africa, where the type has been built under licence by Atlas Aircraft; the type has a secondary air-to-air capability, and is fitted Thomson-CSF Aida II ranging radar instead of the same company's Cyrano IV search radar, and can carry the Matra Super 530 air-to-air missile; other additions are a SFIM inertial platform, a Thomson-CSF laser rangefinder and Doppler navigation; the type entered service in 1975, and has been sold to Ecuador and Libya in addition to South Africa

Mirage F1B: two-seat operational training version of the Mirage F1C with the fuselage lengthened by 0.30 m (11.8 in), the cannon deleted and the internal fuel capacity reduced; the type entered service in 1976 and has been widely exported

Mirage F1C: initial production version of the important Mirage F1 family, which first flew in prototype form in December 1966; the type is designed primarily for the all-weather interception role with monopulse Cyrano IV radar and the Super 530 missile, a combination that offers snap-up/snap-down capability at all operating altitudes; the type has been widely exported in Europe, Africa, the Middle East and South America

Mirage F1C-200: designation of a small number of French Mirage F1Cs modified with an inflight-refuelling probe for rapid overseas deployments

Mirage F1CR-200: designation of the French air force's combat capable reconnaissance version of the Mirage F1C, first flown in November 1981 and retaining radar plus associated armament, but fitted internally with the advanced SNAR navigation system (Cyrano IVM-R radar, Uliss 47 inertial navigation system and ESD computer), conventional Omera 33 vertical and Omera 40 panoramic cameras, Thomson CSF Raphael side-looking airborne radar, and SAT Super Cyclope WCM 2400 infra-red linescan; the type can also carry a Harold underfuselage reconnaissance pod, or a new SLAR pod, or a Nora potronic reconnaissance pod or a Thomson-CSF electronic intelligence pod, and is fitted as standard with an inflight-refuelling probe

Mirage F1E: designation of a subsequently abandoned all-weather version with advanced nav/attack systems and the SNECMA M53 bleed-turbojet, and now applied to the improved Mirage F1C for export with upgraded electronics such as Cyrano IV radar modified for terrain avoidance, air-to-surface ranging and look-down capability; the type also has a Sagem-Kearfott inertial platform, an EMD/Sagem digital central computer and an improved head-up display; the type entered service in 1976 and has achieved significant sales success

Dassault-Breguet Mirage 2000C
(France)
Type: interceptor and air-superiority fighter with powerful secondary attack capability
Accommodation: pilot seated on a Martin-Baker F10Q ejector seat
Armament (fixed): two 30-mm DEFA 554 cannon with 125 rounds per gun
Armament (disposable): this is carried on nine hardpoints, one under the fuselage rated at 1800 kg (3,968 lb), four under the wing roots each rated at 400 kg (882 lb), and four under the wings, the inner pair rated at 1800 kg (3,968 lb) each and the outer pair at 300 kg (661 lb) each, to a maximum weight of 6300 kg (13,889 lb); the underfuselage hardpoint can accommodate one Super 530 air-to-air missile, or one AS.30L air-to-surface missile, or four 250-kg (551-lb) bombs, or one 250-kg (551-lb) laser-guided bomb, or one 1000-kg (2,205-lb) free-fall or laser-guided bomb, or one 400-kg (882-lb) cluster bomb, or 18 BAP 100 anti-runway or BAT 120 anti-vehicle bombs, or one Belouga cluster bomb, or one grenade-launcher pod, or one reconnaissance pod; the under-root hardpoints can take four 250-kg (551-lb) bombs, or three 250-kg (551-lb) laser-guided bombs, or four 400-kg (882-lb) cluster bombs, or 12 BAP 100 or BAT 120 bombs, or four Belouga cluster bombs, or four Type 531 grenade-launchers, or one laser-marking pod, or two CC 421 30-mm cannon pods; the inner underwing hardpoints can each accept a Super 530 air-to-air missile, or four 250-kg (551-lb) bombs, or one 1000-kg (2,205-lb) bomb, or one 400-kg (882-lb) cluster bomb, or one Belouga cluster bomb, or one Type 531 grenade-launcher, or one 68-mm (2.68-in) rocket-launcher pod (F1 with 36, or F2 with six, or LR155 with 18 rockets), or one AS.30L air-to-surface missile, or one AM.39 Exocet anti-ship missile; and the outer underwing hardpoints can each lift one Matra 550 Magic air-to-air missile, or one 250-kg (551-lb) bomb, or one F4 launcher with 18 68-mm (2.68-in) rockets, or one ECM pod

Electronics and operational equipment: communication and navigation equipment, plus Thomson-CSF/ESD RDI pulse-Doppler radar (Thomson-CSF RDM multi-role radar in first 50 French and most export aircraft), Sagem Uliss 52 inertial platform, ESD central digital computer, Thomson-CSF VE-130 head-up display, Thomson-CSF Serval-B radar-warning receiver, Thomson-CSF VMC-180 head-down display, Thomson-CSF/ESD electronic countermeasures equipment and Thomson-CSF laser designator and marked target seeker

Powerplant and fuel system: one 9000-kg (19,840-lb) afterburning thrust SNECMA M53-5 bleed-turbojet (to be replaced in later production aircraft by the 9700-kg/21,385-lb M53-P2), and a total internal fuel capacity of 3800 litres (835 Imp gal) plus provision for two 1700-litre (374-Imp gal) underwing drop-tanks and one 1200-litre (264-Imp gal) under-fuselage tank; provision for inflight-refuelling

Performance: maximum speed more than 2350 km/h (1,460 mph) or Mach 2.2 at 12000 m (39,370 ft), and 1110 km/h (690 mph) or Mach 0.9 at sea level with bombs; initial climb rate more than 18000 m (59,055 ft) per minute; service ceiling 20000 m (65,614 ft); range more than 1800 km (1,118 miles) with two 1700-litre (374-Imp gal) drop-tanks

Weights: empty 7500 kg (16,534 kg); normal take-off 9500 kg (20,994 lb); maximium take-off 17000 kg (37,478 lb)

Dimensions: span 9.00 m (29 ft 6 in); length 14.35 m (47 ft 1 in); wing area 41.0 m² (441.3 sq ft)

Variants
Mirage 2000B: two-seat operational conversion trainer model with the fuselage lengthened by 0.20 m (7.9 in) to 14.55 m (47.74 ft) for the insertion of the second cockpit

Mirage 2000C: this is rapidly becoming France's most important warplane of the later 1980s, and though bearing a visual similarity to the Mirage III series is a far more advanced aircraft with good radar (possessing an all-altitude search range of 100 km/62 miles), automatic leading-edge slats, relaxed stability and 'fly-by-wire' electrically-signalled controls for optimum performance once the pilot's control inputs have been computer-assessed in relation to aircraft conditions and translated into the appropriate commands; the Mirage 2000 first flew in March 1978, and the type has since been ordered into production as France's main interceptor and air-superiority fighter with secondary reconnaissance, close support and interdiction capabilities; the first production aircraft began to enter service in 1983, and the initial 50 aircraft have the Thomson-CSF RDM non-coherent multi-mission radar, though later machines are to have the RDI pulse-Doppler air interception radar; export variants are the **Mirage 2000EAD** for Abu Dhabi, the **Mirage 2000EM** for Egypt, the **Mirage 2000EGM** for Greece, the **Mirage 2000H** for India, and the **Mirage 2000P** for Peru

Mirage 2000N: dedicated low-altitude penetration fighter based on the airframe of the Mirage 2000B strengthened to cope with the stresses of high-speed (Mach 0.9) low-altitude (60 m/200 ft) flight to deliver the 150-kiloton ASMP stand-off missile; features of this model are the ESD Antilope 5 terrain-following and ground-mapping radar, TRT AHV12 radar altimeter, Thomson-CSF coloured CRT cockpit displays, two Sagem Uliss 52 inertial platforms, specialized electronic countermeasures and an Omera vertical camera

Mirage 2000P: version of the Mirage 2000 for the Peruvian air force, and having slightly lower avionics standards

Mirage 2000R: reconnaissance variant of the Mirage 2000C, the sole production variant being the **Mirage 2000RAD** for the Abu Dhabi with COR-2 or AA-3-38 Harold camera pods

Dassault-Breguet Mystère-Falcon 20F
(France)
Type: light transport aircraft
Accommodation: crew of two on the flightdeck, and up to 14 passengers in the cabin
Armament (fixed): none
Armament (disposable): none
Electronics and operational equipment: communication and navigation equipment, plus provision for weather radar
Powerplant and fuel system: two 4,500-lb (2041-kg) General Electric CF700-2D-2 turbofans, and a total internal fuel capacity of 5180 litres (1,139 Imp gal) in two integral wing and two fuselage tanks
Performance: cruising speed 840 km/h (522 mph) or Mach 0.78 at 10000 m (32,810 ft); service ceiling 12800 m (41,995 ft); range 3300 km (2050 miles) with eight passengers
Weights: empty 7530 kg (16,600 lb); maximum take-off 13000 kg (28,660 lb)
Dimensions: span 16.30 m (53 ft 6 in); length 17.15 m (56 ft 3 in); height 5.32 m (17 ft 5 in); wing area 41.0 m² (441.33 sq ft)

Variants
Mystère-Falcon 20: sold in France as the Mystère and elsewhere as the Falcon, this trim twin-jet aircraft was conceived as a corporate transport and first flew in May 1963; the type is powered by two 4,125-lb (1871-kg) General Electric CF700-2C non-afterburning turbofans, and has sold to air arms (as a VIP transport and for other roles) as well as to civil customers; Mystère-Falcon 20s of various types serve with the air forces of Canada, Morocco and Norway in the electronic warfare and electronic/signals intelligence roles with a variety of specialized equipment fits

Mystère-Falcon 20C: extended-range version of the Mystère-Falcon 20, using the same powerplant but featuring additional fuel

Mystère-Falcon 20D: uprated version with 4,250-lb (1928-kg) CT700-2D turbofans

Mystère-Falcon 20E: further uprated version with 4,500-lb (2041-kg) CF700-2D-2 turbofans

Mystère-Falcon 20F: current production version based on the Mystère-Falcon 20E but with better field performance resulting from improved high-lift devices; the **Falcon ST** is a systems trainer version with the Cyrano II radar of the Dassault-Breguet Mirage IIIE

Mystère-Falcon 20G: developed version with 5,440-lb (2468-kg) Garrett AFT3-6-2C turbofans and 5770 litres (1269 Imp gal) of internal fuel, developed specifically to meet the requirements of the US Coast Guard for a medium-range surveillance aircraft, 41 such aircraft being planned under the designation **HU-25A Guardian** with a maximum take-off weight of 15200 kg (33,510 lb) for a range of 4170 km (2,590 miles); the type has accommodation for a flightcrew of two, plus a systems officer and two observers in the cabin, as well as provision for three passengers; the avionics suite includes search radar, forward-looking infra-red, steerable TV with laser illumination and optional side-looking airborne radar

Mystère-Falcon 200: this is a hybrid derivative of the Mystère-Falcon 20F with the powerplant of the Mystère-Falcon 20G; the aircraft is in French service as the Gardian resources protection type with Thomson-CSF Varan radar, and a Crouzet tactical navigation/plotting system

Dassault-Breguet Super Etendard
(France)
Type: carrierborne strike fighter
Accommodation: pilot seated on a Martin-Baker CM4A ejector seat
Armament (fixed): two 30-mm DEFA S53 cannon with 125 rounds per gun
Armament (disposable): this is carried on one underfuselage hardpoint, rated at 600 kg (1,323 lb) and four underwing hardpoints, the inner pair rated at 1100 kg (2,425 lb) each and the outer pair at 450 kg (992 lb) each, to a maximum weight of 2100 kg (4,630 lb); the underfuselage hardpoint can accommodate two 250-kg (551-lb) bombs, or one 600-litre (132-Imp

gal) drop-tank, or one buddy refuelling pod or one reconnaissance pod; the inner underwing hardpoints can accept two 250- or 400-kg (551- or 882-lb) free-fall or retarded bombs, or two LR150 launchers each with 18 68-mm (2.68-in) rockets, or two 625- or 1100-litre (137- or 242-Imp gal) drop-tanks, or one AM.39 Exocet anti-ship missile to starboard and one 1,100-litre (242-Imp gal) drop-tank to port; and the outer wing hardpoints can lift two Matra 550 Magic air-to-air missiles, or two 250- or 400-kg (551- or 882-lb) free-fall or retarded bombs, or two LR150 rocket pods; French aircraft are also fitted for the AN 52 15-kiloton nuclear bomb, and will receive provision for the ASMP nuclear stand-off missile

Electronics and operational equipment: communication and navigation equipment, plus Thomson-CSF/ESD Agave lightweight multi-function radar, Sagem-Kearfott ETNA inertial platform, Thomson-CSF VE-120 head-up display, Thomson-CSF Type BF radar-warning receiver, Crouzet Type 66 air-data computer and Crouzet Type 97 navigation display and armament control system

Powerplant and fuel system: one 5000-kg (11,023-lb) thrust SNECMA Atar 08K-50 turbojet, and a total internal fuel capacity of 3270 litres (719 Imp gal) in integral wing and rubber fuselage tanks, plus provision for 2800 litres (616 Imp gal) in drop-tanks; provision for inflight-refuelling

Performance: maximum speed about 1065 km/h (662 mph) or Mach 1.0 at 11000 m (36,090 ft), and 1180 km/h (733 mph) or Mach 0.96 at sea level; initial climb rate 6000 m (19,685 ft) per minute; service ceiling 13700 m (44,950 ft); combat radius 850 km (528 miles) with AM.39 Exocet missile

Weights: empty 6500 kg (14,330 lb); normal take-off 9450 kg (20,835 lb); maximum take-off 12000 kg (26,455 lb)

Dimensions: span 9.60 m (31 ft 6 in); length 14.31 m (46 ft 11.5 in); height 3.86 m (12 ft 8 in); wing area 24.80 m² (267.0 sq ft)

Variants
Etendard IVM: now serving only in the training role, the Etendard (standard) was the French navy's basic attack fighter until the advent of the Super Etendard; the type first flew in July 1956 and, powered by a 4400-kg (9,700-lb) non-afterburning SNECMA Atar 08B turbojet, can carry (in addition to its two inbuilt 30-mm cannon) some 1360 kg (3,000 lb) of stores on four underwing hardpoints; one of the type's principal limitations was the lack of search radar (the narrow nose holding only an Aïda ranging radar) which limited to the type to IR-homing missiles and free-fall ordnance

Etendard IVP: photo-reconnaissance version of the Etendard IVM without armament but carrying five Omera cameras (three in the nose and two in the ventral position previously occupied by cannon ammunition); the type can also carry Douglas-designed 'buddy' refuelling pods for the support of other Etendards

Super Etendard: this is an upgraded Etendard variant, though the design and structure have needed considerable development to secure true transonic performance; the type first flew in October 1974 and began to enter service in June 1978 with the advanced nav/attack system necessary for the accurate launch of the AM.39 Exocet anti-ship missile and other ordnance; the type can also carry a Thomson-CSF DB 3141 electronic countermeasures pod or a reconnaissance pod on the centreline

Super Etendard NG: this is an improved version of the Super Etendard for service in early 1990s; the main changes are to the avionics, which now include Thomson-CSF cockpit displays, a Sagem inertial navigation system and ESD Anemone search radar; the type also has provision for the carriage of the ASMP nuclear-armed cruise missile

Dassault-Breguet/Dornier Alpha Jet
(France/West Germany)

Type: advanced jet trainer and battlefield close-support/reconnaissance aircraft

Accommodation: pupil and instructor seated in tandem on Martin-Baker AJRM4 (French aircraft) or MBB-built Stencel SIII-S3AJ (German aircraft) ejector seats

Armament (fixed): one 27-mm Mauser or 30-mm DEFA cannon in a detachable underfuselage pod with 150 rounds

Armament (disposable): this is carried on one underfuselage hardpoint (generally used for the cannon pod) and four underwing hardpoints, to a maximum of 2500 kg (5,511 lb); some 36 basic weapon configurations have been qualified for the training and operational roles; the underfuselage hardpoint can take, as an alternative to the cannon pod, one 250-kg (551-lb) bomb or one 400-kg (882-lb) modular bomb; the underwing hardpoints can accept an enormous variety of alternatives, typical stores being the F4 pod for 18 68-mm (2.68-in) rockets, the LAU-3B/A and LAU-61/A pods each for 19 2.75-in (69.85-mm) rockets, the CEM 1 multi-store carrier with 36 68-mm (2.68-in) rockets and various small bombs, the 125-kg (276-lb) free-fall bomb, the 250-kg (551-lb) free-fall or retarded bomb, the Mk 82 500-lb (227-kg) bomb, the Snakeye 500-lb (227-kg) retarded bomb, the Belouga cluster bomb, the BL755 cluster bomb, the Durandal runway-destroying bomb, the CC 420 30-mm cannon pod with 180 rounds, the Matra 550 Magic air-to-air missile, the AGM-65 Maverick air-to-surface missile, and a reconnaissance pod with three Omera 61 and one Omera 40 panoramic cameras

Electronics and operational equipment: communication and navigation equipment, plus (French

aircraft) Thomson-CSF 902 weapon-aiming computer, or (German aircraft) Kaiser/VDO KM 808 head-up display, Litef LDN Doppler navigation and Elettronica ECM
Powerplant and fuel system: two 1350-kg (2,976-lb) thrust SNECMA/Turboméca Larzac 04-C5 turbofans, and a total internal fuel capacity of 1900 litres (418 Imp gal) in three fuselage and three integral wing tanks, plus provision for two 310-litre (68-Imp gal) drop-tanks
Performance: maximum speed 1005 km/h (624 mph) or Mach 0.85 at 3050 m (10,000 ft), and 1000 km/h (621 mph) or Mach 0.82 at sea level; initial climb rate 3420 m (11,220 ft) per minute; service ceiling 14630 m (48,000 ft); combat radius 425 km (264 miles) on a lo-lo-lo sortie with gun pod and underwing stores; ferry range 2940 km (1,827 miles) with external fuel
Weights: empty 3345 kg (7,374 lb) as a trainer, or 3515 kg (7,749 lb) for attack; normal take-off 5000 kg (11,023 lb) as a trainer; maximum take-off 8000 kg (17,637 lb) for attack
Dimensions: span 9.11 m (29 ft 10.75 in); length 12.29 m (40 ft 3.75 in) as a trainer, and 13.23 m (43 ft 5 in) as an attack aircraft; height 4.19 m (13 ft 9 in); wing area 17.50 m² (188.4 sq ft)

Variants
Alpha Jet A: designed in the late 1960s and early 1970s as an advanced trainer and light attack type to replace such types as the Aérospatiale Magister and Aeritalia G91, the Alpha Jet was a joint Franco-German programme that resulted in the first flight during October 1973 of the precursor of an important series; the Alpha Jet A is the West German light attack model, with a pointed nose and probe for the accommodation of various air-data sensors; this model first flew in April 1978 and began to enter West German service in March 1979; Dornier is currently developing an advanced version of this aircraft as a conversion and weapons trainer with a revised cockpit featuring a head-up display for the front-seat pupil (video-relayed to the instructor in the rear seat), two cathode ray tube displays and a single control/display unit two 1440-kg (3,175-lb) thrust Larzac 04-C20 turbofans, and the ability to carry more modern weapons such as the AIM-132 ASRAAM, AGM-88 HARM and projected-weapon dispensers
Alpha Jet E: designation of the French advanced trainer model, which has a secondary light attack capability; the type has a rounded nose and began to enter French service in the summer of 1978; since then the type has scored useful export sales as the **Alpha Jet B** for Belgium, **Alpha Jet C** for Ivory Coast, Qatar and Togo, **Alpha Jet MS-1** for Egypt and **Alpha Jet N** for Nigeria
Alpha Jet NGEA: this is an improved close-support and limited air-combat version of the Alpha Jet E developed by Dassault-Breguet for the export market with an advanced nav/attack system comprising a Sagem Uliss 81 inertial platform, a Thomson CSF VE-110C head-up display, a Thomson-CSF TMV 630 laser rangefinder, and ESD Digibus multiplex digital databus and other improvements; the weapons can include two Matra 550 Magic dogfighting air-to-air missiles and free-fall ordnance or rocket-launchers on the four underwing hardpoints, in addition to the podded underfuselage cannon; the type first flew in April 1982 and serves with at least three air arms, the Egyptian aircraft being designated **Alpha Jet MS-2**
Dassault-Breguet Lancier: under this designation the French company has developed a radically improved single-seat version of the Alpha Jet NGEA optimized for the close support and anti-ship roles with Larzac 04C-20 tubofans plus specialist avionics and weapons; the Lancier (lancer) has a longer nose for the considerably enhanced avionics fit (including air-to-air/air-to-surface Thomson-CSF Agave radar, forward-looking infra-red in an undernose blister fairing, a Thomson-CSF VE-130 wide-angle head-up display behind the new single-piece windscreen, and a CP 2084 central computer); the type first flew in 1985 and exhaustive trials and weapon qualifications are likely to result in sales later in the 1980s; the type is being qualified with weapons such as the Matra 550 Magic air-to-air missile and the Aérospatiale AM.39 anti-ship missile, as well as laser-guided bombs and a podded 30-mm DEFA cannon

de Havilland Canada DHC-4A Caribou
(Canada)
Type: STOL tactical transport
Accommodation: crew of two on the flightdeck, plus up to 32 troops or 8,740 lb (3965 kg) freight in the hold
Armament (fixed): none
Armament (disposable): none
Electronics and operational equipment: communication and navigation equipment
Powerplant and fuel system: two 1,450-hp (1082-kW) Pratt & Whitney R-2000-D5 Twin Wasp radial piston engines, and a total internal fuel capacity of 690 Imp gal (3137 litres) in 10 wing cells
Performance: maximum speed 216 mph (347 km/h) at 6,500 ft (1980 m); cruising speed 182 mph (293 km/h) at 7,500 ft (2285 m); initial climb rate 1,355 ft (413 m) per minute; service ceiling 24,800 ft (7560 m); range 242 miles (390 km) with maximum paylaod
Weights: empty 18,260 lb (8283 kg); maximum take-off 28,500 lb (12298 kg)
Dimensions: span 95 ft 7.5 in (29.15 m); length 72 ft 7 in (22.12 m); height 31 ft 9 in (9.67 m); wing area 912.0 sq ft (84.7 m²)

Variants
DHC-4 Caribou: first flown in July 1958, this tactical transport continued the role specialization of earlier

DHC aircraft, but offered considerably more payload without significant loss of STOL capability; the variant has a maximum take-off weight of 26,000 lb (11793 kg)

DHC-4A Caribou: improved DHC-4 with a 2,500-lb (1134-kg) increase in maximum take-off weight, and the variant that amounted to all but 23 of the 307 Caribou transports built

de Havilland Canada DHC-5D Buffalo
(Canada)
Type: STOL utility transport
Accommodation: crew of three on the flightdeck, plus up to 41 troops, or 34 paratroops, or 24 litters and six passengers, or 18,000 lb (8164 kg) of freight in the cabin
Armament (fixed): none
Armament (disposable): none
Electronics and operational equipment: communication and navigation equipment
Powerplant and fuel system: two 3,133-shp (2336-kW) General Electric CT64-820-4 turboprops, and a total internal fuel capacity of 1,755 Imp gal (7978 litres) in two integral wing tanks in the inner wing panels and 20 bag tanks in the outer wing panels
Performance: maximum speed 290 mph (467 km/h) at 10,000 ft (3050 m) for an assault mission at a weight of 38,950 lb (17667 kg); cruising speed 261 mph (420 km/h) at 10,000 ft (3050 m) for transport; initial climb rate 1,820 ft (555 m) per minute for transport; service ceiling 27,000 ft (8380 m) for transport; range 691 miles (1112 km) with maximum payload, or 2,038 miles (3280 km) with maximum fuel and no payload
Weights: empty 25,160 lb (11412 kg) with allowances for electronics etc; normal take-off 41,000 lb (18597 kg) for STOL assault mission; maximum take-off 49,200 lb (22316 kg) for transport mission
Dimensions: span 96 ft 0 in (29.26 m); length 79 ft 0 in (24.08 m); height 28 ft 8 in (8.73 m); wing area 945.0 sq ft (87.8 m²)

Variants
DHC-5A Buffalo: developed from the DHC-4 as a turboprop-powered tactical and utility transport in response to a US Army requirement, the Buffalo continued DHC's reputation for rugged STOL transports when the type first flew in April 1964; the DHC-4 had a payload of 11,200 lb (5080 kg), but only four trials aircraft for the US Army were built before the company switched production to the improved DHC-5A variant with CT64-810-1 turboprops and payload increased to 13,500 lb (6124 kg); the Canadian Armed Forces took 15 of the type, most being converted subsequently for the maritime patrol and search-and-rescue roles
DHC-5D Buffalo: main production version (the DHC-5B and DHC-5C having been proposals for Indian aircraft with CT64-P4C and Rolls-Royce Dart RDa.12 turboprops respectively) with greater power and payload

Douglas EA-3B Skywarrior
(USA)
Type: electronic reconnaissance aircraft
Accommodation: crew of three on the flightdeck, and a mission crew of four in the cabin replacing the bomb-bay
Armament (fixed): none
Armament (disposable): none
Electronics and operational equipment: communication and navigation equipment, plus ECM gear (ALR-40 surveillance recording and ALR-63 frequency-measuring equipment), side-looking airborne radar (SLAR), forward-looking radar, chaff dispensers and infra-red sensors
Powerplant and fuel system: two 10,500-lb (4763-kg) thrust Pratt & Whitney J57-P-10 turbojets, and a total internal fuel capacity of (KA-3B) of 5,025 US gal (19025 litres) in integral wing and two fuselage tanks; provision for inflight-refuelling
Performance: maximum speed 610 mph (982 km/h) at 10,000 ft (3050 m); cruising speed 520 mph (837 km/h); service ceiling 41,000 ft (12495 m); range 2,900 miles (4665 km)
Weights: empty 39,409 lb (17876 kg); normal take-off 70,000 lb (31751 kg); maximum take-off 82,000 lb (37195 kg)
Dimensions: span 72 ft 6 in (22.10 m); length 76 ft 4 in (23.27 m); height 22 ft 9.5 in (6.95 m); wing area 812.0 sq ft (75.44 m²)

Variants
EA-3B Skywarrior: the most important of the three Skywarrior versions left in US Navy service, the EA-3B is the electronic reconnaissance aircraft derived from the A-3B carrierborne attack aircraft, and first flew in December 1968; the type has the ALR-40 surveillance recording equipment and the LAR-63 frequency-measuring receiver
EKA-3B Skywarrior: designation of the US Navy Reserve for the combined electronic countermeasures training and inflight-refuelling roles; the former role uses the ALQ-76 noise jammer
ERA-3B Skywarrior: electronic 'aggressor' version to test US Navy electronic counter-countermesures training and equipment
KA-3B Skywarrior: US Navy Reserve inflight-refuelling tanker model

EMBRAER EMB-111 Bandeirulha
(Brazil)
Type: maritime reconnaissance aircraft
Accommodation: crew of two on the flightdeck, plus a mission crew of three in the cabin and provision for paratroops or emergency supplies up to 1309 kg (2,886 lb)

Armament (fixed): none
Armament (disposable): an assortment of stores can be accommodated on four underwing hardpoints; these stores generally comprise flares and smoke grenades, but alternatives are eight 5-in (127-mm) HVAR rockets, or four launchers each will seven 2.75-in (69.85-mm) FFAR rockets
Electronics and operational equipment: communication and navigation equipment, Litton LN-33 inertial navigation system, AIL SPAR-1 (APS-128) search radar able to pick up low-profile targets in disturbed sea at a range of 60 miles (96 km), and communication and navigation equipment
Powerplant and fuel system: two 750-shp (559-kW) Pratt & Whitney Canada PT6A-34 turboprops, and a total internal fuel capacity of 2586 litres (569 Imp gal) in four integral wing and two non-jettisonable wingtip tanks
Performance: maximum speed 404 km/h (251 mph) at 3050 m (10,000 ft); cruising speed 347 km/h (216 mph) at 3050 m (10,000 ft); initial climb rate 402 m (1319 ft) per minute; service ceiling 8230 m (27,000 ft); range 2075 m (1,290 miles) with maximum payload, or 2725 km (1,695 miles) with maximum fuel
Weights: empty 3403 kg (7,502 lb); maximum take-off 7000 kg (15,432 lb)
Dimensions: span 15.96 m (52 ft 4.5 in) over tiptanks; length 14.83 m (48 ft 7.9 in); height 4.74 m (15 ft 6.5 in); wing area 29.00 m² (312.2 sq ft)

Variants
EMB-110 Bandeirante: this is the base model of the series, designed as a general-purpose transport in the 1960s with a payload of 15 passengers and a powerplant of two 680-shp (507-kW) Pratt & Whitney Canada PT6A-27 turboprops; the Bandeirante (pioneer) first flew in October 1968, and began to enter civil and military service in 1970
EMB-110A Bandeirante: navaid calibration version of the EMB-110
EMB-110B Bandeirante: photographic survey version with Doppler navigation and an electrically-operated ventral door over the hatch for a battery of cameras
EMB-110C Bandeirante: improved transport version with 15 seats
EMB-110K Bandeirante: improved transport with two 750-shp (559-kW) PT6A-34 turboprops and the fuselage stretched by 0.85 m (2.79 ft) to permit the carriage of a 1880-kg (4,125-lb) payload in the light freighting role; freight is loaded through an upward-opening door in the rear fuselage
EMB-110P1SAR Bandeirante: search-and-rescue version of the EMB-110P1 quick-change passenger/cargo transport
EMB-111A/A Bandeirulha: land-based maritime reconnaissance version of the EMB-110 series, visually identifiable from the transport series by its tip tanks and large nose radome for the search radar

EMBRAER EMB-312 Tucano
(Brazil)
Type: basic trainer
Accommodation: pupil and instructor seated in tandem on Martin-Baker BR8LC lightweight ejector seats
Armament (fixed): none
Armament (disposable): this is carried on four underwing hardpoints, all stressed to 250 kg (551 lb); the inner pair can each accommodate an MS10-21/22-10A gun pod with a single 12.7-mm (0.5-in) machine-gun with 350 rounds, and all hardpoints can accept a 250-lb (113-kg) Mk 81 bomb, or an LM-37/7A rocket-launcher pod with seven 37-mm (1.46-in) rockets, or an LM-70/7 rocket-launcher pod with seven 70-mm (2.75-in) rockets
Electronics and operational equipment: communication and navigation equipment
Powerplant and fuel system: one 750-shp (559-kW) Pratt & Whitney Canada PT6A-25C turboprop, and a total internal fuel capacity of 694 litres (153 Imp gal) in two integral wing tanks, plus provision for the carriage of two 330-litre (73-Imp gal) underwing ferry tanks
Performance: maximum speed 458 km/h (254 mph) at 4115 m (13,500 ft); cruising speed 441 km/h (274 mph) at 3050 m (10,000 ft); initial climb rate 579 m (1,900 ft) per minute; service ceiling 8750 m (28,700 ft); range 1916 km (1,190 miles), or 3555 km (2,209 miles) with ferry tanks
Weights: empty 1790 kg (3,046 lb); normal take-off 2550 kg (5,622 lb); maximum take-off 3175 kg (7,000 lb)
Dimensions: span 11.14 m (36 ft 6.5 in); length 9.86 m (32 ft 4.25 in); height 3.40 m (11 ft 2 in); wing area 19.40 m² (208.82 sq ft)

Variant
EMB-312 Tucano: designed as a basic trainer with turboprop powerplant for a combination of performance, economy of operation and 'jet' handling, the Tucano (toucan) first flew in August 1980 and entered service in 1983, and is proving a successful contender in South American and world markets in the flying and weapon training roles; to meet British requirements for the **Tucano T.Mk 1** version built under licence by Shorts, the type has been upgraded with a Garrett TPE 331 turboprop for enhanced performance

ENAER T-35 Pillan
(Chile)
Type: basic and intermediate-level trainer
Accommodation: pupil and instructor
Armament (fixed): none
Armament (disposable): this is accommodated on two underwing hardpoints, which can each take a 0.5-in (12.7-mm) machine-gun pod, a four- or seven-tube

rocket-launcher, or a 250-lb (113-kg) Mk 81 bomb
Electronics and operational equipment: communication and navigation equipment
Powerplant and fuel system: one 300-hp (224-kW) Avco Lycoming AEIO-540-H1K5 flat-six piston engine, and a total internal fuel capacity of 291.5 litres (64 Imp gal) in two integral wing tanks
Performance: maximum speed 310 km/h (193 mph) at sea level; cruising speed 298 km/h (185 mph) at 2680 m (8,800 ft); initial climb rate 465 m (1,525 ft) per minute; service ceiling 5820 m (19,100 ft); range 1270 km (789 miles) at maximum cruising speed
Weights: empty 930 kg (2,050 lb); maximum take-off 1315 kg (2,900 lb)
Dimensions: span 8.81 m (28 ft 11 in); length 7.97 m (26 ft 1.75 in); height 2.34 m (7 ft 8.25 in); wing area 13.64 m² (146.8 sq ft)

Variant
T-35 Pillan: first flown during 1981 in the USA after development by Piper from the PA-28 Dakota and PA-32 Saratoga series of lightplanes, the T-35 is a useful trainer and counter-insurgency aircraft intended for local manufacture by the developing Chilean aerospace industry; the type has also been developed as the Turbo-Pillan into the 1048-kg (2,310-lb) **Aucan** with a 420-shp (313-kW) Allison 250B-17D turboprop, which offers significantly improved performance (including a maximum speed of 367 km/h; 228 mph) and payload, especially in the armed role

European Helicopter Industries EH.101
(Italy/UK)
Type: multi-role helicopter
Accommodation: normally four for the anti-submarine role
Armament (fixed): none
Armament (disposable): this is carried on two hardpoints (one on each side of the fuselage) each able to accommodate two Stingray or A 244S anti-submarine torpedoes, or anti-ship missiles such as the AM.39 Exocet, AGM-84 Harpoon, Sea Eagle and Sea Skua
Electronics and operational equipment: communication and navigation equipment, plus (on Royal Navy anti-submarine helicopters) Ferranti Blue Kestrel search radar, dunking sonar and sonobuoys used in conjunction with the Marconi AQS-903 acoustic processing system, inertial navigation system, Global Positioning System, Doppler navigation, various electronic and infra-red countermeasures systems
Powerplant and fuel system: three 1,800-shp (1342-kW) General Electric CT7-6 turboshafts, and a total internal fuel capacity of ? Imp gal (? litres)
Performance: cruising speed 184 mph (296 km/h); ferry range 1150 miles (1850 km); endurance 5 hours on station with maximum weapon load
Weights: empty 15,500 lb (7031 kg); maximum take-off 28,660 lb (1300 kg)
Dimensions: main rotor diameter 61 ft 0 in (18.59 m); length overall, rotors turning 75 ft 1.6 in (22.90 m); height overall 21 ft 4 in (6.50 m); main rotor disc area 2,922.5 sq ft (271.5 m²)

Variant
EH.101: designed as a collaborative venture between Westland and Agusta to replace the Sea King variants in service with the British and Italian navies, the EH.101 was scheduled to fly before the end of 1986 in prototype form to validate the concept of a weapon/sensor platform more capable and more powerful than the Sea King, yet possessing smaller overall dimensions; the type is also being developed in utility form with a payload of 13,435 lb (6094 kg), and future plans call for use of the Rolls-Royce/Turboméca RTM322 turboshaft for greater power and thus greater performance; the use of two MIL 1553 databuses offers maximum flexibility in retrofitting additional equipment of the most modern types

Fairchild C-119G Flying Boxcar
(USA)
Type: tactical transport aircraft
Accommodation: crew of four on the flightdeck, and up to 62 troops or freight in the hold
Armament (fixed): none
Armament (disposable): none
Electronics and operational equipment: communication and navigation equipment, plus weather radar
Powerplant and fuel system: two 3,500-hp (2611-kW) Pratt & Whitney R-4360-20 radial piston engines, and a total internal fuel capacity of ? Imp gal (? litres) in wing tanks
Performance: maximum speed 281 mph (452 km/h) at 18,000 ft (4585 m); initial climb rate 1,010 ft (308 m) per minute; service ceiling 23,900 ft (7285 m); range 1,770 miles (2849km) with maximum fuel
Weights: empty 39,800 lb (18053 km); maximum take-off 74,000 lb (33566 kg)
Dimensions: span 109 ft 3 in (33.30 m); length 86 ft 6 in (26.36 m); height 26 ft 4 in (8.03 m); wing area 1,400.0 sq ft (130.06 m²)

Variant
C-119G Flying Boxcar: developed after World War II from the C-82 Packet series, the C-119 has a wider fuselage (and thus greater usable volume for freight) and greater span combined with more power; the C-119G is the only model still in limited service, and is a version of the C-119F with Aero-Products rather than Hamilton Standard propellers

Fairchild C-123B Provider
(USA)
Type: STOL tactical transport

Accommodation: crew of two on the flightdeck, and up to 61 troops, or 50 litters with six seated casualties and six attendants, or freight in the cabin
Armament (fixed): none
Armament (disposable): none
Electronics and operational equipment: communication and navigation equipment, plus weather radar
Powerplant and fuel system: two 2,500-hp (1865-kW) Pratt & Whitney R-2800-99W radial piston engines and two 2,850-lb (1293-kg) thrust General Electric J85-GE-17 turbojets, and a total internal fuel capacity of ? Imp gal (? litres) in wing tanks
Performance: maximum speed 245 mph (394 km/h); cruising speed 205 mph (330 km/h); initial climb rate 1,150 ft (351 m) per minute; service ceiling 29,000 ft (8840 m); range 1,470 miles (2366 km) with maximum fuel
Weights: empty 29,900 lb (13562 kg); maximum take-off 60,000 lb (27216 kg)
Dimensions: span 110 ft 0 in (33.53 m); length 76 ft 3 in (23.92 m); height 34 ft 1 in (10.39 m); wing area 1,223.0 sq ft (113.62 m^2)

Variants
C-123B Provider: developed from the XG-18 cargo glider, which led to the YC-122 light assault transport model and the XG-20 assault glider, the surprisingly long lived C-123 was originally designed by the Chase Aircraft Company and first flown in October 1949; Chase was taken over by Kaizer-Frazer in 1949, and after some pre-production C-123B aircraft had been built the full production contract was awarded to Fairchild, which acquired rights to the design and undertook further development of the C-123 series; the C-123B production model has 2,500-hp (1865-kW) Pratt & Whitney R-2800-99W Double Wasp radials and can carry 60 troops, or 50 litters and six sitting wounded plus six attendants
C-123K Provider: designation of C-123s after the installation of larger wheels (with an anti-skid braking system) and of two underwing pods (each containing one 2,850-lb/1293-kg General Electric J85-GE-17 turbojet) for better performance

Fairchild Republic A-10A Thunderbolt II
(USA)
Type: close-support and anti-tank aircraft
Accommodation: pilot seated on a Douglas ACES II ejector seat
Armament (fixed): one 30-mm General Electric GAU-8/A Avenger rotary-barrel cannon in the forward fuselage with 1,174 rounds
Armament (disposable): this is carried on three underfuselage and eight underwing hardpoints; of the underfuselage hardpoints, which can carry stores on only the centreline or on the two flanking positions at any time, the centreline hardpoint is rated at 5,000 lb (2268 kg) and the two flanking hardpoints each at 3,500 lb (1587 kg); the inboard pair of underwing hardpoints is each rated at 3,500 lb (1587 kg); the next pair outboard is each rated at 2,500 lb (1134 kg); and the four outer hardpoints are each rated at 1,000 lb (454 kg); the maximum external load with reduced fuel is 16,000 lb (7258 kg), reducing to 14,340 lb (6505 kg) with a maximum fuel load; typical loads are 28 500-lb (227-kg) Mk 82 free-fall or retarded bombs, six 2,000-lb (907-kg) Mk 84 general-purpose bombs, eight BLU-1 or BLU-27/B incendiary bombs, 20 Rockeye II cluster bombs, 16 CBU-52 or CBU-71 bomb dispensers, six AGM-65 Maverick air-to-surface missiles, Mk 82 and Mk 84 laser-guided bombs, Mk 84 electro-optically guided bombs, two SUU-23/A 20-mm cannon pods, and several other stores
Electronics and operational equipment: communication and navigation equipment, plus Kaiser head-up display, weapon-delivery package, AAS-35 'Pave Penny' laser designator pod, ALR-46(V) radar-warning receiver, ALQ-119 and Westinghouse ALQ-131 electronic counter-measures pods, and the LANTIRN night/adverse-weather navigation and targeting system
Powerplant and fuel system: two 9,065-lb (4111-kg) thrust General Electric TF34-GE-100 turbofans, and a total internal fuel capacity of 1,606 US gal (6080 litres) in two fuselage and two wing tanks, plus three 600-US gal (2271-litre) drop-tanks; provision for inflight-refuelling
Performance: maximum speed 439 mph (706 mk/h) at sea level in clean condition; cruising speed 387 mph (623 km/h) at 5,000 ft (1525 m); initial climb rate 6,000 ft (1830 m) per minute; combat radius 288 miles (463 km) on a close air support sortie with a 1.7-hour loiter; ferry range 2,455 miles (3950 km) against a 58-mph (93-km/h) wind
Weights: empty 24,960 lb (11322 kg); maximum take-off 50,000 lb (22680 kg)
Dimensions: span 57 ft 6 in (17.53 m); length 53 ft 4 in (16.26 m); height 14 ft 8 in (4.47 m); wing area 506.0 sq ft (47.01 m^2)

Variant
A-10A Thunderbolt II: first flown in May 1972 and entering service in late 1975, the A-10A is slow by modern standards of battlefield aircraft, but is prodigiously strong, highly manoeuvrable, well armoured and filled with the structural and system redundancies necessary for battlefield survival, and also possesses considerable endurance and an excellent weapons load; the type's primary failing is its restriction to clear-weather operations, though this limitation will be reduced from 1987 with the adoption of the LANTIRN (Low-Altitude Navigation and Targeting Infra-Red for Night) system, which comprises navigation and targeting pods, the former with terrain-following radar and forward-looking infra-red, and

the latter with a large-aperture forward-looking infrared, an automatic tracker and a laser ranger/designator) to provide the pilot's new wide-angle head-up display with a thermal image of the land ahead and all necessary targeting information for the use of 'smart' weapons such as the IR and Laser Maverick missiles

FMA IA 58A Pucará
(Argentina)
Type: two-seat counter-insurgency aircraft
Accommodation: pilot and co-pilot in tandem on Martin-Baker Mk AP06A ejector seats
Armament (fixed): two 20-mm Hispano-Suiza HS-804 cannon with 270 rounds each, and four 7.62-mm (0.3-in) FN-Browning machine-guns with 900 rounds each
Armament (disposable): a total of 1620 kg (3,571 lb) can be carried on one underfuselage and two underwing hardpoints, the former rated at 1000 kg (2,205 lb) and the latter each at 500 kg (1102 lb); typical loads include two Martin Pescador air-to-surface missiles, 12 125-kg (276-lb) bombs, six Alkan 530 grenade-dispenser pods, gun pods and drop-tanks
Electronics and operational equipment: communication and navigation equipment, plus an AN/AWE-1 stores programmer and a Matra 83-4-3 sight
Powerplant and fuel system: two 729-kW (978-shp) Turboméca Astazou XVIG turboprops, and a total internal fuel capacity of 1260 litres (277 Imp gal) in two fuselage and two wing tanks, plus provision for a 1130-litre (249-Imp gal) auxiliary tank under the fuselage and two 300-litre (66-Imp gal) tanks under the wings
Performance: maximum speed 500 km/h (311 mph) at 3000 m (9,845 ft); cruising speed 480 km/h (298 mph) at 6000 m (19,685 ft); initial climb rate 1080 m (3,545 ft) per minute; service ceiling 10000 m (32,810 ft); combat radius 350 km (217 miles) on a hi-lo-hi sortie; ferry range 3040 km (1,889 miles)
Weights: empty 4037 kg (8,900 lb); maximum take-off 6800 kg (14,991 lb)
Dimensions: span 14.50 m (47 ft 6.75 in); length 14.25 m (46 ft 9 in); height 5.36 m (17 ft 7 in); wing area 30.30 m² (326.1 sq ft)

Variants
IA 58A Pucara: this basic model of this Argentine counter-insurgency and light attack aircraft first flew in August 1969, and began to enter service in 1974
IA 58B Pucara Bravo: this is a simple development with two 30-mm DEFA 553 cannon with 140 rounds per gun in place of the IA 58A's two 20-mm Hispano Suiza HS-804 cannon
IA 58C Pucara: designation of IA 58A aircraft converted to single-seat configuration with a nose-mounted armament of one 30-mm DEFA 554 cannon with 250 rounds and a Saab TGS2 sight; the payload/range balance is some 30 per cent better than that of the IA 58A

IA 66 Pucara: first flying in 1980, this is an experimental/development model with two 1,000-shp (746-kW) Garrett TPE331 turboprops

FMA IA 63 Pampa
(Argentina)
Type: basic and advanced jet trainer
Accommodation: pupil and instructor seated in tandem on Martin-Baker Mk 8 lightweight ejector seats
Armament (fixed): none
Armament (disposable): five hardpoints (one underfuselage and four underwing) are planned for the proposed export model, which will be able to double as a light attack aircraft
Electronics and operational equipment: communication and navigation equipment
Powerplant and fuel system: one 3,500-lb (1588-kg) thrust Garrett TFE731-2-2N turbofan or one 2,900-lb (1315-kg) thrust Pratt & Whitney Canada PT15D-5 turbofan (the former being installed in the first three and the latter in the last of the four prototypes for evaluative purposes), and a total internal fuel capacity of 980 litres (216 Imp gal) in integral tanks, plus provision for 400 litres (88 Imp gal) of auxiliary fuel
Performance: maximum speed 820 km/h (510 mph) or Mach 0.75 at 9000 m (29,525 ft); initial climb rate 1500 m (4,920 ft) per minute; service ceiling 14000 m (45,925 ft); range 2500 km (1,550 miles) with auxiliary fuel for ferrying
Weights: normal take-off 3490 kg (7,695 lb); maximum take-off 4650 kg (10,251 lb)
Dimensions: span 9.686 m (31 ft 9.25 in); length 10.928 m (35 ft 10.25 in); height 4.28 m (14 ft 0.5 in); wing area 15.63 m² (168.2 sq ft)

Variant
IA 63 Pampa: designed with the assistance of Dornier in West Germany, this is a simple two-seat trainer with light attack capability, and the type first flew in 1983; orders have already been placed by the home country and other South American services

Fokker F.27 Maritime
(Netherlands)
Type: maritime patrol aircraft
Accommodation: crew of two or three on the flight-deck, and a mission crew of two to six in the cabin
Armament (fixed): none
Armament (disposable): none
Electronics and operational equipment: communication and navigation equipment, plus Litton LTN-72 inertial navigation system, IDC air-data computer, Litton APS-504(V)2 panoramic search radar and Sperry SPZ-600 automatic flight control system
Powerplant and fuel system: two 2,320-shp (1730-kW) Rolls-Royce Dart RDa.7 Mk 536-7R turboprops, and a total internal fuel capacity of 6450 litres

(1,639 Imp gal) with 2310 litres (508 Imp gal) in an optional centre-section tank, plus provision for two 938-litre (206-Imp gal) non-jettisonable underwing tanks
Performance: cruising speed 465 km/h (289 mph) at 6100 m (20,015 ft); initial climb rate 442 m (1,450 ft) per minute; service ceiling 8990 m (29,495 ft); range 5000 km (3,107 miles) with maximum internal and external fuel, or 1850-km (1,150 miles) as a transport with a 4536-kg (10,000-lb) load
Weights: empty 13314 kg (29,352 lb); normal take-off 20410 kg (45,000 lb); maximum take-off 21320 kg (47,500 lb)
Dimensions: span 29.00 m (95 ft 2 in); length 23.56 m (77 ft 3.5 in); height 8.50 m (27 ft 11 in); wing area 70.00 m² (753.5 sq ft)

Variants
F.27 Mk 400M: dedicated military transport version of the civil F.27 Friendship Mk 400 airliner; compared with the civil model the Mk 400M has a large cargo door, and (on each side) inflight-openable doors for the despatch of paratroops, of which 46 can be carried; alternative loads are 24 litters and nine attendants, or 6025 kg (13,283 lb) of freight; the type can also be fitted out as a photo-survey aircraft, and other marks of the Friendship series are used in military markings mainly for VIP transport; the basic F.27 or later Fokker 50 airframe is also proposed as the basic of two electronic warfare machines, namely the **Kingbird** airborne early warning aircraft and the **Sentinel** electronic surveillance aircraft; the former would have AWG-9 surveillance radar and an electronic support measures system, while the latter would feature the Motorola APS-135(V) side-looking airborne radar, cameras and an automatic communications intelligence system
F.27 Maritime: maritime patrol version of the standard Friendship airliner, and differing from this latter mainly in extra fuel capacity and the specialized role equipment suiting the stype for coastal patrol, maritime reconnaissance, fishery protection, search-and-rescue, offshore resources protection etc with an endurance of 10 to 12 hours at patrol speeds between 277 and 333 km/h (172 and 207 mph) at an altitude of 460 m (1,510 ft)
F.27MPA Maritime Enforcer: basically the armed version of the F.27 Maritime designed for the anti-ship and anti-submarine roles with an avionics and armament system integrated by Marconi of the UK, and including a digitally-processed version of the APS-504 radar, the AQS-902 sonobuoy acoustic processing system, electronic support measures systems and a central tactical system; 3930 kg (8,664 lb) of weapons are carried on two underfuselage and six underwing hardpoints, typical loads managed by the Alkan stores management system being two or four torpedoes or two anti-ship missiles

Maritime Enforcer Mk 2: this is a considerably developed version of the F.27MPA using many features of the 'F.27 Mk 2' airliner, namely the Fokker 50; the Maritime Enforcer Mk 2 has basically the same airframe as the F.27MPA but with aerodynamic refinements, flight-deck improvements and power provided by two 2,400-shp (1790-kW) Pratt & Whitney Canada PW124 turboprops driving six-blade propellers for greater economy of fuel, reduced exterior and interior noise levels, and reduced vibration; the avionics have been considerably updated (especially in terms of electronic support measures and tactical co-ordination), and the weapons capability is increased to the extent that the Maritime Enforcer Mk 2 can carry up to eight torpedoes and/or depth bombs, or two or four AGM-84 Harpoon or AM.39 Exocet anti-ship missiles, or a mixed load of torpedoes and missiles; maximum take-off weight is 21565 kg (47,542 lb) and typical endurance 14 hours, and the aircraft can cruise at high or low level depending upon mission requirements (translating as a mission radius of 2200 km/1,367 miles with a weapon load of 1800 kg/3,968 lb); maximum range is 6800 km (4,225 miles)

Fuji (Bell) UH-1B (Japan): see Bell UH/1B/E Iroquois

Fuji (Bell) UH-1H (Japan): see Bell UH-1H Iroquis

Fuji T-1A
(Japan)
Type: intermediate trainer with secondary light attack capability
Accommodation: pupil and instructor seted in tandem on ejector seats
Armament (fixed): one 0.5-in (12.7-mm) Colt Browning M2 machine-gun in the nose
Armament (disposable): this is carried on two underwing hardpoints, and can comprise two AIM-9 Sidewinder air-to-air missiles, or two gun pods, or two 340-kg (750-lb) bombs, or two rocket-launcher pods for 2.75-in (69.85-mm) rockets, or napalm tanks
Electronics and operational equipment: communication and navigation equipment
Powerplant and fuel system: one 4,000-lb (1814-kg) thrust Rolls-Royce (Bristol) Orpheus BOr.4 Mk 805 turbojet, and a total internal fuel capacity of ? litres (? Imp gal) in fuselage and wing tanks, plus provision for two ? litre (? Imp gal) drop-tanks
Performance: maximum speed 925 km/h (575 mph) at 6095 m (20,000 ft); cruising speed 620 km/h (385 mph) at 9145 m (30,000 ft); initial climb rate 1980 m (6,495 ft) per minute; service ceiling 15850 m (52,000 ft); range 1300 km (805 miles) on internal fuel, and 1950 km (1,210 miles) on internal and external fuel
Weights: empty 2420 kg (5335 lb); maximum take-off 5000 kg (11,023 lb)

Dimensions: span 10.50 m (34 ft 5 in); length 12.12 m (39 ft 9 in); height 4.08 m (13 ft 4 in); wing area 22.22 m² (239.2 sq ft)

Variants
T-1A: in looks resembling the North American F-86 fighter of the 1950s, this trim intermediate flying and weapons trainer was developed in the 1950s as the T1F, Japan's first operational jet-powered aircraft flew in prototype form during January 1958
T-1B: version with the 1200-kg (2,646-lb) Ishikawa-jima-Harima J3-IHI-3 turbojet and slightly lower performance
T-1C: designation of T-1Bs converted with the 1400-kg (3,086-lb) J3-IHI-7 turbojet

General Dynamics F-16A Fighting Falcon
(USA)
Type: multi-role air-combat and attack fighter
Accommodation: pilot seated on a Douglas ACES II ejector seat
Armament (fixed): one 20-mm General Electric M61A1 Vulcan rotary-barrel cannon in the port wing/fuselage fairing with 500 rounds
Armament (disposable): this is carried on one underfuselage, six underwing and two wingtip hardpoints; the underfuselage hardpoint is rated at 2,200 lb (998 kg), and from inboard outwards the members of each pair of underwing hardpoints are each rated at 4,500 lb (2041 kg), 3,500 lb (1587 kg) and 700 lb (318 kg); the wingtip hardpoints are each rated at 425 lb (193 kg) and are designed for AIM-9 Sidewinder air-to-air missiles; the maximum disposable load is 20,450 lb (9276 kg), and typical loads include six AIM-9 Sidewinder air-to-air missiles (to be replaced in the later 1980s by eight AIM-120A AMRAAMs), four Mk 84 2,000-lb (907-kg) bombs, 19 Mk 82 1,000-lb (454-kg) free-fall or retarded bombs, four guided bombs, 13 CBU-series bomb dispensers, six AGM-66 Maverick air-to-surface missiles, two AGM-88 HARM anti-radiation missiles, and a number of other weapons and stores
Electronics and operational equipment: communication and navigation equipment, plus Westinghouse APG-66 pulse-Doppler range and angle track radar (with look-down and look-up ranges of 35 miles/56 km and 46 miles/74 km respectively), Dalmo Victor ALR-69 radar-warning receiver, Sperry central air-data computer, Singer-Kearfott SKN-2400 (modified) inertial navigation system, Marconi head-up display, Kaiser radar electro-optical display, Delco fire-control computer, two Tracor ALE-40 chaff/flare launchers, Martin-Marietta AAS-35 'Pave Penny' laser-tracker pod, Martin-Marietta LANTIRN war/attack system, Westinghouse ALQ-119 and ALQ-131 ECM pods and other electronic counter-measures equipment
Powerplant and fuel system: one 25,000-lb (11340-kg) afterburning thrust Pratt & Whitney F100-PW-200 turbofan, and a total internal fuel capacity of 1,047 US gal (3962 litres) in wing and fuselage tanks, plus provision for one 300-US gal (1136-litre) and two 370-US gal (1400-litre) or 600-US gal (2271-litre) drop-tanks; provision for inflight-refuelling
Performance: maximum speed more than 1,320 mph (2124 km/h) or Mach 2 at 40,000 ft (12190 m); and 915 mph (1472 km/h) or Mach 1.2 at sea level; initial climb rate 50,000 ft (15240 m) per minute with AAM armaments; service ceiling more than 50,000 ft (15240 m); combat radius more than 575 miles (925 km); ferry range more than 2,145 miles (3890 km) with internal and external fuel
Weights: empty 17,780 lb (8065 kg); normal take-off 25,647 lb (11633 kg); maximum take-off 35,400 lb (16057 kg)
Dimensions: span 31 ft 0 in (9.45 m) over missile rails; length 49 ft 4.9 in (15.09 m); height 16 ft 8.5 in (5.09 m); wing area 300.0 sq ft (27.87 m²)

Variants
F-16A Fighting Falcon: the F-16 is one of the Western world's most important combat aircraft, a comparatively light air-combat fighter with potent secondary attack/strike capability; the need for such a fighter was first truly appreciated during the Vietnam War, in which high-performance but comparatively clumsy US fighters were often caught at a disadvantage by the nimbler Soviet and Chinese fighters used by the North Vietnamese; in 1971, therefore, the USAF launched its Light-Weight Fighter programme, designed to produce experimental aircraft offering moderate performance but very high levels of manoeuvrability through the use of relaxed stability and a computer-assisted 'fly-by-wire' flight-control system; the YF-16 prototype first flew in February 1974 and, with the competing Northrop YF-17, showed markedly superior agility than contemporary fighters; in 1975 the USAF decided to procure the F-16 as a first-line type, production deliveries beginning in 1978; the type has since proved itself an extremely potent short/medium-range combat aircraft
F-16B Fighting Falcon: two-seat combat-capable operational conversion trainer derivative of the F-16A; this variant is also proposed as an 'Advanced Wild Weasel' type with the ALR-46 threat-warning system, ALQ-119 defensive jammer and offensive missiles such as the AGM-45, AGM-78 and AGM-88 types
F-16C Fighting Falcon: version of the F-16A delivered from July 1984 within the context of the Multinational Staged Improvement Program; this variant has APG-68 radar (featuring a programmable digital processing system, increased search range, track-while-scan capability and high-resolution ground mapping), an enlarged tailplane, a revised cockpit with two head-down coloured CRT multi-function

displays and a wide-angle head-up display, wiring for the AIM-120 AMRAAM, and provision for items such as the LANTIRN (Low-Altitude Navigation and Targeting Infra-Red for Night) pod system, JTIDS (Joint Tactical Information Distribution System), GPS (Global Positioning System) and ASPJ (Airborne Self-Protection Jammer); the type also has improved weapons capability for types such as the AGM-65 Maverick and General Electric 30-mm cannon pods; the Fighting Falcon has also been developed with private and service funding (notably as the F-16/101 with the 28,000-lb/12701-kg General Electric F110 afterburning turbofan evolved as the F101DFE from the engine used in the Rockwell B-1B bomber, the F-16/AFTI with control-configured flying surfaces in the Advanced Fighter Technology Integration programme, and the F-16XL cranked-arrow wing version with very considerably enhanced range and weapon payload capabilities [twice the payload over nearly 50% greater range] as offered to the USAF as a strike fighter with the designation F-16F), so there are still many ways in which the Fighting Falcon can yet be developed as an operational type

F-16D Fighting Falcon: two-seat combat-capable conversion trainer derivative of the F-16C

F-16N Fighting Falcon: version ordered by the US Navy as a dissimilar combat training aircraft, based on the F-16C but engined with the F110 afterburning turbofan and fitted with naval avionics plus APG-66 radar; the variant has no internal cannon

General Dynamics F-111F
(USA)
Type: variable-geometry all-weather strike and attack aircraft
Accommodation: crew of two seated side-by-side in a McDonnell Douglas escape capsule with a 40,000-lb (18144-kg) thrust rocket motor
Armament (fixed): one 20-mm General Electric M61A1 Vulcan rotary-barrel cannon and 2,028 rounds (optional) in the weapons bay
Armament (disposable): this is carried in an internal weapons bay and on four or six underwing hardpoints; the internal weapons bay can carry two (only one if cannon pack is fitted) 2,100-lb (953-kg) B43 nuclear free-fall bombs, and the maximum ordnance load is 31,500 lb (14288 kg); weapons that can be carried on the underwing hardpoints are the M117 750-lb (340-kg) retarded bomb, Mk 84 2000-lb (907-kg) free-fall or retarded bomb, Mk 82 500-lb (227-kg) free-fall or retarded bomb, SUU-30 fragmentation-bomb dispenser, CBU-30 and CBU-38 bomb dispensers, Mk 20 Rockeye cluster bomb, GBU-8 and GBU-15 2,000-lb (907-kg) glide bombs, GBU-10 2,000-lb (907-kg) laser-guided bomb and GBU-12 500-lb (227-kg) laser-guided bomb, plus other stores such as the AGM-65A/B/D Maverick air-to-surface missile

Electronics and operational equipment: communication and navigation equipment, plus General Electric APQ-144 attack and navigation radar, Texas Instruments APQ-128 terrain-following radar, Litton inertial navigation system, GPL Doppler navigation, IBM ASQ-133 digital fire-control system, General Electric ASG-25 optical display sight, Dalmo Victor ALR-62 radar-warning receiver, Cincinnati Electronics AAR-44 infra-red warning receiver, Sanders ALQ-137 ECM receiver, Sanders ALQ-94 noise deception jammer, Ford AVQ-26 'Pave Tack' target-acquisition and designation pod, and optional ECM fits such as the Westinghouse ALQ-119 and ALQ-131 pods

Powerplant and fuel system: two 25,100-lb (11385-kg) afterburning thrust Pratt & Whitney TF30-PW-100 turbofans, and a total internal fuel capacity of 16,120 lb (7312 kg) in wing and fuselage tanks, plus provision for two underwing drop-tanks; provision for inflight-refuelling

Performance: maximum speed 1,650 mph (2,655 km/h) or Mach 2.5 at high altitude, and 915 mph (1473 km/h) or Mach 1.2 at sea level; service ceiling 60,000 ft (18290 m); range more than 2,925 miles (4707 km) with maximum internal fuel

Weights: empty 47,481 lb (21537 kg); maximum take-off 100,000 lb (45360 kg)

Dimensions: span 63 ft 0 in (19.20 m) spread and 31 ft 11.4 in (9.74 m) swept; length 73 ft 6 in (22.40 m); height 17 ft 1.4 in (5.22 m); wing area 525.0 sq ft (48.77 m^2) spread and 657.3 sq ft (61.07 m^2) swept

Variants
F-111A: popularly known as the 'Aardvark', the F-111 was the world's first operational variable-geometry combat aircraft, the type having been schemed as a politically-inspired single-design solution to the needs of the USAF for a strike fighter and of the US Navy for a fleet defence fighter (the latter being the vastly overweight F-111B cancelled after the building of five pre-production and two production aircraft); the F-111A prototype first flew in December 1964 and after a protracted development programme began to enter service in October 1967 though there were still many problems with the advanced airframe and its complex avionics (including APQ-113 attack and APQ-110 terrain-following radars); this initial model is powered by two 18,500-lb (8392-kg) Pratt & Whitney TF30-P-3 afterburning turbofans and has Mk 1 avionics; such is the importance of the F-111 in US service that an extremely costly Avionics Modernization Program is being undertaken to upgrade 120 F-111A/E, 57 FB-111A, 166 F-111D/F and 38 EF-111A aircraft with more capable General Electric attack radar, Texas Instruments terrain-following radar, and a host of navigation and digital control systems

EF-111A Raven: designation of 42 electronic warfare aircraft produced by Grumman as F-111A conversions; the type has the ALQ-99E improved tactical jamming system of the Grumman EA-6B Prowler computer-assisted and repackaged for one-man operation, and developing sufficient power to overwhelm the world's most intense radar defences; the type has a canoe fairing over the bomb bay and a pod at the top of the fin; the Raven can thus support tactical aircraft as an escort, in the penetration role or as a stand-off jammer; maximum take-off weight is 89,000 lb (40370 kg), and among the extra avionics items are the ALQ-137(V)4 self-protection electronic countermeasures system, the ALR-62(V)4 terminal threat-warning system, the ALR-23 radar countermeasures receiver system, the ALE-28 chaff dispenser system and (for possible retrofit) the ALQ-131 jammer system in pods under the wings; the type may be retrofitted to carry the AGM-88 HARM radiation-homing missile

FB-111A: strategic bomber version of the basic design for the USAF's Strategic Air Command; powered by two 20,350-lb (9231-kg) TF30-P-7 afterburning turbofans, this variant has strengthened landing gear, longer-span wings (70.0 ft/21.34 m spread and 33.92 ft/10.34 m swept), a maximum take-off weight of 114,300 lb (51846 kg) and APQ-144 attack radar for the delivery of 37,500 lb (17010 kg) of weapons (six free-fall nuclear weapons, or six AGM-69A SRAMs or 42 750-lb/340-kg bombs) over considerable ranges; this version has six (occasionally eight underwing hardpoints for maximum external payload when limited sweep is used, and further upgrading includes the ALQ-137 electronic countermeasures equipment

F-111C: designation of 24 hybrid aircraft for Australia, combining the wings of the FB-111A with the fuselage, empennage, powerplant and avionics of the F-111A; four aircraft have been modified to **RF-111C** reconnaissance configuration with high/low-level cameras, TV and infra-red linescan equipment

F-111D: improved version of the F-111A with 19,600-lb (8890-kg) TF30-P-9 afterburning turbofans, Mk 2 avionics including head-up displays, APQ-130 attack radar and APQ-180 Doppler navigation

F-111E: improved F-111A with modified air inlets, inertial navigation system and attack radar

F-111F: definitive tactical variant based on the F-111D with a revised wing structure with six underwing hardpoints, strengthened landing gear and considerably uprated powerplant, but also incorporating the best avionics features of the F-111E and FB-111A

Government Aircraft Factories Mission Master
(Australia)
Type: STOL utility aircraft
Accommodation: crew of one or two on the flight-deck, plus up to 14 passengers in the cabin or a freight load of 3,350 lb (1520 kg)
Armament (fixed): none
Armament (disposable): a total of 2,000 lb (907 kg) can be carried on four optional underwing hardpoints, each rated at 500 lb (227 kg); loads can include light bombs, gun pods and rocket-launcher pods
Electronics and operational equipment: communication and navigation equipment
Powerplant and fuel system: two 400-shp (298-kW) Allison 250-B17B turboprops, and a total internal fuel capacity of 1,804 lb (818 kg) in flexible bag tanks or 1,726 lb (783 kg) in self-sealing tanks, plus an extra 582 lb (264 kg) in optional wingtip integral tanks
Performance: maximum speed 193 mph (309 km/h) at sea level; cruising speed 161 mph (259 km/h) at sea level; initial climb rate 1,460 ft (445 m) per minute; service ceiling 22,000 ft (6705 m); range 688 miles (1074 km)
Weights: empty 4,714 lb (2150 kg); maximum take-off 8,500 lb (3855 kg)
Dimensions: span 54 ft 0 in (16.46 m); length 41 ft 2.4 in (12.56 m); height 18 ft 1.5 in (5.52 m); wing area 324.0 sq ft (30.10 m²)

Variants
Nomad: this is the basic STOL commercial model that has also been bought by some air arms; the type is available as the N22 with accommodation for 12/13 passengers or an equivalent freight load, and as the N24 with the fuselage stretched by 3.75 ft (1.14 m) to carry 17 passengers or equivalent freight
Mission Master: utility military transport derived from the N22 series
Search Master B: coastal patrol model based on the Mission Master but fitted with Bendix RDR 1400 search radar to scan 60 degrees to each side of the flightpath
Search Master L: improved version of the Search Master B with Litton APS-504 radar with its larger antenna in a 'guppy' radome under the nose for 360-degree search coverage; in a typical sortie the Search Master L can search an area 325% greater than the Search Master B

Grumman A-6E/TRAM Intruder
(USA)
Type: carrierborne all-weather strike and attack aircraft
Accommodation: pilot and systems operator seated in echelon on Martin-Baker GRU7 ejector seats
Armament (fixed): none
Armament (disposable): this is carried on one underfuselage and four underwing hardpoints, each rated at 3,600 lb (1633 kg) up to a maximum weight of 18,000 lb (8165 kg); typical loads are 30 Mk 82 500-lb (227-kg) free-fall or retarded bombs, five Mk 83 1,000-lb (454-kg) bombs, five Mk 84 2,000-lb (907-kg) bombs, four AGM-45 Shrike anti-radiation miss-

iles, two AGM-84 Harpoon anti-ship missiles, and LAU-series launcher for 2.75- and 5-in (69.85- and 127-mm) rockets; additions to the armoury in the immediate future are the AGM-88 HARM anti-radiation missile, the AGM-65 Maverick air-to-surface missile and the Walleye guided bomb; the type can also carry three variable-yield B28, 1-megaton B43, 10-kiloton B57 or megaton-range B61 nuclear weapons

Electronics and operational equipment: communication and navigation equipment, plus Norden APQ-148 multi-mode radar (navigation, target identification and tracking, airborne moving target indication/AMTI, and terrain-clearance and terrain-avoidance using APQ-92 search and APQ-112 tracking components), IBM ASQ-133 digital navigation/attack computer system, Kaiser AVA-1 multi-mode electronic display, Litton ASN-92 inertial navigation system, CNI (communication, navigation and identification) system, Philco-Ford 'Pave Knife' laser-designator pod, an undernose TRAM (Target Recognition and Attack Multi-sensor) package with infra-red and laser sensors, Philco-Ford 'Pave Knife' laser designator pod, and options such as Tracor ALE-40 chaff/flare dispensers and ALQ-41 or ALQ-100 electronic countermeasures pods

Powerplant and fuel system: two 9,300-lb (4218-kg) thrust Pratt & Whitney J52-P-8B turbojets, and a total internal fuel capacity of 2,385 US gal (9028 litres) in wing and fuselage tanks, plus provision for four 300-US gal (1136-litre) drop-tanks; provision for inflight-refuelling

Performance: maximum speed 644 mph (1037 km/h) at sea level; cruising speed 474 mph (763 km/h) at optimum altitude; initial climb rate 7,620 ft (2323 m) per minute; service ceiling 42,400 ft (12925 m); range 1,011 miles (1627 km) with maximum payload, or 2,735 miles (4401 km) for ferrying with maximum internal and external fuel

Weights: empty 25,630 lb (11626 kg); maximum take-off 60,400 lb (27397 kg)

Dimensions: span 53 ft 0 in (16.15 m); length 54 ft 9 in (16.69 m); height 16 ft 2 in (4.93 m); wing area 528.9 sq ft (49.1 m²)

Variants

EA-6A Intruder: this is the strike support version of the A-6A Intruder, the initial production version of this important carrierborne medium attack aircraft whose prototype first flew in April 1960; the A-6A entered service in February 1963 with two 8,500-lb (3856-kg) Pratt & Whitney J52-P-6 turbojets; the only A-6A variant left in service is the EA-6A, which retains a partial attack capability but is used solely for the support role as a tactical jamming aircraft with 30 antennae (the ALH-6 and ALQ-86 systems) to detect, locate and classify enemy radars which can then be jammed by the onboard ALQ-31 noise, ALQ-53 track-breaking and ALQ-76 noise jammers

KA-6D Intruder: 'buddy' refuelling version produced by converting A-6A aircraft with TACAN navigation equipment and a hose-and-drogue refuelling kit in the rear fuselage; the type first flew in May 1966, and its capabilities include the transfer of 3,150 US gal (11924 litres) of fuel immediately after take-off or 2,250 US gal (8517 litres) at a radius of 288 miles (463 km); later aircraft were converted from A-6E standard and can carry five 480-US gal (1817-litre) drop tanks rather than the five 360-US gal (1363-litre) tanks of the earlier aircraft

A-6E Intruder: definitive mid-life attack model which entered service in 1972 with advanced avionics (including a multi-mode radar and an IBM computer), a Conrac stores management system and more powerful engines; surviving A-6As were brought up to this standard, numbers being swelled by new construction; further enhancement of capabilities has been ensured by the evolution of the **A-6E/TRAM** upgraded version with a stabilized Hughes TRAM (Target Recognition and Attack Multi-sensor) electro-optical chin turret with forward-looking infra-red and a laser tracker/designator; the variant first flew in March 1974 to enter service in 1979, and this is now the production standard, to which surviving aircraft are being upgraded; other features are the addition of the Carrier Airborne Inertial Navigation System for automatic carrier landings and provision for fire-and-forget and/or laser-homing weapons; 50 A-6E aircraft were additionally modified to carry six AGM-84 Harpoon anti-ship missiles, and this capability is being added to conversions and new-build aircraft

A-6F Intruder: this is the version due to enter production in 1989 for service in the 1990s; the variant is powered by two General Electric F404-GE-400 non-afterburning turbofans for greater fuel economy and range, and the avionics suite is being considerably enhanced in an all-digital type with coloured CRT cockpit displays and high-resolution synthetic-aperture radar; the weapons capability includes stand-off air-to-surface missile and beyond-visual-horizon air-to-air missile capabilities

Grumman EA-6B Prowler
(USA)

Type: carrierborne electronic countermeasures aircraft

Accommodation: crew of four (pilot, navigator and two electronics officers) seated on Martin-Baker GRUEA 7 ejector seats

Armament (fixed): two AGM-88 HARM radiation-homing missiles can be carried under the wings

Armament (disposable): none

Electronics and operational equipment: communication and navigation equipment, plus the Raytheon ALQ-99 tactical jamming system, in which surveil-

lance receivers in the fintop pod pick up enemy emissions and pass them to the central digital computer for display and recording before their identities, bearings and jamming set-on sequences are analyzed automatically or manually, and any of the five (one underfuselage and four underwing) self-powered noise jamming transmitters actuated; each pod covers one of seven frequency bands, and contains two powerful jammers
Powerplant and fuel system: two 11,200-lb (5,080-kg) thrust Pratt & Whitney J52-P-408 turbojets, and a total internal fuel capacity of 2268 US gal (8585 litres) in wing and fuselage tanks, plus provision for droptanks at the expense of jamming pods; provision for inflight-refuelling
Performance: maximum speed 623 mph (1002 km/h) at sea level; cruising speed 481 mph (774 km/h) at optimum altitude; initial climb rate 10,030 ft (3057 m) per minute; service ceiling 38,000 ft (11580 m); combat radius 332 miles (535 km) a 1-hour loiter; ferry range 2,022 miies (3254 km)
Weights: empty 32,162 lb (14588 kg); normal take-off 54,461 lb (24703 kg) in stand-off jamming configuration; maximum take-off 65,000 lb (29483 kg)
Dimensions: span 53 ft 0 in (16.15 m); length 59 ft 10 in (18.24 m); height 16 ft 3 in (4.95 m); wing area 528.9 sq ft (49.1 m²)

Variant
EA-6B Prowler: developed from the EA-6A Intruder, the EA-6B first flew is May 1968 as the prototype of a powerful stand-off electronic warfare aircraft for the US Navy; the airframe is based on that of the A-6A but stretched by 4.5 ft (1.37 m) and strengthened for operations at higher weights as a result of extra fuel capacity and the addition of the ALQ-99 radar detection, location, classification and jamming system, whose two operators are located in a side-by-side cockpit aft of the standard cockpit for the pilot and navigator; production aircraft were delivered from January 1971, and this type remains a key component of the US Navy's attack capability thanks to a succession of improvement programmes which have improved the original capability (based on the AYA-6 computer) to deal only with single emitters in four frequency bands, to the EXCAP (EXpanded CAPability) standard of 1973 for the jamming of radars in eight frequency bands, the ICAP (Improved CAPability) standard of 1977 with digitally-tuned receivers and computer-controlled systems for the jamming of several emitters forming a weapon system, the ICAP-2 standard of 1983 with the AYA-14 computer (four times the memory and three times the speed of the AYA-6) for the jamming in nine frequency bands of several weapon systems forming a defence complex, the DECM (Defensive Electronic Counter-Measures) standard and the ACAP (Advanced CAPability) standard of the late-1980s with an Amecom (Litton) receiver/processor group operating in 10 frequency bands for improved jamming of communications; APS-130 advanced navigation radar is to be installed in the later 1980s, and other improvements may include the Airborne Self-Protection Jammer, Joint Tactical Information Distribution System and Global Positioning System

Grumman E-2C Hawkeye
(USA)
Type: carrierborne and land-based early-warning aircraft
Accommodation: crew of two on the flightdeck, and a tactical team of three (combat information centre officer, air-control officer and radar operator) in the cabin
Armament (fixed): none
Armament (disposable): none
Electronics and operational equipment: communication and navigation equipment, plus General Electric APS-125 search radar (with a range against small targets of 115 miles/185 km) in a Randtron APA-171 rotodome, Litton ALR-59 passive detection system, Hazeltine APA-172 control-indicator group, Litton ASN-92 carrier aircraft inertial navigation system, Conrac air-data computer, Litton L-304 computer system, APN-153(V) Doppler navigation and other systems
Powerplant and fuel system: two 4,910-ehp (3663-ekW) Allison T56-A-425 turboprops, and a total internal fuel capacity of 1,860 US gal (7041 litres) in wing tanks; provision for inflight-refuelling
Performance: maximum speed 374 mph (602 km/h); cruising speed 365 mph (587 km/h); service ceiling 30,800 ft (9390 m); combat radius 200 miles (320 km) with a 4-hour patrol; ferry range 1,605 miles (2583 km)
Weights: empty 37,945 lb (17211 kg); maximum take-off 51,817 lb (23503 kg)
Dimensions: span 80 ft 7 in (24.56 m); length 57 ft 6.75 in (17.54 m); height 18 ft 3.75 in (5.58 m); wing area 700.0 sq ft (65.03 m²)

Variants
E-2B Hawkeye: this is the improved version of the E-2A Hawkeye initial production version, which first flew in prototype form during October 1960 and entered service in January 1964 to provide the US Navy with an exceptional airborne early warning capability through the use of General Electric APS-96 radar with its antenna in a rotating rotodome above the fuselage; the E-2B was introduced in 1969 by converting E-2As first with the Litton L-304 digital computer and then with the more capable APS-120 radar which added an overland capacity to the APS-96's basic overwater capability; the type also added provision for inflight-refuelling
E-2C Hawkeye: definitive early warning version of

the Hawkeye, which began to enter service in November 1973 after the first flight of a prototype in January 1970; this variant has General Electric/Grumman APS-125 radar able to detect aircraft at ranges of 230 miles (370 km) even in ground clutter and able to perform the simultaneous detection and tracking of more than 250 ship and aircraft targets while controlling 30 or more interceptions at the same time; the radar incorporates electronic counter-countermeasures capability, but the latest E-2Cs have the advanced APS-138 radar with low sidelobes and active-element arrays, and also feature a passive detection system (initially the Litton ALR-59 electronic support measures system and from 1980 the improved ALR-73 electronic support measures system) for the automatic detection, plotting and identification of electronic emitters in a high-density environment at ranges up to 500 miles (805 km); the provision of data-link equipment allows the secure transmission/receipt of information between E-2Cs and other aircraft or surface vessels; from 1988 the standard radar will be an improved version of the APS-138, namely the APS-139 with enhanced capability to detect slow-moving and indeed stationary targets such as warships; and from the early 1990s new-production aircraft will be delivered with the new APS-145 radar, while older aircraft will be retrofitted with this system; the APS-145 is under current development as a system able to operate effectively over normal rather than featureless terrain

TE-2C Hawkeye: training version of the E-2C

C-2A Greyhound: carrier onboard delivery version of the E-2 without the radar and rotodome and provided with a portly fuselage accessed by a ventral ramp; the type first flew in November 1964 and 19 aircraft were delivered in the 1960s, while another 32 are scheduled for delivery in the second half of the 1980s; the type has a maximum take-off weight of 54,354 lb (24654 kg) and has a payload of 39 troops, or 28 passengers, or 20 litters plus four attendants, or 15,000 lb (6804 kg) of freight

Grumman F-14A Tomcat
(USA)
Type: carrierborne variable-geometry multi-role fighter
Accommodation: pilot and systems officer seated in tandem on Martin-Baker GRU7A rocket-assisted ejector seats
Armament (fixed): one 20-mm General Electric M61A1 Vulcan rotary-barrel cannon with 675 rounds in the port side of the forward fuselage
Armament (disposable): this is carried on four underfuselage points and on two hardpoints under the inner portions of the wings, up to a maximum weight of 14,500 lb (6577 kg); the four underfuselage points can each accommodate one AIM-7 Sparrow or AIM-54 Phoenix air-to-air missile; the underwing hardpoints can each take four AIM-9 Sidewinder air-to-air missiles, or alternatively two AIM-9 Sidewinder and one AIM-7 Sparrow or AIM-54 Phoenix air-to-air missile; bombs can be carried with the aid of special racks
Electronics and operational equipment: communication and navigation equipment, plus Hughes AWG-9 weapon control system with pulse-Doppler radar able to detect targets at ranges in excess of 195 miles (315 km), track 24 simultaneously and attack six at varying altitudes simultaneously; Northrop AXX-1 TCS (Television Camera Set) for long-range identification of targets, Kaiser AVG-12 head-up display, ALR-45 radar-warning receiver, Goodyear ALE-39 chaff dispenser, and provision for ALQ-100 and ALQ-162 electronic countermeasures, and the TARPS reconnaissance pod
Powerplant and fuel system: two 20,900-lb (9480-kg) afterburning thrust Pratt & Whitney TF30-P-412A turbofans, and a total internal fuel capacity of 2,430 US gal (9206 litres) in wing and fuselage tanks, plus provision for two 267-US gal (1011-litre) external tanks under the inlet trunkings; provision for inflight-refuelling
Performance: maximum speed 1,545 mph (2486 km/h) or Mach 2.34 at high altitude; cruising speed 633 mph (1019 km/h); initial climb rate more than 30,000 ft (9145 m) per minute; service ceiling more than 50,000 ft (15240 m); range about 2,000 miles (3219 km) with maximum internal and external fuel
Weights: empty 40,104 lb (18191 kg); normal take-off 58,539 lb (26553 kg); maximum take-off 74,348 lb (33724 kg)
Dimensions: span 64 ft 1.5 in (19.54 m) spread and 38 ft 2.5 in (11.65 m) swept; length 62 ft 8 in (19.10 m); height 16 ft 0 in (4.88 m); wing area 565.0 sq ft (52.49 m^2) spread

Variants
F-14A Tomcat: first flown in December 1980, the Tomcat is without doubt the world's most powerful fighter, offering an unrivalled combination of performance (in terms principally of speed and range) with dogfighting, short-, medium- and long-range armament combined with the appropriate avionics and sensors; the type owes its career to the failure of the General Dynamics F-111B naval version of the F-111 strike fighter, a programme in which Grumman had been heavily involved; Grumman's proposal for an F-111B replacement drew heavily on this experience, and used the same primary weapon system (the AIM-54A Phoenix long-range air-to-air missile and the Hughes AWG-9 radar) in a sophisticated airframe with automatically-programmed variable-geometry wings, small moving foreplanes, a two-seat cockpit, considerable fuel and two afterburning turbofans; the type began to enter service in October 1972, and the principal service problem has been continued problems of reliability in the engines

F-14A/TARPS Tomcat: designation of the interim reconnaissance version of the F-14A necessitated by the retirement of the North American RA-5C Vigilante; this version is fitted under the fuselage with the Tactical Air Reconnaissance Pod System containing optical and infra-red sensors

F-14C Tomcat: originally known as the F-14A(Plus), this version is due to enter service in the later 1980s as an interim type between the F-14A and definitive F-14D; the type's main modification is the adoption of a new powerplant in the form of two 29,000-lb (13154-kg) General Electric F110-GE-400 afterburning turbofans, though other updated items are planned including Westinghouse APG-71 radar with programmable digital signal processing

F-14D Tomcat: considerably revised and upgraded model for service in the 1990s with two F110 afterburning turbofans and a completely revised digital rather than analog avionics suite based on the AYK-14 central computer and matched to the long-range AIM-54C and medium-range AIM-120 AMRAAM air-to-ar missiles; the type will also have a new cockpit with CRT displays and Martin-Baker NACES ejector seats, while other significant improvements will be an infra-red search and tracking system, the Joint Tactical Information Distribution System for secure voice and data transmission, the ALQ-165 Airborne Self-Protection Jammer system linked with the ALR-67 threat-warning system, and an ASN-130 laser-gyro inertial platform

Grumman OV-1D Mohawk
(USA)
Type: multi-sensor observation aircraft
Accommodation: pilot and systems operator seated side-by-side on Martin-Baker Mk J5 ejector seats
Armament (fixed): none
Armament (disposable): provision for bombs, rocket-launchers or 7.62-mm (0.3-in) Minigun pods on two underwing hardpoints
Electronics and operational equipment: communication and navigation equipment, plus AAS-4 infra-red reconnaissance or APS-94 side-looking airborne radar (SLAR) systems, two KA-60C 180-degree cameras and one KA-76 serial-frame camera, and provision for electronic countermeasures pods on the underwing hardpoints
Powerplant and fuel system: two 1,400-shp (1044-kW) Avco Lycoming T53-L-701 turboprops, and a total internal fuel capacity of 297 US gal (1125 litres) in wing and fuselage tanks, plus provision for two 150-US gal (567-litre) drop-tanks
Performance: maximum speed 289 mph (465 km/h) on a SLAR mission at 10,000 ft (3050 m), or 305 mph (409 km/h) on an IR mission at 10,000 ft (3050 m); initial climb rate 3,618 ft (1103 m) per minute with SLAR equipment; service ceiling 25,000 ft (7620 m); range 944 miles (1520 km) on a SLAR mission, and 1,011 miles (1627 km) on an IR mission
Weights: empty 12,054 lb (5468 kg) and 11,757 lb (5333 kg) for SLAR and IR missions respectively; normal take-off 15,741 lb (7140 kg) and 15,544 lb (7051 kg) for SLAR and IR missions respectively; maximum take-off 18,109 lb (8214 kg) and 17,912 lb (8125 kg) for SLAR and IR missions respectively
Dimensions: span 48 ft 0 in (14.63 m); length 41 ft 0 in (12.50 m) excluding SLAR pod or 44 ft 11 in (13.69 m) with SLAR pod; height 12 ft 8 in (3.86 m); wing area 360.0 sq ft (33.44 m²)

Variants
OV-1D Mohawk: all earlier variants of the Mohawk in service with the US Army for battlefield reconnaissance have now been upgraded to this standard with strengthened structure and more powerful engines, and offering the reconnaissance capabilities of the OV-1B (APS-94 side-looking airborne radar) and OV-1C (AAS-4 infra-red surveillance system) in a single airframe; the original OV-1A had a span of 42.0 ft (12.80 m) and first flew in April 1959, but the current type offers considerably improved capabilities, the choice between SLAR or IR primary sensors being complemented by an updated vertical panoramic camera for photographic reconnaissance, improved inertial navigation, radiological monitoring and electronic countermeasures

RV-1D Mohawk: electronic reconnaissance model of the OV-1D fitted with the ALQ-133 'Quick Look II' tactical electronic reconnaissance system

EV-1E Mohawk: designation of OV-1Bs converted for the electronic surveillance role with the ALQ-133 radar target locator system packaged in underfuselage and wingtip fairings; the type also has other advanced electronic systems

Grumman S-2E Tracker
(USA)
Type: shipboard and land-based anti-submarine aircraft
Accommodation: crew of two on the flightdeck, and a tactical crew of two in the fuselage
Armament (fixed): none
Armament (disposable): this is carried in a lower-fuselage weapons bay and on six underwing hardpoints; the weapons bay can accommodate two homing torpedoes, or four 385-lb (175-kg) depth charges; the underwing hardpoints can carry 250-lb (113-kg) bombs, or 5-in (127-mm) rockets, or 7.62-mm (0.3-in) Minigun pods or AS.12 air-to-surface missiles
Electronics and operational equipment: communication and navigation equipment, plus APS-38 search radar with its antenna in a retractable ventral 'dustbin', magnetic anomaly detection equipment with its sensor in a retractable tail 'sting', Julie active acoustic ranging equipment with 60 echo-sounding

charges, Jezebel (AQA-3) passive acoustic search equipment, and 32 sonobuoys in the rear of the engine nacelles
Powerplant and fuel system: two 1,525-hp (1138-kW) Wright R-1820-82WA radial piston engines, and a total internal fuel capacity of 728 US gal (2755 litres) in wing and fuselage tanks
Performance: maximum speed 265 mph (426 km/h) at sea level; cruising speed 150 mph (241 km/h) at 1,500 ft (455 m); initial climb rate 1,390 ft (425 m) per minute; service ceiling 21,000 ft (6400 m); range 1,300 miles (2095 km)
Weights: empty 18,750 lb (8505 kg); maximum take-off 29,150 lb (13222 kg)
Dimensions: span 72 ft 7 in (22.12 m); length 43 ft 6 in (13.26 m); height 16 ft 7 in (5.05 m); wing area 496.0 sq ft (46.08 m²)

Variants
S-2A Tracker: this was the US Navy's first dedicated carrierborne anti-submarine aircraft, and flew in XS2F-1 prototype form during December 1952; the type has a narrower fuselage than later variants, the dimensions of the aircraft including a length of 42.0 ft (12.80 m) and a span of 69.67 ft (21.25 m); maximum take-off weight is 26,300 lb (11929 kg); there is also a trainer version designated **TS-2A Tracker**, and the basic model serves in Canada with the designation **CP-141**; the S-2A is at best obsolescent, and many surviving aircraft have been converted into liaison or training aircraft
S-2E Tracker: definitive anti-submarine version produced by conversion of S-2Ds with yet more modern sensors and processing equipment; the S-2D was itself a much developed version of the S-2B (which introduced 'Julie' active acoustic ranging and 'Jezebel' passive long-range search systems) with a wider fuselage, greater span, increased fuel capacity and accommodation for 32 sonobuoys in the engine nacelles
S-2F Tracker: designation of S-2Bs converted to S-2E standard
S-2G Tracker: designation of S-2Es modified by Martin-Marietta with enhanced electronics suiting the type for service on the US Navy's carriers of the 1960s and 1970s pending deliveries of Lockheed S-3 Vikings

Gulfstream Aerospace Gulfstream III
(USA)
Type: multi-role transport aircraft
Accommodation: crew of two or three on the flight-deck, and up to 19 passengers in the cabin
Armament (fixed): none
Armament (disposable): none
Electronics and operational equipment: communication and navigation equipment, plus weather radar
Powerplant and fuel system: two 11,400-lb (5171-kg) thrust Rolls-Royce Spey RB.163 Mk 511-8 turbofans, and a total internal fuel capacity of 4,192 US gal (15868 litres) in integral wing tanks
Performance: cruising speed 581 mph (935 km/h); service ceiling 45,000 ft (13720 m); range 4,842 miles (7792 km)
Weights: empty 38,000 lb (17236 kg); maximum take-off 68,200 lb (30935 kg)
Dimensions: span 77 ft 10 in (23.72 m); length 83 ft 1 in 925.32 m); height 24 ft 4.5 in (7.43 m); wing area 934.6 sq ft (86.83 m²)

Variants
Gulfstream III: developed as a high-performance corporate transport on the basis of the preceding Gulfstream II with a longer fuselage and greater-span wing of supercritical section fitted with winglets, the Gulfstream III first flew in December 1979 and offers excellent range as well as good payload; the type has thus been bought by some air arms as a VIP transport (the USAF designation being **C-20A**), and Denmark took three modified for offshore patrol with Texas Instruments APS-127 surveillance radar and other specialized avionics under the designation **SMA-3**
Gulfstream IV: further improved model with the fuselage lengthened by 4 ft 6 in (1.37 m) to seat a maximum of 19 passengers, or by 18 ft 6 in (5.64 m) for the accommodation of 24 passengers in the **Gulfstream IV-B**; these two variants are powered by two 12,420-lb (5634-kg) thrust Rolls-Royce Tay Mk 610-8 non-afterburning turbofans
SRA-1: this is the surveillance and reconnaissance aircraft derivative of the basic type with Goodyear UPD-8 synthetic-aperture radar or Motorola SLAMMR (Side-Looking Airborne Multi-Mode Radar) in an underfuselage pod, a Rank/Optical KS-146 mirror-lens camera, optional electronic suport measures equipment and underwing hardpoints for advanced weapons such as the AGM-65 Maverick and AGM-84 Harpoon missiles and the GBU-series 'Paveway' laser-guided bomb

Hanzhong Y-8 (China): see Antonov An-12BP 'Cub'

Harbin H-5 (China): see Ilyushin Il-28 'Beagle'

Harbin Z-5 (China): see Mil Mi-4 'Hound-A'

Harbin Z-8 (China): see Aérospatiale SA 321G Super Frelon

Harbin Z-9 Zaitun (China): see Aérospatiale SA 365F Dauphin II

Helibras HB 315B Gaviao (Brazil): see Aérospatiale SA 315B Lama

Helibras HB 350 Esquilo (Brazil): see Aérospatiale AS 350B Ecureuil

Hindustan Aeronautics Ltd Ajeet
(India)
Type: lightweight interceptor and attack aircraft
Accommodation: pilot seated on a Martin-Baker GF4 lightweight ejector seat
Armament (fixed): two 30-mm Aden Mk 4 cannon with 90 rounds per gun
Armament (disposable): this is carried on four underwing hardpoints, the inner pair each capable of accepting a 500-lb (227-kg) bomb, or a BTV cluster bomb, or a BL755 cluster bomb, or a CBLS-200/IA cluster bomb, or a pod for Soviet 55-mm (2.17-in) rockets, or a Type 122 pod for 18 68-mm (2.68-mm) rockets, and the outer pair each capable of taking a Type 122 pod, or a Soviet pod, or a 30-Imp gal (136-litre) drop-tank
Electronics and operational equipment: communication and navigation equipment, plus a Ferranti F 195R/3 ISIS weapon sight
Powerplant and fuel system: one 4,500-lb (2041-kg) thrust Rolls-Royce Orpheus Mk 701-01 turbojet, and a total internal fuel capacity of 297 Imp gal (1350 litres) in nine fuselage and two integral wing tanks, plus provision for 60 Imp gal (273 litres) of external fuel
Performance: maximum speed 635 mph (1022 km/h) or Mach 0.96 at 39,000 ft (11890 m), and 685 mph (1,102 mph) or Mach 0.9 at sea level; climb to 39,000 ft (11890 m) in 6 minutes 2 seconds from brakes-off; service ceiling 45,000 ft (13715 m); combat radius 107 miles (172 km) on a lo-lo attack sortie with two 500-lb (227-kg) bombs
Weights: empty 5,085 lb (2307 kg); normal take-off 7,800 lb (3538 kg); maximum take-off 9,200 lb (4173 kg)
Dimensions: span 22 ft 1 in (6.73 m); length 29 ft 8 in (9.04 m); height 8 ft 1 in (2.46 m); wing area 136.6 sq ft (12.69 m²)

Variants
Ajeet: developed on the basis of the Hawker Siddeley (Folland) Gnat lightweight fighter, the Ajeet is a 'Gnat Mk 2' with features such as integral wing tanks (freeing the underwing hardpoints for armament) and greater longitudinal stability; the first aircraft flew in March 1975 and the type was built in small numbers
Ajeet Trainer: this is a simple advanced and operational conversion trainer derivative of the Ajeet, the fuselage being stretched by 1.4 m (4.6 ft) to accommodate the second cockpit; there is provision for the cannon to be removed, allowing internal fuel capacity to be increased by 273 litres (60 Imp gal)

Hindustan Aeronautics Ltd Cheetah (India): see Aérospatiale SA 315B Lama

Hindustan Aeronautics Ltd Chetak (India): see Aérospatiale SA 316B Alouette III

Hindustan Aeronautics Ltd HF-24 Marut Mk 1
(India)
Type: ground-attack fighter
Accommodation: pilot seated on a Martin-Baker Mk S4C ejector seat
Armament (fixed): four 30-mm Aden Mk 2 cannon in the nose with 120 rounds per gun, and a retractable Matra 103 pack of 50 68-mm (2.68-in) rockets in the lower fuselage
Armament (disposable): this is carried on four underwing hardpoints, each rated at 1,000 lb (454 kg), up to a maximum weight of 4,000 lb (1814 kg); typical loads are four 1,000-lb (454-kg) bombs, or four Matra 116 launchers each with 18 68-mm (2.68-in) rockets, or clusters of T-10 3-in (76-mm) rockets, or napalm tanks
Electronics and operational equipment: communication and navigation equipment, plus a Ferranti ISIS gyro gunsight
Powerplant and fuel system: two 4,850-lb (2200-kg) thrust HAL-built Rolls-Royce (Bristol) Orpheus Mk 703 turbojets, and a total internal fuel capacity of 549 Imp gal (2491 litres) in integral wing and fuselage tanks, plus provision for four 100-Imp gal (454-litre) drop-tanks and an 88-Imp gal (400-litre) auxiliary tank in place of the fuselage rocket pack
Performance: maximum speed 675 mph (1,086 km/h) or Mach 1.02 at 40,000 ft (12190 m); climb to 40,000 ft (12190 m) in 9.33 minutes; combat radius (HF-24 Marut Mk 1T) 898 miles (1,445 km) on a high-altitude interception sortie
Weights: empty 13,658 lb (6195 kg); normal take-off 19,734 lb (6951 kg); maximum take-off 24,085 lb (10925 kg)
Dimensions: span 29 ft 6.25 in (9.00 m); length 52 ft 0.75 in (15.87 m); height 11 ft 9.75 in (3.60 m); wing area 301.4 sq ft (28.00 m²)

Variants
Marut Mk 1: first flown in December 1961, this was India's first indigenously designed jet combat aircraft, though hopes of development to Mach 2 capability were dashed by lack of a suitable powerplant; in the event the type was produced in comparatively small numbers as a useful ground-attack aircraft for service from 1968
Marut Mk 1T: combat-capable operational conversion trainer derivative of the Marut Mk 1, the single-seater's retractable rocket pack being deleted to make room for the second cockpit; the type first flew in April 1970

Hindustan Aeronautics HJT-16 Kiran Mk 1A
(India)
Type: basic trainer, weapons trainer and light attack aircraft
Accommodation: pupil and instructor seated side-by-side on Martin-Baker H4HA ejector seats

Armament (fixed): none
Armament (disposable): this is carried on two underwing hardpoints, each capable of accepting one 500-lb (227-kg) bomb, or one HAL pod with two 7.62-mm (0.3-inn) FN machine-guns, or one pod with seven 68-mm (2.68-in) rockets, or one 50-Imp gal (227-litre) drop-tank
Electronics and operational equipment: communication and navigation equipment, and two Mk IIIB reflector gunsights
Powerplant and fuel system: one 2,500-lb (1134-kg) thrust Rolls-Royce Viper Mk 11 turbojet, and a maximum internal fuel capacity of 250 Imp gal (1137 litres) in fuselage, wing and centre section tanks
Performance: maximum speed 432 mph (695 km/h) at sea level, and 427 mph (688 km/h) at 30,000 ft (9145 m); cruising speed 201 mph (304 km/h); climb to 30,000 ft (9145 m) in 20 minutes; service ceiling 30,000 ft (9145 m); combat radius 506 miles (815 km)
Weights: empty 5,645 lb (2561 kg); normal take-off 8,050 lb (3651 kg); maximum take-off 9,335 lb (4234 kg)
Dimensions: span 35 ft 1.25 in (10.70 m); length 34 ft 9 in (10.60 m); height 11 ft 11 in (3.635 m); wing area 204.5 sq ft (19.00 m²)

Variants
Kiran Mk 1: this unarmed basic trainer first flew in September 1964 and began to enter service in 1968
Kiran Mk 1A: designation of late-production Kiran Mk 1s with provision for armament on two underwing hardpoints for the weapons training and light attack roles
Kiran Mk 2: developed version available from 1983 (after a first flight during July 1976) in the weapons training, light attack and counter-insurgency roles; this variant is powered by a 3,400-lb (1542-kg) Rolls-Royce Orpheus Mk 701-01 non-afterburning turbojet for greater performance and agility, and has four underwing hardpoints for 2,000 lb (907 kg) of weapons or external fuel; the type also has two inbuilt 7.62-mm (0.3-in) machine-guns with 250 rounds per gun

Hughes (McDonnell Douglas Helicopters) AH-64A Apache
(USA)
Type: battlefield and anti-tank helicopter
Accommodation: co-pilot/gunner and pilot seated in tandem
Armament (fixed): one 30-mm Hughes M230E1 Chain Gun cannon with 1,200 rounds, mounted in a flexible turret under the fuselage
Armament (disposable): this is carried on four hardpoints under the stub wings, and can comprise 16 Hellfire anti-tank missiles in four four-round clusters, or four launchers each with 19 2.75-in (69.85-mm) rockets, or combination of the two types up to a maximum weight of 3,880 lb (1760 kg)
Electronics and operational equipment: communication and navigation equipment, plus Singer-Kearfott Doppler navigation, Hughes Black Hole infra-red suppression system, Martin Marietta Target Acquisition and Designation System (TADS), Pilot's Night-Vision System (PNVS), Honeywell Integrated Helmet And Display Sighting System (IHADSS), Teledyne fire-control computer, CPG stabilized sight with forward-looking infra-red (FLIR) sensor, laser ranger and laser designator
Powerplant and fuel system: two 1,536-shp (1145-kW) General Electric T700-GE-700 turboshafts, and a total internal fuel capacity of 375 US gal (1419 litres) in two crash-resistant fuel cells in the fuselage, plus provision for external tanks
Performance: maximum speed 192 mph (309 km/h); cruising speed 182 mph (293 km/h); initial climb rate 2,880 ft (878 m) per minute; service ceiling 20,500 ft (6,250 m); range 380 miles (611 km) on internal fuel
Weights: empty 10,268 lb (4657 kg); normal take-off 13,825 lb (6271 kg); maximum take-off 17,650 lb (8006 kg)
Dimensions: main rotor diameter 48 ft 0 in (14.63 m); length 49 ft 1.5 in (14.97 m); height (rotor head) 13 ft 10 in (4.22 m); main rotor disc area 1,809.5 sq ft (168.11 m²)

Variants
AH-64A Apache: first flown in September 1975, the Apache has become the US Army's most important anti-tank and battlefield helicopter; for considerable size, complexity and cost, the Apache is a highly capable machine offering advanced sensors and considerable performance for the carriage and accurate delivery of a heavy ordnance load; particular features of the design are high survivability, good crew and primary systems protection, and avionics such as TADS (for the optical, thermal or TV acquisition of targets that can then be laser-ranged and laser-designated) and PNVS (for nap-of-the-earth flight profiles under all weather conditions by day and night); the manufacturer is also proposing that the type be retrofitted (under a three-stage programme) with the Global Positioning System, Doppler navigation and improved versions of the TADS and PNVS; current improvements are centred on the T700-GE-701 turboshaft for better performance under 'hot-and-high' conditions, and (from the 100th aircraft) composite-construction main rotor blades
AH-64B Apache: improved version for delivery in the 1990s with voice controls and a 'fly-by-light' fibre-optical control system using a side-arm control stick; the type will also feature improved armament, including a pair of AIM-9 Sidewinder missiles for self defence over the battlefield; the company is also proposing versions for the US Marine corps and US Navy, the first basically similar to the US Army AH-

64B and the latter being optimized for the defence of surface battle groups with target-acquisition radar plus Penguin or Harpoon anti-ship missiles

Hughes (McDonnell Douglas Helicopters) Model 300C
(USA)
Type: utility light helicopter
Accommodation: pilot and up to two passengers on a bench seat
Armament (fixed): none
Armament (disposable): none
Electronics and operational equipment: communication and navigation equipment
Powerplant and fuel system: one 190-hp (142-kW) Avco Lycoming HIO-360-D1A flat-four piston engine, and a total internal fuel capacity of 27.3 US gal (103.5 litres) in a fuselage tank, plus provision for a 19-US gal (72-litre) auxiliary fuselage tank
Performance: cruising speed 94 mph (151 km/h) at sea level; initial climb rate 750 ft (229 m) per minute; service ceiling 10,200 ft (3110 m); range 230 miles (370 km)
Weights: empty 1,050 lb (476 kg); maximum take-off 2,050 lb (930 kg)
Dimensions: main rotor diameter 26 ft 10 in (8.18 m); length overall with rotors turning 30 ft 10 in (9.40 m); height (rotor head) 8 ft 9 in (2.67 m); main rotor disc area 565.5 sq ft (52.5 m²)

Variants
Model 200 Utility: a light utility helicopter intended for communications, training and observation, the two-seat Model 200 series first flew in October 1956 as the **Model 269** initial version and secured considerable sales success in the civil market before being revised as the Model 200 Utility refined variant and as the **Model 200 Deluxe** up-market variant; the type was procured as a trainer by the US Army under the designation **TH-55A Osage** with the 180-hp (134-kW) Avco Lycoming HIO-360 B1A engine, and a similar version was produced under licence in Japan as the **Kawasaki TH-55J** as a flying trainer
Model 300: three-seat production version of the Model 269B development variant, and available from 1967 with a quieter tail rotor
Model 300C: improved model with 45% better payload and available from the 1969; the type was also produced under licence in Italy as the **Breda-Nardi NH-300C**
Model 300QC: version of the Model 300C with noise emission reduced by 75%; this variant has a maximum take-off weight of 1,925 lb (873 kg)

Hughes (McDonnell Douglas Helicopters) OH-6A Cayuse
(USA)
Type: light observation and utility helicopter
Accommodation: pilot and co-pilot, and provision for converting the freight volume for the carriage of four troops
Armament (fixed): none
Armament (disposable): provision for an armament package (on the port side of the fuselage) with one M134 7.62-mm (0.3-in) Minigan or one M75 40-mm grenade-launcher
Electronics and operational equipment: communication and navigation equipment
Powerplant and fuel system: one 317-shp (236-kW) Allison T63-A-5A turboshaft de-rated to 215 shp (160 kW) for continuous running, and a total internal fuel capacity of 63 US gal (240 litres) in two fuselage bladder tanks, plus provision for various internal and external auxiliary tanks
Performance: cruising speed 150 mph (241 km/h) at sea level; initial climb rate 1,840 ft (561 m) per minute; service ceiling 15,800 ft (4815 m); range 413 miles (665 km) with standard fuel, and 1,560 miles (2511 km) with auxiliary fuel
Weights: empty 1,156 lb (524 kg); normal take-off 2,400 lb (1089 kg); maximum take-off 2,700 lb (1225 kg)
Dimensions: main rotor diameter 26 ft 4 in (8.03 m); length (fuselage) 23 ft 0 in (7.01 m); height 8 ft 1.5 in (2.48 m); main rotor disc area 544.63 sq ft (50.6 m²)

Variants
OH-6A Cayuse: first flown in February 1963 as the Model 369 and flown in the US Army's Light Observation Helicopter competition as the OH-6, this type was ordered into production against requirements for some 4,000 aircraft and began to enter service in September 1966; service reports were highly enthusiastic, but production was cancelled after the delivery of 1,434 aircraft because of a declining production rate and increasing procurement costs; the reopened LOH competition was then won by the Bell Model 205, which entered service as the OH-58A Kiowa
OH-6C Cayuse: US Army experimental model based on 'The Quiet One' development version with the 400-shp (298-kW) Allison 250-C20, a five-blade main rotor and an acoustic blanket round the engine
OH-6D Cayuse: proposed development to meet the US Army's Advanced Scout Helicopter programme
Model 500: commercial derivative with the 278-shp (207-kW) Allison 250–C18A turboshaft, and in the export version, limited military applications such as medevac
Model 500C: 'hot and high' version of the Model 500 with the 400-shp (298-kW) Allison 250-C20 turboshaft and a main rotor 26 ft 4 in (8.03 m) in diameter; the type is built under licence by Kawasaki in Japan as the **Kawasaki (Hughes) Model 500C**, and by RACA in Argentina as the **RACA (Hughes) Model 500C**
Model 500D: export version of the OH6C test heli-

copter with a number of quieting features such as a slow-turning five-blade main rotor and four-blade tail rotor driven by a 420-shp (313-kW) Allison 250-C20B turboshaft; the main rotor has a diameter of 26 ft 5 in (8.05 m), and the rotor hub has a low-drag 'coolie hat' fairing; the type is built under licence in Japan as the **Kawasaki (Hughes) Model 500D** for use by the Japan Ground Self-Defense Force as the **OH-6D**
Model 500E: developed version of the Model 500D with a more pointed nose for better streamlining and increased cabin volume
Model 500M: dedicated military export variant based on the OH-6A but fitted with the Allison 250-C18A turboshaft derated to 275 shp (207 kW); examples in Spanish service are used in the light anti-submarine role with Texas Instruments ASQ-81 magentic anomaly detector and an armament of two Mk 44 or Mk 46 torpedoes; the type is also built under licence in Argentina as the **RACA (Hughes) 500M**, in Italy as the **BredaNardi NH-500M** (and **NH-500MC** with 'hot-and-high' features) and in Japan as the **Kawasaki (Hughes) 500M**, which serves with the Japanese forces as the **OH-6J**
Model 500MD Defender: uprated military variant based on the Model 500D (420-shp/313-kW Allison 250-C20B turboshaft, five-blade main rotor, T-tail and other improvements); the type can be used for scouting, liaison, logistic support, training, casevac, anti-submarine warfare and battlefield support; the type has been built under licence in Italy as the **BredaNardi NH-500MD** and in South Korea as the **KAL (Hughes) 500MD**; the type has been developed in a number of subvariants such as the **Model 500MD Scout Defender** (battlefield reconnaissance and light attack with Black Hole Ocarina infra-red suppression, a nose-mounted stabilized sight and a wide assortment of light armament provision including rockets and a gun up to 30-mm calibre), the **Model 500MD Quiet Advanced Scout Defender** (similar to the Scout Defender but with noise reduction features and a mast-mounted sight for 'nap-of-the-earth' flight profiles), the **Model 500MD/TOW Defender** (a dedicated anti-tank version with four BGM-71 TOW missiles and appropriate sight), the **Model 500MD/MMS-TOW Defender** (as the Model 500MD/TOW but with a mast-mounted sight on a pylon 2 ft/0.61 m above the rotor head), the **Model 500MD/ASW Defender** (a dedicated anti-submarine variant with search radar, ASQ-81 magnetic anomaly detector and two Mk 44 or Mk 46 torpedoes), and the **Model 500MD Defender II** (an upgraded model fitted with two Stinger missiles for self-defence, and fitted with items such as forward-looking infra-red and the APR-39 radar-warning receiver
Model 500MG Defender: improved multi-role military export version with the revised nose profile, advanced avionics and the 420-shp (313-kW) Allison 250-C20B turboshaft

Model 530F Lifter: 'hot-and-high' version of the Model 500D with the fuselage of the model 500E and a 650-shp (485-kW) Allison 250-C30 turboshaft derated to 430 shp (317 kW)
Model 530MG Defender: military version of the Model 530F with an advanced cockpit, mast-mounted sight and a removable beam with provision for a wide assortment of weapons; this version has a main rotor diameter of 27.5 ft (8.4 m), a maximum take-off weight of 3,550 lb (1610 kg), a maximum speed of 150 mph (241 km/h) and a range of 275 miles (443 km); the accommodation is a pilot and up to five passengers or an equivalent weight of cargo
Nightfox: low-cost nocturnal surveillance helicopter available in Model 500MG and Model 530MG versions and fitted with FLIR
Paramilitary MG Defender: low-cost patrol and rescue helicopter available in Model 500MG and Model 530MG versions

ICA-Brasov IAR-316B Alouette III (Romania): see Aérospatiale SA 316B Alouette III

ICA-Brasov IAR-317 Airfox (Romania): see Aérospatiale SA 316B Alouette III

ICA-Brasov IAR-330 Puma (Romania): see Aérospatiale SA 300L Puma

Ilyushin Il-14M 'Crate'
(USSR)
Type: short-range transport aircraft
Accommodation: crew of two on the flightdeck, and up to 36 passengers or freight in the cabin
Armament (fixed): none
Armament (disposable): none
Electronics and operational equipment: communication and navigation equipment
Powerplant and fuel system: two 1417-kW (1,900-hp) Shvetsov ASh-82T radial piston engines, and a total internal fuel capacity of 1,330 Imp gal (6500 litres) in eight wing, centre-section and fuselage tanks
Performance: cruising speed 350 km/h (217 mph) at 3000 m (9,845 ft); service ceiling 7400 m (24,280 ft); range 400 km (249 miles) with maximum payload, or 1750 km (1,087 miles) with maximum fuel
Weights: empty 12600 kg (27,778 lb); maximum take-off 17500 kg (38,581 lb)
Dimensions: span 31.70 m (104 ft 0 in); length 22.31 m (73 ft 2.25 in); height 7.80 m (25 ft 7 in); wing area 100.0 m² (1,076.43 sq ft)

Variants
Il-14 'Crate': designed as an improved Ilyushin Il-12 to succeed the Lisunov Li-2 (the Soviet version of the Douglas DC-3/C-47 series) as the nation's standard feederliner and tactical transport, the Il-14 flew in prototype form during 1952 and began to enter service in 1954; several variants were built for civil and military use, typical being the **Il-14P** for 18/26

passengers, the **Il-14M** lengthened by 1.0 m (3.3 ft) for 24/28 passengers, the **Il-14T** freight version of the Il-14P and the **Il-14G** freight version of the I1-14M
Il-14 'Crate-C': electronic warfare version first seen in 1979 and fitted with a number of unspecified equipment items for service with the Polish air force

Ilyushin Il-18D 'Coot'
(USSR)
Type: medium transport
Accommodation: crew of five on the flightdeck, and up to 120 passengers or 13500 kg (29,762 lb) of freight in the cabin
Armament (fixed): none
Armament (disposable): none
Electronics and operational equipment: communication and navigation equipment, plus weather/navigation radar
Powerplant and fuel system: four 3169-ekW (4,250-ehp) Ivchenko AI-20M turboprops, and a total internal fuel capacity of up to 30000 litres (6,600 Imp gal) in one centre-section, two inner-panel integral and 20 outer-panel bag tanks
Performance: maximum speed 675 km/h (419 mph) at 8000 m (26,245 ft); cruising speed 625 km/h (388 mph) at 8000 m (26,245 ft); service ceiling 10,000 ft (32,810 ft); range 3700 km (2,300 miles) with maximum 13,500-lb (29,762-lb) payload, and 6,500 km (4040 miles) with maximum fuel
Weights: empty 35,000 lb (77,160 lb); maximum take-off 64000 kg (141,093 lb)
Dimensions: span 37.40 m (122 ft 8.5 in); length 35.90 m (117 ft 9 in); height 10.17 m (33 ft 4 in); wing area 140.0 m² (1,507.0 sq ft)

Variants
Il-18 'Coot': first flown in July 1957, the Il-18 was designed as a medium-capacity turboprop airliner for the USSR's internal and European routes, and was produced in a number of variants with revised seating arrangements and powerplants; the basic Il-18 accommodated 75 passengers and had four 2983-ekW (4,000-ehp) Kuznetsov NK-4 turboprops, the **Il-18B** was similar but accommodated 84 passengers, the **Il-18V** appeared in 1961 with accommodation for between 90 and 100 passengers, the **Il-18D** had AI-20M turboprops and seating for 122 passengers in a lengthened fuselage, the **Il-18E** was similar to the Il-18D with reduced fuel capacity, and the **Il-18T** was a pure freighter model; many example of this series passed into military service as transports, freighters and VIP aircraft
Il-20 'Coot-A': taken into Soviet service during the mid-1970s, this is a dedicated electronic intelligence derivative of the Il-18; the airframe is substantially unaltered, but under the forward fuselage is a canoe fairing some 10.25 m (33.63 ft) long and 1.15 m (3.77 ft) side, probably housing a side-looking airborne radar; other modifications are a pair of box fairings on the sides of the fuselage just aft of the flightdeck, and 10 blade antennae (eight under and two over the fuselage)

Ilyushin Il-28 'Beagle'
(USSR)
Type: light bomber
Accommodation: pilot, bombardier/navigator and rear gunner in separate positions
Armament (fixed): two 23-mm NR-23 cannon with 100 rounds per gun in the lower nose, and two 23-mm NR-23 cannon with 225 rounds per gun on a flexible mounting in the rear turret
Armament (disposable): this is carried in a weapons bay in the fuselage under the wing, and the maximum load of 3000 kg (6,614 lb) can be made up of free-fall bombs, the usual combinations being four 500-kg (1,102-lb) or eight 250-kg (551-lb) bombs
Electronics and operational equipment: communication and navigation equipment, plus all-weather bombing radar with its antenna in a ventral radome, and tail-warning radar
Powerplant and fuel system: two 2700-kg (5,952-lb) thrust Klimov VK-1A turbojets, and a total internal fuel capacity of 7908 litres (1,740 Imp gal) in wing tanks, plus provision for two 333-litre (73.25-Imp gal) wingtip auxiliary tanks
Performance: maximum speed 900 km/h (559 mph) at 4500 m (14,765 ft), and 800 km/h (497 mph) at sea level; cruising speed 770 km/h (478 mph) at 10000 m (32,810 ft); climb to 10000 m (32,810 ft) in 18 minutes; service ceiling 12300 m (40,355 ft); range 2180 km (1,350 miles) with a 1000-kg (2,205-lb) bombload, or 2400 km (1,490 miles) with maximum fuel
Weights: empty 13000 kg (26,455 lb); normal take-off 18400 kg (40,565 lb); maximum take-off 21200 kg (46,737 lb)
Dimensions: span 21.45 m (70 ft 4.5 in); length 17.65 m (57 ft 11 in); height 6.70 m (22 ft 0 in); wing area 60.8 m² (654.4 sq ft)

Variants
Il-28 'Beagle': though obsolete (the prototype having flown in 1948) in its basic light bomber role, the Il-28 remains in fairly widespread service with Soviet allies
Il-28R 'Beagle': three-seat tactical reconnaissance variant of the Il-28 series, the erstwhile bomb bay being used to accommodate three or five cameras plus some 12 to 18 photo-flash bombs; some aircraft have an electronic reconnaissance package with a radome under the rear fuselage
Il-28T 'Beagle': designation of the torpedo bomber version, the bomb bay being able to accommodate one large or two small torpedoes, or (in an alternative role) a few mines
Il-28U 'Mascot': two-seat operational conversion

trainer with a solid nose and two stepped cockpits
Harbin H-5: designation of the Chinese-built copy of the Il-28, versions of the basic bomber being the **HZ-5** tactical reconnaissance aircraft and the **HJ-5** operational conversion trainer

Ilyushin Il-38 'May-A'
(USSR)
Type: maritime reconnaissance and anti-submarine aircraft
Accommodation: (estimated) crew of three or four on the flightdeck, and a mission crew of eight or nine in the cabin
Armament (fixed): none
Armament (disposable): weapons such as depth charges, homing torpedoes and perhaps missiles are carried in a large bay in the lower fuselage; details of weapon types and total weapon load are not available
Electronics and operational equipment: communications and navigation equipment, plus search radar in an undernose radome, magnetic anomaly detection (MAD) gear in a tail 'sting', sonobuoys and onboard computing and analysis equipment
Powerplant and fuel system: four 3169-ekW (4,250-ehp) Ivchenko AI-20M turboprops, and a total internal fuel capacity of 30000 litres (6,600 Imp gal) in one centre-section, two inner-panel integral and 20 outer-panel bag tanks
Performance: maximum speed 645 km/h (401 mph); cruising speed 645 km/h (401 mph) at 8,230 m (27,000 ft); range 7240 km (4,500 miles)
Weights: empty about 38000 kg (83,775 lb); maximum take-off about 61000 kg (134,480 lb)
Dimensions: span 37.40 m (122 ft 8.5 in); length 39.60 m (129 ft 10 in); height 10.17 m (33 ft 4 in); wing area 140.0 m² (1,507.0 sq ft)

Variants
Il-38 'May-A': this is a fairly radical development of the Il-18 as a dedicated anti-submarine aircraft; the type was first flown in 1966 or 1967 and features a longer fuselage with the wing set relatively farther forward than on the Il-18, presumably to maintain the centre of gravity position with the installation of heavy mission equipment in the forward fuselage; the design philosophy is exactly the same as that used by the Americans to develop the Lockheed P-3 Orion from the Electra airliner, but there is little doubt that the Soviet aircraft is inferior in terms of electronic capability in the maritime reconnaissance and anti-submarine roles
Il-38 'May-B': designation of a variant identified in 1984, and notable for a second large radome (under the weapon bay); it is likely that this version is intended for mid-course missile guidance, and thus lacks anti-submarine capability

Ilyushin Il-76M 'Candid'
(USSR)
Type: heavy transport
Accommodation: crew of seven including two freight-handlers for a payload of up to 40000 kg (88,183 lb); alternatively, up to 140 troops can be accommodated
Armament (fixed): two 23-mm NR-23 cannon on a flexible mounting in the rear turret
Armament (disposable): none
Electronics and operational equipment: communication and navigation equipment, plus weather/navigation radar, computer-controlled automatic flight-control system, and sophisticated freight-handling system
Powerplant and fuel system: four 12000-kg (26,455-lb) thrust Soloviev D-30KP turbofans, and a total internal fuel capacity of some 81830 litres (18,000 Imp gal) in inner- and outer-panel integral wing tanks
Performance: maximum speed 850 km/h (528 mph); cruising speed 800 km/h (497 mph); service ceiling 15500 m (50,855 ft); range 5000 km (3,107 miles) with maximum payload, and 6700 km (4,163 miles) with maximum fuel
Weights: empty about 75000 kg (165,334 lb); maximum take-off 170000 kg (374,780 lb)
Dimensions: span 50.50 m (165 ft 8 in); length 46.59 m (152 ft 10.5 in); height 14.76 m (48 ft 5 in); wing area 300.0 m² (3,229.2 sq ft)

Variants
Il-76 'Candid': first flown in March 1971, this is the Soviet replacement for the Antonov An-12 series in both the civil freighting and military transport roles; the upswept rear fuselage has a ramp and rear doors for ease of access to the unobstructed hold, which measures 20.0 m (65.6 ft) in length (24.5 m/80.4 ft including the ramp), 3.46 m (11.35 ft) in width and 3.4 m (11.15 ft) in height and can handle freight with the aid of roller panels in the floor and two overhead winches; the **Il-76T** version has additional fuel (in a centre-section tank) and higher operating weights; the **Il-76TD** has improved D-30KP-1 turbofans plus further increases in fuel capacity and operating weights; specifically military versions are the **Il-76M** and **Il-76MD** equivalent to the Il-76T and Il-76TD respectively but with a tail turret (often without its two 23-mm NR-23 cannon) and other military equipment; the type is extremely capable, its design, landing gear, high-lift devices and powerful engines bestowing good field performance even under adverse conditions
Il-76 'Mainstay': entering service in the mid-1980s, this is the airborne early warning version of the Il-76 freighter, with a lengthened forward fuselage and tactical compartment for a mission crew deriving data from the large surveillance radar with its antenna in a rotodome above the fuselage; full production is allowing the phasing-out (or relegation to secondary

areas) of obsolescent Tupolev Tu-126 'Moss' AEW aircraft

Il-76 'Midas': this is the inflight-refuelling tanker version of the series, a three-point hose-and-drogue type with considerable extra fuel filling tanks located in the erstwhile cargo hold and providing Soviet tactical and long-range aircraft with significantly improved refuelling capabilities

Israel Aircraft Industries AMIT and Improved Fougas (Israel): see Aérospatiale CM.170-1 Magister

Israel Aircraft Industries 201 Arava
(Israel)
Type: STOL tactical transport
Accommodation: crew of one or two on the flightdeck, and up to 24 troops, or 16 paratroops and two despatchers, or 2350 kg (5,181 lb) of freight in the cabin
Armament (fixed): provision for one 0.5-in (12.7-mm) machine-gun in a special pack on each side of the forward fuselage, and for a flexible rear-firing machine-gun
Armament (disposable): two pods each containing six 81-mm (3.19-in) rockets, each mounted on a short pylon projecting from the side of the fuselage below the fixed machine-gun installation
Electronics and operational equipment: communication and navigation equipment, plus a number of optional items suiting the type for maritime reconnaissance, or for electronic warfare with the aid of pallet-mounted Elint and ESM packages
Powerplant and fuel system: two 750-shp (559-kW) Pratt & Whitney Canada PT6A-34 turboprops, and a total internal fuel capacity of 1663 litres (366 Imp gal) in four integral wing tanks, plus provision for a pair of 1022-litre (225-Imp gal) auxiliary tanks in the hold
Performance: maximum speed 326 km/h (203 mph) at 3050 m (10,000 ft); cruising speed 319 km/h (198 mph) at 3050 m (10,000 ft); initial climb rate 393 m (1,290 ft) per minute; service ceiling 7620 m (25000 m); range 280 km (174 miles) with maximum payload, or 1056 km (656 miles) with maximum fuel
Weights: empty 4000 kg (8,818 lb); maximum take-off 6805 kg (15,002 lb)
Dimensions: span 20.96 m (68 ft 9 in); length 13.03 m (42 ft 9 in); height 5.21 m (17 ft 1 in); wing area 43.68 m² (470.2 sq ft)

Variants
IAI 201 Arava: this is the military transport version of the IAI 101B civil commuter (18 passengers or some 1815 kg/4,000 lb of freight) and IAI 102 civil utility (20 passengers, freight, VIP transport etc) aircraft; the type is slow and short-ranged, but rugged and possessed of good short-field performance thanks to its STOL design; the IAI 201 can carry 24 troops, or 16 paratroops or freight, but is generally used for maritime patrol or electronic warfare, in the latter role fitted with a palletized suite comprising Elta EL/K-1250 and L-8310 communications intelligence receivers, one Elta L-8312 computer and Elta L-8200 series spot and barrage noise jammers
IAI 202: version with the fuselage pod lengthened from 9.33 to 10.23 m (30.61 to 33.56 ft), tip winglets (which can also be retrofitted to the IAI 201) and extra fuel in a 'wet' wing for the carriage of 30 troops, or 20 paratroops, or 12 litters and five attendants, or 2500 kg (5,511 lb) of freight over a range of 630 km (392 miles) on the power of two 750-shp (559-kW) PT6A-36 turboprops

Israel Aircraft Industries 1124N Seascan
(Israel)
Type: coastal patrol aircraft
Accommodation: crew of two on the flightdeck, and mission crew in the cabin
Armament (fixed): none
Armament (disposable): various stores can be carried on two pylons attached to the sides of the fuselage up to a maximum weight of 2180 kg (4,806 lb)
Electronics and operational equipment: communication and navigation equipment, plus Litton APS-504(V)2 360° search radar, Global GNS-500A navigation system and specialized mission gear
Powerplant and fuel system: two 3,700-lb (1678-kg) thrust Garrett TFE731-3-1G turbofans, plus a total internal fuel capacity of 5390 litres (1,186 Imp gal) in integral wing tanks and one rear-fuselage tank
Performance: maximum speed 870 km/h (541 mph) up to an altitude of 5900 m (19,355 ft); cruising speed 740 km/h (460) at 12500 m (41,010 ft); initial climb rate 1525 m (5,000 ft) per minute; service ceiling 13725 m (45,030 ft); range 4445 km (2,762 miles) with five passengers and long-range tank
Weights: empty 5760 kg (12,698 lb); maximum take-off 10660 kg (23,500 lb)
Dimensions: span 13.65 m (44 ft 9.5 in) over the tiptanks; length 15.93 m (52 ft 3 in); height 4.81 m (15 ft 9.5 in); wing area 28.64 m² (308.26 sq ft)

Variant
IAI 112N Seascan: this is the coastal patrol version of the IAI 1124 Westwind executive transport; the type has a low-altitude range of 2555 km (1588 miles), allowing the Seascan to cover a 268056-km² (103,496-sq mile) search area in the course of a 6.5-hour patrol at an altitude of 915 m (3,000 ft); at an altitude of 13720 m (45,000 ft) the patrol endurance is 8 hours and range 4633 km (2,878 miles); IAI is currently planning an armed version for the anti-ship and anti-submarine roles with the appropriate sensors and weapons such as the Gabriel Mk III missile and acoustic-homing torpedoes

Israel Aircraft Industries Dagger (Israel): see Israel Aircraft Industries Kfir-C2

Israel Aircraft Industries Kfir-C2
(Israel)
Type: interceptor and ground-attack aircraft
Accommodation: pilot seated on a Martin-Baker IL10P ejector seat
Armament (fixed): two 30-mm DEFA 552 cannon with 140 rounds per gun
Armament (disposable): this is carried on five under-fuselage and four underwing hardpoints up to a maximum weight of 5775 kg (12,731 lb); for the interception role two Rafael Shafrir 2 or Python 3 air-to-air missiles are carried under the outer wings; for ground-attack duties the aircraft can carry a wide variety of stores, these including the Luz-1 air-to-surface missile, AGM-45 Shrike anti-radiation missile, AGM-65 Maverick air-to-surface missile, HOBOS guided bomb, IMI rocket-launcher pods, napalm bombs, up to four 1,000-lb (454-kg) bombs, up to 10 500-lb (227-kg) free-fall or runway-cratering 'dibber' bombs, and other stores of an increasingly sophisticated nature
Electronics and operational equipment: communication and navigation equipment, plus MBT ASW-41 control-augmentation system, MBT ASW-42 stability-augmentation system, Elbit S-8600 (licence-built Singer-Kearfott) multi-mode navigation system and Rafael Mahat weapon-delivery system (or IAI WDNS-141 weapon-delivery and navigation system), Tamam central air-data computer, Elta EL/M-2001B ranging radar, Israel Electro-Optics head-up display and automatic gunsight, and ECM equipment (internal and podded)
Powerplant and fuel system: one 17,900-lb (8119-kg) afterburning thrust General Electric J79-J1E turbojet, and a total internal fuel capacity of 3243 litres (713 Imp gal) in five fuselage and four wing integral tanks, plus provision for up to five drop tanks of 500-, 600-, 1,300- or 1,700-litre (110-, 132-, 286- or 374-Imp gal) capacity
Performance: maximum speed over 2,440 km/h (1,516 mph) or Mach 2.3 at 11000 m (36,090 ft), and 1390 km/h (864 mph) or Mach 1.1 at sea level; initial climb rate 14000 m (45,930 ft) per minute; service ceiling 17680 m (58,000 ft) in stable flight in air-combat configuration, with zone-climb capability to 22860 m (75,000 ft); combat radius 345 km (214 miles) for an intercept mission, or 768 km (477 miles) on a hi-lo-hi ground-attack mission with bombs, fuel and AAMs
Weights: empty 7285 kg (16,060 lb); normal take-off 9390 kg (20,701 lb) for interception or 14670 kg (32,341 lb) for ground-attack; maximum take-off 16200 kg (35,714 lb)
Dimensions: span 8.22 m (26 ft 11.5 in) for wing, and 3.73 m (12 ft 3 in) for canard; length 15.65 m (51 ft 4.5 in) including probe; height 4.55 m (14 ft 11.25 in); wing area 34.80 m² (374.6 sq ft) for wing, and 1.66 m² (17.87 sq ft) for canard foreplanes

Variants
Kfir-C1: first flown in 1973, this initial model of the Kfir (lion cub) was in essence the airframe of the Dassault-Mirage III/5 series mated to the General Electric J79 afterburning turbojet and fitted with Israeli electronics; the type was designed after the manufacturer had gained experience with the **IAI Nesher** (eagle), itself merely an unlicensed copy of the Dassault Mirage IIICJ, produced mainly for Israeli service but later exported as the **Dagger**; the Kfir-C1 entered only limited production, a mere two squadrons being equipped with the type from 1974 pending the introduction of more advanced derivatives with much improved combat capability
Kfir-C2: introduced in 1976, this is a much developed version of the Kfir-C2 designed to keep the type viable against all conceivable threats well into the 1990s; the type is distinguishable from the Kfir-C1 by its dogtoothed outer wing panels, small undernose strakes and, most importantly of all, swept delta canard foreplanes; the result is a warplane with formidable combat capabilities plus short-field performance thanks to the sustained manoeuvrability and control effectiveness resulting from the aerodynamic developments
Kfir-TC2: designation of the combat-capable two-seat conversion trainer variant of the Kfir-C2; the type was first flown in February 1981, and the main distinguishing feature of the type is the visibility-improving droop of its nose section, which is lengthened by 0.84 m (2.75 ft) to accommodate the second cockpit
Kfir-C7: definitive single-seat version introduced in 1983; this is based on the Kfir-C2, but uses a specially-adapted version of the J79-GE-J1E with some 1,000 lb (454 kg) additional afterburning thrust in combat situations; the type has a number of advanced features, including capability for the carriage and use of 'smart' weapons, a revised cockpit with more sophisticated electronics and HOTAS (Hands On Throttle And Stick) controls, and provision for inflight-refuelling; maximum take-off weight is increased by 1540 kg (3,395 lb), but combat radius and (more importantly) thrust-to-weight ratio are improved to a marked degree
Kfir-TC7: tandem two-seat operational conversion trainer variant of the Kfir-C7, developed along lines similar to those that led to the Kfir-TC2; however, it is likely that the Kfir-TC7 is more than a mere conversion trainer, for the provision of the full electronic suite may mean that the type is designed to undertake the two-seat electronic warfare role in addition to its more mundane tasks

Israel Aircraft Industries Lavi
(Israel)
Type: multi-role air-combat and ground-attack fighter

Accommodation: pilot seated on a Martin-Baker IL10LD zero/zero ejector seat
Armament (fixed): not known
Armament (disposable): this is carried on five hardpoints (one under the fuselage and four under the wings) to a maximum weight of 7000 kg (15,432 lb) or more of conventional bombs (free-fall or retarded), cluster bombs units, dispenser weapons, air-to-air missiles, air-to-surface missiles and drop tanks
Electronics and operational equipment: communication and navigation equipment, plus Elta EL/M-2035 multi-role pulse-Doppler radar and a comprehensive suite of Israeli-developed (Elta, Elbit, Elisra and Tadiran) electronic support measures systems, electronic countermeasures systems, a central computer, a head-up display and three head-down displays, a stores-management system and other items
Powerplant and fuel system: one 9400-kg (20,723-lb) afterburning thrust Beth Shemeth-built Pratt & Whitney PW1120 afterburning turbofan, and a total internal fuel capacity of 3330 litres (733 Imp gal) in integral tanks, plus provision for drop tanks carrying a total of 5095 litres (1,121 Imp gal); provision for inflight-refuelling
Performance: maximum speed 1965 km/h (1,222 mph) or Mach 1.85 at optimum altitude, or 997 km/h (619 mph) at low level with eight 340-kg (750-lb) bombs and two air-to-air missiles; initial climb rate not revealed; service ceiling not revealed; combat radius on a combat air patrol 1850 km (1,150 miles) or on a lo-lo-lo mission 450 km (280 miles)
Weights: empty about 7030 kg (15,498 lb); maximum take-off more than 17000 kg (37,478 lb)
Dimensions: span 8.78 m (28 ft 9.7 in); length 14.57 m (47 ft 9.6 in); height 4.78 m (15 ft 8.2 in); wing area 33.05 m² (355.76 sq ft)

Variant
Lavi: an advanced single-seat interceptor and ground-attack aircraft with STOL field performance, the Lavi (young lion) is being developed in Israel with the aid of US funding and expertise in areas of advanced structures; the need of an indigenously produced multi-role aircraft has been clear to the Israelis for some time, and the strategic importance of the Lavi is considerable, and the installation of Israeli electronics in an airframe built in Israel and powered by an Israeli licence-produced American turbofan results in a highly capable aircraft able to undertake the roles of older types whose maintenance is dependent on continued overseas support

Israel Aircraft Industries Nesher (Israel): see Israel Aircraft Industries Kfir-C2

KAL (Hughes) Model 500 MD (South Korea): see Hughes (McDonnell Helicopters) OH-6A Cayuse

Kaman SH-2F Seasprite
(USA)
Type: anti-submarine, anti-ship missile defence search-and-rescue, and utility helicopter
Accommodation: pilot, co-pilot and sensor operator
Armament (fixed): none
Armament (disposable): one or two Mk 46 torpedoes
Electronics and operational equipment: communication and navigation equipment, plus Canadian Marconi LN-66HP surveillance radar with its antenna in an undernose radome, ASQ-81 magnetic anomaly detection (MAD) gear with a towed 'bird', ALR-54 radar-warning receiver (ALR-66 being retrofitted), Teledyne ASN-123 tactical navigation system with cathode-ray tube display, APN-182 Doppler navigation, SSQ-41 passive sonobuoys and SSQ-47 active sonobuoys (being replaced by DIFAR and DICASS respectively)
Powerplant and fuel system: two 1,350-shp (1007-kW) General Electric T58-GE-8F turboshafts, and a total internal fuel capacity of 276 US gal (1046 litres) in fuselage tanks, plus provision for 120 US gal (454 litres) of external fuel
Performance: maximum speed 165 mph (265 km/h) at sea level; cruising speed 150 mph (241 km/h); initial climb rate 2,440 ft (744 m) per minute; service ceiling 22,500 ft (6860 m); range 422 miles (679 km) with maximum fuel
Weights: empty 7,040 lb (3193 kg); normal take-off 12,800 lb (5805 kg); maximum take-off 13,500 lb (6123 kg)
Dimensions: main rotor diameter 44 ft 0 in (13.41 m); length with rotors turning 52 ft 7 in (16.03 m); height (rotor head) 13 ft 7 in (4.14 m); main rotor disc area 1,520.5 sq ft (141.26 m²)

Variant
SH-2F Seasprite: first flown in July 1959 as the single-engined HU2K-1, the Seasprite was redesignated UH-2 in 1962, and the sole version now in service in the SH-2F variant introduced in 1972 as the LAMPS (Light Airborne Multi-Purpose System) anti-submarine, anti-ship missile defence, search-and-rescue, and utility transport helicopter for service on the US Navy's smaller surface combatants; the two-engine configuration had been pioneered by the UH-2C variant, and the SH-2F still offers capabilities unmatched in any helicopter of comparable size; the variant can carry a slung load of 4,000 lb (1814 kg)
SH-2G Seasprite: planned version with two General Electric T700-GE-401 turboshafts driving a new main rotor with composite-construction blades

Kamov Ka-25 'Hormone-A'
(USSR)
Type: anti-submarine, missile-support and utility helicopter
Accommodation: pilot and co-pilot side-by-side on

the flightdeck, and a mission crew of three in the cabin; alternatively, 12 passengers or freight can be carried
Armament (fixed): none
Armament (disposable): this is carried in a bay under the fuselage, and can consist of homing torpedoes, nuclear depth charges and, in the latest variant with a long box-like container under the fuselage, wire-guided torpedoes
Electronics and operational equipment: communication and navigation equipment, plus 'Big Bulge' search radar in an undernose radome, 'Tie Rod' electro-optical sensor, either dunking sonar or a towed magnetic anomaly detection (MAD) 'bird' and (optionally) a container of sonobuoys mounted externally; the number of other antennae and blisters indicates that other systems are also carried
Powerplant and fuel system: two 671-kW (900-shp) Glushenkov GTD-3 turboshafts (early aircraft) or two 738-kW (990-shp) Glushenkov GTD-3BM turboshafts (later aircraft), and a total internal fuel capacity of ? litres (? Imp gal), plus provision for two external fuel tanks, one carried on each side of the fuselage
Performance: maximum speed 210 km/h (130 mph); cruising speed 195 km/h (121 mph); service ceiling 3500 m (11,480 ft); range 400 km (249 miles) with internal fuel, and 650 km (404 miles) with internal and external fuel
Weights: empty about 4765 kg (10,505 lb); maximum take-off about 7500 kg (16,534 lb)
Dimensions: rotor diameter (each) 15.74 m (51 ft 8 in); length (fuselage) 9.75 m (32 ft 0 in); height (to top of rotor head) 5.37 m (17 ft 7.5 in); total rotor disc area 389.2 m² (4,189.0 sq ft)

Variants
Ka-25 'Hormone-A': introduced to service in 1965, the Ka-25 is clearly derived from the Ka-20 prototype revealed in 1961 and itself developed on the conceptual basis of the Ka-15/Ka-18 series; the 'Hormone-A' is the dedicated anti-submarine version of the series
Ka-25 'Hormone-B': this is the missile support version of the Ka-25 family, and is fitted with A-346Z data-link equipment for the provision of targeting data and the mid-course updating of long-range anti-ship missiles such as the SS-N-3 'Shaddock' and SS-N-12 'Sandbox' launched by cruiser-sized surface units; the variant is distinguishable from the 'Hormone-A' by the domed undersurface of its 'Short Horn' nose radome and the provision of a radar with a cylindrical radome under the rear of the cabin
Ka-25 'Hormone-C': the utility and search-and-rescue variant of the family, the 'Hormone-C' lacks the offensive avionics and armament of the 'Hormone-A', and often carries a winch and other specialist equipment

Kamov Ka-27 'Helix-A'
(USSR)
Type: anti-submarine helicopter
Accommodation: (probable) pilot and co-pilot on the flightdeck, and mission crew of two or three in the cabin
Armament (fixed): none
Armament (disposable): this is carried in a bay in the lower fuselage, and probably comprises homing torpedoes or nuclear depth charges
Electronics and operational equipment: communication and navigation equipment, plus search radar in an undernose radome, electro-optical sensor, dunking sonar or a towed magnetic anomaly detection (MAD) 'bird' and sonobuoys in a box on each side of the fuselage; the 'Helix-A' doubtless carries other equipment including onboard analysis and computing gear
Powerplant and fuel system: two 1659-kW (2,225-shp) Isotov TV3-117V turboshafts, and a total internal fuel capacity of ? litres (? Imp gal)
Performance: maximum speed 260 km/h (162 mph); cruising speed 230 km/h (143 mph); service ceiling 6000 m (19,685 ft); range 800 km (497 miles)
Weights: empty 6100 kg (13,448 lb); maximum take-off 12600 kg (27,778 lb)
Dimensions: rotor diameter (each) about 15.90 m (52 ft 2 in); length (fuselage) about 11.30 m (37 ft 1 in); height (to top of rotor head) about 5.50 m (18 ft 0.5 in); total rotor disc area about 198.56 m² (2,137.4 sq ft)

Variants
Ka-27 'Helix-A': this important type can be regarded as a modern equivalent to the Ka-25 series; introduced in the early 1980s, it has considerably greater power and slightly greater dimensions for much enhanced performance and payload within an airframe still able to fit into the same hangar as the Ka-25; the 'Helix-A' is the dedicated anti-submarine version, and can also carry a slung load of 5000 kg (11,023 lb) or a squad of Naval Infantry in the cabin
Ka-27 'Helix-B': this is the dedicated Naval Infantry assault transport verson, able to carry 16 fully-equipped men
Ka-27 'Helix-C': this is the general-purpose member of the Ka-27 series, and can be used in roles as diverse as search-and-rescue and vertical replenishment of warships at sea

Kamov Ka-? 'Hokum'
(USSR)
Type: battlefield air-combat and close-air support helicopter
Accommodation: pilot and weapons officer seated side-by-side
Armament (fixed): probably one 23- or 30-mm cannon

Armament (disposable): this is carried on four underwing and two wingtip hardpoints to an unknown maximum weight of rocket-launchers, anti-tank missiles, air-to-surface missiles and air-to-air missiles
Electronics and operational equipment: communication and navigation equipment, plus a wide assortment of sensors such as low-light-level TV, forward-looking infra-red, laser ranger and marked-target seeker etc
Powerplant and fuel system: two ?-kW (?-shp) turboshafts, and a total internal fuel capacity of ? litres (? Imp gal)
Performance: maximum speed 350 km/h (217 mph); combat radius 250 km (155 miles)
Weights: empty not known; maximum take-off about 5500 kg (12,125 lb)
Dimensions: rotor diameter, each 18.20 m (59 ft 8.5 in); length overall 16.00 m (52 ft 6 in); height 5.40 (17 ft 8.6 in); rotor disc area, total 520.3 m² (5,600.75 sq ft)

Variant
Ka-? 'Hokum': designed with the co-axial twin main rotors typical of Kamov practice, the 'Hokum' is an advanced battlefield helicopter of which few details are available; in combination with a slim fuselage and retractable landing gear, the rotor design offers a high degree of agility and speed, while the elimination of the tail rotor offers the possibility of a shorter fuselage for reduced battlefield visibility and vulnerability; it is likely that the type will be tasked with the anti-helicopter escort role in conjunction with offensive operations by Mil Mi-24s and Mi-28s, and when it enters service in the later 1980s the 'Hokum' will provide the Soviets with a genuine combat helicopter type with advanced capabilities

Kawasaki C-1
(Japan)
Type: STOL tactical transport
Accommodation: crew of four on the flightdeck, and one loadmaster plus 60 troops, or 45 paratroops, or 36 litters and attendants, or a freight load of 11900 kg (26,235 lb) in the cabin
Armament (fixed): none
Armament (disposable): none
Electronics and operational equipment: communication and navigation equipment, plus optional weather radar
Powerplant and fuel system: two 14,500-lb (6577-kg) thrust Mitsubishi-built Pratt & Whitney JT8D-M-9 turbofans, and a total internal fuel capacity of 15,200 litres (3,344 Imp gal) in four integral wing tanks
Performance: maximum speed 805 km/h (500 mph) at 7620 m (25,000 ft); cruising speed 655 km/h (407 mph) at 10670 m (35,000 ft); initial climb rate 1065 m (3,600 ft) per minute; service ceiling 11580 m (38,000 ft); range 3355 km (2,085 miles) with a 2200-kg (4,850-lb) payload and maximum fuel, or 1300 km (808 miles) with a 7900-kg (17,416-lb) load
Weights: empty 24300 kg (53,571 lb); normal take-off 38700 kg (85,317 lb); maximum take-off 45000 kg (99,206 lb)
Dimensions: span 30.60 m (100 ft 4.75 in); length 29.00 m (95 ft 1.75 in); height 9.99 m (32 ft 9.25 in); wing area 120.5 m² (1,297.1 sq ft)

Variants
C-1: developed as a medium-capacity STOL transport to replace Japan's ageing Curtiss C-46 Commando transports, the C-1 was a co-operative effort co-ordinated by Kawasaki, and the first prototype flew in November 1970, with deliveries of 24 aircraft made between 1974 and 1978, plus that of another three in 1981; five aircraft have 4732 litres (1,041 Imp gal) of additional tankage in the centre section for greater range
C-1 KAI: electronic warfare version of the basic C-1 tactical transport, fitted with the XJ/ALQ-5 suite with eight nose/tail antennae, three underfuselage antennae and two lateral nose antennae

Kawasaki (Boeing Vertol) KV 107/II-4
(Japan/USA)
Type: medium-lift transport helicopter
Accommodation: crew of two on the flightdeck, and up to 26 troops or 15 litters in the cabin
Armament (fixed): none
Armament (disposable): none
Electronics and operational equipment: communication and navigation equipment
Powerplant and fuel system: two 932-kW (1,250-shp) Ishikawjima-Harima-built General Electric CT58-IHI-110-1 turboshafts, and a total internal fuel capacity of 1324 litres (291 Imp gal) in the fuselage sponson fairings
Performance: maximum speed 255 km/h (158 mph) at sea level; cruising speed 240 km/h (149 mph) at 1525 m (5,000 ft); initial climb rate 463 m (1,520 ft) per minute; service ceiling 4570 m (15,000 ft); range 175 km (109 miles) with a 3000-kg (6,614-lb) payload
Weights: empty 4870 kg (10,736 lb); normal take-off 8620 kg (19,005 lb); maximum take-off 9705 kg (21,396 lb)
Dimensions: rotor diameter (each) 15.24 m (50 ft 0 in); length (fuselage) 13.59 m (44 ft 7 in); height (to rear rotor head) 5.09 m (16 ft 8.5 in); total disc area 364.6 m² (3,924.7 sq ft)

Variants
KV 107/II-3: the Japanese manufacturer of the Boeing Vertol Model 107 bought first a licence manufacturing agreement and then world rights in the late 1950s and early 1960s, and has since developed the Model 107 basic design for a number of specific applications; the KV 107/II-3 is the mine countermeasures

version with long-range tanks, an external hook and provision for the sweeping and (occasionally) the retrieval of mines

KV 107/II-4: this is the tactical version of the series with a strengthened floor for the carriage of light vehicles and other heavy loads

KV 107/II-5: this is the long-range search-and-rescue variant, with tankage for 3785 litres (833 Imp gal) of fuel and specialized rescue equipment; the version delivered to Sweden is the **HKP 4C**, which has different avionics and two Rolls-Royce Gnome H.1200 turboshafts

KV 107/IIA-3: uprated version of the KV 107/II-3 with 1044-kW (1,400-shp) CT58-IHI-140-1 turboshafts

KV 107/IIA-4: uprated version of the KV 107/II-4 with 1044-kW (1,400-shp) CT58-IHI-140-1 turboshafts

KV 107/IIA-5: uprated version of the KV 107/II-5 with 1044-kW (1,400-shp) CT58-IHI-140-1 turboshafts

KV 107/IIA-SM-1: fire-fighting and crash rescue variant for Saudi Arabia with 606 litres (1033 Imp gal) of additional standard tankage and provision for a 1893-litre (416-Imp gal) internal ferry tank

KV 107/IIA-SM-2: casevac variant for Saudi Arabia; this has the same fuel provisions as the KV 107/IIA-SM-1

Kawasaki (Hughes) TH-55J (Japan): see Hughes (McDonnell Douglas Helicopters) Model 300C

Kawasaki (Hughes) Model 500M (Japan): see Hughes (McDonnell Douglas Helicopters) OH-6A Cayuse

Kawasaki (Lockheed) P-2J
(Japan)
Type: anti-submarine and maritime patrol aircraft
Accommodation: crew of two on the flightdeck, and a tactical crew of 10 in forward (seven-man) and aft (three-man) compartments
Armament (fixed): none
Armament (disposable): none
Electronics and operational equipment: communication and navigation equipment, plus APN-178B-N Doppler radar, APS-80-N search radar, APA-125-N radar indicator, HLR-101 electronic support measures, HSQ-101 magnetic anomaly detection (MAD) gear, ASA-20B 'Julie' recorder, AN/AQA-5-A 'Jezebel' recorder, HQA-101 active sonobuoy indicator, ARR-52A(V) sonobuoy receiver, and HSA-116 integrated data display system and digital data processor
Powerplant and fuel system: two 2282-ekW (3,060-ehp) Ishikawajima-Harima-built General Electric T64-IHI-10E turboprops and two 1550-kg (3,417-lb) thrust Ishikawajima-Harima J3-IHI-7D turbojets, and a total internal fuel capacity of 12947 litres (2,848 Imp gal) in integral wing and port wingtip tanks
Performance: cruising speed 400 km/h (249 mph); initial climb rate 550 m (1,800 ft) per minute; service ceiling 9145 m (30,000 ft); range 4450 km (2,765 miles) with maximum fuel
Weights: empty 19275 kg (42,495 lb); maximum take-off 34020 kg (75,000 lb)
Dimensions: span 29.78 m (97 ft 8.5 in) without tiptanks, length 29.23 m (95 ft 10.75 in); height 8.93 m (29 ft 3.5 in); wing area 92.9 m² (1,000.0 sq ft)

Variants
P-2J: this is a Japanese development of the Lockheed P-2H Naptune with a stretched fuselage, additional power (in the form of two underwing turbojet engines) and a high percentage of Japanese electronics; first flown in July 1966 for service from late 1969, the type is now being phased out in favour of the Lockheed P-3C Orion, the type whose niche in the Japanese inventory the P-2J was designed to fill as a cheaper alternative
EP-2J: designation of P-2Js converted to electronic intelligence aircraft with the HLR-105 and HLR-106 Elint systems
UP-2J: designation of P-2Js converted for the electronic countermeasures and the drone launching and control roles

Kawasaki XT-4
(Japan)
Type: basic and advanced flying trainer, and light attack aircraft
Accommodation: pupil and instructor seated in tandem on Stencel SIII-S03ER zero/zero ejector seats
Armament (fixed): none
Armament (disposable): this is carried on five hardpoints (one under the fuselage and four under the wings) to a maximum weight of about 2000 kg (4,409 lb) bombs, rocket-launchers, gun pods and (later) air-to-air and air-to-surface missiles
Electronics and operational equipment: communication and navigation equipment, plus a Shimadze/Kaiser head-up display, a laser-gyro attitude and heading reference system and (possibly later) search radar
Powerplant and fuel system: two 3325-kg (7,330-lb) thrust Ishikawajima-Harima F3-IHI-30 turbofans, and a total internal fuel capacity of 2270 litres (499 Imp gal) in integral tanks, plus provision for two 455-litre (100-Imp gal) drop tanks
Performance: maximum speed 955 km/h (593 mph) at optimum altitude; initial climb rate 3000 m (9,845 ft) per minute; service ceiling 15000 m (49,215 ft); range 1300 km (808 miles)
Weights: empty 3700 kg (8157 lb); maximum take-off 7500 kg (16,534 lb)
Dimensions: span 9.90 ft m (32 ft 6 in); length 13.00

m (42 ft 7.8 in); height 4.60 m (15 ft 1 in); wing area 21.60 m² (232.51 sq ft)

Variant
XT-4: developed indigenously by Fuji, Kawasaki and Mitsubishi, and first flown in 1985 with indigenously-developed turbofan engines, the XT-4 is designed as a basic and advanced trainer, but its five external hardpoints clearly mark it as capable of the light attack as well as weapons training roles

Lockheed C-5A Galaxy
(USA)
Type: strategic heavy transport aircraft
Accommodation: crew of five on the flightdeck, plus 15 relief personnel and 75 troops on the upper deck, and provision for 270 troops or 220,967 lb (100,228 kg) of freight in the hold
Armament (fixed): none
Armament (disposable): none
Electronics and operational equipment: communication and navigation equipment, plus weather and mapping radar
Powerplant and fuel system: four 41,000-lb (18598-kg) thrust General Electric TF39-GE-1 turbofans, and a total internal fuel capacity of 49,000 US gal (185480 litres) in 12 integral wing tanks; provision for inflight-refuelling
Performance: maximum speed 571 mph (919 km/h) at 25,000 ft (7620 m); cruising speed 553 mph (890 km/h) at 25,000 ft (7620 m); initial climb rate 1,800 ft (550 m) per minute; service ceiling 34,000 ft (10360 m); range 3,750 miles (6035 km) with maximum payload, and 7,990 miles (12860 km) for ferrying with max fuel
Weights: empty 337,937 lb (153285 kg); maximum take-off 769,000 lb (348810 kg)
Dimensions: span 222 ft 8.5 in (67.88 m); length 247 ft 10 in (75.54 m); height 64 ft 1.5 in (19.85 m); wing area 6,200.0 sq ft (576.0 m²)

Variants
C-5A Galaxy: designed as a strategic airlifter able to operate into and out of tactical airstrips, the Galaxy first flew in June 1968 and, though it never achieved the payload/range figures of the overly ambitious specification, it soon acquired an enviable reputation as a heavy-lift transport for items of outsize equipment such as artillery, armoured fighting vehicles and complete missile systems; the type features a multiple-wheel landing gear for low ground loadings, and a fuselage with rear ramp and lifting nose for the through-loading/unloading of freight, which is handled by powered cargo systems; in an effort to meet the US Air Force's payload/range requirements the manufacturer designed the wings to reduced fatigue limits, resulting in service problems, so the entire fleet of C-5As is being re-engined with 43,000-lb (19504-kg) thrust TF39-GE-1C turbofans with 51,155 US gal (193640 litres) of fuel, and rewinged to a higher standard, resulting in an increase in maximum take-off weight to 837,000 lb (379663 kg), and in 2-*g* payload to 275,000 lb (124740 kg)
C-5B Galaxy: designation of 50 new aircraft being built to supplement the 77 upgraded C-5As, to the same basic standard but with such refinements as triple inertial navigation systems, colour weather radar, and improved fatigue and corrosion resistance; the C-5B can carry a 261,000-lb (118390-kg) payload over a range of 3,400 miles (5472 km) after take-off at a maximum weight of 837,000 lb (379663 kg)

Lockheed C-130H Hercules
(USA)
Type: medium tactical transport
Accommodation: crew of four or five on the flightdeck, and up to 92 troops, or 64 paratroops, or 74 litters and two attendants, or 43,400 lb (19686 kg) of freight in the hold
Armament (fixed): none
Armament (disposable): none
Electronics and operational equipment: communication and navigation equipment, plus weather radar and inertial navigation system
Powerplant and fuel system: four 4,508-shp (3362-kW) Allison T56-A-15 turboprops, and a total internal fuel capacity of 6,960 US gal (26344 litres) in six integral wing tanks, plus provision for two 1,360-US gal (5146-litre) underwing tanks; provision for inflight-refuelling
Performance: maximum speed 384 mph (618 km/h); cruising speed 375 mph (603 km/h); initial climb rate 1,900 ft (579 m) per minute; service ceiling 42,900 ft (13075 m); range 2,487 miles (4002 km) with maximum payload, or 4,721 miles (7600 km) with maximum fuel and a 16,078-lb (7293-kg) payload
Weights: empty 76,780 lb (34287 kg); normal take-off 155,000 lb (70310 kg); maximum take-off 175,000 lb (79380 kg)
Dimensions: span 132 ft 7 in (40.41 m); length 97 ft 9 in (29.79 m); height 38 ft 3 in (11.66 m); wing area 1,745.0 sq ft (162.12 m²)

Variants
C-130A Hercules: first flown in YC-130 prototype form during August 1954, the Hercules tactical transport is the mainstay of Western air arms' transport capabilities, and was responsible for what is now the standard configuration for military airlifters, namely a portly fuselage with an unobstructed hold of regular rectangular section accessed at truckbed height by an air-openable rear ramp/door under an unswept tail unit, and supported on the ground by multi-wheel landing gear with main units retracting into fuselage blisters and in the air by a high-mounted wing; the C-130A was the first production version, and entered

service in December 1956 with 3,750-shp (2796-kW) T56-A-1A turboprops driving three-blade propellers for a maximum take-off weight of 116,000 lb (52616 kg); machines of this basic type was converted into **C-130A-II** electronic reconnaissance aircraft, **AC-130A Spectre** gunships (with a side-firing armament of four 20-mm M61A1 cannon and four 7.62-mm/0.3-in Miniguns), **DC-130A** drone launch and control aircraft, **JC-130A** missile-tracking aircraft for use over the Atlantic test range, **NC-130A** special test aircraft, and RC-130A photo-reconnaissance aircraft

C-130B Hercules: improved production model with increased fuel capacity and 4,050-shp (3021-kW) T56-A-7/7A turboprops driving four-blade propellers for a maximum take-off weight of 135,000 lb (61235 kg); variants of this model were the **C-130B-II** (later **RC-130B**) electronic reconnaissance aircraft, the **HC-130B** search-and-rescue aircraft, the **JC-130B** recovery aircraft for satellite data capsules, the **KC-130B** inflight-refuelling tankers and the **WC-130B** weather reconnaissance aircraft

C-130D Hercules: ski-equipped version for arctic and antarctic service

C-130E Hercules: third production version of the transport series, in essence a version of the C-130B with two permanent underwing tanks as well as increased internal capacity; variants of the transport version were the **AC-130E Spectre** gunship (redesignated **AC-130H Spectre** after the retrofitting of T56-A-15 turboprops and the addition of inflight-refuelling capability), the **DC-130E** drone launch and control aircraft, the **EC-130E ABCCC** special forces command and control aircraft fitted with the USC-15 battlefield command capsule in the hold (redesignated **EC-130H** after the retrofitting of T56-A-15 turboprops), the **EC-130E 'Coronet Solo II'** electronic surveillance aircraft for the Air National Guard, the HC-130E crew recovery aircraft, the **MC-130E** special forces insertion and extraction aircraft (redesignated **C-130H(CT)** after the retrofitting of T56-A-15 turboprops and more advanced avionics), and the **WC-130E** weather reconnaissance aircraft

C-130F Hercules: US Navy equivalent of the C-130B; the **KC-130F** is a US Marine Corps inflight-refuelling tanker based on the C-130F but featuring 4,910-shp (3661-kW) T56-A-16 turboprops, and another variant is the ski-equipped **LC-130F** for antarctic use

C-130G Hercules: US Navy equivalent of the C-130E but powered by T56-A-16 turboprops; the four aircraft were revised as VLF aircraft with the TACAMO II package (for relay of communications with submerged submarines) under the designation **EC-130G**

C-130H Hercules: uprated version of the C-130E with airframe/avionics modifications and T56-A-15 turboprops flat-rated to 4508 shp (3362 kW); variants of this major production model were the **DC-130H** drone launch and control aircraft, the **EC-130H**

'Compas Call' communications jamming aircraft with the special 'Compass Call' equipment package, the **HC-130H** rescue and recovery aircraft with spacecraft re-entry tracking radar and Fulton recovery gear (redesignated **JHC-130H** after conversion for the aerial recovery of satellite data capsules), the **KC-130H** inflight-refuelling tanker, the **PC-130H** maritime patrol aircraft (originally designated **C-130H-MP** and fitted with search radar, side-looking airborne radar, forward-looking infra-red, low-light-level TV, infra-red linescan and imaging infra-red sensors), the **VC-130H** VIP transport and the **WC-130H** weather reconnaissance aircraft

C-130H-30 Hercules: designation of the C-130H derivative with the fuselage of the L-100-30 civil model, which features a stretch in overall length to 112.75 ft (34.37 m) to make possible the carriage of 128 troops, or 98 paratroops, or 97 litters plus attendants, or an equivalent freight load

C-130K Hercules: derivative of the C-130H with T56-A-15 turboprops and other detail modifications for British service, in which the type is known as the **Hercules C.Mk 1**, one being converted for weather reconnaissance as the **Hercules W.Mk 2** and 20 with the stretched fuselage of the L-100-30 as the **Hercules C.Mk 3**

HC-130N Hercules: advanced spacecraft and satellite data capsule recovery aircraft based on the C-130H

HC-130P Hercules: inflight-refuelling tanker derivative of the C-130H for the refuelling of helicopters and the aerial recovery of aircrew

EC-130Q Hercules: advanced version of the EC-130G for the TACAMO (TAke Charge And Move Out) role associated with communications relay between the national command authority and submerged SSBNs; the aircraft have the TACAMO III, IV or IVB equipment suites

KC-130R Hercules: version of the KC-130H for the US Marine Corps

LC-130R Hercules: ski-equipped derivative of the C-130H for the US Navy

RC-130S Hercules: redesignation of JC-130A aircraft retrofitted with high-intensity lighting for nocturnal search-and-rescue missions

C-130 AEW: under this designation Lockheed is proposing an airborne early warning aircraft based on the airframe/powerplant combination of the current military Hercules fitted with a palletized tactical compartment fed with data from the two antennae (mounted in bulbous nose and tail fairings) of the GEC APY-920 radar developed for the BAe Nimrod AEW.Mk 3; for a cost considerably more modest than those of current AEW aircraft, this proposal offers 360-degree surveillance to a radius of 230 miles (370 km) from an altitude of 27,000 ft (8230 m); on-station patrol time is planned as 12 hours

L-100: civil transport version of the basic C-130 with 4,050-shp (3020-kW) Allison 501-D22 turboprops;

several of these and later aircraft have been sold to smaller air arms
L-100-20: version of the L-100 with the fuselage stretched by 8.33 ft (2.54 m) and powered by 4,508-shp (3362-kW) Allison 501-D22A turboprops
L-100-30: derivative of the L-100-20 with the fuselage stretched another 6.67 ft (2.03 m)

Lockheed C-141B StarLifter
(USA)
Type: logistics transport aircraft
Accommodation: crew of four on the flightdeck, and up to 94,525 lb (42877 lb) of freight in the hold
Armament (fixed): none
Armament (disposable): none
Electronics and operational equipment: communication and navigation equipment, plus weather and mapping radar and inertial navigation systems
Powerplant and fuel system: four 21,000-lb (9526-kg) thrust Pratt & Whitney TF33-P-7 turbofans, and a total internal fuel capacity of 23,600 US gal (89335 litres) in integral wing tanks; provision for inflight-refuelling
Performance: cruising speed 566 mph (910 km/h); initial climb rate 2,920 ft (890 m) per minute; service ceiling 41,600 ft (12680 m); range 2,935 miles (4725 km) with maximum payload, or 6,390 miles (10280 km) for ferrying with maximum fuel
Weights: empty 148,120 lb (67186 kg); maximum take-off 343,000 lb (155580 kg)
Dimensions: span 159 ft 11 in (48.74 m); length 168 ft 3.5 in (51.29 m); height 39 ft 3 in (11.96 m); wing area 3,228.0 sq ft (299.88 m²)

Variant
C-141B StarLifter: first flown in 1963, the StarLifter was the USAF's first turbofan-powered airlifter, and designed in response to a requirement for a long-range strategic partner to the short/medium-range C-130 series; production of the initial C-141A series amounted to 284 aircraft, and these began to enter service in April 1965, soon building a good reputation for reliability and structural strength; however, the principal limitation of the type was soon discovered to be its volume-limited freight capacity, meaning that the hold in the 145-ft (44.20-m) fuselage was generally full long before the designed payload weight of 70,847 lb (32136 kg) had been reached; in March 1977, therefore, Lockheed flew the prototype of a version with the fuselage stretched 13.33 ft (4.06 m) in front of the wing and 10.0 ft (3.05 m) aft of the wing; this allowed the carriage of 13 rather than 10 standard pallets; it was thus decided to revise all 270 surviving aircraft to this C-141B standard, an inflight-refuelling capacity being added at the same time; the last conversion was redelivered in June 1982, the programme having added the equivalent of 87 new C-141As to the fleet without additional crew requirements or anything like the purchase cost of the new aircraft

Lockheed CP-140 Aurora
(USA)
Type: long-range anti-submarine and patrol aircraft
Accommodation: crew of three on the flightdeck, and a mission crew of eight in the cabin
Armament (fixed): none
Armament (disposable): this is carried in a lower-fuselage weapons bay capable of taking 4,800 lb (2177 kg) of stores on eight stations or 6,350 lb (2427 kg) on three stations, and on 10 underwing hardpoints, the inner six each rated at 2,450 lb (1111 kg), the next pair outboard each at 1,450 lb (658 kg) and the outboard pair each at 611 lb (277 kg); among the weapon types which can be carried are depth bombs, mines, torpedoes and missiles such as the AGM-84 Harpoon anti-ship missile
Electronics and operational equipment: communication and navigation equipment, plus APS-116 search radar, ASQ-502 magnetic anomaly detection (MAD) gear, OR-89/AA forward-looking infra-red (FLIR), OL-82 acoustic data processor, Litton LN-33 inertial navigation system, APN-208 Doppler navigation, Univac AYK-10 navigation and tactical computer system with RD-348. ASQ-147 and ASA-82 displays, ALR-47 electronic support measures, secure data-link equipment and cameras; there is also provision for side-looking airborne radar (SLAR)
Powerplant and fuel system: four 4,910-dhp (3661-kW) Allison T56-A-14LFE turboprops, and a total internal fuel capacity of 22,410 US gal (34826 litres) in one fuselage and four integral wing tanks
Performance: maximum speed 455 mph (732 km/h) at optimum altitude; cruising speed 432 mph (695 km/h); initial climb rate 2,890 ft (881 m) per minute; service ceiling 28,250 ft (8610 m); combat radius 1,150 miles (1850 km) with an 8.2-hour patrol; ferry range 5,180 miles (8336 km)
Dimensions: span 99 ft 8 in (30.37 m); length 116 ft 10 in (35.61 m); height 33 ft 8.5 in (10.29 m); wing area 1,300.0 sq ft (120.77 m²)

Variant
CP-140 Aurora: first flown in March 1979, the Aurora is basically the airframe of the P-3 Orion fitted with the tactical electronics of the S-3A Viking

Lockheed F-19
(USA)
Type: reconnaissance and attack 'stealth' aircraft
Accommodation: pilot seated on an ejector seat
Armament (fixed): none
Armament (disposable): this is carried in a lower-fuselage weapons bay to an unknown weight (probably comparatively small) of specialized anti-radiation and comparable weapons

Electronics and operational equipment: communication and navigation equipment, plus (probably) a number of passive sensors and target-acquisition systems
Powerplant and fuel system: probably two 12,000-lb (5443-kg) thrust General Electric F404 non-afterburning turbofans, and a total internal fuel capacity of ? US gal (? litres); provision for inflight refuelling
Performance: highly classified, but probably including low supersonic speed at altitude and a combat radius of about 500 miles (805 km)
Weights: ('guesstimated') empty 15,000 lb (6804 kg); maximum take-off 30,000 lb (13608 kg)
Dimensions: ('guesstimated') span 33 ft 0 in (10.06 m); length 56 ft 0 in (17.07 m); height 12 ft 0 in (3.66 m); wing area 550.0 sq ft (51.1 m²)

Variant
F-19: virtually nothing is known of this important machine, the world's first 'stealth' aircraft to reach operational status; the US Air Force denies the very existence of the design, but it seems likely that development was undertaken on the conceptual basis of the Lockheed A-12/SR-71 series after a number of Israeli setbacks with US aircraft pitted against Soviet surface-to-air weapon systems in the 1973 'Yom Kippur' War; the type is designed to penetrate enemy airspace to detect and then destroy high-value SAM systems and their associated radar systems, relying on low visual, electro-magnetic and infra-red signatures (derived from a blended fuselage/wing design with minimal reflective angles, inward-canted twin fins, flush inlets and semi-submerged nozzles) to achieve penetration of hostile airspace; all data are highly conjectural, being designed to indicate the aircraft's probabilities rather than actualities; the type first flew in about 1977, and it is believed that a modest number is in service with the US Air Force

Lockheed F-104G Starfighter
(USA)
Type: multi-role fighter
Accommodation: pilot seated on a Martin-Baker Mk GQ7 (F) ejector seat
Armament (fixed): one 20-mm General Electric M61A1 Vulcan rotary-barrel cannon with 750 rounds in the nose
Armament (disposable): this is carried on one underfuselage and four underwing hardpoints, and on two wingtip missile rails, up to a maximum weight of 4,310 lb (1955 kg): the wingtip rails are used for the carriage only of AIM-9 Sidewinder air-to-air missiles; the five hardpoints can be used for a variety of stores including the Mk 84 2,000-lb (907-kg) bomb, the Mk 83 1,000-lb (454-kg) free-fall or retarded bomb, the MLU-10/B 750-lb (340-kg) mine, the AGM-12 Bullpup air-to-surface missile, the LAU-6 launcher with seven 2.75-in (69.85-mm) rockets, the LAU-3 launcher with 19 2.75-in (69.85-mm) rockets, the LAU-10 launcher with four 5-in (127-mm) rockets and other weapons including nuclear stores
Electronics and operational equipment: communication and navigation equipment plus Autonetics F15A NASARR fire-control radar system, General Electric ASG-14 optical sight system, Honeywell MH-97 automatic flight-control system and a Litton LN-3 navigation system
Powerplant and fuel system: one 15,800-lb (7167-kg) afterburning thrust General Electric J79-GE-11A or MAN/Turbo-Union J79-MTU-J1K turbojet, and a total internal fuel capacity of 896 US gal (3392 litres) in five fuselage bag tanks, and provision for one 225-US gal (849-litre) centreline, two 195-US gal (740-litre) underwing and two 170-US gal (645-litre) wingtip drop-tanks; provision for inflight-refuelling
Performance: maximum speed 1,450 mph (2333 km/h) or Mach 2.2 at 36,000 ft (10970 m); cruising speed 610 mph (981 km/h); initial climb rate 55,000 ft (15765 m) per minute; service ceiling 58,000 ft (17680 m); range 1,550 miles (2495 km)
Weights: empty 14,900 lb (6758 kg); normal take-off 21,690 lb (9838 kg); maximum take-off 28,779-lb (13054-kg)
Dimensions: span 21 ft 11 in (6.68 m); length 54 ft 9 in (16.69 m); height 13 ft 6 in (4.11 m); wing area 196.1 sq ft (18.22 m²)

Variants
F-104A Starfighter: evolved in response to a Korean War US Air Force requirement for a fast-climbing interceptor, the Starfighter was designed with a barrel-like fuselage containing the avionics, pilot, afterburning turbojet and fuel, and with unswept but diminutive flying surfaces of exceptionally thin thickness/chord ratio; the XF-104 prototype first flew in March 1954 on the power of a 10,200-lb (4627-kg) Wright XJ65-W-6, but four years of testing and evaluation followed before the F-104A entered service with a fuselage lengthened by 5.5 ft (1.68 m) and power provided by the 14,800-lb (6713-kg) General Electric J79-GE-3A afterburning turbojet; a few of these elderly fighters (armed with a single M61A1 20-nmm cannon and two AIM-9 Sidewinder air-to-air missiles) remain in service as interceptors with the Taiwanese air force
F-104D Starfighter: two-seat conversion trainer variant of the F-104C tactical strike version, and powered by the 15,800-lb (7167-kg) J79-GE-7 turbojet; the **F-104DJ** was the version for Japan
F-104G Starfighter: the earlier Starfighters had not been very successful in US service, but with the development of the F-104G variant the type became a major commercial and military success, for this multi-role fighter was accepted for widespread European service; features of the design were a strengthened structure, an enlarged empennage and other aerody-

namic improvements, advanced avionics, superior weapons capability (in terms of quantity and type), and a more powerful engine; 1,127 of this model were built, mostly in Europe, the number being swelled by 200 basically similar **CF-104** aircraft (J79-OEL-7 turbojet and inflight-refuelling capability) built under licence in Canada by Canadair and 210 **F-104J** aircraft mostly built under licence in Japan by Mitsubishi

RF-104G Starfighter: tactical reconnaissance variant of the F-104G with three cameras

TF-104G Starfighter: two-seat operational conversion trainer of the F-104G but retaining full combat capability; the **CF-104D** was the Canadian equivalent powered by the J79-OEL-7 turbojet

Lockheed P-3C Orion
(USA)
Type: long-range anti-submarine aircraft
Accommodation: crew of five on the flightdeck, and a mission crew of five in the cabin
Armament (fixed): none
Armament (disposable): this is carried in a lower-fuselage weapons bay and on 10 underwing hardpoints; the weapons bay can accommodate a maximum of 7,250 lb (3289 kg) of weapons, typical loads being one 2,000-lb (907-kg) Mk 25, Mk 39, Mk 55 or Mk 56 mine, or three 1,000-lb (454-kg) Mk 36 or Mk 52 mines, or three Mk 57 depth bombs, or eight Mk 54 depth bombs, or eight Mk 43, Mk 44 or Mk 46 torpedoes, or four torpedoes and two Mk 101 nuclear depth bombs; the underwing load of 12,750 lb (5783 kg) can include six, 2,000-lb Mk 25, Mk 39, Mk 55 or Mk 56 mines on the innermost hardpoints, and 1,000- or 500-lb (454- or 227-kg) mines or rockets (single or podded) on the outer hardpoints; provision is also made for the carriage of AGM-84 Harpoon anti-ship missiles under the wings
Electronics and operational equipment: communication and navigation equipment, plus APS-115 search radar with nose and tail antennae, ASQ-81 magnetic anomaly detector with its sensor in the tail 'sting', AAS-36 forward-looking infra-red, AXR-13 low-light-level TV, 84 sonobuoys plus associated ARR-72 receivers and AQA-7 indicator sets, ASQ-114 digital computer, AYA-8 data processor, ASA-66 tactical consoles two Litton LTN-72 inertial navigation systems, APN-227 Doppler navigation, ALQ-78 podded electronic support measures (being replaced by ALQ-77 wingtip ESM) and cameras
Powerplant and fuel system: four 4,910-shp (3661-kW) Allison T56-A-14 turboprops, and a total internal fuel capacity of 9,200 US gal (34826 litres) in one fuselage and four integral wing tanks
Performance: maximum speed 473 mph (761 km/h) at 15,000 ft (4570 m); cruising speed 378 mph (608 km/h) at 25,000 ft (7620 m); initial climb rate 1,950 ft (594 m) per minute; service ceiling 28,300 ft (8625 m); combat radius 1,550 mile (2494 km) with 3-hour patrol, or 2,383 miles (3835 km) with no loiter
Weights: empty 61,490 lb (27892 kg); normal take-off 135,000 lb (61235 kg); maximum take-off 142,000 lb (64410 kg)
Dimensions: span 99 ft 8 in (30.37 m); length 116 ft 10 in (35.61 m); height 33 ft 8.5 in (10.29 m); wing area 1,300-0 sq ft (120.77 m²)

Variants
P-3A Orion: first flown in prototype form in November 1959 as a development of the airframe/powerplant combination of the L-188 Electra turboprop airliner (though with the fuselage shortened by 7.33 ft/2.24 m), the P-3A entered service in August 1962 with 4,500-shp (3356-kW) T56-A-10W turboprops, underwing hardpoints for a variety of stores, and a substantial weapons bay built into a bulged lower fuselage; the type no longer serves in its maritime patrol and anti-submarine roles, though in-service derivatives (all produced by conversion) are the **CP-3A** transport, the **RP-3A** oceanographic reconnaissance aircraft, and the **VP-3A** staff transport
P-3B Orion: this second production series is powered by the 4,910-shp (3661-kW) T56-A-14 without the water/alcohol injection of the P-3A's powerplant; the type was fitted with the Deltic tactical processing system rather than the suite derived from that of the P-2 Neptune used in the initial P-3A aircraft; in-service modification has kept the anti-submarine capability of the type up to high standards, and export models have customer-dictated variations on the basic theme (the Australian aircraft using the Marconi AQS-901 acoustic processing system and the New Zealand aircraft having the Boeing UDACS display and control consoles)
P-3C Orion: current production model, featuring the same basic airframe and powerplant as the P-3B but incorporating the A-NEW avionics suite with new sensors and controls, and based on the Univac ASQ-114 computer system; from 1975 the designation **P-3C Update I** was used for new-production aircraft with increased computer memory, more sensitive acoustic processing equipment and upgraded navigation equipment; a year later the **P-3C Update II** was introduced with an infra-red detection system and provision for AGM-84 Harpoon anti-ship missiles; the **P-3C Update III** was introduced in 1985 with an IBM Proteus acoustic signal-processing system and a new ARS-3 sonobuoy signal receiver allowing the aircraft to plot the position of any buoy without flying over it; for service in the late 1980s the **P-3C Update IV** is being developed with further enhanced processing capability and more sensitive acoustic sensors; the service designation **P-3D Orion** will probably be allocated to this last variant, which may be powered by General Electric GE27 turboprops
RP-3D Orion: designation of one P-3C reconfigured as a special research aircraft

WP-3D Orion: designation of two P-3Cs reconfigured for weather reconnaissance

EP-3E Orion: designation of 12 aircraft (10 P-3As and two EP-3Bs) reworked as electronic surveillance/countermeasures aircraft; this variant has canoe fairing above and below the fuselage, and a ventral radome forward of the wing; in the electronic fit are the ALQ-110 radar signal-gathering system, the ALD-8 radio direction finder, the ALR-52 frequency-measuring receiver, the ALQ-78 electronic support measures system, the ALQ-171 electronic intelligence system, and the ALR-60 radio recorder; the type's maximum take-off weight is 142,000 lb (64411 kg)

P-3F Orion: designation of six Iranian aircraft, basically similar to the P-3C

P-3 AEW&C: under this designation Lockheed is proposing the development of surplus P-3B airframes as airborne early warning and control aircraft with a tactical compartment in the fuselage fed with data from an overhead rotodome for the antenna of the powerful APS-138 radar; the concept offers potential customers with a comparatively cheap AEW aircraft, while the attractions for current Orion operators are reduced spares holdings, reduced crew and maintenance training requirements, and a considerable background of operating experience with the basic airframe/powerplant combination

L-188 Electra: this was the airliner from which the Orion series was developed, and was not a great commercial success after a series of three crashes broke public confidence in the type; the type is still in service, often as a freighter, and several have been acquired by smaller air arms

Lockheed S-3A Viking
(USA)

Type: carrierborne anti-submarine aircraft
Accommodation: crew of two on the flightdeck, and tactical crew of two in the cabin, all seated on Douglas Escapac 1-E ejector seats
Armament (fixed): none
Armament (disposable): this is carried in a split weapons bay in the underside of the fuselage, and on two underwing hardpoints, to a maximum of 7,000 lb (3175 kg); the weapons bay can accommodate four 1,000-lb (454-kg) destructors, or four Mk 46 torpedoes, or four Mk 82 1,000-lb (454-kg) bombs, or two Mk 57 depth bombs, or four Mk 54 depth bombs, or four Mk 53 1,000-lb (454-kg) mines; the underwing hardpoints can carry Mk 52, Mk 55 or Mk 56 mines, or Mk 20-2 cluster bombs, or Mk 36 destructors, or LAU-10A/A launchers each with four 5-in (127-mm) rockets, or LAU-61/A launchers each with 19 2.75-in (69.85-mm) rockets, or LAU-68/A launchers each with seven 2.75-in (69.85-mm) rockets, or LAU-69/A launchers each with 10 2.75-in (69.85-mm) rockets
Electronics and operational equipment: communication and navigation equipment, plus Texas Instruments APS-116 search radar, Texas Instruments ASQ-81 magnetic anomaly detector with its sensor in the tail 'sting', OR-89 forward-looking infra-red, 60 sonobuoys and associated OL-82 processing system and separate sonobuoy reference system, AYK-10 digital central computer, ALR-87 electronic support system, ASN-92(V) inertial navigation system and APN-200 Doppler navigation
Powerplant and fuel system: two 9,275-lb (4207-kg) thrust General Electric TF34-GE-2 turbofans, and a total internal fuel capacity of 1,900 US gal (7192 litres) in integral wing tanks, and provision for two 300-US gal (1137-litre) drop-tanks; provision for inflight-refuelling
Performance: maximum speed 518 mph (834 km/h); cruising speed 426 mph (686 km/h); initial climb rate more than 4,200 ft (1280 m) per minute; service ceiling more than 35,000 ft (10670 m); range more than 2,300 miles (3701 km) for combat, and more than 3,455 miles (5560 km) for ferrying
Weights: empty 26,650 lb (12088 kg); normal take-off 42,500 lb (19277 kg); maximum take-off 52,540 lb (23832 kg)
Dimensions: span 68 ft 8 in (20.93 m); length 53 ft 4 in (16.26 m); height 22 ft 9 in (6.93 m); wing area 598.0 sq ft (55.56 m²)

Variants

S-3A Viking: delivered from October 1973 after a first flight in January 1973, the S-3A is the US Navy's mainstay carrierborne anti-submarine aircraft, and a classic example of how to pour a quart into a pint pot; the capacious fuselage holds the flight and tactical crews in considerable comfort, together with the mass of computerized mission avionics, while the wings hold the substantial fuel load and support the two economical non-afterburning turbofans

KS-3A Viking: designation of a sole inflight-refuelling variant

US-3A Viking: designation of six carrier onboard delivery aircraft

S-3B Viking: designation of the upgraded S-3A variant produced by conversion of existing aircraft; the modification programme includes a Sanders OL-82A acoustic signal-processing system, improved electronic support measures for greater cover, a Hazeltine ARR-78 sonobuoy receiver system with enhanced reference capabilities, a Texas Instruments APS-116 radar with inverted synthetic-aperture techniques and other improvements, and provision for AGM-84 Harpoon anti-ship missiles

Lockheed SR-71A
(USA)

Type: strategic reconnaissance aircraft
Accommodation: pilot and reconnaissance-systems operator seated in tandem on ejector seats

Armament (fixed): none
Armament (disposable): none
Electronics and operational equipment: communication and navigation equipment, plus a number of classified reconnaissance sensors
Powerplant and fuel system: two 32,500-lb (14742-kg) afterburning thrust Pratt & Whitney JT11D-20B bleed-turbojets, and a total internal fuel capacity of more than 12,000 US gal (45425 litres) in integral tanks; provision for inflight-refuelling
Performance: maximum speed 2,250 mph (3620 km/h) or Mach 3.4 at high altitude; cruising speed 1,980 mph (3186 km/h) or Mach 3 at high altitude; service ceiling 100,000 ft (30480 m); range 2,980 miles (4800 km) on internal fuel
Weights: empty 60,000 lb (27216 kg); maximum take-off 170,000 lb (77,111 kg)
Dimensions: span 55 ft 7 in (16.94 m); length 107 ft 5 in (32.74 m); height 18 ft 6 in (5.64 m); wing area 1,800.0 sq ft (167.3 m²)

Variants
SR-71A: this is the world's highest-performance aircraft, and began to enter service in 1966 after a evolutionary design process from the A-11 drone-carrying reconnaissance aircraft and YF-12 interceptor; the aircraft is built largely of titanium to withstand the extreme heat generated by air friction at Mach 3 even at high altitude, and the type's advanced aerodynamics are matched by a potent powerplant centred on the two afterburning bleed-turbojets burning special JP-7 fuel carried by dedicated Boeing KC-135Q inflight-refuelling tankers for maximization of the type's prodigious range; at Mach 3 the engines produce only 18 per cent of the powerplant's total thrust, the remainder being generated by inlet suction (54 per cent) and the special nozzles at the rear of the multiple-flow nacelles (28 per cent); the reconnaissance systems carried by the 'Blackbird' (so-called for its heat-radiating black finish) are highly classified, but it is known that optical, infra-red and electronic sensors are carried in special bays in the wing/fuselage chine fairings, and can survey more than 100,000 sq miles (259000 km²) per hour
SR-71B: designation of the two-seat operational conversion trainer variant
SR-71C: designation of one SR-71A converted to SR-71B standard

Lockheed TR-1A
(USA)
Type: high-altitude tactical reconnaissance aircraft
Accommodation: pilot seated on an ejector seat
Armament (fixed): none
Armament (disposable): none
Electronics and operational equipment: communication and navigation equipment, plus optical, infra-red and electro-magnetic (including Hughes ASARS and UPD-X SLAR) reconnaissance systems in fuselage bays and two wing-mounted pods; advanced electronic countermeasures
Powerplant and fuel system: one 17,000-lb (7711-kg) thrust Pratt & Whitney J75-P-13B turbojet, and a total internal fuel capacity of about 1,175 US gal (4448 litres) in four wing tanks
Performance: cruising speed more than 430 mph (692 km/h) at more than 70,000 ft (21650 m); service ceiling 90,000 ft (27430 m); range more than 3,000 miles (4830 km)
Weights: empty 15,100 lb (6849 kg); maximum take-off 40,000 lb (18143 kg)
Dimensions: span 103 ft 0 in (31.39 m); length 63 ft 0 in (19.20 m); height 16 ft 0 in (4.88 m); wing area about 1,000.0 sq ft (92.9 m²)

Variants
U-2CT: two-seat operational conversion trainer variant (with stepped separate cockpits) of the U-2C electronic intelligence version of the initial U-2 series
U-2R: this was the final production model of the U-2 high-altitude reconnaissance aircraft series, which first flew as a prototype in August 1955 as the realization of the concept of marrying a sailplane-type airframe to a highly reliable non-afterburning turbojet for long-range cruise at very high altitudes; the U-2 series entered service in the second half of the 1950s and went through a number of variants before the U-2R was developed in the mid-1960s to overcome the basic U-2 series' mismatch of engine and airframe; thus the U-2R resembled its predecessors externally, but internally was a virtually new and completely restressed design carefully matched to the J75 engine introduced in the U-2B as replacement for the U-2A's Pratt & Whitney J57-P-7 or J57-P-57A of 10,500- or 11,200-lb (4763- or 5080-kg) thrust respectively; produced in 1969 and 1970, the U-2R has greater structural strength, an increased-span wing and much increased fuel capacity, and carries five 70-mm cameras for high-resolution photo-reconnaissance; large pods can be mounted on the wings for additional reconnaissance equipment, and this feature is carried over to the TR-1A series
TR-1A: introduced in the early 1980s as an advanced development of the U-2R, the TR-1A carries more advanced avionics, and though designated as a tactical reconnaissance type it serves mainly in the strategic role (14 of the 24 aircraft), only 10 being configured for the tactical role with the Lockheed Precision Location Strike System for the stand-off detection, classification and localization of hostile radar systems, whose position is then data-link relayed to friendly attack forces for immediate attack; the planned equipment fit may not finally be realized as a result of technical and budgetary problems
TR-1B: designation of the two-seat conversion trainer variant

Lockheed TriStar K.Mk 1
(USA/UK)
Type: strategic airlifter and inflight-refuelling aircraft
Accommodation: crew of three on the flightdeck, plus cabin accommodation for 204 passengers or 98,110 lb (44500 kg) of freight on pallets
Armament (fixed): none
Armament (disposable): none
Electronics and operational equipment: communication and navigation equipment of advanced types, and a twin Mk 17T hose-and-drogue unit in the rear fuselage
Powerplant and fuel system: three 50,000-lb (22680-kg) thrust Rolls-Royce RB.211524B4 turbofans, and a total internal fuel capacity of 213,240 lb (96724 kg) in integral wing tanks plus 100,060 lb (45387 kg) in two fuselage tanks; provision for inflight-refuelling
Performance: maximum cruising speed 600 mph (964 km/h) at 35,000 ft (10670 m); long-range cruising speed 552 mph (889 km/h) at 33,000 ft (10060 m); range with maximum payload 4,835 miles (7780 km)
Weights: empty 242,864 lb (110163 kg); maximum take-off 540,000 lb (245000 kg)
Dimensions: span 164 ft 6 in (50.09 m); length excluding probe 164 ft 2.5 in (50.05 m); height 55 ft 4 in (16.87 m); wing area 3,541.0 sq ft (329.0 m²)

Variants
TriStar K.Mk 1: this most capable heavy airlifter/inflight-refuelling tanker has been developed to meet the Royal Air Force's continued strategic needs, the type being converted from four ex-British Airways L-1011-500 TriStars by Marshalls of Cambridge with underfloor tanks for transfer fuel, and above-floor accommodation for passengers or freight; the type has a 38,710-Imp gal (175974-litre) maimum fuel load, and can offload some 15,321 Imp gal (69650 litres) of fuel, weighing 124,000 lb (56246 kg), at a radius of 2,647 miles (4260 km through either of the twin Mk 17T hose-and-drogue units fitted in the ventral position; the type first flew in July 1985 for service in 1986, and future developments include provision of full freighting/refuelling capabilities with a pair of Mk 32 hose-and-drogue units under the wings
Tristar KC.Mk 1: designation of two ex-British Airways aircraft converted for the tanker/freight role with additional tanage and HDU arrangement as on the TriStar K.Mk 1, and provision for the installation of 194 seats
TriStar K.Mk 2: designation of three ex-Pan Am aircraft converted to a standard similar to that of the TriStar K.Mk 1 but with about 10,000 lb (4536 kg) less fuel

McDonnell Douglas A-4M Skyhawk II
(USA)
Type: carrierborne and land-based light attack bomber
Accommodation: pilot seated on a Douglas Escapac 1-3C lightweight ejector seat
Armament (fixed): two 20-mm Mk 12 cannon with 200 rounds per gun in the wing roots
Armament (disposable): this is carried on one underfuselage hardpoint, rated at 3,500 lb (1588 kg), and on four underwing hardpoints, the inner pair each rated at 2,250 lb (1021 kg) and the outer pair each at 1,000 lb (454 kg), to a maximum of 9,155 lb (4153 kg); a great variety of weapon loads can be carried, including nuclear bombs, the Mk 84 2,000-lb (907-kg) bomb, the Mk 88 1,000-lb (454-kg) free-fall or retarded bomb, the Mk 82 500-lb (227-kg) free-fall or retarded bomb, the Mk 81 250-lb (113-kg) free-fall or retarded bomb, the LAU-3/A launcher with 19 2.75-in (69.85-mm) rockets, the LAU-10/A launcher with four 5-in (127-mm) rockets, gun pods with either one M61A1 20-mm cannon or two conventional 20-mm cannon, the AN-M-series armour-piercing bomb, the BLU-series napalm bomb, the CBU-series bomb and mine dispenser, torpedoes, depth charges, the AGM-12 Bullpup air-to-surface missile, the AGM-45 Shrike anti-radiation missile, the AIM-9 Sidewinder air-to-air missile and various other stores such as demolition and fragmentation bombs
Electronics and operational equipment: communication and navigation equipment, plus Marconi AVQ-24 head-up display, Hughes ASB-19 Angle-Rate Bombing System, Texas Instruments AJB-3 loft bombing system, APN-153(V)7 ground-mapping and terrain-avoidance radar, ASN-41 inertial navigation system. Itek ALR-45 radar-warning receiver, Magnavox ALR-50 SAM-warning system, Tracor ALE-39 chaff/flare dispenser, and optional ECM such as the Sanders ALQ-100 deception electronic countermeasures and Sanders ALQ-130 tactical communications jammer
Powerplant and fuel system: one 11,200-lb (5080-kg) Pratt & Whitney J52-P-408 turbojet, and a total internal fuel capacity of 800 US gal (3028 litres) in fuselage and integral wing tanks, plus provision for one 150-, 300- or 400-US gal (568-, 1136- or 1514-litre) drop-tank under the fuselage and one 150- or 300-US gal (568- or 1136-litre) drop-tank under each wing; provision for inflight-refuelling
Performance: maximum speed 685 mph (1102 km/h) or Mach 0.9 at sea level, or 646 mph (1040 km/h) with 4,000-lb (1814-kg) bombload; initial climb rate 10,300 ft (3140 m) per minute; range 2,000 miles (3220 km) with maximum fuel
Weights: empty 10,465 lb (4747 kg); normal take-off 24,500 lb (11113 kg); maximum take-off 27,420 lb (12437 kg)
Dimensions: span 27 ft 6 in (8.38 m); length 40 ft 4 in (12.29 m) excluding probe; height 15 ft 10 in (4.57 m); wing area 260.0 sq ft (24.16 m²)

Variants

A-4F Skyhawk: developed in the early 1950s under the leadership of Ed Heinemann, the Skyhawk first flew as the XA4D-1 prototype during June 1954 and soon displayed outstanding speed and manoeuvrability even with a substantial warload; the design team had achieved the near miraculous of combining all the US Navy's design requirements and payload into a diminutive airframe powered by a 7,200-lb (3266-kg) Wright J65 non-afterburning turbojet and possessing only half the maximum take-off weight originally specified; the production A4D-1 (later A-4A) entered service in October 1956 with the 7,700-kg (3493-kg) J65-W-4 turbojet and could carry 5,000 lb (2268 kg) of disposable ordnance in addition to its two inbuilt 20-mm cannon; the oldest variant still in service is the A-4F powered by the 9,300-lb (4218-kg) Pratt & Whitney J52-P-8A non-afterburning turbojet (the engine type introduced on the A4D-5/A-4E variant) and having extra avionics in a dorsal hump

TA-4F Skyhawk: two-seat trainer version of the A-4F; there are also a few examples of the **EA-4F Skyhawk** in service as electronic warfare trainers with the ALE-41 chaff dispenser

A-4H Skyhawk: version of the A-4E for Israel with a square-topped vertical tail, revised avionics, an inbuilt armament of two 30-mm DEFA cannon with 150 rounds per gun, a braking parachute and the J52-P-8A turbojet

TA-4H Skyhawk: two-seat operational conversion trainer variant of the A-4H for Israel, which has modified a small number for the electronic warfare role with Elta chaff and jammer pods

TA-4J Skyhawk: two-seat training version for the US Navy, similar to the TA-4F but with reduced operational capability (only one 20-mm cannon, often omitted) and powered by the 8,500-lb (3856-kg) J52-P-6 turbojet

A-4K Skyhawk: version of the A-4F for New Zealand but fitted with the braking parachute of the A-4H

TA-4K Skyhawk: two-seat operational conversion trainer variant of the A-4K

A-4KU Skyhawk II: version for Kuwait, generally similar to the A-4M

TA-4KU Skyhawk II: two-seat operational conversion trainer variant of the A-4KU

A-4M Skyhawk II: much improved version of the basic series for the US Marine Corps, featuring the square-top tail of the A-4H, a relocated braking parachute, much enhanced avionics, and the considerably more powerful J52-P-408A turbojet

OA-4M Skyhawk II: designation of TA-4F aircraft converted (for the US Marine Corps' forward air control role) to the basic standard of the A-4M and, though fitted for armament, generally carrying none as fuel capacity and loiter time are more important; the type also carries special communications equipment

A-4N Skyhawk II: version of the A-4M for Israel, but featuring the 30-mm cannon armament of the A-4H, upgraded electronics and an Elliott head-up display

A-4P Skyhawk: designation of ex-US Navy A-4B aircraft refurbished and upgraded for the Argentine air force

A-4PTM Skyhawk: designation of ex-US Marine Corps refurbished and updated by Grumman for Malaysia with modern avionics, two additional underwing hardpoints, a braking parachute, provision for two AIM-9 Sidewinder self-defence air-to-air missiles (and later for AGM-65 Maverick air-to-surface missiles), and the hughes Angle-Rate Bombing System; some aircraft are being modified to a two-seat operational conversion trainer standard similar to the TA-4F by the insertion of a lengthened cockpit section

A-4Q Skyhawk: designation of ex-US Navy A-4B aircraft refurbished and upgraded for the Argentine navy

A-4S Skyhawk: designation of ex-US Navy A-4B aircraft refurbished and much improved for Singapore with items such as APQ-145 mapping and ranging radar, 30-mm Aden cannon, more advanced electronics, and the 8,100-lb (3674-kg) J65-W-20 turbojet; local development is in hand for the further upgrading of these aircraft with advanced avionics and General Electric F404 turbofan engines

A-4S-1 Skyhawk: uprated A-4S with electronic improvements and higher-rated hardpoints

TA-4S Skyhawk: two-seat operational conversion trainer variant of the A-4S with two separate cockpits

McDonnell Douglas AV-8A Harrier (USA): see British Aerospace Harrier GR.Mk 3

McDonnell Douglas AV-8B Harrier II (USA): see McDonnell Douglas/British Aerospace AV-8B Harrier II

McDonnell Douglas F-4E Phantom II (USA)
Type: all-weather multi-role fighter
Accommodation: pilot and systems operator seated in tandem on Martin-Baker Mk H7 ejector seats
Armament (fixed): one 20-mm General Electric M61A1 Vulcan rotary-barrel cannon with 640 rounds
Armament (disposable): this is accommodated on four recessed stations under the fuselage (each capable of accepting an AIM-7 Sparrow air-to-air missiles) and five hardpoints (one under the ventral fuselage and four under the wings); to a maximum of 16,000 lb (7257 kg); the fuselage hardpoint can accept nuclear stores (B28, B43, B57 or B61 bombs) up to a weight of 2,170 lb (968 kg) or up to 3,020 lb (1371 kg) of conventional stores (general-purpose bombs, retarded bombs, bomb dispensers, fire bombs, smoke bombs, flares, rocket launchers, SUU-16/A or SUU-23/A gun pods or a spray tank); the four wing points can accommodate a maximum of four AGM-45

Shrike anti-radiation missiles, or four AGM-62 Walleye guided bombs, or four AIM-9 Sidewinder or two AIM-7 Sparrow air-to-air missiles on the inner points, or a maximum of 12,980 lb (5888 kg) of assorted conventional and nuclear stores, rocket-launchers, gun pods, spray tanks, ECM pods or tow targets

Electronics and operational equipment: communication and navigation equipment, plus Hughes APQ-120 fire-control radar, AJB-7 bombing system; ASQ-91 weapon-release system, General Electric ASG-26A computing sight, Northrop ASX-1 TISEO (Target Identification System Electro-Optical), ASN-63 inertial navigation system, CKP-92A central computer and Itek APR-36 radar-warning receiver, plus provision for podded electronic countermeasures such as the ALQ-101, ALQ-119, ALQ-130 and ALQ-131 systems and for the Sanders ALQ-140 infra-red countermeasures system

Powerplant and fuel system: two 17,900-lb (8119-kg) afterburning thrust General Electric J79-GE-17A turbojets, and 1,855 US gal (7022 litres) of internal fuel in integral wing tanks and seven fuselage tanks, plus provision for one 600-US gal (2271-litre) and two 370-US gal (1400-litre) drop tanks; provision for inflight-refuelling

Performance: maximum speed (clean) 1,430 mph (2301 km/h) or Mach 2.17 at 36,000 ft (10975 m); cruising speed 570 mph (917 km/h) with stores; initial climb rate (clean) 49,800 ft (15180 m) per minute; service ceiling (clean) 58,750 ft (17905 m); combat radius 712 miles (1145 km) on an interdiction sortie; ferry range 1,978 miles (3184 km)

Weights: empty 30,328 lb (13757 kg); normal take-off 41,487 lb (18818 kg); maximum take-off 61,795 lb (28030 kg)

Dimensions: span 38 ft 7.5 in (11.77 m); length 63 ft 0 in (19.20 m); height 16 ft 5.5 in (5.02 m); wing area 530.0 sq ft (49.24 m²)

Variants
RF-4B Phantom II: oldest surviving operational variant of the legendary Phantom II series, the RF-4B first flew in March 1965 as a tactical day/night photo-reconnaissance aircraft (derived from the F-4B fighter) for the US Marine Corps; the type is unarmed and provided with a lengthened nose for forward and oblique cameras (radar and infra-red reconnaissance equipment also being carried); the Phantom II was conceived in the early 1950s as the McDonnell AH-1 strike and attack aircraft, but finally built as a dual-role fleet defence and attack fighter for the US Navy, first flying as the XF4H-1 in May 1958 with two 14,800-lb (6713-kg) General Electric J79-GE-3A afterburning turbojets; the type was planned round a missile (fighter) or missile/free-fall bomb (attack) armament, a radar fire-control system removing the need (for the first time in a naval fighter) for surface radar assistance; performance was thus optimized for climb rate, speed and range; the pre-production and interim-production F4H-1F (later F-4A) began to enter service in 1958 with two J79-GE-2 turbojets (pending delivery of the proposed J79-GE-8 turbojets), four semi-active radar-homing Sparrow III missiles and APQ-50 or improved APQ-72 radar; the following F4H-1 (later F-4B) was the first definitive production variant from 1961 with J79-GE-8 turbojets, APQ-72 radar, provision for 16,000 lb (7257 kg) of disposable stores, a missile armament of four Sparrow IIIs plus four IR-homing Sidewinders and a bulged cockpit canopy

F-4C Phantom II: this is the limited-change equivalent of the F-4B for the US Air Force with larger wheels, inflight-refuelling capability, APQ-100 radar, several system changes and 17,000-lb (7711-kg) J79-GE-15 turbojets; the type first flew in May 1963, and several of the aircraft were later passed on to Spain with the revised designation **F-4C(S)** after the aircraft had been refurbished

RF-4C Phantom II: US Air Force tactical reconnaissance version of the F-4C, first flown in production from during May 1964 and generally equivalent to the RF-4B with APQ-99 forward-looking radar, APQ-102 side-looking airborne radar, AAS-18A infra-red linescan and a combination of forward and oblique cameras

F-4D Phantom II: improved variant of the F-4C for the US Air Force, with avionics tailored to USAF needs and fitted with APQ-109 radar; the type first flew in December 1965

F-4E Phantom II: major production version for the US Air Force and incorporating features found wanting in early models during operations over Vietnam; the type flew in prototype form during August 1965 and in production form during June 1967; notable features are an inbuilt cannon (the 20-mm M61A1 Vulcan six-barrel 'Gatling' type), leading-edge slots on the tailplane and (added retrospectively in place of blown leading-edge flaps) slatted outer wing panels for increased combat manoeuvrability, a remodelled nose with reduced-diameter APQ-120 radar, additional fuel capacity bestowed by the introduction of a seventh fuel cell in the rear fuselage, and J79-GE-17 turbojets; the type was also produce for Japan under the designation **F-4E(J)**, and many of these are being updated as close-support aircraft under the designation **F-4E(J)KAI** with Westinghouse APG-66J pulse-Doppler radar, a Kaiser/VDO head-up display, an inertial navigation system, and an electronic countermeasures system; the type can also be used in the anti-ship role with two ASM-1 missiles; Israel also operates a small number of **F-4E(S)** special reconnaissance aircraft of the type modified by General Electric with the HIAC-1 photo-reconnaissance system in a drastically revised nose; the converted aircraft have been further revised in the powerplant for high-altitude high-speed flight for long-range strategic and operational reconnaissance

RF-4E Phantom II: tactical reconnaissance version of the F-4E with the same basic characteristics as the RF-4C, and first flown in September 1970; the type was also built for Japan under the designation **RF-4E(J)**

F-4F Phantom II: variant of the F-4E optimized for the air-superiority role in the hands of the West German air force; the type lacks the seventh fuselage fuel cell, slotted tailplane and air-to-surface weapons capability, but has APQ-100 radar, outer wing slats and Sparrow armament; the type is about to undergo a modernization programme with APG-65 lookdown/shoot-down fire-control radar with AEG displays, AIM-120A AMRAAM missiles, MIL 1553 databus and Honeywell ring laser inertial navigation system

F-4G 'Advanced Wild Weasel': this is currently the most important Phantom II variant, the EF-4E prototype first flying in December 1976 on the basis of an F-4E aircraft converted for the defence-suppression role; the concept was pioneered in Vietnam with North American F-100 Super Sabre, Republic F-105 Thunderchief and EF-4C/D 'Wild Weasel' aircraft and proved itself to be both feasible and tactically important; the F-4G conversions thus have no internal cannon, which is replaced by the APR-38 system, which has a multitude of aerials for the detection, classification, identification and localization of enemy radars which are then engaged with AGM-45 Shrike and AGM-88 HARM anti-radiation missiles; other elements of the 'Advanced Wild Weasel' suite are the APR-36 or APR-37 radar-homing and warning systems, and the ALQ-119-12, ALQ-119-14 or ALQ-131 defensive jammer pods

F-4J Phantom II: this was the US Navy's and US Marine Corps' second-generation Phantom, fitted with 17,900-lb (8119-kg) J79-GE-8 or J79-GE-10 turbojets and first flown in prototype form during June 1965 for service from mid-1966; the type has APG-59 pulse-Doppler radar with the AWG-10 fire-control system, the AJB-7 bombing system, seven fuselage fuel tanks, the slotted tailplane, drooping ailerons, and provision for automatic carrier landings; a small number were later transferred to the UK with minimum modifications under the designation **F-4J(UK)**

F-4K Phantom II: under this designation the manufacturer developed a version of the F-4J for British carrierborne requirements; extensive redesign of the fuselage was required for the installation of two 20,515-lb (9305-kg) Rolls-Royce Spey RB.168-25R Mk 202/203 afterburning turbofans, and other modifications were an extending nosewheel leg for increased incidence at take-off, a folding radome for the AWG-11 fire-control radar system, and reduced anhedral on the tailplane; the type entered service at the beginning of 1967 as the **Phantom FG.Mk 1**, and the aircraft were subsequently transferred to the Royal Air Force

F-4M Phantom II: designation of the F-4K variant for the Royal Air Force, with which the type entered service early in 1968 as the **Phantom FGR.Mk 2**; the type lacks the extending nosewheel leg and folding nose of the F-4K, while other modifications are the AWG-12 radar fire-control system, Sky Flash air-to-air missile and flush-fitting ventral EMI reconnaissance pod capability, a Ferranti inertial navigation system, an unslotted tailplane and, as a retrofit, an electronic countermeasures system in the top of the fin

F-4N Phantom II: designation of F-4B aircraft upgraded structurally and in terms of avionics to F-4J standard; the first such conversion flew in 1972

F-4S Phantom II: designation of F-4J aircraft converted from 1977 to an improved standard with a strengthened structure, the AWG-10A radar fire-control system, leading-edge slats on the outer panels of the wings and J79-GE-10B turbojets; systems improvements include the installation of the APR-43 radar-warning receiver in place of the original APR-32 system

Modernized F-4 Phantom II: Boeing has suggested a programme for maintaining the military viability of the F-4E (in particular) with a radical update centred on the use of two 20,600-lb (9344-kg) Pratt & Whitney PW1120 afterburning turbofans, an underfuselage conformal tank with an additional 1,100 US gal (4164 litres) of fuel plus an ALE-40 chaff/flare dispenser in its rear, and modern avionics based on the MIL1553 databus to include (to customer requirement) Hughes APG-65 multi-mode radar, GEC wide-angle head-up display, GEC air-data computer, Sperry head-down displays and Honeywell laser ring-gyro inertial navigation system; Israel is interested in this possibility, and was in 1986 flight testing an F-4E fitted with one PW1120 and one standard J79

McDonnell Douglas F-15C Eagle
(USA)

Type: air-superiority and attack fighter

Accommodation: pilot seated on a Douglas ACES II ejector seat

Armament (fixed): one 20-mm General Electric M61A1 Vulcan rotary-barrel cannon with 940 rounds in the upper edge of the starboard inlet

Armament (disposable): this is carried on special positions (AIM-7 Sparrow air-to-air missiles) and on three underfuselage and two underwing hardpoints, up to a maximum of 16,000 lb (7258 kg) or 23,600 lb (10705 kg) with FAST packs; the missile load is normally four AIM-7F Sparrow and four AIM-9L Sidewinder medium- and short-range air-to-air missiles; other stores can include the Mk 84 2,000-lb (907-kg) free-fall or guided (laser, infra-red or electro-

optical homing), Mk 83 1,000-lb (454-kg) free-fall bomb, Mk 82 500-lb (227-kg) free-fall or retarded bomb, BLU-27 860-lb (390-kg) napalm bomb, CBU-24 bomb dispenser, CBU-42 bomb dispenser, CBU-49 bomb dispenser, CBU-52 bomb dispenser, Mk 20 Rockeye cluster bomb, LAU series rocket launchers for 2.75-in (69.85-mm) rockets and other weapons
Electronics and operational equipment: communication and navigation equipment, plus Hughes APG-63 pulse-Doppler search and tracking radar, McDonnell Douglas head-up display, IBM central computer, Northrop ALQ-135 internal electronic countermeasures system, Loral ALR-56 radar-warning receiver, Magnavox electronic warfare warning system and other items
Powerplant and fuel system: two 23,950-lb (10864-kg) afterburning thrust Pratt & Whitney F100-PW-100 turbofans, and a total internal fuel capacity of 2,020 UK gal (7646 litres), plus provision for 1,463 US gal (5537 litres) in two conformal FAST (Fuel And Sensor Tactical) packs and three 600-US gal (2271-litre) drop-tanks; provision for inflight-refuelling
Performance: maximum speed more than 1,650 mph (2655 km/h) or Mach 2.5 at high altitude; climb to 40,000 ft (12190 m) in 1 minute; service ceiling 60,000 ft (18290 m); absolute ceiling 100,000 ft (30480 m); range more than 2,878 miles (4631 km) without FAST packs but with droptanks, or more than 3,450 miles (5560 km) with maximum fuel
Weights: empty 27,000 lb (12247 kg); normal take-off 44,560 lb (20212 kg) for interception mission; maximum take-off 68,000 lb (30845 kg)
Dimensions: span 42 ft 9.75 in (13.05 m); length 63 ft 9 in (19.43 m); height 18 ft 5.5 in (5.63 m); wing area 608.0 sq ft (56.5 m²)

Variants
F-15A Eagle: designed during the late 1970s to reflect the air combat lessons of the Vietnam War, the F-15 series is intended primarily for the air-superiority role, but has an excellent secondary capability in the attack role; the type flew in prototype form during July 1972, and amongst its key features are a high thrust-to-weight ratio for exceptional climb rate and altitude performance, advanced avionics and a head-up display, and superb aerodynamic design for low drag and sustained manoeuvrability; the type began to enter service in November 1974
F-15B Eagle: combat-capable two-seat operational conversion trainer derivative of the F-15A, at first designated TF-15A and flown in July 1973
F-15C Eagle: upgraded single-seat model delivered from 1979 with increased internal fuel capacity, a programmable signal processor, other avionic improvements, and provision for FAST (Fuel And Sensor Tactical) packs; these last are conformal pallets that can be attached to the outside of the inlets to carry additional fuel, sensors such as low-light-level TV and electronic countermeasures; these low-drag pallets degrade performance by no appreciable degree, and are stressed for the tangential carriage of missiles and free-fall ordnance; the Japanese version of this model is the **F-15J**
F-15D Eagle: combat-capable two-seat operational conversion trainer derivative of the F-15C, and delivered from mid-1979; the Japanese version of this model is the **F-15DJ**
F-15E Eagle: based on the F-15D, this is an advanced all-weather interdictor and strike aircraft, the adoption of the two-seat formula marking a radical shift in US Air Force thinking, which has hitherto opted for single-seat aircraft in this role even for areas with such problematical weather as Europe; the manufacturer paved the way for the change with the development of the private-venture F-15 Enhanced Eagle optimized for the European theatre, and the F-15E was selected in preference to the General Dynamics F-16E Fighting Falcon in March 1984 for delivery from early 1987; the type is designed to carry the LANTIRN (Low-Altitude Navigation and Targeting Infra-Red for Night) podded nav/attack system, the front cockpit being equipped with a wide-angle head-up display for the presentation of navigation/combat information, and the rear cockpit having head-down cathode ray tube displays for radar, forward-looking infra-red, digital map and threat-warning systems; FAST tanks will be standard for the carriage of an additional 9,000 lb (4082 kg) or more of fuel and of tangentially-mounted weapons; the total external weapons load will be more than 24,250 lb (11000 kg) of the weapons already qualified for the F-15 series, plus the AIM-120 AMRAAM air-to-air missile, AGM-65 Maverick air-to-surface missile, and AGM-88 HARM anti-radiation missile; the type's radar capability is considerably better than that of the basic F-15 series (though the Multi-Stage Improvement Program began to add features of these developments to the F-15C from June 1985) through installation of the APG-70 radar (a much improved development of the original APG-63 system), which has a track-while-scan capability in the air-to-air mode and a synthetic aperture mode for ground mapping and the acquisition of ground targets by day or night; other avionics improvements are an upgraded central computer system, an advanced programmable stores-management system and more modern electronic countermeasures; further development could include canard foreplanes and 2D vectoring nozzles for enhanced manoeuvrability and to allow operations from damaged runways

McDonnell Douglas F/A-18A Hornet
(USA)
Type: carrierborne and land-based attack fighter
Accommodation: pilot seated on a Martin-Baker US10S ejector seat

Armament (fixed): one 20-mm General Electric M61A1 Vulcan rotary-barrel cannon with 570 rounds
Armament (disposable): this is carried on nine hardpoints (one under the fuselage, two under the nacelles, two under the inner wings, two under the outer wings and two at the wingtips for missiles and rated respectively at 3,300 lb/1497 kg, 2,500 lb/1134 kg each, 2,350 lb/1066 kg each, 1,300 lb/590 kg each and 600 lb/272 kg each), up to a maximum weight of 17,000 lb (7711 kg); typical loads are six AIM-7 Sparrow or AIM-9 Sidewinder air-to-air missiles, 12 AGM-65 Maverick air-to-surface missiles, five Mk 84 2,000-lb (907-kg) bombs, 17 Mk 83 1,000-lb (454-kg) bombs, 34 Mk 82 500-lb (227-kg) bombs, four Mk 84 guided bombs, 16 LAU-3/A launchers each with 19 2.75-in (69.85-mm) rockets, 34 Mk 20 Rockeye II cluster bombs, 17 CBU-58 bomb dispensers, and several other weapons such as the BL-755 cluster bomb and JP233 airfield-denial weapon
Electronics and operational equipment: communication and navigation equipment, plus Hughes APG-65 multi-mode air-to-air and air-to-surface tracking radar, Kaiser head-up display, three Kaiser head-down displays, Itek ALR-67 radar-warning receiver, Sanders ALQ-126B electronic countermeasures system, Litton inertial navigation system, General Electric flight-control system with two AYK-14 digital computers, and provision for a Martin-Marietta ASQ-173 laser spot tracker and Ford AAS-38 FLIR
Powerplant and fuel system: two 16,000-lb (7258-kg) afterburning thrust General Electric F404-GE-400 turbofans, and a total internal fuel capacity of 1,650 US gal (6251 litres), plus provision for three 315-US gal (1192-litre) drop-tanks; provision for inflight-refuelling
Performance: maximum speed more than 1,190 mph (1915 km/h) or Mach 1.8 at high altitude; service ceiling 50,000 ft (15240 m); combat radius more than 460 miles (740 km) on a fighter mission, or 633 miles (1019 km) on an mission radius; ferry range 2,300 miles (3700 km)
Weights: empty 23,050 lb (10455 kg); normal take-off 33,585 lb (15234 kg) for a fighter mission; maximum take-off 49,224 lb (22328 kg) for an attack mission
Dimensions: span 37 ft 6 in (11.43 m) over missile rails; length 56 ft 0 in (17.07 m); height 15 ft 3.5 in (4.66 m); wing area 400.0 sq ft (37.16 m^2)

Variants
F/A-18A Hornet: first flown in November 1979 for service from late 1983, the F/A-18A is a dual-role fighter and attack aircraft designed to replace both the McDonnell Douglas F-4 Phantom II multi-role fighter and Vought A-7 Corsair II attack aircraft in service with the US Navy and US Marine Corps; the origins of the type lie with the Northrop YF-17, unsuccessful contender with the General Dynamics YF-16 in the US Air Force's Light-Weight Fighter competition; the type has been considerably reworked by the McDonnell Douglas/Northrop partnership to produce the multi-capable carrierborne F/A-18A, which has already shown its significant capabilities in the apparently incompatible fighter and attack roles; further improvement in range may result from a McDonnell Douglas-funded programme to develop a conformal external fuel tank carried in the dorsal position, and adding only 300 lb (136 kg) to empty weight yet adding capacity for an additional 3,000 lb (1361 kg) of fuel for a total of 13,865 lb (6289 kg); from 1988 the type will be available in a night-attack version with forward-looking infra-red and an improved head-up display
RF-18A: designation of the dedicated tactical and operational reconnaissance variant of the F/A-18A with a pallet for cameras (including the KA-99) and AAD-5 infra-red linescan equipment in a bulged nose no longer fitted with the 20-mm cannon
F/A-18B Hornet: designation of the combat-capable two-seat operational conversion variant trainer variant of the F/A-18A, originally designated **TF/A-18A** and possessing 6% less internal fuel capacity
F/A-18C Hornet: improved tactical version first flown in 1986 with internal ASPJ (Airborne Self-Protection Jammer), a Martin-Baker NACES ejector seat, data-bus-linked small computers rather than one large mission computer,and provision for reconaissance equipment and the AIM-120 AMRAAM air-to-air missile and AGM-65F IIR Maverick air-to-surface missile
F/A-18D Hornet: two-seat version of the F/A-18C equivalent to the TF/A-18A operational conversion and proficiency trainer
CF-18 Hornet: designation of the F/A-18A version for Canada with a different instrument landing system and intended for land-based operations; a basically similar aircraft is being bought by Australia
EF-18A: designation of the F/A-18A for Spain
EF-18B: designation of the F/A-18B for Spain

McDonnell Douglas KC-10A Extender
(USA)
Type: inflight-refuelling tanker with secondary airlift cabability
Accommodation: crew of three on the flightdeck, and various seating options or up to 169,370 lb (76825 kg) of freight in the cabin
Armament (fixed): none
Armament (disposable): none
Electronics and operational equipment: communication and navigation equipment, plus weather and mapping radar and inertial navigation system
Powerplant and fuel system: three 52,500-lb (23814-kg) thrust General Electric CF6-50C2 turbofans, and a total internal fuel capacity of 54,750 US gal (207251 litres) in integral wing tanks and fuselage bladder

tanks; provision for inflight-refuelling
Performance: maximum speed 610 mph (982 km/h) at 25,000 (7620 m); cruising speed 564 mph (908 km/h); initial climb rate 2,900 ft (884 m) per minute; service ceiling 33,400 ft (10180 m); range 4,370 miles (7032 km) with maximum payload, or 11,500 miles (18507 km) with maximum fuel
Weights: empty 240,065 lb (108891 kg) as a tanker; maximum take-off 590,000 lb (267620 kg)
Dimensions: span 165 ft 4.4 in (50.41 m); length 181 ft 7 in (55.35 m); height 58 ft 1 in (17.70 m); wing area 3,958.0 sq ft (367.7 m²)

Variant
KC-10A Extender: this is an advanced tanker and cargo aircraft derivative of the McDonnell Douglas DC-10-30CF convertible freighter in fairly widespread service use; the procurement of the type was desired by the US Air Force to provide a more capable aircraft than the Boeing KC-135 for the support of long-range deployments of tactical aircraft; the Extender suffered from severe political problems, but after a first flight in July 1980 and service deliveries from March 1981 the type has fully proved itself; the main advantages of the Extender over the Stratotanker are its considerably greater volume of transfer fuel, its own considerable range, its ability to serve as navigational mothership to smaller aircraft, the high-capacity refuelling boom which can also carry a drogue adapter, and the ability of the hold to carry a substantial load of spares, equipment and groundcrew for the support of overseas deployments; seven tanks in the lower hold accommodate 18,075 US gal (88421 litres) of transfer fuel, which can be supplemented from the Extender's standard tankage to make possible the transfer of 30,000 US gal (113562 litres) of fuel at a radius of 2,200 miles (3540 km); out to a range of 5,150 miles (8288 km) the KC-10A has cargo capacity comparable to that of the Lockheed C-5 Galaxy, but between that range and 11,500 miles (18507 km) the payload/range figures are better than those of the Galaxy

McDonnell Douglas T-45A Goshawk (USA): see British Aerospace Hawk T.Mk 1A

McDonnell Douglas/British Aerospace AV-8B Harrier II
(USA/UK)
Type: V/STOL close-support aircraft
Accommodation: pilot seated on a Stencel ejector seat (US aircraft) or a Martin-Baker Mk 10 ejector seat (British aircraft)
Armament (fixed): one 25-mm General Electric GAU-12/A rotary-barrel cannon system (with the gun in one underfuselage pod and 300 rounds in the other) on US aircraft, or two Aden 25-mm cannon in British aircraft
Armament (disposable): this is carried on one underfuselage hardpoint (rated at 1,000 lb/454 kg) and six underwing hardpoints (the inner pair each rated at 2,000 lb/907 kg, the centre pair each at 1,000 lb/454 kg and the outer pair each at 630 lb/286 kg), to a maximum of 7,000 lb (3175 kg) for VTO or 17,000 lb (7711 kg) for STO; typical loads are four AIM-9 Sidewinder air-to-air missiles, or four AGM-65 Maverick air-to-surface missiles, or up to 16 500-lb (227-kg) bombs, or up to 12 CBU series cluster bombs, or 10 'Paveway' laser-guided bombs, or 10 fire bombs, or 10 rocket pods, or two cannon pods
Electronics and operational equipment: communication and navigation equipment, plus Hughes ASB-19 Angle-Rate Bombing System, Smiths head-up display, head-down display, IBM weapon-delivery system, Litton ASN-130A dispenser, and provision for the Westinghouse/ITT ALQ-155 Airborne Self-Protection Jammer (in a converted ALQ-131 pod) and ALQ-164 jammer pods
Powerplant and fuel system: one 22,000-lb (9979-kg) thrust vectored-thrust Rolls-Royce Pegasus 11-21 (F402-RR-406) turbofan, and a total internal fuel capacity of 1,125 US gal (4260 litres), plus provision for four 300-US gal (1136-litre) drop-tanks; provision for inflight-refuelling
Performance: maximum speed 615 mph (988 km/h) or Mach 0.93 at 36,000 ft (10975 m), and 673 mph (1083 km/h) or Mach 0.88 at sea level; initial climb rate 14,715 ft (4485 m) per minute; service ceiling more than 50,000 ft (15240 m); combat radius 172 miles (282 km) with 12 Mk 82 Snakeye bombs after STO; ferry range 3,310 miles (5382 km) with four droptanks
Weights: empty 12,750 lb (6,783 kg); normal take-off 22,950 lb (10410 kg); maximum take-off 29,750 lb (13494 kg)
Dimensions: span 30 ft 4 in (9.25 m); length 46 ft 4 in (14.12 m); height 11 ft 7.75 in (3.55 m); wing area 230.0 sq ft (21.37 m²)

Variants
AV-8B Harrier II: this is a US development of the British Harrier V/STOL close support and reconnaissance aircraft with a number of significant improvements for enhanced operational capability, developed largely by BAe's US licensee, McDonnell Douglas, after US Marine Corps experience with the AV-8A Harrier; among these improvements are a wing (and portions of the fuselage and tail unit) of graphite epoxy composite construction, a larger wing of supercritical section (fitted with larger flaps, drooping ailerons and leading-edge root extensions for additional manoeuvrability), extra hardpoints for greater offensive load, underfuselage lift improvement devices (large-area gun pod/strake units and a retractable barrier forward of the pods to boost trapped-gas lift for vertical take-off), strengthened

landing gear, larger inlets for greater power, and a redesigned forward fuselage to provide the pilot with better fields of vision; the type flew in YAV-8B prototype form in November 1978, and service deliveries began in 1983; the AV-8B is also to be provided with a night-attack capability by the installation of equipment similar to that under development for the McDonnell Douglas F/A-18A Hornet (forward-looking infra-red and an improved head-up display)
TAV-8B: designation of the combat-capable two-seat operational conversion trainer variant of the AV-8B with a fuselage stretch of 4 ft (1.22 m) to allow the incorporation of a larger cockpit section for a weight penalty of 1,325 lb (601 kg)
Harrier GR.Mk 5: designation of the British equivalent to the AV-8B assembled in the UK and featuring British electronic countermeasures, a Ferranti inertial navigation system, a Ferranti moving map display, a panoramic camera in the nose, and provision for different weapons (including two AIM-9 Sidewinder air-to-air missiles on small pylons forward of the wing-mounted outrigger wheels)

MBB PAH-1
(West Germany)
Type: anti-tank helicopter
Accommodation: pilot and weapons operator seated side-by-side, and provision for three passengers on a bench seat in the rear of the cabin
Armament (fixed): none, though an HBS 202 20-mm cannon or an Emerson Flexible Turret System may be fitted
Armament (disposable): this is carried on two outriggers, one on each side of the cabin carrying three Euromissile HOT anti-tank missiles
Electronics and operational equipment: communication and navigation equipment, plus one APX 397 gyro-stabilized missile sight and Singer APS-128 Doppler navigation
Powerplant and fuel system: two 420-shp (313-kW) Allison 250-C20B turboshafts, and a total internal fuel capacity of 580 litres (128 Imp gal) in bladder tanks under the cabin floor
Performance: cruising speed 220 km/h (137 mph) at sea level; initial climb rate 540 m (1,770 ft) per minute; service ceiling 4250 m (13,950 ft); range 575 km (357 miles)
Weights: empty 1913 kg (4,217 lb); maximum take-off 2400 kg (5291 lb)
Dimensions: main rotor diameter 9.84 m (32 ft 3.4 in); length (fuselage) 8.81 m (28 ft 11 in); height 3.00 m (9 ft 10 in); main rotor disc area 76.05 m^2 (818.6 sq ft)

Variants
BO 105C: this was the initial production version of MBB's highly capable five-seat utility helicopter, powered by 400-shp (298-kW) Allison 250-C20 turboshafts and bought by many military operators; the type first flew in February 1967; this and other basically civil variants have been assembled under licence in Indonesia, the Philippines and Spain, many going to military customers in roles such as maritime patrol and search-and-rescue
BO 105CB: standard version since 1975 with 420-shp (313-kW) Allison 250-C20B turboshafts, which are also used on the **BO 105CBS** subvariant with the cabin lengthened by 0.25 m (9.8 in) for greater passenger and freight capacity; Sweden uses a version of the BO 105CB in the anti-tank role with an outfit of eight TOW missiles and a stabilized M65 roof sight
BO 105LS: uprated version for 'hot-and-high' operations with 550-shp (410-kW) Allison 250-C28C turboshafts and the cabin of the BO 105CBS; this type has a maximum take-off weight of 2600 kg (5,732 lb) with a slung load
BO 105M: this is the liaison and observation version for the West German army, by which the type is designated **VBH**; the primary differences between this military model and the civilian series are the uprated dynamic system, crash-resistant fuel tanks and seats, and high-impact landing gear
BO 105P: this is the **PAH-1** anti-tank version for the West German army

Meridionali CH-47C (Italy): see Boeing Vertol CH-47D Chinook

Mikoyan-Gurevich MiG-15bis 'Fagot'
(USSR)
Type: fighter
Accommodation: pilot seated on an ejector seat
Armament (fixed): one 37-mm N-37 cannon with 40 rounds and two 23-mm NR-23 cannon with 80 rounds per gun in the nose
Armament (disposable): this is carried on two underwing hardpoints, up to a maximum weight of 500 kg (1,102 lb) of bombs or rockets
Electronics and operational equipment: communication and navigation equipment, plus a gyro sight
Powerplant and fuel system: one 3170-kg (6989-lb) thrust Klimov VK-1A turbojet, and a total internal fuel capacity of 1565 litres (344 Imp gal) in wing and rear fuselage tanks, plus provision for two 250- or 600-litre (55- or 132-Imp gal) underwing tanks
Performance: maximum speed 1076 km/h (669 mph) at sea level and 1044 km/h (649 mph) at 3000 m (9,845 ft); initial climb rate 3500 m (11,480 ft) per minute; service ceiling 15550 m (51,015 ft); range 1860 km (1,156 miles) with maximum fuel
Weights: empty 3400 kg (7,495 lb); normal take-off 4960 kg (10,934 lb); maximum take-off 5785 kg (12,756 lb)
Dimensions: span 10.08 m (33 ft 0.75 in); length 11.05 m (36 ft 3.25 in); height 3.40 m (11 ft 1.75 in); wing area 20.60 m^2 (221.7 sq ft)

Variants
MiG-15 'Fagot': first flown in December 1947, the MiG-15 is one of the most important aircraft of all time, and played a crucial part in the air campaign over the Korean War during the early 1950s; the type is now thoroughly obsolete, but still used by a number of smaller air arms as a combat type or for advanced training; this initial model is powered by the 2270-kg (5,005-lb) Klimov RD-45F non-afterburning turbojet and has an armament of two 23-mm cannon
MiG-15bis 'Fagot': improved version first flown in 1949 and serving in slightly greater numbers for its additional weapons and performance capabilities
MiG-15UTI 'Midget': two-seat trainer version, still widely used for advanced and operational training

Mikoyan-Gurevich MiG-17F 'Fresco-C'
(USSR)
Type: fighter
Accommodation: pilot seated on an ejector seat
Armament (fixed): one 37-mm N-37D cannon with 40 rounds and two 23-mm NR-23 cannon with 80 rounds per gun
Armament (disposable): this is carried on two underwing hardpoints, up to a maximum weight of 500 kg (1,102 lb); typical loads are four pods each with eight 55-mm (2.17-in) rockets, or two 250-kg (551-lb) bombs
Electronics and operational equipment: communication and navigation equipment, plus a gyro sight
Powerplant and fuel system: one 3400-kg (7,495-lb) afterburning thrust Klimov VK-1F turbojet, plus a total internal fuel capacity of 1410 litres (310 Imp gal) in two fuselage tanks, plus provision for two 240- or 400-litre (53- or 88-Imp gal) drop-tanks
Performance: maximum speed 1145 km/h (711 mph) or Mach 0.97 at 3000 m (9,845 ft); initial climb rate 3900 m (12,795 ft) per minute; service ceiling 16600 m (54,460 ft); range 1470 km (913 miles)
Weights: empty 4100 kg (9,040 lb); normal take-off 5340 kg (11,773 lb); maximum take-off 6700 kg (14,770 lb)
Dimensions: span 9.63 m (31 ft 7 in); length 11.26 m (36 ft 11.25 in); height 3.35 m (11 ft 0 in); wing area 22.60 m² (243.3 sq ft)

Variants
MiG-17F 'Fresco-C': the MiG-17 was developed in the 1950s as a much-improved MiG-15, and was authorized for full-scale production in June 1951, deliveries beginning in 1952; the MiG-17 retained the VK-1 non-afterburning turbojet of the MiG-15bis and has now disappeared from service, but was followed by the MiG-17F with the VK-1F afterburning turbojet, wider airbrakes and an exposed jetpipe; the type was also built in Poland as the **LIM-5P**
MiG-17PF 'Fresco-D': all-weather version of the MiG-17F day fighter with RP-5 Izumrud radar and an armament of three 23-mm cannon (supplemented in the out-of-service MiG-17PFU 'Fresco-E' derivative by four AA-1 'Alkali' beam-riding air-to-air missiles)
Shenyang J-5: designation of the MiG-17F built in China, and exported with the designation F-5
Shenyang J-5A: designation of the MiG-17PF built in China, and exported with the designation **F-5A**
Shenyang JJ-5: designation of the Chinese-developed operational trainer version, combining the forward fuselage of the MiG-15UTI with the rest of the airframe of the J-5A; the type has ranging radar in the upper portion of the inlet lip, armament of one 23-mm cannon, powerplant of one 2700-kg (5,952-lb) Wopen-5D (VK-1A) turbojet, a maximum take-off weight of 6215 kg (13,700 lb) and a length of 11.50 m (37.73 ft); the type has been exported as the **FT-5** and **F-5T**

Mikoyan-Gurevich MiG-19SF 'Farmer-C'
(USSR)
Type: fighter and ground-attack aircraft
Accommodation: pilot seated on an ejector seat
Armament (fixed): three 30-mm NR-30 cannon with 80 rounds per gun
Armament (disposable): this is carried on two underwing hardpoints, and can consist of two AA-2 'Atoll' air-to-air missiles, or two 212-mm (8.35-in) rockets, or two packs each with eight 55-mm (2.17-in) rockets, or two 125- or 250-kg (227- or 551-lb) bombs
Electronics and operational equipment: communication and navigation equipment, and a reflector sight
Powerplant and fuel system: two 3250-kg (7,165-lb) afterburning thrust Tumansky RD-9BF turbojets, and a total internal fuel capacity of 2170 litres (477 Imp gal) in fuselage tanks, plus provision for two 200-, 300-, 400-, 800 or 1520-litre (44-, 66-, 88-, 176- or 344-Imp gal) drop-tanks
Performance: maximum speed 1450 km/h (901 mph) or Mach 1.35 at 10000 m (32,810 ft); cruising speed 950 km/h (590 mph) at 10000 m (32,810 ft); initial climb rate 6900 m (22,635 ft) per minute; service ceiling 17900 m (58,725 ft); range 1390 km (863 miles) on internal fuel
Weights: empty 5170 kg (11,397 lb); normal take-off 7400 kg (16,314 lb); maximum take-off 8900 kg (19,621 lb)
Dimensions: span 9.20 m (30 ft 2.25 in); length 12.60 m (41 ft 4 in) excluding probe; height 3.90 ft (12 ft 9.5 in); wing area 25.0 m² (269.1 sq ft)

Variants
MiG-19SF 'Farmer-C': vying with the North American F-100 Super Sabre for the historical niche as the world's first supersonic operational aircraft, the MiG-19 resulted from an appreciation that the MiG-17 was little more than an upgraded MiG-15, and that a new design was required for truly supersonic performance;

the result was the conceptually advanced MiG-19, which flew in definitive prototype form during September 1953 on the power of two Mikulin AM-5 turbojets, earlier prototypes having experimented with a single Lyulka AL-5, or twin Klimov VK-7Fs or twin AM-5s; the type began to enter service in 1954 as the MiG-19F 'Farmer-A' with AM-5F afterburning turbojets and as the MiG-19PF 'Farmer-B' with radar, but these types were withdrawn because of high accident rates; the first major production version was thus the MiG-19S (later redesignated MiG-19SF) with an all-moving tailplane, Tumansky RD-9B turbojets, a revised control system and a gun armament of three long-barrel 30-mm cannon in place of the mixed 23-and 37-mm battery carried by the earlier aircraft

MiG-19PF 'Farmer D': limited all-weather development of the MiG-19SF with RP-5 Izumrud radar and an armament of two 30-mm cannon; the MiG-19PM was similar but carried 'Scan Odd' radar, four AA-1 'Alkali' beam-riding air-to-air missiles and no cannon

Mikoyan-Gurevich MiG-21MF 'Fishbed-J'
(USSR)
Type: multi-role fighter
Accommodation: pilot seated on a KM-1 zero/zero ejector seat
Armament (fixed): one 23-mm GSh-23L twin-barrel cannon with 200 rounds in a belly pack
Armament (disposable): this is carried on four underwing hardpoints, up to a maximum weight of about 1500 kg (3,307 lb); for the interception role, typical loads are two AA-2 'Atoll' air-to-air missiles plus two AA-2-2 'Advanced Atoll' air-to-air missiles or two UV-16 57 launchers each with 16 55-mm (2.17-in) rockets, or four AA-2-2 'Advanced Atoll' air-to-air missiles, or two AA-2-2 'Advanced Atoll' air-to-air missiles and two drop-tanks; for the ground-attack role, typical loads are four UV-16-57 launchers each with 16 rockets, or two 500-kg (1102-lb) and two 250-kg (551-lb) bombs, or four S-24 240-mm (9.45-in) air-to-surface rockets
Electronics and operational equipment: communication and navigation equipment, plus 'Jay Bird' fire-control radar, head-up display, Sirena 3 radar-warning receiver, ARL-S data-link and provision for an underfuselage reconnaissance pod (generally with three cameras, one infra-red linscanner and one chaff dispenser), ECM jammer pods and wingtip ECM pods
inlet centrebody and a gyro gunsight
Powerplant and fuel system: one 6600-kg (14,550-lb) afterburning thrust Tumansky R-13-300 turbojet, and a total internal fuel capacity of 2600 litres (572 Imp gal) in fuselage tanks, plus provision for up to three (one carried on a dedicated underfuselage hardpoint) 490-litre (108-Imp gal) drop-tanks
Performance: maximum speed 2230 km/h (1,385 mph) or Mach 2.1 at 11000 m (36,090 ft) and 1300 km/h (807 mph) or Mach 1.06 at sea level; service ceiling about 15250 m (50,030 ft); combat radius 370 km (230 miles) on a hi-lo-hi sortie with four 250-kg (551-lb) bombs; range 1100 km (683 miles) 'clean' on internal fuel, or 1800 km (1,118 miles) for ferrying with three drop-tanks
Weights: empty 5580 kg (12,302 lb); normal take-off 8200 kg (18,077 lb) with four AA-2 missiles; maximum take-off 9400 kg (20,723 lb) with two AA-2 missiles and three drop-tanks
Dimensions: span 7.15 m (23 ft 5.5 in); length 15.76 m (51 ft 8.5 in) including probe; height 4.10 m (13 ft 5.5 in); wing area 23.0 m² (247.6 sq ft)

Variants
MiG-21F 'Fishbed-C': this was the first large-scale production variant of the MiG-21, a type of great military importance in its day, and still serving in large numbers after development into a capable though limited multi-role fighter; the type was designed in response to a 1953 requirement for a short-range clear-weather interceptor, and developed through a series of prototypes with different wing planforms and sweeps but all powered by Tumansky RD-9E or R-11 turbojets; the tailed delta configuration was adopted in 1956 as offering the best combination of desired attributes, and the pre-production E-6 flew in 1957; full production began in 1958 with the MiG-21 'Fishbed-B' powered by the 5100-kg (11,243-lb) R-11 fed by 2340 litres (515 Imp gal) of fuel, and possessing an armament of two 30-mm cannon; from 1959 this type was supplanted by the MiG-21F with 5750-kg (12,676-lb) R-11, provision for a centreline tank of 490-litre (108-Imp gal) capacity and an armament of one 30-mm cannon plus either two AA-2 'Atoll' air-to-air missiles or two UV-16-57 rocket pods; the type was also built in Czechoslovakia without significant modification, and in China as the Xian J-7
MiG-21PF 'Fishbed-D': designation of the 1960 model with an inlet of increased diameter (to allow the incorporation of a larger centrebody with R1L 'Spin Scan-A' interception radar for limited all-weather capability), greater power from the 5950-kg (13,117-lb) R-11F turbojet and 2850 litres (627 Imp gal) of internal fuel to overcome in part the problems of chronically short range; late-production variants on this basic model were the **MiG-21PFS** with blown flaps and rocket-assisted take-off capability, the NATO-designated **'Fishbed-E'** with broader-chord vertical tail surfaces and provision for a GP-9 underfuselage gun pod with twin-barrel GSh-23L 23-mm cannon, and the **MiG-21FL** export model with R2L 'Spin Scan-B' radar and the 6200-kg (13,668-lb) R-11-300 turbojet, but without blown flaps; this last model was built under licence in India
MiG-21PFM 'Fishbed-F': full-production model

incorporating the various improvements of earlier models, and thus having blown flaps, the R-11-300 turbojet, broader-chord vertical tail and R2L radar; the type also introduced a conventional cockpit enclosure, with fixed windscreen plus side-opening canopy in place of the earlier models' single-piece front-hinged unit

MiG-21PFMA 'Fishbed-J': a straightforward development of the MiG-21PFM in most respects, this marked the start of the MiG-21 series' second generation as a multi-role fighter; the type introduced a radical enlargement of the dorsal fairing for reduced drag plus extra avionics and revised tankage for 2600 litres (572 Imp gal) of internal fuel, this reduction being offset by provision for three drop tanks with a total capacity of 1470 litres (323 Imp gal); other improvements were the KM-1 zero/zero ejector seat, more powerful 'Jay Bird' radar, and four rather than two underwing hardpoints for the carriage of AA-2-2 'Advanced Atoll' as well as AA-2 'Atoll' air-to-air missiles; early MiG-21PFMAs had provision for an external GP-9 gun pod, but later aircraft were fitted with an internal GSh-23L cannon installation; the export version was the **MiG-21M**, which was built under licence in India

MiG-21R 'Fishbed-H': tactical reconnaissance version of the MiG-21PFMA with three cameras in a bay just aft of the nosewheel, and also able to carry an underfuselage pod (for optical or infra-red equipment) and tip-mounted electronic countermeasures equipment comparable to that of the 'Fishbed-K'; a derivative of this type was the **MiG-21RF** derived from the MiG-21MF but otherwise similar to the MiG-21R

MiG-21MF 'Fishbed-J': much improved version of the MiG-21PFMA with the 6600-kg (14,550-lb) R-13-300 turbojet offering greater power for lower installed weight; another modification was the addition of a small rear-view mirror in a neat fairing above the windscreen; an improved version of the MiG-21MF was designated **'Fishbed-K'** by NATO with an enlarged dorsal fairing for greater fuel capacity and yet further enhanced aerodynamics; the subvariant also has provision for detachable electronic countermeasures pods at the wing and fin tips

MiG-21bis 'Fishbed-L': third-generation development of the MiG-21 series with a re-engineered airframe for greater strength but reduced weight, updated avionics and a further enlarged dorsail fairing to allow an increase to 2900 litres (638 Imp gal) of internal fuel

MiG-21bisF 'Fishbed-N': definitive third-generation model with the 7500-kg (16,535-lb) Tumansky R-25 turbojet and further improvements to the avionics, in this instance characterized by the 'bow-and-arrow' probe on the nose for air-data sensors necessary for the accurate delivery of air-to-surface ordnance; the type can carry AA-8 'Aphid' as well as AA-2 'Atoll' air-to-air missiles, and licence production has been undertaken in India

MiG-21U 'Mongol-A': designation of the two-seat conversion trainer based on the MiG-21F but with the larger mainwheels and tyres introduced on the MiG-21PF

MiG-21US 'Mongol-B': designation of the two-seat conversion trainer with broader-chord vertical tail and flap blowing; the instructor also has a retractable periscope for improved vision during take-offs and landings

MiG-21UM 'Mongol-B': designation of the two-seat conversion trainer derived from the MiG-21MF and R-13 engine, but otherwise similar to the MiG-21US

Mikoyan-Gurevich MiG-23MF 'Flogger-B'
(USSR)
Type: variable-geometry air-combat and multi-role fighter
Accommodation: pilot seated on a KM-1 zero/zero ejector seat
Armament (fixed): one 23-mm GSh-23L twin-barrel cannon in a fuselage belly pack with some 400 rounds
Armament (disposable): this is carried on five hardpoints (one under the fuselage, one under each engine inlet duct and one under each fixed inner-wing panel) for a maximum of some 3000 kg (6,614 lb) of disposable stores including FAB-500 (1,102-lb) bombs, UV-16-57 (2.17-in) rocket-launcher pods and air-to-air missiles such as the AA-2 'Atoll', AA-2-2 'Advanced Atoll', AA-7 'Apex' and AA-8 'Aphid'; a typical load in the air-combat role is six AA-7 missiles (four on twin launchers under the inlet ducts and two under the inner wings)
Electronics and operational equipment: communication and navigation equipment, plus 'High Lark' radar (with a search range of 85 km/53 miles and a tracking range of 55 km/34 miles), an undernose laser ranger, Sirena 3 radar-warning receiver, SRO-2 IFF, Doppler radar, 'Swift Rod' instrument landing system, and unspecified electronic countermeasures equipment
Powerplant and fuel system: one 11500-kg (25,353-lb) afterburning thrust Tumansky R-29 turbojet, and a total internal fuel capacity of 5750 litres (1,265 Imp gal), plus provision for three 800-litre (176-Imp gal) drop tanks on the underfuselage and outer wing hardpoints
Performance: maximum speed 2450 km/h (1,522 mph) or Mach 2.3 at high altitude, and 1365 km/h (848 mph) or Mach 1.1 at sea level; service ceiling about 18300 m (60,040 ft); combat radius with a centreline tank on a hi-lo-high mission 950 km (590 miles); ferry range with three drop tanks 2500 km (1,553 miles)
Weights: empty about 11000 kg (24,250 lb); normal take-off about 16000 kg (35,274 lb); maximum take-off about 18900 kg (41,667 lb)

Dimensions: span spread 14.25 m (46 ft 9 in) and swept 8.17 m (26 ft 9.5 in); length with probe 18.15 m (59 ft 6.5 in); height 4.35 m (14 ft 4 in); wing area 27.26 m² (293.4 sq ft)

Variants
MiG-23M 'Flogger-B': this was the first production variant of the MiG-23 air-combat fighter, one of the USSR's most important tactical aircraft; the type was schemed in the late 1950s and early 1960s as a variable-geometry successor to the MiG-21 with modest field requirements but considerably better payload/range figures with higher-performance radar and advanced weapons; the type flew in two basic prototype forms (the E-230 tailed delta with a Lyulka AL-7F-1 main engine and a battery of Kolesov lift engines for V/STOL capability, and the E-231 variable-geometry type with an AL-7F-1 for STOL capability) during the first half of the 1960s, the E-231 proving superior in all important respects; the new Tumansky R-27/R-29 series of afterburning turbojets was planned for the production model, but pending deliveries of these powerful engines the pre-production aircraft were completed with the AL-5F-1 turbojet for operational evaluation, resulting in the MiG-23S and MiG-23SM (with two and four hardpoints respectively), both designated 'Flogger-A' by NATO; the first large-scale production model was thus the MiG-23M which appeared in 1971 with the 10200-kg (22,485-lb) R-27 afterburning turbojet; because this engine is lighter and shorter than the AL-5 series, the airframe had to be revised quite considerably, the wings being located farther forward (with greater sweep on the inner portions), the chord of the outer wings being extended to create a large dogtooth, and the rear fuselage being shortened; the radar of this model was initially a less capable 'High Lark' set, though the full-performance equipment has since been retrofitted
MiG-23MF 'Flogger-B': upgraded production model delivered from 1975 with the R-29 afterburning turbojet and the full-capability version of the 'High Lark' radar; this type has limited look-down/shoot down capability
MiG-21U 'Flogger-C': designation of the tandem two-seat operational conversion trainer version of the MiG-23MF with full weapons capability but reduced fuel and the smaller 'Jay Bird' radar
MiG-23UM 'Flogger-C': designation of a later two-seat operational conversion trainer based on the MiG-23M with the R-27 turbojet and possibly capable of doubling in the electronic warfare role
MiG-23 'Flogger-E': export version of the MiG-23M with lower standards of equipment (including the 'Jay Bird' radar with a search range of 29 km/18 miles and a tracking range of 19 km/12 miles), lacking the laser ranger and Doppler navigation, and fitted only for the AA-2 'Atoll' air-to-air missile

MiG-23BN 'Flogger-F': derivative of the MiG-23MF, introduced in 1974 and originally believed to by only for export but since adopted in large numbers by the USSR and its Warsaw Pact allies; the type is a hybrid with MiG-23 and MiG-27 features, and thus capable of the air combat and attack roles; the model is fitted with the MiG-27 nose without radar but offering the better-protected pilot (external armour is added to the sides of the forward fuselage) a considerably enhanced view forward and downward, and fitted with a laser ranger and marked-target seeker, terrain-avoidance radar, and Doppler navigation; the type also has the R-29B turbojet, larger tyres for operations from semi-prepared airstrips, additional electronic countermeasures, and an infra-red seeker; the type can carry 4500 kg (9,921 lb) of ordnance on five hardpoints (one under the fuselage, two under the inlet trunks and two under the fixed inner wings) each fitted with tandem racks
MiG-23MF 'Flogger-G': dedicated air-combat fighter version of the MiG-23MF with simpler avionics (including a lighter radar and a revised undernose sensor blister), a smaller dorsal fin and a revised nosewheel leg
MiG-23BN 'Flogger-H': version of the MiG-23BN with a small avionics blisters on each side of the lower fuselage just forward of the nosewheel doors
MiG-23 'Flogger-K': designation of an aerodynamically improved version with a smaller ventral fin and dogtoothed inboard leading edges on the wing gloves

Mikoyan-Gurevich MiG-25 'Foxbat-A'
(USSR)
Type: interceptor fighter
Accommodation: pilot seated on a KM-1 zero/zero ejector seat
Armament (fixed): none
Armament (disposable): this is carried on four underwing hardpoints, and generally comprises four AA-6 'Acird' air-to-air missiles, or two AA-7 'Apex' and two AA-8 'Aphid' air-to-air missiles
Electronics and operational equipment: communication and navigation equipment, plus 'Fox Fire' radar (with a lock-on range of 90 km/56 miles), Sirena 3 radar-warning receiver, unspecified electronic countermeasures and continuous-wave illuminators for the rading-homing air-to-air missiles
Powerplant and fuel system: two 12300-kg (27,116-lb) afterburning thrust Tumansky R-31 turbojets, and a total internal fuel capacity of 17900 litres (3,938 Imp gal) in fuselage, inlet saddle and integral wing tanks
Performance: maximum speed 2975 km/h (1,849 mph) or Mach 2.8 at high altitude with four AA-6 'Acrid' air-to-air missiles, and 1,040 km/h (646 mph) at sea level with four AA-6 'Acrid' air-to-air missiles; initial climb rate 12480 m (40,945 ft) per minute;

service ceiling 24400 m (80,050 ft); combat radius 1130 km (702 miles)
Weights: empty about 20000 kg (44,092 lb); maximum take-off 37425 kg (82,507 lb)
Dimensions: span 13.95 m (45 ft 9 in); length 23.82 m (78 ft 1.75 in); height 6.10 m (20 ft 0.25 in); wing area 56.83 m² (611.7 sq ft)

Variants
MiG-25 'Foxbat-A': developed specifically to deal with the North American B-70 Valkyrie Mach 3 strategic bomber, which was then cancelled, the MiG-25 is the world's highest-performance interceptor though limited in this performance to non-manoeuvring flight; the type's avionics are also of somewhat elderly vintage; first flown in April 1965 as the E-266 prototype and research aircraft, the MiG-25 began to enter service in 1970
MiG-25R 'Foxbat-B': this is the basic operational reconnaissance version of the MiG-25 with a maximum take-off weight of 33400 kg (73,633 lb) and unkinked wings reduced in span to 13.40 m (44.0 ft); this version has a maximum speed of Mach 3.2 and a combat radius of 1100 km (694 miles), and carries a small side-looking airborne radar and six cameras in the nose (one vertical and five oblique), a small 'Jay Bird' radar in the nose forward of the SLAR installation, Doppler and ground-mapping radar in ventral radomes, and electronic counter-countermeasures equipment
MiG-25U 'Foxbat-C': two-seat operational conversion trainer using a new nose section with a separate trainee cockpit
MiG-25R 'Foxbat-D': variant of the 'Foxbat-B' with a larger SLAR and nose cameras, and probably fitted for the strategic role with electronic intelligence equipment
MiG-25MP 'Foxbat-E': designation of 'Foxbat-As' converted to an updated standard for limited lookdown/shoot-down capability at a lower-altitude cruise; the type has a new pulse-Doppler search-and-track radar and a new undernose sensor blister, and the engines are uprated R-31F units each with an afterburning thrust of 14000 kg (30,864 lb); the type has an internal gun, and the radar is matched to six of the new AA-9 snap-down air-to-air missiles

Mikoyan-Gurevich MiG-27 'Flogger-D'
(USSR)
Type: variable-geometry ground-attack aircraft
Accommodation: pilot seated on an KM-1 zero/zero ejector seat
Armament (fixed): one 23-mm six-barrel cannon with about 700 rounds in a ventral package
Armament (disposable): this is carried on seven hardpoints (one on the centreline rated at 1000 kg/2,205 lb, two under the rear fuselage each rated at 500 kg/ 1,102 lb, one under each inlet rated at 1000 kg/2,205 lb, and one under each wing glove rated at 500 kg/ 1,102 lb) to a maximum of 4000 kg (8,818 lb) or more; typical weapons are the AA-2 'Atoll' and AA-8 'Aphid' air-to-air missiles, AS-7 'Kerry' and AS-14 air-to-surface missiles, AS-9 anti-radiation missile, 500- and 250-kg (1,102- and 551-lb) bombs, and launchers for 16 and 32 55-mm (2.17-in) rockets; the type can also carry two free-fall nuclear weapons
Electronics and operational equipment: communication and navigation equipment, plus head-up display, laser rangefinder and marked-target seeker, terrain-avoidance radar, Doppler navigation, 'Odd Rods' IFF, Sirena 3 radar-warning receiver, and internal/external electronic countermeasures
Powerplant and fuel system: one 11500-kg (25,352-lb) afterburning thrust Tumansky R-29-300 turbojet, and a total internal fuel capacity of 5750 litres (1,265 Imp gal), plus provision for three 800-litre (176-Imp gal) drop-tanks
Performance: maximum speed 1807 km/h (1,123 mph) or Mach 1.7 at high altitude, and 1345 km/h (836 mph) or Mach 1.1 at sea level; service ceiling 16000 m (52,495 ft); combat radius 390 km (240 miles) on a lo-lo-lo sotie with two AA-2 air-to-air missiles, four 500-kg (1,102-lb) bombs and one 800-litre (176-Imp gal) drop tank; ferry range 2500 km (1,553 miles)
Weights: empty 10790 kg (23,788 lb); normal take-off 15500 kg (34,170 lb); maximum take-off 20100 kg (44,313 lb)
Dimensions: span spread 14.25 m (46 ft 9 in) and swept 8.17 m (26 ft 9.7 in); length 16.00 m (52 ft 5.9 in); height 4.50 m (14 ft 9.2 in); wing area, spread 27.25 m² (293.33 sq ft)

Variants
MiG-27 'Flogger-D': derived from the MiG-23 as a dedicated ground-attack aircraft, the MiG-27, has a new forward fuselage with fixed rather than variable-geometry air inlets (and two-position nozzles) plus a 'duck' nose without radar; this nose section has a fair degree of armour protection and seats the pilot relatively higher than in the MiG-23, offering him far better fields of vision to the front, sides and below over the flattened nose; the nose itself its fitted with a laser ranger and marked-target seeker, and with terrain-avoidance radar; the weapon capability is similar to that of the MiG-23BN in terms of weapons carried, but somewhat greater in terms of weight, and the type has a new six-barrel cannon optimized for the ground-attack role
MiG-27 'Flogger-J': revised version produced from 1981 with sensors that differ in detail from the earlier model's fit, and provision for two gun pods (each containing one 23-mm GSh-23L twin-barrel cannon) with the barrels angled for oblique downward fire

Mikoyan-Gurevich MiG-29 'Fulcrum'
(USSR)

Type: air-superiority and attack fighter

Accommodation: pilot seated on a zero/zero ejector seat

Armament (fixed): probably one 30-mm twin-barrel cannon with ? rounds in the port leading-edge root extension

Armament (disposable): this is carried on six underwing hardpoints up to a weight of ? kg (? lb), and can comprise AA-10 medium-range and/or AA-11 short-range air-to-air missiles, bombs, rockets pods and other stores including electronic warfare pods and drop tanks

Electronics and operational equipment: communication and navigation equipment, plus a pulse-Doppler nose radar, head-up display, infra-red sensor (with search and tracking functions), Sirena 3 radar-warning receiver, 'Odd Rods' IFF and other systems

Powerplant and fuel system: two 8300-kg (18,298-lb) afterburning thrust Tumansky R-33D turbofans, and a total internal fuel capacity of 4000 kg (8,818 lb) in integral tanks; provision for inflight-refuelling

Performance: maximum speed 2335 km/h (1,450 mph) or Mach 2.2 at altitude and 1300 km/h (805 mph) or Mach 1.06 at sea level; service ceiling more than 15000 m (49,215 ft); combat radius 1150 km (715 miles)

Weights: (estimated) empty 7825 kg (17,251 lb); maximum take-off 16500 kg (36,376 lb)

Dimensions: span 11.50 m (37 ft 8.75 in); length including probe 17.20 m (56 ft 5.2 in); height 4.40 m (14 ft 5.2 in); wing area 35.5 m² (382 sq ft)

Variant

MiG-29 'Fulcrum': entering service in 1985, the MiG-29 is a dual-role fighter resembling the McDonnell Douglas F-15 Eagle but sized like the McDonnell Douglas F/A-18 Hornet and, like that latter fighter, optimized for the air combat role with attack a powerful secondary capability; the type clearly possesses great agility, and a genuine look-down/shoot-down capability is offered by the combination of a 40-km (25-mile) range pulse-Doppler radar and the new AA-10 snap-down air-to-air missile

Mikoyan-Gurevich MiG-31 'Foxhound'
(USSR)

Type: long-range all-weather interceptor fighter

Accommodation: pilot and radar operator seated in tandem on zero/zero ejector seats

Armament (fixed): none

Armament (disposable): this is carried on four under-fuselage and four underwing hardpoints, and generally comprises four AA-8 and four AA-9 air-to-air missiles

Electronics and operational equipment: communication and navigation equipment, plus a high-power pulse-Doppler nose radar, 'Odd Rods' IFF, Sirena 3 radar-warning receiver, and other electronic systems

Powerplant and fuel system: probably two 14000-kg (30,864-lb) thrust Tumansky R-31F afterburning turbojets, and a total internal fuel capacity of ? litres (? Imp gal) in integral tanks, plus drop tanks

Performance: maximum speed 2555 km/h (1,588 mph) or Mach 2.4 at altitude; combat radius 1500 km (932 miles)

Weights: (estimated) empty 21825 kg (48,115 lb); maximum take-off 41150 kg (90,719 lb)

Dimensions: (estimated) span 14.00 m (45 ft 11.2 in); length 25.00 m (82 ft 0.25 in); height 6.10 m (20 ft 0.2 in); wing area 58.00 m² (624.33 sq ft)

Variant

MiG-31 'Foxhound': a much revised development of the MiG-25MP with more powerful variants of the R-31 engine, a lengthened two-seat fuselage, leading-edge root extensions, and revised inlets and nozzles, the MiG-31 is optimized for interception of intruders at lower altitudes in conjunction with a more capable look-down radar and the new radar-homing AA-9 missile (of which four are carried in tandem pairs under the central fuselage), so offering the very real possibility of detecting and destroying low-level intruders such as penetration bombers and even cruise missiles; the 'Foxhound' is in every respect an extremely capable interceptor characterized by advanced weapons, avionics, sensors and propulsion

Mil Mi-2 'Hoplite' (USSR): see WSK-PZL Swidnik (Mil) Mi-2 'Hoplite'

Mil Mi-4 'Hound-A'
(USSR)

Type: utility helicopter

Accommodation: crew of two on the flightdeck, and up to 11 passengers or 1740 kg (3,836 lb) of freight in the cabin

Armament (fixed): optional 7.62-mm (0.3-in) or 12.7-mm (0.5-in) machine-gun under the fuselage

Armament (disposable): provision for rocket-launcher pods

Electronics and operational equipment: communication and navigation equipment

Powerplant and fuel system: one 1268-kW (1,700-hp) Shvetsov ASh-82V radial piston engine, and a total internal fuel capacity of 1,000 litres (220 Imp gal) in the fuselage, plus provision for one 500-litre (110-Imp gal) auxiliary tank in the cabin or carried externally

Performance: maximum speed 210 km/h (130 mph) at 1500 m (4,920 ft); cruising speed 160 km/h (99 mph); service ceiling 5500 m (18,050 ft); range 400 km (249 miles) with 10 passengers

Weights: empty 5270 kg (11,618 lb); normal take-off 7350 kg (16,204 lb); maximum take-off 7800 kg (17,196 lb)

Dimensions: main rotor diameter 21.00 m (68 ft 11 in); length (fuselage) 16.80 m (55 ft 1 in); height 4.40 m (14 ft 5.25 in); main rotor disc area 346.36 m² (3,728.3 sq ft)

Variants
Mi-4 'Hound-A': first flown in May 1952, the 'Hound' is a conventional but still useful general-purpose helicopter of the piston-engined variety; the type can carry 14 troops or a light vehicle loaded through the clamshell rear doors; the slung load capability is 1300 kg (2,866 lb)
Mi-4 'Hound-B': anti-submarine version with under-nose search radar, magnetic anomaly detector (using a towed 'bird'), and sonobuoys
Mi-4 'Hound-C': electronic countermeasures version equipped to jam enemy battlefield communications
Harbin Z-5: designation of the Chinese-built version

Mil Mi-6A 'Hook'
(USSR)
Type: heavy transport and assault helicopter
Accommodation: crew of five, and up to 65 passengers or 41 litters and two attendants, or 12000 kg (26,455 lb) of freight in the cabin
Armament (fixed): optional 12.7-mm (0.5-in) machine-gun in the nose
Armament (disposable): none
Electronics and operational equipment: communication and navigation equipment
Powerplant and fuel system: two 4100-kW (5,500-shp) Soloviev D-25V (TV-2BM) turboshafts, and a total internal fuel capacity of 7910 litres (1,740 Imp gal) in 11 fuselage tanks, plus provision for 4375 litres (962 Imp gal) in two external tanks and for 4375 litres (962 Imp gal) in two cabin ferry tanks
Performance: maximum speed 300 km/h (186 mph); cruising speed 250 km/h (155 mph); service ceiling 4500 m (14,765 ft); range 620 km (385 miles) with an 8000-kg (17,637-lb) payload, or 1450 km (900 miles) for ferrying with maximum fuel
Weights: empty 27240 kg (60,055 lb); normal take-off 40500 kg (82,285 lb); maximum take-off 42500 kg (93,700 lb) for VTO
Dimensions: main rotor diameter 35.00 m (114 ft 10 in); length (fuselage) 33.18 m (108 ft 10.5 in); height 9.86 m (32 ft 4 in); main rotor disc area 962.1 m² (10,356.4 sq ft)

Variants
Mi-6 'Hook': first flown in September 1957, the 'Hook' was for its time the world's largest helicopter, and is still a prodigious machine capable of lifting a slung load of 8000 kg (17,637 lb) as an alternative to a load of troops, freight and light vehicles in a hold accessed by clamshell rear doors
Mi-6A 'Hook': late-production model with a number of detail improvements

Mil Mi-8 'Hip-C'
(USSR)
Type: transport helicopter
Accommodation: crew of two or three on the flight-deck, and up to 32 passengers or 4000 kg (8,818 lb) of freight in the cabin
Armament (fixed): one 12.7-mm (0.5-in) machine-gun in the nose
Armament (disposable): this is carried on a twin rack on each side of the cabin, and can comprise a total of 128 55-mm (2.17-in) rockets in four suspended packs, or other weapons such as four AT-2 'Swatter' anti-tank missiles
Electronics and operational equipment: communication and navigation equipment
Powerplant and fuel system: two 1267-kW (1,700-shp) Isotov TV2-117A turboshafts, and a total fuel capacity of 1870 litres (411 Imp gal) in one fuselage and two external tanks, plus provision for 1830 litres (403 Imp gal) of auxiliary fuel in two cabin tanks
Performance: maximum speed 260 km/h (161 mph) at 1000 m (3,280 ft); cruising speed 180 km/h (112 mph); service ceiling 4500 m (14,760 ft); range 445 km (276 miles) with maximum freight load or 1200 km (745 miles) for ferrying with maximum fuel
Weights: empty 6625 kg (14,605 lb); maximum take-off 12000 kg (26,455 lb)
Dimensions: main rotor diameter 21.29 m (69 ft 10.5 in); length (fuselage) 18.17 m (59 ft 7.5 in); height 5.65 m (18 ft 6.5 in); main rotor disc area 356.0 m² (3,832.0 sq ft)

Variants
Mi-8 'Hip-C': basically a turbine-engined development of the Mi-4 for civil and military applications, and first flown in 1961 as the 'Hip-A' with a four-blade main rotor driven by a single Soloviev turboshaft transmission-limited to 2013 kW (2,700 shp), the Mi-8 was clearly a highly capable machine, and the second prototype (first flown in September 1962) pioneered the twin-turboshaft powerplant, the definitive five-blade main rotor being evaluated in the 'Hip-B' prototype; the 'Hip-C' was thus the first production model, and has proved itself in every way an excellent machine with the capacity to lift a slung load of 3000 kg (6,614 lb); features found only on the military version are Doppler navigation and provision for a considerable weight and diversity of armament
Mi-8 'Hip-D': this is a battlefield communications relay helicopter distinguishable by its additional antennae and external box fairings
Mi-8 'Hip-E': this is the advanced battlefield version of the 'Hip' series, and is generally reckoned to be the world's most heavily armed helicopter, with provision for six packs containing a total of 192 55-mm (2.17-in) rockets plus four AT-2 'Swatter' anti-tank missiles, complemented by a 12.7-mm (0.5-in) flexible machine-gun in the nose

Mi-8 'Hip-F': this is the export version of the 'Hip-E' with six AT-3 'Sagger' missiles in place of the 'Swatters' of the Soviet type

Mi-8 'Hip-G': this is a battlefield communications relay helicopter slightly different from the 'Hip-D'

Mi-8 'Hip-J': this is an electronic warfare derivative of the basic series with a number of unspecified systems characterized by external aerials

Mi-8 'Hip-K': this is an advanced electronic countermeasures variant with antennae on the sides of the rear fuselage and box fairings on the cabin sides; the type has no Doppler navigation

Mi-14 'Haze-A': this is the land-based anti-submarine variant with retractable landing gear, a boat hull and stabilizing sponsons for ditching capability; the type has 1417-kW (1,900-shp) TV3-117 or 1641-kW (2,200-shp) TV3-117MT turboshafts and an estimated maximum take-off weight of 13000 kg (28,660 lb); operational equipment includes search radar in an undernose radome, and magnetic anomaly detection (using a towed 'bird') plus a weapons load of torpedoes and/or depth charges; the type can also be used in the search-and-rescue role with armament replaced by a rescue winch and other mission equipment

Mi-14 'Haze-B': mine countermeasures version of the 'Haze-A'

Mi-17 'Hip-H': this is a much uprated and updated version of the Mi-8 with the dynamic system of the Mi-14 main rotor and the tail rotor relocated to the port side of the fin; the cabin is enlarged slightly to allow carriage of the greater payload made possible by the improved powerplant; the type began to enter service in 1981 as a utility and assault helicopter well-suited to 'hot-and-high' operations, and it is likely that the type will be further developed as a tactical helicopter; maximum take-off weight is 13000 kg (28,660 lb) and the type can carry a standard payload of 4000 kg (8,818 lb)

Mil Mi-14 'Haze' (USSR): see Mil Mi-8 'Hip'

Mil Mi-17 'Hip' (USSR): see Mil Mi-8 'Hip'

Mil Mi-24 'Hind-A'
(USSR)
Type: armed assault helicopter
Accommodation: crew of four, and up to eight troops in the cabin
Armament (fixed): one 12.7-mm (0.5-in) machine-gun on a flexible mount in the nose
Armament (disposable): this is carried on four underwing hardpoints and two wingtip twin rails, and usually comprises four UV-32 launchers each with 32 55-mm (2.17-in) rockets on the underwing hardpoints, and four AT-2 'Swatter' anti-tank missiles on the tip installations; the maximum external weapons load is 1275 kg (2,811 lb)

Electronics and operational equipment: communication and navigation equipment
Powerplant and fuel system: two 1641-kW (2,200-shp) Isotov TV3-117 turboshafts, and a maximum internal fuel capacity of ? litres (? Imp gal) in fuselage tanks
Performance: maximum speed about 320 km/h (199 mph)
Weights: empty about 6500 kg (14,330 lb); normal take-off 10000 kg (22,046 lb)
Dimensions: main rotor diameter 17.00 m (55 ft 9 in); length (overall) 17.00 m (55 ft 9 in); height (overall) 4.25 m (14 ft 0 in); main rotor disc area 227.0 m^2 (2,443.3 sq ft)

Variants

Mi-24 'Hind-A': developed in the mid-1960s as a high-performance multi-role helicopter, the Mi-24 has been the subject of contested evaluations in the West, in relation largely to the tactical capability of the gunship models; the type is based on the dynamic system of the Mi-14, in this application married to a new fuselage of slender lines to enhance performance and battlefield survivability; the 'Hind-A' was the second production model, the initial **'Hind-B'** having been produced in only small numbers, probably as a pre-production batch, for service from 1973; this 'Hind-B' is distinguishable by its straight wings with only four hardpoints, and by the location of the tail rotor on the starboard side in the fin

Mi-24 'Hind-C': version of the 'Hind-A' without the nose gun and wingtip missile launcher rails

Mi-24 'Hind-D': much altered version used in the dedicated gunship role with a new forward fuselage featuring stepped cockpits for the gunner (nose) and pilot (behind and slightly above the gunner); the gunner controls a four-barrel 12.7-mm (0.5-in) heavy machine-gun in a turret under the nose, while the wing-mounted armament is similar to that of the 'Hind-A'; the extensive sensor fit for the accurate firing of air-to-surface ordnance includes an air data probe, low-light-level TV, radar and a laser tracker; the maximum weapons load is 1500 kg (3,307 lb), and though the cabin can carry an eight-man infantry squad it is likely that only one man and reload weapons are carried for battlefield rearming; though the type clearly possesses considerable speed and offensive capability, Western analysts point out that the type is large and relatively unmanoeuvrable for the gunship role

Mi-25 'Hind-D': export version of the Mi-24 'Hind-D' with avionics of a reduced standard

Mi-24 'Hind-E': improved version of the 'Hind-D' with provision for AT-6 'Spiral' instead of AT-2 'Swatter' anti-tank missiles

Mi-24 'Hind-?': version of the 'Hind-E' with a revised nose armament of one twin-barrel cannon in a pack fixed on the starboard side of the nose

Mil Mi-25 'Hind-D' (USSR): see Mil Mi-24 'Hind-A'

Mil Mi-26 'Halo'
(USSR)
Type: heavy transport helicopter
Accommodation: crew of four on the flightdeck, and a loadmaster and up to 20 passengers, or a loadmaster and 20000 kg (44,092 lb) of freight in the hold
Armament (fixed): none
Armament (disposable): none
Electronics and operational equipment: communication and navigation equipment
Powerplant and fuel system: two 8504-kW (11,400-shp) Lotarev D-136 turboshafts, and a total internal fuel capacity of ? litres (? Imp gal) in two feeder and eight underfloor fuselage tanks
Performance: maximum speed 295 km/h (183 mph); cruising speed 255 km/h (158 mph); service ceiling 4500 m (14,760 ft); range 800 km (497 miles) with maximum fuel
Weights: empty 28200 kg (62,170 lb); normal take-off 49500 kg (109,127 lb); maximum take-off 56000 kg (123,457 lb)
Dimensions: main rotor diameter 32.00 m (105 ft 9 in); length (fuselage) 33.73 m (110 ft 8 in); height (rotor head) 8.055 m (26 ft 5.25 in); main rotor disc area 804.25 m² (8,657 sq ft)

Variant
Mi-26 'Halo': first flown in 1979, the 'Halo' is the world's largest production helicopter, and highly capable heavy-lift type scaled up from the Mi-6 but with proportionally more power to drive an eight-blade main rotor; as with other such Soviet helicopters the large hold is accessed by clamshell rear doors and a ramp for the loading of vehicles and artillery

Mil Mi-28 'Havoc'
(USSR)
Type: air-combat and gunship helicopter
Accommodation: weapons operator and pilot seated in tandem in separate stepped cockpits
Armament (fixed): probably one 30-mm cannon in an underfuselage turret with ? rounds
Armament (disposable): this is carried on four hardpoints under the stub wings, and is believed to include eight modified AT-6 'Spiral' anti-tank missiles in two four-tube boxes under the inner hardpoints, and four air-to-air missiles in two twin launchers under the outer hardpoints
Electronics and operational equipment: communication and navigation equipment, plus a variety of unknown sensors in the nose, as well as electronic countermeasures systems
Powerplant and fuel system: probably two 1641-kW (2,200-shp) Isotov TV3 turboshafts, and a total internal fuel capacity of ? litres (? Imp gal)
Performance: maximum speed 355 km/h (221 mph); combat radius 240 km (149 miles)
Weights: empty 8000 kg (17,637 lb); maximum take-off 11000 kg (24,250 lb)
Dimensions: main rotor diameter 17.00 m (55 ft 9.3 in); fuselage length 17.40 m (57 ft 1 in); main rotor disc area 226.98 m² (2,443.3 sq ft)

Variant
Mi-28 'Havoc': entering service in 1985 or 1986, the 'Havoc' seems to have confirmed Western doubts about the battlefield viability of the Mi-24 gunship models, for while this new machine is clearly derived from earlier Mil helicopters (the powerplant of the Mi-24 and the rotor system of the Mi-8), it has adopted the US practice of a much slimmer and smaller fuselage for increased manoeuvrability and reduced vulnerability over the modern high-technology battlefield; the 'Havoc' thus bears a passing resemblance to the Hughes AH-64A Apache in US Army service, and amongst its features are infra-red suppression of the podded engines, infra-red decoys, upgraded armour, electro-optical sighting and targeting systems for use in conjunction with the undernose 30-mm cannon and disposable weapons (including air-to-air missiles) carried on the stub wings' four hardpoints, and (possibly) millimetre-wavelength radar; the type clearly possesses an air-combat capability against other battlefield helicopters

Mitsubishi F-1
(Japan)
Type: close-support fighter
Accommodation: pilot seated on a Daiseru-built Weber ES-7J ejector seat
Armament (fixed): one 20-mm JM61A1 Vulcan rotary-barrel cannon with 750 rounds
Armament (disposable): this is carried on one underfuselage, four underwing and two wingtip hardpoints up to a maximum 2780 kg (6,129 lb); among the stores options available are four AIM-9 or AAM-1 air-to-air missiles (carried as two on the outer hardpoints and two on the wingtips), or two ASM-1 (Type 80) anti-ship missiles, or four JLAU-3/A launchers each with 19 2.75-in (69.85-mm) rockets, or four RL-7 launchers each with seven 2.75-in (69.85-mm) rockets, or four RL-4 launchers each with four 5-in (127-mm) rockets, or up to 12 250-lb (113-kg) or 500-lb (227-kg) bombs, or eight 750-lb (340-kg) bombs
Electronics and operational equipment: communication and navigation equipment, plus Mitsubishi Electric J/AWG-11 air-to-air and air-to-surface radar, Mitsubishi Electric (Thomson-CSF) head-up display, Mitsubishi Electric J/AWG-1 fire-control system and bombing computer, Ferranti 6TNJ-F inertial navigation system, Lear Siegler 5010BL attitude and heading reference system, Tokyo Keiki APR-4 radar-warning receiver, and provision for electronic countermeasures pods

Powerplant and fuel system: two 7,070-lb (3207-kg) afterburning thrust Rolls-Royce/Turboméca Adour Mk 801A turbofans, and a total internal fuel capacity of 3823 litres (841 Imp gal) in seven fuselage tanks, plus provision for three 821-litre (180-Imp gal) drop-tanks

Performance: maximum speed 1700 km/h (1,056 mph) or Mach 1.6 at 10970 m (36,000 ft); initial climb rate 10670 m (35,000 ft) per minute; service ceiling 15240 m (50,000 ft); combat radius 555 km (345 miles) on a hi-lo-hi sortie with two ASM-1s and one drop-tank; ferry range 2600 km (1,616 miles) with maximum internal and external fuel

Weights: empty 6358 kg (14,017 lb); normal take-off 9860 kg (21,737 lb); maximum take-off 13675 kg (30,148 lb)

Dimensions: span 7.88 m (25 ft 10.25 in); length 17.84 m (58 ft 6.25 in); height 4.28 m (14 ft 4.25 in); wing area 21.18 m² (228.0 sq ft)

Variants

F-1: this trim close-support and anti-shipping attack fighter was derived from the T-2 supersonic trainer, and bears more than a passing resemblance to the SEPECAT Jaguar, with which it shares a twin Adour powerplant; the type first flew as a T-2 conversion in June 1975, the primary modification being the plating over of the rear cockpit to provide volume for an inertial navigation system, bombing computer and electronic countermeasures system

T-2: designed as Japan's first supersonic aircraft, the T-2 first flew in prototype form during July 1971 and entered service as an unarmed flying trainer; the **T-2A** is the weapons training and light attack equivalent, this having the armament potential of the F-1 but without the inertial navigation, bombing and electronic countermeasures systems

Mitsubishi (Sikorsky) HSS-2 (Japan): see Sikorsky S-61B (SH-3D Sea King)

Myasishchev M-4 'Bison-C'
(USSR)

Type: maritime reconnaissance aircraft

Accommodation: crew of four or five on the flight-deck, and a variable mission crew in the fuselage

Armament (fixed): two 23-mm NR-23 cannon each in one manned rear turret and two remotely-controlled dorsal and ventral barbettes

Armament (disposable): up to 4500 kg (9,921 lb) of stores can be carried in the underfuselage weapons bays

Electronics and operational equipment: communication and navigation equipment, plus 'Puff Ball' missile-guidance radar, 'Bee Hind' tail-warning and gunlaying radar, and a wide assortment of mission electronics with antennae on the nose, along the fuselage and at the wingtips

Powerplant and fuel system: four 9500-kg (20,943-lb) thrust Mikulin AM-3D turbojets, or (according some authorities) four 13000-kg (28,660-lb) thrust Soloviev D-15 turbojets, and a total internal fuel capacity of ? litres (? Imp gal); provision for inflight refuelling

Performance: maximum speed 900 km/h (559 mph) at altitude; service ceiling 15000 m (49,215 ft); range 11000 km (6,835 miles) with 4500-kg (9,921-lb) payload

Weights: empty 90000 kg (198,413 lb); maximum take-off 210000 kg (462,963 lb)

Dimensions: span 52.50 m (172 ft 3 in); length 53.40 m (175 ft 2.33 in); height 14.10 m (46 ft 3 in); wing area 320.0 m² (3,444.0 sq ft)

Variants

M-4 'Bison-A': conceived as a long-range strategic bomber, the M-4 first flew in late 1953 and entered service in 'Bison-A' form with four 8700-kg (19,180-lb) Mikulin AM-3D turbojets during 1955; the type has a defensive armament of six 23-mm cannon in three positions (one manned in the tail and two remotely controlled above and below the fuselage) plus at least 15000 kg (33,060 lb) of free-fall weapons; most of the bombers have been revised since the early 1960s as inflight-refuelling tankers with D-15 engines plus extra fuel and a hosereel units in the erstwhile bomb bay, and additional tankage in the front and rear portions of the fuselage

M-4 'Bison-B': this is the maritime version with an observation nose and ventral blister, and equipped for multi-sensor reconnaissance; the type also has an inflight-refuelling capability, and by the 1980s the type was known to have at least 12 electronic intelligence receivers and (possibly) the capability for mid-course updating of anti-ship missiles

M-4 'Bison-C': updated version of the 'Bison-B' with large radar (believed to be 'Puff Ball' for AS-2 missile guidance) in a lengthened nose, as well as reconnaissance equipment

Nanchang A-5 'Fantan-A'
(China)

Type: attack aircraft

Accommodation: pilot seated on a Martin-Baker PKD10 ejector seat

Armament (fixed): two 23-mm NR-23-2 cannon with ? rounds per gun

Armament (disposable): this is accommodated in a small bomb-bay (deleted in favour of extra fuel on later aircraft) and on eight hardpoints (four under the fuselage and four under the wings, the two outer underwing points normally carrying drop-tanks) to a maximum of 2000 kg (4,409 lb); the bomb-bay can accept four 250-kg (551-lb) bombs or a single 20-kiloton nuclear weapon; the fuselage hardpoints can each lift a single 250-kg (551-lb) bomb, and the underwing points can each accept a single 250-kg (551-lb)

bomb, or one AAM (R.550 Magic, AIM-9 Sidewinder or PL-2 or PL-7 versions of the AA-2 'Atoll') or a pod for eight unguided rockets (Russian S-5 55-mm/2.17-in or Chinese 90-mm/3.54-in weapons); the maximum weapon weight is 2000 kg (4,409 lb)
Electronics and operational equipment: communication and navigation equipment, plus an optical sight
Powerplant and fuel system: two 3750-kg (8,267-lb) afterburning thrust Wopen-6 (Tumansky R-9BF-811) turbojets, and a total internal fuel capacity of 3700 litres (814 Imp gal), plus provision for two 760-litre (167-Imp gal) drop-tanks
Performance: maximum speed 1435 km/h (890 mph) or Mach 1.35 at high altitude and 1160 km/h (722 mph) or Mach 0.95 at sea level; initial climb rate 6000 m (19,685 ft) per minute; service ceiling 16000 m (52,495 ft); combat radius 370 km (230 miles) on a lo-lo-lo sortie with four 250-kg (551-lb) bombs and two drop-tanks; ferry range 1850 km (1,150 miles)
Weights: empty 6200 kg (13,670 lb); normal take-off 10700 kg (23,590 lb); maximum take-off 12000 kg (26,455 lb)
Dimensions: span 10.20 m (33 ft 5 in); length 15.65 m (51 ft 4.1 in); height 4.50 m (14 ft 9.5 in)

Variants
A-5I 'Fantan-A': this is a drastic but basically conventional Chinese development of the MiG-19/Shenyang J-6 series with slightly larger overall dimensions, a comparable powerplant and a completely redesigned forward fuselage to make possible the installation of a small bomb bay; to this end the previous nose inlet has been replaced by two lateral inlets; the original Chinese scheme was apparently to fit turbofan engines and nose radar when these became available, making the A-5 an able medium-range tactical strike aircraft with free-fall nuclear weapons; the type has now been kept as a conventional attack aircraft and fighter-bomber with turbojets, the bomb bay being used for 70% more fuel; the **A-5II** is an export variant for North Korea
A-5III 'Fantan-A': designation of the export version with revised avionics and (on later aircraft) an additional pair of hardpoints for self-defence air-to-air missiles; the aircraft in Pakistani service have been considerably upgraded by the addition of Western avionics
A-5M 'Fantan-A': designation of an improved version under development with the aid of Aeritalia, which is responsible for advanced avionics based on those of the AMX and including a head-up display

North American F-1000D Super Sabre
(USA)
Type: interceptor and fighter-bomber
Accommodation: pilot seated on an ejector seat
Armament (fixed): four 20-mm M39 cannon with 200 rounds per gun in the fuselage

Armament (disposable): this is carried on six underwing hardpoints up to a maximum weight of 7,500 lb (3402 kg), and can comprise air-to-air and air-to-surface missiles, bombs and rocket-launchers
Electronics and operational equipment: communication and navigation equipment, plus APG-25(V) gun-tracking radar, radar gunsight and APR-26(V) radar-warning receiver
Powerplant and fuel system: one 17,000-lb (7711-kg) afterburning thrust Pratt & Whitney J57-P-21A turbojet, and a total internal fuel capacity of 2088 US gal (7,905 litres) in wing and fuselage tanks, plus provision for two 450-US gal (1,705-litre) drop-tanks; provision for inflight-refuelling
Performance: maximum speed 865 mph (1392 km/h) or Mach 1.31 at 35,000 ft (10670 m); cruising speed 565 mph (909 km/h) at 36,000 ft (1970 m); initial climb rate 16,000 ft (4875 m) per minute; service ceiling 45,000 ft (13715 m); combat radius 530 miles (853 km); ferry range 1,500 miles (2411 km) with maximum fuel
Weights: empty 21,000 lb (9525 kg); maximum take-off 34,830 lb (15800 kg)
Dimensions: span 38 ft 9 in (11.81 m); length 54 ft 3 in (16.54 m) including probe; height 16 ft 2.75 in (4.96 m); wing area 385.0 sq ft (35.77 m^2)

Variants
F-100A Super Sabre: this was the US Air Force's first supersonic fighter, making its initial flight in May 1953 and entering service in September 1954; this early variant is powered by one 15,000-lb (6804-kg) J57-P-7 or 16,000-lb (7257-kg) J57-P-39 afterburning turbojet; the type was designed as a pure fighter, and small numbers continue in this role with Taiwan
F-100C Super Sabre: much improved aerodynamically with a taller fin and extended wings, this variant is optimized for the fighter-bomber role with the J57-P-21 afterburning turbojet, eight underwing hardpoints and inflight-refuelling capability
F-100D Super Sabre: attack derivative of the F-100C with flapped wings, electronic countermeasures, a low-altitude bombing system and provision for four AIM-9 Sidewinder air-to-air missiles
F-100F Super Sabre: tandem two-seat operational conversion trainer with two 20-mm cannon and provision for 6,000 lb (2722 kg) of external stores; the fuselage is lengthened by 3.0 ft (0.91 m)

North American T-28B Trojan
(USA)
Type: training aircraft
Accommodation: pupil and instructor seated in tandem
Armament (fixed): none
Armament (disposable): none
Electronics and operational equipment: communication and navigation equipment

Powerplant and fuel system: one 1,425-hp (1063-kW) Wright R-1820 radial piston engine, and a total internal fuel capacity of ? Imp gal (? litres) in fuselage tanks
Performance: maximum speed 343 mph (522 km/h); cruising speed 310 mph (499 km/h) at 30,000 ft (9145 m); range 1,060 miles (1706 km)
Weights: empty 6,424 lb (2914 kg); maximum take-off 8,486 lb (3849 kg)
Dimensions: span 40 ft 1 in (12.22 m); length 33 ft 0 in (10.06 m); height 12 ft 8 in (3.86 m); wing area 268.0 sq ft (24.9 m²)

Variants
T-28A: designed as successor to the same company's T-6 Texan/Harvard series of basic trainers, the T-28 featured more advanced lines, a considerably upgraded powerplant (the 800-hp/597-kW Wright R-1300-1 radial piston engine in this US Air Force model), a frameless canopy and retractable tricycle landing gear; the prototype first flew in September 1949 and proved an immediate success
T-28B Trojan: version for the US Navy with considerably more power and revised avionics
T-28C Trojan: arrester-hooked version for the T-28B
T-28D: ground-attack and counter-insurgency aircraft developed from the T-28A but fitted with the 1,425-hp (1063-kW) R-1820-56S radial and six underwing hardpoints for gun pods, napalm tanks, bombs and rocket-launchers; the type was also developed as the **AT-28D** attack trainer; and in France many T-28As were modified to T-28D standard under the name **Fennec**, these aircraft having two 0.5-in (12.7-mm) machine-guns pods and four 136-kg (300-lb) bombs for use in the close-support role

Northrop F-5A Freedom Fighter
(USA)
Type: lightweight tactical fighter
Accommodation: pilot seated on a rocket-assisted ejector seat
Armament (fixed): two 20-mm Colt-Browning M39 cannon in the nose with 280 rounds per gun
Armament (disposable): this is carried on one underfuselage and four underwing hardpoints and on two wingtip missile-launcher rails, up to a maximum weight of 4,400 lb (1996 kg); the wingtip rails and underwing hardpoints can each carry one AIM-9 Sidewinder air-to-air missile, the underfuselage hardpoint can accommodate one 2,000-lb (907-kg) Mk 84 bomb, and the underwing hardpoints can lift bombs, rocket-launcher pods, gun pods or AGM-12 Bullpup air-to-surface missiles
Electronics and operational equipment: communication and navigation equipment, plus a Norsight optical sight and control equipment for the AGM-12 Bullpup missile when appropriate

Powerplant and fuel system: two 4,080-lb (1850 kg) afterburning thrust General Electric J85-GE-13 turbojets, and a total internal fuel capacity of 583 US gal (2207 litres) in two tanks, plus provision for one 150-US gal (568-litre) centreline, two 150-US gal (568-litre) underwing and two 50-US gal (189-litre) external tanks; provision for inflight-refuelling
Performance: maximum speed 925 mph (1,489 km/h) or Mach 1.4 at 35,000 ft (10970 m); cruising speed 640 mph (1,031 km/h) or Mach 0.97 at 36,000 ft (10670 m); initial climb rate 28,700 ft (8750 m) per minute; service ceiling 50,500 ft (15390 m); combat radius 195 miles (314 km) with maximum payload; ferry range 1,612 miles (2594 km) with maximum internal and external fuel
Weights: empty 8,085 lb (3667 kg); maximum take-off 20,677 lb (9379 kg)
Dimensions: span 25 ft 3 in (7.70 m); length 47 ft 2 in (14.38 m); height 13 ft 2 in (4.01 m); wing area 170.0 sq ft (15.79 m²)

Variants
F-5A Freedom Fighter: designed as the N-156F private venture, the F-5A was adopted as part of the USA's military support programme for important allies and thus supplied in considerable numbers to NATO countries and other allies; keys to the design were useful weapons capability plus small size and limited (though supersonic) performance for low unit cost and easily achieved servicing and airfield requirements; the type first flew in July 1959 and the first production machine followed in October 1963
CF-5A: designation of the F-5A version built in Canada for the Canadian Armed Forces with 4,300-lb (1950-kg) Orenda J85-CAN-15 afterburning turbojets and a number of detail modifications including manoeuvring flaps
NF-5A: designation of the CF-5A built in Canada for the Netherlands
RF-5A Freedom Fighter: designation of the reconnaissance version of the F-5A with four KS-92 cameras in a modified nose
SF-5A: designation of the F-5A version built for the Spanish air force by CASA
SRF-5A: designation of the RF-5A version built for the Spanish air force by CASA
F-5B Freedom Fighter: tandem two-seat operational conversion trainer version of the F-5A, which entered service in April 1964; the type is similar to the F-5A apart from the lengthened cockpit and lack of nose guns
NF-5B: designation of the Canadair-built version of the F-5B for the Netherlands
SF-5B: designation of the CASA-built version of the F-5B for the Spanish air force
CF-5D: designation of the Canadair-built version of the F-5B for the Canadian Armed Forces
T-38A Talon: this is the US Air Force's highly

capable supersonic basic and advanced flying trainer based on the N-156T version of the Nothrop N-156 basic design; the type first flew in April 1959 and began to enter service in March 1961; fitted with two 3,850-lb (1746-kg) J85-GE-5A afterburning turbojets, the Talon has a maximum take-off weight of 12,03 lb (5485 kg) and a maximum speed of Mach 1.3 at 36,000 ft (10975 m); the length is 46,375 ft (14.14 m)

Northrop F-5E Tiger II
(USA)
Type: lightweight tactical fighter
Accommodation: pilot seated on a rocket-assisted ejector seat
Armament (fixed): two 20-mm Colt-Browning M39A2 cannon with 280 rounds per gun in the nose
Armament (disposable): this is carried on one underfuselage and four underwing hardpoints and on two wingtip missile rails, up to a maximum of 7,000 lb (3175 kg); weapons which can be carried include the Mk 84 2,000-lb (907-kg) bomb, the Mk 83 1,000-lb (454-kg) bomb, the Mk 82 500-lb (227-kg) free-fall or retarded bomb, the Mk 36 1,000-lb (454-kg) destructor, the BLU-1, -27 and -32 napalm bombs, the CBU-24, -49, -52 and -58 bomb dispensers, the LAU-3 launcher with 19 2.75-in (69.85-mm) rockets, the LAU-68 launcher with seven 2.75-in (69.85-mm) rockets, the SUU-20 bomb and rocket pack, a 30-mm cannon pod on the centreline, AGM-65 Maverick air-to-surface missiles, various guided bombs and AIM-9 Sidewinder air-to-air missiles
Electronics and operational equipment: communication and navigation equipment, plus Emerson APQ-153 or APQ-159(V) multi-mode radar, General Electric ASG-29 optical sight, and options such as a Litton inertial navigation system, Itek ALR-46 radar-warning receiver, Northrop ALQ-171 conformal electronic countermeasures, and Tracor ALE-40 chaff/flare dispenser
Powerplant and fuel system: two 5,000-lb (2267-kg) afterburning thrust General Electric J85-GE-21 turbojets, and a total internal fuel capacity of 670 US gal (2538 litres) in three fuselage tanks, and provision for three 275-US gal (1041-litre) drop-tanks or one 275-US gal (1041-litre) and two 150-US gal (568-litre) drop-tanks, plus two 50-US gal (189-litre) tiptanks; inflight-refuelling capability
Performance: maximum speed 1,083 mph (1743 km/h) or Mach 1.64 at 36,000 ft (10970 m); and 708 mph (1139 km/h) or Mach 0.93 at sea level; cruising speed 645 mph (1038 km/h) or Mach 0.98 at 36,000 ft (10970 m); initial climb rate 34,500 ft (10515 m) per minute; service ceiling 52,000 ft (15850 m); combat radius 190 miles (305 km) with two Sidewinders and 6,300-lb (2857-kg) dropload; ferry range 2,314 miles (3720 km) with maximum fuel
Weights: empty 9,723 lb (4410 kg); maximum take-off 24,722 lb (11214 kg)

Dimensions: span 26 ft 8 in (8.13 m); length 47 ft 4.75 in (14.73 m); height 13 ft 4.5 in (4.08 m); wing area 186.0 sq ft (17.3 m²)

Variants
F-5E Tiger II: this is essentially an upgraded version of the F-5A series with better performance and weapon capability derived from the installation of more powerful engines; other modifications are enhanced manoeuvrability (derived from better aerodynamics and a greater thrust-to-weight ratio), increased fuel capacity, reduced field requirements and an integrated fire-control system (though still without radar); the type flew in prototype form in March 1969 and began to enter service in April 1973; apart from overseas sales success, the type has found favour with the US forces as a dissimilar air combat training aircraft as its performance and agility are comparable to those of the MiG-21 series
RF-5E Tigereye: reconnaissance derivative of the F-5E with the forward fuselage modified and lengthened by 8 in (0.203 m) to accept a wide variety of optical and infra-red tactical reconnaissance equipment
F-5F Tiger II: this is the tandem-two-seat operational conversion derivative of the F-5E, with the fuselage lengthened by 42.5 in (1.08 m) to allow the insertion of the second seat; the type has only one inbuilt 20-mm cannon, but otherwise retains the full combat capability of the F-5E

Northrop T-38A Talon (USA): see Northrop F-5A Freedom Fighter

Nurtanio NC-212 (Indonesia): see CASA C-212 Series 200 Aviocar

Panavia Tornado IDS
(West Germany/Italy/UK)
Type: variable-geometry multi-role combat aircraft
Accommodation: pilot and systems operator seatedin tandem on Martin-Baker Mk 10A ejector seats
Armament (fixed): two 27-mm IWKA-Mauser cannon with 360 rounds per gun
Armament (disposable): this is carried on three underfuselage and four swivelling underwing hardpoints, to a weight of about 9000 kg (19,840 lb); among the weapons which can be carried are the AIM-9B/AIM-9L Sidewinder air-to-air missile, the Kormoran anti-ship missile, the AGM-65 Maverick air-to-surface missile, the AS.30 air-to-surface missile, the GBU-15 TV-guided bomb, the 'Paveway' laser-guided bomb (used in conjunction with a 'Pave Spike' pod), the Sea Eagle anti-ship missile, the Mk 83 1,000-lb (454-kg) free-fall or retarded bomb, the Mk 13/15 1,000-lb (454-kg) free-fall or retarded bomb, the Mk 82 500-lb (227-kg) free-fall or retarded bomb, the Matra 250-kg (551-lb) free-fall or retarded bomb, the BLU-1B 750-lb (340-kg) fire bomb, the BL755 600-lb (272-kg) cluster bomb, the JP233 airfield-attack munitions-dispenser, the MBB MW-1 battlefi-

eld-attack munitions-dispenser, the LAU-51A rocket-launcher pod, and the LR-25 rocket-launcher pod
Electronics and operational equipment: communication and navigation equipment, plus Texas Instruments pulse-Doppler multi-mode radar with a ground-mapping function and terrain-following sub-unit, Smiths/Teldix/OMI head-up and head-down displays, Ferranti FIN 1010 inertial navigation system, Plessey 72 Doppler navigation, Litef Spirit 3 digital central computer, Microtecnica air-data system, Ferranti laser rangefinder and marked-target seeker, Marconi-Selenia stores-management system, radar-warning receiver (Elettronica in Italian aircraft, Marconi ARI.18241/1 in British aircraft, and Itek/AEG Telefunken in West German aircraft), and options such as Westinghouse ASQ-153(V) 'Pave Spike' laser pod, data-link pod, MBB reconnaissance pod, BAe reconnaissance pod, and electronic countermeasures pods such as the British Marconi ARI.23246/1 Sky Shadow, Italian/West German Elettronica/AEG-Telefunken EL/73, and Italian Elettronica ELT/553
Powerplant and fuel system: two 16,800-lb (7620-kg) afterburning thrust Turbo-Union RB.199-34R Mk 103 turbofans, and a total internal fuel capacity of 6400 litres (1,408 Imp gal) in integral fuselage and wing tanks, plus provision for 1500-litre (330-Imp gal) drop-tanks; provision for inflight-refuelling
Performance: maximum speed more than 2337 km/h (1,453 mph) or Mach 2.2 at 11000 m (36,090 ft), and more than 1480 km/h (920 mph) or Mach 1.21 at sea level; climb to 9145 m (30,000 ft) in less than 2 minutes from brakes-off; service ceiling more than 15000 m (49,210 ft); combat radius 1390 km (863 miles) on a hi-lo-hi sortie with heavy weapons load, and about 3890 km (2,420 miles) for ferrying
Weights: empty 14090 kg (31,063 lb); normal take-off 21400 kg (47,178 lb) with maximum internal fuel but no stores; maximum take-off 27215 kg (60,000 lb)
Dimensions: span spread 13.91 m (45 ft 7.5 in) and swept 8.60 m (28 ft 2.5 in); length 16.72 m (54 ft 10.25 in); height 5.95 m (28 ft 2.5 in); wing area about 25 m² (269 sq ft)

Variants
Tornado IDS: this is one of the most important combat aircraft to have been developed in recent years, and forms the main interdiction and strike strength available to the British, West German and Italian air forces; the type has also secured useful Middle Eastern export orders; the type was designed as a high-performance type with fly-by-wire control system and advanced avionics for extremely accurate navigation and safe flight at supersonic speeds and very low levels in all weathers for pinpoint day/night first-pass attacks on a variety of targets; the type also possesses good STOL characteristics through its variable-geometry wings, powerful afterburning turbofan engines and extensive high-lift devices; the use of a Texas Instruments radar with terrain-following capability allows the Tornado to fly at Mach 1.2+ at an altitude of only 200 ft (60 m), so reducing the warning time available to the defences to organize themselves; the design was originated in the 1960s as the Multi-Role Combat Aircraft (able to undertake the close air support, battlefield interdiction, long-range interdiction, counter-air attack, air-superiority, interception and air defence, reconnaissance and naval strike roles), and the first prototype flew in April 1974 for service delivery from 1980; the type is known to the Royal Air Force as the **Tornado GR.Mk 1**, an operational conversion trainer derivative being the **Tornado GR.Mk 1(T)**; recent developments in British aircraft have been the adoption of the ALARM anti-radiation missile, and this requires the upgrading of the Tornado's avionics system with a MIL 1553B databus, improved computer and revised missile-control system

Tornado ADV: this is the air-defence variant developed in the UK for the Royal Air Force and designed for the long-range interception of intruders detected by airborne early warning aircraft; to allow the semi-recessed carriage of four Sky Flash air-to-air missiles under the fuselage, the aircraft is stretched to 18.06 m (59.25 ft), this having the useful by-products of 909 litres (200 Imp gal) more internal fuel and finer aerodynamic lines for reduced transonic drag and hence superior acceleration; the rest of the armament comprises two or four AIM-9L Sidewinder short-range air-to-air missiles, and a single 27-mm cannon; the primary equipment changes are the adoption of twin inertial platforms, the updating of the main computer, the installation of advanced cockpit displays, the provision of automatic wing sweep and flap/slat scheduling, and the replacement of the Texas Instruments multi-mode radar by Marconi/Ferranti Foxhunter pulse-Doppler interception radar; this has considerable resistance to electronic countermeasures, and possesses a track-while-scan capability to a range of 115 miles (185 km); power is provided by two analog-controlled 16,290-lb (7675-kg) RB.199 Mk 103 turbofans for an additional 7% afterburning thrust; maximum take-off weight is more than 27270 kg (60,120 lb), and the combat patrol duration is 2 hours 20 minutes at a radius of 375 miles (603 km) without inflight-refuelling; the Tornado ADV is known in the RAF as the **Tornado F.Mk 2** or, with digitally-controlled RB.199 Mk 104 afterburning turbofans and lengthened jetpipes for greater afterburning thrust, with automatic wing sweep control, and with other detail modifications, as the **Tornado F.Mk 3**; the type first flew in October 1979 and began to enter service in 1986

Tornado ECR: this is the electronic combat and reconnaissance variant under development for the West German (and possibly Italian) air force; this model will have an emitter location system (possibly the APR-38 system) in place of the internal cannon, a low- and medium-altitude reconnaissance system, a data-link system and MBB-developed electronic intelligence and jammer pods; the use if a MIL 1553 databus and other avionics improvements also provide capability for the carriage and launch of the AGM-88 HARM anti-radiation missile

Pilatus PC-7 Turbo-Trainer
(Switzerland)
Type: multi-role trainer
Accommodation: pupil and instructor seated in tandem
Armament (fixed): none
Armament (disposable): this is carried on six underwing hardpoints, the inner pair each rated at 250 kg (551 lb), the central pair each rated at 250 kg (551 lb), the central pair each at 160 kg (353 lb) and the outer pair at 110 kg (243 lb), up to a maximum of 1040 kg (2,293 lb); a wide variety of weapons (gun pods, rocket pods and light bombs) can be accommodated for weapons training
Electronics and operational equipment: communication and navigation equipment, and provision for a reconnaissance pod under the wings
Powerplant and fuel system: one 650-shp (485-kW) Pratt & Whitney Canada PT6A-25A turboprop, flat-rated to 550 shp (410 kW) at sea level, and a total internal fuel capacity of 474 litres (104 Imp gal) in outer-wing integral tanks, plus provision for two 152- or 240-litre (33.5- or 53-Imp gal) drop-tanks
Performance: maximum speed 500 km/h (311 mph); cruising speed 410 km/h (255 mph) at 6100 m (20,025 ft); initial climb rate 610 m (2,000 ft) per minute; service ceiling 9755 m (32,005 ft); range 1200 km (745 miles) in aerobatic configuration, and 2630 km (1,634 miles) with underwing stores and fuel
Weights: empty 1330 kg (2,932 lb); normal take-off 1900 kg (4,188 lb); maximum take-off 2700 kg (5,952 lb)
Dimensions: span 10.40 m (34 ft 1.5 in); length 9.775 m (32 ft 0.75 in); height 3.21 m (10 ft 6.5 in); wing area 16.60 m² (179.7 sq ft)

Variants
PC-7 Turbo-Trainer: first flown in prototype form during April 1966 with a 550-shp (410-kW) Pratt & Whitney Canada PT6A turboprop as the P-3-06 (subsequently redesignated P-3B) derivative of the piston-engined P-3 trainer, this is a useful multi-role trainer with secondary attack capability; the type began to enter service in 1978 and substantial sales have been made to third-world air forces
PC-9 Turbo-Trainer: developed in 1983 and 1984 as an unsuccessful contender in the competition to find a replacement for the Jet Provost trainer in British service, the PC-9 derivative of the PC-7 offers a maximum speed of 460 km/h (286 mph) amongst other performance increments provided by a 950-shp (708-kW) PT6A-62 turboprop and a slight reduction in span; other modifications are a stepped cockpit with Martin-Baker Mk 11 zero/zero ejector seats, and revised airbrakes, ailerons and landing gear doors

PZL Mielec TS-11 Iskra-bis DF
(Poland)
Type: close-support and reconnaissance trainer
Accommodation: pilot seated on a lightweight ejector seat
Armament (fixed): one 23-mm NR-23 cannon with ? rounds
Armament (disposable): this is carried on four underwing hardpoints, up to a weight of about 400 kg (882 lb), possible stores including 7.62-mm (0.3-in) machine-gun pods, launcher pods for eight 55-mm (2.17-in) rockets, and 100-kg (220-lb) bombs
Electronics and operational equipment: communication and navigation equipment, plus an optical sight and reconnaissance equipment
Powerplant and fuel system: one 1100-kg (2,425-lb) thrust IL SO-3W turbojet, and a total internal fuel capacity of 1400 litres (308 Imp gal) in two fuselage and two integral wing tanks
Performance: maximum speed 770 km/h (478 mph) at 5000 m (16,405 ft); cruising speed 600 km/h (373 mph); initial climb rate 1164 m (3,820 ft) per minute; service ceiling 11000 m (36,090 ft); range 1260 km (783 miles) with 1200 litres (264 Imp gal) of fuel
Weights: empty 2560 kg (5,644 lb); normal take-off 3243 kg (7,150 lb) with 570 litres (125.5 Imp gal) of fuel; maximum take-off 3840 kg (8,465 lb)
Dimensions: span 10.06 m (33 ft 0 in); length 11.15 m (36 ft 7 in); height 3.50 m (11 ft 5.5 in); wing area 17.50 m² (188.4 sq ft)

Variants
Iskra-bis A: this was the initial production version of the Iskra (spark) two-seat primary and advanced trainer; the type first flew in February 1960 and began to enter service in March 1963 with the 780-kg (1,720-lb) HO-10 turbojet and two underwing hardpoints for weapons training; from 1964 production aircraft has the more powerful SO-1 or improved SO-3 turbojet
Iskra-bis B: derivative of the Iskra-bis A with four underwing hardpoints and originally designated Iskra 100
Iskra-bis C: single-seat reconnaissance derivative of the Iskra-bis B and originally called the Iskra 200, with the rear cockpit replaced by additional fuel tankage, and reconnaissance equipment carried in the lower fuselage aft of the cockpit
Iskra-bis D: this was the final production version of the original Iskra series before production ceased in 1979, and is in essence an improved Iskra-bis B with greater weapons versatility
Iskra-bis DF: this version resulted in a resumption of

production in 1982, and is a combination of Iskra-bis C and D features, being a single-seat reconnaissance and close-support aircraft with four hardpoints and three cameras (in the lower fuselage and the bottom of the two inlets)

RACA (Hughes) Model 500 (Argentina): see Hughes (McDonnell Douglas Helicopters) OH-6A Cayuse

Rockwell B-1B
(USA)
Type: variable-geometry long-range strategic bomber and missile-carrier
Accommodation: crew of four on the flightdeck, all seated on Douglas ACES ejector seats
Armament (fixed): none
Armament (disposable): this is carried in three lower-fuselage weapons bays, up to a maximum weight of 64,000 lb (29030 kg), and on eight underfuselage hardpoints, up to a maximum weight of 28,000 lb (12701 kg); in the strategic nuclear role, the weapons bays can accommodate 24 B61 or B83 bombs, or 12 B28 or B43 bombs, or eight AGM-86B air-launched cruise missiles, or 24 AGM-69 short-range attack missiles, while the underfuselage hardpoints can accept 14 B43, B61 or B83 bombs, or eight B28 bombs, or 14 AGM-86B air-launched cruise missiles or AGM-69 short-range attack missiles; in the conventional role, the weapon bays can lift 24 Mk 84 2,000-lb (907-kg) bombs or 128 Mk 82 500-lb (227-kg) bombs, while the underfuselage hardpoints can take 14 Mk 84 2,000-lb (907-kg) bombs or 44 Mk 82 500-lb (227-kg) bombs
Electronics and operational equipment: communication and navigation equipment, plus Westinghouse ALQ-153 tail-warning system and extremely advanced offensive and defensive avionics suites; the Offensive Avionics System is co-ordinated by Boeing on the basic of the same company's OAS for the B-52 bomber and comprises Westinghouse APQ-164 multi-mode attack, ground-mapping and terrain-following radar, Singer-Kearfott inertial navigation, Teledyne Ryan APN-218 Doppler navigation, Northrop NAS-26 astro-inertial navigation; IBM avionic control units and computers and Sperry offensive display sets; the Defensive Avionics System is co-ordinated by Eaton-AIL as the ALQ-161 with quickly reprogrammable digital computers and Sanders display units with Raytheon phased-array antennae and Northrop jammers for the near-instantaneous detection, location, analysis and jamming of hostile emissions
Powerplant and fuel system: four 30,000-lb (13608-kg) afterburning thrust General Electric F101-GE-102 turbofans, and a total internal fuel capacity of 24,000 US gal (90850 litres) in fuselage and integral wing tanks, plus provision for auxiliary fuel in the weapons bays; provision for inflight-refuelling

Performance: maximum speed 825 mph (1330 km/h) or Mach 1.25 at high altitude and 600 mph (966 km/h) or Mach 0.79 for penetration at 200 ft (61 m); range about 7,455 miles (12000 km)
Weights: empty about 172,000 lb (78019 kg); maximum take-off 477,000 lb (216365 kg)
Dimensions: span 136 ft 8.5 in (41.67 m) spread and 78 ft 2.5 in (23.84 m) swept; length 147 ft 0 in (44.81 m); height 34 ft 0 in (10.36 m); wing area about 1,950.0 sq ft (181.2 m^2)

Variant
B-1B: entering service from 1986 as the US Air Force's primary strategic bomber to replace the Boeing B-52 in the penetration role, the B-1B is the result of a protracted and convoluted development history from the time the US Air Force issued its November 1969 requirement for a high-level bomber with dash capability of Mach 2.2+ in the high-speed delivery of free-fall and stand-off weapons; the Rockwell submission was accepted as the B-1 in 1970, and the full-scale design and development of the initial production version was soon under way, this B-1A being a complex and highly advanced variable-geometry type with General Electric F101 turbofans and fully variable inlets; the prototype first flew in December 1974 and the flight test programme moved ahead without undue delay; however, in June 1977 President Carter decided to scrap the programme in favour of cruise missile development, and it was only with the inauguration of President Reagan that matters began to look up again, the new administration deciding during October 1981 on the procurement of 100 B-1B bombers in the low-level high-subsonic penetration role with fixed inlets and revised nacelles (reducing maximum speed to Mach 1.25) but a strengthened airframe and landing gear for operation at high weights with nuclear and conventional weapons over very long ranges; other changes were concerned with reduction of the type's already low radar signature, S-shaped ducts with streamwise baffles shielding the face of the engine compressors, and radar absorbent materials being used in sensitive areas to reduce reflectivity; the second and fourth B-1As were used from March 1983 to flight-test features of the B-1B, which first flew in September 1984 with the advanced offensive and defensive electronic systems; from the ninth aircraft the type is built with revised weapons bays, the forward bay having a movable bulkhead allowing the carriage of 12 AGM-86B ALCMs internally, as well as additional fuel tanks and SRAMs; the initial fit is eight ALCMs carried internally plus another 14 externally; when the Northrop ATB (Advanced-Technology Bomber, probably to be designated B-2) enters service in the early 1990s as the USAF's penetration 'stealth' bomber, the B-1B will be relegated increasingly to the stand-off and conventional bombing roles

Rockwell OV-10A Bronco
(USA)

Type: multi-role counter-insurgency aircraft

Accommodation: crew of two seated in tandem on LW-3B ejector seats

Armament (fixed): two 7.62-mm (0.3-in) machine-guns with 500 rounds per gun in each sponson

Armament (disposable): this is carried on one underfuselage and four undersponson hardpoints, the former rated at 1,200 lb (544 kg) and each of the latter at 600 lb (272 kg) to a maximum weight of 4,000 lb (1814 kg); a very wide variety of stores can be accommodated on these hardpoints, including the Mk 88 1,000-lb (454-kg) bomb, Mk 82 500-lb (227-kg) free-fall or retarded bomb, Mk 81 250-lb (113-kg) free-fall or retarded bomb, Mk 82 and Mk 81 Snakeye bombs, Mk 77 firebomb, LAU-3/A, LAU-10/A, LAU-32/A, LAU-59/A, LAU-60/A, LAU-61/A, LAU-68/A and LAU-69/A launchers for 2.75-in (69.85-mm) rockets, SUU-11A/A 7.62-mm (0.3-in) Minigun pod, GPU-2/A 20-mm cannon pod and CBU-55/B cluster bomb; there is also provision for one AIM-9 Sidewinder air-to-air missile under each wing

Electronics and operational equipment: communication and navigation equipment, plus an optical sight

Powerplant and fuel system: two 715-shp (533-kW) Garrett T76-G-416/417 turboprops, and a total internal fuel capacity of 252 US gal (954 litres) in five wing tanks, plus provision for one 150-US gal (568-litre) drop-tank

Performance: maximum speed 281 mph (452 km/h) at sea level; initial climb rate 2,600 ft (790 m) per minute; service ceiling 24,000 ft (7315 m); combat radius 228 miles (367 km) with maximum payload; ferry range 1,382 miles (2,224 km) with maximum fuel

Weights: empty 6,893 lb (3127 kg); normal take-off 9,908 lb (4494 kg); maximum take-off 14,444 lb (6552 kg)

Dimensions: span 40 ft 0 in (12.19 m); length 41 ft 7 in (12.67 m); height 15 ft 2 in (4.62 m); wing area 291.0 sq ft (27.03 m²)

Variants

OV-10A Bronco: a much-underestimated type designed for the battlefield reconnaissance and counter-insurgency roles, the Bronco first flew in July 1965 in response to the US Marine Corps' Light Armed Reconnaissance Aircraft requirement, and this YOV-10A prototype was powered by two 600-shp (447-kW) Garrett T76-G-6/8 turboprops; the type was generally satisfactory and entered production as the OV-10A with a 10-ft (3.05-m) increase in span and more powerful T76-G-10/12 turboprops; the first production aircraft flew in August 1976, and since that time the aircraft has performed admirably for the USMC and US Air Force in the battlefield reconnaissance, forward air control and helicopter escort roles

OV-10B Bronco: target-towing version of the OV-10A for West Germany, later supplemented by **OV-10B(Z)** aircraft with a 2,950-lb (1330-kg) General Electric J85-GE-4 turbojet pod-mounted above the centre section for better performance

OV-10C Bronco: version of the OV-10A for Thailand

OV-10D Bronco: these are OV-10A aircraft of the USMC revised for the night surveillance and observation roles with 1,040-shp (776-kW) T76-G-420/421 turboprops (with infra-red suppression kits) and uprated armament provision (up to a maximum weight of 3,600 lb/1633 kg, and including laser-guided weapons) used in conjunction with the chin-mounted AAS-37 forward-looking infra-red sensor and a laser designator; this AAS-37 system can also be used with the M97 20-mm cannon turret that can be fitted in place of the standard weapon fit; other modifications in this important type are the APR-38 radar-warning receiver and the ALE-39 chaff dispenser

OV-10E Bronco: version of the OV-10A for Venezuela

OV-10F Bronco: version of the OV-10A for Indonesia

Rockwell T-2C Buckeye
(USA)

Type: general-purpose trainer

Accommodation: pupil and instructor seated in tandem on rocket-assisted LS-1 ejector seats

Armament (fixed): none

Armament (disposable): this is carried on two underwing hardpoints, up to a maximum weight of 640 lb (290 kg), and can comprise several types of practice bomb and rocket-launcher pod

Electronics and operational equipment: communication and navigation equipment, plus an optical sight

Powerplant and fuel system: two 2,950-lb (1339-kg) thrust General Electric J85-GE-4 turbojets, and a total internal fuel capacity of 691 US gal (2616 litres) in one fuselage, two wing and two wingtip tanks

Performance: maximum speed 522 mph (840 km/h) at 25,000 ft (7620 m); initial climb rate 6200 ft (1,890 m) per minute; service ceiling 40,400 ft (12315 m); range 1,047 miles (1685 km)

Weights: empty 8,115 lb (3680 kg); maximum take-off 13,179 lb (5977 kg)

Dimensions: span 38 ft 1.5 in (11.62 m) over tiptanks; length 38 ft 3.5 in (11.67 m); height 14 ft 9.5 in (4.5 m); wing area 255.0 sq ft (23.69 m²)

Variants

T-2B Buckeye: designed as the North American NA-264 to meet a 1956 US Navy requirements for a multi-role trainer, the Buckeye used proved components and features of other North American (later

Rockwell) aircraft to speed the development of the T2J-1 (from 1962 the T-2A) initial model powered by a single 3,400-lb (1542-kg) Westinghouse J34-W-36 non-afterburning turbojet; though the type proved successful, it was clear that greater power and a twin-engine layout would offer better performance and greater reliability, and there thus appeared in August 1962 the T-2B with two 3,000-lb (1361-kg) Pratt & Whitney J60-P-6 turbojets
T-2C Buckeye: definitive twin-engine model with General Electric J85 turbojets for the first flight of a production aircraft in December 1968
T-2D Buckeye: T-2C variant for Venezuela
T-2E Buckeye: T-2C variant for Greece with a secondary attack capability though the provision of six underwing hardpoints for a maximum load of 3,500 lb (1587 kg) of disposable stores

Saab J 35F Draken
(Sweden)
Type: all-weather fighter and attack aircraft
Accommodation: pilot on a Saab 73SE-F ejector seat
Armament (fixed): one 30-mm Aden M/55 cannon with 100 rounds
Armament (disposable): this is carried on nine hardpoints (three under the fuselage and three under each wing with each underfuselage and inboard underwing unit rated at 500 kg/1,102 lb and the four under the outer wings each at 100 kg/220 lb) to a maximum of 2900 kg (6,393 lb); weapons that can be carried include the Rb24 Sidewinder air-to-air missile, Rb27 and Rb28 Falcon air-to-air missiles, and a variety of bombs and rockets
Electronics and operational equipment: communication and navigation equipment, plus Ericcson UAP 13102 (PS-01/A) or UAP 13103 (PS-011/A) fire-control radar, Hughes S72N infra-red system, Hughes/Saab-Scania S7B fire-control system, Saab BT9 bombing system, data-link system (associated with the STRIL-60 national air-defence network), integrated nav/attack system, radar-warning receiver and provision for external eletrcnic countermeasures such as the Pod KA and Pod 70 systems
Powerplant and fuel system: one 8000-kg (17,637-lb) afterburning thrust Svenska Flygmotor RM6B turbojet (licence-built Rolls-Royce Avon with SFA-designed afterburner), and a total internal fuel capacity of 4000 litres (880 Imp gal) plus provision for 1275- and 500-litre (280- and 110 Imp gal) drop tanks; no provision for inflight refuelling
Performance: maximum speed 2125 km/h (1,320 mph) or Mach 2.0 at high altitude; initial climb rate 10500 m (34,450 ft) per minute; service ceiling 20000 m (65,615 ft); hi-lo-hi combat radius 635 km (395 miles) with external warload; ferry range 3250 km (2,020 miles)
Weights: empty 8250 kg (18, 188 lb); maximum take-off 12270 kg (27,050 lb) on a defence mission or 15000 kg (33,060 lb) on an attack mission
Dimensions: span 9.40 m (30 ft 10 in); length 15.35 m (50 ft 4 in); height 3.89 m (12 ft 9 in); wing area 49.20 m² (529.6 sq ft)

Variants
Sk 35C Draken: the Draken (dragon) was Europe's first genuinely supersonic fighter and designed in the first half of the 1950s; the type was a very great achievement in aerodynamic and propulsion terms, the double-delta wing providing good area and considerable fuel volume with comparatively little drag so that a single Avon turbojet with locally-designed afterburner provided sufficient thrust for high performance; the type first flew in prototype form during October 1955, and production J 35A aircraft with the 7000-kg (15,432-lb) Svenska Flygmotor RM6B engine (licence-built Avon) began to enter service in March 1960, being followed by the J 35B with data-link equipment (for use in association with the Swedish STRIL-60 integrated air-defence network), collision-course radar and greater armament; the oldest version left is service is the Sk 35C, a tandem two-seat operational conversion trainer derived from J 35A airframes but without provision for armament
J 35D Draken: much improved interceptor with the more powerful RM6C afterburning turbojet, enlarged air inlets, more advanced avionics and a zero/zero ejector seat
J 35F Draken: definitive interceptor model with Hughes pulse-Doppler radar (for use with semi-active radar-homing Hughes Falcon air-to-air missiles in place of the earlier versions' Sidewinders) and generally upgraded avionics
J 35J Draken: designation of J 35F Drakens reworked to a more modern standard with improvements to the radar, IR, IFF and navigation systems, plus strenthened wings for greater warloads with an additional hardpoint under each inner wing panel
Saab 35X: designation of the export version of the Draken with 30% greater internal fuel capacity, two 30-mm Aden M/55 cannon, and strengthening for operation at higher weights to a maximum of 16000 kg (32,275 lb) including a warload of up to 4500 kg (9,921 lb); the version for Denmark is the **Saab 35XD** with 11 hardpoints, produced in **F-35** radarless ground-attack, **RF-35** tactical reconnaissance and **TF-35** two-seat trainer subvariants, recently upgraded to a more modern standard with a Marconi head-up display, Ferranti 105D laser ranger, Lear-Siegler nav/attack system and Singer-Kearfott inertial navigation system; the version for Finland is the **Saab 35XS**, delivered with the designation **J35XS** and generally similar to the J 35F but with two cannon; Finland has also bought ex-Swedish aircraft, namely the **J35BS** radarless operational trainer (ex-J 35B), the **J35F**

interceptor (ex-J 35F) and the **J35C** two-seat trainer (ex-Sk 35C); ex-Swedish J 35D aircraft are currently being refurbished for sale to Austria

Saab 105Ö
(Sweden)
Type: trainer and light attack aircraft
Accommodation: pupil and instructor seated side-by-side on ejector seats
Armament (fixed): none
Armament (disposable): this is carried on six underwing hardpoints, the inner pair each rated at 275 kg (610 lb) the centre pair each at 454 kg (1,000 lb) and the outer pair each at 275 kg (610 kg), up to a maximum weight of 2000 kg (4,409 lb); typical loads are two 454-kg (1,000-lb) and four 227-kg (500-lb) bombs, or four 227-kg (500-lb) bombs and two 30-mm cannon pods, or two 7.62-mm (0.3-in) Minigun pods and four 227-kg (500-lb) napalm tanks, or 12 135-mm (5.3-in) rockets, or six launchers each with four 5-in (127-mm) rockets, or 18 75-mm (2.95-in) rockets, or two RB 05 air-to-surface missiles and two RB 24 Sidewinder air-to-air missiles and two 7.62-mm (0.3-in) Minigun pods
Electronics and operational equipment: communication and navigation equipment, plus optical sights
Powerplant and fuel system: two 2,850-lb (1293-kg) thrust General Electric J85-GE-17B turbojets, and a total internal fuel capacity of 2050 litres (451 gal) in two fuselage and two integral wing tanks, plus provision two 500-litre (110-Imp gal) drop-tanks
Performance: maximum speed 970 km/h (603 mph) at sea level; climb to 10000 m (32,810 ft) in 4.5 minutes; service ceiling (13000 m (42,650 ft); range 2400 km (1,491 miles)
Weights: empty 2550 kg (5,534 lb); maximum take-off 6500 kg (14,330 lb)
Dimensions: span 9.50 m (31 ft 2 in); length 10.50 m (34 ft 5 in); height 2.70 m (8 ft 10 in); wing area 16.30 m^2 (175.46 sq ft)

Variants
Saab 105: this attractive high-wing multi-role aircraft was developed as a private venture and intended for a variety of civil and military tasks; the prototype first flew in June 1963 with two 745-kg (1,642-lb) Turboméca Aubisque non-afterburning turbofans; the type was soon ordered for the Swedish air force in three variants, namely the **Sk 60A** trainer and liaison aircraft with conventional seating for four, the **Sk 60B** light attack aircraft with accommodation for two on ejector seats, and the **Sk 60C** dual-role light attack and reconnaissance aircraft with ejector seats for two and a camera installation in the nose
Saab 105Ö: improved trainer and light attack version for Austria with a new and more potent powerplant, and a strengthened wing for greater underwing stores capability

Saab-Scania AJ 37 Viggen
(Sweden)
Type: all-weather attack aircraft
Accommodation: pilot seated on a Saab-Scania rocket-assisted ejector seat
Armament (fixed): none
Armament (disposable): this is carried on three underfuselage and four underwing hardpoints (plus another pair of underwing hardpoints if required), up to a maximum weight of 6000 kg (13,228 lb); the centreline hardpoint is normally fitted with an Aden 30-mm cannon pod or a drop-tank; the two outer underfuselage hardpoints can each accommodate a Bofors launcher for six 135-mm (5.3-in) rockets, or an RB 24 Sidewinder air-to-air missile, or an RB 28 Falcon air-to-air missile; the inner pair of underwing hardpoints can each accept a Bofors launcher for six 135-mm (5.3-in) rockets, or one RB 04E anti-ship missile, or one RB 05 air-to-surface missile, or one RB 75 Maverick air-to-surface missile, or one RB 24 Sidewinder air-to-air missile; and the outer pair of underwing hardpoints can each take one RB 24 Sidewinder air-to-air missile; an alternative is 16 assorted (including Virgo fragmentation) bombs
Electronics and operational equipment: communication and navigation equipment, plus L. M. Ericsson PS-37A search and attack radar, Saab-Scania CK-37 central computer, Phillips air-data computer, Marconi head-up display, Decca Type 72 Doppler navigation, SATT radar-warning receiver, Svenska Radio radar display and electronic countermeasures systems
Powerplant and fuel system: one 11800-kg (26,015-lb) afterburning thrust Volvo Flygmotor RM8A (licence-built and modified Pratt & Whitney JT8D-22) turbofan, and a total internal fuel capacity of about 5700 litres (1,254 Imp gal) in four fuselage and two wing tanks, plus provision for one centreline drop-tank
Performance: maximum speed 2125 km/h (1,320 mph) or Mach 2.0 at 12000 m (39,370 ft) and more than 1335 km/h (830 mph) or Mach 1.1 at sea level; climb to 10000 m (32,810 ft) in less than 1 minute 40 seconds; service ceiling about 15200 m (49,870 ft); combat radius at least 1000 km (621 miles) on a hi-lo-hi sortie with external weapons, and 500 km (311 miles) on a lo-lo-lo sortie with external weapons
Weights: empty about 11800 kg (26,015 lb); normal take-off 15000 kg (33,069 lb); maximum take-off 20500 kg (45,194 lb)
Dimensions: span 10.60 m (34 ft 9.25 in), and canard 5.45 m (17 ft 10.5 in); length 16.30 m (53 ft 5.75 in) including probe; height 5.80 m (19 ft 0.25 in); wing area 46.00 m^2 (495.1 sq ft), and canard foreplanes 6.20 m^2 (66.74 sq ft)

Variants
AJ 37 Viggen: the Viggen (thunderbolt) is one of

the most advanced warplanes in the world, custom-designed by Saab-Scania to the particular requirement of the Swedish air force for an integrated weapon system needed as a J 35 Draken successor with high performance, great versatility and STOL capability from dispersed sites using lengths of road for runways; extensive research confirmed that the canard configuration offered the best possibilities in conjunction with advanced high-lift devices, an integral thrust reverser (the first such installation on an afterburning turbofan, in this instance developed by Flygmotor on the basis of a US civil turbofan) actuated by compression of the nosewheel leg after a high-sink no-flare landing; the comprehensive avionics are vital to effective operation in conjunction with the national air-defence network, and also provide electronic countermeasures and an instrument landing system; the first prototype flew in February 1967, and the initial production aircraft was an AJ 37 attack fighter than flew in February 1971 for service in the middle of the year

SF 37 Viggen: designation of the single-seat all-weather overland reconnaissance version of the AJ 37, with a revised nose section accommodating one infra-red and six optical cameras, plus an infra-red sensor, an electronic countermeasures registration system and an air-data camera (to record altitude, position, course etc), the whole reconnaissance system being controlled by the aircraft's central computer; underwing loads can include two RB 24 Sidewinder self-defence air-to-air missiles and two electronic countermeasures pods (one active and one passive), while drop tanks can be carried on the underfuselage hardpoints; the first SF 37 flew in May 1973 and the type entered service in April 1977

SH 37 Viggen: designation of the single-seat all-weather maritime reconnaissance version of the AJ 37, with a revised nose section accommodating a surveillance radar and electronic countermeasures registration system, together with the same type of data camera as carried by the SF 37; under the wings the type can carry two RB 24 Sidewinders and two electronic countermeasures pod (one active and one passive), while the underfuselage hardpoints can accommodate a drop tank on the centreline, a nocturnal reconnaissance pod on the port hardpoint, and a long-range camera pod or FFV Red Baron infra-red reconnaissance pod on the starboard hardpoint; the first SH 37 flew in December 1973 and the type entered service in the second half of 1975

Sk 37 Viggen: designation of the tandem two-seat operational conversion trainer variant of the AJ 37 with reduced fuel capacity and a revised forward fuselage for the accommodation of the separate instructor's cockpit behind and above that of the pupil; the type also has a slightly taller vertical tail, increasing height by 10 cm (3.94 in); the Sk 37 is fully combat-capable, with the same armament provision as the AJ 37, and began to enter service in the second half of 1972 after the type's first flight in July 1970

Saab-Scania JA 37 Viggen
(Sweden)
Type: all-weather interceptor fighter with secondary attack capability
Accommodation: pilot seated on a Saab-Scania rocket-assisted ejector seat
Armament (fixed): one 30-mm Oerlikon KCA long-range high-velocity cannon with 150 rounds, in an underfuselage pack offset to port
Armament (disposable): this is carried on three underfuselage and four underwing hardpoints, up to a maximum weight of 6000 kg (13,228 lb); the same weapons as carried on the AJ 37 version can be accommodated for the attack mission, but for the air-defence role in conjunction with Sweden's STRIL-60 air-defence network the JA 37 normally carries up to six RB 24/74 Sidewinder or RB 71 Sky Flash air-to-air missiles, or a mix of these two types
Electronics and operational equipment: communication and navigation equipment, plus L. M. Ericsson PS-46/A pulse-Doppler long-range search and attack radar, Singer-Kearfott SKC-2037 digital central computer, Garrett LD-5 air-data computer, Singer-Kearfott KT-70L inertial platform, Smiths head-up display, Decca Type 72 Doppler navigation, SATT radar-warning receiver, Svenska Radio radar display and ECM systems, and Honeywell/Saab-Scania SA07 digital automatic flight-control system
Powerplant and fuel system: one 12750-kg (28,108-lb) afterburning thrust Volvo Flygmotor RM8B (developed version of the licence-built Pratt & Whitney JT8D-22) turbofan, and a total internal fuel capacity of about 5700 litres (1,254 Imp gal) in four fuselage and two wing tanks, plus provision for a centreline drop-tank
Performance: maximum speed more than 2125 km/h (1,320 mph) or Mach 2.0 at 12000 m (39,370 ft), and 1470 km/h (913 mph) or Mach 1.2 at sea level; climb to 10000 m (32,810 ft) in less than 1 minute 40 seconds from brakes-off; service ceiling more than 15200 m (49,870 ft); combat radius 1000 km (621 miles) on a hi-lo-hi sortie with external stores, and 500 km (311 miles) on a lo-lo-lo sortie with external stores
Weights: normal take-off 17000 kg (37,478 lb); maximum take-off 20500 kg (45,194 lb)
Dimensions: span 10.60 m (34 ft 9.25 in) and canards 5.45 m (17 ft 10.5 in); length 16.40 m (53 ft 9.75 in) including probe; height 5.90 m (19 ft 4.25 in); wing area 46.00 m² (495.1 sq ft) and canard foreplanes 6.20 m² (66.7 sq ft)

Variant
JA 37 Viggen: this is the dedicated interceptor version (with secondary attack capability) of the Viggen family, but is sufficiently different from its

predecessors to be treated separately; the airframe has been revised (combining the fuselage and wings of the AJ 37 with the tail of the Sk 37) and fitted with the considerably more powerful RM8B afterburning turbofan for enhanced performance in the interception role under command of the national air-defence network; most importantly, however, the JA 37 is fitted with a new generation of electronics and with a revised armament configuration, including a fixed gun in the form of the potent Oerlikon KCA; this fires 0.36-kg (0.79-lb) shells at the rate of 1,350 rounds per minute with a muzzle velocity of 1050 m (3,445 ft) per second, producing after a flat trajectory of 1500 m (4,920 ft) as much penetrative power as the British Aden and French DEFA cannon (of the same calibre) develop at the muzzle; the first JA 37 flew in November 1977, and the type entered service in 1979

Saab-Scania JAS 39 Gripen
(Sweden)
Type: all-weather fighter, attack and reconnaissance aircraft
Accommodation: pilot seated on a Martin-Baker S10LS zero/zero ejector seat
Armament (fixed): one 27-mm Mauser BK27 cannon plus ? rounds in the port side of the nose
Armament (disposable): this is carried on four hardpoints under the wings and on two wingtip missile rails up to a maximum weight of some 6500 kg (14,330 lb), and can include RB Sidewinder, RB 71 Sky Flash and other advanced air-to-air missiles, RB 75 Maverick air-to-surface missiles, RBS 15F anti-ship missiles, bombs of various 'smart' and 'dumb' types, unguided rocket pods, and various reconnaissance and/or electronic countermeasures systems such as the BOZ-100 chaff pod and Lake 200 jammer pod
Electronics and operational equipment: communication and navigation equipment, plus Ericsson pulse-Doppler search and acquisition radar, podded FLIR, Hughes head-up display, three SRA head-down displays (one for flight data, one for ground mapping and one for radar/FLIR data), digital central computer used in conjunction with the Lear Siegler fly-by-wire flight-control system, laser inertial navigation system, and many electronic countermeasures and electronic support measures systems
Powerplant and fuel system: one 8165-kg (18,000-lb) Volvo Flygmotor/General Electric RM12 afterburning turbofan, and a total internal fuel capacity of ? litres (? Imp gal) in integral tanks
Performance: (estimated) maximum speed 2128 km/h (1,323 mph) or Mach 2 at high altitude
Weights: empty not revealed; normal take-off 8000 kg (17,637 lb); maximum take-off about 11350 kg (25,022 lb)
Dimensions: span 8.00 m (26 ft 3 in); length 14.00 m (45 ft 11.2 in); height not revealed; wing area not revealed

Variant
JAS 39 Gripen: designed from 1980 as a successor to the Viggen in the fighter, attack and reconnaissance roles, the Gripen (griffon) is further proof of the Swedish aerospace industry's extraordinary ability to design advanced combat aircraft tailored to Sweden's peculiar political, climatic and operational requirements; the basic design (a canard delta configuration) is developed from that of the Viggen, but the adoption of composite materials for some 30% of the airframe increases strength while reducing weight and making possible the effective use of the RM12 afterburning turbofan developed jointly by Flygmotor and General Electric on the basis of the latter's F404J; other features are the triplex 'fire-by-wire' flight-control system for enhanced manoeuvrability, one wide-angle head-up and three head-down displays (flight information, computer-generated terrain map with tactical overlay, and computer-controlled radar/infra-red sensor information); these and other features offer performance and capabilities (including STOL and dispersed-site operations) comparable or superior to those of the 15000-kg (33,068-lb) Viggen in an 8000-kg (17,637-lb) aircraft; the first prototype is scheduled to fly in 1987 for service in 1992

SEPECAT Jaguar S
(France/UK)
Type: close support and reconnaissance aircraft
Accommodation: pilot seated on a Martin-Baker Mk 9B Mk II ejector seat
Armament (fixed): two 30-mm Aden cannon with 150 rounds per gun
Armament (disposable): this is carried on one underfuselage hardpoint (rated at 2,500 lb/1134 kg) and two underwing hardpoints (the inner pair each rated at 2,500 lb/1134 kg and the outer pair each at 1,250/567 kg), up to a maximum weight of 10,500 lb (4763 kg); a wide diversity of stores can be carried, but typical loads are eight 1,000-lb (454-kg) free-fall or retarded bombs, 11 500-b (227-kg) free-fall retarded bombs, six BL755 cluster bombs, six JP233 anti-airfield bombs and launchers for 19 2.75-in (70-mm) or 36 68-mm (2.68-in) rockets; the Mk 13/18 'Paveway' laser-guided bomb, AJ.168 Martel air-to-surface missile and AIM-9 Sidewinder air-to-air missile can also be carried
Electronics and operational equipment: communication and navigation equipment, plus Smiths head-up display, Ferranti FIN1064 digital inertial and weapon-aiming system, Marconi-Elliott MCS 920 digital central computer, Ferranti Type 105/106 laser rangefinder and marked-target seeker, Marconi ARI.18223 radar-warning receiver, and options such as the Philips/Matra Phima chaff dispenser, Westinghouse ALQ-101 electronic countermeasures pod and BAe reconnaissance pod

Powerplant and fuel system: two 8,040-lb (3647-kg) afterburning thrust Rolls-Royce/Turboméca Adour Mk 104 turbofans, and a total internal fuel capacity of 924 Imp gal (4200 litres) plus provision for three 264-Imp gal (1200-litres) drop-tanks; provision for inflight-refuelling
Performance: maximum speed 1,056 mph (1700 km/h) or Mach 1.6 at 36,000 ft (10970 m) and 840 mph (1350 km/h) or Mach 1.1 at sea-level; climb to 30,000 ft (9145 m) in 1 minute 30 seconds; service ceiling 46,000 ft (14020 m); lo-lo-lo combat radius 334 miles (537 km) with external warload but no drop tanks; ferry range 2,190 miles (3525 km)
Weights: empty 15,430 lb (7000 kg); normal take-off 24,150 lb (10955 kg); maximum take-off 34,610 lb (15700 kg)
Dimensions: span 28 ft 6 in (8.69 m); length 55 ft 2.5 in (16.83 m) including probe; height 16 ft 0.5 in (4.89 m); wing area 260.27 sq ft (24.18 m^2)

Variants
Jaguar A: the Jaguar resulted from separate British and French efforts in the early 1960s to develop a two-seat advanced/operational trainer and a single-seat close-support and attack aircraft; the countries decided to work on a collaborative basis, and from the Breguet Br.121 project developed the Jaguar, which first flew in two-seat prototype form in September 1968; the Jaguar A is the French single-seat attack version, first flying in March 1969 for delivery up to the end of 1981; this type is powered by 3315-kg (7,305-lb) Adour Mk 102 afterburning turbofans, and carries all the weapons in the specification above, plus the 15-kiloton AN 52 free-fall nuclear weapon; the last 30 of France's 160 aircraft were completed with provision for the AS.30L laser-homing missile and associated ATLIS II designator pod
Jaguar B: this is the Royal Air Force's **Jaguar T.Mk 2** operational conversion trainer version; only 38 examples were built, the first flying in August 1971, and the main change is the lengthening of the fuselage to 17.53 m (57.51 ft) to accommodate the second cockpit; the type has the complete weapons capability of the Jaguar S (Jaguar GR.Mk 1)
Jaguar E: this is the French air force's two-seat trainer, first flown in September 1968 and built to the extent of 40 aircraft with Adour Mk 102 turbofans for delivery from May 1972
Jaguar S: this is the British **Jaguar GR.Mk 1** single-seat close support aircraft first flown in October 1969 and built to the extent of 165 aircraft; delivered with Adour Mk 102 turbofans, from 1978 the survivors were upgraded with Adour Mk 104s
Jaguar International: first flown in August 1976, this is an improved multi-role type based on the Jaguar A/S single-seater but fitted with 3900-kg (8,598-lb) Adour Mk 804s or (in later aircraft) 4205-kg (9,270-lb) Adour Mk 811s plus revised avionics and weapons; aircraft of this type have been sold to Ecuador, India, Nigeria and Oman, India also building the type under licence; the Jaguar International has two overwing hardpoints for air-to-air missiles such as the Matra 550 Magic or AIM-9 Sidewinder, and the underwing hardpoints can accept anti-ship missiles such as the AGM-84 Harpoon, AM.39 Exocet and Kormoran; the Indian aircraft have the most advanced avionics, including Thomson-CSF Agave multi-mode radar (in some aircraft) with target indication for anti-ship missiles on the digital HUDWAS head-up display and weapon-aiming system, Sagem inertial navigation and Ferranti COMED 2045 combined map and electronic display

Shenyang J-5 (China): see Mikoyan-Gurevich Mig-17F 'Fresco-C'

Shenyang J-6C 'Farmer'
(China)
Type: fighter, attack and reconnaissance aircraft
Accommodation: pilot on a Martin-Baker PKD10 ejector seat
Armament (fixed): three 30-mm NR-30 cannon with ? rounds per gun
Armament (disposable): this is accommodated on four underwing hardpoints, the outer pair normally being used for drop-tanks, to a maximum weight of 500 kg (1,102 lb); typical loads are four air-to-air missiles, or eight 212-mm (8.35-in) rockets, or two 250-kg (551-lb) bombs
Electronics and operational equipment: communication and navigation equipment, plus a gyro sight
Powerplant and fuel system: two 3250-kg (7,165-lb) afterburning thrust Wopen-6 (Tumansky R-9BF-811) turbojets, and a total internal fuel capacity of 2170 litres (477 Imp gal) in two main and two smaller fuselage tanks, plus provision for two 760- or 1140-litre (167- or 251-Imp gal) drop-tanks
Performance: maximum speed 1540 km/h (957 mph) or Mach 1.45 at 11000 m (36,090 ft) and 1340 km/h (832 mph) or Mach 1.09 at sea level; cruising speed 950 km/h (590 mph); initial climb rate more than 9145 m (30,000 ft) per minute; service ceiling 17900 m (58,725 ft); combat radius 685 km (426 miles) with two 760-litre (167-Imp gal) drop-tanks; ferry range 2200 km (1,366 miles)
Weights: empty 5670 kg (12700 lb); normal take-off 7545 kg (16,634 lb); maximum take-off 8965 kg (19,764 lb)
Dimensions: span 9.20 m (30 ft 2.25 in); length 12.60 m (41 ft 4 in) excluding nose probe; height 3.88 m (12 ft 8.75 in); wing area 25.00 m^2 (269.1 sq ft)

Variants
J-6 'Farmer': this is the Chinese-built version of the Mikoyan-Gurevich MiG-19SF 'Farmer-C', produced after the signature of a licence agreement in January

1958 with deliveries beginning in December 1961; since that time the type has been built in large numbers and also exported with the designation **F-6**, impressing customers with the excellence of the finish and the great attention to detail; the aircraft is technically obsolete, but still a formidable close-range air-combat adversary as a result of its great agility and powerful close-range gun armament (three 30-mm cannon)

J-6A: the Chinese equivalent of the MiG-19PF with two 30-mm cannon (in the wing roots) and limited interception radar

J-6B: the Chinese equivalent of the MiG-19PM 'Farmer-D' with two 30-mm cannon, interception radar and radar-homing air-to-air missiles

J-6C: Chinese development of the J-6 with the brake parachute relocated to a bullet fairing at the base of the rudder

J-6Xin: Chinese development of the J-6A with Chinese radar in a sharp-tipped radome on the splitter plate rather than Soviet radar in the inlet centrebody

JJ-6: Chinese trainer development equivalent to (but not identical with) the MiG-19UTI; the forward fuselage is lengthened by 0.84 m (33.07 in) forward of the wing to provide volume for the insertion of a tandem-seat cockpit; armament comprises a single 30-mm cannon, and the type is exported with the designation FT-6

JZ-6: Chinese version of the MiG-19R reconnaissance aircraft with the fuselage cannon replaced by a camera installation

Shenyang J-8 'Finback'

(China)
Type: air-superiority fighter and close-support aircraft
Accommodation: pilot seated on an ejector seat
Armament (fixed): two 30-mm NR-30 cannon each with ? rounds in the underside of the fuselage
Armament (disposable): this is carried on four underwing hardpoints to an unknown maximum weight of air-to-air missiles, bombs, rocket pods and drop tanks
Electronics and operational equipment: communication and navigation equipment, plus 'Odd Rods' IFF and a radar-warning receiver
Powerplant and fuel system: two 6200-kg (13,668-lb) thrust Wopen WP-7 afterburning turbojets, and a total internal fuel capacity of ? litres (? Imp gal), plus provision for at least four 1140-litre (251-Imp gal) drop tanks
Performance: maximum speed about 2500 km/h (1,553 mph) or Mach 2.35 at high altitude; Initial climb rate 12000 m (39,370 ft) per minute; service ceiling 18000 m (59,055 ft); maximum range 1850 km (1,150 miles)
Weights: empty 12000 kg (26,455 lb); maximum take-off 19000 kg (41,888 lb)
Dimensions: (estimated) span 10.00 m (32 ft 9.7 in); length 19.00 m (62 ft 4 in); height 5.20 m (17 ft 0.7 in); wing area 40.0 m² (430.57 sq ft)

Variants

J-8I 'Finback': revealed in September 1984, the long-expected J-8 is a radical Chinese development of the Xian J-7 (Mikoyan-Gurevich MiG-21) along the lines of the Mikoyan-Gurevich E-152A experimental aircraft, about which the Chinese received limited data in the late 1950s; compared with the J-7 the airframe has been scaled up and a twin-engine powerplant installed; the programme has been long in the development, work having started in the early 1960s and accelerated once the success of the Q-5 had proved the basic soundness of such developments; the type uses two Chinese-built versions of the Tumansky R-11 afterburning turbojet to secure genuine Mach 2 performance, the engines being fed through a single large nose inlet with a translating centrebody

J-8II 'Finback': upgraded version revealed in 1986 with a revised nose in which the central inlet with small radar bullet is replaced by a large nose radome and lateral inlets just forward of the wing leading edges; the type has a span of 9.30 m (30 ft 6.1 in) and a length of 21.50 m (70 ft 6.5 in), and at a maximum take-off weight of 17800 kg (39,242 lb) possesses a maximum speed of Mach 2.2 at altitude; the type is offered for export as the **F-8II**, and China is seeking the participation of Western companies in the provision of avionics and medium-range semi-active homing missiles

Shijiazhung Y-5 (China): see Antonov An-2 'Colt'

Shin Meiwa PS-1

(Japan)
Type: STOL anti-submarine flying-boat
Accommodation: crew of five on the flightdeck, and a mission crew of five in the cabin
Armament (fixed): none
Armament (disposable): this is carried in an internal weapons bay, two underwing pods and two wingtip hardpoints; the weapons bay can carry four 150-kg (331-lb) anti-submarine bombs, the pods can each accommodate two homing torpedoes, and the wingtip hardpoints can each support three 5-in (127-mm) rockets
Electronics and operational equipment: communication and navigation equipment, plus APS-80N search radar with its antenna in the thimble nose, AQS-101B dunking sonar, ASQ-10A magnetic anomaly detection (MAD) gear, AQA-3 'Jezebel' passive sonar with 20 sonobuoys, 'Julie' active sonar ranging equipment with 12 charges, APN-153 Doppler radar, AYK-2 navigation computer, N-OA-35/HSA tactical plotting group and other items
Powerplant and fuel system: four 2285-kW (3,064-

shp) Ishikawajima-built General Electric T64-IHI-10 turboprops, and a total internal fuel capacity of 17900 litres (2,387 Imp gal) in two fuselage and five wing tanks
Performance: maximum speed 547 km/h (340 mph) at 1525 m (5,000 ft); cruising speed 315 km/h (196 mph) at 1525 m (5,000 ft) on two engines; initial climb rate 690 m (2,264 ft) per minute; service ceiling 9000 m (29,530 ft); range 2170 km (1,348 miles) for a normal mission, or 4745 km (2,948 miles) for ferrying
Weights: empty 26300 kg (58,000 lb); normal take-off 36000 kg (79,365 lb); maximum take-off 43000 kg (94,797 lb)
Dimensions: span 33.14 m (108 ft 8.75 in); length 33.50 m (109 ft 11 in); height 9.715 m (31 ft 10.5 in); wing area 135.8 m² (1,461.8 sq ft)

Variants
PS-1: one of the few modern flying-boats in service anywhere in the world, the PS-1 is a thoroughly capable machine with excellent anti-submarine sensors capable of deployment with the aircraft floating or in the air; the type has good STOL performance and can operate in rough water
US-1: search-and-rescue variant of the PS-1 with retractable wheeled landing gear to turn the type into an amphibian; the type has a crew of nine, and can carry 20 passengers or 12 litters
US-1A: version of the US-1 with uprated turboprops for better performance

Shorts Tucano T. Mk 1 (UK): see EMBRAER EMB-312 Tucano

SIAI-Marchetti S.211
(Italy)
Type: basic trainer and light attack aircraft
Accommodation: pupil and instructor seated in tandem on Martin-Baker Mk 8 ejector seats
Armament (fixed): none
Armament (disposable): this is carried on four under-wing hardpoints, the inner pair each rated at 300 kg (661 lb) and the outer pair each at 150 kg (331 lb), up to a maximum weight of 600 kg (1,323 lb); on these hardpoints a wide variety of external loads can be accommodated, typical combinations being four SIAI pods each with one or two 7.62-mm (0.3-in) FN machine-guns, or four SIAI pods each with one 12.7 mm (0.5-in) FN-Browning machine-gun, or two pods each with one 20-mm cannon, or four Aerea AL-18-50 launchers each with 18 51-mm (2-in) rockets, or four Matra F2 launchers each with six 68-mm (2.68-in) rockets, or four LAU-32/A launchers each with seven 2.75-in (69.85-mm) rockets, or Aerea AL-6-80 launchers each with six 81-mm (3.19-in) rockets, or two Matra 155 launchers each with 18 68-mm (2.68-in) rockets, or two SNORA RWK-020 launchers each with 12 81-mm (3.19-in) rockets, or four 150-kg (331-lb) bombs, or two 300-kg (661-lb) bombs or napalm tanks, or two photographic reconnaissance pods each with four cameras and infra-red linescan equipment
Electronics and operational equipment: communication and navigation equipment, plus provision for Doppler radar, attack radar, head up display, radar-warning receiver and ECM to customer specifications
Powerplant and fuel system: one 2,500-lb (3,134-kg) thrust Pratt & Whitney Canada JT15D-4C turbofan, and a total internal fuel capacity of 800 litres (176 Imp gal) in one fuselage and integral wing tanks, plus provision for two 350-litre (77-Imp gal) drop-tanks
Performance: cruising speed 667 km/h (414 mph) at 7620 m (25,000 ft); initial climb rate 1310 m (4,300 ft) per minute; service ceiling 12800 m (42,000 ft); combat radius 555 km (345 miles) on a hi-lo-hi sortie with four rocket-launchers; ferry range 2485 km (1,544 miles) with maximum fuel
Weights: empty 1615 kg (3,560 lb); normal take-off 2500 kg (5,511 lb) as a trainer; maximum take-off 3000 kg (6,614 lb) as an attack aircraft
Dimensions: span 8.43 m (27 ft 8 in); length 9.31 m (30 ft 6.5 in); height 3.80 m (12 ft 5.5 in); wing area 12.60 m² (135.63 sq ft)

Variant
S.211: first flown in April 1981, the S.211 is a lightweight trainer with good secondary attack capability

SIAI-Marchetti SF.260TP
(Italy)
Type: trainer and tactical support aircraft
Accommodation: pilot and co-pilot seated side-by-side, plus provision for a passenger to the rear of the cockpit
Armament (fixed): none
Armament (disposable): this is carried on four underwing hardpoints, up to a maximum weight of 300 kg (661 lb); a large number of alternative loads can be accommodated, typical installations being two SIAI pods each with one or two 7.62-mm (0.3-in) FN machine-guns and 500 rounds; two Aerea AL-8-70 launchers each with eight 2.75-in (69.85-mm) rockets; two LAU-32/A launchers each with seven 2.75-in (69.85-mm) rockets; two Aerea AL-18-50 launchers each with 18 51-mm (2-in) rockets; two Aerea AL-8-68 launchers each with eight 68-mm (2.68-in) rockets; two Aerea AL-6-80 launchers each with six 81-mm (3.19-in) rockets; two 125-kg (276-lb) general-purpose bombs; two 120-kg (265-lb) fragmentation bombs; or two photographic-reconnaissance pods; or a large assortment of training stores
Electronics and operational equipment: communication and navigation equipment, plus an optical sight and provision for a reconnaissance pod
Powerplant and fuel system: one 350-shp (261-kW) Allison 250-B17C turboprop, and a total internal fuel capacity of 243 litres (53.5 Imp gal) in two integral

wing and two non-jettisonable wingtip tanks, plus provision for two 80-litre (17.6-Imp gal) drop-tanks
Performance: maximum speed 380 km/h (236 mph) at sea level; cruising speed 370 km/h (230 mph) at 3050 m (10,000 ft); initial climb rate 661 m (2,170 ft) per minute; service ceiling 8535 m (28,000 ft); range 950 km (590 miles) at 4570 m (15,000 ft)
Weights: empty 795 kg (1,753 lb); normal take-off 1200 kg (2,646 lb) as a trainer; maximum take-off 1300 kg (2,866 lb) as an armed aircraft
Dimensions: span 8.35 m (27 ft 4.75 in) over tiptanks; length 7.40 m (24 ft 3.25 in); height 2.41 m (7 ft 11 in); wing area 10.10 m² (108.7 sq ft)

Variants
SF.260M: first flown in October 1970 as the military trainer derived from the F.250 prototype of 1964, the SF.260M has proved highly successful as a primary and basic trainer in Italian and third-world service
SF.260W Warrior: this is the armed weapons trainer and tactical support derivative of the SF.260M; the type is powered by a 260-hp (194-kW) Avco Lycoming O-540-E4A5 piston engine, and first flew in May 1972; the Warrior is similar in all basic ways to the SF.260TP apart from the engine installation, which results in a length of 7.10 m (23.29 ft); the type has a maximum take-off weight of 1300 kg (2,866 lb) and a maximum speed of 305 km/h (190 mph) at sea level
SF.260TP: this is the turboprop-powered derivative of the SF.260W and first flew in July 1980, deliveries beginning in 1982; SF.260M and SF.260W aircraft can be upgraded to this standard with a company-developed conversion kit

Sikorsky S-55 (H-19 Chickasaw)
(USA)
Type: general-purpose helicopter
Accommodation: crew of two on the flightdeck, and up to 10 troops, or six litters, or freight in the cabin
Armament (fixed): none
Armament (disposable): none
Electronics and operational equipment: communication and navigation equipment
Powerplant and fuel system: one 800-hp (596-kW) Wright R-1300-13 radial piston engine, and a total internal fuel capacity of ? US gal (? litres) in fuselage tanks
Performance: maximum speed 112 mph (180 km/h); cruising speed 91 mph (146 km/h); initial climb rate 1,020 ft (311 m) per minute; range 360 miles (580 km)
Weights: empty 5,250 lb (2381 kg); maximum take-off 7,900 lb (3583 kg)
Dimensions: main rotor diameter 53 ft 0 in (16.15 m); length (fuselage) 42 ft 3 in (12.88 m); height 13 ft 4 in (4.06 m); main rotor disc area 2,206.2 sq ft (204.95 m²)

Variants
S-55: first flown in November 1949, the S-55 was a pioneer of utility helicopter operations, entering service with the US Navy as the HO4S, with the US Marine Corps as the HRS, with the US Air Force as the H-19, and with the US Army as the H-19 Chickasaw, powered by the Wright R-1300 or Pratt & Whitney R-1340 radial piston engines; small numbers remain in service with third-world air arms
S-55T: introduced in 1971 with a 650-shp (485-kW) Garrett TSE331-3U-303 turboshaft, this version has been produced in small numbers by conversion of piston-engined helicopters, an increase of 900 lb (408 kg) being secured by the reduction in powerplant weight

Sikorsky S-58A (CH-34A Choctaw)
(USA)
Type: general-purpose helicopter
Accommodation: crew of two on the flightdeck, and up to 18 troops, or eight litters or freight in the cabin
Armament (fixed): none
Armament (disposable): none
Electronics and operational equipment: communication and navigation equipment
Powerplant and fuel system: one 1,525-hp (1137-kW) Wright R-1820-84 radial piston engine, and a total internal fuel capacity of 306.5 US gal (1159 litres) in internal tanks
Performance: maximum speed 122 mph (196 km/h) at sea level; cruising speed 97 mph (156 km/h); initial climb rate 1,100 ft (335 m) per minute; service ceiling 9,500 ft (2895 m); range 247 miles (400 km) with maximum fuel, or 182 miles (293 km) with maximum payload
Weights: empty 7,750 lb (3515 kg); normal take-off 13,000 lb (5900 kg); maximum take-off 14,000 lb (6350 kg)
Dimensions: main rotor diameter 56 ft 0 in (17.07 m); length (fuselage) 46 ft 9 in (14.25 m); height 15 ft 11 in (4.85 m); main rotor disc area 2,463.0 sq ft (228.8 m²)

Variants
S-58A: this was the basic designation of the S-58 civil and military general-purpose helicopter, which first flew in prototype form during March 1954 as a far more capable machine than the preceding S-55 series thanks to the use of a considerable more powerful engine and a rethought fuselage with larger cabin; the type was produced for the US services as the US Army's CH-34 Choctaw, the US Marine Corps' UH-34 Seahorse and US Navy's SH-34 Seabat, and the type was also widely exported; most survivors conform generally to the standard described above
S-58T: under this designation many S-58s have been revised with a turboshaft powerplant, namely the 1,875-shp (1398-kW) Pratt & Whitney Canada PT6T

Turbo Twin Pac with consequent improvements in reliability and payload, the latter deriving from the turboshaft's considerably lighter weight

Sikorsky S-61B (SH-3D Sea King)
(USA)
Type: anti-submarine operator
Accommodation: crew of two on the flightdeck, and two sensor operators in the cabin
Armament (fixed): none
Armament (disposable): provision on the sponson struts for 840 lb (381 kg) of weapons including depth bombs and torpedoes
Electronics and operational equipment: communication and navigation equipment, plus Bendix ASQ-13 dunking sonar and APN-130 Doppler navigation
Powerplant and fuel system: two 1,400-shp (1044-kW) General Electric T58-GE-10 turboshafts, and a total internal fuel capacity of 840 US gal (3180 litres) in three fuselage tanks
Performance: maximum speed 166 mph (267 km/h); cruising speed 136 mph (219 km/h); initial climb rate 2,200 ft (667 m) per minute; service ceiling 14,700 ft (3200 m); range 625 miles (1005 km) with maximum fuel
Weights: empty 11,685 (5382 kg); normal take-off 18,626 lb (8449 kg); maximum take-off 20,500 lb (9300 kg)
Dimensions: main rotor diameter 62 ft 0 in (18.90 m); length (fuselage) 54 ft 9 in (16.69 m); height (rotor head) 15 ft 6 in (4.72 m); main rotor disc area 3,019.0 sq ft (280.5 m^2)

Variants
SH-3A Sea King: the Sikorsky S-61 was designed in the later 1950s to a US Navy requirement for an anti-submarine helicopter combining, for the first time in such a machine, the hunter and the killer roles previously requiring two helicopters; Sikorsky's design featured a boat hull with outrigger sponsons for emergency waterborne stability and to provide accommodation for the retracted main landing gear units, far greater payload and reliability being ensured by the use of a powerful twin-turboshaft powerplant; the prototype of the **S-61B** naval helicopter first flew in March 1959, and the HSS-2 Sea King (from 1962 the SH-3A Sea King) began to enter service in September 1961 with 1,250-shp (932-kW) General Electric T58-GE-8B turboshafts, ASQ-13 dunking sonar and up to 840 lb (381 kg) of weapons; the type was also produced in licence in Japan as the essentially similar **Mitsubishi (Sikorsky) S-61B**, which is designated **HSS-2** in Japanese naval service; subvariants of the SH-3A are the **HH-3A** VIP transport and the **CH-124** for Canada
SH-3D Sea King: designation of the improved anti-submarine version with more powerful engines; the type is also built under licence in Italy as the Agusta (Sikorsky) ASH-3D; the only current variant of the basic model is the **VH-3D** VIP transport; the Japanese licence-built version is the **HSS-2A**
HH-3E Jolly Green Giant: this is the US Air Force's combat search-and-rescue variant evolved during the Vietnam War for the recovery of aircrew downed in combat areas or behind enemy lines; the variant was developed from the CH-3C and CH-3E transport helicopters (now both out of service) produced as the **S-61R** variant with retractable tricycle landing gear, a completely revised fuselage and a rear ramp for the straight-in loading of freight; the HH-3E was produced with two 1,500-shp (1119-kW) T58-GE-5 turboshafts, armour, an armament of two 7.62-mm (0.3-in) Miniguns, external fuel for greater range, an inflight-refuelling probe and specialist rescue equipment
HH-3F Pelican: US Coast Guard search-and-rescue version of the HH-3E without armour or armament, but with advanced search radar; the type is also built under licence in Italy as the ASH-3F
SH-3G Sea King: designation of SH-3As converted into utility helicopter with removable anti-submarine equipment
SH-3H Sea King: designation of the multi-role version of the SH-3G with upgraded anti-submarine equipment (including ASQ-13B dunking sonar, active/passive sonobuoys and Texas Instruments ASQ-81 magnetic anomaly detection with a towed 'bird') plus specialist equipment (including Canadian Marconi LN-66 high-performance radar and electronic support measures) for the fleet missile defence role, primarily the detection and localization of incoming anti-ship missiles; the Japanese licence-built version is the HSS-2B, and that produced for home and export sales by Agusta is the ASH-3H
S-61A: designation of the amphibious transport version able to carry 26 troops, or 15 litters, or 12 VIPs, or a substantial freight load
S-61A-4 Nuri: version of the S-61 for Malaysia with additional fuel and able to carry 31 passengers
S-61D-4: SH-3D equivalent for Argentina

Sikorsky S-65A (CH-53D Sea Stallion)
(USA)
Type: heavy assault helicopter
Accommodation: crew of three on the flightdeck, and up to 64 troops, or 24 litters and four attendants, or freight in the cabin
Armament (fixed): none
Armament (disposable): none
Electronics and operational equipment: communication and navigation equipment
Powerplant and fuel system: two 3,925-shp (2927-kW) General Electric T64-GE-413 turboshafts, and a total internal fuel capacity of 630 US gal (2384 litres) in two fuselage tanks
Performance: maximum speed 196 mph (315 km/h)

at sea level; cruising speed 173 mph (278 km/h); initial climb rate 2,180 ft (664 m) per minute; service ceiling 21,000 ft (6400 m); range 257 miles (413 km)
Weights: empty 23,485 lb (10653 kg); maximum take-off 36,400 lb (16510 kg)
Dimensions: main rotor diameter 72 ft 3 in (22.02 m); length (fuselage) 67 ft 2 in (20.47 m) excluding probe; height (rotor head) 17 ft 1.5 in (5.22 m); main rotor disc area 4,070.0 sq ft (378.1 m^2)

Variants
CH-53A Sea Stallion: developed as the **S-65** under a US Navy requirement for an assault transport helicopter for the US Marine Corps, the Sea Stallion first flew in prototype form during October 1964, and deliveries to the US Marine Corps began in 1966; the type is powered by two 2,850-shp (2125-kW) General Electric T64-GE-6 turboshafts, and the load can include one 105-mm (4.13-in) howitzer and ammunition, or 38 fully-equipped troops; all but 32 were built with provision to tow a minesweeping sled
RH-53A: designation of 15 CH-53As converted into dedicated minesweepers for the US Navy with 3,925-shp (2927-kW) T64-GE-413 turboshafts
HH-53B Super Jolly: designation of the US Air Force combat search-and-rescue variant, fitted out with the same provisions as the HH-3E Jolly Green Giant (including a retractable inflight-refuelling probe and all-weather flight instrumentation) and powered by two 3,080-shp (2297-kW) T64-GE-3 turboshafts
HH-53C Super Jolly: upgraded version of the HH-53B for the US Air Force, fitted with 3,925-shp (2927-kW) T64-GE-7 turboshafts for greater performance and payload
CH-53D Sea Stallion: improved version of the CH-53A for the US Marine Corps with greater power for higher performance and increased payload (55 troops) over short ranges; most of the type have provision to tow a minesweeping sled
RH-53D: first flown in October 1972, this is a dedicated minesweeping version of the CH-53D for the US Navy; the type is powered by two 4,380-shp (3266-kW) T64-GE-415 turboshafts, has a 500-US gal (1893-litre) fuel tank in each sponson, and can tow the Mk 103 mechanical, Mk 104 acoustic, Mk 105 magnetic and Mk 106 magnetic/acoustic sweeps; as well as the SPU-1 Magnetic Orange Pipe for dealing with shallow-water magnetic mines; two 0.5-in (12.7-mm) heavy machine-guns are carried for the detonation of any mines brought to the surface
CH-53E Super Stallion: this is a radical development of the basic type, developed as the Sikorsky S-80 series and first flown in prototype form during March 1974 for service in 1981; the type has a seven- rather than six-blade main rotor with a diameter of 79.0 ft (24.08 m) and driven by three rather than two turboshafts, in this instance 4,380-shp (3266-kW) T64/GE-416 units; this much improved dynamic system allows an increase in maximum take-off weight to 73,500 lb (33339 kg) for freight loads of 30,000 lb (13608 kg) carried internally or 36,000 lb (16329 kg) slung externally, or alternatively of 55 fully-equipped troops or 30,000 lb (13608 kg) of freight in a fuselage lengthened to 73.33 ft (22.35 m)
MH-53E Sea Dragon: US Navy mine-countermeasures version of the CH-53E with greater fuel capacity (an additional 1,000 US gal/3785 litres) in larger sponsons) and provision for inflight-refuelling; further improvements are planned for the series, including composite-construction rotor blades and main rotor hub, night-vision equipment, infra-red suppressors for uprated engines, and other operational improvements
CH-53G: version of the Sea Stallion series for West Germany with T64-GE-7 turboshafts
HH-53H Super Jolly: improved version of the HH-53C for nocturnal SAR with an AAQ-10 infra-red sensor, APQ-158 terrain-following radar, and much-improved navigational capability

Sikorsky S-70 (UH-60A Black Hawk)
(USA)
Type: combat assault helicopter
Accommodation: crew of two on the flightdeck, and up to 11 troops, or four litters, or freight in the cabin
Armament (fixed): provision for one or two 7.62-mm (0.3-in) M60 machine-guns in the cabin firing through the doorways
Armament (disposable): this is carried on four hardpoints (two under each of the stub wings of the optional External Stores Support System and each rated at 400 lb/181 kg) to a maximum of 1,600 lb (726 kg); the usual stores are 16 AGM-114 Hellfire anti-tank missiles, four M56 mine dispensers, or four launchers for 2.75-in (69.85-mm) rockets
tank missiles
Electronics and operational equipment: communication and navigation equipment, plus Singer Doppler navigation, E-Systems/Loral APR-39 radar-warning receiver, M130 chaff dispenser and Sanders ALQ-144 infra-red jammer
Powerplant and fuel system: two 1,560-shp (1151-kW) General Electric T700-GE-700 turboshafts, and a total internal fuel capacity of 354 US gal (1340 litres) in two fuselage tanks, plus provision for two 230-US gal (871-litre) and two 450-US gal (1703-litre) external tanks
Performance: maximum speed 184 mph (296 km/h) at sea level; cruising speed 167 mph (269 km/h); service ceiling 19,000 ft (5790 m); range 373 miles (600 km) on internal fuel, or 1380 miles (2220 km) with maximum internal and external fuel
Weights: empty 10,624 lb (4819 kg); normal take-off 16,260 lb (7375 kg); maximum take-off 20,250 lb (9185 kg)

Dimensions: main rotor diameter 53 ft 8 in (16.36 m); length (fuselage) 50 ft 0.75 in (15.26 m); height (rotor head) 12 ft 4 in (3.76 m); main rotor disc area 2,261.0 sq ft (210.05 m²)

Variants
UH-60A Black Hawk: designed as the **S-70** to meet a Utility Tactical Transport Aircraft System requirement issued by the US Army for a Bell UH-1 replacement (and thus being able to carry a full 11-man infantry squad and its equipment), the Black Hawk first flew in October 1974 and beat a Boeing Vertol competitor for selection as the production model in late 1976; the type began to enter service in 1979, and is a highly capable and versatile helicopter able to carry a slung load of 8,000 lb (3631 kg); there have been in-service problems with the transmission, but a new gearbox is being developed to improve reliability and increase maximum take-off weight to 26,450 lb (11997 kg), which will allow the carriage of a greater assortment of external loads as well as improved armament
EH-60A Black Hawk: electronic warfare version of the UH-60A designed for the jamming of battlefield communications with the 1,800-lb (817-kg) ALQ-151 'Quick Fix II' package
HH-60A Night Hawk: US Air Force combat search-and-rescue variant of the UH-60A with the dynamic system and rescue winch of the SH-60B; the avionics of this important type include advanced radar (with terrain-following, terrain-avoidance and ground-mapping capabilities), forward-looking infra-red, and multi-function cockpit and helmet displays; other equipment includes stub wings for two 230-US gal (871-litre) external tanks, a retractable inflight-refuelling probe, two side-mounted machine-guns and other items; interim types are the **Credible Hawk** with additional fuel and inflight refuelling capability but not fitted with FLIR or terrain-following radar, and the **Pave Hawk** development of the Credible Hawk with 'Pave Low III' FLIR
SH-60B Seahawk: this is the maritime **S-70L** version produced to meet the US Navy's requirement for a Light Airborne Multi-Purpose System III to take over from the Sikorsky SH-3 Sea King series on destroyers and larger frigates as complement to the lighter Kaman SH-2F Seasprite LAMPS I helicopter carried by smaller surface vessels; first flown in December 1979 for service in 1983, the SH-60B has a folding main rotor, a relocated twin-wheel rear landing gear unit and a deck-recovery system, is powered by two 1,690-shp (1261-kW) T700-GE-401 turboshafts, and has a maximum take-off weight of 21,884 lb (9926 kg) with an armament of two Mk 46 anti-submarine torpedoes and an electronics fit comprising Texas Instruments APS-124 search radar in a chin radome, Texas Instruments ASQ-81 magnetic anomaly detection on a starboard-side pylon, electronic support measures equipment in two lateral chin fairings, a data-link system for real-time control from the parent ship and for mid-course updating of RGM-84 Harpoon anti-ship missiles, and a port-side 25-round pneumatic launcher for 125 sonobuoys
SH-60F Seahawk: version of the SH-60B designed as the inner-zone anti-submarine helicopter carried by aircraft-carriers, and fitted with Bendix ASQ-13F dunking sonar for this important task; the type is planned for service in 1988
SH-60J Seahawk: designation of the licence-built Japanese version of the SH-60B basically similar to the US helicopter but featuring items of specifically Japanese nature such as a ring-laser gyro
VH-60 Seahawk: VIP transport version of the SH-60 for use by the US president and other high-ranking officials
S-70A: designation of the export version of the UH-60 basic type
S-70B: designation of the export version of the SH-60 basic type, the Royal Australian Navy having ordered the **S-70B-2** type for its RAWS (Role-Adaptable Weapon System) requirement
S-70C: utility version available in civil and military applications, and optimized for adequate operation in 'hot-and-high' conditions

Soko G-2 Galeb (Yugoslavia): see Soko J-1 Jastreb

Soko G-4 Super Galeb
(Yugoslavia)
Type: two-seat basic trainer and light attack aircraft
Accommodation: pupil and instructor seated in tandem on Martin-Baker J8 ejector seats
Armament (fixed): one 23-mm GSh-23L two-barrel cannon and 200 rounds in a removable ventral pod
Armament (disposable): this is carried on four underwing hardpoints to a maximum weight of 1200 kg (2,646 lb), and can include bombs, cluster bombs, anti-personnel/anti-tank bomblet dispensers, napalm tanks, rocket-launchers and one reconnaissance pod
Electronics and operational equipment: communication and navigation equipment, plus an optical sight and provision for a reconnaissance pod
Powerplant and fuel system: one 4,000-lb (1814-kg) thrust Rolls-Royce Viper Mk 632 non-afterburning turbojet, and an internal fuel capacity of 1720 litres (378 Imp gal) in integral tanks plus provision for drop-tanks
Performance: maximum speed 910 km/h (565 mph) at 6000 m (19,685 ft); initial climb rate 1500 m (4,920 ft) per minute; absolute ceiling 15000 m (49,215 ft); lo-lo-lo combat radius 300 km (186 miles)
Weights: empty 3250 kg (7,165 lb); maximum take-off 6330 kg (13,955 lb)
Dimensions: span 9.88 m (32 ft 5 in); length 11.86 m (38 ft 11 in); height 4.28 m (14 ft 0.5 in); wing area 19.50 m (209.9 sq ft)

Variant
G-4 Super Galeb: first flown in July 1978, the Super Galeb bears no more relationship to the G-2 Galeb than an identity of role, being an altogether more advanced aircraft with marked similarities to the BAe Hawk; the type began to enter service in 1983, and offers a capable light attack facility to the standard training role

Soko J-1 Jastreb
(Yugoslavia)
Type: light attack aircraft
Accommodation: pilot seated on an HSA (Folland) Type 1-B lightweight ejector seat
Armament (fixed): three 0.5-in (12.7-mm) Colt-Browning machine-guns with 135 rounds per gun in the nose
Armament (disposable): this is carried on eight underwing hardpoints; the inner pair are able to accept two 250-kg (551-lb) bombs, or two clusters of smaller bombs, or two 150-litre (33-Imp gal) napalm tanks, or two launchers each with 12 55-mm (2.17-in) rockets; and the outer six are each able to lift one 127-mm (5-in) rocket
Electronics and operational equipment: communication and navigation equipment, plus an optical sight
Powerplant and fuel system: one 3,000-lb (1361-kg) thrust Rolls-Royce (Bristol) Viper Mk 531 turbojet, and a total internal fuel capacity of 975 litres (214 Imp gal) in two fuselage tanks, plus provision for two 275-litre (60.5-Imp gal) jettisonable wingtip tanks
Performance: maximum speed 820 km/h (510 mph) at 6000 m (19,685 ft); cruising speed 740 km/h (460 mph) at 5000 m (16,405 ft); initial climb rate 1260 m (4,135 ft) per minute; service ceiling 12000 m (39,375 ft); range 1520 km (945 miles) with maximum fuel
Weights: empty 2820 kg (6,217 lb); maximum take-off 5100 kg (11,243 lb)
Dimensions: span 11.68 m (38 ft 4 in) over tiptanks; length 10.88 m (35 ft 8.5 in); height 3.64 m (11 ft 11.5 in); wing area 19.43 m² (209.14 sq ft)

Variants
G-2A Galeb: this was Soko's first design, the programme being launched in 1957 to provide the Yugoslav air force with a capable yet simple turbojet-powered trainer; the Galeb first flew in May 1961 and immediately impressed Western analysts with its similarity to the Aermacchi M.B.326 series; production started in 1963 and the type entered service with the Yugoslav air force powered by the 2,4500-lb (1134-kg) Viper Mk 22-6 non-afterburning turbojet; the pupil and instructor are seated in tandem but unstepped Folland Type 1-B lightweight ejector seats, and the Galeb has a maximum speed of 812 km/h (505 mph) at 6200 m (20,350 ft)
G-2AE Galeb: designation of the export model delivered from 1974 to Libya and Zambia with a number of detail modifications and improvements
J-1 Jastreb: this is the single-seat light attack derivative of the G-2, with a strengthened airframe and uprated powerplant for higher performance and greater payload
J-1-E Jastreb: export version of the J-1 with updated equipment and other improvements
RJ-1 Jastreb: tactical reconnaissance version of the J-1 with cameras in the forward fuselage, and in the forward portion of the tiptanks
RJ-1-E Jastreb: export version of the RJ-1 with various combinations of Vinten cameras and flash equipment for day and night photo-reconnaissance
TJ-1 Jastreb: operational conversion trainer version of the J-1

Soko J-22 Orao-B/CNIAR IAR-93B
(Yugoslavia/Romania)
Type: close-support aircraft
Accommodation: pilot seated on a Martin-Baker RU10J ejector seat (Romanian aircraft) or Martin-Baker YU10J ejector seat (Yugoslav aircraft)
Armament (fixed): two 23-mm Gsh-23L twin-barrel cannon with 200 rounds per gun
Armament (disposable): this is carried on one underfuselage and four underwing hardpoints, up to a maximum weight of 2500 kg (5,511 lb); typical loads are five FAB-250 250-kg (551-lb) bombs, or five FAB-500 500-kg (1,102-lb) bombs, or 12 FAB-100 100-kg (220-lb) bombs, or 12 RAB-120 120-kg (265-lb) fragmentation bombs, or clusters of RAB-16 16-kg (35-lb) fragmentation bombs, or clusters of ZAB-45 45-kg (99-lb) incendiary bombs, or DPT-150 150-kg (331-lb) cluster bombs, or PLAB-150-L 150-kg (331-lb) napalm bombs, or 55-mm (2.17-in) or 127-mm (5-in) rocket-launcher pods; it is unknown what types of guided weapon can be carried
Electronics and operational equipment: communication and navigation equipment, plus 'head-up display, radar-warning receiver and provision for a reconnaissance pod
Powerplant and fuel system: two 5,000-lb (2268-kg) afterburning thrust Rolls-Royce Viper Mk 633-47 turbojets, and a total internal fuel capacity of 3280 litres (722 Imp gal) in integral wing tanks, plus provision for three 540-litre (119-Imp gal) drop-tanks
Performance: maximum speed 1160 km/h (721 mph) or Mach 0.95 at sea level; cruising speed 730 km/h (453 mph) at 7000 m (22,965 ft); initial climb rate 3960 m (12,992 ft) per minute; service ceiling 12500 m (41,010 ft); combat radius 650 km (404 miles) on a hi-lo-hi sortie with internal fuel and external weapons
Weights: empty 5900 kg (13,007 lb); normal take-off 8600 kg (18,959 lb); maximum take-off 101000 kg (22,266 lb)
Dimensions: span 9.62 m (31 ft 6.75 in); length 14.90 m (48 ft 10.6 in); height 4.45 m (14 ft 7.25 in); wing area 26.00 m² (279.86 sq ft)

Variants
J-22 Orao-A/IAR-93A: a simple yet effective light attack aircraft well suited to the operational requirements and capabilities of the two sponsor nations, this light attack and close support aircraft was designed from 1970 under the Yugoslav name Orao and the Romanian designation IAR-93; two prototypes were flown on 31 October 1974 (one in each country), and the type entered service in the late 1970s with 4,000-lb (1814-kg) Viper Mk 632-41R non-afterburning turbojets and a fuel capacity of 3080 litres (678 Imp gal) in wing bag tanks; there is also a two-seat version of this initial model with fuel capacity reduced to 2700 litres (594 Imp gal)

J-22 Orao-B/IAR-93B: full-production version of the series with Viper Mk 633-47 afterburning turbojets and integral tanks for the fuel; this variant has also been developed as a two-seater for operational conversion

Sukhoi Su-7BMK 'Fitter-A'
(USSR)
Type: ground-attack fighter
Accommodation: pilot seated on a KM-1 ejector seat
Armament (fixed): two 30-mm NR-30 cannon with 70 rounds per gun in the wing roots
Armament (disposable): this is carried on six hardpoints, the two under the fuselage and the two under the inner wings each rated at 500 kg (1,102 lb) and the pair under the outer wings each at 250 kg (551 lb), up to a nominal weight of 2500 kg (5,511 lb) reduced to 1000 kg (2,205 lb) when drop-tanks are carried under the fuselage; typical weapons are the 750- and 500-kg (1,653- and 1,102-lb) bombs, the S-24 250-kg (551-lb) runway-piercing bomb, various types of nuclear stores, napalm tanks, and the UV-16-57U launcher with 16 55-mm (2.17-in) rockets
Electronics and operational equipment: communication and navigation equipment, plus 'High Fix' ranging radar in the inlet centrebody, Sirena 3 radar-warning receiver and ASP-5PF gyro sight
Powerplant and fuel system: one 10000-kg (22,046-lb) afterburning thrust Lyulka AL-7F-1 turbojet, and a total internal fuel capacity of 2940 litres (647 Imp gal) in fuselage and integral wing tanks, plus provision for two 600-litre (132-Imp gal) drop-tanks under the fuselage and two 900-litre (198-Imp gal) ferry tanks under the wings
Performance: maximum speed 1700 km/h (1,055 mph) or Mach 1.6 at 11000 m (36,090 ft), and 1350 km/h (840 mph) or Mach 1.1 at sea level; initial climb rate about 9120 m (29,920 ft) per minute; service ceiling 15150 m (49,700 ft); combat radius 345 km (215 miles) with two drop-tanks; ferry range 1450 km (901 miles) with maximum fuel
Weights: empty 8620 kg (19,004 lb); normal take-off 12000 kg (26,455 lb); maximum take-off 13500 kg (29,762 lb)

Dimensions: span 8.93 m (29 ft 3.5 in); length 17.37 m (57 ft 0 in) including probe; height 4.57 m (15 ft 0 in); wing area 31.5 m^2 (339.1 sq ft)

Variants
Su-7B 'Fitter-A': an aircraft of almost unbelievable battlefield toughness and accuracy, the Su-7 was designed in the early 1950s as the Soviet counter to the North American F-100 Super Sabre and McDonnell F-101 Voodoo, and first flew in 1955 after development through a number of prototype aircraft; the type was ordered into production in 1958 and entered service in 1959 as the Su-7B tactical fighter with the 9000-kg (19,841-lb) Lyulka AL-7F afterburning turbojet, a powerful but very thirsty engine that requires the use of drop tanks on at least two of the four (soon increased to six) hardpoints to secure even the minimum acceptable combat radius

Su-7BKL 'Fitter-A': version optimized for operation from poor airfields and featuring redesigned landing gear with low-pressure tyres (requiring bulged doors over the nose unit) and a steel skid outboard of each main wheel, as well as twin braking parachutes and more efficient brakes

Su-7BM 'Fitter-A': uprated Su-7B with the 10000-kg (22,046-lb) AL-7F-1 afterburning turbojet and detail modifications such as a relocated pitot boom, twin ducts above the central fuselage, a Sirena 3 radar-warning receiver, a zero/zero KM-1 ejector seat, improved air data sensors to provide input for a fire-control system based on a ballistic computer, and higher-velocity cannon

Su-7BMK 'Fitter-A': version combining features of the Su-7BKL and Su-7BM, and possessing improved avionics and an additional pair of hardpoints behind the main landing gear units

Su-7U 'Moujik': two-seat operational conversion trainer variant of the Su-7B

Su-7UM 'Moujik': two-seat operational conversion trainer variant of the Su-7BM

Su-7UMK 'Moujik': two-seat operational conversion trainer variant of the Su-7BMK

Sukhoi Su-15 'Flagon-F' (USSR): see Sukhoi Su-21 'Flagon-F'

Sukhoi Su-20 'Fitter-K'
(USSR)
Type: variable-geometry ground-attack fighter
Accommodation: pilot seated on a KM-1 ejector seat
Armament (fixed): two 30-mm NR-30 cannon with 70 rounds per gun
Armament (disposable): this is carried on eight hardpoints (two tandem pairs under the fuselage with each unit ratyed at 500 kg/1,102 lb, and two under each wing the inner pair each rated at 500 kg/1,102 lb and the outer pair each at 750 kg/1,653 lb) to a maximum weight of 4000 kg (8,818 lb); various types of tactical nuclear weapons can be carried, but more common

loads are two or four AS-7 'Kerry' air to surface missiles, or six 500-kg (1,102-lb) and two 250-kg (551-lb) bombs, or eight 240-mm (9.45-in) anti-runway rockets, or eight launchers each with 16 or 32 55-mm (2.17-in) rockets

Electronics and operational equipment: communication and navigation equipment, plus SRD-5M 'High Fix' ranging radar, terrain-avoidance radar, head-up display, 'Odd Rods' IFF, Sirena 3 radar-warning receiver, Doppler navigation, and laser ranger and marked-target seeker

Powerplant and fuel system: one 11200-kg (24,961-lb) afterburning thrust Lyulka AL-21F-3 turbojet, and a total internal fuel capacity of 4550 litres (1,000 Imp gal) plus provision for four 800-litre (176-Imp gal) drop tanks; no provision for inflight-refuelling

Performance: maximum speed 2369 km/h (1,472 mph) or Mach 2.2 at 11000 m (36,090 ft) and 1287 km/h (800 mph) or Mach 1.05 at sea level; initial climb rate 14950 m (49,050 ft) per minute; service ceiling 18000 m (59,055 ft); lo-lo-lo combat radius 500 km (311 miles) with 2000-kg (4,409-lb) warload

Weights: empty 10900 kg (24,030 lb); maximum take-off 19500 kg (42,990 lb)

Dimensions: span spread 14.00 m (45 ft 11.2 in) and swept 10.60 m (34 ft 9.5 in); length 19.20 m (63 ft 0 in) including probe; height 5.35 m (17 tf 6.6 in); wing area spread 40.10 m² (431.65 sq ft) and swept 37.20 m² (400.4 sq ft)

Variants

Su-17 'Fitter-C': first flown in 1966 as the Su-7IG 'Fitter-B' prototype derived from the fixed-wing Su-7, the Su-17 is a radical development of the 'Fitter' family with variable-geometry outer wing panels for improved field performance and better range with useful weapons load; the solution offered a significant improvement of capabilities without the cost and potential technical problems of a new design with variable geometry applied to the entire wing planform, but the Su-17 entered only limited service with the 7000-kg (15,432-lb) Lyulka AL-21F-1 afterburning turbojet and additional fuel

Su-17MK and Su-20 'Fitter-C': definitive early version of the variable-geometry 'Fitter' series derived from the Su-17 but fitted with the 11200-kg (24,691-lb) AL-21F-3) afterburning turbojet and much improved avionics; the type is used by the USSR (Su-17MK) and by Warsaw Pact countries and favoured export customers (Su-20) with reduced avionics standards

Su-20MK 'Fitter-D': important derivative of the 'Fitter-C' with the nose lemgthened by 0.25 m (9.85 in), a small blister radome under the inlet for terrain-avoidance radar, a laser ranger and marked-target seeker in the inlet centrebody, a head-up display, a radar-warning receiver, an internal chaff/flare dispenser and (most significantly) the ability to deliver tactical nuclear weapons

Su-17UM 'Fitter-E': tandem two-seat operational conversion trainer varient, based on the Su-17MK but having a lengthened and drooped nose for improvrd visibility and a widened dorsal fairing for additional fuel; the type has only the wing-root cannon; the export version is the **Su-22U 'Fitter-E'** with the Tumansky R-29 turbojet

Su-22 'Fitter-F': export derivative of the Su-17MK for third-world clients, and distinguishable by its larger dorsal fin and enlarged spine; the type is powered by the 11500-kg (25,353-lb) Tumansky R-29BS-300 afterburning tubojet, but has generally inferior avionics in comparison with Soviet and Warsaw Pact varients

Su-17UM 'Fitter-G': two-seat operational conversion trainer similar to the 'Fiter-E' but distinguishable by its taller tail, small ventral fin, an enlarged dorsal spine and the provision of a laser ranger and marked-target seeker in the inlet centrebody

Su-22BM 'Fitter-H': Lyulka-powered single-seater based on the 'Fitter-C' and 'Fitter-D' but incorporating tne refinements of the 'Fitter-E' (including the down-tilted forward fuselage), updated avionics and an additional hardpoint under the inner portion of each wing for the carriage of two air-to-surface missiles

Su-22BKL 'Fitter-J': third-world export version of the 'Fitter-H' with the rear fuselage and powerplant of the 'Fitter-F' but less advanced avionics and weapons (the air-to-air missile fit being limited to AA-2 'Atoll' and AA-2-2 'Advanced Atoll' weapons); there is also a two-seat operational conversion trainer development of the 'Fitter-J', but as yet no designation or NATO reporting name has been revealed

Su-20 'Fitter-K': latest Lyulka-powered version based on the'Fitter-H' and intended for service with the USSR and Warsaw Pact, with additional avionics and electronic countermeasures to permit effective operation in areas of intense air and ground defences; this varient is based on the 'Fitter-H' but distinguishable by the enlargement of the cooling air inlet at the base of the fin

Sukhoi Su-21 'Flagon-F'
(USSR)

Type: all-weather interceptor fighter
Accommodation: pilot on a KM-1 ejector seat
Armament (fixed): none
Armament (disposable): this is carried on six hardpoints (two under the fuselage each rated at 500kg/1,102 lb, and two under each wing rated at 350 kg/772 lb) to a maximum of 1500 kg (3,307 lb); typical weapons are the AA-2 'Atoll', AA-2-2 'advanced Atoll', AA-3 'Anab', AA-3-2 'Advanced Anab 'and AA-8 'Aphid' air-to-air missiles, and pod with 23-mm GSh-23L twin-barrel cannon

Electronics and operational equipment: communication and navigation equipment, plus 'Improved Skip Spin' or 'Twin Scan' fire-control radar, visual search system, 'Odd Rods' IFF, Sirena 3 radar-warning system and several electronic countermeasures systems
Powerplant and fuel system: two 11200-kg (24,691-lb) afterburning thrust Lyulka AL-21F-3 turbojets, and a total internal fuel capacity of about 8000 litres (1,760 Imp gal) plus provision for two 800- or 600-litre (2176- or 132-Imp gal) drop tanks
Performance: (estimated) maximum speed about 2655 km/h (1,650 mph) or Mach 2.5 at 11000 m (36,090 ft); initial climb rate 13700 m (44,950 ft) per minute; service ceiling 20000 m (65,615 ft); hi-hi-hi combat radius 725 km (450 miles); ferry range 2250 km (1,398 miles)
Weights: (estimated) empty 12250 kg (27,006 lb); normal take-off 18000 kg (39,683 lb); maximum take-off 20000 kg (44,092 lb)
Dimensions: (estimated) span 10.50 m (34 ft 5.4 in); length 22.00 m (72 ft 2.1 in); height 5.00 m (16 ft 5 in); wing area 35.70 m² (384.3 sq ft)

Variants
Su-15U 'Flagon-C': this is the combat-capable operational conversion trainer variant of the 'Flagon' family, derived from the 'Flagon-D' initial-production single-seater but fitted with two separate cockpits; the Su-15 series was derived from the Su-11 after a 1962 requirement for genuine Mach 2.5 interceptors, with two afterburning turbojets drawing air through lateral inlets (instead of one afterburning turbojet with a single nose inlet), nose-mounted radar and updated avionics; the prototype probably flew in 1964, and the pre-production 'Flagon-A' of the late 1960s had Su-11 wings of 9.3-m (30.5-ft) span and two 10000-kg (22,046-kg) Lyulka AL-21F-1 afterburning turbojets
Su-15F or Su-15MF 'Flagon-D': first true production model, based on the 'Flagon-A' but having greater-span wings with compound sweep and deeper inlets, probably signifying the adoption of AL-21F-3 afterburning turbojets
Su-21 'Flagon-E': improved model introduced in 1973 with 11200-kg (24,691-lb) AL-21F-3 afterburning turbojets, twin nosewheels and 'Twin Scan' air-interception radar
Su-21 'Flagon-F': definitive single-seat model introduced from 1975 with a low-drag ogival rather than conical radome for a larger radar providing limited look-down/shoot-down capability; as with earlier models of the series, the facts about this highly effective fighter are uncertain
Su-21U 'Flagon-G': two-seat operational conversion trainer based on the 'Flagon-F' with a periscope-fitted rear cockpit behind the standard cockpit

Sukhoi Su-24 'Fencer-C'
(USSR)
Type: variable-geometry attack and interdiction aircraft
Accommodation: pilot and weapons officer seated side-by-side on KM-1 ejector seats
Armament (fixed): one 30-mm rotary-barrel cannon with ? rounds (port underfuselage fairing) and possibly one 30-mm single-barrel cannon with ? rounds (starboard underfuselage fairing)
Armament (disposable): this is carried on eight hardpoints (four under the fuselage as one tandem pair and two flanking units, one under each wing glove and one under each wing, each rated at 1500 kg/3,307 lb except the underwing hardpoints which are each rated at 1000 kg/2,205 lb) to a maximum weight of 11000 kg (24,250 lb); the type can carry tactical nuclear weapons, but more common weapons are AA-2 'Atoll', AA-7 'Apex' and AA-8 'Aphid' air-to-air missiles, AS-7 'Kerry', AS-9, AS-10 'Karen', AS-11, AS-12 and AS-14 'Kedge' air-to-surface missiles, and 1000-, 750-, and 500-kg (2,205-, 1,653- and 1,102-lb) bombs of various types
Electronics and operational equipment: communication and navigation equipment, plus pulse-Doppler multi-mode radar, twin terrain-following radars, head-up display, head-down displays, radar and infrared warning receivers, inertial navigation system, Doppler navigation, laser rangefinder and marked-target seeker, and a variety of electronic countermeasures and target acquisition/designation systems
Powerplant and fuel system: two 11200-kg (24,691-lb) afterburning thrust Lyulka AL-21F-3 turbojets, and a total internal fuel capacity of about 13000 litres (2,860 Imp gal) plus provision for two 3000-litre (660-Imp gal) drop tanks; provision for inflight-refuelling
Performance: (estimated) maximum speed 2320 km/h (1,441 mph) or Mach 2.18 at 11000 m (36,090 ft) and 1470 km/h (913 mph) or Mach 1.2 at sea level; service ceiling 16500 m (54,135 ft); lo-lo-lo combat radius 320 km (199 miles) with 8000-kg (17,637-lb) warload, or hi-lo-hi combat radius 1300 km (805 miles) with 3000-kg (6,614-lb) warload and two drop tanks; ferry range 6450 km (4,008 miles)
Weights: (estimated) empty 19000 kg (41,887 lb); normal take-off 29000 kg (63,933 lb); maximum take-off 41000 kg (90,388 lb)
Dimensions: (estimated) span spread 17.50 m (57 ft 5 in) and swept 10.50 m (34 ft 5.5 in); length 21.29 m (69 ft 10 in) excluding probe; height 6.00 m (19 ft 8 in); wing area spread 47.0 m² (505.9 sq ft) and swept 42.5 m² (457.5 sq ft)

Variants
Su-24 'Fencer-A': apparently developed as the Soviet counterpart to the General Dynamics F-111 series in the long-range interdiction role, the Su-24 was the USSR's first true variable-geometry combat aircraft,

initially flying in prototype form during 1969 and entering service during 1974 as the 'Fencer-A' initial production model; since that time the aircraft has displayed range and weapon capabilities that make it likely that its wartime role would be unrefuelled interdiction raids launched from Eastern European bases against targets of operational importance in Western Europe; the type has the latest in Soviet avionics, allowing blind first-pass attacks in all weathers by day and night

Su-24 'Fencer-B': revised version of the 'Fencer-A' with a bullet fairing at the base of the fin for a braking parachute

Su-24 'Fencer-C': introduced in 1981, this variant is based on the 'Fencer-B' but has a radar-warning receiver system with antennae on the fin and on the inlet lips, a revised air-data probe, an underfuselage aerial (possibly for mid-course missile-update purposes) and modified tailpipes, the last suggesting that the variant is possibly powered by two uprated AL-21F-3 turbojets in place of the original two models' 10000-kg (22,046-lb) AL-21F-1 turbojets

Su-24 'Fencer-D': variant similar to the 'Fencer-C' but with a retractable inflight-refuelling probe, a broader-chord lower portion to the fin, the nose section forward of the cockpit lengthened by 0.75 m (29.5 in) possibly indicating a new radar type, glove pylons integral with the large overwing fences, and an undernose blister window/fairing

Su-24 'Fencer-E': electronic warfare variant designed to supplant the Yak-28 'Brewer-E' in the escort and stand-off roles

Su-24 'Fencer-F': reconnaissance variant with internal side-looking airborne radar, infra-red linescanner and cameras

Sukhoi Su-25 'Frogfoot'
(USSR)
Type: tactical close-support aircraft
Accommodation: pilot seated on an ejector seat
Armament (fixed): one 30-mm rotary barrel cannon with ? rounds in the lower fuselage
Armament (disposable): this is carried on 10 underwing pylons up to a maximum weight of 4500 kg (9,921 lb) of bombs, rocket-launcher pods, air-to-air and air-to-surface missiles, electronic countermeasures pods etc
Electronics and operational equipment: communication and navigation equipment, plus a laser ranger and marked-target seeker, a head-up display, electronic and infra-red countermeasures, and other items
Powerplant and fuel system: probably two 5100-kg (11,244-lb) thrust Tumansky R-13-300 turbojets, and a total internal fuel capacity of about 5000 kg (11,023 lb), plus provision for at least two drop tanks on the innermost pair of underwing hardpoints
Performance: maximum speed 880 km/h (547 mph) at optimum altitude; combat radius with a 4000-kg (8,818-lb) payload 550 km (342 miles)
Weights: empty 9500 kg (20,944 lb); normal take-off 16000 kg (35,273 lb); maximum take-off 19200 kg (42,329 lb)
Dimensions: span 15.50 m (50 ft 10 in); length 14.50 m (47 ft 7 in); height not revealed; wing area 37.60 m² (404.74 sq ft)

Variant
Su-25 'Frogfoot': following the path pioneered by the US Air Force with its AX programme that led to the adoption of the Fairchild Republic A-10A Thunderbolt II battlefield attack and close-support aircraft, the USSR moved mor slowly during the definition and design phases during the early and mid-1970s, the resulting Su-25 being comparable in many respects to the A-10A but lighter, considerably higher powered and possessing markedly better performance; the type was developed with great care, pre-production types seen in Afghanistan from 1982 having a small cannon installation and a less refined avionics/weapon fit, but from 1984 the full-production standard has been achieved with a new large-calibre cannon and complex avionics; the type probably carries two air-to-air missiles on the outermost pair of underwing hardpoints for self defence

Sukhoi Su-27 'Flanker'
(USSR)
Type: air-superiority fighter
Accommodation: pilot seated on a KM-1 zero/zero ejector seat
Armament (fixed): one 30-mm six-barrel cannon with ? rounds in a wing root installation
Armament (disposable): this is carried on 10 hardpoints up to a maximum weight of about 6000 kg (13,228 lb) of missiles and free-fall ordnance, and can include eight air-to-air missiles, the normal fit being six AA-10 active-homing medium-range missiles, or four AA-10s and four AA-8 'Aphid' IR-homing short-range missiles
Electronics and operational equipment: communication and navigation equipment, plus pulse-Doppler nose radar (with a search range of 240 km/149 miles and a tracking range of 185 km/115 miles), a head-up display, electronic countermeasures, Sirena 3 radar-warning receiver and other items
Powerplant and fuel system: two 13600-kg (29,982-lb) afterburning thrust turbofans, and a total internal fuel capacity of 7000 kg (15,432 lb) in integral tanks
Performance: maximum speed 2495 km/h (1,550 mph) or Mach 2.35 at high altitude; initial climb rate 18300 m (60,040 ft) per minute; service ceiling more than 15000 m (49,215 ft); combat radius 1500 km (932 miles)
Weights: empty 17700 kg (39,021 lb); normal take-off 20000 kg (44,092 lb); maximum take-off 28800 kg (63,492 lb)

Dimensions: span 14.50 m (47 ft 6.9 in); length 21.00 m (68 ft 10.75 in); height 5.50 m (18 ft 0.5 in); wing area 46.5 m² (500.5 sq ft)

Variant
Su-27 'Flanker': developed during the 1970s as a direct counterpart to the US Air Force's McDonnell Douglas F-15 Eagle air-superiority fighter, the Su-27 is a capable and high-performance type, and almost certainly the USSR's first genuine look-down/shoot-down fighter with its pulse-Doppler radar and an offensive complement of four active-homing air-to-air missiles; the type began to enter service in 1985

Transall C.160
(France/West Germany)
Type: tactical transport
Accommodation: crew of three on the flightdeck, and up to 93 troops, or 88 paratroops, or 62 litters and four attendants, or a 16000 kg (35,273 lb) freight load
Armament (fixed): none
Armament (disposable): none
Electronics and operational equipment: communication and navigation equipment, including Omera ORB 37 weather radar
Powerplant and fuel system: two 6,100-shp (4549-kW) Rolls-Royce Tyne RTy.20 Mk 22 turboprops, and a total internal fuel capacity of 28050 litres (6,170 Imp gal) in four integral wing tanks and one optional 9000-litre (1,980-Imp gal) centre-section tank; provision for inflight-refuelling
Performance: maximum speed 513 km/h (319 mph) at 4875 m (16,000 ft); initial climb rate 396 m (1,300 ft) per minute; service ceiling 8230 m (27,000 ft); range 1853 km (1,150 miles) with maximum payload, or 8858 km (5504 miles) for ferrying with centre-section tank
Weights: empty 29000 kg (63,935 lb); maximum take-off 51000 kg (112,435 lb)
Dimensions: span 40.00 m (131 ft 3 in); length 32.40 m (106 ft 3.5 in); height 11.65 m (38 ft 2.75 in); wing area 160.00 m² (1,722.3 sq ft)

Variants
C.160: this Franco-German venture was conceived in the late 1950s as a twin-engine tactical transport to replace the Nord Noratlas and Douglas C-47; the design was of the standard transport configuration pioneered by the Lockheed C-130 (large-volume hold accessed by a rear ramp/door under an upswept tail, high wing with turboprop engines, and multi-wheel main landing gear retracting into fuselage blister fairings); the prototype first flew during March 1963, and from 1967 to 1972 production aircraft were built for France (**C.160F**), West Germany (**C.160D**), Turkey (**C.160T**) and South Africa (**C.160Z**); by the mid-1970s the French decided that additional aircraft would be useful and the production line was reopened for 25 more aircraft to an updated standard with new-generation avionics and (to meet the requirements of France's overseas commitments) additional tankage in the centre section plus an inflight-refuelling probe; the revised model first flew in 1981, deliveries beginning in the same year; 10 of the aircraft are outfitted an inflight-refuelling tankers with a hose-and-drogue unit in the starboard main landing gear blister, and another five can be so converted at short notice with the aid of kits; there exist plans for the development of other variants such as the C.160S maritime surveillance aircraft (with search radar and other role equipment), the C.160SE electronic surveillance and intelligence aircraft (Thomson-CSF Varan radar, forward-looking infra-red, side-looking airborne radar, and electronic and signals intelligence systems), and the C.160AAA airborne early warning aircraft (with the radar and electronics of the BAe Nimrod AEW.Mk 3 as well as electronic support measures pods at the wingtips)
C.160 Astarte: designation of four C.160s specially outfitted with a Collins package of hardened electronics to serve in the airborne communication-relay role between the French nuclear deterrent forces and the national command authorities

Tupolev Tu-16 'Badger-A'
(USSR)
Type: medium bomber and maritime reconnaissance aircraft
Accommodation: crew of six
Armament (fixed): two 23-mm NR-23 cannon each in dorsal and ventral barbettes, and in the radar-laid rear turret, plus one 23-mm NR-23 fixed cannon in the starboard side of the nose of aircraft without a radome
Armament (disposable): this is carried in a weapons bay in the lower part of the fuselage, up to a maximum of 9000 kg (19,841 lb) of free-fall ordnance
Electronics and operational equipment: communication and navigation equipment, plus mapping radar and other systems
Powerplant and fuel system: two 9500-kg (20,944-lb) thrust Mikulin AM-3M turbojets, and a total internal fuel capacity of about 44900 litres (9,877 Imp gal) in fuselage and wing tanks, plus provision for underwing auxiliary tanks; provision for inflight-refuelling
Performance: maximum speed 990 km/h (615 mph) at 6000 m (19,685 ft); cruising speed 850 km/h (530 mph) at optimum altitude; initial climb rate 1250 m (4,100 ft) per minute; service ceiling 12300 m (40,355 ft); combat radius 2900 km (1,802 miles) without inflight-refuelling
Weights: empty 37200 kg (82,010 lb); maximum take-off 75800 kg (167,108 lb)
Dimensions: span 32.93 m (108 ft 0.5 in); length 34.80 m (114 ft 2 in); height 14.00 m (45 ft 11.25 in); wing area 164.65 m² (1,772.3 sq ft)

Variants
Tu-16 'Badger-A': for its period a remarkable type that still serves in large numbers with great utility in a number of roles, the Tu-16 was designed in the late 1940s and first flew early in 1952 as the Tu-88 prototype; the type began to enter service in late 1953 or early 1954 as a strategic bomber with free-fall nuclear weapons; most of the surviving aircraft in Soviet service have been converted for other roles, the same designation being retained for two inflight-refuelling tanker derivatives (one with additional tanks in the erstwhile bomb bay and a hose trailed from the starboard wingtip to meet a receptacle on the port wingtip of the receiving aircraft, and the other with an underfuselage hose-and-drogue unit)
Tu-16 'Badger-B': variant of the 'Badger-A' for the carriage, aerial launch and guidance of two AS-1 'Kennel' anti-ship missiles on underwing hardpoints; the type carried Komet III guidance radar, and most aircraft have now been rebuilt to 'Badger-G' standard
Tu-16 'Badger-C': anti-shipping variant with one AS-2 'Kipper' missile carried semi-recessed under the fuselage; the nose is revised to eliminate the bombardier's position and nose gun in favour of 'Puff Ball' acquisition radar and A-39Z missile-guidance equipment; in recent years the type has been reworked to carry an alternative load of two AS-6 'Kingfish' missiles under the wings; overall length is 36.60 m (120 ft 1 in)
Tu-16 'Badger-C (Modified)': 'Badger-C' variant designed to carry two AS-6 'Kingfish' anti-ship missiles on underwing pylons; AS-2 capability is retained by this variant
Tu-16 'Badger-D': maritime reconnaissance and electronic intelligence variant with a row of three ventral blisters for receiver antennae and a comprehensive suite of multi-sensor reconnaissance equipment; the type has the same nose as the 'Badger-C' but without the missile-guidance equipment
Tu-16 'Badger-E': designation of 'Badger-A' bombers rebuilt for optical reconnaissance with a revised bomb bay to accommodate additional fuel and various pallet-mounted reconnaissance systems
Tu-16 'Badger-F': this is a 'Badger-E' subvariant with two electronic intelligence pods (probably broad-band receivers) mounted under the wings
Tu-16 'Badger-G': based on the 'Badger-A', this is an advanced missile platform for two AS-5 'Kelt' anti-ship or stand-off missiles under the wings; the nose is similar to that of the 'Badger-A' but reworked to accommodate 'Short Horn' navigation and bombing radar
Tu-16 'Badger-G (Modified)': subvariant of the 'Badger-G' designed to carry AS-6 'Kingfish' rather than AS-5 missiles
Tu-16 'Badger-H': dedicated electronic warfare variant based on the 'Badger-A' but intended for the escort or stand-off role with the bomb bay revised to accommodate broad-band receiver/analysis and jammer equipment; the type also dispenses up to 9000 kg (19,841 lb) of chaff, flares and electronic decoys from a ventral tube and other ejectors; the type is distinguishable by two teardrop radomes (one forward and one aft of the payload bay), two blade antennae and a hatch afat of the payload bay
Tu-16 'Badger-J': dedicated electronic warfare variant comparable to the 'Badger-H' but with improved receiver/analysis and jammer capability (spot and barrage types, resulting in a large ventral canoe fairing) at the expense of dispensing capability
Tu-16 'Badger-K': long-range electronic intelligence variant with additional fuel and receiving antennae in two teardrop radomes (one in the weapons bay and the other just ahead of it) and four small strut-mounted pods on the underside of the fuselage just ahead of the rear radome
Tu-16 'Badger-L': apparently a variant based on the 'Badger-A' but fitted with a thimble radome on the nose and a new chin radar characterized by a rotary strip antenna in a flat dish radome
Xian H-6: designation of the 'Badger-A' model produced in China for the strategic bomber role
Xian H-6D: maritime patrol and anti-ship variant equivalent to the 'Badger-C/G' with two underwing C-601 anti-ship missiles; the type is also known as the **H-6 IV**

Tupolev Tu-22 'Blinder-A'
(USSR)
Type: supersonic reconnaissance bomber
Accommodation: crew of three seated in tandem on ejector seats
Armament (fixed): one 23-mm NR-23 cannon in tail barbette
Armament (disposable): this is carried in a weapons bay in the lower fuselage, and comprises up to 10000 kg (22,046 lb) of free-fall conventional or nuclear weapons
Electronics and operational equipment: communication and navigation equipment, plus 'Down Beat' nav/attack radar, 'Fan Tail' gunlaying radar, Sirena 3 radar-warning receiver, 'Odd Rods' IFF and several electronic countermeasures systems
attack radar and chaff dispensers
Powerplant and fuel system: two 12250-kg (27,006-lb) afterburning thrust turbojets, and a total internal fuel capacity of about 45450 litres (10,000 Imp gal)*Performance:* maximum speed 1480 km/h (920 mph) or Mach 1.4 at 12200 m (40,025 ft); service ceiling 18300 m (60,040 ft); combat radius 3100 km (1,926 miles) with weapons and 400-km (249-mile) supersonic dash on internal fuel; ferry range 6500 km (4039 miles) on internal fuel
Weights: empty 40000 kg (88,183 lb); maximum take-off 83900 kg (184,965 lb)

Dimensions: span 27.70 m (90 ft 10.5 in); length 40.53 m (132 ft 11.5 in); height 10.67 m (35 ft 0 in); wing area 145.00 m² (1,560.8 sq ft)

Variants
Tu-22 'Blinder-A': this supersonic bomber and maritime patrol aircraft began development in 1955 as the Tu-105, the prototype first flying during 1959; the type was seen initially as replacement for the Tu-16 in the role of high-altitude penetration against increasingly sophisticated Western defences; entering service in 1961, the initial 'Blinder-A' variant is a reconnaissance bomber with cameras in the rear of the main landing gear fairings (protruding aft of the wing trailing edges and a 'trademark' of large Tupolev aircraft) and side-looking airborne radar or infra-red linescan equipment between the nose radar and the crew compartment
Tu-22 'Blinder-B': basic missile-carrying variant with one AS-4 'Kitchen' weapon semi-recessed under the fuselage for use with the 'Down Beat' radar that replaces the 'Short Horn' type of the 'Blinder-A'; this variant also introduced an inflight-refuelling capability by means of a probe above the nose
Tu-22 'Blinder-C': dedicated maritime reconnaissance variant in which the weapons bay is occupied by a substantial reconnaissance package with six or seven cameras, side-looking airborne radar, infra-red linescan equipment, electronic intelligence gear and electronic countermeasures equipment
Tu-22 'Blinder-D': operational conversion trainer variant with a cockpit for the instructor added behind and above the standard cockpit; this model appeared in 1968, at about the time modifications to the inlet design indicated that a new engine type had been introduced; although not confirmed, it is believed that initial models were powered by Mikulin AM-3 afterburning turbojets, these giving way to Kolesov VD-7 or VD-7F engines later in the production run, which was comparatively short as the type was not deemed a great success because of its limited performance (especially in speed and range)

Tupolev Tu-26 'Backfire-B'
(USSR)
Type: variable-geometry medium bomber and maritime reconnaissance aircraft
Accommodation: crew of four, probably seated on ejector seats
Armament (fixed): two 23-mm NR-23 cannon in a rear barbette
Armament (disposable): this is carried in a lower-fuselage weapons bay and on four hardpoints (one under each wing glove and one under each inlet duct for tandem triple ejector racks) to a maximum of 12000 kg (26,455 lb) of free-fall ordnance, or one semi-recessed AS-4 'Kitchen' air-to-surface missile, or two AS-6 'Kingfish' air-to-surface missiles

Electronics and operational equipment: communication and navigation equipment, plus 'Down Beat' nav/attack radar, 'Fan Tail' gunlaying radar, Sirena 3 radar-warning receiver, 'Odd Rods' IFF and a number of electronic countermeasures systems
Powerplant and fuel system: two 20000-kg (44,092-lb) afterburning thrust turbofans, and a total internal fuel capacity of about 60000 litres (13,200 Imp gal); provision for inflight-refuelling
Performance: maximum speed 2125 km/h (1,320 mph) or Mach 2 at high altitude, and 1100 km/h (684 mph) or Mach 0.9 at sea level; service ceiling more than 16000 m (52,495 ft); combat radius 5470 km (3,400 miles) with typical weapon load on a hi-lo-hi sortie with a 320-km (199-mile) supersonic dash and another 320 km (199 miles) at sea level; ferry range 12000 km (7,457 miles)
Weights: empty 54000 kg (119,048 lb); maximum take-off 122500 kg (270,062 lb)
Dimensions: span 34.45 m (113 ft 0 in) spread and 26.21 m (86 ft 0 in) swept; length 42.00 m (137 ft 10 in); height 10.06 m (33 ft 0 in); wing area about 170.0 m² (1,829.9 sq ft) spread

Variants
Tu-22M 'Backfire-A': this powerful bomber, missile-carrier and reconnaissance aircraft is clearly derived from the Tu-22 in the same way as the Sukhoi Su-17 was developed from the Su-17, with pivoting outer wings for improved field performance and better range in the cruising regime without detriment to dash performance with the wings fully swept; in the case of the Tu-22M the revision was extended further, the engine installation being modified from two afterburning turbojets pod-mounted on each side of the vertical tail surfaces to two afterburning turbofans in a conventional lateral installation with variable-geometry inlets; the revised type featured a number of other differences, and the Tu-136 prototype first flew in 1969, allowing a small number of Tu-22M 'Backfire-A' aircraft to enter service by 1973; like the Tu-22, the type proved disappointing in range, largely because of the drag of the main landing gear pod fairings, and only a few pre-production aircraft (possibly Tu-22 conversions rather than new-build machines) were built
Tu-26 'Backfire-B': definitive version of the basic design, considerably revised and re-engineered for better performance and offensive capability; the only parts of the Tu-22M retained were the vertical tail, fuselage shell and inner wing structure, the rest of the aircraft being new and featuring main landing gear units retracting inwards into centre section bays rather than rearwards into trailing-edge pods
Tu-26 'Backfire-C': revised version introduced in 1983 with ramp inlets; this offers higher dash performance, and may mark the introduction of a new engine type; the Tu-26 is used by the strategic bomber force

for offensive operations with free-fall weapons and cruise missiles, and by the naval air force for electronic reconnaissance and anti-shipping attack with air-to-surface missiles

Tupolev Tu-28P 'Fiddler-B'
(USSR)
Type: long-range all-weather interceptor
Accommodation: crew of two seated in tandem on ejector seats
Armament (fixed): none
Armament (disposable): this is carried on four underwing hardpoints, and comprises two IR-homing AA-5 'Ash' and two radar-homing AA-5 'Ash' air-to-air missiles
Electronics and operational equipment: communication and navigation equipment, plus 'Big Nose' attack radar
Powerplant and fuel system: two 11000-kg (24,250-lb) afterburning thrust Lyulka AL-21F-3 turbojets, and a total internal fuel capacity of 13000 kg (28,660 lb
Performance: maximum speed 1850 km/h (1,150 mph) or Mach 1.75 at 11000 m (36,090 ft); initial climb rate 7500 m (24,605 ft) per minute; service ceiling 20000 m (65,615 ft); range 4990 km (3,101 miles) with maximum fuel
Weights: empty 24500 kg (54,012 lb); normal take-off 40000 kg (88183 lb); maximum take-off 45000 kg (99,206 lb)
Dimensions: span 18.10 m (59 ft 11.75 in); length 27.20 m (89 ft 3 in); height 7.00 m (23 ft 0 in); wing area 80.0 m² (861.1 sq ft)

Variant
Tu-28P 'Fiddler-B': this is the world's largest fighter, a truly enormous aircraft for its role and designed for the unusual long-range patrol requirements of the USSR's great northern reaches, which lack any abundance of airfields; the type originated as the Tu-102 prototype for a planned series of multi-role fighter and reconnaissance aircraft that did not in the event enter production; the prototype flew in 1960 or 1961, and was followed by a small number of Tu-128 'Fiddler-A' pre-production aircraft which, like the Tu-102, had canted ventral fins; the type was then developed as the Tu-28P 'Fiddler-B' pure interceptor with high performance, considerable range and a powerful missile armament allied to the 100-km (62-mile) range 'Big Nose' search and target-illumination radar

Tupolev Tu-95 'Bear-A'
(USSR)
Type: long-range strategic heavy bomber
Accommodation: crew of 10
Armament (fixed): two 23-mm NR-23 cannon each in remotely-controlled dorsal and ventral barbettes and in the manned rear turret
Armament (disposable): this is carried in a weapons bay in the lower fuselage, and comprises up to 20000 kg (44,092 lb) of free-fall conventional or nuclear weapons
Electronics and operational equipment: communication and navigation equipment, plus 'Short Horn' attack radar with its antenna in a chin radome, 'Bee Hind' tail warning and gunlaying radar, A-322Z Doppler navigation and several electronic countermeasures system
Powerplant and fuel system: four 11033-ekW (14,795-ehp) Kuznetsov NK-12MV turboprops, and a total internal fuel capacity of 72980 litres (16,540 Imp gal) in wing tanks; provision for inflight-refuelling
Performance: maximum speed 850 km/h (528 mph) or Mach 0.78 at 12500 m (41,010 ft); service ceiling about 13400 m (43,965 ft); range with 11340-kg (25,000-lb) bombload 12550 km (7,798 miles); ferry range 17500 km (10,874 miles) on internal fuel
Weights: empty 86000 kg (189,594 lb); maximum take-off 152400 kg (335,979 lb)
Dimensions: span 51.10 m (167 ft 7.75 in); length 47.50 m (155 ft 10 in); height 12.12 m (39 ft 9.25 in); wing area 310.5 m² (3,342.3 sq ft)

Variants
Tu-95 'Bear-A': the Tu-95 series of bombers, reconnaissance aircraft and (eventually) missile-carriers first flew in prototype form during late 1954 and began to enter service in 1955; the type was developed as a long-range strategic bomber able to carry two thermonuclear bombs or 20000 kg (44,092 lb) of conventional bombs over intercontinental ranges, as was conceived as a technically less-demanding fallback for the conceptually more advanced Myasishchev M-4 'Bison'; in the event the Tupolev aircraft entered service as the **Tu-20** and has proved far superior to the M-4 in terms of versatility and range, and despite its propeller propulsion it achieves jet speeds; the 'Bear-A' was the original bomber version and is still in service
Tu-95 'Bear-B': missile-carrier variant with the AS-3 'Kangaroo' in a semi-recessed position under the fuselage; the nose is revised to accommodate the 'Crown Drum' high-definition search radar and A-336Z missile-guidance system; from 1962 the aircraft were revised with inflight-refuelling probes, and some have been modified subsequently for the strategic reconnaissance role with a blister fairing on the starboard side of the rear fuselage
Tu-95 'Bear-C': introduced in about 1963, this is a dedicated maritime reconnaissance and electronic intelligence derivative of the 'Bear-B'; the type has the search radar and inflight-refuelling nose of the 'Bear-B', but no offensive capability as the weapons

bays are reworked for an additional 19000 litres (4,179 Imp gal) of fuel and electronic warfare equipment characterized by two lateral blister fairings on the rear fuselage and a number of ventral radomes

Tu-95 'Bear-D': introduced in about 1966, this is a maritime multi-sensor reconnaissance and missile-support version produced by converting 'Bear-A' aircraft; there are a number of detail differences between aircraft, but most have 'Mushroom' radar in a chin radome, 'Big Bulge' radar in a ventral radome, blister fairings on both sides of the rear fuselage, pods at the tip of each tailplane half, anti-flutter masses (probably containing the antennae for a Sirena 3 radar-warning system), a longer inflight-refuelling probe, and 'Box Tail' rear-warning and tail-turret fire-control radar; for the missile-support part of its task the 'Bear-D' carries A-346Z digital data-link equipment; in the mid-1970s some 'Bear-Ds' were revised with a long fairing (with four aerials) in place of the tail turret

Tu-95 'Bear-E': introduced in the later 1960s, this is a multi-sensor maritime reconnaissance aircraft produced by converting 'Bear-As' with inflight-refuelling capability plus (in the rear weapons bay) additional fuel tankage and a conformal pallet for six or seven cameras and other sensors (probably side-looking airborne radar and infra-red linescan equipment) and (in the forward weapons bay) electronic equipment characterized by two external blister fairings; the type also has the lateral rear-fuselage blister strakes of the 'Bear-C' and 'Bear-D'

Tu-142 'Bear-F': the Tu-142 designation is believed to apply to new-build aircraft produced from the late 1960s in response to the continued demand for this highly versatile long-range type; the Tu-142 has a number of engineering and equipment improvements compared with the original series, allowing take-off at a maximum weight of some 188000 kg (414,462 lb) attributable in part to considerably greater fuel capacity; fuselage length is increased to 49.50 m (162.4 ft) and the protruding flaps of the earlier variants are replaced by narrower-chord units in a completely restressed wing with stronger landing gear units and larger tyres; the type was designed for the long-range anti-submarine role and began to enter service in 1972; the rear fuselage accommodates a crew rest area (with galley and toilet) as well as stowage and launchers for sonobuoys, magnetic anomaly detection equipment is housed in a pod at the top of the fin, the gun armament is reduced to the two-gun rear turret, a smaller ventral radar is fitted, and there is provision for a smaller chin radar

Tu-95 'Bear-G': designation of 'Bear-B' and 'Bear-C' aircraft reworked as launch platforms for the AS-4 'Kitchen' missile

Tu-142 'Bear-H': designation of new-build aircraft used as launch platforms for four AS-15 cruise missiles, carried in pairs under the inboard wing panels and in the lower-fuselage weapons bay; the fuselage has been considerably refined, the ventral barbette being eliminated, and external ECM and ESM equipment being replaced by internal units, the type has a maximum take-off weight of 154000 kg (339,512 lb) and an unrefuelled radius of 8285 km (5,148 miles)

Tu-95 'Bear-J': older aircraft reworked as long-range relay aircraft fitted with VLF radio for communication with submerged nuclear submarines

Tupolev Tu-142 'Bear': see Tupolev Tu-95 'Bear-A'

Tupolev Tu-? 'Blackjack'
(USSR)

Type: variable-geometry supersonic bomber

Accommodation: crew probably of four, seated on ejector seats

Armament (fixed): none

Armament (disposable): this is carried in two lower-fuselage weapons bays and on hardpoints under the wing gloves to a maximum generally quoted in the West at about 16500 kg (36,376 lb) of bombs and/or missiles (including up to eight AS-15 air-launched cruise missiles)

Electronics and operational equipment: communication and navigation equipment, plus attack radar, electronic support measures, electronic countermeasures systems and other items

Powerplant and fuel system: four 23000-kg (50,705-lb) afterburning thrust turbofans of unknown designation, and a total internal fuel capacity of ? kg (? lb); provision for inflight-refuelling

Performance: (estimated) maximum speed 2230 km/h (1,386 mph) or Mach 2.1 at high altitude; long-range cruising speed 955 km/h at 13500 m (44,290 ft); hi-lo-hi combat radius 7300 km (4,536 miles)

Weights: (estimated) empty 118000 kg (260,140 lb); maximum take-off 270000 kg (595,238 lb)

Dimensions: (estimated) span, spread 52.00 m (170 ft 7 in) and swept 33.75 m (110 ft 9 in); length 50.65 m (166 ft 2 in); height 13.75 m (45 ft 1.3 in); wing area 360.0 m² (3,875.1 sq ft)

Variant

Tu-? 'Blackjack': conceptually akin to the Rockwell B-1B in being a variable-geometry supersonic penetration bomber designed for the strategic role, the 'Blackjack' is due to enter service in 1988 after a first flight in 1982; the Soviet type is somewhat larger than the B-1B, and the unrefuelled combat radius (estimated at 7300 km/4,536 miles) is posited on the assumption of high-level subsonic cruise, low-level transonic penetration and high-level supersonic attack; all data are highly speculative, and it seems likely that the type probably possesses far greater payload than that with which it has generally been credited

Tupolev Tu-126 'Moss'
(USSR)
Type: airborne warning and control system aircraft
Accommodation: crew of 12
Armament (fixed): none
Armament (disposable): none
Electronics and operational equipment: communication and navigation equipment, plus 'Flat Jack' search and tracking radar with its antenna in an 11.0-m (36-ft) rotodome above the rear fuselage, and several other systems including computerized data-processing, data-link communications and other aspects relating to the AWACS role
Powerplant and fuel system: four 11185-ekW (15,000-ehp) Kuznetsov NK-12MK turboprops, and a total internal fuel capacity of 75000 litres (16,498 Imp gal) in wing and fuselage tanks; provision for inflight-refuelling
Performance: maximum speed 850 km/h (528 mph) at high altitude; cruising speed 650 km/h (404 mph) at high altitude; service ceiling 11000 m (36,090 ft); range 12550 km (7,800 miles) on internal fuel, sufficient for an endurance of more than 20 hours
Weights: empty about 105000 kg (231,481 lb); maximum take-off 175000 kg (385,802 lb)
Dimensions: span 51.10 m (167 ft 7.75 in); length 57.30 m (188 ft 0 in); height 16.05 m (52 ft 8 in); wing area 311.1 m² (3,348.8 sq ft)

Variant

Tu-126 'Moss': adapted from the airframe of the obsolescent Tu-114 airliner (itself developed from the Tu-95 bomber), the 'Moss' was the USSR's first airborne early warning and control system aircraft; the type first flew in 1967 or 1968 to enter service in 1971, and the 'Flat Jack' surveillance radar (with its antenna in an 11-m/36-ft) rotodome above the rear fuselage, is believed to be of only limited capability, effective performance being secured only over water; the type has inbuilt IFF capability, and the tactical compartment in the fuselage is data-linked to ground stations and supporting fighters

Vought A-7E Corsair II
(USA)
Type: shipboard attack aircraft and tactical fighter
Accommodation: pilot seated on a McDonnell Douglas Escapac ejector seat
Armament (fixed): one 20-mm General Electric M61A1 Vulcan rotary-barrel cannon with 1,000 rounds in the port side of the fuselage
Armament (disposable): this is carried on two underfuselage and six underwing hardpoints, those under the fuselage each rated at 500 lb (227 kg), the two inboard underwing hardpoints at 2,500 lb (1134 kg) each, and the four outer underwing hardpoints at 3,500 lb (1587 kg) each, up to a maximum weight of 15,000 lb (6804 kg) or more; among the weapons that can be carried are various types of free-fall or retarded bombs (general-purpose, napalm and fuel-air explosive), Rockeye cluster bombs, CBU-series bomb dispensers, mines, destructors, LAU-series rocket-launchers, up to four AIM-9 Sidewinder air-to-air missiles, up to four AGM-45 Shrike or AGM-88 HARM anti-radiation missiles, up to four AGM-65 Maverick air-to-surface missiles, up to four AGM-84 Harpoon anti-ship missiles, up to six Walleye I or four Walleye II or two Walleye II ER guided bombs, or nuclear stores, or various types of laser- or TV-guided bombs
Electronics and operational equipment: communication and navigation equipment, plus Texas Instruments APQ-126(V) multi-mode nav/attack radar, AAR-42 forward-looking infra-red pod, ASN-91 inertial navigation system, ASN-190 Doppler navigation, ASU-99 projected-map display, Itek ALR-45 radar-warning receiver, Magnavox ALR-50 SAM-launch warning system, Sanders ALQ-126 electronic countermeasures, Loral APR-43 tactical radar-warning system, Tracor ALE-39 chaff/flare dispenser, and provision for podded items such as Westinghouse ALQ-119 and ALQ-131 electronic countermeasures, Xerox ALQ-123 infra-red countermeasures, Sanders ALQ-126 deception electronic countermeasures, Eaton-AIL ALQ-130 tactical communications jammer and Northrop ALQ-162 radar jammer
Powerplant and fuel system: one 15,000-lb (6804-kg) thrust Allison TF41-A-2 (licence-built Rolls-Royce Spey) turbofan, and a total internal fuel capacity of 1,500 US gal (4678 litres) in fuselage and integral wing tanks, plus provision for 1,200 US gal (4542 litres) of external fuel in four drop-tanks; provision for inflight-refuelling
Performance: maximum speed 690 mph (1110 km/h) or Mach 0.9 at sea level, and 645 mph (1038 km/h) or Mach 0.86 at 5,000 ft (1525 m) with 12 Mk 82 500-lb (227-kg) bombs; combat radius 715 miles (1151 km) on a hi-lo-hi sortie; range 2,280 miles (3669 km) on internal fuel, and 2,860 miles (4603 km) with maximum internal and external fuel
Weights: empty 19,127 lb (8676 kg); maximum take-off 42,000 lb (19050 kg)
Dimensions: span 38 ft 9 in (11.80 m); length 46 ft 1.5 in (14.06 m); height 16 ft 0.75 in (4.90 m); wing area 375.0 sq ft (34.83 m²)

Variants

TA-7C Corsair II: this is the oldest variant of the Corsair family still in service, a tandem-two seat operational conversion trainer based on the A-7B and A-7C single-seaters and powered by the 13,400-lb (6078-kg) Pratt & Whitney TF30-P-408 non-afterburning turbofan; the type has the operational capabilities of the A-7E; the Corsair II is the US Navy's standard medium attack aircraft, and was derived aerodynamically from the supersonic Vought F-8 Crusader to save time when the US Navy issued an urgent requirement for a carrierborne medium attack aircraft in the early 1960s; the prototype first flew during September 1965 and the Corsair II entered service during October 1966 in the A-7A initial production form with the 11,350-lb (5148-kg) TF30-P-6 turbofan for great range with a substantial load of disposable ordnance; following the A-7A were the A-7B with the 12,200-lb (5534-kg) TF30-P-8 or TF40-P-408 turbofan and A-7C with the latter engine and the avionics of the A-7E; a version in very limited service is the **ETA-7C Corsair II**, an electronic training subvariant with the ALE-41 chaff dispenser system

A-7D: tactical fighter for the US Air Force with the 14,500-lb (6577-kg) Allison TF41-A-1 turbofan (derived from the Rolls-Royce Spey), M61A1 internal cannon, inflight-refuelling capability, and an advanced all-weather nav/attack system; the type was later retrofitted with automatic manoeuvring flaps and the 'Pave Penny' laser tracker

A-7E Corsair II: definitive attack, close support and interdiction variant for the US Navy, based on the A-7D but using the TF41-A-2 engine; later aircraft were equipped with a forward-looking infra-red pod under the starboard wing for enhanced adverse-weather and nocturnal attack capability

A-7H: land-based version of the A-7E for the Greek air force, denavalized but retaining wing-folding capability

TA-7H: two-seat operational conversion trainer derivative of the A-7H for the Greek air force

A-7K: two-seat operational conversion trainer derivative of the A-7D for the Air National Guard

EA-7L Corsair II: electronic warfare conversions of TA-7C two-seaters

A-7P: designation of A-7As refurbished for the Portuguese air force with TF40-P-408 engines and the avionics of the A-7E

Vought F-8E Crusader
(USA)
Type: shipboard fighter
Accommodation: pilot seated on an ejector seat
Armament (fixed): four 20-mm Colt-Browning M39 cannon with 144 rounds per gun in the forward fuselage
Armament (disposable): this is carried on the fuselage sides (four AIM-9 Sidewinder air-to-air missiles or eight 5-in/127-mm rockets) and on two underwing hardpoints, up to a maximum weight of 5,000 lb (2268 kg); typical underwing loads are two Mk 84 2,000-lb (907-kg) bombs, or two Mk 83 1,000-lb (454-kg) bombs, or four Mk 82 500-lb (227-kg) bombs, or 12 Mk 81 250-lb (113-kg) bombs, or 24 5-in (127-mm) rockets; the French F-8E(FN) version carries two Matra R530 air-to-air missiles on the sides of the fuselage
Electronics and operational equipment: communication and navigation equipment, plus APQ-94 search and fire-control radar, AAS-115 infra-red sensor and other systems
Powerplant and fuel system: one 18,000-lb (8165-kg) afterburning thrust Pratt & Whitney J57-P-20 turbojet, and a total internal fuel capacity of 1,400 US gal (5296 litres) in fuselage and wing tanks; provision for inflight-refuelling
Performance: maximum speed 1,135 mph (1,827 km/h) or Mach 1.72 at 36,000 ft (10970 m); cruising speed 560 mph (901 km/h) at 40,000 ft (12,190 m); initial climb rate about 21,000 ft (6400 m) per minute; service ceiling 58,000 ft (17680 m); combat radius 600 miles (966 km)
Weights: normal take-off 28,000 lb (12700 kg); maximum take-off 34,000 lb (15420 kg)
Dimensions: span 35 ft 8 in (10.87 m); length 54 ft 6 in (16.61 m); height 15 ft 9 in (4.80 m); wing area 375.0 sq ft (34.84 m²) sq ft

Variants
F-8E(FN) Crusader: resulting from a 1952 US Navy requirement for a carrierborne supersonic air-superiority fighter, this type first flew in March 1955 as the XF8U-1 with the 14,800-lb (6713-kg) Pratt & Whitney J57-P-11 afterburning turbojet; it entered service in March 1957 as the F8U-1 (later F-8A) with an armament of four 20-mm cannon plus a retractable rocket pack (Sidewinder air-to-air missiles being added later) and powered by the 16,200-lb (7348-kg) J57-P-4A; the type underwent considerable development, the F-8E(FN) being the last production model, derived from the F-8E with blown flaps and other high-lift improvements for operation from French aircraft-carriers after delivery in January 1965

F-8H Crusader: designation of the modernized and refurbished F-8D for the Philippines, fitted with the 18,000-lb (8165-kg) J57-P-20 engine, blown flaps and updated avionics

Westland Commando (UK): see Westland Sea King HAS.Mk 5

Westland Lynx HAS.Mk 2
(UK)
Type: anti-submarine and light anti-ship helicopter
Accommodation: crew of two on the flightdeck, and up to two mission crew in the cabin
Armament (fixed): none
Armament (disposable): this is carried on a pylon on each side of the cabin, and comprises two Mk 44, Mk 46 or Stingray torpedoes, or two Mk 11 depth charges for the anti-submarine role; for the anti-ship role the armament comprises four Sea Skua sea-skimming missiles or four AS.12 air to-surface missiles
Electronics and operational equipment: communication and navigation equipment, plus Ferranti ARI.5979 Seaspray search radar, Texas Instruments ASQ-81 magnetic anomaly detector with a towed

'bird', Racal MIR-2 'Orange Crop' electronic support measures equipment, Decca 71 Doppler navigation and (in AS.12-fitted helicopters) SFIM APX-M334 gyro-stabilized sight
Powerplant and fuel system: two 900-shp (671-kW) Rolls-Royce Gem 2 turboshafts, and a total internal fuel capacity of 202 Imp gal (918 litres) in five fuselage tanks, plus provision for 180 Imp gal (818 litres) of auxiliary fuel in two cabin tanks
Performance: cruising speed 144 mph (232 km/h) at sea level; initial climb rate 2,170 ft (661 m) per minute; combat radius 58 miles (93 km) with two torpedoes and 2-hour loiter; ferry range 650 miles (1046 km) with maximum internal and external fuel
Weights: empty 6,040 lb (2740 kg); maximum take-off 10,500 lb (4763 kg)
Dimensions: main rotor diameter 42 ft 0 in (12.802 m); length (fuselage) 39 ft 1.3 in (11.92 m); height (rotor head) 10 ft 6 in (3.20 m); main rotor disc area 1,385.44 sq ft (128.71 m²)

Variants
Lynx AH.Mk 1: this is the major land-based variant of the Lynx helicopter, which first flew in prototype form during March 1971 and began to enter service in 1977, and possesses better performance than the naval Lynx as it has the same basic powerplant but a reduced maximum take-off weight of 10,000 lb (4536 kg); other major divergences from the naval Lynx are skid landing gear, a slightly longer fuselage (39.57 ft/ 12.06 m), and a different combination of avionics and weapons; the Lynx has been qualified for a wide assortment of weapons, but in its basic anti-tank role with the British army the type is fitted with launchers for eight BGM-71 TOW heavyweight anti-tank missiles guided with the aid of a stabilized roof-mounted M65 sight; the cabin can accommodate reload TOW missiles, or eight fully-equipped troops, or a Milan ground-launched anti-tank missile crew with launcher and other equipment, or casualties, or freight
Lynx HAS.Mk 2: this is the baseline naval Lynx model, an advanced anti-submarine and anti-ship helicopter suitable for deployment on small surface vessels; the type has non-retractable but castoring tricycle landing gear, a folding tail and a naval avionics/weapons fit; as an alternative to the primary roles, the Lynx HAS.Mk 2 can be used for the carriage of 10 troops, or of 2,000 lb (907 kg) of freight carried internally, or 3,000 lb (1361 kg) carried externally
Lynx Mk 2: French navy version of the Lynx HAS.Mk 2 with Alcatel DUAV 4 dunking sonar and Omera-Segid ORB 31-W search radar
Lynx HAS.Mk 3: improved British naval model with 1,120-shp (835-kW) Gem 41-1 turboshafts
Lynx Mk 4: French navy version of the Lynx HAS.Mk 3
Lynx AH.Mk 7: improved model for the British army with Gem 41 turboshafts and a tail rotor rotating in the opposite direction to that of the Lynx AH.Mk 1; this version has better low-level hovering and manoeuvring capabilities, facilitating nap-of-the-earth anti-tank operations
Lynx Mk 21: Brazilian navy version of the Lynx HAS.Mk 2
Lynx Mk 23: Argentine navy version of the Lynx HAS.Mk 2
Lynx Mk 25: Dutch navy version of the Lynx HAS.Mk 2 used mainly for search-and-rescue, and designated **UH-14A** in that service
Lynx Mk 27: Dutch navy version of the Lynx HAS.Mk 3 used for the anti-submarine role with Alcatel dunking sonar, and designated **SH-14B** in that service
Lynx Mk 28: version of the Lynx AH.Mk 1 for the Qatari police with Gem 41-1 turboshafts and special equipment, including emergency flotation gear
Lynx Mk 80: Danish navy version of the Lynx HAS.Mk 3 used for anti-submarine and maritime patrol work
Lynx Mk 81: Dutch navy version of the Lynx HAS.Mk 3 with magnetic anomaly detection equipment for the anti-submarine role, and designated **SH-14C** in that service
Lynx Mk 86: Norwegian air force version of the Lynx HAS.Mk 3 with Gem 41-2 turboshafts, non-folding tail and specialist equipment for the coastal search-and-rescue role
Lynx Mk 87: Argentine navy version of the Lynx HAS.Mk 3 with Gem 41-2 turboshafts
Lynx Mk 88: West German navy version of the Lynx HAS.Mk 3 with ASQ-18 dunking sonar and designed for the anti-submarine role
Lynx Mk 89: Nigerian navy version of the Lynx HAS.Mk 3 with Gem 43-1 turboshafts and RCA Primus radar for use in the search-and-rescue role
Lynx-3: much improved dedicated anti-tank development with 1,346-shp (1004-kW) Gem 60 turboshafts, an advanced-technology main rotor, and a slightly lengthened fuselage housing two pilots and a deployable anti-tank missile team, the missile team's reloads, and pylon-mounted weapons such as Hellfire, HOT or TOW anti-tank missiles, and Stinger air-to-air missiles

Westland Sea King HAS.Mk 5
(UK)
Type: anti-submarine helicopter
Accommodation: crew of two on the flightdeck, and a tactical crew of two in the cabin
Armament (fixed): none
Armament (disposable): this is carried on the sides of the fuselage, and comprises four Mk 46 homing torpedoes or four Mk 11 depth charges
Electronics and operational equipment: communication and navigation equipment, plus Marconi

ARI.5991 Sea Searcher radar in a dorsal radome, Decca 71 Doppler navigation, Plessey Type 195 dunking sonar, Ultra Electronic mini sonobuoys, Marconi LAPADS (Lightweight Acoustic Processing and Display System) tactical system and Louis Newmark Mk 31 automatic flight control system

Powerplant and fuel system: two 1,660-shp (1238-kW) Rolls-Royce Gnome H.1400-1 turboshafts, and a maximum internal fuel capacity of 800 Imp gal (3636 litres) in fuselage tanks

Performance: cruising speed 129 mph (208 km/h) at sea level; initial climb rate 2,020 ft (616 m) per minute; service ceiling 4,000 ft (1220 m) on one engine; range 764 miles (1230 km) with standard tankage

Weights: empty 13,672 lb (6201 kg) for anti-submarine role; maximum take-off 21,000 lb (9525 kg)

Dimensions: main rotor diameter 62 ft 0 in (18.90 m); length (fuselage) 55 ft 9.75 in (17.01 m); height (rotor head) 15 ft 6 in (4.72 m); main rotor disc area 3,019.1 sq ft (280.47 m^2)

Variants

Sea King HAS.Mk 2: derived from the Sikorsky S-61B (SH-3) series for the US Navy, the licence-built Westland Sea King is a better-equipped helicopter designed for autonomous rather than ship-controlled anti-submarine operations by virtue of its onboard tactical compartment fed with data from the Plessey Type 195 dunking sonar, the Ekco AW391 radar (with its antenna in a dorsal radome) and the Marconi AD580 Doppler navigation system; the type first flew in British form during May 1969, and entered service during the same year as the Sea King HAS.Mk 1 with two Rolls-Royce Gnome H.1400 turboshafts; the Sea King HAS.Mk 2 is basically similar but powered by 1,660-shp (1238-kW) Gnome H.1400-1 turboshafts and incorporates features developed for Australia's Sea King Mk 50

Sea King AEW.Mk 2: designation of Sea King HAS.Mk 2s converted into airborne early warning helicopters with the Thorn-EMI Search Water radar, its antenna in a pressurized radome on a swivelling arm on the starboard side of the fuselage; the arm and radome are turned to the rear for carrier operations and cruising flight, then swivelled down to the vertical position below the fuselage for patrol operations; the type also has the Racal MIR-2 'Orange Crop' electronic support measures system

Sea King HAR.Mk 3: dedicated search-and-rescue derivative of the Sea King HAS.Mk 2 for the Royal Air Force; the type entered service in 1977 and has additional avionics and a cabin outfitted for the carriage (in addition to the two rescue crew) of 19 survivors, or six litters, or two litters and 11 survivors

Sea King HC.Mk 4: assault transport version of the Commando Mk 2 for the Royal Marines with the folding tail and main rotor of the Sea King series

Sea King HAS.Mk 5: much upgraded anti-submarine and search-and-rescue variant for the Royal Navy with advanced sensors and data-processing capability in a cabin enlarged by the 6.5-ft (1.98-m) rearward movement of the rear bulkhead; the Sea Searcher radar has twice the range of the AW391 type as well as better discrimination against electronic countermeasures, and the LAPADS allows faster and more accurate processing of data from the dunking sonar and sonobuoys

Sea King Mk 41: search-and-rescue variant for the West German navy with Gnome H.1400 turboshafts

Sea King Mk 42: anti-submarine variant for the Indian navy with Gnome H.1400 turboshafts

Sea King Mk 42A: improved anti-submarine variant for the Indian navy with Gnome H.1400-1 turboshafts and haul-down gear

Sea King Mk 42B: developed anti-submarine and anti-ship variant for the Indian navy with the Super Searcher surveillance radar, the Marconi AQS-902 acoustic processing system and provision for Sea Eagle anti-ship missiles

Sea King Mk 42C: upgraded version of the Sea King Mk 42B for the Indian navy

Sea King Mk 43: search-and-rescue variant for Norwegian air force with Gnome H.1400 turboshafts

Sea King Mk 43A: improved search-and-rescue variant for the Norwegian air force with Gnome H.1400-1 turboshafts

Sea King Mk 45: anti-submarine and anti-ship variant for the Pakistani navy with Gnome H.1400 turboshafts and provision for AM.39 Exocet missiles

Sea King Mk 47: anti-submarine variant for the Egyptian navy with Gnome H.1400-1 turboshafts

Sea King Mk 48: dual-role VIP transport/search-and-rescue variant for the Belgian air force with Gnome H.140-1 turboshafts

Sea King Mk 50: anti-submarine variant of the Sea King HAS.Mk 1 for the Australian navy with Gnome H.1400-1 turboshafts and Bendix ASQ-13B dunking sonar

Sea King Mk 50A: improved anti-submarine variant of the Sea King Mk 50 for the Australian navy

Commando Mk 1: this is a land-based derivative of the Sea King series that first flew in September 1973 with Gnome H.1400 turboshafts, non-retractable wheeled landing gear, no sponsons, and provision for a wide variety of armament; the type is operated by the Egyptian air force in the trooping role with accommodation for 21 men in the cabin

Commando Mk 2: uprated version of the Commando Mk 1 with Gnome H.1400-1 turboshafts and able to carry 27 men

Commando Mk 2A: version of the Commando Mk 2 for the Qatari air force

Commando Mk 2B: VIP derivative of the Commando Mk 2 for the Egyptian air force

Commando Mk 2C: VIP derivative of the Commando Mk 2A for the Qatari air force
Commando Mk 2E: electronic intelligence and jamming version for the Egyptian air force, fitted with the Selenia IHS-6 Elint and jamming system

Westland Wasp HAS.Mk 1
(UK)
Type: light anti-submarine helicopter
Accommodation: crew of two seated side-by-side in the cockpit, and provision for up to four passengers
Armament (fixed): none
Armament (disposable): this is carried on pylons projecting one on each side of the fuselage, and comprises two Mk 44 torpedoes, depth charges or AS.11 air-to-surface missiles
Electronics and operational equipment: communication and navigation equipment, plus APX-Bézu 260 stabilized sight for the missile installation
Powerplant and fuel system: one 710-shp (530-kW) Rolls-Royce (Blackburn) Nimbus Mk 503 turboshaft, and a total internal fuel capacity of 155 Imp gal (755 litres) in a fuselage tank
Performance: maximum speed 120 mph (193 km/h) at sea level; initial climb rate 1,440 ft (439 m) per minute; service ceiling (12,200 ft (3720 m); range 270 miles (435 km) with four passengers
Weights: empty 3,452 lb (1566 kg); maximum take-off 5,600 lb (2495 kg)
Dimensions: main rotor diameter 32 ft 3 in (9.83 m); length (fuselage) 30 ft 4 in (9.24 m); height (rotor head) 8 ft 11 in (2.72 m); main rotor disc area 816.9 sq ft (75.89 m^2)

Variants
Scout AH.Mk 1: this is the British army's version of the Wasp, with a 685-shp (511-kW) Nimbus Mk 102 turboshaft and skid landing gear; the type has better performance than the Wasp as it is lighter; the Scout can be armed with a flexible cannon or machine-gun, rocket-launcher pods or wire-guided missiles such as the AS.11 or AS.12, but is generally used as a five-seat utility helicopter; the type first flew as the Saunders-Roe P.531 in July 1958, and the Scout began to enter service in 1963
Wasp HAS.Mk 1: naval version with greater power, castoring quadricycle landing gear, a folding tail and main rotor blades, and naval avionics and armament

WSK-PZL Swidnik (Mil) Mi-2 'Hoplite'
(Poland/USSR)
Type: utility light helicopter
Accommodation: crew of one on the flightdeck, and up to 10 passengers, or four litters and one attendant, or two litters and two seated casualties, or 700 kg (1,543 lb) of freight in the cabin
Armament (fixed): none
Armament (disposable): this can be carried on stub pylons located one on each side of the fuselage for two rocket-launcher pods or four anti-tank missiles
Electronics and operational equipment: communication and navigation equipment, plus a radar-warning receiver
Powerplant and fuel system: two 298- or 335-kW (400- or 450-shp) Polish-built Isotov GTD-350P turboshafts, and a total internal fuel capacity of 600 litres (131 Imp gal) in one fuselage tank, plus provision for two 238-litre (52-Imp gal) external tanks on the sides of the fuselage
Performance: cruising speed 200 km/h (124 mph) at 500 m (1,640 ft); initial climb rate 270 m (885 ft) per minute; service ceiling 4000 m (13,125 ft); range 170 km (105 miles) with maximum payload, or 440 km (273 miles) with maximum internal fuel, or 797 km (495 miles) with internal and external fuel
Weights: empty 2365 kg (5,213 lb); normal take-off 3550 kg (7,826 lb); maximum take-off 3700 kg (8,157 lb)
Dimensions: main rotor diameter 14.50 m (47 ft 6.9 in); length (fuselage) 11.40 m (37 ft 4.75 in); height (to rotor head) 3.75 m (12 ft 3.5 in); main rotor disc area 165.13 m^2 (1,777.5 sq ft)

Variant
Mi-2: this is basically a turboshaft-powered derivative of the Mil Mi-1, two small turboshafts replacing the single piston engine of the earlier machine for enhanced reliability and greater power at reduced weight; the development was undertaken by the parent design bureau in the USSR, the prototype first flying in September 1961; thereafter final development and production was switched to Poland, where the first series machine was flown in 1965; early aircraft had two 298-kW (400-shp) turboshafts, but in 1974 the 336-kW (450-shp) turboshafts became the production norm; the type can undertake the whole range of military tasks, and an 800-kg (1,764-lb) slung load can be lifted

Xian H-6 (China): see Tupolev Tu-16 'Badger-A'

Xian F-7M Airguard
(China)
Type: interceptor and multi-role fighter
Accommodation: pilot seated on an ejector seat
Armament (fixed): two 30-mm Type 30-1 cannon with 60 rounds per gun
Armament (disposable): this is carried on four underwing hardpoints to a maximum weight of 2000 kg (4,409 lb) of PL-2, PL-2A, PL-7 or Matra 550 Magic air-to-air missiles, rocket-launchers (each containing 18 55-mm/2.17-in or seven 90-mm/3.54-in rockets) and free-fall weapons such as 50-, 100-, 250- and 500-kg (110-, 220-, 551- and 1,102-lb) bombs
Electronics and operational equipment: communication and navigation equipment, and Type 226 ranging radar, Type 956 head-up display and weapon-

aiming system, and other items
Powerplant and fuel system: one 6100-kg (13,448-lb) afterburning thrust Wopen-7B turbojet, and a maximum internal fuel capacity of 2385 litres (525 Imp gal) plus provision for one 800- or 500-litre (176- or 110-Imp gal) centreline drop tank and two 500-litre (110-Imp gal) underwing drop-tanks
Performance: maximum speed 2180 km/h (1,355 mph) or Mach 2.05 at high altitude; initial climb rate 10800 m (35,433 ft) per minute; service ceiling 18200 m (59,710 ft); combat radius 600 km (373 miles); ferry range 2230 km (1,386 miles)
Weights: empty 5145 kg (11,343 lb); maximum take-off 8900 kg (19,621 lb)
Dimensions: span 7.154 m (23 ft 5.6 in); length without probe 13.945 m (45 ft 9 in); height 4.103 m (13 ft 5.5 in); wing area 23.00 m² (247.6 sq ft)

Variants
J-7: this is the Chinese-built copy of the MiG-21F series, produced in limited numbers during the 1960s but then suspended because of technical problems on the production line; the type re-entered production in the late 1970s to meet renewed Chinese demand, and this new-production version has features of the 'Fishbed-E' and 'Fishbed-J' Soviet fighters as well as the basic airframe of the 'Fishbed-C'; greater performance is secured by the use of a WP-7B turbojet, which offers an afterburning thrust of 6100 kg (13,448 lb), and greater offensive punch by the addition of a second 30-mm Type 30-1 cannon in the fuselage; each gun has 60 rounds of ammunition, and further capability is provided by two underwing hardpoints each able to carry one PL-2 air-to-air missile, a pod with 18 55-mm (2.17-in) rockets, or one 250- or 500-kg (551- or 1,102-lb) bomb; improved avionics include a Type 222 ranging radar in a fully-variable inlet centrebody, and an SM-3A optical sight; the type has also been exported as the **F-7**, which can be fitted to carry the latest variants of the AIM-9 Sidewinder air-to-air missile used in conjunction with a retrofitted head-up display
F-7M Airguard: much improved export model developed in the mid-1980s with an additional pair of underwing hardpoints (for the carriage of two PL-7 air-to-air missiles or two 48-litres/106-Imp gal drop tanks), relocated probes, a revised cockpit with new canopy and ejector seat, and a number of other Chinese developments such as the Type 226 ranging ranger with improved resistance to electronic counter-measures (the GEC Sky Ranger is offered as an alternative); most important, however, is the inclusion of a mass of Western avionics such as the GEC Type 956 HUDWAS head-up display and weapon-aiming system, air data computer, radar altimeter and multi-mode radios

Xian Y-7 (China): see Antonov An-24RT 'Coke'

Yakovlev Yak-28P 'Firebar'
(USSR)
Type: all-weather fighter
Accommodation: crew of two seated in tandem on ejector seats
Armament (fixed): none
Armament (disposable): this is carried on two underwing hardpoints, and normally comprises one IR-homing AA-3 'Anab' and one radar-homing AA-3 'Anab' air-to-air missile; two auxiliary underwing hardpoints can be fitted for the carriage of two AA-2 'Atoll' or AA-8 'Aphid' air-to-air missiles
Electronics and operational equipment: communication and navigation equipment, plus 'Skip Spin' attack radar in the nose and a radar-warning receiver in the tail, and other systems
Powerplant and fuel system: two 6200-kg (13,668-lb) afterburning thrust turbojets (probably Tumansky R-11 units), and a total internal fuel capacity of ? litres (? Imp gal) in fuselage and wing tanks, plus provision for two 1000-litre (220-Imp gal) slipper tanks
Performance: maximum speed 1180 km/h (733 mph) or Mach 1.1 at 10670 m (35,000 ft); initial climb rate 8520 m (27,955 ft) per minute; cruising speed 920 km/h (571 mph) at optimum altitude; service ceiling 16750 m (55,000 ft); combat radius 900 km (559 miles) on a hi-hi sortie; ferry
Weights: empty 13600 kg (29,982 lb); normal take-off 15875 kg (34,998 lb); maximum take-off 20000 kg (44,092 lb)
Dimensions: span 12.95 m (42 ft 6 in); length 23.00 m (75 ft 5.5 in); height 3.95 m (12 ft 11.5 in); wing area 37.6 m² (404.7 sq ft)

Variants
Yak-28R 'Brewer-D': the Yak-28 series has served the Soviet air forces well over a long period, the basic type having proved itself capable of considerable development and versatility; the type was developed from the twin-engined Yak-25 all-weather interceptor with more sharply swept flying surfaces and twin wheels on each unit of the bicycle landing gear, whose main unit was moved aft to allow the incorporation of a weapon bay in the lower fuselage; the type first flew in 1960, and the series was developed for a number of tactical roles, those not in service being the Yak-28 'Brewer-A' light bomber prototype, the Yak-28I 'Brewer-B' strike bomber, and the Yak-28I 'Brewer-C' strike bomber with features of the Yak-28P such as a lengthened fuselage and longer engine inlets; the 'Brewer-D' is a multi-sensor reconnaissance aircraft introduced in the late 1960s for the carriage of optical, side-looking radar or infra-red linescan equipment on interchangeable pallets in the erstwhile weapons bay
Yak-28E 'Brewer-E': this is the electronic escort version of the series, introduced in the late 1960s and crewed by a pilot plus a mission officer, the latter

operating the electronic systems and associated countermeasures jammers (both spot and barrage) and dispensers

Yak-28P 'Firebar': this was the original production model, introduced in 1962 and featuring 'Skip Spin' radar in a solid nose; this radar provides tracking to a range of 60 km (37.3 miles) and tracking to shorter ranges; from 1967 the original short nose and short air inlets were replaced by longer units with lower drag, increasing overall length from 21.70 m (71.19 ft) to 23.00 m (75.46 ft) and boosting transonic performance

Yak-28U 'Maestro': two-seat operational conversion trainer variant with two separate cockpits

Yakovlev Yak-38 'Forger-A'
(USSR)
Type: shipboard V/STOL combat aircraft
Accommodation: pilot seated on an ejector seat
Armament (fixed): none
Armament (disposable): this is carried on four underwing hardpoints, up to a maximum weight of about 3600 kg (7,937 lb); typical stores are launchers for 55-mm (2.17-in) rockets, AA-2 'Atoll' or AA-8 'Aphid' air-to-air missiles, air-to-surface missiles, bombs and gun pods containing the 23-mm GSh-23L twin-barrel cannon
Electronics and operational equipment: communication and navigation equipment, plus ranging radar, head-up display, 'Odd Rods' IFF, Sirena 3 radar-warnig receiver, and various electronic countermeasures systems
Powerplant and fuel system: one 8160-kg (17,989-lb) thrust Lyulka AL-21F vectored-thrust non-afterburning turbojet and two 3750-kg (7,870-lb) thrust Kolesov ZM non-afterburning lift turbojets, and a total internal fuel fuel capacity of about 2900 litres (639 Imp gal) plus provision for two 600-litre (132-Imp gal) drop-tanks
Performance: maximum speed 1110 km/h (627 mph) or Mach 0.95 at high altitude, and 1125 km/h (700 mph) or Mach 0.8 at sea level; initial climb rate 4500 m (14,765 ft) per minute; service ceiling 12000 m (39,370 ft); combat radius 370 km (230 miles) on a hi-lo-hi sortie with maximum weapon load, or 240 km (150 miles) on a lo-lo-lo sortie with maximum weapon load
Weights: empty 7385 kg (16,281 lb); maximum take-off 11700 kg (25,794 lb) for VTO or 13000 kg (28,660 lb) for STO
Dimensions: span 7.32 m (24 ft 0.2 in); length 15.50 m (50 ft 10.3 in); height 4.37 in (14 ft 4 in); wing area 18.50 m^2 (199.14 sq ft)

Variants
Yak-38 'Forger-A': developed from the Yak-36 VTOL prototype, the Uak-38 is a carrierborne tactical aircraft with V/STOL capability; the type first flew in the early 1970s and entered service in 1976, and though a limited type designed to provide operational experience with such aircraft, it provides Soviet helicopter carriers and aircraft-carriers with useful interception and attack capabilities when out of reach of land-based air defences
Yak-38UV 'Forger-B': tandem two-seat operational conversion trainer variant with the fuselage lengthened to 17.68 m (58.0 ft) to accommodate the pupil's cockpit

Table 3.1 *Flying training aircraft*

Name (country)	powerplant	speed	range	maximum weight	span
Aérospatiale Epsilon (France	1 × 300-hp (224-kW) Lycoming AEIO-540 piston engine	380 km/h (236 mph)	1110 km (690 miles)	1250 kg (2,755 lb)	7.92 m (25 ft 11.75 in)
Aerotec A-122A Uirapurú (Brazil)	1 × 160-hp (119-kW) Lycoming O-320 piston engine	227 km/h (147 mph)	800 km (497 miles)	840 kg (1,825 lb)	8.50 m 27 ft 10.75 in)
Beech Model 24 Sundowner (USA)	1 × 180-hp (134-kW) Lycoming O-360 piston engine	147 mph (237 km/h)	735 miles (1183 km)	2,450 lb (1111kg)	32 ft 9 in (9.98 m)
British Aerospace Bulldog T.Mk 1 (UK)	1 × 200-hp (149-kW) Lycoming IO-360 piston engine	150 mph (241 km/h)	620 miles (1000 km)	2,350 lb (1066 kg)	33 ft 0 in (10.06 m)
CASA 223 Flamingo (Spain/West Germany)	1 × 200-hp (149-kW) Lycoming IO-360 piston engine	245 km/h (152 mph)	1150 km (715 miles)	1050 kg (2,315 lb)	8.28 m (27 ft 2 in)
Cessna Model 152 (USA)	1 × 110-hp (82-kW) Lycoming O-235 piston engine	127 mph (204 km/h)	794 miles (1278 km)	1,670 lb (757 kg)	32 ft 8.5 in (9.97 m)
Cessna Model 172 (USA)	1 × 210-hp (157-kW) Continental IO-360 piston engine	153 mph (246 km/h)	1,010 miles (1625 km)	2,550 lb (1156 kg)	35 ft 10 in (10.02 m)
de Havilland Canada DHC-1 Chipmunk Mk 20 (Canada)	1 × 145-hp (108-kW) de Havilland Gipsy Major 10 piston engine	138 mph (222 km/h)	280 miles (451 km)	2,014 lb (914 kg)	34 ft 4 in (10.46 m)

Table 3.1 *Flying training aircraft–continued*

ENAER T-35 Pillan (Chile)	1 × 300-hp (224-kW) Lycoming AEIO-540 piston engine	310 km/h (193 mph)	1270 km (789 miles)	1315 kg (2,900 lb)	8.81 m (28 ft 11 in)
Fuji KM-2B (Japan)	1 × 340-hp (254-kW) Lycoming IGSO-480 piston engine	367 km/h (228 mph)	965 km (600 miles)	1542 kg (3,400 lb)	10.004 m (32 ft 10 in)
FFA AS 202/18A Bravo (Switzerland)	1 × 180-hp (134-kW) Lycoming AEIO-360 piston engine	240 km/h (149 mph)	965 km (600 miles)	1050 kg (2,315 lb)	9.75 m (31 ft 11.75 in)
Heliopolis Gomhouria 2 (Egypt)	1 × 145-hp (108-kW) Continental C-145 piston engine	220 km/h (136 mph)	790 km (491 miles)	800 kg (1,764 lb)	10.60 m (34 ft 9 in)
Hindustan Aeronautics Ltd HPT-32 (India)	1 × 260-hp (194-kW) Lycoming AEIO-540 piston engine	157 mph (253 km/h)	435 miles (700 km)	2,665 lb (1209 kg)	31 ft 2 in (9.50 m)
ICA IAR-823 (Romania)	1 × 290-hp (216-kW) Lycoming IO-540 piston engine	310 km/h (193 mph)	1800 km (1118 miles)	1500 kg (3,307 lb)	10.00 m (32 ft 9.75 in)
Lockheed T-33A (USA)	1 × 5,400-lb (2450-kg) Allison J33-A-35 turbojet	600 mph (965 km/h)	1,275 miles (2050 km)	15,061 lb (6832 kg)	38 ft 10.5 in (11.65 m)
Nanchang CJ-6 (China)	1 × 213-kW (285-hp) Huosai-6A piston engine	286 km/h (178 mph)		1400 kg (3,088 lb)	10.70 m (35 ft 1.25 in)
NZ Aerospace CT/4 (New Zealand)	1 × 210-hp (157-kW) Continental IO-360 piston engine	178 mph (286 km/h)	884 miles (1422 km)	2,400 lb (1088 kg)	26 ft 0 in (7.92 m)
North American T-6 Texan (USA)	1 × 550-hp (409-kW) Pratt & Whitney R-1340 piston engine	212 mph (341 km/h)	870 miles (1400 km)	5,617 lb (2550 kg)	42 ft 0 in (12.80 m)
Piaggio P.149D (Italy)	1 × 270-hp (201-kW) Lycoming GO-480 piston engine	312 km/h (194 mph)	980 km (609 miles)	1680 kg (3,700 lb)	11.12 m (36 ft 5.8 in)
RFB Fantrainer 600 (West Germany)	1 × 600-shp (447-kW) Allison 250-C30 turboshaft	430 km/h (267 mph)	1930 km (863 miles)	2300 kg (5,070 lb)	9.70 m (31 ft 10 in)
Saab 91B Safir (Sweden)	1 × 190-hp (142-kW) Lycoming O-435 piston engine	275 km/h (171 mph)	1075 km (668 miles)	1220 kg (2,690 lb)	10.60 m (34 ft 9.25 in)
Saab MFI 17 Supporter (Sweden)	1 × 200-hp (149-kW) Lycoming IO-360 piston engine	236 km/h (146 mph)	1120 km (696 miles)	1200 kg (2,645 lb)	8.85 m (29 ft 0.5 in)
UTVA-75 (Yugoslavia)	1 × 180-hp (134-kW) Lycoming IO-360 piston engine	220 km/h (136 mph)	2000 km (1,242 miles)	960 kg (2,116 lb)	9.73 m (31 ft 11 in)
Valmet L-70 Vinka (Finland)	1 × 200-hp (149-kW) Lycoming AEIO-360 piston engine	240 km/h (149 mph)	1015 km (630 miles)	1250 kg (2,756 lb)	9.85 m (32 ft 3.75 in)
Zlin Z-526 (Czechoslovakia)	1 × 119-kW (160-hp) Walter 6-III Minor piston engine	245 km/h (152 mph)	650 km (404 miles)	975 kg (2,150 lb)	10.60 m (34 ft 9 in)

SECTION 4
MISSILES

STRATEGIC MISSILES

DF-2 (CSS-1)
(China)
Type: single-stage medium-range ballistic missile
Dimensions: diameter 1.60 m (5.25 ft); length 22.80 m (74.80 ft)
Weights: total round 26000 kg (57,319 lb); post-boost vehicle not known
Propulsion: one liquid-propellant rocket delivering ?-kg (?-lb) thrust
Range: 1200 km (745 miles)
CEP: 2780 m (3,040 yards)
Warhead: one 15-kiloton nuclear (fission) or conventional RV
Launch: hot type from pad
Guidance: radio-updated inertial

Variant
DF-2: designated **CSS-1** in the US terminology for Chinese weapons, this simple weapon was China's first strategic missile and was introduced in 1970; it is based on the technology of the Soviet SS-3 'Shyster', itself based generally on the German Peenemünde A-4 (V-2) of World War II; the type remains in limited Chinese service (some 50 such weapons), but has only very limited value

DF-3 (CSS-2)
(China)
Type: single-stage intermediate-range ballistic missile
Dimensions: diameter 2.46 m (8.07 ft); length 20.60 m (67.59 ft)
Weights: total round 27000 kg (59,524 lb); post-boost vehicle not known
Propulsion: one storable liquid-propellant rocket delivering ?-kg (?-lb) thrust
Range: 3200 km (1,988 miles)
CEP: 1390 m (1,520 yards)
Warhead: one 200-kiloton nuclear or 2-megaton thermonuclear RV
Launch: hot type from pad
Guidance: radio-updated inertial

Variant
DF-3: designated **CSS-2** in the US terminology for Chinese weapons, this missile began to enter service in 1971, and is based on Soviet technology; some 60 such missiles are is service, and an alternative estimate puts range at 2700 km (1,678 miles) with the larger 2-megaton warhead

DF-4 (CSS-3)
(China)
Type: two-stage limited-range intercontinental ballistic missile
Dimensions: diameter 2.46 m (8.07 ft); length 26.80 m (87.93 ft)
Weights: total round 50000 kg (110,229 lb); post-boost vehicle not known
Propulsion: (first stage) one storable liquid-propellant rockets delivering ?-kg (?-lb) thrust and (second stage) one storable liquid-propellant rocket delivering ?-kg (?-lb) thrust
Range: 5000 km (3,107 miles) for early model, and 6960 km (4,325 miles) for later model
CEP: 930 m (1,017 yards)
Warhead: one 3-megaton thermonuclear RV for early model, and three or four 200-kiloton MRVs for later model
Launch: hot type from silo
Guidance: inertial

Variant
DF-4: designated **CSS-3** in the US terminology for Chinese weapons, this type began to enter service in 1978, and was soon developed into a longer-range variant with a MRV or MIRV payload; only a very few such missiles are in service

DF-5 (CSS-4)
(China)
Type: two-stage intercontinental ballistic missile
Dimensions: diameter 3.35 m (10.99 ft); length 32.50 m (106.63 ft)
Weights: total round 200000 kg (440,917 lb); post-boost vehicle not known
Propulsion: (first stage) four storable liquid-propellant rockets each delivering 70000-kg (153,321-lb) thrust and (second stage) one storable liquid-propellant rocket delivering ?-kg (?-lb) thrust
Range: 10000 km (6,214 miles)
CEP: 930 m (1,017 yards)
Warhead: one 4-megaton thermonuclear RV
Launch: hot type from silo
Guidance: inertial

Variants
DF-5: designated **CSS-4** in the US terminology for Chinese weapons, this ICBM began to enter very limited service in 1981, and has also been developed as a satellite launcher
DF-6: improved DF-5 (also designated CSS-4 by the USA) carrying a 5-megaton thermonuclear warhead

over a range of 13000 km (8,078 miles), and believed to be the genuine production version of the type

DF-7 (CSS-5)
(China)
Type: three-stage intercontinental ballistic missile
Dimensions: not known
Weights: not known
Propulsion: not known
Range: not known
CEP: not known
Warhead: up to 10 ?-kiloton MIRVs
Launch: hot type from a silo
Guidance: inertial

Variant
DF-7: known in US terminology as the **CSS-5,** this is China's latest and most capable ICBM with an advancedtype of warhead; the launch system is based on that of the CZ-3 (Long March 3) satellite launcher; DF stands for Dong Feng (east wind)

HY-1 (CSS-N-2)
(China)
Type: two-stage submarine-launched ballistic missile
Dimensions: diameter 1.50 m (4.92 ft); length 10.00 m (32.81 ft)
Weights: total round 14000 kg (30,864 lb); post-boost vehicle not known
Propulsion: (first stage) one solid-propellant rocket delivering ?-kg (?-lb) thrust, and (second stage) one solid-propellant rocket delivering ?-kg (?-lb) thrust
Range: 2700 km (1,678 miles)
CEP: 2800 m (3,062 yards)
Warhead: one 1-megaton thermonuclear RV
Launch: submarine tube
Guidance: inertial

Variant
HY-1: designated **CSS-N-2** in the US terminology for Chinese weapons, this is a limited-service SLBM pending the widespread adoption of the JL-1

HY-2 (CCS-N-4)
(China)
Type: two-stage submarine-launched ballistic missile
Dimensions: diameter 2.30 m (7.54 ft); length 12.80 m (41.99 ft)
Weights: total round 20000 kg (44,092 lb); post-boost vehicle not known
Propulsion: (each stage) one solid-propellant rocket delivering ?-kg (?-lb) thrust
Range: 3200 km (1,988 miles)
CEP: 1850 m (2,023 yards)
Warhead: one 1-megaton thermonuclear RV
Launch: submarine tube
Guidance: inertial

Variant
JL-1: designated **CSS-N-4** in the US terminology for Chinese weapons, this missile has undergone a lengthy development programme and began to enter service in 1983 on board China's very slowly increasing number of missile submarines; JL stands for Ju Lang (giant wave)
JL-2: much improved version with a 2-megaton warhead currently under development

Aérospatiale SSBS S-3
(France)
Type: two-stage intermediate-range ballistic missile
Dimensions: diameter 1.50 m (4.92 ft); length 13.80 m (45.28 ft)
Weights: total round 25800 kg (56,878 lb); post-boost vehicle not revealed
Propulsion: (first stage) one SEP 902 Hérisson (P16) solid-propellant rocket delivering 55000-kg (121,252-lb) thrust and (second stage) one SEP Rita II (P6) solid-propellant rocket delivering 32000-kg (70,547-kg) thrust
Range: 3000 km (1,864 miles)
CEP: 830 m (908 yards)
Warhead: one 1.2-megaton thermonuclear RV
Launch: hot type from silo
Guidance: Sagem/EMD inertial

Variant
SSBS S-3: designed as a successor to the SSBS S-2 during the mid- and late 1970s, the S-3 mounts a higher-performance second stage on the first stage of the S-2, and began to enter service in 1980 in France's 18 IRBM silos; the improved TN-61 warhead is hardened against high-altitude nuclear explosions, and carries a new generation of penetration aids

Aérospatiale MSBS M-4
(France)
Type: three-stage submarine-launched ballistic missile
Dimensions: diameter 1.92 m (6.30 ft); length 11.05 m (36.25 ft)
Weights: total round 35073 kg (77,323 lb); post-boost vehicle not revealed
Propulsion: (first stage) one SEP 401 (P10) solid-propellant rocket delivering 71000-kg (156,526-lb) thrust, (second stage) one SEP 402 (P6) solid-propellant rocket delivering 30000-kg (66,138-lb) thrust and (third stage) one SEP 403 solid-propellant rocket delivering 7000-kg (15,432-lb) thrust
Range: 4500 km (2,796 miles) with TN-70 warheads and 5000+ km (3,105+ miles) with TN-71 warheads
CEP: 460 m (503 yards)
Warhead: six 150-kiloton TN-70 or miniaturized TN-71 thermonuclear MIRVs
Launch: submarine tube
Guidance: Sagem/EMD inertial

Variant
M-4: becoming operational in 1985, the M-4 is of a wholly new French SLBM generation designed of considerably greater weight and dimensions for significantly deeper underwater launch depth and improved range with a MIRV bus hardened against electro-magnetic pulse and carrying penetration aids; range is improved with the miniaturized TN-71 warheads, and each system is designed to cover a target area 350 km (217.5 miles) long by 150 km (93.2 miles) wide

Aérospatiale MSBS M-20
(France)
Type: two-stage submarine-launched ballistic missile
Dimensions: diameter 1.50 m (4.92 ft); length 10.40 m (34.12 ft)
Weights: total round 20055 kg (44,213 lb); post-boost vehicle not revealed
Propulsion: (first stage) one SEP 904 (P10) solid-propellant rocket delivering 45000-kg (99,206-lb) thrust and (second stage) one SEP Rita 11 (P6) solid-propellant rocket delivering 32000 kg (70,547-lb) thrust
Range: 3000 km (1,864 miles)
CEP: 930 m (1,017 yards)
Warhead: one 1.2-megaton MR-61 thermonuclear RV
Launch: submarine tube
Guidance: Sagem/EMD inertial

Variant
MSBS M-20: this medium-weight SLBM entered service in 1977 as successor to the first-generation M-2, itself an updated M-1; the TN-60 warhead is carried in a modified Rita 11/P6 second stage and is supported by penetration aids; the RV is hardened against the effects of high-altitude nuclear explosions

SS-4 'Sandal'
(USSR)
Type: single-stage medium-range ballistic missile
Dimensions: diameter 1.60 m (5.25 ft); length 21.00 m (68.90 ft)
Weights: total round 27000 kg (59,524 lb); post-boost vehicle 1350 kg (2,976 lb)
Propulsion: one liquid-propellant rocket delivering ?-kg (?-lb) thrust
Range: 2000 km (1,243 miles)
CEP: 2400 m (2,625 yards)
Warhead: one 1-megaton thermonuclear or HE RV
Launch: hot type from pad with reload capability when fired from unhardened sites
Guidance: inertial

Variant
SS-4: a development of the SS-3 'Shyster', the SS-4 entered service in 1958 and is now being phased out on favour of the SS-20; another estimate of CEP puts the figure at 2000 m (2,187 yards)

SS-11 'Sego'
(USSR)
Type: two-stage lightweight intercontinental ballistic missile
Dimensions: diameter 2.50 m (8.20 ft); length 20.00 m (65.62 ft)
Weights: total round 45000 kg (99,205 lb); post-boost vehicle (Model 1) 1000 kg (2,205 lb) or (Model 3) 1135 kg (2,502 lb)
Propulsion: (first stage) one storable liquid-propellant rocket delivering ?-kg (?-lb) thrust and (second stage) one storable liquid-propellant rocket delivering ?-kg (?-lb) thrust
Range: (Model 1) 8800 km (5,470 miles) or (Model 3) 9600 km (5,965 miles)
CEP: (Model 1) 1400 m (1,530 yards) or (Model 3) 1110 m (1,215 yards)
Warhead: (Model 1) one 950-kiloton thermonuclear RV or (Model 3) three 250-kiloton thermonuclear MRVs
Launch: hot type from hardened silo, with limited reload capability
Guidance: inertial

Variants
SS-11 Model 1: this highly capable light ICBM was introduced in 1966 and is a third-generation light ICBM built in very substantial numbers, the Model 1 being distinguished by its single RV and modest range
SS-11 Model 2: experimental model of the late 1960s with a single 1-megaton RV and advanced penetration aids
SS-11 Model 3: second operational variant, with longer range and three MRVs; deployed from 1973
SS-11 Model 4: development model of the late 1970s with three or six low-yield MIRVs

SS-13 'Savage'
(USSR)
Type: three-stage lightweight intercontinental ballistic missile
Dimensions: diameter 1.70 m (5.58 ft); length 20.00 m (65.62 ft)
Weights: total round 34000 kg (74,955 lb); post-boost vehicle 680 kg (1,499 lb)
Propulsion: (first stage) liquid-propellant rocket delivering ?-kg (?-lb) thrust, (second stage) liquid-propellant rocket delivering ?-kg (?-lb) thrust and (third stage) liquid-propellant rocket delivering ?-kg (?-lb) thrust
Range: 8000 km (4,970 miles)
CEP: 1850 m (2,025 yards)
Warhead: one 600-kiloton nuclear RV
Launch: hot type from silo
Guidance: inertial

Variant
SS-13: deployed from 1969 onwards, the SS-13 is a third-generation light ICBM comparable to the US Minuteman but built and deployed only in small numbers; other estimates of the type suggest a 750-kiloton warhead carried in a 450-kg (992-lb) post-boost vehicle over a range of 10000 km (6,214 miles)

SS-17 'Spanker'
(USSR)
Type: two-stage lightweight intercontinental ballistic missile
Dimensions: diameter 2.50 m (8.20 ft); length 24.00 m (78.74 ft)
Weights: total round 65000 kg (143,298 lb); post-boost vehicle (Models 1 and 3) 2740 kg (6,041 lb) or (Model 2) 2730 kg (6,019 lb)
Propulsion: (first stage) storable liquid-propellant rocket delivering ?-kg (?-lb) thrust and (second stage) storable liquid-propellant rocket delivering ?-kg (?-lb) thrust
Range: (Models 1 and 3) 10000 km (6,214 miles) or (Model 2) 11000 km (6,835 miles)
CEP: (Model 1) 440 m (480 yards), (Model 2) 425 m (465 yards) or (Model 3) 350 m (385 yards)
Warhead: (Models 1 and 3) four 750-kiloton nuclear MIRVs or (Model 2) one 6-megaton thermonuclear RV
Launch: cold type from a hardened silo, offering a reload capability
Guidance: inertial

Variants
SS-17 Model 1: this fourth-generation lightweight ICBM was introduced in 1975 and is designated **RS-16** by the Soviets; the type is comparable in many performance regards to the SS-11, but offers the decided strategic advantage of cold launch and thus the opportunity for rapid reload of the silo
SS-17 Model 2: introduced in 1977, this variant differs from the Model 1 in having a single medium-yield RV and slightly longer range
SS-17 Model 3: introduced in the early 1980s this variant differs from the Model 1 in having an improved guidance system for reduced CEP; the accuracy of the type enhances the SS-17 family's capability against US missile silos and thus makes it a powerful aspect of the counterforce balance between the USA and USSR

SS-18 'Satan'
(USSR)
Type: two-stage heavyweight intercontinental ballistic missile
Dimensions: diameter 3.00 m (9.84 ft); length 35.00 m (114.83 ft)
Weights: total round 225000 kg (496,032 lb); post-boost vehicle (Model 1) 7560 kg (16,667 lb), (Models 2 and 4) 7590 kg (16,733 lb) or (Model 3) 7500 kg (16,534 lb)
Propulsion: (first stage) storable liquid-propellant rocket delivering ?-kg (?-lb) thrust and (second stage) storable liquid-propellant rocket delivering ?-kg (?-lb) thrust
Range: (Model 1) 12000 km (7,455 miles), (Models 2 and 4) 11000 km (6,835 miles) or (Model 3) 16000 km (9,940 miles)
CEP: (Models 1 and 2) 425 m (465 yards), (Model 3) 350 m (385 yards), or (Model 4) 260 m (285 yards)
Warhead: (Model 1) one 27-megaton thermonuclear RV, (Model 2) eight to 10 900-kiloton thermonuclear MIRVs, (Model 3) one 20-megaton thermonuclear RV or (Model 4) 10 500-kiloton thermonuclear MIRVs
Launch: cold type from a hardened silo, offering a reload capability
Guidance: inertial

Variants
SS-18 Model 1: introduced in 1974 with the Soviet designation **RS-20**, this fourth-generation heavyweight ICBM is the largest missile so far deployed, and a truly prodigious weapon with range, accuracy and warhead making it primarily effective as a counterforce weapon against missile silos and buried command/communications centres
SS-18 Model 2: following the Model 1 in 1976, this variant has a computer-controlled post-boost bus with eight of 10 MIRVs
SS-18 Model 3: introduced in 1977, this variant offers greater range and accuracy than the Model 1 with a slightly smaller but still enormously formidable warhead
SS-18 Model 4: introduced in 1979, this variant is derived from the Model 2 but has greater accuracy and an improved post-boost bus carrying up to 14 MIRVs (usually 10 real weapons and four decoys, plus penetration aids)
SS-18 Model 5: introduced in the mid-1980s, this new variant is still relatively unknown, but is believed to have a range of 9000 km (11,806 miles) with a post-boost bus carrying 10 750-kiloton thermonuclear MIRVs, probably with great accuracy than the Model 4

SS-19 'Stiletto'
(USSR)
Type: two-stage lightweight intercontinental ballistic missile
Dimensions: diameter 2.75 m (9.02 ft); length 22.50 m (73.82 ft)
Weights: total round 78000 kg (171,958 lb); post-boost vehicle (Model 1) 3420 kg (7,540 lb), (Model 2) 3180 kg (7,011 lb) or (Model 3) 3410 kg (7,518 lb)
Propulsion: (first stage) storable liquid-propellant rocket delivering ?-kg (?-lb) thrust and (second stage)

storable liquid-propellant rocket delivering ?-kg (?-lb) thrust
Range: (Model 1) 9600 km (5,965 miles) or (Models 2 and 3) 10000 km (6,215 miles)
CEP: (Model 1) 390 m (425 yards), (Model 2) 260 m (285 yards) or (Model 3) 280 m (305 yards)
Warhead: (Models 1 and 3) six 550-kiloton nuclear MIRVs, or (Model 2) one 10-megaton thermonuclear RV
Launch: hot type from a hardened silo, with limited reload capability
Guidance: inertial

Variants
SS-19 Model 1: introduced in 1975 as a lightweight ICBM successor (together with the SS-17) to the SS-11, this fourth-generation missile is designated RS-18 in Soviet terminology; the type features the same type of advanced guidance as the SS-17 and SS-18, the onboard computer either correcting the course to remove any deviations from the planned norm, or alternatively generating a new course if this is more efficient; like the SS-17 and SS-18, the SS-19 has the range, accuracy and warhead to make it primarily a counterforce weapon against US missile silos and against buried command and communications centres
SS-19 Model 2: introduced in 1978, this variant has a single medium-yield RV and much improved accuracy
SS-19 Model 3: introduced in 1980, this variant reverts to a MIRVed payload, but with better accuracy and range than the Model 1

SS-20 'Saber'
(USSR)
Type: land-mobile two-stage intermediate-range ballistic missile
Dimensions: diameter 1.70 m (5.58 ft); length 16.00 m (52.49 ft)
Weights: total round 25000 kg (55,115 lb); post-boost vehicle not known
Propulsion: (first stage) solid-propellant rocket delivering ?-kg (?-lb) thrust and (second stage) solid-propellant rocket delivering ?-kg (?-lb) thrust
Range: (Models 1 and 2) 4000 km (2,485 miles) or (Model 3) 7400 km (4,600 miles)
CEP: 425 m (465 yards)
Warhead: (Model 1) one 650-kiloton nuclear or 1.5-megaton thermonuclear RV, (Model 2) three 150-kiloton nuclear MIRVs, or (Model 3) one 50-kiloton nuclear RV
Launch: cold type from wheeled transporter/erector/launcher, with reload capability
Guidance: inertial

Variants
SS-20 Model 1: based on the upper two stages of the SS-16 lightweight ICBM, and designed to replace the SS-4 and SS-5 missile systems in the intermediate role, the SS-20 began to enter service in 1977 and is a formidable system offering the advantages of good accuracy with tactical land mobility; the warhead is believed to be a 650-kiloton type, though many reports indicate a 1.5-megaton type; this variant may have entered only limited service
SS-20 Model 2: also introduced in 1977, this model offers comparable performance but increased tactical capability through the use of a MIRVed warhead
SS-20 Model 3: increased-range variant with a single low-yield warhead; possibly not operational

SS-24 'Scalpel'
(USSR)
Type: three-stage intercontinental ballistic missile
Dimensions: diameter 2.50 m (8.20 ft); length 21.00 m (68.90 ft)
Weights: total round 100000 kg (220,459 lb); post-boost vehicle 3625 kg (7,992 lb)
Propulsion: (first stage) solid-propellant rocket delivering ?-kg (?-lb) thrust, (second stage) solid-propellant rocket delivering ?-kg (?-lb) thrust, and (third stage) solid-propellant rocket delivering ?-kg (?-lb) thrust
Range: 12000+ km (7,455 miles)
CEP: 200 m (220 yards)
Warhead: up to 10 350-kiloton nuclear MIRVs
Launch: cold type from superhardened silos or special rail cars, offering a reload capability
Guidance: stellar-inertial

Variant
SS-24: entering service in the mid-1980s, the SS-24 is the Soviet counterpart to the US Peacekeeper, and may also be developed for a mobile basing system, probably using special rail cars hidden in the tunnels of those parts of the US rail network without overhead wires; the type offers a formidable combination of range, accuracy and warhead

SS-25 'Sickle'
(USSR)
Type: three-stage lightweight intercontinental ballistic missile
Dimensions: diameter 1.70 m (5.58 ft); length 19.00 m (62.34 ft)
Weights: total round 37000 kg (81,570 lb); post-boost vehicle 725 kg (1,598 lb)
Propulsion: (first stage) solid-propellant rocket delivering ?-kg (?-lb) thrust, (second stage) solid-propellant rocket delivering ?-kg (?-lb) thrust, and (third stage) solid-propellant rocket delivering ?-kg (?-lb) thrust
Range: 9000+ km (5,590+ miles)
CEP: 200 m (220 yards)
Warhead: one 550-kiloton nuclear RV or three/four 150-kiloton nuclear MIRVs

Launch: cold type from hardened silo or wheeled transport/erector/launcher vehicle, offering a reload capability
Guidance: stellar-inertial

Variant
SS-25: according to the Soviets this is an upgraded version of the SS-13 rather than a new design (hence the Soviet designation **RS-12M**), but the result is a road-mobile weapon with great accuracy and the offensive capability of the US Minuteman lightweight ICBM

SS-X-26
(USSR)
Type: three-stage intercontinental ballistic missile
Dimensions: not known
Weights: not known
Propulsion: not known
Range: not known
CEP: not known
Warhead: up to 10 ?-kiloton MIRVs
Launch: cold-launch from a hardened silo, offering reload capability
Guidance: inertial

Variant
SS-X-26: due to begin flight trials in 1986 or 1987, this is believed to be the prospective replacement for the SS-18, and is thus a massive solid-propellant weapon

SS-X-27
(USSR)
Type: three-stage intercontinental ballistic missile
Dimensions: not known
Weights: now known
Propulsion: not known
Range: not known
CEP: not known
Warhead: not known
Launch: cold type from a hardened silo, offering reload capability
Guidance: inertial

Variant
SS-X-27: apparently designated RS-20M by the Soviets in an attempt to convince the world that it is an SS-18 derivative, this is a truly enormous prototype solid-propellant weapon entered flight test in 1986 and is clearly designed as successor to the SS-18

SS-X-28
(USSR)
Type: two-stage intermediate-range ballistic missile
Dimensions: not known
Weights: not known
Propulsion: not known
Range: not known
CEP: not known
Warhead: one ?-kiloton nuclear RV
Launch: cold type from wheeled transporter/erector/launcher, with reload capability
Guidance: inertial

Variant
SS-X-28: currently in flight test, this is believed to be a high-mobility IRBM possibly first thought to be designated **SS-20 Model 5**

SS-N-5 'Serb'
(USSR)
Type: two-stage submarine-launched ballistic missile
Dimensions: diameter 1.42 m (4.66 ft); length 12.90 m (42.32 m)
Weights: total round 17000 kg (37,477 lb); post-boost vehicle not known
Propulsion: (first stage) one solid-propellant rocket delivering ?-kg (?-lb) thrust and (second stage) one solid-propellant rocket delivering ?-kg (?-lb) thrust
Range: 1400 km (870 miles)
CEP: 2800 m (3,060 yards)
Warhead: one 1-megaton thermonuclear RV
Launch: submarine tube
Guidance: inertial

Variant
SS-N-5: now serving in the theatre nuclear role on board 13 'Golf II' class missile submarines, the SS-N-5 missile is comparable in many respects to the original version of the US Polaris SLBM and began to enter service in 1964 as the Soviets' first underwater-launched ballistic missile; data on the weapon are still scanty, and it is not known with certainty if the type uses solid-propellant or storable liquid-propellant rockets; by current standards the range and accuracy are poor

SS-N-6
(USSR)
Type: single-stage submarine-launched ballistic missile
Dimensions: diameter 1.80 m (5.91 ft); length 10.00 m (32.81 ft)
Weights: total round 18900 kg (41,667 lb); post-boost vehicle 680 kg (1,499 lb)
Propulsion: one storable liquid-propellant rocket delivering ?-kg (?-lb) thrust
Range: (Model 1) 2400 km (1,490 miles) or (Models 2 and 3) 3000 km (1,865 miles)
CEP: 1850 m (2,023 yards)
Warhead: (Model 1) one 700-kiloton nuclear RV, (Model 2) one 650-kiloton nuclear RV, or (Model 3) two 350-kiloton nuclear MRVs
Launch: submarine tube
Guidance: inertial

Variants
SS-N-6 Model 1: introduced in 1970, this is a hybrid second/third-generation SLBM using technology and components derived from the SS-11 'Sego'; the warhead size and comparatively poor CEP dictate the weapon's targeting against area targets; for many years it was thought that the NATO reporting name **'Sawfly'** applied to this weapon, but this is apparently not the case
SS-N-6 Model 2: introduced in 1973, this version trades throw-weight for range, the reduction of 50 kilotons in warhead yield providing an additional 600 km (373 miles) of range to provide the launch submarines with greater operating area
SS-N-6 Model 3: introduced in 1974, this variant combines the range of the Model 2 with a double warhead configuration for maximum effect against cities and other area targets

SS-N-8
(USSR)
Type: two-stage submarine-launched ballistic missile
Dimensions: diameter 1.65 m (5.41 ft); length 12.90 m (42.32 ft)
Weights: total round 20400 kg (44,974 lb); post-boost vehicle 680 kg (1,499 lb)
Propulsion: (first stage) on storable liquid-propellant rocket delivering ?-kg (?-lb) thrust, and (second stage) one storable liquid-propellant rocket delivering ?-kg (?-lb) thrust
Range: (Model 1) 7800 km (4,845 miles) or (Model 2) 9100 km (5,655 miles)
CEP: (Model 1) 1410 m (1,540 yards) or (Model 2) 1550 m (1,695 yards)
Warhead: one 800-kiloton nuclear RV
Launch: submarine tube
Guidance: stellar-inertial

Variants
SS-N-8 Model 1: introduced in 1971 aboard 'Delta I' class SSBNs, the SS-N-8 is a fourth-generation SLBM with good range but only moderate CEP despite the use of two stellar fixes to update the inertial guidance
SS-N-8 Model 2: introduced in 1977, this variant is a developed version carrying the same warhead over greater range

SS-N-17 'Snipe'
(USSR)
Type: two-stage submarine-launched ballistic missile
Dimensions: diameter 1.65 m (5.41 ft); length 11.06 m (36.29 ft)
Weights: total round not known; post-boost vehicle 1135 kg (2,502 lb)
Propulsion: (first stage) one solid-propellant rocket delivering ?-kg (?-lb) thrust and (second stage) one solid-propellant rocket delivering ?-kg (?-lb) thrust
Range: 3900 km (2,425 miles)
CEP: 1400 m (1,530 yards)
Warhead: one 1-megaton thermonuclear RV
Launch: submarine tube
Guidance: stellar-inertial

Variant
SS-N-17: introduced in 1977 on the sole 'Yankee II' class SSBN, the SS-N-17 is the first Soviet SLBM known to have solid-propellant rockets; it is also believed that the type has post-boost propulsion for manoeuvring in space, yet the CEP is still too great for the missile's use in anything but the countervalue role

SS-N-18 'Stingray'
(USSR)
Type: two-stage submarine-launched ballistic missile
Dimensions: diameter 1.80 m (5.91 ft); length 14.10 m (46.26 ft)
Weights: total round 25000 kg (55,115 lb); post-boost bus not known
Propulsion: (first stage) one storable liquid-propellant rocket delivering ?-kg (?-lb) thrust, and (second stage) one storable liquid-propellant rocket delivering ?-kg (?-lb) thrust
Range: (Models 1 and 3) 6500 km (4,040 miles) or (Model 2) 8000 km (4,970 miles)
CEP: (Models 1 and 3) 1410 m (1,540 yards) or (Model 2) 1550 m (1,695 yards)
Warhead: (Model 1) three 200-kiloton nuclear MIRVs, (Model 2) one 450-kiloton nuclear RV, or (Model 3) seven 200-kiloton nuclear MIRVs
Launch: submarine tube
Guidance: stellar-inertial

Variants
SS-N-18 Model 1: designated **RSM-50** by the Soviets, this fourth-generation SLBM entered service in 1976 aboard 'Delta III' class SSBNs and was the first Soviet SLBM with a MIRVed warhead
SS-N-18 Model 2: introduced in 1979, this variant offers greater range with a payload reduced to one RV
SS-N-18 Model 3: introduced in 1979, this variant took over from the Model 1 with a load of seven smaller MIRVs

SS-N-20 'Sturgeon'
(USSR)
Type: three-stage submarine-launched ballistic missile
Dimensions: diameter 2.00 m (6.56 ft); length 15.00 m (49.2 ft)
Weights: not known
Propulsion: (first stage) one solid-propellant rocket delivering ?-kg (?-lb) thrust, (second stage) one solid-propellant rocket delivering ?-kg (?-lb) thrust) and

(third stage) one solid-propellant rocket delivering ?-kg (?-lb) thrust)
Range: 8300 km (5,160 miles)
CEP: better than 500 m (547 yards)
Warhead: between six and nine 100/200-kiloton nuclear MIRVs
Launch: submarine tube
Guidance: stellar-inertial

Variant
SS-N-20: introduced in 1981 as the primary armament of the huge 'Typhoon' class SSBN, the SS-N-20 entered development in 1973 as a fifth-generation SLBM of good range and advanced capabilities; the range of the missile allows launch submarines to operate from Soviet sanctuary areas and under the arctic ice cap and still hit targets anywhere in the USA

SS-N-23 'Skiff'
(USSR)
Type: two-stage submarine-launched balistic missile
Dimensions: diameter 2.00 m (6.56 ft); length 13.60 m (44.62 ft)
Weights: not known
Propulsion: (first stage) one solid-propellant rocket delivering ?-kg (?-lb) thrust, and (second stage) one solid-propellant rocket delivering ?-kg (?-lb) thrust
Range: 8300 km (5,160 miles)
CEP: 560 m (612 yards)
Warhead: seven 150-kiloton nuclear MIRVs
Launch: submarine tube
Guidance: stellar-inertial

Variant
SS-N-23: introduced in 1985 aboard the 'Delta IV' class SSBN, this is a capable fifth-generation SLBM notable for its good range and low CEP

Martin-Marietta LGM-25C Titan II
(USA)
Type: two-stage heavyweight intercontinental ballistic missile
Dimensions: diameter 10.00 ft (3.048 m); length 103.00 ft (31.39 m)
Weights: total round 330,000 lb (151049 kg); post-boost vehicle 8,300 lb (3765 kg)
Propulsion: (first stage) two Aerojet LR87-AJ-5 storable liquid-propellant rockets each delivering 216,000-lb (97975-kg) thrust, and (second stage) one Aerojet LR91-AH-5 storable liquid-propellant rocket delivering 100,000-lb (45360-kg) thrust
Range: 9,325 miles (15000 km)
CEP: 1,425 yards (1305 m)
Warhead: one 9-megaton W53 thermonuclear warhead and penetration aids carried in a single General Electric Mk 6 RV
Launch: hot type from a hardened silo
Guidance: AC/IBM inertial

Variant
LGM-25C Titan II: last of the USA's heavyweight liquid-propellant ICBMs to remain in service (until October 1987), the Titan II was introduced in 1962 as successor to the HGM-25A Titan I; the guidance systems were extensively upgraded in 1980 and 1981 to provide effective life into the late 1980s

Boeing LGM-30G Minuteman III
(USA)
Type: three-stage lightweight intercontinental ballistic missile
Dimensions: diameter 6.07 ft (1.85 m); length 59.71 ft (18.20 m)
Weights: total round 78,000 lb (335380 kg); post-boost vehicle (Mk 12) 2,400 lb (1087 kg) or (Mk 12A) 2,535 lb (1150 kg)
Propulsion: (first stage) one Thiokol M55A1 (TU-120) solid-propellant rocket delivering 202,600-lb (91900-kg) thrust, (second stage) one Aerojet SR18-AJ-1 solid-propellant rocket delivering 60,625-lb (27500-kg) thrust, and (third stage) one Aerojet/Thiokol SR73-AJ/AG-1 solid-propellant rocket delivering 33,800-lb (15332-kg) thrust
Range: 8,700 miles (14000 km)
CEP: 300 yards (275 m) with Mk 12 RV or 240 yards (220 m) with Mk 12A RV
Warhead: three 170-kiloton W62 nuclear MIRVs in a single General Electric Mk 12 RV, or three 335-kiloton W78 nuclear MIRVs in a single Mk 12A RV, both with penetration aids
Launch: hot type from a hardened silo
Guidance: Rockwell Autonetics inertial

Variants
LGM-30G Minuteman III Model 1: this lightweight ICBM was introduced in 1970 as the third-generation companion to the second-generation Minuteman II, with an improved third stage and MIRVed payload, whose Mk 12 bus contains both chaff- and decoy-dispensing penetration aids
LGM-30G Minuteman III Model 2: introduced in 1979, this upgraded model has an improved Mk 12A RV with higher-yield warheads and more advanced penetration aids, made possible by miniaturization of the warheads and RV components

Boeing LGM-30F Minuteman II
(USA)
Type: three-stage lightweight intercontinental ballistic missile
Dimensions: diameter: 6.00 ft (1.83 m); length 59.71 ft (18.20 m)
Weights: total round 72,810 lb (33027 kg); post-boost vehicle 1,610 lb (730 kg)

Propulsion: (first stage) one Thiokol M55A1 (TU-120) solid-propellant rocket delivering 202,600-lb (91900-kg) thrust, (second stage) one Aerojet SR18-AJ-1 solid-propellant rocket delivering 60,000-lb (27216-kg) thrust, and (third stage) one Hercules M57A1 solid-propellant rocket delivering 17,650-lb (8006-kg) thrust
Range: 7,775 miles (12510 km)
CEP: 400 yards (365 m)
Warhead: one 1.2-megaton W56 thermonuclear warhead carried in an Avco Mk 11C Model 4 RV with Tracor Mk 1A penetration aids
Launch: hot type from a hardened silo
Guidance: Rockwell Autonetics inertial

Variant
LGM-30F Minuteman II: this second-generation lightweight ICBM entered service in 1966, swiftly replacing the LGM-30A and LGM-30B variants of the original Minuteman I type; the Minuteman II is a simple upgrading of the LGM-30B with more advanced guidance (including an eight-target selection capability and chaff-dispensing penetration aids) and greater range; eight of the 450 current rounds are fitted as launch vehicles for the Emergency Rocket Communications System to provide a short-term pre-launch link between the US National Command Authorities and the US strategic forces should the USA's satellite communications system be knocked out

Martin-Marietta LGM-118A Peacekeeper
(USA)
Type: four-stage heavyweight intercontinental ballistic missile
Dimensions: diameter 7.67 ft (2.337 m); length 70.87 ft (21.60 m)
Weights: total round 195,000 lb (88452 kg); post-boost vehicle 7,935 lb (3600 kg)
Propulsion: (first stage) one Thiokol solid-propellant rocket delivering 570,000-lb (258552-kg) thrust, (second stage) one Aerojet solid-propellant rocket delivering 335,000-lb (151956-kg) thrust, (third stage) one Hercules solid-propellant rocket delivering 77,000-lb (34927-kg) thrust, and (fourth stage) one Rockwell hypergolic liquid-propellant rocket delivering ?-lb (?-kg) thrust
Range: 8,700 miles (14000 km)
CEP: 65/100 yards (60/90 m)
Warhead: 10 300-kiloton MIRVs in an Avco RV carried in a Rockwell RS-34 bus
Launch: cold type from a hardened silo, offering a reload capability
Guidance: Rockwell Autonetics/Honeywell/Northrop inertial

Variant
LGM-118A Peacekeeper: conceived as a replacement for the LGM-25 and LGM-30 series, the Peacekeeper is a fourth-generation heavyweight ICBM, and was greatly troubled by political opposition during its relatively smooth development life; the Peacekeeper, previously known as the **MX**, began to enter service during 1986 in upgraded Minuteman silos because political pressure prevented any of the proposed mobile basing methods; the type's accuracy makes it a potent counterforce weapon

Lockheed UGM-27C Polaris A3TK Chevaline
USA/UK)
Type: two-stage submarine-launched ballistic missile
Dimensions: diameter 4.50 ft (1.372 m); length 32.29 ft (9.84 m)
Weights: total round 35,000 lb (15876 kg); post-boost vehicle 1,500 lb (680 kg)
Propulsion: (first stage) one Aerojet A3P solid-propellant rocket delivering 80,000-lb (36288-kg) thrust, and (second stage) one Hercules solid-propellant rocket delivering ?-lb (?-kg) thrust
Range: 2,950 miles (4750 km)
CEP: 1,015 yards (930 m)
Warhead: three 60-kiloton nuclear MRVs plus an unknown number of decoys and penetration aids
Launch: submarine tube
Guidance: General Electric/MIT/Hughes/Raytheon Mk 2 inertial

Variant
UGM-27C Polaris A3TK Chevaline: this is now the only version of the Polaris SLBM left in service, the missiles having been updated in the 'Chevaline' programme of the 1970s to carry, instead of the original three British 200-kiloton warheads, several (but generally assumed to be three) British MRVs each with a yield believed to be 60 kilotons and capable of a lateral separation of 45 miles (72 km); the warheads are hardened against electro-magnetic pulse and fast radiation, and the warhead bus also contains chaff penetration aids and several (perhaps three) decoys

Lockheed UGM-73A Poseidon C3
(USA)
Type: two-stage submarine-launched ballistic missile
Dimensions: diameter 6.17 ft (1.88 m); length 34.00 ft (10.36 m)
Weights: total round 65,000 lb (29484 kg); post-boost vehicle 3,300 lb (1497 kg)
Propulsion: (first stage) two Thiokol/Hercules solid-propellant rockets each delivering ?-lb (?-kg) thrust, and (second stage) two Hercules solid-propellant rockets each delivering ?-lb (?-kg) thrust
Range: 2,485 miles (4000 km) with 14 MIRVs or 3,230 miles (5200 km) with 10 MIRVs
CEP: 605 yards (553 m)
Warhead: 10 Mk 3 MIRVs each with a W68 40/50-

kiloton nuclear warhead, or 14 Mk 3 MIRVs each with a W76 100-kiloton nuclear warhead
Launch: submarine tube
Guidance: General Electric/MIT/Hughes/Raytheon Mk 4 inertial

Variant
UGM-73A Poseidon C3: introduced in 1970 as successor to the Polaris SLBM, the Poseidon marked an eightfold increase in target-devastation capability at the same range as its predecessor, the Poseidon's real advantages being much improved CEP and a MIRVed payload (the first in a US SLBM)

Lockheed UGM-96A Trident I C4
(USA)
Type: three-stage submarine-launched ballistic missile
Dimensions: diameter 6.17 ft (1.88 m); length 34.00 ft (10.36 m)
Weights: total round 73,000 lb (33113 kg); post-boost vehicle 3,000+ lb (1361+ kg)
Propulsion: (first stage) one Thiokol/Hercules solid-propellant rocket delivering ?-lb (?-kg) thrust, (second stage) one Hercules solid-propellant rocket delivering ?-lb (?-kg) thrust, and (third stage) one UTC-CSD solid-propellant rocket delivering ?-lb (?-kg) thrust); there is also a powered post-boost bus
Range: 4,230 miles (6810 km)
CEP: 500 yards (547 m)
Warhead: eight Mk 4 MIRVs each with a W76 100-kiloton nuclear warhead
Launch: submarine tube
Guidance: Mk 5 stellar-inertial

Variants
UGM-96A Trident I C4: introduced in 1979 as the primary armament of the 'Ohio' class SSBNs and also carried in a number of converted 'Benjamin Franklin' and 'Lafayette' class SSBNs, the Trident I is essentially the Poseidon SLBM with a third stage for much increased range and a very much more advanced warhead based on the Mk 4 MIRV, the bus being manoeuvrable in space for maximum accuracy; should the Soviets upgrade their SLBM capabilities to a marked degree, the USA plans to develop the Mk 5 Evader MARV for installation on the Trident I; this payload can be manoeuvred independently in space and during re-entry to offer maximum target scope while adversely affecting the defence's ability to achieve interceptions
Trident II D5: due to enter flight test in 1987, this is a much improved development for use in the 'Ohio' class SSBN; the type has a diameter of 6 ft 10.66 in (2.10 m) and a length of 44 ft 7 in (13.59 m), and a launch weight of 70,550 lb (32001 kg); the first two stages have graphite-epoxy rather than Kevlar casings but contain the same motors as the Trident I, while the third stage has a United Technologies motor; the payload comprises 10 to 15 Mk 5 RVs each with one 335-kiloton W78 warhead; guidance is entrusted to the Mk 6 stellar-inertial system offering a CEP as low as 130 yards (120 m); the Royal Navy's version will probably have eight British-designed warheads in a US-provided manoeuvring bus

CRUISE MISSILES

Aérospatiale ASMP
(France)
Type: air-launched theatre/strategic cruise missile
Dimensions: diameter 0.96 m (37.8 in); length 5.38 m (17.65 ft); span not known
Weights: total round 900 kg (1,984 lb); warhead 100/150-kiloton nuclear
Propulsion: one integral solid-propellant booster rocket delivering ?-kg (?-lb) thrust, and one ramjet sustainer delivering ?-kg (?-lb) thrust
Performance: speed Mach 4; range 300 km (186 miles)
CEP: not revealed
Guidance: inertial and terrain following

Variant
ASMP: designed as a medium-range air-launched cruise missile for carriage by Dassault-Breguet Mirage IVP bombers (after conversion from Mirage IVA standard with an armament of one 60/70-kiloton AN 22 free-fall nuclear weapon), by Dassault-Breguet Mirage 2000N low-altitude strike aircraft and by Dassault-Breguet Super Etendard strike fighters, the ASMP entered service in 1986 and provides the French with a stand-off nuclear capability against large targets such as railway yards, main bridges, and command and communications centres

Boeing AGM-86B
(USA)
Type: air-launched strategic cruise missile
Dimensions: diameter 27.3 in (0.693 m); length 20.75 ft (6.325 m); span 12.00 ft (3.658 m)
Weights: total round 2,825 lb (1281.4 kg); warhead

270 lb (123 kg) 200-kiloton W80-1 nuclear
Propulsion: one Williams Research F107-WR-100 turbofan delivering 600-lb (272-lb) thrust
Performance: speed 500 mph (805 km/h); range 1,950 miles (3140 km)
CEP: between 30 and 100 ft (9 and 30.5 m)
Guidance: Litton inertial with McDonnell Douglas DPW-23 TERCOM (TERrain COntour Matching) update

Variant
AGM-86B: entering service in 1981, this air-launched cruise missile is one of the US Air force's most important weapons, some 95 Boeing B-52G bombers being configured to carry 12 such weapons apiece (six on each of two underwing stations) plus four AGM-69A SRAMs and four free-fall nuclear weapons in the bomb bay; work is under way on modifying 95 B-52H bombers to the same standard, with provision on both Stratofortress marks for the weapons-bay carriage of the Common Strategic Rotary Launcher for eight AGM-86B and/or AGM-69A SRAM missiles; it is estimated that 85% of the strategic areas more than 225 miles (360 km) from Soviet airspace, with the SRAMs available for defence suppression should the bombers have to penetrate Sovet airspace; the new Rockwell B-1B supersonic penetration bomber can carry eight AGM-86Bs on a rotary launcher in the weapons bay, plus another 14 under the wings; though 4,348 AGM-86Bs were planned, only 1,715 have been built as newer cruise missiles are planned with greater capabilities at a time the Soviets are strengthening their capabilities against air-launched cruise missiles

General Dynamics BGM-109G Tomahawk
(USA)
Type: ground-launched strategic cruise missile
Dimensions: diameter 21.00 in (0.533 m); length 21.00 ft (6.40 m); span 8.33 ft (2.54 m)
Weights: total round 2,650 lb (1202 kg); warhead 270 lb (123 kg) 200-kiloton W84 nuclear
Powerplant: one Atlantic Research solid-propellant booster rocket delivering 7,050-lb (3198-kg) thrust, and one Williams Research F107-WR-100 turbofan delivering 600-lb (272-kg) thrust
Performance: speed 500 mph (805 km/h); range 1,555 miles (2502 km)
CEP: 64 ft (20 m)
Guidance: Litton inertial plus McDonnell Douglas DPW-23 TERCOM (TERrain COntour Matching) update and Digital Scene-Matching Area Correlation terminal homing

Variants
BGM-109A Tomahawk: the BGM-109 series is the land- and ship/submarine-launched equivalent to the AGM-86B, and the type began to enter service in 1983; the BGM-109A naval variant is known as the **TLAM-N (Tactical Land Attack Missile – Nuclear)**, a ship- and submarine-launched encapsulated version with the W80 Mod 0 warhead, a range of 1,555 miles (2502 km) and a CEP of 305 yards (280 m) with inertial and TERCOM guidance

BGM-109B Tomahawk: this is also known as the **TASM (Tactical Anti-Ship Missile)** designed for ship and submarine launch, and carries a 1,000-lb (454-kg) HE warhead (derived from that of the AGM-12 Bullpup air-to-surface missile) over a range of 280 miles (450 km); terminal guidance is provided by an active radar seeker derived from that of the RGM-84A Harpoon anti-ship missile, and this commands the missile into a pop-up/dive attack to strike the target on its vulnerable upper surfaces

BGM-109C Tomahawk: this is also known as the **TLAM-C (Tactical Land Attack Missile – Conventional)**, and is essentially the airframe of the BGM-109A with the warhead of the BGM-109B and DSW-15(V) DSMAC terminal guidance for great accuracy over a range of 925 miles (1490 km)

BGM-109G: this is also known as the **GLCM (Ground-Launched Cruise Missile)**, and is the strategic version of the series deployed extensively in Europe in the theatre nuclear role; the weapon is carried in a four-round transporter/erector/launcher vehicles designed for rapid deployment (in times of crisis and war) away from Soviet-targeted base areas; a group of four TELs is supported by two wheeled launch control centres to provide launch and targeting data, and to provide the crews with NBC protection

AGM-109H Tomahawk: proposed medium-range air-launched airfield attack cruise missile for the US Air Force with the Teledyne Continental J402-CA-401 engine and a conventional warhead

AGM-109K Tomahawk: proposed medium-range air-launched land attack and sealane control cruise missile for the US Air Force with the J402-CA-401 engine and the WDU-25A/B conventional warhead

AGM-109L Tomahawk: proposed medium-range air-launched land/sea attack cruise missile for the US Navy with the J402-CA-401 engine and the WDU-7B conventional warhead

General Dynamics Advanced Cruise Missile
(USA)
Type: air-launched long-range cruise missile
Dimensions: not known
Weights: not known
Propulsion: one Williams Research F112 turbofan delivering ?-lb (?-kg) thrust
Performance: speed high subsonic; range not known
CEP: not known
Guidance: not known

Variant
Advanced Cruise Missile: although this important weapon may enter service before 1990, very little is

known of its highly classified development; the type is designed for use on B-52, B-1 and B2 bombers, and key features are longer range than the AGM-86B combined with a high measure of 'stealth' technology to reduce the missile's observability

AS-15 'Kent', SS-C-4 and SS-N-21
(USSR)
Type: air-, land- and submarine-launched tactical/strategic cruise missile
Dimensions: diameter 0.533 m (21.00 in); length 6.90 m (22.64 ft); span 3.45 m (11.32 ft)
Weights: total round 1500 kg (3,307 lb); warhead 150-kiloton thermonuclear or ?-kg (?-lb) HE
Propulsion: one turbojet delivering ?-kg (?-lb) thrust
Performance: speed not known; range 2750 or 3000 km (1,710 or 1,865 miles)
CEP: 45 m (150 ft)
Guidance: inertial with terrain-following update

Variants
AS-15: this is the air-launched version of the USSR's standard cruise missile, which began to enter service during the mid-1980s in a number of forms; the AS-15 is currently operated by the Tupolev Tu-95 'Bear-H', and will equip the new Tupolev 'Blackjack' supersonic variable-geometry bomber due in service during 1988; the weapon compares favourably to the US AGM-86 and BGM-109 types
SS-C-4: entering service in 1985, this is the land-based variant of the family, and is based in wheeled transporter/erector/launchers
SS-N-21: due to enter service in the second half of the 1980s, this is the encapsulated tube-launched submarine variant, and presumably offers a choice of targeting and warhead options to provide Soviet attack submarines with a wide variety of attack choices

SS-NX-24
(USSR)
Type: ship- and submarine-launched cruise missile
Dimensions: (estimated) diameter 1.25 m (49.2 in); length 12.0 m (39.37 ft); span 5.9 m (19.36 ft)
Weights: not known
Propulsion: not known
Performance: speed supersonic; range not known
CEP: not known
Guidance: not known

Variant
SS-NX-24: little is known of this new Soviet cruise missile, probably designed for the delivery of a nuclear warhead against high-value naval targets; is also being developed for a land-based application

SURFACE-TO-SURFACE MISSILES

Avibras SS-70
(Brazil)
Type: surface-to-surface battlefield missile
Dimensions: diameter 0.33 m (13.0 in); length not known
Weights: not known
Propulsion: one solid-propellant rocket delivering ?-kg (?-lb) thrust
Range: 70 km (43 miles)
CEP: not known
Warhead: HE
Launch: wheeled transporter/erector/launcher
Guidance: inertial

Variant
SS-70: little is known of this new Brazilian weapon, but it may be a development of the X-40 rocket, which can carry a 146-kg (322-lb) payload over a range of 68 km (42 miles)

Avibras SS-300
(Brazil)
Type: surface-to-surface battlefield missile
Dimensions: diameter 0.93 m (36.6 in); length not known
Weights: not known
Propulsion: one solid-propellant rocket delivering ?-kg (?-lb) thrust
Range: 300 km (186 miles)
CEP: not known
Warhead: HE
Launch: wheeled transporter/erector/launcher
Guidance: inertial

Variant
SS-300: due to enter flight test in 1987, this Brazilian weapon offers modest battlefield capability with a conventional warhead, though a submunition-dispenser type may later be developed

CPMIEC Model M
(China)
Type: single-stage surface-to-surface theatre/battlefield missile
Dimensions: diameter 1.0 m (39.37 in); length 9.1 m (29.86 ft)

Weights: total round 6200 kg (13,668lb); warhead ?-kg (?-lb)
Propulsion: one solid-propellant rocket delivering ?-kg (?-lb) thrust
Range: 600 km (373 miles)
CEP: not known
Warhead: HE/fragmentation
Launch: wheeled transporter/erector/launcher
Guidance: inertial

Variant
Model M: this Chinese weapon is being developed for several theatre and battlefield roles with HE/fragmentation and submunition-dispenser warheads, though a tactical nuclear warhead is also a possibility; ranges between 200 and 600 km (124 and 373 miles) are possible

Aérospatiale Hades
(France)
Type: surface-to-surface theatre/battlefield missile
Dimensions: diameter 0.60 m (1.97 ft); length 8.00 m (26.25 ft); span 1.50 m (4.92 ft)
Weights: not revealed
Propulsion: one SNPE/SEP solid-propellant rocket delivering ?-kg (?-lb) thrust
Range: 350 km (217 miles)
CEP: not revealed
Warhead: 10/25-kiloton variable-yield nuclear
Launch: trailer-mounted vertical box
Guidance: inertial

Variant
Hades: under development for service in the late 1980s or early 1990s, the Hades is designed as the longer-range successor to the Pluton

Aérospatiale Pluton
(France)
Type: surface-to-surface battlefield missile
Dimensions: diameter 0.55 m (1.80 ft); length 7.64 m (25.07 ft)
Weights: total round 2350 kg (5,181 lb); warhead 350 or 500 kg (772 or 1,102 lb)
Propulsion: one SEP/SNPE/Aérospatiale dual-thrust solid-propellant rocket delivering ?-kg (?-lb) thrust
Range: 10/120 km (6.2/75 miles)
CEP: 330 m (360 yards)
Warhead: 15-kiloton nuclear or 25-kiloton AN 51 nuclear
Launch: box on converted AMX-30 MBT chassis
Guidance: SFENA strap-down inertial

Variant
Pluton: designed in the late 1960s, the Pluton began to enter French service in 1974 and is a capable battlefield weapon with fair accuracy for use primarily against operational-level targets such as transport centres and follow-on troop concentrations

Ching Feng
(Taiwan)
Type: surface-to-surface battlefield missile
Dimensions: diameter 0.60 m (23.6 in); length 7.0 m (22.97 ft)
Weights: total round 1400 kg (3,086 lb); warhead not known
Propulsion: one solid-propellant rocket delivering ?-kg (?-lb) thrust
Range: 120 km (75 miles)
CEP: not known
Warhead: HE
Launch: not known
Guidance: an unknown radar-based system

Variant
Ching Feng: this Taiwanese battlefield system appears to be based conceptionaly on the US Lance, though the fins are somewhat smaller and a radar guidance system is used

FROG-3
(USSR)
Type: two-stage surface-to-surface battlefield rocket
Dimensions: diameter (warhead) 0.55 m (21.65 in) and (rocket) 0.40 m (15.75 in); length 10.50 m (34.45 ft)
Weights: total round 2250 kg (4,960 lb); warhead 450 kg (992 lb) HE
Propulsion: one non-jettisonable solid-propellant booster rocket delivering ?-kg (?-lb) thrust, and one solid-propellant sustainer rocket delivering ?-kg (?-lb) thrust
Range: 40 km (25 miles)
CEP: 400 m (440 yards)
Warhead: originally provided with interchangeable 200-kiloton nuclear and HE warheads, only the latter now being available to the Soviet allies still using the system
Launch: converted chassis of the PT-76 light amphibious tank
Guidance: none

Variant
FROG-3: the oldest of the FROG (Free Rocket Over Ground) series still in service, the FROG-3 was introduced in 1957, and now possesses little real value as its CEP is so great

FROG-5
(USSR)
Type: two-stage surface-to-surface battlefield rocket
Dimensions: diameter 0.55 m (21.65 in); length 9.10 m (29.86 ft)
Weights: total round 3000 kg (6,614 lb); warhead 450 kg (992 lb) HE

Propulsion: one non-jettisonable solid-propellant booster rocket delivering ?-kg (?-lb) thrust, and one solid-propellant sustainer rocket delivering ?-kg (?-lb) thrust
Range: 55 km (34 miles)
CEP: 400 m (440 yards)
Warhead: originally provided with interchangeable 200-kiloton nuclear and HE warheads, only the latter now being available to the Soviet allies still using the system
Launch: converted chassis of the PT-76 light amphibious tank
Guidance: none

Variant
FROG-5: introduced in the late 1950s or early 1960s, this is a development of the FROG-3 with a revised propulsion system, the body of the rocket being increased in diameter to that of the FROG-3's warhead, in the process being slightly shortened; like the FROG-3, the FROG-5 is now elderly and of only the smallest operation value

FROG-7
(USSR)
Type: surface-to-surface battlefield rocket
Dimensions: diameter 0.55 m (21.65 in); length 9.10 m (29.86 ft)
Weights: total round 2300 kg (5,071 lb); warhead 550 kg (1,213 lb) nuclear and HE, or 390 kg (860 lb) chemical
Propulsion: one solid-propellant rocket delivering ?-kg (?-lb) thrust
Range: 11/70 km (6.8/43 miles)
CEP: 450/700 m (490/765 yards) depending on range
Warhead: 10-, 100- or 200-kiloton nuclear, or HE or chemical
Launch: ZIL-135 8 × 8 transporter/erector/launcher
Guidance: none

Variant
FROG-7: introduced in 1965, this is the last of a series of FROG (Free Rocket Over Ground) weapons, and is being replaced by the longer-ranged and more accurate SS-21

SS-1C 'Scud-B'
(USSR)
Type: surface-to-surface battlefield missile
Dimensions: diameter 0.84 m (2.76 ft); length 11.40 m (37.40 ft)
Weights: total round 6370 kg (14,043 lb); warhead 1000 kg (2,205 lb)
Propulsion: one storable liquid-propellant rocket delivering ?-kg (?-lb) thrust
Range: 80/180 km (50/112 miles) with nuclear warhead, or 80/280 km (50/174 miles) with HE or chemical warhead
CEP: 930 m (1,015 yards) at maximum range, reducing with shorter range
Warhead: 40/100-kiloton variable-yield nuclear, or HE, or chemical
Launch: MAZ-537 8 × 8 transporter/erector/launcher
Guidance: inertial

Variants
SS-1B 'Scud-A': introduced in 1957, this original operational model is no longer in front-line service with the Warsaw Pact, and is a considerably less capable weapon than its successors, having a weight of 4400 kg (9,700 lb) and being able to carry a 40-kiloton (now HE) warhead over a range of 130 km (81 miles) after launch from its TEL, a converted JS-III heavy tank chassis
SS-1C 'Scud-B': introduced in 1964, this is an altogether more capable weapon
SS-1C 'Scud-C': introduced in about 1970, this variant has a maximum range boosted to 450 km (280 miles) by a better propellant load, but maximum-range CEP increases to 1100 m (1,205 yards) as a consequence

SS-12 'Scaleboard'
(USSR)
Type: single-stage short-range ballistic missile
Dimensions: diameter 1.05 m (3.45 ft); length 11.25 m (36.90 ft)
Weights: total round 8800 kg (19,400 lb); warhead 1250 kg (2,756 lb)
Propulsion: one solid-propellant rocket delivering ?-kg (?-lb) thrust
Range: 220/800 km (138/497 miles)
CEP: 480 m (525 yards)
Warhead: 800-kiloton nuclear
Launch: MAZ-537 8 × 8 transporter/erector/launcher
Guidance: inertial

Variants
SS-12: introduced in 1969, this weapon provides the Soviet army with long-range strategic/operational capability at front (army group' level
SS-12M: originally designated **SS-22** in Western terminology, this improved SS-12 began to enter service in 1977, and has different ranges (220 to 880 km/138 to 547 miles) combined with reduced CEP (320 m/350 yards) and a 550-kiloton nuclear warhead that can be exchanged for a chemical or cluster-munition warhead

SS-21 'Scarab'
(USSR)
Type: surface-to-surface battlefield missile
Dimensions: 0.46 m (1.51 ft); length 9.44 m (30.84 ft)

Weights: not known
Propulsion: one solid-propellant rocket delivering ?-kg (?-lb) thrust
Range: 14/120 km (8.75/75 miles)
CEP: 50/100 m (55/110 yards)
Warhead: 10- or 100-kiloton nuclear, chemical or cluster munition
Launch: ZIL-167 8 × 8 transporter/erector/launcher
Guidance: inertial

Variant
SS-21: introduced in 1976, this system is designed to replace the FROG series, and is deployed at front (army group) level as a primary means of removing NATO defensive positions slowing or halting the main axes of Soviet advance; the type is known to the Soviets as the **Tochka**

SS-23 'Spider'
(USSR)
Type: short-range ballistic missile
Dimensions: not known
Weights: not known
Propulsion: one solid-propellant rocket delivering ?-kg (?-lb) thrust
Range: 80/500 km (50/311 miles)
CEP: 280 m (305 yards)
Warhead: 200-kiloton nuclear, or chemical or cluster munition
Launch: MAZ 8 × 8 transporter/erector/launcher
Guidance: inertial

Variant
SS-23: introduced in 1980, this is the Soviet replacement for the 'Scud' series, and offers significantly reduced time into action as well as better range and CEP

Douglas MGR-1B Honest John
(USA)
Type: surface-to-surface battlefield missile
Dimensions: diameter 2.50 ft (0.76 m); length 24.84 ft (7.57 ft)
Weights: total round 4,710 lb (2136 kg); warhead 1,500 lb (580 kg) nuclear or HE, or 1,243 lb (564 kg) chemical
Propulsion: one solid-propellant rocket delivering ?-lb (?-kg) thrust
Range: 4.5/23 miles (7.2/37 km)
CEP: 910 yards (830 m)
Warhead: 2-, 20- or 40-kiloton W31 nuclear, or HE or cluster munition
Launch: 6 × 6 truck transporter/erector/launcher
Guidance: none

Variant
MGR-1B Honest John: introduced in 1960 to replace the 1953-vintage MRG-1A initial version, the MGR-1B is an obsolescent system combining a powerful warhead and good cross-country mobility with the accuracy and range of tube artillery, but in the non-nuclear role offering limited offensive capability only at considerable cost

Martin-Marietta MGM-31B Pershing II
(USA)
Type: two-stage short-range ballistic missile
Dimensions: diameter 3.28 ft (1.00 m); length 34.45 ft (10.50 m)
Weights: total round 16,000 lb (7257 kg); warhead 650 lb (295 kg)
Propulsion: (first stage) one Hercules solid-propellant rocket delivering ?-lb (?-kg) thrust, and (second stage) one Hercules solid-propellant rocket delivering ?-lb (?-kg) thrust
Range: 1,125 miles (1810 km)
CEP: 22/49 yards (20/45 m)
Warhead: 5/50-kiloton selectable-yield W85 nuclear
Launch: M656 truck/trailer-mounted transporter/erector/launcher
Guidance: Singer-Kearfott inertial plus Goodyear RADAG terminal homing

Variants
MGM-31A Pershing IA: introduced in 1969 as a long-range battlefield interdiction missile, the Pershing IA is derived from the 1962-vintage MGM-31 Pershing I deployed on the M474 tracked launch vehicle; the Pershing IA changes to the M656 truck/trailer TEL, is 34.78 ft (10.60 m) long, weighs 10,140 lb (4600 kg), is powered by one Thiokol MX105 solid-propellant booster rocket delivering 26,750-lb (12134-kg) thrust and one Thiokol XM106 solid-propellant sustainer rocket delivering 15,560-lb (7058-kg) thrust, and carries a 1,650-lb (748-kg) 60- or 400-kiloton W50 airburst nuclear warhead over a range of 100/460 miles (161/740 km) with a CEP of 400 yards (365 m) under the guidance of a Bendix inertial system
MGM-31B Pershing II: introduced in 1985, the Pershing II is a modular upgrading of the Pershing IA with far greater range and other system improvements, the most important of which is the RADAG (RADar Area Guidance) terminal homing, which reduces the CEP to so low a figure that a much smaller warhead (air-burst or earth-penetrating) can be carried; the range and extreme accuracy of the weapon mean that even with a small warhead the type has operational and indeed limited strategic capabilities within the European theatre

Vought MGM-52C Lance
(USA)
Type: single-stage battlefield support missile
Dimensions: diameter 1.83 ft (0.56 m); length 20.25 ft (6.17 m)
Weights: total round 3,373 lb (1530 kg) with nuclear

warhead or 3,920 lb (1778 kg) with conventional warhead; 467 lb (212 kg) nuclear or 1,000 lb (454 kg) conventional
Propulsion: two Rocketdyne dual-thrust storable liquid-propellant rockets each delivering ?-lb (?-kg) thrust
Range: 3/75 miles (4.8/121 km)
CEP: 500 yards (455 m)
Warhead: 1-, 10- or 50-kiloton selectable-yield W70-1/2/3 nuclear, 0.5-kiloton enhanced-radiation W70-4 nuclear, or M251 cluster munition
Launch: M752 tracked transporter/erector/launcher

Guidance: E-Systems/Sys-Donner/Arma simplified inertial

Variants
MGM-52C Lance: introduced in 1972 as replacement for the Sergeant and Honest John weapons, the Lance is a highly capable battlefield missile able to deliver an assortment of alternative warheads, including the W70-4 'neutron' type and the M251 cluster munition type with 836 0.95-lb (0.43-kg) anti-personnel/anti-materiel bomblets to saturate a circle of 900-yard (820-m) diameter

AIR-TO-SURFACE MISSILES

CITEFA ASM-2 Martin Pescador
(Argentine)
Type: air-to-surface tactical missile
Dimensions: diameter 0.2185 m (8.60 in); length 2.94 m (9.65 ft); span 0.73 m (28.74 in)
Weights: total round 140 kg (308.6 lb); warhead 40 kg (88/2 lb) FRAG-HE
Propulsion: one solid-propellant rocket delivering ?-kg (?-lb) thrust
Performance: speed Mach 1+; range 2500/9000 m (2,735/9,845 yards)
Guidance: radio command

Variant
ASM-2 Martin Pescador: this is a simple radio command-guided missile which entered service in 1979

Aérospatiale AS.11 (France): see Aérospatiale SS.12 anti-tank missile

Aérospatiale AS.12 (France): see Aérospatiale AS.12 anti-ship missile

Aérospatiale AS.20
(France)
Type: air-to-surface tactical missile
Dimensions: diameter 0.25 m (9.84 in); length 2.60 m (8.53 ft); span 0.80 m (31.50 in)
Weights: total round 143 kg (315.3 lb); warhead 30 kg (66 lb) impact-fused FRAG-HE
Propulsion: one SNPE/Aérospatiale Aspic solid-propellant booster rocket delivering ?-kg (?-lb) thrust, and one SNPE/Aérospatiale Icare solid-propellant sustainer rocket delivering ?-kg (?-lb) thrust
Performance: speed Mach 1.7; range 7000 m (7,655 yards)
Guidance: radio command

Variant
AS.20: developed from the radio-commanded AA.20 air-to-air missile, the simple but effective AS.20 was introduced in 1961 and remains in limited service, mainly as a training round for the AS.30 series

Aérospatiale AS.30 Laser
(France)
Type: air-to-surface tactical missile
Dimensions: diameter 0.34 m (13.4 in); length 3.65 m (11.98 ft); span 1.00 m (3.28 ft)
Weights: total round 520 kg (1,146.4 lb); warhead 240 kg (520 lb) impact- or delay-fused HE or SAP-HE
Propulsion: one SNPE/Aérospatiale two-stage solid-propellant rocket delivering ?-kg (?-lb) thrust
Performance: speed Mach 1.5; range 11250 m (12,305 yards); CEP 2 m (6.6 ft)
Guidance: Thomson-CSF Ariel semi-active laser homing

Variants
AS.30: introduced in 1960 to provide French Dassault Mirage IIIE fighter-bombers with a moderate stand-off capability, the AS.30 is essentially a scaled-up AS.20; use of this missile allowed the launch aircraft to come no closer than 3000 m (3,280 yards) to the target, the operator than radio-commanding the 3.885-m (12.75-ft) missile to the visual line of sight with a CEP of less than 10 m (33 ft)
AS.30TCA: introduced in 1964, this is a semi-automatic command to line of sight development, obviating the need for the operator to track both the missile and the target, a severe tactical disadvantage for the AS.30 model
AS.30 Laser: introduced in 1983, this much-improved

model is designed for use with the ATLIS (Automatic Tracking Laser Illuminator System) pod

Matra BGL 400
(France)
Type: laser-guided tactical glide bomb
Dimensions: diameter 0.40 m (15.75 in); length 3.20 m (10.50 ft); span 1.35 m (4.43 ft)
Weights: total round 475 kg (1,047 lb); warhead 400 kg (882 lb) bomb
Propulsion: none
Performance: speed high subsonic; range 10000 m (10,935 yards)
Guidance: Thomson-CSF TMV 630 EBLIS semi-active laser homing

Variants
BGL 400: this is derived in concept from the US 'Paveway' series, being a 400-kg (882-lb) HE bomb provided with cruciform wings/fins on a rearward extension of the body and fitted with a seeker/control section on the nose; the laser seeker is based on the Ariel seeker of the AS.30L missile, and the target can be designed from the ground or by an aircraft using the ATLIS designator system
BGL 1000: version based on the French standard 1000-kg (2,205-lb) HE bomb, with a diameter of 0.50 m (19.7 in), a length of 4.00 m (13.12 ft) and a span of 1.60 m (5.25 ft)

Saab-Bofors RB 05A
(Sweden)
Type: air-to-surface tactical missile
Dimensions: diameter 0.30 m (11.8 in); length 3.60 m (11.81 ft); span 0.80 m (31.50 in)
Weights: total round 305 kg (672.4 lb); warhead not revealed
Propulsion: one Volvo Flygmotor VR-35 dual-thrust prepackaged liquid-propellant rocket delivering 2500-kg (5,511-lb) thrust in the boost phase and 500-kg (1,102-lb) thrust in the sustain phase
Performance: speed Mach 1+; range 9000 m (9,845 yards); CEP less than 10 m (33 ft)
Guidance: radio command

Variant
RB 05A: introduced in the early 1970s, the RB 05A is a fairly large but capable air-to-surface missile intended for land and sea targets; the missile is launched at an altitude between 20 and 50 m (65 and 165 ft) and then climbs to some 400 m (1,315 ft) as the operator guides it to the target, where the FRAG-HE warhead is detonated by a proximity fuse

AS-7 'Kerry'
(USSR)
Type: air-to-surface tactical missile
Dimensions: 0.30 m (11.8 in); length 3.50 m (11.48 ft); span not known
Weights: total round 1200 kg (2,646 lb); warhead 100 kg (220 lb) FRAG-HE
Propulsion: one solid-propellant rocket delivering ?-kg (?-lb) thrust
Performance: speed Mach 0.6; range 11100 m (12,140 yards); CEP not known
Guidance: radio command, or beam riding

Variant
AS-7: this was the USSR's first genuine air-to-surface missile even though it first appeared only in the late 1970s; earlier Soviet types had been modified from air-to-air missiles, but the AS-7 is a more capable weapon roughly comparable to the AGM-12 Bullpup

AS-9
(USSR)
Type: anti-radiation tactical missile
Dimensions: diameter 0.50 m (19.69 in); length 6.00 m (19.69 ft); span not known
Weights: not known
Propulsion: one liquid-propellant rocket delivering ?-kg (?-lb) thrust
Performance: speed Mach 0.8; range 80/90 km (49.7/55.9 miles)
Guidance: passive radar homing

Variants
AS-9: virtually nothing is known of this long-range anti-radar missile, somewhat unusually powered by a storable liquid-propellant rocket and designed for use on Soviet tactical aircraft
AS-11: the weapon originally thought to have borne this designation is now known as the AS-14, and the AS-11 may thus be a variant of the AS-9 with a different homing system
AS-12 'Kegler': virtually nothing is known of this weapon, but some reports suggest that it may be another AS-9 variant with a different homing system

AS-10 'Karen'
(USSR)
Type: air-to-surface tactical missile
Dimensions: diameter 0.30 m (11.81 in); length 3.00 m (9.84 ft)
Weights: not known
Propulsion: one solid-propellant rocket delivering ?-kg (?-lb) thrust
Performance: speed Mach 0.8; range 10000 m (10,935 yards)
Guidance: semi-active laser homing

Variant
AS-10: this laser-homing weapon forms an important part of the inventory for the Mikoyan-Gurevich MiG-27 and Sukhoi Su-17 attack fighters, and for the Sukhoi Su-24 interdictor

AS-11 (USSR): see AS-9

AS-12 'Kegler' (USSR): see AS-9

AS-14 'Kedge'
(USSR)
Type: air-to-surface tactical missile
Dimensions: not known
Weights: not known
Propulsion: solid-propellant rocket delivering ?-kg (?-lb) thrust
Performance: speed not known; range 40 km (25 miles)
Guidance: electro-optical homing

Variant
AS-14 'Kedge': virtually nothing is known of this tactical air-launched missile

BAe/Marconi ALARM
(UK)
Type: anti-radiation tactical missile
Dimensions: diameter 8.66 in (0.22 m); length 13.91 ft (4.24 m); span 28.33 in (0.72 m)
Weights: total round 385 lb (175 kg); warhead not revealed
Propulsion: one ROF Nuthatch solid-propellant two-stage rocket delivering ?-lb (?-kg) thrust
Performance: not revealed
Guidance: Marconi passive radar seeking

Variant
ALARM: based aerodynamically and structurally on the Sky Flash air-to-air missile, the ALARM (Air-Launched Anti-Radiation Missile) is due to enter service in the late 1980s as the primary weapon of British tactical aircraft operating against hostile radars; the weapon is designed to zoom to 40,000 ft (12190 m) after a low-level launch in the vicinity of any possible target, thereupon descending nose down under a small drogue parachute as the broad-band seeker searches for hostile emissions; once a target has been selected, the drogue is discarded and the motor re-ignited for a supersonic dive 'down the throat' of the hostile radar system

BAe/Matra Martel and ARMAT
(UK/France)
Type: (AJ.168) air-to-surface tactical missile or (AS.37) anti-radiation tactical missile
Dimensions: diameter 15.75 in (0.40 m); length (AJ.168) 12.70 ft (3.87 m) or (AS.37) 13.52 ft (4.12 ft); span 47.25 in (1.20 m)
Weights: total round (AJ.168) 1,213 lb (550 kg) or (AS.37) 1,168 lb (530 kg); warhead 330 lb (150 kg) proximity- or impact-fused FRAG-HE
Propulsion: one Hotchkiss-Brandt/SNPE Basile solid-propellant booster rocket delivering ?-kg (?-lb) thrust, and one Hotchkiss-Brandt/SNPE Cassandre solid-propellant sustainer rocket delivering ?-kg (?-lb) thrust
Performance: speed Mach 2; range 60 km (37.3 miles) from a high-altitude launch declining to 30 km (18.6 miles) from a low-altitude launch
Guidance: (AJ.168) Marconi TV command or (AS.37) EMD AD.37 passive radar seeker

Variants
AJ.168 Martel: introduced in the late 1960s, the British version of the Martel uses command guidance with a small TV camera in the nose of the missile for the accurate placement of the substantial warhead
AS.37 Martel: the French version of the Martel is designed for anti-radiation use with a broad-band passive seeker which is locked to the required frequency before the weapon is fired
ARMAT: developed by Matra on the basis of the Martel airframe, this is France's new anti-radiation missile, differing from the AS.37 that it is replacing by having a span of 1.20 m (3.94 ft), a weight of 545 kg (1,201 lb), a range of 80+ km (50+ miles) and a more modern and capable ESD passive radar seeker

Martin-Marietta AGM-12C Bullpup
(USA)
Type: air-to-surface tactical missile
Dimensions: diameter 18.00 in (0.457 m); length 13.58 ft (4.14 m); span 48.00 in (1.22 m)
Weights: total round 1,785 lb (810 kg); warhead 1,000-lb (454-kg) impact-fused FRAG-HE
Propulsion: one Thiokol LR62-RM-2/4 prepackaged liquid-propellant rocket delivering ?-lb (?-kg) thrust
Performance: speed Mach 1.75; range 17,600 yards (16095 m)
Guidance: radio command

Variant
AGM-12C Bullpup: this is the only version of this obsolescent air-to-surface missile still in service, and though limited in terms of its guidance system (requiring the aircraft to loiter in the vicinity of the target until the weapon impacts) it still packs a powerful punch in areas of reduced air-defences

Texas Instruments AGM-45A Shrike
(USA)
Type: anti-radiation tactical missile
Dimensions: diameter 8.00 in (0.203 m); length 10.00 ft (3.048 m); span 36.00 in (0.914 m)
Weights: total round 390 lb (176.9 kg); warhead 145 lb (65.8 kg) impact- and proximity-fused FRAG-HE
Propulsion: one Rocketdyne Mk 39 or Aerojet Mk 53 solid-propellant rocket delivering ?-lb (?-kg) thrust
Performance: speed Mach 2; range 28.9 miles (46.5 km); CEP adequate so long as the target continues to emit
Guidance: Texas Instruments passive radar seeking

Variant
AGM-45A Shrike: introduced in 1963, the Shrike was the USA's first tactical anti-radiation missile; the missile was produced in 10 blocks, and there are at least 13 different seeker units tailored to individual radar types; this has proved a tactical disadvantage (requiring specific tuning before take-off and thus preventin attacks on targets of opportunity), as does the seeker's lack of inbuilt memory (resulting in a ballistic trajectory if the emitter shuts down)
AGM-45B Shrike: improved AGM-45A with the Aerojet Mk 78 solid-propellant rocket

Hughes/Martin-Marietta AGM-62A Walleye I
(USA)
Type: electro-optically guided tactical glide bomb
Dimensions: diameter 12.50 in (0.318 m); length 11.33 ft (3.454 m); span 45.50 in (1.156 m)
Weights: total round 1,130 lb (512.6 kg); warhead 825 lb (374.2 kg) shaped-charge HE
Propulsion: none
Performance: speed high subsonic; range 1,975/32,400 yards (1805/29625 m) depending on launch altitude; CEP between 15 and 20 ft (4.6 and 6.1 m)
Guidance: optical imaging with nose-mounted TV

Variants
AGM-62A Walleye I: developed in the mid-1960s by the Naval Weapons Center, the Walleye I entered service in 1966 and was made by Hughes and Martin-Marietta; the type was designed for stand-off attacks from heights up to 35,000 ft (10670 m) against high-contrast buildings and bridges, whose optical images are easy to pick up and lock into the nose-mounted gyro-stabilized TV camera, allowing the launch aircraft to drop the weapon and depart the area as the missile automatically homes onto its target; the operator in the launch aircraft monitors the progress of the missile and can break into the automatic homing system to redirect the weapon if necessary
AGM-62A Walleye I ER/DL: this is the extended-range model of the late 1970s with slightly reduced length but fitted with data link equipment so that the weapon can be dropped in the general direction of the target and have its guidance system locked onto the clear image of the target (by the departing operator) only when the weapon has approached to moderately close range

Martin-Marietta AGM-62A Walleye II
(USA)
Type: electro-optically guided tactical glide bomb
Dimensions: diameter 18.00 in (0.457 m); length 13.25 ft (4.039 m); span 51.00 in (1.295 m)
Weights: total round 2,340 lb (1061.4 kg); warhead 1,900 lb (861.8 kg) shaped-charge HE
Propulsion: none
Performance: speed high subsonic; range 1,975/49,300 yards (1805/45080 m) depending on launch altitude; CEP between 15 and 20 ft (4.6 and 6.1 m)
Guidance: optical imaging with nost-mounted TV

Variants
AGM-62A Walleye II: developed to overcome the Walleye I's limitations against hard targets, the Walleye II was introduced in 1974 and has a considerably larger warhead (derived from that of the Mk 84 free-fall bomb) and a revised seeker with a smaller optical gate for increased accuracy; built by Martin-Marietta, the Walleye II requires a target with sharp visual contrasts for good lock-on before launch
AGM-62A Walleye II ER/DL: introduced in 1976, this Extended-Range/Data-Link model was developed so that greater range can be ensured; the missile is dropped from high altitude (so extending range to 65,500 yards/59895 m) and is locked onto the target late in the flight using the two-way data-link system to the launch aircraft, where the operator monitors the TV image in the weapon's guidance system

Hughes AGM-65 Maverick
(USA)
Type: multi-role air-to-surface tactical missile
Dimensions: diameter 12.00 in (0.305 m); length 8.17 ft (2.489 m); span 28.30 ft (0.719 m)
Weights: total round (AGM-65A/D) 463 lb (210 kg) or (AGM-65E/F) 633.6 lb (287.4 kg); warhead (AGM-65A/D) 125 lb (56.7 kg) shaped-charge HE or (AGM-65E/F) 300 lb (136.1 kg) penetrating FRAG-HE
Propulsion: one Thiokol SR109-TC-1 (TX-481) dual-thrust solid-propellant rocket delivering ?-lb (?-kg) thrust
Performance: speed subsonic; range 985/26,400 yards (900/24140 m); CEP (AGM-65A/D) 5 ft (1.5 m) or (AGM-65E/F) less than 5 ft (1.5 m)
Guidance: (AGM-65A and AGM-65B) TV imaging, (AGM-65C and AGM-65E) semi-active laser homing, or (AGM-65D and AGM-65F) infra-red imaging

Variants
AGM-65A Maverick: introduced in 1972 as a US Air Force weapon, the Maverick series is the smallest fully-guided air-to-surface missile in the US inventory, and one of the West's most important weapons of the type, largely because in all its versions it is a fire-and-forget type; the AGM-65A is the initial TV imaging version (often known as the **TV Maverick**), and suffers the tactical disadvantage that the low magnification of its nose-mounted camera forces the pilot of the launch aircraft to fly close to the target to secure lock on before missile launch and subsequent automatic attack
AGM-65B Maverick: improved TV imaging model

with double the image magnification of the AGM-65A to overcome the earlier version's tactical disadvantages; the type is often known as the **Scene-Magnification Maverick**

AGM-65C Maverick: highly capable development using laser homing compatible with the 'Pave Knife', 'Pave Penny', 'Pave Spike', 'Pave Tack' and several non-US designation systems; the weapon was known as the **Laser Maverick**, but was cancelled before reaching full service with the US Marine Corps in the close support role against targets generally designated by ground-based laser

AGM-65D Maverick: imaging infra-red version for the USAF with twice the lock-on range of the AGM-65A/B types, and capable of operation in adverse weather conditions and at night; the type is known as the **IR Maverick** and is designed for use in conjunction with FLIR (Forward-Looking Infra-Red) systems or with the LANTIRN (Low-Altitude Navigation and Targeting Infra-Red for Night) pod carried by Fairchild Republic A-10 attack aircraft and General Dynamics F-16 fighters in the European theatre, or indeed with the APR-38 radar-warning system of aircraft such as the McDonnell Douglas F-4 Phantom II 'Wild Weasel II' defence-suppression fighter

AGM-65E Maverick: this model was introduced in 1985 as successor to the cancelled AGM-65C with a less costly laser seeker but a more powerful 250-lb (113-kg) warhead with impact or delay fuses

AGM-65F Maverick: this is a version for the US Navy combining the airframe and guidance of the AGM-65D with the warhead and fuse of the AGM-65D plus software changes to optimize the weapon's capabilities against warship targets

AGM-65G Maverick: this is the USAF equivalent of the AGM-65F, and is designed for the destruction of 'hard' tactical targets

Boeing AGM-69A SRAM
(USA)
Type: air-to-surface strategic/tactical missile
Dimensions: diameter 17.50 in (0.445 m); length (internal carriage) 14.00 ft (4.267 m) or (external carriage) 15.83 ft (4.826 m); span 30.00 in (0.762 m)
Weights: total round 2,240 lb (1016 kg); warhead 170-kiloton W69 nuclear
Propulsion: one Lockheed SR75-LP-1 two-stage solid-propellant rocket delivering ?-lb (?-kg) thrust
Performance: speed Mach 3.5; range 100/137.5 miles (160.9/221.3 km) from a high-altitude launch declining to 35/50 miles (56.3/80.5 km) from a low-altitude launch; CEP 500 yards (457 m)
Guidance: Singer-Kearfott KT-76 inertial plus terrain-avoidance radar altimeter

Variant
AGM-69A SRAM: introduced in 1972 as part of the primary armament for Boeing B-52 heavy bombers (20 SRAMs or eight SRAMs and four free-fall nuclear weapons) and General Dynamics FB-111 medium bombers (six SRAMs or two SRAMs and four free-fall nuclear weapons), the Short-Range Attack Missile began development in 1964 as replacement for the AGM-28 Hound Dog; the role envisaged for the new weapon was stand-off attack against primary defensive installations; the type is now tasked with defence suppression tasks in association with the launch of AGM-86B air-launched cruise missiles and the delivery of gravity nuclear weapons; the SRAM has four basic attack profiles: semi-ballistic from launch to impact, altimeter-supervised terrain-following, ballistic pop-up from screening features followed by terrain-following, and combined terrain-following and inertial; any of these profiles can include preprogrammed 180-degree course changes, and the warhead can be set for ground- or air-burst; it is likely that surviving missiles will be re-engined with a new Thiokol liquid-propellant rocket

AGM-69B SRAM: upgraded AGM-69A

SRAM II: Boeing is developing a new SRAM missile only about two-thirds the size of the current AGM-69A and using an airframe of largely composite construction as well as a guidance system based on the latest VHSIC (Very High Speed Integrated Circuit) technology; the B-1B will be able to carry 20 SRAM IIs internally when the type enters service in 1992

General Dynamics AGM-78B Standard ARM
(USA)
Type: anti-radiation tactical missile
Dimensions: diameter 13.5 in (0.343 m); length 15.00 ft (4.572 m); span 43.00 in (1.092 m)
Weights: total round 1,356 lb (615.1 kg); warhead 214.7 lb (97.4 kg) FRAG-HE
Propulsion: one Aerojet Mk 27 Mod 4 dual-thrust solid-propellant rocket delivering ?-lb (?-kg) thrust
Performance: speed Mach 2.5; range 70+ miles (112.6 + km); CEP good even if the target ceases emission
Guidance: Maxson Electronics passive radar seeking

Variants
AGM-78A Standard ARM: derived from the RIM-66 Standard surface-to-air missile, the US Navy's AGM-78A was introduced in 1968 as a longer-range complement to the AGM-45A Shrike, and uses the same no-memory seeker with its attendant tactical limitations

AGM-78B Standard ARM: improved model with a gimballed wide-band seeker plus memory; this requires no pre-launch tuning (enabling targets of opportunity to be engaged) and keeps the missile on its committed course even if the emitter shuts down

AGM-78C Standard ARM: version of the AGM-78B for the US Air Force

AGM-78D Standard ARM: final version (with the

AGM-78D2 subvariant) marked by further-improved seeker capabilities

Texas Instruments AGM-88A HARM
(USA)
Type: anti-radiation tactical missile
Dimensions: diameter 10.00 in (0.254 m); length 13.68 ft (4.171 m); span 44.00 in (1.118 m)
Weights: total round 796 lb (361.1 kg); warhead 145 lb (65.8 kg) FRAG-HE
Propulsion: one Thiokol/Hercules YSR113-TC-1 solid-propellant rocket delivering ?-lb (?-kg) thrust
Performance: speed Mach 3+; range 46+ miles (74+ km); CEP very good
Guidance: Texas Instruments passive radar seeking

Variant
AGM-88A HARM: the High-speed Anti-Radiation Missile began to enter US service in 1982 as successor to the tactically-limited AGM-45 Shrike and the costly (also weighty) AGM-78 Standard ARM; the type offers good passive radiation homing (with detection from the launch aircraft's Itek ALR-45 or McDonnell Douglas APR-38 radar-warning receivers) and high speed so that the hostile emitter has minimum close-down time; the HARM can be used in any of three modes: self-protection mode when the launch aircraft's radar-warning receiver detects a hostile emitter and programmes the missile before launch, pre-briefed mode with the missile fired blind towards a possible target with the seeker searching in flight and commanding a self-destruct if no target is discovered, and target-of-opportunity mode when the seeker of the unfired missile detects and locks onto an emitter

Rockwell AGM-114A Hellfire
(USA)
Type: air-launched anti-tank tactical missile
Dimensions: diameter 7.00 in (0.178 m); length 5.84 ft (1.78 m); span 13.00 in (0.33 m)
Weights: total round 95 lb (43.1 kg); warhead 20 lb (9.1 kg) HEAT
Propulsion: one Thiokol TX-657 solid-propellant rocket delivering ?-lb (?-kg) thrust
Performance: speed high subsonic; range 6,500 yards (5945 m)
Guidance: Rockwell semi-active laser homing

Variant
AGM-114A Hellfire: this is an impressive weapon opening up the possibility of fire-and-forget helicopter attacks against armoured formations, the missile having proved itself able to home on the reflections of tanks laser-illuminated by aerial or ground-based designators; the target need only be illuminated once the missile has been launched, thereby reducing the time available for the target to undertake defensive measures and also reducing the launch platform's vulnerability to countermeasures
AGM-114B Hellfire: version of the AGM-114A for the US Navy and US Marine Corps with a different motor and homing guidance provided by three different seeker modules
AGM-114C Hellfire: US Army version of the AGM-114B without the safe arm feature
RBS 17: due to enter service by 1988, this is á joint US and Swedish development of the Hellfire for coastal defence purposes with the Swedish army; the missile has been fitted with a Bofors delayed-action blast/fragmentation warhead suitable for attacks on landing craft and assault hovercraft, and with revisions to the seeker head weighs a total of 48 kg (106 lb) and has a length of 1.625 m (5.44 ft); the missile is transported in a special container which has a loaded weight of 71 kg (156.5 lb), and is fired from a light tripod launcher with the aid of a laser designator

Motorola AGM-122A Sidearm
(USA)
Type: air-to-surface anti-radiation tactical missile
Dimensions: diameter 5 in (127 mm); length 9.42 ft (2.87 m); span 24.8 in (0.63 m)
Weights: total round 185 lb (83.9 kg); warhead not known
Propulsion: one Rocketdyne Mk 36 Mod 2 solid-propellant rocket delivering ?-lb (?-kg) thrust
Performance: speed Mach 2; range 4,000 yards (3660 m)
Guidance: Motorola passive radar seeking

Variant
AGM-122A Sidearm: currently in flight test, the Sidearm is a low-cost self-protection anti-radiation weapon for helicopters and light tactical aircraft produced by the conversion of in-storage semi-active radar-homing AIM-9C Sidewinder air-to-air missiles

Emerson Electric AGM-123A Skipper 2
(USA)
Type: air-to-surface rocket-boosted air-to-surface tactical glide bomb
Dimensions: diameter not revealed; length 14.1 ft (4.3 m); span not revealed
Weights: total round 1,283 lb (582 kg); warhead not revealed
Propulsion: one Rocketdyne Mk 78 solid-propellant rocket delivering ?-lb (?-kg) thrust
Performance: speed not revealed; range 34 miles (55 km)
Guidance: semi-active laser homing

Variant
AGM-123A Skipper 2: this simple yet highly effective tactical weapons uses off-the-shelf components (the Mk 83 1,000-lb/454-kg 'iron bomb', the rocket motor

of the AGM-45 Shrike and the guidance-control system of the 'Paveway II' series guided bomb) to produce a stand-off weapon of great accuracy for US Navy aircraft

Rockwell HOBOS/KMU-353A/B
(USA)
Type: electro-optically guided Mk 84 2,000-lb (907-kg) tactical glide bomb
Weight: total round 2,100 lb (952.6 kg)
Performance: range between 1,650 and 26,750 yards (1510 and 24460 m) depending on launch altitude; CEP 20 ft (6 m)

Variants
HOBOS: the HOming BOmb System was developed by Rockwell during the 1960s in parallel with the Paveway I series, though this type uses electro-optical rather than laser guidance, at first with a TV image-contract tracker and then with more advanced TV and infra-red seekers for enhanced nocturnal and adverse-weather capability; the three kits are the **KMU-353A/B** and **KMU-390/B** image-contrast systems, and the **KMU-359/B** infra-red system, each comprising a nose-mounted guidance section and (connected by strakes along the body of the bomb) a tail-mounted control section with fins; the packages can be added to the 2,000-lb (907-kg) Mk 84 GP bomb (KMU-353A/B and KMU-359/B) and the 3,000-lb (1361-kg) M118E1 demolition bomb (KMU-390/B)
GBU-15: this is an improved version of the HOBOS concept, developed by Rockwell on the basis of the Modular Glide Weapon System as a means of extending yet farther the range of the HOBOS concept; the core weapon is the 2,000-lb (907-kg) Mk 84 bomb or SUU-54 dispenser bomb (carrying 1,800 BLU-63 and/or BLU-86 anti-tank bomblets), to which is fitted a guidance package (with either a TV or an imaging infra-red seeker), a control package, a data-link package and a cruciform-wing aerodynamic package; the guidance package of the **GBU-15(V)1/B** initial version is of the TV type using the AXQ–14 data-link data-link system, while that of the later **GBU-15(V)2/B** model uses the imaging infra-red guidance of the AGM-65D Maverick; this weapon weighs 2,450 lb (1111 kg), and has ranges between 1,650 yards (1,510 m) and 51 miles (82 km) depending on launch altitude, the CEP being in the order of 20 ft (6 m)
AGM-130A: version of the GBU-15 weighing 2,917 lb (1323 kg) with a Mk 84 bomb as payload and powered by Hercules solid-propellant rocket for a range of 15 miles (24 km) from a low-altitude launch; the type is controlled by the same type of TV imaging infra-red guidance used in the GBU-15 series, and is also being developed to carry boosted kinetic-energy penetrators as a submunition load; the role of the missile is low-altitude stand-off attack against heavily-defended targets
AGM-130B: comparable to the AGM-130A but carrying an SUU-54 airfield-attack submunitions-dispenser warhead (15 boosted kinetic-energy penetrators and 75 British-designed HB876 area-denial submunitions) and having an overall weight of 2,560 lb (1161 kg)
AGM-130C: comparable to the AGM-130A but carrying the BLU-109/B penetrating warhead

Texas Instruments Paveway I/KMU-351A/B
(USA)
Type: laser-guided 2,000-lb (907-kg) Mk 84 tactical glide bomb
Weight: total round 2,100 lb (952.6 kg)
Performance: range between 1,650 and 20,000 yards (1510 and 18290 m) depending on launch altitude; CEP 27 ft (8 m)

Variants
Paveway I: introduced in the late 1960s, the Paveway I designation covers a series of add-on laser-homing kits (marked-target seekers and associated control surfaces developed by Texas Instruments) that can be added to standard low- or high-drag bombs for maximum accuracy when used in conjunction with an air- or ground-based laser designator; the kits are the **KMU-342/B** (750-lb/340-kg M117 demolition bomb), the **KMU-351A/B** (2,000-lb/907-kg Mk 84 high/low-drag GP bomb), the **KMU-370B/B** (3,000-lb/1361-kg M118E1 demolition bomb), the **KMU-388A/B** (500-lb/227-kg Mk 82 high/low-drag GP bomb), the **KMU-420/B** (500-lb/227-lb Rockeye Mk 20 Mod 2 cluster bomb with 358 1.1-lb/0.5-kg anti-tank bomblets) and the **KMU-421B** (2,000-lb/907-kg SUU-54B or Pave Storm I cluster bomb with 1,800 1.1-lb/0.5-kg anti-tank bomblets); the advantage of the system is that the launch aircraft needs no modification and can leave the target area (unless it is the laser-designator aircraft); the system works at night and in poor visibility, and the minimum cloud base for successful operation is 2,500 ft (760 m); the airborne designators most commonly associated with the series are three podded systems, the AVQ-26 'Pave Tack', AVQ-23 'Pave Spike' and LANTIRN (Low-Altitude Navigation and Targeting Infra-Red for Night)
Paveway II: entering service in 1978, this is an improved series based on the same basic concept but with a folding wing aerofoil group added at the tail (for extra manoeuvrability and additional lateral range) and an updated electronics package; the weapons within this important basic designation are the **GBU-10E/B** and **GBU-10F/B** (both based on the 2,000-lb/907-kg Mk 84 high/low-drag GP bomb), the **GBU-12D/B** and **GBU-12F/B** (both based on the 500-lb/227-kg Mk 82 high/low-drag GP bomb), the **GBU-16B/B** and **GBU-16C/B** (both based on the 1,000-lb/

454-kg Mk 83 low-high-drag GP bomb) and the **Mk 13/18UK** (British 1,000-lb/454-kg Mk 13/18 GP bomb)
Paveway III: entering service in 1987, this is an improved version of the Paveway II for release at low level and thus fitted with high-lift folding wings; the type has the service designation **GBU-24** and also features an improved seeker and more advanced microprocessor technology
AGM-123A Skipper: powered version combining the 1,000-lb (454-kg) Mk 83 bomb with a Paveway II infra-red seeker and the propulsion package of the AGM-45 Shrike anti-radiation missile; the role of the missile is low-altitude stand-off attack against heavily-defended targets

ANTI-SHIP MISSILES

CPMIEC C-601
(China)
Type: air-launched medium-range anti-ship missile
Dimensions: diameter not revealed; length 7.38 m (24.21 ft); span 2.40 m (7.87 ft)
Weights: total round 2440 kg (5,379 lb); warhead not revealed
Propulsion: one solid-propellant booster rocket delivering ?-kg (?-lb) thrust, and one sustainer turbojet delivering ?-kg (?-lb) thrust
Performance: speed 1100 km/h (684 mph); range not known
Guidance: monopulse active radar

Variant
C-601: little is known of this useful anti-ship missile, which began to enter service in 1982 as the primary armament of the H-6D anti-shipping version of the Chinese-built 'Badger' bomber that can carry two of the type on underwing hardpoints; the C-601 is clearly derived from the Soviet SS-N-2 'Styx' weapon, and in the US standard terminology is known (together with its derivatives) as the **CSS-N-1**
FL-1: this is the ship-launched version of the C-601, and after launch cruises at an altitude of 30 m (100 ft) for an attack with its 513-kg (1,131-lb) HE warhead; FL stands for Fei Lung, or flying dragon
FL-7: much-developed version of the FL-1 intended for use against medium-sized naval targets and fired singly or in salvoes; the type has a length of 6.6 m (21.65 ft) and a weight of 1800 kg (3,968 lb), and uses a combination of jettisonable solid-propellant booster under the fuselage and an internal liquid-propellant rocket for Mach 1.4 flight out to a range of 32 km (20 miles) at a cruise height between 50 and 100 m (150 and 330 ft)
HY-2: this is the coast-launched version of the C-601 launched from a wheeled single-rail launcher or from a tracked chassis; this basic weapon comes in three forms as the HY-2 with active radar homing and a cruise altitude of 100 m (330 ft), the HY-2A with passive IR homing and a cruise altitude of 30 m (100 ft), and the HY-2G with active radar homing and a radar altimeter for constant cruising height to a range of 95 km (59 miles); the type has been exported in a variant designated **'Silkworm'** by NATO
HY-4: this is a longer-range version of the HY-2 with the fuselage lengthened to allow the insertion of a turbojet engine and its fuel together with a ventral air inlet; there are probably air- and ship-launched versions of this model
SY-1: simplified version of the FL-1 with a mid-course cruise altitude of 100 to 300 m (330 and 985 ft) and a less advanced homing system

CPMIEC C-801
(China)
Type: ship-launched medium-range anti-ship missile
Dimensions: diameter not revealed; length 5.2 m (17.06 ft); span 1.0 m (3.28 ft)
Weights: total round about 1000 kg (2,205 lb); warhead ?-kg (?-lb) semi-armour piercing HE
Propulsion: one solid-propellant booster rocket delivering ?-kg (?-lb) thrust, and one solid-propellant sustainer rocket delivering ?-kg (?-lb) thrust
Performance: speed Mach 1.2
Guidance: active radar

Variant
C-801: introduced to Chinese service in 1983, the C-801 is an advanced anti-ship missile of which little is known with certainty; the type is said to be compatible with the fire-control system used with the Chinese HY-2 and FL-1 derivatives of the Soviet SS-N-2 'Styx' missile; this is certainly an advanced anti-ship cruise missile, and can also be launched from aircraft, shore batteries and surfaced submarines; the type cruises at a preset altitude of 20 or 30 m (66 or 98 ft); the weapon is also known as the **YJ-6**, YJ standing for Ying Ji, or eagle strike

Aérospatiale AS.12
(France)
Type: helicopter-launched short-range anti-ship tactical missile
Dimensions: diameter 0.21 m (8.25 in); length 1.87 m (6.14 ft); span 0.65 m (2.13 ft)
Weights: total round 76 kg (168 lb); warhead 28.4 kg (62.6 lb) HE
Propulsion: one solid-propellant rocket delivering ?-kg (?-lb) thrust
Performance: speed low subsonic; range 8000 m (8,750 yards)
Guidance: wire-guided command to line of sight

Variants
AS.12: introduced in 1960 as the air-launched version of the SS.12M surface-to-surface tactical missile, the AS.12 has been made in very large numbers for more than 30 countries, but is now obsolescent because of its small size (suiting it for effective use against only submarines and fast attack/patrol craft), low speed and poor range
SS.12M: naval version of the surface-to-surface SS.12, associated with a gyro-stabilized sight suiting the type to installation as the armament of attack craft and other small naval platforms

Aérospatiale AS.15TT
(France)
Type: air-launched short-range anti-ship tactical missile
Dimensions: diameter 0.185 m (7.3 in); length 2.16 m (7.09 ft); span 0.564 m (1.85 ft)
Weights: total round 96 kg (212 lb); warhead 29.7 kg (65.5 lb) HE
Propulsion: one SNPE/Aérospatiale solid-propellant rocket delivering ?-kg (?-lb) thrust
Performance: speed Mach 0.82; range 16000 m (17,500 yards)
Guidance: (height) Thomson-CSF preprogrammed descending cruise to a height determined by radar altimeter and (bearing) radar command

Variant
AS.15TT: entering service in the mid-1980s, the AS.15TT was financed by Saudi Arabia and is designed as successor to the AS.12; the missile is controlled in bearing by the launch helicopter's Thomson-CSF Agrion radar, and in height by a programme that lets the missile down to wavetop height for the seaskimming approach and attack
MM.15: ship-launched variant designed to arm fast attack craft

Aérospatiale MM.38 Exocet
(France)
Type: ship-launched anti-ship tactical missile
Dimensions: diameter 0.45 m (13.75 in); length 5.21 m (17.09 ft); span 1.004 m (3.29 ft)
Weights: total round 750 kg (1,653 lb); warhead 165 kg (364 lb) proximity and impact delay-fused GP1 FRAG-HE
Propulsion: one SEP Epervier solid-propellant booster rocket delivering ?-kg (?-lb) thrust, and one SEP Eole V solid-propellant sustainer rocket delivering ?-kg (?-lb) thrust
Performance: speed Mach 0.93; range 4.5/45 km (2.8/28 miles)
Guidance: inertial plus TRT RAM.01 radar altimeter, and EMD ADAC active radar terminal homing

Variants
MM.38 Exocet: the most widely produced anti-ship missile of Western origins, the Exocet was designed in the late 1960s and began to enter service in 1974; since that date the missile has seen extensive operational use (notably in the Falklands war of 1982 and in the current war between Iraq and Iran); the missile is launched towards the target on data provided by the launch ship's sensors and fire-control system, and cruises at low altitude until some 10 km (6.2 miles) from the anticipated target position, when the monopulse active seeker head is turned on, the target acquired and the terminal phase initiated at the one of three heights preselected at launch with sea state in mind; residual fuel adds to the effects of the warhead detonation, which has in itself proved somewhat troublesome and unreliable
AM.38 Exocet: limited-production helicopter-launched version of the MM.38 with a 1-second ignition delay to avoid damage to the launch platform
AM.39 Exocet: full-production aircraft-launched version of the MM.38 with a revised propulsion arrangement (SNPE Dondor booster and SNPE Hélios sustainer) in a shorter body (4.69 m/15.39 ft), reducing weight to 652 kg (1,437 lb) but increasing range to 50/70 km (31/43.5 miles) depending on launch altitude
SM.39 Exocet: full-production submarine-launched version of the MM.38; this version is launched from a submarine torpedo tube in a 5.80-m (19.03-ft) powered and guided VSM capsule weighing 1350 kg (2,976 lb) complete with missile; the capsule is driven to the surface at a shallow angle by its own solid-propellant booster, and on breaking the surface blows off its nose cap before initiating a gas generator to eject the missile for an otherwise conventional attack profile
MM.40 Exocet: much improved version of the ship-launched missile; the rocket engines have been improved, and the use of a lightweight container/launcher allows the type (despite the greater missile weight) to be carried by smaller vessels, or alternatively in larger numbers by larger launch platforms;

the dimensions of the MM.40 version include a length of 5.78 M (18.96 ft) and a span of 1.135 m (3.72 ft), and the total-round weight is 850 kg (1,874 lb) for a range of 65+ km (40.4+ miles)

Aérospatiale/MBB ANS
(France/West Germany)
Type: air-launched anti-ship tactical missile
Dimensions: diameter 0.35 m (13.78 in); length 5.70 m (18.70 ft); span not revealed
Weights: total round 950 kg (2,094 lb); warhead 180 kg (397 lb) HE
Propulsion: one MBB integral rocket-ramjet delivering ?-kg (?-lb) thrust
Performance: speed Mach 2+; range 6/100+ km (3.7/62.1+ miles)
Guidance: strapdown inertial and ESD Super ADAC active radar terminal homing

Variant
ANS: the Anti-Navire Supersonique (supersonic anitship) missile is currently under development by Aerospatiale and MBB for service in the early 1990s as successor to the Exocet and Kormoran air-launched missiles; the weapon features advanced propulsion for very high attack speed (so reducing the time available to the target for the implementation of countermeasures), and the terminal guidance will probably incorporate a home-on-jam mode

MBB Kormoran
(West Germany)
Type: air-launched medium-range anti-ship tactical missile
Dimensions: diameter 0.344 m (13.5 in); length 4.40 m (14.44 ft); span 1.00 m (39.37 in)
Weights: total round 600 kg (1,323 lb); warhead 165 kg (364 lb) delay-fused radial HE
Propulsion: two SNPE Pradès solid-propellant booster rockets each delivering ?-kg (?-lb) thrust, and one SNPE Eole IV solid-propellant sustainer rocket delivering ?-kg (?-lb) thrust
Performance: speed Mach 0.95; range 37 km (23 miles)
Guidance: SFENA-Bodenseewerk strap-down inertial plus TRT radar altimeter, and Thomson-CSF RE576 active radar terminal homing

Variants
Kormoran: developed from the Nord AS.34 missile projected in France, the Kormoran entered West German naval service in 1977; the missile cruises at a height of 30 m (100 ft) to the approximate location of the target, then descends to wave-top height for the preset passive or active radar attack
Kormoran 2: this improved weapon has enhanced capability against electronic countermeasures, greater range and a 220-kg (485-lb) warhead

IMI Gabriel Mks I and II
(Israel)
Type: ship-launched medium-range anti-ship tactical missile
Dimensions: diameter 0.34 m (13.4 in); length (Mk I) 3.35 m (11.00 ft) or (Mk II) 3.42 m (11.22 ft); span (Mk I) 1.38 m (4.53 ft) or (Mk II) 1.34 m (4.40 ft)
Weights: total round (Mk I) 431 kg (950 lb) or (Mk II) 522 kg (1,151 lb); warhead 180 kg (397 lb) impact and proximity delay-fused HE
Propulsion: one solid-propellant booster rocket delivering ?-kg (?-lb) thrust, and one solid-propellant sustainer rocket delivering ?-kg (?-lb) thrust
Performance: speed (Mk I) Mach 0.65 or (Mk II) Mach 0.7; range (Mk I) 21 km (13 miles) or (Mk II) 36 km (22.4 miles)
Guidance: inertial plus radar altimeter, and optical or semi-active radar terminal homing

Variants
Gabriel Mk I: developed in the 1960s, this somewhat short-ranged anti-ship missile nevertheless has good capabilities (especially in terms of its resistance to electronic countermeasures), the cruise and approach phases of its flight being akin to those of the Gabriel Mk III, though the sea-skimming attack is made under optical or semi-active radar control
Gabriel Mk II: extended-range version of the Mk I with a longer body and more propellant
Hsiung Feng: Taiwanese version of the Gabriel incorporating a number of local modifications and components
Skorpioen: South African version of the Gabriel incorporating local components

IMI Gabriel Mk III
(Israel)
Type: ship-launched medium-range anti-ship tactical missile
Dimensions: diameter 0.34 m (13.4 in); length 3.81 m (12.50 ft); span 1.34 m (4.40 ft)
Weights: total round 558 kg (1,230 lb); warhead 150 kg (331 lb) impact and proximity delayed-fuse HE
Propulsion: one solid-propellant rocket delivering 3600-kg (7,937-lb) thrust
Performance: speed Mach 0.73; range 60+ km (37.3+ miles)
Guidance: inertial plus radar altimeter, and active radar terminal homing

Variants
Gabriel Mk III: derived from the earlier Gabriel antiship missiles, the ship-launched Gabriel Mk III introduces a frequency-agile active radar seeker, though the optical and semi-active radar homing systems of the Gabriel Mks I and II can also be used; the weapon can thus have three guidance modes, namely fire-and-forget, fire and update via a data-link from a targeting

helicopter or similar, and fire and command using the launch vessel's radar for better targeting data; the missile cruises at 100 m (330 ft) and then descends to 20 mm (66 ft) for the approach to the target, the actual attack being made at a present height of 1.5, 2.5 or 5 m (4.9, 8.25 or 13.1 ft) depending on the sea state

Gabriel Mk IIIA/S: air-launched 40-km (25-mile) range version developed for use by Israeli attack aircraft; this model weighs 558 kg (1,230 lb), its dimensions include a length of 3.84 m (12.60 ft) and a span of 1.10 m (3.61 ft), and transonic speed; the three operating modes are radar-set range-and-bearing launch, manually-set range-and-bearing launch, and bearing-only launch

Gabriel Mk IIIA/S ER: extended-range version of the Gabriel Mk IIIA/S with a longer sustainer rocket to provide a range of 60 km (37.3 miles); both the Gabriel Mk IIIA/S and Mk IIIA/S ER have the same guidance options as the basic Gabriel Mk III as detailed for the Gabriel Mk IIIA/S (see above); this model has a weight of 600 kg (1,323 lb)

Gabriel Mk IV: updated version with a length of 4.7 m (15.42 ft) and powered by a small turbojet for a range of 200 km (124 miles); there will doubtless be provision for mid-course update of the inertial navigation system of this variant, whose development is anticipated for the later 1980s

OTO Melara/Matra Otomat
(Italy/France)
Type: coast- or ship-launched anti-ship tactical missile
Dimensions: diameter 0.46 m (18.1 in); length 4.82 m (15.81 ft); span 1.19 m (3.90 ft)
Weights: total round 770 kg (1,698 lb); warhead 210 kg (463 lb) delay-fused SAP-HE
Propulsion: two jettisonable Hotchkiss-Brandt/SNPE solid-propellant booster rockets each delivering 3500-kg (7,717-lb) thrust, and one Turboméca TR 281 Arbizon sustainer turbojet delivering 400-kg (882-lb) thrust
Performance: speed Mach 0.9; range 6/60+ km (3.7/37.3+ miles) when fired as a singleton, or 6/80+ km (3.7/49.7+ miles) when fired in salvo
Guidance: autopilot plus radar altimeter, and (Otomat) Thomson-CSF 'Col Vert' or (Otomat 2) SMA active radar terminal homing

Variants
Otomat: developed for shipboard or coastal operation from a self-contained container-launcher, the Otomat is a capable air-breathing anti-ship missile; the weapon cruises at 250 m (820 ft) before descending to 20 m (66 ft) in the closing stages of the attack, when the active radar system is turned on; the Thomson-CSF 'Col Vert' system causes the missile to climb to 175 m (575 ft) and the dive on the vulnerable upper surfaces of the target

Otomat 2 Teseo: version of the basic Otomat fitted with SMA active radar terminal homing and the TG-2 command guidance system so that a mid-course update can be fed into the missile guidance package from a helicopter or other source; this allows the missile to operate effectively to a maximum-fuel range of 180+ km (112 miles) before delivering a sea-skimmer attack

Otomach 2: supersonic version projected for development in the late 1980s with a version of the Alfa-Romeo AR318 turbojet, a range of 100+ km (62.1+ miles) and a 100-kg (220-lb) warhead; the type will also have a more advanced seeker

Sistel Sea Killer
(Italy)
Type: ship- and helicopter-launched anti-ship tactical missile
Dimensions: diameter 0.206 m (8.11 in); length (Sea Killer Mk 1) 3.50 m (11.48 ft) or (Sea Killer Mk 2) 4.70 m (14.42 ft); span 0.99 m (38.98 in)
Weights: total round (Sea Killer Mk 1) 170 kg (374.8 lb) or (Sea Killer Mk 2) 300 kg (661 lb); warhead (Sea Killer Mk 1) 35 kg (77.2 lb) HE or (Sea Killer Mk 2) 70 kg (154 lb) proximity- and impact-fused SAP-HE
Propulsion: (Sea Killer Mk 1) one solid-propellant rocket delivering ?-kg (?-lb) thrust, or (Sea Killer Mk 2) one SEP solid-propellant booster rocket delivering 4000-kg (8,818-lb) thrust and one SEP solid-propellant sustainer rocket delivering 100-kg (220-lb) thrust
Performance: speed (Sea Killer Mk 2) 1080 km/h (671 mph); range (Sea Killer Mk 1) 10000 m (10,935 yards) or (Sea Killer Mk 2) 25 km (15.5 miles)
Guidance: (Sea Killer Mks 1 and 2) Sistel radar beam-riding plus radar altimeter or (in heavy ECM conditions) radio command

Variants
Sea Killer Mk 1: a lightweight anti-ship weapon designed for use from fast attack craft, the Sea Killer Mk 1 uses the Contraves Sea Hunter Mk 2 radar for guidance and is a sea-skimming missile now no longer in service in the surface-to-surface role

Sea Killer Mk 2: much upgraded tandem-motor version with longer range and a more capable warhead, the Sea Killer Mk 2 is associated with the Sea Hunter Mk 4 radar of larger surface vessels; the missile is currently being evaluated with a seeker head derived from that of the considerably larger Otomat anti-ship missile

Mariner: designation of a shipboard weapon system using the Sea Killer Mk 2, and essentially the helicopter-carried Marte system revised for shipboard application

Marte: designation of the helicopter weapon system using the Sea Killer missile and SMA APQ-706 radar

Mitsubishi Type 80 (ASM-1)
(Japan)
Type: air-launched anti-ship tactical missile
Dimensions: diameter 0.335 m (13.19 in); length 4.00 m (13.12 ft); span 1.20 m (47.25 in)
Weights: total round 610 kg (1,345 lb); warhead 250 kg (551 lb) HE
Propulsion: one Nissan Motors solid-propellant rocket delivering ?-kg (?-lb) thrust
Performance: speed 1100 km/h (684 mph); range 80 km (49.7 miles)
Guidance: Japan Aviation Electronics strapdown inertial plus TRT radar altimeter, and Mitsubishi Electric active radar terminal homing

Variant
Type 80: introduced in 1981 as the primary anti-ship armament of the Mitsubishi F-1 attack fighter, the **ASM-1** is a simple yet effective anti-ship missile of the fire-and-forget type, and is notable for its comparatively high speed

Kongsberg Penguin Mk III
(Norway)
Type: air- and ship-launched medium-range anti-ship tactical missile
Dimensions: diameter 0.28 m (11.02 in); length 3.17 m (10.40 ft); span 1.00 m (3.28 ft)
Weights: total round 360 kg (794 lb); warhead 121 kg (267 lb) impact delay-fused SAP-HE
Propulsion: one Raufoss Ammunisjons solid-propellant rocket delivering ?-kg (?-lb) thrust
Performance: speed Mach 0.9; range 40 km (25 miles) from a surface launch and 60 km (37.3 miles) from an air launch
Guidance: Kongsberg Vapenfabrikk inertial and Kongsberg Vapenfabrikk infra-red terminal homing

Variants
Penguin Mk I: the Western world's first anti-ship missile, the Penguin Mk I was conceived in the early 1960s and entered service in 1972 as part of Norway's defence against maritime invasion, and as such was optimized for good performance in the country's peculiar coastal waters after launch from fast attack craft; the result is a missile with infra-red terminal homing, treated as a round of ammunition and launched on information supplied by the launch platform's sensors and fire-control system; this ship-launched version has a two-stage rocket and height control by a pulsed-laser altimeter; the weight is 340 kg (750 lb), the dimensions include a length of 2.96 m (9.71 ft) and a span of 1.40 m (459 ft), and the performance figures include a speed of Mach 0.8 and a range of 20 km (12.4 miles); surviving missiles are being upgraded to **Penguin Mk I Mod 7** standard with the seeker of the Penguin Mk II Mod 3
Penguin Mk II: improved Mk I with range boosted to 30 km (18.6 miles) and a weight of 340 kg (750 lb) in the ship-launched **Penguin Mk II Mod 3** (also used by Sweden with the designation **RBS 12**), which has a length of 2.95 m (9.68 ft) and a span of 1.40 m (4.59 ft); the type entered service in the early 1980s, and surviving missiles are being upgraded to **Penguin Mk II Mod 5** standard with enhanced seeker performance; the basic type has since been developed for helicopter launch as the folding-wing **Penguin Mk II Mod 7** (US designation **AGM-119B**) with a weight of 385 kg (849 lb) and a length of 3.02 m (9.91 ft); the variant incorporates a number of Penguin Mk III improvements (especially in the seeker and signal processor), and has a new two-stage motor and fully digital electronics, unlike the earlier Mk II models which have analog electronics
Penguin Mk III: air-launched development (also capable of ship launch) with a longer body, shorter-span wings, a single-stage rocket and a radar altimeter; the type has the US designation **AGM-119A**, and is a highly capable weapon programmed to fly a circuitous approach to the target via a waypoint; as with the earlier versions of the missile, the use of infra-red terminal homing gives the target virtually no warning of the missile's imminent arrival, so reducing the time available for countermeasures

Skorpioen (South Africa): see IMI Gabriel Mks I and II

Saab RB 04E
(Sweden)
Type: air-launched anti-ship tactical missile
Dimensions: diameter 0.50 m (19.69 in); length 4.45 m (14.60 ft); span 1.97 m (6.45 ft)
Weights: total round 616 kg (1,358 lb); warhead 300 kg (661 lb) proximity and impact delay-fused FRAG-HE
Propulsion: one IMI two-stage solid-propellant rocket delivering ?-kg (?-lb) thrust
Performance: speed high subsonic; range 32 km (20 miles) from a high-altitude launch
Guidance: Saab autopilot, and Philips Elektronikindustrier active radar terminal homing

Variant
RB 04E: the only current version of this important anti-ship missile still left in service, the RB 04E was derived from the RB 04D with updated avionics, a strengthened structure and reduced-span wings; particular features of the type are its fully autonomous operation after launch, and the advanced nature of its electronics (especially in the field of electronic counter-countermeasures)

Saab RB 08A
(Sweden)
Type: ship- and coast-launched medium/long-range anti-ship tactical missile

Dimensions: diameter 0.66 m (25.98 in); length 5.72 m (18.77 ft); span 3.01 m (9.88 ft)
Weights: total round 1215 kg (2,679 lb); warhead 250 kg (551 lb) FRAG-HE
Propulsion: two jettisonable solid-propellant booster rockets each delivering ?-kg (?-lb) thrust, and one Turboméca Marboré IID turbojet delivering 400-kg (882-lb) thrust
Performance: speed Mach 0.85; range 250 km (155 miles)
Guidance: autopilot plus TRT radar altimeter, and Thomson-CSF active radar terminal homing

Variant
RB 08A: developed by French and Swedish interests on the basis of the Nord CT-20 target drone and introduced in 1967, the RB 08A is a large and powerful anti-ship missile whose considerable range capability requires the use of active radar terminal homing

Saab RBS 15M
(Sweden)
Type: ship-launched medium/long-range anti-ship tactical missile
Dimensions: diameter 0.50 m (19.7 in); length 4.35 m (14.27 ft); span 1.40 m (4.59 ft)
Weights: total round 780 kg (1,720 lb); warhead not revealed
Propulsion: two jettisonable Saab solid-propellant booster rockets each delivering ?-kg (?-lb) thrust, and one Microturbo TRI 60-2 sustainer turbojet delivering 377-kg (831-lb) thrust
Performance: speed high subsonic; range 100 km (62 miles)
Guidance: autopilot plus radar altimeter, and Philips Elektronikindustrier active radar terminal homing

Variant
RBS 15M: entering service in the second half of the 1980s, the RBS 15M is a potent ship-launched anti-ship missile fired (after onboard programming using data derived from the ship's sensors and fire-control system) with the aid of two jettisonable solid-propellant rockets to fly at medium or low altitude (the former capability being essential in the Swedish archipelago) to the point at which the active seeker is turned on, whereupon the missiles drops to sea-skimming height for the attack
RBS 15F: air-launched version weighing 598 kg (1,318 lb) as no booster rockets are needed, and possessing a range in the order of 150 km (93.2 miles)

Hsiung Feng (Taiwan): see IMI Gabriel Mks I and II

AS-2 'Kipper'
(USSR)
Type: air-launched long-range anti-ship tactical missile
Dimensions: diameter 0.90 m (35.4 in); length 10.00 m (32.81 ft); span 4.90 m (16.08 ft)
Weights: total round 4200 kg (9,250 lb); warhead ?-kiloton nuclear or 1000 kg (2,205 lb) HE
Propulsion: one turbojet (possibly Lyulka AL-5) delivering 5000-kg (11,023-lb) thrust
Performance: speed Mach 1.2; range 185 km (115 miles) from a high-altitude launch
Guidance: autopilot with capability for mid-course radio update, and active radar terminal homing

Variant
AS-2 'Kipper': introduced in 1960 as the primary anti-ship weapon of the Tupolev Tu-16 'Badger-C', the AS-2 is obsolescent by modern standards because of its size and lack of electronic sophistication

AS-3 'Kangaroo'
(USSR)
Type: air-launched long-range anti-ship tactical/operational and area-attack strategic missile
Dimensions: diameter 1.85 m (72.8 in); length 14.90 m (48.88 ft); span 9.15 m (30.02 ft)
Weights: total round 11000 kg (24,250 lb); warhead 800-kiloton/1-megaton thermonuclear
Propulsion: one turbojet (possibly Tumansky R-11) delivering 5100-kg (11,244-lb) thrust
Performance: speed Mach 1.8; range 650 km (405 miles) from a high-altitude launch
Guidance: autopilot and radio command

Variant
AS-3 'Kangaroo': this massive weapon entered service in 1961 as a strategic weapon carried by Tupolev Tu-95 'Bear-B' aircraft, and was intended as an area attack weapon without any form of terminal guidance; the warhead also makes the weapon useful against tactical/operational naval targets such as carrier battle groups

AS-4 'Kitchen'
(USSR)
Type: air-launched long-range anti-ship tactical/operational and area-attack strategic missile
Dimensions: diameter 0.90 m (35.4 in); length 11.30 m (37.07 ft); span 3.00 m (9.84 ft)
Weights: total round 5900 kg (13,007 lb); warhead 200/350-kiloton thermonuclear or 1000 kg (2,205 lb) HE
Propulsion: one liquid-propellant rocket delivering ?-kg (?-lb) thrust
Performance: speed Mach 3.5; range 460 km (286 miles) from a high-altitude launch reducing to 300 km (186 miles) from a low-altitude launch
Guidance: inertial and active or passive (see below) terminal homing

Variant
AS-4 'Kitchen': introduced in 1965 and used by

Tupolev Tu-22 'Blinder' and Tu-26 'Backfire' supersonic bombers, the AS-4 is the USSR's first multi-role missile, the type being available with no terminal homing and a thermonuclear warhead as a strategic attack weapon, or with active radar or passive infra-red or anti-radar terminal homing as an anti-ship weapon with conventional or thermonuclear warhead, the flight profile in the latter case being a high altitude cruise followed by a devastating dive; the type is now tasked generally with operations against US surface battle groups, which are operationally and strategically important targets with high electromagnetic and thermal signatures

AS-5 'Kelt'
(USSR)
Type: air-launched long-range anti-ship tactical missile
Dimensions: diameter 0.90 m (35.4 in); length 8.60 m (28.22 ft); span 4.60 m (15.09 ft)
Weights: total round 3000 kg (6,614 lb); warhead 1000 kg (2,205 lb) HE
Propulsion: one liquid-propellant rocket delivering ?-kg (?-lb) thrust
Performance: speed Mach 1.2; range 230 km (143 miles) from a high-altitude launch reducing to 180 km (112 miles) from a low-altitude launch
Guidance: autopilot and active or passive radar terminal homing

Variant
AS-5 'Kelt': introduced in 1966 as replacement for the AS-1 'Kennel' in the air-launched role, the AS-5 arms primarily the Tupolev Tu-16 'Badger-G' bomber, and has the same basic aeroplane configuration as its predecessor, an HE warhead and more advanced avionics including two terminal homing options

AS-6 'Kingfish'
(USSR)
Type: air-launched long-range anti-ship tactical/operational and area-attack strategic missile
Dimensions: diameter 0.90 m (35.4 in); length 10.00 m (32.81 ft); span 2.50 m (8.20 ft)
Weights: total round 5000 kg (11,023 lb); warhead 350-kiloton nuclear or 1000 kg (2,205 lb) HE
Propulsion: one solid-propellant rocket delivering ?-kg (?-lb) thrust
Performance: speed Mach 3; range 560 km (348 miles) from a high-altitude launch reducing to 250 km (155 miles) from a low-altitude launch
Guidance: inertial, and active radar or passive anti-radar terminal homing

Variant
AS-6 'Kingfish': introduced in the 1970s as a complement to the AS-4 'Kitchen', and generally carried in pairs by Tupolev Tu-16 'Badger-C (Mod)' and 'Badger-G (Mod)' and by Tu-26 'Backfire-B' aircraft, the AS-6 has the same targeting and homing methods as the AS-4, but greater range and a cruise altitude of 18000 m (59,055 ft) before the high-supersonic dive onto the target

SS-N-2A/B 'Styx'
(USSR)
Type: ship-launched medium-range anti-ship tactical missile
Dimensions: diameter 0.75 m (29.53 in); length 6.30 m (20.67 ft); span 2.75 m (9.02 ft)
Weights: total round 3000 kg (9,843 lb); warhead 500 kg (1,102 lb) impact delay-fused HE
Propulsion: one jettisonable solid-propellant booster rocket delivering ?-kg (?-lb) thrust, and one storable liquid-propellant sustainer rocket delivering ?-kg (?-lb) thrust
Performance: speed Mach 0.9; range 46 km (28.6 miles)
Guidance: autopilot with capability for mid-course radio update, and active radar terminal or radio command terminal homing

Variants
SS-N-2A 'Styx': entering service in 1958, this seminal but large anti-ship missile is obsolescent but still in widespread service, despite its limitations when faced by electronic countermeasures or effective point defences; the type has a preset cruise altitude of up to 300 m (985 ft)
SS-N-2B 'Styx': entering service in 1965, this updated missile has folding wings for better shipboard stowage, and may have infra-red terminal homing
SS-N-2C 'Styx': entering service in the early 1970s, this version of the SS-N-2B has longer range (80 km/50 miles) and updated avionics, the latter allowing a seaskimming approach to the target; the extended range of this weapon makes mid-course guidance by a helicopter all but essential if the type's full range capability is to be used; the variant is sometimes known as the **SS-N-2 (Mod)**

SS-N-3B/C 'Shaddock'
(USSR)
Type: ship- and surfaced submarine-launched long-range anti-ship tactical/operational missile
Dimensions: diameter 0.86 m (33.86 in); length 19.90 m (65.29 ft); span 2.10 m (6.89 ft)
Weights: total round 4500 kg (9,921 lb); warhead 350-kiloton nuclear or 1000 kg (2,205 lb) HE
Propulsion: two jettisonable solid-propellant booster rockets each delivering ?-kg (?-lb) thrust, and one sustainer turbojet or ramjet delivering ?-kg (?-lb) thrust
Performance: speed Mach 1.4; range 460 km (286 miles)

Guidance: autopilot with capacity for mid-course radio update, and active radar terminal homing

Variants
SS-N-3A 'Shaddock': introduced in 1962 as a successor to the NATO-designated SS-N-3C strategic missile, the SS-N-3A is the 'Shaddock' variant designed for launch from surfaced submarines, its primary targets being US carrier battle groups with the nuclear warhead; the long range of the weapon makes mid-course guidance update a key necessity on ranges of more than 200 km (124 miles); alternative estimates give nuclear warhead yields of 800 kilotons and 1 megaton
SS-N-3B 'Shaddock': this is the surface vessel companion to the SS-N-3A, and was introduced at the same time
SS-N-3C 'Shaddock': introduced in 1960, this was the initial member of the 'Shaddock' series, and designed as an area-attack weapon against US strategic targets with autopilot control but no terminal homing system
SS-C-1B 'Sepal': introduced in 1962, this is the coastal defence version of the 'Shaddock', container-mounted on an 8 × 8 truck

SS-N-7
(USSR)

Type: submarine-launched medium-range anti-ship tactical/operational missile
Dimensions: diameter 0.55 m (21.66 in); length 6.70 m (21.98 ft); span not known
Weights: total round 3500 kg (7.716 lb); warhead 200-kiloton nuclear or 500 kg (1,102 lb) HE
Propulsion: one solid-propellant rocket delivering ?-kg (?-lb) thrust
Performance: speed Mach 0.95; range 55 km (34 miles)
Guidance: autopilot and active radar terminal homing

Variant
SS-N-7: designed for underwater launch through the tubes of a submarine, the SS-N-7 was designed as a weapon against US carrier battle groups, and began to enter service in 1968; the type cruises at a height of some 30 m (100 ft)

SS-N-9 'Siren'
(USSR)

Type: ship- and submarine-launched medium/long-range anti-ship tactical/operational missile
Dimensions: diameter 0.55 m (21.66 in); length 9.20 m (30.18 ft); span 2.50 m (8.20 ft)
Weights: total round 3000 kg (6,614 lb); warhead 200-kiloton nuclear or 500 kg (1,102 lb) HE
Propulsion: one solid-propellant rocket delivering ?-kg (?-lb) thrust
Performance: speed Mach 0.9; range 110 km (68 miles)

Guidance: autopilot with capability for radio mid-course update, and combined active radar and infra-red terminal homing

Variants
SS-N-9 'Siren': introduced in 1969 on board light missile corvettes, and then developed for encapsulated launch from the tubes of submerged submarines, the SS-N-9 is similar to the SS-N-7 in concept and operation, though its range makes necessary the use of third-party targeting for effective maximum-range accuracy; the use of two terminal homing modes dictates a cruise altitude of 75 m (245 ft)
SS-N-22: improved version of the SS-N-9 introduced in 1982 for use on board the Soviet 'Sovremenny' class destroyers; the type is thought to have a third-party-targeted range of 150 km (93 miles), and the advanced avionics are believed to confer a true sea-skimming capability as well as a 'home-on-jam' terminal mode in conditions of electronic countermeasures; like the SS-N-9, the SS-N-22 is associated with 'Band Stand' shipborne radar

SS-N-12 'Sandbox'
(USSR)

Type: ship-launched long-range anti-ship tactical/operational missile
Dimensions: diameter 0.86 m (33.86 in); length 10.70 m (35.10 ft); span 2.50 m (8.20 ft)
Weights: total round 5000 kg (11,023 lb); warhead 350-kiloton nuclear or 1000 kg (2,205 lb) HE
Propulsion: two jettisonable solid-propellant booster rockets delivering ?-kg (?-lb) thrust, and one sustainer turbojet or ramjet delivering ?-kg (?-lb) thrust
Performance: speed Mach 2.5; range 560 km (348 miles)
Guidance: autopilot with capability for mid-course radio update, and active radar terminal homing

Variant
SS-N-12 'Sandbox': this potent missile is believed to be an updated version of the SS-N-3 series, with longer range and more advanced avionics so that a sea-skimming attack profile can be flown; like the 'Shaddock' series, the SS-N-12 is of aeroplane configuration with folding wings for shipboard stowage, and requires mid-course updating for effective use over its maximum range

SS-N-19
(USSR)

Type: ship- and submarine-launched long-range operational/tactical anti-ship missile
Dimensions: diameter not known; length about 10.0 m (32.8 ft); span not known
Weights: total round not known; warhead ?-kg (?-kg (?-lb) 500-kiloton thermonuclear

Propulsion: solid-propellant rocket delivering ?-kg (?-lb) thrust
Performance: speed not known; range 550 km (342 miles)
Guidance: inertial, with capability for mid-course update by helicopter or aircraft data-link

Variant
SS-N-19: very little is known of this weapon, which entered service in 1980 and is carried by the 'Kirov' class battle-cruiser and the 'Oscar' class nuclear-powered cruise missile submarine

SS-C-2B 'Samlet'
(USSR)
Type: coast-defence anti-ship tactical missile
Dimensions: diameter 1.20 m (47.24 in); length 7.00 m (22.97 ft); span 5.00 m (16.40 ft)
Weights: total round 3000 kg (6,614 lb); warhead not known
Propulsion: one jettisonable solid-propellant rocket delivering ?-kg (?-lb) thrust, and (probably) one RD-500 sustainer turbojet delivering 1600-kg (3,527-lb) thrust
Performance: speed high subsonic; range 200 km (124 miles)
Guidance: autopilot plus altimeter, and active radar terminal homing

Variant
SS-C-2B 'Samlet': this obsolescent coast-defence missile is derived from the now-gone AS-1 'Kennel' air-launched anti-ship missile, and closely related to the out-of-service SS-C-2A 'Salish', from which it differs in having more advanced avionics, including a larger nose radome; the type is associated with 'Sheet Bend' long-range radar, and may also be capable of acepting mid-course guidance to make effective use of its long range

BAe Sea Eagle
(UK)
Type: air-launched medium/long-range anti-ship tactical missile
Dimensions: diameter 15.75 in (0.40 m); length 13.58 ft (4.14 m); span 3.94 ft (1.20 m)
Weights: total round 1,300 lb (590 kg); warhead not revealed
Propulsion: one Microturbo TRI 60-1 turbojet delivering 367-kg (787-lb) thrust
Performance: speed Mach 0.9+; range 30/60 miles (48.3/96.6 km) depending on launch altitude
Guidance: autopilot plus radar altimeter, and Marconi active radar terminal homing

Variants
Sea Eagle: entering service in the mid-1980s, the Sea Eagle is derived from the airframe of the Martel air-to-surface missile but fitted with a turbojet engine for greater range, and was designed for aircraft and helicopter launch platforms; the weapon has good speed and range, and cruises at low altitude to reduce the chance of the target spotting it visually or electro-magnetically; the active seeker is of an advanced type, and the large warhead (weighing some 400 lb/181 kg) is effective against most targets
Sea Eagle SL: currently under development, this ship-launched derivative of the Sea Eagle is fitted with boost rockets and carried in a lightweight container/launcher

BAe Sea Skua
(UK)
Type: air-launched short-range anti-ship tactical missile
Dimensions: diameter 8.75 in (0.222 m); length 9.33 ft (2.85 m); span 24.4 in (0.62 m)
Weights: total round 325 lb (147 kg); warhead 77 lb (35 kg) HE
Propulsion: one BAJ Vickers two-stage solid-propellant rocket delivering ?-lb (?-kg) thrust
Performance: speed Mach 0.9+; range 22,000 yards (20115 m)
Guidance: autopilot plus radar altimeter, and Marconi semi-active radar terminal homing

Variant
Sea Skua: entering premature service in 1982, the Sea Skua lightweight helicopter-launched anti-ship missile has proved an excellent weapon; the auto-pilot/altimeter cruise phase of the flight is followed by semi-active radar homing on the reflections of the Ferranti Seaspray radar carried by the Westland Lynx, though the adoption of other types as launch platforms will necessitate alterations to this system

McDonnell Douglas RGM-84A Harpoon
(USA)
Type: ship-launched medium/long-range anti-ship tactical missile
Dimensions: diameter 13.5 in (0.343 m); length 15.06 m (4.58 m); span 3.00 ft (0.914 m)
Propulsion: one Aerojet solid-propellant booster rocket delivering 14,550-lb (6600-kg) thrust, and one Teledyne CAE J402-CA-400 sustainer turbojet delivering 680-lb (308-kg) thrust
Weights: total round 1,498 lb (679 kg); warhead 500 lb (227 kg) proximity and impact delayed-action penetration/blast HE
Performance: speed Mach 0.85; range 68 miles (110 km)
Guidance: Lear-Siegler or Northrop strap-down inertial plus Honeywell APN-194 radar altimeter, and Texas Instruments PR-53/DSQ-28 active radar terminal homing

Variants

RGM-84A Harpoon: rapidly becoming the Western world's most important ship-launched anti-ship missile, the Harpoon was conceived in the late 1960s as a capable but comparatively cheap weapon emphasizing reliability rather than outright performance in all respects but electronic capability and range, where the use of a turbojet rather than a rocket in the sustainer role pays handsome dividends; the missile can be fired in range and bearing mode, allowing the late activation of the active radar as a means of reducing the chances of the missile being detected through its own emissions, or in the bearing-only mode for earlier activation of the radar where the precise location of the target is not available at missile launch time; if no target is found after the low-level approach, the missile undertakes a preprogrammed search pattern, and acquisition of the target is followed in the 60-mile (97-km) range Block I initial-production missiles by a steep pop-up climb and dive onto the target's more vulnerable upper surfaces; later Block IB and Block IC missiles have greater range and the capability for a sea-skimming rather than pop-up attack, and the Block IC type has enhanced electronic counter-countermeasures capability plus the ability to fly a dog-leg course via three preprogrammed waypoints; currently under development is the Block II type with range increased to more than 120 miles (193 km), plus variable flight profiles and increased capability against electronic countermeasures

AGM-84A Harpoon: air-launched version of the RGM-84A without the booster rocket, reducing length to 12.58 ft (3.84 m) and weight to 1,160 lb (526 kg)

UGM-84A Sub-Harpoon: submarine-launched version installed in a capsule for tube firing in a manner similar to that of the SM.39 version of the Exocet

RGM-84C Harpoon: much improved model incorporating the flight profile capabilities of the RGM-84A Blocks IB, IC and II missiles, and entering service in the second half of the 1980s; the type is also available in submarine- and air-launched versions as the **UGM-84C Sub-Harpoon** and **AGM-84C Harpoon** respectively

Texas Instruments Sea Ray
(USA)
Type: air-, land- and ship-launched short-range tactical anti-ship missile
Dimensions: diameter 8 in (0.203 m); length 9.75 ft (2.97 m); span not revealed
Weights: total round 400 lb (181 kg); warhead 150 lb (68 kg) FRAG-HE
Propulsion: one Aerojet Mk 78 solid-propellant rocket delivering ?-kg (?-lb) thrust
Performance: speed not revealed; range 4.35 miles (7 km)
Guidance: semi-active laser homing

Variant
Sea Ray: this weapon is under development as a private venture by Texas Instruments to the order of an undisclosed customer (possibly South Korea) using the powerplant of the AGM-45A Shrike anti-radiation missile and the seeker system of the Paveway III series laser-homing bomb; the Sea Ray flies a ballistic trajectory until some 10 seconds from anticipated impact, when the seeker is activated; the weapon is associated with the Sea Tiger fire-controlled system, and could enter service in 1987 or 1988

ANTI-SUBMARINE MISSILES

DDS Ikara
(Australia)
Type: ship-launched anti-submarine tactical weapon system
Dimensions: span 5.00 ft (1.52 m); length 11.22 ft (3.42 m); height 5.15 ft (1.57 m)
Weights: total round varies according to payload; payload typically 507 lb (230 kg) Mk 46 lightweight torpedo
Propulsion: one Murawa two-stage solid-propellant rocket delivering ?-lb (?-kg) thrust
Performance: speed Mach 0.8; range 15 miles (24 km)
Guidance: autopilot plus radar altimeter, and radio command

Variants
Ikara: developed in Australia as a major anti-submarine weapon, the Ikara is a small aircraft-configured missile launched (on data provided by the ship's sonar systems via the fire-control system) towards the anticipated position of the target, the missile being fed with upgraded information in flight so that it can paradrop its homing torpedo payload (initially the US Mk 44, then the Mk 46 and recently a number of European weapons such as the Marconi

Stingray and Whitehead A 184) over the target for autonomous continuance of the attack while the missile clears the area and crashes

Ikara GWS 40: British version of the system with target data provided through the ship's Action Data Automation Weapons System

Branik: version for Brazil with a dedicated missile tracking and guidance system

Super Ikara: being developed by the Australian Department of Defence Support and BAe, the Super Ikara is a much updated and improved version of the basic concept matched to the performance of modern sonar detection systems; the launch vehicle is to be boosted by a solid-propellant rocket before an air-breathing sustainer engine takes over for flight to a range of 60 miles (96 km), a range that will almost certainly dictate the provision of mid-course update

Latécoère Malafon Mk 2
(France)
Type: ship-launched anti-submarine tactical weapon system
Dimensions: diameter 0.65 m (25.6 in); span 3.30 m (10.83 ft); length 6.15 m (20.18 ft)
Weights: total round 1500 kg (3,307 lb); payload 540 kg (1,190 lb) L4 homing torpedo
Propulsion: two SNPE Vénus solid-propellant booster rockets each delivering ?-kg (?-lb) thrust
Performance: speed 830 km/h (516 mph); range 13000 m (14,215 yards)
Guidance: autopilot plus radar altimeter, and radio command

Variant
Malafon Mk 2: introduced in 1966, the production Malafon Mk 2 operates in a manner similar to the Ikara, but glides under radio control after the initial boost phase of the flight, dropping its nose-mounted torpedo payload some 800 m (875 yards) from the target's anticipated position by decelerating the airframe with a braking parachute; the torpedo then falls into the water and continues the attack autonomously

FRAS-1
(USSR)
Type: ship-launched anti-submarine tactical weapon system
Dimensions: diameter 0.70 m (27.56 in); length 6.20 m (20.34 ft); span 1.30 m (51.2 in)
Weights: total round 800 kg (1,764 lb); warhead 15-kiloton nuclear (fission) or 450-mm (17.7-in) homing torpedo
Propulsion: one solid-propellant rocket delivering ?-kg (?-lb) thrust
Performance: speed Mach 1+; range 30 km (18.6 miles)
Guidance: none

Variant
FRAS-1: associated with the SUW-N-1 twin launcher (with 20-round magazine) found on many large Soviet surface vessels, the FRAS-1 rocket entered service in 1967 and is believed to be derived from the FROG series of artillery rockets; the weapon is launched on data supplied by the launch ship's sonar via the appropriate fire-control system, and operates in a manner similar to that of the American ASROC system

SS-N-14 'Silex'
(USSR)
Type: ship-launched anti-submarine tactical weapon system
Dimensions: diameter 0.55 m (21.65 in); length 7.60 m (24.93 ft); span 1.10 m (43.3 in)
Weights: total round 1000 kg (2,205 lb); warhead 2.5-kiloton nuclear (fission) or 450-mm (17.7-in) homing torpedo
Propulsion: one solid-propellant rocket delivering ?-kg (?-lb) thrust
Performance: speed Mach 0.95; range 7.4/55 km (4.6/34.2 miles)
Guidance: radio command

Variant
SS-N-14: though now rightly appreciated as an anti-submarine weapon in the mould of the Ikara and Malafon (though of considerably superior performance), the 'Silex' was at first thought by the West to be a surface-to-surface tactical missile, the Soviets having waged a subtle but convincing disinformation campaign between 1968 (when the first launchers went to sea) and 1974 (when the missile was introduced); the weapons are carried in quadruple launchers, though the 'Kirov' class battle-cruisers each have a twin launcher with 26 rounds, and guidance is effected with the aid of 'Head Light' or 'Eye Bowl' radars after the missile has been launched on information provided by the ship's sonar system

SS-N-15
(USSR)
Type: submarine-launched anti-submarine tactical weapon system
Dimensions: diameter 0.533 m (21.00 in); length 6.50 m (21.33 ft)
Weights: total round 1900 kg (4,189 lb); warhead 15-kiloton nuclear (fission)
Propulsion: one solid-propellant rocket delivering ?-kg (?-lb) thrust
Performance: speed Mach 1.5; range 37 km (23 miles)
Guidance: none

Variant
SS-N-15: introduced in 1972, the SS-N-15 is believed

to be based on the American SUBROC system and is very similar in its concept and method of operation; the type is deployed aboard most Soviet nuclear attack submarines, and on the 'Tango' class patrol submarine

SS-N-16
(USSR)
Type: submarine-launched anti-submarine tactical weapon system
Dimensions: diameter 0.618 m (24.33 in); length 6.50 m (21.33 ft)
Weights: total round 2150 kg (4,740 lb); warhead 450-mm (17.7-in) homing torpedo
Propulsion: one solid-propellant rocket delivering ?-k (?-lb) thrust
Performance: speed Mach 1.5; range 55 km (34.2 miles)
Guidance: inertial

Variant
SS-N-16: introduced in the mid-1970s as an updated, inertially-guided and longer-ranged development of the SS-N-15 concept, the SS-N-16 is of greater diameter than its precursor, and paradrops its homing torpedo payload into the water over the anticipated target position; most Soviet nuclear-powered attack and modern conventionally-powered patrol submarines carry the weapon

Boeing/Gould ASW-SOW
(USA)
Type: submarine-launched anti-submarine tactical weapon system
Dimensions: diameter 21.00 in (0.533 m); length 21.00 ft (6.40 m)
Weights: total round 2,700 lb (1224.7 kg); warhead 800 lb (362.9 kg) Mk 50 Barracuda homing torpedo or nuclear depth charge
Propulsion: one Hercules solid-propellant rocket delivering ?-lb (?-kg) thrust
Performance: speed Mach 1.5+; range 63.3/103.6 miles (101.9/166.7 km)
Guidance: none

Variant
ASW-SOW: due to enter service in the 1990s, the Anti-Submarine Warfare – Stand-Off Weapon is the successor to SUBROC, offering greater range and full compatibility with the latest digital Mk 117 fire-control system, which programmes the weapon and fires it towards the target area; the encapsulated missile is launched through a standard torpedo tube, and its rocket motor is ignited only after the weapon has reached the surface, wrap-round fins unfolding as the missile leaves the capsule to fly a ballistic trajectory, probably discarding its burned-out rocket and leaving the warhead to continue into the anticipated target area

Honeywell RUR-5A ASROC
(USA)
Type: ship-launched anti-submarine tactical weapon system
Dimensions: diameter 12.75 in (0.324 m); length (Mk 44 payload) 15.00 ft (4.57 m) or (Mk 46 payload) 14.79 ft (4.51 m); span 33.25 in (0.845 m)
Weights: total round (Mk 44 payload) 957 lb (434 kg), (Mk 46 payload) 1,073 lb (487 kg) or (W44 payload) 825 lb (374 kg); warhead (Mk 44 payload) 425 lb (192.8 kg), (Mk 46 payload) 508 lb (230.4 kg) or (W44 payload) 260 lb (117.9 kg) 1.5-kiloton Mk 17 nuclear depth charge
Propulsion: one Naval Propellant Plant solid-propellant rocket delivering 11,000-lb (4990-kg) thrust
Performance: speed Mach 0.8; range 2,025/11,600 yards (1850/10,605 m)
Guidance: none

Variants
RUR-5A ASROC: introduced in 1962 aboard surface units of the US Navy, the Anti-Submarine ROCket is a ballistic weapon aimed and launched from the Mk 16 octuple dedicated launcher or Mks 10 and 26 twin SAM/ASROC launchers on information supplied by the ship's sonar system via the underwater weapons fire-control system; the weapon flies ballistically, the rocket being jettisoned when exhausted to leave the payload to continue alone to the target's anticipated position for an autonomous attack; only the US Navy has the Mk 17 depth charge with the W44 nuclear warhead
ASROC(VL): updated version currently under development for service in the 1990s for vertical launch from the Mk 41 launcher unit able to accommodate ASROC, BGM-109 Tomahawk, RGM-84 Harpoon and RIM-66/67 Standard missiles; the weapon is boosted vertically, aligned onto the correct trajectory, discards the booster and follows a ballistic trajectory to the impact area

Goodyear UUM-44A SUBROC
(USA)
Type: submarine-launched anti-submarine tactical rocket system
Dimensions: diameter 21.00 in (0.533 m); length 22.00 ft (6.71 m)
Weights: total round 4,000 lb (1814 kg); warhead 650 lb (294.8 kg) 5-kiloton W55 nuclear (fission)
Propulsion: one Thiokol TE-260G solid-propellant rocket delivering ?-lb (?-kg) thrust
Performance: speed Mach 1.5; range 35 miles (56.3 km)
Guidance: Singer-Kearfott SD-510 inertial

Variant
UUM-44A SUBROC: introduced in 1965 as part of the anti-submarine weapon inventory of US nuclear-

powered attack submarines (which generally carry four or six of these SUBmarine ROCkets), the UUM-44A is a tube-launched weapon fired on the data supplied by the boat's BQQ-2 sonar suite via the analog Mk 113 fire-control system; the rocket is ignited under the water, forcing the missile up and out of the water to fly towards the target's anticipated position, above which the nuclear warhead is released (by explosive bolts and retro-thrust deceleration of the missile) to fall into the water and detonate at the preset depth with a lethal radius of 5,300/7,000 yards (4845/6400 m); the type is being phased out of service as older boats retire, for the system is incompatible with the newer boats' digital Mk 117 fire-control system

SURFACE-TO-AIR MISSILES

CMPIEC HM-5 Red Tassel: see SA-7 'Grail' (USSR)

CPMIEC HQ-2 (China): see SA-2 'Guideline' (USSR)

CPMIEC HQ-61
(China)
Type: land- and ship-launched short-range tactical surface-to-air missile
Dimensions: diameter 0.286 m (11.26 in); length 3.99 m (13.09 ft); span 1.166 m (45.9 in)
Weights: total round 300 kg (661 lb); warhead not revealed
Propulsion: one solid-propellant rocket delivering ?-kg (?-lb) thrust
Performance: speed Mach 3; range 3/10 km (1.86/6.2 miles); altitude limits 0/8000 m (0/26,245 ft)
Guidance: strapdown inertial and CW semi-active radar homing

Variant
HQ-61: this Chinese-developed SAM bears a resemblance to the American Sparrow series, but is larger and heavier to the US weapon; in both services the weapon is associated with a twin launcher, that on ships being located above the below-decks rotary magazine, and that on land being based on a 6 × 6 chassis; both types are used in conjunction with a Chinese radar based on the Soviet 'Flat Face'

Ayn as-Sakr (Egypt): see SA-7 'Grail' (USSR)

DTCN/Matra Masurca Mk 2 Mod 3
(France)
Type: ship-based two-stage medium-range area-defence tactical surface-to-air missile
Dimensions: diameter (missile) 0.406 m (16.00 in) and (booster) 0.57 m (22.44 in); length (missile) 5.38 m (17.65 m) and (booster) 3.32 m (10.89 ft); span (missile) 0.77 m (30.31 in) and (booster) 1.50 m (59.06 in)
Weights: total round 2098 kg (4,625 lb); warhead 100 kg (220 lb) proximity- and impact-fused FRAG-HE
Propulsion: one SNPE Polka solid-propellant booster rocket delivering 34780-kg (76,675-lb) thrust, and one SNPE Jacée solid-propellant sustainer rocket delivering 2423-kg (5,342-lb) thrust
Performance: speed Mach 3; range 50 km (31 miles); altitude limits 30/2300 m (100/75,460 ft)
Guidance: semi-active radar homing

Variant
Masurca Mk 2 Mod 3: the only version of this weapon now in service, the Mascurca Mk 2 Mod 3 introduced semi-active radar rather than radar beam-riding sight guidance; the type is used on a twin-rail launcher and fed from a 48-round magazine, the associated systems being a 3D surveillance radar, a weapon-direction system and two DRBC 51 missile-control radars

Matra R.440
(France)
Type: land-based point-defence tactical surface-to-air missile
Dimensions: diameter 0.15 m (5.9 in); length 2.89 m (9.48 ft); span 0.54 m (21.25 in)
Weights: total round 85 kg (187.4 lb); warhead 15 kg (33 lb) proximity-fused focalized FRAG-HE
Propulsion: one SNPE Lens solid-propellant rocket delivering 4850-kg (10,962-lb) thrust
Performance: speed Mach 2.3; range 500/13000 m (545/14,215 yards) against helicopters and non-manoeuvring targets, or 500/8500 m (545/9,295 yards) against manoeuvring targets, or 500/6500 m (545/7,110 yards) against low-altitude and sea-skimming targets; altitude limits 40/3600 m (130/11,810 ft)
Guidance: Thomson-CSF radio command with infra-red/radar gathering and tracking

Variant
R.440: entering service in 1971, the R.440 SAM is designed for use in the **Cactus** and **Crotale** land-based SAM systems, and is designed for low-altitude close-range engagements; the same basic missile is also

used in the **Naval Crotale** system, a four- or eight-round system (**Naval Crotale 4B** or **Naval Crotale 8B**) designed for the short-range air defence of warships against medium-altitude, low-altitude and sea-skimming attacks; the Naval Crotale system uses radar acquisition of targets, though any such threats below an altitude of 50 m (165 ft) are handled by a TV tracking system; for adverse-weather operations there is also the **Naval Crotale EDIR** system (also known as **Naval Crotale 8S**) with SAT passive infra-red tracking of the target and the missile; the IR tracker has a maximum range of 20 km (12.4 miles); there is also a **Naval Crotale 8MS** modular system for installation on craft down to a displacement of 200 tons and featuring a lightweight eight-round launcher and fire-director turret that can also be used for gunfire control

Matra R.460
(France)
Type: land-based point-defence tactical surface-to-air missile
Dimensions: diameter 0.156 m (6.14 in); length 3.15 m (10.33 ft); span 0.59 m (23.23 in)
Weights: total round 100 kg (220.5 lb); warhead 14-kg (30.9-lb) proximity-fused focalized FRAG-HE
Propulsion: one SNPE Lens dual-thrust solid-propellant rocket delivering ?-kg (?-lb) thrust
Performance: speed Mach 2.5; range 500/12000 m (545/13,125 yards); altitude limits 15/6000 m (50/19,685 ft)
Guidance: Thomson-CSF radio command with infra-red/radar gathering and tracking

Variant
R.460: a version of the R.440 with longer range and greater speed, the R.460 entered service in 1982, and is designed for used with the **Shahine**, **SICA** AFV-based battlefield SAM system

Matra Mistral
(France)
Type: land- and ship-based point-defence tactical surface-to-air missile
Dimensions: diameter 0.90 m (3.54 in); length 1.80 m (5.91 ft); span 0.19 m (7.48 in)
Weights: total round 17 kg (37.5 lb); warhead 3.0 kg (6.6 lb) proximity- and impact-fused FRAG-HE
Propulsion: SEP solid-propellant booster and sustainer rockets delivering ?-kg (?-lb) thrust
Performance: speed Mach 2.6; range 500/6000 m (545/6,560 yards); altitude limits not revealed
Guidance: SAT passive infra-red homing

Variant
Mistral: entering service in 1986, this lightweight SAM is designed for land and naval use, the former comprising the **SATCP** system (a pedestal-mounted containerized missile together with the target-acquisition and fire-control unit, able to detect aircraft at 6000 m/6,560 yards and helicopters at 4000 m/4,375 yards), and the latter in the **SADRAL** system designed as point defence for larger warships; the SADRAL system uses a remotely-controlled six-round launcher also mounting the target-acquisition TV camera, though longer-range acquisition is undertaken by the parent ship's radar, with an optronic system as back-up for the two main acquisition channels

Euromissile Roland 1 and 2
(France/West Germany)
Type: land-based point-defence tactical surface-to-air missile
Dimensions: diameter 0.16 m (6.3 in); length 2.40 m (7.87 ft); span 0.50 m (19.7 in)
Weights: total round 63 kg (139 lb); warhead 6.5 kg (14.3 lb) proximity- and impact-fused FRAG-HE
Propulsion: one SNPE Roubaix solid-propellant booster rocket delivering 1600-kg (3,527-lb) thrust, and one SNPE Lampyre solid-propellant sustainer rocket delivering 200-kg (441-lb) thrust
Performance: speed Mach 1.6; range 500/6300 m (545/6,890 yards); altitude limits 20/3000 m (65/9,845 ft)
Guidance: infra-red gathering then optical (Roland 1) or radar (Roland 2) semi-automatic radio command to line of sight

Variants
Roland 1: designed from the early 1960s by the antecedents of MBB (West Germany) and Aérospatiale (France) working as Euromissile, the Roland 1 was intended as a clear-weather low-altitude surface-to-air missile for use by France on the AMX-30R converted MBT chassis and for export; the type entered service in 1981
Roland 2: produced in parallel with the Roland 1 for the West Germany army, the Roland 2 is designed for all-weather operation on the chassis of the Marder MICV; the type is in limited US service as the **MIM-115 Roland**, produced under licence in the USA by Hughes and Boeing; the missile is also used in the **Shelter Roland** and **FlaRakRad** systems, the former being designed for static use and the latter being based on a 10-tonne 8 × 8 truck for the defence of airfields in association with one Roland Co-ordination Centre for each tactical unit of four to eight fire units
Roland 3: introduced in 1984, the Roland 3 overcomes some of the earlier models' performance shortcomings by introducing a more powerful motor to boost maximum speed to Mach 1.9 and maximum range to 8000 m (8,750 yards); otherwise the missile is closely akin to the Roland 2, and designed like that weapon for all-weather engagements with radar guidance

IMI Barak 1
(Israel)
Type: land- and ship-based point-defence tactical surface-to-air missile
Dimensions: diameter 0.17 m (6.7 in); length 2.17 m (7.12 ft); span 0.68 m (26.77 in)
Weights: total round 86 kg (189.6 lb); warhead 22 kg (48.5 lb) FRAG-HE
Propulsion: one solid-propellant triple-thrust rocket delivering ?-kg (?-lb) thrust
Performance: speed Mach 1.7; range 500/10000 m (550/10,935 yards); altitude limits not revealed
Guidance: Elta AMDRS radar command with optical back-up

Variant
Barak 1: developed for the point-defence of small warships, the Barak 1 can be carried in an eight-round 2500-kg (5,511-lb) launcher on a mounting derived from that of the TCM-20 AA gun mounting, or in a 1300-kg (2,866-lb) eight-round vertical launcher; as the missile is of the radar command type all-weather operation is ensured, and the automatic operation of the system gives capability against multiple attack by aircraft or seaskimming missiles; the land-based version is known as **ADAMS** (Air Defence Advanced Missile System), and comprises two vertical banks of five launchers on an 8 × 8 LAV vehicle

Selenia Aspide 1A
(Italy)
Type: ship- and ground-based medium-range tactical surface-to-air missile
Dimensions: diameter 0.203 m (8.00 in); length 3.70 m (12.14 ft); span 0.80 m (31.5 in)
Weights: total round 220 kg (485 lb); warhead 35 kg (77 lb) proximity- and impact-fused FRAG-HE
Propulsion: one SNIA-Viscosa solid-propellant rocket delivering ?-kg (?-lb) thrust
Performance: speed Mach 4; range 50/100 km (31/62 miles); altitude limits not revealed
Guidance: Selenia semi-active radar homing

Variant
Aspide 1A: this is basically similar to the Aspide air-to-air missile, in this instance differing only in the cropping of the wings to fit the quadruple or octuple launcher of the **Albatros** system; the system works in conjunction with the RTN 10X tracking radar and RTN 12X target-illumination radar; the **Albatros Mk 1** system uses a single director, and the **Albatros Mk 2** system two directors; the land-based equivalent of the Albatros system is the **Spada** system designed for the protection of small strategic areas against saturation attack; the system is based on a sextuple launcher for Aspide missiles supported by search and interrogation radars, tracking and illuminating radars and control units, all integrated into a high-level air-defence system able to deal with single or multiple, sequential or saturation attacks at all levels in conditions of heavy electronic countermeasures

Toshiba Type 81
(Japan)
Type: land-based point-defence tactical surface-to-air missile
Dimensions: diameter 0.16 m (6.3 in); length 2.70 m (8.86 ft); span 0.60 m (23.6 in)
Weights: total round 100 kg (220 lb); warhead not revealed
Propulsion: one Nissan Motors solid-propellant rocket delivering 8400-kg (18,519-lb) thrust
Performance: speed Mach 2.4; range ?/10000 m (?/10,935 yards); altitude limits ?/3000 m (?/9,845 ft)
Guidance: Toshiba inertial and infra-red terminal homing

Variant
Type 81: developed with great delays as the **Toshiba Tan-SAM** from the 1960s, the Type 81 entered Japanese service only in 1981; the type was designed for land-mobility, and the four-round launcher is carried on the rear of a Type 73 6 × 6 truck, two launchers being accompanied by a fire-control vehicle with a phased-array pulse-Doppler surveillance and acquisition radar

Bofors RBS 70 Rayrider
(Sweden)
Type: land-based point-defence tactical surface-to-air missile
Dimensions: diameter 0.106 m (4.17 in); length 1.32 m (4.4.33 ft); span 0.32 m (12.6 in)
Weights: total round 15 kg (33.07 lb); warhead 1 kg (2.2 lb) proximity- and impact-fused FRAG-HE
Propulsion: one Bofors solid-propellant booster rocket delivering ?-kg (?-lb) thrust, and one IMI solid-propellant sustainer rocket delivering ?-kg (?-lb) thrust
Performance: speed supersonic; range ?/5000 m (?/5,470 yards); altitude limits 0/3000 m (0/9,845 ft)
Guidance: laser beam-riding

Variants
RBS 70 Rayrider: developed in the late 1960s and early 1970s, the RBS 70 comprises a stand (comprising a post with tripod legs), a sight unit and 24-kg (52.9-lb) container with preloaded missile; once fired, the missile is gathered into the sight unit's field of vision and then rides the sight's laser straight to the target, where the detonating warhead fills its lethal volume with heavy metal pellets; the empty container can then be replaced by a full one for another engagement; the type can also be mounted on the rear of trucks or on the turret rings of light AFVs

RBS 70 ARMAD: updated version comprising an APC-compatible turret housing the RBS 70 system (on an elevating arm), seven missiles and an Ericsson HARD 3D surveillance radar
RBS 70 SLM: navalized version with a redesigned stand and gyro-stabilized optics
RBS 70M Nightrider: night-capable version using the same basic technology as the RBS 70 Rayrider, but featuring an updated missile with a larger sustainer rocket and heavier warhead used on a remotely-controlled launcher complete with TV and IR cameras and the laser transmitter

Tien Kung (Taiwan): see Raytheon MIM-104A Patriot (USA)

ABM-1B 'Galosh'
(USSR)
Type: land-based three-stage exoatmospheric strategic anti-ballistic missile missile
Dimensions: diameter 2.57 m (8.43 ft); length 19.80 m (64.96 ft); span not known
Weights: total round 32700 kg (72,090 lb); warhead 3- or 5-megaton thermonuclear
Propulsion: (first stage) storable liquid-propellant rocket delivering ?-kg (?-lb) thrust, (second stage) storable liquid-propellant rocket delivering ?-kg (?-lb) thrust, and (third stage) storable liquid-propellant rocket delivering ?-kg (?-lb) thrust
Performance: speed hypersonic; range ?/740 km (?/460 miles); altitude limits not known
Guidance: radar command

Variant
ABM-1B 'Galosh': introduced in the early 1960s as the world's first ABM missile, the 'Galosh' (which has the Soviet designation **UR-96**) is designed for exoatmospheric interception of ballistic missiles, which it destroys with the electromagnetic pulse and fast radiation effects of its substantial nuclear warhead; the type remains a truly formidable system that the Soviets are updating with new radar and command systems; the most important radars are five phased-array warning radars, and 11 'Hen House' long-range warning radars supported by 'Dog House' and 'Cat House' medium-range battle-management radars in support of the 'Try Add' engagement radars

ABM-2
(USSR)
Type: land-based exoatmospheric strategic anti-ballistic missile missile
Dimensions: not known
Weights: not known
Propulsion: now known
Performance: not known
Guidance: semi-active radar and active radar terminal homing

Variant
ABM-2: originally known by the temporary designation **SH-04** (derived from the test site at Sary Shagan where the type was first observed), the ABM-2 is believed to be an advanced development of the ABM-1B 'Galosh'

ABM-3 'Gazelle'
(USSR)
Type: land-based exoatmospheric anti-ballistic missile missile
Dimensions: not known
Weights: not known
Propulsion: now known
Performance: not known
Guidance: semi-active radar and active radar terminal homing

Variant
ABM-3: this is a highly advanced ABM missile used in conjunction with 'Flat Twin' and 'Pawn Shop' phased-array radars; virtually nothing in the way of detail is known of this advanced complement to the ABM-1 'Galosh'; the type was formerly known in the West as the **SH-08**; it is possible, however, that the designation refers to a versatile silo-launch system able to handle an SH-04 exoatmospheric or SH-08 endoatmospheric missile

SA-1 'Guild'
(USSR)
Type: land-based medium-range tactical/operational surface-to-air missile
Dimensions: diameter 0.70 m (27.56 in); length 12.00 m (39.37 ft); span 1.10 m (43.31 in)
Weights: total round 3200 kg (7,055 lb); warhead not known
Propulsion: one liquid-propellant rocket delivering ?-kg (?-lb) thrust
Performance: speed supersonic; range ?/32 km (?/20 miles); altitude limits not known
Guidance: radar command

Variant
SA-1 'Guild': the world's second operational surface-to-air missile when it entered service in 1954 as a high-altitude destroyer of strategic bombers, the 'Guild' was for its time a stupendous technical achievement and is still a reserve weapon in the USSR; the missile itself is capable, but the real efficiency of the system derives largely from the associated 'Yo-Yo' radar, which is very powerful and can tracks some 30 targets simultaneously

SA-2 and SA-N-2 'Guideline'
(USSR)
Type: land- and ship-based two-stage medium-range tactical surface-to-air missile

Dimensions: diameter (missile) 0.50 m (19.69 in) and (booster) 0.70 m (27.56 in); length (typical overall) 10.70 m (35.10 ft); span (missile) 1.70 m (5.58 ft) and (booster) 2.20 m (7.22 ft)
Weights: total round 2300 kg (5,071 lb); warhead 130 kg (286.6 lb) proximity- and impact-fused FRAG-HE
Propulsion: one jettisonable solid-propellant booster rocket delivering ?-kg (?-lb), and one solid-propellant sustainer rocket delivering ?-kg (?-lb) thrust
Performance: speed Mach 3.5; range 50 km (31 miles); altitude limits not known
Guidance: radar command

Variants
SA-2 'Guideline': introduced in 1958 as an aerodynamic development of the SA-1 and designed for the same high-altitude bomber-destroying role, the 'Guideline' was for some two decades the most important surface-to-air missile in the world and produced in a number of variants on the basic **V750VK** theme; the type is associated with the 'Fan Song' fire-control radar, additional warning and height information being supplied by other radars; the 'Guideline' is currently operational in **SA-2B**, **SA-2C**, **SA-2D**, **SA-2E** and **SA-2F** versions, the last being capable of engagements as low as 100 m (330 ft)
HQ-2: Chinese-built derivative of the SA-2, accorded the designation **CSA-1** in the US terminology for Chinese weapons; the type has followed a development comparable to that of the Soviet original, the latest being the **HQ-2J** associated with Red Flag 2A radar

SA-3 and SA-N-1 'Goa'
(USSR)
Type: land- or ship-based two-stage medium-range area-defence tactical surface-to-air missile
Dimensions: diameter (missile) 0.46 m (18.1 in) and (booster) 0.701 m (27.6 in); length overall 6.70 m (21.98 ft); span (missile) 1.22 m (4.00 ft) and (booster) 1.50 m (4.92 ft)
Weights: total round (SA-N-1A) 946 kg (2,085.1 lb) or (SA-N-1B) 950 kg (2,094.4 lb); warhead 10-kiloton nuclear or 60 kg (132 lb) proximity-fused FRAG-HE
Propulsion: one jettisonable solid-propellant booster rocket delivering ?-kg ((?-lb) thrust, and one solid-propellant sustainer rocket delivering ?-kg (?-lb) thrust
Performance: speed Mach 2.1; range 6/22 km (3.7/13.7 miles); altitude limits 50/15250 m (165/50,030 ft)
Guidance: (SA-3 and SA-N-1A) radar command or (SA-N-1B) semi-active radar terminal homing

Variants
SA-3A 'Goa': introduced in 1961 as a short-range SAM intended primarily for operation against low-flying targets, the 'Goa' has the Soviet designation **S-125 Pechora** and is a small two-stage missile at first carried on a two-rail launcher on the back of 6 × 6 trucks; more recently the Yugoslavs have introduced a three-rail launcher, and the Soviets have in turn produced a four-rail type; warning and target acquisition are provided by 'Flat Face' radar, and guidance is entrusted to 'Low Blow' radar
SA-N-1A 'Goa': navalized version of the land-based SA-3A, the SA-N-1A was introduced in 1961 as the USSR's first fully-deployed surface-to-air missile; the type is used on a twin launcher supplied from a 16-round magazine, and is controlled with the aid of 'Peel Group' radar
SA-N-2B 'Goa': improved version with the 'Peel Group' radar modified to provide semi-active radar guidance

SA-4 'Ganef'
(USSR)
Type: land-based medium/long-range area-defence tactical surface-to-air missile
Dimensions: diameter 0.90 m (35.43 in); length 8.80 m (28.87 ft); span (wings) 2.30 m (7.74 ft) and (tail) 2.60 m (8.53 ft)
Weights: total round 1800 kg (3,968 lb); warhead 100 to 130 kg (220 to 286.6 lb) proximity-fused HE
Propulsion: four jettisonable solid-propellant booster rockets each delivering ?-kg (?-lb) thrust, and one sustainer ramjet delivering ?-kg (?-lb) thrust
Performance: speed Mach 2.5; range 9.3/72 km (5.8/44.7 miles); altitude limits 1100/24000 m (3,610/78,740 ft)
Guidance: radar command plus semi-active radar terminal homing

Variant
SA-4 'Ganef': introduced in the early 1960s with the Soviet designation **3M8 Krug**, the 'Ganef' is widely used by the Warsaw Pact forces as the rear-area SAM component within the Soviet doctrine of front-line air defence in depth; the weapons are carried in pairs on a hydraulically-operated ramp/turntable on a tracked chassis derived from that of the GMZ minelayer, and are generally salvo-fired in pairs in conjunction with the 'Long Track' surveillance, 'Thin Skin' height-finding and 'Pat Hand' fire-control radars; the 'Ganef' has been produced in four versions, one of them optimized for a lower effective ceiling (down to 300 m/985 ft) and another (with a longer nose for a length of 9.00 m/29.53 ft) for greater range and ceiling; some reports attribute a weight of 2500 kg (5,511 lb)

SA-5 'Griffon' and 'Gammon'
(USSR)
Type: land-based single-stage long-range tactical/operational surface-to-air missile
Dimensions: diameter (missile) 0.85 m (33.46 in); length 10.60 m (34.78 ft); span 2.90 m (9.51 ft)
Weights: total round not known; warhead ?-kg (?-lb) nuclear or HE

Propulsion: four jettisonable solid-propellant booster rockets each delivering ?-kg (?-lb) thrust, and one liquid-propellant sustainer rocket delivering ?-kg (?-lb) thrust
Performance: speed Mach 4+; range ?/300 km (?/186 miles); altitude limits ?/29000 m (?/95,145 ft)
Guidance: probably radio command and semi-active radar terminal homing

Variant
SA-5 Gammon': introduced in the early 1960s the SA-5 (which has the Soviet designation **S-200**) is the 'high' component of the SA-3/SA-5 'low/high' air-defence mix (three SA-3 battalions to two SA-5 battalions for the protection of targets wthin the USSR); the type has frequently been identified with the massive and often-displayed surface-to-air missile allocated the reporting name 'Griffon' by Western analysts, but this is now believed to have been a prototype weapon revealed by the Soviets for misinformation purposes; no photograph of the SA-5 has been revealed publicly in the West, but it is known that the weapon was designed for the interception of medium- and high-altitude targets (aircraft, short/medium-range missiles and, under some circumstances, ballistic missiles); the type is thought to have the option of HE or low-yield nuclear warheads (in a terminal stage that may be powered), and the associated radar is the 'Square Pair' system for target tracking and missile guidance, target detection being provided by 'Back Net' or 'Bar Lock' radars, and height-finding by a 'Side Net' radar; the type is now thought to be obsolescent in the face of the latest Western electronic countermeasures; it is possible that the type has been produced in three versions as the **SA-5A** initial-production version first deployed in 1960 with a conventional warhead, the **SA-5B** first deployed in 1970 witha nuclear warhead, and the **SA-5A** initial-production version first deployed in 1960 with a conventional warhead, the **SA-5B** first deployed in 1970 with a nuclear warhead, and the **SA-5C** first deployed in 1975 with optional HE or nuclear warheads and enhanced terminal manoeuvring capability

SA-6 'Gainful'
(USSR)
Type: land-based medium-range tactical surface-to-air missile
Dimensions: diameter 0.335 m (13.2 in); length 6.20 m (20.33 ft); span 1.52 m (5.00 ft)
Weights: total round 550 kg (1,212 lb); warhead 80 kg (176.4 lb) proximity- and impact-fused FRAG-HE
Propulsion: one integral solid-propellant rocket-ramjet delivering ?-kg (?-lb) thrust
Performance: speed Mach 2.8; range 4/30 km (2.5/18.7 miles) at low altitude or 4/60 km (2.5/37.3 miles) at high altitude; altitude limits 100/18000 m (330/59,055 ft)
Guidance: radar command plus semi-active radar terminal homing

Variants
SA-6 'Gainful': though a potent weapon when introduced to Soviet service in the mid-1960s and used by the Arab armies in the 1973 war with Israel, the single-stage SA-6 (Soviet designation **9M9**) is now obsolescent through lack of capability against Western electronic countermeasures; the missiles are carried on triple launchers mounted on a chassis derived from that of the ZSU-23-4 quadruple 23-mm AA gun mounting, and are used in conjunction with the 'Long Track' surveillance radar, 'Thin Skin' height-finding radar and 'Straight Flush' fire-control radar; though the type has not been upgraded electronically, system improvements have considerable enhanced performance compared with that of early models, which had a maximum range of 22000 m (24,060 yards) and a ceiling of 9000 m (29,530 ft); the system is designed **ZRK Kub** in the USSR
SA-6B Mod 1 'Gainful': designation of an interim type using the combined launcher/guidance radar vehicle of the SA-11 'Gadfly' system with a three-rail SA-6 fit

SA-7A and SA-N-5 'Grail'
(USSR)
Type: man-portable point defence tactical surface-to-air missile
Dimensions: diameter 0.07 m (2.75 in); length 1.30 m (4.27 ft)
Weights: total round 9.2 kg (20.3 lb); warhead 2.5 kg (5.5 lb) impact-fused FRAG-HE
Propulsion: one solid-propellant booster rocket delivering ?-kg (?-lb) thrust, and one solid-propellant sustainer rocket delivering ?-kg (?-lb) thrust
Performance: speed Mach 1.5; range ?/3600 m (3,935 yards); altitude limits 45/1500 m (150/4,920 ft)
Guidance: IR homing

Variants
SA-7A 'Grail': developed in the early 1960s, the 'Grail' is known to the Soviets as the **Strela** and is a simple weapon used by two-man teams (one carrying the firing unit and a missile, and the other carrying a reload missile) for short-range air defence of front-line units; operations soon revealed that the type was prone to chasing decoys, and was also ineffective against the faster type of tactical target
SA-7B 'Grail': improved model with an IFF system and a more powerful motor, increasing speed to Mach 1.7 and ceiling to 4800 m (15,750 ft); the type has also been seen on a four-round launcher carried on light vehicles
SA-N-5: naval version of the SA-7B and known to the Soviets as the **9M32M Strela 2M**, this point-

defence weapon is used in a four-round pedestal mount (with lockers containing four or eight reloads) on small surface combatants, auxiliaries and amphibious warfare ships

Ayn as-Sakr: Egyptian development of the SA-7B with a number of improvements (Rank Pullin night sight, Thomson-CSF IFF, Thomson-Brandt digital electronics in place of the Soviet analog electronics, and Teledyne seeker) to increase tactical capabilities with a range of 4400 m (4,810 yards and a ceiling of 2400 m (7,875 ft)

HN-5 Red Tassel: designation of the SA-7 made in China; the original version was the HN-5 capable only of pursuit engagement of jet aircraft (but head-on engagement of helicopters), while the improved **HN-5A** has a more powerful warhead, greater detection range by use of a cooled seeker, and better discrimination against alternative IR sources; the **HN-5C** is the version used on a mobile system introduced by the Chinese in 1986 with eight ready-to-launch missiles (four missiles on each side of the powered fire-control turret and its IR tracker, TV monitor and laser rangefinder) and eight reloads in the hull of the HRB-320 cross-country vehicle

SA-8A and SA-N-4 'Gecko'
(USSR)
Type: land- and ship-based point-defence tactical surface-to-air missile
Dimensions: diameter 0.21 m (8.25 in); length 3.20 m (10.50 ft); span 0.64 m (25.2 in)
Weights: total round 190 kg (419 lb); warhead 40 to 50 kg (88 to 110 lb) proximity-fused HE
Propulsion: one solid-propellant dual-thrust rocket delivering ?-kg (?-lb) thrust
Performance: speed Mach 2; range 1600/12000 m (1,750/13,125 yards); altitude limits 50/13000 m (165/42,650 ft)
Guidance: radar or optical command plus semi-active radar and/or infra-red terminal homing

Variants
SA-8A 'Gecko': introduced to Soviet service in the early 1970s, the single-stage SA-8 system features four ready-to-fire missiles on rail launchers carried on a 6 × 6 chassis believed to be derived from that of the ZIL-167 truck; the launcher system has a 'Land Roll' radar system with a surveillance radar and twin target tracking/missile guidance radars so that a salvo of two missiles can be fired for optimum hit probability; an optical system (including a low-light-level TV) is boresighted with the missile radars for use in conditions in which electronic countermeasures make radar command impossible or problematical; the system is designated **ZRK Romb** (square) in the USSR
SA-8B 'Gecko': upgraded missile (possibly the same as that of the naval SA-N-4 system) first seen in 1980 with a revised launcher arrangement of three containers (rather than two rails) on each side of the radar group); this missile is longer than the original and has greater capabilities, but details are lacking
SA-N-4 'Gecko': naval version of the SA-8B, the SA-N-4 entered service in the early 1970s and is associated with a retractable twin launcher (fed by an 18-round magazine) and the 'Pop Group' radar fire-control system; the missile can also be used in the surface-to-surface role over any range to the radar horizon

SA-9 'Gaskin'
(USSR)
Type: land-based point-defence tactical surface-to-air missile
Dimensions: diameter 0.12 m (4.72 in); length 2.00 m (6.56 ft); span 0.30 m (11.8 in)
Weights: total round 30+ kg (66+ lb); warhead not known
Propulsion: one solid-propellant booster rocket delivering ?-kg (?-lb) thrust, and one solid-propellant sustainer rocket delivering ?-kg (?-lb) thrust
Performance: speed Mach 1.5+; range ?/8000 m (?/8,750 yards); altitude limits ?/4000 m (?/13,125 ft)
Guidance: infra-red homing

Variant
SA-9 'Gaskin': this weapon entered service with the Soviet forces in the late 1960s, and was at first thought in the West to be derived from the SA-7, though more recent thinking attributes the ancestry of the 'Gaskin' to the AA-2 air-to-air missile; the single-stage weapon is normally carried as two container-boxed pairs (one pair on each side of a turret-like launch position) on a converted BRDM-2 4 × 4 reconnaissance vehicle; the weapon has severe tactical limitations in that it is restricted to clear-weather operations, and the lack of any remote-sensing apparatus entails the operator having to traverse the whole turret continuously in his search for targets

SA-10 'Grumble' and SA-N-6
(USSR)
Type: land- and ship-based medium-range area-defence tactical surface-to-air missile
Dimensions: diameter 0.45 m (17.72 in); length 7.00 m (22.97 ft); span 1.20 m (3.94 ft)
Weights: total round 1500 kg (3,307 lb); warhead 100 kg (220 lb) proximity-fused FRAG-HE
Propulsion: one solid-propellant rocket delivering ?-kg (?-lb) thrust
Performance: speed Mach 6; range 9.5/65 km (5.9/40.4 miles); altitude limits 10/30500 m (33/100,065 ft)
Guidance: semi-active radar for the cruise phase and active radar terminal homing

Variants

SA-10 'Grumble': entering service in 1980, the single-stage SA-10 is a highly capable weapon at first associated (**SA-10A** version) with three different tower-mounted surveillance radars and fixed launch sites for the interception of cruise missiles; the type has since been developed (as the **SA-10B** version) for mobile operations, this high-speed missile being deployed on four-round transporter/erector/launchers located on 8 × 8 trucks, with the associated planar-array radar located on a similar chassis

SA-N-6: entering service in 1978, this is the naval version of the SA-10, designed for vertical launch from underdeck silos each fed by an eight-round rotary magazine; mid-course target illumination is provided by the launch ship's 'Top Dome' radar

SA-11 'Gadfly' and SA-N-7
(USSR)

Type: land- and ship-based two-stage short/medium-range area-defence tactical surface-to-air missile
Dimensions: diameter 0.40 m (15.75 in); length 5.60 m (18.37 ft); span 1.20 m (3.94 ft)
Weights: total round 650 kg (1,433 lb); warhead 90 kg (198 lb)
Propulsion: one jettisonable solid-propellant booster rocket delivering ?-kg (?-lb) thrust, and one solid-propellant sustainer rocket delivering ?-kg (?-lb) thrust
Performance: speed Mach 3; range 3/30 km (1.86/18.6 miles); altitude limits 30/15000 m (100/49,215 ft)
Guidance: radio command plus semi-active monopulse radar terminal homing

Variants

SA-11 'Gadfly': introduced in the early 1980, this two-stage missile is the Soviet replacement for the SA-6 'Gainful', a system of severe tactical limitation as each four-vehicle launch battery (12 missiles) has only one 'Straight Flush' guidance radar, limiting the battery to single-target engagement capability; the SA-11 system uses four rather than three rails on each launch vehicle, which is also provided with its own guidance radar (probably a derivative of the 'Front Dome' radar used in the SA-N-7 naval version); the system thus offers considerably enhanced performance in terms of multiple target-engagement capability, and also features greater resistance to electronic countermeasures

SA-N-7: naval version of the SA-11, the SA-N-7 was introduced in 1981 and is intended to replace the SA-N-1 'Goa' series, and the standard fit is two single launchers with wix to eight 'Front Dome' target-illuminating radars for the engagement of multiple targets

SA-12 'Gladiator'
(USSR)

Type: ground-based medium-range tactical/operational surface-to-air missile
Dimensions: 0.70 m (27.56 in); length 7.00 m (22.97 ft); span 1.30 m (4.27 ft)
Weights: not known
Propulsion: one- or two-stage solid-propellant rocket delivering ?-kg (?-lb) thrust
Performance: speed high supersonic; range ?/100 km (?/62.1 miles); altitudes limits 30/30000 m (100/98,425 ft)
Guidance: active radar homing

Variant

SA-12 'Gladiator': introduced in 1986, the SA-12 is designed as successor to the SA-4 'Ganef', and is a very capable vertical-launch weapon (mounted on a tracked vehicle) able to tackle targets over every altitude band at short and medium ranges; the type is associated with a phased-array radar able to deal with several targets simultaneously, and it is believed that the SA-12 may have limited capability against tactical and theatre ballistic missiles

SA-13 'Gopher'
(USSR)

Type: land-based point-defence tactical surface-to-air missile
Dimensions: diameter 0.12 m (4.72 in); length 2.20 m (7.22 ft); span 0.40 m (15.75 in)
Weights: total round 55 kg (121 lb); warhead 6 kg (13.2 lb) FRAG-HE
Propulsion: one solid-propellant rocket delivering ?-kg (?-lb) thrust
Performance: speed Mach 2; range 500/10000 m (550/10,935 yards); altitude limits 10/10000 m (33/32,810 ft)
Guidance: infra-red homing

Variant

SA-13 'Gopher': a new weapon of which little is known, the one-stage SA-13 (known to the Soviets as the **Strela 10**) is successor to the SA-9 in the role of point defence of Soviet ground forces, and is based on a four-round launcher (two containers on each side of the central ranging radar) on the chassis of MT-LB tracked carrier

SA-14 'Gremlin'
(USSR)

Type: man-portable point defence tactical surface-to-air missile
Dimensions: not known
Weights: not known
Propulsion: one solid-propellant rocket delivering ?-kg (?-lb) thrust

Performance: speed supersonic; range ?/7000 m (?/7,655 yards); altitude limits not known
Guidance: laser beam-riding

Variant
SA-14 'Gremlin': this system was introduced in the mid-1980s as replacement for the elderly and limited SA-7 series, and virtually nothing is known of the type; the missile can be distinguished by its conical nose, which projects from the front of the revised launcher

SA-X-15 and SA-N-9
(USSR)
Type: short/medium-range surface-to-air missile
Dimensions: (estimated) 0.60 m (23.6 in); length 3.50 m (11.48 ft); span not known
Weights: not known
Propulsion: solid-propellant rocket delivering ?-kg (?-lb)
Performance: speed not known; range ?/16000 m (?/17,500 yards); altitude limits 18/18000 m (60/59,055 ft)
Guidance: radar command

Variant
SA-X-15: virtually nothing is known of this verticall-launched short-range defence missile, which is probably under development as the replacement for the SA-8 'Gecko' on the basis of the SA-N-9 system
SA-N-9: ship-launched missile first seen in the form of the vertical launchers carried on the 'Udaloy' class, *Novorossiysk* ('Kiev' class) and *Frunze* ('Kirov' class) ships, which were initially thought to be for the SA-NX-8 missile; the SA-N-9 is designed as successor to the SA-N-4 'Goa' series, and is matched to the 'Top Mesh/Top Plate' 3D surveillance and 'Cross Sword' guidance radars; the missiles are accommodated in an eight-round underdeck magazine that rotates to bring a missile into the single launch position under the hatch

SA-N-3 'Goblet'
(USSR)
Type: ship-based medium-range area-defence tactical/operational surface-to-air missile
Dimensions: diameter 0.70 m (27.6 in); length 6.40 m (21.00 ft); span 1.70 m (5.58 ft)
Weights: total round 540 kg (1,190 lb); warhead 25-kiloton nuclear or 150 kg (331 lb) proximity-fused FRAG-HE
Propulsion: one solid-propellant two-stage rocket delivering ?-kg (?-lb) thrust
Performance: speed Mach 2.8; range (early version) 6/30 km (3.7/18.6 miles) or (later version) 6/55 km (3.7/34 miles); altitude limits 90/24500 m (295/80,380 ft)
Guidance: radar command

Variant
SA-N-3 'Goblet': at first thought to be a derivative of the SA-6 'Gainful', this powerful SAM entered Soviet service in 1967 as partial successor to the SA-N-1 'Goa' system; the type is fired from a twin launcher, and the radar fire-control system is the 'Head Light' system; the missile can also be used in the surface-to-surface role over ranges as far as the radar horizon

Bristol/Ferranti Bloodhound Mk 2
(UK)
Type: land-based medium-range area-defence tactical surface-to-air missile
Dimensions: diameter 21.5 in (0.546 m); length 25.42 ft (7.75 m); span 9.29 ft (2.83 m)
Weights: total round 5,070 lb (2300 kg); warhead not revealed
Propulsion: four Royal Ordnance jettisonable solid-propellant booster rockets each delivering ?-kg (?-lb) thrust, and two Rolls-Royce (Bristol) Thor sustainer ramjets each delivering ?-kg (?-lb) thrust
Performance: speed Mach 3.6; range 50+ miles (80+ km); altitude limits 325 to 75,500 ft (100 to 23000 m)
Guidance: Ferranti semi-active radar homing

Variant
Bloodhound Mk 2: introduced in 1964 as replacement for the Bloodhound Mk 1 which had entered service in 1958, the Bloodhound Mk 2 is air-transportable and land-mobile, the complete system comprising four launchers, a target-illuminating radar and a launch control post, the latter two using data provided by a separate surveillance radar; the large HE warhead is detonated by a proximity fuse

Shorts Blowpipe
(UK)
Type: man-portable point-defence tactical surface-to-air missile
Dimensions: diameter 3.00 in (0.076 m); length 4.56 ft (1.39 m); span 10.9 in (0.274 m)
Weights: total round 24.5 lb (11.1 kg); warhead 4.85 lb (2.2 kg) proximity- and impact-fused shaped-charge and blast HE
Propulsion: one Royal Ordnance two-stage solid-propellant rocket delivering ?-kg (?-lb) thrust
Performance: speed Mach 1.5; range ?/4,400+ yards (?4025+ m); altitude limits 0/6,600 ft (0/2010 m)
Guidance: radio command to line of sight after infra-red gathering

Variant
Blowpipe: a capable weapon proved in combat against targets as difficult as high-speed crossing aircraft, the Blowpipe comes in a sealed container, the whole unit weighing 28.7 lb (13 kg) and being clipped to a 19.6-lb (8.9-kg) aiming unit; the operator

acquires the target visually with his monocular sight, fires the missile and then guides it to the target with a thumb controller before discarding the empty 4.59-ft (1.40-m) launcher tube and clipping on another round; the Blowpipe can also be installed on a pedestal mount (land and naval applications) and on the turret ring of light AFVs

Shorts Javelin
(UK)
Type: man-portable point-defence tactical surface-to-air missile
Dimensions: diameter not revealed; length 4.56 ft (1.39 m); span not revealed
Weights: total round not revealed; warhead not revealed
Propulsion: one Royal Ordnance dual-thrust solid-propellant rocket delivering ?-lb (?-kg) thrust
Performance: speed supersonic; range ?/2.5 miles (?/4+ km); altitude limits not revealed
Guidance: semi-automatic command to line of sight

Variants
Javelin: this is a development of the Blowpipe designed to deal more effectively with battlefield targets such as combat helicopters, which can launch their anti-tank missiles at ranges of more than 2.5 miles (4 km); to provide the necessary range a more powerful sustainer is fitted, and destructive capability is enhanced by the provision of a new-pattern warhead; targeting at greater range is aided by the use of semi-automatic command to line of sight guidance, requiring the operator merely to keep the target centred in his sight
Sea Javelin: proposed naval development of the land-based Javelin designed to provide warships ranging in size from fast attack craft to destroyers with a point defence SAM capability; launch tubes for five containerized missiles are mounted on a Laurence-Scott LS30B stabilized weapon platform for operation by a single man, an alternative system being integration with the ship's sensors and fire-control equipment

Shorts Sea Cat
(UK)
Type: ship-based point-defence tactical surface-to-air missile
Dimensions: diameter 7.5 in (0.191 m); length 4.86 m (1.48 m); span 25.6 in (0.65 m)
Weights: total round 150 lb (68 kg); warhead 22 lb (10 kg) proximity and impact delay-fused FRAG-HE
Propulsion: one IMI dual-thrust solid-propellant rocket delivering ?-kg (?-lb) thrust
Performance: speed Mach 0.9; range ?/6000 yards (?/5485 m); altitude limits 100/3,000 ft (30/915 m)
Guidance: radio command to line of sight with optical gathering and tracking

Variants
Sea Cat GWS 20: introduced to service in the early 1960s, the Sea Cat is a lightweight optically-controlled point-defence missile well-suited to the requirements of small ships and small navies; the missile is generally carried on quadruple launchers, though a lightweight triple mounting is available for smaller vessels, and the operator acquires his target through binoculars before firing the missile, acquiring it visually and guiding it to the target; the latest version of the Sea Cat is being developed with anti-radiation homing, though it is unclear whether this is intended for anti-ship or anti-aircraft use
Sea Cat GWS 21: Sea Cat system linked with a radar fire-control system for blind-fire capability; the radar fire-control systems used by navy's other than the Royal Navy include the Contraves Sea Hunter, San Giorgio NA9 and Hollandse Signaalapparaten WM-40 series
Sea Cat GWS 22: Sea Cat system linked with the MRS3/GWS 22 gun/missile fire-control system with Types 903 and 904 radars for blind-fire capability
Sea Cat GWS 24: Sea Cat system linked with a Marconi closed-circuit TV fire-control system
Tigercat Mk 1: land-mobile point-defence version on a triple launcher carried by a two-wheel trailer
Tigercat Mk 2: lightweight version of the Tigercat Mk 1 with solid-state electronics and general lightening of the system

Bae Sea Dart
(UK)
Type: ship-based two-stage medium-range area-defence tactical surface-to-air missile
Dimensions: diameter 16.5 in (0.42 m); length 14.44 ft (4.40 m); span 36.00 in (0.914 m)
Weights: total round 1,210 lb (549 kg); warhead not revealed
Propulsion: one jettisonable IMI solid-propellant booster rocket delivering 35,275-lb (16000-kg) thrust, and one variable-thrust Rolls-Royce Odin sustainer ramjet
Performance: speed Mach 3.5; range ?/50+ miles (?/80+ km); altitude limits 100/82,000 ft (30/24995 m)
Guidance: GEC/Sperry semi-active radar homing

Variants
Sea Dart GWS 30: this is a highly capable third-generation naval surface-to-air missile designed for all-altitude engagements at long range, and also for limited surface-to-surface use at ranges out to 18 miles (29 km); the GWS 30 fire-control system is associated with the Type 909 missile-guidance radar after the ship's Type 965 radar has provided long-range warning and the designation of the target in three dimensions to the twin-rail launcher and Type 909 radars; the proximity-fused FRAG-HE warhead is capable of dealing with all aerial targets

Lightweight Sea Dart: this version is suitable for smaller warships, and comprises a fixed launcher (for containerized Sea Dart rounds), a small fire-control system and Marconi 805SD tracking/illuminating radar

BAe Sea Wolf
(UK)
Type: ship-based point-defence tactical surface-to-air missile
Dimensions: diameter 7.1 in (0.18 m); length 6.25 ft (1.91 m); span 22.00 in (0.56 m)
Weights: total round 176 lb (79.8 kg); warhead 29.5 lb (13.4 kg) proximity- and impact-fused FRAG-HE
Propulsion: one Bristol/RPE Blackcap solid-propellant rocket delivering ?-kg (?-lb) thrust
Performance: speed Mach 2+; range ?/7,000 yards (?/6400 m); altitude limits 15/10,000 ft (5/3000 m)
Guidance: Marconi radio and semi-automatic command to line of sight with radar and/or infra-red missile and target tracking

Variants
Sea Wolf GWS 25: designed during the 1960s and introduced in 1979, the Sea Wolf is a highly capable point-defence system capable of dealing with aircraft, missiles and even shells; the complete **Sea Wolf GWS 25 Mod 0** equipment (sextuple launcher, automatic reloader, 30-round magazine, fire-control system, Marconi-Elliott TV tracker, Type 910 radar etc) is both bulky and weighty (making it difficult to instal the system on ships of less than 2,500-ton displacement), and the desire to instal two such systems on the double-ended 'Type 22' class frigates dictated the eventual dimensions and displacement of the ships; the later **Sea Wolf GWS 25 Mod 3** makes use of the Marconi 805SW (Type 911) radar for a considerably lighter installation suitable for ships down to 1,000-ton displacement
Sea Wolf VM40: this is a lightweight system devised by BAe and Hollandse Signaalapparaten using the HSA STIR tracking radar and a converted four-round Sea Cat launcher with automatic reloading or (considerably lighter) four disposable plastic or long-life metal launch tubes
Containerized Sea Wolf VM40: containerized version of the Sea Wolf VM40 system in two standard containers suitable for installation on warships or auxiliaries
Vertical-Launch Sea Wolf GWS 26: this system is being developed for British warships and for export, the type offering faster response time, greater range (11,000 yards/10060 m) and reduced system weight; current thinking advocates a 32-cell launcher designed by IMI or developed by Vickers from the Martin-Marietta VLS for the US Navy
Vertical-Launch Sea Wolf GWS 27: private-venture development by BAe and Marconi for autonomous fire-and-forget capability with enhanced electronic counter-countermeasures capability (through the installation of an active radar seeker operating in conjunction with SACLOS or inertial mid-course guidance) over an increased maximum range of 6.2 miles (10 km)

BAe Rapier
(UK)
Type: land-based point-defence tactical surface-to-air missile
Dimensions: diameter 5.25 in (0.133 m); length 7.35 ft (2.24 m); span 15.0 in (0.381 m)
Weights: total round 94 lb (42.6 kg); warhead 1.1 lb (0.5 kg) impact-fused SAP HE
Propulsion: one IMI Troy dual-thrust rocket delivering ?-kg (?-lb) thrust
Performance: speed Mach 2+; range 900/7,100+ yards (825/6490 m); altitude limits 0/10,000 ft (0/3050 m)
Guidance: Marconi/Decca/Barr & Stroud microwave semi-automatic command to line of sight with optical gathering and tracking

Variants
Rapier: one of the most important land-mobile SAMs yet to appear, the Rapier was conceived in the 1960s as a point-defence weapon whose extreme accuracy would ensure impact with the target, so doing away with the need for a large warhead with proximity and impact fuses; the weapon began to enter service in 1971 and has since been developed into a number of operating forms all based on the same missile; the basic clear-weather Rapier system comprises a four-round launcher on a trailer, a generator and an optical tracker, the missile being slaved to the operator's line of sight; the basic system can also be updated with a surveillance radar and improved capability against electronic countermeasures; other systems using the missile are **Tracked Rapier** (with the fire unit and optical/thermal-imaging tracker on a converted M548 tracked carrier), **Rapier Laserfire** (a pallet-mounted system of greatly reduced cost and marginally reduced capability, comprising a surveillance radar, Racal/Ferranti laser tracker and missile guidance system, control post, command link and four-round launcher) and **Rapier Blindfire** (Marconi Blindfire surveillance radar, four-round launcher and automatic command to line of sight control system); the weapon is still being improved, largely with digital rather than analog electronics; the latest version is the **Rapier Darkfire** using an electro-optical guidance system based on infra-red tracking of missiles fired from a six- rather than four-round launcher for enhanced tactical capability by day and night
Rapier Mk 2: updated missile under development for the **Rapier 2000** system, which is based on three

towed vehicles designed for minimum radio-active contamination and maximum resistance to electromagnetic impulse effects; one vehicle has eight ready-to-launch Rapiers and an electro-optical tracking system, one has Plessey 3D surveillance and target-acquisition radar, and one has improved Blindfire 2000 tracking and fire-control radar; the revised missile has a Royal Ordnance Thermopylae rocket motor for a range of 5 miles (8 km), and comes in two forms as the **Rapier Mk 2A** with a FRAG-HE warhead and 'smart' proximity fuse (for the engagement of RPVs and missiles) and the **Rapier Mk 2B** with an impact-fused hollow-charge warhead (for the engagement of aircraft), and the fire-control system is capable of the simultaneous engagement of two targets with two missiles

Shorts Tigercat (UK): see Shorts Sea Cat

Martin-Marietta/Oerlikon ADATS (USA/Switzerland): see Martin-Marietta/Oerlikon ADATS anti-tank missile

General Dynamics RIM-2D(N) Terrier
(USA)
Type: ship-based medium-range area-defence tactical/operational surface-to-air missile
Dimensions: diameter (missile) 13.5 in (0.343 m) and (booster) 18.00 in (0.457 m); length (missile) 13.50 ft (4.115 m) and (booster) 25.83 ft (7.874 m); span (missile) 3.50 ft (1.07 m)
Weights: missile 1,180 lb (535.2 kg); booster 1,820 lb (825.5 kg); warhead 1-kiloton W45 nuclear
Propulsion: one Allegany Ballistics solid-propellant booster rocket delivering ?-lb (?-kg) thrust, and one Allegany Ballistics solid-propellant sustainer rocket delivering ?-lb (?-kg) thrust
Performance: speed Mach 3; range ?/23 miles (?/37 km); altitude limits 500/80,000 ft (150/24385 m)
Guidance: radar beam-riding

Variants
RIM-2D(N) Terrier: introduced in 1958, this nuclear-armed version of the RIM-2D dedicated air-defence SAM was developed to provide US surface combatants with a nuclear-armed surface-to-air missile with powerful surface-to-surface capability out to the radar horizon; most other variants of the extensive Terrier family have been or are being replaced in the surface-to-air role by the RIM-66/RIM-67 Standard series; associated with the Terrier is the Mk 10 twin launcher and magazines holding 40, 60 or 80 reload rounds
RIM-2F Terrier: introduced in 1963, this conventionally-armed SAM has an overall length of 26.17 ft (7.98 m), a weight of 3,090 lb (1402 kg) and propulsion comprising an Atlantic Research Mk 30 Mod 2 solid-propellant booster rocket and a Naval Ordnance Station Indian Head Mk 12 Mod 1 (later Thiokol Mk 70 Mod 1) solid-propellant sustainer rocket; the variant uses semi-active radar homing

Raytheon RIM-7M Sea Sparrow
(USA)
Type: ship-based point-defence tactical surface-to-air missile
Dimensions: diameter 8.00 in (0.203 m); length 13.08 ft (3.99 m); span 40.00 in (1.02 m)
Weights: total round 503 lb (228.2 kg); warhead 88 lb (40 kg) proximity- and impact-fused FRAG-HE
Propulsion: one Hercules Mk 58 Mod 4 two-stage solid-propellant rocket delivering ?-lb (?-kg) thrust
Performance: speed Mach 3+; range ?/24,250 yards (?/22175 m); altitude limits 25/50,000 ft (8/15240 m)
Guidance: semi-active radar homing

Variants
RIM-7E5 Sea Sparrow: developed in the early 1960s from the AIM-7E Sparrow air-to-air missile, the RIM-7E5 was adopted in 1969 as the US Navy's Basic **Point Defense Missile System** together with the Mk 25 octuple launcher located on a converted 3-in (76-mm) gun mounting; associated with this system are the analog Mk 115 fire-control system and the Mk 51 manually-controlled director and illuminator; a version of the same missile, the AIM-7E2 air-to-air missile with reduced minimum engagment range plus enhanced manoeuvrability, was adopted for the **Canadian Sea Sparrow System**, which uses a four-round launcher associated with the Hollandse Signaalapparaten WM-22/6 fire-control system with monopulse track while scan, moving target indication search, pulse-Doppler tracking and continuous-wave target illumination
RIM-7F Sea Sparrow: the missile is derived from the AIM-9F with improved capabilities against low-flying targets, and is compatible with the NATO **Sea Sparrow System** and its Mk 29 octuple launcher, digital fire-control system, and powered tracker and illuminator
RIM-7H5 Sea Sparrow: RIM-7E5 reconfigured for compatibility with the Mk 29 launcher and its associated electronics
RIM-7M Sea Sparrow: improved missile based on the AIM-7M with its monopulse seeker, and adopted in 1983
Sparrow HAWK: company-funded development of the Sparrow SAM installed for land applications on the launcher of the NIM-23 HAWK system, the result being a nine- rather than three-round launcher with more rapid reload capability than with the considerably heavier HAWK missiles

Western Electric MIM-14B Nike Hercules
(USA)
Type: ground-based two-stage long-range tactical/operational surface-to-air missile
Dimensions: diameter 31.50 in (0.80 m); length overall 41.00 or 41.50 ft (12.50 or 12.65 m); span 6.17 ft (1.88 m)

Weights: total round 10,415 or 10,710 lb (4720 or 4858 kg); warhead 1,120 lb (508 kg) 2-, 20- or 40-kiloton interchangeable W31 nuclear or FRAG-HE
Propulsion: four jettisonable Hercules solid-propellant booster rockets each delivering ?-lb (?-kg) thrust, and one Thiokol M30A1 solid-propellant sustainer rocket delivering ?-lb (?-kg) thrust
Performance: speed Mach 3.65; range ?/90+ miles (?/145+ km); altitude limits ?/150,000 ft (?/45720 m)
Guidance: radar command

Variant
MIM-14B Nike Hercules: introduced in 1958 as successor to the first-generation Nike Ajax long-range surface-to-air missile, the Nike Hercules is still in service though obsolescent; its replacement in the US inventory is the NIM-104 Patriot; key features of the Nikes Hercules' design are its range and altitude capability, the capable General Electric HIPAR radar, and the powerful warhead (only the USA having fielded the nuclear type)
MIM-14C Nike Hercules: improved MIM-14B

Raytheon MIM-23B Improved HAWK
(USA)
Type: land-based medium-range area-defence tactical surface-to-air missile
Dimensions: diameter 14.00 in (0.356 m); length 16.79 ft (5.12 m); span 4.00 ft (1.22 m)
Weights: total round 1,398 lb (634 kg); warhead 120+ lb (54.4+ kg) proximity-fused FRAG-HE
Propulsion: one Aerojet M-22E8 dual-thrust solid-propellant rocket delivering ?-lb (?-kg) thrust
Performance: speed Mach 2.5; range ?/25 miles (?/40 km); altitude limits 100/49,000 ft (30/17985 m)
Guidance: Raytheon semi-active radar homing

Variants
MIM-23A HAWK: baseline model introduced as the Homing All-the-Way Killer land-mobile air-defence missile during 1960
MIM-23B Improved HAWK: improved model introduced in the 1970s with an upgraded engine, new guidance package and larger warhead; the entire HAWK system is effective but cumbersome, being designed for limited battlefield mobility by wheeled transport in several parts, the most important being the three-round launcher on a two-wheel trailer (three launchers to a battery), the pulse-Doppler acquisition radar, the continuous-wave acquisition radar, the range-only radar, the continuous-wave illuminator and the battery control centre; there are various improvement standards for the MIM-23B system, based largely on improved radar capability at low levels and the adoption of electro-optical tracking systems
MIM-23C Improved HAWK: upgraded MIM-23B
MIM-23D Improved HAWK: upgraded MIM-23C
MIM-23E Improved HAWK: variant of the MIM-23C with improved guidance in a multi-jamming environment
MIM-23F Improved HAWK: variant of the MIM-23D with improved guidance in a multi-jamming environment
M727 SP HAWK: self-propelled version mounted on an M548 tracked carrier chassis
NOAH: this acronym stands for NOrway Adapted HAWK, and indicates a development of the basic weapon using Hughes LASR 3D radar to replace the HAWK battery's current array of search radars; other advantages of the LASR are good results in clutter conditions, and high resistance to electronic countermeasures

General Dynamics RIM-24B Improved Tartar
(USA)
Type: ship-based medium-range area-defence tactical surface-to-air missile
Dimensions: diameter 13.50 in (0.343 m); length 15.50 ft (4.724 m); span 24.00 in (0.609 m)
Weights: total round 1,325 lb (601 kg); warhead not revealed
Propulsion: one Aerojet Mk 27 Mod 2/3 dual-thrust solid-propellant rocket delivering ?-lb (?-kg) thrust
Performance: speed Mach 1.8; range ?/20 miles (?/32.2 km); altitude limits 59/70,000 ft (15/21335 m)
Guidance: semi-active radar homing

Variant
RIM-24B Improved Tartar: introduced in 1963, this is the only version of the RIM-24 Tarter series remaining in service; it is a comparatively light medium-range weapon (with a dual-thrust rocket to obviate the need for a booster) designed for use as the primary weapon of destroyers and lighter vessels, as the secondary armament of cruisers; the basic role is the interception of low- and medium-level attackers with a FRAG-HE warhead, though the type can engage high-altitude attackers and has a surface-to-surface capability out to the radar horizon; the type is associated with the Mk 11 twin launcher (plus a magazine for 42 reload rounds) and Mks 13 and 22 single launchers (40 and 16 reload rounds respectively), and with the SPG-51 missile-guidance radar, the combination generally being operated in conjunction with the SPS-52 search radar for early warning

General Dynamics FIM-43A Redeye
(USA)
Type: man-portable point-defence tactical surface-to-air missile
Dimensions: diameter 2.75 in (0.70 m); length 4.00 ft (1.22 m); span 5.40 in (0.14 m)
Weights: total round 18 lb (8.2 kg); warhead not disclosed
Propulsion: one Atlantic Research M115 dual-thrust

solid-propellant rocket delivering ?-lb (?-kg) thrust
Performance: speed supersonic; range ?/3,700 yards (?/3385 m); altitude limits are revealed
Guidance: infra-red homing

Variant
FIM-43A Redeye: developed from the late 1950s in an effort to provide US infantry units with man-portable surface-to-air capability, the Redeye entered service only in 1969 after a protracted development programme, and has never proved entirely successful; the type is notable for its comparatively short range and its restriction to pursuit-only engagements (after the target has launched its attack); the missile is carried in a sealed container, and this is clipped to the sight unit for an engagement; the empty container is discarded after firing and replaced by another round
FIM-43B Redeye: upgraded FIM-43A with the M110 solid-propellant rocket
FIM-43C Redeye: upgraded FIM-43B with the M115 solid-propellant rocket and revised launcher
FIM-43D Redeye: upgraded FIM-43C

General Dynamics RIM-66A/B Standard SM-1 MR
(USA)
Type: ship-based medium-range area-defence tactical surface-to-air missile
Dimensions: diameter 13.50 in (0.343 m); length (RIM-66A) 14.67 ft (4.47 m) or (RIM-66B) 15.50 ft (4.724 m); span 36.00 in (0.914 m)
Weights: total round (RIM-66A) 1,276 lb (578.8 kg) or (RIM-66B) 1,342 lb (608.7 kg); warhead not revealed
Propulsion: one Aerojet/Hercules Mk 56 Mod 0 dual-thrust solid-propellant rocket delivering ?-lb (?-kg) thrust
Performance: speed Mach 2+; range (RIM-66A) ?/28.75 miles (?/46.25 km) or (RIM-66B) ?/41.6 miles (?/67 km); altitude limits (RIM-66A) ?/50,000 ft (?/15240 m) or (RIM-66B) ?/62,500 ft (?/19050 m)
Guidance: semi-active radar homing

Variants
RIM-66A Standard SM-1 MR: designed from the early 1960s as replacement for the Tartar short/medium-range area-defence SAM, the Standard SM-1 MR series reflected a growing realization in the US Navy that the primary threat faced in Soviet air attack was saturation of the American warships' target-illumination capabilities as the Terrier and Tartar systems require a dedicated illuminator for each missile; the RIM-66A (notable as the first US Navy missile with all-solid-state electronics, though still of the analog type) entered service in 1968, and is generally associated with the Mk 11/12 twin launcher (42 missiles), Mk 13 single launcher (40 missiles), Mk 22 single launcher (16 missiles) and Mk 26 single launcher (24, 44 or 64 missiles); the type has a horizon-limited surface-to-surface capability, and a proximity- and impact-fused HE warhead
RIM-66B Standard 1-MR: improved RIM-66A with upgraded motor for increases of 45% in range and 25% in maximum engagement altitude, but otherwise having the capabilities of the RIM-66A

General Dynamics RIM-67A Standard SM-1 ER
(USA)
Type: ship-based two-stage medium/long-range area-defence tactical surface-to-air missile
Dimensions: diameter 13.50 in (0.343 m); length 26.17 ft (7.98 m); span 36.00 in (0.914 m)
Weights: total round 1,360 lb (616.9 kg); warhead not revealed
Propulsion: one jettisonable Naval Propellant Plant Mk 12 Mod 1 solid-propellant booster rocket delivering ?-lb (?-kg) thrust, and one Atlantic Research Mk 30 Mod 2 solid-propellant sustainer rocket delivering ?-lb (?-kg) thrust
Performance: speed Mach 2.5+; range ?/46 miles (?/74 km); altitude limits ?/80,000 ft (?/24385 m)
Guidance: semi-active radar homing

Variant
RIM-67A Standard SM-1 ER: produced in parallel with the RIM-66A and RIM-66B as a longer-range weapon to replace the Terrier and Talos systems of the previous generation, the RIM-67A has a different propulsion arrangement (a booster stage being appended to the base of the missile proper) for greater range and a higher altitude limit; the ER (Extended Range) versions of the Standard missile are associated with the Mk 10 twin launcher and a magazine for 40, 60 or 80 missiles depending on the size of the parent ship

General Dynamics RIM-66C Standard SM-2 MR
(USA)
Type: ship-based medium-range area-defence tactical surface-to-air missile
Dimensions: diameter 13.50 m (0.343 m); length 15.50 ft (4.724 m); span 36.00 in (0.914 m)
Weights: total round 1,342 lb (608.7 kg); warhead not revealed
Propulsion: one Aerojet/Hercules Mk 56 Mod 0 dual-thrust solid-propellant rocket delivering ?-lb (?-kg) thrust
Performance: speed Mach 2+; range ?/56 miles (?/74 km); altitude limits ?/80,000 ft (?/24385 m)
Guidance: semi-active radar homing

Variants
RIM-66C Standard SM-2 MR: introduced to match the AEGIS weapon system (SPY-1A phased-array radar, Mk 99 fire-control system, Mk 1 weapons-control system and Mk 1 command and decision

system), the SM-2 series is an electronic development of the SM-1 type with digital electronics including a monopulse homing receiver (instead of the SM-1's conical-scan semi-active homing) with greater resistance to electronic countermeasures, an inertial reference unit for mid-course guidance, and a data-link for mid-course updating and thus optimization of the flight profile by making possible a more energy-efficient trajectory to increase range; in the SM-2 MR model this produces a 60% increase in range compared with that of the SM-1 MR
Standard SM-2 MR(N): version under development with a W81 low-yield nuclear warhead, providing the US Navy with a capable tactical/operational weapon against high-threat attacks on major units (together with a powerful surface-to-surface capability out to the radar horizon, or beyond it with the aid of mid-course update from a targeting helicopter)

General Dynamics RIM-67B Standard SM-2 ER
(USA)
Type: ship-based two-stage long-range area-defence tactical surface-to-air missile
Dimensions: diameter 13.50 in (0.343 m); length 27.00 ft (8.23 m); span 36.00 in (0.914 m)
Weights: total round 3,058 lb (1387.1 kg); warhead not revealed
Propulsion: one jettisonable Naval Propellant Plant Mk 12 Mod 1 solid-propellant booster rocket delivering ?-lb (?-kg) thrust, and one Atlantic Research Mk 30 Mod 2 solid-propellant sustainer rocket delivering ?-lb (?-kg) thrust
Performance: speed Mach 2.5+; range ?/92 miles (?/148 km); altitude limits ?/100,000 ft (?/30480 m)
Guidance: semi-active radar homing

Variants
RIM-67B Standard SM-2 ER: this is the extended-range equivalent of the RIM-66B, and has the same electronic improvements for optimum trajectory, in this case boosting range by 100%
Standard SM-2 ER(N): version under development with a W81 low-yield nuclear warhead, and offering the same type of capabilities as the SM-2 MR(N) though having greater need of mid-course guidance if the full range capability of the missile is to be used in the surface-to-surface role

Ford MIM-72C/R Improved Chaparral
(USA)
Type: man-portable point-defence tactical surface-to-air missile
Dimensions: 5.12 in (0.13 m); length 9.54 ft (2.91 m); span 24.76 in (0.629 m)
Weights: total round 190 lb (86.3 kg); warhead 24.7 lb (11.2 kg) proximity-fused M250 FRAG-HE
Propulsion: one Bermite Mk 50 Mod 0 solid-propellant rocket delivering ?-lb (?-kg) thrust

Performance: speed supersonic; range ?/3.7 miles (?/6 km); altitude limits 1,150/10,000 ft (350/3050 m)
Guidance: Ford Aerospace infra-red homing

Variants
MIM-72A Chaparral: this was the initial Sidewinder AAM-derived missile used in the M48 Chaparral system in the late 1960s; the weapon weighs 187 lb (85 kg), and the system was designed to provide US Army land formations with low-level air defence to the end of the 20th century
MIM-72B Chaparral: improved model developed in the 1970s with a more powerful warhead, more sensitive General Electric/Raytheon seeker and better fuse; the weight of this model is 190 lb (86.3 kg), and the system is used in conjunction with the MPQ-94 surveillance radar, the warned operator traversing the four-round turret to acquire the target visually and fire a missile
MIM-72C/R Chaparral: version of the late 1970s with a more discriminating Ford Aerospace seeker, smokeless propellant to reduce the chance of the launcher being detected, and a forward-looking infra-red sensor for adverse-weather and nocturnal operation
MIM-72D Improved Chaparral: experimental development of the MIM-72C with an improved warhead and directional Doppler fuse; the type was developed further as the **MIM-72E** and **MIM-72F** versions
MIM-72C Sea Chaparral: this is the naval version of the system, used only by the Taiwanese naval for the point-defence of major units; the missile is identical with the land-based weapon, and the launcher incorporates a minimum of change
Bodensee Shorad: this West German development is a trailer-based short-range air defence system using the AIM-9L Sidewinder air-to-air missile on a four-round launcher with a remotely-controlled TV camera, and is designed for the point defence of West German airfields

General Dynamics FIM-92A Stinger
(USA)
Type: man-portable point defence tactical surface-to-air missile
Dimensions: diameter 2.75 in (0.07 m); length 5.00 ft (1.52 m)
Weights: total round 22.3 lb (10.1 kg); warhead 6.6 lb (3.0 kg) proximity-fused FRAG-HE
Propulsion: one Atlantic Research dual-thrust solid-propellant rocket delivering ?-lb (?-kg) thrust
Performance: speed Mach 2+; range ?/5,500 yards (?/5030 m); altitude limits ?/15,750 ft (?/4800 m)
Guidance: General Dynamics (Pomona) infra-red homing

Variants
FIM-92A Stinger: the Stinger was designed as

successor to the Redeye with all-aspect engagement capability, greater performance and manoeuvrability, an IFF system and enhanced resistance to electronic countermeasures; the weapon was introduced in 1981 and has proven generally successful; the missile and launcher weigh 30 lb (13.6 kg), or 33.3 lb (15.1 kg) with the IFF system and battery
Stinger-POST: this advanced model currently under development has a Passive Optical Scanning Technique seeker to provide better discrimination between the target and the ground for low-altitude engagements

Raytheon MIM-104A Patriot
(USA)
Type: land-based medium-range area-defence tactical/operational surface-to-air missile
Dimensions: diameter 16 in (0.406 m); length 17.00 ft (5.18 m); span 3.00 ft (0.914 ft)
Weights: total round 2,200 lb (998 kg); warhead not revealed
Propulsion: one Thiokol TX-486-1 solid-propellant rocket delivering ?-lb (?-kg) thrust
Performance: speed Mach 3; range ?/37.3 miles (?/60 km); altitude limits ?/78,750 ft (?/24000 m)
Guidance: radar command (track via missile) and Raytheon semi-active radar homing

Variants
MIM-104A: entering service in 1986 after a protracted development, the Patriot is a highly capable medium-range air-defence weapon designed to replace the Improved HAWK and Nike Hercules; each launcher system has four containerized missiles on a semi-trailer, and up to eight (normally five) launchers can be controlled by the MSQ-104 engagement centre working with the MPQ-53 surveillance and target-acquisition radar
Tien Kung I: this is a Taiwanese surface-to-air missile based conceptually on the Patriot; no details have been released, but it is known that the type entered production in 1986

Tien Kung II: radical development of the Tien Kung I slated for service entry in about 1990; the type has integral rocket/ramjet propulsion for high speed and long range

General Dynamics RIM-116A RAM
(USA)
Type: ship-based point-defence tactical surface-to-air missile
Dimensions: diameter 5 in (0.127 m); length 9.17 ft (2.79 m); span 17.25 in (0.438 m)
Weights: total round 159 lb (72.1 kg); warhead 22.4 lb (10.2 kg) proximity- and impact-fused FRAG-HE
Propulsion: one Bermite/Hercules Mk 36 Mod 7 solid-propellant rocket delivering 3,000-lb (1361-kg) thrust
Performance: speed Mach 3; range ?/10,300 yards (?/9420 m); altitude limits not revealed
Guidance: General Dynamics semi-active radar and infra-red terminal homing

Variant
RIM-116A RAM: designed as a missile counterpart to the Phalanx 20-mm close-in gun system, the Rolling Airframe Missile is intended as a point-defence system for frigates and attack craft; the missile uses proved components wherever possible (the warhead, fuse and motor of the AIM-9 Sidewinder, and the infra-red seeker of the FIM-92 Stinger for example), together with a rolling airframe and passive radar seeker for mid-course guidance and to ensure that the terminal homing system is accurately aligned towards the end of the flight; current launcher options are the EX-31 with 24 tubes on a Phalanx gun mounting for training and elevation (West German and Danish vessels), and five-round inserts fitted into the upper two cells of the Mk 16 octuple ASROC launcher or of the Mks 25 and 29 octuple Sea Sparrow launchers (US vessels)

AIR-TO-AIR MISSILES

IAE MAA-1 Piranha
(Brazil)
Type: short-range air-to-air missile
Dimensions: not revaled
Weights: not revealed
Propulsion: one solid-propellant rocket delivering ?-kg (?-lb) thrust

Performance: not revealed
Guidance: infra-red homing

Variant
MAA-1 Piranha: another good example of the considerable strides being made by the Brazilian armaments industry, this simple yet capable air-to-air

missile began development in 1977 and entered service in 1986

CATIC PL-2
(China)
Type: short-range air-to-air missile
Dimensions: diameter 0.127 m (5 in); length 2.99 m (9.81 ft); span 0.528 m (20.79 in)
Weights: total round 76 kg (167.5 lb); warhead 11.3 kg (24.9 lb) IR proximity-fused FRAG-HE
Propulsion: one solid-propellant rocket delivering ?-kg (?-lb) thrust
Performance: not known, but probably comparable to that of the AA-2 'Atoll'
Guidance: infra-red homing

Variants
PL-2: this is essentially the Chinese version of the Soviet AA-2 'Atoll', itself derived from early models of the AIM-9 Sidewinder; the PL-2 has an uncooled seeker and is thus equivalent to the AIM-9B and restricted to pursuit interceptions at medium and high altitudes
PL-2A: improved pursuit-course missile with greater discrimination against hot spots as a result of its thermo-electric cooling system for the seeker, allowing operations closer to the ground; the missile is basically similar to the AIM-9E conversion of the AIM-9B

CATIC PL-5B
(China)
Type: short-range air-to-air missile
Dimensions: diameter 0.127 m (5 in); length 2.89 m (9.48 ft); span 0.657 m (25.87 in)
Weights: total round 85 kg (187.4 lb); warhead ? kg (? lb) IR proximity-fused continuous-rod or radio proximity-fused FRAG-HE
Propulson: one solid-propellant rocket delivering ?-kg (?-lb) thrust
Performance: speed not revealed; range 16 km (10 miles)
Guidance: infra-red homing

Variant
PL-5B: with double delta control surfaces and 'off-boresight' rather than 'all-aspect' engagement capability (equivalent in some respects to the AIM-9L but with a lead-sulphide seeker cooled by compressed air), this is a useful air-to-air missile reflecting China's continued faith in the basic design, and has considerable agility for snap-up and snap-down engagements

CATIC PL-7
(China)
Type: short-range dogfighting air-to-air missile
Dimensions: not revealed
Weights: not revealed
Propulsion: one solid-propellant rocket delivering ?-kg (?-lb) thrust
Performance: not revealed
Guidance: infra-red homing

Variant
PL-7: this is an 'all-aspect' development of the PL-2/5 series with a continuous-rod warhead and an indium-antimonide seeker cooled by liquid nitrogen; in appearance it is similar to the Matra R550 Magic, and is probably designed as a dogfighting missile

Matra R530
(France)
Type: medium-range air-to-air missile
Dimensions: diameter 0.263 m (10.35 in); length (semi-active radar version) 3.284 m (10.77 ft) or (IR-homing version) 3.198 m (10.49 ft); span 1.103 m (43.4 in)
Weights: total round (semi-active radar version) 192 kg (423.3 lb) or (IR-homing version) 193.5 lb (426.6 lb); warhead 27 kg (59.5 lb) proximity and impact delay-fused continuous-rod HE or FRAG-HE
Propulsion: one Hotchkiss-Brandt/SNPE Antoinette or SNPE Madelain dual-thrust solid-propellant rocket delivering ?-kg (?-lb) thrust
Performance: speed Mach 2.7; range 18000 m (19.685 yards)
Guidance: EMD AD26 semi-active radar homing or SAT AD3501 infra-red homing

Variant
R530: introduced in the early 1960s as armament for the Dassault Mirage III and later used on the Dassault-Breguet Mirage F1, the R530 has been developed in semi-active radar and IR homing versions, neither of which has proved particularly successful in combat

Matra Super R530F
(France)
Type: medium/long-range air-to-air missile
Dimensions: diameter 0.263 m (10.35 in); length 3.54 m (11.61 ft); span 0.90 m (35.4 in)
Weights: total round 250 kg (551 lb); warhead 30+ kg (66+ lb) proximity-fused FRAG-HE
Propulsion: one Thomson-Brandt/SNPE Angéle dual-thrust solid-propellant rocket delivering ?-kg (?-lb) thrust
Performance: speed Mach 4.6; range 35 km (21.75 miles)
Guidance: EMD Super AD26 semi-active radar homing

Variants
Super R530F: developed from the R530 series in the early 1970s as armament for the Dassault-Breguet Mirage F1, this high-speed missile is matched to the

Cyrano IV radar, and has snap-up/snap-down capabilities against aircraft 7000 m (22,695 ft) above/below the launch aircraft
Super R530D: improved version matched to the RDI and RDM pulse-Doppler radars of the Dassault-Breguet Mirage 2000 series; this version has an improved motor for a range of up to 60 km (37.3 miles), and the altitude differential in the snap-up/snap-down mode is 9000 m (29,530 ft); the missile can deal with Mach 3 targets at a height of 24000 m (78,740 ft) and, considerably more importantly, with targets flying fast and low in ground clutter conditions

Matra R550 Magic 1
(France)
Type: short-range dogfighting air-to-air missile
Dimensions: diameter 0.157 m (6.18 in); length 2.75 m (9.02 m); span 0.66 m (25.98 in)
Weights: total round 89.8 kg (198 lb); warhead 12.5 kg (27.6 lb) proximity and impact delay-fused FRAG-HE
Propulsion: one SNPE Roméo single-stage solid-propellant rocket delivering ?-kg (?-lb) thrust
Performance: speed Mach 3; range 320/10000 m (350/10,935 yards)
Guidance: SAT AD3601 infra-red homing

Variants
R550 Magic 1: developed in the late 1960s and early 1970s as a French counterpart to the US AIM-9 Sidewinder missile, the Magic is designed for all-aspect (except head-on) engagement at altitudes up to 18000 m (59,055 ft), the engagement envelope being reduced above this altitude; the Magic can be launched by an aircraft manoeuvring at +6g
R550 Magic 2: entering service in the mid-1980s, this improved version has a genuine all-aspect engagement capability, an improved motor, greater structural strength and better aerodynamic controls for enhanced dogfight manoeuvrability, and a seeker that can be slaved to the launch aircraft's radar to provide optimum launch conditions

Matra MICA
(France)
Type: short/medium range air-to-air missile
Dimensions: diameter not revealed; length 3.10 m (10.17 ft); span not revealed
Weights: total round 110 kg (242.5 lb); warhead ? kg (? lb) FRAG-HE
Propulsion: one SNPE solid-propellant rocket delivering ?-kg (?-lb) thrust
Performance: speed not revealed; range ?/60 km (37.3 miles)
Guidance: strapdown inertial plus infra-red or semi-active radar homing

Variant
MICA: due to enter service in the 1990s as France's premier air-to-air missile, the MICA is similar in appearance to the US Standard series SAM (though of course much smaller) and is designed to be launched on the basis of the parent aircraft's primary sensor, cruising under strapdown inertial guidance before the seeker in turned on, the IR type being used for shorter-range air combat and the semi-active radar type being used for longer-ranged interception

BBG AIM-132 ASRAAM
(West Germany/UK)
Type: short-range air-to-air missile
Dimensions: diameter 0.168 m (6.6 in); length 2.73 m (8.96 ft); span not revealed
Weights: total round 100 kg (220.5 lb); warhead 10 kg (22 lb) FRAG-HE
Propulsion: one solid-propellant dual-thrust rocket delivering ?-kg (?-lb) thrust
Performance: speed not revealed; range 300/15000 m (330/16,405 yards)
Guidance: strapdown inertil plus infra-red homing

Variant
AIM-132: this Advanced Short-Range Air-to-Air Missile is under development by BAe and Bodenseewerk Gerätetechnik, and is due to enter service in the early 1990s; the weapon uses body-lift aerodynamics, and is controlled by four rear fins

Rafael Python 3
(Israel)
Type: short-range dogfighting air-to-air missile
Dimensions: diameter 0.16 m (6.3 in); length 3.00 m (9.84 ft); span 0.86 m (33.86 in)
Weights: total round 120 kg (264.6 lb); warhead 11 kg (24.25 lb) proximity and impact delay-fused FRAG-HE
Propulsion: one double-base solid-propellant rocket delivering ?-kg (?-lb) thrust
Performance: speed Mach 3; range 500/15000 m (545/16,405 yards)
Guidance: infra-red homing

Variant
Python 3: developed in the late 1970s and early 1980s as the Shafrir 3, the Python is an advanced development of the basic Shafrir with an improved motor offering greater range, a revised seeker with greater sensitivity for all-aspect engagement capability, and aerodynamic improvements for enhanced manoeuvrability

Rafael Shafrir 2
(Israel)
Type: short-range dogfighting air-to-air missile
Dimensions: diameter 0.16 m (6.3 in); length 2.47 m

(8.10 ft); span 0.52 m (20.47 in)
Weights: total round 93 kg (205 lb); warhead 11 kg (24.25 lb) proximity and impact delay-fused FRAG-HE
Propulsion: one double-base solid-propellant rocket delivering ?-kg (?-lb) thrust
Performance: speed Mach 2.5; range 500/5000 m (545/5,470 yards)
Guidance: infra-red homing

Variant
Shafrir 2: derived from the preproduction Shafrir 1, the Mk 2 version is a capable short-range AAM that has performed well in combat with a 65/70 'kill' percentage

Selenia Aspide
(Italy)
Type: medium-range air-to-air missile
Dimensions: diameter 0.203 m (8 in); length 3.70 m (12.14 ft); span 1.00 m (39.37 in)
Weights: total round 220 kg (485 lb); warhead 33 kg (72.75 lb) proximity and impact delay-fused FRAG-HE
Propulsion: one SNIA-Viscosa single-stage solid-propellant rocket delivering ?-kg (?-lb) thrust
Performance: speed Mach 4; range 70 km (43.5 miles) from a high-altitude launch dropping to 44 km (27.3 miles) from a low-altitude launch
Guidance: Selenia semi-active radar homing

Variant
Aspide: developed conceptually from the Raytheon AIM-7 Sparrow, the Aspide is like the American missile in being a multi-role type (surface-to-air and air-to-air), and entered air-to-air service with the Italian air force in 1978; the type arms the Aeritalia F-104S (and will arm the Panavia Tornado), and has greater low-altitude range than the Sparrow, together with much improved guidance and electronic counter-countermeasures

Mitsubishi AAM-1 (Japan): see Ford Aerospace/Raytheon AIM-9L Sidewinder

Armscor V3B
(South Africa)
Type: short-range dogfighting air-to-air missile
Dimensions: diameter 0.127 m (5 in); length 2.944 m (9.66 ft); span 0.53 m (20.86 in)
Weights: total round 73.4 kg (161.8 lb); warhead 10+ kg (22.05+ lb) proximity and impact delay-fused FRAG-HE
Propulsion: one double-base solid-propellant rocket delivering ?-kg (?-lb) thrust
Performance: speed Mach 3; range 300/5000 m (330/5,470 yards) at high altitude or 300/2000 m (330/2,185 yards) at low altitude
Guidance: infra-red homing

Variants
V3A: designed during the early 1970s, this South African dogfighting missile was introduced to service in 1975, and was the world's first AAM used with a helmet-mounted sight for initial (and rapid) designation of the target aircraft
V3B: improved model introduced in 1979 with greater seeker and helmet sight acquisition angles, and an improved motor
V3 Kukri: export model introduced in 1982

AA-2-2C/D 'Advanced Atoll'
(USSR)
Type: short-range air-to-air missile
Dimensions: diameter 0.12 m (4.72 in); length (AA-2-2C) 2.80 m (9.17 ft) or (AA-2-2D) 3.10 m (10.17 ft); span 0.53 m (20.87 in)
Weights: total round (AA-2-2C) 70 kg (154.3 lb) or (AA-2-2D) 75 kg (165.3 lb); warhead 6 kg (13.2 lb) impact delayed fuse FRAG-HE
Propulsion: one solid-propellant rocket delivering ?-kg (?-lb) thrust
Performance: speed Mach 2.5; range 8000 m (8,750 yards)
Guidance: (AA-2-2C) infra-red homing or (AA-2-2D) semi-active radar homing

Variants
AA-2A 'Atoll': developed with the Soviet designation **K-13A** from the AIM-9B Sidewinder missile (using intelligence information and examples of the missile captured in 1958 and later years), the first-generation AA-2A infra-red homing missile began to enter Soviet service in the mid-1960s; this variant has a range of 5700 m (6,235 yards)
AA-2B 'Atoll': produced in parallel with the AA-2A, this is the semi-active radar homing version of the first-generation series, and has a range of 8000 m (8,750 yards)
AA-2-2A 'Advanced Atoll': second-generation infra-red missile with greater reliability and range
AA-2-2B 'Advanced Atoll': second-generation semi-active radar homing missile with greater reliability and range

AA-3A/B 'Anab'
(USSR)
Type: short/medium-range air-to-air missile
Dimensions: diameter 0.28 m (11.02 in); length (AA-3A) 4.10 m (13.45 ft) or (AA-3B) 4.00 m (13.12 ft); span 1.30 m (51.18 in)
Weights: total round 275 kg (606.3 lb); warhead 30 kg (66 lb) proximity- and impact-fused FRAG-HE
Propulsion: one solid-propellant rocket delivering ?-kg (?-lb) thrust
Performance: speed Mach 2.5; range (AA-3A) 20 km (12.4 miles) or (AA-3B) 30 km (18.6 miles)
Guidance: (AA-3A) infra-red homing or (AA-3B) semi-active homing

Variants
AA-3A 'Anab': this was the USSR's second-generation air-to-air missile, and was the first all-weather weapon of this type in the Soviet inventory; the type is generally associated with Sukhoi and Yakovlev fighters

AA-3B 'Anab': following the Soviet practice of developing air-to-air missiles in infra-red and semi-active radar versions for maximum kill probability under all operational conditions, this is the semi-active radar homing version, generally associated with Sukhoi and Yakovlev fighters; as with the AA-3A, the type is being supplanted by the AA-7 'Apex'

AA-5A/B 'Ash'
(USSR)
Type: medium-range air-to-air missile
Dimensions: diameter 0.305 m (12.00 in); length (AA-5A) 5.21 m (17.09 ft) or (AA-5B) 5.50 m (18.04 ft); span 1.30 m (4.27 ft)
Weights: total round 390 kg (860 lb); warhead not known
Propulsion: one solid-propellant rocket delivering ?-kg (?-lb)thrust
Performance: speed supersonic; range (AA-5A) 20 km (12.5 miles) or (AA-5B) 65 km (40.4 miles)
Guidance: (AA-5A) infra-red homing or (AA-5B) semi-active radar homing

Variants
AA-5A 'Ash': developed in the second half of the 1950s as primary armament of the Tupolev Tu-28P long-range interceptor, the AA-5A remains in limited service despite its technical obsolescence
AA-5B 'Ash': semi-active radar version of the AA-5A associated with the 'Big Nose' radar of the Tu-28P

AA-6A/B 'Acrid'
(USSR)
Type: long-range air-to-air missile
Dimensions: diameter 0.40 m (15.75 in); length (AA-6A) 5.80 m (19.03 ft) or (AA-6B) 6.29 m (20.64 ft); span 2.25 m (7.38 ft)
Weights: total round (AA-6A) 750 kg (1,653.4 lb) or (AA-6B) 800 kg (1,763.7 lb); warhead 90 kg (198.4 lb) proximity- and impact-fused FRAG-HE
Propulsion: one solid-propellant rocket delivering ?-kg (?-lb) thrust
Performance: speed Mach 4.5; range (AA-6A) 25 km (15.5 miles) or (AA-6B) 70 km (43.5 miles)
Guidance: (AA-6A) infra-red homing or (AA-6B) semi-active radar homing

Variants
AA-7A 'Acrid': developed in the 1960s as a long-range bomber-destroyer at medium and high altitudes, the AA-6 entered service in the early 1970s as the primary armament of the Mikoyan-Gurevich MiG-25 and Sukhoi Su-21
AA-6B 'Acrid': semi-active radar homing version of the AA-6A, the Soviet practice being a ripple of two missiles (one of each homing type) to generate maximum kill probability

AA-7A/B 'Apex'
(USSR)
Type: medium-range air-to-air missile
Dimensions: diameter 0.223 m (8.78 in); length (AA-7A) 4.20 m (13.78 ft) or (AA-7B) 4.50 m (15.09 ft); span 1.40 m (4.59 ft)
Weights: total round (AA-7A) 300 kg (661 lb) or (AA-7B) 320 kg (705.5 lb); warhead 50 kg (88 lb) proximity- and impact-fused FRAG-HE
Propulsion: one solid-propellant rocket delivering ?-kg (?-lb) thrust
Performance: speed Mach 3.5; range (AA-7A) 20 km (12.5 miles) or (AA-7B) 55 km (34.2 miles)
Guidance: (AA-7A) infra-red homing or (AA-7B) semi-active radar homing

Variants
AA-7A 'Apex': introduced in the mid- to late 1970s, this third-generation Soviet air-to-air missile can be regarded as comparable to the US AIM-7E Sparrow III but with greater manoeuvrability; the type is generally associated with the Mikoyan-Gurevich MiG-23S and MiG-25E, and is believed to be optimized for medium- and low-level interceptions; the Soviet designation is **R-23T**
AA-7B 'Apex': semi-active radar homing version of the AA-7A; the Soviet designation is **R-23R**

AA-8A/B 'Aphid'
(USSR)
Type: short-range dogfighting air-to-air missile
Dimensions: diameter 0.12 m (4.72 in); length (AA-8A) 2.15 m (7.05 ft) or (AA-8B) 2.35 m (7.71 ft); span 0.40 m (15.75 in)
Weights: total round (AA-8A) 55 kg (121.25 lb) or (AA-8B) 60 kg (132.3 lb); warhead 6 kg (13.2 lb) proximity- and impact-fused FRAG-HE
Propulsion: one solid-propellant rocket delivering ?-kg (?-lb) thrust
Performance: speed Mach 3; range (AA-8A) 10000 m (10,935 yards) or (AA-8B) 15000 m (16,405 yards)
Guidance: (AA-8A) infra-red homing or (AA-8B) semi-active radar homing

Variants
AA-8A 'Aphid': introduced in the mid-1970s as the replacement for the AA-2 series and as the short-range companion to the medium-range AA-7, the extremely capable AA-8 is used by most modern Soviet fighters, and is notable for its high degree of manoeuvrability; the Soviet designation is **R-60T**

AA-8B 'Aphid': semi-active radar homing version of the AA-8A; the Soviet designation is **R-60R**

AA-9
(USSR)
Type: medium-range air-to-air missile
Dimensions: not known
Weights: not known
Propulsion: one solid-propellant rocket delivering ?-kg (?-lb) thrust
Performance: speed not known; range 45 km (28 miles) from a high-altitude launch or 20 km (12.5 miles) from a low-altitude launch
Guidance: semi-active radar homing

Variant
AA-9: deployed on Mikoyan-Gurevich MiG-29 and MiG-31 fighters, the AA-9 was developed from the early 1970s as a weapon able to deal with low-altitude penetrations of Soviet airspace, particularly by US cruise missiles; trials have confirmed that the AA-9 can snap-down at least 6000 m (19.685 ft) to engage such target flying at an altitude as low as 50 m (165 ft); US sources credit the missile with a range of 18.5/23.2 km (11.5/14.4 miles) in a head-on low-altitude engagement, or of 8.4/9.25 km (5.2/5.75 miles) in a tail-on low-altitude engagement, or of 39.8/45.4 km (24.75/28.2 miles) in a high-altitude head-on engagement

AA-10 'Alamo'
(USSR)
Type: medium-range air-to-air missile
Dimensions: not known
Weights: not known
Propulsion: one solid-propellant rocket delivering ?-kg (?-lb) thrust
Performance: speed supersonic; range 20.4/35 km (12.66/21.9 miles)
Guidance: active radar homing

Variant
AA-10: introduced in 1984, this is an all-aspect snap-down missile matched to the radars of new-generation Soviet fighters capable of 'look-down/shoot-down' engagements

AA-11 'Archer'
(USSR)
Type: short-range air-to-air missile
Dimensions: not known
Weights: not known
Propulsion: one solid-propellant rocket delivering ?-kg (?-lb) thrust
Performance: speed supersonic; range not known
Guidance: infra-red homing

Variant
AA-11 'Archer': this is the latest Soviet short-range AAM, and forms a key part of the inventory for aircraft such as the Mikoyan-Gurevich MiG-29

BAe Red Top
(UK)
Type: short-range air-to-air missile
Dimensions: diameter 8.75 in (0.223 m); length 10.88 ft (3.32 m); span 37.75 in (0.908 m)
Weights: total round 330 lb (149.7 kg); warhead not revealed
Propulsion: one solid-propellant two-stage rocket delivering ?-lb (?-kg) thrust
Performance: speed Mach 3; range 13,200 yards (12070 m)
Guidance: infra-red homing

Variant
Red Top: introduced in 1974, the Red Top is essentially an upgraded and rationalized Firestreak missile, and possesses an all-aspect engagement capability plus a somewhat larger FRAG-HE warhead than that of the Firestreak

BAe Firestreak
(UK)
Type: short-range air-to-air missile
Dimensions: diameter 8.75 in (0.223 m); length 10.46 ft (3.188 m); span 29.4 in (0.747 m)
Weights: total round 300 lb (137 kg); warhead 50 lb (22.7 kg) proximity-fused FRAG-HE
Propulsion: one solid-propellant two-stage rocket delivering ?-lb (?-kg) thrust
Performance: speed Mach 2+; range 7,000 yards (6400 m)
Guidance: infra-red homing

Variant
Firestreak: introduced in 1958, this missile is obsolete but still retained in small numbers; the type is limited to pursuit-type engagements

BAe/Marconi Sky Flash (UK): see Raytheon/General Dynamics AIM-7M Sparrow

Hughes AIM-4F Falcon
(USA)
Type: short/medium-range air-to-air missile
Dimensions: diameter 6.6 in (0.168 m); length 7.17 ft (2.18 m); span 24.00 in (0.61 m)
Weights: total round 150 lb (68 kg); warhead 28.65 lb (13 kg) proximity-fused HE
Propulsion: one Thiokol M46 two-stage solid-propellant rocket delivering 4,440-lb (2014-kg) thrust
Performance: speed Mach 4; range 12,350 yards (11295 m)
Guidance: Hughes semi-active radar homing

Variants
AIM-4C Falcon: this elderly AAM still serves with the Swedish and Swiss air forces with the designations **RB 28** and **HM58** respectively
AIM-4F Falcon: introduced in 1959, this is a much-improved model matched to the Convair F-106A interceptor and based on a longer body (with greater-span wings) fitted with a two-stage motor
AIM-4G Falcon: version of the AIM-4F with infra-red guidance, decreasing length by 5 in (0.127 m)

Raytheon/General Dynamics AIM-7M Sparrow
(USA)
Type: medium-range air-to-air missile
Dimensions: diameter 8 in (0.203 m); length 12.07 ft (3.68 m); span 40 in (1.02 m)
Weights: total roudn 514 lb (233 kg); warhead 86 lb (39 kg) proximity and impact delayed-fuse FRAG-HE
Propulsion: one Hercules Mk 58 or Aerojet Mk 65 dual-thrust solid-propellant rocket delivering ?-lb (?-kg) thrust
Performance: speed Mach 4; range 43.5 miles (70 km) from a high-altitude launch reducing to 27 miles (43.5 km) from a low-altitude launch
Guidance: Raytheon Advanced Monopulse Seeker semi-active radar homing

Variants
AIM-7E Sparrow: introduced in 1962 as the first version of the Sparrow III to enter large-scale production, this variant remains in limited service; the type is powered by a Rocketdyne Mk 38 or Aerojet Mk 52 single-stage motor, weighs 452 lb (205 kg), has a range of 28 miles (44 km) with its 66-lb (30-kg) warhead, and is guided by Raytheon continuous-wave semi-active radar guidance; the variant is built in Japan as the **AIM-7J Sparrow**, and a limited-production short-range version is the **AIM-7E2 Sparrow**
AIM-7F Sparrow: redesigned version introduced in 1977 on the McDonnell Douglas F-15 and now carried also by the McDonnel Douglas F/A-18 Hornet; the variant has a much larger engagement envelope through the adoption of solid-state electronics and a dual-thrust motor; like the AIM-7E it is 12.00 ft (3.66 m) long, but weighs 503 lb (228 kg) and has a 86-lb (39–kg) warhead; the use of continuous-wave and pulse-Doppler guidance considerably enhanced look-down/shoot-down capability, even at the longer ranges of the variant
AIM-7M Sparrow: introduced in 1982 as an interim type pending availability of the AIM-120 AMRAAM, this variant has a new monopulse seeker offering performance comparable with that of the British Sky Flash derivative, especially in the longer-range look-down/shoot-down mode
BAe Sky Flash: British development of the AIM-7E introduced in 1977; this model has a Marconi monopulse semi-active radar homing system, a 66-lb (30-kg) warhead, a total weight of 424 lb (192.3 kg) and a range of 31 miles (50 km); the type is also used by Sweden with the local designation **RB 71**
BAe Sky Flash 90: proposed designation for an advanced development proposed by the UK and Sweden with a Marconi active radar seeker system and a Royal Ordnance Hoopoe solid-propellant rocket instead of the Aerojet motor of the current Sky Flash; the new motor is designed to provide greater speed, range and terminal manoeuvrability

Ford Aerospace/Raytheon AIM-9L Sidewinder
(USA)
Type: short-range dogfighting air-to-air missile
Dimensions: diameter 5 in (0.127 m); length 9.35 ft (2.85 m); span 24.8 in (0.63 m)
Weights: total round 188 lb (85.3 kg); warhead 22.5 lb (10.2 kg) proximity-fused FRAG-HE
Propulsion: one Bermite/Hercules Mk 36 Model 8 solid-propellant rocket delivering ?-lb (?-kg) thrust
Performance: speed Mach 2.5; range 11 miles (17.7 km)
Guidance: Bodensee Gerätetechnik ALASCA infra-red homing

Variants
AIM-9B Sidewinder: though still in the operational inventory of a few nations, this venerable version of the Sidewinder series entered service in 1956, and powered by a Naval Propellant Plant Mk 17 single-stage motor the 167-lb (76-kg) missile achieves a range of 4000 yards (3660 m); the **AIM-9B/FWG.2** is an improved European-built model with greater seeker sensitivity, and the **AIM-9E** a US-built improvement with a Ford cooled seeker, again for greater sensitivity; the Japanese **Mitsubishi AAM-1** is virtually a copy of the AIM-9E, though it is slightly smaller (2.60 m/8.53 ft long) and lighter (70 kg/154.3 lb) than the American weapon
AIM-9G Sidewinder: introduced in 1967, this is an improved version of the US Navy's 1965 **AIM-9D Sidewinder** (the real precursor of the latest Sidewinder variants) with a Rocketdyne Mk 36 solid-propellant rocket and the Sidewinder Expanded Acquisition Mode (SEAM) seeker for improved target acquisition capability
AIM-9H Sidewinder: introduced in 1970, this 195-lb (88.5-kg) missile is a further improved AIM-9G with a Bermite- of Hercules-built Mk 36 Mod 6 motor, an improved seeker providing enhanced all-weather capability, and other modifications to improve manoeuvrability
AIM-9J Sidewinder: developed from the AIM-9E by converting obsolescent AIM-9Bs (**AIM-9J2**) and making new missiles (**AIM-9J1** and the product-improved **AIM-9J3**), this variant appeared in the

early 1970s and features a Thiokol Mk 17 solid-propellant motor, a cooled seeker and some solid-state electronics; this model weighs 172 lb (78 kg), has a length of 10.07 ft (3.07 m) and a span of 20 in (0.559 m), and can reach a range of 15,850 yards (14495 m); the type also introduced foreplanes of revised shape in a successful attempt to increase the missile's manoeuvrability

AIM-9K Sidewinder: upgraded AIM-9H

AIM-9L Sidewinder: much improved version combining the latest Sidewinder airframe (featuring foreplanes of yet further refined shape for increased manoeuvrability) with the Bodensee Gerätetechnik seeker developed for the abortive Viper AAM, offering an excellent all-aspect seeker capability combined with a dogfighting airframe

AIM-9M Sidewinder: US-made version of the European AIM-9l with a weight of 189.6 lb (86 kg) as a result of using a closed-cycle cooler for improved seeker sensitivity (especially a low altitudes) and a 'smokeless' Bermite/Hercules Mk 36 Mod 9 motor for reduced visual signature

AIM-9N Sidewinder: designation of AIM-9B and AIM-9E missiles upgraded to an improved AIM-9J standard

AIM-9P Sidewinder: US-made export model (either new-build or upgraded existing weapons) to AIM-9L standard with a weight of 170 lb (77.1 kg); there are three subvariants in the form of the **AIM-9P1** with an active optical proximity fuse, the **AIM-9P2** with 'smokeless' motor and the **AIM-9P3** combining AIM-9P1 and AIM-9P2 features with a new warhead of reduced thermal sensitivity and greater shelf life

AIM-9R Sidewinder: model under development for production in the late 1980s with an improved seeker for greater acquisition range and better resistance to countermeasures

Hughes AIM-26B Super Falcon
(USA)
Type: medium-range air-to-air missile
Dimensions: diameter 11.4 in (0.29 m); length 6.79 ft (2.07 m); span 24.4 in (0.62 m)
Weights: total round 262 lb (119 kg); warhead 40 lb (18 kg) proximity fused HE
Propulsion: one Thiokol M60 solid-propellant rocket delivering 5,625-lb (2552-kg) thrust
Performance: speed Mach 2; range 10,500 yards (9600 m)
Guidance: Hughes semi-active radar homing

Variant
AIM-26B Super Falcon: introduced as the primary weapon of the Convair F-106 Delta Dart, the Super Falcon still serves with Sweden and Switzerland under the designations **RB 27** and **HM55** respectively

Hughes AIM-54A Phoenix
(USA)
Type: long-range air-to-air missile
Dimensions: diameter 15 in (0.38 m); length 13.15 ft (4.008 m); span 36.4 in (0.925 m)
Weights: total round 985 lb (447 kg); warhead 132 lb (60 kg) proximity and impact delayed-fuse FRAG-HE
Propulsion: one Rocketdyne Mk 47 Mk 0 single-stage solid-propellant rocket delivering ?-lb (?-kg) thrust
Performance: speed Mach 4.3; range 2.4/85 miles (3.86/137 km)
Guidance: DSO-26 semi-active homing for the cruise phase, and active radar terminal homing

Variants
AIM-54A Phoenix: designed for the abortive General Dynamics F-111B swing-wing naval fighter, the Phoenix was brought to fruition (together with the accompanying Hughes AWG-9 pulse-Doppler fire-control system) in the Grumman F-14 fleet-defence fighter; the type began to enter service in 1973, and has a ceiling of 81,400 ft (24810 m)

AIM-54C Phoenix: introduced in the early 1980s, this considerably improved missile has digital rather than analog electronics, a strapdown inertial reference system, solid-state radar and very much improved electronic counter-countermeasures capability; the weight of this variant is 1,008 lb (457.2 kg), the speed Mach 5, the ceiling 100,000 ft (30490 m) and the range 2.2/92 miles (3.5/148 km); as with the AIM-54A, this legend range has been considerably exceeded in service, especially when the F-14 launch aircraft is supported by a Grumman E-2C Hawkeye warning and control system aircraft

Hughes AIM-120A AMRAAM
(USA)
Type: medium-range air-to-air missile
Dimensions: diameter 7.00 in (0.178 m); length 11.98 ft (3.65 ft); span 20.7 in (0.526 m)
Weights: total round 335 lb (152 kg); warhead 45 lb (20.4 kg) proximity and impact delay-fused FRAG-HE
Propulsion: one solid-propellant rocket delivering ?-lb (?-kg) thrust
Performance: speed supersonic; range 45 miles (72 km) from a high-altitude launch declining to 34 miles (55 km) from a low-altitude launch
Guidance: inertial and active radar homing

Variant
AIM-120A AMRAAM: designed in the late 1970s and early 1980s for a service debut in the mid- to late 1980s as replacement for the AIM-7 Sparrow, the Advanced Medium-Range Air-to-Air Missile is a fire-and-forget BVR (Beyond-Visual-Range) weapon

544 MISSILES

with the size and weight of the AIM-9 Sidewinder but capabilities and performance better than those of the considerably larger Sparrow; the missile cruises on a preprogrammed course (with the capability for mid-course radio update) until its reaches the vicinity of the target, when it activates its active radar seeker, which also possesses a 'home-on-jam' mode

Vought ASAT
(USA)
Type: air-launched anti-satellite missile
Dimensions: diameter (first stage) 17.5 in (0.445 m) and (second stage) 19.8 in (0.53 m); length 17.72 ft (5.40 m); span not revealed
Weights: total round 2,632 lb (1194 kg); warhead not revealed
Propulsion: (first stage) one Lockheed SR75-LP-1 solid-propellant rocket delivering ?-lb (?-kg) thrust, and (second stage) one Altair III solid-propellant rocket delivering ?-lb (?-kg) thrust
Performance: speed hypersonic; range 620-mile (1000-km) orbital altitude
Guidance: infra-red homing

Variant
ASAT: one of the world's most advanced weapons, the ASAT is being tested during the mid-1980s for deployment under modified McDonnell Douglas F-15 fighters in the early 1990s; the type is based on the propellant section of the Boeing AGM-69A SRAM missile with a high-velocity second stage carrying the Vought 1005 Miniature Vehicle, which measures 12 by 13 in (305 by 330 mm) and is fired once the vehicle's telescopic infra-red guidance system has locked onto the target satellite, the MV being designed to shatter the target with the kinetic energy of a 30,680-mph (49375-km/h) impact

ANTI-TANK-MISSILES

CITEFA Mathogo
(Argentina)
Type: ground-launched short/medium-range anti-tank missile
Dimensions: diameter 0.102 m (4.20 in); length 0.998 m (3.27 ft); span not known
Weights: total round 11.3 kg (24.9 lb); warhead not known
Propulsion: one solid-propellant booster, rocket delivering ?-kg (?-lb) thrust, and one solid-propellant sustainer rocket delivering ?-kg (?-lb) thrust
Performance: speed 325 km/h (202 mph); range 350/2100 m (385/2,295 yards)
Guidance: wire command to line of sight

Variant
Mathogo: introduced in the late 1970s or early 1980s, this is a simple first-generation weapon reminiscent in concept and layout (but not in dimensions and weights) to the Swedish Bantam weapon; the operator can control four missiles each separated from the firing position by 50-m (165-ft) cables; the weight of the missile and launcher is 19.5 kg (64 lb)

Avibras MSS-1 (Brazil): see MBB Mamba

NORINCO Red Arrow 8
(China)
Type: ground-launched short/medium-range anti-tank missile
Dimensions: diameter 0.12 m (4.72 in); length 0.875 m (34.4 in); span 0.32 m (12.6 in)
Weights: total round 11.2 kg (24.7 lb); warhead not revealed
Propulsion: one solid-propellant rocket delivering ? kg (? lb) thrust
Performance: speed 865 km/h (538 mph); range 100/3000 m (110/3,280 yards)
Guidance: wire semi-automatic command to line of sight

Variants
Red Arrow 8: this indigenously-designed Chinese weapon began to enter service in the mid-1980s, and may be regarded as China's counterpart to the British Swingfire and US TOW systems; the missile has wrap-round fins and is carried in a container that doubles as the launch tube when installed on the head of the fire tripod; the weight of the loaded tube is 24.5 kg (54 lb), and of the launch tripod 23 kg (50.7 lb) increased by 24 kg (52.9 lb) top-section control equipment; the weapon has a HEAT warhead with an armour penetration capability of 800 mm (31.5 in), and a hit probability of 90% is claimed; the Red Arrow 8 can also be used on a pedestal mount on 4 × 4 vehicles and tracked APCs

NORINCO Red Arrow 73 (China): see AT-3 'Sagger'

Aérospatiale Eryx
(France)
Type: man-portable short-range light anti-tank missile
Dimensions: diameter 0.16 m (6.3 in); length 0.905 m (2.97 ft); span not revealed
Weights: total round 11 kg (24.25 lb); warhead 3.6 kg (7.94 lb) HEAT
Propulsion: one solid-propellant rocket delivering ?-kg (?-lb) thrust
Performance: speed 72 km/h (44.7 mph); range ?/600 m (?/655 yards)
Guidance: wire semi-automatic command to line of sight

Variant
Eryx: this is a lightweight anti-tank weapon selected for the French army in 1986; the type has an extremely small motor resulting in very low speed, but this tactical limitation is offset by the type's minimum backblast and ability to be fired from inside a building; the warhead can penetrate 900 mm (35.4 in) of armour

Aérospatiale SS.11
(France)
Type: ground-, vehicle- or helicopter-launched medium-range anti-tank missile
Dimensions: diameter 0.164 m (6.46 in); length 1.20 m (3.94 ft); span 0.50 m (19.7 in)
Weights: total round 29.9 kg (65.9 lb); warhead not revealed
Propulsion: one two-stage solid-propellant rocket delivering ?-kg (?-lb) thrust
Performance: speed 685 km/h (426 mph); range 500/3000 m (545/3,280 yards)
Guidance: wire command to line of sight

Variants
SS.11: entering service in 1956, the SS.11 was the first of the modern anti-tank and battlefield missiles, and though obsolescent is still in fairly widespread service; the type is launched from a rail, and can carry Type 140AC anti-tank warhead capable of penetrating 600 mm (23.62 in) of armour, the Type 140AP02 delay-action anti-tank/anti-personnel warhead capable of penetrating 100 mm (0.4 in) of armour, and the Type 140AP59 anti-personnel fragmentation warhead
SS.11B1: improved model of 1962 with transistorized electronics in the control unit
AS.11: air-launched version of the SS.11 series, generally associated with converted utility helicopters fitted with the necessary sight system
Harpon: 1967 model with semi-automatic command to line of sight guidance

Euromissile HOT
(France/West Germany)
Type: ground-, vehicle- or helicopter-launched long-range heavy anti-tank missile
Dimensions: diameter 0.165 m (6.5 in); length 1.25 m (4.18 ft); span 0.312 m (12.28 in)
Weights: total round 23.5 kg (51.8 lb); warhead 6 kg (13.2 lb) HEAT
Propulsion: one SNPE Bugéat solid-propellant booster rocket delivering ?-kg (?-lb) thrust, and one SNPE Infra solid-propellant sustainer rocket delivering 24-kg (53-lb) thrust
Performance: speed 865 km/h (537 mph); range 75/4250 m (82/4,650 yards)
Guidance: SAT/Eltro wire semi-automatic command to line of sight

Variants
HOT: entering service in the early 1970s, the HOT (Haute subsonique Optiquement téléguidé tiré d'un Tube, or high-subsonic optically-guided tube-launched) is a powerful anti-tank weapon designed for static ground use, or for installation in armoured fighting vehicles and helicopters; the missile can be used from single or multiple tube launchers, and can penetrate more than 800 mm (31.5 in) of armour; the type's primary disadvantage is a modest speed, which means that the missile takes some 17 seconds to reach its maximum range even when launched from a helicopter which thus has to remain exposed for this time
HOT 2: this improved version of the basic weapon features much-enhanced armour penetration as a result of its larger warhead, and is also faster than its predecessor with all the consequent tactical advantages of this fact
HOT Commando ATLAS: introduced in 1986, this is the Affût de Tir Léger Au Sol (ground-based lightweight fire equipment) version designed specifically for launch from the ground (using a tripod-mounted launcher) or from a light vehicle; the HOT 2 missile is used with the launcher developed for the SA 342 helicopter and the Thomson-CSF Castor infra-red tracker designed for the Renault VAB carrier; the standard anti-tank warhead can penetrate 1300 mm (51.2 in) of armour at maximum range, and a new multi-purpose warhead is being developed with a 150-mm (5.9-in) diameter HEAT charge able to defeat 350 mm (13.78 in) of armour and a casing able to fill a radius of 25 m (82 ft) with lethal 1,000 fragments

Euromissile MILAN
(France/West Germany)
Type: man-portable ground- or vehicle-launched short/medium-range anti-tank missile
Dimensions: diameter 0.90 m (3.54 in); length 0.769 m (2.52 ft); span 0.265 m (10.43 in)
Weights: total round 6.65 kg (14.66 lb); warhead 2.98 kg (6.57 lb) HEAT

Propulsion: one SNPE Artus dual-thrust solid-propellant rocket delivering ?-kg (?-lb) thrust
Performance: speed 720 km/h (447 mph); range 25/2000 m (28/2,185 yards)
Guidance: SAT/Eltro wire semi-automatic command to line of sight

Variants
MILAN: an advanced second-generation missile, the MILAN (Missile d'Infanterie Leger ANti-char, or light infantry anti-tank missile) began to enter service in the mid-1970s and was designed as the man-portable counterpart to the heavyweight HOT from the same stable; the weapon is tube fired from a single launcher on the ground or located on a pedestal, or from multiple launchers located on light vehicles; the warhead can penetrate more than 650 mm (25.6 in) of armour; like the larger HOT, the MILAN suffers the tactical disadvantages of modest speed, though this failing is in part mitigated by shorter range
MILAN 2: this improved version of the basic weapon features much-improved armour penetration as a result of its larger (115- rather than 103-mm/453- rather than 4.06-in diameter) K115 warhead with extended probe for optimum stand-off detonation distance, and is also faster than its predecessor

MBB Mamba
(West Germany)
Type: man-portable short/medium-range anti-tank missile
Dimensions: diameter 0.12 m (4.72 in); length 0.955 m (3.13 ft); span 0.40 m (15.75 in)
Weights: total round 11.2 kg (24.7 lb); warhead 2.7 kg (5.95 lb) HEAT
Propulsion: one solid-propellant dual-thrust rocket delivering ?-kg (?-lb) thrust
Performance: speed 505 km/h (314 mph); range 300/2000 m (330/2,185 yards)
Guidance: wire command to line of sight

Variants
Mamba: introduced in 1972, the Mamba in an upgraded Cobra 2000 with a single rocket providing lift as well as propulsion; the operator can control a total of 12 missiles via a junction box, and the warhead can penetrate 500 mm (19.69 in) of armour; the type is also being evaluated with a fibre-optic guidance system
Avibras MSS-1: this is a Brazilian anti-tank missile apparently identical with the Mamba externally, but possessing a launch weight of 17.2 kg (38 lb)

MBB Bö 810 Cobra 2000
(West Germany)
Type: man-portable short/medium-range anti-tank missile
Dimensions: diameter 0.10 m (3.94 in); length 0.95 m (3.12 ft); span 0.48 m (18.90 in)
Weights: total round 10.3 kg (22.7 lb); warhead 2.7 kg (5.95 lb) HEAT or anti-tank shrapnel
Propulsion: one solid-propellant booster rocket delivering ?-kg (?-lb) thrust, and one solid-propellant sustainer rocket delivering ?-kg (?-lb) thrust
Performance: speed 305 km/h (190 mph); range 400/2000 m (435/2,185 yards)
Guidance: wire command to line of sight

Variant
Cobra 2000: designed in the late 1950s as a first-generation man-portable missile, the Cobra 2000 has proved a successful export weapon; the missile is simply placed on the ground facing the target, and launched with the aid of a non-jettisonable booster by an operator who can be laterally displaced from the weapon by as much as 70 m (230 ft) and co-ordinate eight missile via a junction box; the HEAT warhead is able to penetrate 500 mm (19.69 in) of armour, and the anti-tank shrapnel type 350 mm (13.78 in)

IMI Mapats
(Israel)
Type: ground- and vehicle-launched long-range heavy anti-tank missile
Dimensions: diameter 0.148 m (5.83 in); length 1.450 m (4.76 ft) with probe extended; span not revealed
Weights: total round 18.5 kg (40.78 lb); warhead 3.6 kg (7.94 lb) HEAT
Propulsion: one solid-propellant rocket delivering ?-kg (?-lb) thrust
Performance: speed 1135 km/h (704 mph); range ?/4000 m (?/4,375 yards
Guidance: laser beam-riding

Variant
Mapats: known colloquially as the 'Togger' because it contains features from the US TOW and Soviet 'Sagger' anti-tank missiles, the Mapats is a highly capable weapon with advanced guidance and capable of penetrating 800 mm (31.5 in) of multi-layer armour; the weapon is supplied as a round in a fibre-glass container which serves as the clip-on launch tube, the missile-container assembly weighting 29 kg (63.9 lb) and measuring 1.50 m (4.92 ft) in length

Kawasaki Type 64
(Japan)
Type: ground-, vehicle- and helicopter-launched short/medium-range anti-tank missile
Dimensions: diameter 0.12 m (4.72 in); length 1.00 m (3.28 ft); span 0.60 m (23.6 in)
Weights: total round 15.7 kg (34.6 lb); warhead not revealed
Propulsion: one Daicel/Nippon Oil two-stage solid-propellant rocket delivering ?-kg (?-lb) thrust

Performance: speed 305 km/h (190 mph); range 350/1800 m (385/1,970 yards)
Guidance: Nihon Electric wire command to line of sight

Variant
Type 64: introduced in 1965 and otherwise known as the **KAM-3D**, this is a simple, slow yet effective missile now being replaced by the Type 79 weapon; the system requires a three-man crew, and the operator can control one or more laterally-separated missiles from a central firing position; the missile is launched from a small ground support/baseplate

Kawasaki Type 79
(Japan)
Type: ground- and vehicle-launched long-range heavy anti-tank missile
Dimensions: diameter 0.152 m (6.00 in); length 1.565 m (5.13 ft); span 0.332 m (13.1 in)
Weights: total round 33 kg (72.75 lb); warhead 1.9 kg (4.2 lb) HEAT or SAP-HE
Propulsion: one Nihon Yushi solid-propellant booster rocket delivering ?-kg (?-lb) thrust, and one Daicel solid-propellant sustainer rocket delivering ?-kg (?-lb) thrust
Performance: speed 720 km/h (447 mph); range ?/4000 m (?/4,375 yards)
Guidance: Nihon Electric wire semi-automatic command to line of sight

Variant
Type 79: also known as the **KAM-9D** and introduced in 1980, this is a powerful weapon modelled conceptually on the American TOW system, and able to deal effectively with tanks (the missile's armour penetration being 500+ mm/19.67+ in) and landing craft; the containerized missile weighs 42 kg (92.6 lb) and is generally fired from a tripod launcher, two missiles being allocated to each laterally-separated firing unit; low speed is the type's worst disadvantage

Bofors Bantam
(Sweden)
Type: man-portable short/medium-range anti-tank missile
Dimensions: diameter 0.11 m (4.33 in); length 0.85 m (2.79 ft); span 0.40 m (15.75 in)
Weights: total round 11.5 kg (25.35 lb) including container; warhead 1.9 kg (4.2 lb) HEAT
Propulsion: one two-stage solid-propellant rocket delivering ?-kg (?-lb) thrust
Performance: speed 305 km/h (190 mph); range 300/2000 m (330/2,185 yards)
Guidance: wire command to line of sight

Variant
Bantam: this is a first-generation anti-tank missile (carried in and launched from a small container) with simple guidance and low cruising speed, the operator being afforded a measure of protection by a separation of up to 120 m (130 yards) between the launcher and the control position; the weapon was adopted in 1963, and its warhead can penetrate 500 mm (19.68 in) of armour

Bofors RBS 56 Bill
(Sweden)
Type: ground-, vehicle- and helicopter-launched medium-range anti-tank missile
Dimensions: diameter 0.15 m (5.91 in); length 0.90 m (2.95 ft); span 0.41 m (16.14 in)
Weights: total containerized round 16 kg (35.3 lb); warhead not revealed
Propulsion: one Royal Ordnance solid-propellant rocket delivery ?-kg (?-lb) thrust
Performance: speed 720 km/h (447 mph); range 150/2000 m (165/2,185 yards)
Guidance: wire semi-automatic command to line of sight

Variant
RBS 56 Bill: entering service in the later 1980s, this is a highly advanced missile fired from a container tube fitted with a sight unit and mounted on a tripod; the key to the Bill's capabilities is the shaped-charge warhead, which is angled to fire 30 degrees downwards from the missile centreline as the weapon overflies the target's most vulnerable upper surfaces, thus allowing the gas/vaporized metal jet to burn through the target tank's thinnest armour

Kun Wu (Taiwan): see AT-3 'Sagger'

AT-1 'Snapper'
(USSR)
Type: man-portable ground- or vehicle-launched medium-range anti-tank missile
Dimensions: diameter 0.14 m (5.51 in); length 1.13 m (3.71 ft); span 0.78 m (30.7 in)
Weights: total round 22.25 kg (49.1 lb); warhead 5.25 kg (11.5 lb) HEAP
Propulsion: one solid-propellant rocket delivering ?-kg (?-lb) thrust
Performance: speed 320 km/h (199 mph); range 500/2300 m (545/2,515 yards)
Guidance: wire command to line of sight

Variant
AT-1 'Snapper': known to the Soviets as the **3M6 Shmell**, this first-generation anti-tank missile entered service in the early 1960s, and it notable for its considerable size and slow flight; the warhead can penetrate 380 mm (14.96 in) of armour

AT-2 'Swatter'
(USSR)

Type: vehicle- or helicopter-launched medium/long-range anti-tank missile

Dimensions: diameter 0.132 m (5.2 in); length 1.14 m (3.74 ft); span 0.66 m (26.00 in)

Weights: total round ('Swatter-A') 26.5 kg (54.8 lb), ('Swatter-B') 29.5 kg (65 lb) or ('Swatter-C') 32.5 kg (71.65 lb); warhead not known

Propulsion: one solid-propellant rocket delivering ?-kg (?-lb) thrust

Performance: speed 540 km/h (336 mph); range ('Swatter-A') 500/3000 m (545/3,280 yards), ('Swatter-B') 500/3500 m (545/3,825 yards) or ('Swatter-C') 500/4000 m (545/4,375 yards)

Guidance: ('Sagger A' and 'Sagger-B') radio command to line of sight and infra-red terminal homing, or ('Sagger-C') radio semi-automatic command to line of sight and infra-red terminal homing

Variants

AT-2 'Swatter-A': a first-generation vehicle-launched anti-tank missile with countermeasures-prone radio command guidance, the 'Swatter' was introduced in the mid-1960s and is known to the Soviets as the **PTUR-62 Falanga**; the HEAP warhead can penetrate 480 mm (18.9 in) of armour

AT-2 'Swatter-B': increased-range model with an improved motor and a more capable warhead able to penetrate 510 mm (20.08 in) of armour

AT-2 'Swatter-C': helicopter-launched version for the Mil Mi-8 and Mi-24 with semi-automatic command to line of sight guidance and yet more range

AT-3 'Sagger'
(USSR)

Type: man-portable ground-, vehicle- and helicopter-launched medium/long-range anti-tank missile

Dimensions: diameter 0.119 m (4.69 in); length 0.883 m (2.90 ft); span 0.46 m (18.1 in)

Weights: total round 11.29 kg (24.9 lb); warhead 3 kg (6.6 lb) HEAP

Propulsion: one solid-propellant dual-thrust rocket delivering ?-kg (?-lb) thrust

Performance: speed 430 km/h (267 mph); range 300/3000 m (330/3,280 yards)

Guidance: wire command to line of sight

Variants

AT-3 'Sagger-A': introduced in about 1970, the AT-3 is a more capable and versatile weapon that the AT-2, with a countermeasures-proof wire guidance system and more powerful motor for greater (but still comparatively low) speed; the type is designated **PTUR-64 Malyutka** in the USSR, and its warhead can penetrate 410 mm (16.14 in) of armour; Egypt is revising its weapons with a new warhead and semi-automatic command to line of sight guidance, probably based on that of the Euromissile MILAN

AT-3 'Sagger-B': introduced in 1973, this upgraded model has an improved motor for greater speed and thus reduced flight time

AT-3 'Sagger-C': introduced in the late 1970s, this is a much developed version with semi-automatic command to line of sight guidance

Red Arrow 73: Chinese version of the AT-3, with a length of 0.868 m (2.85 ft) and a span of 0.393 m (15.47 in)

Kun Wu: unlicensed Taiwanese development of the AT-3 with a nose section of revised shape

AT-4 'Spigot'
(USSR)

Type: man-portable short/medium-range anti-tank missile

Dimensions: not known

Weights: not known

Propulsion: one solid-propellant rocket delivering ?-kg (?-lb) thrust

Performance: speed 540 km/h (336 mph); range 25/2000 m (27/2,185 yards)

Guidance: wire semi-automatic command to line of sight

Variant

AT-4 'Spigot': introduced in the late 1970s, this is a second-generation anti-tank missile using technologies 'gleaned' from the West; the type is fired from a tube mounted on a light tripod, and the HEAT warhead can penetrate 600 mm (23.62 in) of armour

AT-5 'Spandrel'
(USSR)

Type: vehicle- and helicopter-launched long-range heavy anti-tank missile

Dimensions: diameter 0.155 m (6.1 in); length 1.30 m (4.27 ft); span not known

Weights: total round 12 kg (26.5 lb); warhead not known

Propulsion: one solid-propellant rocket delivering ?-kg (?-lb) thrust

Performance: speed 720 km/h (447 mph); range 100/4000 m (110/4,375 yards)

Guidance: wire semi-automatic command to line of sight

Variant

AT-5 'Spandrel': introduced in about 1980, the AT-5 is a powerful second-generation vehicle-launched weapon using technology 'gleaned' from such Western types as the Euromissile HOT; the HEAT warhead can penetrate 750 mm (29.53 in) of armour

AT-6 'Spiral'
(USSR)
Type: helicopter-launched long-range heavy anti-tank missile
Dimensions: not known
Weights: 32 kg (70.55 lb); warhead not known
Propulsion: one solid-propellant dual-thrust rocket delivering ?-kg (?-lb) thrust
Performance: speed not known; range 100/7000 m (110/7,655 yards)
Guidance: radio command to line of sight (or possibly laser beam-riding)

Variant
AT-6 'Spiral': so far seen only on the Mil Mi-24 attack helicopter, the AT-6 is the first Soviet third-generation anti-tank missile, its good range being matched by excellent armour penetration qualities (about 800 mm/31.5 in)

AT-7
(USSR)
Type: man-portable short-range lightweight anti-tank missile
Dimensions: not known
Weights: not known
Propulsion: one solid-propellant rocket delivering ?-kg (?-lb) thrust
Performance: not known
Guidance: wire semi-automatic command to line of sight

Variant
AT-7: this Soviet anti-tank weapon is designed for infantry use, and is comparable in many respects to the US M47 Dragon system, but details are lacking; the type has the Soviet designation **Metis** (mongrel)

AT-8
(USSR)
Type: tube-launched medium-range anti-tank missile
Dimensions: diameter 0.125 m (4.92 in); length not known; span not known
Weights: not known
Propulsion: one solid-propellant rocket delivering ?-kg (?-lb) thrust
Performance not known
Guidance: semi-active laser homing

Variant
AT-8: this weapon is fired from the main armament fo the T-64, T-72 and T-80 series main battle tanks of the Soviet army, and is apparently used in two forms with a shaped-charge warhead for anti-tank use and with a FRAG-HE warhead for anti-helicopter use; the Soviet designation is **Kobra**

BAe Swingfire
(UK)
Type: vehicle-launched or crew-portable medium/long-range heavy anti-tank missile
Dimensions: diameter 6.7 in (0.17 m); length 3.50 ft (1.07 m); span 14.7 in (0.373 m)
Weights: total round 60 lb (27 kg); warhead 15.4 lb (7 kg) HEAT
Propulsion: one IMI solid-propellant two-stage rocket delivering ?-lb (?-kg) thrust
Performance: speed 415 mph (668 km/h); range 165/4 yards (150/4000 m)
Guidance: wire command to line of sight

Variants
Swingfire: introduced to service in 1969, the Swingfire is a ground- or vehicle-launched (FV432 armoured personnel carrier or Striker tracked vehicle) weapon, and though it is comparatively elderly it remains useful for its powerful warhead (able to penetrate 31.5 in/800 mm of armour), range and ability for the operator to remain under cover while flying the missile; the type can also be fitted with a thermal imager for night or adverse-weather operations, and the latest development has replaced earlier-generation electronics with micro-miniaturized components for greater reliability and maintainability
Beeswing: pallet-mounted version that can be carried on any light truck and fired from this or removed by a three-man crew for ground operation

Martin-Marietta/Oerlikon ADATS
(USA/Switzerland)
Type: vehicle-launched air-defence and anti-tank missile
Dimensions: diameter 0.152 m (6.00 in); length 2.05 m (6.73 ft); span not revealed
Weights: total round 112 lb (50.8 kg); warhead 26.5+ lb (12+ kg) HEAT
Propulsion: one Hercules solid-propellant rocket delivering ?-lb (?-kg) thrust
Performance: speed Mach 3+; range 8,750 yards (8000 m)
Guidance: laser beam-riding

Variant
ADATS: one of the most interesting weapons to be developed in recent years, the ADATS (Air Defence Anti-Tank System) missile has been produced in the USA by Martin-Marietta within the context of the Oerlikon-integrated turret for an eight-tube launcher/fire-control system that can be installed on armoured personnel carriers and the like to produce a highly capable dual-role air-defence and anit-tank system; the 360-degree traverse turret has a Contraves Italiana surveillance radar, forward-looking infra-red sensor, TV tracker, laser ranger, fire-control system

and missile-guidance laser in its centre, with the two four-tube elevating missile launchers to its sides; eight reload missiles are carried in the hull; the anti-tank performance includes an armour penetration figure of 900 mm (35.4 in) using an impact fuse, and the AA performance includes an effective range of 5000 m (5,470 yards) using a proximity fuse

Ford Aerospace MGM-51C Shillelagh
(USA)
Type: gun-launched anti-tank missile
Dimensions: diameter 6 in (0.152 mm); length 3.75 ft (1.143 m); span 11.50 in (0.292 m)
Weights: total round 59 lb (26.8 lb); warhead 15 lb (6.8 kg) shaped-charge HE
Propulsion: one Amoco Chemicals/Hercules solid-propellant sustainer rocket delivering ?-lb (?-kg) thrust
Performance: speed Mach 3.45; range 5,685 yards (5200 m)
Guidance: Ford Aerospace infra-red semi-automatic command to line of sight

Variant
MGM-51C Shillegah: the MGM-51A version of this pioneering gun-launched missile began to enter service in 1967, and the MGM-51C is the definitive version; the type has a minimum effective guided range of 1,250 yards (1145 m), for it only at this range that the missile climbs into the guidance beam and is guided to the target

Hughes BGM-71A TOW
(USA)
Type: vehicle- or helicopter-launched medium/long-range heavy anti-tank missile
Dimensions: diameter 6.00 in (0.152 m); length 3.85 ft (1.174 m); span 13.5 in (0.343 m)
Weights: total round 49.6 lb (22.5 kg); warhead 8.6 lb (3.9 kg) HEAP
Propulsion: one Hercules K41 solid-propellant two-stage rocket delivering ?-lb (?-kg) thrust
Performance: 700 mph (1127 km/h); range (pre-1976 models) 70/3,280 yards (65/3000 m) or (post-1976 models) 70/4,100 yards (65/3750 m)
Guidance: Emerson Electric wire semi-automatic command to line of sight

Variants
BGM-71A TOW: entering service in 1970, the TOW (Tube-launched Optically-tracked Wire-guided) missile is the West's most important anti-tank weapon, and is a heavyweight type designed for vehicle- or helicopter-borne launchers; the type has proved its capabilities in several wars, and the warhead has shown itself able to penetrate 600 mm (23.62 in) of armour
BGM-71B Extended-Range TOW: semi-official designation of the basic weapon produced from 1976 with greater range than the original model
BGM-71C Improved TOW: introduced in the early 1980s in response to the development of Soviet tanks with improved armour, this interim model features a larger-diameter warhead (5 in/127 mm) with a telescoping nose probe that extends in flight to ensure a perfect stand-off distance (15 in/381 mm) for the detonation of the shaped-charge warhead; this model has a length of 5.10 ft (1.777 m) with the probe extended, and weighs 56.65 lb (25.7 kg); armour penetration with the new warhead increases to 700 mm (27.56 in)
BGM-71D TOW 2: introduced in 1983 as the standard US weapon for tackling the latest Soviet tanks at long range, this model has an improved motor and a 13.2-lb (6-kg) warhead of 6-in (152-mm) diameter, the nose probe ensuring optimum stand-off distance for penetration of 800 mm (31.5 in) of armour; this variant weighs 61.95 lb (28.1 kg); though this model can be fired from the original analog-electronics launcher, it is designed for use with an improved digital-electronics launcher fitted with thermal as well as optical sights, and with a more advanced guidance package; Hughes is also proposing four developments of this powerful weapon as the **TOW 2A** with an improved direct-attack warhead, the **TOW 2B** top-attack version with a millimetre-wave radar or electro-optical fuse, the **TOW 2D** of which no details have been released, and the **TOW 2N** with wireless command, supersonic speed and a range of 10,000 yards (9145 m)

McDonnell Douglas/Raytheon FGM-77A Dragon
(USA)
Type: man-portable short/medium-range anti-tank missile
Dimensions: diameter 5.00 in (0.127 m); length 2.44 ft (0.744 m); span 13.00 in (0.33 m)
Weights: total round 13.6 lb (6.2 kg); warhead 5.4 lb (2.45 kg) HEAT
Propulsion: 60 solid-propellant side thrusters each delivering 265-lb (120-kg) thrust
Performance: speed 225 mph (362 km/h); range 80/ 1,100 yards (75/1005 m)
Guidance: Raytheon wire semi-automatic command to line of sight

Variant
FGM-77A Dragon: developed from the mid-1960s as an anti-tank and bunker-busting weapon, the Dragon entered service in the early 1970s and has since seen considerable action in various parts of the world; the whole system (missile, fibreglass container/launcher, support stand and sight unit) weighs only 30.4 lb (13.8 kg) but is both bulky and tall, and suffers the tactical disadvantages of considerable backblast and visibility;

once the missile has been fired the empty tube is discarded and the sight (now enhanced with a thermal imager) plus support stand clipped to a fresh round

Rockwell AGM-114A Hellfire: see Rockwell AGM-114A Hellfire air-to-surface missile

Index

1
LAND WEAPONS

5.5-in Medium Gun 96
17-pdr Gun 97
20/1 mm M75 114
20/3 mm M55 114
25-pdr Gun 97
40L60 110
40L70 109
105-mm Howitzer C1 99
122-mm SP howitzer 43
130-mm MRS 121
320-mm rocket launcher 117
2S1 78
2S3 77
2S7 77
2S9 78
4K 4FA series 40
4K 7FA series 21, 39
5PZF Gepard 102
53T2 Tarasque 107
76T2 Cerbère 107

A-19 93
ACL-APX 127
ACMAT TPK 4.20 48
  VBL 48
AIFV series 65
AML series 32
AMX-10 series 30, 44
AMX-13 22
AMX-13 DCA 101
AMX-30 series 6
AMX-32 4
AMX-40 4
AMX-48 Leclerc 4
AMX VCI series 45
ASTROS SS-30 117
  SS-40 117
  SS-60 117
ASU-57 27
ASU-85 27
AT105 Saxon series 63
AT-P Series 60

AVIBRAS ASTROS SS-30 117
  ASTROS SS-40 117
  ASTROS SS-60 117
  FGT108 116
  SBAT-70 116
  SBAT-127 116
Abbot 78
Abrams 18
Alvis FV101 Scorpion series 37
  FV102 Striker 63
  FV103 Spartan 63
  FV104 Samaritan 63
  FV105 Sultan 63
  FV106 Samson 63
  FV107 Scimitar 37
FV601 Saladin 37
  FV603 Saracen series 62
  Stormer 62
  Streaker 63
American Locomotive M7 Priest 82
Armored Infantry Fighting Vehicle series 65
Armoured Vehicle General Purpose 42
Armoured Bridgelayer (Vickers) 18
Armoured Recovery Vehicle (Vicker)s 18
Armscor G5 87
  G6 76
  Valkyri 121
Arrow 108
Artemis 30 108
Autocamiones BMR-600 series 54

B-10 129
B-11 129
B-300 127
BAe Tracked Rapier series 105
BDX 41

BK 710 25
BLG-60 13
BLR 54
BM-13 124
BM-14 124
BM-21 124
BM-24 119, 123
BM-25 123
BM-27 123
BMD-1 57
BMD-20 123
BMP-1 series 57
BMP-2 56
BMR-600 54
BN Constructions SIBMAS 40
BOV/20-mm 106
BRAVIA Chaimite series 52
BRDM-1 series 36
BRDM-2 series 35
BREM-1 11
BS-3 94
BTR-40 series 60
BTR-50 series 58
BTR-60 series 58
BTR-70 series 57
BTR-80 58
BTR-152 series 59
Bandkanon 1A 77
Basic Vulcan 114
Battle Tank Mk 5 23
Beherman Demoen BDX 41
Bergepanzer 3 7
Bergepanzer Greif 21
Berliet VXB-170 46
Bernardini XLP-10 22
  X1A Carcara 21
Blindicide 126
Bofors 40L60 110
  40L70 109
  m/39 89
  Tp 4140 89
  FH-77 88

554 INDEX/LAND WEAPONS

PV-1110 128
Bandkanon 1A 77
Karin 89
Stridsvagn 103 9
Bradley 64
Breda 40L70 109
Brückenlegepanzer 68 10
Buffalo 48

CCVL 23
CITEFA Model 77 82
 SAPBA-1 116
 SLAM-Pampero 116
Cactus 101
Cadillac (General Motors) M24
 Chaffee 25
 M41 Walker Bulldog 24
 M42 106
 M108 81
Cadillac Gage Commando
 series 68
 Commando Ranger 70
 Commando Scout 38
 Stingray 24
Carcara 21
Cardoen VTP-2 128
Carl Gustav 128
Cascavel 29
Cerbère 107
Ch-26 95
Chaffee 25
Chaimite series 52
Challenger 14
Chaparral 66
Char de Depannage 23
Char Poseur de Pont 23
Chieftain 15
Chrysler (General Dynamics) M48
 Patton 19
Cleveland Army Tank Plant M109
 series 80
 M992 81
Commando series 68
Commando Ranger 70
Commando Scout 38
Condor 49
Cougar 42
Creusot-Loire AMX-13 22
 AMX VCI series 45
 Char de Depannage 23
 Char Poseur de Pont 23
 Mk 61 74
 Mk F3 74
Crotale 101

D-1 91
D-3/L-10 122
D-20 91
D-30 92
D-44 95
D-48 94
D-74 92
DAF PW series 52
 YP-408 series 52

DANA 73
Daewoo KIFV 65
Daimler FV701 Ferret series 38
Detroit Arsenal M52 82
Detroit Tank Plant/American
 Locomotive M47 20
Diana 111
Dragon 300 70

E-3/L-21 122
EE-T1 Osorio 3
EE-T2 Osorio 3
EE-3 Jararaca 29
EE-9 Cascavel 29
EE-11 Urutu 41
EBR series 30
EEFA Bourges M-50 75
EMC 81 32
ENASA VEC 35
ERC series 31–32
ENGESA EE-T1/2 Osorio 3
 EE-3 Jararaca 29
 EE-9 Cascavel 29
 EE-11 Urutu 41
Eland 32
Entpannungspanzer 65 10
Euromissile Roland/Marder 102

FFV Carl Gustav 128
 Miniman 127
FGT108 116
FH-70 85
FH-77 88
FH105 99
FH155 99
FIROS 6 120
FIROS 25 120
FK 20-2 109
FM Model 1968 126
FMC Armored Infantry Fighting
 Vehicle 65
 CCVL 23
 LVTP7 series 71
 M2 Bradley 64
 M3 Bradley 64
 M48 Chaparral 66
 M59 68
 M75 68
 M106 66
 M113 series 66
 M125 66
 M163 Vulcan 66
 M548 67
 M577 67
 M688 67
 M752 67
 M806 66
 M901 Improved TOW
  Vehicle 67
FUG 33
FV101 Scorpion series 37
FV102 Striker 63
FV103 Spartan 63
FV104 Samaritan 63

FV105 Sultan 63
FV106 Samson 63
FV107 Scimitar 37
FV432 Series 61
FV433 Abbot 78
FV601 Saladin 37
FV603 Saracen series 62
FV701 Ferret series 38
FV721 Fox 37
FV1609 series 64
FV4002 Centurion AVLB 16
FV4003 Centurion AVRE 16
FV4006 Centurion ARV 15
FV4018 Centurion BARV 16
FV4017 Centurion 16
FV4201 Chieftain 15
FV4030/2 Shir 15
FV4030/4 Challenger 14
FV4204 Chieftain ARV 15
FV4205 Chieftain AVLB 15
Fahd 43
Federal Construction
 Brückenlegepanzer 68 10
 Entpannungspanzer 65 10
 M35 90
 M46 89
 M50 90
 M57 90
 Pz 61 10
 Pz 67 10
Ferret 38
Fiat/OTO Melara Tipo 6614 51
 Tipo 6616 34
Field Mount 90-mm 83
Firefly (Sherman) 21
Fox 37
Fuchs 48

G-3/L-8 122
G2 96
G5 87
G6 76
GAI 111–112
GBF Diana 111
GBI 111
GC 45 82
GCT 74
GDF-001/002 110
GHN 45 83
GIAT 53T2 Tarasque 107
 76T2 Cerbère 107
 ACL-APX 127
 AMX-10 series 30, 44
 AMX-13 22
 AMX-13 DCA 101
 AMX-30 series 6
 AMX-32 4
 AMX-40 4
 AMX-48 Leclerc 4
 GCT 74
 Shahine, SICA 6
 TR 84
GKN Sankey AT105 Saxon
 series 63

INDEX/LAND WEAPONS

FV432 series 61
'Gadfly' 104
'Gainful' 104
'Ganef' 104
'Gecko' 104
'Gopher' 105
General Dynamics M1
  Abrams 18
  M48 Patton 19
  M60 18
  M728 CET 19
General Motors (Cadillac) M41
  Walker Bulldog 24
  M24 Chaffee 25
General Motors Canada LAV-25 42
  Armored Vehicle General Purpose 41–42
  Cougar 42
  Grizzly 41
  Husky 42
Gepard 102
Grenadier 55
Grizzly 41

H-1 43
HAI Artemis 30 108
Hagglund & Soner PS 701/R 102
  Ikv 91 26
  Lvrbv 701 102
  Pbv 302 series 54
  Pvrbv 551 27
Hellcat 29
Henschel Hanomag (Thyssen Henschel) JPK 90 26
  Jagdpanzer Jaguar 1 25
  Jagdpanzer Jaguar 2 26
  Jagdpanzer Kanone 26
Humber FV1609 series 64
Hunting LAW 80 129
Husky 42

IAFV 51
IMI B-300 127
  BM-24 119
  LAR-160 119
  MAR-290 119
  SMAW 127
IMR 13
Improved Tornado 54
Improved TOW Vehicle 67
Infanterikanonvagn (Ikv) 91 26
Ingersoll (Borg-Warner) LVTP5 series 72
Instalza M-65 127
Israeli Ordnance Corps
  Merkava 8

JPK 90 26
JPz 4-5 26
Jackson 28
Jagdpanzer Jaguar 1 25
Jagdpanzer Jaguar 2 26
Jagdpanzer Kanone 26
Jagdpanzer Rakete 25

Jaguar 1 25
Jaguar 2 26
Jararaca 29

K-63 42
KH178 87
KH179 87
KIFV 65
KM-900 series 51
KS-12 113
KS-18 113
KS-19 112
KS-30 113
Kadar Fahd 43
Karin 89
Khalid 15
Kia Machine Tool KH178 87
  KH179 87
Komatsu Type 60 26
  Type 82 35
  Type 87 35
Komatsu/Japan Steel Works Type 74 76
Kongsberg FK 20-2 109
Krauss-Maffei/Contraves 5PZF
  Gepard 102
Krauss-Maffei/Krupp MaK
  Bergepanzer 3 7
  Leopard 1 7
  Leopard 2 6
Kuka Arrow 108
Kung Feng III 122
Kung Feng IV 122
Kung Feng VI 123
Kürassier 21

L-33 75
L118 Light Gun 96
L119 Light Gun 96
L127 Light Gun 97
LAR-160 119
LARS-2 119
LAV-25 42
LAW 80 129
LRAC 89-STRIM 127
LVTP5 series 72
LVTP7 series 71
Lvrbv 701 102
Lanze 127
Leclerc 4
Leopard 1 7
Leopard 2 6
Light Gun 96
Luchs 33

m/36 110
m/39 89
M-30 93
M-37/10 84
M-50 75
M-55 126
M-60 series APC 73
M-60 series gun 84
M-61/37 84
M-65 127

M-68 86
M-71 86
M-74 84
M-980 series 73
M1 Abrams 18
M2 halftrack 71
M2 Bradley 64
M3 halftrack 71
M3 Bradley 64
Me series 47
M4 mortar-carrier 71
M4 Sherman 20
M5 halftrack 71
M7 Priest 82
M9 halftrack 71
M13 71
M14 71
M15 71
M16 71
M17 71
M18 Hellcat 29
M18 launcher 129
M20 3.5-in bazooka 130
M20 75-mm recoilless gun 129
M21 71
M24 Chaffee 25
M27 130
M32 ARV 21
M35 90
M36 Jackson 28
M40 130
M41 Walker Bulldog 24
M42 106
M44 series 81
M46 89
M47 20
M48 76-mm gun 100
M48 Chaparral 66
M48 Patton 19
M50 90
M51 118
M52 82
M52/55 83
M53 12.7-mm AA 107
M53 100-mm gun 83
M53/59 101
M53/70 101
M55 0.5-in AA gun 115
M55 20-mm AA gun 115
M56 100
M57 90
M59 155-mm gun 98
M59 APC 68
M59A 126
M60 18
M63 Plaman 125
M65 99
M72 130
M74 ARV 21
M75 20-mm AA gun 115
M75 APC 68
M77 Oganj 125
M101 99

M102 99
M106 66
M107 series 79
M108 series 81
M109 series 80
M110 series 79
M113 series 66
M114 98
M115 97
M116 99
M117 114
M118 114
M125 66
M139 99
M163 Vulcan 66
M167 Vulcan 114
M198 98
M548 67
M577 67
M578 79
M688 67
M752 67
M806 66
M901 Improved TOW Vehicle 67
M728 CET 19
M992 81
M1937 152-mm gun/howitzer 91
M1938 122-mm howitzer 93
M1939 37-mm AA gun 113
M1939 85-mm AA gun 113
M1942 76-mm gun 95
M1943 152-mm howitzer 91
M1944 85-mm AA gun 113
M1944 100-mm gun 94
M1946 130-mm gun 91
M1955 122-mm gun 92
M1955 152-mm gun/howitzer 91
M1955 180-mm gun 90
M1956 130-mm gun 92
M1966 76-mm howitzer 95
M1967 APC 42
M1967 76-mm howitzer 95
M 1970 APC 42
M1973 SPG/H 77
M1974 SPH 78
M1975/BM-21 125
M1975 SPG 77
M1976/BM-21 125
M1977 ARV 12
M 18 series 85
MAR-290 119
MCV-80 Warrior series 61
MECAR MPA 75 126
    RL-83 Blindicide 126
    Field Mount 90-mm 83
MK 20 Rh 202 107
ML-20 91
MLRS
MOWAG SPY 35
    Grenadier 55
    Improved Tornado 55
    Piranha 55
    Roland 56

MPA 75 126
MT-55 13
MT-LB series 59
MTU 14
MTU-20 13
Main Battle Tank Mk 1 17
    Mk 3 17
    Mk 7 17
Marder 48
Massey Harris M44 81
Merkava 8
Miniman 127
Mitsubishi Type 61 MBT 9
    Type 67 AEV 9
    Type 67 AVLB 9
    Type 70 ARV 9
    Type 73 APC 51
    Type 74 MBT 9
    Type 75 measurer 52
    Type 75 SPH 76
    Type 78 ARV 9
    Type 88 MBT 8
    Type S 60 series 52
Mk 61 74
Mk F3 74
Model 1968 126
Model 77 82
Model 77 DANA 73
Model 839 86
Model 845 86
Modele 50 85
Modello 56 86
Multiple-Launch Rocket System 125

NHV-1 43
NHV-4 43
NIMDA Shoet 50
NM-116 25
NORINCO H-1 43
    Type 63 42
NORINCO-Vickers NVH-1 43
    NHV-4 43
Nobel PZF 44 Lanze 127

OF-40 8
OT-62 59
OT-64 series 43
OT-65 34
OT-66 50
OTO Melara IAFV 51
    OF-40 8
    Modello 56 86
    Palmaria 75
Oerlikon GAI 111–112
    GBF Diana 111
    GBI 111
    GDF-001 110
    GDF-002 110
    RWK-014 122
Oganj 125
Osorio 3

PSZH-IV 50
PS 701/R 102

PT-76 23
PV-1110 128
PW series 52
PZF 44 Lanze 127
Pbv 302 series 54
Pvrbv 551 26
Pz 61 10
Pz 67 10
Pacific Car and Foundry M107 series 79
    M110 series 79
    M578 79
Palmaria 75
Panga 38
Panhard AML series 32
    EBR series 30
    M3 series 47
    VBL series 32–33
    VCR series 46
    Buffalo 48
Panzeraufklärungsradargerät 49
Patton 19
Pegaso 3560 series 54
'Pig' series 64
Piranha 55
Plaman 125
Porsche Wiesel 33
Priest 82

R-6/B-2 121
R-50 series 88
RAM V-1 34
RAM V-2 34
RAMTA RAM V-1 34
    V-2 34
    TCM-20 108
    TCM Mk 3 108
RL-83 Blindicide 126
ROF FV721 Fox 37
    FV4017 Centurion 16
    FV4021 Chieftain 14
    FV4030/2 Shir 14
    FV4030/4 Challenger 14
    FV4204 Chieftain ARV 15
    FV4205 Chieftain AVLB 15
    Panga 38
RPG-2 128
RPG-7 128
RPG-16 128
RPG-18 128
RPU-14 124
RWK-014 122
Radarpanzer TUR 48
Rapier 67, 105
Ratel series 53
Reinosa R-50 series 88
Renault EMC 81 32
    ERC Sagaie series 32
    VAB series 45
    VBC 90 30
    VCAC series 46
    Sagaie 2 31
Rheinmetall MK 20 Rh 202 107
Rheinmetall/OTO Melara/Vickers

FH-70   85
Rhino   20
Roland   56
Roland/Marder   102
Royal Ordnance SP122   73

S-23   90
S-60   113
SA-4 'Ganef'   104
SA-6 'Gainful'   104
SA-8 'Gecko'   104
SA-11 'Gadfly'   104
SA-13 'Gopher'   105
SAKR-18   118
SAKR-30   118
SAPBA-1   116
SARPAC   126
SB 155   88
SBAT-70   116
SBAT-127   116
SD-44   94
SIBMAS   40
SISU XA-180   43
SK105 Kurassier   21
SKOT-2 series   43
SKP-5   14
SLAM/Pampero   116
SM-4-1   92
SMAW   127
SNIA FIROS 6   120
   FIROS 25   120
SO-120   78
SP122   73
SPG-9   129
SPY   35
SRC International GC 45   82
Saint Etienne LRAC 89–
   STRIM   127
Saladin   37
Samaritan   63
Samson   63
Sandock-Austral Ratel series   53
Santa Barbara BLR   54
   SB 155   88
   Teruel   121
Saracen series   62
Saxon series   63
Schützenpanzer Neu M-1966
   Marder   48
Scimitar   37
Scorpion series   37
Shahine, SICA   6
Sherman   20
Sherman Firefly   21
Shir   15
Shoet   50
Shorland   38
Short Brothers Shorland   38
Soltam L-33   75
   M-68   86
   M-71   86
   Model 839   86
   Model 845   86
Spähpanzer Luchs   33

Spartan   63
Steyr 4K 74A series   40
   4K 7FA series   21, 39
SK105 Kürassier   21
Stingray   24
Stormer   62
Streaker   63
Stridsvagn 103   9
Striker   63

T-12   93
T-34/85 MBT   14
T-34/122 SPH   77
T-54 MBT   13
T-55 MBT   13
T-62 MBT   12
T-64 MBT   11
T-72 MBT   11
T-80 MBT   12
T65   99
Tp 4140   89
TAM   3
TAMSE/Thyssen Henschel
   TAM   3
   VCT series   39
TCM-20   108
TCM Mk 3   108
TELAR 1   105
TELAR 2   105
TM 125   49
TM 170   49
TOPAS-2   59
TPK 4.20   48
TR   84
Tampella M-37/10   84
   M-60   84
   M-61/37   84
   M-74   84
Tarasque   107
Teruel   121
Thomson-Brandt SARPAC   126
Thomson-CSF/Matra Cactus   101
   Crotale   101
Thyssen Henschel TM 125   49
   TM 170   49
   UR-416   50
   Condor   49
   Panzeraufklärungsradargerät   49
   Spähpanzer Luchs   33
   Transportpanzer 1 Fuchs   48
Thyssen Henschel/Krupp MaK
   Radarpanzer TUR   48
   Schützenpanzer Neu M-1966
      Marder   48
Timoney Mk 6   41
Tipo 6614   51
Tipo 6616   34
Tracked Rapier   67, 105
Transportpanzer 1 Fuchs   48
Type 51 bazooka   130
Type 52 recoilless gun   129
Type 54 76-mm gun   95
Type 54 122-mm howitzer   93

Type 54 152-mm howitzer   91
Type 55 37-mm AA gun   113
Type 55 57-mm gun   96
Type 56 14.5-mm AA gun   114
Type 56 85-mm AA gun   113
Type 56 85-mm gun   95
Type 56 launcher   128
Type 56 recoilless gun   129
Type 58 14.5-mm AA gun   114
Type 59 57-mm AA gun   113
Type 59 100-mm AA gun   113
Type 59 100-mm gun   94
Type 59 130-mm gun   92
Type 59 MBT   3
Type 60 122-mm gun   92
Type 60 APC   26
Type 60 light tank   23
Type 61 MBT   9
Type 62 light tank   22
Type 63 37-mm AA gun   113
Type 63 APC   42
Type 63 light tank   22
Type 63 MRS   118
Type 63 SPAAG   10
Type 64 light tank   25
Type 66 152-mm gun/
   howitzer   91
Type 67 AEV   9
Type 67 AVLB   9
Type 67 MRS   120
Type 69 launcher   128
Type 69-I MBT   3
Type 69-II MBT   3
Type 69-II Mk B MBT   3
Type 70 ARV   9
Type 70 MRS   117
Type 73 APC   51
Type 74 MBT   9
Type 74 SPH   76
Type 75 measurer   52
Type 75 MRS   120
Type 75 SPH   76
Type 78 ARV   9
Type 80 MBT   3
Type 80 SPAAG   3
Type 81 MRS   118
Type 82 CP   35
Type 84 AVLB   3
Type 87 APC   35
Type 88 MBT   8
Type 653 APC   3
Type S 60 series   52

UR-416   50
Urutu   41

VAB   45
VAP   119
VBL series   32–33, 48
VCAC series   46
VCR series   46
VCT series   39
VEC   35
VTP-2   42
VXB-170   46

558 INDEX/WARSHIP CLASSES

Valiant 17
Valkyr 41
Valkyri 121
Value Engineered Abbot 79
Verne Dragoon 300 70
Vijayanta 17
Vickers FV433 Abbot 78
  Armoured Bridgelayer 18
  Armoured Recovery
    Vehicle 18
  Main Battle Tank Mk 1 17
    Mk 3 17
    Mk 7 17

Valiant 17
Valkyr 41
Vijayanta 17
Vickers/FMC Battle Tank Mk 5 23
Vought Multiple-Launch Rocket
  System 125
Vulcan 66, 114

WP-8 121
WPT-34 14
WZT-1 14
Walker Bulldog 24
Warrior series 61
Wiesel 33

X1A 21
XA-180 43
XLP-10 22

YP-408 series 52
YPR-765 series 65

ZIS-2 96
ZIS-3 95
ZPU-1 114
ZPU-2 114
ZPU-4 114
ZSU-23-4 103
ZSU-57-2 103
ZU-23 113

2

WARSHIP CLASSES

'A69' 241
'Agosta' 159
'Akizuki' 228
'Akula' 147
'Al Mansur' 295
'Al Siddiq' 294
'Al Wafi' 295
'Alia' 312
'Allen M.Sumner' 229
'Allen M.Sumner (FRAM II)' 229
'Alligator' 317
'Almirante' 202
'Almirante Pereira da Silva' 268
'Alpha' 147
'Alpino' 268
'Amatsukaze' 203
'Amazon' 242
'Anchorage' 318
'Andrea Doria' 191
'Annapolis' 269
'Anshan' 203
'Arleigh Burke' 204
'Asashio' 159
'Asheville' 310
'Assad' 286
'Audace' 204
'Audace (Improved)' 205
'Austin' 318
'Ayanami' 230

'BDC' 320
'Badr' 286
'Bainbridge' 192
'Baleares' 251
'Baptista de Andrade' 269
'Barbel' 159
'Batral' 319
'Belknap' 192
'Benjamin Franklin' 137
'Bergamini' 269
'Berk' 270
'Blue Ridge' 320
'Boxer' 206
'Bremen' 242
'Broad-Beam Leander' 256
'Broadsword' 205

'Bronstein' 270
'Brooke' 243
'Brooke Marine 32.6-m' 294
'Brooke Marine 37.5-m' 295

'C65' 206
'C70/AA' 207
'C70/ASW' 207
'California' 193
'Cannon' 271
'Carpenter (FRAM I)' 231
'Casma' 174
'Charles F. Adams' 208
'Charles F. Adams
  (Improved)' 209
'Charles Lawrence' 271
'Charlie I' 143
'Charlie II' 144
'Charlie III' 144
'Chengdu' 243
'Chikugo' 272
'Churchill' 147
'Claud Jones' 273
'Clemenceau' 181
'Colbert' 194
'Colossus' 182
'Comandante Joao Belo' 273
'Commandant Rivière' 244
'Constitucion' 295
'Coontz' 211
'Cormoran' 295
'Cornwall' 206
'County' 209
'Crosley' 271

'D20' 232
'D60' 234
'Daphné' 160
'Daring' 210
'Delfinen' 161
'Delta I' 137
'Delta II' 138
'Delta III' 138
'Delta IV' 138
'Descubierta' 245
'D'Estienne d'Orves' 241
'Dolfijn' 161

'Draken' 162
'Duke' 245
'Dvora' 296

'E-71' 267
'Echo' 144
'Echo I' 148
'Enterprise' 182
'Esmeraldas' 287
'Ethan Allen' 148
'Exeter' 222

'F67' 211
'F70' 251
'F80' 261
'F2000' 246
'Farragut' 211
'Fatahillah' 246
'Fearless' 321
'Flagstaff 2' 296
'Fletcher' 231
'Forrestal' 183
'Foxtrot' 162
'Friesland' 232
'Frosch I' 321
'Frosch II' 321

'Garcia' 274
'Garibaldi' 183
'Gearing (FRAM I)' 233
'Gearing (FRAM II)' 235
'Glavkos' 174
'Glenard P. Lipscomb' 149
'Glover' 275
'Godavari' 247
'Golf II' 146
'Grisha I' 287
'Grisha III' 287
'Groningen' 247
'Guppy I' 163
'Guppy II' 163
'Guppy IIA' 164
'Guppy III' 164

'HDP 1000' 288
'Hai Lung' 181
'Halifax' 212
'Halland' 212

INDEX/WARSHIP CLASSES

'Halland (Modified)' 236
'Hamburg' 213
'Han' 149
'Haruna' 213
'Hatakaze' 214
'Hatsuyuki' 214
'Hatsuyuki (Improved)' 215
'Hauk' 296
'Helsinki' 297
'Hermes' 184
'Heroj' 165
'Holland' 236
'Hotel II' 139
'Hotel III' 139
'Huangfen' 306
'Huchuan' 297
'Hugin' 297

'Impavido' 216
'Independence' 184
'Inhauma' 248
'Invincible' 185
'Iowa' 190
'Iroquois' 236
'Ishikari' 249
'Isuzu' 275
'Ivan Rogov' 322
'Iwo Jima' 322

'Jacob van Heemskerck' 249
'Jaguar' 298
'Jeanne d'Arc' 185
'Jerong' 305
'Jiangdong' 249
'Jianghu I' 250
'Jianghu II' 250
'Jianghu III' 250
'Jiangnan' 275
'Joao Coutinho' 276
'John F. Kennedy' 185
'Juliett' 146

'Kaman' 299
'Kanin' 216
'Kara' 194
'Kartal' 298
'Kashin' 216
'Kashin II' 217
'Kashin (Modified)' 217
'Kebir' 295
'Kidd' 218
'Kiev' 186
'Kildin (Modified)' 219
'Kilo' 165
'Kirov' 191
'Kitty Hawk' 187
'Knox' 251
'Köln' 276
'Komar' 298
'Koni' 253
'Kortenaer' 253
'Kotlin' 237
'Kotlin (Modified)' 237
'Krasina' 195
'Kresta I' 195

'Kresta II' 196
'Krivak I' 252
'Krivak II' 252
'Krivak III' 253
'Kynda' 197

'LSM 1' 323
'LST 1-1152' 323
'La Combattante II' 299
'La Combattante III' 300
'Lafayette' 139
'L'Inflexible' 140
'Lazaga' 302
'Lazaga (Modified)' 302
'Le Redoutable' 142
'Leahy' 197
'Leander Batch 1' 277
'Leander Batch 2' 255
'Leander Batch 3' 256
'Leopard' 277
'Long Beach' 198
'Los Angeles' 150
'Luda' 219
'Lupo' 257
'Lürssen FPB-57' 301
'Lürssen TNC-45' 303

'MV400' 306
'Mackenzie' 278
'Madina' 246
'Maestrale' 258
'Majestic' 187
'Makut Rajakumarn' 278
'Manchester' 223
'Matka' 305
'Meko 200' 259
'Meko 360H2' 220
'Midway' 188
'Mike' 151
'Minegumo' 238
'Minerva' 288
'Ming' 166
'Minister' 312
'Mirka I' 279
'Mirka II' 279
'Mol' 313
'Muntenia' 220
'Murasame' 238

'Näcken' 167
'Najin' 279
'Nanuchka I' 289
'Nanuchka II' 289
'Nanuchka III' 289
'Narvhalen' 166
'Narwhal' 151
'Newport' 324
'Niels Juel' 260
'Nilgiri' 260
'Nimitz' 189
'Niteroi' 221
'Niteroi (Modified)' 221
'November' 152

'Oberon' 167
'Obuma' 280

'October' 306
'Ohio' 142
'Oliver Hazard Perry' 261
'Osa I' 306
'Osa II' 306
'Oscar' 145
'Oslo' 262
'Ouragan' 324

'P400' 307
'PF103' 281
'PR72' 308
'PSMM Mk 5' 309
'Papa' 145
'Parchim' 289
'Patra' 307
'Pauk' 290
'Peder Skram' 263
'Pegasus' 308
'Perdana' 300
'Permit' 156
'Perth' 208
'Petya I' 280
'Petya I (Modified)' 280
'Petya II' 280
'Petya II (Modified)' 280
'Polnochny' 325
'Porpoise' 167
'Poti' 290
'Potvis' 161
'President' 281
'Principe de Asturias' 189
'Province' 309

'Rahmat' 282
'Raleigh' 326
'Ramadan' 310
'Ratanakosin' 290
'Ratcharit' 310
'Reshef' 311
'Resolution' 142
'Restigouche (Improved)' 282
'Riga' 283
'River' 283
'Romeo' 168
'Ropucha' 326
'Rothesay' 284
'Rubis' 152

'S60' 161
'S70' 158
'SAM Kotlin' 221
'SNA72' 152
'Saam' 265
'Saar 2' 304
'Saar 3' 304
'Saar 4' 311
'Saar 4.5' 312
'Saar 5' 291
'Salisbury' 284
'Sauro' 169
'Sava' 169
'Shanghai II' 312
'Shanghai III' 312
'Shanghai IV' 312

'Sheffield' 222
'Shershen' 313
'Shirane' 223
'Sierra' 152
'Sjöormen' 170
'Skate' 153
'Skipjack' 153
'Skory' 239
'Skory (Modified)' 239
'Sleipner' 291
'Snögg' 314
'Sohang' 299
'Soju' 307
'Sovremenny' 223
'Sparviero' 314
'Spica I' 314
'Spica II' 315
'Spica III' 315
'Spica-M' 315
'Spruance' 224
'Stenka' 316
'Stockholm' 315
'Storm' 316
'Sturgeon' 154
'Suffren' 225
'Sutjeska' 170
'Sverdlov' 198
'Swiftsure' 155

'T47' DD 240
'T47' DDG 225
'TCD90' 327
'TR1700' 171
'Tachikaze' 226
'Takatsuki' 226
'Tango' 171
'Tarantul I' 292
'Tarantul II' 292
'Tarawa' 327
'Thetis' 292
'Thomaston' 328
'Thresher' 156
'Ticonderoga' 199
'Toti' 171

'Trafaalgar' 156
'Tribal' DD 236
'Tribal' FF 285
'Tromp' 227
'Truxtun' 200
'Tuima' 307
'Turunmaa' 292
'Turya' 316
'Type 051' 219
'Type 053J' 249
'Type 12' 266
'Type 12 (Modified)' 284
'Type 21' 242
'Type 22' 205
'Type 23' 245
'Type 41' 277
'Type 41/61' 285
'Type 42' 222
'Type 61' 284
'Type 81' 285
'Type 82' 201
'Type 120' 276
'Type 122' 242
'Type 143A' 303
'Type 143B' 303
'Type 148' 300
'Type 205' 172
'Type 206' 172
'Type 207' 173
'Type 209/0' 174
'Type 209/1' 174
'Type 209/2' 175
'Type 209/3' 176
'Type 209/4' 176
'Type 210' 176
'Type 420' 292
'Type 540' 177
'Type 101A' 213
'Type 1100' 174
'Type 1200' 174
'Type 1300' 175
'Type 1400' 176
'Type 1500' 176
'Type 2400' 177

'Type FS1500' 263
'Typhoon' 142
'Tzu Chiang' 296

'Udaloy' 227
'Ula' 176
'Ulsan' 264
'Upholder' 177
'Uzushio' 178

'V-28' 248
'Valiant' 147
'Van Speijk' 265
'Vastergotland' 178
'Victor I' 157
'Victor II' 157
'Victor III' 158
'Virginia' 201
'Vittorio Veneto' 202
'Vosper Mk 1' 293
'Vosper Thornycroft Mk 3' 293
'Vosper Thornycroft Mk 5' 265
'Vosper Thornycroft Mk 7' 266
'Vosper Thornycroft Mk 9' 294
'Vosper Thornycroft Mk 10' 221

'Wasp' 328
'Whidbey Island' 329
'Whidbey Island (Modified)' 329
'Whiskey' 179
'Whitby' 266
'Wielingen' 267
'Willemoes' 317
'Wuhan' 146

'Xia' 143

'Yamagumo' 240
'Yankee I' 143
'Yankee II' 143
'Yubari' 267
'Yubari (Improved)' 267
'Yuushio' 179

'Zeeuleuw' 180
'Zwaardvis' 181

3

AIRCRAFT

35 (Saab) 456
37 (Saab) 457
39 (Saab) 459
105 (Saab) 457

A 109 353
A 129 353
A-4 431
A-5 448
A-6 406
A-7 477
A-10 401
A-37 397
AB.204 354
AB.205 366

AB.206 367
AB.212 355
AB.412 355
AH-1 373
AH-64 413
AIDC AT-TC-3 356
  T-CH-1 357
AMIT Fouga 347
AMX 343
AS 332 350
AS 350 352
ASH-3 356
AT-26 344
AT-28 450
AT-TC-3 356

AV-8A/C 378
AV-8B 437
Aeritalia F-104S Starfighter 341
  G91R 341
  G91Y 342
  G222 342
Aeritalia/Aermacchi/EMBRAER
  AMX 343
Aermacchi M.B.326 343
  M.B.339 and M.B.339 Veltro
  2 345
Aero L-29 Delfin 345
  L-39 Albatros 346
Aérospatiale AS 332 Super
  Puma 350

INDEX/AIRCRAFT

AS 350 Ecureuil 352
CM.170 Magister 346
HH-65 Dolphin 353
SA 315 Lama 348
SA 316 Alouette III 348
SA 318 Alouette II Astazou 347
SA 319 Alouette III Astazou 349
SA 321 Super Frelon 349
SA 330 Puma 350
SA 341/342 Gazelle 351
SA 360/366 Dauphin 352
SE 313 Alouette II 347
Agusta A 109 353
 A 129 Mangusta 354
Agusta (Bell) AB.204 354
 AB.205 366
 AB.206 367
 AB.212 355
 AB.412 Grifone 355
Agusta (Sikorsky) ASH-3 356
Airguard 481
Airtech CN-235 387
Ajeet 412
Albatros 346
Alizé 389
Alouette II 347
Alouette II Astazou 347
Alouette III 348
Alouette III Astazou 349
Alpha-XH1 349
Alpha Jet 396
An-2 357
An-12 357
An-14 358
An-22 359
An-24 359
An-26 359
An-28 358
An-30 360
An-32 360
An-72 360
An-124 360
Andover 375
Antonov An-2 'Colt' 357
 An-12 'Cub' 357
 An-14 'Clod' 358
 An-22 'Cock' 359
 An-24 'Coke' 359
 An-26 'Curl' 359
 An-28 'Cash' 358
 An-30 'Clank' 360
 An-32 'Cline' 360
 An-72 'Coaler' 360
An-124 'Condor' 360
Apache 413
Arava 418
Atlantic 389
Atlantique 389
Atlas Alpha-XH1 349
 Cheetah 391
 Impala 344
Aucan 400
Aurora 426

Aviocar 386
Aviojet 386

B-1 454
B-52 368
BN-2 384
BO105 438
'Backfire' 474
'Badger' 472
Bandeirante 399
Bandeirulha 396
'Beagle' 416
'Bear' 475–476
Beech C-12 Huron 361
 King Air 362
 Maritime Patrol 200 362
 Super King Air 361
 T-34 Mentor/Turbine Mentor 362
 U-21 Ute 361, 362
Bell AH-1G/Q/R/S HueyCobra 363
 AH-1J SeaCobra 364
 AH-1T Improved SeaCobra 364
 AH-1W SuperCobra 364
 OH-58 Kiowa 366
 TH-57 SeaRanger 367
 UH-1 Iroquois 'Huey' 365
 Model 204 365
 Model 205 366
 Model 206 JetRanger/LongRanger 367
 Model 212 366
 Model 214 367
Bell/Boeing V-22 Osprey 368
Be-12 368
Beriev Be-12 (M-12) 'Mail' 368
'Bison' 448
Black Hawk 465
'Blinder' 473
Boeing B-52 Stratofortress 368
 C-18 375
 C-135 Stratolifter 370
 E-3 Sentry 374
 E-4 374
 E-6 374
 E-18 375
 EC-135 370
 KC-135 Stratotanker 369
 KC-707 375
 KE-3 375
 RC-135 371
 VC-137 375
 Model 707 375
Boeing Vertol CH-46 Sea Knight 374
 CH-47 Chinoock 374
 CH-113 Labrador/Voyageur 374
'Brewer' 482
British Aerospace HS 748 Military Transport 375
 VC10 383

Andover 375
Buccaneer 375
Canberra 376
Coastguarder 375
Harrier 377
Hawk 378
Hunter 379
Jet Provost 382
Lightning 380
Nimrod 380
Sea Harrier 381
Shackleton 382
Strikemaster 382
Victor 384
Britten-Norman BN-2 Defender/Islander/Trislander 384
Bronco 455
Buccaneer 375
Buckeye 455
Buffalo 398

C.160 472
C-1 422
C-2 408
C-5 434
C-12 361
C-18 375
C-20 411
C-101 386
C-119 400
C-123 400
C-130 424
C-135 370
C-141 426
C-212 386
CAC Model 206 367
CASA C-101 Aviojet 386
 C-212 Aviocar 386
CASA/Nurtanio CN-235 387
CH-46 374
CH-47 374
CH-113 374
CH-118 366
CH-135 366
CH-136 367
CL-41 385
CL-215 385
CM.170 346
CN-235 387
CNIAR IAR-93 467
CP-140 426
CT-114 385
Canadair CL-41 Tebuan 385
 CL-215 385
 CT-114 Tutor 385
Canberra 376
'Candid' 417
Caribou 397
'Cash' 358
Cayuse 414
Cessna A-37 Dragonfly 387
 O-2 388
 OA-37 388
 T-37 387

INDEX/AIRCRAFT

T-48 388
Model 337 388
Cheetah (Atlas) 391
  (HAL) 348
Chetak 349
Chickasaw 463
Chinook 374
Choctaw 463
'Clank' 360
'Cline' 360
'Clod' 358
'Coaler' 360
Coastguarder 375
'Cock' 359
'Coke' 359
'Colt' 357
Commando 480
'Condor' 360
Convair F-106 Delta Dart 389
'Coot' 416
Corsair II 477
'Crate' 416
Crusader 478
'Cub' 357
'Curl' 359

DHC-4 397
DHC-5 388
Dagger 419
Dassault-Breguet Alizé 389
  Atlantic/Atlantique 389
  Etendard 396
  Gardian 395
  Guardian 395
  Lancier 397
  Mirage III 390
  Mirage IV 391
  Mirage 5 392
  Mirage F1 393
  Mirage 2000 394
  Mystère-Falcon 20 395
  Mystère-Falcon 200 395
  Super Etendard 395
Dassault-Breguet/Dornier Alpha Jet 396
Dauphin 352
de Havilland Canada DHC-4 Caribou 397
  DHC-5 Buffalo 398
Defender (BN) 384
  (Hughes/McDonnell Douglas) 414
Delfin 345
Delta Dart 389
Dolphin 366
Douglas EA/EKA/ERA/KA-3 Skywarrior 398
Dragonfly 397
Draken 456

E-2 408
E-3 374
E-4 374
E-6 374

E-18 375
EA-3 398
EA-6 407
EC-135 370
EF-111 406
EH.101 400
EKA-3 398
EMB-110 398
EMB-111 398
EMB-312 399
EMB-326 344
EMBRAER AT-26 Xavante 344
  EMB-110 Bandeirante 398
  EMB-111 Bandeirulha 398
  EMB-312 Tucano 399
  EMB-326 344
ENAER Aucan 400
  T-35 Pillan 399
Eagle 434
Ecureuil 352
Electra 429
Etendard 396
European Helicopter Industries EH.101 400
Extender 436

F.27 402
F-1 447
F-4 432
F-5 (Northrop) 450
  (Shenyang) 439
F-7 481
F-8 (Shenyang) 461
  (Vought) 478
F-14 409
F-15 434
F-16 404
F-19 426
F-100 449
F-104 341, 427
F-106 389
F-111 405
F/A-18 434
FB-111 405
FMA IA 58 Pucara 402
  IA 63 Pampa 402
  IA 66 Pucara 402
FTMA 388
'Fagot' 438
Fairchild C-119 Flying Boxcar 400
  C-123 Provider 400
Fairchild Republic A-10 Thunderbolt II 401
'Fantan' 448
'Farmer' 439
'Fencer' 470
Fennec 450
'Fiddler' 475
Fighting Falcon 404
'Firebar' 482
'Fishbed' 440
'Fitter' 468–469
'Flagon' 469

'Flanker' 471
'Flogger' 441, 443
Flying Boxcar 400
Fokker F.27 Maritime/Maritime Enforcer 402
'Forger' 483
'Foxbat' 442
'Foxhound' 444
Freedom Fighter 450
'Fresco' 439
'Frogfoot' 471
Fuji T-1 403
Fuji (Bell) UH-1 365, 366
'Fulcrum' 444

G-2 467
G-4 466
G91 341-342
G222 342
Galeb 467
Gardian 395
Gaviao 348
Gazelle 351
General Dynamics EF-111 Raven 406
  F-16 Fighting Falcon 404
  F/FB-111 405
Goshawk 379
Government Aircraft Factories Nomad/Mission Master/Search Master 406
Greyhound 408
Grifone 355
Gripen 459
Grumman A-6 Intruder 406
  C-2 Greyhound 408
  E-2 Hawkeye 408
  EA-6 Prowler 407
  F-14 Tomcat 409
  OV-1 Mohawk 410
  S-2 Tracker 410
Guardian 395
Gulstream III/IV 411
Gulfstream Aerospace C-20 411
  SMA-3 411
  SRA-1 411
Gulfstream III/IV 411

H-5 417
H-6 473
HAL Cheetah 348
  Chetak 349
HB 315 348
HF-24 412
HH-65 353
HJ-5 417
HJT-16 412
HS 748 375
HU-25 395
HZ-5 417
'Halo' 447
Hanzhong Y-8 358
Harbin H-5 417
  HJ-5 417

INDEX/AIRCRAFT

HZ-5 417
Z-5 445
Z-8 350
Z-9 Zaitun 353
Harrier 377
Harrier II 437
'Havoc' 447
Hawk 378
Hawkeye 408
'Haze' 446
Helibras HB 315 Gaviao 348
   HB 350 Esquilo 352
'Helix' 421
Hercules 424
'Hind' 446
Hindustan Aeronautics Ltd HF-24
   Marut 412
   HJT-16 Kiran 412
   Ajeet 412
'Hip' 445
'Hokum' 421
'Hook' 445
'Hoplite' 481
'Hound' 444
'Hormone' 421
Hornet 435
'Huey' 365
HueyCobra 363
Hughes (McDonnell Douglas
   Helicopters) AH-64
   Apache 413
   OH-6 Cayuse 414
   TH-55 Osage 414
   Model 200/269/300 414
   Model 500/530 Defender 414
Hunter 379
Huron 371

IA 58 402
IA 63 402
IA 66 402
IAI AMIT Fouga 347
   Arava 418
   Dagger 419
   Improved Fouga 347
   Kfir 419
   Lavi 419
   Nesher 419
   Seascan 418
IAR-93 467
IAR-316 349
IAR-317 347
IAR-330 350
ICA-Brasov IAR-316 Alouette
   III 349
   IAR-317 Airfox 349
   IAR-330 Puma 350
Il-14 415
Il-18 416
Il-20 416
Il-28 416
Il-38 417
Il-76 417
Ilyushin Il-14 'Crate' 415

Il-18 'Coot' 416
Il-20 'Coot' 416
Il-28 'Beagle' 416
Il-38 'May' 417
Il-76 'Candid' 417
Il-76 'Mainstay' 418
Il-76 'Midas' 418
Impala 344
Improved Fouga 347
Improved SeaCobra 364
Intruder 406
Iroqois 365
Isfahan 368
Iskra 453
Islander 384

J-1 467
J-5 439
J-6 460
J-7 482
J-8 461
J-22 467
JJ-6 460
JZ-6 460
Jaguar 459
Jastreb 467
Jet Provost 382
JetRanger 367
Jolly Green Giant 464

KA-3 398
KC-10 436
KC-135 369
KC-707 375
KE-3 375
KS-3 429
KV 107 422
Ka-? 421
Ka-25 420
Ka-27 421
Kaman SH-2 Seasprite 420
Kamov Ka-25 'Hormone' 420
   Ka-27 'Helix' 421
   Ka-? 'Hokum' 421
Kawasaki C-1 422
   TH-55 414
   XT-4 423
Kawasaki (Boeing Vertol) KV
   107 422
Kawasaki (Lockheed) P-2 423
Kfir 419
King Air 362
Kiowa 366
Kiran 412

L-29 345
L-39 346
L-188 429
LIM-5 439
Labrador 374
Lama 348
Lancier 397
Lavi 419
Lightning 380
Lockheed C-5 Galaxy 424

C-130 Hercules 424
C-141 StarLifter 426
CP-140 Aurora 426
F-19 426
F-104 Starfighter 427
KS-3 429
L-188 Electra 429
P-3 Orion 428
S-3 Viking 429
SR-71 429
TR-1 430
U-2 430
US-3 430
TriStar 431
LongRanger 367
Lynx 478

M-4 448
M-12 368
M.B.326 343
M.B.339 345
MBB BO105/PAH-1/VBH 438
'Maestro' 483
'Mail' 368
'Mainstay' 417
Mangusta 354
Maritime/Maritime Enforcer 402
Maritime Patrol 200 362
Marut 412
'May' 417
McDonnell Douglas A/EA/OA/TA-4
   Skyhawk II 431
   F/RF-4 Phantom II 432
   F-15 Eagle 434
   F/A-18 and RF-18 Hornet 434
   KC-10 Extender 436
McDonnell Douglas/British
   Aerospace AV-8B Harrier
   II 437
McDonnell Douglas Helicopters
   AH-64 Apache 413
   OH-6 Cayuse 414
   Model 200/269/300 414
   Model 500/530 Defender 414
Mentor 362
'Midas' 418
'Midget' 439
MiG-15 439
MiG-17 439
MiG-19 439
MiG-21 440
MiG-23 441
MiG-25 442
MiG-27 443
MiG-29 444
MiG-31 444
Mikoyan-Gurevich MiG-15 'Fagot'
   and 'Midget' 439
MiG-17 'Fresco' 439
MiG-19 'Farmer'
MiG-21 'Fishbed' and
   'Mongol' 440
MiG-23 'Flogger' 441
MiG-25 'Foxbat' 442

INDEX/AIRCRAFT

MiG-27 'Flogger' 443
MiG-29 'Fulcrum' 444
MiG-31 'Foxhound' 444
Mi-2 481
Mi-4 444
Mi-6 445
Mi-8 445
Mi-14 446
Mi-17 446
Mi-24 446
Mi-25 446
Mi-26 447
Mi-28 447
Mil Mi-2 'Hoplite' 481
   Mi-4 'Hound' 444
   Mi-6 'Hook' 445
   Mi-8 'Hip' 445
   Mi-14 'Haze' 446
   Mi-17 'Hip' 446
   Mi-24 'Hind' 446
   Mi-25 'Hind' 446
   Mi-26 'Halo' 447
   Mi-28 'Havoc' 447
Military Transport 375
Mirage III 390
Mirage IV 391
Mirage 5 392
Mirage F1 393
Mirage 2000 394
Mission Master 406
Mitsubishi F-1 447
   T-2 448
Model 200/269/300 414
Model 500/530 414
Model 707 375
Mohawk 410
'Mongol' 441
'Moss' 477
'Moujik' 468
Myasishchev M-4 'Bison' 448
Mystère-Falcon 20 395
Mystère-Falcon 200 395

Nanchang Q-5 'Fantan' 448
Nesher 419
Night Hawk 466
Nimrod 380
Nomad 406
North American AT-28 450
   F-100 Super Sabre 449
   T-28 Trojan 449
   Fennec 450
Northrop F-5A Freedom
   Fighter 450
   F-5E Tiger II 451
   T-38 Talon 450
Nuri 464

OH-6 414
OV-1 410
OV-10 455
Orao 467
Orion 428
Osage 414
Osprey 368

P-2 423
P-3 428
PAH-1 438
PC-7 453
PC-9 453
PS-1 461
PZL Mielec TS-11 Iskra 453
Pampa 402
Panavia Tornado 451
Pelican 464
Phantom II 432
Pilatus PC-7/9 Turbo-Trainer 453
Pillan 399
Provider 400
Prowler 407
Pucara 402
Puma 350

Q-5 448

Raven 406
Rockwell B-1 454
   OV-10 Bronco 455
   T-2 Buckeye 455

S.211 462
S-2 410
S-3 429
S-55 463
S-58 463
S-61 464
S-65 464
S-70 465
SEPECAT Jaguar 459
SF.260 462
SH-2 420
SIAI-Marchetti S.211 462
   SF.260 Warrior 462
SMA-3 411
SR-71 429
SRA-1 411
Saab 35 Draken 456
   37 Viggen 457
   39 Gripen 459
   105 457
Scout 481
Sea Dragon 465
Sea King (Sikorsky) 464
   (Westland) 479
Sea Knight 374
Sea Harrier 381
Sea Stallion 464
SeaCobra 364
Seahawk 466
SeaRanger 367
Search Master 406
Seascan 418
Seasprite 420
Sentry 374
Shackleton 382
Shenyang F-5 439
   F-8 461
   J-5 'Fresco' 439
   J-6 'Farmer' 460
   J-8 'Finback' 461

JJ-6 460
JZ-6 460
Shin Meiwa PS-1 461
   US-1 461
Shorts Toucan 399
Sikorsky S-55 (H-19
   Chickasaw) 463
   S-58 (CH-34 Choctaw) 463
   S-61 (SH-3 Sea King/
   Pelican) 464
   S-65 (CH-53 Sea Stallion/Super
   Jolly/Super Stallion) 464
   S-70 (UH-60 Black Hawk/Night
   Hawk/Seahawk) 465
Skyhawk II 431
Soko G-2 Galeb 467
   G-4 Super Galeb 466
   J-1 Jastreb 467
   J-22 Orao 467
Starfighter 341, 427
StarLifter 426
Stratofortress 368
Stratolifter 370
Stratotanker 369
Strikemaster 382
Su-7 468
Su-15 470
Su-17 469
Su-20 469
Su-21 469
Su-22 469
Su-24 470
Su-25 471
Su-27 471
Sukhoi Su-7 'Fitter/Moujik' 468
   Su-15 'Flagon' 470
   Su-17 'Fitter' 469
   Su-20 'Fitter' 469
   Su-21 'Flagon' 469
   Su-22 'Fitter' 469
   Su-24 'Fencer' 470
   Su-25 'Frogfoot' 471
   Su-27 'Flanker' 471
SuperCobra 364
Super Etendard 395
Super Jolly 465
Super King Air 361
Super Puma 350
Super Sabre 449
Super Stallion 465

T-1 403
T-2 (Mitsubishi) 448
   (Rockwell) 455
T-28 449
T-35 399
T-38 450
T-45 379
T-48 388
T-CH-1 357
TH-55 414
TH-57 367
TR-1 430
TS-11 453

INDEX/MISSILES 565

Talon 450
Tebuan 385
Thunderbolt II 401
Tiger II 451
Tigereye 451
Tomcat 409
Tornado 451
Tracker 410
Transall C.160 472
Trislander 384
TriStar 431
Trojan 449
Tucano 399
Tu-? 476
Tu-16 472
Tu-22 473
Tu-26 474
Tu-28 475
Tu-95 475
Tu-95 475
Tu-142 476
Tu-126 477
Tupolev Tu-? 'Blackjack' 476
   Tu-16 'Badger' 472
   Tu-22 'Blinder' 473
   Tu-26 'Backfire' 474

Tu-28 'Fiddler' 475
Tu-95 'Bear' 475
Tu-142 'Bear' 476
Tu-126 'Moss' 477
Turbine Mentor 362
Tutor 385

U-2 430
U-21 361, 362
UH-1 365
US-1 461
US-3 430
Ute 362

V-22 368
VBH 438
VC10 383
VC-137 375
Veltro 2 345
Victor 384
Viggen 457
Viking 429
Vought A-7 Corsair II 477
   F–8 Crusader 478
Voyageur 374

WSK-PZL Swidnik (Mil) Mi-2 'Hoplite' 481

Wasp 481
Warrior 462
Westland Commando 480
   Lynx 478
   Scout 481
   Sea King 479
   Wasp 481

XT-4 423
Xavante 344
Xian F-7 Airguard 481
   H-6 473
   J-7 482
   Y-7 359

Y-7 359
Y-8 358
Yak-28 482
Yak-38 483
Yakovlev Yak-28 'Brewer/Firebar/Maestro' 482
   Yak-38 'Forger' 483

Z-5 445
Z-8 350
Z-9 353
Zaitun 353

4

MISSILES

AA-2 'Atoll' 539
AA-2-2 'Advanced Atoll' 539
AA-3 'Anab' 539
AA-5 'Ash' 540
AA-6 'Acrid' 540
AA-7 'Apex' 540
AA-8 'Aphid' 540
AA-9 541
AA-10 'Alamo' 541
AA-11 'Archer' 541
AAM-1 542
ABM-1 'Galosh' 524
ABM-2 524
ABM-3 'Gazelle' 524
ADAMS 523
ADATS 549
AGM-12 504
AGM-45 504
AGM-62 505
AGM-65 505
AGM-69 506
AGM-78 506
AGM-84 517
AGM-86 496
AGM-88 507
AGM-114 507
AGM-119 513
AGM-122 507
AGM-123 507
AGM-130 508
AIM-4 541
AIM-7 542

AIM-9 542
AIM-26 543
AIM-54 543
AIM-120 543
AIM-132 538
ALARM 504
AM.38 510
AM.39 510
AMRAAM 543
ANS 511
ARMAT 504
AS.11 545
AS.12 510
AS.15 510
AS.20 502
AS.30 502
AS.30 Laser 502
AS.30 TCA 502
AS-2 'Kipper' 514
AS-3 'Kangaroo' 514
AS-4 'Kitchen' 514
AS-5 'Kelt' 515
AS-6 'Kingfish' 515
AS-7 'Kerry' 503
AS-9 503
AS-10 'Karen' 503
AS-11 503
AS-12 'Kegler' 503
AS-14 'Kedge' 504
AS-15 'Kent' 498
ASAT 544
ASM-1 513

ASM-2 502
ASMP 496
ASRAAM 538
ASROC 520
ASW-SOW 520
AT-1 'Snapper' 547
AT-2 'Swatter' 548
AT-3 'Sagger' 548
AT-4 'Spigot' 548
AT-5 'Spandrel' 548
AT-6 'Spiral' 549
AT-7 549
AT-8 549
'Acrid' 540
'Advanced Atoll' 539
Advanced Cruise Missile 497
Aérospatiale AM.38 Exocet 510
   AM.39 Exocet 510
   AS.11 545
   AS.12 510
   AS.15 510
   AS.20 502
   AS.30 502
   AS.30 Laser 502
   AS.30 TCA 502
   ASMP 496
   MM.38 Exocet 510
   MM.40 Exocet 510
   MSBS M-4 488
   MSBS M-20 489
   SM.39 Exocet 510
   SS.11 545

INDEX/MISSILES

SSBS S-3 488
Eryx 545
Hades 499
Pluton 499
Aérospatiale/MBB ANS 511
'Alamo' 541
Albatros 523
'Anab' 539
'Apex' 540
'Aphid' 540
'Archer' 541
Armscor V3 539
   Kukri 539
'Ash' 540
Aspide 539
Aspide 1A 523
'Atoll' 539
Avibras SS-70 498
   SS-300 498
Ayn as-Sakr 527

BAe Firestreak 541
   Rapier 531
   Red Top 541
   Sea Dart 530
   Sea Eagle 517
   Sea Skua 517
   Sea Wolf 531
   Sky Flash 542
   Swingfire 549
BAe/Marconi ALARM 504
BAe/Matra Martel 504
BBG AIM-132 ASRAAM 538
BGL 400 503
BGL 1000 503
BGM-71 5
BGM-109 497
Bantam 547
Barak 523
Bill 547
Bloodhound 529
Blowpipe 529
Bö 810 Cobra 2000 546
Bodensee Shorad 535
Boeing AGM-69 SRAM 506
   AGM-86 496
   LGM-30 Minuteman II/III 494
Boeing/Gould ASW-SOW 520
Bofors Bantam 547
   RBS 56 Bill 547
   RBS 70 Rayrider 523
Branik 519
Bristol/Ferranti Bloodhound 529
Bullpup 504

C-601 509
C-801 509
CATIC PL-2 537
   PL-5 537
   PL-7 537
CITEFA ASM-2 Martin
   Pescador 502
   Mathogo 544

CPMIEC C-601 509
   C-801 509
   FL-1 509
   FL-7 509
   HQ-61 521
   HY-2 509
   HY-4 509
   SY-1 509
   YJ-6 509
   Model M 498
CSA-1 525
CSS-1 487
CSS-2 487
CSS-3 487
CSS-4 487
CSS-5 488
CSS-N-1 509
CSS-N-2 488
CSS-N-4 488
Cactus 521
Ching Feng 499
Cobra 2000 546
Crotale 521

DDS Branik 519
   Ikara 518
   Super Ikara 519
DF-2 487
DF-3 487
DF-4 487
DF-5 487
DF-6 487
DF-7 488
DTCN/Matra Masurca 521
Douglas MGR-1 Honest
   John 501
Dragon 550

Emerson Electric AGM-123
   Skipper 2 507
Eryx 545
Euromissile HOT 545
   MILAN 545
   Roland 522
Exocet 510

FGM-77 550
FIM-43 533
FIM-92 535
FL-1 509
FL-7 509
FRAS-1 519
FROG-3 499
FROG-5 499
FROG-7 500
Falcon 541
Firestreak 541
Ford MIM-72 Improved
   Chaparral 535
Ford Aerospace MGM-51
   Shillelagh 550
Ford Aerospace/Raytheon AIM-9
   Sidewinder 542

GBU-10 508
GBU-12 508
GBU-15 508
GBU-16 508
GBU-24 508
Gabriel 511
'Gadfly' 528
'Gainful' 526
'Galosh' 524
'Gammon' 525
'Ganef' 525
'Gaskin' 527
'Gazelle' 524
'Gecko' 527
General Dynamics AGM-78
   Standard ARM 506
   BGM-109 Tomahawk 497
   FIM-43 Redeye 533
   FIM-92 Stinger 535
   RIM-2 Terrier 532
   RIM-24 Improved Tartar 533
   RIM-66/67 Standard 534–535
   RIM-116 RAM 536
   Advanced Cruise Missile 497
'Gladiator' 528
'Goa' 525
'Goblet' 529
Goodyear UUM-44 SUBROC 520
'Gopher' 528
'Grail' 526
'Gremlin' 528
'Griffon' 525
'Grumble' 527
'Guideline' 524
'Guild' 524

HARM 507
HN-5 Red Tassel 527
HOBOS 508
HOT 545
HQ-2 525
HQ-61 521
HY-1 488
HY-2 509
HY-4 509
Hades 499
Harpoon 517
Hellfire 507
Honest John 501
Honeywell RUR-5 ASROC 520
Hsiung Feng 511
Hughes AGM-65 Maverick 505
   AIM-4 Falcon 541
   AIM-26 Super Falcon 543
   AIM-54 Phoenix 543
   AIM-120 AMRAAM
   BGM-71 TOW 550
Hughes/Martin-Marietta AGM-62
   Walleye I 505

IAE MAA-1 Piranha 536
IMI Barak 523
   Gabriel 511

INDEX/MISSILES

Mapats 546
Ikara 518
Improved Chaparral 535
Improved HAWK 533
Improved Tartar 533

JL-1 488
JL-1 488
Javelin 530

KAM-3 547
KAM-9 547
'Kangaroo' 514
'Karen' 503
Kawasaki Type 64 546
   Type 79 547
'Kedge' 504
'Kegler' 503
'Kelt' 515
'Kent' 498
'Kerry' 503
'Kipper' 514
'Kitchen' 514
Kongsberg Penguin 513
Kormoran 511
Kukri 539

LGM-25 494
LGM-30 494
LGM-118 495
Lance 501
Latécoère Malafon 519
Lockheed UGM-27 Polaris 495
   UGM-73 Poseidon 495
   UGM-96 Trident 496

M727 SP HAWK 533
MAA-1 536
MBB Bö 810 Cobra 2000 546
   Kormoran 511
   Mamba 546
MGM-31 501
MGM-51 550
MGM-52 501
MGR-1 501
MICA 538
MILAN 545
MIM-14 532
MIM-23 533
MIM-72 535
MIM-104 536
MIM-115 522
MM.38 510
MM.40 510
MSBS M-4 488
MSBS M-20 489
Magic 538
Malafon 519
Mamba 546
Mapats 546
Mariner 512
Marte 512
Martel 504

Martin Pescador 502
Martin-Marietta AGM-12
   Bullpup 504
   AGM-62 Walleye II 505
   LGM-25 Titan II 494
   LGM-118 Peacekeeper 495
   MGM-31 Pershing 501
Martin-Marietta/Oerlikon
   ADATS 549
Masurca 521
Mathogo 544
Matra ARMAT 504
   BGL 400 503
   BGL 1000 503
   MICA 538
   R530 537
   R550 Magic 538
   R.440 521
   R.460 522
   Mistral 522
   Super R530 537
Maverick 505
McDonnell Douglas RGM-84
   Harpoon 517
McDonnell Douglas/Raytheon
   FGM-77 Dragon 550
Mitsubishi AAM-1 542
   ASM-1 513
   Type 80 513
Mk 13/18UK 509
Minuteman II/III 494
Mistral 522
Model M 498
Motorola AGM–122 Sidearm 507

NOAH 533
NORINCO Red Arrow 8 544
   Red Arrow 73 548
Naval Crotale 522
Nightrider 524
Nike Hercules 532

OTO Melara/Matra Otomat 512
Otomat 512

PL-2 537
PL-5 537
PL-7 537
Patriot 536
Paveway I/II/III 508
Peacekeeper 495
Penguin 513
Pershing 501
Piranha 536
Pluton 499
Polaris 495
Poseidon 495
Python 3 538

R530 537
R550 537
R.440 521
R.460 522

RAM 536
RB 04 513
RB 05 503
RB 08 513
RBS 12 513
RBS 15 514
RBS 17 507
RBS 56 547
RBS 70 523
RGM-84 517
RIM-2 532
RIM-7 532
RIM-24 533
RIM-66/67 534-535
RIM-116 536
RS-16 490
RS-20 490
RSM-50 493
RUR-5 520
Rafael Python 3 538
Shafrir 2 538
Rapier 531
Rayrider 523
Raytheon MIM-23 Improved
   HAWK 533
   MIM-104 Patriot 536
   RIM-7 Sea Sparrow 532
Raytheon/General Dynamics AIM-
   7 Sparrow 542
Red Arrow 8 544
Red Arrow 73 548
Red Top 541
Redeye 533
Rockwell AGM-114 Hellfire 507
   AGM-130 508
   GBU-15 508
   HOBOS 508
Roland 522

SA-1 'Guild' 524
SA-2 'Guideline' 524
SA-3 'Goa' 525
SA-4 'Ganef' 525
SA-5 'Griffon/Gammon' 525
SA-6 'Gainful' 526
SA-7 'Grail' 526
SA-8 'Gecko' 527
SA-9 'Gaslin' 527
SA-10 'Grumble' 527
SA-11 'Gadfly' 528
SA-12 'Gladiator' 528
SA-13 'Gopher' 528
SA-14 'Gremlin' 528
SA-X-15 529
SA-N-1 'Goa' 525
SA-N-2 'Guideline' 524
SA-N-3 'Goblet' 529
SA-N-4 'Geclo' 527
SA-N-5 'Grail' 526
SA-N-6 'Grumble' 527
SA-N-7 'Gadfly' 528
SA-N-9 529
SADRAL 522

INDEX/MISSILES

SATCP 522
SH-04 524
SH-08 524
SM.39 510
SRAM 506
SS.11 545
SS-1 'Scud' 500
SS-4 'Sandal' 489
SS-11 'Sego' 489
SS-12 'Scaleboard' 500
SS-13 'Savage' 489
SS-17 'Spanker' 490
SS-18 'Satan' 490
SS-19 'Stiletto' 490
SS-20 'Saber' 491
SS-21 'Scarab' 500
SS-22 500
SS-23 'Spider' 501
SS-24 'Scalpel' 491
SS-25 'Sickle' 491
SS-70 498
SS-300 498
SS-C-1 'Sepal' 516
SS-C-2 'Samlet' 517
SS-C-4 498
SS-N-2 'Styx' 515
SS-N-3 'Shaddock' 515
SS-N-5 'Serb' 492
SS-N-6 492
SS-N-7 516
SS-N-8 493
SS-N-9 'Siren' 516
SS-N-12 'Sandbox' 516
SS-N-14 'Silex' 519
SS-N-15 519
SS-N-16 520
SS-N-17 'Snipe' 493
SS-N-18 'Stingray' 493
SS-N-19 516
SS-N-20 'Sturgeon' 493
SS-N-21 498
SS-N-22 516
SS-N-23 'Skiff' 494
SS-NX-24 498
SS-X-26 492
SS-X-27 492
SS-X-28 492
SSBS S-3 488
SUBROC 520
SY-1 509
Saab RB 04 513
　RB 08 513

RBS 15 514
Saab-Bofors RB 05 503
'Saber' 491
'Sagger' 548
'Samlet' 517
'Sandal' 489
'Sandbox' 516
'Satan' 490
'Savage' 489
'Scaleboard' 500
'Scarab' 500
'Scalpel' 491
'Scud' 500
Sea Cat 530
Sea Chaparral 535
Sea Dart 530
Sea Eagle 517
Sea Javelin 530
Sea Killer 512
Sea Ray 518
Sea Skua 517
Sea Sparrow 532
Sea Wolf 531
'Sego' 489
Selenia Aspide 539
　Aspide 1A 523
'Sepal' 516
'Serb' 492
'Shaddock' 515
Shafrir 2 538
Shahine, SICA 522
Shorad 535
Shorts Blowpipe 529
　Javelin 530
　Sea Cat 530
　Sea Javelin 530
　Tigercat 530
Shrike 504
'Sickle' 491
Sidearm 507
Sidewinder 542
'Silex' 519
'Siren' 516
Sistel Sea Killer/Mariner/
　Marte 512
'Skiff' 494
Skipper 2 507
Skorpioen 511
Sky Flash 542
'Snapper' 547
'Snipe' 493
Spada 523
'Spandrel' 548

'Spanker' 490
Sparrow 542
'Spider' 501
'Spigot' 548
'Spiral' 549
Standard 534–535
Standard ARM 506
'Stiletto' 490
Stinger 535
'Stingray' 493
'Sturgeon' 493
'Styx' 515
Sub-Harpoon 517
Super 530 537
Super Falcon 543
Super Ikara 519
'Swatter' 549
Swingfire 549

TOW 550
Tan-SAM 523
Terrier 532
Texas Instruments AGM-45
　Shrike 504
　AGM-88 HARM 507
　Paveway I/II/III 508
　Sea Ray 518
Tigercat 530
Titan II 494
Tomahawk 497
Toshiba Tan-SAM 523
　Type 81 523
Trident 496
Type 64 546
Type 79 547
Type 80 513
Type 81 523

UGM-27 495
UGM-73 495
UGM-84 517
UGM-96 496
UUM-44 520

V3 539
Vought ASAT 544
　MGM-52 Lance 501

Walleye I/II 505
Western Electric MIM-14 Nike
　Hercules 532

YJ-6 509